Morphing Wing Technologies

Morphing Wing Technologies

Large Commercial Aircraft and Civil Helicopters

Editors-in-Chief

Antonio Concilio, Ignazio Dimino

Adaptive Structures Division, The Italian
Aerospace Research Centre,
Capua (CE), Italy

Leonardo Lecce, Rosario Pecora

Department of Industrial Engineering, Aerospace Division,
University of Naples "Federico II",
Naples, Italy

ELSEVIER

Butterworth-Heinemann
An imprint of Elsevier

Butterworth-Heinemann is an imprint of Elsevier
The Boulevard, Langford Lane, Kidlington, Oxford OX5 1GB, United Kingdom
50 Hampshire Street, 5th Floor, Cambridge, MA 02139, United States

Library of Congress Cataloging-in-Publication Data
A catalog record for this book is available from the Library of Congress

British Library Cataloguing-in-Publication Data
A catalogue record for this book is available from the British Library

ISBN: 978-0-08-100964-2

For information on all Butterworth-Heinemann publications
visit our website at https://www.elsevier.com/books-and-journals

Publisher: Matthew Deans
Acquisition Editor: Carrie Bolger
Editorial Project Manager: Carrie Bolger
Production Project Manager: Kiruthika Govindaraju
Cast Designers: the Editors-in-Chief and the Publisher's Team

Typeset by SPi Global, India

To our beloved wives, Maria Grazia, Valeria, Anna, Deniz

Contents

SECTION 3 MORPHING SKINS

SECTION 4 SYSTEMS DESIGN

Contributors

Frederico Afonso
University of Lisbon, Lisbon, Portugal

Alessandro Airoldi
Politecnico di Milano, Milano, Italy

Salvatore Ameduri
The Italian Aerospace Research Centre, CIRA SCpA, Capua (CE), Italy

Gianluca Amendola
The Italian Aerospace Research Centre, CIRA SCpA, Capua (CE), Italy

Gennady A. Amiryants
Federal State Unitary Enterprise Central Aerohydrodynamic Institute Named After Professor N.E. Zhukovsky (TSAGI), Zhukovsky, Russia

Francesco Amoroso
University of Naples "Federico II", Naples, Italy

Alexandre Antunes
EMBRAER SA, Sao Jose dos Campos, Brazil

Alfonso Apicella
Leonardo S.p.A., Pomigliano D'Arco (NA), Italy

Gianvito Apuleo
Piaggio Aerospace, Villanova d'Albenga (SV), Italy

Maurizio Arena
University of Naples "Federico II", Naples, Italy

Uwe T.P. Arnold
ZF Luftfahrttechnik GmbH, Calden, Germany

Silvestro Barbarino
Sikorsky, A Lockheed Martin Company, Stratford, CT, United States

Marco Bellucci
MARE Engineering, Napoli, Italy

Paolo Bettini
Politecnico di Milano, Milano, Italy

Robert Blackwell
Sikorsky, A Lockheed Martin Company, Stratford, CT, United States

Ruxandra M. Botez
École de Technologie Supérieure, ETS, Montreal, QC, Canada

Miguel Á. Castillo Acero
Aernnova Group, Madrid, Spain

Vasily Chedrik
Federal State Unitary Enterprise Central Aerohydrodynamic Institute Named After Professor N.E. Zhukovsky (TSAGI), Zhukovsky, Russia

Alexander Chedrik
Federal State Unitary Enterprise Central Aerohydrodynamic Institute Named After Professor N.E. Zhukovsky (TSAGI), Zhukovsky, Russia

Alexander Chevagin
Federal State Unitary Enterprise Central Aerohydrodynamic Institute Named After Professor N.E. Zhukovsky (TSAGI), Zhukovsky, Russia

Monica Ciminello
The Italian Aerospace Research Centre, CIRA SCpA, Capua (CE), Italy

Antonio Concilio
The Italian Aerospace Research Centre, CIRA SCpA, Capua (CE), Italy

Alessandro De Gaspari
Politecnico di Milano, Milano, Italy

Federico Martín de la Escalera
Aernnova Engineering Division S.A., Madrid, Spain

Luca Angelo Di Landro
Politecnico di Milano, Milano, Italy

Ignazio Dimino
The Italian Aerospace Research Centre, CIRA SCpA, Capua (CE), Italy

José Lobo do Vale
University of Lisbon, Lisbon, Portugal

Antoine Dumont
ONERA, Meudon, France

Roman Efimov
Federal State Unitary Enterprise Central Aerohydrodynamic Institute Named After Professor N.E. Zhukovsky (TSAGI), Zhukovsky, Russia

Sergio Esposito
Leonardo S.p.A., Pomigliano D'Arco (NA), Italy

Yasser Essa
Aernnova Engineering Division S.A., Madrid, Spain

Rolf Evenblij
Technobis Fibre Technologies, Alkmaar, The Netherlands

Alessandro Gilardelli
Politecnico di Milano, Milano, Italy

André Gratias
Fraunhofer ENAS, Chemnitz, Germany

Teodor L. Grigorie
École de Technologie Supérieure, ETS, Montreal, QC, Canada; University of Craiova, Craiova, Romania

Generoso Iannuzzo
Leonardo S.p.A., Pomigliano D'Arco (NA), Italy

Fanil Ishmuratov
Federal State Unitary Enterprise Central Aerohydrodynamic Institute Named After Professor N.E. Zhukovsky (TSAGI), Zhukovsky, Russia

Vladimir Kulesh
Federal State Unitary Enterprise Central Aerohydrodynamic Institute Named After Professor N.E. Zhukovsky (TSAGI), Zhukovsky, Russia

Innokentiy Kursakov
Federal State Unitary Enterprise Central Aerohydrodynamic Institute Named After Professor N.E. Zhukovsky (TSAGI), Zhukovsky, Russia

Ksenia Kuruliuk
Federal State Unitary Enterprise Central Aerohydrodynamic Institute Named After Professor N.E. Zhukovsky (TSAGI), Zhukovsky, Russia

Svetlana Kuzmina
Federal State Unitary Enterprise Central Aerohydrodynamic Institute Named After Professor N.E. Zhukovsky (TSAGI), Zhukovsky, Russia

Fernando Lau
University of Lisbon, Lisbon, Portugal

Thomas H. Lawrence
Sikorsky, A Lockheed Martin Company, Stratford, CT, United States

Leonardo Lecce
University of Naples "Federico II", Naples, Italy

Grace Lima
EMBRAER SA, Sao Jose dos Campos, Brazil

Peter F. Lorber
Sikorsky, A Lockheed Martin Company, Stratford, CT, United States

Andreas Lühring
Fraunhofer IFAM, Bremen, Germany

Alexander Lysenkov
Federal State Unitary Enterprise Central Aerohydrodynamic Institute Named After Professor N.E. Zhukovsky (TSAGI), Zhukovsky, Russia

Marco Magnifico
University of Naples "Federico II", Naples, Italy

Vladimir Malenko
Federal State Unitary Enterprise Central Aerohydrodynamic Institute Named After Professor N.E. Zhukovsky (TSAGI), Zhukovsky, Russia

Victor Malyutin
Federal State Unitary Enterprise Central Aerohydrodynamic Institute Named After Professor N.E. Zhukovsky (TSAGI), Zhukovsky, Russia

Mihir P. Mistry
The Boeing Company, Chicago, IL, United States

Christof Nagel
Fraunhofer IFAM, Bremen, Germany

Maria Chiara Noviello
University of Naples "Federico II", Naples, Italy

Felipe Odaguil
EMBRAER SA, Sao Jose dos Campos, Brazil

Antonio Pagano
The Italian Aerospace Research Centre, CIRA SCpA, Capua (CE), Italy

Damiano Pasini
McGill University, Montreal, QC, Canada

Modesto Pecora
The Italian Aerospace Research Centre, CIRA SCpA, Capua (CE), Italy

Rosario Pecora
University of Naples "Federico II", Naples, Italy

Fabian Peter
RWTH Aachen University, Aachen, Germany

Mikhail Pronin
Federal State Unitary Enterprise Central Aerohydrodynamic Institute Named After Professor N.E. Zhukovsky (TSAGI), Zhukovsky, Russia

Francesco Rea
University of Naples "Federico II", Naples, Italy

Sergio Ricci
Politecnico di Milano, Milano, Italy

Lorenzo Rossi
Leonardo S.p.A., Pomigliano D'Arco (NA), Italy

Salvatore Russo
Leonardo S.p.A., Pomigliano D'Arco (NA), Italy

Giuseppe Sala
Politecnico di Milano, Milano, Italy

Andrey Saprykin
Federal State Unitary Enterprise Central Aerohydrodynamic Institute Named After Professor N.E. Zhukovsky (TSAGI), Zhukovsky, Russia

Oliver Schorsch
Fraunhofer IFAM, Bremen, Germany

Martin Schueller
Fraunhofer ENAS, Chemnitz, Germany

Sergey Shalaev
Federal State Unitary Enterprise Central Aerohydrodynamic Institute Named After Professor N.E. Zhukovsky (TSAGI), Zhukovsky, Russia

Stefan Storm
Airbus Group Innovations, Ottobrunn, Germany

Tobias Strobl
3DSE Management Consultants, Muenchen, Germany

Eike Stumpf
RWTH Aachen University, Aachen, Germany

Afzal Suleman
University of Lisbon, Lisbon, Portugal; University of Victoria, Victoria, BC, Canada

Viktor Timokhin
Federal State Unitary Enterprise Central Aerohydrodynamic Institute Named After Professor N.E. Zhukovsky (TSAGI), Zhukovsky, Russia

Maurizio Verrastro
Leonardo S.p.A., Caselle (TO), Italy

Andrea Vigliotti
The Italian Aerospace Research Centre, CIRA SCpA, Capua (CE), Italy

William A. Welsh
Sikorsky, A Lockheed Martin Company, Stratford, CT, United States

Matthew L. Wilbur
US Army Research Laboratory, Adelphi, MD, United States

Mikhail Zichenkov
Federal State Unitary Enterprise Central Aerohydrodynamic Institute Named After Professor N.E. Zhukovsky (TSAGI), Zhukovsky, Russia

Editor-in-Chief Biographies

Antonio Concilio (Section 6 Editor, Chapters 1, 15, 17, 18, & 22) took his degree in aeronautics engineering with honors at the University of Napoli "Federico II" (Italia) in 1989. There, he was awarded his PhD in aerospace engineering in 1995. In 2007, he completed the ECATA Master in Aerospace Business Administration at ISAE-Supaero, Toulouse, France. Since 1989, he has worked as a researcher at the Italian Aerospace Research Centre (Italia), where he is currently the Head of the Adaptive Structures Division. Between 1995 and 2000, he was head of the area of methods in vibroacoustics. In 2000, he was appointed Head of the new Smart Structures Laboratory, later integrated with the Vibration and Acoustics Lab (2002). Between 2012 and 2014, he was the National Relationship Manager at CIRA. In 2014, he was appointed head of the newly founded Adaptive Structures Division, one of ten scientific areas of the Italian Aerospace Research Centre. He has represented the company in a number of specific liaisons with external firms. Strongly involved in social activities, he is a co-founder and has been the Secretary of the ex-Alumni Association of the Aerospace Engineers at the University of Naples Federico II (AIAN), and a member of the Italian Association for Aeronautics and Astronautics (AIDAA). He is associated with MP4EVER, a non-profit organization for supporting people with eating disorders and their families. He is affiliated with the Spartacus Rugby Social Club.

Since 2005, Dr. Concilio is Lecturer at the Doctorate School at "Federico II" University (SCUDO) on smart structures: theory and applications. Since 2004, he has held courses on the same subject for postgraduate students from national programs funded by the Italian Ministry for Education and Research and Regione Campania. In 2007, he held a course on smart structures within a master in systems engineering, jointly organized by the University of Missouri-Rolla and the Federico II University. His main research interests are vehicles' noise and vibration control for increased comfort and devices protection, smart structural systems design, and morphing of aircraft wing and control surfaces. He was scientifically responsible for CIRA of several EU and nationally funded research contracts on smart structures and noise and vibration control. He was the European scientific coordinator of the IDEA PACI project (Identification of an Aircraft Passenger Comfort Index, 1998–2001) and the National Project Manager of the PON SMART project (Design and Realization Methods of Intelligent Systems for the Monitoring and Control of Aeronautic and Aerospace Structures, 2002–06). Recently, he had been leading an international team for developing, manufacturing, and testing an adaptive trailing edge device (ATED) within the EU SARISTU project (Smart Aircraft Intelligent Structures, 2011–15). He is author of about 170 scientific papers, presented at conferences (132) or published in specialized Journals (36).

Ignazio Dimino (Section 7 Editor, Chapters 15, 17, 18, 21, & 22) graduated with honors in aeronautical engineering at the University of Palermo, Italy, in 2004. He earned his PhD in aeronautics from the Imperial College of London under the supervision of Prof. Ferri Aliabadi. He was also visiting researcher at the Center of Acoustics and Vibration of Penn State University in 2008. Since 2004, his research has focused on the fundamentals of adaptive structures, including wing morphing, active noise and vibration control and smart materials. He currently holds the position of scientist at the Department of Adaptive Structures of the Italian Aerospace Research Centre (CIRA) and project manager in the field of smart structures.

Dr. Dimino was project manager at CIRA of the recently successfully completed SARISTU project (Smart Aircraft Intelligent Structures) dealing with the design, manufacturing, and testing of a morphing wing trailing edge device, and responsible of the actuation and control systems, aircraft performance, and safety. In 2015, he participated in the wind tunnel test campaign in Russia of the first true-scale morphing wing ever delivered in Europe. He also joined the Italian/Canadian team of the CRIAQ MDO-505 project supervising the design and testing of a morphing aileron.

In 2012, in the framework of Clean Sky 1 Green Regional Aircraft (GRA) ITD, he supervised the aeroacoustic test campaign of advanced high lift devices tailored to both open-rotor and turbo-prop aircraft configurations carried out in the subsonic wind tunnel at INCAS (Romania). From 2016, he is responsible of the Morphing Technology stream of the Clean Sky 2 Green Regional Aircraft IADP, and board member of Airgreen 2 PMO managing the technologies review and demonstrations. He is leading also the international team for the development and integration of full scale morphing devices (Droop Nose, Morphing Flap and Adaptive Winglet) for on-ground and in-flight test demonstration.

Dimino is author of over 50 peer-reviewed papers on adaptive structures, structural dynamics, and vibro-acoustic control, and co-author of two books. He is also independent reviewer for several international journals dealing with smart, green and integrated applications and patents. In 2014, Dr. Dimino received the Best Paper Award at the 3AF/CEAS conference, "Greener Aviation: Clean Sky Breakthroughs and Worldwide Status," in Brussels. In 2016, he served as program committee member and chairman of the second Greener Aviation conference in Brussels, organized by the 3AF *Association Aéronautique et Astronautique de France*. He has supervised more than ten master's degree and two PhD students. He is president of the Giuseppe Mario Bellanca association in Sciacca, promoting the value of the aviation genius Bellanca who designed first enclosed-cabin monoplane in the United States in 1922.

Leonardo Lecce (Section 1 Editor, Chapter 1) graduated in 1971 with honors in aeronautical engineering at the University of Naples Federico II and has spent his academic career at the same university. He retired from his position as full professor of aerospace structures, on September 1st, 2016. From 2000 to 2006, he was the Aeronautical Engineering Department Chair. He has supervised more than 250 graduation and 20 doctoral (PhD) theses. He has been a member of the EU Expert Commission for the evaluation of research proposals many times. He has taken up appointment as a member of the Scientific Committee at the Italian Aerospace Research Centre (CIRA) many times, too. He is a member of the Board of the Italian branch of the Advisory Council for Aeronautics Research in Europe (ACARE), and since 2006, he is a member of the Executive Committee of the European Association of Structural Health Monitoring. Founder of the ex-Alumni Association of the Aerospace Engineers at the University of Naples Federico II (AIAN), he was its President for many years. In 2013, he was named President of the Italian Association of Aeronautics and Astronautics (AIDAA), after having directed the Naples Chapter since 2010. Now is CEO of the company Novotech Advanced Aerospace Technology, S.r.L.

Prof. Lecce's main research interests include development of structures and systems integrating innovative smart materials, damage detection and health monitoring of structures, morphing of aircraft wing and control surfaces, and recently, prediction and control of noise and vibration in transportation systems. He was scientifically responsible for the Federico II University for several EU and nationally funded research contracts. He was the European Scientific Coordinator of the MADAVIC (Magnetostrictive Actuators for Damage Analysis and Vibration Control, 5th Framework Programme, FP) and MESEMA (Magnetoelastic Energy Systems for Even More Electric Aircraft, 6th FP) projects. Within the 7th FP (2007–13), he was the coordinator of Airgreen (a company association made of nine partners), associate member of the Alenia-managed Green Regional Aircraft Consortium (GRA), a part of the EU Joint Technology Initiative (JTI), Clean Sky Integrated Technological Demonstrator (ITD).

The results of his research activities are reported in more than 250 papers published in national and international journals and conference proceedings.

Rosario Pecora (Section 5 Editor, Chapters 10, 14, 15, 17, 18, & 22) was born in Napoli, Italy, on August 24, 1977. In 2002, he earned his master's degree, with honours, in aeronautical engineering from the University of Naples "Federico II" (Italy); here he also earned his PhD in transport engineering in 2005. From 2002 until 2011, he cooperated with several aeronautical companies (Vulcanair S.p.A., Tesi S.a.S., OMA SUD S.p.A., DEMA S.p.A., Piaggio AeroIndustries S.p.A., Bombardier Aircraft, Leonardo Aircraft, S.p.A.) mainly in the fields of loads, aeroelasticity, and structural design. In this framework, he also covered leading roles in the certification of new or derived aircraft according to EASA airworthiness requirements (VLA, Parts 23 and 25).

From 2005 until 2011, he also worked as a post-doc in the Aerospace Engineering Department of the University of Naples "Federico II." During this period, he lectured on aircraft structures with reference to FEM-based structural design and structures dynamics; he was appointed technical responsible for several national and European research projects dealing with design and validation of conventional and unconventional (smart) aircraft structures. Among these projects, he worked to the development of innovative UAVs (CAPECON, 2005-2006), active vibration dampers based on magnetorheological fluids (TELLUS, 2006-2008), hybrid box configurations for next generation regional aircraft wings (ATR-UNINA joint project, 2010-2011), shape memory alloys actuators and smart architectures for wing trailing edge morphing (Alenia AerMacchi-UNINA joint project, 2006-2008). The results of the Alenia AerMacchi-UNINA joint project were successfully patented both in Europe and in the United States (EP EP2147856 B1 and US P 8348201 B2).

In December of 2011, he was appointed senior researcher at the University of Naples "Federico II" and responsible for the academic course on aircraft structures stability, both positions he holds today. Starting in 2011, his ever-increasing responsibilities continued to grow in large research programs representing his institution at national, European and trans-European levels. He worked as primary responsible for all the activities assigned to the University of Naples in the framework of SARISTU (2011-2015), CRIAQ MDO-505 (2013-2015), CleanSky (2008-today), and Tivano (2014-today) projects. In 2012, he was nominated head of the Smart Structures Laboratory of the University of Naples "Federico II," and in 2015, he was appointed institutional scientific responsible for Airgreen2, a European cooperative project running under the CleanSky2 platform. In April of 2017, he achieved the national scientific recognition of his technical and scientific maturity, and consequently became eligible for the role of associate professor. In the midst of intense research and academic work, evidenced by nearly one-hundred scientific contributions to conferences and technical journals, he continues to maintain close relationships with relevant national industries, and also offers technical support for the certification of new products as an EASA-approved compliance verification engineer for structures and flutter.

Biographies

CO-EDITOR BIOGRAPHIES

Ferri (Fraydon) M.H. Aliabadi (Section 3 Editor) is Chair of Aerostructures and heads the Department of Aeronautics at the Imperial College London. He is also the Zaharoff Professor in aviation. He is a fellow of the Royal Aeronautical Society. Prior to joining the Imperial College in 2005, he was professor of computational mechanics and the Director of Aerospace Engineering at Queen Mary University of London (1997–2004) and reader and head of the Damage Tolerance Division at WIT, Southampton (1987–97). He has around five hundred publications in the areas of computational structural mechanics, acoustics, fracture mechanics and fatigue, contact mechanics, optimization, materials modelling and boundary element methods, including two hundred fifty journal publications and fifty authored or edited books. Professor Aliabadi is the editor of the International Journal of Multiscale Modelling and serves on editorial boards of several other journals. He is editor-in-chief of the book series, *Computer and Experimental Methods in Structures and Advances in Fracture.*

Ruxandra Botez (Section 4 Editor, Chapter 12) received her bachelor's degree from the Faculty of Aircraft Design in Bucharest, Romania, in 1984. Ruxandra became an aircraft design engineer and her project concerned gust loads studies on the IAR-99 military aircraft. She worked until 1987 as an engineer at ICA-Brasov in Romania, the manufactory for Puma and Alouette helicopters. She received her MScA degree from Ecole Polytechnique in 1989, completing her master's thesis on semi-empirical dynamic stall methods for helicopters, a subject of interest to Bell Helicopter Textron. Ruxandra obtained her PhD from McGill University in 1994 in fluid-structure interactions and nonlinear dynamics. Ruxandra worked as a postdoctoral fellow in helicopters control at Auburn University in the United States until 1995. She worked at Bombardier Aerospace as aeroservoelasticity task leader on the Active Control Technology (ACT) project until 1998, when she began work at ETS as an associate professor, becoming full professor in 2003. Dr. Botez founded and is currently head of the Research Laboratory in Active Controls, Avionics and AeroServoElasticity (LARCASE). LARCASE is equipped with three major aerospace infrastructures: an aircraft flight research simulator from CAE, Inc., for the Cessna Citation X, a subsonic blow-down wind tunnel Price-Païdoussis, and an unmanned aerial system, UAS-S4, from Hydra Technologies in Mexico. The LARCASE team consists of more than 45 students and researchers per semester. Dr. Botez collaborates on various research projects with aerospace companies Bombardier, Bell Helicopter Textron, CMC Electronics-Esterline, CAE, Inc., NASA Dryden Flight Research Center, Presagis, Thales Avionics, Alenia, and many other companies, research centers, and universities. She has been the Canada Research Chair Holder Level 1 in Aircraft Modeling and Simulation Technologies since 2011. Dr. Botez and her team have received more than thirty awards. Dr. Botez' research achievements have been the subject of 75 media articles. Dr. Botez and her team have published 110 papers in referred journals, 260 papers in conference proceedings, and 5 book chapters, by invitation. Dr. Botez has been featured as a keynote speaker at 20 conferences, and she is editor-in-chief of the INCAS Bulletin.

Sergio Ricci (Section 2 Editor, Chapter 6) earned his MS in aeronautical engineering in 1986 from Politecnico di Milano, where he was also awarded his PhD in aerospace engineering in 1989. He started his university career as an assistant professor in 1993, and he has been an associate professor since 2003. Over the years, his teaching has mainly focused on preliminary design of airplanes, fundamentals of measurements, and graduate-level experimental methods for airplane structures. His main research interests relate to aeroelastic analysis and multidisciplinary optimization of aircraft, fast structural sizing, morphing aircraft, numerical and experimental structural dynamics and aeroelasticity, and aeroelastic analysis of aircraft at the conceptual design level (Prof. Sergio Ricci is the founder and leader of the open source project NeoCASS: Next generation Conceptual Aero Structural Sizing www.neocass.org). During his career, Prof. Ricci has been PI for various national and international research programs, including EU-FP7(JTI) Project "GLAMOUR – Gust Load Alleviation techniques assessment on wind tUnnel MOdel of advanced Regional aircraft (2014–2016)," EU-FP7 Project "NOVEMOR – NOVel AIR VEhicle configurations: from fluttering wings to MORphing flight (2011–2014)," EU-FP7 Project "SARISTU – Smart Intelligent Aircraft Structures (2011–2015)," EU-FP7(JTI) Project "AWAHL – Advanced Wing And High-Lift Design (2011–2012)," for Politecnico di Milano in cooperation with FOI (Sweden) and ASCO (Belgium): 129'595 EURO, EU-FP7 Project "FFAST – Future Fast Aeroelastic Simulation Technologies (2010–2013)," "IDHEAL – IDentification of HElicopter cAbin vibration Levels (2011–2013)," EU-FP6 Project "SimSAC – Simulating Aircraft Stability And Control Characteristics for Use in Conceptual Design (2008–2011)," EU-FP5 Project "3AS – Active Aeroelastic Aircraft Structures (2005–2007)." He is a member of many international committees and author of more than one hundred twenty publications presented at conferences or published in international journals.

Fabio Semperlotti (Section 8 Editor) is an assistant professor in the School of Mechanical Engineering at Purdue University and a member of the Ray W. Herrick Laboratories there. Before joining Purdue, Dr. Semperlotti was a member of the faculty in the Aerospace and Mechanical Engineering Department at the University of Notre Dame in the United States (2011–15). He received his MS in aerospace engineering in 2000 and his MS in astronautic engineering in 2002 from the University of Rome, "La Sapienza," in Italy. Later, he completed his doctoral studies at the Pennsylvania State University (Penn State), where in 2009 he received a PhD in aerospace engineering. Prior to joining Penn State, Dr. Semperlotti served as structural engineer for several European aerospace industries, including the French Space Agency (CNES), where he worked on space launch systems and satellite platforms.

Dr. Semperlotti directs the Purdue University Structural Health Monitoring and Dynamics Laboratory, where he conducts research related to structures and materials including structural dynamics and wave propagation, acoustic metamaterials, structural health monitoring, and smart and adaptive structures. He has authored over 60 technical publications including both conference proceedings and journal articles. His research has been supported by the US National Science Foundation, US Air Force, US ARMY, DARPA, and industrial sponsors. Dr. Semperlotti is the recipient of the National Science Foundation CAREER award (2015) and of the Air Force Office of Scientific Research Young Investigator Program (YIP) (2015).

CONTRIBUTOR BIOGRAPHIES

Frederico Afonso (Chapter 4) completed his MSc in aerospace engineering in 2011 and his PhD in aerospace engineering in 2015, both at Instituto Superior Técnico, University of Lisbon, in Portugal. During his PhD studies, he participated in the EU 7th Framework Program Project NOVEMOR, the scope of which was to analyze and optimize novel and morphing aircraft. He is currently a postdoctoral researcher at Instituto Superior Técnico. His main research interests include multidisciplinary design optimization, nonlinear aeroelasticity, morphing structures, preliminary and conceptual aircraft design. He is currently studying modular aircraft within the scope of the Portugal 2020 Framework Project FLEXCRAFT. He is author of 6 scientific papers published in international peer-reviewed journals and 22 conference papers.

Alessandro Airoldi (Chapter 9) has been an associate professor since 2016 at the Department of Aerospace Science and Technology of Politecnico di Milano, having previously been an assistant professor beginning in 2007. He is currently professor of Aeronautical Structures, and he has been a lecturer on the subject of space structures and aerospace materials and technology. He is author of more than 80 scientific and technical publications, mainly focused on numerical modelling of structures and materials for aerospace applications and on the development of structural solutions based on innovative and smart materials.

Salvatore Ameduri (Chapters 10, 13, 16 & 20), after defending his PhD thesis (Univ. of Naples "Federico II") on smart structures, entered CIRA in 2002 in the Smart Structures and Materials Department. He is the reference person of the Unit of Sensors and Actuators for Adaptive Structures, author of about 60 journal and conference papers on smart structures, morphing, noise & vibration, author and co-author of two international patents (one for a drop nose and another for a SMA variable chamber TE), reviewer for scientific journals such as JIMSS and IOP, and recipient of the Best Presentation award at the ICMAE Conference in 2014. Additionally, Dr. Ameduri has played coordination and scientific roles for many projects such as: Graphene-Polymeric Spray Sensor for Shape Recognition of Super-Deformable Structures, GRAPSS CIRA Internal Project, project manager; AIRGREEN 2 (CS2 Program 2015–21), technical coordinator; SARISTU (FP7 2011-2015), responsible for estimating cross-related effects of morphing concepts; SADE, "Smart High Lift Devices for Next-Generation Wings" (FP7 2019-2012), droop nose designer.

Gianluca Amendola (Chapters 11, 18 & 22) was born in Avellino, Italy, on July 29th, 1987. He graduated with his bachelor's degree in 2009, and his master's degree in 2012, both with 110/110 *magna cum laude*, in aeronautical engineering at the University of Napoli "Federico II." In 2012, he was involved in the framework of the EU Clean Sky project on the design of a morphing trailing edge device created to improve aircraft aerodynamic performance. Since 2014, he is a researcher at the Department of Smart Structures of the Italian Aerospace Research Centre (CIRA). He is involved in the EU SARISTU project (Smart Aircraft Intelligent Structures) dealing with the design and testing of a morphing wing trailing edge. He was involved in experimental characterization of the morphing trailing edge device. Dr. Amendola is working in the Clean Sky 2 (Airgreen) WP 2.1.2 for the design of a morphing winglet for flight tests and in the WP 2.1.4 of Clean Sky 2 (Airgreen) project for the development of a meta-sensor (software sensor) for active wing load distribution control. Dr. Amendola is active in the field of aerospace design with innovative hypersonic deployable re-entry heat shields (MINI-IRENE, Italian Re-entry NacElle), with focus on the deployment kinematic mechanism. In 2016, he successfully completed his PhD at the University of Naples (Department of Industrial Engineering—Aerospace Division) in the framework of the CRIAQ MD0505 project, a cooperation between CIRA, Alenia, University of Naples, and a Canadian university (ETS) and industry (Bombardier). His research pertains to the design, manufacturing and testing of a morphing aileron for next-generation regional aircraft.

Gennady A. Amiryants (Chapter 23) Education: Moscow Aviation Institute (1960); Moscow Institute of Physics and Technology (postgraduate, PhD, 1968). Affiliations and Functions: Engineer of Central Aerohydrodynamics Institute, TsAGI (1960), Leading Engineer, TsAGI (1966); Professor: Methods and Experience of Contemporary Aircrafts Static Aeroelasticity Problems Solving, TsAGI (1982); Head of Department, TsAGI (1988). Experienced in static aeroelasticity, use of elasticity, active aeroelastisity concepts, multidisciplinary theoretical and experimental investigations, and smart (morphing) adaptive structures. Dr. Amiryants is author of more than 100 scientific papers and patents, as well as eight books about aviation history and the history of science.

Francesco Amoroso (Chapters 15, 17 & 18) earned his degree with honors in aerospace engineering at the University of Napoli "Federico II" (Italy) in 2004. There, he earned his PhD in aerospace engineering in 2009. Dr. Amoroso has been an assistant professor of turbomachinery structural design and a lecturer of advanced aircraft structures at the same university since 2011. He has worked for many companies (ALENIA AERMACCHI, AVIO SPA, PININFARINA) and research centers (CIRA, CNR, ONERA), as well as for European academies (Trinity College of Dublin, Universität des Saarlandes - LPA). His research focuses on experimental aeroacoustics, structures dynamics, and the design of conventional and smart structures. He has held leading roles in national and international research projects (MESEMA, COMFORT, SEFA, COSMA, CleanSky GRA, CleanSky 2), and his research has been published in nearly 50 papers related to conferences and scientific journals.

Alexandre Pequeno Antunes (Chapter 6) was born in Rio de Janeiro, Brazil, on May 13, 1974. He earned his degree in mechnical engineering at the Federal University of Rio de Janeiro in 1998. In 2000, he completed his master's degree in aeronautics at the aeronautical institute of technology, and then in 2014, he earned his doctorate there in computer science. Since 2000, he has been at Embraer as a product development engineer and, during this time, he has worked in the Aeronautical Department, the Air Management System Department, and the Advance Conceptual Design Department, and for the last 6 years has been a part of the Research and Development Department. There, he served as project manager of the MDO group for 4 years and has most recently been responsible for the aerodynamic activities of the Laminar Flow Project. He is a senior computational fluid dynamics engineer with experience in performing external and internal analysis/design of many aircrafts' components. He participated in wind tunnel and flight test campaigns for many developed products from Embraer. His main research interests are aerodynamic design of wings and high-lift systems, conceptual aircraft design, and multidisciplinary optimization.

Mr. Alfonso Apicella (Chapter 20) Leonardo S.p.A., Aircraft Division—Structures Analysis and Methods Dept. Mr. Apicella has more than 18 years of experience managing European projects. From 2002 to 2007, he has been European leader and focal point of various research projects ADPRIMAS, SMAAC, ADMIRE, MUSCA 2004–07. Mr. Apicella is responsible for the management and development of European research, FUBACOMP, WELAIR, DIALFAST, ITOOL, AEROMAG, ALCAS, TANGO. From 2005 to the present, he has been advisor for the preparation of European large projects MAAXIMUS, JTI Clean Sky GRA LWC, SARISTU.

Gianvito Apuleo (Chapter 2) earned his master's degree in aerospace engineering in 2004 at the Second University of Naples "S.U.N." (Italy). After several experiences between 2004 and 2009 as a stress analyst in AIRBUS Deutschland and EADS for both civil and military projects, he joined the research division (PHT) at Piaggio Aero Industries in 2009 as Senior Structural Stress Engineer. His work focuses, in particular, on composite structures analysis for Piaggioaerospace industrial programs and the development of stress analysis methodologies and manufacturing processes in research projects for new products. He has been involved with several EU research projects such as WASIS, INFUCOMP, ESPOSA, TEST-X-TRANSPORT, CESAR, and CLEAN SKY in developing and testing aircraft composite structures. He is the Work Area Leader of the "More Affordable Composite Structures" WP in SAT (Small Air Transport) framework within the CLEAN SKY 2 project.

Maurizio Arena (Chapter 18) is working towards his PhD in the Department of Industrial Engineering of University of Napoli "Federico II" (Italy), where he graduated with honors in aerospace engineering in 2015 with a thesis focused on the aeroservoelastic stability assessment of a morphing trailing edge within the SARISTU European project. His research focuses on the design and validation of smart systems for next generation aircraft. In this context, several innovative solutions for optimizing aircraft weight and operative costs and enhancing flight performance and internal comfort have been developed.

Uwe T.P. Arnold (Chapter 24) received his diploma in aeronautical engineering in 1988 from the University of Braunschweig, Germany. He worked as a research assistant at the Institute of Flight Mechanics, where he finished his PhD thesis on blade instabilities of stoppable rotors. Thereafter, he was awarded a postdoctoral fellowship for a research tenure at the Department of Mechanical and Aerospace Engineering at Princeton University in the United States. Returning to Germany in 1996, Dr. Arnold joined ZF Luftfahrttechnik GmbH, a German company specializing in R&D, production, and MRO of dynamic helicopter components. There, he is responsible for R&D and serves as the head of the design organization. He has published 30 scientific papers and articles, holds more than 10 patents, and has received multiple awards for his work. He is also a lecturer at the Academic Training Center of the German Armed Forces.

Silvestro Barbarino (Chapter 24) earned his MS in aerospace engineering from University of Napoli "Federico II" (Italy) in 2006; there, he was also awarded his PhD in aerospace engineering in 2010. Silvestro joined Sikorsky, a Lockheed Martin company, in 2014 as a senior engineer in the Rotor Structures group and is currently a data scientist analyst within the iNtelligent Technologies, Analytics, and Sustainability (iTAS) branch. Previously, he has been a research associate and adjunct faculty at Rensselaer Polytechnic Institute, a postdoctoral scholar at Penn State University and a research officer at Swansea University (UK). Silvestro's expertise includes topics such as aerospace morphing structures, smart materials, fatigue analysis of metals, nonlinear structural dynamics, analytics, and artificial intelligence. He is author of more than 45 technical papers, reports, and book chapters, and he has been awarded patents in both the EU and the United States. He is a member of AIAA, ASME, and AHS.

Marco Bellucci (Chapter 15) earned his degree in materials engineering *cum laude* at the University of Naples Federico II in 1998, and he has always been interested in the field of numerical modelling for engineering design and research. He also has substantial experience in the development and management of research and design activities, with expertise in the most common CAE simulation methods and tools used in industrial and research applications. From 1998 to 2004, he worked as a researcher at the Italian Public Research Center, dealing with numerical modelling with implicit/explicit methods and in the framework of thermo-elastic analysis, structural nonlinear analysis, crash simulation and biomechanics, numerical modelling, and calibration of unconventional materials. In 2001 he founded Mare Engineering, a company providing consultant services to SME and large enterprises. From 2001 to 2005, he worked as chairman of Mare, while from 2005 forward, he has also been responsible for CAE. In 2012, he joined Step Sud Mare (a subsidiary company of Mare Engineering founded in 2012) where he currently works as responsible of CAE and R&D activities. During his career, he has written and managed more than thirty regional, national, and European R&D projects.

Paolo Bettini (Chapter 9), received his PhD in aerospace engineering at Politecnico di Milano in 2008, and he has been an assistant professor there since 2013. Presently, he is a lecturer in aerospace technologies and materials at the Department of Aerospace Science and Technology. His research focuses on smart materials (FO, PZT, and SMA) and their integration on smart structures, especially on the development of health monitoring systems and composite structures with morphing capabilities. Other activities involve the design of high sensitivity sensors based on fiber Bragg grating technology. He is the author of more than sixty papers printed in international journals or presented at international conferences.

Robert Blackwell (Chapter 24) joined the Sikorsky Aircraft Division of United Technologies in 1970 after receiving his bachelor's and master's degrees from Rensselaer Polytechnic Institute. During his time at Sikorsky, he has worked on the UH-60 series helicopters, the Comanche helicopter, the X2 Technology Demonstrator aircraft, and several R&D efforts focusing on advanced rotor designs. He has authored several publications concerning advanced rotor concepts and is a past chairman of the American Helicopter Society Dynamics Committee. He currently teaches the dynamics portion of the Penn State University rotary wing short comprehensive short course in rotary wing technology. He was named a Sikorsky Technical Fellow in 2003 and an American Helicopter Society Technical Fellow in 2012.

Miguel Á. Castillo Acero (Chapter 19), PhD in aerospace engineering (2016) and master's in aerospace engineering (1990), Escuela Técnica Superior Ingenieros Aeronáuticos, Universidad Politécnica Madrid (Spain). Dr. Acero completed postgraduate studies in science and technology management (1992) at the Universidad Carlos III, in Getafe, Spain. Over the span of his 27-year aerospace analysis engineering career, starting in 1990, Dr. Acero has held various positions with C.A.S.A. (Getafe, Spain), Boeing (Seattle, United States), Hispano Suiza (Le Havre, France), and Aernnova, formerly Gamesa Aeronautica (since 2000). Dr. Acero is the author of aerospace-related patents and scientific articles in aerospace specialized journals.

Alexander Chedrik (Chapter 23) studied from 2003 to 2009 at the Moscow Institute of Physics and Technology (MIPT) at the Department of Aeromechanics and Aircrafts in Zhukovsky (Russia). In June of 2009, he graduated from the MIPT with a master's degree in applied mathematics and physics. From 2009 to 2013, he did his postgraduate studies at the Central Aerohydrodynamic Institute (TsAGI, Zhukovsky) on the strength and thermal conditions of aircraft. Since 2009, he has worked as a junior scientific researcher in TsAGI. He focuses on static aeroelasticity and is author of about 10 publications which have been presented at conferences or published in scientific journals.

Vasily Chedrik (Chapter 23) earned his degree in aeronautics engineering from the Moscow Institute of Physics and Technology (MIPT) in 1983; there he was also awarded his PhD in structural optimization in 1990. In 1994, he became a head of the Multidisciplinary Structural Design Optimization Division at the Central Aerohydrodynamic Institute (TsAGI). He is responsible for the development and application of advanced methods for structural and multidisciplinary optimization and is an expert of the Russian Aviation Register. He is author/co-author of more than 100 scientific papers on different fields of static and dynamic strength of aircraft. He is a lecturer at MIPT on theory of elasticity and plasticity.

Alexander Chevagin (Chapter 23) earned his degree in aerothermodynamics at the Moscow Institute of Physics and Technology (State University) in 1977. Since 1977, he has worked as an engineer at Central Aerohydrodynamic Institute (TsAGI), where he is currently the head of the Propulsion Department. There he was awarded his PhD in aero- and thermodynamics of aircraft in 1987. He is a lecturer at the Department of Aeromechanics and Flight Engineering of the Moscow Institute of Physics and Technology in aerodynamics of propulsion systems. He is also author of more than 150 scientific papers, presented at conferences or published in specialized journals.

Monica Ciminello (Chapter 11 & 17) earned her master's degree in physics at "Federico II" University of Naples with a specialization in cybernetics for computer vision applications. From 2006 to 2009, she produced several peer-reviewed journal publications on semi-active piezo-shunt control systems and vibro-acoustic coupled systems modelling, and earned a PhD in aerospace engineering at "Federico II" University of Naples and mechanical engineering at Conservatoire National des Artes et Metiers (CNAM) of Paris, earning a mention of excellence. From 2011, she has been a permanent researcher at the Italian Aerospace Research Center in the Adaptive Structures Department specializing in the design and integration of smart sensors. She has worked on the design of a sensor system network based on FBG optical extensimeter for shape reconstruction of an adaptive trailing edge in the Saristu project spearheaded by Airbus. She is currently responsible for the design of a SHM system for skin-stringer debonding of a wing box using a distributed Rayleigh scattering optical technique and principle component analysis (SMAF project). She is also the project focal point of the optical sensor integration for SHM of pressure vessel (VECEP project leaded by AVIO).

Alessandro De Gaspari (Chapter 6) earned his MS in aerospace engineering in 2006 at Politecnico di Milano, where he was also awarded his PhD in aerospace engineering in 2010, with his thesis on optimal design of morphing wings based on compliant structures. Since 2007, he has worked at the Department of Aerospace Science and Technology of Politecnico di Milano as a researcher. He is a structural engineer with experience in structural dynamics, aeroelasticity, multidisciplinary design optimization, wind tunnel testing, active aeroelastic control, and morphing structures. His teaching focuses on finite element modelling and design for undergraduate students. He is developer of two computation codes: a parametrization framework for the design and modelling of aircraft, and a software for the synthesis of compliant mechanisms and adaptive compliant structures. Dr. Gaspari has participated in several European projects including: EU-FP7(JTI) Project "GLAMOUR – Gust Load Alleviation techniques assessment on wind tUnnel MOdel of advanced Regional aircraft (2014–2016)," EU-FP7 Project "NOVEMOR – NOVel AIR VEhicle configurations: from fluttering wings to MORphing flight (2011–2014)," EU-FP7 Project "SARISTU – Smart Intelligent

Aircraft Structures (2011–2015)," and EU-FP5 Project "3AS – Active Aeroelastic Aircraft Structures (2005–2007)." Currently he is involved in CLEAN SKY 2 with a main focus on WP2.1.2 (Morphing Structures) and WP2.1.4 (Load Control & Alleviation). He was a reviewer for several international journals, including *Journal of Aircraft (JA)*, *Journal of Neurocomputing (Neucom)*, and *The International Journal of Aerospace Engineering (IJAE)*, and he is author of more than 30 scientific papers presented at conferences or published in international journals.

Federico Martin De La Escalera (Chapter 19), PhD in mining engineering (2015), Polytechnic University of Madrid, master's in finite element technologies and engineering applications (1995), UNED (National University of Distance Education), has 25 years' experience as an analysis engineer for the aerospace industry, starting in 1995 at EADS C.A.S.A. Space (San Fernando de Henares, Spain), Ariane V launcher groups, Gamesa Aeronautica since 2001. Author of various aerospace publications including mechanical and materials scientific articles in aerospace specialized journals. He has 15 years' experience as a lecturer in several universities, with 5 years in the Department of Applied Mathematics, School of Mines, Polytechnic University, and 10 years in the Carlos III University Continuum Mechanics and Structures Theory Department.

Luca Di Landro (Chapter 9) has been an associate professor since 2005 at the Department of Aerospace Science and Technology of Politecnico di Milano, and was previously an associate and then assistant Professor in the Chemistry and Materials Department starting in 1990. He is currently professor of aerospace technologies and materials and of polymer technologies. He is author of more than 200 papers and congressional communications mainly focused on polymers and composites for aerospace applications and on physical, chemical, and mechanical characterization of materials.

Antoine Dumont (Chapter 5) earned his master's degree in mechanical engineering at the French Institute for Advanced Mechanics (IFMA) in Clermont-Ferrand (France) in 2006. In 2005, while a research assistant under the direction of Prof. Joseph Katz at the University of San Diego in the United States, he discovered a true passion for aerodynamics. This led him to prepare a doctoral thesis on aerodynamic shape optimization applied to helicopter rotor blades at ONERA in collaboration with the University of Poitiers (France). Following his PhD defense in 2010, he joined a research team working in aerodynamics for civil aircraft applications at ONERA. His main research interests are aerodynamic shape optimization to reduce drag of civil aircraft, multidisciplinary design optimization (mainly aero-structural optimization), and the development of adjoint methods in CFD. He was part of the EU SARISTU project (Smart Aircraft Intelligent Structures, 2011–15) where he demonstrated the potential of optimization techniques to find adapted morphing shapes that reduce wing drag along the cruise segment. He is author of 2 scientific papers published in international journals and 13 conference papers.

Roman Efimov (Chapter 23) earned his degree in rocket propulsion systems at the Omsk State Technical University in 2007. Since 2007, he has worked as an engineer in wind tunnel T-104 at Central Aerohydrodynamic Institute (TsAGI). In 2008, he began work as a designer of aerodynamic models in the Design-Engineering department, where he is currently head of group. He is author of seven scientific papers, presented at conferences or published in specialized journals.

Sergio Esposito (Chapter 20), Leonardo Company, Viale dell'Aeronautica snc, Pomigliano d'Arco (Naples)-Italy. His aeronautic engineering degree was attained at "Federico II" Univerisity – Naples in 26/03/1998, with a score of 110/110 *cum laude*. At Leonardo company, he worked in Alenia Aeronautica as a job shopper (Ergon Line) from March 1999 to March 2001, and was employed in the microscopy laboratory as an SEM (Scanning Electron Microscope) technician. He was then confirmed in April 2001 and took charge of the Non-Destructive Laboratory in 2003. Since 2005, he has been responsible for the Optical/Electronic Microscopy Laboratory. Later he managed reconciliation activities for the B787 horizontal stabilizer at Foggia Plant. He subsequently became responsible for metallic materials technology in M&P Department, until 2014. In 2015, he assumed responsibility for metal & assembly technologies in the MAR&D Department.

Yasser Essa (Chapter 19), Mechanical Engineer, MSc, PhD, has 20 years' experience in the development of different technological and industrial projects such as KC390: Birdstrike simulation of inboard and outboard flap structure, led and funded by EMBREAR Aircraft Company, advanced HLFC HTP design (CS2), modified cost effective fiber-based structures with improved multi-functionality and performance (MODCOMP, CS2), Smart Intelligent Aircraft Structures (SARISTU, FP7), ELECTRICAL (FP7), Cost-Effective Small AiRcraft (CESAR, FP6), Advanced design and configurations of future aircraft and Rear End Optimization, ICARO (CDTI-Spain), A400M program (Airbus), etc. His research field is numerical and experimental testing of composite materials under static and dynamic behavior. A lecturer in different universities, Dr. Essa has two patents and is the author of numerous journal papers.

Rolf Evenblij (Chapter 11) graduated with a BSc in physics-photonics from Delft/Rijswijk University of Technology in 1994. Currently employed as program manager in aerospace & space, and automotive testing at Fibre Technologies in Alkmaar. After a brief period as a system engineer for system and software development for fiber optic sensing systems, he managed various development and implementation projects for all relevant markets including aerospace. Starting 2 years ago, he became involved in program management specifically targeting aerospace/space and automotive testing, focusing on business development and strategic company management. As program manager, he has been involved in various aerospace projects (e.g., SARISTU, TAPAS, and ALGESMO) where miniature and reliable fiber-sensing devices are utilized for applications including damage and impact detection, shape sensing of morphing structures, helicopter blade monitoring, and load monitoring, etc.

Alessandro Gilardelli (Chapter 9) is a researcher in the Department of Aerospace Science and Technology of Politecnico di Milano, where he earned his master's degree in aeronautical engineering in 2015. His focus is on application of corrugated skin and auxetic chiral topologies for morphing structures.

André Gratias (Chapter 22), graduated from the Chemnitz University of Technology in 2010 with a degree in information and communication technology. He is currently employed as a researcher at the Fraunhofer Institution for Electronic Nanosystems in Chemnitz. His area of expertise is the design, development and test of embedded and microcontroller-based electrical systems for actuation control and sensor data acquisition. He has contributed to several European projects in the area of aeronautics, including development of an actuator control for an adaptive trailing edge within the EU SARISTU project (Smart Aircraft Intelligent Structures, 2011–15). Other involvements are AFLoNext and Clean Sky 2 in aeronautics projects, as well as PBSA in the area of space.

Teodor Lucian Grigorie (Chapter 12) received a BsC and a MsC in aerospace engineering from the Faculty of Electrical Engineering, University of Craiova, Romania, in 1998 and 1999, respectively. In 2006, he earned his PhD at the Faculty of Aerospace Engineering, Politechnica University of Bucharest in the aerospace engineering field. Between 2008 and 2009, he was a postdoctoral fellow at the University of Quebec, Ecole de Technologie Supérieure in Montreal, Canada, with which he has collaborated in research projects since 2002. His research concerns navigation systems, micro- and nano-inertial sensors, aircraft board equipment and systems, automatic control systems (including intelligent control systems), and morphing aircraft structures.

Generoso Iannuzzo (Chapter 20) Leonardo S.p.A., Aircraft Division—Head of Airframe Division. Education, Università degli Studi di Napoli Federico II, mechanical engineering, with experience in aircraft airframe design and knowledge of methods and processes gained through collaboration on ATR, AM-X, MD95, and A380 programs since 1989. He was Lead of Stress Department from July 2009 to January 2013 and presently is head of the Airframe Division.

Fanil Ishmuratov (Chapter 23) graduated from Moscow Institute of Physics and Technology in 1975, and received his PhD in aerospace engineering at Central Aerohydrodynamic Institute (TsAGI) in 1985. He was responsible for the development and application of advanced computational aeroelasticity methods at aeroservoelasticity division of TsAGI. Since 1999, he has worked as a Head of the Static Aeroelasticity Division within the Strength Standards, Loads and Aeroelasticity Department at TsAGI. He received his second doctoral degree in multidisciplinary analysis of aeroelasticity characteristics in 2012 at TsAGI. He is an expert of the Russian Aviation Register. He is author/co-author of more than 100 scientific papers on aeroelasticity and structural dynamics, presented at conferences or published into specialized journals.

Vladimir Kulesh (Chapter 23) graduated from Moscow Institute of Physics and Technology (State University, Russia) in 1972. There, he also received his PhD in physical and mathematical sciences in 1979. Since 1972, he has worked as a researcher at Central Aerohydrodynamic Institute (Russia), where he obtained a doctoral degree in technical science in 1992, and currently is the head of the Optical and Physical Methods of Research Division. Since 1991, he has been a lecturer at the Department of Aeromechanics and Flight Engineering at Moscow Institute of Physics and Technology. He is author of more than 100 scientific papers presented at conferences or published in specialized journals.

Innokentiy Kursakov (Chapter 23) earned his degree in applied mathematics and physics with honors at the Moscow Institute of Physics and Technology (State University) in 2006. Since then, he has worked as a junior researcher at Central Aerohydrodynamic Institute (TsAGI), where he is currently the senior researcher in the Propulsion Department. There he was awarded his PhD in aero- and thermodynamics of aircraft in 2010. Since 2008, he has been a lecturer at the Department of Aeromechanics and Flight Engineering of the Moscow Institute of Physics and Technology in Applied Computational Fluid Dynamics. He is author of more than 40 scientific papers, presented at conferences or published in specialized journals.

Ksenia Kuruliuk (Chapter 23) graduated from Moscow Institute of Physics and Technology (State University, Russia), the Department of Aeromechanics and Flight Engineering in 2012. Since 2012, she has worked as a researcher at Central Aerohydrodynamic Institute (Russia) at Optical and Physical Methods of Research Division. Since 2013, she is a postgraduate student at Central Aerohydrodynamic

Institute working on a dissertation on an optical videogrammetry method for measuring deformation field in strength and wind tunnel tests. She is co-author of about 20 scientific papers presented at conferences or published in specialized journals.

Svetlana Kuzmina (Chapter 23) is the leading research fellow for the Dynamic Aeroelasticity Division within the Strength Standards, Loads and Aeroelasticity Department at Central Aerohydrodynamic Institute (TsAGI). She graduated from Moscow Institute of Physics and Technology, receiving her PhD in aerospace engineering in 1976. She has been working at TsAGI in the Aeroelasticity Department since 1971. She is responsible for the development and application of advanced computational aeroelasticity and unsteady aerodynamics methods. Dr. Kuzmina is an author of over 60 technical papers, presentations at scientific conferences, and reports in unsteady aerodynamics, aeroelasticity and structural dynamics. She is a Member of the IFASD Scientific Committee.

Fernando Lau (Chapter 4) graduated in engineering physics in 1992 from the Instituto Superior Técnico of University of Lisbon in Portugal. There he was also awarded his PhD in aerospace engineering in 2000. Since then, he has been a researcher at the Institute of Mechanical Engineering (IDMEC), developing research in the areas of aeroacoustics, adaptive structures, aircraft conceptual design, and multidisciplinary aircraft optimization. He has authored 20 articles in international journals and presented more than 30 articles in national and international conferences. He co-supervised two PhD theses and more than 20 MSc theses. He was an evaluator of European projects in FP7 and is an evaluator of the Portuguese National Innovation Agency (ADI), as well as several international journals. He has participated in several European projects (SNAAP, APIAN, FRIENDCOPTER, ARTIMA, 3AS, SMORPH and NOVEMOR), and currently he is the principal investigator of FLEXCRAFT under the national Portugal 2020 Framework. At present, he is the vice coordinator of the Masters in Aerospace Engineering program and a member of the Pedagogical Council of Instituto Superior Técnico.

Thomas Lawrence (Chapter 24) received his master's degree in aeronautical engineering from MIT in 1976. He worked for 35 years in the field of helicopter handling qualities, flight simulation, and flight tests. He has presented papers to the AHS Forum and has given lectures by invitation to the AHS, AIAA, and SAE. Tom was a member of the small team that developed and flew the Sikorsky X2 to a speed of 250 knots, the team subsequently being awarded the Collier Trophy. The X2 was later inducted into the Smithsonian Air and Space Museum in Washington, DC.

Grace Lima (Chapter 6) (MSc Engineering) was born in Guarulhos, Brazil, on January 2nd, 1980. She graduated as an aeronautical engineer from ITA (Aeronautical Institute of Technology, Brazil) in 2003, being one of the first women to graduate from this renowned institution. There she also earned her master's degree in aeronautics engineering in 2010. Since 2004, she has been working in several companies in the aeronautics industry, and her major fields of professional concentration are aerodynamics and flying qualities. At Embraer, she worked on the Aerodynamics Team in executive aircraft programs like Phenom 100 and Phenom 300, as well as in the defense program KC 390, helping develop the aerodynamic data bank for these aircrafts and also performing several transonic and subsonic CFD analysis. At Mectron, she worked on the development of ACAUAN (MK-82 guided bomb) and the development of the fifth generation Missile A-Darter (Mectron/Denel Dynamics), a South African project co-funded by the Brazilian Government. She also participated in wind tunnel tests and flight test campaigns during all these projects. Recently, she has been working in the R&D Group at Embraer studying static aeroelastic effects in aerodynamics and flying qualities characteristics of airplanes with high aspect ratio wings.

Peter Lorber (Chapter 24) received his BS from Cornell University (applied and engineering physics, 1979), and his SM (1981) and PhD (aerodynamics, 1984) from the Massachusetts Institute of Technology. He has been employed at United Technologies Research Center and Sikorsky, a Lockheed Martin company, since 1984, and is currently a fellow in Flight Sciences. He has investigated helicopter aerodynamics, aero-acoustics, and active rotor and flow control, and has been responsible for scale model aeromechanics testing in support of multiple helicopter development programs. He has published over 65 technical papers and reports, and has contributed to 12 US patents. He is a senior member of the AIAA and a Technical Fellow of the AHS International.

Andreas Lühring (Chapter 7) is a graduate engineer in chemical engineering (University of Applied Sciences Emden) and successfully completed his diploma thesis at the Fraunhofer Institute for Manufacturing Technology and Advanced Materials (IFAM), Bremen, Germany, in 2001. Since 2005, he has worked as a researcher at the Adhesives and Polymer Chemistry Division at the Fraunhofer Institute for Manufacturing Technology and Advanced Materials (IFAM), Bremen, Germany. His research focuses on the formulation of adhesives for special applications and the development of fast curing processes.

Alexander Lysenkov (Chapter 23) earned his degree in applied mathematics and physics at the Moscow Institute of Physics and Technology (State University) in 2001. There he graduated from postgraduate studies in 2004. Since 2004, he has worked as a researcher at Central Aerohydrodynamic Institute (TsAGI), where he held the positions of Senior Researcher, Head of Group, and now is the Head of the Computational Fluid Dynamics Division of the Propulsion Department. There he was awarded his PhD in aero- and thermodynamics of aircraft in 2004. Since 2005, he has been a lecturer at the Department of Aeromechanics and Flight Engineering of the Moscow Institute of Physics and Technology in Methods of Grids Generation for Aerodynamic Tasks. He is author of more than 40 scientific papers presented at conferences or published in specialized journals.

Marco Magnifico (Chapter 17) earned his master's degree in aerospace engineering with honors at the University of Napoli "Federico II" (IT) in 2012. There, he earned his PhD in aerospace, naval, and quality engineering in 2016. Since 2014, he has worked as a freelance designer and maker in additive manufacturing. He worked for a year in the aeronautical industry, where he gained experience in the certification of very light aircraft, and since the end of 2016, he is involved in the design and manufacturing of prototypes and wind tunnel models at Eurotech Sas (IT), in which he is responsible for sensor systems installation.

Vladimir Malenko (Chapter 23) earned his degree in aircraft engineering at the Moscow Aviation Institute (Technical State University) in 1966. Since 1959, he has worked at the Propulsion Department of Central Aerohydrodynamic Institute (TsAGI). From 1964 to 1988, he worked in the Experimental Section of the Rescue Systems, and from 1988 to 2013, he was head of the TsAGI T-104 wind tunnel. He has 20 certificates of invention and is author of four scientific papers presented at conferences or published in specialized journals.

Victor Malyutin (Chapter 23) studied from 1995 to 2001 at the Moscow Institute of Physics and Technology (MIPT) in the Department of Aeromechanics and Flight Engineering in Zhukovsky, Russia. In June of 1999, he graduated from the MIPT with a master's degree in applied mathematics and physics. From 2001 to 2005, he did his postgraduate studies at the Moscow Institute of Physics and Technology

(MIPT, Zhukovsky) in the speciality area of strength and thermal conditions of aircraft. Since 2001, he has worked as a scientific researcher in TsAGI. His professional interests range from analytical-experimental investigations of static aeroelasticity, design and fabrication of special structures and equipment, and stiffness tests. He is author of about 20 publications presented at conferences or published in scientific journals.

Mihir Mistry (Chapter 24) earned his PhD in aerospace engineering from the Pennsylvania State University in University Park, Pennsylvania. His graduate research focus was on the application of smart structures to rotorcraft. During his PhD studies, he was funded as a National Defense Engineering and Science Graduate (NDSEG) fellow. After graduation, he worked at the Rensselaer Polytechnic Institute (RPI) in Troy, New York, as a visiting researcher focusing on various smart structures-related projects. Subsequently, he joined the Boeing Company in 2013 as a flying qualities engineer at their Ridley Park, Pennsylvania, location where his focus has primarily been in the area of modelling and simulation for various programs.

Christof Nagel (Chapter 7) earned his degree in physics at the University of Halle "Martin Luther" (Germany) in 1995. He was awarded his PhD in applied materials science with great honors at the University of Kiel "Christian Albrecht" (Germany) in 1999. Since 2000, he has worked as a researcher at the Materials Science and Engineering Division at the Fraunhofer Institute for Manufacturing Technology and Advanced Materials (IFAM), Bremen, Germany. His research fields are the strength, fatigue, and damage tolerance of adhesively bonded structures.

Maria Chiara Noviello (Chapter 15) was born in Cerreto Sannita, Italy, on January 6th, 1991. On April 7th, 2015, she graduated with honors in aerospace engineering at the University of Naples "Federico II" with a master's thesis about the experimental shape reconstruction of a morphing wing trailing edge prototype. She is currently a PhD candidate in industrial engineering. Her research is mainly focused on the aeroelastic stability assessment of adaptive lifting surfaces in nominal and failure conditions. Recent activities regard the control system design for a morphing wing flap demonstrator. She is author of three scientific papers presented at conferences.

Felipe Issamu Kitadani Odaguil (Chapter 6) was born in São Paulo, Brazil, on April 24, 1987. He graduated in mechatronics engineering from the University of São Paulo in 2010. In 2011, he joined the Engineering Specialization Program in a partnership between EMBRAER and the Aeronautical Institute of Technology in São José dos Campos. Since 2012, he has worked at EMBRAER as a product development engineer at the Multidisciplinary Design Optimization team with emphasis in external computational aerodynamics. In 2013, he completed his master's in aerospace engineering at the Aeronautical Institute of Technology.

Antonio Pagano (Chapter 25) earned his degree in aeronautics engineering at the University of Napoli "Federico II" (Italia) in 1990. Since then, he has been working at the Italian Aerospace Research Center (Italia), gaining more than 25 years of experience in the field of unsteady aerodynamics, mainly applied to the design/analysis of rotorcraft configurations (helicopters and tiltrotors) and propeller-driven aircraft. His main fields of interest are the development of CFD solvers for rotorcraft analyses, the design of software environments for code integration and multidisciplinary simulations, and the aerodynamic shape optimization of rotary wings. Presently, he is working for the Technologies Integration and Aeronautic Demonstrators department.

Damiano Pasini (Chapter 8) has been a faculty member of McGill University since 2004, and is now a professor of Mechanical Engineering. He received his master's in civil engineering (1995) and architectural engineering (2000) from the University of Pavia, a PhD (2003) in structural mechanics from the University of Bristol, and a Marie-Curie Postdoctoral Fellowship (2003–04) from the Institute of Structural Mechanics of DLR (German Aerospace Center) in Braunschweig. His research expertise revolves around mechanical metamaterials with specific focus on micro-manufactured porous materials, as well as fiber-steered laminates. He is fully engaged in understanding their mechanics, introducing reliable predictive models, and using them to develop manufactured materials with unprecedented properties. Besides being a member of the McGill Institute for Advanced Materials, the McGill Institute for Aerospace Engineering, and the Center for High Performance Polymer and Composite Systems, he holds cross-appointments in the departments of bioengineering and biomedical engineering at McGill. He serves on the editorial board of Scientific Reports and the Journal of Bionic Engineering.

Modesto Pecora (Chapter 14) was born in Salerno, Italy, on May the 2nd, 1949. He earned his degree in aeronautical engineering at the University of Naples "Federico II" in 1974. From 1975 to 1995, he worked at Aeritalia (currently known as Leonardo Aircraft); in the framework of the international cooperation between Aeritalia and Aerospatiale, he was responsible for ATR42 and ATR72 certification with reference to loads and flutter domains. He left Aeritalia in 1995 and joined CIRA (Italian Aerospace Research Center), where he was responsible for the loads and aeroelasticity laboratory until his retirement in September 2015. During his forty-year carrier, he has been a contract professor of aeroelasticity at the Second University of Naples (SUN) and held PhD courses on elastic aircraft dynamics at the University of Naples "Federico II"; as load and flutter specialist, he has been involved in the certification process of several general aviation (AP68TP, P68100, P68 Spartacus, VF600W, P166) and ultralight aircraft (Stiletto, Squalo, Takana, P92, P96, BS115). He is currently a consultant for Blackshape Aircraft S.p.A., where he is responsible for aeroelastic stability analyses and serves as a compliance verification engineer (CVE) for loads.

Fabian Peter (Chapter 3) earned his diploma in aerospace engineering at the RWTH Aachen University, Germany, in 2011. He worked as researcher at the Institute of Aerospace Systems of RWTH Aachen University in the field of systems design, overall aircraft assessment, and aircraft design software development until 2016. During this time, he was involved in several projects focusing on technology integration and assessment on an overall aircraft level (e.g., EU FP7 project SARISTU). Since 2016, he is the lead of the research focus area Systems and Aircraft Technologies at the Research Institute Bauhaus Luftfahrt e.V., in Munich, Germany.

Mikhail Pronin (Chapter 23) is the head of the Flutter Division at Strength Standards, Loads, and Aeroelasticity Department at Central Aerohydrodynamic Institute, named after Professor N.E. Zhukovsky (TsAGI). He graduated from Moscow Institute of Physics and Technology in 2010. He has been working at TsAGI in the Aeroelasticity and Structural Dynamics Department since 2008. Mikhail Pronin is responsible for flutter modeling in subsonic/transonic wind tunnels, GVT, and aircraft certification.

Francesco Rea (Chapter 10), born in Naples (Italy) on August 9th, 1990 and graduated with honors in aerospace engineering (MSc) on February 5th, 2015 with a master's thesis on the aeroelastic behavior of a morphing aileron. He is currently a PhD candidate in industrial engineering at the University of Naples

"Federico II." He is currently working on the structural design of morphing high-lift devices in compliance with industrial standards and applicable airworthiness requirements, covering all phases from the assessment of preliminary structural layout to full-scale ground static testing. He is author of three scientific papers presented at conferences.

Lorenzo Rossi (Chapter 20) Leonardo S.p.A., Aerostuctures Division—Product Engineering. Mechanical Engineer, PhD in chemical engineering, materials, and production, involved in Leonardo Aerostructures manufacturing research & development in the field of composite materials manufacturing and assembly processes for aeronautical components. He has been involved in European projects and many other national research activities in collaboration with major Italian companies and universities active in the aerospace field (JTI, CleanSky, LOCOMACHS, MAAXIMUS, SARISTU, ALCAS). He has been also involved in the development and support of major Leonardo production programs: ATR 787 (fuselage sections, Stabilizers), Boeing 787 (fuselage sections, circumferential frame and shear tie, horizontal stabilizers), Bombardier C Series (vertical and horizontal stabilizers), and Airbus A380 (fuselage sections).

Salvatore Russo (Chapter 20) Leonardo S.p.A., Aircraft Division—Structures Analysis and Methods Department. Education: Master of Science in Aeronautics Engineering, University of Naples "Federico II," Napoli, Italy. Collaborated on several national and European projects (e.g., SARISTU, IASS, COALESCE, ITOOl), with experience in numerical simulation, FE analysis, & mechanical tests of composite/metallic materials and structures. Experience includes 2002–04 work as a fuselage designer with the B787 Program structural analysis, 2004–06 as a R&D engineer for the Integrated Tool for Simulation of Textile Composites (ITOOL) project: technical & scientific responsibilities for Alenia Aeronautica. Other experience includes 2007–14 stress engineer, COALESCE project, technical & scientific responsibility for Alenia Aeronautica, Crash Analysis Innovative Technologies for Aeronautic Structures in hybrid composite/aluminum materials (SALINA) project, technical & scientific responsibilities for Alenia Aeronautica, design, development, and validation of new advanced composite material for regional fuselage aircraft wing innovative processes and structural architectures (ASIA) project, technical & scientific responsibilities for Alenia Aeronautica, 2014–16 Airframe Smart Intelligent Aircraft Structures (SARISTU) project, technical & scientific responsible for Alenia Aermacchi, and Advanced Methods for Certification and Verification (CERVIA) project, technical & scientific responsibilities for Alenia Aermacchi.

Giuseppe Sala (Chapter 9) graduated in 1982 in aeronautical engineering and he became a full professor in 2001. Presently, he lectures on aerospace technologies and materials I and II at the Department of Aerospace Science and Technology of Politecnico di Milano and science and engineering of materials at the doctoral school of the same university. He is director of the Aerospace Science and Technology Department and effective member of the Senato Accademico. He is the author of more than 150 printed papers in the field of advanced materials and technologies. Giuseppe Sala has taken part in research programs financed by MIUR, ASI, CNR, and the EU. Research activities were performed in cooperation with Italian and foreign universities (Atlanta, Chicago, London, Monaco, Stockolm, Helsinki, Tolose, Pechin, Shangai), research centers (CIRA, CISE, ENEL, JRC) and important companies operating in the high technology and innovative materials fields.

Andrey Saprykin (Chapter 23) earned his degree in hydroaerodynamics at the Moscow Aviation Institute (Technical State University). Since 2009, he has worked as an engineer of experimental research of the propulsion and propeller systems group at Central Aerohydrodynamic Institute (TsAGI), where he is currently the head of this group.

Oliver Schorsch (Chapter 7) earned his degree in chemistry at the University of Bremen, Germany, in 2001. He was awarded his PhD in chemistry with great honor at the University of Bremen, Germany, in 2004. Since 2005, he has worked as a researcher in the Adhesives and Polymer Chemistry Division at the Fraunhofer Institute for Manufacturing Technology and Advanced Materials (IFAM), Bremen, Germany. His research focuses on the development of elastomers for low-temperature applications and the development of structural adhesives.

Martin Schueller (Chapter 22), graduated in 2007 with a degree in engineering for microtechnology/mechatronics. He is currently finishing his PhD with a thesis on actuators for active flow control. He is employed as a research fellow at the Fraunhofer Institution for Electronic Nanosystems in Chemnitz, Germany, and leads the Flow Control Actuators and Systems group. His areas of expertise are microfluidics and actuators for active flow control. He contributed to several national (Germany: Cluster of Excellence MERGE) and international projects in the area of aeronautics (Clean Sky 1, AFLoNext, SARISTU, Clean Sky 2) and space (PBSA).

Sergey Shalaev (Chapter 23) is the principal specialist for the Dynamic Aeroelasticity Division within the Strength Standards, Loads and Aeroelasticity Department at Central Aerohydrodynamic Institute (TsAGI). He graduated from Moscow Aviation Technology Institute in 1970. He has been working at TsAGI in the Strength Standards, Loads and Aeroelasticity Department since 1970. Sergey Shalaev is responsible for flutter modelling of all types of flying vehicles and flutter tests in subsonic/transonic wind tunnels.

Stefan Storm (Chapter 26) received his diploma in physical engineering from the University of Applied Sciences of Munich, Germany, in 1993. Stefan fulfilled the mandatory community service requirement in the field of environmental awareness raising at Helmholtz Zentrum in Munich, Germany, from 1993 until 1995. Within the framework of the Bavarian sponsorship program BIT, he worked as a trainee at IABG in the Space Test Center in Ottobrunn, Germany, from 1995 until 1996. Since 1996, he has been employed as a research & development engineer in the Aeromechanics Systems Team at Airbus Group Innovations (formerly Daimler Benz Forschung) in Ottobrunn, Germany. There he has been involved in various projects in the field of active vibration control, morphing structures and ice protection systems in close connection with the business units airbus, helicopters, defense and space. Stefan presented dedicated results at various international conferences, published articles in numerous scientific papers, and has filed more than 30 patents on behalf of the Airbus Group.

Tobias Strobl (Chapter 3) graduated from the Technical University of Munich, Germany, in aerospace engineering in 2011. He was awarded his doctorate in aerospace engineering at the Technical University of Munich, Germany, in 2016. Tobias was working as a research & development engineer with Airbus Group Innovations in Ottobrunn, Germany from 2011 until 2015. Since 2015, he has worked as a consultant in product development and innovation management with 3DSE Management Consultants in Munich, Germany. Tobias is author of various scientific papers, has made numerous conference contributions, and filed a number of patents on behalf of the Airbus Group.

Eike Stumpf (Chapter 3) earned his diploma in aerospace engineering at the Technical University of Berlin, Germany, in 1996. He worked at BMW-Rolls-Royce Aeroengines and Airbus. At the German Aerospace Center (DLR), he started as a PhD student in 1997, and after receiving his PhD at RWTH Aachen University, he acted as assistant to Prof. Szodruch and head of the Technology Assessment & Scenario Technique Department. In 2011, he became Head of Institute and Chair of Aerospace Systems at RWTH Aachen University.

Afzal Suleman (Chapter 4) is a professor and Canada Research Chair in Aerospace Systems at the University of Victoria. He is currently the Director of the Center for Aerospace Research on Unmanned Air Systems at the University of Victoria, where he collaborates with Boeing USA, Bombardier Canada, and Department of Defense in Canada. He is also an adjunct faculty member at the Technical University of Lisbon (Instituto Superior Tecnico) in Lisbon, Portugal. The focus of Afzal's research is on multidisciplinary design optimization of next generation, environmentally efficient, and novel aircraft configurations. He leverages unmanned air systems technology for experimental validation and airworthiness evaluation. He was the coordinator of the EU-funded project on Novel Air Vehicle Configurations from fluttering Wings to Morphing Flight (NOVEMOR), and has participated in EU projects (ARTIMA, 3AS, SMORPH). He has authored/co-authored over 100 scientific journal publications and over 200 conference papers, and holds 2 patents. He has also supervised 30 PhD and 64 MSc theses in diverse areas such as aeroelasticity, aeroacoustics, multidisciplinary design optimization, advanced composite structures, unmanned air systems, and small satellites.

Viktor Timokhin (Chapter 23) graduated from Bauman Moscow State Technical University in 1969 and held several positions at Central Aerohydrodynamic Institute (TsAGI) and the Design Department of Mil Helicopter Plant. He is an author of a series of software packages on strength and aeroelasticity of aviation structures. He also is an advanced user of MSC software on static and dynamic structural analysis. Now he is a leading engineer at the Strength Standards, Loads and Aeroelasticity Department, TsAGI. He is author/co-author of more than 30 scientific papers, presented at conferences or published in specialized journals.

José Vale (Chapter 4) graduated in aerospace engineering from Instituto Superior Técnico, University of Lisboa in 2003, where he completed his MSc degree in morphing aircraft in 2007 and proceeded with his studies on morphing technologies assessment toward the completion of his PhD. He obtained his PhD degree in 2012. During his MSc and near the end of his PhD studies, he participated as a researcher in EU collaborative projects of the Eurocores S3T (project SMorph) and 7th Framework Programme (project NOVEMOR). Following his PhD completion, he continued his research in the scope of the NOVEMOR project on the subject of Novel and Morphing Aircraft Analysis and Optimization. He is currently a research associate in the Aerospace Research Group at IST, Lisbon, continuing the research and development of computational and experimental models for morphing technologies, Fluid-Structure Interaction (FSI) and Non-Linear Aeroelasticity (NLAE) of high aspect ratio wings. He is also involved as a researcher in a Portuguese National Funded Project (Project FLEXCRAFT) with focus on modular aircraft. Dr. Vale has published 8 papers in international peer-reviewed journals and 23 peer-reviewed conference proceedings.

Maurizio Verrastro (Chapter 21) earned his degree in aeronautics engineering at Politecnico di Torino University in 2007. He has worked since 2007 in an aeronautics environment covering different positions, from Testing and Quality employee to Logistic Support specialist. From 2010, he began work as a consultant for the Alenia System Safety department. From 2012, he became part of the Alenia Aermacchi (now Leonardo Aircraft Division) safety office and worked for both military A/Cs certification aspects and for research programs' safety activities.

Andrea Vigliotti (Chapter 8) received a master's degree in aerospace engineering from the University of Naples with a thesis on the active control of noise in enclosures. He also received a PhD in mechanical engineering from the McGill University (Montreal) with a thesis on the nonlinear homogenization of materials with a repetitive microstructure. He was recipient of the Newton International Fellowship, awarded by the Royal Society (UK). Since March 2015, he has been the head of the Innovative Materials Laboratory at the Italian Aerospace Research Centre in Capua (Italy).

William A. Welsh (Chapter 27) received his BS in mechanical engineering from the State University of Buffalo (New York) in 1975 and his MS from Cornell University in 1981. He works at Sikorsky, a Lockheed Martin company, in Stratford Connecticut, United States, as a senior technical fellow in the area of dynamics. Sikorsky is a manufacturer of helicopters including the UH-60 M Black Hawk and the S-76D and S-92 helicopters. During his career at Sikorsky, he has served as a member and supervisor of the dynamics group, which is responsible for helicopter vibration and stability. He has authored or co-authored 15 technical papers and is the holder of 24 patents. Mr. Welsh is an American Helicopter Society International Technical Fellow as well as a member of the American Institute of Aeronautics and Astronautics.

Matthew Wilbur (Chapter 24) has a BS in aeronautical engineering from the Embry-Riddle Aeronautical University in Daytona Beach, Florida, and a MS in structural engineering from the George Washington University in Washington, DC. He works as a researcher for the US Army Research Laboratory (ARL) Vehicle Technology Directorate at the NASA Langley Research Center in Hampton, Virginia. His duties include analytical investigation of advanced rotor system concepts, and the planning, preparation, and execution of wind-tunnel tests of aeroelastically similar, model-scale helicopter rotor systems in the Langley Transonic Dynamics Tunnel (TDT). Mr. Wilbur received the 2001 ARL Achievement Award for Engineering for his efforts to demonstrate the use of embedded piezoelectric-fiber actuators to impart active-twist control on helicopter rotor blades, a field of study that has proven application for rotorcraft vibration and noise reduction. He is the author of more than 60 technical papers.

Mikhail Zichenkov (Chapter 23) graduated from Moscow Institute of Physics and Technology and received his PhD in aerospace engineering in 1991. He has been working at TsAGI in Strength Standards, Loads and Aeroelasticity Department since 1979. Dr. Mikhail Zichenkov is responsible for the development and application of advanced methods in experimental aeroservoelasticity. He is an author of over 90 technical papers, presentations at scientific conferences, and reports in aeroelasticity and structural dynamics. He is a Member of the IFASD Scientific Committee.

Foreword 1

During several centuries, all interested enthusiasts who developed ideas for flying vehicles tried to copy nature while studying the flight of birds and hoped to achieve the dream of mankind "flying like the birds" by using flexible adaptive structures. So, morphing structures are as old as the dream of flying.

Sir John Cayley in the 18th century was the first to postulate the use of a fixed wing structure, to separate the forces of lift, drag, and thrust and use an independent thrust producer.

Most of the aircraft which are flying today are still using these principles from Cayley, i.e., a fixed wing component to produce lift, a separate engine (or several) to produce thrust, and a fuselage to accommodate the payload.

The development of the jet engine for civil aircraft application after World War II enabled commercial aircraft to fly higher and faster. But by optimizing the wing for the high subsonic cruise Mach numbers at around Ma 0.8, a new challenge appeared. Takeoff and landing conditions became more severe and afforded a much higher lift coefficient for takeoff and landing in order to keep the takeoff and landing speeds reasonable and manageable for pilots.

Besides a good aerodynamic performance in cruise conditions, the lift coefficient for takeoff and landing had to be increased by a factor of 4–5 in order to achieve the required good landing performances. These high lift coefficients for landing could only be achieved with very sophisticated high-lift systems, which had slats and flaps to increase the wing camber and wing surface and enable the air flow around to stay attached up to fairly high angles of attack. At that time, we were not yet speaking of morphing structures. It was the pure optimization of the aircraft operation in high- and low-speed conditions which required this additional complexity at the front and rear parts of the wing.

These movables for high lift had to be simple, robust, and reliable, and it was the task of the aircraft design engineers to ensure the operability of these movables in all weather conditions and within the defined operational flight envelope.

In the 1980s, Airbus was developing the A300–600 as a follow-on aircraft from the former A300 B4 by using some proven technologies from the A310 aircraft.

In this context, the design engineers realized that the camber of the wing could be slightly increased to improve cruise performance. This was easily feasible by adjusting the rigging of the flap for cruise flight conditions. During this design exercise, the aerodynamic specialists proposed moving the flap in small steps during the cruise flight phase to achieve an optimum L/D at any time in cruise flight. This concept was called "variable camber wing," and it was intended to move the flap during the cruise phase slightly (some tenths of a degree), further downwards or upwards depending on the actual aircraft weight.

This proposal has been investigated thoroughly as it promised another half to 1% improvement of fuel consumption in a 2000 nm flight. But this would have also meant adapting the flight control system and the corresponding control laws. After investigating all the critical failure cases, the program management decided not to introduce the "variable camber concept" and to stay with the existing high-lift design. This was the first discussion about whether to use a "morphing wing geometry" during the cruise phase.

The adaptable wing concept in cruise has been investigated several times since the 1980s in the Airbus design office. For the A330 and A340 program, where both aircraft have a nearly identical wing geometry, the variable camber concept has been studied intensively, as both aircraft types are operating at slightly different Cl during cruise (MTOW are quite different for both aircraft!)

Main arguments at that time (the beginning of the 1990s) were again about the additional complexity of the system architecture and additional maintenance cost for the operators. So, variable camber for adaptive wings was a subject of research and also an important point for the wing design for nearly all Airbus aircraft. But over the years, as the methods of predicting wing flexibility in cruise and cruise loading, and actual twist of the wing under 1 g loads, became much more accurate, the need for an adaptable camber control during cruise conditions became less obvious. The approximately 1% additional increase in cruise performance has never been dominant compared to the additional system complexity and the related increased maintenance cost.

The discussion with the operator has often confirmed the decision of the aircraft manufacturer to keep the wing structure fixed and the necessary high-lift system as simple as possible.

Today, a lot of new materials are being developed. Also, sensor and control technologies are constantly improving, which can be seen in the miniaturization and integration of sensor and controllers into very small units. These new technologies are opening up the discussion of reconsidering and using morphing structures in the aircraft design.

The following articles in this book provide a very good and vast overview about all different research aspects for morphing structures like aerodynamic requirements, new materials, new actuators with robust and precise positioning of wing movables and surfaces, new mechanical deformation architectures, new system architectures to optimize aircraft operation during all flight phases, and manufacturing aspects. Certification issues will be the last element in the development phase.

The research field for morphing structures is very wide, and a lot of new technologies will be required and are available today to convince the industry aircraft design engineers to adopt new concepts for their next new aircraft design. The performance gains have to be very clearly demonstrated in wind tunnel tests, ground tests, and finally, flight tests, and the overall benefit of morphing concepts has to include the additional complexity, weight, and cost issues of the system.

I believe that there is now a good chance to develop the morphing structures in such a way that the additional weight and complexity are small compared to the potential gain in performance, and that this technology will have a chance to be integrated into future new aircraft designs.

Dieter Schmitt

Former Vice-President, Research and Technology at Airbus in Toulouse,
Toulouse, France

Foreword 2

Morphing studies have fascinated me since the very first moment I heard about them; that interest grew thanks to the enthusiasm of some colleagues of mine with whom I discussed the matter, not only for the early time, but also for the undeniable charm of the topic. Morphing philosophy aims at using the most recent technological attainments to build complex structures which are able to perceive what is happening around or within them, and modify their behavior accordingly.

In aeronautics, morphing solutions will make aircraft increasingly similar to birds, more than they have ever been. This new approach will realize the dream of wings that can change their geometry as a function, for instance, of the inner stress field, thus achieving optimal efficiency and the highest safety level, as a beautiful bird can do instinctively.

The *artifex* Daedalus built wax wings for himself and his son, Icarus, to escape the gilded prison of the Cretan labyrinth, with the terrible outcomes the myth tells us. Leonardo picked up the challenge, as did many others throughout history, "[...] che per queste ragioni potrai conoscere l'uomo con le sue congegnate e grandi ale, facendo forza contro alla resistente aria e vincendo, poterla soggiogare e levarsi sopra di lei [...]" (for these reasons, you will be able to know the man who wears large and elaborated wings, to overcome the air drag and finally rise upon it).

I recognize a unique engineering power in the capability to integrate innovative materials, distributed actuator and sensor networks, response and analysis models, and software into one single application. In my opinion, however, "morphing" is more a philosophy than a technology; it brings to mind non-static solutions from mechanical and physical behavior, solutions that are able to dynamically overcome variable external conditions. In fact, this approach may apply not only to aircraft, but also to many other mechanical and general products. I wonder how people's lives could change if many commonly used objects were designed and realized from a morphing perspective. Moving from this consideration, I asked the colleagues of mine that work on this subject in aeronautics to organize a congress and set up a continuous forum for meeting and discussing with other specialists of different educational disciplines, for instance by founding an open scientific association, focused on topics like "a morphing life." According to my ideas, this could be an important accelerator for exporting this approach to many fields of knowledge.

In the meantime, by proposing some of the most important recent scientific investigations in aeronautics worldwide, this volume is already a valuable contribution to disseminate and spread this new technological approach. The colleagues that are herein involved in the preparation of the single chapters are some of the major international experts on the addressed topics. The simultaneous presence of researchers that operate inside universities, research centers, and industries, ensure a vision merging knowledge and innovation capability, essential nowadays for technology development and advancement. The articles are mainly dedicated to the aeronautic segment, but approaches and methods for solving selected practical problems are applicable to other sectors, too. The introductory chapter, dedicated to the history of morphing techniques and their current and future evolution is important for understanding the long maturation of this concept and its continuous effort to mimic nature (biomimicry). Nature always does It better than artificial human efforts! The technical chapters are a progressive and exhaustive guide to every single aspect related to the adaptive structures: design methods and numerical simulations, materials, sensors, actuators, some relevant applications (case studies), etc.

A wonderful synthesis appears in the section dedicated to the systems where everything is combined and melded: therein, logic and matter are each at the service of the other.

As president of the Aerospace District of Campania and former president of the Italian Aerospace Research Center, I cannot but underline with full satisfaction the strong presence of professionals who operate in my own country and region. With their own work, they demonstrate the extraordinary potential of knowledge and innovation herein available, together with the strong international network they have created.

My wholehearted thanks go to all the editors and authors of this book; I had the pleasure to work with many of them, and I truly appreciate that they asked me to draw up these short and non-exhaustive considerations. I am sure the portion of future that is in their hands will be better than the current time.

Luigi Carrino
President, Aerospace District of Campania, Capua (CE), Italy

Preface

As we look backward in time, what has been realized sounds like something special: 80 people from universities, research centers, and industries throughout the world, including Europe, Russia, and North and South America, worked together on this project; this not a common thing. No more than 5 years ago, morphing wing technology appeared strange, even weird to most. It is trivial, then, to conclude that those who cooperated on the matter felt that a dramatic change had been happening in the engineering approach to aircraft design. Further, they felt altogether that the ongoing transformation had been leading to applicable results, even from an industrial realization point of view. In the aeronautical field, airworthiness authorities demand a huge process of qualification, standardization, and verification. Nothing can leave the ground and fly without a deep investigation of any possible failure scenario, error, or other events that can result in unsafe operating conditions. Therefore, the idea that creating aircraft wings that could change their shape while bearing the usual operational loads seemed available within the limits of the technology that is accessible or will be accessible in the very next future was, itself, noteworthy. The simple awareness of this idea should have implied a revolution of the minds and appeared simply astonishing in itself.

A consideration that appears even more surprising is that engineers and scientists have worked on the matter herein presented for a long time, since the early stages of aircraft evolution. If we observe the first aircraft concepts, we realize that the wing was not thought of—in principle—as a rigid element with an immutable shape. Later, it was forced to be conceived of and constructed as a stiff structure, due to the increasingly challenging flight speed requirements and consequently increasing aerodynamic loads. Nevertheless, this immediately appeared to be a clear limitation on generic aircraft operations: an aircraft would never have landed, taken off, or been controlled in flight with a fixed-geometry wing! The need for flaps, slats, aerodynamic brakes, ailerons, and all the other movable surfaces placed all along the wing, was related to that reality. The structural system became a disordered collection of different subsystems, in turn becoming more and more complex from the design, realization, assembly, maintenance, and control points of view. That was the price to be paid for the inability to construct a wing that could change its shape, depending on the current necessity. Among the few, even if outstanding, examples that tried to partially overcome those limitations were the F-14 "Tomcat" and the Concorde (70s–80s). It is remarkable that scientists and aeronautical engineers, irrespective of actual technological status, never stopped aiming at such a dream. Patents, research, and simple ideas had continuously enriched the bibliography from the Wrights' flyer to present day, and, presumably, will continue doing so in the future. This research was also accompanied by some remarkable experiences, such as the F111 variable shape wing in the 1980s.

It was only around 2000, however, that morphing wing technology had a push forward, related to the discovery (or, better, the "re-discovery") of so-called "smart materials." They were intrinsic transformers of molecular energy into mechanical energy and promised to enable the creation of extremely compact and powerful actuators. This perspective gave a dramatic impulse to research in the field. Ironically, as science history shows, it is currently understood that traditional technology is definitely able to realize shape-changing wings without expecting the "new breakthrough material technology" to reach satisfactory readiness levels. This point has been widely demonstrated in research programs like Mission Adaptive Compliant Wing (MACW), Smart Intelligent Aircraft Structures (SARISTU), and the most recent CleanSky2. Those achievements, and many others that have been widely cited in this book, demonstrate how the world's diffuse knowledge finally allowed scientists to approach the design of a "morphing wing."

This book proposes a logical scheme that simulates the realization of a morphing wing mainly intended for large commercial aircraft and rotorcraft, including nothing but assessed processes and tools.

Moving from a historical survey that shows how the idea of creating adaptive aerodynamic surfaces has always accompanied aircraft design since the very first flight, the book starts with a section (Section 1) dedicated to the current worldwide status of morphing technologies, industrial development expectations, and what is already available in terms of "flying systems." This short and basic premise paves the way to presentation of an engineering approach to properly deal with morphing wing technology. In order to correctly design an adaptive wing system, the requirements should be defined in terms of reference aircraft and desired targets, which must then be translated into applicable shape variations of the wing sections. In this way, the geometrical needs are emphasized, namely indicating the wing region most likely to be affected by morphing, and the expected shape modifications. On the basis of the achieved results, and making reference to the envisaged structural architectures, further optimization can be performed, sketching the layout of the structural arrangements enabling shape variation (Section 2). As an adaptive skin realization, one of the major challenges in properly completing a morphing structure is finding different materials or architectural concepts that can be used to match the specifications, and these difficulties are covered; in detail, polymeric, composites, and alternative approaches are presented. These offer potential solutions for those applications requiring anisotropic stiffness for bending and in-plane deformation, without suffering large out-of-plane displacements under aerodynamic and inertial loads. The viability of, and prospects for, some of the most promising applications are also given by highlighting the most critical factors affecting their performance (Section 3).

The next step is to break down the reference system structure into its main subsystems, basically meaning the sensor, actuation, and control architectures. The issues related to installation of such active components in conventional wing structures, both metal and composite, are discussed, along with issues associated with the number of required channels, the installation process, and interface management (Section 4). A certain prospect of the devices that are available on the market is presented, favoring the ones particularly suitable to integration purposes and those that exhibit desirable properties like compactness, lightness, and reliability, and likelihood of certification. Adaptive structures spawn completely new paradigms concerning modeling strategies and aeroelastic approaches. For instance, they undergo large displacement fields that, in turn, give rise to peculiar deformed shapes while maintaining classical limitations for strain and stress values, and display many more degrees of freedom, producing increased modal density. These characteristics affect and modify many aspects of the traditional aeronautical engineering process, like simulation activity, design criteria assessment, and interpretation of the dynamic response, including the use of codes and elaboration on the relative output (Section 5). Having established guidelines for the design of these devices, based upon the individuation of the main components, the assessment of the respective interfaces, and the formalization of the relative project processes, a section dedicated to the practical realization of selected morphing devices is introduced. This part incorporates both structural and aerodynamic studies, and proposes some of the most promising and potentially flyable solutions for performance improvement of large commercial aircraft. An adaptive leading edge, a trailing edge, and a winglet device are herein considered subcomponents of a generic morphing wing (Section 6).

Next, the problem of moving to the real manufacture of an adaptive structural system is faced. That step is not trivial and, once more, introduces novelties into classical industrial practices. For instance, a kinematic system needs a number of checks during its assembly, in order to avoid the insurgence of residual stresses. Furthermore, a variable architecture implies the use of a large number of parts, with all the related issues connected to safety, maintenance, operation, and so on. Manufacture and assembly processes, verification procedures, and fault & hazard analysis (FHA) are performed as the basis for application of these devices to real aircraft. A specific section is then dedicated to exploration of the most relevant and common difficulties in moving from CAD to real morphing system handcraft, and the necessary experimentation process for validating their behavior (Section 7). About this latter point, a large and exhaustive chapter is completely devoted to the main results achieved during different wind tunnel test campaigns on full-scale demonstrators. Having dealt with the airplane characteristics and problems, the final unit aims to disclose the status of these technologies with reference to rotary wing systems. It wants to answer the possible question, "What about the helicopters?" Moving from an industrial view of the current perspectives on implementation of morphing technology onto rotorcraft, a study is presented on the envisaged application of adaptive blades for tilting rotor machines. Other aspects concerning the current use of smart structures and their impact on de-icing and the active noise and vibration control are discussed in the final section (Section 8).

The Editors-in-Chief:
Antonio Concilio
Ignazio Dimino
Leonardo Lecce
Rosario Pecora

SECTION 1

Introduction

Editor:

Leonardo Lecce
University of Naples "Federico II", Naples, Italy

List of Chapters

CHAPTER

1 HISTORICAL BACKGROUND AND CURRENT SCENARIO

Antonio Concilio*, Leonardo Lecce†

The Italian Aerospace Research Centre, CIRA SCpA, Capua (CE), Italy University of Naples "Federico II", Naples, Italy†*

CHAPTER OUTLINE

Morphing Wing Technologies. https://doi.org/10.1016/B978-0-08-100964-2.00001-0

NOMENCLATURE

ACARE	Advisory Council for Aeronautics Research in Europe
ADIF	Adaptiver Flügel (adaptive wing)
AFRL	Air Force Research Laboratory (United States)
AG2	Air Green 2
AS	aerospace
ATED	adaptive trailing edge device
ATM	air traffic management
CFD	computational fluid dynamics
CHANGE	combined morphing assessment SW using flight envelope data and mission based morphing prototype wing development
CIRA	Centro Italiano Ricerche Aerospaziali (I) (Italian Aerospace Research Center)
CRIAQ	Consortium de Recherche et Innovation Synergétiques en Aérospatiale (Quebec, CAN) (Consortium for Aerospace Multidisciplinary Research and Innovation)
CSIR	Council for Scientific and Industrial Research (SA)
DARPA	Defense Advanced Research Projects Agency (United States)
DGLR	Deutsche Gesellschaft für Luft-und Raumfahrt (D) (German Agency for Aeronautical and Aerospace Research)
DLR	Deutsche Luft-und Raumfahrt (D) (German Aeronautic and Aerospace)
DOF	degree of freedom
EADS	European Aeronautic Defence and Space Company
EMBRAER	Empresa Brasileira de Aeronautica (BR) (Aeronautic Brazilian Industry)
EREA	Association of European Research Establishments in Aeronautics
ESB	expansion shoulder bump

ETS	École de Technologie Supérieure (Quebec, CAN) (High Technology School)
EU	European Union
EUROMART	European Measures on Aerospace Research and Technology
FBG	fibre bragg's grating
FHA	failure harshness analysis
FFA	Flygtekniska Försöksanstalten (S) (Swedish Institute for Aeronautic Research)
FO	fibre optics
FOI	Totalförsvarets Forskningsinstitut (S) (Swedish Defence Research Agency)
GRA	Green Regional Aircraft
IMG4	Industry Management Group 4 (merging EUROMART, Engine, Equipment, ATM)
INASCO	Integrated Aerospace Sciences Corporation
INCAS	Institutul National de Cercetari Aerospatiale (ROM) (National Institute for Aerospace Research)
INTA	Instituto Nacional de Técnica Aeroespacial (E) (National Institute for Aerospace Technology)
IST	Instituto Superior Técnico (P) (High Technological Institute)
JTI	joint technology initiative
JW	joint wing
KVE	Kok & Van Engelen (NL)
MAS	morphing aircraft structures
MACW	mission adaptive compliant wing
MAW	mission adaptive wing
MBB	Messerschmitt-Bölkow-Blohm (D)
MDOF	multiple degrees of freedom
MSMA	magnetic shape memory alloys
NASA	National Aeronautics and Space Administration (United States)
NOVEMOR	Novel Air Vehicle Configurations: From Fluttering Wings to Morphing Flight
NRC	National Research Council (CAN)
ONERA	The Office National d'Etudes et de Recherches Aérospatiale (F) (French National Aerospace Research Centre)
PZT	Pb-Zr-Ti materials (lead zirconium titanate; piezoelectric materials)
RBM	root bending moment
RSB	rear shoulder bump
SADE	Smart High Lift Devices for Next Generation Wings
SARISTU	Smart Intelligent Aircraft Structures
SDOF	single degree of freedom
SFWA	smart fixed wing aircraft
SMA	shape memory alloys
SME	small medium enterprise
TDT	Transonic Dynamics Tunnel (United States)
TRL	technology readiness level
ЦАГИ (TSAGI)	Центральный аэрогидродинамический институт (RUS) (Central Aerohydrodynamic Institute)
UAV	unmanned aerial vehicle
UCAV	unmanned combat aerial vehicle
VCCTEF	variable camber continuous trailing edge flap
WT	wind tunnel

1 INTRODUCTION

Morphing is a recurrent word in the aeronautical and aerospace (AS) scenario, as by 2016. In the scientific bibliography, a lot has been written about this topic and its related technologies. In this introduction, we try to assess and define the fundamentals of the concept with specific reference to aeronautics.

Morph is a word deriving from ancient Greek, μορφή or "morfé," and indicates shape, figure, or other synonyms. Therefore, morphing is something related to shaping, the modification of the original aspect of a certain entity. When used in aeronautics, it is mainly referred to changes in the wing configuration. In general, morphing is one of the branches of the novel technological discipline of the "smart structures," meaning to release fully integrated devices into a structural architecture so to give that system, additive capabilities to adapt to external changes or variation of needs while preserving the classical characteristic of "bearing loads." Often, smart structures are often confused with "adaptive structures" or "intelligent structures." Truly, there are some differences in the three concepts and a sort of hierarchy can be drafted. All the structures that can be allowed to modify themselves, in any of their characteristic feature, is said to be "adaptive." Classical aircraft systems examples are the wing of the F14 "Tomcat" [1], able to alter their dihedral, or the nose of the Concorde [2], able to rotate around the pitch axis. Among them, the ones that allow fully integrated actuators and sensors systems, can be called "smart." Therefore, smart herein indicates the full integration of the active devices inside the structural skeleton. If a smart or even an adaptive structure is provided with a sort of "intelligence" that we can herein define as the capability of processing the information coming from the outside (sensors) and the inside (background knowledge, preprogrammed or acquired), the structure can be called "intelligent." Graphically, the concept is represented in Fig. 1.

Morphing structural systems may then be simply adaptive or smart or even intelligent, depending on their intrinsic capabilities and even aspect. In this book, smart intelligent structures are basically referred to as the active elements (actuator and sensor systems) are fully integrated in the structural body and connected to functions for elaborating the signals caught from the environment, combining them with the basic instructions, and transmitting the resultant information as commands to properly modify its current characteristics.

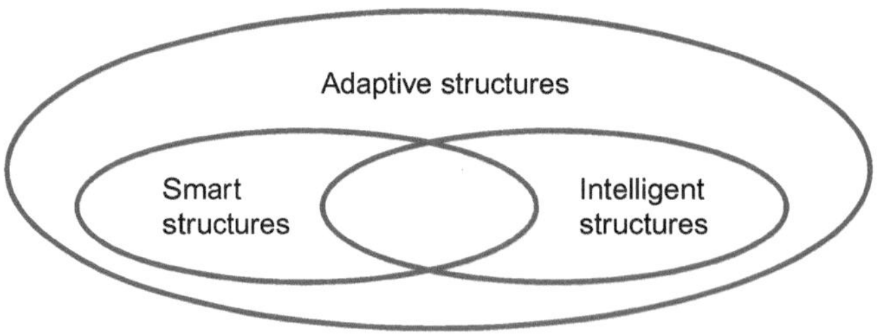

FIG. 1

Conceptual representation of adaptive, smart, intelligent structures domain.

Dealing with commercial transport airplanes, the wing is the structural component most sensitive for flight performance. The wing shall support the aircraft mission in each condition: take-off, landing, cruise, maneuver, approach, climb, (just to cite some of them) … and even taxing, when they are a sort of cumbersome and dangerous ballast for the vehicle. Furthermore, the aircraft may travel at full or almost empty weight, at high or low altitude, at high or low speed, into very different environmental conditions (depending for instance on the latitude, the season, and the climate). Yet, the aircraft is designed with reference to a single condition (the design point), individuated as the best match among all the possible missions it is envisaged to run. The best design translates itself into the optimal geometrical shape, governing aerodynamics. It is trivial to assume that a wing that could modify its form to face at the highest extent the different conditions that can be met during its mission, could improve the performance of the aircraft significantly. On the other side, nature knows very well this problem, having given birds remarkable adaptability. In Fig. 2 [3–8], a number of very nice pictures clearly illustrate some of the different phases of flight, with great emphasis on the extremely different wing configurations. Going deeper into details, furthermore, we easily realize a single mission (for instance, diving) is not associated to a single condition, simply because it can occur with different modalities (velocity, angle) and even different external situations (lateral or front wind, temperature and so on). Then, the conclusion may be drawn that even for a single referred standard mission, a set of geometries (or a geometry envelope) should be better referred to.

FIG. 2

Bird Flight (from the top left to the right): loitering [3], approach (1) [4], diving [5], landing [6], approach (2) [7], take-off [8].

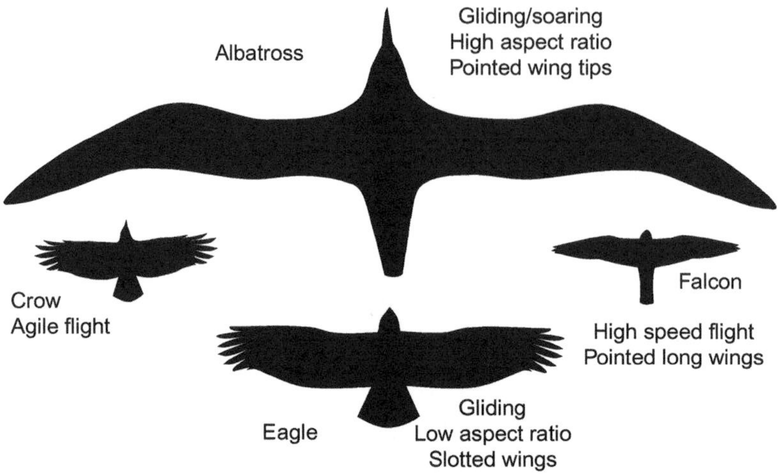

FIG. 3

Wing shapes for different bird species and their relative aim [9].

However, with this great adaptability, the solution is not universal as the many species of birds witness, Fig. 3. This again arises from another simple consideration that an even adaptable wing could not embrace all the possible conditions, including bearable weight. The resulting structural systems shall be different, both in terms of mass (linked to extension and stability) and rigidity (linked to the necessity of guaranteeing a prescribed shape under the already cited external and internal loads) while damping can be considered constant for similar architectures. It should be accepted that an adaptive wing itself would maintain a limited design envelope, anyway significantly larger than a single design point. The fact is different conditions vary for a quantity of parameters that is impossible to think a unique solution for all. As a concluding remark, we can state that morphing is a means to achieve a specific answer to certain wing aerodynamic performance increase, regarding a defined target. For the process to be properly accomplished, the specifications shall be assessed exhaustively in order that this innovative approach may address the technologist towards a very powerful solution.

It is worthwhile to discuss the generated load, perhaps the most important parameter, structurally speaking. Generated load is associated with aerodynamic and inertial forces. The different configurations should be able to sustain external and internal stresses while modifying their external aspect. This leads to "the paradox of morphing structures": maintaining bearing property with minimal deviations in terms of displacement from the original configuration, while changing its shape (in other words, ensuring large deformability characteristics) [10]. The solution to this paradox, reconciling these two conflicting requirements, is key to choosing the structural layout.

2 COMPONENTS OF A WING MORPHING STRUCTURAL SYSTEM

This book deals with wing morphing structural systems. In detail, these complex devices are applied to wings, enabling limited changes of their geometry and preserving their capability to bear external and internal loads. The morphing system should be able to deform itself, as it is sufficiently rigid to resist

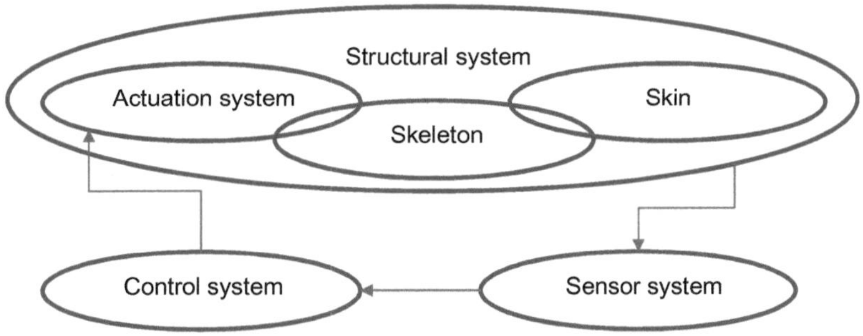

FIG. 4

Schematics of a generic adaptive structure.

the action of several forces. Of course, these general principles can be specialized in many ways when the real application occurs. Here, we try to define the main guidelines in detail, and although a generic morphing system can be specialized when its application is entirely defined, five main components are identified:

- Structural skeleton.
- Actuator system.
- Skin.
- Sensor system.
- Control system.

A typical scheme of a morphing structure may be seen in Fig. 4, where the main elements are reported with their logical collocation and links.

2.1 STRUCTURAL SKELETON

The structural skeleton is the body of the system and is the part that shall sustain the upcoming load. A solution that is generally adopted to overcome the cited paradox of smart structures, is to rely on the actuator system to absorb the incoming forces. The logic is simple. For the structure to become deformable, a kinetic chain or a quasikinetic chain is defined. Usually, in the applications that are herein presented, "quasi" means that the skin concurs to the structural characteristic of the system [11]. Because of that, the main role of the structural skeleton is to create suitable transmission patterns for the stress and strain field, so that it can be distributed uniformly inside the system, alleviating the actuator load. The structural skeleton also preserves the shape of the system for any deformed configurations. It hosts the actuation system, completed with its kinematics, and is designed to respect the connections with the rest of the structure. An alternative configuration is based on compliant architectures, which are aimed at giving rise to continuous deformation fields under the action of loads. Even in this case, to minimize the strain and stress field peaks inside the material, it is necessary to arrange the layout in such a way the deformation is uniformly distributed within the different components as the structure is conceived as a continuous domain instead of a set of blocks. The same fact, however, leads to other drawbacks, like a significant larger flexibility.

2.2 ACTUATION SYSTEMS

Following these assumptions, the actuation system plays different roles in the set-up of the morphing device. It is connected to the structural skeleton and, in general, it has a structural role. In fact, if the structural skeleton is a pure kinematic chain, all the loads are transferred through its actuation arms. This is nothing new, as the usual flaps, slats, ailerons, and so on are working on this principle. The real novelty is that the system is fully integrated within the main body and, because it must guarantee complex deformed shapes, with some continuity, it is made of several actuators. Now this is a conflicting point. On one hand, it is necessary to keep a large number for a better load distribution and a proper actuation. On the other hand, safety and maintenance cost requirements would keep that number low to minimize overall impact. This point is currently open and represents one of the most important design challenges. We should assume a typical mean pressure load is around thousands N/sqm, which in turn leads to, for usual dimension of adaptive devices, momentum per unit of surface equal to hundreds N/m. These are the target values the actuation devices should accomplish and can greatly increase when the camber or attitude changes. Again, it is worthwhile to consider that the system should bear the current resultant load without giving rise to excessive aeroelastic deformations, which could make the applied displacement field vanish.

From an operational point of view, we should clearly point out that the system, aimed at bearing the operational loads, is made of both the skeleton and the actuation chains. In fact, the whole collection of their components absorbs the generated forces and make up the stress space domain [12]. Although it is opportunistic to separate the skeleton from the actuation system design, the two parts should be optimized together in the final refinement, for instance to minimize the weight.

2.3 SKIN

Another component that also has a structural functionality but is usually designed apart, is the skin [12]. The skin needs to face extremely high strains (magnitude of 1% to 10% for typical configurations and typical structures), because of dealing with morphing and variable-shape structures, while withstanding the operational loads (again, the smart structures paradox comes back), in order to ensure a smooth profile to the complete system. The ideal skin has a zero-elastic modulus in the tangential plane with respect to its deployment and an infinite resistance along its local normal direction. Speaking of aeronautical structures, it must satisfy typical demanding environmental requirements, such as being exposed to temperatures varying between −50°C and 80°C and to assure a constant behavior along a significant number of cycles.

Although current literature presents some concepts that do not have the use of external skin, leaving the skeleton naked and in the form of a sort of "snake skeleton", these solutions do not seem viable from an industrial point of view. The common trend is reducing the noise emission from gaps (for instance, between flaps and the main body) and the associated efficiency reduction. Skin characteristics have a direct impact on the sensor system. In fact its nature may determine if a direct or indirect measurement may take place. The peculiarities of its architecture could in fact create obstacles to sensors installations. Moreover, the large occurring strain level is the point that mainly determines the sensor network layout. The cited strain levels are hardly recoverable with standard devices and the use of some architectural modulation or special modulation devices may be necessary. Of course, where applicable, the first is the most convenient, while the second brings with itself a cost increment. In Fig. 5 [13,15,16],

FIG. 5

A pictorial version of the mimetic robot, shown in the movie "Terminator2" [13,14] and some fantasy representation of living metal [15,16].

a pictorial version of the idea realized in the very famous science fiction movie "Terminator 2" [14] is shown, where a robot shows the clear capability in adapting its body and its skin covering to different situations, exploiting the idea of a sort of "living metal" or "mimetic alloy."

2.3.1 Sensor system

As in the case of the actuator system, a sensor system for integrated adaptive (smart) structures should be completely embodied. This aspect may be less dramatic than in the case of actuators, because of the current availability of many miniaturized and performant sensors (also developed per the MEMS [17] philosophy, for instance). Indeed, this design phase poses a challenge because of the large number of devices necessary to monitor the reference domain, and the need to arrange them all so they are accessible for maintenance. The large number of elements required leads in turn to a severity for the cabling and the cabling route; when the different weight contributions are considered, it is not guaranteed that the actuator system is the heaviest. The necessity of having a large sensor network, both in terms of extension and sensible points number gives rise to interesting specific fall-out researches, ranging from the use of multiplexed fiber optics [18] to wireless systems [19]. The first devices use common Fiber Bragg Gratings (FBG) [20] as an alternative to the standard strain gauges (a single fiber may typically host 10 FBG and more). However, curvature limits in the deployment of such devices and other related aspects, limit the use of this technology. The second option hides difficulties as well: the network needs an "invisible" path to transmit the information (light travels along straight lines), sensors should be fed (and the solution is again the use of cables) and so on. This latter point may be overcome with "energy harvesting" technology solutions, adding complexities to the complete architecture. An interesting platform that is taking increasing attention is the use of Rayleigh or Brillouin fiber optic sensors [21,22], that allow placing a virtual infinite number of sensors over a single fiber (these techniques are always referred to as "distributed sensing"). Again, drawbacks are also present in this case, but interesting potential is arising.

2.4 CONTROL SYSTEM

The control system integrates information coming from the sensor network and arranges it to be suitable for the actuator system. However, some difficulties may arise from response time as the control system has to interface with several inputs and outputs (usually, inputs are much larger than outputs).

A 50 Hz bandwidth may be sufficient for most of the possible problems (typical control or maneuver surface response below 20 Hz) for aeronautical structural systems that this book refers to (large commercial aircraft, business, and general aviation aircraft). Both feedforward and feedback architectures should be considered. The first envisages the use of smart systems like added traditional ones, demanding a continuous or stepped intervention of the pilot. The other performs better but requires continuous acquisition from the sensor network, elaboration of data and commands to the actuator system. For large, high-speed systems, this can in turn lead to instability problems that can be limited by using well-assessed techniques, however deteriorating the performance levels [23,24].

2.5 CABLING

As mentioned earlier, both actuator and sensor systems bring with themselves a major complication related to cables [25]. For their envisaged extension (overall length), the associated weight is not negligible and may be of the same magnitude of the related systems. Surely, sensor cables may be heavier than the sensors in themselves. Weight, however, is not the larger part of the problem. A major complication comes from their deployment and, therefore, the necessity of foreseeing appropriate paths. A complication comes from the fact that the structure is morphing, then deformable and experiencing large displacement. The placement should consider the possibility that cables can be truncated, after the action of the movable structural system and, then, apart from the traditional cure in fixating the wires, additional attention should be paid in preserving additional mobility and avoiding any contact with the mobile parts. A schematic and qualitative vision of the expected relative importance of the different components (namely, structure, actuator and sensor system, skin, and cables) is shown in Fig. 6.

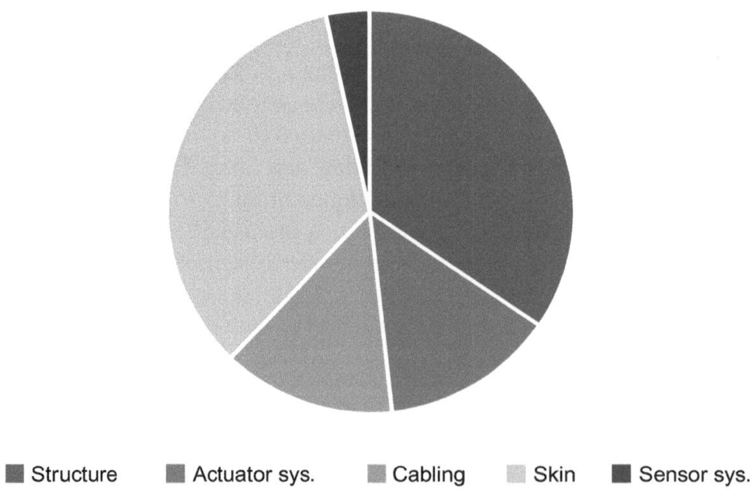

FIG. 6

Relative importance of the components weight in adaptive structures (qualitative representation).

FIG. 7

An adaptive structural system like a mechanical machine.

2.6 ASSEMBLY

Finally, once the different components are defined and designed, they should be assembled. This is not a trivial task because of the many items involved in the process [26]. They are also deformable, and are not easily handled. Besides, because additional complexities are added to the structural system through different devices, maintenance is clearly more sophisticated. Special and added inspection windows should be inserted into the construction, to make sure that any additive and active element can be accessible. Therefore, special tools and rigs should be considered as part of the design process. For the structural system, the assembly process is innovative and should be clearly suited, as it is intuitive, for each considered configuration. Once again, it is found that a specific morphing structure responds to a specific problem and undergoes a specific assembly route. On the other side, it is realized that, having transformed a typical load-bearing, static structure into something that modifies itself in its shape and characteristics while maintaining its original duty to sustain the action of external and internal forces, the structural system has been modified into a mechanical machine, Fig. 7.

3 THE MAIN CHALLENGES

3.1 SKINS

In the aeronautical field, shape is the most important aspect, being directly linked to aerodynamic performance. In fact, aerodynamics is strongly influenced by geometry, either from a macroscopic (system aspect) or a microscopic (surface roughness, for instance) point of view. This is why skin occupies a major place in the components that are crucial for the set-up of any aeronautical morphing system. It has the twofold objective to preserve the inner structural devices and maintain an adequate smoothness as morphing occurs. This target shall be accomplished in the operative environment, ensuring a constant behavior over the prescribed range while resisting over enough cycles, so to not dramatically affect maintenance and operability costs. A material that can be a suitable candidate for this scope shall also prove its resistance to chemical attackers, like skydrol, the aeronautic benchmark. Some other issues should also be considered, specific for different parts of the aircraft. If leading edges are spoken about, for instance, bird impact resistance should be taken into account; other upper skin parts should also be strong enough to meet hail impact endurance requirements and so on. A simple geometrical consideration is that for ideal morphing, skin deformation should occur just in the desired direction or, in each direction according to a specific input. This in turn would mean Poisson's module of the ideal skin morphing material is 0. Last but not least, the candidate material should show an excellent workability to match the demanding smoothness requirements.

From these statements, it cannot be a surprise that the choice and the realization of a suitable skin is not trivial. Foil metals cannot be considered because they cannot be compliant with the necessary deformations. Analogously, carbon composites, arranged through traditional layouts, should also be excluded because of their very high stiffness. Fiberglass-based materials present some interesting potentialities, yet maintaining the classical drawbacks of a significant in-plane stiffness. Rubbers and foams are instead very sensible to temperature variations and undergo important aging phenomena. Related fatigue life is still a nonexplored matter for the topics of direct interest, herein. So, the conclusion is pretty simple: skin materials for morphing applications should explore nonconventional architectures. In literature, there are so many examples of exotic proposals, ranging from auxetic [27,28] to chiral [29–31], Fig. 8 [30,31], to micro-structured materials [32,33]. However, none of the proposed solutions can be said to be completely acceptable for the final implementation. The application of smart materials could somehow reduce the problem. shape memory alloys (SMA) like NiTi shows the peculiar capability of nonpresenting "residual stains" even if they are solicited to very high strains (larger than 10%!), per their super- or pseudo-elastic properties [34]. In this case, however, two drawbacks arise: the high mean Young modulus (close to the aluminium one), that would increase the actuation morphing forces, and the fatigue life which is not so long for standard architectures. Indeed, a possible driver can be related to the ideal performance: an optimal skin should present zero normal, in plane resistance (morphing plane) and infinite resistance in the orthogonal direction (the load direction). Then, the choice should be addressed toward composite materials, arranged into a very particular way. The solution to this problem, simply stated, has not been found yet.

Once again, an easy prophecy is to state that, whatever the solution, it will be targeted to a specific application and will not be applicable for any request. Unless the skin itself has "adaptable" characteristics and can "morph" its inner nature (once again, we are speaking about T1000 skin).

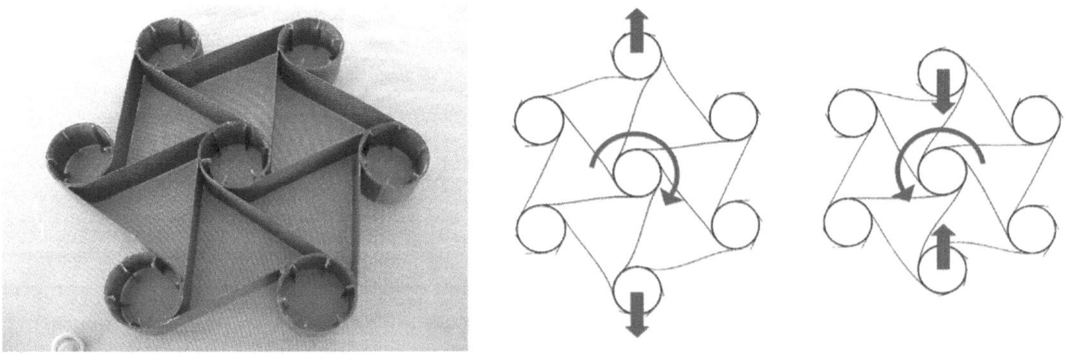

FIG. 8

Example of a chiral architecture and its working modes [30,31].

3.2 ACTUATION SYSTEMS

The actuation system is one of the most important keys in the development of a smart architecture aimed at aeronautic morphing. Again, there are conflicting requirements that hardly match each other. In the case of the actuation system, in order for it to command complex geometries, it seems necessary to refer to a distribution of several active devices. In turn, this can be unacceptable from a safety point of view, because the possibility of failure increases as the number of devices does. On the other side, one can say that many actuators reduce the possibility of dramatic breakages because of the explicit redundancy. However, this statement holds only if a dedicated failure harshness analysis (FHA) study is properly carried out and assessed, being the answer not so trivial [35]. For instance, if an actuator breaks or the general actuator system fails, blocking the device completely in a certain position, this could lead to important consequences on the whole system, affecting the stress field over the whole structure. The most obvious way to solve this problem is to make the structure heavier, and this can hardly be acceptable. Even many actuators and many failure risks have a direct impact on maintenance and one of the possible ways to reduce those dangers is to reduce inspection intervals. Again, the cost increases [36]. A point apart should be dedicated to the volume of the single actuation unit (motor). In fact, because the load is always the same, for a certain mission, reducing the number of actuators means to increase the power effectiveness of each of them. This in turn means that each actuator is larger and, then, its volume could not fit inside the structure and specific nacelles should be designed for it. This also makes the airflow dirtier and the overall resulting architecture heavier. On the other hand it should also be considered that the dimensional ratio between power and actuator volume grows as provided power increases. In synthesis, actuation system architecture complexity is a multifaceted topic, the solution of which is not simple. A summary of the questions is reported in Table 1.

SMAs raises the possibility of using smart materials to replace the actuation system and solve some of the related issues. These materials show very high energy density: for instance, a 1 mm^2 diameter wire is able to produce force of hundreds kg! And, as said, strains around the tenth of the unit. Of course, resulting displacement depends on the initial length. Herein it is briefly recalled that a smart material may be defined as a substance that is able to transform a form of energy (in input) into mechanical energy (in output) by their intrinsic characteristics. More in general, a smart material is something that modifies its mechanical peculiarities after an energy input. This transformation occurs without any necessity of complex and complicated external apparatuses. Because these materials have the intrinsic property to make that conversion, the system is intrinsically safe (the material cannot stop working!). On the other side, there are many aspects related to fatigue life and aging that should be faced. For sure, the higher the deformation is used (up to 10% and more), the shorter is the expected life, which reaches millions of cycles for values around 0.5% [37]. If this holds for static behavior, the coupling with extremely vibrating environment of a general aircraft is not completely documented and understood

Table 1 Main Qualitative Figures of Different Architectures of Morphing Actuation Systems

Actuator Systems	Geometry	Failure Risks	Inspection Times	Overall Weight	Unit Volume
Few	Limited	Low	Long	Small	Large
Many	Expanded	High	Short	Large	Small

yet. In fact, in this case it should be considered the superposition of millions or billions of cycles of small amplitude but characterized by very high frequencies. The change of behavior with time is also another open point.

SMA are indeed used for implants in the human body, but in that case, they figure as a static addition, without any kind of action (aortic stents, bone patches and extensions, and so on) [38]. Furthermore, nowadays it is preferred to elaborate actuators [39] that use the SMAs as "motor," while the generation of force or displacement is filtered through a devoted architecture. This kind of strategy, in spite of some configuration difficulties, brings many advantages like the one to preserve the "active" element, facilitate the connections (SMA are very hard to be attached to other elements), etc. This approach is becoming more and more common, above all for devices performing simple operations. For instance, several firms have lately marketed many pin-puller devices [40].

The use in the aeronautic field has been slowed until now mainly because the evolution of the classical electromechanical actuators has brought impressive feasibility levels to extremely compact devices [41]. Therefore, these last are preferred with respect to devices not yet fully characterized in operational conditions. It is believed a new perspective will be disclosed as architectures with many devices would take place to answer aircraft manufacturer needs. For the reason already discussed, the use of a very dense actuator network is unacceptable because of the unavoidable increase of the maintenance operations interval and the related costs. In any case, the use of SMAs can be imagined only for quasistatic devices like flaps, slats, or aerodynamic brakes. In fact, because of their own characteristic to convert heat energy into mechanical energy, the process is very slow (around the unit of Hertz) and is unacceptable for control devices like ailerons, rudders, and others. These limitations could be dropped as Magnetic SMA are used, Fig. 9 [42], having larger bandwidths, but also prominent nonlinear behavior [43,44]. However, these systems would lead to heavier architectures and there would be several issues related to their compatibility with the other electrical systems on-board.

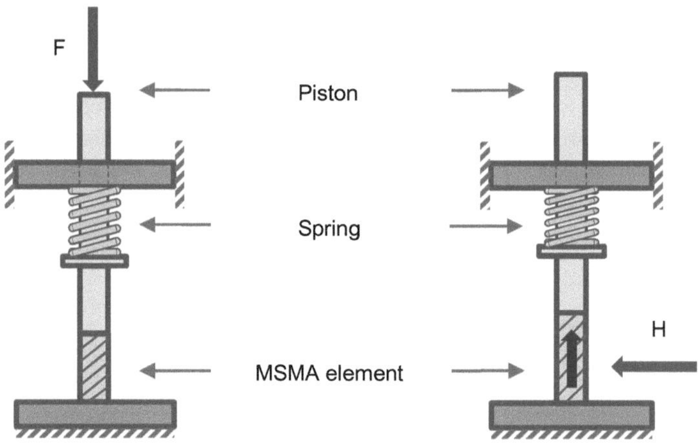

FIG. 9

A magnetic SMA concept [42].

Another possibility to increase the frequency response of a smart actuator system is the use of piezoelectric actuators [45]. As the SMA or the magnetic shape memory alloys (MSMA) they have capability to produce very high forces with very low strains (around hundreds of microstrain or 0.01%, two or three magnitudes lower). On the other hand, their bandwidth is impressive, up to 1 MHz or, more likely, several hundreds of kHz. Stroke limitations and considerable weight do not make these devices good candidates as aerodynamic surface actuation systems. Indeed, the high voltage operation (hundreds of V) is another showstopper to their implementation. They are classically applied as micro-positioners for electron microscopes, as ultrasonic generators and voltage transformers. Regarding aeronautical applications, there is a great deal of interesting research on their use as compact actuator devices for noise and vibrations control, whether for passenger's comfort increase or the protection of electronic cards and other sensible devices [46,47].

While magnetorheological fluids can be associated to dissipative systems or to the improvement of performance of active systems [48], without being the main driver component, magneto-strictive and electro-strictive materials are able to generate large forces and large displacements, with interesting bandwidths. However, as for the MSMA, these materials lead easily to heavier architectures that make their application extremely improbable for aeronautical structures. This possibility is made even weaker for their typical nonlinear behavior and association to strong electric or magnetic fields that may lead to incompatibilities with the instrumentation and communications on-board.

3.3 SENSOR SYSTEMS

The important consequences in operation and maintenance following the issue of complex actuation networks, are amplified with the implementation of the sensor network [49]. In fact, as the structural system has virtually infinite degrees of freedom, an infinite number of active devices is needed. The usual, traditional sensors are cut out of the possible choices. It does not really matter of their number, but the mass of cables necessary to connect them. If it is considered that a pair of (or three) cables is needed for each device, considering signal transmission and feeding, it easily comes out that great complications arise for at least two aspects: mass penalty and layout arrangement. It is always worthwhile to remind that a morphing structure is something that is made by relatively moving parts, and cables are exposed to risks of being truncated. Added mass is definitely another issue, in the same way as the necessity of using shielded cables to avoid interference with the on-board instrumentation. From the maintenance and the operability point of view [50], finally, several sensors implies a higher probability of local failures and a harder detectability. This would in turn cause an increase of the inspection and intervention times, not to talk about the layout complexity, Fig. 10 [51–56].

At least one of these points may be alleviated by the implementation of so-called energy harvesting [57]. This technology, takes advantage of other "environmental" sources of energy, like structural vibrations and tries to convert that into electrical energy to feed the sensors locally. The generated energy is very low but is sufficient to make a single sensor work. Definitely, this approach is not aimed at saving energy but reducing the amount and the mass of cables running through the aircraft structure and, then, the associated electrical network complexity. Many studies are available by now and, currently, some system is working [58,59]. So, this is a mature approach, even though each configuration has to be investigated separately because of different architecture peculiarity needing different solutions. All these considerations, trivial enough, lead to two different options to overcome the immanent difficulty related to cables installation.

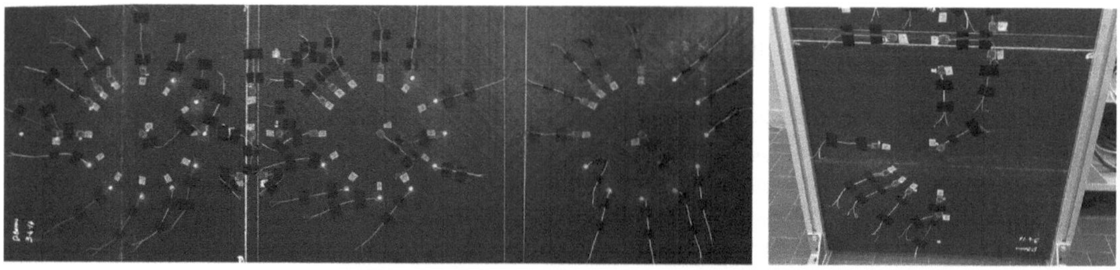

FIG. 10

Load-monitoring sensor technologies to estimate the structural health of composite elements (wing panel) [50–56].

The first, and the most obvious maybe, is the use of fiber optics (FO) [18] instrumented for instance with FBG, a sort of a hardware-printed narrow-band filter. They have the unique property to guarantee the possibility of deploying several sensible devices on a single fiber. There are commercial instruments that, with different technologies, guarantee the presence of thousands of measure points per meter! Brillouin, Rayleigh, or Raman scatter technique-based systems, for instance, allow having a virtual infinite number of sensors on fibers: a 10 m long fiber may host something like a thousand sensors! This achievement is attained with a limited increase of complexity (and cost) of the interrogator system and the use of common telecommunication fibers (largely cheaper than the ones implementing FBG). However, these approaches have the drawback of necessarily increasing the number of acquisition points, then slowing down the elaboration times. In fact, all the input coming from the sensible points necessarily have to be acquired, processed, and eventually canceled. Recently, quasicontinuous FBG printing and interrogation techniques have been set-up, trying to exploit this technology toward continuous monitoring capabilities. For now, applications are limited but they are expected to expand. Other commercial systems combine more fibers together in order to get direct information about the resulting shapes. However, in this case, the number of cables could even increase. As usual, the optimal solution does not exist but shall be assessed case by case. Generally speaking, it could be stated that for simple geometries, discrete FBG may be preferred, because the expected information is regular. For regions where high density acquisitions are needed, the Brillouin technique may be preferred. An alternative could be represented by the recent technique implementing quasicontinuous FBG distribution. For very extended domains, like bridges, rails, and so on, Rayleigh or Raman techniques could be exploited. For complex domains, where limited points may be acquired in order to reconstruct the shape, again FBG may be used. Finally, for isolated, hardly accessible points, single standard sensors may be preferred, essentially for the easier installation, while guaranteeing a high reliability level (years of experience in operating them). Opportunity reasons suggested the authors to avoid referring to any specific commercial system in this book. However, the interested reader can easily find a quantity of commercial systems through a fast survey on the Net.

Cabling can be substantially reduced by the implementation of wireless sensor networks [19]. In such a configuration, data cables could be avoided. The other contribution, linked to supply, could also be skipped through the same approach. This technique is pretty common in literature and uses at times targeted distributed antennas to convey the necessary energy [60]. Communication interference is safe if specific bandwidth is accessed, devoted for these applications. Dedicated protocols prevent the superposition of information coming from the different sensors (for instance, each assigned to an exclusive channel). Apparently, the system seems perfect! However, some important drawbacks occur with real implementations… as usual, there are not only advantages. Because light runs along a straight line, a reserved path should be built to guarantee the transmission from point to point. Intermediate antennas (relays) can be placed in strategic points to facilitate the transmission. These steps increase the background noise and deteriorate the signal quality and the information it transmits. Also, the system complexity and its cost increase. Antennas for sensors feeding may be effectively printed on the structure. There are in this case at least two difficulties. First, the structural system is made of several parts, therefore, exhibiting a certain quantity of discontinuities and, then, the path may not be so immediate. Second, the printed antenna should always ensure a certain performance level; as deterioration of the film is unavoidable, aging may heavily drive the maintenance plans. A wireless system, may then result in a more expensive, more complex, and heavier system than the standard ones. Much study and research is currently being carried out to overcome the negative aspects of such architectures [61,62].

4 BACK TO THE PAST

4.1 THE WRIGHT'S FLYER

The first powered aircraft flew in 1903. The Wright's Flyer [63] was very different from a modern airplane. It was a biplane without any movable surfaces for maneuver, attitude control, and so on. Its flight was ensured by leverage systems that allowed the pilot to modify the geometrical aspect of the bottom wing (camber, twist, curvature). In other words, the Wright Brothers' airplane had morphing wings. Of course, no one knew about morphing at that time, so this evidence was not publicized a lot. The most common comment (at least, the most common the authors of this chapters have heard) about this point is that, at that time and in that specific case, velocities were so limited that the aerodynamic loads were in turn restrained and, it was then possible to fly with extremely flexible (deformable) wings. As time went by, aircraft velocity increased, imposing the use of namely rigid structures to face larger and larger external forces. This is a well-known and appears to be a reasonable circumstance. The conclusion however may be different: flexible wings did not become obsolete with respect to new challenges. They were discarded because no one knew how to design a deformable wing while still capable of bearing larger external loads. Per this vision, the so-called classical aeronautic structures were born because of our incapacity of designing morphing wings, capable to resist huge loads. Rigid structures seemed the most obvious and easy way to face new difficulties. A beautiful virtual representation of the Wright's Flyer (including the possibility of acting on commands of the plane!) may be found on the web at http://www.pbs.org/wgbh/nova/space/pilot-wright-brothers-flyer.html. Fig. 11 [60], shows the schematic of the plane, as illustrated in the Wright Brothers' patent ("The Flying Machine"). The picture clearly shows the deformed shape, assumed by the wing under the control commands.

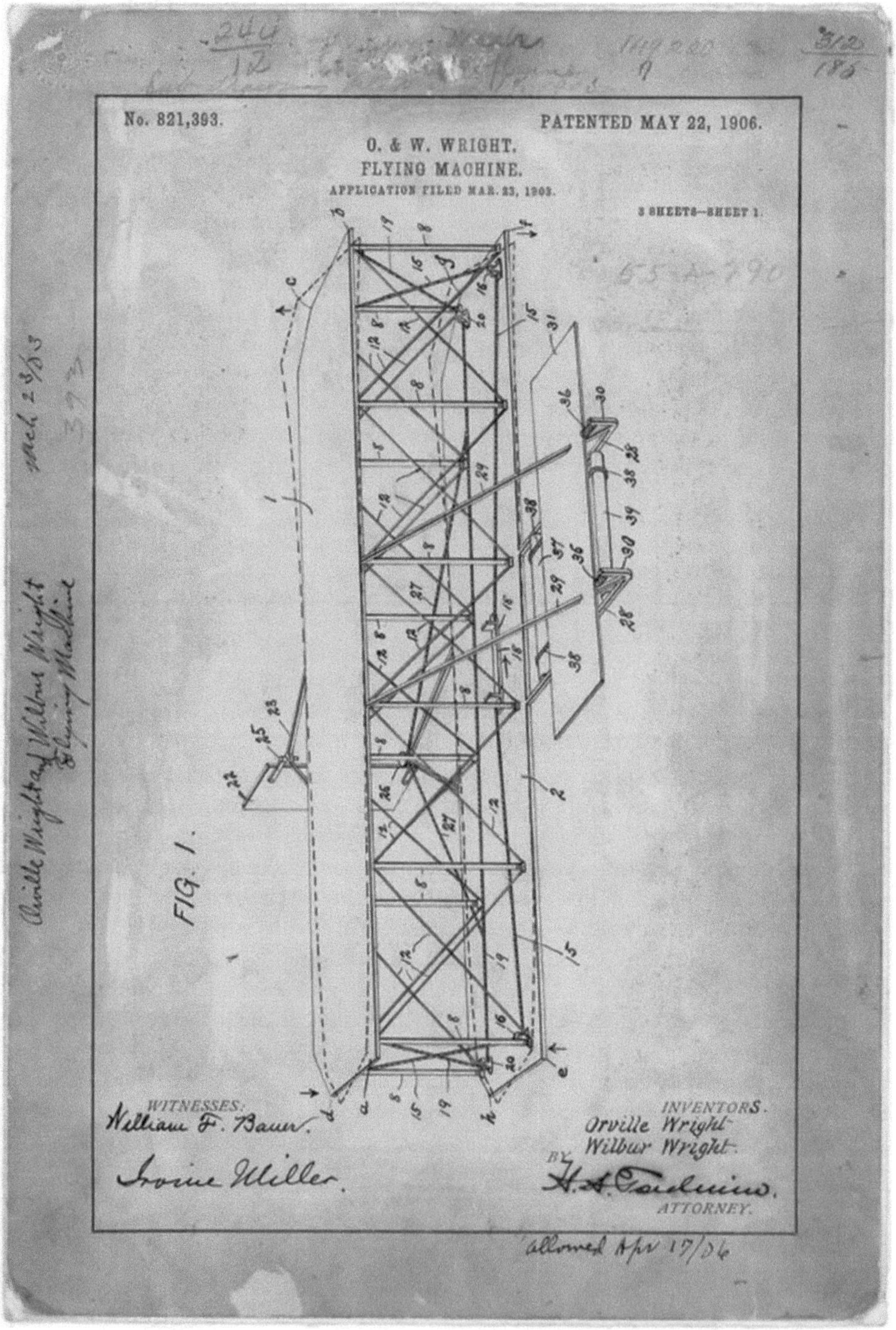

FIG. 11

An image taken by the patent, issued to the Wright Brothers on May 22, 1906 (the Flying Machine) [60].

4.2 PLANE AND THE LIKE FOR AEROPLANES

Indeed, morphable wings never did go out from the aeronautical designers' mind... at least not for some of them. Per the authors best knowledge, the first patent ever released about morphing wings was dated October 15, 1915, and attributed to a Dutch Engineer, Alexander Albert Holle, resident in Thames Ditton, England, who applied for his invention in Germany (a wonderful precursor of the European integration) and, then, in the United States, Fig. 12 [64,65]. For people investigating the matter of morphing wings, the American patent foreword is of a certain impact: "*I* (...omitted...) *have invented certain new and useful improvements in and relating to planes and the like for aeroplanes* (airfoils, ed.) *of which the following is a specification...*". And, continuing in the text: "*This invention* (...omitted...) *consists of an improved construction which has for its object to extend the range of speeds of the machine* (the aeroplane, ed.) *in the horizontal and vertical planes, to extend its range of angles of ascent and descent, to extend its vertical range or maximum attainable altitude, to increase its efficiency as measured by its capacity for carrying a maximum load with a minimum expenditure of power, to improve its stability and maneuvering qualities, and lastly to balance the forces acting on the surfaces of the aerofoil so as to reduce to a minimum the effort necessary for its control.*"

It is questionable that the proposed devices could really accomplish all those objectives. If a contemporary scientist were asked about general morphing purposes, he or she could definitely state the same sentence. By looking at the invention sketches, it is easy to comprehend how much attention the author paid to preserve the structural characteristics of the system, introducing kinematic systems that would be capable of bearing loads while deforming. Truly, these arrangements look very similar to what have been recently proposed by a number of scientists throughout the world, as an alternative to "compliant architectures."

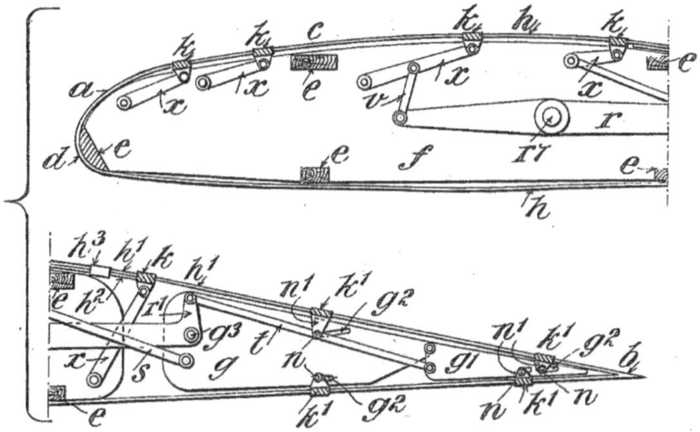

FIG. 12

A sketch of Holle's adaptive systems for modifying leading and trailing edges [64,65].

4.3 THE PARKER'S WING

In a report dated 1920, not much later than the Wright's Flyer first flight, the American designer H.F. Parker reported the use of "variable camber wings" for his mono, bi and tri-planes [66]. In a few years, velocities increased so much that wing structures were already much stiffer than the first ones. Similarly to Holle, Parker reported how much a deformable wing may be of certain advantage with respect to a rigid one. In this case, Parker does not refer to morphing devices installed on the wing. His problem was dealing with variable speeds, particularly for landing and flight in quote where, at least for fighters, there was the need of accessing a wide velocity range. The question was then how to access broader speed ranges without suffering excessive penalizations for the variation of L/D ratio at different angles of attack and overcoming the limits of the stall angle. The introduction of a variable camber wing could again answer those needs. In his text, Parker cites the possibility of modifying the wing surface. This is fascinating with respect to the current researches in extensible, telescopic wings. One of his conclusions is also interesting; he states that "mechanical difficulties… are so serious that there does not seem any prospect of their being overcome in the near future," so giving the idea he did instead expect something in the *far* future.

To Parker's eyes, variable camber was possible already in his times, even if the expected advantages were not so large, yet sufficient enough to provide a satisfactory solution of the defined problem. The achieved solution aimed at modifying the wing geometry by exploiting the aerodynamic forces. Parker's wing is then a compliant system, aeroelastically tailored, different from Holle's proposal. In this case, continuous variations of the geometrical wing characteristics are envisaged, achieved by means of a proper arrangement of the internal structure. It is worthwhile to note that this duality between kinematic and compliant systems has survived until today! The mechanism consists of placing a front spar at the leading edge and a rear spar at roughly 2/3 of the chord. The wing portion behind the rear spar is rigid, while the rest is flexible enough. The force distribution makes the wing to camber at large angles of attack (low velocity) while it stands in a streamline (minimum drag) at small angles of attack (high velocity). To make variations happen in this configuration, the ribs are allowed to slide over the rear spar, Fig. 13. Going into detail, Parker's solution is even more intriguing. For bi-plane configurations, for instance, he chose to have a rigid wing (the bottom), while morphing was devoted to the upper aerodynamic surface. In this way, he combined a virtual surface enlargement to curvature. For the sake of truth, it should be recalled that as reported by the editors of his historical report, "*At the time this wing was designed it was Mr. Parker's belief that the wing would be automatic in operation. Subsequent examination indicates that this is not true, at least for the rib as now designed. Means for flexing the wing mechanically are not discussed.*" In other words, this first compliant system would have only worked if operated by external mechanics. Indeed, this is the current trend.

4.3.1 An earlier vision

The reader should now be sufficiently accustomed with the idea that the adaptive wing concept was born at the same time of the modern aviation, soon after or with the Wright's flight itself. Indeed, things are once more a bit different. Going backwards, we encounter Clement Ader. As reported by T. Weisshaar in a wonderful retrospective article [67], "*Early aviation enthusiasts watched birds soar, changing wing shape as they dove and loitered. On the basis of his observations, in 1890, the French*

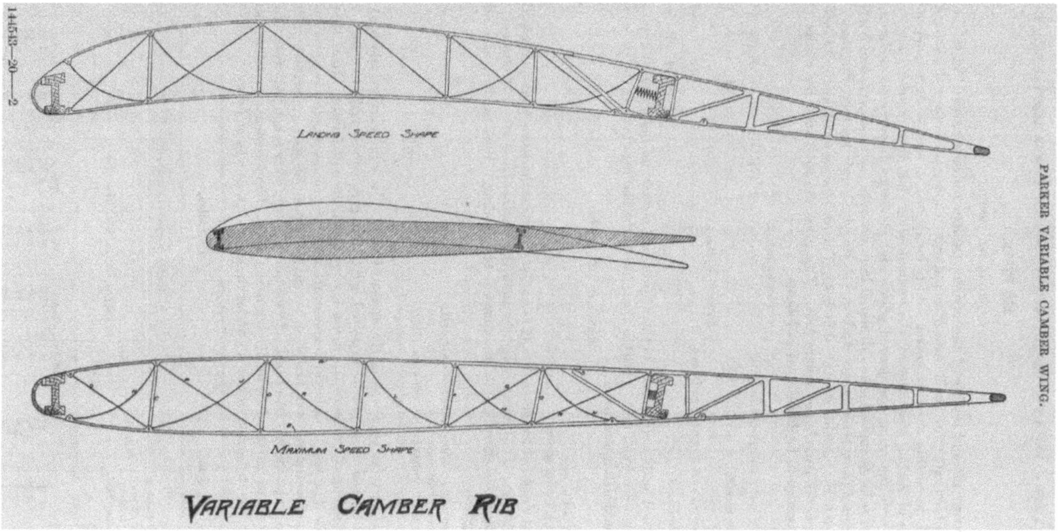

FIG. 13

The Parker's variable camber wing [66].

aviation pioneer Clement Ader (…omitted…) *developed ideas for the future of aviation warfare and described them in a monograph published in 1909.*" In that paper [68], the following impressively modern statements can be read: "*Whatever category airplanes might belong to… their wings must be articulated in all their parts and must be able to fold up completely… When advances in aircraft design and construction permit, the frames will fold and the membranes will be elastic in order to diminish or increase the bearing surfaces at the wish of the pilot… Their wings will be… bird-type, long and narrow, with a minimum of surface and a heavy load for square meter… the wings will be adjustable, so that in flight they can be reduced by a half or a third or even less… The wings will be extendable in flight and their surface will be increased or decreased at will. These airplanes will be characterized by their agility but will also be of solid construction… I propose to make some of them of metal, following experiments and plans already made. Because these airplanes will be stored in great numbers, their wings will fold up completely with great ease…*"

As we look at the contemporary aircraft and the latest tendencies, these statements appear impressively up-to-date. Without looking at the near or far future, some airplanes already have the requirements: the already cited F-14 sweeps its wings; air-carriers aircraft have foldable wings for a better stowage… Above all, the planes are predominantly made of metal (or composite… and also plastic) and all of them have command surfaces, specifically suited for a particular maneuver (flaps, slats, ailerons…). The Ader's book also correctly anticipates the well-known paradox of smart structures, envisaging the advent of agile (transformable) but solid systems. A concept plane by Ader is still shown nowadays at the Museum "*des Arts et Métiers*" (Arts and Professions) in Paris (F), Fig. 14 [69]. For complete information, the patent N.205155, issued by France on August 11, 1890, protects this concept; this example is the first fully documented idea of a heavier-than-air air navigation system.

FIG. 14

Clément Ader's Avion III (Éole) as displayed at the Musée des Arts et Métiers in Paris [69].

5 MODERN TIMES

5.1 NASA STUDIES

Since those early times, many other speculative studies followed. Yet, as the operational velocities increased and the wing became stiffer and stiffer, the necessary mechanical complications to attain morphable wings moved away from the scientific and technological community interest for this approach. It can be once more stated that morphing was abandoned because the available technologies were not advanced enough to achieve that performance. In other words, we can affirm wings were born to be deformable, but the limits of the knowledge led to simpler, engineered solutions that could overcome the problem by simply neglecting it. Modern scientific attainments finally allowed coming back to the original nature of the wing.

As commercial aircraft performance seemed to be levelling, for example, arriving to low transonic cruise flight, the search for new possibilities to further improve their capabilities, again moved to "novel" solutions. Not surprisingly, morphing was one. In 1980, the Preliminary Design Department of Boeing, under NASA commissioned research, published an interesting report, where the potential of variable wing geometry was discussed [70]. In detail, the study concerned with the potential of introducing systems to vary the wing camber continuously in cruise, to optimize selected aerodynamic reference parameters to reduce fuel consumption and operating costs. It is relevant to observe that the possibility of modifying the wet aerodynamic surface was completely ignored. Indeed, technological difficulties were still hard to overcome at those times (and

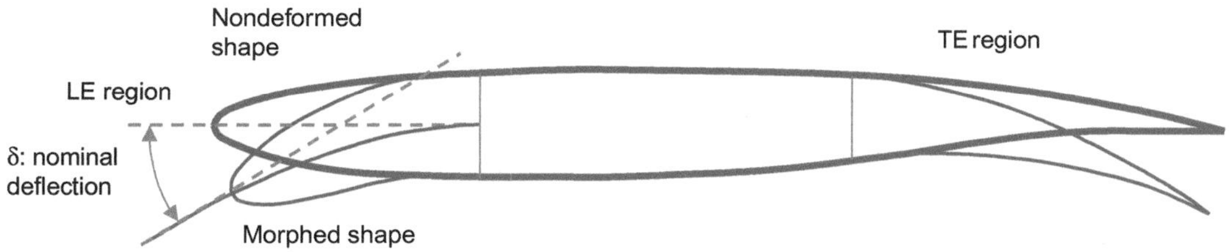

FIG. 15

The NASA concept and the deflection angle definition for morphed structures, 1980 [70].

nowadays). It is also worthwhile to mention that the authors took into account that aircraft already had several devices to allow for configuration changes as the flight condition shifted. For instance, classical devices like leading and trailing edge flaps, slats or both consent to optimally face low-speedtake-off and landing operations.

In that study, variable camber concepts explored shape-altering systems to deflect and smoothly contour leading and trailing edges during different flight conditions, Figs. 15 and 16. Three different aspects of the problem were discussed: the design of systems that would have allowed achieving variable wing geometries; the aerodynamic gains associated to the implementation of those capabilities at certain selected stages (particularly take-off, cruise and landing), and the actual possibility of integrating those devices within the existing transport aircraft systems. Applications for intercontinental (10,000 km) and domestic missions (4,000 km) were discussed. It is a pity that the document indicated the NASA reviewers and did not indicate the names of the authors, real scientific pioneers, limited to recall their affiliation (Boeing). The poor bibliography refers to just three former publications: in particular, it is believed that a 1978 AIAA paper by Levinsky E.S. and Palko R.L., *Semispan Wind Tunnel Test of a Computer-ControlledSelf-Optimizing Flexible Technology Wing* surely inspired them.

For the sake of completeness, it is worthwhile to cite another, former NASA report, published in 1977, which referred to morphable leading and trailing edges as well [71]. Again, it focused on the improvement of commercial aircraft performance and suggested peculiar concept systems to modify their shape. Despite the real applicability of those ideas, it is clear evidence that morphing did neverstop getting attention. Besides, the therein referred bibliography, discloses that attention was paid to both civil and military aircraft in the previous period (1968–77) and that former attempts to modulate the aircraft performance were conducted by investigating the possibility of exploiting the capabilities of the existing devices (mainly, flaps).

5.2 DGLR STUDIES

Not too long after that important and historical publication dated 1980, another work was issued, even more impressive from the social impact point of view. Airbus was born in 1970 and its first aircraft had its maiden flight in 1972 (A300). In 1984, DGLR (the former DLR) and Messerschmitt-Bolkov-Blohm (MBB, which merged into Airbus) published in German a sort of answer to the Boeing-NASA study, concerning what they called "current trends" in the "commercial aircraft" design [72]. The study was

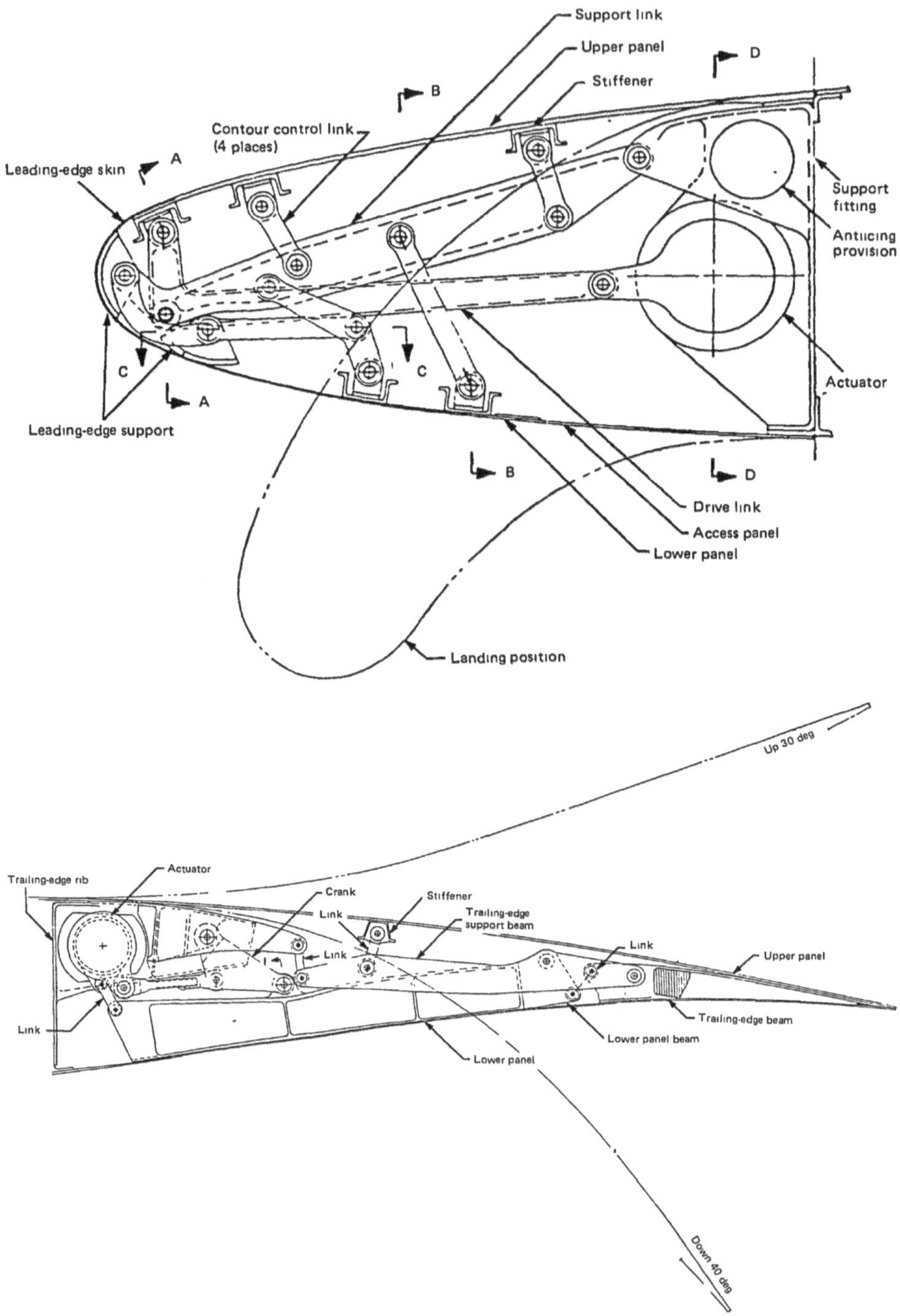

FIG. 16

The leading edge and the trailing edge mechanisms for the NASA concept, 1980 [70].

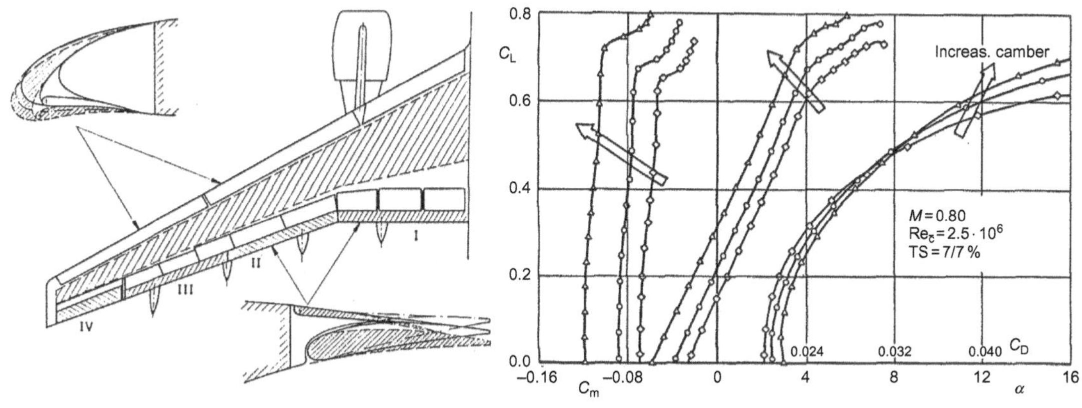

FIG. 17

Architectural sketches and predicted camber-led performance increase, following DGLR-MBB study, 1984 [72,73].

translated in English and published as a NASA Technical Memorandum in 1986 [73]. The study starts with a detailed presentation of the requirements applicable for the launch of novel passenger planes seen from the industrial point of view. This step is very interesting because it allows a direct figure of the expected benefits in the operation of such an innovative machine, as perceived by the manufacturer and seller at those times. Aerodynamic advances are outlined together with their expected impacts. Fig. 17 shows the sketch of a typical contemporary commercial aircraft wing, including a hypothetical layout and early concept of the adaptive systems, together with a synthetic preliminary forecast of the supposed gains.

In detail, the work focused on the possible implementation and the overall integration of

- Transonic, variable camber aerofoils for the control of the efficiency aerodynamic parameters and the buffeting alleviation, including load control.
- Transonic technologies for shock boundary layer control.
- Passive and active laminarization of transonic aerofoils.
- New propulsion systems, including prop-fan systems.

The publication concluded with some suggestions about the development of the concepts inside National or European projects. According to their vision, the authors clearly stated how the attainment of those ambitious objectives could be achieved only through a structured cooperation among universities, research centers and industries. Some years later, the European Union (EU) assessed their own strategic research programs and independent European associations were constituted, grouping different stakeholder categories, like the aeronautical industries (IMG4), the research centers (EREA), and the universities (European Aeronautics Science Network, EASN). Several investigation activities were then launched among the ones that were foreseen in that article. Nowadays, many others are currently being developed, in both Europe and overseas.

5.3 THE MISSION ADAPTIVE WING

The F-111 Mission Adaptive Wing (MAW) joint research program among Boeing, USAF, and NASA started in the early 1980s [74]. It proposed the use of adaptive leading and trailing edge surfaces to provide ideal wing shapes for selected flight conditions. Key points were the attainment of a smooth profile with hinge-less devices, avoiding any interruption of the geometrical continuity as it happens in standard flaps and ailerons. In detail, a General Dynamics F-111 fighter was retrofitted with wings whose camber could be actively modified by hydraulic actuators. Conventional rigid-link mechanisms and fiberglass flex-panels were also used. The achieved results are impressive: the aircraft actually had 59 flights between 1985 and 1989, Fig. 18 [74], and allowed direct measurements of the aerodynamic benefits that were achieved. Variable sweep capabilities permitted maintaining high aerodynamic efficiency for all investigated speeds, outstanding conventional leading and trailing-edge systems (slats and flaps). Deflections up to 10 and 18 degrees. were respectively implemented, for both cruise and maneuver conditions [75–78].

However, drawbacks in the mechanical design resulted in severe weight penalties and a significant increase of the overall system complexity. These difficulties almost completely vanished the attained aerodynamic benefits and generated serious doubts on the evolution of the concept [79]. After the first phase, investigations continued for more years, shifting the attention towards the system integration and the aircraft controllability [79]. Simplifications were introduced, in the attempt to make the architecture more appealing for real employment [78]. A digital control device ensured preselected changes to the wing shape. The system was provided with four automatic control modes, combining together the deflection of the leading and the trailing edge: maneuver camber control; cruise camber control; maneuver load control; and maneuver alleviation. These modes were respectively related to attain the maximum aerodynamic efficiency, the maximum speed (for defined altitude and power setting), the highest load factor, and the (partial) reduction of the gust effects.

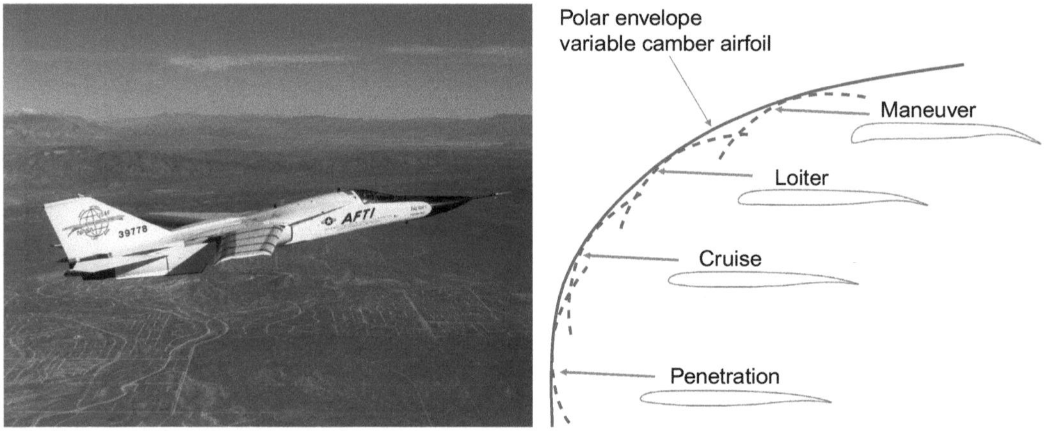

FIG. 18

The F-111 MAW in flight (left) [74] and morphing philosophy (right, an authors representation of the results, shown in Ref. [75]).

5.4 FURTHER NASA STUDIES

None of the cited studies did lead to any actual application. Available technologies, complexity, cost, weight, and operability penalties, associated to the difficulties, associated to the physical implementation of the individuated systems, prevented their material development. Over the years, however, new possibilities arose and the thematic had a new push. In 1999, NASA published a systematic work on the estimated benefits of variable geometry wing camber control for transport aircraft [80]. In this case, analytical investigations were carried out. Aerodynamic performance models allowed estimating optimal performance benefits of the variable camber. The most important result was that by considering the complete wing trailing edge surfaces, drag could be reduced at virtually all points in the flight envelope, Fig. 19 [80].

Dealing with jet engines, the focus was given to lift-to-drag ratio. It dealt with the symmetric use of ailerons and flaps to achieve those benefits. Experimental evidence, following devoted wind tunnel (WT) tests, were used for performance predictions. For diagram points not originally included in the investigation, simple interpolation was carried out. The research introduced an original algorithm for computing optimal surface deflections for given conditions. The results showed that variable camber systems for a commercial aircraft using a simple trailing-edge device could lead to gains exceeding 10%, especially for nonstandard conditions. In standard cruise regime, the benefits were estimated around 1%–3%. The report refers to contemporary studies carried out at those times in Europe by Airbus (MBB), trying to apply adaptive systems to A-330 and 340 series [81].

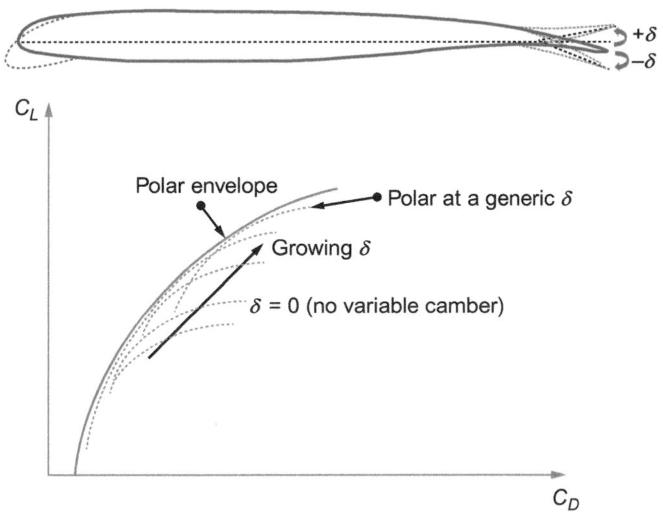

FIG. 19

Conceptual vision of the expected polar variations following the introduction of variable chamber systems (single flap) [80].

6 RECENT ACTIVITIES—UNITED STATES

6.1 ADAPTIVE WING REBORN: SMAs

Sometimes technology perspectives change in unexpected ways. Often, once they have been modified, things change forever. As we have known till this point, the wing was born as a flexible element, able to deform itself to accomplish the upcoming necessities, in accordance to different flight stages. As aircraft started flying faster and faster, these basic capabilities were lost, because the accessible technologies did not allow maintaining them. Indeed, many studies (and patents) did follow each other but were not applied in reality because of the difficulties related to costs, weight, room occupation and, above all, operability (maintenance and production). Despite the advantages being very clear, the solution seemed unreachable. The unpredicted push came from material technology.

SMAs with macroscopic effects were discovered in 1959 [82–85]. Truly, SMAs had been known before for a long time, but to significantly lower quantitative levels. Moreover, the related materials were expensive and rare (gold-cadmium, for instance) [86]. Nickel titanium compounds exhibited around 10% strain recovery with a potential of generating forces of tens of kilograms per square millimeter (wire section). The working characteristic is very simple. Once the material is stretched beyond its (pseudo) plastic limit and a residual strain is detected after the load removal, direct heating is able to completely recuperate that deformation [86]. Some industrial applications started in the early 1970s (cryogenic tube joints) [87] after coming to the attention of several scientists from different educations. The high biocompatibility favored medical developments (for instance, aortic stents) [88]. At the beginning of the 90s, someone had the idea of using these wonderful metallic materials as compact actuators for morphing wings [89].

6.2 THE DARPA SMART WING PROGRAM

The guidelines of the Defence Advanced Research Projects Agency (DARPA) program "Smart Wing," carried out in cooperation with the Air Force Research Laboratory (AFRL) and NASA, looked at the use of SMA to overcome former difficulties experienced in the development of adaptive wings. The Northrop Grumman Corporation led the project, awarded on September 1994 [90]. It was arranged into two different phases: the first ranging between January 1995 and February 1999, mainly devoted to technology development and ground tests [91], and the second ranging between January 1997 and November 2001, especially focused on technology refinement and WT tests [92]. Fig. 20 [93] shows an artistic vision of the deployment of the adaptive command surfaces onto a real Northrop-Grumman unmanned combat aerial vehicle (UCAV). The original picture (by the US Navy [93]) was in fact modified to show the possible location of the morphing surfaces onto that aircraft, per the cited research program.

As a smart material, SMAs allow intrinsic transformation of a form of energy (heat, in this case) into mechanical energy and then permit achieving extremely compact active systems: the material itself is the actuator. This peculiarity has impressive impacts on necessary room and design complexity. Implementation costs and reliability are also positively affected, being the actuation capability an intrinsic property of the material, that cannot stop working. Operation and maintainability do not show specific criticalities. An issue is associated with fatigue life, because Ni-Ti alloys show limited lifecycles for large strain activations (10^5 for 2% and 10^7 for 0.5% recovery). Another issue is

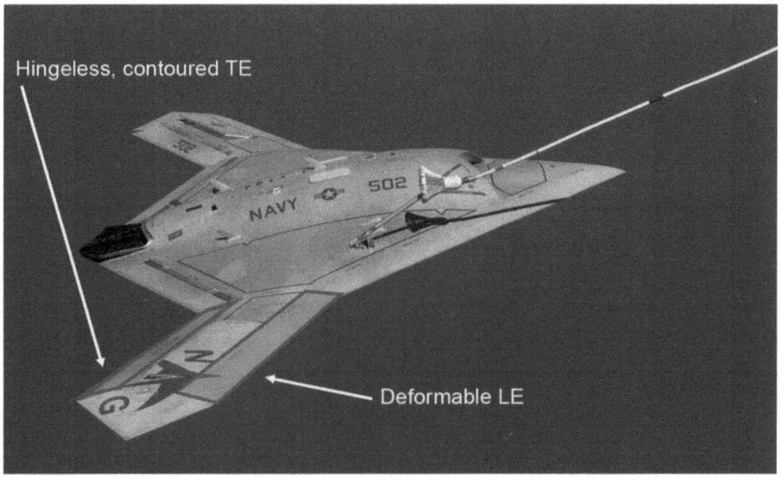

FIG. 20

A pictorial modification of a US Navy picture [93], representing the possible layout of smart devices on an actual NG UCAV.

associated to its dynamic response: because they are activated with temperature, slow processes can be faced. In other words, quasistatic phenomena can be targeted. Trim control systems, typically having around 20 Hz bandwidths, are therefore skipped out [92].

Despite confining the use of SMA actuators within the defined boundaries (static applications and limited strain recovery), the design solutions did not take off. Indeed, many other details seemed critical to overcome. The major question regarded the improper managing of the "smart structure paradox." The deformable structure should have been designed to be soft, then largely exposed to external loads. The mentioned study showed that the complete actuator system should have concurred and driven the load absorption. The research team therefore addressed their attention to different command devices, focalizing in the end towards the use of eccentric actuators or "eccentuators" [94]. These objects, implementing standard technology, could deform the structure and participate to absorb the external loads. The conclusion was amazing: after all, standard technology could be used to set-up morphing wings!

In detail, the "Smart Wing" program targeted the realization of integrated hinge-less wing systems, capable of replacing the standard control surfaces to allow attaining optimal aerodynamic shapes for different flight conditions. The keyword is "integrated," i.e., the system should have been hidden into the main body (smart structure), to have a minimum impact on the aerodynamics and generating an almost perfectly clean shape; "hinge-less," i.e., the surface should have smoothly changed its shape. The team constructed a scaled WT model, representative of a contemporary military aircraft wing (F18 E/F) [91–95], tested at the Transonic Dynamics Tunnel (TDT) in the NASA Langley Research Center. The morphing systems were made of wing elements actuated by SMA tendons, placed at the trailing edge where deflections up to 10 degrees were recorded. In addition, SMA torque tubes allowed achieving up to 5 degrees of twist. Assuming steady conditions, around 10% performance improvement was predicted by exploiting WT test results and numerical predictions.

Several reasons made the research team move towards a different aircraft in the second part of the activity. A subsonic UCAV was chosen mainly because

- The wing section was thicker, allowing an easier integration of the actuation systems.
- The smaller size of the referenced aircraft permitted testing larger scale models (1/3 to 1/6).
- The technology transition was expected to be faster and simpler for unmanned vehicles.

Again, it is worthwhile to focus on two substantives of the statements above (integration benefits were already discussed). "Larger scale models" means that the integrated actuation system size would have been closer to reality; in fact, this kind of system is generally nonscalable because the embedding possibility itself and their performance depend on available room or, equivalently, on allowed dimensions and accessible layout. "Technology transition" means that the developed systems, even after successful tests, should have undergone further investigations and advances before they could be installed on real crafts.

The final WT model was designed to host both conventional and innovative actuators, on the left and right span, respectively. Both SMA actuators and ultrasonic piezoelectric motors were used [92]. In the first case, a direct actuation was implemented (SMA devices acted directly on the target structural elements to be deformed). In the second case, the motors drove kinematic systems that in turn forced the desired deformation and displacement fields to the structure. This last choice allowed improving the stiffness of the overall system and the bandwidth. The investigators concluded that the achieved results were significant mainly because they could be exploited to full-size aircraft without any significant change in the architecture. The proved system can be ranked TRL 5. The investigators added significant statements in the summary of their investigation, to achieve a full maturity, issues linked to power supply, cost, system reliability, and the overall system integration should have been considered. Even today, together with maintenance and operability, these same items are key issues. Until today, none of the tried concepts arrived to a real production aircraft; nevertheless, it can be stated that the DARPA Smart Wing Program paved the path for many other investigations to come. It is not too far from reality to affirn that, since then, any other research on smart wing systems has followed the same schematic approach.

An important fall out of the DARPA Smart Wing Program was the birth of NextGen Aeronautics. It was founded by the main investigator of that research, Dr. Jay Kudva, one of the major references in the world of integrated adaptive wings. Since 2003, NextGen has participated in several projects on adaptive wing concept developments, sponsored by DARPA and other US agencies. NextGen role in the field of adaptive structures has been often acknowledged in many occasions [96]. NextGen has expanded interest from AS to a wide range of disciplines, including multifunctional materials and structures, sensing and control systems, and integrated antennas, while maintaining their capability to follow the project from the design to the manufacturing and testing of demonstrators and prototypes, a necessary characteristic when dealing with innovative, and sometimes breakthrough concepts.

6.3 THE DARPA MORPHING AIRCRAFT STRUCTURES PROGRAM

On the wake of the previous success, DARPA launched a new program in the period 2003–07, called Morphing Aircraft Structures (MAS), which was aimed at flying UAV with variable wings for military applications [97,98]. The aerodynamic surfaces were requested to change their sweep, chord, and area. The project investigated concurrent concepts carried out by three companies (NextGen Aeronautics,

Lockheed-Martin, and Raytheon Missiles Corporation) on three different aircraft types. The general objective was to combine low-speed loiter with high-speed dash capabilities of different aircraft systems that have typical conflicting requirements as one limits the other. This consideration is schematically shown in Fig. 21, an elaboration of a former bibliographical sketch [68]. By acting on the global wing geometry variation, the performance may be expanded and the design diagram can be enlarged (an essential schematic representing the concept carried out by Lockheed-Martin is reported at the bottom of the same Fig. 21 [68]). Basic requirements included responsiveness (time critical deployment, in order to properly respond to unpredictable crisis), agility (ability to run fast after air and ground targets), and persistence (ability to loiter on selected areas for long time).

NextGen proposed a substantial in-plane shape change and surface area variation to modify the wing from a high-aspect-ratio loiter to a reduced-wing-area for transonic, low altitude dash shape [67–103]. A kinematic system operated to open and close the wing, whose inner body to the trailing edge was made of a sort of fan-type fabric, Fig. 22 [103,104]. The Lockheed-Martin concept was aimed at reducing a substantial portion of the wing area during low altitude, transonic dash portion of its mission, while expanding in loitering. The idea involved the use of advanced skin materials, at least locally, to guarantee a certain surface smoothness when the wing was folded or deployed [67–105]. In the end, the concept of Raytheon implemented the development of telescopic wing to increase the range capabilities of generic missiles. It demonstrated the most challenging concept because it faced the highest load per unit of surface (around 1 ton per square meter), while the available room for installation of the actuators and the support structure was extremely reduced [67–107]. After a first development phase, the project addressed its focus towards the two airplane concepts. Both passed a further evaluation phase and were tested in the NASA Langley Transonic Dynamics Wind Tunnel (TDT).

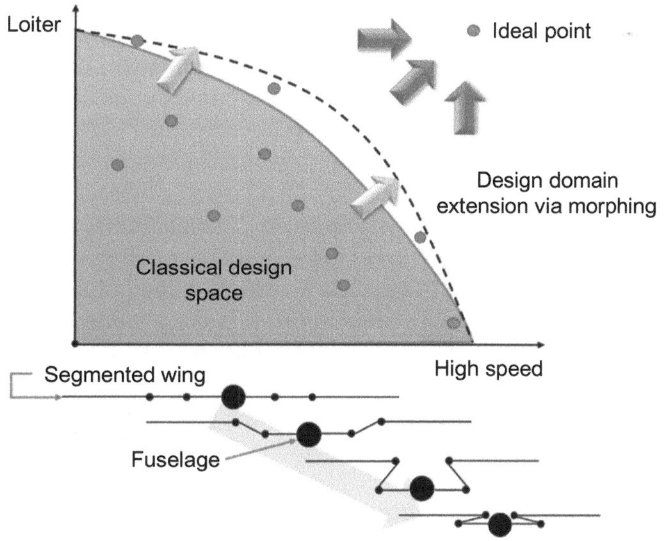

FIG. 21

Qualitative picture showing characteristic points combining max velocities and loitering times [68].

FIG. 22

A NextGen Aeronautics concept for a UAV morphing wing [103,104].

DARPA MAS Program, Courtesy of NextGen Aeronautics.

The WT models included the complete adaptive system (power units, actuator systems, control, and sensor systems), able to respond to the prescribed specifications in terms of time response and geometry variations [67–105].

7 RECENT ACTIVITIES—EUROPE

7.1 ADIF

In 1995, the European community started their first applied investigations on adaptive structural wings. The ADIF (Adaptiver Flügel) project may be considered the turning point, marking the start of this kind of research on morphing in Germany, at the DLR research center [108–111]. This first activity came out of other former numerical aerodynamic investigations, concerning with the alleviation of buffet phenomena. Previous works did in fact show that deploying a properly contoured bump just under a shock wave could lead to achieve a nearly isentropic compression, attenuating the intensity of the shock itself and reducing the associated impedance [112,113]. The further inclusion of an adaptive trailing edge gave the envisaged system some capabilities of flow control, aimed at supporting the functionality of the previous device [112,113], Fig. 23. Once again, it is worthwhile to state that each problem has a specific solution; in fact, an adaptive trailing edge may be used for a great deal of applications that must be specified in order so it can properly accomplish the assigned objectives within the prescribed conditions (boundaries, loads, and so on). The referenced transonic wing system should host an adaptive bump (top box skin) and allow trailing edge variations. At that stage, the study focused on two-dimensional airfoils.

Results show this conceptual 2-component device could effectively be used in a wide range of velocities and corresponding angles of attack. In detail, the flow control associated to the use of the adaptive trailing edge was effective at low speeds, while the bump behaved very well at high speeds [109]. The use of their combination compensated and the achieved performance improvement was

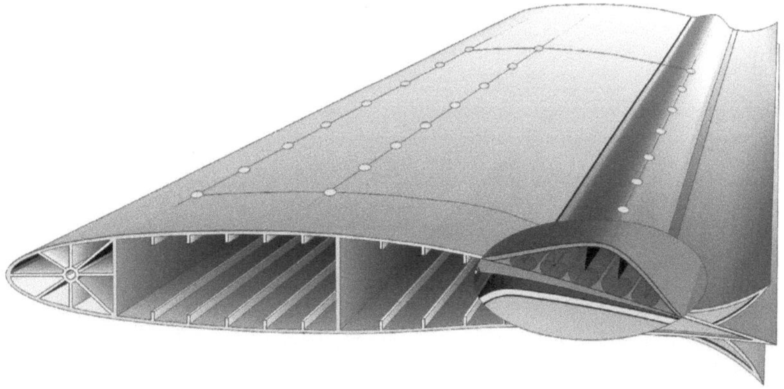

FIG. 23

ADIF (DLR): the Adaptive Airfoil concept, combination of variable camber and bump systems.

Courtesy of DLR.

almost flat in the considered ranges. Indeed, it could be noted as the effects of the two subcomponents were positively amplified when they acted together [109]. Two different architectures were implemented to attain the desired trailing edge shape variations. The so-called "finger concept" substituted rigid with flexible ribs, realized by a continuous series of hinged plates (like a finger or a snake skeleton). The various elements were then linked together to reduce the degrees of freedom to two per rib for limiting the necessary number of actuators. The other architecture used a closed belt, transferring the actuator stroke to inner laminae, vertical to the chord, inducing in turn a certain deformation to the modified structure. Both concepts were validated in lab tests. In this very early stage, both kinematic and compliant philosophies were tried. This peculiar and important aspect was not very evidenced in those first papers, as it was too early with respect to technology challenge!

7.2 CLEAN SKY

In 2010, Europe launched a major technology activity, intended in the end to address all the research topics in the aeronautical field [114]. Clean Sky was split into different subprograms, one of them specifically directed to adaptive structures (smart fixed wing aircraft, SFWA). Among the others, the Green Regional Aircraft (GRA) dedicated a room in two segments for developing adaptive systems: the first addressed Low Weight aspects while the second targeted the transmitted noise reduction (Low Noise) [115,116]. In detail, among the explored technologies, GRA-Low Weight considered the use of integrated sensor systems for the assessment of structural monitoring networks, capable of providing information that allowed the reduction of established safety factors in the structural design, that in turn would have decreased the overall weight [117]. GRA-Low Noise focused on the development of adaptive architectures for reducing the radiated noise in landing and take-off, a great deal of which produced by the gaps that are present in the high-lift devices like flaps and slats. In order to validate the numerical predictions in the very complex scenario of airflows running into those openings, intensive experimental campaigns were performed, including large activities in WTs (most of them were carried out at the facilities of the INCAS research centre, Romania), Fig. 24.

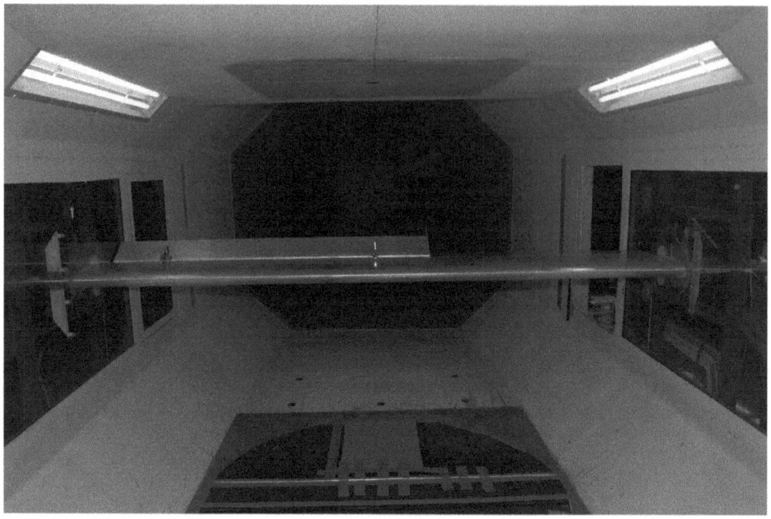

FIG. 24

A picture taken during WT tests at INCAS, JTI-GRA Low Noise [118].

Courtesy of INCAS.

This comprehensive and intensive study involved many partners spanning across Europe and started with spread preliminary investigations [119,120]. The summary of that research, published in the annual conference of the Green Regional Aircraft Program, was awarded with a special prize by the EU (Greener Aviation Award) [118]. The project aimed at designing specific conformable surfaces for minimizing acoustic emissions. WT tests however did not use adaptive mechanical systems, but rigid shaped airfoils and additional wing components, simulating different stages of morphing. This step was necessary in order to have an experimental evaluation of the proposed technology benefits, before moving to the executive design and the manufacture of real-size demonstrators. Later, these systems were actually realized and are currently in the final phase of demonstration (2017). In particular, morphing technologies were selected among many others as "feasible to be implemented in perspective on real aircraft." They did in fact pass an internal selection where many innovative systems were evaluated after several parameters, including availability of the key technologies, the perspective advantages and the actual feasibility of the on-board.

8 CURRENT SCENARIO

8.1 AIRBUS—SARISTU (SMART INTELLIGENT AIRCRAFT STRUCTURES)

In 2010, the major research ever carried out in Europe on adaptive structures was devised. With a 51 million Euro budget and more than 60 partners throughout the Old Continent (and more), it resulted in a reference for all the future activities on the subject. Airbus led the SARISTU program (www.saristu.eu), which was articulated into several subprojects or "Application Scenarios." Each of them targeted at developing specific technologies to be integrated in two full-scale demonstrators, managed in the so-called "Integration Scenarios." Within those domains, an advanced wing section and an advanced fuselage section were realized, each implementing specific novel systems [121,122]. Concerning the adaptive wing activities, a 5.5 m span wing section hosted a complete set of three morphing devices, in turn realized by different teams.

An adaptive leading edge aimed at increasing take-off and landing performance, including antiicing apparatus and tested for bird impact, was designed by a group led by the Deutsche Luft und Raumfahrt (DLR) [123]. They therein exploited their own experience, matured in several years of continuous investigations since the previously mentioned ADIF. The system implemented a single electromechanical actuator that forced mechanical deformation of a composite skin. It combined high bending resistance with reasonable in-plane deformability, taking advantage of the anisotropy properties. The European Aircraft and Defence Systems—Innovation Work (EADS-IW), led a team aimed at developing an innovative adaptive winglet system [124]. It targeted the possibility of giving the aircraft an effective device for gust alleviation. It implemented a classical, large band electromechanical actuator driving a single-hinged (plain) flap, in turn connected to the main body by means of a special rubber ensuring surface continuity while being compliant with the aeronautical standards. The system was proved to have a bandwidth of several tens of Hz.

The Italian Aerospace Research Centre (CIRA) guided a group focused on the design and the realization of an adaptive trailing edge device (ATED), aimed at improving the aircraft performance in cruise, in terms of fuel consumption reduction, or equivalently, range increase [125]. A set of flyable electrical actuators was used, compact enough to be entirely integrated into the main body in spite of the limited available room. They drove a single degree of freedom (SDOF) kinematics, made of 3-hinge ribs and related structural components (stiffeners and spars) devoted to ensure the prescribed shape. Also in this case, a suited deformable skin was implemented to guarantee aerodynamic surface continuity, this time combining metal and special rubbers.

Finally, the wing box hosting the installation of the three adaptive devices, was completely instrumented with a fiber optic system, making up a dense sensor network for structural health monitoring objectives. This structure was entirely made of carbon composite. Alenia Aermacchi (today the Aeronautics Division of Leonardo Finmeccanica) drove the consortium that conceived and manufactured this core element [126]. All the devices were designed for sustaining real flight conditions for the referenced aircraft they were developed for, and realized according to the best practice, common to the aeronautical industry. The assembled wing, Fig. 25, underwent WT tests at TsAGI, the Russian Aeronautical Research Centre, where the three systems proved their capabilities in low speed flow ($M = 0.3$), April–June 2015.

8.2 BOEING—ADAPTIVE WING

Around the same period, another project took place in the United States similar to SARISTU. Surprisingly enough, the addressed architecture for the trailing edge system looked like the same concept (3-hinge based ribs); in Fig. 26, the authors' interpretation of that concept is shown, per what was found in the bibliography [127–130]. However, some important differences emerged. The system is arranged with three independent actuators on each hinge line to realize a full 3-DOF architecture, different from the 1-DOF system proposed in SARISTU. At the beginning of the project, three different kind of actuation systems were suggested: the first being based on classical electromechanical motors, the

FIG. 25

The SARISTU wing inside the Wind Tunnel Facility at TsAGI (RUS), WT T-104.

Courtesy of TsAGI.

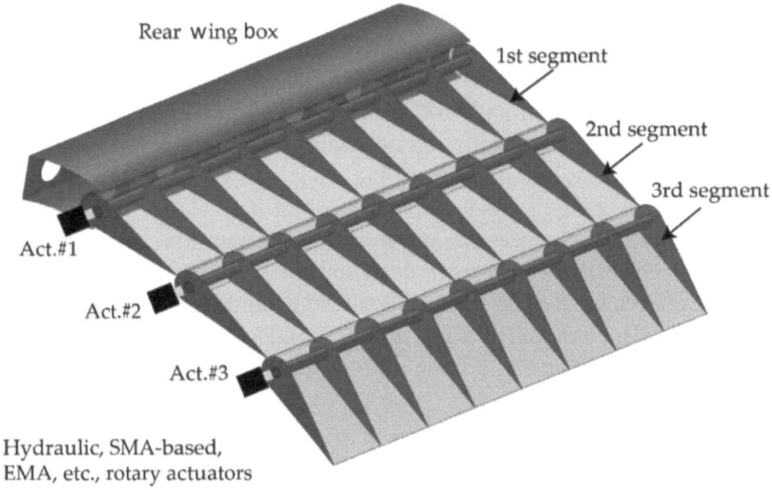

FIG. 26

The novel Boeing concept [127–130].

second one implementing hydraulic devices and the final one using advanced materials, namely SMAs, realizing an active torque tube. Public declarations by Boeing presented the three options as complementary and placed at different stations along the chord [129,130]. In particular, SMA rods present a good ratio between developed power and total weight, yet limited from the point of view of the well-known slow response times. Hydraulic motors show high speeds and small size with respect to the produced power while they are heavy, need dedicated hydraulic lines and require valves. Finally, electromechanical systems show some penalties in terms of size with respect to the former but are rather simple, very reliable and fast. Furthermore, they are more affordable and coherent with the current trend of "all electric aircraft." For this kind of application, according to the authors' opinion, there are several reasons indicating that both hydraulic and SMA-based systems should not be used for a real (short-term) industrial development. In fact, weight, complexity, consistency, and controllability reasons push in this direction; furthermore, SMA-based actuators are not considered mature enough yet.

The concept, named variable camber continuous trailing edge flap (VCCTEF) was once again aimed at optimizing lift to drag ratio (*L/D*) [127], finalized to compensate the fuel weight reduction during flight. The project mentioned the possibility to achieve twist control and reported the sound benefits an adaptive wing can lead to [130]. An extensive presentation of the approach may be found in [131]. The wing is divided in two main sections: up to and after the kink. The inner wing device is more devoted to augment polar characteristics (lift over drag), while segmented adaptive flaps on the outer region are instead aimed at twist control. In the bibliography, there is little information concerning the skin implemented into the proposed architecture to allow a continuous, no-gap deformation [131,132]. It is deployed both on bottom and top surfaces in the inner region, while segments are also installed between each chord-wise segment at the outer region, to guarantee span-wise geometrical continuity. It is known that skin is a very crucial aspect in the design of these objects and this could somehow justify such a lack of details. From various papers concerned with these investigations, it is clear that flutter analysis is perceived to play a fundamental role when dealing with reduced stiffness, increased mass, and increased degree of freedom (DOF) systems [133–136]. In general, this kind of wing architectures should undergo a systematic aeroelastic investigation, not only concerned with flutter (dynamic instabilities).

8.3 FLEXSYS AND GULFSTREAM

Since its very first approach, the FlexSys concept tried to exploit the flexibility of the wing underlying structure to favor the attainment of prescribed shapes. In Fig. 27 (top left) [137], the authors' pictorial representation of that idea is shown. The strategy aimed at distributing the effort of a limited number of actuators through selected and defined stress-strain transmission paths. Instead of referring to a complex active kinematic system network, structural compliance characteristics were considered. The compliant structure considered therein and in other publications [79], may be defined as a load-bearing mechanism, optimized to distribute point driven strains to produce a significant shape change of the target body.

The study was referred to as mission adaptive compliant wing (MACW) to indicate a step forward with respect to the MAW, previously referenced [79]. The design was developed using dedicated SW tools to draw the architectural topology, able to undergo a global deformation while avoiding high-stress concentrations. In other words, the configuration could evenly distribute the internal tensions. Summarizing, the main key of this development is perhaps the use of classical actuation systems, applied to a compliant internal structure and seamless flexible skins. This allows achieving a very thorough simplicity level. Per the investigators, the structural system was also designed to allow nominal locking in any attained form, while withstanding the external operational loads. Indeed, this assumption may hold from a static point of view but should undergo significant further analyses to

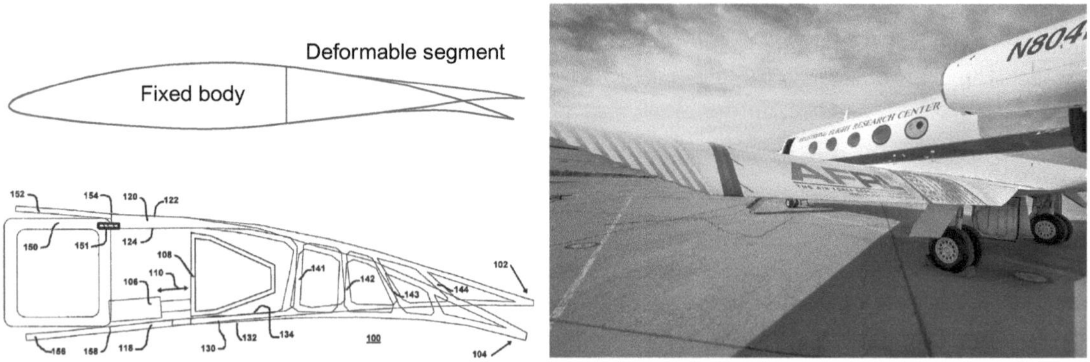

FIG. 27

FlexSys System: from the concept [137] (top left) to the Patent [138] (bottom left) until taxing [139] and flight tests (right).

verify its complex dynamic response, a direct consequence of its augmented elastic properties. The proposed technology also allows linear variation of the imposed deflection along the span to implement enforced twist for load alleviation, like root bending moment (RBM) reduction. The proposed device did well behave under the action of the design external forces, reaching +40 to −10 degrees deflections without exceeding material limits [79]. The system also showed the interesting advantage to absorb significantly less power if compared to a flap with the same characteristics (increase of C_l). Smoothness of surface permitted achieving additive levels of drag resistance reduction, while the complete removal of joints and seams made the system less compliant to icing and less vulnerable to debris. Finally, laminarity was preserved during the shape variation process, permitting to attain important increases of performance levels. These additive effects are characteristic of the morphing structures with respect to the classical kinematic architectures. Unfortunately, because of the high classification level of this activity, little detailed information is available and some statements cannot be verified by independent reviews, either at qualitative or quantitative levels.

Some additional details can be seen through one of the applied patents, placed in 2008 [138], Fig. 27 (bottom left), and a paper dealing with preliminary flight tests, when a compliant wing section was mounted as a passenger experiment on-board of the Scaled Composite White Night [140]. Recently (2015), FlexSys morphing flaps underwent flight testing under supervision of Boeing and the US AFRL on a Gulfstream III [139,141]. The next step is aimed at testing similar kinds of compliant flap tabs on an Air Force-operated Boeing KC-135, which could be the last step before industrial production [142,143]. It may be expected that the involved companies could give rise to spin-off joint ventures for addressing large scale commercial applications. Morphing technology for certain specific challenges seems to offer the same possibility of retrofit to current aircraft like winglet devices a few years ago. Lately, some relevant issues have been arisen with respect to the actual certification possibility of these devices, mostly with respect to the aeroelastic impact [144].

8.3.1 Other relevant projects: Change and NOVEMOR

In Europe, the CHANGE project (Combined Morphing Assessment Software Using Flight Envelope Data and Mission Based Morphing Prototype Wing Development, 2012–15) developed morphing leading [145] and trailing [145] edges for an unmanned aircraft vehicle (UAV) with the further capability to modify its span [146]. These variations were aimed at facing different flight phases: camber for take-off and landing, shortening or enlarging span for diving or loitering, and so on. The activities focused on a specific wing so that, the reciprocal influence of the different devices could be checked from both the structural (elastic interference) and the aerodynamic (flow interference) point of view. An interesting attainment was the setup of a sort of a virtual iron bird (a devoted simulation software, including both structural and aerodynamic behavior). Such a tool was envisaged to provide the scientists, involved with morphing, with a tool able to define the most suitable wing shape for a given mission [147,148]. Per this output, and considering the selected actuation strategy, the software provided the optimal devices layout. Experimental tests proved the correctness of the selected approach that included the development of a morphable skin [149]. In the project website, detailed information is available, including a wide record of publications (http://change.tekever.com). The consortium included nine partners and was led by Tekever, a Portuguese company with its core business in communication and UAV development.

The NOVEMOR project (Novel Air Vehicle Configurations: From Fluttering Wings to Morphing Flight, 2011–15) aimed at investigating novel aircraft configurations implementing innovative lifting concepts and morphing wing solutions to enable cost-effective air transportation [150]. The Instituto Superior Técnico of Lisboa led the consortium, consisting of seven partners (universities, research centers and one major industry, EMBRAER). They designed an active camber wing for a concept regional jet based on compliant devices, with the target of reducing the aerodynamic drag and the radiated noise, and some potential to retrofit conventional wings. A point of the followed approach was to include the morphing devices at the very early stage of the project, then conditioning the design process from the beginning. This is one of the first cases where morphing was not seen as an adjustment to an already established configuration but as an intrinsic part of the system development. Central to this research activity was also the definition of a complete aircraft model, by EMBRAER, as a benchmark for estimating selected potential benefits of the planned smart devices.

The main characteristics of the referenced regional jet were then available including geometry, aerodynamic results, flight envelope, flying qualities, weight and balance calculations, performance analyses, and engine performance. On that basis, a first estimation of the expected benefits, associated to the inclusion of some adaptive devices was assessed, for instance in terms of fuel consumption reduction for typical missions. The work followed two main technological lines: the application of morphing concepts to the reference plane and the study of joined-wing (JW) configurations, including the installment of selected morphing mechanisms. Many concepts were analyzed: an adaptive wingtip for attitude control improvement and fuel consumption reduction, adaptive leading and trailing edge devices, a telescopic wing, sweep and planform morphing systems. Further details may be found at the official website of the project, http://www.novemor.eu/novemor/, and the official website of the European Community Research Development and Information Service (CORDIS), http://cordis.europa.eu/result/rcn/55978_en.html.

The main attainments are summarized in the final project publishable report, stored at the internet address previously mentioned. Unfortunately, all the developed concepts and the associated tests were devoted to scaled objects but the wingtip device that was full size. Therefore, there is a generalized issue concerning the size of the developed systems and the referenced structure. It is not always easy to scale such devices to real aircraft proportions. Of course, this limitation does not apply to SW tools that can be used on different scales. In detail, the development of the morphing architectures was jointly conducted by DLR, mainly focusing on the leading-edge wingtip, and the Polytechnic of Milano, taking care of the leading and trailing edge devices in the main body (inner region). A certain emphasis should be given to the engineering tools development, apart from the design, manufacture and test of the proposed system [151]. The test campaign occurred at the WTs of the University of Bristol and the Polytechnic of Milano, demonstrating the goodness of the established approach for designing compliant morphing systems [152,153]. This research continued after the end of the project and proved to be applicable for general design of morphing aerodynamic surfaces [154].

Among the others, the NOVEMOR project dealt with the development of extremely challenging devices as telescopic wing architectures, applied on a small remotely piloted vehicle (RPV), or planform and sweep geometry variable systems, addressing a reduced size regional aircraft Joint Wing model. These ideas were too complex to foresee a close perspective application on real planes. Further high speed WT tests were conducted at the facilities of CSIR, the Council for Scientific and Industrial Research (RSA), Fig. 28. In this case, rigid models were used, featuring morphing shapes occurring at different steps of the transformation. The structural effect was then lost but the information was useful to validate computational fluid dynamics (CFD) codes. NOVEMOR recognized the morphing trailing edge as the device showing greater potential for reducing the fuel consumption in all flight segments; the morphing leading edge instead resulted more effective for low speeds. Concerning this point, a relevant issue regarded the difficulty in assessing the actual advantages related to the employment of morphing devices in real operations. This is mainly due to the lack of data that can be used to validate the estimates. As a final consideration, EMBRAER recognized morphing as a design approach that could lead to significant benefits, compliant to the established European and International goals. It was also realized that a major issue for the effective implementation of such a technology is represented by the relatively small size of regional jet wings so that any envisaged solution should combine morphing and classical systems.

8.3.2 Future projects

Many projects in Europe started after SARISTU. Currently (2016), the second phase of the already mentioned JTI-GRA, the thematic Innovative Aircraft Demonstrator Platform (IADP), have assigned major contracts mainly devoted to develop morphable systems on adaptive wings. They are

FIG. 28

Wind tunnel transonic tests at CSIR.

concerned with the design and realization of adaptive leading edge, trailing edge, winglet and wing-tip with innovative architectures while trying to extend their applicability range. One of the drivers is in fact the reduction of the available systems complexity, to come to cheaper, more reliable and sustainable devices. Both WT and flight tests are forecast at the end of the studies, namely around 2020. Similar trends occur inside the technology stream Integrated Technology Demonstrator Air-frame (ITD), transversal to the products line and keeping some of the heritage from the previous SFWA Program [155–157].

Meanwhile, the EU research funding program Horizon 2020 (H2020) explicitly refers to "morphing" as one of the research lines for many of the listed topics [158]. On this theme, all the major industries are active in the Old Continent, including and above all Airbus. In this timeframe, it seems that particular attention is devoted to the structural optimization, from both the point of view of the topology and the reduction of the number of the system parts. In the United States, Boeing activities are continuing the already referenced projects like MACW and VCCTEF. CLEEN (Continuous Lower Energy, Emissions and Noise) is a multitasked research, aimed at improving technologies for greener transportation, sponsored among the others by the Federal Aviation Administration (FAA) and the International Civil Aviation Organization (ICAO) [159,160]. It covers very different topics, merging fuel, engine, and aerodynamic surfaces research. Concerning this latter, the activities are (once again) especially concentrated on adaptive trailing edges development. Indeed, it is not possible to talk properly about morphing devices, because the implemented system is a plain flap. Nevertheless, Boeing applied for a patent, released in 2015 [161].

Currently (2016), DARPA is focusing on the development of smart drivers and the assessment of suitable design environments for these new technologies as in the late funding opportunity Complex Adaptive System Composition and Design Environment (CASCADE). The goal of this program is to exploit novel modelling capabilities for a deeper understanding of component interactions and a unified view of the complex system behavior. It also aims at developing a formal language for designing sophisticated adaptive systems [162]. On these topics, there are explicit interests of practically all major aircraft companies worldwide. Bombardier recently closed an investigation on these themes, inside a cooperation between Italy and Canada (MDO-505, Morphing Architectures for Aerodynamic Performance Improvement, http://www.criaq.aero/Projets_de_recherche/detail_projet/12409213096). EMBRAER just come out of the successful project NOVEMOR and is heavily involved with smart on-board architectures like structural health monitoring systems [163]. In the same way, many others medium-small transport aircraft manufacturers are approaching the adaptive structures technology.

The sound sensation of adaptive wings is entering the common dictionary in the aeronautical and AS world and, in a short period, it is expected that morphing characteristics will be common on any kind of aircraft.

9 THE TRADITION AT THE UNIVERSITY OF NAPOLI AND CIRA

Traditionally, Italy has been very active in the field of smart structures and materials. The first work on the subject was published in 1988 [164], as the first cooperation in these matters between the University of Napoli and CIRA started, soon after the release of the first historical papers by Edberg and Crawley and de Luis (1987) [165,166]. These investigations did concern with noise and vibration control.

9.1 ADAPTIVE AIRFOIL

The first application of smart materials on an aeronautical airfoil for increasing its performance started a bit later, in 1997. The occasion was the transnational project Adaptive Airfoil, conceived inside the EREA group (Association of the European Research Establishments in Aeronautics) and involved five of the major research centers on the continent [167]. It was officially launched in January 1998 as the first cooperative research in Europe on this topic and was parallel to the studies carried out in Germany (DLR, Braunschweig) in the same period [108–111]. Both of them followed the results of ongoing research projects in aerodynamics (Euroshock I & II) that showed how localized geometrical modifications could lead to interesting improvements in terms of drag reduction and buffet effect alleviation [112,113]. Combined deformations (bumps) on the upper side of the airfoil around the leading edge zone (expansion shoulder bump, ESB), and close to shock foot respectively, towards the trailing edge region (rear shoulder bump, RSB), could significantly affect the aerodynamic field.

In "Adaptive Airfoil," the objective was to develop a mechanical system able to statically and dynamically shape the box skin of a transonic wing profile. The project had the participation of the Italian, German, Swedish, Spanish and French aerospace research centers (CIRA, DLR Gottingen, FOI—the former FFA, INTA, and ONERA) [168]. On these topics, CIRA funded a PhD activity at the University of Napoli [169], mainly focusing on the design of a structural system commanded by and integrated with piezoelectric ceramic actuators. The targets, envisaged to be applied to a quasitwo-dimensional structure, gave rise to a lot of interesting research topics, whose results and basic hypotheses intersected each other. For the involved investigators, this was the first evidence of the complexity of an aeronautical adaptive structure. Each specialization had its specific role [168]. Aerodynamics was mainly aimed at defining the shapes, able to attain the target benefits, together with the issue of proper models of the coupling between structural deformations and derived flow. Moving from the aerodynamic specifications, structural mechanics should have defined a structure able to achieve the desired shape under the combined loads deriving from the external pressure and the internal actuator actions. Furthermore, the conceived systems should have been able to produce the desired strain field in operational conditions.

Here was the adaptive structure paradox, appeared for the first time to the engaged investigators: while the construction should have been flexible enough to be driven by command forces, it should have been strong enough to resist the external loads. On the other side, the actuator system itself should have been able to lead the complex device, winning both its stiffness and the aerodynamic action. Stated in this way, it follows the actuators were trivially called to withstand at least part of the derived stress. This is the reason why the actuator system is generally referred to and not just "actuator;" it has to be in fact considered as a complex architecture aimed at bearing the external and internal tensions, and not as a mere force generator. What seems sound now, was not entirely understood at those times. This fact caused the project to achieve only a partial success. The system showed good performance in lab tests but excessive static and dynamic (i.e., vibrations) deformation would have occurred when the prototype had been exposed to an actual transonic flow. This uncertainty linked to its actual behavior, once in the flow of a certain transonic facility, brought the team to overcome the final functionality tests.

The sensor system was thought to be integrated into the structure in order to minimally interfere with the measured parameters, characteristic of the investigated phenomenon, and targeted to read both the aerodynamic (pressure) and the structural parameters (strains). Finally, the control system (here imagined for a feedback action on the oscillating shock wave—buffet) was thought robust enough

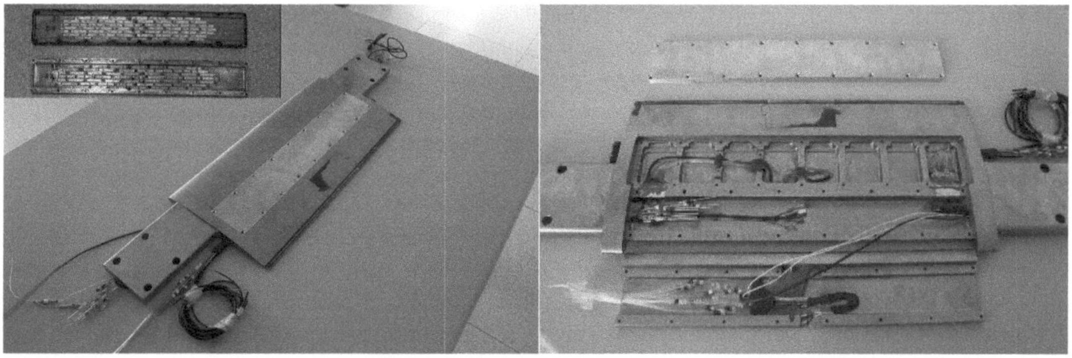

FIG. 29

The Adaptive Airfoil experimental prototype (left), integrated with kulites, static pressure sensors, (right) and piezo patches (top left, bottom of the adaptive bump, light gray in the picture).

to overcome the uncertainties related to the modelled physics. It is worthwhile to underline that the aforementioned peculiarities of each subsystem have been confirmed in any further investigation related to morphing, up to now. Finally, the team released a complex technological demonstrator, Fig. 29, incorporating complete actuator and sensor systems. It was entirely manufactured and assembled at CIRA. The uniqueness of this realization is also on its size: the whole object (a WT model) had a 20 cm chord, a 30 cm span, and a thickness of few centimeters. Static bump deformation (the light segment in the picture) was reached through a pneumatic differential pressure, inflating the metallic skin like a balloon. Two different piezoelectric based devices were implemented to generate dynamic oscillations. The first system (not reported in the figure and designed by INTA) used a devoted kinematics to amplify the displacements of piezoelectric stacks and acted as a 2D span-wise point force on a selected line of the bump, pulling and pushing it inward and outward, respectively. The second architecture applied piezoelectric patches, directly attached to the skin, to actuate the necessary strain to achieve the target displacements (picture at the top left in Fig. 29). This configuration was settled for the RSB while the ESB was figuratively implemented through rigid insertions [170–174].

9.2 THE HINGE-LESS WING

The former experience somehow demonstrated that piezoelectric ceramics could not have been applied to real aeronautical structures for guaranteeing the necessary strains and forces for morphing. Simply, the energy input was too small for the investigated physics. On the other side, this kind of smart material is well behaved for attenuating noise and vibration, whose physics is told of the "small disturbances." On the wake of the innovative ideas carried out by DARPA in the Smart Wing Program, attention was moved to SMAs. This material is the perfect complement to piezoelectric; it is able to give rise to high levels of force combined with high levels of strain but confined into the low frequency range (few Hertz at the most), only then convenient for static or quasistatic applications.

Nevertheless, there are many aerodynamic surfaces on a typical wing that do not necessitate a wide bandwidth to operate well: flaps, slats, and aerodynamic drag systems are usually activated in seconds. Control surfaces like ailerons instead need a higher band, around 20 Hz and could not be affected by such applications. It is relevant to remark that the choice to use Smart Materials (PZT or SMA) relies mainly on the possibility of realizing very compact and simple actuator systems. This fact envisages the generation of devices that could have a minimal impact on the shape of the aircraft, avoiding the use of classical nacelles to mask the huge actuation mechanisms for flaps, slats, and so on, necessary from the aerodynamic point of view. Moving from these considerations, the research objective was addressed to the realization of architectures, able to substitute the current mechanisms, installed on-board of wings.

CIRA then gave rise to an exploratory investigation on the use of SMAs for structural shaping (morphing was not a word that was so used at those times) in cooperation with the National Research Council of Italy (CNR). The "Hingeless Wing" project started in 2002 [175,176], while "Adaptive Airfoil" drifted to its conclusion during the same year. The investigation started from a standard aeronautical structure, to find some solution to make it "adaptable," or allow it to be able to change its configuration. The scientists working on the subject immediately understood that a common architecture was too stiff to permit the occurrence of suitable deformations (suitable to the prescribed aims). Therefore, they concluded that an adaptive structure could only be designed from the very beginning as a novel system, responding to different requirements than a standard one [177,178].

This statement seems trivial nowadays but it was not that common in those years. The obvious consequence was to guess that SMA or similar inner actuation systems, could only deform continuous structures if these latter were "flexible" enough [179]. In spite of this final deduction, the research consented to develop many kinds of very compact and original actuator devices, implementing SMA wires into appropriate arrangements to amplify their response in terms of displacement or force. In Fig. 30, the sketch of one of those original systems is shown [177,178]. The adopted layout allowed attaining the same performance (driven vertical displacement of a target elastic structure) than the one produced by classical longitudinal devices (direct actuation), achieving a substantial reduction of employed smart material mass (per unit of length). A novel active torque tube [180] and a tuneable dynamic panel [181–183], both based on an SMA-type actuation, were other remarkable output of this preliminary investigation.

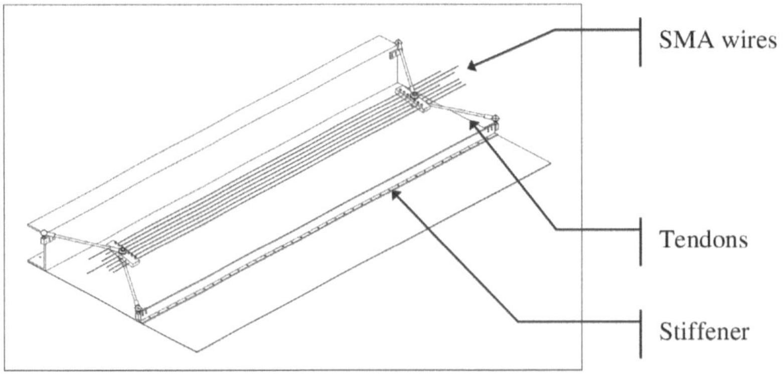

FIG. 30

SMA actuator system architecture [177,178].

9.3 SMARTFLAP

In 2006, Alenia (today, the Aeronautics Division of Leonardo Finmeccanica), assigned a contract to the University of Napoli and CIRA for the development of novel flaps for the current regional propeller aircraft, driven by smart materials (SMAs). This was a keystone for the activity of the entire research group, for the first time involved in the re-design of an existing component of a real structure. From the previous experiences, it was understood that the project could not produce a definitive answer to the issued problem, but it could be the occasion to better understand many former concepts. The first point was to define a system able to be deformed and bear external loads. On the wake of what was commonly done on the operative aircraft, it was decided to rely on the actuator architecture itself that should have been considered as a structural system (in a traditional flap, for instance, the load is born by the stiff rod that connects the motors to the aerodynamic surface) [184,185]. Since then, the group referred to these kind of devices as "structural actuators." However, because SMA elements should have been continuously driven and would have been completely part of the main structure, the resulting overall system would have had additive elastic properties. In turn, these would have strongly affected the dynamic response of the original structure. While the first point (continuous guidance) moved the investigators towards a specific and stiff device (the real actuator system), the second point (the actual embedment of the realized actuator system into the structural body) led to two separate architectures.

The actuator layout resulted in a very stiff arc, strong enough to afford the external aerodynamic loads and, at the same time, able to accomplish the SMA action [186,187]. SMAs ribbons, connected to the extremal points of the structural titanium arcs, drove its deformation that should have been maintained under the action of the aerodynamic loads. The resulting stress field would have been generated by both these contributions, internal and external excitations. In lab, they were simulated through a devoted weight arrangement. The designed system demonstrated to face these combined forces well and was successfully patented [188]. In spite of the stiffness, however, relevant dynamic characteristics were shown, resulting in significant modal response variations. A simple combination of these devices permitted to realize an "active rib." Statically, it proved to be a powerful force generator, achieving displacements that were completely in line with the specifications, Fig. 31 [189]. Meanwhile, a

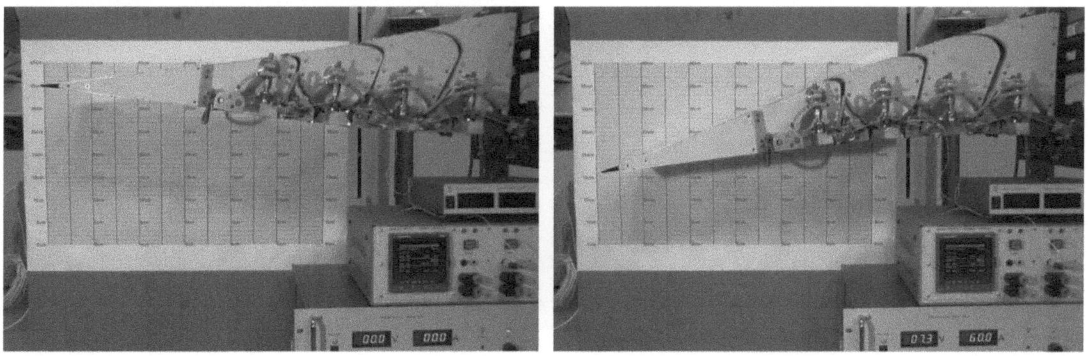

FIG. 31

Elastic morphing ribs, mounted on a dummy demonstrator [189].

dynamic analysis confirmed its complexity (in terms of eigenfrequencies and related vectors) and prospected the sound necessity of carrying out devoted aeroelastic investigations. The single-bay flap, around 20 cm span and 30 cm chord, underwent an extensive modal campaign, showing remarkable results despite its geometrical extension limits. In particular, a strong coupling between bending and torsional modes was highlighted; it would have had potential consequences on aeroelastic instabilities and, led to serious constraints in the design process [190]. A further investigation was then dedicated to the assessment of a proper design methodology, taking into account the complexity of the structural layout [191].

A way to minimize the impact of the augmented elasticity (that could never be completely avoided), was to transform a classical aeronautical structure into a hinged kinematics, driven by the same actuator. The resulting design was compact and effective, Fig. 32 [192]. The effect on dynamics was reduced and the static elastic response to the external loads was well diminished. However, using a smaller number of devices in the available room that was allowed (for the referred configuration, they were the only load-bearing elements), the structural capabilities were also decreased, leading to lower safety factors. Again, functional lab tests did verify the morphing capability of the proposed trailing-edge structure, either loaded or unloaded, fully integrating the complete actuator mechanical kinematics within the structure. The required energy level was coherent with the theoretical estimates, a sign of having realized efficient mechanics. At this stage, the performed activities were addressed to verify the feasibility of assessing a morphing architecture implementing SMA actuators. Installation, production, maintenance, fatigue behavior, and, above all, certification of these kinds of devices are nowadays still an issue, (2016). However, the achieved results allowed expressing a first positive opinion on the reliability of the general concept and let the researchers move towards other, more assessed actuation devices while respecting weight and volume limitations. In fact, it was understood that in order to increase the TRL of the demonstrated concept, it was necessary to implement commercial products. Finally, the investigated solutions evidenced the strong difference between compliant and kinematic architectures. The first one showed better integration possibilities and either a wide choice of layouts but complex dynamic responses. The second one proved to give rise to a stiffer system, leading to important gains in terms of modal and aeroelastic characteristics.

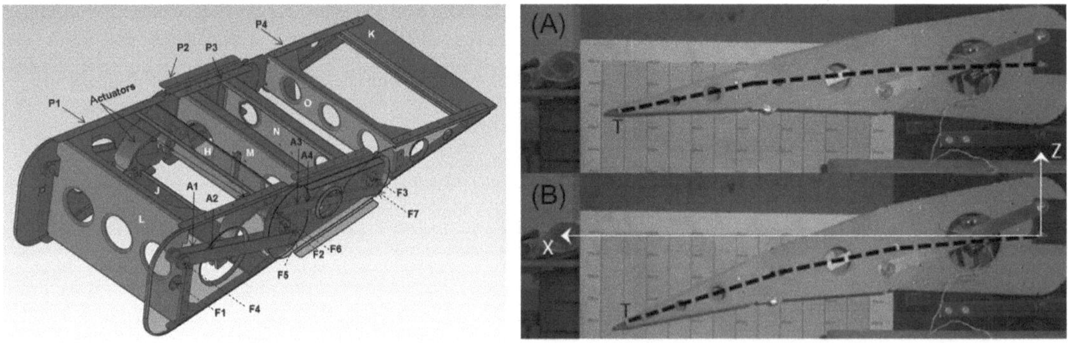

FIG. 32

Morphing trailing edge, actuated by an inner SMA-based device: structural schematics (left); operation example (right) [192].

9.4 SADE

The European project SADE (Smart High Lift Devices for Next Generation Wings, 2008–12), aimed at developing adaptive structural technologies for augmenting laminar wing capabilities, reducing noise emissions and high-lift qualities. DLR led the working group, made of industries (Airbus and Piaggio), research centers and universities. In that research, special attention was devoted to the leading-edge segment. The activities addressed the use of electric motors, transmitting motion to the wing skin through dedicated internal architectures. The issue of realizing such a skin, able to resist the external loads, while guaranteeing a sufficient deformability required the involvement of many scientific and technological disciplines. Another relevant point, then followed by future projects on structural morphing, was the demonstration on full-scale elements and a devoted investigation on the aeroelastic effects. In fact, adaptive devices are tightly correlated to their own size. Volume and stiffness requirements strongly condition their design and performance. Systems showing great potentialities at a certain scale, could completely fail if enlarged or reduced. It is therefore definitely necessary to act on the real article dimensions.

CIRAs role was focused on the development of a compliant mechanism for implementing a droop nose effect. The architecture was based on a metal skin, deformed by a kinematic system and in turn driven by a single motor, Fig. 33. The adopted configuration permitted achieving the large specified displacements without overcoming the material limits and without buckling. The deformations could be modulated through an internal architecture, controlling the structural stiffness for governing shape variations and nose retractions. An important feature was its capability to enable compliant movements by taking advantage of the external aerodynamic forces [193,194]. Special FE tools were assessed for designing the kinematics while considering the action of both the internal and the pressure loads. A heuristic optimization process was set-up to assess the architecture topology and to size the structural elements, whose admissible domain was constrained by the material characteristics and the required geometry envelope. CIRA activities concluded with the manufacture of the proposed full-scale device and the experimental verification of its functionality. Attained shape variations and global displacements were measured while monitoring the insurgent skin stresses. Finally, results permitted to refine the numerical models, previously released [195,196]. It is worthwhile to note that, holding the

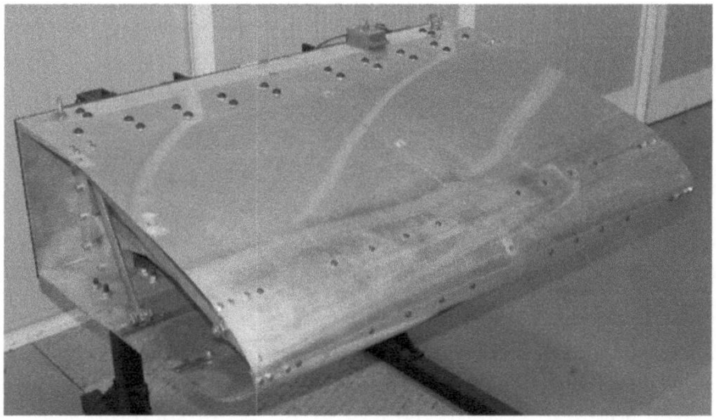

FIG. 33

The CIRA full-scale droop nose demonstrator, released inside the European project SADE.

differences between scaled and real devices, the design process used a downsized prototype as a preliminary benchmark for testing the proposed concept. In this case, the investigation focused on the selected architecture and the way of actuating deformations, neglecting the local strain levels [197]. The proposed system was patented and the related process was entirely endorsed by the European project [198].

9.5 CLEAN SKY—JTI-GRA—LOW NOISE

Led by Leonardo Finmeccanica (the former Alenia Aermacchi) in the general frame of JTI (joint technology initiative), the European cooperation program in Aeronautics, the GRA (Green Regional Aircraft) has been one of the industrial development platforms. It has involved more than 30 stakeholders, including engine manufacturers, system engineering companies, research centers, universities, and small and medium enterprises (SME). Directed to concepts for improving the environmental impact of medium size and medium range aircraft, the program has faced problems concerning the reduction of the aircraft weight and its energy consumption. This general requirement forced the necessity of increasing the aerodynamic performance of the plane. In its first phase, GRA was structured in different subprograms, each targeting a specific topic. In that frame, CIRA and the University of Napoli have developed suitable morphing concepts inside the low weight configuration (LWC, 2008–16).

In detail, the activities were aimed at designing a more efficient flap device in terms of produced acoustic emissions during take-off and landing. The adaptive structural system was conceived as an additional component to be applied to the existing flap without modifying the basic geometrical envelop, with the twofold objective of reducing the necessary operational deflection angles by attaining a more efficient aerodynamic shape and diminishing the overall complexity by limiting the flap series extension (for instance, allowing the transition from a double to a single slot configuration). While cooperating at the level of requirements for refinement and the general concept assessment, the groups of the University of Napoli and CIRA worked on distinct architectures. CIRAs idea was based on a semicompliant layout: the single ribs were divided into blocks, each connected by "X-springs," properly sized to bear the aerodynamic loads and allow the desired deflections [199–201]. Integrated SMA wires actuated the whole system, powered by an original heat device, conceived and realized by the Integrated Aerospace Sciences Corporation (INASCO) inside a thematic call (an external project sponsored by the EU in the frame of JTI-GRA). A simple 2-bay demonstrator was manufactured and tested to validate the concept and verify the predicted structural performance, Fig. 34 [202]. The system realized by the University of Naples relied on a configuration where the basic rib was divided into three chord-wise blocks, hinged and driven by a rotational motor through dedicated kinematics. Correspondent chord-wise blocks were connected span-wise to attain an adequate torsional stiffness. This device underwent laboratory tests to verify its behavior with and without external loads, Fig. 35 [203].

Those activities were carried out inside a technological innovation frame where many other proposals concurred towards the same goals. They included the development of novel aerodynamic surfaces, innovative active flow control devices, and so on. All such systems were finally evaluated vs. their estimated performance (by lab test measurements), the actual use of available and assessed technologies and their real perspective applicability (including issues related to integration and maintainability). Both the proposed adaptive systems overcame the established threshold and were classified at the very first positions. The University of Napoli architecture ranked absolute first and was endorsed with an additional budget to come to a full-scale laboratory system (more than 3 m span for half a meter

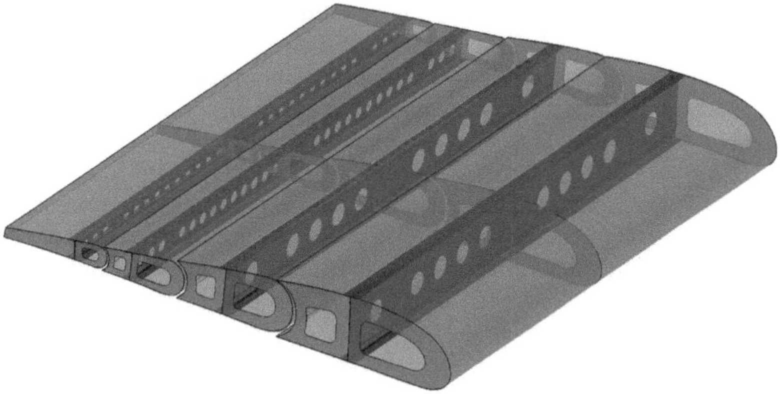

FIG. 34

CIRA adaptive flap demonstrator: CAD sketch [202].

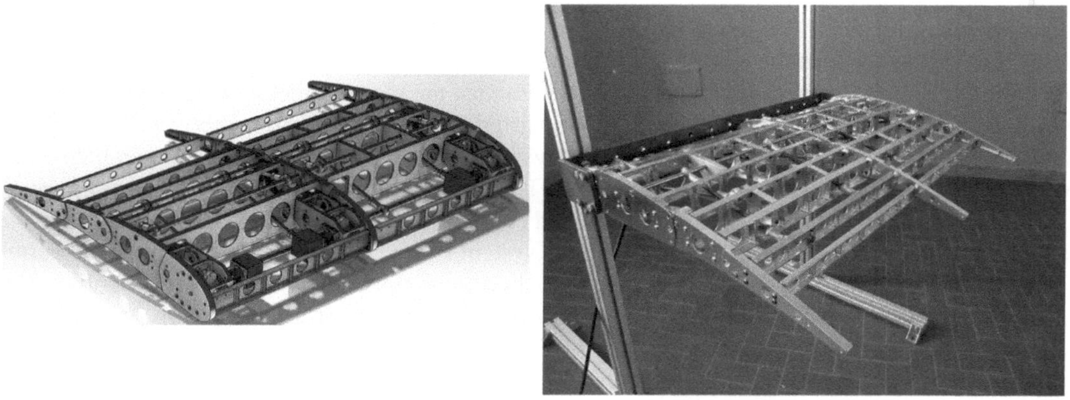

FIG. 35

University of Napoli adaptive flap demonstrator: CAD rendering and lab demonstrator [203].

chord). The final configuration included an additive system to allow rigid deflections (like a classical flap) while ensuring the morphing movement (via the aforementioned three-hinge mechanism) in order to attain a sort of augmented high-lift device [204]. The system was finally released during 2016. It gained a nomination for the "Best Clean Sky Project Award in 2016, as one of the best examples of successful activities performed within the Clean Sky Programme."

9.6 EU—SARISTU

The European project SARISTU (Smart Intelligent Aircraft Structures, 2011–15) was a large industrial project involving more than 50 partners across the continent for a budget of over 50 million Euro [121]. Divided into a technological development phase and an integration phase, it was devoted at assessing critical systems for the implementation of morphing, intelligent sensing, and multifunctional materials over the wing and the fuselage of medium size commercial aircraft. Herein, CIRA and the University of Napoli were deeply involved in the assessment of a wing morphing trailing edge to minimize the deviations from the current optimal layout during a typical mission, compensating the weight reduction following the fuel use. This special task was carried out together with other valuable partners. Industries were involved for the specifications issue (Airbus and Alenia Aermacchi, now the Aeronautics Division of Leonardo Finmeccanica), the exploitation of the assessed configurations to other aircraft models (IAI), the manufacture of dummy test articles (ATARD) and the release of the demonstrator for final lab and WT tests (Aernnova). The other research centers (Fraunhofer Institute with its institutes IFAM and ENAS, and ONERA) developed morphing skins, the electronic architecture and suitable aerodynamic models for the performance prediction and the optimization of the wing shape at each stage of morphing, respectively. Mare Engineering, INASCO and Innowattech took care of detailed FEM simulations, also addressed to stress analysis, the support to the sensor network development, and the preliminary aeroelastic simulations, in the order [205].

The architectural layout was defined, following the specifications issued by Alenia and Airbus, with the support of the Polytechnic of Milano [206]. Extended over 70% of the wing span, the width of the adaptive device was fixed to 10% of the mean aerodynamic chord (MAC). It generally resulted in a small percentage of the aft region of the wing, the rear spar being located at around 70% of the local chord. The region between the wing box aft spar and the active trailing edge movable part was familiarly called "dead box" (fixed region) and was used to host the motors and the main body of the cable system in the manufacturing phase. The ATED consisted of four main items, namely the structural, the actuator, the sensor, and the skin system. It was driven by a devoted control architecture, implementing both a feedforward and feedback logic [207,208]. The first configuration applied a classical joystick-type action, where the kinematics followed a direct external command. The second configuration was meant to approach the real operation. In flight, the system is thought to get information on the current weight or, more properly, on the instantaneous fuel consumption and adapt the wing camber to the optimal shape for achieving minimum. The integration with the aircraft flight systems was not considered at this stage and the feedback action moved from a representation of the instantaneous form. The sensor system gathered and elaborated information for reconstructing the actual trailing edge deformation as a product of the combined action of the internal actuators and the aerodynamic loads [209].

The structural system core consisted of deformable kinematic ribs, each made of subsequent hinged blocks in a typical finger-like outline. Mechanical cross-links connected the various pieces to attain a SDOF architecture. Longitudinal stiffeners (minor spars) were deployed to increase the torsional characteristics of the structure and to provide a support to the morphing skin, Fig. 36 [210,211]. This SDOF mechanism was constrained by the actuator system, designed to bear loads, as the corresponding classical flap or aileron devices do. It was completely integrated into the referenced geometry so that a clean aerodynamic shape resulted, without the necessity of including any further cover (nacelle) [212–214]. The real structural system was then made of the combination of the structural skeleton and the actuation apparatus. It is worthwhile to underline that this architecture was designed according to Certification Specification N.25 (CS-25) regulations, in order to get a product compliant with industrial requirements at the largest extent as possible, though not yet commercial. The stress analysis and the experimental characterization campaigns confirmed the robustness of the chosen configuration and evidenced large margins of improvement [125,215,216].

The sensor network was completely integrated in the referenced geometry as well. It was based on the use of fiber optic devices (Fiber Bragg's Gratings, FBG), arranged to get information on the

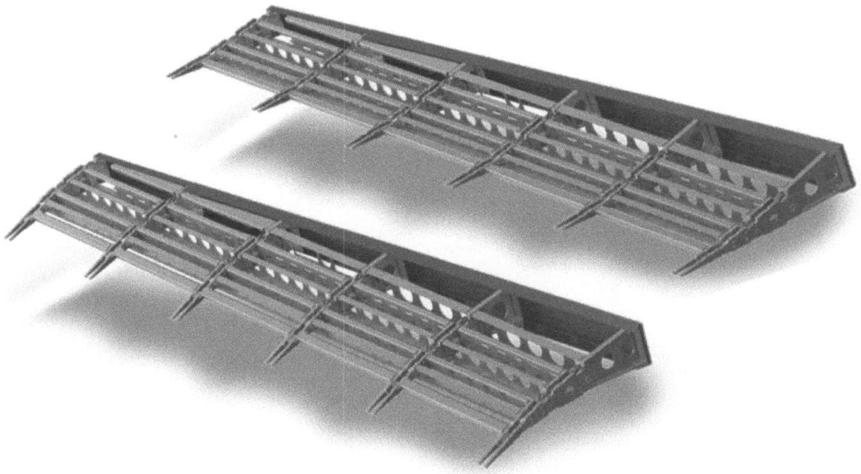

FIG. 36

The ATED kinematics [210,211].

longitudinal and the transversal strains. Because of the high deformation levels related to a morphing structure, special layouts were necessary to match those values and the use of commercial products, whose range is generally limited to 1%–2% [217–219]. The Dutch firms KVE and Technobis supported the ATED partnership in the realization of this component [220], the so-called "sliding beam." Aiming at the generalization of the proposed concept, this specific activity led to interesting fallouts as the proposal of an innovative sensor system, implementing the same basic idea [221,222]. A targeted study was then carried out with the aim of verifying the applicability of other kind of transducer technologies, with the main aim to further reduce the cumbersome problem of the cabling deployment. Wireless architectures, specifically based on the use of graphene sensors were investigated [223,224]. Structure, actuator, sensor components were basically standard items, generally commercial, arranged into a novel and integrated (smart) configuration. The skin system was instead an innovative and breakthrough element. In that case, several constraints should have been considered. For instance, the resulting surface had to undergo deformations larger than 5% for several cycles (magnitude 10^5; the system was supposed to operate a single cycle per flight, equivalent to the fuel consumption history) while maintaining its mechanical properties at the target temperature interval, ranging between −80°C and 50°C [225,226].

The ATED design was supported by several parallel studies. Starting from the initial assumptions, aerodynamic computations looked for the optimal shape and the best distribution of the hinges and the geometrical architecture [206]. Special attention was paid to the static and dynamic behavior of the morphing skin under the aerodynamic loads. Aeroelastic studies were necessary to characterize the system behavior, having an increased number of DOF, different elastic characteristics (both distributed and concentrated) and additional mass contributions [227]. The manufacture and assembly of such a complex object was not an easy task as a direct consequence of having transformed a passive structure in an articulated machine, Fig. 37. The construction experience in itself was a project apart. The team faced the trivial fact that once the design philosophy had been changed, the manufacture and

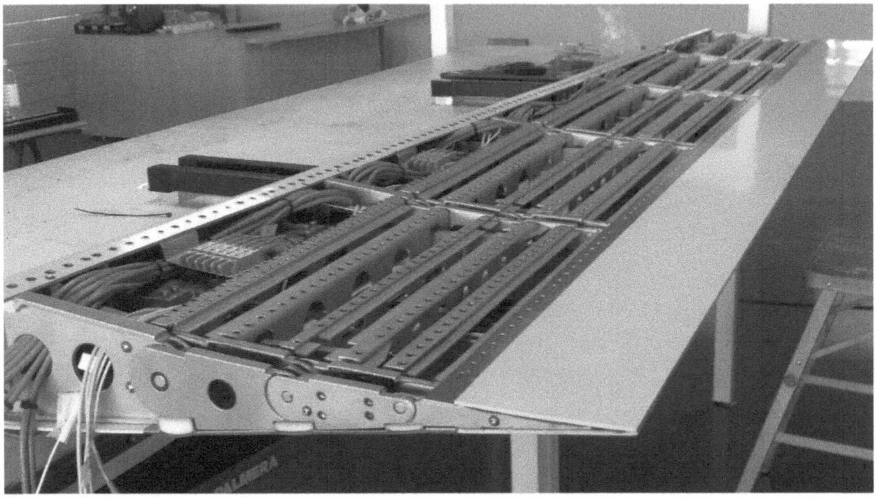

FIG. 37

The adaptive trailing edge device (ATED) structure like a mechanical machine.

the assembly strategy should have been changed in turn. Dedicated tools and rigs to accomplish the integration needs were then imagined and realized [228]. To tackle this additional difficulty, a preliminary task was introduced, for producing, testing, and characterizing simplified versions of the ATED at increasing level of difficulty [215]. In this way, the partnership gained essential basic confidence with the matter, to be spent during the final release activities. Similarly, preliminary WT tests were conducted at the University of Napoli on a reduced span model, to get a preliminary knowledge of the system behavior under aerodynamic loads and installation issues [228].

Already at this stage, the study considered impacts on the maintenance, the operations, and the safety through a devoted FHA study [229]. The promised benefits in terms of fuel consumption were estimated larger (around 3%) than the weight penalty associated to its final realization (around 1%). Open points are the correlated fuel cost savings at each flight versus the additive periodical maintenance costs required [230–232]. An interesting exploitation study was also addressed, to verify the prospect to expand the application of the investigated concept to other classes of aircraft like business jets [183]. Finally, the ATED was installed onto the complete wing mock-up at the Alenia Aermacchi plants in Pomigliano D'Arco—I and then moved to the WT facilities of Tsagi in Zhukovsky, Moscow Region—RUS, where the final test campaign took place; there, it and the overall adaptive assembly proved their full functionality.

9.7 THE ADAPTIVE AILERON

Following the enthusiastic results achieved with the ATED, the team moved to a new challenge. The ATED original configuration did refer to the standard layout of the flap, leaving apart the aileron region. There are several reasons to leave that part unchanged. The most relevant is the aileron has a crucial function in flight and its failure would lead to potentially catastrophic scenarios. Because of the limited size of the referred zone, the integration of an active system would have furthermore interested the aileron volume completely and, for the same reason, there would have been major impacts on the components installation (above all, the actuators). Finally, because the aeroelastic response of that control surface is critical, the strong effects following its substantial modification (again, important variations of stiffness and mass) should have been deeply studied. On the other hand, the former studies showed that tip regions (the farthest from the wing root) give the most significant contribution in the process to attain the optimal aerodynamic shape for compensating the weight reduction, following the fuel consumption in cruise. This was the reason why it was chosen to investigate the possibility to extend the developed technology to the aileron. The activity was proposed within a joint Italian-Canadian research activity fostered by the Consortium de Recherche et d'Innovation en Aérospatiale au Québec (CRIAQ) and participated by the Canadian branches of Thales Aerospace and Bombardier AS (http://en.etsmtl.ca/Unites-de-recherche/LARCASE/Recherche-et-innovation/Projets).

The project aimed at realizing a full-scale demonstrator of a tip wing section for investigating the capability to ensure a certain level of flow control and aerodynamic performance increase through the the combined action of an adaptive wing box and a morphing trailing edge. The Canadian team (École de Technologie Supérieure, ETS, and the National Canadian Research Council, NRC) was in charge of the first device definition, while the Italian group (University of Napoli and CIRA), targeted the preparation of a shape variable 4-bay aileron. All the previously listed problems were evident at the very first approach [233,234]. The volume limitations forced the designers to change the strategy, followed for the ATED. Instead of having a couple of actuators acting on each rib, the architectural layout was specialized for each single bay. A single actuator drove each of the two inner ribs (aileron root), while the outer two bays (aileron tip) were driven by a single motor, deployed at the third, middle rib. In other words, the last two segments were made of two slave and one master ribs, commanded by a single active device. Calculations proved that this configuration could bear the specified loads, both internal (actuation forces) and external (aerodynamic pressure) [235,236]. The aeroelastic studies confirmed the system reliability, demonstrating the absence of related instabilities in the design conditions [237]. It should be remarked that the system design referred to WT test and not to flight parameters; therefore, low speeds were considered, limited to $M = 0.2$.

This important limitation should not minimize the success of the experiment: in fact, this activity allowed testing several new solutions, impossible to try otherwise. Among those, the most relevant are the presence of a single actuator per rib in the inner region, the choice to have a shaft configuration to drive three ribs at the outer region with a single motor, the complete absence of a stretchable skin and, above all, the coexistence of the classical aileron motor and the morphing actuator systems. It was necessary in fact to safeguard the basic aileron functionality while ensuring deformability characteristics. The former device, part of the reference configuration, acted on the main shaft; the design aimed at preserving its characteristic dynamic response for flight control purposes. The latter devices acted on the ribs to properly morph the shape. The requirement was then to maintain a certain aileron shape for the functional deflections necessary for roll control, yet giving it the possibility of assuming different static cambers. Concerning the skin, the possibility of avoiding the use of a heavy and sophisticated compliant solution, was checked. Then, the single hinged blocks were designed in order that they could slide one into the other, like a meniscus, preserving a certain geometrical regularity. This possibility was strongly correlated to manufacture tolerances and assembly precision: in fact, small deviations would have had significant impacts on the kinematics behavior [238,239]. As usual, advantages and disadvantages try to compensate each other.

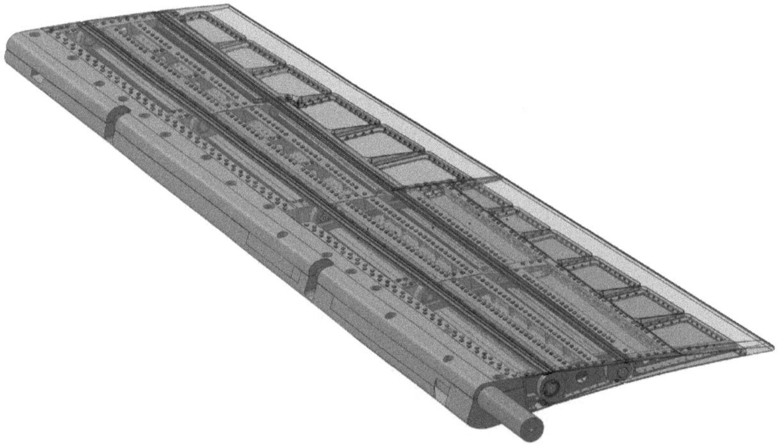

FIG. 38

The adaptive aileron—CAD.

The manufactured system was assembled onto a wing section model and tested in a WT at the NRC facilities in Ottawa (CAN). A schematic of the morphing aileron is shown in Fig. 38. On the same model, the adaptive wing box was also installed, designed, and manufactured by the Canadian team. The adaptive aileron device proved its functionality in the designed flow conditions [240]. The project also took advantage of previous studies carried out on the adaptive morphing trailing edge regarding the sensor and control equipment. An exploitation of the formerly produced concepts on the integrated sensor network was also tried. In this case, they were limited to theoretical predictions, yet they produced interesting results [241].

9.7.1 The AG2 project (JTI-GRA2) and the next future

On the wake of the successes of JTI-GRA—Low Noise, SARISTU and Adaptive Aileron, CIRA was called to coordinate an international working group within the second frame of Clean Sky, JTI—GRA, aimed at developing selected innovative technologies for the new generation of regional aircraft. The team focused their attention on morphing, structural monitoring, and novel aerodynamic surfaces for improved flow control (JTI-CS2-2014-CPW01-REG-01-02; Advanced Wing for Regional Aircraft—Technologies Development, Design and Manufacturing for FTB#1).

Design specifications were assigned by the Aeronautic Division of Leonardo Finmeccanica, with the final target of testing the planned systems in WT and in flight. The research on the adaptive structures is directly coordinated by CIRA and is articulated in four more tasks. The first one is devoted to the release of an improved adaptive trailing edge system and is in charge of a team led by the University of Napoli. Polytechnic of Milano is responsible for the second one, dealing with the set-up of a novel adaptive droop nose concept for the laminar flow preservation and the hyper lift generation, and the third one, aimed at developing an innovative wingtip for load alleviation purposes. The last one concerns the arrangement of an adaptive winglet and is coordinated by CIRA itself. Like SARISTU, this project (AG2, 2015–22) is focused on a medium-range, medium-size turboprop aircraft and is focused on the expansion or refinement of already assessed technologies. For instance, the adaptive trailing edge is installed on the flap; it will maintain its original functions while assuming different cambers in cruise, and will give some additional advantages in take-off and landing. In the same way, the adaptive winglet and the innovative wingtip should prove advantages for gust alleviation and load control operations Fig. 39.

Currently (2016), the University of Napoli and CIRA are very active in several research proposals, within Europe and National frames. The two partners are devoting more and more attention to the improvement of the developed prototypes and the possibility of facing flight tests. The available morphing architecture is ready for real aircraft, apart from the due actions of adapting its layout to the actual configuration of the selected machine and performing the integration with the other on-board systems. Flight tests are expected to take place by 2020 and should consider several evolutions of the original concept. The first step regards a strong topology optimization of the architecture, and the objective is to come to a 30% reduction of the current weight. The simplification of the device is the second target, aimed at reducing the parts number by 50%.

The concept evolution may also involve a reduction of the DOF that would in turn generate a deterioration of the perspective performance. This operation needs careful analyses before it is implemented. While this progress is necessary to increase the potential fly-ability of the architecture, there is a growing demand for more capable systems, consequently including kinematics with augmented degrees of freedom (for instance, refer to https://www.youtube.com/watch?v=2NfGBKFBRpQ). The possibility to exploit some compliant concepts at a certain extent will also be explored. At this stage, it is possible to think that future morphing system architectures may have a hybrid nature, trying to combine the benefit of the two approaches. Within this scenario, a step forward should involve the

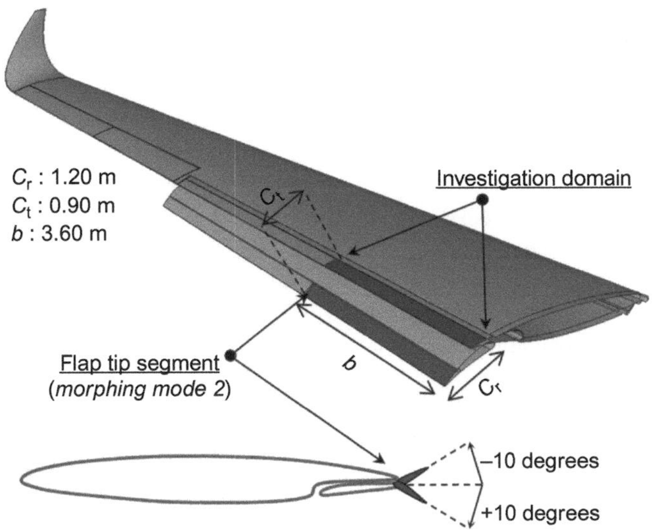

FIG. 39

AG2; conceptual vision of an adaptive flap trailing edge.

migration of current actuation systems to smart material-based architectures. Open issues, related to time response, load bearing ability, and their intrinsic behavior should then be assessed (fatigue life, durability, aging, and so on). It is also easy to anticipate that to get systems that are more robust, much effort should be devoted to the improvement of the skin surface. There are many opportunities on the table, planned to be verified by systematic lab and in-flight experiments.

10 FUTURE PERSPECTIVES

10.1 SAFE DESIGN

Structural kinrematic adaptive systems bring important deals with respect to safety [242]. First of all, because of the higher number of parts, the probability of failure increases for a complex system. A key point is then the capability of reducing the number of parts, classically a good index of the system fragility (or, on the contrary, its inverse is a good measure of the system robustness). Classical kinematic solutions are therefore consolidated on one side and are complex on the other. Fully compliant systems are instead structural components, forced to continuous deformation along an established path by relatively simple actuation architectures. This philosophy was addressed by the joint teams of Flexsys, NASA, and Gulfstream. It has shown its major disadvantage in being strongly exposed to very severe aeroelastic behavior by increasing its eigenmodes [243]. It should also be added that mechanical devices are more rigid and their behavior is better known in literature and in the aeronautical field, as a simple evolution of the flap, aileron, slat, etc. It is also worthwhile to discuss mass contribution. In principle, it shall be higher for kinematic chains than compliant structures. Still in this condition, the modal density is expected to be larger for the latter than the former, meaning very low rigidity levels for those architectures. This can also be an issue for the static response.

Motors themselves are worth some dedicated statements. A kinematic actuation requires distributed active devices to achieve the required performance. A compliant architecture could rely on a smaller number of actuators because the driven structure is conceived as a continuous system. Please note that, the larger the actuators number, generally, the higher the stiffness. Now, from the point of view of maintenance, costs, operability, and hazards, the larger the number of motors, the higher the risk of failure, and the higher the complete system price. It should be added that a system with many devices can sometimes be attributed with a certain level of redundancy and safety (as each component is less significant with respect to the whole). In any case, the reduction of the installed motors may be a design indication for the future. A way to accomplish this achievement is the use of a shaft architecture, for a compliant as well as a kinematic structural system, Fig. 40 [36].

Generally, from the certification point of view, a kinematic system is more acceptable than a compliant system because it derives from very large experience on similar architectures. In other words, it is already in the *forma mentis* of the engineers and regulation public officers. Compliant systems have a long way to go before they can be fully accepted. This same statement applies to manufacture (tolerances) and assembly (jigs and tools) processes, as well as maintenance (inspections and repairs). On the other side, if the kinematics is complex, the associated costs will stay high, in spite of the acceptable levels of safety. However, when all these contributions are combined and subjected to a qualitative metric, the results show a substantial equivalence of the two philosophies. A subjective evaluation is herein reported as an application exercise, performed

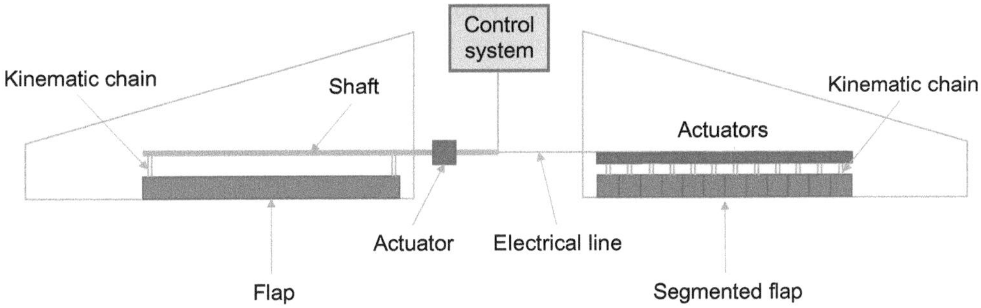

FIG. 40

Conceptual comparison b/w conventional shafted (left) and distributed (right) actuation layouts.

Table 2 Comparison Between Compliant and Kinematic Architecture Complexity. Measures: L, low; M, medium; H, high. Score: 1–5.

Char	DOF	K	Parts	M	MoD	Ael	Shaft	Cert	FHA	Main	Man	Asse
Com	H	L	*L*	*L*	H	H	*H*	H	H	H	L/M	H
Kin	*M*	*L/M*	H	M/H	*M*	*M*	M	*M*	*M/H*	*M/H*	*L*	*M/H*
μ=:27	1	1	*5*	*5*	1	1	*5*	1	1	1	4	1
μ:=31	*3*	2	1	2	*3*	*3*	3	*3*	2	2	*5*	2

Ael, *aeroelastic behavior criticality;* Asse, *assembly complexity;* Cert, *certification issue;* Char, *characteristic;* Com, *compliant system;* DOF, *degrees of freedom;* FHA, *failure and hazard analysis;* K, *stiffness;* Kin, *kinematic system;* M, *mass;* Main, *maintenance complexity;* Man, *manufacture complexity;* MoD, *modal density;* Parts, *parts number;* Shaft, *shaft insertion possibility;* μ, *merit.*

by the authors per their personal experience and feelings, Table 2. Furthermore, both the resulting figures of merit are around 2.5 on a scale 1 to 5, meaning the two architectures are far from a good 4/5 rank. It is herein understood as the DLR solution for their adaptive droop nose (SADE, SARISTU) results in a very smart compromise between compliant (structural skin) and kinematic (motor system) configurations.

10.2 SKINS AND FILLERS

Kinematic and compliant architectures both rely on very peculiar skins. These components should undergo significant loads, exhibit large strains and be exposed to harsh aeronautical environment (for instance, a temperature range between −50°C and 80°C). Their in-plane stiffness can dramatically affect the actuator size, because it constitutes an extra-load to overcome. A good question is whether the use of such a skin may be omitted, thereby removing a lot of practical complications. Unfortunately, the easy answer is no (at least, currently), because the implementation of piecewise architectures lead to even more complications, also different from aerodynamic penalties. Gaps give rise to higher radiated noise and higher relevance to tolerance in manufacturing and fabrication, affecting the flow quality for the separation among the parts and the interaction with the air itself. This last point does not imply a louder sound radiation but the insurgence of broadband structural vibrations. The use of continuous (piece-wise?) skins should then be somehow addressed. Because morphing deals with large deformations, it is almost trivial to find out that some solutions rely on the development of polymeric materials, suited for the referred environment [11–252]. Some concerns remain about the industrial fabrication process, the resistance to fatigue, the aging behavior, the cost, and the weight. Another point is the compliance to vertical loads. Because a polymer is generally soft, easily stretchable, and homogeneous, it also deforms along the vertical directions, causing bubbles that can have an impact on its performance. This bubbling phenomenon is connected to the pressure gradient between the internal (dynamic) and the external (static) pressure. In turn, the effect may be either static, producing a wake enlargement, or dynamic, inducing flow oscillations. In the first case, the effects should be confined, because the flow is already separated at the trailing edge region. The second effect can usually be neglected if the displacement amplitude is limited to few mm.

Fiberglass is a material that satisfies high strain requirements and shows reasonable and stable mechanical properties [123–243,246–254]. It well behaves under load and has good characteristics of aging and fatigue life. It is a common aircraft material and its density is acceptable. However, as a structural material, its stretching requires high forces. Because they can only be generated by the inner actuators, the device size increases, regardless of the external aerodynamic pressure. On the other hand, because fiberglass is a composite, its structural response can be opportunely modelled. The best (ideal) choice is something that exhibits infinite bending vs. zero normal stiffness. It is tough to reach this performance but may be a main driver. Specified applications could make the difference: leading edge technologies should consider further resistance needs for bird strike, while trailing edge and wing box should take into account hail effects. Both considerations lead to thickness and, then, weight increase. Another important requirement regarding morphing loads is the transversal elongation or deformation occurring in the material plane direction (by the inner actuator system). Poisson's coupling may generate some difficulties because stretch or elongation is generally desired along a selected direction, giving rise to large stresses at the transversal, constrained borders. Auxetic materials may be a viable solution to this issue [27–243,245–255]. They have zero or negative Poisson's ratio, so that they cannot be transversally ripped while being loaded along a specific line. Furthermore, they show several interesting properties that could produce applications well beyond the AS sector, Fig. 41 [256]. Their basic characteristics lead to unusual and interesting behaviors, as the one illustrated in Fig. 42 [257], where the material tries to compact itself, becoming denser under an external pressure. The picture is a personal authors' interpretation of a concept reported in literature.

In general, material systems can be structures at micro or macro scale, in order to have auxetic capabilities. A special kind of such a product is the one based on chiral architectures [28–243,245–258]. Of course, zero Poisson's module value is better for morphing and can be attained for instance by creating a weighted mixture of positive and negative materials. Applications may range from skins to fillers, aimed at sealing possible gaps generated by the morphing architectures (as in the case of naked meniscus systems). Traditionally, special plasters like Flexcore [259,260] are associated to this concept. In this case, the bulk material is a macroscopic arrangement of laminae, repeatedly bent, that offers a good shape resistance but no longitudinal stiffness. Their appearance resembles typical honeycomb architectures. Among the other relevant properties, auxetic materials also have the peculiarity to exhibit large displacements while limiting the overall stress levels. In fact, these latter distribute

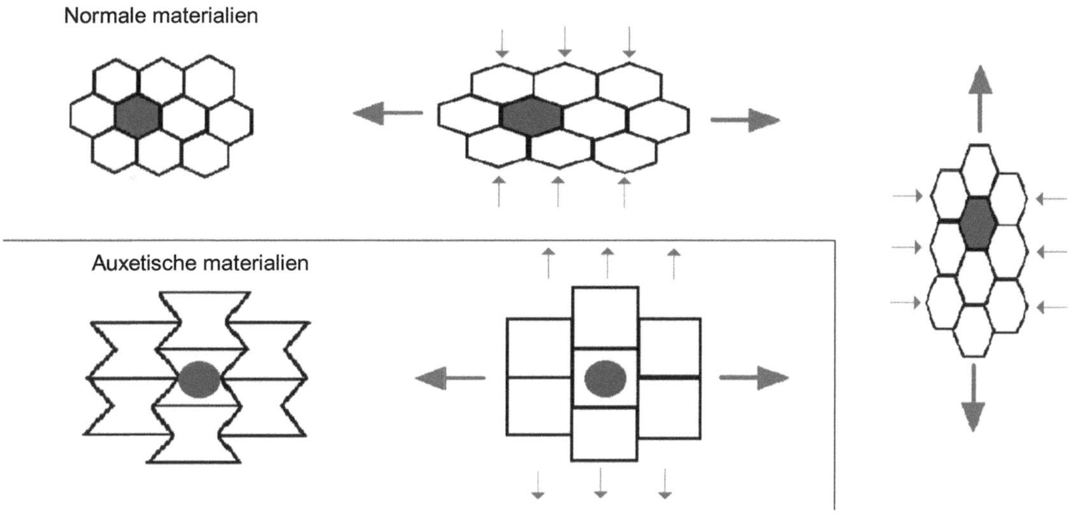

FIG. 41

Auxetic material and its behavior vs a traditional material (left) [256].

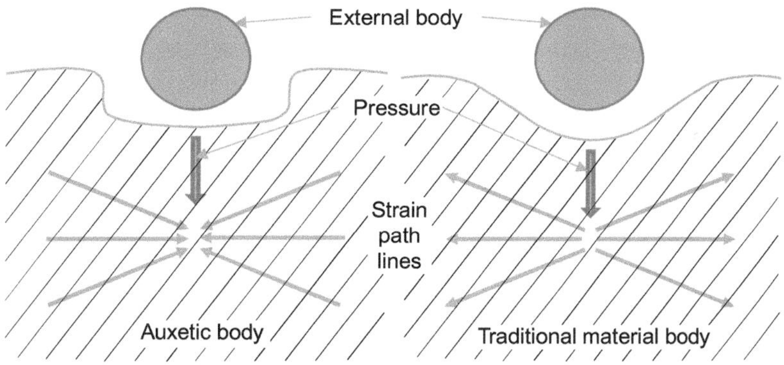

FIG. 42

Auxetic behavior under a pressure force (schematic) [257].

uniformly as a direct consequence of the geometrical topology. This peculiarity is almost essential for developing appropriate morphing structures, where the most important challenge is to face high tension and strain values. Currently, special attention is devoted to the manufacturing process of these systems, which still pose several issues for the industrial implementation [261]. It is worthwhile to mention that these arrangements, because of their intrinsic porous nature, have several limitations in the use as skins and are mainly employed as fillers.

10.3 DIRECT ACTUATION: THE USE OF SMART MATERIALS

When the idea arose of systematically developing new, adaptive structural systems, capable of changing their shape and their functionality as required by the mission or the actual conditions of the aircraft, specific attention was immediately devoted to actuators. In fact, it was easy to imagine that each desired motion would have needed a dedicated motor. To have a fully deformable wing, these motors should have been completely embedded in the structure and strong enough to counteract the external loads and deform the architectural layout. In the case of a distributed actuation network, it would not have been feasible to construct several nacelles and covers to limit the aerodynamic flow perturbations. The renewed presence of SMAs was seen as a sort of philosopher's stone, able to solve that (and many others) innovative design quest [34]. However, engineers and scientists soon realized that available technology already provided actuators that were compact enough to fit into the investigated bodies. It was finally discovered that smart structural systems no longer needed smart materials [92]. A lot has been done since that moment; however, as ambition grows, so does the need for more and more challenging perspectives. After it was demonstrated that adaptive structural systems could be made with standard technology, researches went back to smart materials to face a thousand of other problems.

Since 1997, when the "Adaptive Wing Program" by DARPA was launched [90–92], it had been clear that SMA could not have been implemented as they were, but should have been better arranged into proper driving systems. SMA, as part of the structure, should have had load-bearing properties to sustain external and internal forces. These considerations still hold and may be considered as a main guide to the use of smart materials. An exception is with the case where no or small forces are present, as space deployable systems or equilibrated hinges. SMAs and even piezoelectric ceramics (usually indicated as PZT, from their classical chemical composition: lead, zirconium, and titanium) give the possibility to produce very compact actuators, capable of generating exceptional high force and displacement levels (SMAs) or of providing broadband frequency responses (piezo) [45]. They offer the possibility of a direct actuation, with enormous impacts on the actuation systems complexity and safety. Simplicity of the arrangement has beneficial consequences on the overall weight. Furthermore, the certification process is easier because the actuator capabilities do not depend on the architectural arrangement but on the material properties. Energy conversion works at the level of material and will work until it breaks. Lifetime and aging may be critical issues. A short summary of the expected characteristics of smart actuation systems (not just smart actuators) are shown in Table 3, as an immediate reference while devoted literature should be addressed for a better estimate of the parameters.

Table 3 Comparison Between Smart and Classical Actuator Systems Implementation Characteristics. Measures: L, low; M, medium; H, high. Score: 1–5.

Char	Effic	En den	Parts	M	K	Cert	FHA	Main	Man	Life
Smart	L	*H*	*L*	*L*	M	H	M/H	*L*	*L*	M
Class	*H*	M	H	M	*H*	*L/M*	*L/M*	M	M/H	*H*
μ:33	1	*5*	*5*	*3*	3	1	2	*5*	*5*	3
μ:33	*5*	3	1	1	*5*	*4*	*4*	3	2	*5*

Cert, *certification issues;* Char, *characteristic;* Class, *classical actuator system;* Eff, *efficiency;* En den, *energy density;* FHA, *failure and hazard analysis;* K, *stiffness;* Life, *lifetime;* M, *mass;* Main, *maintenance complexity;* Man, *manufacture complexity;* Parts, *parts number;* Smart, *smart actuator system;* μ, *merit.*

In the table, classical vs. smart actuation systems were considered. The use of traditional devices has a relevant impact for certification and FHA analyses. The table should be intended as a photograph more than as a future perspective, shot at the time this article was written (2016). Surprisingly enough, a substantial equivalence for the two systems results. It is easily predictable to see an increasing presence of smart actuation systems soon, as some weak points will be overcome with the operational practice and knowledge about their behavior will be improved. A few smart systems are already available on the market. Some of them are used in very harsh environments like space, for specific functions of satellites or rockets. A leading example is the family of pin-pullers for stage detachment, initially conceived at Penn State University [262] and then developed by several industrial firms like Cedrat Technologies [263], Arquimea [264], TiNi AS, Technosprings, and many others. The concept of the working mechanism is simple: when necessary, a pin maintaining contact between two components, is pulled by an SMA device. This latter is activated by heating and wins the resistance of a precompressed mechanical spring. Activation time is confined to around 1 ms. The device architecture is extremely easy and made up of a small number of parts. Consequently, flight and space certification are facilitated and the kinematics itself is practically trouble-free. An example of these devices is shown in Fig. 43, following an example reported in literature [262].

10.4 WIRELESS, DISTRIBUTED SENSING

The high number of DOF, nearly infinite, is an immediate problem to manage when dealing with structural systems, A suitable sensor network aimed at monitoring morphing architectures should have a relevant extension and considerable density, suitable of catching the various shapes that are achieved during operation. This fact implies the deployment of a huge quantity of transducers that can reveal needed information on a suitable number of points. Accelerometers are widely used in common practice, both in lab and in flight, but have some features that suggest that other kinds of detection elements should be employed. Generally speaking, strain devices are more appropriate than displacement (velocity, acceleration) sensors because they avoid rigid motion contributions (coming from taxing, maneuver, and so on) that can disturb measurement output. Usually, they are also lighter. Moving from the former to the latter is just a matter of installation. Examples are strain gauges or optical fiber-based devices (Fiber Bragg's Gratings for instance, FBG).

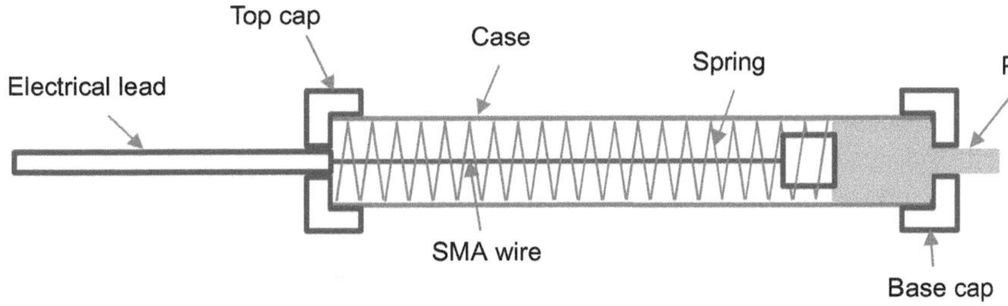

FIG. 43

A classical pin-puller schematic [262].

However, the main problem is not related to the transducer itself but to the associated cabling. Any sensible element should have feeding and signal connections, which means three or four wires per element, or only one or two in the case of optical fibers. In this latter case, the number can be further reduced for their multiplexing capability or, in other words, the possibility of deploying more detection points along the same line. However, installation concerns could strongly limit this benefit, mainly linked to the necessity of avoiding excessive small curvatures along the placement path of the fiber, which would dramatically affect the light transmission capability. Acquisition and elaboration data systems may definitely be equivalent for the different configurations. At least two major impacts should be considered on the global performance evaluation. The first one is linked to the cabling mass, layout and installation, into hardly accessible zones (in a common aircraft, there are many places that are unreachable). The other one regards maintenance, both ordinary and extraordinary (for example, as a consequence of a fault). For a standard smart or adaptive structural system, the estimated weight of cables can reach roughly 5% of the total. Penalties at level of manufacturing and inspection time can also be significant and should be considered.

These difficulties can partially be removed using wireless systems [19]. They can drastically reduce the number of wires, with important benefits in terms of complexity reduction, weight savings, and maintenance operations. Unfortunately, current drawbacks prevent the extensive use of these architectures in common practice. Primarily, the sensors must be fed. Placing cables to meet this intrinsic necessity would imply the failure of the concept. At least two solutions stand out among many possibilities. The first consists of creating small, localized harvesting energy systems that use other types of allowable energy, to satisfy the necessity of each sensor [50–59]. For instance, energy can be derived from the vibration field, characteristic of any structure in any operational condition. The added complexity by the set-up of a self-standing system, apparently uncorrelated with the rest of the architecture is well paid back. Because the required energy is small, another possibility is to feed the devices through antenna or equivalent systems [265]. It is important to indicate that this problem is common in many technological fields, mechanical or not.

In this case, the same antenna can be used to send recorded signals to the central acquisition and elaboration units [223,224]. The signals cannot be sent to the final receivers directly, because they also need a specific transmission path [266]. Then, the antenna system should fit into the target architecture in a way that avoids massive structural obstacles. Its use implies a transmitter must be made for each sensor, unless it is naturally radiating. Furthermore, several smaller antennas should be deployed as a single device cannot run all over the structural system. Each of them would refer to intermediate receivers and could be equipped with some signal preprocessing functions (filtering, digitalization) to transmit a preliminary, cleaner signal to the main station, Fig. 44. Interference with other on-board instrumentation is something to be faced. In a partial way, this issue is typically solved by using a selected and devoted frequency band. In any case, leaks, ill-functioning, and other similar occurrences should be properly evaluated when the complete system is released and ready to undergo the certification process. Recapping, a wireless system should consist of independent sensors, each fed by an energy harvesting system or antenna and, in any case, a transmitting device. Cabling issues do not vanish but are confined to the sensor region. It is worthwhile to note that, in principle, any kind of physical sensor can be used in a wireless network, and associated electronics is the key to make it a part of a wireless or cabled system.

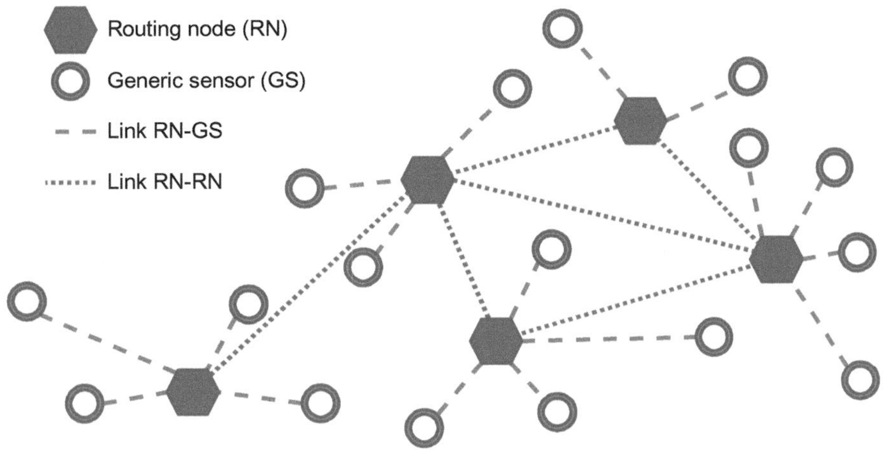

FIG. 44

A wireless sensor network (WSN) schematic (first level, sensors and intermediate receivers).

10.5 CONTROL SYSTEM ARCHITECTURE

To deal with control logic, some assumptions should be made at the beginning. In the case under investigation, it is assumed that morphing improves certain aerodynamic performances of aircraft (lift, drag, lift over drag ratio, and so on) with respect to operational conditions (cruise, take-off, descent, etc.). It is also supposed that static or quasistatic devices are involved. This hypothesis allows neglecting dynamic coupling between the control and main systems, even in the case of feedback logic implementation. Because the aerodynamic performance depends on the achieved elastic geometry, this is the vector parameter to be monitored. With the possibility to change the airplane form to improve behavior, that information is vital to perform an exact, targeted action. Classical on-board systems allow achieving flight mechanics-type info to adequately define the optimal configuration for certain phases of the mission with a high level of confidence. The classical inverse problem of trim, tightly correlated with safety issues, is widely addressed in literature [267,268]. However, this cannot be sufficient for defining the most suitable shaping action, as more information is required. There are many reasons beyond this statement. For instance, the same load can result from several different configurations, not all of which are equally efficient. Because the wing shape changes, different polars may be achieved in an instant and the airplane state is represented on a single point of those curves. Once that point is assessed by a proper morphing action, the working point may be moved onto another line, shifting towards a more suitable characteristic.

In literature, the process of drawing the actual wing shape from signals derived from the available sensor network during its operation is known as shape reconstruction. This term is taken from archaeology [269,270]. Many devices are necessary because structural elasticity (leading in itself to an infinite DOF system) is accompanied by the additional presence of a morphing architecture that makes the structural response change (without speaking about the correlated change of the aerodynamic load). Indeed, infinite structures, each with infinite degrees of freedom, should be talked about. In the case of large civilian commercial aircraft, speeds are usually high ($M \asymp 0.8$) and small perturbations can give rise to important variations. It is possible to imagine a code able to estimate the consequent aerodynamic field around the aircraft and the related forces once the deformation field has been acquired and the geometry has been extrapolated. Alternatively, it is possible to refer to an off-line database where the optimal shapes are stored (derived by high fidelity CFD computation, validated through a number of experimental correlations) and a dedicated software able to estimate the performance variations between actual and measured (extrapolated) forms. In this way, a complete loop between action (morphing) and distributed effect (aerodynamics) is achieved and the morphing command can be properly modulated to adjust the effects. The proposed solutions are good benchmarks but many others can be viable.

This step does not close the loop. In fact, the objective is to impove more practical parameters, like fuel consumption, and not to achieve a specific flow map, or even particular lift or drag values. The above operation should verify that kerosene flux diminishes, for instance through a check on the fuel valve. To achieve a minimum, further iterations can be run. This procedure can easily implemented if the action of a single morphing system is assumed, as implicitly done in the discussed case. The related block diagram is sketched in Fig. 45. In actual practice, simpler solutions can be implemented. For instance, a straight comparison between a reference strain vector (experimentally acquired on the ground or numerically derived through a validated numerical model) and the measured deformations,

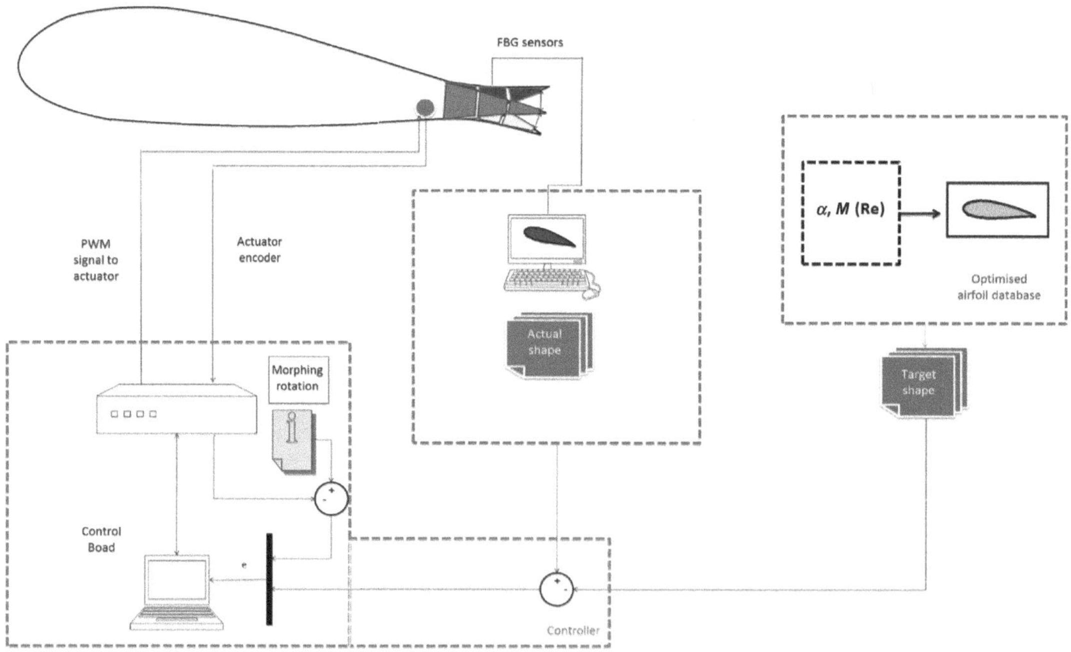

FIG. 45

Logic flow of a morphing control system; conceptual schematic [207].

can take place. Still more directly, a straight loop between selected command parameters (like the imposed camber variation) and macro indicators of the flight situation (like the instantaneous consumption) can be established.

The control system has to take care of many further, sometimes crossed, information if more devices are working simultaneously (concurring to the same or even different objectives) which makes the situation more complicated. In other words, a detailed map of the structural configuration is necessary to move toward an integrated morphing system, which can change the aerodynamic shape of the aircraft as a whole, derived from information acquired through a highly distributed and dense sensor network. An aerodynamic code could estimate the actual load distribution and drive necessary adjustments to the control actions. This step should be verified through a check with the available flight parameters and then iterated till convergence (trying to avoid instabilities).

10.6 CYBERNETICS AND ROBOTICS

Looking at a generic aircraft morphing architecture, whether designed or realized, it is easy to imagine that the natural evolution of these systems is to expand the deformability regions and increase the macroscopic degrees of freedom of the reference structure. It could be stated that this tendency aims at emulating biological flight (like the bird one) more and more, even trying to move beyond that boundary. The control systems should govern the behavior of this futuristic machine, preventing general instabilities or strain excess, while an autonomous system or a human pilot drives it at a macroscopic level. An ornitomorphic or, generally, a biomimetic system is then imagined, commanded by an artificial or a natural intelligence.

Aircraft morphing systems can be classified as Smart Structures [271]. These are defined as integrated load-bearing architectures that can adapt themselves to variations of external environments, enlarging their operational domain while maintaining optimal performance levels. Adjustments follow mutated needs and try to maintain the structure into the best possible configuration while new occurrences arise and evolve with time. Such structural systems exhibit additional DOF that allows continuous transformation. Their design does not refer to a single reference point but to an extended domain. Smart Intelligent Structures may be introduced if the former concept is combined with recent achievements in information technology, signal processing, and cybernetics. A classical definition of cybernetics comes from N. Wiener, that explained the terms as the "scientific study of control and communication in the animal and the machine" (1948) [272]. The etymology of the word comes from Greek κυβερνητική (kybernetiké), meaning governance or guidance. It may be seen as an interdisciplinary matter that links together control systems, electronics and electronic networks, informatics, neuroscience, logics, evolutionary biology and psychology, both human and animal.

Autonomous decision-making and experience-based learning capabilities are traditionally linked to the development of information data processing and management systems that are at the base of biomimetic robot intelligence [273]. Analogously, robotics is the fusion of some specialized mechanic, materials, electric, electronic, systems, and information engineering technology branches, dealing with the design, construction, operation, and application of robots [274]. A robot is a mechanical or virtual artificial agent, as an electro-mechanical machine, driven through the support of devoted computer codes or electronic circuitry. The word is derived from the Czech "robota," meaning heavy duty, in turn descended from the ecclesiastic Slavic "rabota," meaning servitude or servants. Exoskeletons or anthropomorphic autonomous or remote driven mechanical systems are classical examples of

FIG. 46

Two examples of actual Hunter Drones: The J-UCAS Boeing X-45A UCAV (a technology demonstrator, left) [277] and the Northrop Grumman RQ-4B Global Hawk (EuroHawk, right) [278].

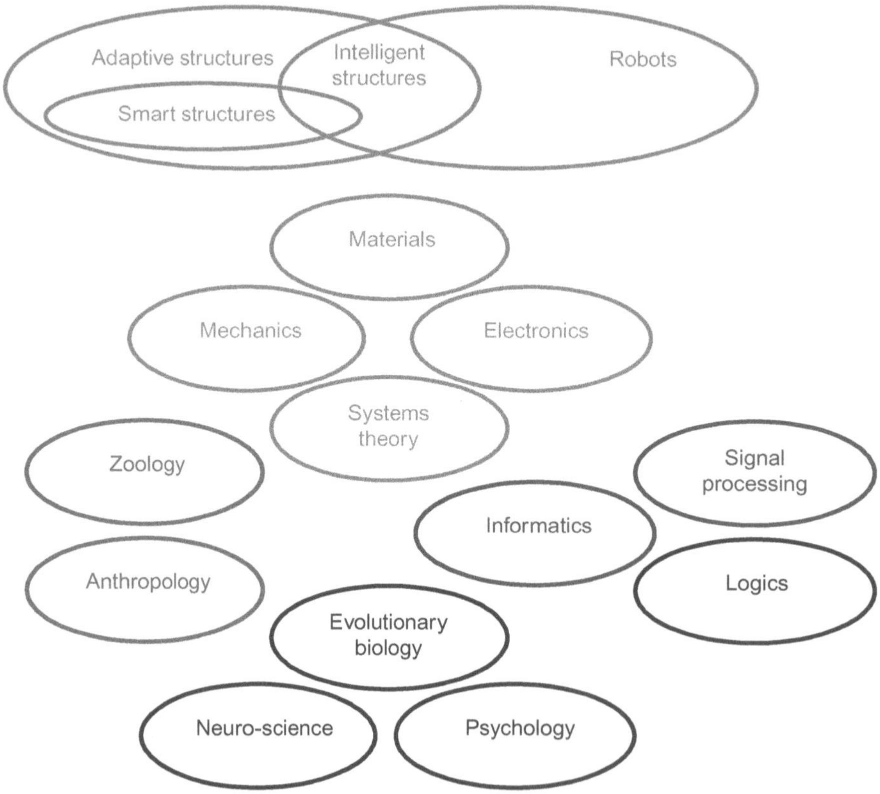

FIG. 47

Morphing systems and involved or potentially involved scientific disciplines.

robots. Computer systems should be included in this definition as main elements of the control and sensor systems, including associated data processing and information feedback.

From these considerations, it can be stated that morphing devices may be classified as peculiar smart structures, in turn seen as a particular class of mechanical system and excellent candidates as part of a new category of robots. Indeed, standard drones, currently available on the mass market, are basic examples of robots. In the authors' opinion, morphing systems are to evolve towards machines capable of multiple degrees of freedom that can resemble the behavior of living flying animals (ornitomorphic), as they move to higher, complex levels. The fact that these systems could be conceived for low to high speeds up to super and hypersonic velocities as well as for space missions, makes the future very close to science fiction. On the other hand, some prototypic versions of these concepts are already available, while a lot of movies try to envision far more prospects. Many of us can remember the terrible Skynet Hunter Killers, as seen in the movie Terminator 3 [275], or other multishape vehicles, like Transformers [276]. When we see some of the current aircraft, Fig. 46 [277,278], the natural question is how far we actually are from that future. In Fig. 47, a schematic vision of a possible classification of morphing systems is shown, together with a nonexhaustive list of the involved disciplines as they are and as they could be, while their architecture is moving towards a robotic approach.

ACKNOWLEDGMENTS

The authors want to address a warm appreciation to all their colleagues that actively supported the writing of this chapter, providing their experience and their knowledge to improve its form and content. Their sincere thanks go especially to Salvatore Ameduri, Gianluca Amendola, Francesco Amoroso, Ignazio Dimino, Monica Ciminello, and Rosario Pecora, who increased the value of this chapter through their valuable experience in the development of morphing systems.

REFERENCES

[1] G.C. Sponsler, E. Dare, D. Gignoux, N.N. Rubin, The F-4 and the F-14, Columbia Research Corporation, 1973. Naval Analysis Programs—Office of Naval Research Grant N.N00014-72-C-0339, Arlington, VA, USA, May.
[2] W.E. Goff, Droop nose, Flight International, Sutton, UK, (August) (1971).
[3] https://en.wikipedia.org/wiki/Bird_flight#/media/File:Rissa_tridactyla_2.jpg. By © Hans Hillewaert, CC BY-SA 4.0, https://commons.wikimedia.org/w/index.php?curid=639330.
[4] https://pixabay.com/it/falcon-wildpark-poing-approccio-1325785/. CC0 Public Domain.
[5] https://en.wikipedia.org/wiki/Northern_gannet#/media/File:Gannet_(Morus_bassana),_Belmont_-_geograph.org.uk_-_529175.jpg, By Mike Pennington, CC BY-SA 2.0.
[6] https://pixabay.com/it/gabbiani-uccelli-volo-degli-uccelli-738756/. CC0 Public Domain.
[7] https://en.wikipedia.org/wiki/Northern_gannet#/media/File:Morus_bassanus_30.jpg. CC BY-SA 3.0, https://commons.wikimedia.org/w/index.php?curid=195569.
[8] https://en.wikipedia.org/wiki/Bird_flight#/media/File:Domestic_Pigeon_Flock.jpg. By Toby Hudson–Own work, CC BY-SA 3.0, https://commons.wikimedia.org/w/index.php?curid=8267094.
[9] https://en.wikipedia.org/wiki/Bird_flight#/media/File:FlightSilhouettes.svg. By L. Shyamal–Own work, CC BY-SA 2.5, https://commons.wikimedia.org/w/index.php?curid=3212520.

[10] Barbarino S., Pecora R., Lecce L., Concilio A., Ameduri S., Calvi E. A. "A novel SMA-based concept for airfoil structural morphing," J. Mater. Eng. Perform. Vol. 8, N.5, pp. 696–705, 2009.
[11] O. Schorsch, A. Lühring, C. Nagel, in: P.C. Wölcken, M. Papadopoulos (Eds.), Elastomer-based skin for seamless morphing of adaptive wings, Smart Intelligent Aircraft Structures (SARISTU)–Proceedings of the Final Project Conference, Springer International Publishing, Cham, Switzerland, 2016, ISBN 978-3-319-22412-1.
[12] R. Pecora, M. Magnifico, F. Amoroso, L. Lecce, M. Bellucci, I. Dimino, A. Concilio, M. Ciminello, in: P.C. Wölcken, M. Papadopoulos (Eds.), structural design of an adaptive wing trailing edge for large aeroplanes, Smart Intelligent Aircraft Structures (SARISTU)—Proceedings of the Final Project Conference, Springer International Publishing, Cham, Switzerland, 2016, ISBN 978-3-319-22412-1.
[13] https://www.flickr.com/photos/25348772@N03/2390063960. Public Domain.
[14] J. Cameron (Producer & Director), "Terminator 2: Judgment Day" (Motion Picture), TriStar Pictures, Culver City, CA, 1991.
[15] http://i.vimeocdn.com/video/522317976_1280x720.jpg. Public Domain.
[16] https://www.flickr.com/photos/criminalintent/3226136197. CC Attribution ShareAlike 2.0 Generic (CC BY-SA 2.0) License @, https://creativecommons.org/licenses/by-sa/2.0/legalcode.
[17] M. Kraft, N.M. White (Eds.), MEMS for Automotive and Aerospace Applications, Woodhead Publishing Series in Electronic and Optic Materials, Woodhead Publishing, Cambridge, ISBN 978-0-85709-118-5, 2013.
[18] G. Rajan (Ed.), Optical Fiber Sensors: Advanced Techniques and Applications, CRC Press, Boca Raton, FL, 2015, ISBN 978-1-48222-825-0.
[19] J. Cao, X. Liu, Wireless Sensor Networks for Structural Health Monitoring, Springer Briefs in Electrical and Computer Engineering, Springer, Basel, 2016.
[20] A. Cusano, A. Cutolo, J. Albert, Fiber Bragg Grating Sensors: Recent Advancements, Industrial Applications and Market Exploitation, Bentham eBooks, Bentham Science Publishers, Sharjah, ISBN 978-1-608-05084-0, 2011.
[21] Iten M., Novel Applications of Distributed Fiber-Optic Sensing in Geotechnical Engineering (ETH-dissertation),Vdf-Hochschulverlag an der ETH, Zurich, 2012, ISBN 978-3-728-13454-7.
[22] J.L. Santos, F. Farahi (Eds.), Handbook of Optical Sensors, CRC Press, Boca Raton, FL, ISBN 978-1-439-86686-3, 2015.
[23] P. Hagedorn, G. Spelsberg-Korspeter (Eds.), Active and Passive Vibration Control of Structures, CISM International Centre for Mechanical Sciences, Springer, Vienna, ISBN 978-3-709-11821-4, 2014.
[24] A. Tewari, Aeroservoelasticity: Modeling and Control, Birkhauser Series in Control Engineering, Springer, New York, NY, ISBN 978-1-493-92368-7, 2015.
[25] M. Tooley, A. Filippone, T.H.G. Megson, M.V. Cook, P.W. Carpenter, E.L. Houghton, D. Wyatt, L.R. Jenkinson, J. Marchman, F. De Florio, et al., Aerospace Engineering e-Mega Reference, Butterworth-Heinemann Elsevier Science, Oxford, ISBN 978-1-856-17576-0, 2009.
[26] S. Jackson, Systems Engineering for Commercial Aircraft: A Domain-Specific Adaptation, Ashgate Publishing Company, Burlington, VT, ISBN 978-1-472-43921-5, 2015.
[27] M. Mariam, N.A. Murtaza, J. Sami, U. Ansari, Review of mechanics and applications of auxetic structures, Adv. Mater. Sci. Eng. 2014 (2014) 17.
[28] Liu Y., Hu H. "A review on auxetic structures and polymeric materials," Sci. Res. Essays, Vol. 5, N.10, pp. 1052–1063, 2010.
[29] S.H. Kang, S. Shan, W.L. Noorduin, M. Khan, J. Aizenberg, K. Bertoldi, Buckling-induced reversible symmetry breaking and amplification of chirality using supported cellular structures, Adv. Mater. 2013 (2013) 29.
[30] Bettini P., Airoldi A., Sala G., Di Landro L., Ruzzene M., Spadoni A. "Composite chiral structures for morphing airfoils: numerical analyses and development of a manufacturing process," Compos. Part B, Vol. 41, N.2, pp. 133–147, 2010.

[31] Airoldi A., Bettini P., Panichelli P., Asaro A., Sala G.. "Development and application of composite chiral honeycomb with auxetic behaviour," Auxetics 2014, 5th International Conference "Auxetics and Other Material Models with Negative Characteristics," 10th International Workshop "Auxetics and Related Systems", Poznan, September 15-19, 2014.
[32] McKnight G. P., Henry C. P., Herrera G., Microstructured Reconfigurable Composite Material, United States Patent 9211690 B1, Priority Date July 25, 2005; Publication Date, December 15, 2015.
[33] A. Vigliotti, D. Pasini, Analysis and design of lattice materials for large chord and curvature variations in skin panels of morphing wings, Smart Mater. Struct. 24 (3) (2015) 11.
[34] L. Lecce, A. Concilio (Eds.), Shape Memory Alloy Engineering: For Aerospace, Structural and Biomedical Applications, Butterworth-Heinemann Elsevier Science, Oxford, ISBN 978-0-080-99921-0, 2014.
[35] Verrastro M., Metge S.. "Morphing wing integrated safety approach and results," Smart Intelligent Aircraft Structures (SARISTU)—Proceedings of the Final Project Conference, Wölcken P. C., Papadopoulos M. (Eds.), Springer International Publishing, Cham, Switzerland, 2016, ISBN 978-3-319-22412-1.
[36] Dimino I., Ciminello M., Concilio A., Pecora R., Amoroso F., Magnifico M., Schueller M., Gratias A., Volovick A., Zivan L. "Distributed actuation and control of a morphing wing trailing edge," Smart Intelligent Aircraft Structures (SARISTU)—Proceedings of the Final Project Conference, Wölcken P. C., Papadopoulos M. (Eds.), Springer International Publishing, Cham, Switzerland, 2016, ISBN 978-3-319-22412-1.
[37] H. Funakubo, J.B. Kennedy, Shape Memory Alloys, Gordon and Breach Science Publishers, New York, NY, ISBN 978-2-881-24136-9, 1987.
[38] T. Yoneyama, S. Miyazaki, Shape Memory Alloys for Biomedical Applications, CRC Press, Boca Raton, FL, ISBN 978-1-845-69344-2, 2009.
[39] Danoiu S., Rotaru P., Degeratu S., Rizescu S., Bizdoaca N. G. "Shape memory alloy-based smart module structure working under intense thermo-mechanical stress," J. Therm. Anal. Calorim., Vol. 118, N.2, pp. 1323–1330, 2014.
[40] Nava N., Collado M., Alvarez F., Cabás R., San Juan J., Patti S., Lautier J. M. "Ultra-low-weight rotary actuator for operation on mars and pin puller mechanism based on a novel shape memory alloy technology," Proceedings of the 41st Aerospace Mechanisms Symposium, Jet Propulsion Laboratory, Pasadena, CA, May 16-18, 2012.
[41] Dimino I., Ciminello M., Concilio A., Gratias A., Schueller M., Pecora R. "Control system design for a morphing wing trailing edge," Proceedings of the 7th ECCOMAS Thematic Conference on Smart Structures and Materials, Smart 2015, Azores, June 3-6, 2015.
[42] M. Grolms, presenting a concept by K. Schlüter, A. Raatz, Braunschweig University of Technology, and Holz B., Saarland University, August 28, http://www.materialsviews.com/optimizing-actuators-with-shape-memory-alloys/, 2012.
[43] Flaga S., Sapinski B. "The properties of the magnetic shape memory alloy actuator," Proceedings of Active Noise and Vibration Control Methods, Krakow–Wojanow, June 6-8, 2011.
[44] De Souza V. F., Savil M. A., Paiva A., Boyd J. G., Lagoudas D. C. "Nonlinear vibration analysis of a magnetic shape memory alloy oscillator," Proceedings of the XVII International Symposium on Dynamic Problems of Mechanics, Natal, February 22-27, 2015.
[45] K. Uchino, Piezoelectric Actuators and Ultrasonic Motors, Electronic Materials: Science \& Technology-Springer Science + Business Media LLC, New York, NY, ISBN 978-0-792-39811-0, 1997.
[46] S.B. Choi, Y.M. Han, Piezoelectric Actuators: Control Applications of Smart Materials, CRC Press, Boca Raton, FL, ISBN 978-1-439-8-1809-1, 2010.
[47] N. Jalili, Piezoelectric-Based Vibration Control: From Macro to Micro/Nano Scale Systems, Springer Science + Business Media LLC, New York, NY, ISBN 978-1-441-90070-8, 2009.
[48] S.B. Choi, Y.M. Han, Magnetorheological Fluid Technology: Applications in Vehicle Systems, CRC Press, Boca Raton, FL, ISBN 978-1-439-85673-4, 2012.

[49] S.C. Mukhopadhyay (Ed.), New Developments in Sensing Technology for Structural Health Monitoring, Lectucameronre Notes in Electrical Engineering, Springer-Verlag, Berlin, Heidelberg, ISBN 978-3-642-21099-0, 2011.

[50] M.H. Tooley, D. Wyatt, Aircraft electrical and electronic systems: principles, operation and maintenance, in: Aircraft Engineering Principles and Practice, Butterworth-Heinemann, Burlington, MA, ISBN 978-0-750-68695-2, 2009.

[51] V. Memmolo, L. Maio, N.D. Boffa, E. Monaco, F. Ricci, Damage detection tomography based on guided waves in composite structures using a distributed sensor network, Opt. Eng. 55 (1) (2016), 011007.

[52] Monaco E., Boffa N. D., Memmolo V., Ricci F., Testoni N., De Marchi L., Marzani A., Hettler J., Tabatabaeipour M., Delrue S., Van Den Abeele K. Methodologies for guided waves based SHM system implementation on composite wing panels: results and perspectives from SARISTU scenario 5, Smart Intelligent Aircraft Structures (SARISTU), Springer International Publishing, Cham, Switzerland, 497–529, 2016.

[53] L. Maio, V. Memmolo, F. Ricci, N.D. Boffa, E. Monaco, R. Pecora, Ultrasonic wave propagation in composite laminates by numerical simulation, Compos. Struct. 121 (2015) 64–74.

[54] Memmolo V., Ricci F., Maio L., Boffa N. D., Monaco E. "Model assisted probability of detection for a guided waves based SHM technique," in Health Monitoring of Structural and Biological Systems 2016, Proc. SPIE 9805, 9805-3 2016.

[55] Memmolo V., Ricci F., Boffa N. D., Maio L., Monaco E. "Statistical recognition of structural health in composites through ultrasonic testing," 19th World Conference on Non-Destructive Testing, 21 (7), July, ndt.net 2016) ISSN 1435-4934.

[56] Memmolo V., Boffa N. D., Monaco E., Ricci F., Pasquino N. "Experimental characterization of a composite structures health monitoring methodology," in 3rd IEEE International Workshop on Metrology for Aerospace, Florence, June 22-23, 2016.

[57] D. Briand, E. Yeatman, S. Roundy (Eds.), Micro Energy Harvesting, Advanced Micro and Nanosystems, Wiley-VCH Verlag, Weinheim, ISBN 978-3-527-67292-9, 2015.

[58] L. Mescia (Ed.), Innovative Materials and Systems for Energy Harvesting Applications, Advances in Environmental Engineering and Green Technologies, Engineering Science Reference-IGI Global, Hershey, PA, ISBN 978-1-466-68254-2, 2015.

[59] Penella M. T., Gasulla M. "A review of commercial energy harvesters for autonomous sensors," Proceedings of the Instrumentation and Measurement Technology Conference–IMTC 2007, Warsaw, May 1-3, 2007.

[60] Maehara D., Akai R., Tran G. K., Sakaguchi K., Sampei S., Araki K., Iwai H. "Experiment on battery-less sensor activation via multi-point wireless energy transmission," Proceedings of the 2013 IEEE 24th International Symposium on Personal Indoor and Mobile Radio Communications Conference (PIMRC), London, September 8-11, 2013.

[61] Zolghadri N., Halling M. W., Barr P. J., "Comparison of wireless and wired structural system identification," Proceedings of the Structures Congress 2014, Boston, MA, April 3-5, 2014.

[62] Trott A., Swanson B. "Wireless structural monitoring for spacecraft," AIAA Space 2012 Conference & Exposition, Pasadena, CA, September 11-13, 2012.

[63] H.S. Wolko (Ed.), The Wright Flyer: An Engineering Perspective, The Smithsonian Institution Press for the National Air and Space Museum, Washington, DC, 1987, ISBN 978-0-87474-979-3.

[64] Holle A. A., Plane and the Like for Aeroplanes, Germany Patent N.DE366693C, Priority Date, October 15, 1915; Publication Date, January 10, 1923.

[65] Holle A. A., Plane and the Like for Aeroplanes, United States Patent N.1225711, Priority Date, Dec. 22nd, 1916; Publication Date, May 8, 1917.

[66] Parker H. F., The Parker variable camber wing, Report No. 77, National Advisory Committee for Aeronautics (NACA), Washington Government Printing Office, Washington, DC, 1920.

[67] Weisshaar T. A., "Morphing aircraft technology—new shapes for aircraft design," Proceedings of the NATO Research Technology Organization Research Applied Vehicle Technology Panel Symposium RTO-MP-AVT-141, 2006, Vilnius, October 2-4, 2006.
[68] T.A. Weisshaar, New Aircraft Systems Concepts–Towards New Horizons in Aeroelasticity, Purdue University, Course Material, Aeroelasticity Course AAE 556, West Lafayette, IN, USA, 2008.
[69] PHGCOM, Own work, CC BY-SA 3.0, http://creativecommons.org/licenses/by-sa/3.0 or GFDL, http://www.gnu.org/copyleft/fdl.html. via Wikimedia Commons.
[70] Brissenden R. F., Heath A. R., Conner D. W., Spearman M. L. (technical monitors), The Boeing Preliminary Design Department, Assessment of variable camber for application to transport aircraft, NASA Contractor Report CR-158930, National Aeronautics and Space Administration Langley Research Center, Hampton, VA, October, 1980.
[71] Ferris J. C., Wind-tunnel investigation of a variable camber and twist wing, NASA technical note TN-D8475, National Aeronautics and Space Administration, Washington, DC, August, 1977.
[72] R. Hilbig, H. Koerner, in: Aerodynamische Entwicklungsrichtungen fuer Verkehrsflugzeuge, Deutsche Gesellschaft fuer Luft-und Raumfahrt (DGLR), DGLR Annual Convention 1984, Hamburg, October 1-3, 1984.
[73] Hilbig R., Koerner H., "Aerodynamic Design Trends for Commercial Aircraft," NASA Technical Memorandum TM-77976, National Aeronautics and Space Administration, Washington, DC 1986 January (translation from the original German version).
[74] Hardy R., "AFTI/F-111 mission adaptive wing technology demonstration program," Proceedings of the 1983 AIAA Aircraft Prototype and Technology Demonstrator Symposium, Air Force Museum Dayton, OH, March 23-24, 1983.
[75] K.L. Bonnema (Ed.), AFTI/F-111 mission adaptive wing briefing to industry, Air Force Wright Aeronautical Laboratories, Wright-Patterson Air Force Base, Dayton, OH, 1988. AFWAL Technical Report, TR-88-3082. October.
[76] Smith J. W., Lock W. P., Payne G. A., "Variable-camber systems integration and operational performance of the AFTI/F-111 mission adaptive wing," NASA Technical Memorandum TM-4370, National Aeronautics and Space Administration, Edwards, CA, April 1992.
[77] Webb L. D., McCain W. E., Rose A., "Measured and predicted pressure distributions on the AFTI-F-111 mission adaptive wing," NASA Technical Memorandum TM-100443, National Aeronautics and Space Administration, Edwards, CA, 1988 November.
[78] Goecke P.S., Webb L. D., Friend E. L., Lokos W. A., "Flight test results from a supercritical mission adaptive wing with a smooth variable camber," NASA Technical Memorandum TM-4415, National Aeronautics and Space Administration, Edwards, CA, November, 1992.
[79] Thornton S. V., "Reduction of structural loads using maneuver load control on the Advanced Fighter Technology Integration (AFTI)/F-111 mission adaptive wing," NASA Technical Memorandum TM-4526, National Aeronautics and Space Administration, Edwards, CA, September, 1993.
[80] Bolonkin A., Gilyard G. B., "Estimated benefits of variable-geometry wing camber control for transport aircraft," NASA Technical Memorandum TM-1999-206586, National Aeronautics and Space Administration, Edwards, CA, October, 1999.
[81] Szodruch J., Hilbig R., "Variable wing camber for transport aircraft," Prog. Aerosp. Sci., Vol. 25, N.3, pp. 297–328, 1988.
[82] W.J. Buehler, in: Nitinol Re-Examination, White Oak Laboratory Alumni Association (WOLAA) Leaf, WOL History Supplement, Winter, Vol. 7, 2006. www.wolaa.org.
[83] Buehler W. J., Wiley R. C., Nickel-Base Alloys, United States Patent N.3174851, Priority Date, Dec. 1st, 1961; Publication Date, March 23, 1965.
[84] Buehler W. J., Gilfrich J. V., Wiley R. C., "Effects of low-temperature phase changes on the mechanical properties of alloys near composition Ti-Ni," J. Appl. Phys., Vol. 1963, N.34, pp. 1475–1477, 1963.

[85] Wang F. E., Buehler W.J., Pickart S.J., "Crystal structure and a unique martensitic transition of Ti-Ni," J. Appl. Phys., Vol. 1965, N.36, pp. 3232–3239, 1965.

[86] L. Lecce, A. Concilio (Eds.), Shape Memory Alloys Engineering, Butterworth-Heinemann, Oxford, ISBN 978-0-080-99920-3, 2014.

[87] Harrison J. D., Hodgson D. E., "Use of Ti-Ni in mechanical and electrical connectors," in Perkins J. (Ed.), "Shape Memory Effects in Alloys," Proceedings of the International Symposium on Shape Memory Effects and Applications, Toronto, May 19-22, 1975.

[88] Bokhari S. W., Vahdat O., Winters R. J., "The first clinical experience with a peripheral, self-expanding-nitinol stent in the treatment of saphenous vein graft disease: angiographic evidence of late expansion," J. Invasive Cardiol., Vol. 2003, N.15, pp. 418–422, 2003.

[89] J.N. Kudva, Overview of the DARPA smart wing project, J. Intell. Mater. Syst. Struct. 15 (2004) 261–267.

[90] Garcia I., Sater J., Fischer G., "Smart material based actuation," Proceedings of the 11th International Conference on Adaptive Structures and Technologies, Nagoya, October 23-26, 2000.

[91] Kudva J. N., Martin C. A., Scherer L. B., Jardine A. P., McGowan A. R., Lake R. C., Sendeckyj G., Sanders B., "Overview of the DARPA/AFRL/NASA smart wing program," Proceedings of the SPIE Conference on Industrial and Commercial Applications of Smart Structures Technologies, Newport Beach, CA, March 1-5, 1999.

[92] Kudva J. N., Sanders B., Pinkerton-Florance J., Garcia E., "The DARPA/AFRL/NASA smart wing program—final overview," Proceedings of the SPIE Conference on Industrial and Commercial Applications of Smart Structures Technologies, San Diego, CA, March 17-21, 2002.

[93] US Navy–US Navy.mil, Public Domain, https://commons.wikimedia.org/w/index.php?curid=39742405.

[94] Musgrove R. G., Eccentric Actuator, United States Patent N.4386761, Priority Date, August 15, 1978; Publication Date, September 1, 1981.

[95] Fleming G. A., Burner A. W., "Deformation measurements of smart aerodynamic surfaces," Proceedings of the SPIE Conference on Optical Diagnostics for Fluids/Heat/Combustion and Photomechanics for Solids, Denver, CO, July 21-23, 1999.

[96] Glotzer S. C. (chair), Committee for a review of adaptive structural materials research that could pose a threat to US National Security, "Adaptive Materials and Structures: A Workshop Report," The National Academy of Sciences, Washington, DC, 2013, ISBN 978-0-309-27271-1.

[97] McGowan A.-M. R., Vicroy D. D., Busan R. C., Hahn A. S., "Perspectives on Highly Adaptive or Morphing Structures," Proceedings of the NATO Research Technology Organization Applied Vehicle Technology Panel Symposium, RTO-MP-AVT-152 2009, Evora, April 20-24, 2009.

[98] Barbarino S., Bilgen O., Ajai M. R., Friswell M. I., Inman D. J. "A review of morphing aircraft," J. Intell. Mater. Syst. Struct., Vol. 22, N.9, pp. 823–877, 2011.

[99] L. Kennet (Ed.), Military Aviation, Air University Press, Maxwell Air Force Base, Montgomery, AL, ISBN 978-1-58566-118-3, 2003. Reprint of the Clement Ader's work, "L'Aviation Militaire," originally published by Berger-Levrault, Paris, 1909.

[100] Ader C., Appareille Ailè pour la Navigation Aerienne, France Patent N.205155, Priority Date, April 19, 1890; Publication Date, August 11, 1890.

[101] https://commons.wikimedia.org/wiki/File%3AAircraft_Fighter_Jet_F-111_AFTI_NASA_0.jpg https://upload wikimedia.org/wikipedia/commons/e/e9/Aircraft_Fighter_Jet_F-111_AFTI_NASA_0.jpg. by NASA photogra-pher [Public domain], via Wikimedia Commons.

[102] Kota S., Osborn R., Ervin G., Maric D., "Mission adaptive compliant wing–design, fabrication and flight test," Proceedings of the NATO Research Technology Organization Research Symposium, Morphing Vehicles, RTO-MP-AVT-168, 2009, Evora, April 20-24, 2009.

[103] Flanagan J. S., Strutzenberg R. C., Myers R. B., Rodrian J. E., "Development and flight testing of a morphing aircraft, the NextGen MFX-1," Proceedings of the 48th AIAA/ASME/ASCE/AHS/ASC Structures, Structural Dynamics, and Materials Conference, 2007, Honolulu, HI, April 23-26, 2007.

[104] Gandhi N., Cooper J., Ward D., Howard D. P., Neal D., Cannon B., "A hardware demonstration of an integrated adaptive wing shape and flight control law for morphing aircraft," Proceedings of the AIAA Guidance, Navigation, and Control Conference, Chicago, IL, August 10-13, 2009.

[105] Ivanco T. G., Scott R. C., Love M. H., Zink S., Weisshaar T. A., "Validation of the Lockheed Martin morphing concept with wind tunnel testing," Proceedings of the 48th AIAA/ASME/ASCE/AHS/ASC Structures, Structural Dynamics, and Materials Conference, 2007, Honolulu, HI, April 23-26, 2007.

[106] Rodriguez A. R., "Morphing aircraft technology survey," Proceedings of 45th AIAA Aerospace Sciences Meeting and Exhibit, Reno, NA, January 8-11, 2007.

[107] Chattopadyay N. C., Jony B., Acharya A., "An analysis on wing morphing," Proceedings of the Global Engineering, Science and Technology Conference 2012, Dhaka, December 28-29, 2012.

[108] Sinapius M., Monner H. P., Kintscher M., Riemenschneider J., "DLR's morphing wing activities within the european network," Proceedings of the 23rd International Congress of Theoretical and Applied Mechanics (ICTAM), Bejing, August 19-24, 2012.

[109] Richter K., "Das DLR-Leitprojekt "Adaptiver Flügel" (ADIF)" DLR web site, http://www.dlr.de/as/desktopdefault.aspx/tabid-333/378_read-2905/.

[110] Ahrend H., Heyland D., Martin W., "Das Leitkonzept Adaptiver Flügel (ADIF)" DGLR Jahrestagung, DGLR-JT97–147, München (D), 1997 (in German).

[111] Bein T., Büter A., Campanile L. F., Heyland D., Monner H. P., "Das Leitkonzept Adaptiver Flügel (ADIF): Zielsetzungen und Bisherige Ergebnisse des DLR auch Sicht der Adaptronik" Proceedings of the Adaptronic Congress 1999, Potsdam, March 3-4, 1999 (in German).

[112] J. Delery, J. Fulker, P. de Matteis, Drag Reduction by Shock and Boundary Layer Control: Results of the Project Euroshock II, Supported by the European Union 1996–1999, Notes on Numerical Fluid Mechanics and Multidisciplinary Design, vol. 80, Springer Science & Business Media, Berlin, Heidelberg, ISBN 978-3-540-45856-2, 2013.

[113] E. Stanewsky, J. Delery, J. Fulker, W. Geissler, Euroshock-Drag Reduction by Passive Shock Control: Results of the Project Euroshock, AER2-CT92–0049, Supported by the European Union, 1993–1995, in: Notes on Numerical Fluid Mechanics, vol. 56, Springer Science & Business Media, Berlin, Heidelberg, ISBN 978-3-322-90711-0, 2013.

[114] L. Ulmann (Director), The European Files: The Aviation Industry Goes Green With Clean Sky, Les Dossiers Europeens, Paris, vol. 18, January, 2010.

[115] Clean Sky, "GRA—Green Regional Aircraft," Clean Sky website, http://www.cleansky.eu/content/page/gra-green-regional-aircraft.

[116] Clean Sky, Annual activity report 2010, Clean Sky website, http://cleansky.eu/sites/default/files/documents/clean-sky-annual-report-2010.pdf.

[117] Lecce L., Monaco E., Concilio A., Casciati F., "A survey of structural health monitoring research in Italy," Proceedings of the 50th AIAA/ASME/ASCE/AHS/ASC Structures, Structural Dynamics, and Materials Conference, Palm Springs, CA, May 4-7, 2009.

[118] Barbarino M., Dimino I., Carozza A., Nae C., Stoica C., Pricop V., Peng S. H., Eliasson P., Grundestam O., Tysell L., Davidson L., Eriksson L. E., Yao H. D., Ben Khelil S., Moens F., Le Garrec T., Mincu D. C., Simon F., Manoha E., Godard J. L., Averardo M. A., "Airframe noise reduction technologies applied to high-lift devices of future green regional aircraft," Proceedings of 3AF/CEAS Conference Greener Aviation: Clean Sky Breakthroughs and Worldwide Status, Bruxelles, March 12-14, 2014 (Best Conference Paper–Greener Aviation Award).

[119] D. Casalino, M. Barbarino, Optimization of a single-slotted lined flap for wing trailing-edge noise reduction, J. Aircr. 49 (4) (2012) 1051–1063.

[120] Eliasson P., Grundestam O., Peng S.-H., Yao H., Davidson L., Erik Eriksson L., "Assessment of high-lift concepts for a regional aircraft in the ALONOCO project," 50th AIAA Aerospace Sciences Meeting Including the New Horizons Forum and Aerospace Exposition, Nashville, TN, January 9-12, 2012.

[121] Woelcken P. C., Papadopoulos M. (Eds.), "Smart Intelligent Aircraft Structures (SARISTU)—Proceedings of the Final Project Conference," Springer International Publishing, Cham, Switzerland, 2016, ISBN 978-3-319-22413-8.

[122] Woelcken P. C., "SARISTU Smart Intelligent Aircraft Structures," Aerodays 2015—Aviation in Europe—Innovating for Growth, London, October 20-23, 2015.

[123] Kintscher M., Kim J., Storm S., Peter F., "Assessment of the SARISTU enhanced adaptive droop nose," Woelcken P. C., Papadopoulos M. (Eds.), "Smart Intelligent Aircraft Structures (SARISTU)—Proceedings of the Final Project Conference," Springer International Publishing, Cham, 2016, ISBN 978-3-319-22413-8.

[124] Wildscheck A., Storm S., Herring M., Drezga D., Korian V., Roock O., "Design, optimization, testing, verification and validation of the wingtip adaptive trailing edge," Woelcken P. C., Papadopoulos M. (Eds.), "Smart Intelligent Aircraft Structures (SARISTU)—Proceedings of the Final Project Conference," Springer International Publishing, Cham, 2016, ISBN 978-3-319-22413-8.

[125] A. Concilio, I. Dimino, R. Pecora, An adaptive trailing edge for large commercial aircraft, in: VII European Congress on Computational Methods in Applied Science and Engineering, ECCOMAS 2016, Greece, Crete Island, 2016.

[126] Carossa G. M., Giuliani M., Johnston A., Altkvist C., Airoldi A., Khodaei Z. S., Aliabadi M. H., "Implementation of morphing, structural health monitoring and nanomaterials on an outer wing box." Woelcken P. C., Papadopoulos M. (Eds.), "Smart Intelligent Aircraft Structures (SARISTU)—Proceedings of the Final Project Conference," Springer International Publishing, Cham, 2016, ISBN 978-3-319-22413-8.

[127] Nguyen N., Kaul U., Lebofsky S., Ting E., Chaparro D., Urnes J., "Development of variable camber continuous trailing edge flap for performance adaptive aeroelastic wing," SAE 2015 AeroTech Congress & Exhibition, Seattle, WA, September 22-24, 2015.

[128] Burdette D., Kenway G. K. W., Lyu Z., Martins J. R. R. A., "Aerostructural design optimization of an adaptive morphing trailing edge wing," 56th AIAA/ASCE/AHS/ASC Structures, Structural Dynamics, and Materials Conference, Kissimmee, FL, January 5-9, 2015.

[129] Warwick G., "NASA, boeing study flexible wing control," Aviation Week and Space Technology Blog "Things with Wings, January 30," http://aviationweek.com/blog/nasa-boeing-study-flexible-wing-control, 2013.

[130] Urnes J., Nguyen N., Dykman J., "Development of Variable Camber Continuous Trailing Edge Flap System," NASA Fundamental Aeronautics Program (FAP) Technical Conference, Cleveland, OH, March 13-15, 2012.

[131] Tal E. A., Coupled Flight Dynamic and Unsteady Aeroservoelastic Modeling and Control of a Flexible Wing Aircraft (PhD thesis), Delft University of Technology, Faculty of Aerospace Engineering, July 17, 2015.

[132] Tal E., Nguyen N., "Unsteady aeroservoelastic modeling of flexible wing generic transport aircraft with variable camber continuous trailing edge flap," 33rd AIAA Applied Aerodynamics Conference, Dallas, TX, June 22-26, 2015.

[133] Nguyen N. T., Ting E., Lebofsky S., "Aeroelastic analysis of wind tunnel test data of a flexible wing with a variable camber continuous trailing edge flap (VCCTEF)," 56th AIAA/ASCE/AHS/ASC Structures, Structural Dynamics, and Materials Conference, Kissimmee, FL, January 5-9, 2015.

[134] Urnes J., Nguyen N., Ippolito C., Totah J., Trinh K., Ting E., "A mission-adaptive variable camber flap control system to optimize high lift and cruise lift-to-drag ratios of future N+3 transport aircraft," 51st AIAA Aerospace Sciences Meeting including the New Horizons Forum and Aerospace Exposition, Dallas, TX, January 7-10, 2013.

[135] Nguyen N., "A framework for adaptive aeroelastic wing shaping control," NASA Advanced Modeling & Simulation (AMS) Seminar Series, April 22, 2014.

[136] Liu Y., Bai J., Livne E., "Robust optimization of variable-camber continuous trailing-edge flap system action using stochastic kriging," 33rd AIAA Applied Aerodynamics Conference, AIAA Aviation, Dallas, TX, June 22-26, 2015.
[137] Kota S., Hetrick J. A., Osborn R., Paul D., Pendleton E., Flick P., Tilmann C., "Design and application of compliant mechanisms for morphing aircraft structures,"SPIE Smart Structures and Materials Conference 2003: Industrial and Commercial Applications of Smart Structures Technologies, San Diego, CA, March 2-6, 2003.
[138] Kota S., Hetrick J. A., Adaptive Compliant Wing and Rotor System, United States Patent N.US7384016B2, Priority Date, March 3, 2003; Publication Date, June 10, 2008.
[139] NASA Website, http://www.nasa.gov/press/2014/november/nasa-tests-revolutionary-shape-changing-aircraft-flap-for-the-first-time, Release 14-308, November 7, 2014.
[140] Hetrick J. A., Osborn R., Kota S., Flick P., Paul D., "Flight testing of mission adaptive compliant wing 48th AIAA/ASME/ASCE/AHS/ASC Structures, Structural Dynamics, and Materials Conference Honolulu, HI, April 23-26, 2007.
[141] NASA, US AFRL, "Adaptive compliant trailing edge flight experiment," RC Soaring Digest, Vol. 31, N.14, pp. 85–86, 2014.
[142] G. Warwick, Flexible Thinking, Aviation Week and Space Technology, New York, NY, 2015, p. 17 December 7-20.
[143] G. Warwick, Morphing Wings Are Still a Long Way From Reality, Aviation Week and Space Technology, New York, NY, 2016, p. 17 January 8-22.
[144] Herrera C. Y., Spivey N. D., Lung S. F., Ervin G., Arbor A., Flick P., "Aeroelastic airworthiness assessment of the adaptive compliant trailing edge flaps," 46th Society of Flight Test Engineers International Symposium, Lancaster, CA, September 14-17; 2015.
[145] Radestock M., Riemenschneider J., Monner H. P., Rose M., "Experimental investigation of a compliant mechanism for an uav leading edge," 7th Eccomas Thematic Conference on Smart Structures and Materials, Smart 2015, Azores, June 3-6, 2015.
[146] Santos P. D. R., Gamboa P. V., Santos P. M. B., Silva J. M. A., "Structural design of a composite variable-span wing," Integrity, Reliability and Failure of Mechanical Systems, Irf 2013, Funchal, June 23-27, 2013.
[147] Albuquerque P. F., Gamboa P. V., Silvestre M. A., "Multidisciplinary and multilevel design methodology of unmanned aerial vehicles using enhanced collaborative optimization," IJMME Vol. 9, N.4, pp. 544–553, 2015.
[148] Ajaj R., Keane A., Beaverstock C., Friswell M., Inman D., "Morphing aircraft: the need for a new design philosophy," 7th Ankara International Aerospace Conference, 2013, Ankara, September 11-13, 2013.
[149] Woods B. K. S., Friswell M. I. "The adaptive aspect ratio morphing wing design concept and low fidelity skin optimization," Proceedings of the ASME 2014 Conference on smart materials, Adaptive Structures and Intelligent Systems, SMASIS 2014, Newport RI, September 8-10, 2014.
[150] Suleman A., Vale J. L., Afonso F., Lau F. P., Ricci S., De Gaspari A., Riccobene L., Cavagna L., Cooper J., Lambert L., Wales C., Cheung R., Nangia R., Rizzi A., Zhang M., Monner H. P., de Kamp B., Vasista S., Morelli M., Parizi J., Odaguil F., de Lima G. R., Antunes A., "Novel air vehicle configurations: from fluttering wings to morphing flight," 11th World Congress on Computational Mechanics, WCCM XI. Barcelona, July 20-25, 2014.
[151] De Gaspari A., Ricci S., Travaglini L., "Aeroelastic analysis of a regional aircraft with morphing wing using PyPAD framework," 4th SCAD, Symposium on Collaboration in Aircraft Design, Toulouse, November 25-27, 2014.
[152] Vasista S., De Gaspari A., Ricci S., Riemenschneider J., Monner H. P., van de Kamp B., "Compliant structures-based wing and wingtip morphing devices," Aircr. Eng. Aerosp. Technol., Vol. 88, N.2, pp. 311–330, 2016.

[153] De Gaspari A., Ricci S., Riccobene L., "Design, manufacturing and wind tunnel test of a morphing wing based on compliant structures," 24th AIAA/AHS Adaptive Structures Conference, SciTech 2016, San Diego, CA, January 4-8, 2016.
[154] De Gaspari A., Ricci S., "Knowledge-based shape optimization of morphing wing for more efficient aircraft," J. Aerosp. Eng., Vol. 2015, Article ID 325724, 19 p., 2015.
[155] JTI-CS2-2015-CPW03-AIR-01-04, Next generation movables for high speed aircraft, CS2-IA Innovation Action, Published October 22, 2015, Deadline February 4 2016.
[156] JTI-CS2-CPW-REG-01-02, Advanced wing for regional A/C–technologies development, design and manufacturing for FTB1," CS2-IA Innovation Action, Publication July 9, 2014, Deadline November 5 2014.
[157] JTI-CS2-CPW-REG-02-01, "Flight physics and wing integration in FTB2," CS2-IA Innovation Action, Publication July 9, 2014, Deadline November 5 2014.
[158] H2020, "Draft horizon 2020 work programme 2016–2017 in the area of transport," https://ec.europa.eu/programmes/horizon2020/sites/horizon2020/files/11.%20SC4_2016–2017_pre-publication.pdf.
[159] Boeing, "Continuous lower energy, emissions and noise (CLEEN) technologies—Boeing program overview," Cleen Consortium Public Session, Atlanta, GA, November 19, 2014.
[160] Akiyama D., "Accelerating advances in environmental performance," ICAO Seminar on the Environment, Montreal, September 9-10, 2014.
[161] M.A. Moser, M.R. Finn, Adaptive Trailing Edge Actuator System and Method, United States Patent Application US2015/0083853 A1, Priority Date September 24, 2013; Publication Date, March 26, 2015.
[162] DARPA, "Complex adaptive system composition and design environment (CASCADE)," Solicitation Number DARPA-BAA-16-11, Agency: Other Defense Agencies, 2015.
[163] SAE, G-11SHM, Structural Health Monitoring and Management Group (Aisc), Guidelines for Implementation of Structural Health Monitoring on Fixed Wing Aircraft, Standard, ARP6461, SAE International, Warrendale, PA, 2013.
[164] Concilio A., About the Use of Piezoelectric Actuators for Controlling Intelligent Structural Elements (Master thesis), University of Napoli "Federico II", Faculty of Aeronautical Engineering, December 15, 1988 (in Italian).
[165] Edberg D. L., "Control of flexible structures by applied thermal gradients," AIAA J. Vol. 25, N.6, pp. 877–883, 1987.
[166] Crawley E. F., de Luis J., "Use of piezoelectric actuators as elements of intelligent structures," AIAA J. Vol. 25, N.10, pp. 1373–1385, 1987.
[167] Geissler W., Koch S., "Adaptive Airfoil", H. Sobieczky (editor), "IUTAM Symposium Transsonicum IV," Proceedings of the IUTAM Symposium, Göttingen, September 2-6, 2002, Springer Science and Business Media, Dordrecht, 2003, ISBN 978-94-010-3998-7.
[168] Concilio A., Andersson B., Geissler W., Gobert J. L., San Millan J., "EREA project: adaptive airfoil," 5th ONERA-DLR Aerospace Symposium, ODAS 2003, Toulouse, June 4-6, 2003.
[169] Ameduri S., Integrated Systems for Improving the Aerodynamic Performance of a Transonic Airfoil (PhD thesis), University of Napoli "Federico II," Faculty of Engineering, March 27, 2002 (in Italian).
[170] Concilio A., Esposito C., Ameduri S., Lecce L., "A smart actuator to transonic wave drag reduction," Actuator 2000, Bremen, June 19-21, 2000.
[171] Ameduri S., Esposito C., Concilio A., "Transonic wave drag reduction through a smart structure system," 2nd International Symposium on Smart Structures and Micro-Techniques, IS3M, Hong Kong, October 19-21, 2000.
[172] Ameduri S., Esposito C., Concilio A., "Active shape airfoil control through composite-piezoceramic actuators," 8th SPIE Annual Symposium on Smart Structures and Materials, Newport Beach, CA, March 4-8, 2001.
[173] Ameduri S., Gianvito A., Concilio A., "An adaptive bump for transonic shock wave drag reduction," Actuator 2002, Bremen, June 10-12, 2002.

[174] Concilio A., Gianvito A., Ameduri S., "Latest experimental results concerning a smart actuator for transonic wave drag reduction," 13th International Conference on Adaptive Structures and Technologies, ICAST 2002, Potsdam, October 7-9, 2002.

[175] A. Concilio, C. Izzo, C. Pascarella, Hingeless Wing (HIWIN): A Technology Proposal for UAV, CIRA, Capua, 2002. Doc. Number CIRA-TN-02-014, January.

[176] A. Vigliotti, A. Concilio, Hingeless Wing (HIWIN): A UAV Technology Project, CIRA, Capua, 2003. Doc. Number CIRA-TN-02-014 Rev. 1, March.

[177] A. Vigliotti, S. Ameduri, A. Concilio, L. De Vivo, Hingeless Wing–Structure & Actuation Architecture Selection Report, CIRA, Capua, 2003. Doc. Number CIRA-TR-03-255, May.

[178] Vigliotti A., Ameduri S., Concilio A., "Evaluation of two structural architectures for a hingeless flap prototype, 14th International Conference on Adaptive Structures and Technologies, ICAST '03, Seoul, October 7-9, 2003.

[179] Antonucci V., Concilio A., Faiella G., Giordano M., Mennella F., Nicolais L., "Shape modification of an adaptive aeronautical profile," Actuator '04, Bremen, June 14-16, 2004.

[180] Antonucci V., Concilio A., Giordano M., Nicolais L., "Torque composite tube with embedded shape memory alloys actuators," 4th International Workshop on Structural Health Monitoring, Stanford, CA, September 15-17, 2003.

[181] Diodati G., Ameduri S., Concilio A., "A SMA embedded anisotropic panel aimed at controlling vibration due to variable regimes sources," 8th Conference on Dynamical Systems Theory and Applications, Łódź, December 12-15, 2005.

[182] Diodati G., Ameduri S., Concilio A., "Adaptive vibration control through an SMA embedded panel," J. Theor. Appl. Mech., Vol. 45, N.4, pp. 919–930, 2007.

[183] Diodati G., Ameduri S., Concilio A., "SMA embedded panel optimized through a genetic approach," J. Intell. Mater. Syst. Struct., Vol. 20, N.-, pp. 1529–1540, 2009.

[184] Ameduri S., Barbarino S., Pecora R., Lecce L., Concilio A., E. Calvi, "A novel SMA implementation oriented to airfoil morphing," The International Conference for Shape Memory and Superelastic Technologies, SMST 2008, Stresa, September 21-25, 2008.

[185] Barbarino S., Smart Morphing Concepts and Applications for Advanced Lifting Surfaces (PhD thesis) University of Napoli "Federico II," Faculty of Engineering, November 27, 2009.

[186] S. Barbarino, R. Pecora, L. Lecce, A. Concilio, S. Ameduri, Airfoil morphing architecture based on shape memory alloys, Proceedings of the ASME Conference on Smart Materials, Adaptive Structures and Intelligent Systems, SMASIS2008, vol. 1, 2008, pp. 729–737.

[187] Barbarino S., Ameduri S., Lecce L., Concilio A., "Wing shape control through an SMA-based device," J. Intell. Mater. Syst. Struct., Vol. 20, N.3, pp. 283–296, 2009.

[188] R. Pecora, G. Iannuzzo, M. Riccio, S. Russo, E. Calvi, L. Lecce, S. Barbarino, A. Concilio, S. Ameduri, Actuator Device Based on a Shape Memory Alloy and a Wing Flap Assembly Fitted with Such an Actuator Device, United States Patent N.US8348201, Priority Date, Jul. 23rd, 2008; Publication Date, January 28, 2010.

[189] Barbarino S., Pecora R., Lecce L., Concilio A., Ameduri S., De Rosa L., "Airfoil structural morphing based on SMA actuator series: numerical and experimental studies," J. Intell. Mater. Syst. Struct. Vol. 22, N.10, pp. 987–1004, 2011.

[190] Mirone G., Concilio A., Siracusa S., Dimino I., Pellegrino A., "Dynamics modification of adaptive aerostructures components" XXIX AIAS National Congress, Maratea, September 7-10, 2010 (in Italian).

[191] Concilio A., Ameduri S., "Influence of structural architecture on linear SMA actuators performance and morphing system layout optimisation," J. Intell. Mater. Syst. Struct., Vol. 25, N.16, pp. 2037–2051, 2014.

[192] Pecora R., Barbarino S., Concilio A., Lecce L., Russo S., "Design and functional test of a morphing high-lift device for a regional aircraft," J. Intell. Mater. Syst. Struct., Vol. 22, N.10, pp. 1005–1023, 2011.

[193] Ameduri S., “A smart leading edge architecture aimed at preserving laminar regime,” 1st EASN Workshop on Aerostructures, Paris, October 7-8, 2010.
[194] Tomassetti G., Ameduri S., Carozza A., “Innovative streamline-flow preserving actuation strategies for wing airfoil nose,” Int. J. Mater. Struct. Integr., Vol. 2, N.4, pp. 437–457, 2011.
[195] Ameduri S., “A leading edge morphing architecture for droop nose effect,” XXII AIDAA National Congress, Napoli, September 9-12, 2013.
[196] Ameduri S., “A SMA based morphing leading edge architecture,” The 5th International Conference on Mechanical and Aerospace Engineering, ICMAE 2014, Madrid, July 18-19, 2014 (Best Conference Oral Presentation Award).
[197] Ameduri S., Concilio A., Daniele E., “A drop nose laboratory demonstrator: experimental characterisation and validation,” 23rd International Conference on Adaptive Structures and Technologies, ICAST 2012, Nanjing, October 11-13, 2012.
[198] S. Ameduri, Device for Modifying the Airfoil Geometry, European Patent N.EP2727825A1, Priority Date October 30, 2012; Publication Date, May 7, 2014.
[199] Ameduri S., Concilio A., “Single slotted flap SMA-based morphing architecture: feasibility study and preliminary design,” 4th International Conference on Experiments/Process/System Modeling/Simulation/Optimization, 4th IC-EpsMsO, Athens, July 6-9, 2011.
[200] Brindisi A., Tiseo B., Ameduri S., Concilio A., “Optimization and integration of shape memory alloy (SMA)-based elastic actuators within a morphing flap architecture,” J. Intell. Mater. Syst. Struct., Vol. 23, N.4, pp. 381–396, 2011.
[201] Ameduri S., Concilio A., Pecora R., “A SMA-based morphing flap: conceptual and advanced design,” Smart Struct. Syst., Vol. 16, N.3, pp. 555–557, 2015.
[202] Ameduri S., Concilio A., Pecora R., D. Karagiannis, “A single slotted morphing flap based on SMA technology,” Smart Struct. Syst., Vol. 17, N.5, pp. 819–835, 2016.
[203] Pecora R., Amoroso F., Amendola G., Concilio A., “Validation of a smart structural concept for wing-flap camber morphing,” Smart Struct. Syst., Vol. 14, N.4, pp. 659–678, 2014.
[204] R. Pecora, F. Amoroso, M. Magnifico, Toward the Bi-Modal Camber Morphing of Large Aircraft Wing Flaps: the Clean Sky Experience, SPIE Smart Structures/NDE, Las Vegas, NA, 2016, pp. 20–24. March.
[205] Concilio A. (Ed.), “Adaptive trailing edge device (ATED): exploitation of morphing—booklet of technologies,” SARISTU Workshop, Chemnitz, December 11, 2014. http://www.cira.it/it/comunicazione/events/file/Booklet%20ATED.pdf.
[206] Carossa G. M., Ricci S., De Gaspari A., Liauzun C., Dumont A., Steinbuch M., “Adaptive trailing edge: specifications, aerodynamics, and exploitation,” Woelcken P. C., Papadopoulos M. (Eds.), “Smart Intelligent Aircraft Structures (SARISTU)—Proceedings of the Final Project Conference,” Springer International Publishing, Cham, 2016, ISBN978-3-319-22413-8.
[207] I. Dimino, A. Concilio, M. Schueller, A. Gratias, An Adaptive Control System for Wing TE Shape Control, SPIE Smart Structures/NDE, San Diego, CA, 2013. March 10-14.
[208] S. Barbarino, S. Ameduri, R. Pecora, Wing chamber control architectures based on SMA: Numerical investigations, Proceedings of SPIE—The International Society for Optical Engineering, vol. 6423, 2007. Article number 64231E.
[209] Ciminello M., Dimino I., Ameduri S., Concilio A., “Sensor network structural integration for shape reconstruction of morphing trailing edge,” 26th International Conference on Adaptive Structures and Technologies, ICAST 2015, Kobe, October 14-16, 2015.
[210] Pecora R., Concilio A., Dimino I., Amoroso F., Ciminello M., “Structural design of an adaptive wing trailing edge for enhanced cruise performance,” 24th AIAA/AHS Adaptive Structures Conference, January 4-8, AIAA SciTech, San Diego, CA, 2016.

[211] R. Pecora, F. Amoroso, M. Magnifico, I. Dimino, A. Concilio, KRISTINA: Kinematic Rib Based Structural System for Innovative Adaptive Trailing Edge, SPIE Smart Structures/NDE, Las Vegas, NA, 2016. March 20-24.
[212] Gratias A., Schueller M., Lipowski M., Dimino I., "Actuator net design for morphing wing structure," Smart System Integration Conference, Amsterdam, March 13-14, 2013.
[213] I. Dimino, D. Flauto, G. Diodati, A. Concilio, R. Pecora, Actuation system design for a morphing wing trailing edge, Recent Pat. Mech. Eng. 7 (2014) 138–148.
[214] I. Dimino, G. Diodati, A. Concilio, A. Volovick, L. Zivan, Distributed Electromechanical Actuation System Design for a Morphing Trailing Edge Wing, SPIE Smart Structures/NDE, Las Vegas, NA, 2016. March. 20-24.
[215] Gratias A., Schueller M., Dimino I., Pecora R., Concilio A., "Static and dynamic test campaign of an adaptive trailing edge device," International Conference on Adaptive Structures and Technologies, ICAST 2014, The Hague, October 6-8, 2014.
[216] Dimino I., Concilio A., Pecora R., "Primary structural components characterization of an adaptive trailing edge device (ATED)," 24th AIAA/AHS Adaptive Structures Conference, January, AIAA SciTech, San Diego, CA, 4–8 2016.
[217] M. Ciminello, A. Concilio, D. Flauto, F. Mennella, FBG Sensor System for Trailing Edge Chord-Wise Hinge Rotation Measurements, SPIE Smart Structures/NDE, San Diego, CA, 2013, pp. 10–14. March.
[218] Ciminello M., Ameduri S., Flauto D., "Design of FBG based-on sensor device for large displacement deformation," International Microwave and Optoelectronics Conference, IMOC 2013, Rio de Janeiro, August 4-7, 2013.
[219] Ciminello M., Concilio A., Ameduri S., Flauto D., Mennella F., "Hinge rotation of a morphing rib using FBG strain sensors," Smart Struct. Syst. Vol. 15, N.6, pp. 1393–1410, 2015.
[220] Evenblij R., Kong F., Koimtzoglou C., Ciminello M., Dimino I., Concilio A., "Shape sensing for morphing structures using fiber bragg grating technology," Woelcken P. C., Papadopoulos M. (Eds.), "Smart Intelligent Aircraft Structures (SARISTU)—Proceedings of the Final Project Conference," Springer International Publishing, Cham, 2016, ISBN 978-3-319-22413-8.
[221] Ciminello M., Bettini P., Ameduri S., Guerreschi E., Cuneo G., "Experimental validation of a sensorized ring based on optical fibers for strain monitoring of morphing structure," 29th Congress of the International Council of the Aeronautical Sciences, ICAS 2014, San Petersburg, September 7-12, 2014.
[222] Ciminello M., Bettini P., Ameduri S., Guerreschi E., Concilio A., Sala G., "Monito-ring: an original fiber optic system for morphing application," J. Intell. Mater. Syst. Struct., Vol. 26, N.18, pp. 2463–2476, 2015.
[223] Rinaldi A., Proietti A., Tamburrano A., Ciminello M., Sarto M. S., "Graphene-based strain sensor array on carbon fiber composite laminate," J. Sens. Stud., Vol. 15, N.12, pp. 7295–7303, 2015.
[224] Rinaldi A., Proietti A., Aquarelli C., Marra F., Tamburrano A., Ciminello M., Sarto M. S., Monaco E., "High sensitive graphene-based strain sensors for SHM of composite laminates," Woelcken P. C., Papadopoulos M. (Eds.), "Smart Intelligent Aircraft Structures (SARISTU)—Proceedings of the Final Project Conference," Springer International Publishing, Cham, 2016, ISBN 978-3-319-22413-8.
[225] Schorsch O., Lühring A., Nagel C., Pecora R., Dimino I., "Polymer based morphing skin for adaptive wings," 7th ECCOMAS Conference on Smart Structures and Materials, SMART 2015, Azores, June 3-6, 2015.
[226] R. Pecora, F. Amoroso, R. Palumbo, M. Arena, G. Amendola, I. Dimino, Preliminary aeroelastic assessment of a large aeroplane equipped with a camber-morphing aileron, in: Proceedings of SPIE Conference on Smart structures and NDE, Portland (Oregon), 2017. Vol. 10166.
[227] Pecora R., Magnifico M., Amoroso F., and E. Monaco, "Multi-parametric flutter analysis of a morphing wing trailing edge," Aeronaut. J., Vol. 118, N. 1207, pp. 1063–1078, 2014.

[228] Y. Essa, F.M. de la Escalera Cutillas, I. Dimino, M. Ciminello, A. Concilio, in: P.C. Woelcken, M. Papadopoulos (Eds.), Manufacturing and testing of smart morphing SARISTU trailing edge, Smart Intelligent Aircraft Structures (SARISTU)—Proceedings of the Final Project Conference, Springer International Publishing, Cham, ISBN 978-3-319-22413-8, 2016.
[229] Dimino I., Concilio A., Pecora R., "Safety and reliability aspects of an adaptive trailing edge device (ATED)," 24th AIAA/AHS Adaptive Structures Conference, AIAA SciTech, San Diego, CA, January 4–8, 2016.
[230] G. Diodati, A. Concilio, S. Ricci, A. De Gaspari, F. Huvelin, A. Dumont, J.-L. Godard, Estimated Performances of an Adaptive Trailing Edge Device Applied to Medium-Size Aircraft Flights, SPIE Smart Structures/NDE, San Diego, CA, 2013. March 10-14.
[231] G. Diodati, A. Concilio, Actuation Needs for an Adaptive Trailing Edge Device Aimed at Reducing Fuel Consumption on a Regional Aircraft, SPIE Smart Structures/NDE, San Diego, CA, 2013. March 10-14.
[232] Diodati G., Actuation System Design and Test for an Adaptive Trailing Edge Morphing Device (PhD thesis), University of Rome "La Sapienza," Department of Mechanical and Aerospace Engineering, December 1, 2016.
[233] M. Arena, M.C. Noviello, F. Rea, F. Amoroso, R. Pecora, G. Amendola, Modal stability assessment for a morphing aileron subjected to actuation system failures: Numerical analysis supported by test evidence, in: Proceedings of 2016 7th International Conference on Mechanical and Aerospace Engineering, ICMAE 2016, 2016, pp. 437–442.
[234] R. Pecora, F. Amoroso, L. Lecce, Effectiveness of wing twist morphing in roll control, J. Aircr. 49 (6) (2012) 1666–1674.
[235] G. Amendola, I. Dimino, R. Pecora, M. Magnifico, Distributed actuation concepts for a morphing aileron device, Aeronaut. J. 120 (1231) (2016) 1365–1385.
[236] Amendola G., Dimino I., Concilio A., Pecora R., Amoroso F., "Actuation system design for a morphing aileron," Appl. Mech. Mater., Vol. 798, N.-, pp. 582–588, 2015, Presented at the International Conference on Aerospace Engineering, ICOAE 2015, St. Petersburg, June 18-19, 2015 (Best Conference Oral Presentation Award).
[237] G. Amendola, I. Dimino, F. Amoroso, R. Pecora, Experimental Characterization of an Adaptive Aileron: Lab Tests and FE Correlation, SPIE Smart Structures/NDE, Las Vegas, NA, 2016. March 20-24.
[238] G. Amendola, I. Dimino, F. Amoroso, R. Pecora, A. Concilio, Preliminary design of an adaptive aileron for next generation regional aircraft, J. Theor. Appl. Mech. 55 (1) (2017) 307–316.
[239] G. Amendola, I. Dimino, A. Concilio, M. Magnifico, R. Pecora, Numerical Design of an Adaptive Aileron, SPIE Smart Structures/NDE, Las Vegas, NA, 2016. March 20-24.
[240] Amendola G., Full-Scale Servo Actuated Morphing Aileron for Wind Tunnel Tests (PhD thesis), University of Napoli "Federico II," Department of Industrial Engineering, Aerospace Division, Mar 31st, 2016.
[241] G. Amendola, M. Ciminello, I. Dimino, S. Ameduri, Real-time monitoring of a variable-camber aileron rib by original strain-angle transducer, J. Intell. Mater. Syst. Struct. 28 (8) (2017) 1023–1035.
[242] Rushby J., "A safety-case approach for certifying adaptive systems," AIAA Infotech@Aerospace Conference, Seattle, WA, April 6-9, 2009.
[243] Herrera C. Y., Lung S. F., Ervin G., Flick P., "Aeroelastic airworthiness assessment of the adaptive compliant trailing edge flaps," 46th International Symposium of the Society of Flight Test Engineers, Lancaster, CA, September 14-17, 2015.
[244] Thill C., Etches J., Bond I., Potter K., Weaver P. "Morphing skins," Aeronaut. J., Vol. 112, N. 1129, pp. 117–139, 2008.
[245] Wright O., Wright W., Flying Machine, US Patent N. US821393 A, Priority Date, Mar. 23rd, 1903; Publication Date, May 22, 1906.

[246] NASA/Ken Ulbrich, http://www.nasa.gov/sites/default/files/flap_0.jpg. [Public domain], via Wikimedia Commons, https://commons.wikimedia.org/wiki/File%3AACTE_taxi_testing_2014-10-31.jpg. https://uploadwikimedia.org/wikipedia/commons/d/de/ACTE_taxi_testing_2014-10-31.jpg.

[247] Arslan P., Kalkan U., Tiraş H., Tunçöz I. O., Yang I., Gürses E., Sahin M., Özgen S., Yaman Y., "Structural analysis of an unconventional hybrid control surface of a morphing wing," 25th International Conference on Adaptive Structures and Technologies, Icast 2014, The Hague, October 6-8, 2014.

[248] NOVEMOR Consortium, "Project final report," EU Grant Agreement Number 285395, Collaborative Project: Level 1, Period Covered: from September 1, 2011 to February 28 2015.

[249] Clean Sky 2 Team, "Clean Sky 2, Joint Technical Programme", a publication of Clean Sky 2 JU, Brussels, March 23, 2015. http://ec.europa.eu/research/participants/data/ref/h2020/other/guide-appl/jti/h2020-guide-techprog-cleansky-ju_en.pdf.

[250] Magnifico M, KRISTINA: Kinematic Rib Based Structural System for Innovative Adaptive Trailing Edges (PhD thesis), University of Napoli "Federico II," Department of Industrial Engineering, Aerospace Division, March 31, 2016.

[251] James T., Menner A., Bismarck A., Iannucci L., "Morphing skins: development of new hybrid materials," 4th SEAS DTC Technical Conference, Edinburgh, July 7-8, 2009.

[252] Chen S., Chen Y., Zhang Z., Liu Y., Leng J., "Experiment and analysis of morphing skin embedded with shape memory polymer composite tube," J. Intell. Mater. Syst. Struct., Vol. 25, N.16, pp. 2052–2059, 2014.

[253] Guiler R. W., Huebsch W. W., "Control of a swept wing tailless aircraft through wing morphing," 26th International Congress of the Aeronautical Sciences, ICAS 2008, Anchorage, AK, September 14-19, 2008.

[254] J.E. Elgersma, Conceptual Layout of Wing Structure Using Topology Optimization for Morphing Micro Air Vehicles in A Perching Maneuver, Air Force Institute of Technology, Department of the Air Force Air University, Wright-Patterson Air Force Base, OH, USA, March 9, 2012.

[255] R.D. Vocke, C.S. Kothera, B.K.S. Woods, E.A. Bubert, N.M. Wereley, One-dimensional morphing structures for advanced aircraft, in: R. Agarwal (Ed.), Recent Advances in Aircraft Technology, InTech Open, Rijeka (HR), 2012, ISBN 978-953-51-0150-5.

[256] https://en.wikipedia.org/wiki/Auxetics#/media/File:Auxetische_Materialien.wiki.png By LZ6387 at German Wikipedia, CC BY-SA 3.0, https://commons.wikimedia.org/w/index.php?curid=1885 9076.

[257] J.N. Grima, D. Attard, R. Gatt, A Novel Chemo.-Mechanical process for making Auxetic foams and for Their Reconversion to Conventional Form, http://www.slideshare.net/asimraza1234/004-foamspptpresentation.

[258] A. Airoldi, M. Crespi, G. Quaranti, G. Sala, Design of a morphing airfoil with composite chiral structure, J. Aircr. 49 (4) (2012) 1008–1019.

[259] Morishima R., Analysis of Composite Wing Structures With a Morphing Leading Edge (PhD thesis), Cranfiled University, Deaprtment of Aerospace Engineering, January 15, 2011.

[260] FAA, Aviation Maintenance Technician Handbook–Airframe, vol. 1, Department of Transportation, Federal Aviation Administration-Flight Standards Service, Oklahoma City, OK, 2012.

[261] Grima J. N., Attard D., Gatt R., Cassar R. N., "A novel process for the manufacture of Auxetic foams and for their re-conversion to conventional form," Adv. Eng. Mater., Vol. 11, N.7, pp. 533–535, 2009.

[262] Cipollo P. M., Surrusco B. S., Bilen S. G., "An electrically actuated pin-puller for space applications using nickel-titanium memory alloys," Frank J. Redd Student Scholarship Competition, Small Satellite Conference, Logan, UT, August 11, 2004.

[263] Hihoud M., Pagès A., Benoit C., Claeyssen F., "BRUCE–electromagnetic actuated pin puller," 13th International Conference on New Actuators, ACTUATOR 2012, Bremen, June 18-20, 2012.

[264] Nava N., Collado M., Cabás R., "New deployment mechanisms based on sma technology for space application," 15th European Space Mechanisms & Tribology Symposium, ESMATS, Noordwijk, September 25-27, 2013.

[265] Bella V., Faraci F. F., Wireless Energy Transfer System and Method for Feeding an Electric Load, United States Patent N.US9071064 B2, Priority Date, December 9, 2009; Publication Date, June 30 2015.

[266] Chang C. W., Kothari A., Jafri A., Koutsonikolas N., Peroulis D., Hu Y. C., "Radiating sensor selection for distributed beamforming in wireless sensor networks," 2008 IEEE Military Communications Conference, MILCOM 2008, San Diego, CA, November 16-19, 2008.
[267] G.W.H. Van Es, Analysis of Aircraft Weight and Balance Related Safety Occurrences, NLR, Amsterdam, 2007. Doc. Number NLR-TP-2007-153. June.
[268] Vancroonenburga W., Verstichela J., Tavernierc K., Vanden Berghe G., "Automatic air cargo selection and weight balancing: a mixed integer programming approach," Transp. Res. E, Special Issue on Modeling, Optimization and Simulation of the Logistics Systems, Vol. 65, N.-, pp. 70–83, 2014.
[269] Akl W., Poh S., Baz A., "Wireless and distributed sensing of the shape of morphing structures," Sensors Actuators A, Vol. 140, N.-, pp. 94–102, 2007.
[270] Igwe P. C., Sangole A. P., Knopf G. K., "Shape morphing and reconstruction using a self-organizing feature map," The 2006 IEEE International Joint Conference on Neural Network, Vancouver, IJCNN 2006, July 16-21, 2006.
[271] A.V. Srinivasan, D.M. McFarland, Smart Structures: Analysis and Design, Cambridge University Press, Cambridge, 2001.
[272] N. Wiener, Cybernetics or Control and Communication in the Animal and the Machine, The MIT Press, Cambridge, MA, ISBN 978-0-262-73009-9, 1961.
[273] Kim J. H., Matson E. T., Myung H., Xu P. (eds), "Robot Intelligence Technology and Applications 2012—An Edition of the Presented Papers from the 1st International Conference on Robot Intelligence Technology and Applications," Springer Science & Business Media, Berlin Heidelberg, 2013, ISBN 978-3-642-37374-9.
[274] Kalkan S., IEEE Staff (ed), "2015 International Conference on Advanced Robotics (ICAR)," IEEE Publishing, New York, NY, 2015, ISBN 978-1-467-37510-8.
[275] M. Kassar, A.G. Vajna, J.B. Michaels, H. Lieberman, C. Wilson (Producers), J. Mostow (Director), "Terminator 3: Rise of the Machines" (Motion Picture), Warner Bros Pictures, Columbia Pictures, 2003.
[276] D. Murphy, T. De Santo, L. di Bonaventura, I. Bryce (Producers), M. Bay (Director), "Transformers" (Motion Picture), DreamWorks Pictures, Paramount Pictures, 2007.
[277] NASA/Dryden Flight Research Center/Jim Ross, http://archive.darpa.mil/j-ucas/X-45/gallery.htm (direct link, alt site), Public Domain, https://commons.wikimedia.org/w/index.php?curid=450653.
[278] J. Herzog, CC BY 4.0, https://commons.wikimedia.org/w/index.php?curid=21567585.

UNIVERSITY OF NAPOLI AND CIRA INTERNATIONAL AWARDS

[1] April 2016, Morphing Flap Project, Nomination for the Best Clean Sky Project Award in 2016.
[2] June 2015, Best Conference Oral Presentation Award, International Conference on Aerospace Engineering, ICOAE 2015, St. Petersburg, June 18-19, 2015.
[3] May 2015 Award: "Premio Venezia," Category "Scientific and Academic" for the CRIAQ MDO 505 project, "Morphing Architectures and Related Technologies for Wing Efficiency Improvement" http://www.italchamber.qc.ca/fr/documents/programme2015.pdf.
[4] July 2014, Best Oral Presentation Award, The 5th International Conference on Mechanical and Aerospace Engineering, ICMAE 2014, Madrid, July 18-19, 2014.
[5] April 2014 Award: 2nd Award for the CRIAQ MDO 505 project "Morphing Architectures and Related Technologies for Wing Efficiency Improvement," at the CRIAQ 7th Forum, http://infoaeroquebec.net/international-project-awarded-at-the-criaq-gala/.
[6] March 2014, Best Conference Paper–Greener Aviation Award, 3AF/CEAS Conference Greener Aviation: Clean Sky Breakthroughs and Worldwide Status, Bruxelles, March 12-14, 2014.

CHAPTER

AIRCRAFT MORPHING—AN INDUSTRY VISION

2

Gianvito Apuleo

Piaggio Aerospace, Villanova d'Albenga (SV), Italy

CHAPTER OUTLINE

NOMENCLATURE

a/c	aircraft
ADL	allowable damage limit
AM	additive manufacturing
b	wing span
C_D	drag coefficient
C_L	lift coefficient
DSG	design service goal
LL	limit load
E	energy of impact

Morphing Wing Technologies. https://doi.org/10.1016/B978-0-08-100964-2.00002-2

g	gravitational acceleration factor
Λ	wing sweep
λ	wing planform taper
MTOW	maximum takeoff weight
OEM	original equipment manufacturer
OEW	aircraft operative empty weight
P	probability of impact
RPAS	remote piloted aircraft system
S	wing wetted area
SRM	structures repair manual
TRL	technological readiness level
W	weight
W_{FUEL}	aircraft fuel weight
$W_{PAYLOAD}$	aircraft payload weight

1 INTRODUCTION

Taking inspiration from birds, the dream of every aeronautical engineer is to design the best possible wing, able to smoothly run in all operating flight conditions. Birds, bats, and other flying creatures can alter lift generated by their wings. In fact, they can adapt their wings to the needed conditions time after time. Wings can be folded tightly to dive for prey, or fully extended to glide to save energy. They are able to modify the camber and wing twist to control their flight.

On the other hand, engineers in aircraft industries still use the "single point design" standard, robust and structured approach, which in itself, still represents the most feasible method practical to apply.

Fortunately, however, today the consolidation of different technologies help to overcome some factual difficulties, that until a few years ago seemed insurmountable, in the manufacturing of the wings, giving greater impetus and credit to the "morphing discipline."

The challenge of morphing in the future will mainly be to convince industry that this kind of technology can bring real advantages in terms of enhanced aerodynamic performance, lower installation impact with respect to traditional control surface systems, reduced weight, and equivalent safety level.

2 CURRENT AIRCRAFT CAPABILITIES

The term "morphing wing" is used for a wide variety of different designs, in which the wing or portions of the wing change shape, by classical control devices (see Fig. 1) or by innovative arrangement of advanced materials and actuators.

Wing morphing, in fact, is not a new concept in aircraft design. The Wright Brothers, for example, used wing warping through structural deformation to control the roll of the aircraft [1], Parker [2] proposed a variable camber wing to increase flight speed. The Wright Flyer was made >100 years ago and used a very simple design, largely of cloth.

Today, aircraft primary and secondary structures are very complex. To have an adaptive wing, that must be: 9 g compliant (a military aircraft for example), free from flutter, fatigue, and damage tolerant,

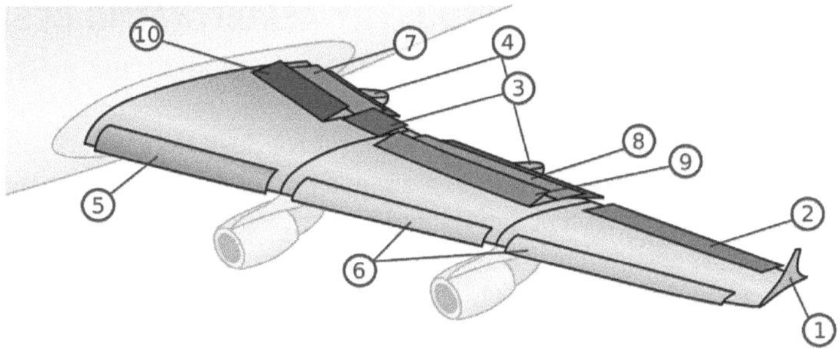

FIG. 1

Wing control device scheme: 1. winglet, 2. outer aileron, 3. inner aileron, 4. flap track fairing, 5. slat, 6. slat, 7. flap, 8. flap, 9. spoiler, 10. spolier.

able to support hard points and still maintain all subsystems within the wing is not easy from industry point of view unless we consider the classic change of shape obtained through standard control devices.

To better understand the current interest of aircraft industries around morphing technology, it is necessary to clarify what the market requests are: military and civil needs are, in most cases, profoundly different. To meet the military's need, it would be perfect to have an aircraft able to perform in all different "permissive" and "not permissive" scenarios.

Obtaining such a machine from industry could overshadow the cost of design, production, and certification, as well as any high operating costs.

Civil aircraft industry is different, because from one side, operators do not have such demanding requests, and from the other side, industry is aware that the selling price and developing time frame are of paramount importance.

2.1 INTEREST OF INDUSTRY

During the last decades, morphing is back again on the table of aircraft industry, mainly for the following reasons:

1. Renewed need to increase efficiency and reduce emissions of commercial aircraft.
2. A more urgent request by different governments to raise the security level of territory using a multirole aircraft.
3. Improved capabilities in compliant structures that foresee simpler, more robust and safety critical shape-changing mechanisms.

For commercial-aircraft designers, the next step in aerodynamic efficiency could come from more slender, lower-drag wings, but these would be flexible and require active control to suppress flutter and alleviate loads. This implies variable-camber trailing edges that are in motion throughout the flight and not just on takeoff and landing.

Another big increase in efficiency could come from low-drag laminar flow, but this requires leading edges that move seamlessly, without gaps to disturb the flow.

2.2 SOME CONSIDERATIONS ABOUT INDUSTRY AERODYNAMIC DESIGN PROCESS

Even if several approaches to morph wings are currently being studied, the most current techniques focus on a single methodology (sweep, dihedral, twist, or span motion). Variable sweep allows a pilot to control the speed of an aircraft, and is particularly useful for dash maneuvers and high speed flight. The dihedral motion refers to the upward angle of a wing; a positive angle affects the lateral stability of an aircraft around its roll axis, whereas a negative angle increases maneuverability. The twist of a wing allows for greater control over aerodynamic forces, helping to maintain a level body. Finally, the wing span affects aspect ratio and wing loading, which enables optimization for specific flight regimes.

Wing morphing concepts are classified into three major types [3]:

1. Planform alternation.
2. Out-of-plane transformation.
3. Airfoil adjustment.

The planform alternation is performed through the wing area resizing by changing parameters including the span, chord length, and sweep angle. The out-of-plane transformation include the wing twist, the chord, and spanwise camber changes. The airfoil adjustment regroup designs that can alter the wing profile with no significant change in the wing camber; the wing thickness control comes under this category (Fig. 2).

Aerodynamic wing design, in common aircraft industry practice, is essentially still based on the idea to optimize, as much as possible, just a single mission scenario. Small aircraft original equipment manufacturers (OEMs), with limited resources in respect to the “big sisters,” try to directly optimize only one flight condition, usually the cruise. For different types of aircraft, having different structures, speed, cruise altitude, operative range, comfort level, weight and cost, in-flight wing geometric variations could allow multirole missions that are impossible without aircraft reconfiguration.

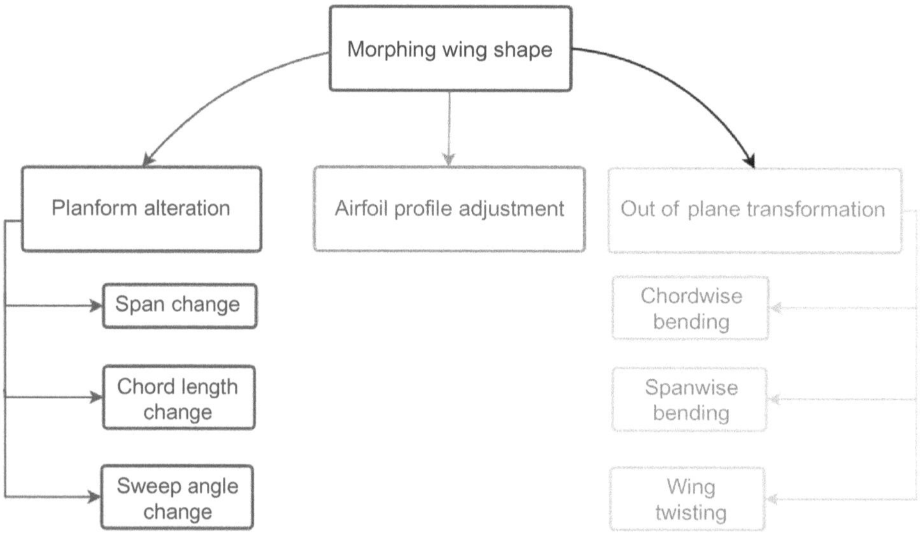

FIG. 2

Morphing wing shape classification [3].

FIG. 3

True wing control device overview.

A rational compromise that tries to mitigate the real difficulty in having a so called "adaptive wing," that ideally tries to incorporate as many morphing wing capabilities as possible is in fact to accept only one optimal condition.

Conventional aircraft (whether civil or military) use mechanisms to discretely change the wing area in different flight configurations. These configurations include takeoff, climb, cruise, and landing. Glenn Curtiss [4] started using hinged flaps as ailerons mainly because the Wrights' continually flexing wooden wing could lead to structural failure. Since then aviation has been wedded to these hinged control surfaces: rudders, elevators, flaps, ailerons, and spoilers all work on this kind of principle, changing the shape of the wing or tail to help the aircraft move in a particular direction (see Fig. 3). The discrete shape change is achieved by extending or retracting flaps, slats, tabs, or ailerons to either modify the wing area and the airfoil camber for additional lift or the aircraft controllability characteristics.

Often, in common practice, several aspects must be considered (i.e., wing deflection, inspection and accessibility problems, regulation rules for safety, actuation, and so on) which make a simple installation quite complicated. Morphing aircraft technology can provide a solution to those problems by developing adaptive structures that can carry out multiple objectives and control aircraft during flight, take-off, and landing.

2.3 EXPECTED PERFORMANCE TARGETS

Adaptive morphing can essentially be used both in flight for maneuvers and to increase performance. This dual aspect will require the morphing to be considered as a "primary" system for certification purposes, and certainly will need to work together with a "fly by wire" system. Expectation and the efforts to obtain these results is remarkable. Particular attention will be made on the reliability of the complete system. For certification purposes, it will be necessary to demonstrate the redundancy of the actuation system, in line with the fail-safe design philosophy.

Several methods exist to increase efficiency of different flight aspects of an aircraft through changing the aerodynamic characteristics of the wing. Firstly, changing the span or the aspect ratio of the aircraft wing alters aircraft lift characteristics and stealth characteristics for military aircraft. Secondly, loitering can be performed more efficiently by changing the airfoil shape through drooping the wings, increasing the airfoil camber, or twisting the wing. Performing any of these changes by morphing during a mission would give increased efficiency in the loiter stage [5].

In particular, Raymer [6] lists five fundamental aerodynamic features that strongly determine aircraft performance.

These parameters are: wing loading (denoted as W/S, the ratio of aircraft weight, W, to wing reference area, S); wing thickness-to-chord ratio (denoted as t/c); wing planform taper (denoted as λ, the ratio of wing tip chord to wing root chord); wing span (denoted as b); and wing sweep (denoted as Λ) [7].

The change of each wing geometric characteristic (i.e., span, aspect ratio, sweep, taper ratio, t/c, camber) has influence on several aircraft parameters, such as:

(1) Maximum speed.
(2) Range.
(3) Landing distance.
(4) Take off distance.
(5) Maneuverability.
(6) State change effectiveness.

It is quite easy to state that main changes in these parameters are essentially affected by variations of span, sweep, and camber. Considering aircraft performance parameters such as range and endurance, it can be shown that these parameters are strongly dependent on C_L/C_D and $C_L^{3/2}/C_D$, respectively. Furthermore, these ratios are directly proportional to wing aspect ratio.

It is thus obvious that an increase in wing aspect ratio will result in a rise of both endurance and range. Therefore, by tailoring the wing geometry through morphing concepts, its lift and drag characteristics can be adjusted to a variety of missions or flight segments.

2.4 MANUFACTURING: NEW MATERIALS AND CONTROLLED INDUSTRIAL PROCESSES

Conventional aircraft operating temperature conditions are usually between +80°C (i.e., standing on the tarmac in desert airfield), and −55°C (i.e., at cruise altitude level). These conditions are already taken into consideration in normal a/c design to validate the safety of all structural parts. Particular attention must be made to the use of composite materials that are generally very sensitive to temperature changes and significantly alter mechanical characteristics. It is obvious, that, external wing

FIG. 4

Example of morphing mechanism inside leading edge.

surfaces or, in general, coating parts that can morph their shape and dimensions must be manufactured with materials suitable for large deformations. This will soon imply an increasingly massive use of composite materials, in particular of elastomers (thermosetting or thermoplastic).

A lot has already been done, especially in the space field, but aeronautical standards are very different, and sometimes very far away, in terms of requirements from them.

A big effort will be necessary for the chemical industry to produce composite materials to reach desired needs. Only the new material generation can be the real "game changer."

For aeronautical standards, it is absolutely mandatory to have a deep production process control, able to always guarantee the same quality in parts production (Fig. 4).

The actual state of art today says that composite material parts are not so easy to manage from this point of view. Therefore a significant step forward in this direction is absolutely required to reach the desired target. New production processes must be assessed in detail, evaluating time, cost, and related risks.

Concerning materials, the last but not the least, is the weight aspect. Light weight structures are always the primary target for aircraft manufacturers. Density of new materials will be another parameter under the magnifying glass.

Another important aspect to take into consideration is the future TRL progress of other parallel technologies like: additive manufacturing (AM), development of low power actuators, and micromechanics with high reliability.

2.5 ASSEMBLY AND QUALITY: AUTOMATION AND INTEGRATED PARTS

The assembly of a complete wing, with systems and control surfaces, is usually quite a complicated and long process. A large part of this process today is still executed "by hand" with specifically trained workers, in small a/c OEM. In the coming future, in which, smart wings with integrated morphing systems will greatly represent the standard, it seems evident to foresee the increase of automation. This will be in parallel with a high precision assembly of the morphing mechanism, which could be supported by the introduction of parts made with AM technologies. Having more and more integrated structures could result in ease of assembly and reduction of time and error arising from putting together several mechanical parts.

Therefore, introducing more automation in the assembly process can guarantee a higher safety level, which is exactly what the aircraft industry always want.

For initial quality, usually a pre in-service inspection is required. Therefore, acceptance criteria are established for structure with porosity, voids, and disbonds to account for initial quality (flaws) developed during the manufacturing process. Damage modes such as porosity, voids, and disbonds are subjected to specified acceptance criteria. This initial quality is intrinsic to the manufacturing process and inspection standards and represent the as-delivered state, and therefore, the structure must meet all requirements of strength, stiffness, safety, and longevity.

The other important aspect to consider is the post in-service inspection, for continued airworthiness purposes. This is usually based on interpretation of the test results, inspection procedures, threshold time, and frequency of inspections are established and published in airplane manuals. A factor is applied to allow for damage to exist over several inspection intervals, depending on the criticality of the structure. (A further factor may be needed to account for scatter revealed in the flaw growth test results.)

2.6 MAINTENANCE: ASSESSED STEPS AND PERSONNEL TRAINING

The awareness of operations and service personnel involved is in general a very important aspect from the industrial point of view. Surely the discussion between a/c manufactures and civil authority will concern some interesting aspects summarized in the following questions:

How to provide awareness training? What is their current level of education? What is the anticipated attention level?

How to ensure they do not act as qualified inspectors? Worried about losing their job if they report their mistake? What can an a/c OEM do to minimize these problems?

Let’s start with main two recommendations:

- For sure a good start can adopt robust design criteria (including, for instance, impact damage resistance).
- Personnel in positions of responsibility first need education on what level of events can cause damage beyond that protected by scheduled maintenance and existing source documentation.

Even if a path to a solution is not unique and probably not yet defined, it is not difficult to indicate a following approach that can indicate what required levels of safety management are needed for structure design, maintenance, and operation awareness.

- Damaging events outside the scope of those considered in design must be of a magnitude that ensures it will be reported (i.e., designs to sufficient impact damage resistance and damage tolerance).
- Simple training is needed to ensure that aircraft service personnel are not to blame with essential “reporting” role of operations.
- Source documentation should reinforce training for line maintenance and structural engineers to realize that disposition of such events is beyond scheduled maintenance and inspection/repair

procedures typically outlined in the structures repair manual (SRM; e.g., exterior visual ADL will not apply).

In addition, it could be useful to follow some conditional inspections and dispositions. For instance:

- The Aircraft Maintenance Manual should contain instructions for conditional inspections to be performed following a collision (both in flight or on ground).
- Exterior instrumented NDI at point of contact and adjacent supporting structure will be needed.
- Interior detailed visual inspection will also help determine the severity of the damage.

It could be that disposition of damage and repair may be beyond the procedures documented in the SRM and that is why additional structural design and process substantiation may be needed (i.e., combination of analysis and tests to address fatigue, damage tolerance, static strength, etc.).

Another two aspects that must be carefully evaluated and could give the "go—no go" for industrialization are: the behavior of used composite material and morphing systems with respect to fuel absorption (i.e., usually the wing is filled with fuel) and galvanic corrosion which usually occurs at interfaces where carbon fibers and metals are in contact.

2.7 SAFETY: ASSESSED METHODS FOR STANDARD ARCHITECTURES

A winning strategy to attract interest from industries to research results, is of course, to correctly assess safety, while considering CS/FAA regulations. This can show main problems to face and solve, in a multidisciplinary complex system such as morphing. In general design philosophy will not change from the latest approach commonly used by aircraft industries: "damage tolerance" together with "fail safe" will lead the process.

Attention must not only be paid to design and analysis of the undamaged structure, but mainly to the assessment of damage influenced by the strength and durability characteristics with particular reference to a/c airframe, within analyses and certification process of composite adaptive parts.

In metallic structures, damage tolerance is demonstrated using fracture mechanics to characterize crack growth under cyclic loading, predict the rate of crack growth in the structure under anticipated service loads, and establish inspection intervals based on realistic damage detection reliability considerations. Since typical CFRP composites have relatively flat S-N curves, and because these damages do not propagate under aircraft operational loading spectra, the previous approach cannot be used to establish inspection plans. Therefore, a no-growth approach is used to demonstrate compliance with damage tolerance requirements for composite primary structures.

In fact, damage tolerance provides a measure of the structural ability to sustain design loads with a level of damage or defect and be able to perform its operating functions [8].

Particularly interesting are the following steps for approaching the discussion between research and industry:

1. Agree on critical technical issues and areas of safety concern for damage tolerance and maintenance of composite adaptive structures.
2. Identify key similarities and differences in methods used for justifying damage capability for both CS23 and CS25 class aircraft composite adaptive structures.
3. Identify the key elements necessary for justifying maintenance inspection and repair procedures for composite aircraft adaptive structures.

4. Identify related content needed to update appropriate approved source (OEM) documentation (Design Manual, SRM, etc.) focused on field safety issues.
5. Identify areas for safety-related standardization of composite damage tolerance, maintenance, and related research needed in the future.

Damage tolerance assessments involve the application of known damage threats to the aircraft structure during its typical service usage and demonstration that this damage will not alter the safe operation of the aircraft prior to detection. Damage tolerance builds upon fail-safe analysis by including determinations of damage growth characteristics and by establishing damage detection methods and inspection plans.

The aircraft industry traditionally uses the "building block" approach to demonstrate the damage tolerance capability of critical fracture airframe structures. In the building block approach, one would run different levels of tests in concert with damage tolerance analysis. Because there are many impact threat events, such as tool drop, runway debris, hail, lighting strikes, etc., in the operation of an aircraft, impact damage is the predominant "defect" to be considered. The minimal environmental deterioration and fatigue damage leaves accidental damage as a primary damage source in composite structures.

Therefore, in parallel, it would also be necessary to develop criteria to identify "adaptive primary structures" for damage tolerance evaluation (i.e., wing, empennage, flap, aileron, spoiler, etc.). Different types of composite adaptive structures will be identified with a particular category to enable its specific program to be developed.

Damage from manufacturing processes, accidental damage from assembly and handling, and accidental damage from in-service operation will be defined.

From each of these sources, the modes (porosity, delamination, puncture, core crush, etc.) of damage will be identified.

For each mode of damage, criteria will be established for determining characteristics such as initial size, threshold of detectability, critical size, and whether growth during spectrum loading will be produced.

The critical modes of loading for composite adaptive structures will be identified for purposes of selecting principal structural elements and for developing appropriate testing requirements. Requirements for static testing of structural elements with various modes of damage and various modes of loading will be defined to provide residual strength information.

Requirements for cyclic testing of structural elements with various modes of damage and various modes of loading will be defined to provide damage growth (or no growth) information (Fig. 5).

Requirements for cyclic testing of important full-scale components will be defined to verify that damage would not grow to a critical size in the full-scale article during the life of the airframe. Requirements for static testing of important full-scale components will be defined to verify that they would carry required loads with any damage growth inflicted from cyclic loading. In addition, appropriate testing will be identified to validate analyses methods used and to demonstrate certification requirements.

Of course, damage threats must be correctly identified, with particular attention to: hail impact, lightning strike, bird strike, tire debris, rotor debris, runway debris, fire, and maintenance process. Unfortunately, today, some threats that are not in common:

- reference standards
- test protocols
- statistical basis

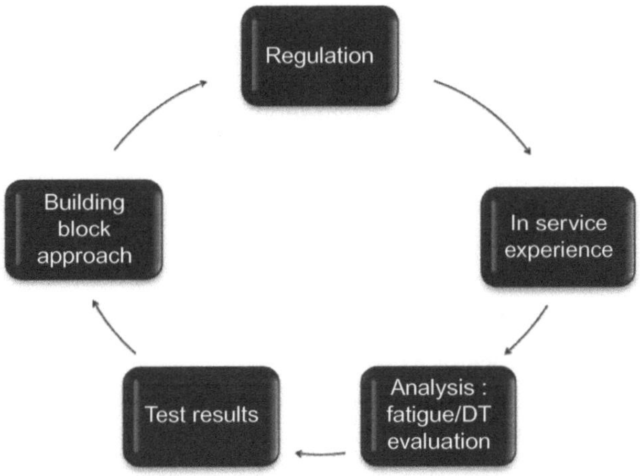

FIG. 5

Damage tolerance approach.

Therefore, investigation is still required in this direction. Damage inspection and detection will be the next step once a common strategy is defined for identification. It will in fact be necessary to define appropriate metrics associated with damage types.

The impact threat for use in the damage tolerance assessment must be defined. All possible low velocity impact sources for the structure in question (e.g., tool drop during maintenance, collision with ground based equipment…) have to be considered. A probabilistic definition can be used for the impact sources defining impact severity (impact energy) and probability rate (per flight hour). The probability density function is expressed in terms of probability rate for occurrence of an impact with energy greater than or equal to a given energy (E_j).

The impact threat must be validated by comparison to in-service damage incidents. Typically, a survey of damage on an in-service aircraft component should be carried out. The severity of such damage needs to be quantified in terms of impact energy. This can be carried out by establishing and applying an appropriate calibration relationship, which relates observed damage dent size to impact energy. It is necessary to show that the highest impact energy seen in-service does not occur more frequently than the impact threat definition to be used in the analysis. This has led to a proposed impact threat of general form:

$$p_j(E \geq E_j) = 10^{-x-\frac{E_j}{15}} \quad \begin{matrix} P_j(E \geq 30\,J) = 10^{-5}/fh \\ P_i(E \geq 90\,J) = 10^{-9}/fh \end{matrix}$$

In this equation, P is the probability (per flight hour) of occurrence of an impact of energy greater than or equal to E_j. The value x relates to the probability (per flight hour) of occurrence of impact events of any energy (10^{-x} per flight hour).

Initial damages (due to impact) must be considered by integrating their influence on the probability of existence of damage at the end of the manufacturing and assembly phases. This can be done through the description of potential impact sources during the different steps of manufacturing and assembly: tool drop, fall of control apparatus, etc. It is known that damage during manufacturing and assembly phases can have a negligible influence regarding the global risk at the last flight hour. Nevertheless the influence of the initial damage on the global risk of failure must be checked. The use of an impact threat derived from in-service threat (based on in-service survey and/or engineering judgment) can inherently cover damage caused by impacts prior to entry into service. Once the impact threat is defined, energy thresholds can be deduced.

The basis of the probabilistic approach is therefore to demonstrate that the inspection program will ensure that the combination of an occurrence of a load k*LL intensity, with the presence of a missed accidental damage reducing the structure strength to k*LL level, remains acceptable. This combination must be extremely improbable (probability $<10^{-9}$ per flight hour).

Therefore this analysis involves the combination of three probabilities function:

- The probability rate of occurrence of external loads.
- The distribution of residual strength of the structure damaged by any given impact energy.
- The maximum probability over design life (DSG) for undetected damage.

3 CURRENT AND EXPECTED NEEDS

Advantages of morphing are paid off (sometime too much) by complexity of both systems and structure, as well as weight increase. Aircraft maximum takeoff weight (MTOW) is a sum of three different weight components:

$$\mathrm{MTOW} = \mathrm{OEW} + W_{\mathrm{FUEL}} + W_{\mathrm{PAYLOAD}}$$

And it is defined as the heaviest weight at which the aircraft has been shown to meet all the airworthiness requirements applicable to it.

Therefore, an increase of morphing systems and structure weight has direct influence on aircraft operative empty weight (OEW). In fact, once MTOW is fixed, an increment of OEW means a fuel quantity reduction.

The aerodynamic advantages coming from the adaptive wing, unfortunately, are not always able to compensate for the reduction of endurance coming from fuel quantity reduction. One of the most famous examples is the F-111 equipped with an adaptive wing (Fig. 6).

Therefore, it is quite clear that industries could hardly accept such disadvantages. Long endurance is one of the issues that become more central with the increasing development of RPAS. It is clear that morphing technology must also answer in this direction for a direct market requirement.

On the other hand for civil markets, we can say that the cost of morphing systems that even highly affect the aircraft sale cost, may not be a binding parameter for the success of the final product, but on condition, have low operating costs. These latter are of primary importance and are carefully evaluated by operators. It is clear that civil markets still play around weight and cost. In the military field, performance prevails on both sale and operative cost.

3.1 TECHNOLOGY TRANSITION

Once aerodynamic performances' "pay-offs" are clear (including benefits in terms of fuel consumption), it is necessary to start discussing about the impact that morphing technologies have on aircraft structures. From an industrial point of view, it is necessary to evaluate correctly when a technology transition is possible. Mainly five factors must be taken into account:

FIG. 6

F-111 with adaptive wing during experimental flight.

1. Stabilized material and processes.
2. Producibility: manufacturing scale-up.
3. Characterized mechanical properties.
4. Predictability of structural performance.
5. Supportability.

- As already previously stated, it is necessary to still invest on material development and screening that can give the possibility of expected benefits.
- Also, manufacturing must be evaluated taking into account the process of scaling-up, acceptance criteria, facility assessment, manufacturing methods development, and producibility.
- Once suitable materials for the application have been found, the certification approach assumes materials for structural components must be qualified (i.e., statistical B basis characterization as a function of temperature, moisture, damage, defect, and fatigue), as well as a developed production process.
- Numerical analysis of new scenarios must show the possibility to correctly predict the structural performance with good correlation to testing.
- For certification purposes, it is absolutely mandatory to demonstrate the reliability of structure (and, in this case, of the system as well). But this is not enough to evaluate the supportability. It is absolutely important to have ease of reparability and maintainability.

3.2 A MISSION CONFIGURABLE WING

An aircraft wing that can mimic a bird wing morphing ability would be able to perform dissimilar missions.

Fighter aircraft or unmanned aerial vehicles could have multiple roles including reconnaissance, fighting and heavy bombing with an increased stealth capability. Commercial aircraft could smoothly

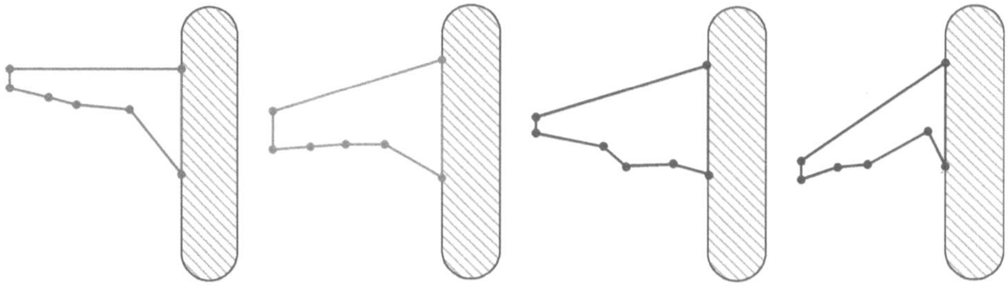

FIG. 7

Some interesting morphing concepts.

adapt wings for high lift during takeoff and landing, fully extending them for optimum cruising, and slightly folding them for an efficient descent (Fig. 7).

3.3 IMPROVED FLAPS AND AILERONS

Adaptive compliant wings able to morph at the position where normally flaps and aileron are installed, could be a great advantage especially for aircraft with a high wingspan that sacrifice performance instead of weight. Probably, in this case, the best examples are Unmanned Aerial Vehicle System (UAS), that could improve performance especially during takeoff and landing as well as overtaking some difficulties in these two phases when weather conditions are not optimal, and reducing required runway length (Fig. 8).

By definition, compliant mechanisms are single-piece flexible structures that deliver the desired motion by undergoing elastic deformation as opposed to jointed rigid body motions of conventional mechanisms [9,10]. This means joint less structures, easy to assemble, monolithic mechanical devices.

Advantages of compliant mechanics include a reduced complexity, zero backlash and wear, submicron accuracy, and embedded actuation and sensing devices.

FIG. 8

Adaptive compliant trailing edge.

Disadvantages could be that power needed by the actuators (electric or pneumatic) to overcome stiffness of structure are too high to take full advantage of the morphing concept.

4 MORPHING AS A SOLUTION

Certainly, many solutions are feasible, although some may be impractical both from the weight point of view, or from the cost point of view. As previously stated, each solution must always be assessed, considering the possibility of inspection and reparability. Let us briefly review some of the key aspects of major interest that is inherent to both wing and control surfaces.

4.1 WING AND CONTROL SURFACE FEASIBLE SOLUTIONS

Table 1 summarizes the wing parts, actuation types and skin types that are involved in morphing feasible solutions.

4.2 SOME SPECIFIC REQUIREMENTS

Without detailed evaluation of test results for each possible solution, it is only possible to list a range of variations of parameters of particular interest, it would be convenient to have a variation of some a/c characteristic.

Table 1 Morphing Concepts, Actuation and Skin Types

Structural Part	Morphed Aspects	Actuation Types	Skin Types
Wing	Sweep	Pneumatic Linear actuator Electric DC motor	Rigid segmented Stretchable
Wing	Span	Pneumatic Linear actuator Servo motor DC motor	Rigid segmented Stretchable
Wing, flap, aileron	Chord	Pneumatic Linear actuator Servo motor Shape memory alloys	Stretchable Flexible
Wing, flap, aileron	Camber	Pneumatic Servo motor Shape memory alloys Lead zirconate titanate Ultrasonic	Rigid segmented Stretchable Flexible
Wing, flap, aileron	Twist	Shape memory alloys Electrical Lead zirconate titanate Pneumatic DC motor	Rigid segmented Stretchable Flexible

It is clear that more flexibility in the configuration change, given the considerable diversity of the type of mission, is always required for military aircraft, whether they are piloted on board or remotely controlled.

For example, it might be useful to have a change in aspect ratio between 150% and 200%, 60%–80% change in area, 40%–60% change in span, while independently controlling the sweep and area, and a wing sweep varying from 15 to 35 degrees. An increase between 30% and 40% in the aerodynamic efficiency from cruise flight to the loiter configuration for surveillance missions would be really advantageous.

With this kind of flexibility for change of aircraft configurations, it could be possible to combine the loiter endurance of a surveillance platform with the high-speed dash capability of an attack aircraft.

The requirements for the weight of the system and structure might be more stringent, as previously explained. They should be contained in the range of 2%–5% compared to an equivalent classic configuration, to avoid placing an undue burden on performance.

5 CONCLUSIONS

Today, morphing of primary structural aircraft parts is neither the industry standard nor a compelling need, but without it, it is impossible to design, manufacture, and sell airplanes.

It is clear that the desire of each aircraft manufacturer, whether large or small, civil or military, will always be able to provide an innovative, secure, and competitive product for the market.

The world around us requires that the horizon of research and industry come closer. The more they will be able to communicate, the more they will be able to exchange vital information to meet our needs.

Ideally, a morphing structure should be able to alter its shape, carry the required load, and be lightweight at the same time. Today it is still difficult to accomplish these three aspects simultaneously. Therefore, the goal of current and future research is to develop particular devices, and to use innovative materials, that can enable morphing through shape-changing airframes whose variation in geometry will significantly influence aerodynamic performance as well as stealth characteristics for military applications. There will be little opportunity that industry will massively invest in this kind of technology without matching these three aspects together while considering the cost and weight limitation.

The innovation derived from the introduction of morphing aircraft part, passes through a combination of other technologies that must be mature enough to be taken into consideration at industry level.

The military sector will provide the greater thrust and implementation. This is especially true because, these technologies can potentially provide very flexible and effective solutions to deal with current requests to raise and strengthen the security and surveillance measures.

As previously mentioned, there will be several challenges in the design of morphing aircraft, in particular:

1. The integrity of compliant structures must be ensured under all circumstances.
2. The system should be designed so that the required actuation force falls within the adequate limits. Thus, the morphing can occur efficiently.
3. The wing skin should provide a smooth and seamless aerodynamic surface while efficiently bearing the aerodynamic loads.

4. The design process should be extended to encompass multiple flight regimes.
5. The control systems should be efficient even with highly coupled control effectors.

Thus, the main challenge faced by morphing technology is to design, fabricate, and operate effective integrated combinations of deformable wing skins, actuators and mechanisms, and flight controls to provide the morphing system designer with the freedom to deal with future diverse, conflicting vehicle mission capabilities [11].

The dream to mimic bird wings is not far, but a lot must still be done to reach that target. To get closer to the objective, it will be necessary to deepen investigations of flexibile wings with anisotropy and nonlinear membrane elasticity to a high degree.

REFERENCES

[1] J.D. Anderson, Introduction to Flight, second ed., McGraw-Hill, Blacklick, OH, USA, 1985.
[2] H. F. Parker, The Parker variable camber wing, Report No. 77.
[3] A.Y.N. Sofla, A.S. Meguid, K.T. Tan, W.K. Yeo, Shape morphing of aircraft wing: status and challenges, Mater. Des. 31 (3) (2010) 1284–1292.
[4] https://en.wikipedia.org/wiki/Glenn_Curtiss.
[5] C.E.S. Cesnik, A Framework for Morphing Capability Assessment, Department of Aerospace Engineering, University of Michigan, Ann Arbor, MI, 2004.
[6] D.P. Raymer, Aircraft Design: A Conceptual Approach, in: AIAA Education Series, AIAA, Washington, DC, USA, 1992.
[7] T.A. Weisshaar, New aircraft systems concepts-towards new horizons in aeroelasticity, NATO Paper RTO-MP-IST-999.
[8] MIL-HDBK-17-3F.
[9] M. I. Friswell, D. Baker, J. E. Herencia, F. Mattioni and P. M. Weaver, Compliant structures for morphing aircraft, 17th International Conference on Adaptive Structures and Technologies Oct. 16 ~ Oct. 19, Taipei, Taiwan.
[10] S. Kota, J. Joo, Zhe Li1, S.M. Rodgers, and J. Sniegowski, Design of Compliant Mechanisms: Applications to MEMS, ISSN 1573-1979, Springer Dordrecht.
[11] Z. You, J. Crabtree, and G. Lederman, Large-Scale Shape Change in Aircraft Wings, Department of Engineering Science, University of Oxford.

CHAPTER

THE DEVELOPMENT OF MORPHING AIRCRAFT BENEFIT ASSESSMENT

3

Fabian Peter, Eike Stumpf
RWTH Aachen University, Aachen, Germany

CHAPTER OUTLINE

NOMENCLATURE

ACTE	adaptive compliant trailing edge
CD	drag coefficient
CL	lift coefficient
C_p^*	critical pressure coefficient
FEM	finite element method
L/D	lift to drag ratio
Ma	Mach number
MAW	mission adaptive wing
NASA	National Aeronautics and Space Administration
RBM	root bending moment
Re	Reynolds number
VCCTEF	variable camber continuous trailing edge flap

1 EXPERIMENTS AS BASIS FOR MORPHING PROGRESS

To understand why performance assessment of morphing aircraft is crucial, it is important to reflect on the history of morphing. In the early days of aviation, the examples in nature, especially the flight of birds, were the focus of aircraft designers. The flight of insects, birds, and bats as we know it would not

Morphing Wing Technologies. https://doi.org/10.1016/B978-0-08-100964-2.00003-4

be possible without morphing. The wing planform and camber of birds change dramatically and continuously with every change in flight conditions [1]. In addition, the structures and materials used in early aviation (e.g., wood and canvas) were somewhat flexible and allowed for a relatively easy deformation. In the 1890s, experiments were conducted to achieve roll control of gliders with wing warping by Montgomery. The best-known example, which is cited in most publications about morphing as an early application, is the Wright Flyer and its first powered flight in 1903. Here the Wright brothers realized roll control by asymmetrical deformation of the outboard wing sections [2]. In a report in 1920, Parker was one of the first to address the challenge that increasing aircraft cruise speeds posed toward the requirement of an acceptable landing speeds. He cited three possible approaches for this problem: the variation of incidence angle of the wing relative to the fuselage, increasing the surface of the wing, and variable camber. He chose a mechanical concept that he described as variable camber. The intent would be to deform a surface (i.e., the upper surface of a biplane) in a way that offers little drag and lift at high speed but high lift with the necessary drag at low speeds. He envisioned a design by which the surface is beneficially deformed by aerodynamic loads of the different flight attitudes integrating flexible elements into the wing structure (see Fig. 1). Such a design would require no external control mechanisms. The design was tested but did not find its way into a commercially successful airplane [3].

Another example of early geometry changes was the telescopic wing presented by Makhonine in 1931. The concept was integrated into an airplane called Makhonine Mak-10, which conducted test flights in 1932 [4].

Since then, morphing has constantly been part of aviation research, but morphing by deformation shifted out of focus with the appearance of rigid structures and materials, which do not respond well to large deformations. However, morphing by the deflection or shifting of rigid elements stayed an essential part of aircraft design. Even the deflection of an aileron or the deployment of a movable landing gear could be interpreted as morphing, making it a part of current civil aviation. However, the general interpretation of the term morphing is a large deformation of parts of the aircraft or continuous deployment of control surfaces. The need to change the geometry of the wing increases with the difference in flight conditions in which an aircraft has to operate.

The most fundamental change is the difference between a high cruise speed to a low landing speed. For this reason, with the increase in aircraft's cruise speed, high lift devices were introduced.

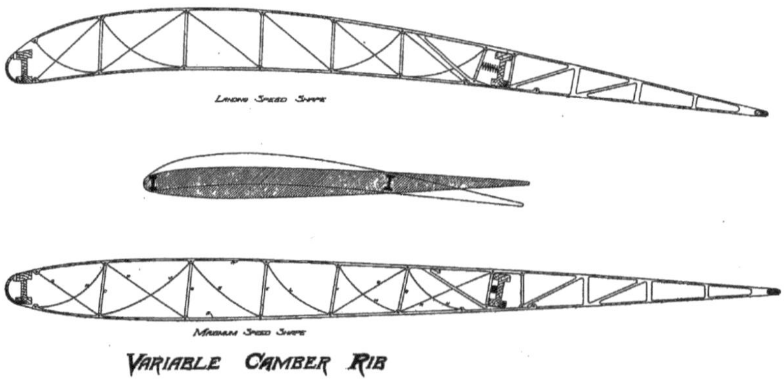

FIG. 1

Parker variable camber wing [3].

FIG. 2

Bell X-5 variable sweep test aircraft [6].

Subsequently, the swept wing enabled speeds approaching and exceeding the sound barrier. The difficult low-speed handling qualities of highly swept wings of military supersonic aircraft called for a change of sweep angle during flight.

The first example for this technology was the Bell X-5 in 1951. Based on a captured Messerschmitt P.1101, whose wing sweep could be altered on the ground, the Bell X-5 could vary its wing sweep angle from 20 to 60 degrees (see Fig. 2). Because the center of pressure moved with variation of the sweep, the wings were mounted on rails in order to move forward, resulting in a complex mechanism. The X-5 had a successful test career and provided important data, building the basis for the introduction of variable sweep in military aircraft [5].

Variable sweep wings found application in several western and eastern military production designs, one of the first being the General Dynamics F-111 Aardvark. With the feasibility of high speeds and the emerging possibilities of transonic wind tunnels and theoretical methods for transonic and supersonic flow calculations, more research was focused on the economics of airfoils at different speeds. This led to large-scale research projects.

2 THE ADVENT OF TRANSONIC METHODS

The F-111 served as a testbed for the National Aeronautics and Space Administration (NASA) research programs for almost 20 years. The first program started in the 1970s and was named Transonic Aircraft Technology (TACT). It focused on the integration of a supercritical wing airfoil and the validation of calculation methods and wind tunnel tests [7]. In the 1980s, the Advanced Fighter Technology Integration (AFTI) program was introduced. Its aim was to increase the efficiency in variable mission and

FIG. 3

General Dynamics F-111 Aardvark with the mission adaptive wing [10].

flight conditions by integration of the mission adaptive wing (MAW) as shown in Fig. 3. For this purpose, the wing of the F-111 was equipped with a smooth variable camber leading and trailing edge [8]. The forward 20% of the wing chord could be uniformly deflected −5 degrees up and 30 degrees down. The aft 40% of chord could be deflected from −7.5 degrees to 25 degrees with a restriction of 1 degrees/foot of span. [9].

The goal of the variable camber application was to achieve lift to drag ratio (L/D) improvement, integral roll control, maneuver load control, direct lift mode (increased lift without increased angle of attack), and gust alleviation. Flight test results comparing the TACT with the AFTI are shown in Fig. 4. For a lift coefficient (CL) of 0.5, as indicated by the dashed line, a reduction of trimmed drag coefficient (CD) of more than 8% compared to the TACT drag was achieved. The increase in weight for the integration of the MAW type leading and trailing edge was stated to be 94 lb. (42.6 kg). The F-111's original wing, which was equipped with a Krueger flap and a single slotted Fowler flap, weighed 9068 lb. (4113.2 kg), showing an increase in weight of about 1%. [11].

Variable sweep wings are still in operation in current military aircraft, such as the Panavia Tornado from 1974. With complex and heavy hinge mechanisms, variable sweep wings never found their way

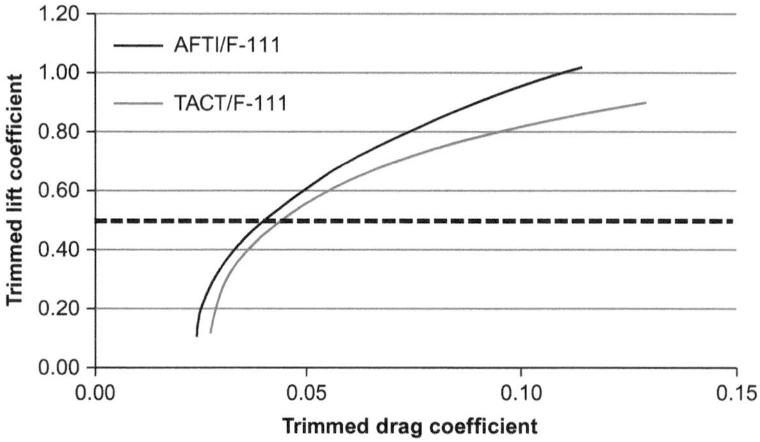

FIG. 4

Lift to drag improvement for mission adaptive wing [11].

into civil aviation and were replaced by delta wings in most military applications. However, they were essential for enabling high speeds. In civil aviation, the methods developed for transonic speed regimes showed the potential of fixed camber change for higher Mach numbers. These methods were applied during the development of the Airbus A310, and then were used to identify potential improvements for the Airbus A300-B4. The A300-B4's wing was developed with theoretical subsonic tools, extensive wind tunnel testing, and experiences made during the development of the Trident and HS 125. A certain amount of rear loading was already applied in the A300-B4's original wing, however, the methods available in the 1980s allowed for further optimization of the airfoil. It was found that an increase in camber of the trailing edge moved the shock aft and increased the area of supersonic flow. Consequently, the camber of the A300-600 was increased. The increase was stronger inboard and smaller outboard. The result on the drag can be described by a rotation of the polar in a CL versus CD diagram similar to Fig. 4. The increase in camber results in a rotation of the polar around a specific point situated on the polar. In order for the change to be beneficial, the point has to be at lower CL than the intended cruise CL, thereby shifting the polar points of higher CL values to smaller CD values. The position of the rotation point is dependent on the Reynolds number (*Re*) and moves to higher CL values with increasing *Re*. This was confirmed by flight tests with a modified A300-B4. In total, the aerodynamic improvements, which also included other changes such as the elimination of boundary layer fences and measurements for parasite drag decrease, amounted to an L/D improvement of 6% [12]. Furthermore, the stronger camber increase inboard changed lift distribution and reduced wing root bending moment (RBM) by 2.4% [13].

The advances in 2D airfoil analysis were also beneficial for the improvements for helicopter blades. Furthermore, the application of camber variation is very attractive for helicopters. This is explained in more detail in a publication of Awani from 1982 [14]. He describes the inherent problem of very different flow conditions for advancing and retreating blade and the resulting necessary compromise for a fixed airfoil selection (see Fig. 5). For the advancing blade, compressibility effects dominate the design choice; for the retreating blade, blade stall is the aspect with the strongest impact on the airfoil choice.

The study by Awani presents an integration of a simple flap geometry in the Vertol VR-7 rotor blade. The flap extents from 50% chord to the trailing edge of the blade and is able to deflect 20 degrees downward. The chord length is a compromise between small deflections and large flap chords with the goal of achieving the desired camber change (Fig. 6).

The analysis was conducted with computational methods and yielded a reduction of necessary power in hovering flight of 2.6%. Helicopter blades are still an area of research for morphing, incorporating piezo elements for actuation and with the long-term goal of reducing the complex and maintenance intensive rotor kinematics and decreasing vibrations and noise (e.g., [16]). In the following, the scope is set on fixed-wing aircraft.

Another NASA program including variable camber was the X-29 (see Fig. 7), which had its first flight in 1984 and the second in 1989. While the most obvious design features were the planes forward-swept main wings and its canards, it also featured discrete variable camber flaperons (see Fig. 8). In comparison to the AFTI smooth variable camber, the discrete variable camber was introduced to the X-29 because of its lower cost. While offering a smooth upper surface at design conditions, its penalties at off-design conditions are described to be minor. In addition to the performance optimization over-large CL regimes, an additional goal for the variable camber application in the X-29 was the usage for roll control [17]. Because of the many highly innovative design features of the X-29, it was difficult to allocate performance improvements to specific technologies [18].

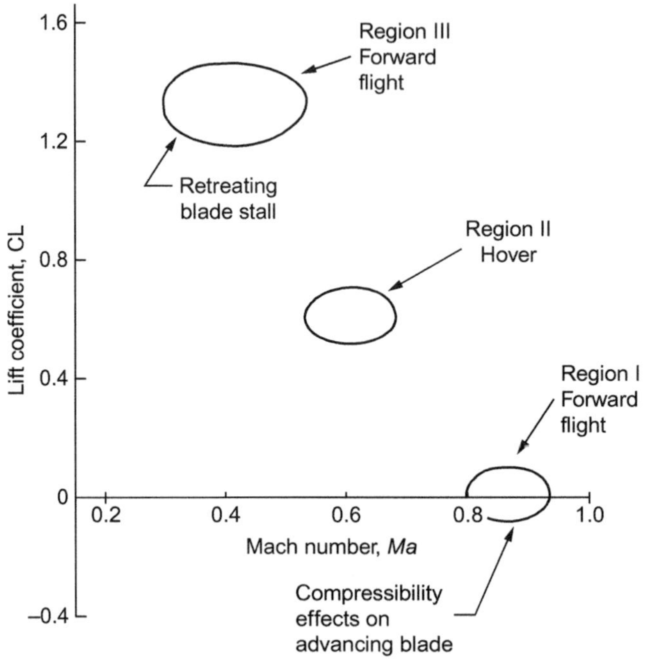

FIG. 5

Influence of lift coefficient and Mach number on rotor aerodynamics [15].

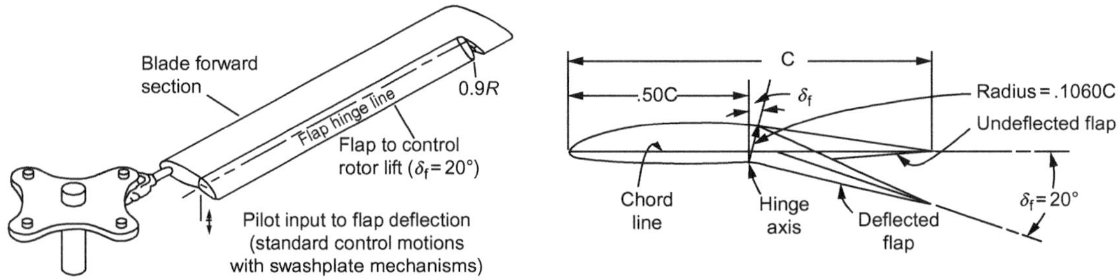

FIG. 6

Helicopter blade variable camber concept [14].

The knowledge gained by research for transonic airfoils and NASA research aircraft featuring variable airfoils showed the potential of variable camber. Consequently, the focus of research in the field of morphing for civil aviation shifted to the topic of variable camber. Fundamental experiments were conducted to improve the understanding of the camber influence and the design of specific variable camber airfoils. The basic phenomenon is that, for constant lift, the minimum drag is situated at higher CL with increasing camber. This can be explained by the increase in rear loading, which reduces local velocities and hence skin friction at the leading edge. Therefore, the relatively sharp optimum of a fixed

FIG. 7

Grumman X-29 [17].

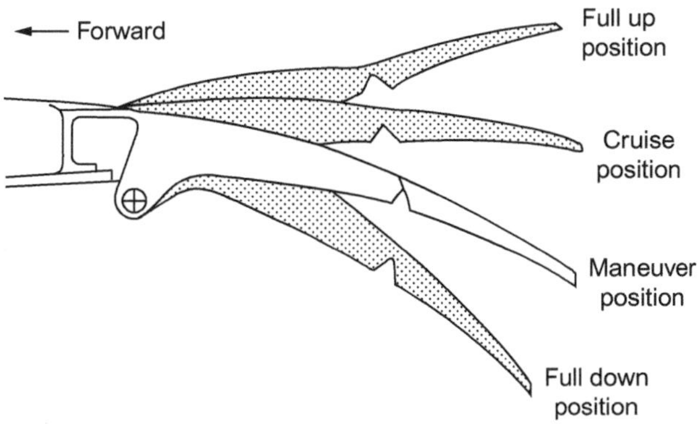

FIG. 8

The X-29's discrete variable camber flaperon [17].

wing L/D versus CL graph can be changed to a rather flat optimum for a variable camber wing. If the improvements are located at higher CL, which is the case in most studies, the buffet onset is also delayed to higher CL. The increase in camber is sensible until trailing edge separation occurs. Also taking into account the relaxation of the design compromise of a fixed wing and restraints for maneuver loads, Szodruch states L/D improvements of 3%–9%, a buffet boundary increase of 12% and a block fuel reduction of 5%. The studies also showed that a continuous camber change is not necessary because most of the improvements are already reached when the camber setting is changed two times during a short-range mission [19].

Greff published an extensive study about the opportunities for variable camber in civil aircraft families in 1990 [20]. The primary motivation for this publication was the application of variable camber on a wing that can be used for medium- and long-range aircraft. The wing of an aircraft family has to be large enough to meet the requirements of the largest family member, which is usually the most stretched version. The specific design tasks in the focus of the Airbus engineers in that period was the common wing for the Airbus A330 and the Airbus A340. For smaller planes of the family, this wing means a design compromise. This compromise is accepted since the wing contributes more than 50% of the development cost, 45% of structural weight, and 60% of the drag. Hence, a common wing as basis for an aircraft family offers large cost savings. The concept of the MAW is too expensive and complex for civil transport aircraft, which spend most of their cruise flight in patterns of relatively similar speed and altitudes. Furthermore, the need for lift enhancement during take-off and landing is larger for civil transport aircraft than for fighter planes, thereby an application of the MAW concept would conflict with high-lift devices needed for transport aircraft. For this reason, variable camber research focused on the use or modification of conventional trailing edge components in order to minimize development effort. The concept introduced by Hilbig [21] and analyzed by Greff makes use of an additional rotation track that enables the Fowler flap to rotate and translate for a limited amount during which no gap occurs between flap and spoiler (see Fig. 9) [20].

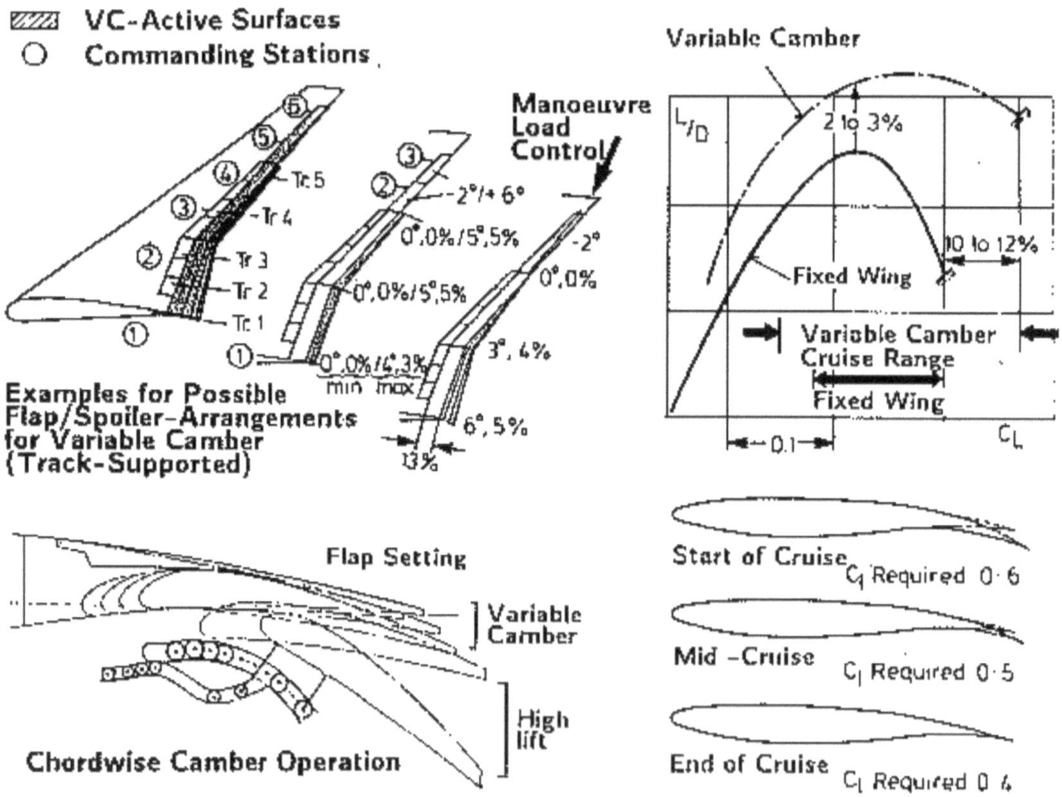

FIG. 9

Greff and Hilbig variable camber concept [20].

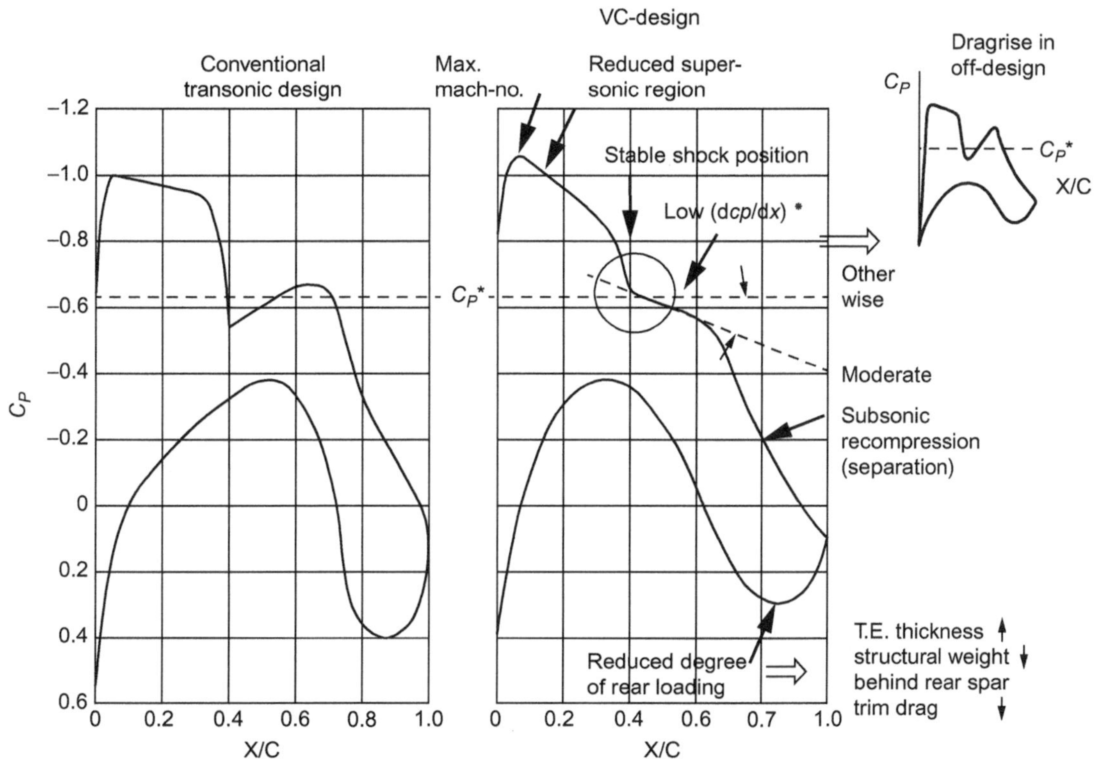

FIG. 10

Greff variable camber airfoil design [20].

Greff states that special attention has been paid to the airfoil design for variable camber application. Highly loaded airfoils with large supersonic flow regions are prone to trailing edge separation, especially if the hinge position of the varied surface is close to the shock position. Also, the regions around the critical pressure coefficient (C_p^*) should exhibit small gradients to guarantee a stable shock for off-design conditions. The rear loading of the lower surface should be decreased to counter the increase in pitching moment. These goals are illustrated in Fig. 10.

A crossover point (where a benefit of the variable camber polars exists) below the design CL would be achieved only if these aspects are taken into account. For this reason, an airfoil was designed specifically for the VC application. The basis for comparison was a highly loaded airfoil designed for a local CL of 0.6 and a local Mach number of 0.74. The VC airfoils design point was chosen at a lower local CL of 0.48 in order to enable the performance increase by deflection for higher CL. The VC airfoil already had a 4% lower minimum drag compared to the basic airfoil. Without varying the flap deflection, the VC airfoil showed higher drag for higher CL than the basic airfoil. This clarifies that the VC specific airfoil has to be operated with deflection for its optimum performance. With flap deflections, the variable camber airfoil showed superior drag characteristic. Taking into account the increased trim drag, because of the increase in rear loading by deflection, an L/D improvement of 4% was determined for the family common wing and for long-range aircraft. As control, a simple table lookup function was proposed, providing the flap deflection at a specific speed, weight, and atmospheric data. Further benefits discussed but not taken into the final assessment were the possible reduction of RBM with a possible manufacturing weight empty reduction of 1.5%–2%, drag decrease by a ~1 degrees reduced wing root setting affecting fuselage drag, and the possible reduction of horizontal and vertical stabilizer area, which is connected to a decrease in wing area. Subsequently, a variable camber benefit of 3%–6% block fuel reduction is stated as expectation for the use in a medium- and long-range family application [20].

Greff did state in his 1990s publication that although a spanwise differential variable camber might be advantageous with respect to load alleviation and offer opportunities to adjust to impacts of engine integration influences, the increase in complexity did not justify an application at that time.

Spillmann concentrated on the application of a spanwise differential camber in a publication in 1992. He noted that birds are able to cope with their very different flight attitudes not only by varying the angle of incidence but also by varying camber across the span. Most of the mission of a civil transport aircraft takes place in level flight, which determines the performance with respect to aerodynamic efficiency. However, the structure of the wing is determined by very brief periods of maneuvers and gusts. In these situations, the aerodynamic optimum lift and load distribution along the span is unfavorable from a structural point of view. If load distribution could be altered, e.g., load shifted toward the wing root, the structural weight of the wing could be reduced. Studies at Cranfield University incorporating this approach resulted in a 10% wing weight reduction for a 150-passenger aircraft, including weight increases of 25% for the flap structures and 10% for the flight control system [22].

Rossow presents a quite dramatic geometry change in his publication from 1991 [23]. He proposes a nose flap and a trailing edge flap that deflect from the upper surface from the airfoil in order to establish and trap a vortex (see Fig. 11). The resulting flow field is stated to achieve CL values up to 6.

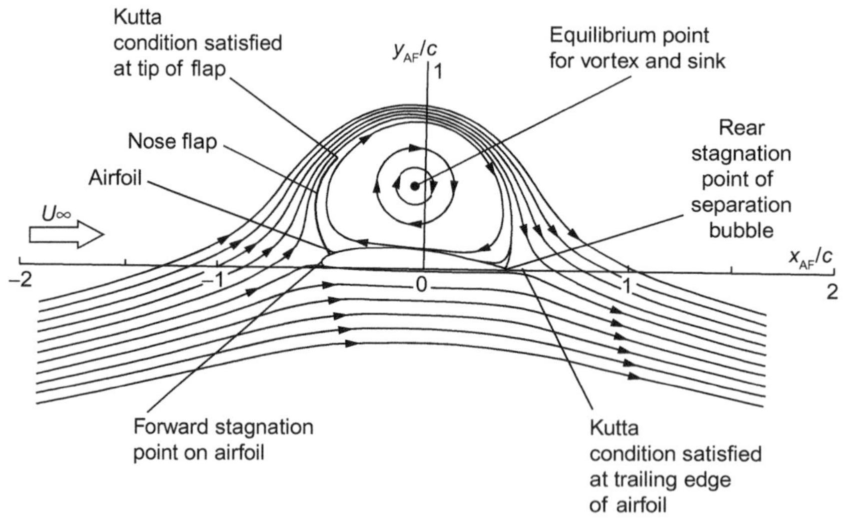

FIG. 11

Lift increase by establishing a trapped vortex [24].

Studies were conducted on a 2D model in a water tunnel and showed that a vortex could be established when fluid was removed through a sink at the center of the vortex. A former publication discussed the use of only one single flap at the front of the airfoil [24]. The aft flap was found to be beneficial to limit the extent of the vortex. Furthermore, with the aft flap, the sink was necessary only to establish the vortex.

3 AUTOMATED METHODS AS ENABLER FOR LARGE SCALE STUDIES

The publications listed thus far were based on methods used for one specific design and presumably a large share of nonautomated work. A very large effort would be required for a study of different geometries. In addition, the studies concentrated on airfoil changes by mostly conventional control surfaces such as Fowler flaps. The availability of inexpensive computation power coupled with several freely available low- and higher-level aerodynamic computational methods led to an increase in aerodynamic research in many fields at the beginning of the 1990s. For morphing, two-dimensional variable camber studies included automatic airfoil geometry optimization. In a study published in 1997, Martins presented a 3D vortex lattice analysis that is coupled with a multivariable function optimizer algorithm. The basis was a swept tapered wing and its airfoil was deformed from the leading edge to the 20% chord and from the 70% chord to the trailing edge. The deformation was restricted by dividing the wing surface with a linear finite element mesh and limiting the deformation of these elements. With constraints considering high-lift performance, an induced drag reduction of 1.8%–2.3% was achieved [25]. Martins did further research, and a publication in 1998 focused on 2D airfoil sections, which were allowed to deform by translation and rotation of a leading and trailing edge segment, and a fixed wing box composed of an unchanged chord section of the airfoil (27.6%–64.5% chord). The kinematics enabling these deformations were not part of the research. The aerodynamic analysis was conducted with a combination of tools, including methods incorporating integral boundary layer, transition, and separation. Compressibility effects were not included, and therefore all studies were conducted at Mach numbers (*Ma*) below the critical *Ma*. The flight conditions for the analysis were set to 25,000 ft and *Ma* 0.575. The evaluated target parameter is the range factor, which is the result of the integrated range equation. For these boundary conditions, an automatic optimization study was conducted. For the case of a sole deformation of the trailing edge, an increase of about 7.03% was realized, and for simultaneous deformation of the leading edge, 24.5% could be achieved. The calculated optimum leading edge contour showed a very strong downward curvature. It was stated that this geometry most likely would be not sensible if wave drag were analyzed. The importance of transonic effects on variable camber was extensively covered in the previously mentioned publications. On its own, the vortex lattice method is not able to incorporate compressibility or viscous effects. This has to be kept in mind for results for transonic flight regimes, which are produced with a pure vortex lattice application [26,27].

The application of 2D and 3D Navier Stokes calculations in an extensive research project allowed the assessment of an innovative trailing edge device concept by Airbus and the German Aerospace Center. In this study, an extended outboard flap replaced the aileron. Additionally, the trailing edges of the flap featured small simple hinge flaps with 10% wing chord or, alternatively, 2% chord split flaps. These devices could be deployed differentially along the span. The stated results of the project are a 7% increase in L/D and a 7% increase in maximum lift. This was achieved by improving the lift distribution and reducing the wave drag for higher *Ma* with a decambering of wing sections.

Furthermore, the wing structure weight was estimated to be reduced by 2% through the application of the trailing edge devices for active load alleviation [28].

4 REINTRODUCTION OF FLEXIBLE MATERIALS

Because of the progress in research of smart materials (e.g., advanced composites and shape memory alloys), flexible wing parts are maturing and becoming more attractive for application in the aviation industry, especially outside the wing box. These structures, also called compliant structures, are often cited to achieve lower weight and lower maintenance cost [29]. Monner published the concept of the adaptive wing project in 1998 [30]. The concept integrates a bump for weakening the shock and a flexible variable camber trailing edge. Weakening the shock reduces wave drag, which is relevant for transonic flight regimes. Because of the curvature at the shock position, the shock is divided into several weaker shocks. One approach is to influence the boundary layer in order to create an artificial bump. This can be realized by a passive system, consisting of holes connected by a ventilation chamber. An active solution with controlled flow through the holes is also conceivable. Monner, however, states that in both cases this approach is prone to congestion of the perforation and therefore is not practical from a technical point of view. The bump for shock control also can be realized by physically deforming the airfoil surface as with the adaptive wing concept. The flexible Fowler flap is realized by connecting chord-wise separated flap segments with revolute and prismatic joints, thereby allowing a smooth deformation with one actuator per span-wise distributed kinematic station. A demonstrator with a span of 3.2 m and a chord of 0.9 m was built and proved the functionality of the bump and the flexible flap. A more detailed description of the flexible Fowler flap is given in Ref. [31].

Different approaches for actuation of a variable camber trailing edge also can be found [32]. Most of the concepts feature finger-like elements with multiple hinges building a framework (e.g., the X-29's discrete variable camber flaperon). One concept, which differs considerably, is the horn concept introduced by Mueller [33]. Here the actuation is realized by rotating rigid elements inside the flap, with a horn or tusk-like shape. When the curved end of these elements is rotated downward, the flexible skin of the flap is deflected. The main advantage is that the bearings carry the largest part of the aerodynamic forces and only a small portion has to be countered when the rigid elements are rotated. The actuation was successfully integrated in a 2.4 m span demonstrator with the outer shape of the outboard flap of an Airbus A340 and three rotational elements. The kinematics were completely integrated into the flap body, which offered the same stiffness as the original flap. The additional weight of the demonstrator was less than 1% compared to the span portion of the original A340 flap. The demonstrator was successfully used to emulate the usage as a roll-control device with deformation cycles equivalent to a flight time of 60,000 h and loads imitating the aerodynamic forces.

Another concept for actuation involves the specific control of the stiffness of structural components of the airfoil. Thereby actuation can either be reduced to very small loads or completely replaced by use of aerodynamic forces. Although the basic idea already was presented in the 1890s by Parker, modern materials can change stiffness properties, e.g., by applying electrical voltage, thereby opening new prospects. An example for this is a study published by Raither in 2015 about an airfoil with variable stiffness skin realized by electrobonded laminate [34].

The introduction of variable camber in small existing aircraft is a sensible step before designing a new wing, although the full potential will not be used. Pfeiffer et al. published a report in 2005 testing

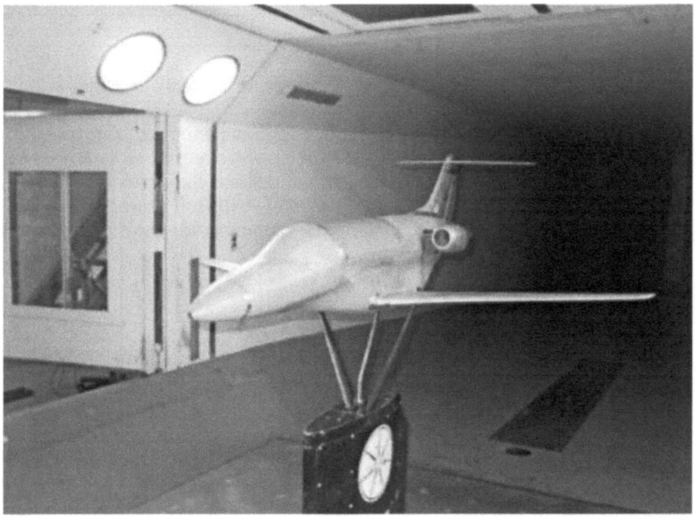

FIG. 12

Beechcraft Premier I Model at the UWAL [35].

different active trailing edge devices on a wind tunnel model of the Beechcraft Premier I [35]. The Beechcraft Premier I is a light business jet that had its first flight in 1998 with a maximum cruise *Ma* of 0.5. An existing wind tunnel model (see Fig. 12) was tested at the University of Washington Aeronautical Laboratory (UWAL).

The four concepts tested were trailing edge wedges, distributed lower surface wedges, a slotted flap, and a cruise tab. Different trailing edge geometries were attached to the wing for each concept. The wind tunnel tests were accompanied by transonic CFD-calculations. The cruise tab was the most beneficial wing modification. The results showed that the optimum deflection of the tab is strongly dependent on *Ma*. A fuel burn reduction of 1.5% was estimated (see Fig. 13). Furthermore, the cruise speed could be increased by 2–4 knots, leading to additional operational benefits.

The kinematic actuation also was investigated. An electrical actuation was judged to be beneficial, because of the existing electrical flap actuation of the Beechcraft Premier I and because of easier realization of locking the position after reaching a target deployment. The integration study into the Premier I is a conservative example because of the small space available in the flap. Further challenges included the tapered outboard flap and the connections of the tab actuation in the flap panel. A kinematic actuation with simple pushrods for the inboard flap and bell cranks on the tapered outboard flap was devised (see Fig. 14). A single interconnected rod could drive the actuators in each flap.

To increase of the credibility of expected benefits, flight tests are an absolute necessity as the unconventional technology of morphing matures. In 2006, the FlexSys Company began flight tests of a morphing airfoil carried by White Knight Aircraft [36]. The goal was to demonstrate the capability of the L/D increase from the improved camber setting as well as increase of laminar flow region for higher CL. For this reason, a natural laminar flow airfoil was selected, one that can reach laminar flow up to 65% chord on the upper airfoil side and 90% chord on the lower. Since the application being considered was on a high-altitude, long-endurance reconnaissance aircraft, *Ma* were relatively low, with a

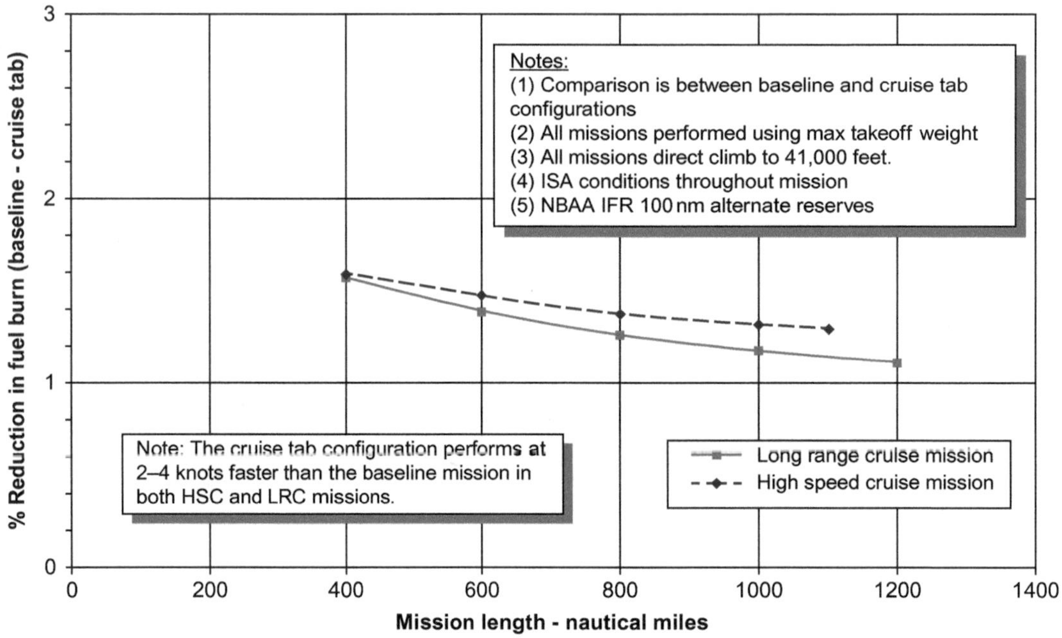

FIG. 13

Beechcraft Premier I cruise tab modification results [35].

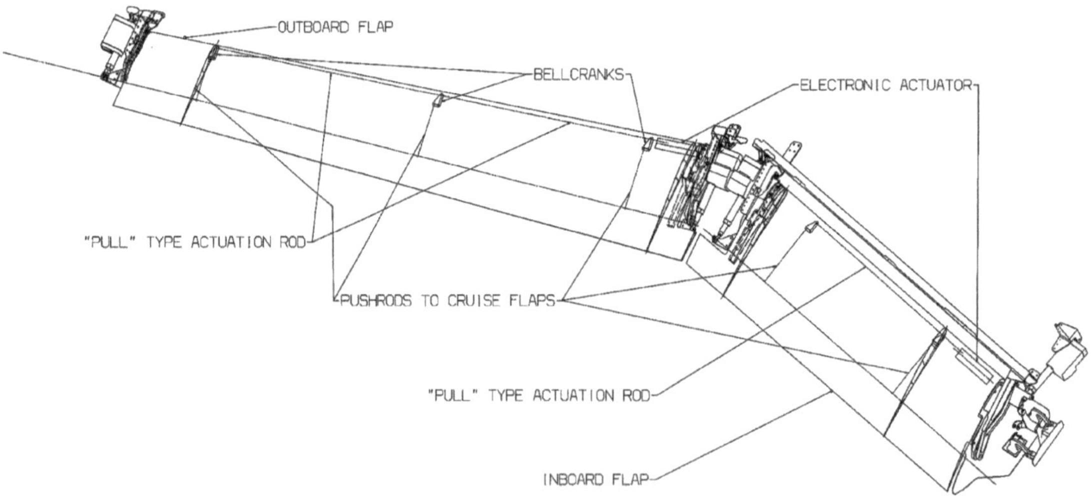

FIG. 14

Beechcraft Premier I cruise tab actuation concept [35].

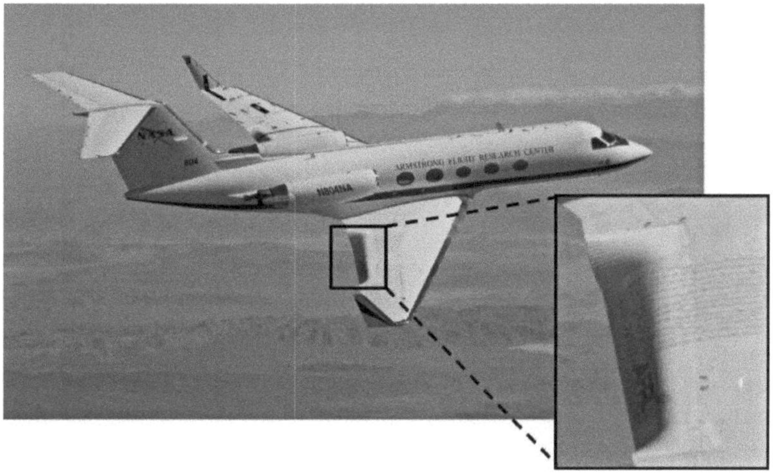

FIG. 15

FlexSys adaptive compliant trailing edge on NASA Gulfstream GIII Aircraft [37].

maximum of 0.55. The variable camber function was realized by a deformable trailing edge extending from 70% to 100% chord, driven by two electrical servomotors inside the flap. The internal structure deforms as a whole, and differential flap setting and actuation speeds of 30 degrees/s are possible. This would enable later use for load alleviation for RBM reduction and gust load alleviation. No more details of the kinematics causing the deformation were given. The tested wing section had a span of 50 in. and a constant 30-in. chord. A wake probe collected the airfoil momentum deficit and hot film sensors determined boundary layer transition location. The results showed that the region of laminar flow could be increased by 10% compared to a fixed trailing edge. FlexSys continued with the morphing trailing concept. In cooperation with NASA, a flight test campaign was planned in order to validate structural effectiveness and airworthiness of an adaptive compliant trailing edge (ACTE). The ACTE was designed to replace the conventional Fowler flap on NASA's subsonic research aircraft testbed, which is based on a Gulfstream G-III (see Fig. 15).

The ACTE consisted of five main elements: the main flap section, a connection spar, the actuation system, and two transition sections connecting the inboard and outboard edges of the flap section to the main wing without a gap. The dimensions of ACTE are given as a 19 ft span and 2 ft chord. After the creation of finite element methods (FEMs) of the ACTE, ground vibration tests were conducted. For the aeroelastic analysis, the joined ACTE and aircraft FEM were coupled with a linear panel method. Flight tests were undertaken from Nov. 2014 to Apr. 2015. The ACTE were not actuated in flight, but on the ground between flights. In order to excite the structure in the relevant frequencies, multiple maneuvers were conducted including roll, pitch, yaw. The collected data showed that the ACTE was airworthy from a structural point of view [37].

The effort of developing morphing trailing edge devices is mainly motivated by the long-term goal of integration in large civil aircraft. Apart from practical experiments, which approach the topic with increasing size of flight tests, the theoretical research focuses on the optimization of concepts and the refinement of the applied methods. One example is NASAs variable camber continuous trailing edge flap (VCCTEF) concept [38] (Fig. 16).

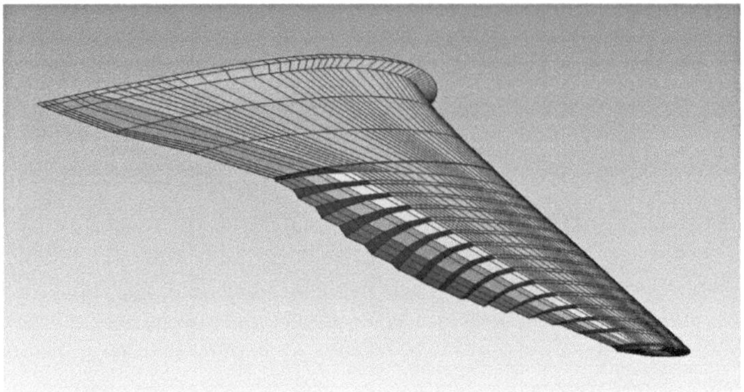

FIG. 16

Variable camber continuous trailing edge flap [39].

Early predictions for the VCCTEF state benefits up to 9.8% for an aircraft comparable in size to a Boeing B757 and aerodynamic analysis with vortex lattice methods [40]. Another example is the research of the University of Milano. Gaspari published several papers presenting the combination of airfoil design optimization methods, including the internal compliant structures to achieve the desired deformation, and aeroelastic analysis [41–43].

5 THE FINAL STEP TO INDUSTRIAL APPLICATION

More research activities are bringing morphing to an industrial level. Integrating a deforming skin to an aircraft component, while maintaining the necessary functions (e.g., ice and lightning strike protection) has been part of the EU research program SARISTU (Smart Intelligent Aircraft Structures) that concluded in 2015 (leading edge [44], trailing edge [45], and morphing winglet [46]). This integration was successfully tested in a wind tunnel. The benefit of the combination of these technologies has realized a block fuel reduction of more than 10% [47]. Research also is focusing on operational aspects for system failures and mission planning [48, 49].

Almost no morphing features have been implemented into current large civil aircraft and conventional control surfaces have been neglected, likely because aircraft manufacturers believe predicted potential is too little, they have low confidence in the quality of the predictions, or they believe the technologies are not adequately mature. Aircraft that already have limited capability for morphing applications likely will make use of such functions as deforming skin. During the development of the Boeing 787, several trailing edge devices were considered, including a variable camber concept patented by Sakurai in 2007 [50]. Subsequently, however, a simple-hinged flap with drooping spoilers was incorporated in the B787, a design that can only facilitate small flap adjustments in flight [51]. The Airbus A350 with its adaptive drop hinge flap would be able to deploy its trailing edge device during cruise without creating a gap [52]. So far, certification activities have focused on load alleviation for take-off and gusts. In general, this is an application of almost conventional high-lift devices for variable camber. The next step could be a stronger segmentation of control devices up to the point where the actuation and kinematics make a further increase uneconomical. This optimum can be found only by increasing the level of assessment from a sole aerodynamic view to the overall aircraft level. Aspects of weight implications because of changes in the system architecture (e.g., kinematics, actuation, power demand) have to be taken into account in order to increase the level of confidence toward the predicted benefits. This is even more crucial if morphing is used for primary control functions because safety aspects and subsequent redundancies must be considered. In addition, the topic of controlling these devices has to be investigated further. Although publications concerning control date back to the 1980s [53], it has to be clarified if additional sensors (e.g., static ports at the trailing edge) are needed. Since deforming morphing surfaces are a big step away from currently produced aircraft, their introduction is most likely to be expected after the application of conventional control surfaces for morphing. The ever-growing pressure on the aircraft manufacturers to maximize revenues and mitigate risks in certification and industrial production of new components pose a big challenge to the introduction of morphing by deformation. Ongoing and advancing research in materials, actuation concepts and flight-testing for morphing by deformation continuously decrease the gap toward an industrial application and thereby decrease the risk for the application by aircraft manufacturers.

REFERENCES

[1] J. Valasek (Ed.), Morphing Aerospace Vehicles and Structures, American Institute of Aeronautics and Astronautics, Chichester, West Sussex/Reston, VA, 2012.

[2] F. Culick, Wright brothers: first aeronautical engineers and test pilots, in: Proc. 45th Annual Symposium The Society of Experimental Test Pilots, Los Angeles, CA, 26–29 September, 2001.

[3] H.F. Parker, The Parker variable camber wing, NACA Technical Report 77, 1920.

[4] S. Spooner, The Makhonine Way, Flight—The Aircraft Engineer and Airship, The Royal Aero Club of the United Kingdom, vol. 20(1220), 1932, pp. 417–418.

[5] D.R. Jenkins, T. Landis, J. Miller, American X-Vehicles: An Inventory—X-1 to X-50, Monographs in Aerospace History No. 31, June 2003.

[6] Y. Gibbs, NASA Armstrong Fact Sheet: X-5 Research Aircraft, https://www.nasa.gov/centers/armstrong/news/FactSheets/FS-081-DFRC.html (retrieved September 14, 2016).

[7] W.D. Painter, L.J. Caw, Design and Physical Characteristics of the Transonic Aircraft Technology (TACT) Research Aircraft, NASA-TM-56048, January 1979, http://www.nasa.gov/centers/dryden/pdf/87897main_H-976.pdf (retrieved March 4, 2016).

[8] R. Hardy, AFTI/F-111 mission adaptive wing technology demonstration program, in: Aircraft Prototype and Technology Demonstrator Symposium, 1983, pp. 121–126.

[9] W.W. Gilbert, Mission adaptive wing system for tactical aircraft, J. Aircr. 18 (7) (1981) 597–602, https://doi.org/10.2514/3.57533.

[10] S. Goecke Powers, L.D. Webb, E.L. Friend, W.A. Lokos, Flight Test Results From a Supercritical Mission Adaptive Wing With Smooth Variable Camber: NASA Technical Memorandum 4415, NASA, Washington, DC, USA, 1992.

[11] R.W. DeCamp, R. Hardy, D.K. Gould, in: Mission-adaptive wing, International Pacific Air and Space Technology Conference and Exposition, vol. 6, SAE International, Warrendale, PA, 1987, pp. 1795–1801.

[12] D. Schmitt, Die aerodynamische Auslegung des Airbus A300-600, in: DGLR (Ed.), DGLR Jahrbuch, 83-123, 1983.

[13] J. Mantel, Leistungssteigernde Modifikationen an transsonischen Tragflügeln von Verkehrsflugzeugen, in: Jahrestagung, DGLR-NR, 82-031, 1982.

[14] A.O. Awani, The Investigation of a Variable Camber Blade Lift Control for Helicopter Rotor Systems, National Aeronautics and Space Administration, NASA Contractor Report, 1982.

[15] L.U. Dadone, Design and Analytical Study of a Rotor Airfoil, National Aeronautics and Space Administration, NASA Contractor Report, 1978.

[16] L.R. Centolanza, E.C. Smith, B. Munsky, Induced-shear piezoelectric actuators for rotor blade trailing edge flaps, Smart Mater. Struct. 11 (1) (2002) 24. http://stacks.iop.org/0964-1726/11/i=1/a=303.

[17] G.A. Trippensee, D.P. Lux, X-29A Forward-Swept-Wing Flight Research Program Status, National Aeronautics and Space Administration, NASA Technical Memorandum, November 1987.

[18] F.A. Johnsen, Sweeping Forward. Developing and flight testing the Grumman X-29A forward Swept Wing Research Aircraft, Library of Congress Cataloging-in-Publication Data, 2013.

[19] J. Szodruch, R. Hilbig, Variable wing camber for transport aircraft. Prog. Aerosp. Sci. 25 (3) (1988) 297–328, https://doi.org/10.1016/0376-0421(88)90003-6.

[20] E. Greff, The development and design integration of a variable camber wing for long/medium range aircraft, Aeronaut. J. 94 (939) (1990) 301–312.

[21] H. Hilbig, H. Wagner, Messerschmitt-Bölkow-Blohm GmbH/Vereinigte Flugtechnische Werke GmbH, Variable wing camber control for civil transport aircraft, in: 14th Congress of the International Council of the Aeronautical Sciences, 1984, pp. 243–248.

[22] J.J. Spillman, The use of variable camber to reduce drag, weight and costs of transport aircraft, Aeronaut. J. 96 (951) (1992) 1–9.

[23] V.J. Rossow, Lift Enhancement by Trapped Vortex, National Aeronautics and Space Administration, NASA Technical Report, 1992.

[24] V.J. Rossow, Lift enhancement by an externally trapped vortex, J. Aircr. 15 (9) (1978) 618–625, https://doi.org/10.2514/3.58416.

[25] A.L. Martins, F.M. Catalano, Aerodynamic optimization study of a mission adaptive wing for transport aircraft, in: 15th Applied Aerodynamics Conference, 1997.

[26] A.L. Martins, R.S. Ribeiro, Aircraft induced drag minimization using a constrained optimization method, in: 20th Congress of the International Council of the Aeronautical Sciences, 1996, pp. 2549–2557.

[27] A.L. Martins, F.M. Catalano, Drag optimization for transport aircraft mission adaptive wing, in: 38th Aerospace Sciences Meeting and Exhibit, 2000.

[28] G. Dargel, H. Hansen, J. Wild, T. Streit, H. Rosemann, K. Richter, Aerodynamische Flügelauslegung mit multifunktionalen Steuerflächen, in: DGLR (Ed.), DGLR Jahrbuch, vol. 1, DGLR, 2002, p. 1605.

[29] R.M. Ajaj, C.S. Beaverstock, M.I. Friswell, Morphing aircraft: the need for a new design philosophy, Aerosp. Sci. Technol. 49 (2013) 154–166, https://doi.org/10.1016/j.ast.2015.11.039.

[30] H.P. Monner, T. Bein, H. Hanselka, E. Breitbach, Design aspects of the adaptive wing—the elastic trailing edge and the local spoiler bump, in: RAS—Multidisciplinary Design and Optimisation Conference, 1998, pp. 15-1–15-9.

[31] H.P. Monner, H. Hanselka, E.J. Breitbach, Development and design of flexible Fowler flaps for an adaptive wing, in: 5th Annual International Symposium on Smart Structures and Materials, SPIE, 1998, pp. 60–70.

[32] I. Dimino, G. Diodati, A. Concilio, A. Volovick, L. Zivan, Distributed electromechanical actuation system design for a morphing trailing edge wing, SPIE Smart Structures/NDE, Las Vegas, Nevada (USA), March 2016. in: Proc. SPIE 9801, Industrial and Commercial Applications of Smart Structures Technologies 2016, 980108, April 16, 2016. https://doi.org/10.1117/12.2219223.

[33] D. Müller, Das Hornkonzept. Realisierung eines formvariablen Tragflügelprofils zur aerodynamischen Leistungsoptimierung zukünftiger Verkehrsflugzeuge, Universität Stuttgart, Stuttgart, Germany, 2000.

[34] W. Raither, E. Furger, M. Zundel, A. Bergamini, P. Ermanni, Variable-stiffness skin concept for camber-morphing airfoils, J. Intell. Mater. Syst. Struct. 26 (13) (2015) 1609–1621, https://doi.org/10.1177/1045389X14546780.

[35] N.J. Pfeiffer, J.G. Christians, Active Tailoring of Lift Distribution to Enhance Cruise Performance, NASA Langley Research Center, Contract Report, April 1, 2005.

[36] J. Hetrick, R. Osborn, S. Kota, P. Flick, D. Paul, Flight testing of mission adaptive compliant wing, in: 48th AIAA/ASME/ASCE/AHS/ASC Structures, Structural Dynamics, and Materials Conference, 2007.

[37] C.Y. Herrera, N.D. Spivey, G. Ervin, P. Flick, Aeroelastic airworthiness assessment of the adaptive compliant trailing edge flaps, in: 46th SFTE Annual International Symposium, 2015.
[38] J. Urnes, N. Nguyen, A mission adaptive variable camber flap control system to optimize high lift and cruise lift to drag ratios of future N+3 transport aircraft, in: 51st AIAA Aerospace Sciences Meeting including the New Horizons Forum and Aerospace Exposition, 2013.
[39] N. Nguyen, A framework for adaptive aeroelastic wing shaping control, Presentation at the Advanced Modeling & Simulation (AMS) Seminar Series, April 22, 2014.
[40] C.A. Ippolito, N.T. Nguyen, J. Totah, K. Trinh, E.B. Ting, Initial assessment of a variable-camber continuous trailing-edge flap system for drag-reduction of non-flexible aircraft in steady-state cruise condition, in: AIAA Infotech@Aerospace (I@A) Conference, 2013.
[41] A. de Gaspari, S. Ricci, A two-level approach for the optimal design of morphing wings based on compliant structures, J. Intell. Mater. Syst. Struct. 22 (10) (2011) 1091–1111, https://doi.org/10.1177/1045389X11409081.
[42] A. de Gaspari, S. Ricci, L. Travaglini, L. Cavagna, A. Antunes, F. Odaguil, G. Lima, Active camber morphing wings based on compliant structures: an aeroelastic assessment, in: 53rd AIAA Aerospace Sciences Meeting, 2015.
[43] A. de Gaspari, S. Ricci, L. Riccobene, Design, manufacturing and wind tunnel test of a morphing wing based on compliant structures, in: 24th AIAA/AHS Adaptive Structures Conference, 2016.
[44] M. Kintscher, J. Kirn, S. Storm, F. Peter, Assessment of the SARISTU Enhanced Adaptive Droop Nose, in: P. C. Wölcken, M. Papadopoulos (Eds.), Smart Intelligent Aircraft Structures (SARISTU), first ed., Springer International Publishing, Cham, Switzerland, 2016, pp. 113–140.
[45] R. Pecora, F. Amoroso, M. Magnifico, I. Dimino, A. Concilio, KRISTINA: kinematic rib-based structural system for innovative adaptive trailing edge, in: Proc. SPIE—The International Society for Optical EngineeringVolume 9801, (2016), Article No. 980107.
[46] A. Wildschek, S. Storm, M. Herring, D. Drezga, V. Korian, O. Roock, Design, optimization, testing, verification, and validation of the wingtip active trailing edge, in: P.C. Wölcken, M. Papadopoulos (Eds.), Smart Intelligent Aircraft Structures (SARISTU), first ed., Springer International Publishing, Cham, Switzerland, 2016, pp. 219–255.
[47] F. Peter, E. Stumpf, G.M. Carossa, M. Kintscher, I. Dimino, A. Concilio, R. Pecora, A. Wildschek, Morphing value assessment on overall aircraft level, in: P.C. Wölcken, M. Papadopoulos (Eds.), Smart Intelligent Aircraft Structures (SARISTU), first ed., Springer International Publishing, Cham, Switzerland, 2016, pp. 859–871.
[48] F.N. Peter, K. Risse, F. Schueltke, E. Stumpf, Variable camber impact on aircraft mission planning, in: 53rd AIAA Aerospace Sciences Meeting, 2015.
[49] I. Dimino, A. Concilio, R. Pecora, Safety and reliability aspects of an adaptive trailing edge device (ATED), in: Proc. 24th AIAA/AHS Adaptive Structures Conference, San Diego, United States, January 4–8 2016.
[50] S. Sakurai, S.J. Fox, K.W. Beyer, D.S. Lacy, P.L. Johnson, S.L. Wells, J.S. Noble, P.T. Meredith, N. V. Huynh, R.R. Christianson, Multi-Function Trailing Edge Devices and Associated Methods, US Patent No. US 7,243,881 B2, 2007.
[51] M. Goldhammer, The Next Decade in Commercial Aircraft Aerodynamics Aircraft Aerodynamics. A Boeing Perspective, Aerodays 2011, European Commission, Madrid, Spain, 2011.
[52] H. Strüber, The aerodynamic design of the A350 XWB-900 high lift system, in: 29th Congress of the International Council of the Aeronautical Sciences, 2014.
[53] J.H. Renken, Mission-adaptive wing camber control systems for transport aircraft, in: 3rd Applied Aerodynamics Conference, 1985.

SECTION 2

Requirements and Performance

Editor:

Sergio Ricci
Polytechnic of Milano, Milano, Italy

List of Chapters

CHAPTER

SPAN MORPHING CONCEPT: AN OVERVIEW

4

José Lobo do Vale*, Frederico Afonso*, Fernando Lau*, Afzal Suleman*,†

University of Lisbon, Lisbon, Portugal University of Victoria, Victoria, BC, Canada†*

CHAPTER OUTLINE

NOMENCLATURE

AC	aerodynamic center
AOA	angle of attack
AR	wing aspect ratio
b	wing span
b_E	extended wing span
b_R	retracted wing span
c	wing chord
CD	drag coefficient
CD_0	base drag coefficient
CD_i	induced drag coefficient
CG	center of gravity
CL	lift coefficient

Morphing Wing Technologies. https://doi.org/10.1016/B978-0-08-100964-2.00004-6

D	Wing drag
e	Oswald efficiency coefficient
F_1	aileron extension factor
F_2	aileron distance factor
I_x	rolling inertia
l_A	distance from wing root to central aileron position
L	wing lift
L/D	lift to drag ratio
MAC	mean aerodynamic chord
p	roll rate
q	dynamic pressure
RBM	wing root bending moment
Re	Reynolds number
RM	rolling moment
$\mathbf{RM}_A$	rolling moment because of aileron deflection
$\mathbf{RM}_S$	rolling moment because of asymmetric span morphing
TOW	take-off weight
W	aircraft weight
Δ**CL**	lift coefficient variation
$\Delta \mathbf{CL}_A$	lift coefficient variation because of aileron deflection
$\Delta \mathbf{L}_A$	lift variation because of aileron deflection
Δx	span variation

1 INTRODUCTION

Span morphing is not a recent concept. The MAK-10 aircraft was the first aircraft with a span morphing wing design and dates from 1931 [1]. The used mechanism was based on a telescopic wing actuated pneumatically, varying span from 13 to 21 m (62%) and area from 21 to 33 m^2 (57%). Another early example of span morphing application was the human-actuated German FS-29 glider from 1972. Since then, only small-scale aircraft have been built with span morphing technology, serving as flight demonstrators of the concept.

Until now, span morphing has not brought substantial benefit to flight performance, partly because the aircraft have been designed for a single mission and are rarely thought of as multiple role vehicles. After the main segment of a flight mission is determined, the design process usually focuses on the flight condition for which the aircraft will be optimized, and the mission itself is adapted to maximize the performance of the aircraft.

Span morphing is potentially advantageous, as are other morphing concepts, for aircraft that significantly change their flight condition in large portions of the mission. The span morphing enables efficient flight at high and low speeds, allowing transition from dash to loiter flight segments with improved levels of performance, and allows changes in the maneuverability of the aircraft.

The designer must be aware of the variety of missions the aircraft has to perform during its lifecycle in order to assess whether span morphing would be beneficial. This evaluation can only be made in regard to design optimization and complete mission calculations. Not only the aerodynamic benefits must be accounted for, but the increase in structural mass and complexity also should play a part in the assessment of the concept.

Although sweep and dihedral angle variations cause the span of the aircraft to change, the span morphing concept discussed here involves actual variation of the length of the wing from root to tip.

In this chapter, Section 2 deals with the effects of span change, starting with a brief description showing the motivation for the use of the concept as a means for aerodynamic drag reduction. Structural effects then are highlighted as the source of depreciation of the aerodynamic benefits, followed by analysis about the stability and control.

Section 3 describes some of the current knowledge about symmetric and asymmetric span morphing. Results are presented for performance parameters in comparison with conventional fixed-wing aircraft performance for small-scale aircraft. Effects of scaling up the concept are also discussed. Asymmetric span morphing as a means for roll control is analyzed and evaluated.

Some of the implementation challenges are discussed in Section 4, in which the most common structural solutions for concept implementation are described and major drawbacks noted.

Finally, a global evaluation of the span morphing concept is done in the conclusions, which reflects the authors' opinion about the feasibility and adequacy of the concept for application in aircraft.

2 EFFECTS OF SPAN INCREASE

Wing span (*b*) is one of the most important geometrical parameters in aircraft design. It is directly related to the wing's lift induced drag [2] because, as span increases, the percentage of inboard wing area airfoils behaving nearly as in two-dimensional flow conditions increases. The downwash in the inboard part of the wing caused by the wing tip vortices is reduced because of the increase in the distance between the vortices and the wing sections near the root.

On the other hand, structural requirements become more severe because of span increase. As lift is distributed by a higher length, the lift bending moment at the wing root is increased, and lead to higher wing structural mass, which can jeopardize the aerodynamic benefits with consequent reduction of the overall benefits or even penalties. Aeroelastic effects and minimum lift requirements play a role in structural design when span is increased.

Another aspect to account for is the effect in maneuverability. Span increase causes rolling inertia to increase but also higher rolling moment production if there is the capability to change the lift near the wing tips (e.g., using ailerons). If the wing is swept, significant changes in the lift pitching moment of the aircraft can be expected, as well as changes in the center of gravity (CG) position, affecting trim conditions and stability.

In this section, the effects of span variation on aerodynamics, structures, stability and control are explored.

2.1 AERODYNAMIC EFFECTS

For the same wing area, an increase in wing span causes a reduction in the wing's mean aerodynamic chord (MAC), increasing the wing's aspect ratio (AR). An increase in AR influences the induced drag coefficient as can be observed in the induced drag coefficient expression given in the following equation (Eq. 1):

$$\mathrm{CD_i} = \frac{CL^2}{\pi ARe}. \tag{1}$$

A reduction in the MAC means a reduction in the local operating Reynolds number (*Re*) of the wing's airfoils which, in general, increases their local base drag coefficient. This effect is more significant at low Reynolds numbers. Although local base drag coefficient might increase, actual local drag decreases because of the wing chord reduction and consequent lower length for boundary layer development. This local base drag reduction is compensated by a higher span length required to maintain wing area, thus making the span increase effect in the base drag nearly null, at least for sufficiently high Reynolds numbers.

Total wing drag, as a result of the integration of the drag components along the span of the wing and expressed as in Eq. (2), is therefore affected by wing span mainly because of its effect in the induced drag coefficient:

$$\mathrm{CD} = \mathrm{CD_0} + \mathrm{CD_i}. \tag{2}$$

Expressions (1) and (2) result from the usual parabolic modeling of the drag as a function of the wing lift and base drag. It can be seen from those expressions that, as the lift coefficient (CL) is reduced, so is the induced drag, therefore increasing the base drag component percentage relatively to the total drag. Inversely, as CL increases, the induced drag component percentage relatively to the total drag increases (see Fig. 1).

Therefore, one can expect more significant aerodynamic benefits from increasing the wing span for operation at high CL values, i.e., at lower flight speeds, in an attempt to maximize the lift to drag ratio (L/D) of the wing throughout the flight envelope.

For lower CL values, i.e., at higher flight speeds, decreasing the wing span without reducing the wing area does not contribute to any drag reduction, because the AR decreases (with the corresponding induced drag increase) and the wing wetted area is maintained, roughly preserving the same base drag except for fairly insignificant Reynolds number increases.

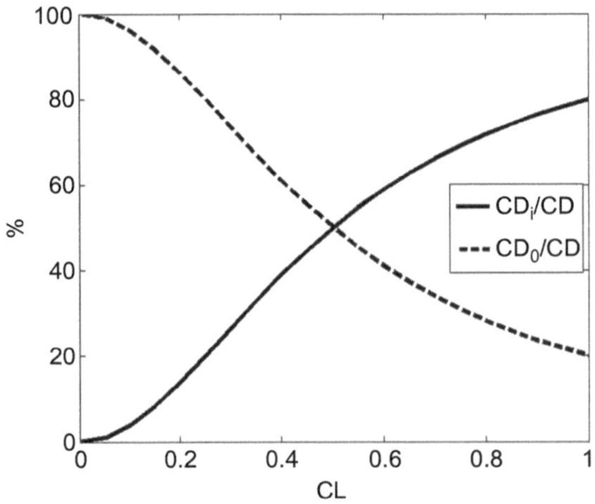

FIG. 1

Typical example of base and induced drag coefficients as functions of the CL given as percentage of the total drag coefficient.

2.2 STRUCTURAL EFFECTS

In terms of structural effects, two main aspects are noticeable if the span is increased while maintaining the wing area (i.e., increase of the AR): The wing root bending moment (RBM) increases, and the maximum thickness of the airfoil sections along the span is reduced because of the chord reduction.

These two effects are both penalizing in terms of structural mass requirements; the loading is more severe for the same amount of lift as span increases, while the load-bearing structure (e.g., wing spars or wing box) resisting area and second moment of inertia cannot be increased efficiently by increasing the height of the structure due to airfoil thickness limitations. Thus, a larger thickness and/or chord-wise extension is required for the wing spars or wing box structures, respectively (Fig. 2).

The structural mass increase is a major drawback that has prevented the wing span morphing development especially in the civil transportation domain.

Another aspect of span increase effects in aircraft structures is the associated increase of wing flexibility, which may result in potentially adverse aeroelastic effects.

In the attempt to limit the wing's structural mass while increasing the span, higher levels of deformation are achieved, namely wing twist, which, if not properly considered, could introduce significant aerodynamic penalties by altering the lift distribution along the span, even in cruise conditions. Additionally, the safety margin for aeroelastic instabilities such as flutter can be reduced when significant deformation occurs in the wing, namely when the vertical tip displacement is a significant percentage of the span of the wing (>30%) [3,4] as depicted in Fig. 3.

Typically, aircraft use enhanced lift devices such as flaps and slats as well as, which are used to reduce the stall speed and consequently the take-off and landing speeds and distances. The increase in AR of a wing, in addition to the previously mentioned chord-wise extension increase of the wing structure, limits the space available to install flaps and slats and limits also their extensions chord-wise, resulting in an increase in the aircraft stall speed and take-off and landing distances. Thus there might be a decrease in operational safety levels associated with span increase.

2.3 STABILITY AND CONTROL EFFECTS

Stability and control are also affected when span is increased. The first and most obvious effect is the increase in rolling inertia. The effect is a quadratic increase with span, because the rolling inerta varies with the square of the distance of the mass elements to the longitudinal axis of the aircraft.

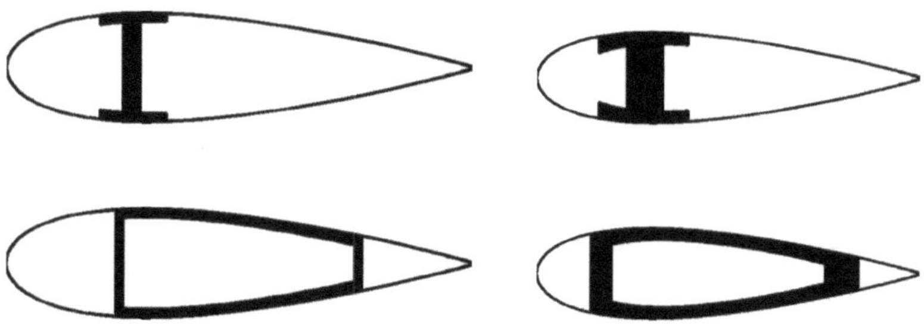

FIG. 2

Spars and wing box structures at a wing section with span increase for the same wing area and lift.

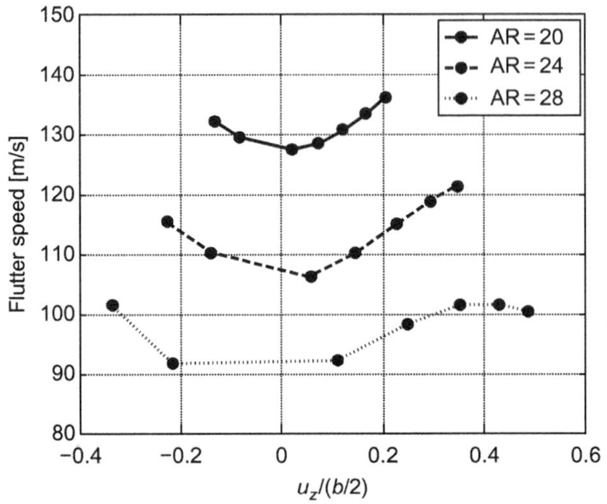

FIG. 3

Flutter speed vs. wing AR for rectangular wings.

Adapted from F. Afonso, Performance Based Multidisciplinary Design Optimization of Energy Efficient and Novel Aircraft Configurations (Ph.D. Thesis), Instituto Superior Técnico, Universidade de Lisboa, Portugal, 2015.

Therefore, if it is not possible to decrease the wing structural mass toward the tips, significant reductions in the achievable roll rate can be expected even if the device producing the rolling moment remains near the wing tip. This occurs because the rolling moment varies linearly with the distance to the aircraft's longitudinal axis, unless the device extension is increased span-wise. This results in loss of maneuverability.

Also contributing to loss of maneuverability (thus to an increased rolling stability) is the aerodynamic damping of the rolling movement, which is increased because of the higher tip rotating speed components, reducing or increasing the local angle of attack (AOA) of the airfoils near the wingtip, and counteracting more intensely the device producing the rolling moment for maneuvering.

If the wing is swept, a span increase is likely to produce a shift in the CG of the aircraft, which will be more significant as the wing's structure mass relative to the aircraft mass increases. In addition, the position of the aerodynamic center (AC) of the wing is also shifted. The shifts in the CG and AC will, in turn, have an effect in trim drag and static stability margin.

Similarly, if the span is increased asymmetrically, the CG and AC will shift out of the symmetry plane of the aircraft and there will be coupling between the longitudinal and lateral movement of the aircraft, which will be more intense as the aircraft becomes more asymmetric. This is the case when span morphing is used for roll control as described in the next section.

3 SPAN MORPHING CONCEPTS AND AIRCRAFT PERFORMANCE

Two major trends for span morphing are being investigated, generally applying the concept to small-scale subsonic unmanned aircraft: exploring the benefits of symmetric span morphing (the span from root to tip of the wing remains the same for the right and left sides of the aircraft) in the longitudinal performance of the aircraft; and exploring asymmetric span morphing (right and left sides of the aircraft with different spans from root to tip of the wing) as a means for roll control (substituting ailerons), while still taking advantage of the beneficial effects of symmetric span morphing.

3.1 SYMMETRIC SPAN MORPHING

Symmetric span morphing is the less complex technology, relating all geometric and inertial variations with only one degree of freedom while maintaining aircraft symmetry at all times. Although longitudinal and lateral dynamic behavior might change because of changes in the total lift point of application or CG position and inertial variations, their level of coupling is similar to that of other aircraft with the same general configuration.

3.1.1 Aerodynamic aspects of span morphing

With span morphing, the wing not only increases its span, but also its lifting area. Therefore, the wing changes not only its AR, which affects the induced drag, but also its wetted area, which affects the base drag component.

While evaluating aerodynamic coefficients for the span morphing wing, one must choose the reference area of the wing and maintain the same reference area independently of the wing configuration being assessed. Fig. 4 shows the results obtained from CFD calculations for four different configurations of a constant airfoil, rectangular, unswept wing with 1 m half-span, from which a rectangular telescopic wing segment is extended; the rectangular telescopic wing has a chord equal to the 80% of that of the inboard segment and increases the wing span up to the 75%. In this case, the reference area is the area of the wing when fully extended [5].

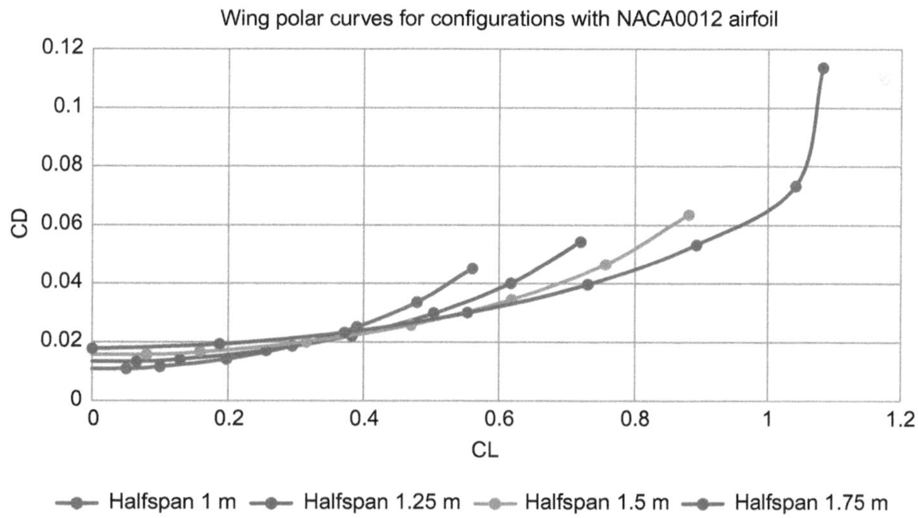

FIG. 4

Polar curves of the telescopic wing with NACA0012 airfoil for different span configurations.

Adapted from J. Vale, A. Leite, F. Lau, A. Suleman, Aero-structural optimization and performance evaluation of a morphing wing with variable span and camber. J. Intell. Mater. Syst. Struct. 22 (2011) 1057–1079, https://doi.org/10.1177/1045389X11416031.

The different polar curves cross each other, showing that for a given CL value, one span configuration is more advantageous than the others. Based on Fig. 4, one can also expect that the optimal variation of span with CL is a continuous function.

For higher CL values, higher span configurations can deliver not only more lift (because of the higher lifting area), but also reduced drag as a consequence of lower AOA requirements and lower induced drag. For lower CL values, the reduction of the wetted area of the wing will decrease its base drag, even if the induced drag is increased because of the AR decrease and AOA increase.

Aerodynamically speaking, it seems reasonable to assume that span morphing can bring significant benefits for an aircraft with significant CL variations during its mission.

Fig. 5 shows the optimum polar curve for the span morphing wing as being composed of segments of different configuration curves chosen for the minimum drag as CL varies. The span morphing wing polar curve tends to be flattened as CL increases when compared to any of the different configuration polar curves. The preferable mechanism for lift production as CL increases seems to be an increase in lifting area instead of AOA. This situation can present other aerodynamic advantages, such as smaller variations of fuselage AOA for a wide CL range of operation, reducing also the fuselage drag component.

For the considered wing and assuming a 100 N weight aircraft, changing from maximum to minimum span results in a stall speed increase of 36% (assumed at a 10 degree AOA) and 62% wing drag reduction at maximum speed (CL < 0.1).

Using span morphing as an alternative to flaps, it was found that in correspondence of a 75% span increase and of a 60% increase in lifting area, a ΔCL of about 0.77 was reached. This is a modest CL

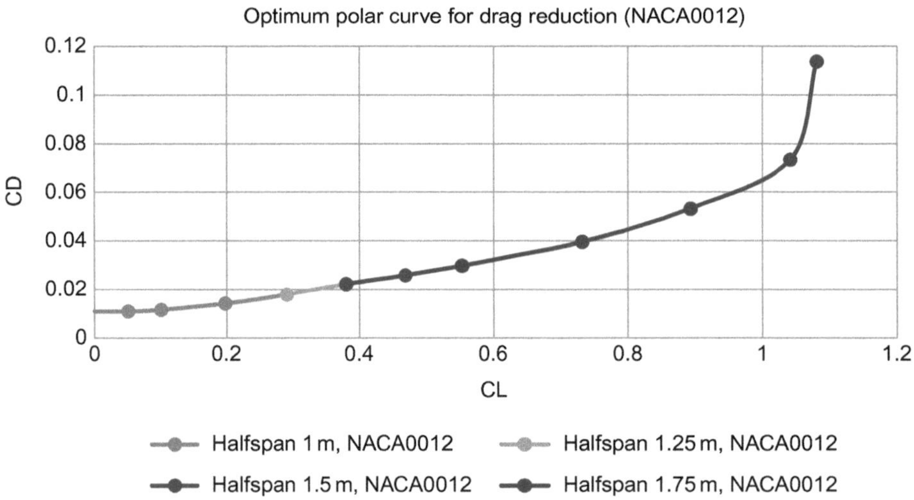

FIG. 5

Optimum polar curve for telescopic wing with NACA0012 airfoil.

Adapted from J. Vale, A. Leite, F. Lau, A. Suleman, Aero-structural optimization and performance evaluation of a morphing wing with variable span and camber. J. Intell. Mater. Syst. Struct. 22 (2011) 1057–1079, https://doi.org/10.1177/1045389X11416031.

increase if compared to what is attainable with Fowler type flap system for take-off and landing. This is also one of the reasons why span morphing is unlikely to be applied to higher scale aircraft (e.g., commercial transport aircraft), for which take-off and landing distances are critical performance parameters.

3.1.2 Structural and performance aspects of span morphing

Although span morphing makes sense aerodynamically, it introduces a mass penalty in the wing structure that is not easy to quantify and depends on the implemented structural solution.

It is difficult to determine which wing dimensions and structure the span morphing wing should be compared to.

For instance, if we consider a loiter mission aircraft with a high span and AR wing, it probably makes sense to compare its wing with a span morphing wing that retracts its wings in dash or cruise to destination and back flight segments. In this case, the mass penalty for the wing structure would be mainly from the actuation mechanism and not so much from the extra material for the wing structure.

If the aircraft is optimized for cruise or dash segments, however, the mass penalty increases, because the actuation system mass has increased due to the material added to enable the wing extension and to withstand the increase in RBM.

It is therefore advisable to compare the benefits in performance for the mass penalties relative to a reference aircraft. This comparison was performed with reference to an aircraft equipped with the previousily described telescopic wing and a conventional fixed wing optimized for cruise at 30 m/s, in correspondence of two different mass penalty values of 10% and 20% of the total aircraft mass. Obtained results are shown in Table 1 for some of the longitudinal performance parameters. It should be noted that besides the span increase capability, these calculations assumed also constant span-wise camber change capability for the morphing wing, which favors the performance increase of the morphing wing [5].

As expected, results show that the optimum fixed-wing aircraft outperforms the span morphing wing aircraft in the design point, and the penalty of the morphing wing increases with the mass penalty. In fact, all the benefits decrease and penalties increase when the mass penalty becomes larger for the morphing wing aircraft. The major penalty of the morphing wing aircraft occurs in rate of climb and climb angle, which is natural because a weight component adds to drag in this flight condition; maximum range is also penalized.

It is upon consideration of the aircraft mission(s) that a decision can be made about the implementation of such span morphing wing concept.

Fig. 6 shows the calculated relative difference between the total energy output of the engine of the morphing wing aircraft and the fixed-wing aircraft for three different climb-cruise mission profiles that were simulated with trajectory tracking algorithms. A negative slope of the curves means that the morphing wing aircraft is consuming less energy during that time period, while a positive slope means the opposite [6].

It can be seen how mission requirements have a strong influence on the energetic consumption of one aircraft relatively to the other. If the aircrafts mission is, for example, 80% of the times closer to the profile of Mission 2, the morphing wing aircraft will not be beneficial; if, however, during the operational lifetime of the aircraft, Missions 1 and 3 are performed a significant number of times, it is probable that the morphing wing aircraft will save energy in the long run.

Table 1 . Performance Parameters Results and Comparison Between Morphing and Optimum Fixed Wing With 10% and 20% Mass Penalty

Wing	Morphing 10%	Morphing 20%	Optimum	Relative Difference 10%	Relative Difference 20%
Span (m)	1–1.7	1–1.7	1	0%–70%	0%–70%
Camber	0–7%	0–7%	4.2% @ root; 0% @ tip		
Area (m^2)	0.52–0.81	0.52–0.81	0.42	23.81%–92.86%	23.81%–92.86%
Max speed (m/s)	56.87	56.80	56.26	1.07%	0.96%
Stall speed (m/s)	12.23	12.71	19.18	−36.24%	−33.71%
Drag @ 20 m/s (N)	9.08	9.39	10.58	−14.15%	−11.21%
Drag @ 30 m/s (N)	15.32	15.65	14.71	4.15%	6.34%
Drag @ 56 m/s (N)	46.67	46.83	48.04	−2.84%	−2.52%
Max climb angle (degree)	22.64	21.12	22.14	2.28%	−4.58%
Max ROC @ 40 m/s (m/s)	9.69	8.97	10.75	−9.91%	−16.54%
Max range (km)	284.70 @ 19 m/s	272.05 @ 20 m/s	284.97 @ 26 m/s	−0.09%	−4.53%
Max range @ 26 m/s (km)	270.25	259.68	284.97	−5.17%	−8.87%
Min glide angle (degree)	3.55	3.55	6.04	−41.26%	−41.26%
Max endurance (min)	251.07	231.3	157.17	59.74%	47.2%

Adapted from J. Vale, A. Leite, F. Lau, A. Suleman, Aero-structural optimization and performance evaluation of a morphing wing with variable span and camber, J. Intell. Mater. Syst. Struct. 22 (2011) 1057–1079, https://doi.org/10.1177/1045389X11416031.

3.1.3 Scalability aspects of span morphing

What about the scalability of the span morphing concept to larger aircraft? Modern full-scale aircraft need the inner space of their wings to accommodate fuel, mechanisms, and structures, which would prevent extension or retraction percentages as high as the ones discussed here (70%) thus reducing relevance of the potential benefits. There is also the matter of actuation power requirements, which for the span morphing increases at a higher rate than the engine power requirements [7]. Eventually, this limits the allowed number of span change cycles if the actuation energy has not to exceed a prescribed amount of the total energy consumption and becomes an important sink of energy, thus impairing the range or endurance of the aircraft.

A study performed on the application of span morphing to a regional transport aircraft [8] with a base half-span of 15.7 and a 3 m half-span extension morphing capability (Fig. 7) shows benefits in off-design flight segments, such as alternate and hold: the magnitude of the benefits increases as the take-

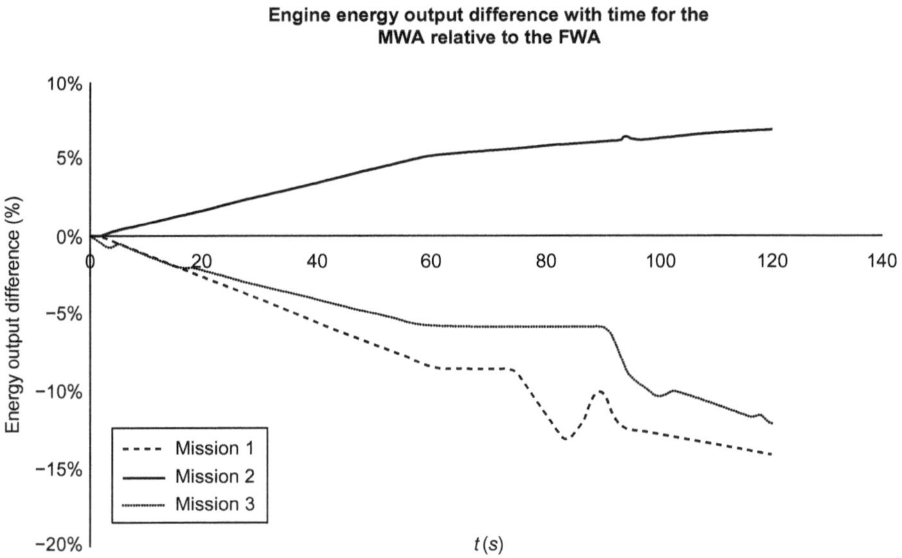

FIG. 6

Engine energy output difference with time for the MWA relative to the FWA for three different climb-cruise missions: Mission 1 [(AOC = 5 degree, $U = 45$ m/s) – ($U = 25$ m/s)]; Mission 2 [(AOC = 10 degree, $U = 34$ m/s) – ($U = 34$ m/s)], and Mission 3 [(AOC = 15 degree, $U = 20$ m/s) – ($U = 45$ m/s)] [6].

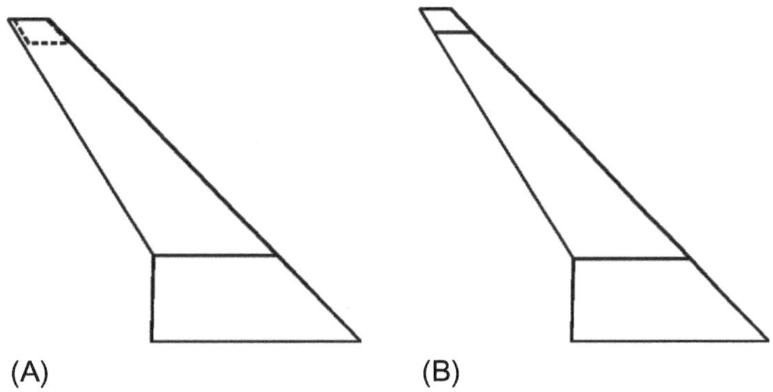

FIG. 7

Span morphing in a regional transport aircraft: (A) fully retracted telescopic wing tip and (B) fully extended telescopic wing.

Adapted from J. Vale, F. Afonso, F. Lau, A. Suleman, Performance based MDO of a joined-wing regional transport aircraft, in: 56th AIAA/ASCE/AHS/ASC Structures, Structural Dynamic, and Material Conference, AIAA Science and Technology Forum and Exposition (SciTech 2015), Kissimmee, FL, USA, January 5–9, 2015, https://doi.org/10.2514/6.2015-0696 (AIAA Paper 2015-0696).

Table 2 Fuel Consumption Savings for the Span Morphing Concept Relatively to the Baseline Configuration of a Regional Transport Aircraft Model for a Typical Mission Profile

	Mission		
Segment	**1 (TOW = 56,863 kg)**	**2 (TOW = 48,826 kg)**	**3 (TOW = 44,826 kg)**
Climb	0.13%	0.25%	0.00%
Cruise	−0.78%	−0.01%	−0.01%
Alternate	5.58%	0.50%	−0.30%
Hold	14.05%	12.64%	5.21%
Overall	4.42%	3.39%	1.35%

Adapted from J. Vale, F. Afonso, F. Lau, A. Suleman, Performance based MDO of a joined-wing regional transport aircraft, in: 56th AIAA/ASCE/ AHS/ASC Structures, Structural Dynamic, and Material Conference, AIAA Science and Technology Forum and Exposition (SciTech 2015), Kissimmee, FL, USA, January 5–9, 2015, https://doi.org/10.2514/6.2015-0696 (AIAA Paper 2015-0696).

off weight (TOW) of the aircraft is increased for the same mission profile, i.e., when aircraft weight in the off-design segments is higher; the magnitude of the benefits decreases while the weight penalty associated with the morphing mechanism increases.

This 3m amount of span morphing in each half wing would be a maximum for this aircraft, given the limits imposed by the fuel tanks, flaps, slats, and ailerons that will have to be considered in the wing design.

Table 2 shows the estimated benefits from span morphing in fuel consumption relative to the baseline aircraft configuration, using low fidelity aerodynamics calculations and assuming no weight penalty for morphing implementation.

Although the accuracy of the fuel consumption calculations (strongly dependent on the accuracy of drag estimation) might be questionable, the trends seem to be according to the expectations based on the actual knowledge of span increase and induced drag reduction: flight segments with lower flight speeds and higher lift requirements (higher aircraft mass) benefit from induced drag reduction caused by the increase in AR and AOA reduction because of the higher lifting area.

Fig. 8 depicts the magnitude of the estimated fuel savings for several weight penalties associated with span morphing implementation. In this case, the increase in weight penalty tends to become a penalty in terms of fuel consumption for lower TOWs, because the span extension brings benefits only in the hold segment and slight penalties in all other segments. As the TOW increases, the benefits extend to the alternate segment, and although the magnitude of the savings decreases as the weight penalty increases, span morphing still compensates in the overall fuel consumption.

Once again, the benefits quantification depends on the aircraft mission profile(s), as was noted for the smaller scale aircraft.

3.2 ASYMMETRIC SPAN MORPHING

Span morphing is difficult to integrate with conventional roll control systems such as ailerons. The chord of the extendable part is significantly reduced by the presence of the aileron in the retracted configuration, and the authority of the aileron is reduced significantly when the wing is fully extended because of the increased roll damping and inertia (Fig. 9). Therefore, either the ailerons need longer span-wise extensions or different options should be considered for roll control with span morphing wings.

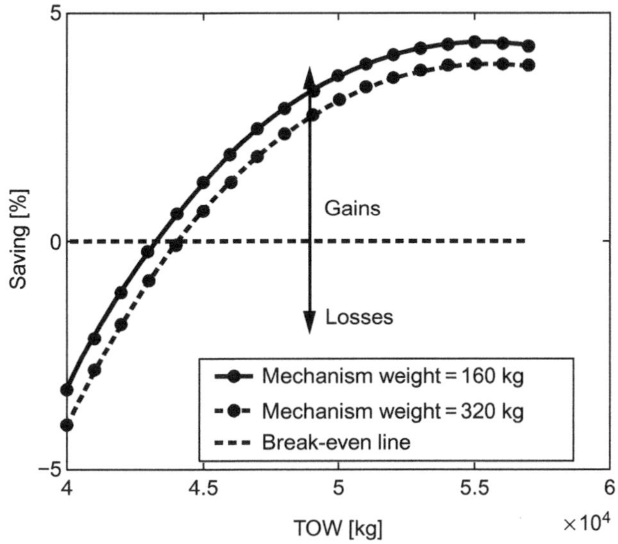

FIG. 8

Effect of mechanism weight penalty on span morphing benefits in fuel consumption for the regional transport aircraft.

Adapted from J. Vale, F. Afonso, F. Lau, A. Suleman, Performance based MDO of a joined-wing regional transport aircraft, in: 56th AIAA/ASCE/AHS/ASC Structures, Structural Dynamic, and Material Conference, AIAA Science and Technology Forum and Exposition (SciTech 2015), Kissimmee, FL, USA, January 5–9, 2015, https://doi.org/10.2514/6.2015-0696 (AIAA Paper 2015-0696).

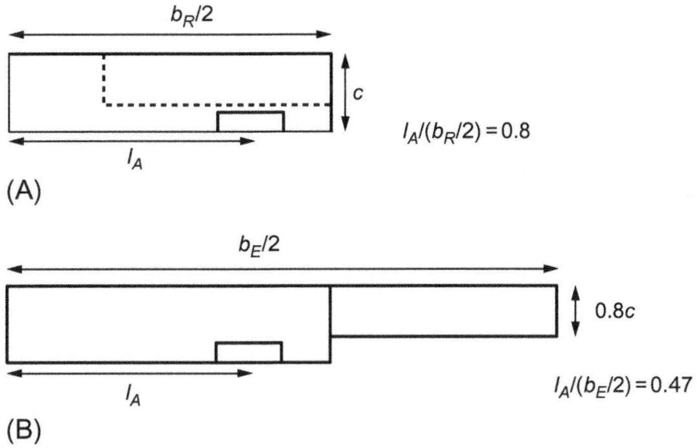

FIG. 9

Extendable part chord reduction and aileron efficiency loss: (A) fully retracted wing and (B) fully extended wing.

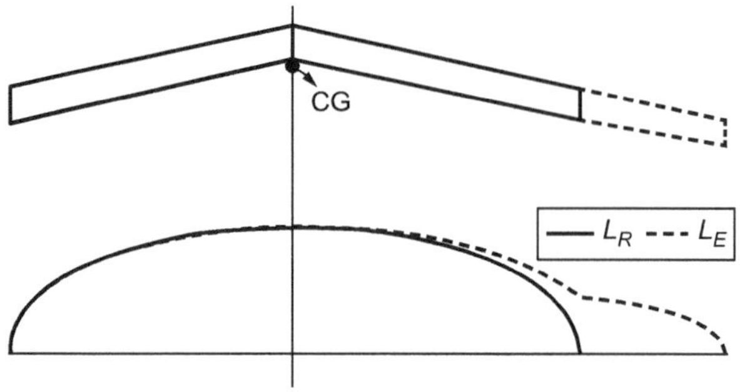

FIG. 10

Changes because of asymmetric span morphing.

Asymmetric span morphing then becomes a natural extension of the symmetric span morphing concept for two main reasons: (1) It is a relatively easy task to add another degree of freedom to a symmetric span morphing mechanism; (2) It replaces aileron devices for roll control.

This concept introduces complexity in the lateral dynamics of the aircraft: There are changes in the CG position; there are changes in the inertia tensor of the aircraft; and there is an asymmetric loading of the wing. All these changes contribute to the coupling of the rolling and yawing motions, and possibly of the pitching movement in case of swept wings (Fig. 10). Control laws need to be developed in order to assure aircraft stability during a rolling maneuver.

3.2.1 Actuation speed requirements

An immediate question that arises from using asymmetric span morphing for roll control is the actuation speed requirement. In order to start rolling the aircraft with a similar rate as that usually obtained with ailerons, the span morphing system must be able to produce a similar rolling moment coefficient.

Fig. 11 shows a simplistic view of the rolling moment generation with ailerons and asymmetric span morphing in a rectangular wing. When using ailerons, the rolling moment is generated by the changes

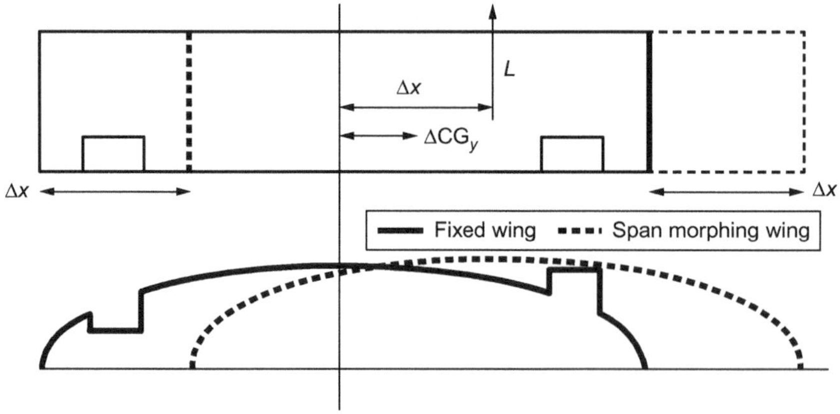

FIG. 11

Rolling moment generation with ailerons and asymmetric span morphing.

in lift along the ailerons span-wise extension, but the span morphing wing generates the rolling moment by shifting the lift distribution to one side of the aircraft.

If we assume a rough approximation of a rectangular lift distribution for a span morphing wing and a constant lift variation along the span-wise extension of the ailerons, we can establish a relationship between the span extension/retraction in each side of the span morphing wing with the dimensions of the wing, dimensions of the ailerons and with aircraft weight.

By assuming the same amount of extension/retraction (Δx), the span morphing rolling moment (RM_S) is:

$$\mathrm{RM}_S = L\Delta x, \tag{3}$$

where L is the total lift produced by the wing. The rolling moment caused by the ailerons (RM_A) is then:

$$\mathrm{RM}_A = 2\Delta L_A l_A, \tag{4}$$

with $\Delta L_A = \Delta \mathrm{CL}_A\, qcF_1 \cdot b/2$ and $l_A = F_2 \cdot b/2$, where $\Delta \mathrm{CL}_A$ is the magnitude of the CL variation due to aileron deflection, q is the dynamic pressure, c is the chord, F_1 and F_2 are factors defining the aileron length and the distance between the central point of the ailerons and the symmetry plane of the aircraft relatively to the half-span, respectively, and b is the wing span.

Setting both moments to be equal and assuming the maneuver starting from leveled static flight ($L = W$, with W being the weight of the aircraft) the following expression can be obtained:

$$\frac{\Delta x}{b} = \frac{\Delta \mathrm{CL}_A F_1 F_2}{\mathrm{CL}}. \tag{5}$$

Expression (5) gives some insight about what to expect in terms of wing span variation in order to achieve rolling moments similar to those produced by ailerons deflection:

- Heavier aircraft require lower span variations;
- Higher deflection, higher aileron length, and higher distance to the symmetry plane correspond to higher span variations, as expected.

As an example, for a cruise CL of 0.5, an aileron length factor F_1 of 0.2, an aileron distance factor F_2 of 0.8 and an aileron deflection causing a $\Delta \mathrm{CL}_A$ of 0.5, an extension/retraction of 8% (16% total wing movement) of the span is required. For a high speed CL of 0.1, the required $\Delta \mathrm{CL}_A$ to achieve the same rolling moment of the previous case is five times lower, if the weight is maintained, so the same 16% total wing movement is required.

If the CL is 0.25 because of an aircraft weight decrease instead of a speed increase, then the total wing movement doubles to 32%. If rolling inertia decrease is coupled to this weight decrease, then the maneuverability might be maintained without increasing wing movement. Nevertheless, actuation speed should be able to vary as significantly as the aircraft weight and as the required span variation, if the rolling moment has to be maintained.

3.2.2 CG position shifts and inertial effects

It should be noted that we did not account for the CG position variation in the previous simplistic analysis. The CG variation will tend to be higher as the structural mass being displaced while varying the span increases. If the wing mass is 10% of the total mass, the CG position (relatively to the aircraft reference frame) will be shifted toward the increasing span wing by 10% of Δx, therefore reducing the rolling moment generated by the same amount.

In addition to this effect, there is an increase in rolling inertia, affecting not only the rolling moment, but also reducing roll acceleration. It will also increase the coupling between lateral modes, as products of inertia appear in the inertia tensor.

Besides the changes in the inertia tensor, the maneuverability is also affected by the variation of the inertia tensor with time, obeying Newton's Second Law, particularly when returning to a steady state.

Assuming only rolling motion, the rolling moment may be expressed as:

$$\mathrm{RM} = I_x\dot{p} + \dot{I}_x p \Leftrightarrow \dot{p} = \frac{\mathrm{RM} - \dot{I}_x p}{I_x}. \qquad (6)$$

After starting a rolling motion to obtain an arbitrary bank angle, the aircraft needs to generate an opposing rolling moment in order to stop banking. To do so, the extended side of the wing needs to be retracted and vice versa. When doing it in an antisymmetric motion, the rolling inertia rate of change is affected only because of the CG shift. If the motion is not antisymmetric, the rolling inertia rate of change primarily contributes to the increase or decrease of the rolling acceleration, depending on the differential rate of change of the span between the two wing sides. The faster the span variation, the more significant the inertia rate of change in the rolling motion of the aircraft; this requires particular attention when developing the control laws associated with aircraft turns.

This also highlights that, similarly to what happens with aileron control, the rolling motion is initiated with a span variation in one direction and stopped with a span variation in the opposite direction.

The energy requirements of such a roll control mechanism, based on what was stated previously, might impair the benefits associated with symmetric span morphing, depending on the number of rolling maneuvers the aircraft performs during the missions.

4 IMPLEMENTATION CHALLENGES

As with other morphing concepts, span morphing presents challenges in achieving significant geometrical changes with acceptable mechanisms' weight and actuation energy. Significant amount of research has been directed toward the internal structure of the span morphing wings, while an equal amount of effort has been directed to the use of new materials able to comply with the strength and flexibility required for the skin of span morphing wings.

4.1 TELESCOPIC WINGS

The first attempts to obtain a span morphing wing involved telescopic wings, whose segments slide out of and into each other.

In general, this type of wing requires that the outer wing can accommodate the smaller segments inside. This implies that the airfoils used in the different segments cannot have arbitrary geometries. In particular, using highly cambered airfoils in the outer segments limits the chord of the inner segments because of geometrical interference. This condition would limit the wing design.

The skin of the inner segments, however, could be rigid, and traditional materials such as aluminum or composite skins could be used.

The load-bearing structure of the telescopic wings varies from:

- Monocoque wing segments supported by the overlapping between consecutive segments with the outer wing shell supporting the total loads. The span variation elements do not carry loads [9].
- Telescopic spars partially supporting the loads of the telescopic segments while acting as an actuated component for span variation [7,10].
- Telescopic wing-box, acting similarily to the telescopic spars with the difference that a wing-box structure is used [11].

The major difficulty with such structures is that, in order to maximize the inner segments area and span, large span-wise extents of the load-bearing outer wing are required, thus leaving little additional space available for structures such as spars and ribs to support the load as the inner segments are deployed; this leads to an increase in skin thickness to compensate for lack of supporting structural elements.

Telescopic wings, in general, also present discontinuities in the chord when transitioning from one segment to the other because of the thickness of the outer segments. These have the effect of increasing drag in small amounts. Alternative designs make use of stretchable skins to avoid this situation, at the expense of higher actuation energy [10,12].

The mechanisms used to cause the motion of the telescopic parts vary from pneumatic [9] to servomechanic, using rack and pinion [13] or belt and pulley [7] actuation elements.

4.2 HINGED STRUCTURES

Hinged structures such as the Zigzag wingbox [14] and other scissors-type mechanisms [15–17] (see Fig. 12) have been proposed as means for actuation and bearing the loads simultaneously, commonly associated with stretchable skins.

The variations in the structural stiffness, due to the structures' internal moving parts, increase the complexity of the design and the probability of failure, and usually require a higher number of actuators compared to the telescopic wing.

The association of these mechanisms with stretchable skins also implies extra actuation energy because of the required deformation of the skin during the span morphing.

4.3 TWIN SPARS

The concept of twin spars is based on two parallel spars running along holes in the wings ribs [12]. Each wing side is connected to one of the spars and complete aircraft symmetry occurs only in fully retracted configuration. As the span is increased, each wing side has one spar in a slightly different chord-wise

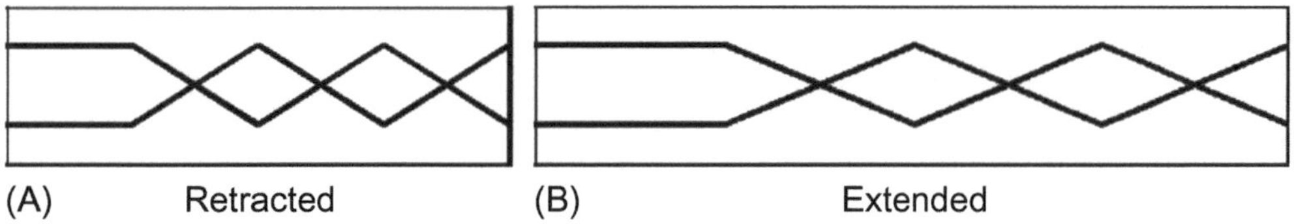

FIG. 12

Scissors type mechanism concept: (A) retracted and (B) extended.

position. Thus, the deformation of the wing will be slightly different from one side to the other, and in the case of a more flexible structure, asymmetries can lead to difficulties in flight control or slight penalties in drag because of extra control surface deflections for trimming.

4.3.1 Elastic skins

The span morphing concept frequently requires an elastic skin, able to achieve strains up to 100%. This is currently possible only with elastomeric skins, which are flexible enough to achieve such strains.

Problems arise when the skin in the lower span configurations is not strained when it should be to withstand the aerodynamic loading while maintaining the airfoil shape. This jeopardizes the aerodynamic performance.

A possible solution is to prestress the skin before bonding it to the wing structure. The stress of the skin helps reducing skin deformation due the aerodynamic loading. As a consequence, the actuation force requirements increase, as will the strain levels of the skin when fully extended. Another possibility is to reinforce an elastomeric skin with fibers of different materials, such as glass or carbon, along the direction perpendicular to the span-wise direction.

Another drawback of elastomeric skins is that they generally preserve the volume, which corresponds to a nearly 0.5 Poisson ratio. A consequence of this is that the chord and thickness of the airfoils between ribs will decrease if not supported by some kind of structure. To tackle this issue, zero Poisson ratio structures have been developed to support the elastomeric skin, while being able to increase the span significantly [12,18].

Durability and damage tolerance is also an issue with flexible skins.

5 CONCLUSIONS

Span morphing may be reasonably used in multiple-role aircraft that require significantly different flight conditions in a mission, or alternatively, are thought to perform significantly different missions.

The concept does not seem advantageous for the substitution of flaps and slats as a mean to reduce stall speed and take-off/landing distances, because it would require a great amount of span variation and lifting area increase in order to produce the same CL variations as the flaps-slats systems.

It does, however, seem advantageous for use in low-speed flight segments, expanding the endurance, and at maximum speed flight segments, increasing the maximum speed, when compared to an aircraft designed for maximum range at an intermediate speed.

The upscaling of the concept presents some difficulties because span morphing greatly reduces the available space inside the wing, which for large-scale aircraft would be prohibitive. Nevertheless, results show a trend of potential benefits with a limited (20%) amount of span increase, for a regional transport aircraft when flown in off-design conditions for a significant part of the mission.

The main drawback is the structural weight penalty, which directly affects range, climb rate and climb angle. Defined multi-mission criteria should be used to evaluate the performance parameters in the context of the group of missions to be performed.

Actuation energy is also a drawback, particularly when scaling up the concept, because it becomes a larger percentage of the engine output energy. This means that either power is drawn from the engine or that another source of energy should exist on board of the aircraft. Either way, it is extra weight degrading the aircraft performance.

From this point of view, asymmetric span morphing for roll control becomes less attractive, because of the higher frequency of actuation compared to symmetric span morphing. Nevertheless, the concept is feasible for roll control although of more complex application. Moreover, in case of swept wings, it increases the coupling between flight modes, not only laterally but also longitudinally.

All of the proposed mechanisms described here present implementation challenges that still need further research. Telescopic wings with rigid skins require optimization in order to minimize the weight of the structure. Hinged structures increase the amount of moving parts and the probability of failure. Twin spars introduce asymmetry in the wing that must be accounted for. Flexible skins require reinforcements or prestress, increasing the actuation energy and power, among other issues.

ACKNOWLEDGMENTS

The authors acknowledge the Portuguese Science Foundation FCT, through IDMEC, under LAETA, Project UID/EMS/50022/2013. This work was partially funded by the projects SMORPH and POCTI-EME-61587-2004. The research presented was also funded by the EU Seventh Framework Programme Project NOVEMOR (Novel Air Vehicle Configurations—From Fluttering Wings to Morphing Flight) under a Grant Agreement No. 285395. Additional support was also provided by the National Science and Engineering Research Council in Canada.

REFERENCES

[1] T. Weisshaar, Morphing Aircraft Technology – New Shapes for Aircraft Design, RTO-MP-AVT-141, Neuilly-sur-Seine, France, 2006.

[2] V. de Brederode, Fundamentos de Aerodinâmica Incompressível, first ed., 1997 (Author's Edition, in Portuguese).

[3] F. Afonso, Performance Based Multidisciplinary Design Optimization of Energy Efficient and Novel Aircraft Configurations (Ph.D. Thesis), Instituto Superior Técnico, Universidade de Lisboa, Portugal, 2015.

[4] F. Afonso, J. Vale, É. Oliveira, F. Lau, A. Suleman, Non-linear aeroelastic response of high aspect-ratio wings in the frequency domain, Aeronautical J. (2017), https://doi.org/10.1017/aer.2017.29.

[5] J. Vale, A. Leite, F. Lau, A. Suleman, Aero-structural optimization and performance evaluation of a morphing wing with variable span and camber, J. Intell. Mater. Syst. Struct. 22 (2011) 1057–1079, https://doi.org/10.1177/1045389X11416031.

[6] J. Vale, F. Lau, A. Suleman, Energy efficiency studies of a morphing unmanned aircraft, J. Aeronaut. Aerosp. Eng. 2 (2013) 122, https://doi.org/10.4172/2168-9792.1000122 (open access).

[7] J. Vale, Development of Computational and Experimental Models for the Synthesis of Span and Camber Morphing Aircraft Technologies (Ph.D. Thesis), Instituto Superior Técnico, Universidade de Lisboa, Portugal, 2012.

[8] J. Vale, F. Afonso, F. Lau, A. Suleman, Performance based MDO of a joined-wing regional transport aircraft. In: 56th AIAA/ASCE/ AHS/ASC Structures, Structural Dynamic, and Material Conference, AIAA Science and Technology Forum and Exposition (SciTech 2015), Kissimmee, FL, USA, January 5–9, 2015, (AIAA Paper 2015-0696). https://doi.org/10.2514/6.2015-0696.

[9] J. Blondeau, J. Richeson, D.J. Pines, Design, development and testing of a morphing aspect ratio wing using an inflatable telescopic spar, In: 44th AIAA/ASME/ASCE/AHS/ASC Structures, Structural Dynamics and Materials Conference, Norfolk, VA, 2004.

[10] K.S.B. Woods, M.I. Friswell, The adaptive aspect-ratio morphing wing: design concept and low fidelity skin optimization, Aerosp. Sci. Technol. 42 (2015) 209–217.
[11] P. Gamboa, P. Santos, Telescopic wing-box for a morphing wing, In: 24th AIAA/AHS Adaptive Structures Conference, San Diego, CA, USA, 2016.
[12] R.M. Ajaj, M.I. Friswell, M. Bourchak, W. Harasani, Span morphing using the GNATSpar wing, Aerosp. Sci. Technol. 53 (2016) 38–46.
[13] J. Mestrinho, P. Gamboa, P. Santos, Design optimization of a variable-span morphing wing for a small UAV, In: 52nd AIAA/ASME/ASCE/AHS/ASC Structures, Structural Dynamics and Materials Conference, 19th AIAA/ASME/AHS Adaptive Structures Conference,Denver, Colorado, 2011.
[14] R.M. Ajaj, et al., The zigzag wingbox for a span morphing wing, Aerosp. Sci. Technol. 28 (2013) 364–375.
[15] P. Aleixo, J. Vale, P. Gamboa, F. Lau, A. Suleman, Design and testing of a morphing wing for an experimental UAV, in: The Applied Vehicle Technology Panel Symposium (AVT-146), 2007.
[16] J.J. Joo, B. Sanders, T. Johnson, M.I. Frecker, Optimal actuator location within a morphing wing scissor mechanism configuration, in: Smart Structures and Materials 2006: Modeling, Signal Processing, and Control, vol. 6166, SPIE, San Diego, CA, 2006.
[17] T. Johnson, M. Frecker, M. Abdalla, Z. Gurdal, D. Lindner, Nonlinear analysis and optimization of diamond cell morphing wings, J. Intell. Mater. Syst. Struct. 20 (2009) 815–824.
[18] R.D. Vocke, C.S. Kothera, B.K. Woods, N.M. Wereley, Development and testing of a span-extending morphing wing, J. Intell. Mater. Syst. Struct. 22 (2011) 879–890.

CHAPTER

5 ADJOINT-BASED AERODYNAMIC SHAPE OPTIMIZATION APPLIED TO MORPHING TECHNOLOGY ON A REGIONAL AIRCRAFT WING

Antoine Dumont
ONERA, Meudon, France

CHAPTER OUTLINE

Morphing Wing Technologies. https://doi.org/10.1016/B978-0-08-100964-2.00005-8

NOMENCLATURE

$\boldsymbol{B(I)}$	Bézier curve of order three given by its four control points P_0, P_1, P_2, P_3
$\boldsymbol{Cst_j}$	jth constraint function of the optimization problem
$\boldsymbol{D}$	bounded domain of fluid
$\boldsymbol{E}$	total energy of the flow (J kg^{-1})
$\boldsymbol{f}$	a function representing either Obj or Cst_j
$\vec{i}$	unit vector in the freestream flow direction
$\boldsymbol{M}$	Mach number
$\boldsymbol{n}$	normal unit vector of Γ of components $(n_x, n_y, n_z)^T$ in R_a
$\boldsymbol{Obj}$	objective function that the optimization process tends to minimize
$\boldsymbol{r}$	perfect gas constant (=287 J K^{-1})
$\boldsymbol{R}$	discrete residual of the flow equations
$\boldsymbol{R_a}$	reference coordinate system
$\vec{V}, V$	flow velocity vector of components $(u, v, w)^T$ with respect to R_a (m s^{-1})
$\boldsymbol{W}$	discrete field of conservative variables
$\boldsymbol{X}$	computational mesh around the rotor
$\boldsymbol{\alpha}$	vector of shape parameter of size N whom components are $(\alpha_1, \ldots, \alpha_N)^T$
$\boldsymbol{\gamma}$	ratio of specific heats
$\boldsymbol{\lambda}$	discrete adjoint vector associated to function f
$\boldsymbol{\rho}$	flow density (kg m^{-3})
$\bar{\bar{\tau}}$	shear stress tensor
$\bar{\bar{\tau}}_R$	Reynolds stress tensor
$\boldsymbol{\Delta H}$	stagnation enthalpy relative to its freestream value (J kg^{-1})
$\boldsymbol{\Delta s}$	variation of entropy relative to its freestream value (J K^{-1} kg^{-1})
$\boldsymbol{\Gamma}$	external surface of volume D
$\boldsymbol{\Phi}$	heat flux vector
$\boldsymbol{\Phi T}$	turbulent heat flux vector
$\mathcal{W}$	vector of conservative variables
∞	subscript for freestream state value

1 INTRODUCTION

In principle, the morphing technology integrated into an aircraft wing concerns an automated shape adaptation to reach the best aerodynamic efficiency given a flying condition. A mechanism, usually inside the wing, adapts the shape following a set of inputs. This process could be seen as an ideal candidate to apply a shape optimization to find the optimal inputs for any flying conditions encountered by the aircraft. This chapter focuses on the resolution of a local shape optimization problem using CFD tools to compute the flow around the wing. The problem is said to be local because it only intends to find a local optimum close to the starting point of the optimization. The addressed methods rely mainly on gradient-based optimization techniques in which, given the sensitivities of the problem functions with respect to the parameters, a direction of improvement is found in which a search for the optimum will start. The process will then iterate until convergence. There are different techniques to compute the sensitivities, but in this case, we chose the adjoint method because of its remarkable ability to compute sensitivities at a constant CPU cost, regardless of the number of shape parameters. In this chapter, the focus will be on an application of adjoint-based aerodynamic optimization performed within the EU project SARISTU, that aimed (in a dedicated task) to design and test a morphing concept based on trailing edge camber variation for a regional aircraft wing. The chapter is organized into five main sections. First, the handling of the CFD mesh deformation taking account of morphing shape changes will be presented. An overview of the ONERA in-house developed tools will follow with reference to the CFD solver *elsA* [1] and the postprocessing chosen in this study that relied upon a far-field drag breakdown based on ONERA *ffd72* software [2]. Then the adjoint method will be introduced, from its equations to their resolution with the *elsA* software. A short introduction to the local shape optimization problem will be presented, with a focus on the feasible directions used in the optimizer DOT [3] chosen for the applications in this work. A test of the method is presented using the classical Rosenbrock's function with the addition of a simple constraint. The chapter ends with the presentation of two optimization process applications to the wing of the SARISTU project where two flying conditions are considered for optimizing the shape of the trailing edge device. Aerodynamic improvement is discussed in the light of the pressure distribution, spanwise wing-loading and far-field drag breakdown comparison with respect to the reference clean wing configuration. Expectations from such morphing technology are finally highlighted at the end of the chapter.

2 HANDLING OF MORPHING SHAPE CHANGES IN A CFD CONTEXT

When considering CFD computations in an aerodesign process, one of the most important issues is the generation of the computational mesh fitting the shape to study. In the special case of morphing technology, it appears very important to produce a deformed mesh following the wing shape changes in a robust manner without impacting the mesh quality. After a short introduction of the context of the study, the whole process of mesh deformation will be presented. It starts from the modeling of the kinematics of the morphing system (which, in this particular case, is dedicated to trailing edge camber variation). Then, from any inputs on the morphing system, a field of displacements on the surface of the wing is generated and then propagated into the three-dimensional computational grid.

2.1 CONTEXT OF THE STUDY

The mechanism considered in this study, based on the results of the SARISTU EU project, can be seen in Fig. 1 as a three-hinged system[1] able to change the local angle of the mechanism parts from one main input (taken, for example, from the translation of a hydraulic actuator). The mechanism presented in Fig. 1 shows one of the elementary systems fitted into the wing that composes the global active trailing edge device (ATED). It is worth noting that this system can be distributed along the span over significant length. This work considers two regions along the wing span for the installation of the morphing system: one on the inboard wing section (before the kink of the wing planform) and a larger one on the outer wing region. The wing regions interested by the morphing trailing edge systems are depicted in Fig. 2. The target of the study is to determine which shapes maximize the aerodynamic efficiency in the design space made of the six hinges angles (three hinge angles per system). For this, it is essential for one single mechanism system to realize any hinge rotation angle defined as a user input and to generate the CFD mesh corresponding to the resulting geometry of the system. This section

[1]For more details on morphing system mechanics, please refer to Chapter 17.

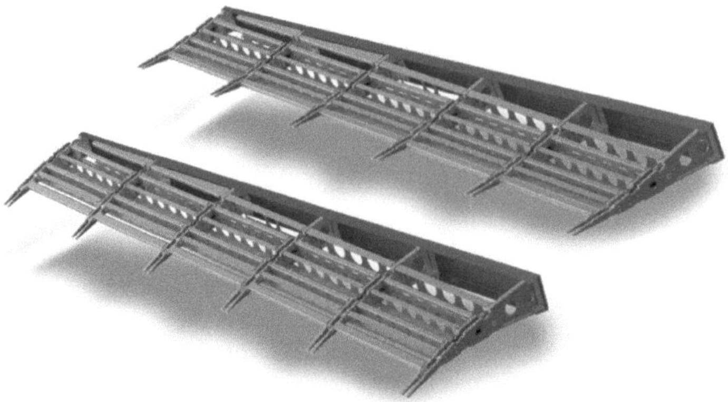

FIG. 1

Morphing mechanism presentation.

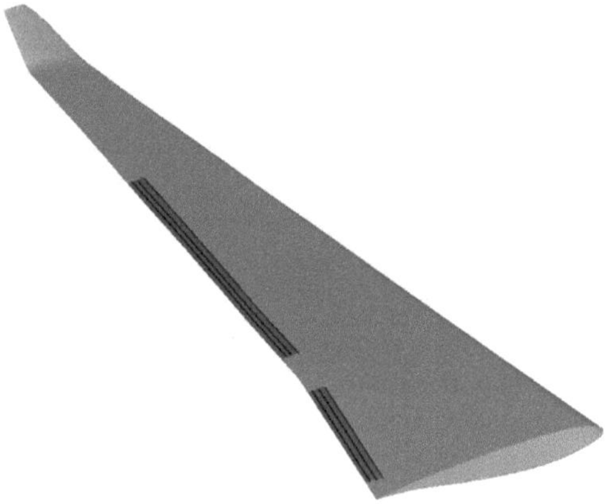

FIG. 2

Implantation of the two trailing edge morphing systems.

will not deal with kinematic design as a system-level mechanical engineering design task. Instead, a predesign capability for such morphing systems is proposed in order to maximize aerodynamic efficiency using CFD in multiblock structured grids for the aerodynamic flow model, as shown in Fig. 3.

The first goal of this section is to illustrate the 2D geometrical model of the ATED shape retained at ONERA to model this mechanism. The second goal is to present the process allowing for the generation of the three-dimensional CFD mesh of the wing, from the 2D ATED model to the three-dimensional surface mesh of the wing and finally to the volume mesh.

2.2 DISCRETE MODEL OF DISPLACEMENT FIELD AT THE TRAILING EDGE

When examining the detailed design of the morphing system in a 2D section as in Figs. 1 and 4, one can see that the mechanism is made of rigid parts that rotate with respect to each other, and of different layers of flexible skin. These skins are made of special materials developed by the Fraunhofer Institute,

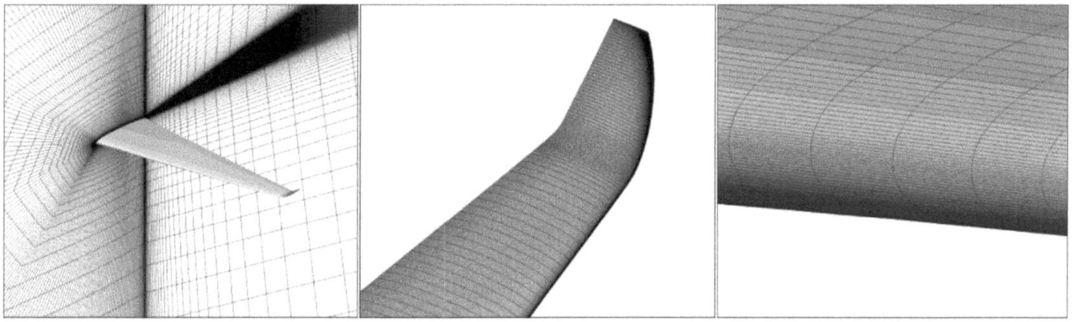

FIG. 3

Multiblock structured mesh of the wing.

FIG. 4

Details description of the skin of the morphing system.

partner of the SARISTU project. The layers play different roles. The external skin, Fig. 4, is made of hard and soft segments; soft segments are positioned across the hinges and are fully stretchable in order to accommodate the deformation induced by morphing (for more technical details on skin arrangement please refer to Chapters 7 and 17).

From this knowledge of the morphing mechanism, it was decided to create a 2D discrete kinematic model enabling the generation of the displacement field at the wing surface mesh due to the three input angles characterizing the rotation in the hinges. This surface mesh displacement represented the input for the 3D volume deformation technique.

For both inner and outer wing trailing edges, a fine discretization of the profile shape, going from the trailing edge over 10% of the mean aerodynamic chord of the wing, was extracted by the CAD using CATIA tools. The two extractions were done in the midsections of the morphing systems in a plane normal to the trailing edge line. An example of such a discretized geometry is presented in Fig. 5 with the upper dotted line representing the suction side, and the lower dotted line representing the pressure side.

Inspired by the detailed definition of the mechanism, the hinge axis position (H1, H2, H3, with H1 being the most upstream hinge) and flexible skin region were placed on a 2D model (for example, for the inner wing system, as shown in Fig. 5). In this study, hinges axes were supposed to be parallel to the trailing edge. Their positions on the wing planform were deduced from drawings of the mechanism. Distances of the hinges from the trailing edge resulted, for H1: 0.320 m, for H2: 0.225 m, and for H3: 0.138 m. Vertical positions of the hinges were defined as the center of the circle that fits into the profile section at this chordwise position, as presented in Fig. 5.

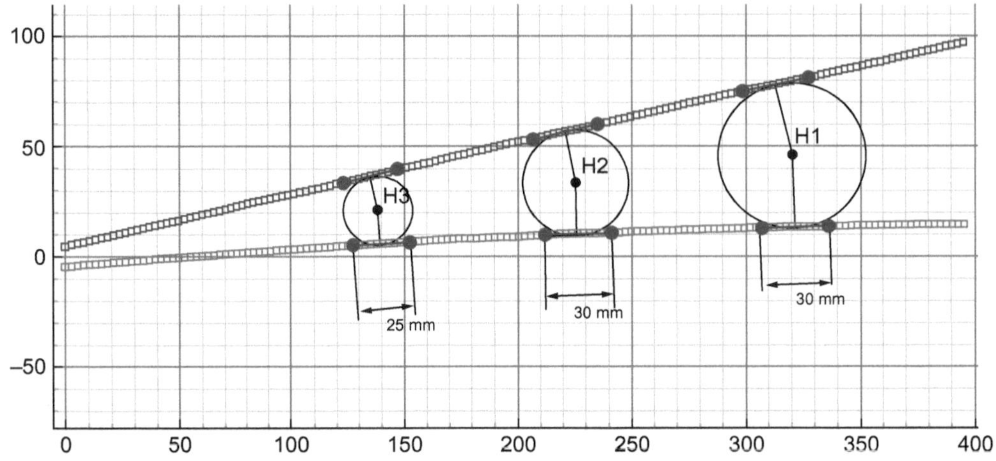

FIG. 5

Definition of flexible and solid regions.

For the flexible skin region close to each hinge, and delimited by two filled dots in Fig. 5, it was assumed an extension of 25 mm across the hinge H3, and of 30 mm across the hinges H1 and H2. It is worth noting that these flexible regions are centered on the intersection point of the skin with the circle included in the wing profile at the given hinge axis position, as presented in Fig. 4, which may be slightly different from the original mechanism. (This work had been done while the mechanism design was still underway.) The 2D model must then deal with two types of displacement: the one made of solid parts (where the thin skin is attached with rivets), and the ones made with thick flexible skin which fills the gap of the mechanism close to the hinges axes.

For all regions outside the flexible ones, the motion corresponds to the composition of rotations around the hinges axes. The strategy is to successively apply rotations to the points defining the geometry, depending on their position about the three-hinge axis. For a point downstream of H3, the geometric transformation will be a rotation around H3, then around H2, and finally around H1. If the point to transform is between H3 and H2, the transformation consists of two rotations, first around H2 then around H1. If the point is between H2 and H1, it is only concerned with one rotation around H1. It is worth noting that, given any set of rotation angles around the hinge axis (during the optimization process for example), only the relative angle variation from the reference ATED shape is considered. By using this strategy, only the initial positions of a hinge axis are used, and no intermediate hinge axis position (for H1 and H2) is necessary or computed, which appeared to be very convenient.

Bezier curves were used to geometrically describe the flexible regions. They can ensure smooth shape evolution while respecting tangency condition at their extremities. It was decided to work with third degree curves to have only four control points to handle. The parametric expression of a curve as a function of the control point position is recalled in Eq. (1):

$$B(t) = P_0(1-t) + 3P_1t(1-t)^2 + 3P_2t^2(1-t) + P_3t^3 \tag{1}$$

Concerning the placement of the control points in the 2D model, Fig. 6 illustrates the build-up process. P_0 and P_3 are placed at the extremities of the flexible zone, and P_1 and P_2 are uniformly distributed in

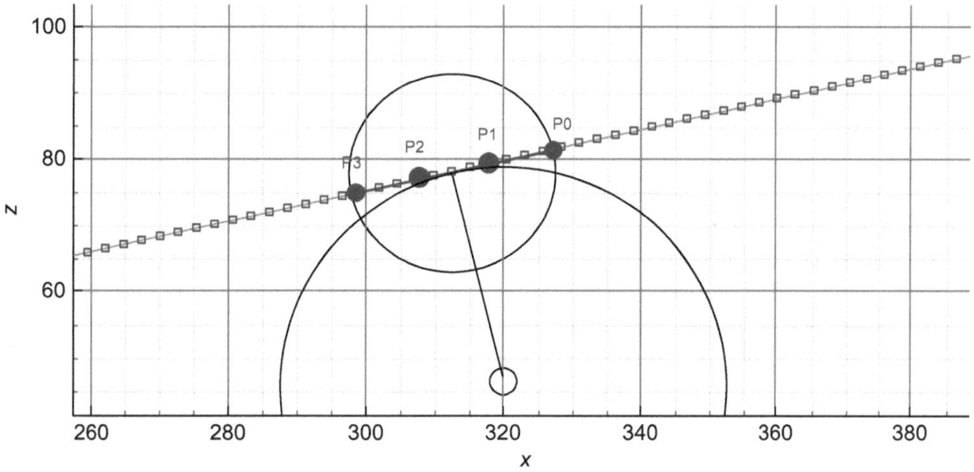

FIG. 6

Bezier control points in a flexible region.

the x axis direction and placed in the z position in order to ensure tangency of the segment P_0P_1 with the rigid part at its right and tangency of P_3P_2 with the rigid part at its left, as can be seen in Fig. 6. Once the control points are set, an association of the initial geometry point M is done following a simple interpolation technique along the x axis:

$$t_M = \frac{x_M - x_{P_3}}{x_{P_3} - x_{P_0}} \tag{2}$$

Once this parametric position is computed on the Bezier curve, it is possible to start the deformation process. The control points are transformed according to the rotation of the hinge with the peculiarity that the first two control points (P_0 and P_1) will rotate as the rigid geometry on the right and the two following points (P_2 and P_3) will rotate as the rigid geometry on the left. Once the hinge rotation motion is applied to the control points, the new position of a point M belonging to the initial geometry is obtained by replacing the initial t_M value, in the Bezier curve equation, considering the new control point locations.

From the displacement of rigid and flexible zones presented above, the result is a 2D displacement field for the fine discretization of the trailing edge geometry extracted from the CAD model. An example is presented in Fig. 7, where each angle of rotation was set at +5 degrees on each hinge axis (when positive angle is applied, it leads to a downward deflection). The position of the axis can be guessed on the deformed shape, but the smoothness and tangency between the rigid and flexible parts are preserved.

Once the 2D model is properly set to give the displacement field in one wing section, given a set of rotation angles for all three hinges, the next step is to use this displacement field to transform the clean wing surface mesh to the corresponding morphing configuration. This process is made node-by-node as presented in Fig. 8. For each point on the surface mesh, called M for example, a plane (P) normal to the trailing edge line and containing the node M is considered. This plane intersects the trailing edge at the point H. A point B is set as a reference on the trailing edge line. In the plane (P), the displacement of the vector $\overrightarrow{HM}$ follows exactly the 2D field displacement

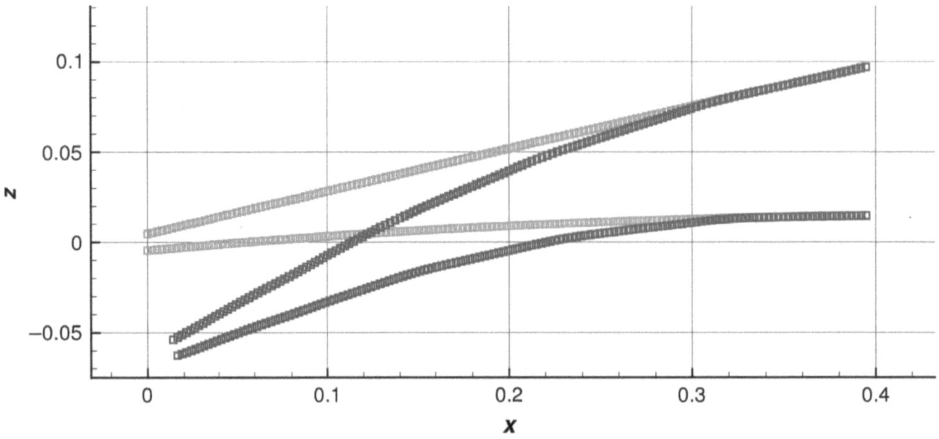

FIG. 7

Example of 2D trailing edge model deformation with a rotation of +5 degrees in all hinges.

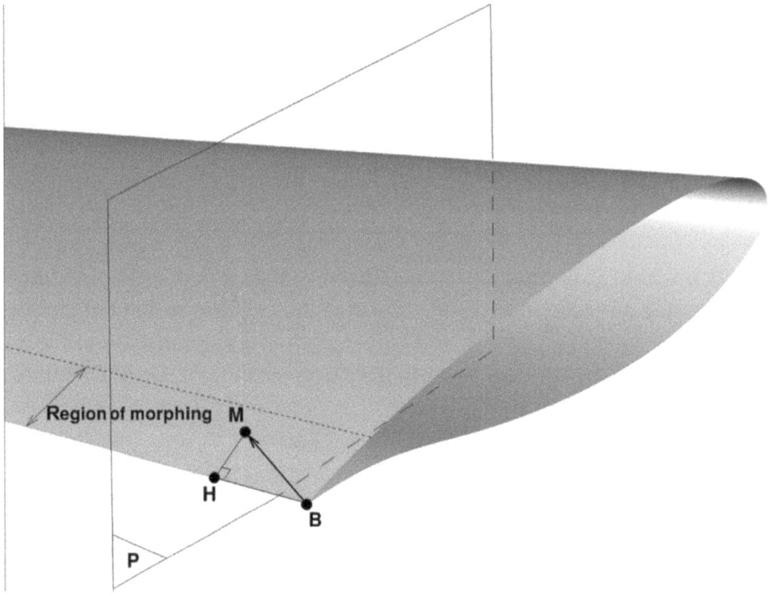

FIG. 8

Strategy for surface mesh deformation.

computed by the 2D model presented above. It is thus straightforward to find the displacement of M, in this case by a linear interpolation method between the two points of the discrete model around M (for example, in terms of the horizontal axis component in plane (P)). All the points outside the morphing area are left undisturbed. The input of the point B on the trailing edge line is useful to control the spanwise extent of the morphing mechanism. In fact, the distance [HB] is very useful to define the length of the mechanism and to represent what happens in the extremities of the real

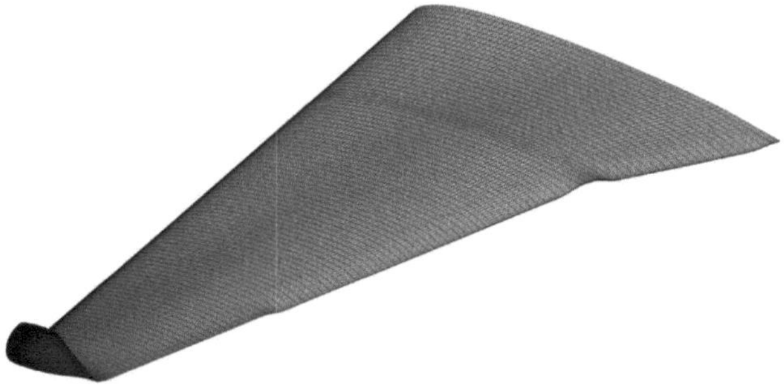

FIG. 9

Surface mesh deformation with an angle of rotation of +5 degrees in all hinges.

mechanism. Coherently to the implemented stretchable skin solution, it has been chosen to define the shape by a linear interpolation technique between the reference shape and the modified one.

The spanwise extent over which the linear interpolation occurs has been set up to 0.4 m on each side of the morphing regions. An example of the process is presented in Fig. 9, where a rotation of +5 degrees is applied for all the hinges of the two morphing mechanisms. For information on the extension of morphing regions, the morphing of the inner wing is applied between $y = -0.1$ m and $y = -3.80$ m. The morphing of the outer wing is applied between $y = -4.15$ m and $y = -11.60$ m (the wing is oriented along negative y axis).

2.3 3D CFD MESH DEFORMATION TECHNIQUE

Having set a way to access to the surface mesh deformation on all computational mesh nodes defining the wing surface, one can now propagate this information into the volume of the 3D CFD mesh. In this case, a process based on a combination of analytical and interpolation techniques has been used. This method is actually used within the ONERA CFD flow solver *elsA* [1] to handle mesh deformation especially in the aeroelastic context in what is called the *elsA*/*Ael* module. The aim of this section is to give to the reader a view on how the process to deform meshes works and to show its use on a morphing application.

From the input of the displacements on the wing surface mesh nodes, boundary conditions (like on the far-field surfaces where no displacement is assumed), and some constraints (for example on the symmetry plane where displacement is constrained into the x-z plane), the present deformation technique produces a new computational grid satisfying all given inputs and conditions. The principle of the mesh deformation technique is first to compute the displacement, thanks to an analytical formulation on all mesh points belonging to each face of each block composing the computational structured CFD grid, then to apply (for all interior points of a given block) a transfinite interpolation technique, taking as boundary conditions the displacements of the nodes of all block faces obtained previously.

The first analytical technique, commonly called "integral method," is based on the following expression:

$$\delta P = \frac{\sum_{M \in \text{all } BC} \frac{\delta M}{PM}}{\eta \sum_{M \in \text{all } BC} \frac{1}{PM}} \tag{3}$$

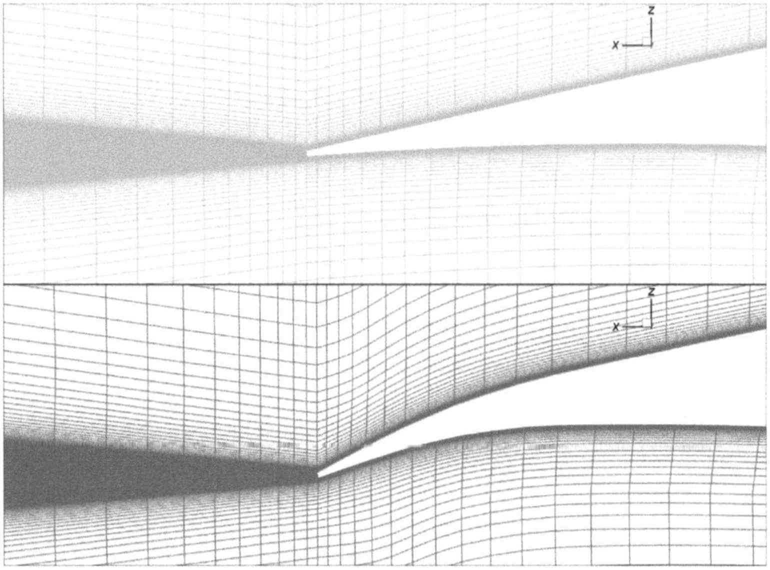

FIG. 10

View of the 3D mesh deformation in the middle section of the inner morphing mechanism with a rotation of +5 degrees in all three hinges.

where M is a point on the physical boundaries where a displacement δM is imposed, P a point on the block boundary where no condition is imposed and for which computed displacement is δP, PM is the distance between these two points, and η a damping function whose value is close to 1 near a wall where a nonzero displacement is imposed and close to ∞ near a fixed boundary. Once the block boundaries displacement is computed, a Trans-Finite Interpolation technique (TFI) is used to compute the interior point displacements. Based on a parametrization with respect to each topological direction of all interior points of a block, the technique uses the displacement of the boundaries to compute the displacement of the interior points based on their position inside the block. A detailed presentation of the complete equations of the TFI technique is out of the scope of this section, and the reader may refer to the handbook on grid generation by Thompson et al. [4] which presents the method in detail.

The application of the method to the 3D mesh around the wing considering the morphing configuration with a rotation of all hinges by +5 degrees is presented in Fig. 10, which shows a cut in a constant y plane on the inner wing. The method proved to be quite robust on this particular application; the only weakness observed is its inability to preserve the initial angle made by the mesh line starting from the surface on the morphing region, but this issue has not been critical on this case.

3 CFD EVALUATION AND FAR-FIELD DRAG ANALYSIS OVER A WING EQUIPPED WITH A MORPHING SYSTEM

Now that the computational grid is able to handle any morphing settings on the wing trailing edge as presented before, let's present the aerodynamic computation and postprocessing tools used for this study. First the RANS finite-volume approach within the ONERA CFD code *elsA* [1] will be introduced.

In a second part, the postprocessing using the far-field drag extraction tools *ffd72* [2] developed in the Applied Aerodynamics Department of ONERA will be illustrated, and its ability to extract a physical drag breakdown from the CFD solution will be shown to provide rich and useful information for aerodynamic design.

3.1 FINITE-VOLUME SOLVER FOR THE RANS EQUATIONS IN *elsA*

The wing is embedded within a sufficiently large fluid domain D bounded by the surface Γ of outward unit normal $n = (n_x, n_y, n_z)^{\mathrm{T}}$ expressed in R_a, a reference coordinate system attached to the wing. The velocity of the flow with respect to R is given by $V = (u, v, w)^{\mathrm{T}}$. The governing flow equations given in their conservative form considering the RANS model are given by:

$$\frac{\partial}{\partial t}\int_{\mathcal{D}} \mathcal{W} dD + \oint_{\Gamma} (F_C - F_D) d\Gamma = 0 \tag{4}$$

where, $\mathcal{W}$ is the vector of conservative variables:

$$W = (\rho, \rho u, \rho v, \rho w, \rho E)^{\mathrm{T}} \tag{5}$$

F_C is the convective flux vector:

$$F_C(W, n) = \begin{pmatrix} \rho V \cdot n \\ \rho u V \cdot n + p n_x \\ \rho v V \cdot n + p n_y \\ \rho w V \cdot n + p n_z \\ \rho E V \cdot n + p V \cdot n \end{pmatrix} \tag{6}$$

and F_D is the diffusive flux vector:

$$F_D(\mathcal{W}, n) = \begin{pmatrix} 0 \\ (\overline{\overline{\tau}} + \overline{\overline{\tau}}_R) \cdot n \\ (\overline{\overline{\tau}} + \overline{\overline{\tau}}_R) \cdot V \cdot n - (\Phi + \Phi_T) \cdot n \end{pmatrix} \tag{7}$$

$\overline{\overline{\tau}}$ and $\overline{\overline{\tau}}_R$ are, respectively, the viscous and turbulent stress tensor and Φ and Φ_T are, respectively, the heat flux and turbulent heat flux vectors.

The RANS equations are discretized using a finite volume method with cell-centered evaluation of conservative flow variables. They are solved iteratively using the elsA CFD code developed at ONERA [1]. Appropriate numerical schemes are selected in order to ensure consistency with the development of the discrete adjoint solver presented in Section 4. F_C is discretized using Roe's flux difference formula [5] and the MUSCL approach with Van Albada limiting function [6]. For the diffusive flux F_D, a centered formula is used; velocity and temperature gradients were evaluated at the center of each cell interface. Classical Boussinesq assumption is used for the Reynolds stress tensor. The turbulent viscosity is calculated with the Spalart-Allmaras model [7]. Convergence to steady state is performed using an implicit time integration algorithm based on a Backward-Euler scheme with locally varying intervals. An adiabatic no-slip condition is applied on the wing surface. A symmetry plane is placed at the more inboard section of the wing, and a nonreflecting condition is used for all other far-field boundaries, where a steady infinite state is imposed outside the boundary surface.

The residual form of the finite volume discretization of the RANS equations is denoted as:

$$R(W,X)=0 \tag{8}$$

where W denotes the cell-centered state variables and X the mesh of the computational domain.

3.2 FAR-FIELD DRAG EXTRACTION TOOL

The far-field drag formulation in *ffd72* is derived from that introduced by Van der Vooren [8]. Details of ONERA's so-called one vector variant can be found in Ref. [2]. This formulation uses the following quantities:

$$\Delta\overline{u}=u_\infty\sqrt{1+2\frac{\Delta H}{u_\infty{}^2}-\frac{2}{(\gamma-1)M_\infty{}^2}\left(\exp\left(\frac{\Delta s}{r}\right)^{\frac{\gamma-1}{\gamma}}-1\right)}-u_\infty \tag{9}$$

$$\vec{f}_i^*=-\rho(u-u_\infty-\Delta\overline{u})\vec{V}-(p-p_\infty)\vec{i} \tag{10}$$

The scalar quantity $\Delta\overline{u}$ has the dimension of a velocity and the vector $\vec{f}_i^*$ that of a stress.

An exact near-field balance reads:

$$D_p=D_{vp}+D_w+D_i+D_{sp} \tag{11}$$

where D_p is the pressure drag, D_{vp}, the viscous pressure drag, D_w, the wave drag, D_i, the induced drag and D_{sp}, the spurious drag. Finally, the viscous drag is defined by:

$$D_v=D_{vp}+D_f \tag{12}$$

where D_f is the friction drag, overall near-field drag is:

$$D_{nf}=D_p+D_f \tag{13}$$

And the physical (i.e., without spurious drag) far-field drag:

$$D_{ff}=D_{vp}+D_w+D_i+D_f \tag{14}$$

Far-field extraction relies on the definition of control volumes V_V, V_W, V_I, for the integration of viscous pressure, wave, and induced drag respectively. The boundary of V_W will be denoted S_W, and the outer boundaries of V_V and V_I (whose inner boundary is the fluid/solid interface S_A), S_V and S_I.

Expressions for the far-field components are:

$$D_{vp}=-\iint_{S_V}\left(\vec{f}_i^*\cdot\vec{n}\right)dS+D_p \tag{15}$$

$$D_w=-\iint_{S_W}\left(\vec{f}_i^*\cdot\vec{n}\right)dS \tag{16}$$

$$D_i=\iint_{S_I}\left(\vec{f}_i^*\cdot\vec{n}\right)dS \tag{17}$$

Spurious drag is defined as:

$$D_{sp}=D_{nf}-D_{ff} \tag{18}$$

As an example, the integration volumes used for wave drag and viscous pressure drag evaluation at cruise condition for the reference wing of the SARISTU project are presented in Fig. 11.

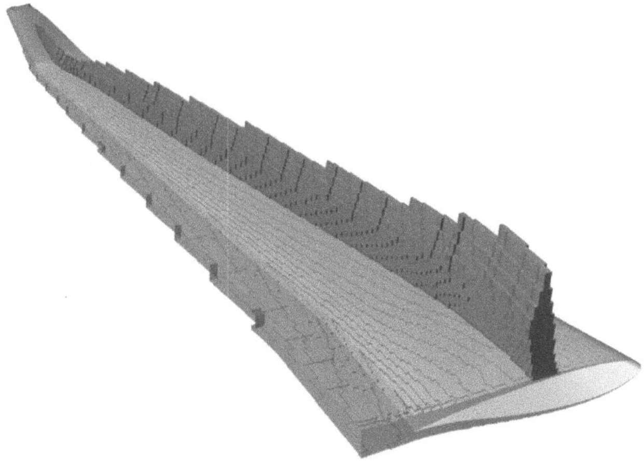

FIG. 11

View of the integration volumes to evaluate wave drag and viscous drag.

4 SENSITIVITY ANALYSIS USING A DISCRETE ADJOINT OF THE RANS EQUATIONS

The wing morphing system is defined by a vector of parameters $\alpha = (\alpha_1, \ldots, \alpha_N)$ representing the hinges rotations and allows modification of the shape of the wing. The local optimization process will explore the design space of size N, to find a vector α^* that will maximize or minimize an objective function (Obj) given a set of constraint functions, of size N_c (Cst_j with $j \in \{1, \ldots, N_c\}$). The next section presents the method used to evaluate the derivatives of a function f, that may represent either Obj or any Cst_j with respect to each α_i, quantity that is central to any gradient-based optimizers as it will presented in Section 5.

4.1 RESIDUAL AND OBJECTIVE FUNCTION DEPENDENCIES

Shape variations of the wing are transferred to the computational mesh and it is assumed that $X(\alpha)$ is a function (C^1). Under the assumptions on R, that the determinant $\det(\partial R/\partial W(W,X))$ is not null, the implicit function theorem defines W as a C^1 function of α. This leads to a more detailed expression for Eq. (8):

$$R(W(\alpha), X(\alpha)) = 0 \tag{19}$$

Taking the example of a function f, that may represent either a combination of aerodynamic coefficients or a coefficient on its own, it will be computed by integration of pressure on the wing surface or on field variables over integration volumes in the fluid as seen in Section 3.2. Thus, f depends only on $X(\alpha)$ and $W(\alpha)$, and this is written as:

$$f(W(\alpha), X(\alpha)) \tag{20}$$

4.2 DISCRETE ADJOINT METHOD IN *elsA*

It is assumed that the residual given by Eq. (19) is verified for all α of the design space. This yields to $dR/d\alpha = 0$. Developing this expression for α_i, gives:

$$\frac{\partial R}{\partial W}\frac{dW}{d\alpha_i} + \frac{\partial R}{\partial X}\frac{dX}{d\alpha_i} = 0 \tag{21}$$

Besides, considering Eq. (20), the gradient of f with respect to α_i, can be expressed as:

$$\frac{df}{d\alpha_i} = \frac{\partial f}{\partial W}\frac{dW}{d\alpha_i} + \frac{\partial f}{\partial X}\frac{dX}{d\alpha_i} \tag{22}$$

Multiplying Eq. (21) by an arbitrary line vector λ^{T}, adding the corresponding expression to Eq. (22) and factorizing yields:

$$\frac{df}{d\alpha_i} = \left(\frac{\partial f}{\partial W} + \lambda^{\mathrm{T}}\frac{\partial R}{\partial W}\right)\frac{dW}{d\alpha_i} + \left(\frac{\partial f}{\partial X} + \lambda^{\mathrm{T}}\frac{\partial R}{\partial X}\right)\frac{dX}{d\alpha_i} \tag{23}$$

In the above equation, the vector λ^{T} is arbitrary, however it is chosen to eliminate the flow sensitivities in Eq. (23), by imposing the following discrete adjoint equation:

$$\frac{\partial f}{\partial W} + \lambda^{\mathrm{T}}\frac{\partial R}{\partial W} = 0 \tag{24}$$

Given λ^{T}, the solution of the adjoint equation, the gradient of f with respect to α_i may be written as:

$$\frac{df}{d\alpha_i} = \frac{\partial f}{\partial X}\frac{dX}{d\alpha_i} + \lambda^{\mathrm{T}}\frac{\partial R}{\partial X}\frac{dX}{d\alpha_i} \tag{25}$$

It is worth noting that the computational cost of the adjoint vector solution presented in Eq. (24) does not depend on the number of shape parameters. However, the computation of the gradients of f depends on the efficiency of the method used to compute the derivatives of the volume mesh with respect to the design parameters. In this work, a finite difference formula of order one is used to compute this quantity:

$$\frac{dX}{d\alpha_i} \approx \frac{X(\alpha_i + \delta\alpha_i) - X(\alpha_i)}{\delta\alpha_i} \tag{26}$$

Hence for each gradient evaluation, N calls to the mesh deformation process are required to compute the mesh sensitivities. The cost is, in our case, low if compared to that of the adjoint equation solution; this because the number of shape parameters is low (seven parameters = six hinges rotation angles + the angle of attack [AoA]).

The inversion of the discrete adjoint equation presented in Eq. (24) is not straightforward as the Jacobian $\partial R/\partial W$ is a large sparse matrix. An iterative quasi-Newton method has been implemented in the *elsA* flow solver [1]:

$$\left(\frac{I}{\delta t} + \frac{\partial R}{\partial W}^{\mathrm{T}(APP)}\right)\left(\lambda^{(m+1)} - \lambda^{(m)}\right) = -\left(\frac{\partial R}{\partial W}^{\mathrm{T}}\lambda^{(m)} + \frac{\partial f}{\partial W}^{\mathrm{T}}\right) \tag{27}$$

where δt is a positive real, the matrix $(\partial R/\partial W)^{\mathrm{T}(APP)}$ is an approximate Jacobian of R. Jacobian approximations and an efficient LU-SSOR implicit solver have been implemented in *elsA* by Peter and Drullion [9]. The expressions of the discrete convective and diffusive flux vectors of Eqs. (6) and (7)

Table 1 Gradients of CL and CD Comparison When Obtained With Finite Differences (FD) and Adjoint Solver With or Without the Eddy Viscosity Differentiated

Gradients CL	FD	Adjoint Frozen *vt*	Variation (%)	Adjoint Diff. *vt*	Variation (%)
θ_{H1}	1.00E−02	9.71E−03	−3.2	9.55E−03	−4.8
θ_{H2}	8.03E−03	7J6E−03	−3.4	7.62E−03	−5.2
θ_{H3}	5.87E−03	5.65E−03	−3.6	5.54E−03	−5.6
θ_{H4}	2.02E−02	1.96E−02	−2.8	1.95E−02	−3.5
θ_{H5}	1.63E−02	1.58E−02	−3.0	1.56E−02	−3.9
θ_{H6}	1.20E−02	1.16E−02	−3.4	1.15E−02	−4.4
AoA	1.28E−01	1.21E−01	−5.5	1.26E−01	−1.1
Gradients CD	**FD**	**Adjoint Frozen *vt***	**Variation (%)**	**Adjoint Diff. *vt***	**Variation (%)**
θ_{H1}	5.21E−04	5.04E−04	−3.2	4.92E−04	−5.5
θ_{H2}	4.16E−04	4.03E−04	−3.0	3.93E−04	−5.6
θ_{H3}	3.03E−04	2.94E−04	−2.8	2.86E−04	−5.6
θ_{H4}	7.92E−04	7.38E−04	−6.9	7.59E−04	−4.2
θ_{H5}	6.34E−04	5.95E−04	−6.2	6.08E−04	−4.1
θ_{H6}	4.64E−04	4.38E−04	−5.5	4.45E−04	−4.1
AoA	6.64E−03	5.09E−03	−23.4	5.34E−03	−19.7

have been differentiated formally. For the differentiation of the viscous flux vector, two types of solutions are possible. The first one consists in neglecting the linearization of the turbulence model in the viscous fluxes, this assumption called "frozen"-eddy viscosity often leads to more robustness of the adjoint solution but could altered sensitivities accuracy. The other approach consists in differentiating the turbulence model, the Spalart-Allmaras model in this case. This approach is more computationally expensive but, most of the times, leads to better sensitivities accuracy.

An example of *CL* and *CD* sensitivities computations considering or not the linearization of the eddy viscosity modeled by the Spalart-Allmaras approach has been done and compared to a finite difference formula of order one. The set of parameters consisted in the six-hinges angle of the morphing systems, plus the AoA considered here as shape parameters (solid rotation of the mesh).

The results are presented in Table 1. The two methods reveal a good behavior in terms of deviation to the finite difference values.

5 LOCAL SHAPE OPTIMIZATION TECHNIQUE

Aerodynamic shape optimization can be formulated as a generic mathematical optimization problem and solved by local optimization techniques. The principle is to find a local optimum of an objective function satisfying a set of constraints. It does not intend to find the global optimum, which may be of very high computational demand depending on the problem. In this section, the mathematical formulation of the problem will be addressed and the focus will be set on the method of feasible direction that will be used in the morphing application of the next section. Here, a two-dimensional constrained function optimization will be considered to illustrate how the method works and, at the same time, demonstrate the capability of the method.

5.1 DEFINITION OF THE PROBLEM

The general statement of the optimization problem, reusing the notations already introduced in Section 4 is given by:

$$\text{Minimize}: \quad Obj(\alpha) \tag{28}$$

$$\text{Subject to}: \quad Cst_j(\alpha) \leq 0 \quad j = 1, \; N_c \tag{29}$$

$$\alpha_i^l \leq \alpha_i \leq \alpha_i^u \quad i = 1, \; N \tag{30}$$

The general way to solve the problem in a local approach is to find a suitable search direction S and then move into this direction to update the α vector according to the following iterative process:

$$\alpha^q = \alpha^{q-1} + \delta^* S^q \tag{31}$$

with q indicating the current iteration number and δ^* the optimal scalar minimizing Obj in the search direction S^q. Different methods exist to define the search direction. In the next section we will focus on the method of feasible direction because it is the one used in the algorithm chosen for this study. The general idea behind this method is to find a search direction that enables to reduce Obj while satisfying the set of active constraints.

5.2 THE METHOD OF FEASIBLE DIRECTIONS

The method of feasible directions relies on a process to find the search directions, which is based on a very practical engineering design sense. At each iteration it tries to follow two main goals: improving Obj by minimizing it and satisfying the constraints imposed on the design space.

Let's start with the first goal and see what it means, in mathematical terms, for the search direction. Trying to minimize Obj at a given design point actually requires that the dot product between ∇Obj and S^q must be positive:

$$\nabla Obj \cdot S^q \leq 0 \tag{32}$$

Every S^q satisfying the above equation is called a *usable direction*, meaning that there is a potential for Obj to be minimized if one moves into such a direction. Because of the presence of the constraints on the design space, not all the usable directions are possible without activating or violating one (or several) of the constraints. Then a condition of "feasibility," in terms of the design constraints must be added. Mathematically, this means that the dot product of S^q with all the gradients of the constraints Cst_j must be negative:

$$\nabla Cst_j \cdot S^q \leq 0 \quad j = 1, \; N_c \tag{33}$$

Hence, to find a successful search direction, a sub-problem of optimization is addressed leading to the so-called *feasible search direction*. The details of the formulation and solution of this sub-problem are out of the scope of this section. The reader can find all the details in the Van der Plaats' textbook on numerical optimization. [10].

5.3 A 2D EXAMPLE: THE ROSENBROCK'S FUNCTION CONSTRAINED BY A DISK

In this example, a classical two-dimensions function was selected to present an application of local optimization solution with the method of feasible directions. The 2D Rosenbrock's function is used here as the objective function, it is given by the formula:

$$Obj(x,y) = \frac{1}{k}\left((a-x)^2 + b(y-x)^2\right) \tag{34}$$

with k, a, and b being constant real parameters. In this example they have been set to $k=3500$, $a=1$, and $b=100$ (Fig. 12).

One constraint is added to the problem: the design point (x,y) must be outside the disk of radius 2.5 whose center is set at $(x_o;y_o)=(2;2)$. This is summarized by the inequality:

$$Cst(x,y) \geq 2.5 \Leftrightarrow \sqrt{(x-2)^2 + (y-2)^2} \geq 2.5 \tag{35}$$

Bounds on variables are also considered:

$$\begin{array}{c} -2 \leq x \leq 2 \\ -2 \leq y \leq 2 \end{array} \tag{36}$$

To illustrate the problem, Fig. 13 presents isolines contours of *Obj* (in flood contours) and *Cst* (in white, labeled line representing constant values of *Cst*) in the design space. This picture reveals important features of the problem. In fact, without constraints, the global optimum of *Obj* is analytically known to be at $(a;a^2)$, therefore (1;1) in our example, which falls into the infeasible region considering the maximum available *Cst* function fixed at 2.5. Hence, independently of the algorithm adopted, the optimization problem will

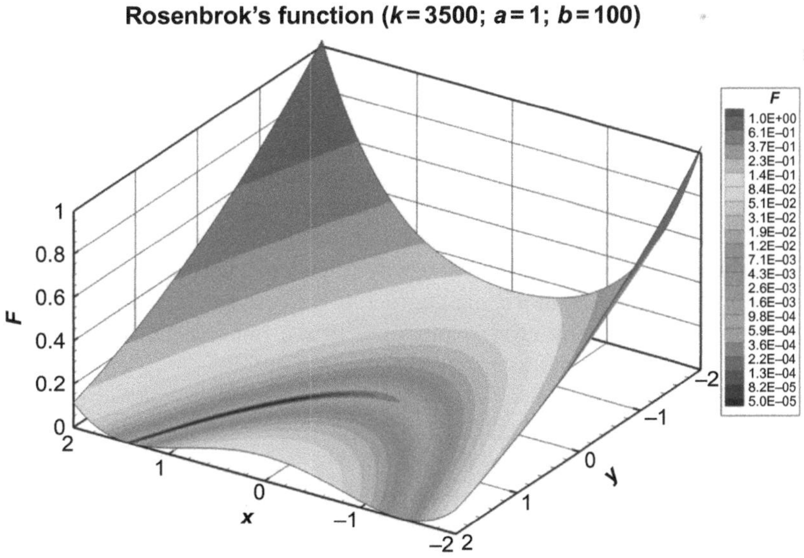

FIG. 12

Rosenbrock function illustration on [–2;2]*[–2;2].

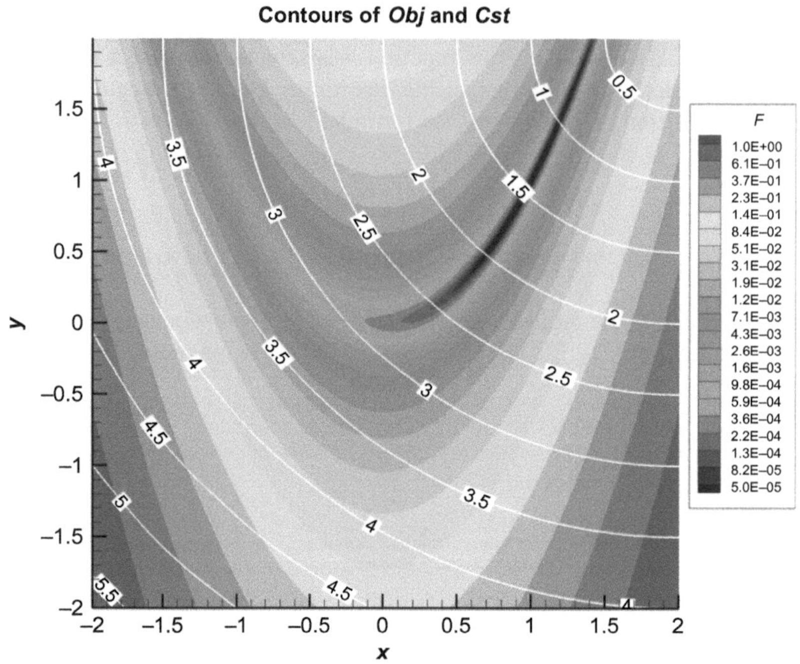

FIG. 13

Contours of *Obj (colored)* and *Cst* (labeled lines in *white*).

stop on the line where *Cst* equals 2.5 and find afterwards the optimum on that particular isoline. From the graph, the optimum of *Obj* satisfying the constraints is very close to $(x^*; y^*) = (0.35; 0.12)$.

In order to illustrate a solution based on the method of feasible directions, the DOT algorithm developed by Van der Plaats has been retained [3]. It uses a Fletcher-Reeves method for unconstrained optimum search and a feasible direction method in constraint search. The software was used through the DAKOTA optimization library [11], which offers the great advantage of proposing a unique interface to several optimization algorithms. The evolution of the algorithm toward the optimum is presented in Fig. 14, where the black dots represent a starting point for a new line search. The optimum is found successfully with 4 line searches that call 23 times the evaluation of the problem functions. This demonstrates the interest of the method and why it has been selected to solve the shape optimization problem addressed within the morphing context. It is worth noting that solving the problem in a global approach would require many more evaluations of the function, which may prove to be very costly when considering CFD computations requiring hours of CPU.

6 AERODYNAMIC SHAPE OPTIMIZATION OF MORPHING SYSTEM: AN APPLICATION WITHIN THE EU PROJECT SARISTU

After having introduced all the components needed to solve a local shape optimization problem, it is time to assemble all these necessary components into one complete process that will be able to exchange data with any gradient-based optimizer. This process will be described here for the wing

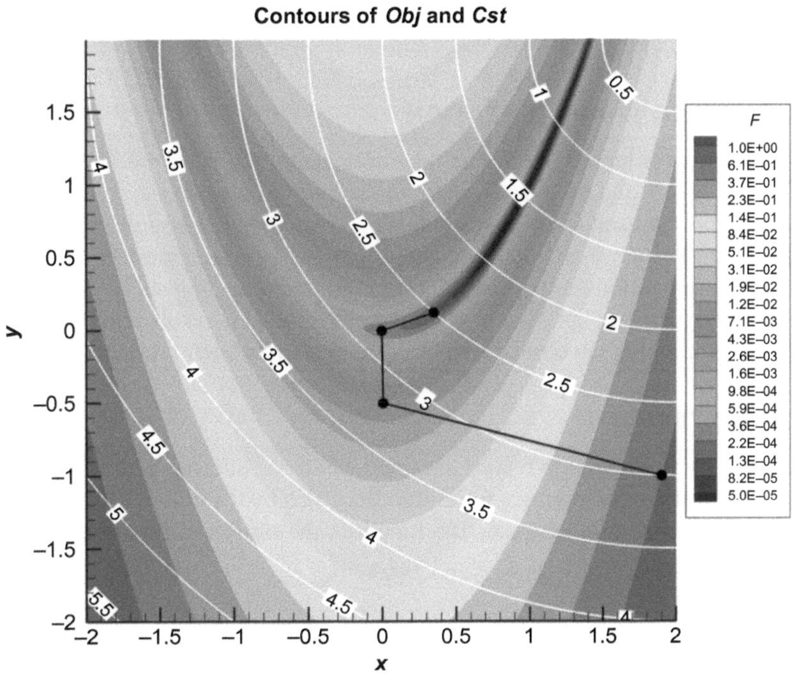

FIG. 14

History of the method of feasible direction with DOT.

optimization of a regional aircraft equipped with two morphing systems on the trailing edge. After having introduced the problem of optimization that is aiming to solve, a presentation of the optimization loop will be done. Then two results of the optimization process will be analyzed carefully. Conclusions and perspectives for the addressed morphing system will be drawn.

6.1 OPTIMIZATION PROBLEM

Given the six rotations modifying the trailing edge of the wing, the objective is to find the best set of parameters that will achieve the best aerodynamic performance at a given flight condition. In this work, the goal is to improve the lift-to-drag ratio (CL/CD) while ensuring a minimum lift coefficient. As presented in Section 2, H1, H2, and H3 are the hinges of the inner wing morphing system, with H1 being the more upstream hinge. H4, H5, and H6 are the hinges on the outer wing morphing system with H4 the more upstream hinge. A boundary constraint on maximum and minimum angle of rotation at each hinge is also considered.

To solve this problem, an iterative gradient-based algorithm has been adopted. Moreover, in order to demonstrate the potential of the ATED in terms of continuous adaptation of aerodynamic performance to the flight condition, two different problems taking into account different constraints in terms of lift coefficient have been considered. For the reference Mach number of 0.75, the first flight point retained is the one corresponding to the cruise aerodynamic condition at CL of 0.52, and the second one is the point yielding the maximum lift-to-drag ratio on the reference wing, which is at a CL of 0.423.

6.2 OPTIMIZATION LOOP PRESENTATION

The optimization loop is composed of two main components: the optimization algorithm presented in the left-hand side of Fig. 15, and the aerodynamic evaluation process on the right-hand side, for which each sub-process has been presented in the previous sections. The open source optimization library tool kit DAKOTA [11] developed initially by SANDIA Lab is used as the optimizer. It covers a wide range of global and local optimization methods. It has been assumed, in this study, to consider a local optimization problem and to use an iterative gradient-based algorithm to solve the problem. As presented before, the DOT algorithm [3] developed by Van der Plaats has been retained.

6.3 FIRST OPTIMIZATION

This section presents the optimization results obtained when considering a maximization of the lift-to-drag ratio under the constraint of having a minimum lift given by the cruise CL. Because the optimizer DOT works with minimizing a given objective function, the objective function is set to:

$$Obj = 100 - CL/CD_{\text{far-field}} \tag{37}$$

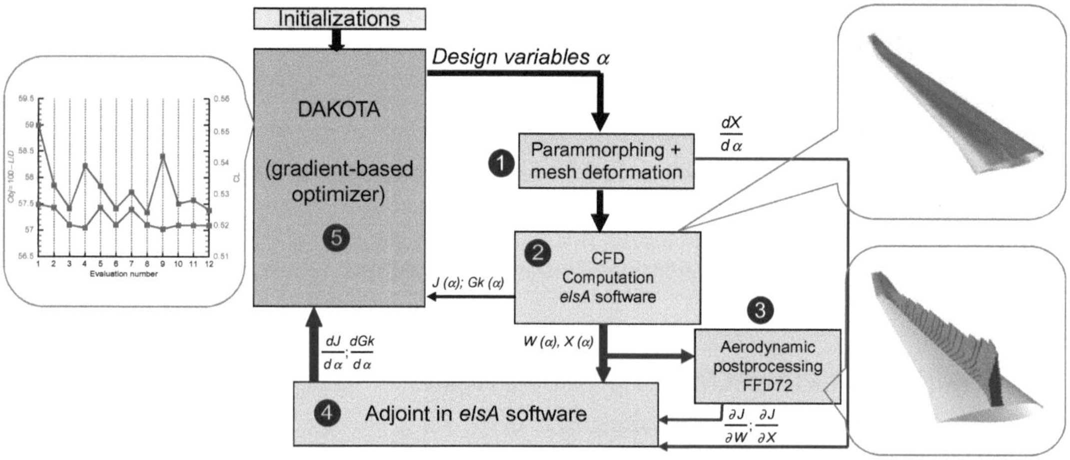

❶ 2D morphing model + mesh deformation process considering the 6 hinges rotations proposed by the optimizer at each iteration of the loop

❷ Flow computation with ONERA elsA software

❸ Aerodynamic performance assessment with Far-Field Extraction ONERA ffd72 software

❹ elsA/Adjoint to compute shape gradient of objective and constraints functions with respect to the 6 parameters when required by the optimizer (blue circuit on the loop).

❺ Dakota Optimization Library: use of DOT to find a local optimum

FIG. 15

Aerodynamic optimization loop.

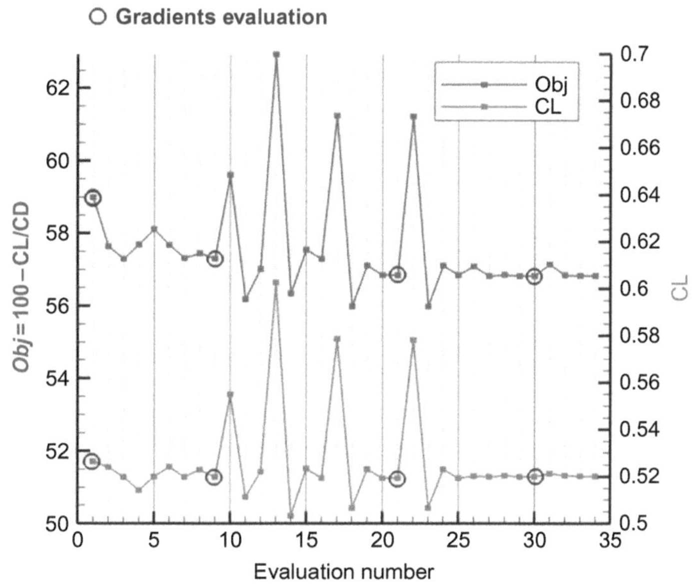

FIG. 16

Optimization history with gradient evaluations steps marked with *circles*.

The starting point was set so that the constraint on the lift was not active (CL > 0.52). This was achieved by setting an increase in the AoA of 0.05 degrees.

The optimizer needed 34 functions evaluations and 4 gradients evaluations (therefore adjoint computations) to reach an optimum. The evolution of the objective and the constraint functions along the process are shown in Fig. 16. The optimum configuration in terms of hinge rotations and angle of attach at CL of 0.52 is given in Fig. 17. It is worth recalling that positive values are associated to an increase of the camber. From there, it appears that in the inner wing ATED the camber is decreased, while it is largely increased in the outer wing. The AoA is reduced at the optimum in comparison to the initial wing, for which the CL of 0.52 is found for an AoA of 1.273 degrees.

The results for the optimum in terms of lift-to-drag ratio, drag components and pitching moment coefficients is given in Table 2. The gain is about +2.6% on lift-to-drag ratio at the CL of 0.52; this corresponds to a far-field drag coefficient reduction by 4.2 counts. When looking at the drag breakdown, it can be seen that the reduction is mainly due to a wave drag reduction, while induced drag stays constant. It is worth noting that due to the camber of the ATED on the outer wing, the pitching moment coefficient has increased of nearly 15% in absolute value. Even if the drag reduction is important for the isolated wing, the final drag reduction of the aircraft may not be as important. In fact, to stabilize an airplane with a higher pitching moment on the main wing, more lift is required on the horizontal tail wing and this induces a higher trim drag.

The polar curves comparison in terms of *L*/*D* as function of CL and CL versus CD in Fig. 18 shows that the increase of lift-to-drag ratio is important for CL equal or higher than 0.52 and tends to decrease for lower CL. The optimum presents even lower *L*/*D* performance below CL = 0.45. This illustrates/confirms the fact that the morphing has a local impact on aerodynamic performance.

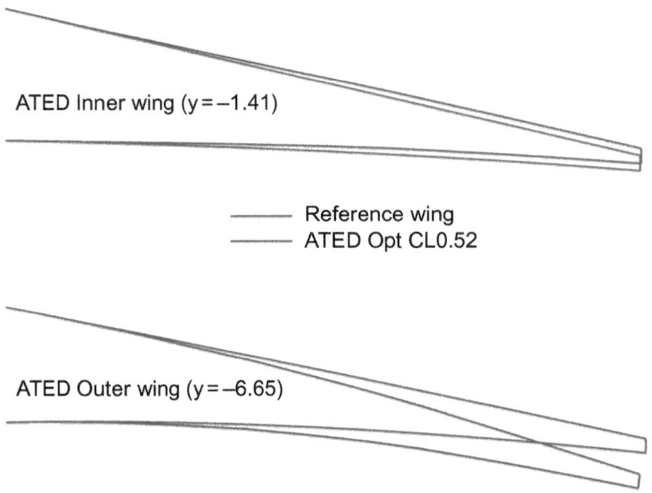

FIG. 17

ATED shape in inner wing *(top)* and outer wing *(bottom)* and value of the six hinges + AoA (degree) at the optimum.

Table 2 Lift-to-Drag Ratio, Drag and Moment Coefficients Evolution After Optimization No. 1

CL = 0.52	Evolution
Lift to drag ratio	*Gain (%)*
$CL/CD_{far\text{-}field}$	2.6
Drag components	*Variation (drag counts)*
$CD_{far\text{-}field}$	−4.2
CD_{wave}	−4.8
$CD_{induced}$	0.0
$CD_{viscous\ pressure}$	0.1
$CD_{friction}$	0.4
Moment coeff.	*Variation (%)*
CM	14.9
Angle of attack	*Variation (degree)*
AoA	−0.595

Fig. 19 presents a Cp comparison in two sections at mid-span of the inner and outer morphing regions for the same lift coefficient of 0.52. In the inner wing section, the rear loading of the profile shows a little reduction. The incidence reduction is clearly visible in the reduction of the high speed on the upper side (leading to an increase of Cp value); this has an impact on the shock intensity. For the outer wing section, the rear loading is very important due to the increase of camber in the ATED region. It appears to play an important role in the reduction of shock intensity. Again, the decrease of incidence induces a reduction of the high speed on the upper side, which is favorable to the shock intensity reduction.

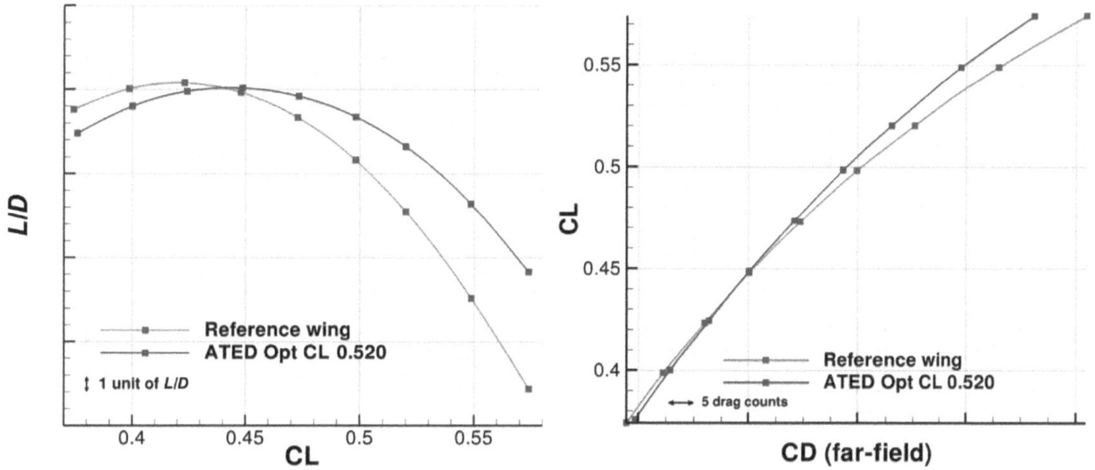

FIG. 18

Polar curves for initial and optimized wing at CL = 0.52.

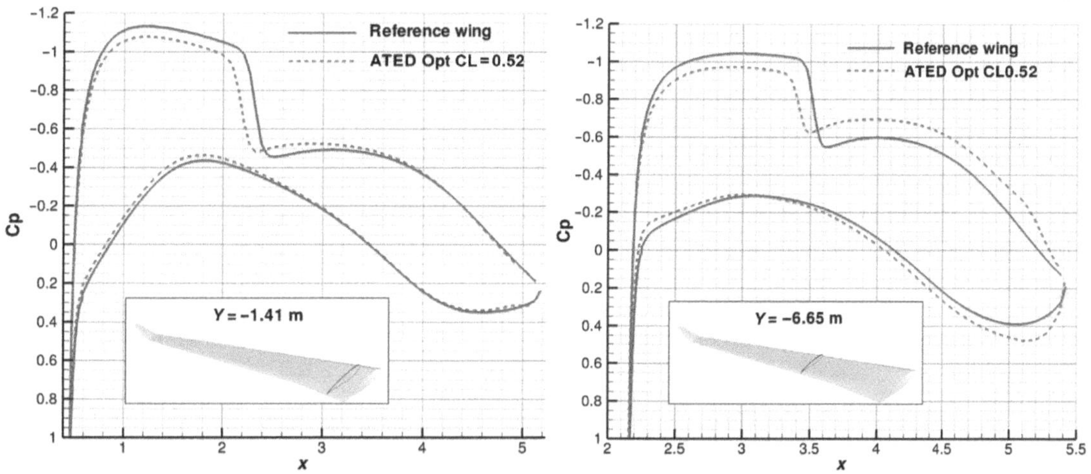

FIG. 19

Cp comparison for inner (left) and outer (right) of initial *(continuous line)* and optimized *(dashed line)* wings.

The spanwise loading comparison at cruise CL of 0.52 is presented in the left of Fig. 20. The general interpretation that camber at trailing edge induces more load is confirmed by the graph. It appears clearly that in the inner wing where the trailing edge camber is reduced, the loading is reduced, and in the outer wing region where the camber is increased, the loading is also increased. It is worth noting that the variation can be quite high, especially in the outer wing region. Another interesting graph is presented in the right of Fig. 20. Using a capability of *ffd72* software, the wave drag distributions for the initial and the optimized wing have been compared on the same diagram, hence the reduction of wave drag can be clearly seen. To conclude, for this first application, the optimization process worked well; it was able to find an optimal solution by adjusting all seven parameters (hinge rotations + AoA) and satisfying the imposed constraints.

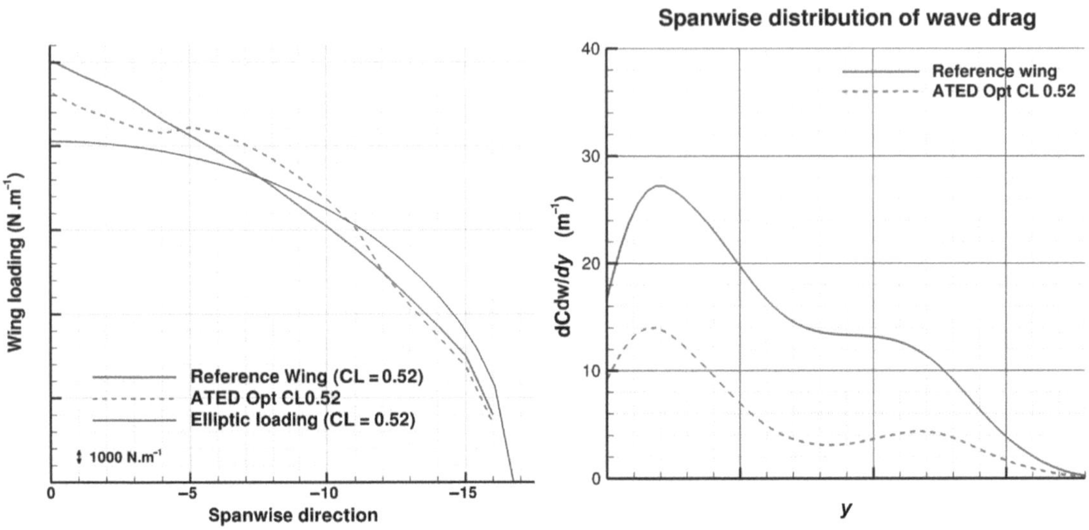

FIG. 20

Spanwise loading (left) and wave drag distribution (right) comparison for initial and optimized wings.

6.4 SECOND OPTIMIZATION

For this optimization, the constraint is modified and the focus is put on the condition of maximum lift-over-drag ratio. Looking at the polar curve of the reference SARISTU wing in Fig. 18, this condition can be found around a CL of 0.423. So this value constitutes the new minimum lift value in the search to minimize the objective: $100 - CL/CD_{\text{far-field}}$.

The optimizer needed 14 evaluations and 2 adjoint computations to reach an optimum. The evolution of the objective and of the constraint along the optimization process is shown in Fig. 21. The starting point was again set so that the constraint in lift was not active ($CL > 0.423$), this was done by increasing the value of the initial AoA by +0.05 degrees.

The optimum configuration in terms of shape and AoA at CL of 0.423 is depicted in Fig. 22. It appears that inner and outer ATED are set to reduce the camber. This is particularly relevant along the inner wing. In the meantime, to maintain the lift coefficient, the AoA is increased from its initial value by 0.5 degrees (that gives a CL of 0.423 on the reference wing).

The results, at the optimum, in terms of lift-to-drag ratio, drag components and pitching moment coefficients is given in Table 3. The gain of +1.3% in lift-to-drag ratio at the CL of 0.423 appears less important than the optimized runs at CL of 0.52. This increase of lift-to-drag ratio is associated to a far field drag coefficient reduction by 1.6 counts. When looking at the drag breakdown, it can be seen that the reduction is mainly due to a relevant decrease of induced drag (−2.5 drag counts (d.c.)) that is balanced by a wave drag increase (+0.8 d.c). One can note that for this flight point, the pitching moment coefficient decreases by the 5% in absolute value; this may have a good impact on the trim drag, for the reason explained in previous section with reference to the total drag of an aircraft configuration equipped with morphing devices.

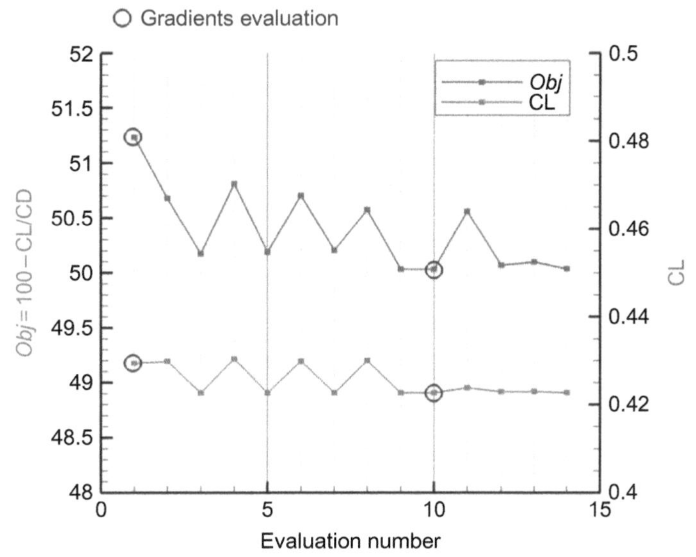

FIG. 21

Optimization history with gradients evaluations *(circles)*.

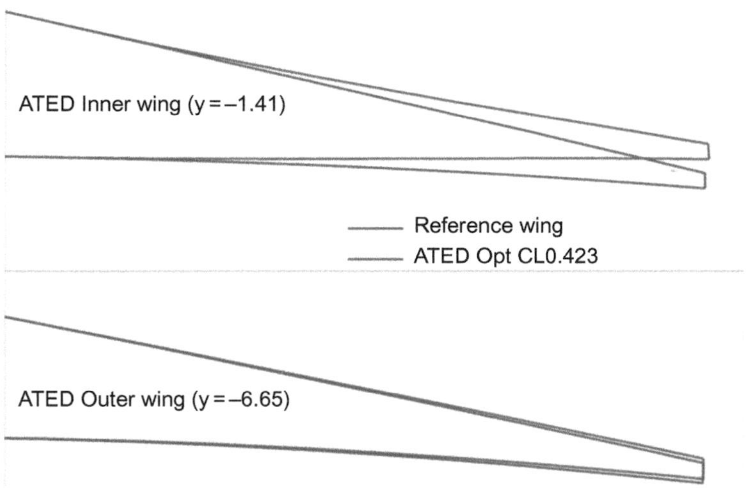

FIG. 22

ATED shape in inner wing *(top)* and outer wing *(bottom)* and value of the six hinges + AoA (degree) at the optimum.

The comparison of polar curves in terms of L/D as function of CL in Fig. 23 reveals that the increase of lift to drag ratio is relevant for CL equal or lower than 0.47. This increase seems to become even higher for CL < 0.42. This confirms again the fact that the morphing has a local impact on aerodynamic performance.

Table 3 Lift-to-Drag Ratio, Drag and Moment Coefficients Evolution After Optimization No. 2

CL = 0.423	Evolution
Lift to drag ratio	*Gain (%)*
$CL/CD_{far\text{-}field}$	1.3
Drag components	*Variation (drag counts)*
$CD_{far\text{-}field}$	−1.6
CD_{wave}	0.8
$CD_{induced}$	−2.5
$CD_{viscous\ pressure}$	0.2
$CD_{friction}$	−0.2
Moment coeff.	*Variation (%)*
CM	−5.0
Angle of attack	*Variation (degree)*
AoA	0.302

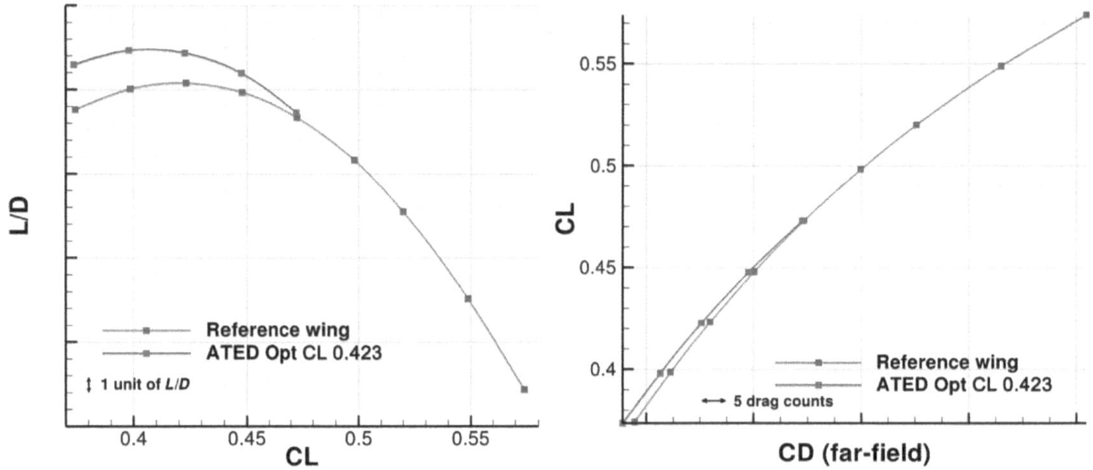

FIG. 23

Polar curves for initial and optimized wing at CL = 0.423.

Fig. 24 presents a comparison of Cp in the mid-span section of the inner and outer morphing regions for the same lift coefficient of 0.423. The increase of the AoA is clearly visible in both sections, by the reinforcement of the high speed flow on the upper side (inducing a decrease of Cp, which becomes more negative). The reduction of the rear loading in both sections leads to an increase of the shock intensity. It is interesting to observe how morphing has worked to avoid a too high increase in wave drag. For this, it is worth noting that the initial shock is less intense in the outer wing than in the inner wing, and that the same behavior is kept on the optimized wing. The decrease of rear loading is high in

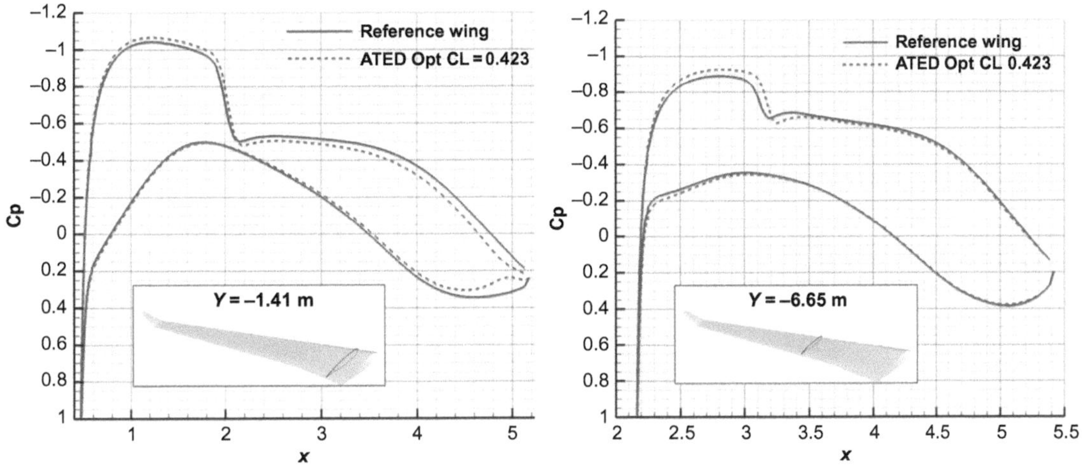

FIG. 24

Cp comparison for inner (left) and outer (right) initial *(continuous line)* and optimized *(dashed line)* wings.

the inner wing section due to the important decrease of camber. In the outer wing section, the tendency has been to reduce the camber but in a very small amount. This induces barely the same rear loading despite the increase of incidence. From the drag breakdown analysis, it can be seen that the optimum shape compensated the increase in wave drag by a stronger reduction of induced drag. This is confirmed by the spanwise loading presented in Fig. 25, which shows this tendency to unload the inner wing in favor of the outer wing, in order to be closer to an elliptical wing loading distribution.

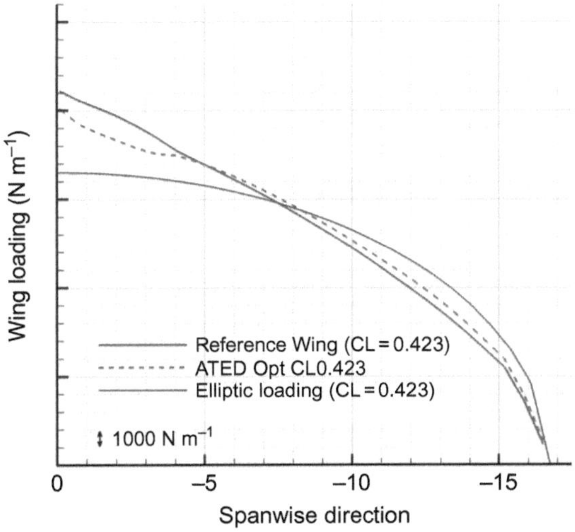

FIG. 25

Spanwise loading comparison for initial and optimized wings.

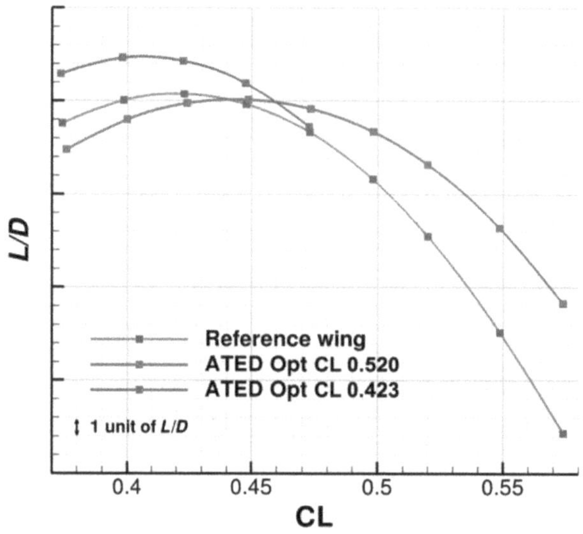

FIG. 26

Lift-to-drag ratio curves comparison for both optimized ATED shape.

6.5 EXPECTATIONS ON MORPHING TECHNOLOGY

For a global summary of the optimizations results, the polar curves of the reference wing and of the two optimized wings are plotted together in Fig. 26. The gains obtained for the optimized wing at the different CL is clearly visible, as seen before, but more interesting is the potential foreseen in the morphing at any CL value. It illustrates that morphing could be really efficient to adapt the wing geometry to the operating flight condition (CL in this case) and enhance its performance all along a mission for which the CL is constantly varying (for example, during the cruise, because of the decrease of mass due to the fuel burnt by the engine and the constrained flight levels).

7 CONCLUSION

A local shape optimization technique involving CFD simulation has been successfully applied to the problem of morphing addressed by the European research project SARISTU with reference to an active trailing edge camber variation of a regional aircraft wing. The main components used to build the optimization process are: a robust mesh deformation technique that handles the trailing edge shape variations, the CFD-based aerodynamic simulations and the adjoint-based sensitivities computations with the CFD/Adjoint solvers of the ONERA *elsA* software, and the ONERA far-field drag extraction software *ffd72*. On the optimizer side, the choice of the gradient-based algorithm DOT inside the Dakota Tool Kit proved to be robust and efficient.

The optimization process was applied to two different flight conditions. The first one corresponds to the actual cruise condition and the second one to the best lift-to-drag ratio condition. For the first optimization, the gain in lift-to-drag ratio is of 2.6%. This gain is obtained using the ATED systems to increase the camber on the outer wing trailing edge while de-cambering the inner wing, which enables a significant reduction of wave drag. For the second optimization the gain in lift-to-drag ratio is of 1.3%. It is obtained by a strong decrease of the camber in the inner wing trailing edge and a moderate decrease in the outer wing trailing edge which enables the reduction of the lift-induced drag.

The proposed optimization process in this study has led to interesting solutions that were analyzed from an aerodynamic point of view. They illustrate clearly the promising use of morphing technology at the trailing edge of a wing to enhance the aerodynamic efficiency of an aircraft along its whole mission. It is worth mentioning that other applications of similar aerodynamic optimization techniques have also been performed at ONERA using the adjoint method [12–16].

REFERENCES

[1] L. Cambier, S. Heib, S. Plot, The ONERA elsA CFD software: input from research and feedback from industry, Mec. Ind. 14 (2013) 159–174.

[2] D. Destarac, Far-Field/Near-Field Drag Balance and Applications of Drag Extraction in C.F.D., VKI Lecture Series CFD Based Aircraft Drag Prediction and Reduction, Hampton, VA, 2003.

[3] G.N. Vanderplaats, DOT User's Manual, Version 4.20, 1995.

[4] J.F. Thompson, B.K. Soni, N.P. Weatherill, Handbook of Grid Generation, CRC Press LLC, Boca Raton, FL, 1999.

[5] P.L. Roe, Approximate riemann solvers, parameters vectors and difference schemes, J. Comput. Phys. 43 (1981) 292–306.

[6] G.D. Van Albada, B. Van Leer, W. Roberts, A comparative study of computational methods in cosmic gas dynamics, Astron. Astrophys. 108 (1982) 76–84.

[7] P.R. Spalart, S.R. Allmaras, A one-equation turbulence model for aerodynamic flows, La recherche aérospatiale 1 (1994) 5–21.

[8] J. Van der Vooren, D. Destarac, Drag/thrust analysis of a jet-propelled transonic aircracft; definition of physical drag components, Aerosp. Sci. Technol. 8 (2004) 545–556.

[9] J. Peter, F. Drullion, Large stencil flux linearization for the simulation of 3D compressible turbulent flows with backward-Euler schemes, Comput. Fluids 36 (2007) 1005–1027.

[10] G.N. Vanderplaats, Numerical Optimization Techniques for Engineering Design with Applications, Mc Graw-Hill, New York, USA, 1984.

[11] Dakota: Design Analysis Kit for Optimization and Terascale Applications, Developed by SANDIA National Laboratories, Albuquerque, New Mexico, http://endo.sandia.gov/DAKOTA/software.html.

[12] I. Salah El Din, G. Carrier, S. Mouton, in: Discrete adjoint method in elsA (Part 2): application to aerodynamic design optimization, 6th ONERA-DLR Aerospace Symposium, Toulouse, France, October 4–6, 2006.

[13] A. Dumont, A. Le Pape, J. Peter, S. Huberson, Aerodynamic shape optimization of hovering rotors using a discrete adjoint of the reynolds-averaged Navier–Stokes equations, J. AHS 56 (2011) 032002.

[14] M. Méheut, A. Arntz, G. Carrier, Aerodynamic Shape Optimizations of a Blended Wing Body Configuration for Several Wing Planforms, AIAA Paper 2012-3122.

[15] G. Carrier, D. Destarac, A. Dumont, M. Méheut, J. Peter, I. Salah El Din, J. Brézillon, M. Pestana, in: Gradient-based aerodynamic optimization with the elsA software, AIAA SciTech Conference, Maryland, USA, January 2014.

[16] A. Dumont, G. Carrier, in: Multi-point aerodynamic optimization of a flexible transport aircraft wing using an aeroelastic adjoint method, 6th European Conference on Computational Fluid Dynamics (ECFD VI), Barcelona, July 2014.

FURTHER READING

[1] J. Peter, Discrete adjoint method in *elsA* (part 1): method/theory, 6th ONERA-DLR Aerospace Symposium, Toulouse, France, October 4th–6th, 2006.

CHAPTER

EXPECTED PERFORMANCES

6

Alessandro De Gaspari*, Sergio Ricci*, Alexandre Antunes†, Felipe Odaguil†, Grace Lima†

Politecnico di Milano, Milano, Italy EMBRAER SA, Sao Jose dos Campos, Brazil†*

CHAPTER OUTLINE

NOMENCLATURE

BMx	wing root bending moment (N m)
CD	drag coefficient
$\mathbf{CD_{wave}}$	wave drag coefficient
$\mathbf{CD_{ind}}$	induced drag coefficient
$\mathbf{CD_{fp}}$	friction and pressure drag coefficient
CL	lift coefficient
Cl	lift coefficient for unit span
Cp	pressure coefficient

Morphing Wing Technologies. https://doi.org/10.1016/B978-0-08-100964-2.00006-X

L/D	aerodynamic efficiency
α_{ind}	induced angle of attack
δ_{LE}	equivalent leading edge deflection (degree)

1 INTRODUCTION

The need for more efficient aircraft able to meet the new challenging requirements defined by US and EU organizations (for example, Horizon 2020) forces researchers to look for more advanced aircraft configurations, based on more efficient aerodynamics and structures combined with more sophisticated flight control systems. Aiming at this global target, a great effort is underway concerning morphing aircraft and especially morphing airfoils, based on their potential capability to optimize selected aircraft performance index during the mission, (for example, fuel burning) by adapting their shape. Nevertheless, the design of this kind of devices does not represent an easy task and would require the availability of ad hoc developed procedures able to tackle the conflicting requirements (such as the high deformability required to change the airfoil shape coupled to the load-carrying capability and low weight).

Looking at great amounts of literature concerning morphing [1–4] it is possible to draw two different conclusions. First, there are many concepts investigated, but there is not clear evidence, for the moment, of which is the most promising concept, or when there is a clear advantage in using it, taking into account many factors such as certification issues, maintainability, fatigue, and potential cost increases. Secondly, there is the need for specific design tools that could assist the engineers in the design, from the conceptual phase, of both the morphing systems and the optimal internal structure guaranteeing the required shape changes with the most efficient use of embedded actuators.

The paper proposes a solution for optimal airfoil morphing design based on a compact approach to describe the airfoil geometry coupled with a two-level optimization procedure. In the first level, the best deformed airfoil configuration is determined to be the most efficient aerodynamic shape which simultaneously limits the energy required to deform the airfoils skin. In the second optimization level, the best internal structural configuration is obtained using a topology optimization tool based on genetic algorithms which synthesize a compliant structure able to adapt itself for matching the optimal shape arising from the first level. Both the topology of the internal load paths and the size of the structural elements are considered during this second level optimization. Once the optimal morphing airfoil is obtained, the finite element models are automatically generated for both the single rib and for a complete wing section as well.

The motivation of the present work lies in studying the benefits in terms of performance and weight saving by the use of morphing devices during few considered flight mission. One of the expected benefits is the possibility of operating in off-design conditions without considerable loss of performance. Improvements in drag reduction, even by a small amount, can result in a remarkable cumulative reduction of atmospheric emissions if one computes the global effect of the entire fleet over years of operation.

The present article introduces the main results obtained by the NOVEMOR project and addresses future developments to raise the morphing devices Technological Readiness Level (TRL).

The performed aerodynamic study is focused on both drag reduction, the resulting fuel consumption reduction, and weight reduction. This chapter summarizes the work executed and presents some comments from the industry's point of view.

Basically, the aerodynamic analysis consisted of the following steps:

- Aerostructural optimization to define a feasible, compliant mechanism and a more detailed lofting;
- Low-fidelity aerodynamic and performance analysis for fuel-saving estimation;
- High-fidelity aerodynamic analysis for validation;
- Weight-saving estimation.

Morphing concepts, based on conformable leading and trailing edges applied to a Reference Aircraft (RA), are presented, along with some important assumptions.

For this conventional configuration, it was assumed that the morphing must not have a big impact on the overall aircraft design, i.e., it should include only small-scale modifications. Therefore, it was decided to keep the wingbox (which is an important component that influences many disciplines) unchanged. Leaving the wingbox untouched offered the possibility of developing a retrofit system, with the potential to be applied in the short term.

The objective of the trailing edge camber concept was to benefit drag at cruise condition by changing lift distribution along the span. Different chord extensions were evaluated, from 5% to the whole flap panel area, and along the span it was considered from the wing root to the aileron, excluding the aileron region. At first, a constant trailing edge deflection was considered, but the benefit of having a variable trailing edge deflection along the span, including upwards deflections, influenced researchers decision to do otherwise. Fig. 1 illustrates the trailing edge concept applied to the RA.

The leading edge concept, on the other hand, was designed to produce maximum CL at subsonic (takeoff and landing) conditions, in substitution of the conventional leading edge high lift devices. A variable chordwise extension was considered due to the front spar limitation.

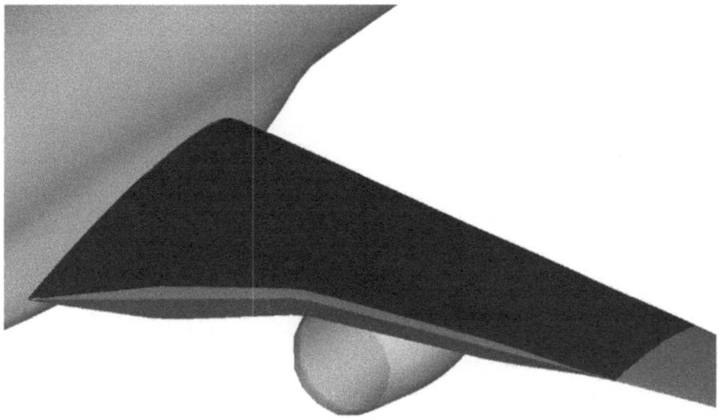

FIG. 1

Morphing wing trailing edge concept applied on the conventional aircraft.

2 THE REFERENCE AIRCRAFT

One of the difficulties during the evaluation of new technologies such as morphing is the lack of realistic application examples. For this reason, NOVEMOR partners decided to develop from scratch a complete, realistic aircraft, potentially to be distributed as an open source model for adoption by the world community, to be used as a platform to test the benefits of different morphing concepts. This key duty was covered by EMBRAER, the only industrial partner of NOVEMOR consortium [5].

The so-called Reference Aircraft (RA) is a typical regional airplane capable of providing the operational flexibility to accomplish different missions at the transonic regime. In particular, the operational requisites for the RA were established considering missions that encompass 600 nm. Nevertheless, the same RA is capable of accomplishing missions up to 2300 nm. The RA was designed to achieve an optimum cruise performance at Mach number 0.78 and lift coefficient, CL, of 0.47. In Table 1, one can see a brief description of the other relevant operational characteristics of the RA.

The active camber concept, where the camber of the wing is adapted during the mission profile, is not a new concept, having been adopted over many years by means of discrete control surfaces like flaps, slats, and ailerons. However, only recently, new structural concepts made it possible to obtain continuous camber variation, generating a smoothing surface able to guarantee better aerodynamic efficiency. Recently, interest in these devices increased due to the possibility of combining them with the Natural Laminar Wing approach, considered potentially a key factor in drag reduction. EU funded research projects, such as SADE and SARISTU, investigated these concepts. The RA is used as a benchmark in order to evaluate the potential benefits that morphing devices can bring in global performances. The use of conformable morphing surfaces to guarantee a smooth chordwise and

Table 1 Reference Aircraft Main Characteristics

Reference Aircraft Characteristics	
Maximum takeoff weight	58,000 kg
Maximum landing weight	53,000 kg
Maximum zero fuel weight	48,000 kg
Basic operating weight	34,000 kg
Maximum payload	14,000 kg
Maximum fuel	18,000 kg
Maximum operating Mach	0.82
Cruise Mach @ 37,000 ft	0.78
Landing Mach @ Sea Level	0.15
Time to climb to FL 350, 600 nm TOW	17 min
Takeoff field length, SL, MTOW	1.49 m
Takeoff field length, SL, 600 nm TOW	1.152 m
Landing field length, SL, MLW	1.626 m
Range 113 PAX @ 100 kg, LRC	4.387 km
Mean aerodynamic chord	3.86 m
Reference wing area	110 m^2
Reference wing span	15.73 m

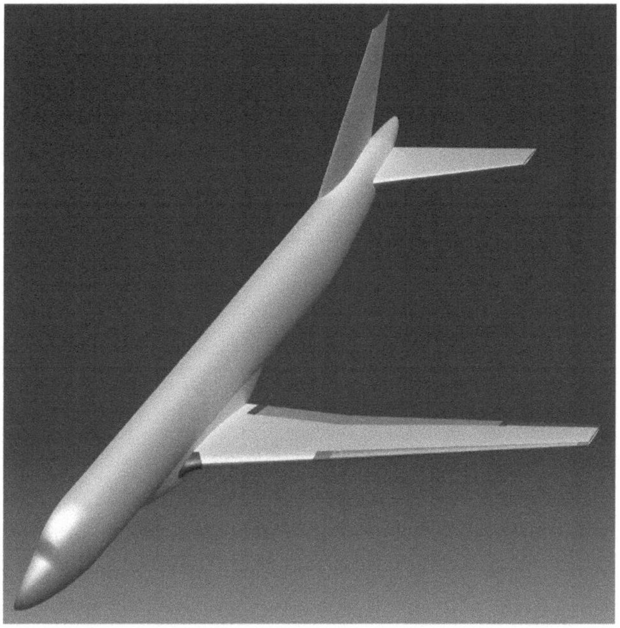

FIG. 2

Reference Aircraft with morphing wing configuration.

spanwise variation of the airfoil camber is being investigated. Fig. 2 shows the morphing configuration adopted during the activities of NOVEMOR.

3 ACTIVE CAMBER USING CONVENTIONAL CONTROL SURFACES

At first, a sequence of preliminary investigations was carried out to identify the potential benefits of active camber as applied to the RA. First, trailing edge morphing was investigated, obtained by moving classical control surfaces like flaps and ailerons, or by simulating the effect of rotating a small portion of airfoil chord. To speed up the calculations, a medium-fidelity approach was adopted based on BLFW code. Researchers changed the boundary condition, imposing a normal velocity on the wing wall, to simulate deflection of the trailing edges. The simulations took into account possible upward and downward deflections of the trailing edge of the wing by a total of 3 degrees (a positive value means a downward deflection). The analyzed configuration corresponds to five panels over the flap region.

The study considered two objectives to be optimized, in this case the drag coefficient, CD, and the bending moment at the root of the wing, BMx. In the first moment, the present morphing analysis was accomplished at the cruise flight condition. The optimizations are performed for the cruise flight condition, and the results are shown below.

3.1 FIVE PANELS OVER THE FLAP REGION

In this optimization, the inboard and outboard flap panels were divided to create five panels, and each of them could move independently. Fig. 3 shows the divisions of the panels for the first study case. In Fig. 4, one can observe the Pareto curve showing the trade-off between the bending moment and the drag coefficient for the viscid analysis. It can be observed that the decrease of the bending moment

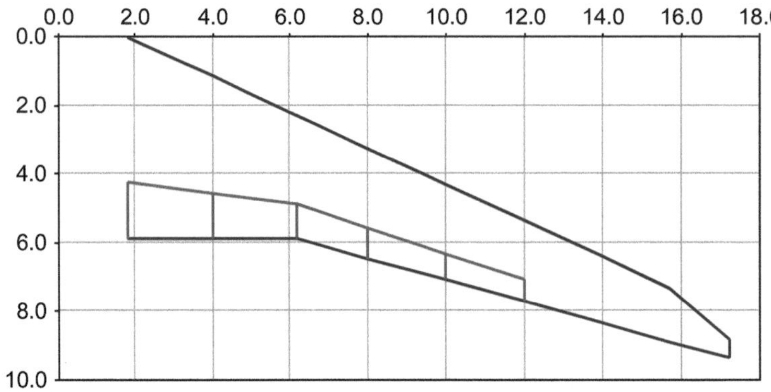

FIG. 3

The five panels over the flap region.

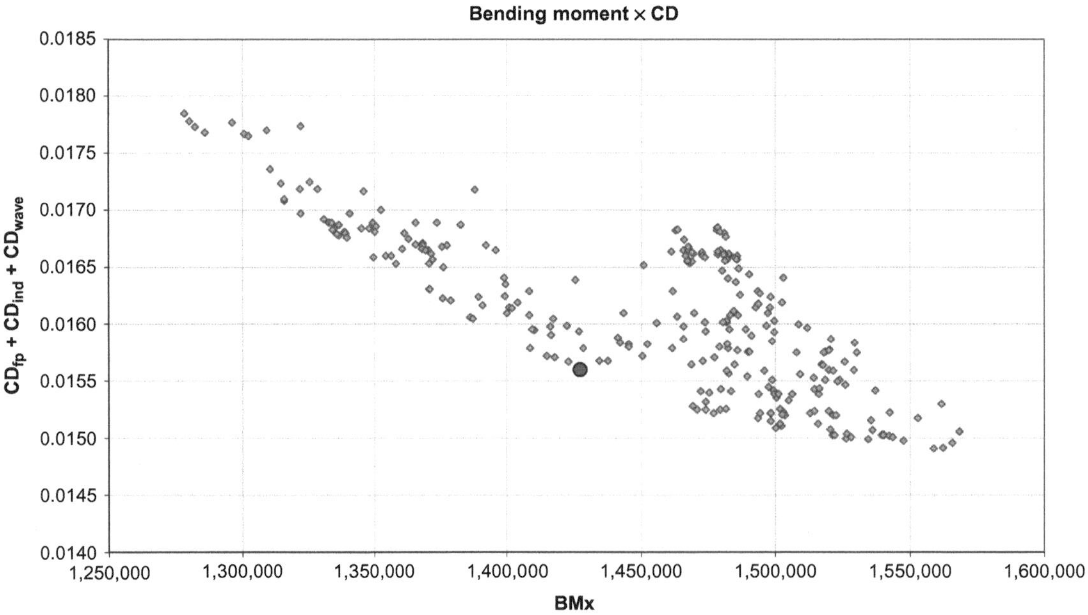

FIG. 4

The Pareto front.

Table 2 Drag Breakdown in the Case of a Five Panel Flap

Drag Counts	Ref.	ID131	Δ
CD_{fp}	68.4	69.4	+1.0
CD_{ind}	83.2	73.9	−9.3
CD_{wave}	4.4	5.7	+1.3
CD_{total}	156.0	149.0	−7.0

leads to an increase in the drag coefficient and vice versa. The right-side solutions have positive deflections along the wing spanwise, and a reduction of 7. drag counts is reached. However, this specific solution leads to an increase in the bending moment of 9.%. On the other hand, the left-side solutions result in an increase in the drag coefficient with a decrease in the bending moment.

The red point in Fig. 4 represents the design point selected as the best compromise between the two objectives. It would be interesting to have an additional analysis for the moment not implemented, in order to define the decrease in structural weight due to the reduction of the bending moment. Table 2 shows the breakdown of the drag components for the RA and the design point. It can be seen that the trailing edge morphing yields a reduction in the induced drag with a small penalty in the wave drag.

4 COUPLED AEROSTRUCTURAL SHAPE OPTIMIZATION

Before the high-fidelity aerodynamic assessment, a knowledge-based (KBE) framework combining parametric geometry representation, prediction of the structural response of morphing skin, and aerodynamic analyses, was used to define the optimal morphing wing shape in terms of mission profile performances [5]. This framework was coupled with a wing shape optimization procedure which combined aerodynamic performances and structural requirements related to deformation of the skins. Afterward, a dedicated code, based on Genetic Algorithm and named SPHERA (Synthesis of comPliant mecHanisms for EngineeRing Applications), was used for optimal design of morphing devices, based on compliant mechanisms able to produce, once actuated, the optimal shape [6,7].

The morphing shape design level represents an aerodynamic optimization which accounts for skin structural constraints such as deformability and axial and bending stresses. It incorporates a new 2.5D aerodynamic tool, based on a 2D/Infinite Swept Wing (ISW) Navier-Stokes method [8], coupled with a 3D panel method for potential flow Calculations developed at Politecnico di Milano [9].

The morphing shape optimization procedure described in Ref. [5] was applied to the shape design of the RA in order to evaluate the global performances and the potential benefits of morphing technologies when implemented into commercial aircraft.

Here, it is important to point out that both the aerodynamic and performance analyses must be coupled. In terms of computational effort, this is quite a demanding approach, because it requires constant creation of the aerodynamic database for each of the morphing configurations and a performance/flight mechanic module to actually fly the aircraft in different missions. Most of the times, this is not affordable in the framework of an optimization procedure. Therefore, the optimal shapes for both the morphing leading and trailing edges are obtained by solving different optimization calculations

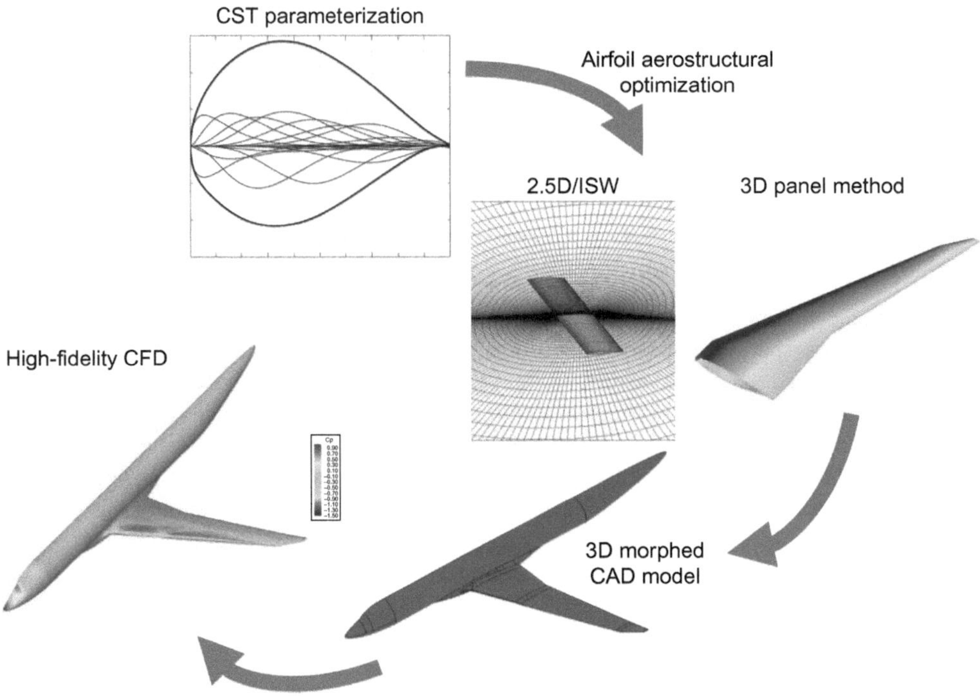

FIG. 5

Aerostructural optimization workflow.

wherein the morphing shape is defined to maximize specific user-defined aerodynamic performance indexes.

The optimal shape is then assessed on a high-fidelity computational fluid dynamics (CFD) tool, which implements a Reynolds-averaged Navier-Stokes (RANS) solver, and is able to trim the complete aircraft along the entire mission profile in order to verify the obtained benefits.

The parametrization of the RA wing follows the optimization sheme of Fig. 5. Different sections were identified in order to reproduce the parametric shape and to construct a mathematical model able to introduce the shape changes into the trailing and leading edge morphing regions.

A new aerodynamic tool inside PHORMA (Parametrical sHapes for aerOdinamic and structural Modeling of Aircraft) allows the addition of 2D/ISW polar curves to the set of identified cross-sections to provide an aerodynamic description of the complete wing model. This could be used to make local corrections to a 3D low-fidelity simulation, based on a panel method, with the computational cost of a reduced number of 2D/ISW flow solutions equal to the number of cross-sections [8]. This technique provides an accurate aerodynamic description of the 3D wing which is very close to the higher-fidelity analyses, greatly reducing the computational cost. Another advantage of using this technique during the optimizations is that it works by separating the three drag components: the induced drag (computed by the Trefftz plane), viscous drag, and pressure drag.

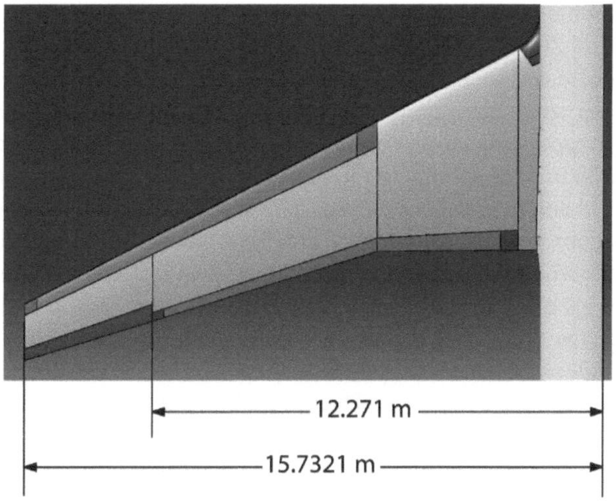

FIG. 6

Leading edge *(between the kink and the wing tip)* and trailing edge *(across the kink)* morphing regions.

After the morphing shape optimization, corresponding CAD models can be generated and used to perform high-fidelity verifications. In this work, optimal results were postprocessed by EMBRAER through adoption of the RANS formulation, in combination with a performance/flight mechanic module to actually fly the aircraft, considering different missions.

Because different design requirements have been provided for each morphing device, different shape optimization problems have been tuned and solved for the morphing leading edge and for the morphing trailing edge. The camber morphing was not applied to the whole wingspan; for the leading edge, it was applied only on the outboard part of the wing due to the engine mounting, and for the trailing edge, it started near the wing root and went up to the aileron. Fig. 6 illustrates the morphing regions, where the darker ends represent the transition areas between the deformed and undeformed sections.

The geometries were evaluated at two flight conditions:

- Subsonic regime (Mach 0.30, Reynolds 6 million): this is a condition to assess the aerodynamics in which high lift is required. The leading edge is expected to present benefits in this condition, because it is a device that could substitute the conventional leading edge high lift devices;
- Transonic regime (Mach 0.78, Reynolds 21 million): this is the typical cruise condition. In this condition, the trailing edge is expected to present a behavior that could be beneficial due to the CL variation expected during cruise.

4.1 MORPHING LEADING EDGE

From a structural point of view, the main leading edge design rule is to keep the circumferential skin length constant in order to avoid axial strains. The amount of droop of the individual sections depends on the maximum curvature change, which is restricted to 25 L/m. Other requirements are related to the spanwise deflection law which must be linear, while three different chordwise configurations were investigated: 10%, 15%, and 20%.

The optimization process was tuned using an objective function that tried to introduce the maximum droop deflection along the wing span, under the skin structural constraint. This may be suitable for a low-speed condition, corresponding, for example, to the landing condition, because alleviation of the bending stress inside the skin corresponds to the increase of the leading edge radius, helping preserve laminar flow for low speed and high lift conditions. The droop deflection to be maximized is the equivalent leading edge deflection δ_{LE}, as defined in Ref. [5]. A set of four identified parametric sections, equally spaced along the outboard region, were used in order to introduce the morphing shape changes along the wingspan. For each section, the optimization process worked on 3, 5, and 7 variables, for a total of 12, 20, and 28 shape optimization variables corresponding to the 10%, 15%, and 20% chordwise morphing configurations. A family of possible shapes has been generated for different values of the shape variation law, as defined in Ref. [5]. The optimal aerodynamic shape was then selected by the 2.5D code.

Corresponding curvature difference functions computed along the morphing leading edge arclength and along the wingspan are reported in Fig. 7.

The constraint on the maximum curvature change along the skin leads to a linear droop angle which varies from 16 degrees at the wing kink to 10 degrees at the wing tip, from 14 to 8.5 degrees, and from 12.5 to 7.5 degrees, for the 10%, 15%, and 20% chordwise configurations, respectively. The shape optimization processes and the following aerodynamic analyses provided the morphing leading edge shapes shown in Fig. 8.

The main effect of the leading edge device is to increase the stall angle, when the trailing edge high lift devices are also deployed. Analyses of single aerodynamic performance indices, performed after the structural shape optimization by the 2.5D code, showed that, for the subsonic flight regime, the morphing leading edge is able to improve the lift-over-drag ratio by the 4% between 0.43 and 1.0 of lift coefficient. This improvement is maximum for the 10% chordwise morphing leading edge, around CL = 0.6, and for the 15% chordwise leading edge, around CL = 0.73.

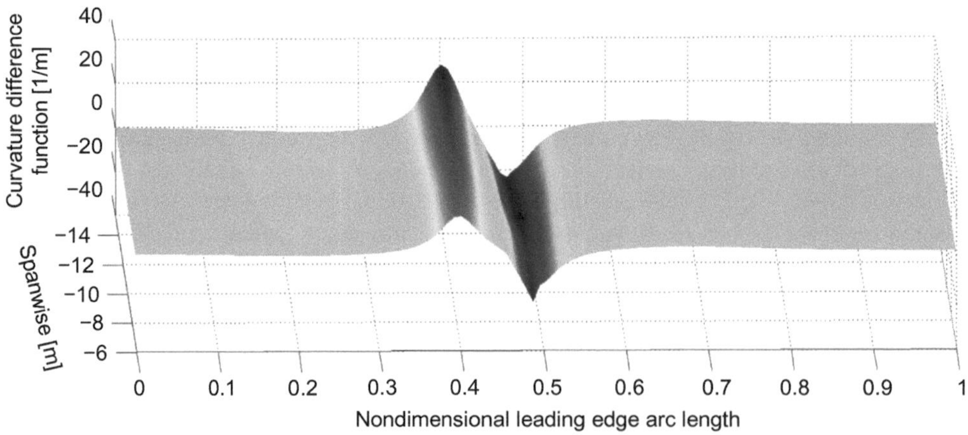

FIG. 7

Morphing leading edge curvature change along the wing span.

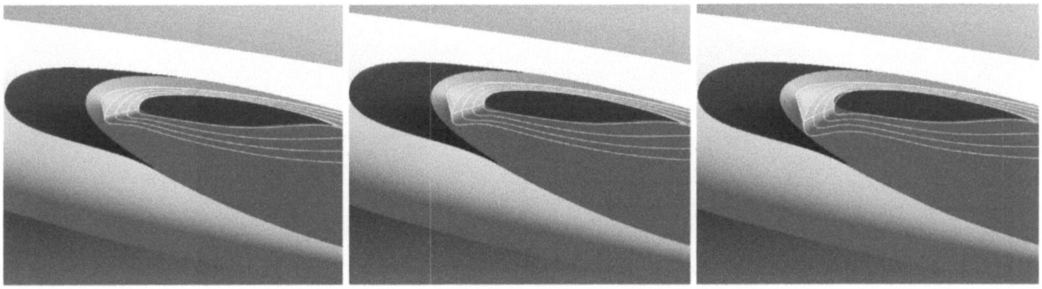

FIG. 8

Parametric CAD model corresponding to 10%, 15%, and 20% chordwise morphing leading edge obtained maximizing droop angle under skin structural constraint.

4.2 MORPHING TRAILING EDGE

A different approach has been followed for the morphing trailing edge, due to the particular configuration and its different structural requirements. Two different chordwise regions were investigated: 5% and 28%. The shape optimization procedure was applied to the wing of the RA to improve its performances by means of a morphing trailing edge optimized for cruise condition. The fitness function is related to the aerodynamic efficiency *L*/*D*, while the optimization process works without spanwise deflection constraints and tries to keep the total CL greater than the reference one required for the trim.

Different structural requirements were adopted for the trailing edge skin: the upper skin length constraint is set up to be constant in order to keep the axial strain equal to zero, while the lower skin length constraint has been relaxed in order to obtain an efficient aerodynamic shape. This means that the lower skin is free to slide along its contour and corresponds to the introduction of a linear slider to avoid axial strain. The constraints related to the skin curvature variations are not taken into account due to small deflections which are bounded between ± 5 degrees, in terms of equivalent deflection δ_{TE}, as defined in Ref. [5].

A set of seven parametric sections along the span, four equally spaced in the inboard region from the wing root to the wing kink section, and four equally spaced in the outboard region from the same kink section to the transition region before the aileron, was identified in order to introduce the morphing shape changes. For each section, the optimization process worked on 4 and 6 variables, for a total of 28 and 42 shape optimization variables, respectively, corresponding to the 5% and 28% chordwise morphing configurations, as defined in Ref. [5].

The aerodynamic models are composed by the structured surface mesh of the wing for the potential flow computations, shown in Fig. 9 together with the corresponding wake, and by the seven structured meshes corresponding to the seven wing sections placed along the span for the 2D/ISW Navier-Stokes analyses. Each of these meshes is a volume mesh extruded along its local sweep angle.

The aerodynamic models are parametrically updated during each optimization step in order to adapt the mesh to the morphing shape changes. The results in terms of polar curves, obtained from the 2D/ISW RANS computations performed on the seven cross-sections, are used to make a sort of "local trim," based on the results obtained from the 3D potential flow simulation. This "local trim" is an

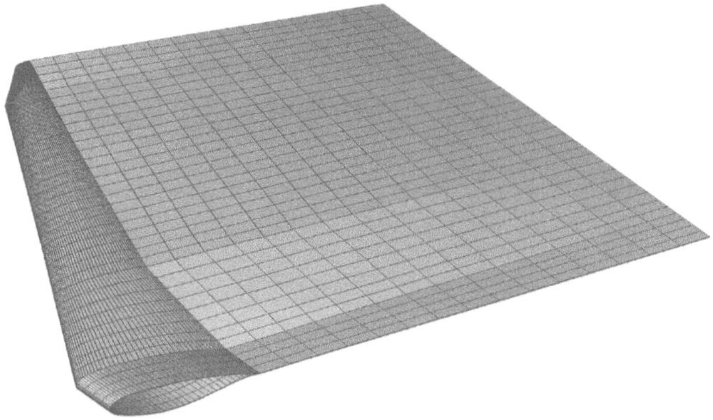

FIG. 9

3D structured mesh around the complete wing model and corresponding wake used for potential flow calculations.

iterative algorithm which runs until the convergence of α_{ind}, computed as the difference between the 3D angle of attack (AoA) adopted in the panel method and the effective angle of attack, computed in the 2D/ISW RANS computations. Once the Cl values of the 2.5D simulations match the local values of the 3D potential flow simulation, corresponding drag components can be computed and integrated in order to calculate the 2.5D polar curves. In particular, the induced drag components come from the 3D potential flow simulation, while the pressure and viscous components come from the integration of the 2D/ISW RANS results corresponding to the local Cl value needed for the trim.

The aerodynamic field corresponding to the morphing shape obtained from the optimization process is reported in Fig. 10, which shows how the morphing trailing edge is able to change its shape to adapt the aerodynamic field along the span. In order to improve wing efficiency, the optimal spanwise deflections reduce the local lift coefficient Cl in the region close to the wing root and increase it in the region near the aileron. The trailing edge deflection allows the control of the Cp distribution and of the shock wave position along the chord, as shown in Figs. 11 and 12 where the 2D/ISW results around two wing section are shown. The mesh of each section is obtained by extruding a 2D structured mesh along the local sweep angle, which is 21.5 degrees for all the sections before the wing kink and 25.2 degrees for the sections placed after the kink.

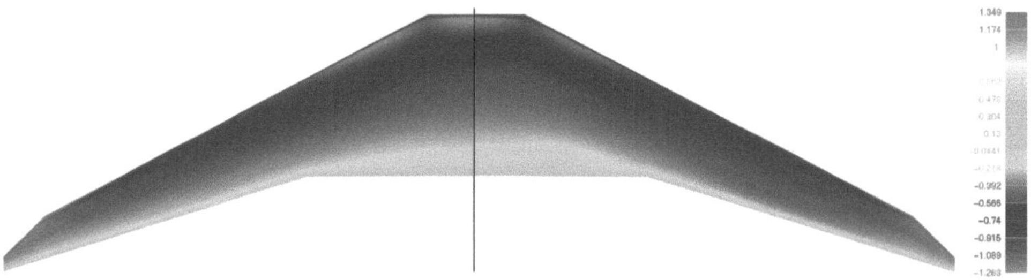

FIG. 10

Comparison between the Cp distribution at cruise Mach over the reference wing (left) and over the wing equipped with the optimal morphing trailing edge (right); results obtained from 3D panel method.

FIG. 11

Comparison between 2D/ISW results in terms of Cp chordwise distributions around the reference (left) and the optimal deformed shape (right) corresponding to the wing inboard section with sweep angle of 21.5 degrees, in high speed conditions.

FIG. 12

Comparison between 2D/ISW results in terms of Cp chordwise distributions around the reference (left) and the optimal deformed shape (right) corresponding to the wing outboard section with sweep angle of 25.2 degrees, in high speed conditions.

The optimal morphing shape is able to reduce the CD for each value of CL, as shown in Fig. 13 (right). In the same Fig. 13 (left), it is possible to see how the whole improvement is due to a combination of effects which reduce the induced drag, while increasing the viscous drag. This variation can be partially attributed to the morphing wing load distribution, which appears in Fig. 14. For this reason, the morphing trailing edge shape obtained by the optimization process has been implemented on the RA, and several high-fidelity CFD computations have been performed on the complete aircraft.

The shape optimization provided morphing trailing edge shapes where the optimal spanwise deflections increase from −3.1 degrees at the wing root to 2.2 degrees at the transition region close to the

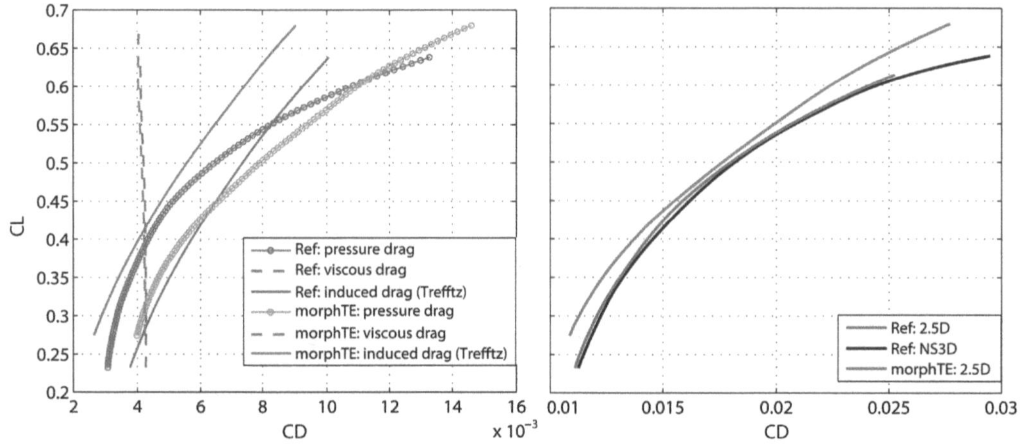

FIG. 13

Comparison between reference and morphing 2.5D polar curves.

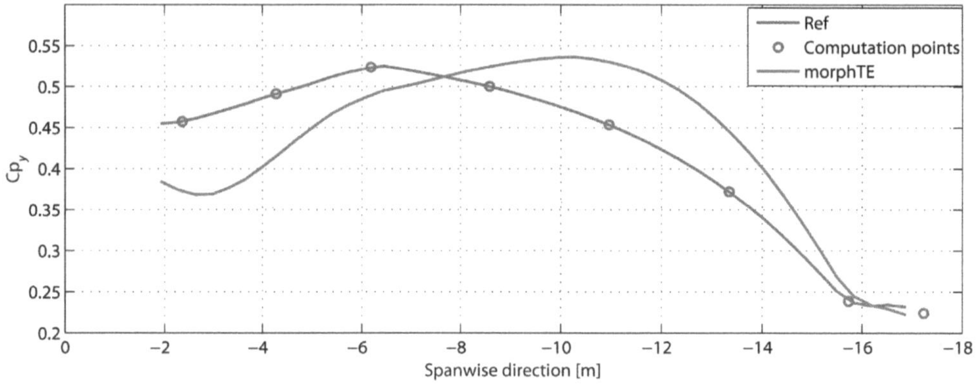

FIG. 14

Comparison between the Cl spanwise distribution at cruise Mach over the reference wing and over the wing equipped with morphing trailing edge.

aileron. Fig. 15 shows the corresponding CAD model, considering 28% of morphing trailing edge chord extension.

5 FUEL SAVINGS

This section presents the medium-fidelity aerodynamic and performance analysis of the morphing concepts adopted to support the down-selection process of the most promising concepts to be further developed.

The aerodynamics of the morphing concepts applied to the conventional transport aircraft was performed with a CFD code that implements a full potential formulation coupled with a boundary layer

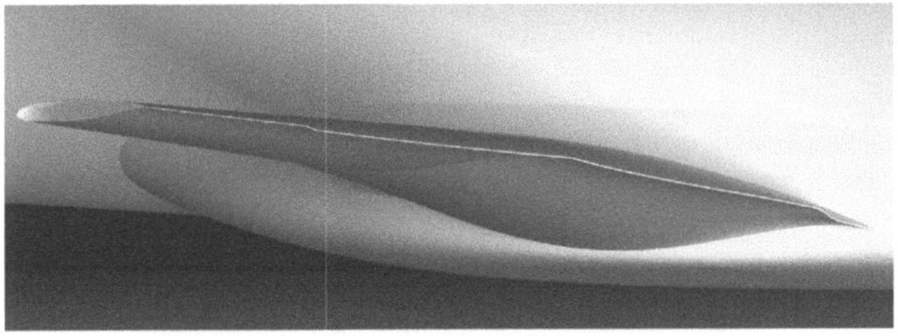

FIG. 15

Parametric CAD model of the 28% chordwise morphing trailing edge for high speed condition under skin structural constraint.

correction. This formulation was chosen due to its low computational costs and the capability to reasonably capture the important physical phenomena.

Each flight condition, i.e., each combination of lift coefficient and Mach number, was evaluated and the results were compared in terms of drag counts. The leading edge morphing could not be assessed at this stage due to limitations of the CFD tool formulation, which is not able to capture the expected benefits.

The aerodynamic results were used to generate an aerodynamic database that, together with an engine deck, was used to assess the performance of the aircraft. The tool used to evaluate the perfomance throughout the mission covers all flight segments, including the reserve fuel needed to fly 100 nm to reach an alternative destination, plus 45 min of loiter. The flight mission profile is illustrated in Fig. 16.

The results were assessed in terms of block fuel, i.e., the fuel consumed to accomplish a certain mission given by a range and a payload. Fig. 17 shows, for many combinations of payload and range, the percentage of block fuel reduction obtained by the trailing edge morphing concept with respect to the RA. During a typical mission for this category of aircraft (600 nm with 12,000 kg of payload), the obtained reduction was in the order of 0.1%.

It is important to emphasize that these results are referred only to the aerodynamic benefits. No additional weight was considered on the basic operating weight (BOW) due to the additional mechanisms.

The morphing of the leading and trailing edge was later analyzed using the RANS formulation.

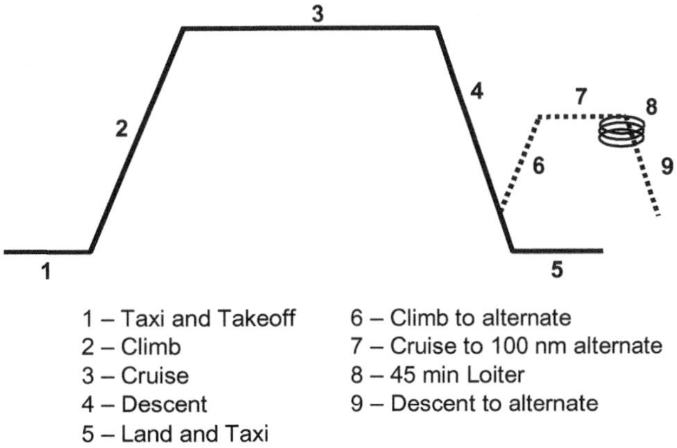

FIG. 16

Typical mission segments.

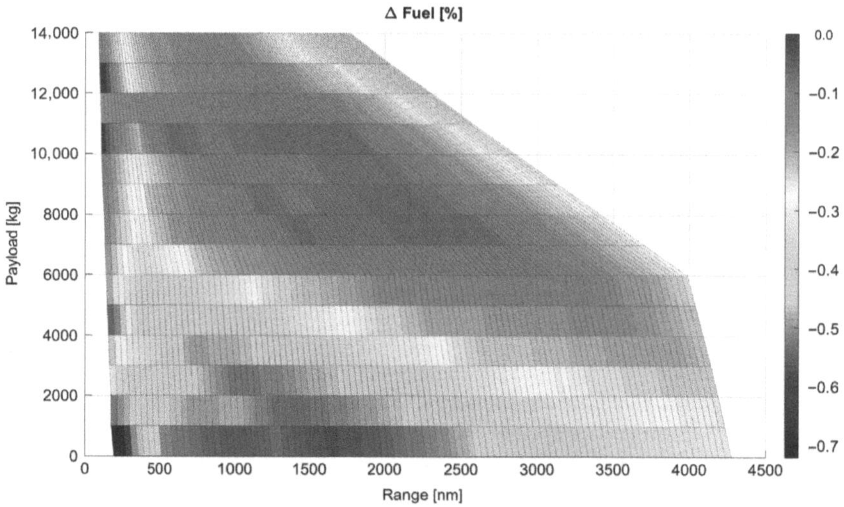

FIG. 17

Block fuel reduction due to morphing of the trailing edge.

6 HIGH-FIDELITY AERODYNAMIC ANALYSIS

The morphing concepts that presented a promising performance benefit were further evaluated with higher fidelity tools. The goal of this high-fidelity analysis was to have more accurate data to validate the methodology.

The versatility of the procedure described in the previous sections for investigating the potential benefits of morphing technologies allows the user to analyze different scenarios. Different generated trailing and leading edge morphing devices, applied to the RA, have been postprocessed through the adoption of the Reynolds Average Navier-Stokes formulation, RANS. This evaluation is carried out for the subsonic and the transonic flight regimes in order to assess the potential benefits such technology can provide in terms of fuel burn and extension of the flight envelope. Indeed, the observed aerodynamic advantages and disadvantages are just part of the feasibility evaluation, which must involve a larger amount of disciplines to push the morphing technology toward its implementation in commercial aircraft. In terms of computational effort, the aerodynamic and performance analysis is quite demanding and required the creation of an aerodynamic database for each of the morphing configurations, as well as the use of a performance/flight mechanics module to actually fly the aircraft considering different missions. The results are presented in the following paragraphs.

6.1 LEADING EDGE MORPHING

This morphing concept has not been evaluated previously because the benefits would not be captured by the medium-fidelity CFD tool. This morphing concept is expected to increase the maximum CL, and to assess the concept, a high-fidelity CFD tool is required to correctly capture the flow characteristics and compare the investigated configurations.

Table 3 Improvements From the MLE Concept

CL	mLE_c10 (%)	mLE_C15 (%)	mLE_C20 (%)
0.6	5	0	3
0.7	3	6	1
0.8	3	5	1
1	1	3	1

Three morphing leading edge geometries were proposed: mLE_c10 with 10% chord extension, mLE_c15 with 15%, and mLE_c20 with 20%, all of them with a maximum deflection compliant with skin structural constraints in terms of maximum allowable deformation. Once the best shapes were selected, they were implemented on the RA 3D model, as shown in Fig. 8, and postprocessed using RANS computations. For the subsonic flight regime, it is possible to observe an improvement in the *L*/*D* ratio for a lift coefficient ranging from 0.43 up to 1.0. Table 3 shows, for the analyzed morphing leading edge concepts, the improvements in the *L*/*D* ratio for different values of the lift coefficient and for the three different values of the chord length.

In order to quantify the fuel savings benefits, a typical flight mission for the RA was considered. This mission comprised a 600 nm, point-to-point distance, with an alternate of 100 nm and a holding phase of 45 min.

The analyses showed that, for the subsonic flight regime, there is an improvement in the lift-over-drag ratio between 0.43 and 1.0 of the lift coefficient. This improvement is about 5% at $CL = 0.6$ for the 10% chordwise extension configuration, and the same value can be achieved at $0.8 \leq CL \leq 0.9$ for the 20% chordwise extension configuration. It was noticed that the holding phase is performed at the lift coefficient falling inside the range in which the 10% chordwise extension configuration can provide benefits in the *L*/*D* ratio. These values agree with the results obtained from the 2.5D/ISW method coupled with the 3D panel method adopted during the optimization process.

The results in Fig. 18 show it is possible to see benefits in terms of either *L*/*D* for CL values over 0.5 or the increase the maximum CL of almost 0.1. This benefit could be used to save fuel in the holding condition and to increase the maximum CL in take-off and landing conditions. The adoption of this morphing device at this flight phase would lead to 62 kg fuel savings.

6.2 TRAILING EDGE MORPHING

From the medium-fidelity analyses and the analyses performed during the shape optimizations, two parameters of the trailing edge were evaluated: the chord extension and the angle of deflection of the morphing surface. The results provided an insight into the range these parameters should have in order to be beneficial, and they were used for generation of the high-fidelity geometries.

Lift-over-drag ratio results, obtained from preliminary studies performed with a spanwise constant deflection at the transonic regime, showed no fuel savings could be achieved, because the advantages can only be observed for CL values different from those corresponding to the trimmed condition at the cruise flight phase.

At first, geometries with a constant trailing edge deflection were analyzed. For each chordwise extension evaluated, a maximum camber deflection was proposed, such that skin structural constraints

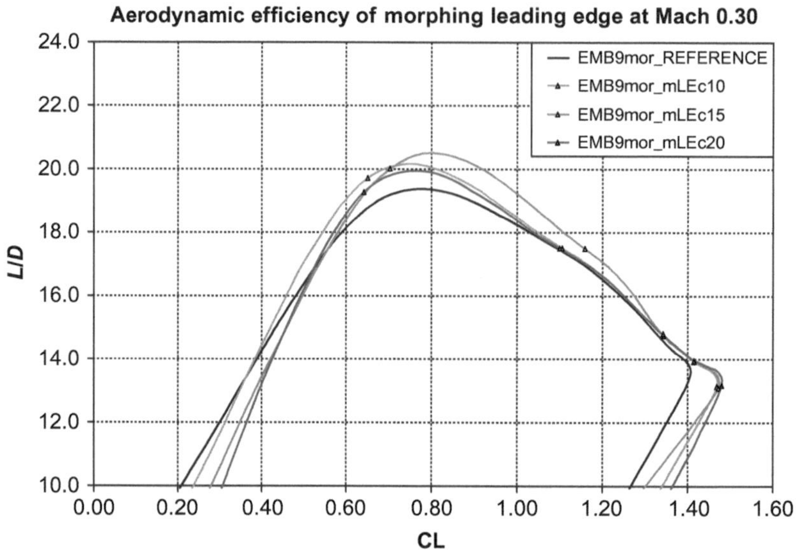

FIG. 18

Aerodynamic efficiency of morphing leading edge at Mach 0.3.

were satisfied. Additionally, an intermediate deflection was also evaluated. Table 4 shows all the geometries analyzed.

Fig. 19 shows a comparison of aerodynamic efficiency (*L*/*D*) over CL. It is possible to see that all morphing geometries present benefits at high CL values, and the higher the deflection, the higher the CL at which there is a benefit for the same chord extension.

Although these configurations led to some improvements in the *L*/*D* coefficient, no fuel savings were achieved. In evaluating the performance analysis for these morphing trailing edge configurations, the MTOW was adopted as take-off weight. Because the CL required to achieve the trimmed condition during the cruise flight phase at different flight levels is below 0.61, there is no justification for adopting a morphing trailing edge working with constant spanwise deflection.

Table 4 Trailing Edge Morphing Geometries

	% Chord	Deflection (Downward) (Degree)
EMB9mor_mTEc5r3_25	5	3.25
EMB9mor_mTEc5r6_5	5	6.50
EMB9mor_mTEc10r2_5	10	2.50
EMB9mor_mTEc10r5_02	10	5.00
EMB9mor_mTEc15r1_75	15	1.75
EMB9mor_mTEc15r3_5	15	3.50
EMB9mor_REFERENCE	0	0.00

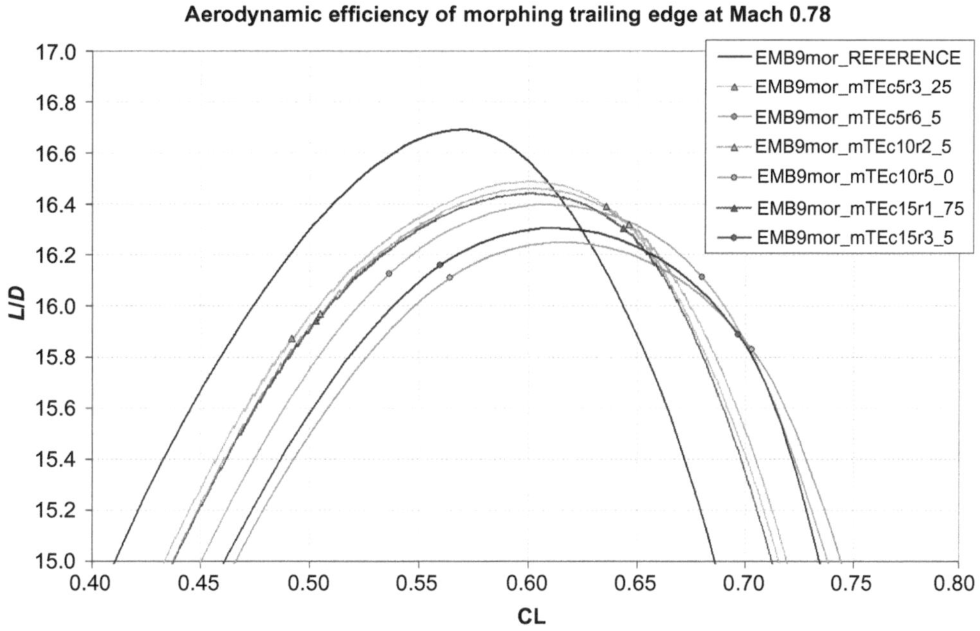

FIG. 19

Aerodynamic efficiency of constant deflection morphing trailing edge at Mach 0.78.

It is important to notice that CL values above 0.61 can be achieved with lower velocities; nevertheless, this implies a decrease in velocity in an off-design condition for current deflections of the morphing trailing edge, thus, implying no guarantee that even the aerodynamic benefits can still be present for these analyzed proposals.

It is possible to see that there is a benefit for values of lift coefficient higher than about 0.61. However, if we look at the mission profile of Fig. 20, it is clear that this is a condition in which the aircraft will rarely be flying. This is a 1500 nm mission in which the aircraft starts at the maximum take-off weight, and therefore, is a condition requiring the highest lift coefficient at cruise for a given altitude and speed.

Despite the not-so-encouraging results obtained at this specific flight condition, there is an indication that other deflections for the morphing trailing edge can provide not only aerodynamic benefits but also fuel savings. Given the condition in which the benefits are present, further geometries were evaluated.

In searching these benefits, the variable trailing edge deflection obtained by the shape optimization process presented in Section 4.2 was investigated. This solution permits the lift distribution to get closer to the elliptical one by giving more freedom to the morphing region.

The results considering two optimal morphing shapes with different chordwise extensions, were analyzed: mTEc5rOptFlap and mTEc28rOptFlap related to 5% and 28% chordwise extension, respectively. Fig. 21 presents the results obtained for the simulations performed by means of the CFD++ code and utilizing the RANS formulation with the Spalart-Allmaras turbulence model.

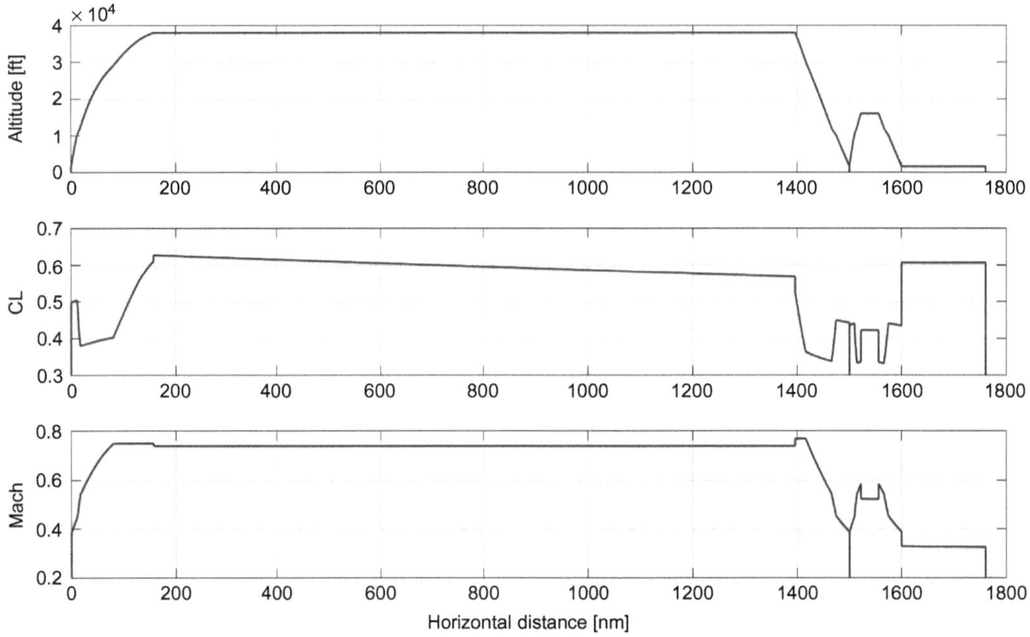

FIG. 20

Flight profile for a 1500 nm mission.

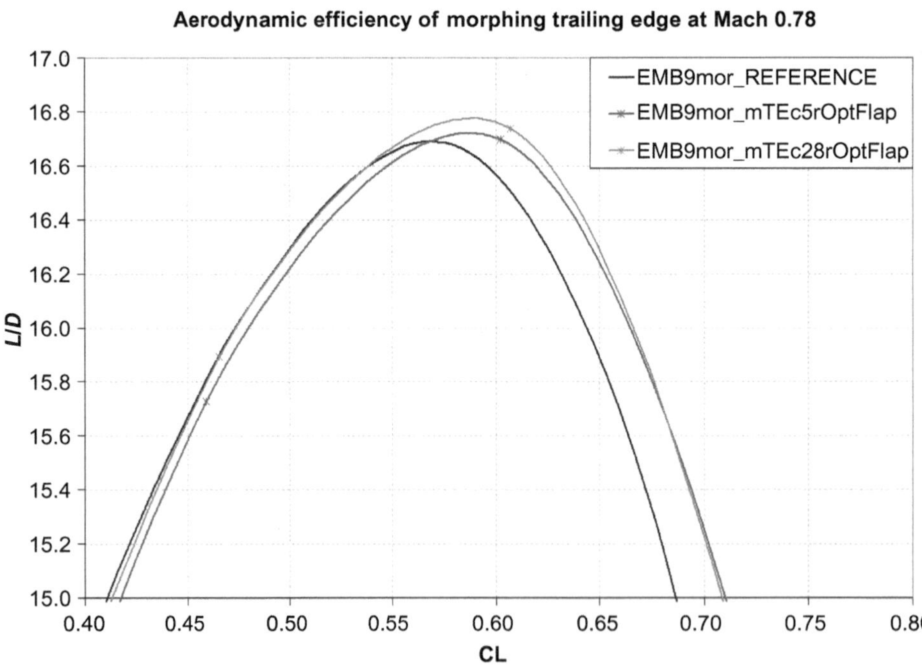

FIG. 21

Aerodynamic efficiency of variable deflection morphing trailing edge at Mach 0.78.

It is worth mentioning that the plotted data consider only the drag components due to the wing, while neglecting the drag contribution of the fuselage; the effect of that aerodynamic component is provided by its volumetric interference over the wing. It is possible to see that the new morphed geometries start to present benefits for lower CL values, and that the 28% chordwise extension configuration shows a slightly higher aerodynamic efficiency. An improvement in the *L*/*D*, with respect to the RA, can be obtained in a range of CL that can encompass different possible missions. This proves the coherence of the results obtained by the 2.5D/ISW method coupled with the 3D panel method adopted during the optimization process.

The above decreasing deflection law in spanwise direction allowed an increase in the *L*/*D* of 2% for $CL = 0.45$ and 5.85 for $CL = 0.55$, which indicates that the potential aerodynamic benefits can be translated into fuel savings. This variation can be partially attributed to the wing loading distribution, which is more triangular than elliptical. This effect, positive from the aerodynamic point of view, moves the center of pressure outboard and increases the root bending moment.

However, the observed aerodynamic advantages are just part of the feasibility evaluation, so during the structural design of the wingbox, the impact of the increased wing root bending moment on the overall wing weight should be duly addressed.

7 WEIGHT SAVING

The assessment of the weight savings can only be accomplished by comparing morphing leading and trailing edge to the conventional wing components performing the same functions, high lift devices for instance. The above-mentioned comparisons make sense once a clear definition of the aircraft architecture, with regard to both concepts, is established. It is necessary to specify whether the morphing device is replacing the high-lift devices or simply being combined with them. This decision demands information only available at a detailed phase of the product design. Nevertheless, once enough aircraft designs implementing both concepts have been performed, the acquired engineering knowledge can assist in judging the best combination. In what follows, the replacement of the high-lift devices with morphing devices has been considered.

It is worth remembering that take-off and landing length requirements determine the maximum lift coefficient demanded to high-lift devices. Fig. 22 shows how the maximum lift coefficient for some aircraft is related to the system complexity of these high-lift devices, which are heavier components. For regional aviation, in particular, the maximum lift requirements are fairly demanding and usually require adoption of systems with at least slat and flap components. Substituting these high-lift systems with morphing devices would require that the same level of maximum lift coefficient has to be assured. Otherwise, field performance would be penalized.

Weight estimation for conventional aircraft (based on geometrics) is a short process due to the vast database from aviation history. However, for aircraft employing new technologies such as morphing, there is a lack of data in the literature, so this is not an easy task. This lack of data makes it a challenge to validate semiempirical methods for the evaluation of the weight of leading and trailing edge morphing devices, especially with reference to the possibility of discarding conventional high-lift devices in favor of morphing devices.

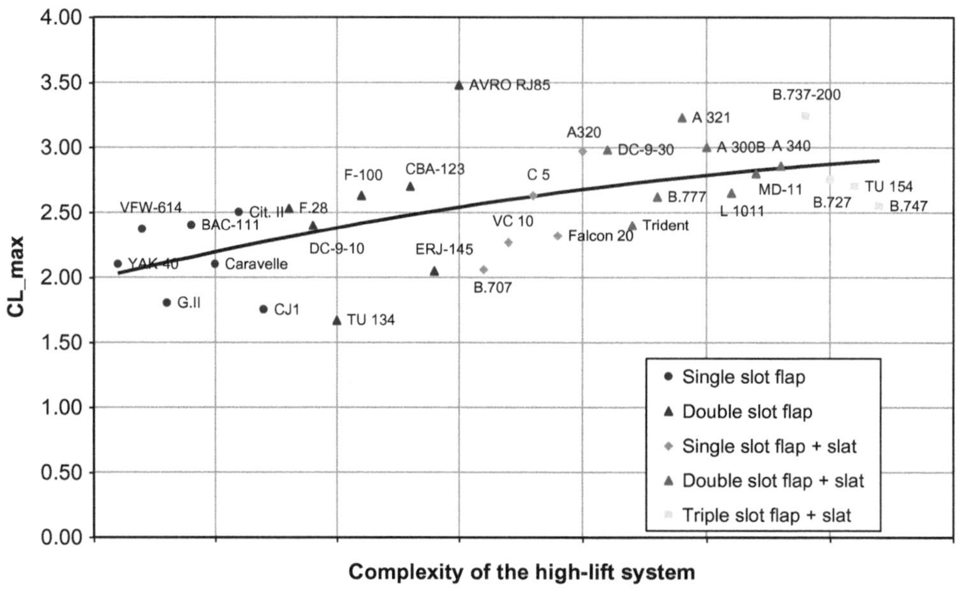

FIG. 22

High-lift systems for different aircraft.

7.1 MORPHING DEVICES

Definition of leading and trailing edge morphing devices accounted not only for their aerodynamic aspects but also structural constraints such as skin deformability and axial and bending stresses. These structural constraints led to deflections of the leading and trailing morphing devices different from those found on high-lift devices. The lower deflection angles provided by the morphing devices result in a lower maximum CL. However, if the morphing devices can be lighter than high-lift devices, and field performance requirements are relaxed, it is possible to adopt them. This trade-off depends on many aspects of the aircraft design, and it must be studied via a multi-disciplinary optimization process. In the present project, such a multidisciplinary optimization (MDO) exercise has not been performed.

Rules of thumb simplify the decision making process in the earlier design phases, and should indicate how much lighter the morphing device is without getting into the details of the design. The present method proposes two metrics. The first one is the based on the weight per length for the leading edge, and the second one is the weight per area for the trailing edge. In order to define these, it is necessary to use data from real aircraft and semiempirical correlations available in the literature.

Fig. 23 shows the structural components that constitute the leading edge of a wing without the slat component. The weight per length for this specific part of the wing is approximately 6.5 kg/m. It is important to keep in mind that this metric is an estimate based on a few analyzed aircraft with a similar structural design concept.

The same sort of metric was constructed for a wing leading edge with a slat component. Fig. 24 shows an example of some structural components from the slat element. The weight per length for this

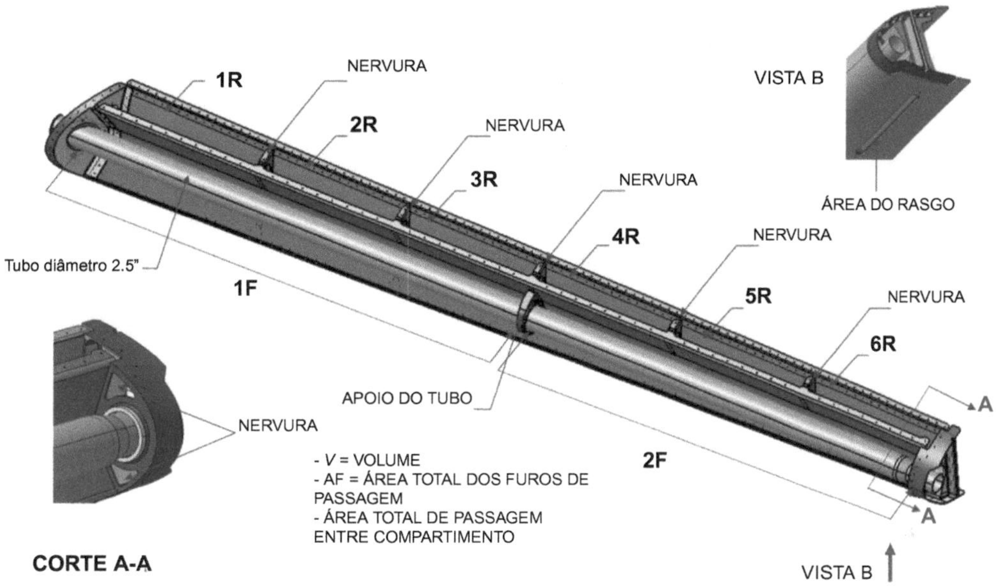

FIG. 23

Structural components of the wing leading edge—no slat.

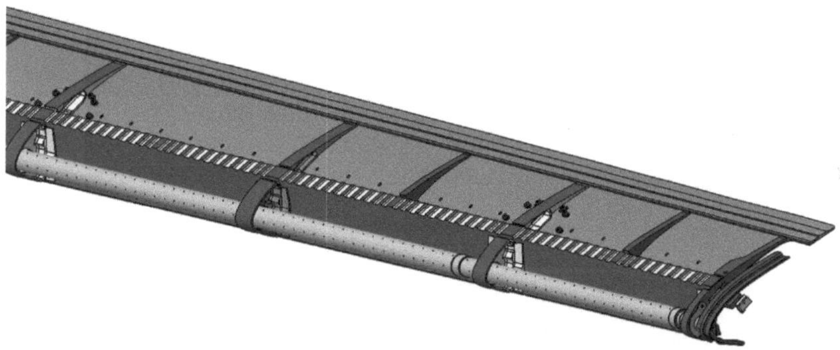

FIG. 24

Structural components of the wing leading edge—with slat.

wing part is 13.0 kg/m, which is heavier than the previous result due to additional movable components. Applying this metric to estimate the leading edge of the RA leads to a total weight of 410 kg. The computed weight for the same RA leading edge resulted in 550 kg. It is important to notice that the elaborated metric is a proposed semiempirical correlation that averages the results from different designs.

In order to compare conventional and nonconventional high lift devices, a morphing device based on compliant mechanisms coupled with conventional actuators was considered in this work. SPHERA

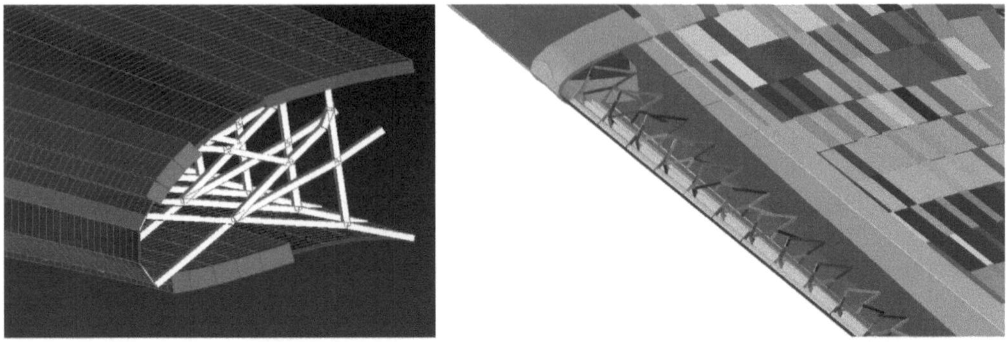

FIG. 25

Structural solution for the leading edge morphing device.

was used for a preliminary design of the leading edge morphing device shown in Fig. 25 and for its integration into the structural model of the reference wing [11]. This component with anti-icing systems has a weight per length of 8.0 kg/m, including compliant mechanisms and morphing skin, which is lower than the conventional solutions, thus demonstrating weight benefits.

The definition of the weight per length for the trailing edge of the high-lift devices is more complicated than for the leading edge. This peculiarity results from the different design solutions that can be adopted for the trailing edge. The design systems that can be employed are:

- Split flap
- Plain flap
- Simple slotted flap
- Single-slotted Fowler flap
- Fixed vane/main double-slotted flap
- Articulating vane/main double-slotted flap
- Main/aft double-slotted flap
- Triple-slotted flap

In the present project, the weight-per-area metric, defined in the NASA RP-4746 document [10], is adopted as the main metric for the trailing edge devices. Table 5 shows the weight-per-area for few of the trailing devices referred to in the NASA document. Employing this metric to compute the weight of the trailing edge devices for the RA provides a total of 1587 kg for a total flap area of 51 m^2. The current weight for the flap devices computed with Embraer's methodology is 1150 kg, a noticeable difference which can be partially attributed to the use of composite materials in the construction of the flap component.

Fig. 26 shows the structural solution of the morphing device concept developed by POLIMI, based on compliant mechanisms [11].

As previously mentioned, the idea is to have some rapid procedure, like the one shown in Fig. 27, to evaluate morphing devices in earlier design phases. However, the weight obtained using the defined metric indicates that a great disparity can be encountered for high-lift devices (around 25%), and the

Table 5 Weight-Per-Area (lb/ft^2) for Different Trailing Edge Concepts

	Single Slotted	Fixed Vane/Main	Articulating Vane/Main	Double Slotted	Triple Slotted	Single Slotted
Support	***Hooked Track***	***Hooked Track***	***Hooked Track***	***Hooked Track***	***Hooked Track***	***Link/ Track***
Flap panels	2.7	3.0	3.5	4.8	5.5	2.7
Supports	3	3.2	3.8	4.7	5.6	1.5
Actuation	2.2	2.2	2.3	2.4	2.5	2.0
Fairing/flap area	(0.45)	(0.45)	(0.50)	(0.55)	(0.60)	(0.05)
Fairing	1.0	1.0	1.2	1.3	1.4	0.1
Total flap	8.9	9.4	10.8	13.2	15.0	6.3

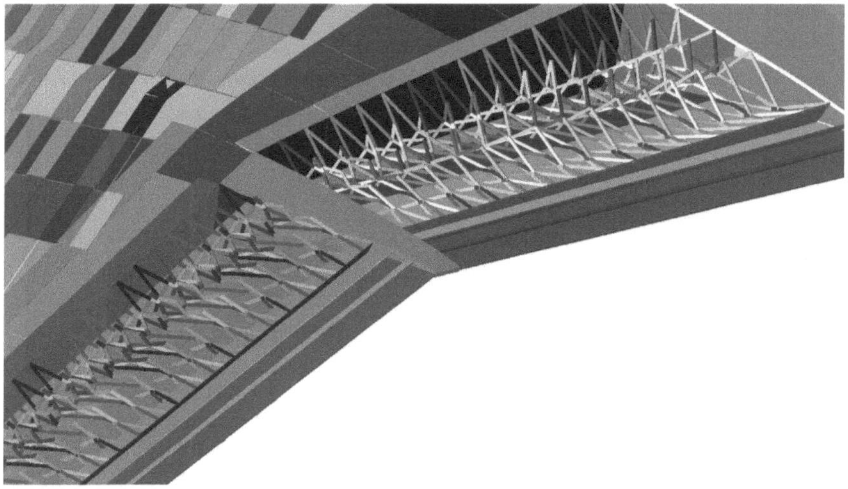

FIG. 26

Structural solution for the flap panel.

weight for morphing devices has not even been validated with a real case. It is clearly necessary to have a more profound validation process before the proposed metric can be used to make conclusions regarding the weight benefits of morphing devices.

8 BENEFIT EXPLOITATION IN THE TRANSPORT AIRCRAFT DESIGN

The studies performed during NOVEMOR project have shown that morphing devices are a promising technology that may help in achieving some environmental goals expected by ACARE. Nevertheless, there is still a long path to cover before this concept can be considered ready to be operated by civil

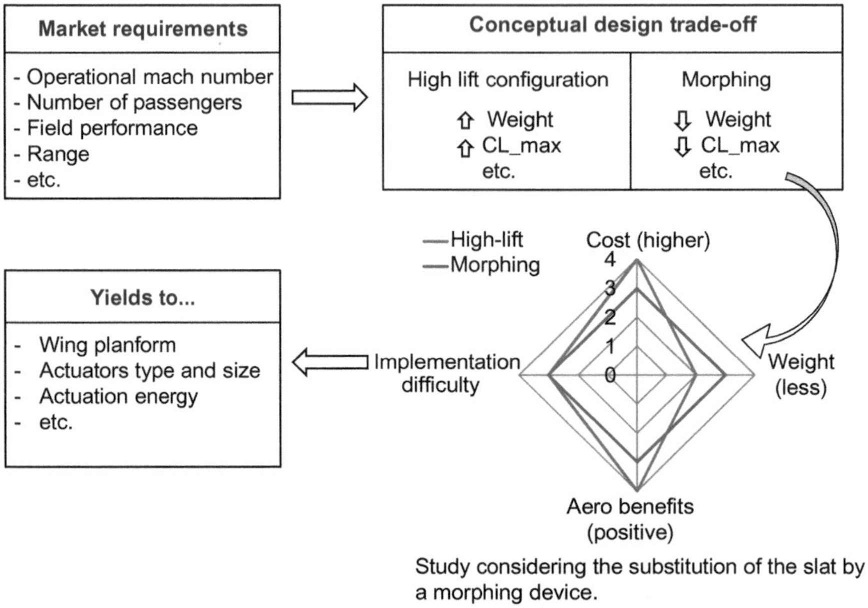

FIG. 27

Decision process to define the employment of morphing devices.

aviation. Some of the key-issues are not intrinsically related to the technology concept itself, but to the economic profit that can be obtained by implementing the technology for regional aircraft. Probably, morphing technology will emerge in the following sequence:

(a) UAV
(b) Business jet
(c) Long-range aircraft
(d) Regional aircraft

In order to have any new technology implemented into an aircraft, the entire system must reach TRL nine, and if a single component doesn't satisfy a requirement, then the technology is not viable. Fig. 28 shows the definition of each TRL level.

It can be observed that the TRL varies as a function of the considered analysis. Based on the overall analysis performed with the results of the NOVEMOR project, it is expected to have TRL nine in approximately 10 years. There are some topics that are more evolved, but as mentioned before, the entire system must have TRL nine. Thus, the most immature topic will delay the entire development, unless efforts are directed towards that specific topic with the lowest TRL in order to accelerate the maturity of the system.

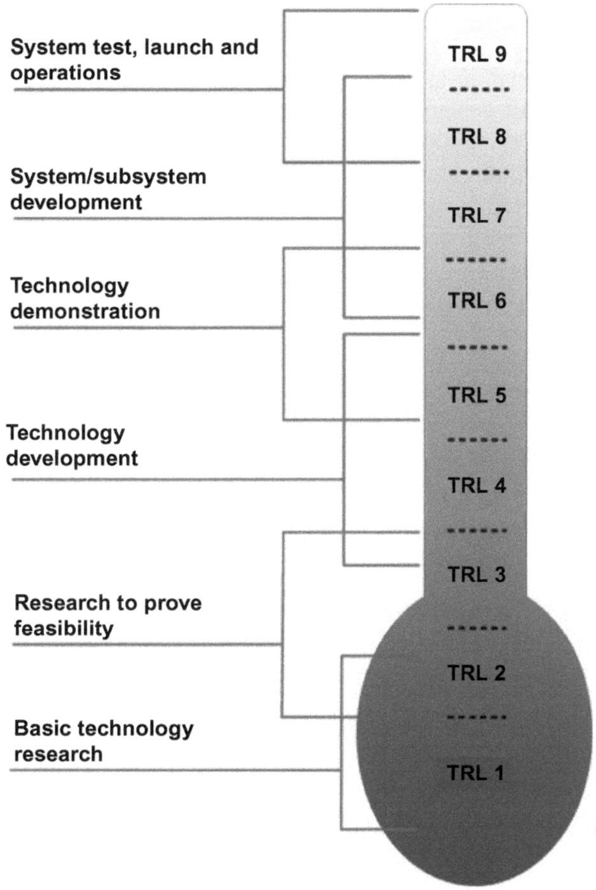

FIG. 28

Definition of the technological readiness level, TRL.

9 CONCLUSIONS

The aerodynamic analyses for the RA and the evaluated morphing devices were performed in accordance with suitably conceived strategy. The definition of the mathematical and numerical formulations considered the need to have the most adequate approach to capture the pertinent physical phenomena at the subsonic and transonic regimes. Indeed, as expected, the transonic regime demanded the use of high-fidelity tools to capture the nonlinearity of the flow.

The aerodynamic scheme, based on the 2.5D/Infinite Swept Wing (ISW) Navier-Stokes method, coupled with a 3D panel method for potential flow calculations, proved to be capable of measuring small differences in the lift and drag coefficients.

Another important conclusion from the aerodynamic and performance assessment of the morphing devices was that the design process must consider the use of such technology since the earlier design phases. The earlier the decision to install a morphing device, the easier it is to rethink the definition of the flight points for the aircraft design. In the present project, the design of the RA design has not considered the possibility of changing the design points to tailor the expected benefits. This fact has led to benefits of the morphing devices at regions of the range-versus-payload diagram that are not so convenient to be operated by the RA. Based on this lesson learned, the following procedure is recommended for future research projects:

(a) define the wing planform,
(b) define the morphing devices: leading edge and trailing edge and the chordwise extension of these devices,
(c) define the maximum allowed deflections based on the structural constraints: assume a load distribution and parametric variations that lead to an efficient lift distribution (consider the aircraft Vn diagram to define the CL values),
(d) obtain the abacus relating the maximum deflection along the wing spanwise direction with the type of lift distribution,
(e) define the flight conditions that satisfy the market survey: nominal flight condition, LR: long range and the FC: fast cruise,
(f) design the reference wing for one of the conditions of item (e) without the use of the morphing devices: the lift loading distribution of the designed wing might be different from those considered on item (c), use engineering judgment to come up with allowable deflections of the morphing devices for the designed wing,
(g) generate an aerodynamic database for the reference design and all the possible morphing deflections defined on item (f),
(h) for each pair (Mach, CL), define the best CD from item (g) and the configuration associated with this value,
(i) use a performance code to fly the aircraft considering the entire payload versus range diagram,
(j) check the regions where the morphing devices provide the benefits,
(k) if this region is not the most recommended under the operation point of view reconsidered the design point selected at item (e) and come back to item (f). (This is an iterative process that must be executed until the reference wing the morphing devices can provide benefits on the expected conditions of the flight envelope.)

The performed study of the NOVEMOR project about the weight benefits does not allow a conclusive position about this issue. The results are not conclusive due to the dispersion. The weight assessment is not simple because it requires evaluation of the configuration at the component level or the use of consolidated database (which is not available for the morphing technology). Thereby, in order to have any sort of weight assessment for the morphing device at earlier design stages, it is necessary to adopt a detailed analysis of the structural solution, mechanisms, and actuators.

ACKNOWLEDGMENTS

The research leading to these results has been partially funded by the European Union's Seventh Framework Programme [FP7/2007-2013] under the grant agreement "NOVEMOR – Novel Air Vehicle Configurations: From Fluttering Wings to Morphing Flight" No. 285395. A special thanks to Filippo Ghirardini from Politecnico di Milano for provided support in the development of the 2.5D/ISW aerodynamic code.

REFERENCES

[1] H.P. Monner, Realization of an optimized wing camber by using formvariable flap structures, Aerosp. Sci. Technol. 5 (2001) 445–455.

[2] J.A. Hetrick, S. Kota, An energy formulation for parametric size and shape optimization of compliant mechanisms, J. Mech. Des. 121 (6) (1999) 229–234.

[3] J. Bowman, B. Sanders, B. Cannon, J. Kudva, S. Joshi, T. Weisshaar, in: Development of next generation morphing aircraft structures, 48th AIAA/ASME/ASCE/AHS/ASC Structures, Structural Dynamics and Materials (SDM) Conference, Honolulu, Hawaii, 2007.

[4] B. Roth, C. Peters, W. Crossley, Aircraft sizing with morphing as an independent variable: motivation, strategies and investigations, 48th AIAA/ASME/ASCE/AHS/ASC Structures, Structural Dynamics and Materials (SDM) Conference, Honolulu, Hawaii, 2007.

[5] A. De Gaspari, S. Ricci, Knowledge-based shape optimization of morphing wing for more efficient aircraft, Int. J. Aerosp. Eng. (2015) 1–19, https://doi.org/10.1155/2015/325724.

[6] A. De Gaspari, S. Ricci, A two-level approach for the optimal design of morphing wings based on compliant structures, J. Intell. Mater. Syst. Struct. 20 (10) (2011) 1091–1111.

[7] S. Vasista, A. De Gaspari, S. Ricci, J. Riemenschneider, H. Monner, B. van de Kamp, Compliant structures-based wing and wingtip morphing devices, Int. J. Aerosp. Eng. 88 (5) (2016), https://doi.org/10.1108/AEAT-02-2015-0067.R1.

[8] S. Ghasemi, A. Mosahebi, E. Laurendeau, in: A two dimensional/infinite swept wing Navier-Stokes solver, Proceedings of the 52nd Aerospace Sciences Meeting (AIAA SciTech), National Harbor, MD, USA, January 2014.

[9] G. Bindolino, P. Mantegazza, L. Visintini, in: A comparison of panel methods for potential flow calculations, VIII Congresso Nazionale AIDAA, Torino, Italy, September 23–27, 1985.

[10] P.K.C. Rudolph, High-Lift Systems on Commercial Subsonic Airliners, NASA Contractor Report 4746, September 1996.

[11] A. De Gaspari, S. Ricci, L. Travaglini, in: Aeroelastic analysis of a regional aircraft with active camber morphing device, Proceedings of the International Forum on Aeroelasticity and Structural Dynamics (IFASD '15), Saint Petersburg, Russia, June–July, 2015.

SECTION 3

Morphing Skins

Editor:

Ferri Aliabadi
Imperial College London, London, UK

List of Chapters

CHAPTER

MORPHING SKIN: FOAMS

7

Oliver Schorsch, Christof Nagel, Andreas Lühring

Fraunhofer IFAM, Bremen, Germany

CHAPTER OUTLINE

NOMENCLATURE

$\sim$	nearly
ATED	Adaptive wing Trailing Edge Device
D	drag
D_i	temperature-dependent material parameter
J	volume ratio
k	material-dependent parameter
K_0	initial bulk modulus
L	lift
n	Poisson's ratio
N	material parameter
N_f	number of cycles to failure
PDMDPS	polydimethyldiphenylsiloxane
PDMS	polydimethylsiloxane
R	displacement ratio
r.h.	relative humidity
TE	trailing edge
U	strain energy density
UV	ultra-violet (radiations)
α_i	temperature-dependent material parameter

Morphing Wing Technologies. https://doi.org/10.1016/B978-0-08-100964-2.00007-1

ε_a	strain amplitude
ε_{a0}	material-dependent parameter
$\overline{\lambda}_i$	deviatoric principal stretch
μ_i	temperature-dependent material parameter

1 INTRODUCTION

From the beginning of aircraft industry, aerodynamic morphing has been used to optimize aerodynamic performances along the flight by changing surface area or camber. Today, entirely mechanical actuation systems are established, where changes of angle/shape or surface area are obtained by moving solid elements. Examples are the movable engine of Osprey helicopters, thrust vectoring of the Harrier, swing wing of fighter aircrafts, as well as slats, flaps and spoilers of conventional aircraft [1].

Although widely used, moving solid elements have many disadvantages. Some solutions are too heavy, complex, or a gapless approach is demanded for reduction of noise and fuel consumption [2].

In contrast to moving solid elements, seamless or gapless morphing is still a research topic. Gapless morphing combines an appropriate actuation system and a skin layer which is able to deform and to carry loads simultaneously. In the past, many actuation systems have been investigated, but up to now there is no single material known yet that solves the mechanical paradox of morphing skins for aviation applications satisfactorily [3].

As shown in many former morphing wing projects, the implementation of large morphing areas leads to heavy and expensive solutions. If small shape or camber changes are requested, one possible solution is the integration of elastomer covered adaptive elements.

The advantages of elastomer based skins are the possibility of a completely seamless approach including outstanding compression and tension properties. On the other hand, elastomers are known for stiffening at low temperature, low abrasion resistance, and fatigue caused by high strain.

A new morphing approach was investigated in the SARISTU project, funded by the European Union's Seventh Framework Programme [4–6]. The work focused on the integration of smaller polymer based morphing elements to established structures like the wing trailing edge and winglet.

The SARISTU Adaptive wing Trailing Edge Device (ATED) was designed with reference to the outer wing of a CS-25 category aircraft. Fraunhofer IFAM developed a polymer based morphing skin to seamlessly cover a multifinger rib architecture enabling conformal and differential airfoil camber morphing. Wing shape is controlled during flight (cruise condition) to compensate weight reduction following fuel burning, by allowing the trimmed configuration to remain optimal in terms of efficiency (*L*/*D* ratio) or minimal drag (*D*). Trailing edge adaptations were investigated to achieve significant benefits in aircraft fuel consumption whose reduction may range from 3% to 5% due to the improved aerodynamic efficiency.

2 DESIGN PRINCIPLES

In this section it is investigated how seamless morphing can be achieved in an active trailing edge device (ATED). The basic idea here is to keep the continuity of the aerodynamic surfaces by the elastic deformation of a solid material, which will be referred to as "morphing skin" hereafter. The morphing

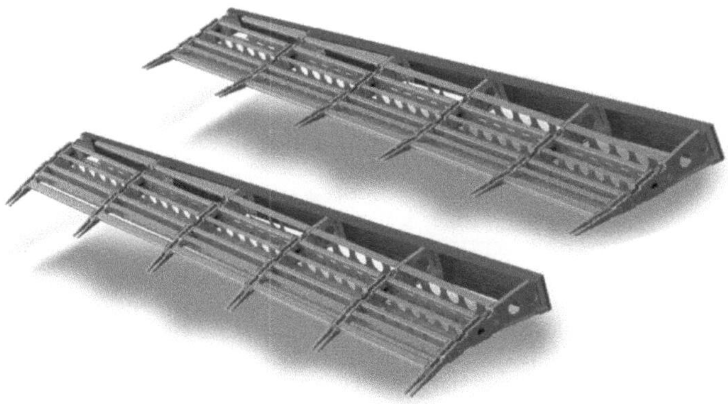

FIG. 1

Primary structure of the ATED [7,8].

skin is to be placed onto a primary substructure, which differs from a conventional structure in that the ribs contain hinges, such that the camber can be changed, as shown in Fig. 1.

A major challenge is to make a highly deformable, yet stable structure, which can be operated at low (−55°C) and high (+80°C) temperatures, and which is stable under fatigue loading conditions and environmental impact. The morphing skin should fill the gap between two stiff parts of the outer aerodynamic surface, one of those remaining in its initial position and the other one moving with respect to it. The morphing skin must be able to follow dimensional changes of the gap evolving between the stiff parts but must also maintain conformity in the deflected configuration for aerodynamic reasons. The morphing skin must withstand movements of about 5 degrees of the trailing edge and relevant air pressure in operation. Furthermore, it should allow for maintenance and repair. This means that the morphing skin should have a removable connection to the aircraft structure. Also, the weight of the morphing skin must be as low as possible. In addition, it is crucial that the actuator capacity needed to deform the morphing skin remains sufficiently low for the whole system to be economic. To obtain sufficient stiffness against air pressure and—at the same time—flexibility for aerodynamic continuity and economic operation is probably the biggest challenge within elastomer based morphing approaches [1].

In order to solve this problem, the morphing skin was segmented, alternating stiff and soft blocks. Thus, only small, soft parts are subject to air pressure and the related deformation remains small because the support length of the soft segments is relatively short. It should, however, not be too short because otherwise the TE movement would generate strains which are critical to the material. Primary structure and kinematics were defined on the base of specific functional constraints, so the space left for the thickness of the morphing skin was limited to values between 4 and 12 mm, depending on the distance from the trailing edge tip. For this reason, a significant thickness variation, which would have been an alternative way for stiffness variation, was impossible. Hence, the design process was focused on the length of the soft segments.

Intermediate results of the design process are shown in Fig. 2. Results of the preliminary design phase were that out-of-plane displacements due to air pressure loading could be kept below 1 mm if the distance of the supporting points were below 50 mm and the stiffness would be in the order of about 150 N/mm per 100 mm span wise edge length and 10 mm skin thickness.

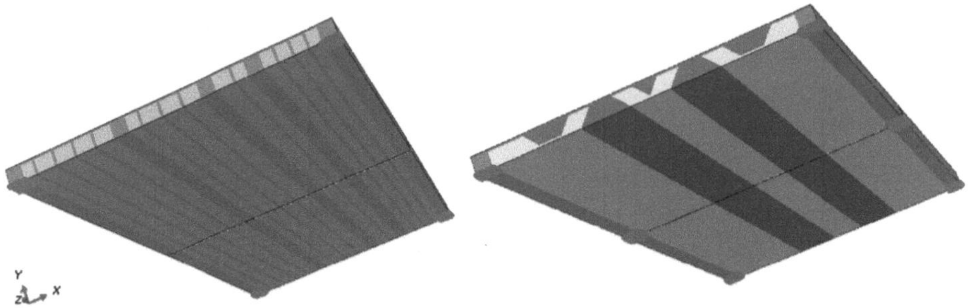

FIG. 2

Design study of morphing skins, representing panels of 200 mm span wise (*z* direction) and 160 mm chord wise (*x* direction) edge length and 10 mm thickness. Alternating soft and hard segments are differently shaded. The top layer, perpendicular to the *y*-direction, is shown in dark.Direction of strain load: *x*, direction of air pressure load: *y*.

A joining concept was needed in order to assemble soft and stiff segments and the top layer. Because of the soft materials involved, and the material mix with stiffness variations over several orders in magnitude, a laminating process was chosen. This is an effective technology to avoid concentrated forces, which would immediately cause failure in soft materials. Because it was also required at a later stage to take off the skin for maintenance of actuators, hinge mechanics, instrumentation, etc., bolts were foreseen in areas where stiff skin segments go onto the ribs of the primary substructure shown in Fig. 1.

3 LOW TEMPERATURE ELASTOMERS

The core material of elastomer based morphing skins has to meet several important requirements. Most importantly, almost constant compression and tension properties are needed as well as constant storage modulus over the whole temperature range of an aircraft wing surface (−55°C to 80°C) and over a very large number of temperature cycles.

Moreover, resistance to abrasion (sand, small stones on runway), environmental resistances (weather, UV radiation) as well as chemical resistance (Skydrol, deicing agent) are essential. The strain within the soft skin segments is of great importance for the material selection as well. It is given by geometry and deflection angle of the morphing trailing edge. Maximum strain should be kept below 10%, otherwise the elastomer would fail due to fatigue after several loading cycles. At least 100,000 cycles are expected to be sufficient for a full-service time of a common aircraft. Finally, high adhesion strength between elastomer skin layers, elastomer foam, and aluminum parts are required to load transmissions between hard and soft segments.

There are several certified elastomers available for aircraft applications. Examples are silicone, EPDM, and polysulfide. All these elastomers are used as sealants, but not for high strain. The main reason is the loss of elasticity as soon as environmental temperatures falls below the glass transition temperature of the elastomer (usually between 45°C and −10°C, Fig. 3).

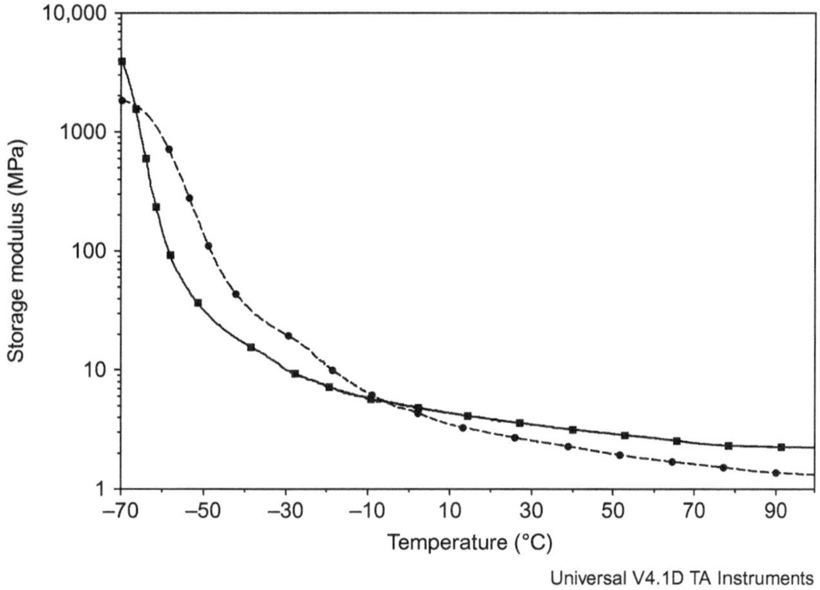

FIG. 3

Dynamic mechanical analysis of EPDM sheeting (Dr. D. Müller, *circle*) and low temperature polyurethane Sikaflex 553 (Sika, *square*) [9].

Silicone elastomers are known to provide excellent elasticity over a wide temperature range. Among the variety of silicone elastomers, polydimethylsiloxane (PDMS) is the most used for technical applications. Fluorosilicone is based on polytrifluoropropyl-methylsiloxane and is used for applications that require fuel or hydrocarbon resistance.

Although glass transition temperatures of these silicones are well below −100°C, common PDMS and fluorinated silicone lose flexibility below 35°C due to cold crystallization of polymer segments. Crystallization can be reduced or even inhibited by substitution of functional groups at polymer side chains. Polydimethyldiphenylsiloxane (PDMDPS) is the most important crystallization hindered silicone. Because of almost constant elasticity between 100°C and −100°C, phenylated silicone is an important encapsulation material for space applications (Fig. 4).

Several commercial low temperature silicones are available, mainly used for the space industry or other low temperature applications. Most of low temperature silicones are either peroxided or platinum catalyzed products. Up to now no commercially available low temperature silicone sheets and foams meet all the material properties for a morphing skin. Furthermore, silicones are known to be very difficult to bond to adhesives.

Reactive formulations can be tailored to improve the needed material properties and to maximize the adhesion strength between all material interfaces of the morphing skin.

Examples for low temperature platinum curing encapsulants are Momentive RTV655 and Dow Corning 3–6121. After curing the two-part formulations, the resulting elastomers show almost constant storage modulus between −60°C and 100°C (see Fig. 5).

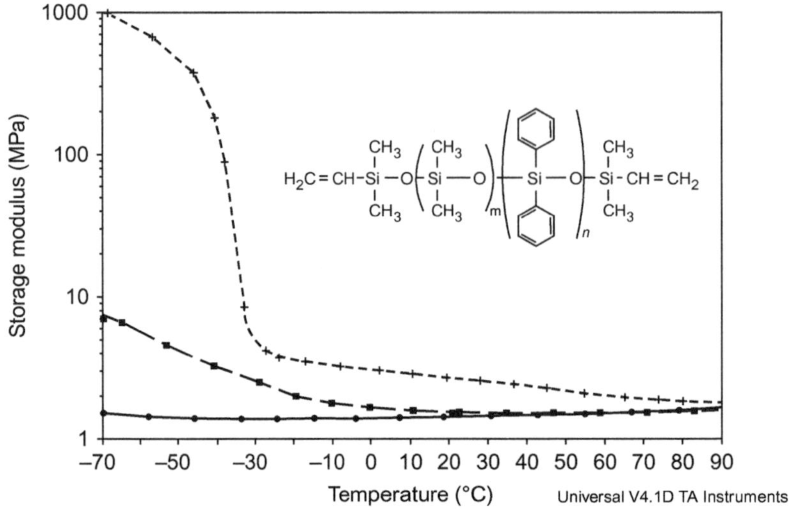

FIG. 4

Dynamic mechanical analysis of common silicone sheeting (MVQ Silicones GmbH, star, $n = 0$) and low temperature silicones with different methyl/phenyl ratio (*m*/*n*), Sylgard 184 (Dow Corning, *square*), Elastosil S690 (Wacker, *circle*) [9].

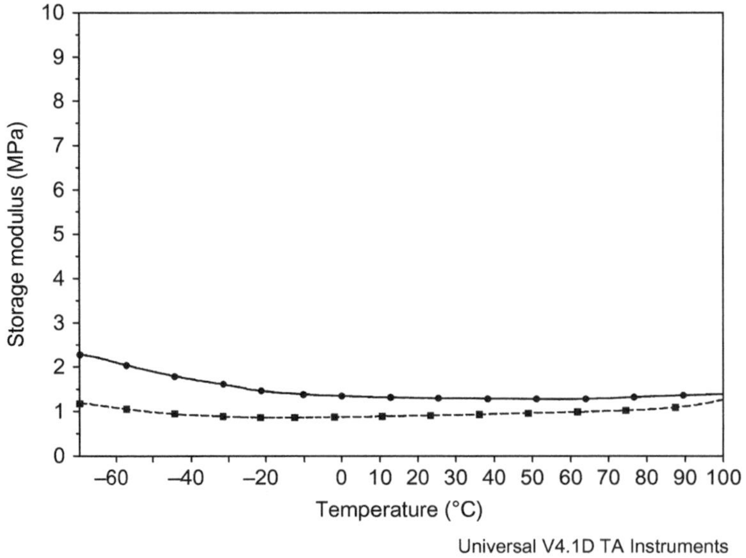

FIG. 5

Dynamic mechanical analysis of two commercial low temperature platinum curing encapsulants, showing almost constant storage modulus between −60°C and 100°C. Dow Corning 3–6121 (*square*) and Momentive RTV655 (*circle*).

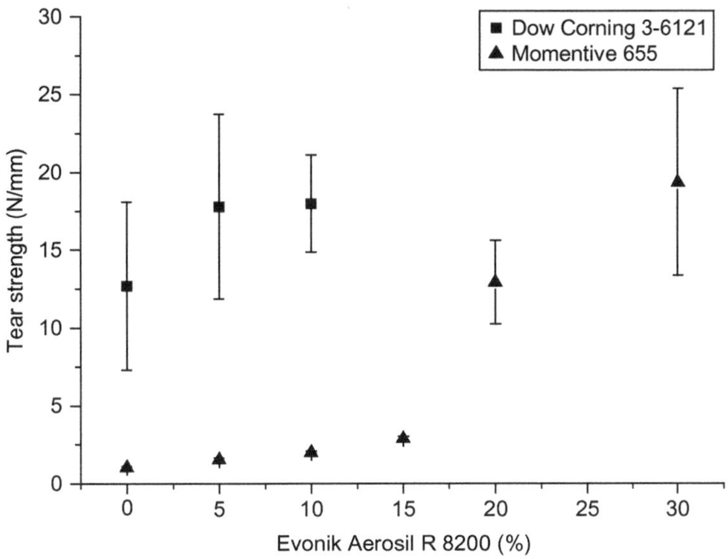

FIG. 6

Effect of fumed silica on tear strength of different low temperature silicones [9].

Unfilled silicone elastomer has poor mechanical properties. Tensile strength, elongation at break, and tear strength can be improved by filler materials. In opposite to pure encapsulant, Momentive 655 Dow Corning 3-6121 already contains an unknown amount of fumed silica.

The effect of fumed silica on tear strength and tensile properties is shown in Figs. 6 and 7. Different levels of Evonik AEROSIL R 8200 were added to Momentive 655 and Dow Corning 3-6121. All formulations were vacuum dispersed by Planetary Centrifugal Vacuum Mixer and cured 1 h at 100°C.

A tear strength test was carried out with DIN 53515 using Graves angle test piece. A tensile test was carried out with DIN 53504 to measure tensile strength and elongation at break.

Mechanical properties can already be improved by low filler concentrations. But >15% fumed silica is needed to increase the resistance to tearing ("locking effect"). On the other hand, additional filler material always increases formulation viscosity that makes processing more difficult or even impossible. Special surface treated fumed silica is available for silicone rubber to reduce this effect.

Due to crosslinking reaction, the formulation viscosity increases after mixing part A and B. Fig. 8 shows the viscosity changes of different encapsulants. The time to double viscosity is defined as pot life. Depending on how to process the liquid formulation, the pot life can be increased by the addition of inhibitors.

Silicone foams are usually manufactured with special foaming equipment. Fig. 9 shows an example of silicone foam, which was formed in an extrusion process based on peroxide crosslinking.

Liquid formulations of reactive silicone are difficult to foam without special foaming equipment. One option is the addition of physical or chemical foaming agents. While chemical foaming agents are thermal reactive components, physical foaming agents consist of encapsulated low-boiling solvents. Both foaming agents release gas after reaching a critical temperature. Those reactive formulations must

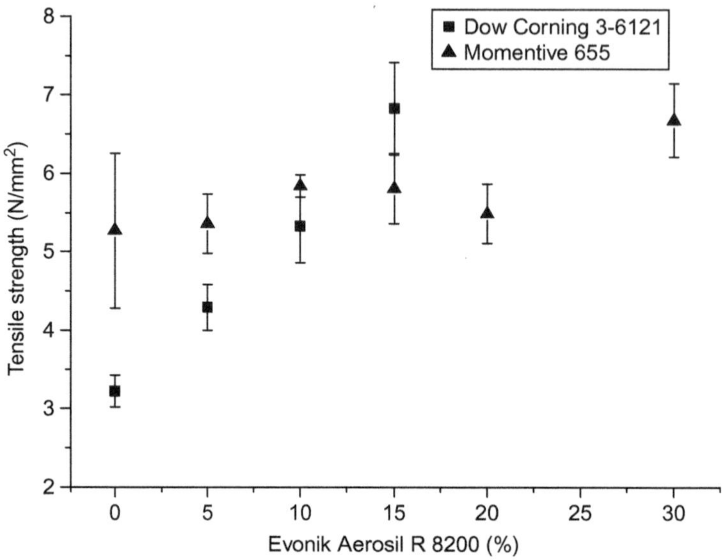

FIG. 7

Effect of fumed silica on tensile strength and elongation at break of different low temperature silicones.

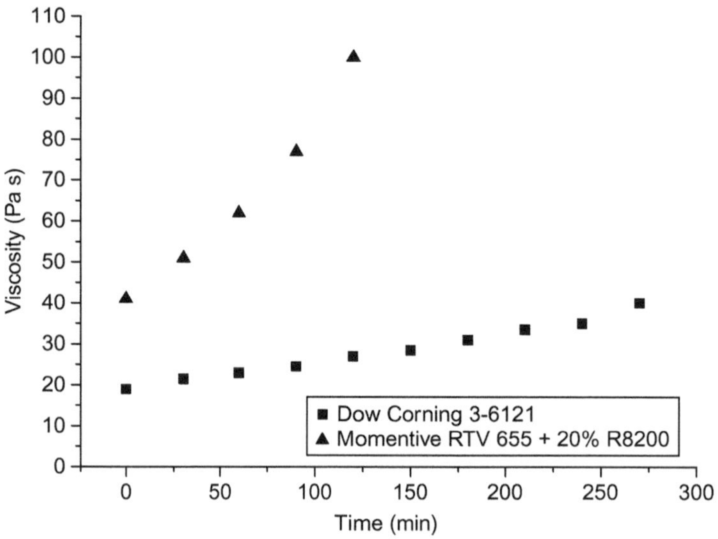

FIG. 8

Rheological measurement of reactive silicone encapsulants (viscosimeter, 4 cm plate 10 1/s @ 23°C, shear time 150–300 s).

FIG. 9

Example of BIW silicone foam extrusion with silicone sheeting on top.

be tailored specifically, since the curing reaction and foaming reaction occur at the same time. The speed of crosslinking reactions can be adjusted by adding inhibitors or crosslinking catalysts.

4 MATERIAL PROPERTIES OF HYPERFLEX

A flexible HYPERFLEX-02 foamed elastomer was formulated specifically for application in morphing devices. Quasistatic tensile and compression tests were performed at −55°C, 23°C, and +80°C at a nominal strain rate of approximately 0.1/s (ISO 527-2 and ISO 604, [10,11]). A total of six specimens were tested under each condition. The material parameters derived from the tests are summarized in Table 1 and the stress-strain data are shown in Fig. 16. Also, the material properties of HYPERFLEX-03 homogeneous, flexible elastomer from [9] were used. Results are summarized in Table 2. It can be seen that both materials behave like soft elastomers over the whole temperature range. No significant stiffening effect is visible. The observed increase in strain at fracture with decreasing temperature is useful because most of the operation cycles will occur in the flight phase, which is at low temperature. Most important results from compression tests were low and temperature insensitive initial elastic moduli and the absence of failure until ~70% or ~80% strain.

Poisson's ratio (ν) of HYPERFLEX-02, as determined by video analysis, was found to be close to 0.2, which is significantly lower as compared to HYPERFLEX-03, but nonzero. This means that there is a transverse contraction for HYPERFLEX-02 parts if loaded in uniaxial tension, but loading of a typical soft segment is biaxial and transverse strain will consequently be even lower. The value measured for HYPERFLEX-03 at +80°C is given in parentheses since it is larger than 0.5. This effect is due to the relatively high value of ν for this material and the sensitivity of the method. A physically meaningful value of 0.5 is obtained if two standard deviations are subtracted from the mean value. The tear strength was measured in accordance with ISO 34 [12] using angle test pieces at a test speed of

Table 1 Results of Quasistatic Tensile Tests on HYPERFLEX-02 at a Strain Rate of ~0.1/s and Indicated Temperatures

Condition (°C)	Initial Elastic Modulus (MPa)			Strain at Fracture (%)			Tensile Strength (N/mm^2)			Tear Strength (N/mm)			Poisson's Ratio		
−55	3.63	±	0.28	94	±	9	1.53	±	0.07	8.74	±	0.94	0.14	±	0.04
23	2.17	±	0.04	81	±	7	1.17	±	0.08	3.91	±	0.26	0.22	±	0.00
80	3.60	±	0.55	59	±	6	0.67	±	0.10	3.00	±	0.16	0.22	±	0.02

Table 2 Results of Quasistatic Tensile Tests on HYPERFLEX-03 at a Strain Rate of ~0.1/s and Indicated Temperatures

Condition (°C)	Initial Elastic Modulus (MPa)			Strain at Fracture (%)			Tensile Strength (N/mm^2)			Tear Strength (N/mm)			Poisson's Ratio		
−55	2.33	±	0.09	309	±	66	10.08	±	1.78	28.94	±	4.71	0.46	±	0.03
23	1.96	±	0.05	173	±	11	5.49	±	0.38	12.90	±	2.70	0.47	±	0.03
80	1.90	±	0.06	140	±	10	4.88	±	0.39	4.93	±	2.00	(0.59	±	0.05)

500 mm/min. Results given in Table 2 show that there is significant tear strength but in comparison to typical sealing elastomers, the measured values tend to be low.

5 PROPERTIES OF BONDED JOINTS

T-joints consisting of aluminum parts bonded with HYPERFLEX02 and coated with HYPERFLEX03 were tested under quasistatic conditions at −55°C, 23°C, and +80°C. The T-joint represents a piece of the finalized skin module (see Fig. 17). It can be loaded in-plane in tension or compression, which represents stretching or compression of the skin module induced by trailing edge movements. Shape and dimensions of the T-joint are shown in Fig. 10. The purpose of the current tests was to confirm stability of adhesion among elastomer and aluminum parts and to provide experimental data for finite element calculations and sizing.

A crosshead speed of 0.03 mm/s was chosen, which corresponds to a nominal strain rate of 0.001/s. All samples were tested until visual damage, which was coincident with a sudden load drop. The recorded load-displacement data are shown until the maximum load in Fig. 11. From these data, the nominal strength, the displacement at maximum load, and the initial stiffness were calculated. The nominal strength was calculated as the maximum load divided by the bonding area. The initial stiffness was defined as the slope of the load-displacement curve in the range from zero to 10% nominal strain. The nominal strain was defined as the measured displacement divided by the initial length.

Test results are summarized in Table 3, showing similar variations as uniaxial test data but lower nominal values. Reasons for the low nominal values are the constraint effect imposed onto the foam layer by the relatively stiff aluminum parts, and locally weak adhesion as indicated by the fracture pattern, Fig. 12.

T-joints were tested under cyclic loading at 23°C using harmonic load-time functions and displacement control. A displacement ratio of $R = -1$ was chosen to simulate upward and downward TE movements. T-joints were clamped such that a clearance of 1 mm was obtained between grip and short limb of the aluminum L-profile on either side. A value of 50% stiffness change was used as a failure condition because displacement control hardly produces total fracture.

Results are shown in Fig. 13 using nominal strain amplitudes, which were calculated as the measured deflection divided by the initial length of the sample (Fig. 10). Statistical data analysis was

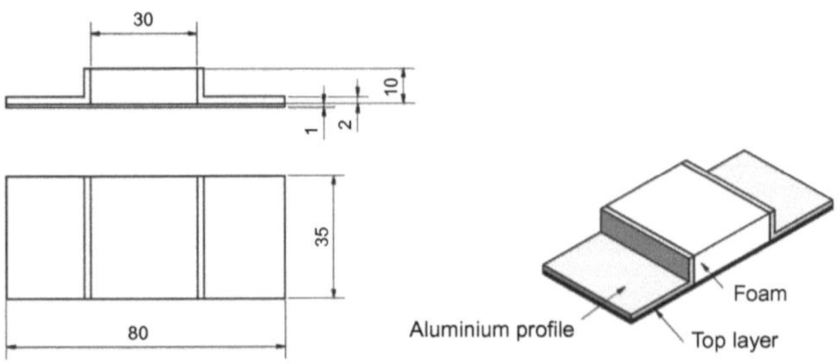

FIG. 10

T-joint.

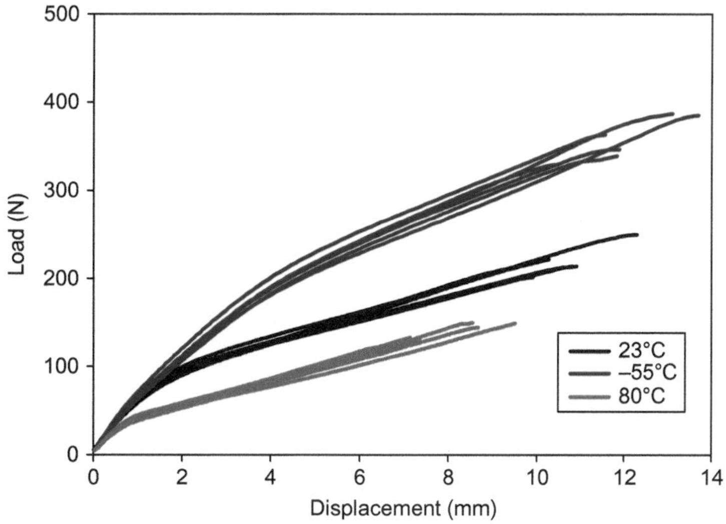

FIG. 11

Load-displacement curves from quasistatic tests of T-joints, tested under quasistatic loading with a loading rate of 0.03 mm/s. Test temperatures are indicated.

Table 3 Results of Quasistatic Tests of T-Joints Consisting of Aluminum Parts and HYPERFLEX, Tested Under Quasistatic Loading With a Nominal Strain Rate of 0.001/s

Condition (°C)	Fracture Load (N)			Nominal Strength (MPa)			Displacement at Fracture (mm)			Initial Stiffness (N/mm)		
−55	358.6	±	23.9	1.09	±	0.10	12.23	±	0.97	70.77	±	6.43
23	219.1	±	16.4	0.64	±	0.04	10.64	±	0.94	69.47	±	5.56
80	138.5	±	11.1	0.40	±	0.02	7.98	±	1.12	52.76	±	5.27

Test temperatures are indicated.

performed in order to estimate the number of cycles to failure at given strain amplitude with statistical significance. A power law has been used to represent the data in Fig. 13 in the form

$$\varepsilon_a = \varepsilon_{a0} \cdot N_f^{-1/k} \tag{1}$$

where ε_a denotes the strain amplitude, N_f denotes the number of cycles to failure, and ε_{a0} and k are material-dependent parameters. Since ε_a acts as controlled variable and N_f as dependent variable, Eq. (1) was resolved for N_f and subsequently least-squares fitted to measured N_f data as function of applied ε_{a0}, assuming a lognormal distribution of measured N_f.

Strain-life data measured at −55°C and +80°C tend to fall below the results at 23°C, specifically at higher amplitudes. Only −55°C and 23°C test data were considered in the fit because no significant contribution to damage was to be expected at 80°C, as effectively, no fatigue loading occurred at that

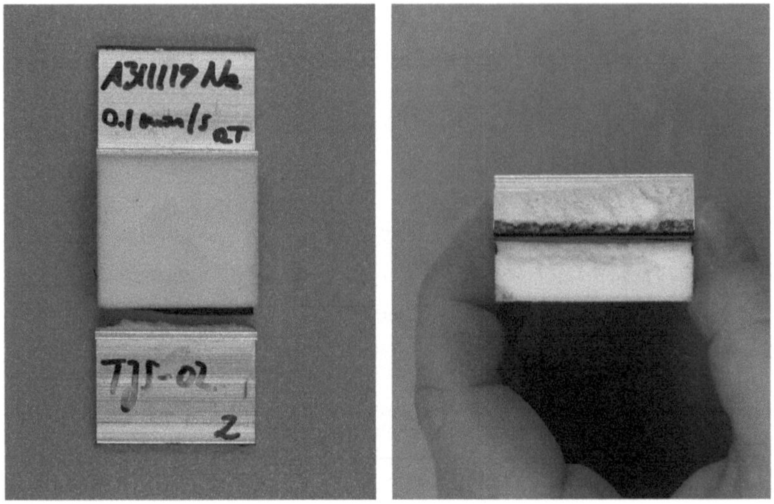

FIG. 12

Fracture surfaces of T-joints consisting of aluminum parts, HYPERFLEX-02, and HYPERFLEX-03, tested under quasistatic loading with a nominal strain rate of 0.06/s.

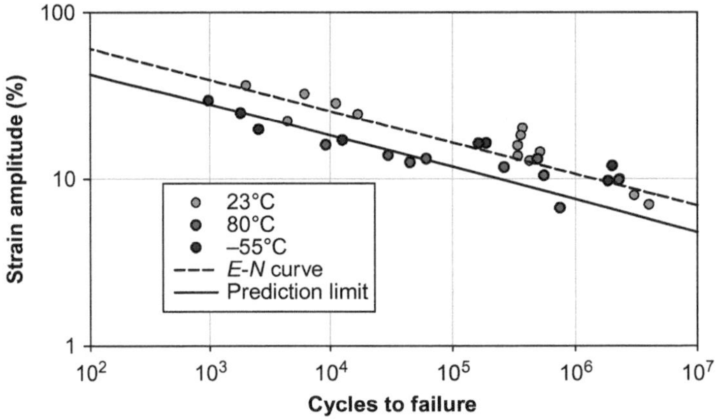

FIG. 13

Strain-life data from a displacement controlled, fully reversed fatigue test of T-joints, consisting of aluminum parts, HYPERFLEX-02 foam and HYPERFLEX03 top layer, at given temperatures and a test frequency of 5 Hz. Dashed and full lines represent the limits for a probability of 50% survival and 90% survival, respectively.

range. The fit curve is shown in Fig. 13 as a dashed line, giving a reasonable representation of the test data. The values of the corresponding fit parameters are $\varepsilon_{a0} = 10^{2.16 \pm 0.13}\%$ and $k = (5.32 \pm 0.57)$. Based on the standard deviation of the regression and a t-distribution for $\log(N_f)$, the lower prediction limit of the E-N curve was calculated for a chosen survival probability of $P = 90\%$ and $n-2$ degrees of freedom using a one-sided tolerance limit following the procedure outlined in ISO 12107 [13]. The lower limit is shown as a full line in Fig. 10. It can be seen that the fatigue life for a chosen probability of survival of 90% is reduced by about one order in magnitude as compared to the dashed *E-N* curve, which corresponds to a probability of survival of 50%.

Visual inspection of the fracture surfaces after completion of the fatigue test indicated that the joints failed predominantly by cohesive fracture within the elastomer material, similarly to Fig. 12. This means that the adhesion between elastomer and aluminum surface is stable against fatigue loads at −55°C, 23°C, and +80°C.

T-joints as described in the preceding paragraphs were exposed to cyclic laboratory aging conditions followed by quasistatic testing. One set of samples was exposed according to simulated "exterior, land, and air" environment (method C in ASTM D 1183-03 [14]), one cycle consisting of 48 h at (71 ±3)°C and <10% relative humidity (r.h.), 48 h at (23 ± 1)°C immersed in water, 8 h at (−57 ± 3)°C and 100% r.h., 64 h at (38.5 ± 2)°C and 100% r.h. A furnace and climate chambers equipped with calibrated temperature and humidity sensors were used. A second set of samples was prepared by immersion in SKYDROL 500-B-4 at room temperature for a certain number of cycles, one cycle equal to 168 h. After completion of one cycle, two samples were taken and tested immediately while the other samples were exposed to the next cycle. This was repeated until the last two samples had passed the last cycle. Test specifications and data evaluation were exactly as described in the section on quasistatic testing.

An overview on the joint properties after exposure to laboratory aging conditions to method C ("exterior, land, and air") in ASTM D 1183-03 is given in Fig. 14. Error bars refer to one standard deviation and full lines represent the mean value over all cycles. Neither strength, nor displacement at maximum load, nor the initial stiffness show systematic variation with the exposure time, indicating that the material was not affected by the environmental attack. Visual inspection of the fracture surfaces indicated that from cycles 1 to 4, joints failed predominantly by mixed cohesive fracture within the elastomer and adhesive failure at the foam-aluminum interface. From cycles 5 to 9, cohesive fracture within the elastomer was the predominant mechanism.

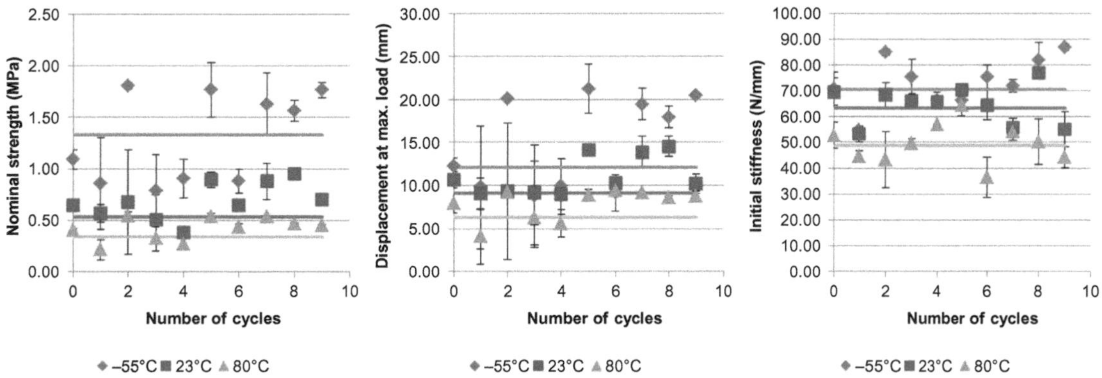

FIG. 14

Mechanical properties of T-joints tested under quasistatic conditions as function of the number of exposure cycles to ASTM D 1183–03C, "exterior, land, and air". Temperatures refer to the test conditions.

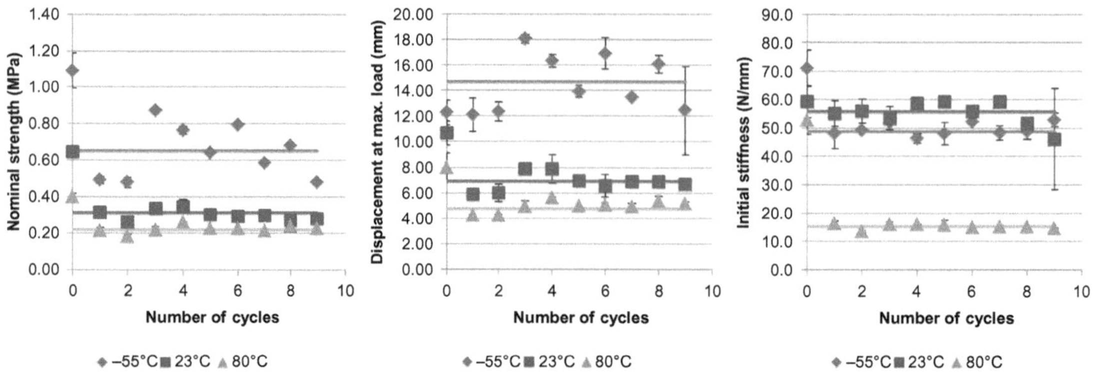

FIG. 15

Mechanical properties of T-joints tested under quasistatic conditions as function of the number of exposure cycles in SKYDROL at room temperature. Temperatures refer to the test conditions.

T-joint test results after immersion in SKYDROL 500-B-4 at room temperature are shown in Fig. 15. The error bars refer to one standard deviation and full lines represent the mean value over all cycles. It can be seen that after an initial drop, strength, displacement, and initial stiffness show no changes with increasing cycle number. This means that the joint properties are initially rapidly changed by immersion in SKYDROL, but stabilize after a couple of cycles. Visual inspection of the fracture surfaces indicated that the joints failed predominantly by cohesive fracture within the elastomer. A small adhesive failure fraction at the foam-aluminum interface was also visible, but it was significantly smaller than the adhesive failure fraction of the T-joints exposed to simulated "exterior, land, and air" conditions.

6 PROPERTIES OF MORPHING SKIN

It should be noted at this point that the stiffness of elastomers decreases with increasing number of load-displacement cycles. It is therefore not adequate for such materials to derive a material model based on the first loading cycle of a quasistatic tensile test. Load-displacement cycles of a T-joint are shown in Fig. 16, where the displacement was increased in each loading cycle. It can be noticed from Fig. 16 that the initial stiffness decreases with increasing number of load cycles, but the load-displacement curve in the current cycle approaches that of the preceding cycle as soon as the maximum imposed to the material in the preceding cycle is exceeded. This is commonly known as Mullins effect [15,16].

An Ogden type material model [17] was used in the form

$$U=\sum_{i=1}^{N}\frac{2\mu_i}{\alpha_i^2}\left(\bar{\lambda}_1^{\alpha_i}+\bar{\lambda}_2^{\alpha_i}+\bar{\lambda}_3^{\alpha_i}-3\right)+\sum_{i=1}^{N}\frac{1}{D_i}\left(J^{el}-1\right)^{2i} \tag{2}$$

with

$$\bar{\lambda}_i=J^{-1/3}\lambda_i \tag{3}$$

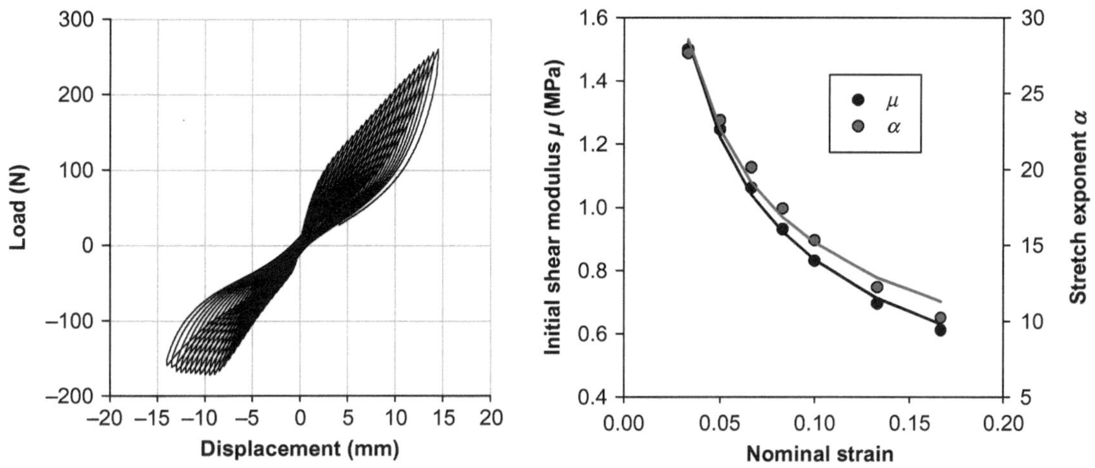

FIG. 16

Full hysteresis of T-joints, measured with increasing strain amplitudes (left), parameters of hyperelastic Ogden potential as function of nominal strain for HYPERFLEX-02 (right).

to describe the material stiffness as a function of the prestrain. Herein, $\overline{\lambda}_i$ refers to the deviatoric principal stretches, λ_i refers to the principal stretches, and J refers to the volume ratio. In Eq. (3), N is a material parameter, and α_i, μ_i, and D_i are temperature-dependent material parameters. The initial shear modulus and bulk modulus are given by $\mu_0 = \Sigma\, \mu_i$ and $K_0 = 2/D_1$. The initial bulk modulus was estimated from the measured Poisson's ratio. For the current material, it was sufficient to use a single term approach of Eq. (2). The values obtained for α and μ are shown in Fig. 16.

The initial shear modulus and the stretch exponent are depicted in Fig. 16 as a function of the nominal strain. A two-parameter power law

$$y = a \cdot \varepsilon^b \tag{4}$$

was fitted to the relation between initial shear modulus and nominal strain, yielding $a = 0.24 \pm 0.01$ and $b = -0.55 \pm 0.02$. The relation between exponent alpha and nominal strain was also fitted by Eq. (4), leading to $a = 4.02 \pm 0.41$ and $b = -0.58 \pm 0.04$. For nominal strains between 2% and 7% appearing in the soft segments of the morphing skin, this means that adequate values of μ and α are between 1 and 2 MPa and between 19 and 39, respectively.

A model of the finalized cross section of the upper and lower skins is shown in Fig. 17. The soft segments are on top and below the hinge lines, which are the regions where the highest deformations are generated. Between the hinge lines, there is almost no deformation generated in the skin due to the relatively high stiffness of the rib segments. Consequently, there was no flexibility needed and stiff segments were placed in those regions. Boundary conditions of the final design were a specified amount of rotation (typically two degrees around hinge lines) and a linearized pressure distribution, pressure values typically ranging from 10,000 to 6000 Pa at the upper skin and from −8000 to 6000 Pa at the lower skin (chord-wise).

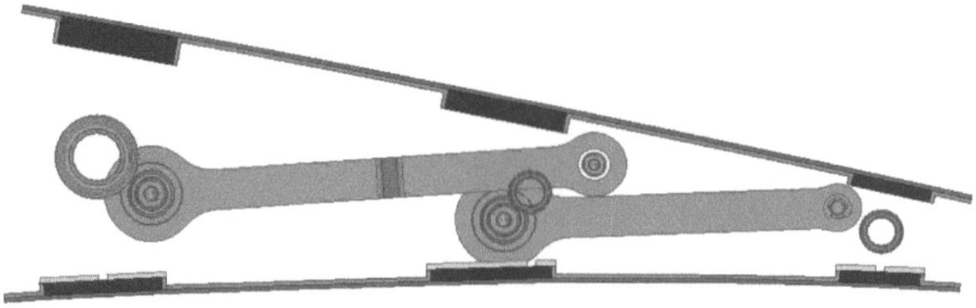

FIG. 17

Finalized geometry of upper (*top*) and lower (*bottom*) morphing skins with actuation mechanism (*middle*). Morphing skins consist of soft segments (*dark shaded*), stiff segments (*bright shaded*), and top layer (*medium shaded*).

The thickness of the soft segments was between 4 and 12 mm and the chordwise length was between 25 and 30 mm, depending on thickness. The variations are due to different amounts of space locally available on the primary structure and due to the variation of the pressure as a function of the chordwise position. The average value of the maximum principal strain at maximum trailing edge displacement and applied pressure load was typically 2% and did not exceed 7%. A typical strain distribution in a soft segment is shown in Fig. 18. The maximum out-of-plane displacements were below 0.5 mm.

The number of actuation cycles at full TE deflection was estimated as $N = 20{,}000$. Based on the maximum value of the maximum principal strain of 7% and the lower limit of the *E-N* curve in Fig. 13 one obtains cycle numbers of $N = \{1E6, 3E5, 6E4\}$ if the probability of failure P_{fail} is set to $\{10\%, 1\%, 0.1\%\}$. If the average maximum principal strain of 2% is considered, which is reasonable because the T-joint contains, due geometric similarity, the same local strain peaks like the morphing skin, one obtains $N = 7E4$ even if P_{fail} is set to a value as low as 1E-6. It can hence be seen that the probability of skin failure due to TE actuation is very low.

The force needed to deform upper and lower skins of the model in the position of maximum displacement at applied pressure was typically 2 kN per 1000 mm span width. This means that only a small part of the actuator load capacity would be necessary to deform the morphing skins in operation.

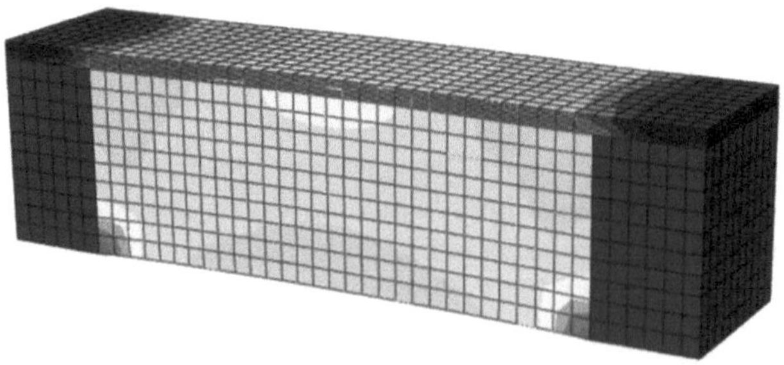

FIG. 18

Strain distribution in a soft segment of the morphing skin.

It is often discussed that morphing materials should have a Poisson's ratio of zero. In the current design, the Poisson's ratio of the foamed Hyperflex-02 material is close to 0.2 which in fact means that the transversal strain is about 0.4% at the average chord-wise strain level of 2%. Because the skin is in a state of planar strain, transversal strain can only be effective in the direction perpendicular to the skin, and the related out-of-plane deformation can be estimated as 50 μm in regions where the skin thickness is 12 mm, which is in fact very low. Out-of-plane deformations found in the model were larger but this effect was related to bending as well as to suction and compression due to air pressure. This indicates that a Poisson's ratio of zero is not so important for the current design.

7 SKIN MANUFACTURING

The design of the SARISTU ATED morphing skin consists of the three main components. The space between the aluminum profiles is filled with low temperature foam. The foam region as well as the aluminum profiles are covered with a 1 mm thin protective layer of a low temperature silicone elastomer (Fig. 19).

The soft segments are located above and under the hinges of the rib architecture and are responsible for skin movement. The lower and upper skin each have three foam segments (Fig. 20).

A new manufacturing process was developed due to the unavailability of appropriate adhesives to bond these materials for low temperature purposes. A core feature of this process is to perform curing, foaming, and adhesive bonding of different reactive silicone formulations and aluminum profiles at the same time. Chemical crosslinking during the curing reaction lead to a strong connection between the foam and the elastomer top layer. To increase the adhesion strength between the silicone and metal interface, the metal profiles were degreased and primered with an adhesion promoter for platinum catalyzed silicones.

In the first step, the liquid reactive rubber formulation is poured into a cast to obtain a 1 mm thin protective layer after curing (Fig. 21).

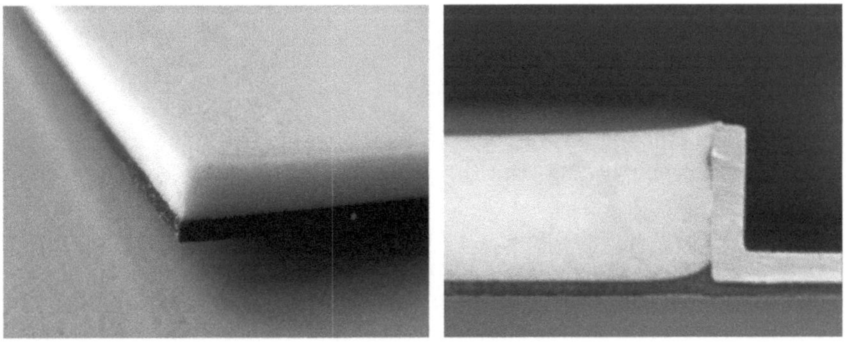

FIG. 19

Aluminum profiles, low temperature silicone foam and a protective layer of low temperature silicone elastomer are the main skin components of the SARISTU ATED.

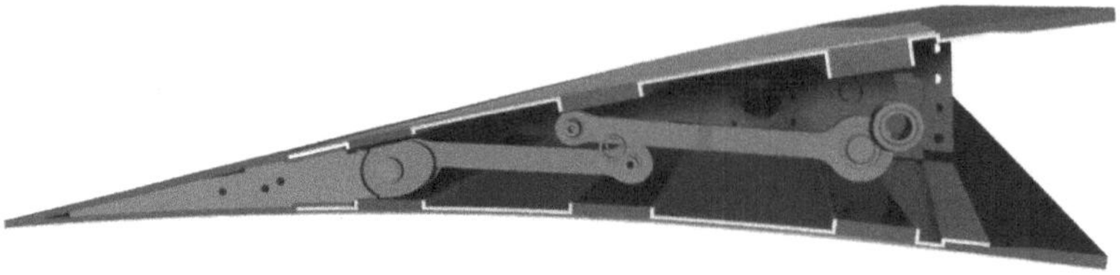

FIG. 20

The morphing skin consists of hard and soft segments. Elastomer foam is used for soft segments which are located above and under the rib hinges. Hard segments are aluminum profiles (*reversed U profiles, top and bottom boundaries*). Hard and soft segments are covered with a thin elastomer layer. The rib kinematic is shown simplified (*the two hinged rods, at the middle of the picture*).

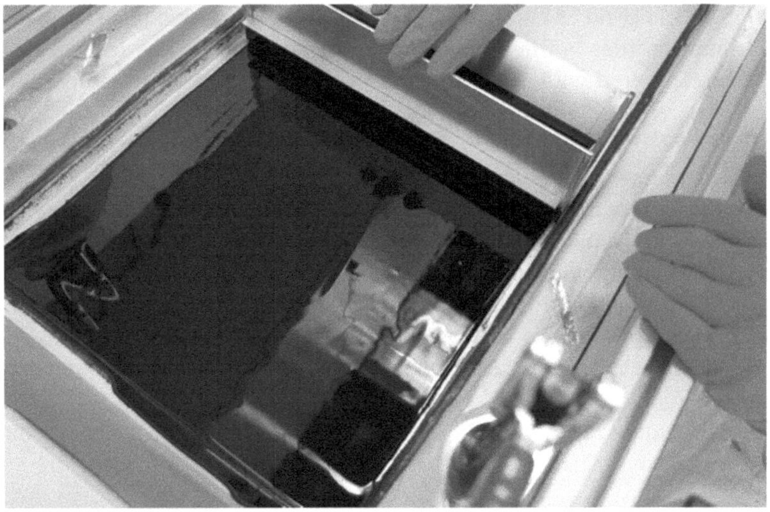

FIG. 21

A liquid rubber formulation is poured into a heatable cast. After curing the rubber leads to a 1 mm thin protective layer.

In the next step, the aluminum profiles are positioned on the still liquid rubber layer. A vacuum process is used to prevent bubbles between the protective rubber layer and the aluminum profiles (Fig. 22).

A reactive rubber formulation containing a foaming agent was filled between the aluminum profiles leading to the soft segments of the morphing skin after curing (Fig. 23).

Fig. 24 shows a sketch of a heating device to manufacture morphing skins up to 2.3 m length. The heating device consists of several heating pads which are separately controlled to obtain a homogenous temperature distribution and can reach temperatures up to 150°C.

After the heating process, the foam segments are cut with an oscillating knife to obtain the required foam thickness (Fig. 25).

FIG. 22

Vacuum bag to prevent bubbles between the protective rubber layer and the aluminum profiles.

FIG. 23

The liquid reactive rubber formulation contains a foaming agent leading to the soft segment between the aluminum profiles after curing.

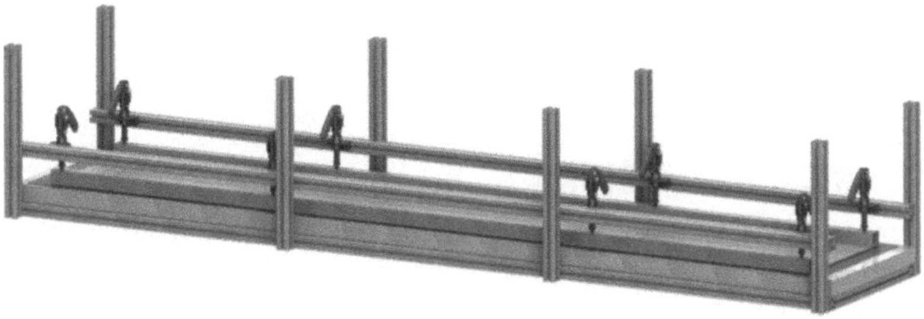

FIG. 24

2.8 m large heating device to process curing, foaming, and adhesive bonding of morphing skin components at the same time [9].

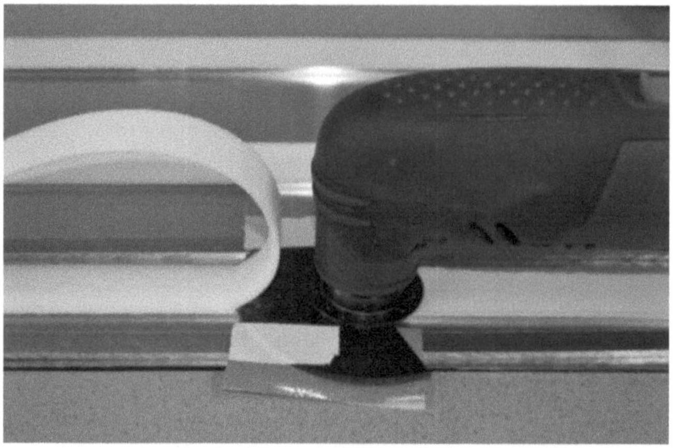

FIG. 25

Cutting the rubber foam segments with an oscillating knife.

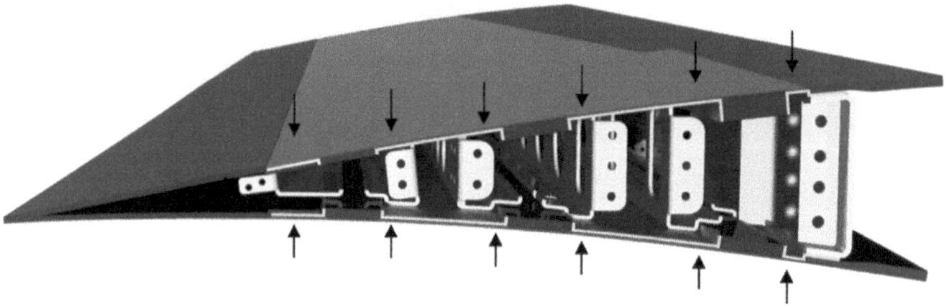

FIG. 26

Upper and lower skin sides are connected to stringers located between the ribs.

The obtained upper and lower skin side was attached to stringers located between the active ribs (Fig. 26) to transmit load in a spanwise direction and to assure attachment/detachment for ease of maintenance and repair.

During the SARISTU project several large skin panels were manufactured and successfully assembled into a true-scale wind tunnel demonstrator (see Fig. 27).

8 SUMMARY AND CONCLUSIONS

In this chapter, a morphing skin concept was investigated and manufactured for an active trailing edge device of an adaptive wing demonstrator. In the current design, geometric gap changes are converted into mechanical strain to maintain a smooth transition between connected parts. The proposed solution was developed up to the physical subcomponent level. The manufacturing technology was demonstrated and adaptive parts were delivered for demonstrator assembly, ground tests, and wind tunnel tests.

FIG. 27

Lower side morphing skins for the SARISTU adaptive trailing edge.

The stiffness measured in cyclic tests on bonded joints was accurately predicted by a hyperelastic material model. It was shown that the elastomeric skin materials have excellent fatigue and aging resistance as well as stable adhesion on aluminum surfaces within aircraft operational temperature range. The presented results suggest the elastic deformation approach based on elastomeric multimaterial design is a potential solution for the morphing paradox to deform and carry loads simultaneously.

The research leading to these results has gratefully received funding from the European Union's Seventh Framework Programme for research, technological development, and demonstration under grant agreement no. 284562.

REFERENCES

[1] C. Thill, J. Etches, I. Bond, K. Potter, P. Weaver, Morphing skins, Aeronaut. J. (March) (2008). Paper no. 3216.

[2] P.K.C. Rudolph, High-Lift Systems on Commercial Subsonic Airliners, NASA Ames Research Center, Moffet Field, CA, 1996.

[3] R. Pecora, F. Amoroso, M. Magnifico, Toward the bi-modal camber morphing of large aircraft wing flaps: the CleanSky experience, in: Proc. SPIE—The International Society for Optical Engineering, Volume 9801, 2016, Article No. 980106.

[4] R. Pecora, F. Amoroso, M. Magnifico, I. Dimino, A. Concilio, KRISTINA: kinematic rib-based structural system for innovative adaptive trailing edge, in: Proc. SPIE—The International Society for Optical Engineering, Volume 9801, 2016, Article No. 980107.

[5] I. Dimino, A. Concilio, R. Pecora, Safety and reliability aspects of an adaptive trailing edge device (ATED), in: Proc. 24th AIAA/AHS Adaptive Structures Conference, San Diego, CA, USA, Jan. 4–8, 2016.

[6] I. Dimino, A. Concilio, M. Schueller, A. Gratias, An adaptive control system for wing TE shape control, in: Proc. SPIE International Conference on Smart Structures 2013, San Diego, CA, USA, Mar. 10–14, 2013.

[7] Pecora, R., Concilio, A.,Dimino, I., Amoroso, F., Ciminello, M., "Structural design of an adaptive wing trailing edge for enhanced cruise performance", 24th AIAA/AHS Adaptive Structures Conference, 2016; San Diego; United States; 4 January 2016-8 January 2016.

[8] I. Dimino, G. Diodati, A. Concilio, A. Volovick, L. Zivan, Distributed electromechanical actuation system design for a morphing trailing edge wing, SPIE Smart Structures/NDE, Las Vegas, Nevada (USA) March 2016. in: Proc. SPIE 9801, Industrial and Commercial Applications of Smart Structures Technologies 2016, Article No. 980108, Apr. 16, 2016. https://doi.org/10.1117/12.2219223.
[9] C. Nagel, A. Fiedler, O. Schorsch, A. Lühring, in: Design, manufacture, and testing of a seamless morphing concept for a smart aircraft wingtip, Proceedings of the 7th ECCOMAS Thematic Conference on Smart Structures and Materials (SMART 2015), Ponta Delgada, Azores, 2015.
[10] ISO 527, Plastics—Determination of Tensile Properties, International Organizationfor Standardization, ISO, Geneva, 2012.
[11] ISO 604, Plastics—Determination of Compressive Properties, International Organization for Standardization, ISO, Geneva, 2002.
[12] ISO 34, Rubber, Vulcanized or Thermoplastic—Determination of Tear Strength—Part 1: Trouser, Angle and Crescent Test Pieces, International Organization for Standardization, ISO, Geneva, 2015.
[13] ISO 12107, Metallic Materials—Fatigue Testing— Statistical Planning and Analysis of Data, International Organization for Standardization, ISO, Geneva, 2012.
[14] ASTM D1183–03, Standard Practices for Resistance of Adhesives to Cyclic Laboratory Aging Conditions, ASTM International, West Conshohocken, PA, 2011.
[15] L. Mullins, Softening of rubber by deformation, Rubber Chem. Technol. 42 (1969) 339.
[16] R.W. Ogden, D.G. Roxburgh, A pseudo–elastic model for the Mullins effect in filled rubber, Proc. R. Soc. Lond. Series A: Math. Phys. Eng. Sci. 455 (1999) 2861–2877.
[17] R.W. Ogden, Recent advances in the phenomenological theory of rubber elasticity, Rubber Chem. Technol. 59 (3) (1986) 361–383.

CHAPTER

THE DESIGN OF SKIN PANELS FOR MORPHING WINGS IN LATTICE MATERIALS

8

Andrea Vigliotti*, Damiano Pasini†

The Italian Aerospace Research Centre, CIRA SCpA, Capua (CE), Italy McGill University, Montreal, QC, Canada†*

CHAPTER OUTLINE

NOMENCLATURE

1PK first Piola-Kirchoff stress tensor
FE finite element
RVE representative volume element

1 INTRODUCTION

With the aim to improve the performance and the efficiency of current aircraft design, a number of research projects have recently addressed the implementation of morphing wing concepts [1–4]. The benefits of morphing wing over the current designs involve the aerodynamic efficiency, the structural weight, and noise emission [5,6].

Although several studies have focused on the design of the wing internal structure [7–10,10a,10b], only a limited number of solutions have been proposed for the skin.

Morphing Wing Technologies. https://doi.org/10.1016/B978-0-08-100964-2.00008-3

The skin of morphing wings is the structural element that poses some of the most difficult challenges in the design of a morphing aircraft [11–15]. Whereas high compliance to reduce the actuation energy is required in the main stretching direction, high stiffness is necessary in the out-of-plane direction to maintain a given shape under the action of the aerodynamic loads. In addition, the skin should be able to undergo in the actuation direction large recoverable deformation, in the order of 20% up to 40%, thus minimizing the deformation in the orthogonal direction. This specification requires the deformation in the cross-direction to be as small as possible for two main reasons, (1) to allow a fine overall control of the wing shape, and (2) to limit the mass of the wing structure. No traditional materials, either metallic or polymeric, could fulfill the conflict existing among these design requirements, but those with a periodic microstructure offer a viable solution. Recent techniques of rapid prototyping allow the manufacturing of components with a lattice architecture from a variety of solid materials, with high precision and acceptable costs [16,17]. By finely tuning their geometric parameters, the structural performance of a lattice material can be optimized to levels that cannot be attained by other traditional materials. The factor that contributes the most is the topology of the unit cell, which can be tailored to maximize desired, as well as conflicting, structural objectives [18,19].

The structural analysis of a component made of a lattice material cannot overlook the geometry of both the unit cell and the macroscopic domain. When the scale of the component is significantly larger than the scale of the unit cell, a direct approach that involves finite element (FE) modeling at the microscopic scale would result in considerably large models. This process would be either unpractical or unfeasible to handle [20,21]. A viable alternative is to resort to multiscale approaches that generally assume a length scale separation between the micro (unit cell) and macro (component) geometry. In this case, because the smaller scale physics can be considered as homogeneous by the larger scale, it is possible to separately model each physical field after defining suitable boundary conditions that allow information to be passed among the different scales. At the larger scale, the component is assumed to consist of homogeneous material with mesh size independent of the unit cell size. Multiscale approaches have been proved successful in delivering accurate results while limiting the computational effort.

In this chapter, we use a multiscale homogenization method that accounts for the geometric nonlinearity of the microstructure and derive the macroscopic constitutive relationship of a skin made out of a lattice material. We first identify the material requirements for the skin of a morphing wing. Then we describe the homogenization method for the analysis of the mechanical properties of lattice materials. Finally, the method is used to analyze in detail the properties of candidate lattice topologies, which in turn generate design property charts for the selection of the most suitable cell topology for a morphing lattice skin.

2 REQUIREMENTS FOR THE SKIN OF A MORPHING WING

In a morphing wing, the airfoil skin has to undergo membrane deformation in the skin-plane, and withstand the aerodynamic pressure in the out-of-plane direction [12,14]. Fig. 1A describes the main requirements of a morphing wing: in camber changes, which impose surface deformation of opposite sign; and in span changes, where the upper and the lower surfaces undergo deformation of the same

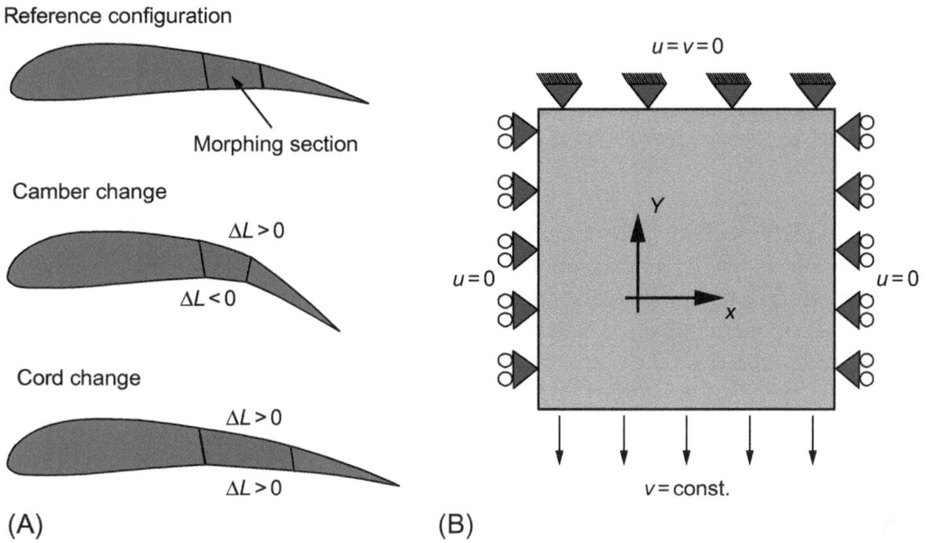

FIG. 1

(A) Camber and span variations achieved via wing morphing. (B) Reference boundary conditions used to determine the material properties of the skin.

sign. As a result, the skin should withstand large tensile and compressive recoverable deformations. In addition, to reduce both the power requirements and the impact of the actuation forces on the rest of the wing structure, very high compliance in the direction of actuation is required for the material of the skin. Another requirement of the actuation is that the skin should have small in-plane Poisson's ratio, a behavior that reduces the loads transmitted to the rest of the wing in the cross-directions.

In summary, the main design requirements for the skin microstructure of a morphing wing can be specified as:

1. Small microscopic deformation, within the elastic limit of the solid material to allow repeated actuations;
2. High in-plane compliance in the direction of actuation to minimize the actuator's power and overall structural mass of the wing;
3. Large recoverable macroscopic elastic deformation to allow the shape changes required by the aerodynamics;
4. High bending stiffness to ensure no shape variation occurs under the action of the aerodynamic pressure;
5. Minimal structural mass for the skin panel;
6. Zero in-plane Poisson's ratio to avoid the occurrence of undesired forces in the span-wise direction, and to reduce the overall structural mass of the wing.

As a reference case, we consider a plate subjected to the boundary conditions shown in Fig. 1B; here the macroscopic strain and stress fields of the plate are uniform and given by.[1]

$$\begin{aligned} G_{22} &= \frac{\Delta L_Y}{L_Y} = \text{const} \\ G_{11} &= G_{21} = G_{12} = 0 \\ P_{33} = P_{23} &= P_{32} = P_{13} = P_{31} = 0 \end{aligned} \tag{1}$$

where P_{ij} and G_{ij} are, respectively, the components of the macroscopic first Piola-Kirchoff (1PK) stress tensor and of the macroscopic displacement gradient.

The above requirements are generally in conflict and unachievable to meet for any homogeneous material. In fact, a polymeric rubber, which is naturally very compliant, could satisfy criteria 1, 3 and 1, but very likely unable to satisfy the other criteria. Another option is given by composite materials that can provide a high level of anisotropy; nevertheless, their tolerance to large deformation is rather limited.

3 A METHODOLOGY FOR NONLINEAR HOMOGENIZATION OF PERIODIC STRUCTURES

In this section, we describe the method that enables the evaluation of the macroscopic constitutive response of a lattice. The scheme relies on the analysis of a representative volume element (RVE) of a lattice, subjected to a given macroscopic deformation field and to periodic boundary conditions. The components of the macroscopic stress tensor are found by calculating the derivative of the deformation energy density of the RVE with respect to the components of the macroscopic deformation gradient. The following briefly summarizes the main step, while a complete description can be found in Refs. [21,22].

A lattice material consists of ordered periodic arrangement of cells obtained by replication of a unit cell along independent tessellation vectors (Fig. 2). In this context, the unit cell is the minimal entity that is capable of generating the entire lattice after replication along the tessellation vectors. Possible RVEs are made out of any $m \times n$ array of contiguous unit cells, where m and n are integers. For the linear case, it can be shown that the estimated macroscopic lattice stiffness is independent from m and n [21]. This applies also to the nonlinear case as long as no bifurcation point is crossed on the load path of the lattice. In the presence of bifurcations, the choice of the RVE influences the behavior of the material model and a detailed analysis should be a priori performed to select the suitable RVE.

To evaluate the equilibrium configuration of the RVE subjected to a given macroscopic deformation, we introduce the following kinematic assumptions:

1. The periodic directions of the lattice change according to the macroscopic displacement gradient;
2. The lattice remains periodic during deformation.

Assumption (1) applies only to the periodic directions, and no further restrictive hypothesis is made on the displacement of the internal points of the RVE, whose configuration is determined by imposing the

[1] We recall that the displacement gradient and the first Piola-Kirchoff tensor are not symmetric, thus $G_{ij} \neq G_{ji}$ and $P_{ij} \neq P_{ji}$.

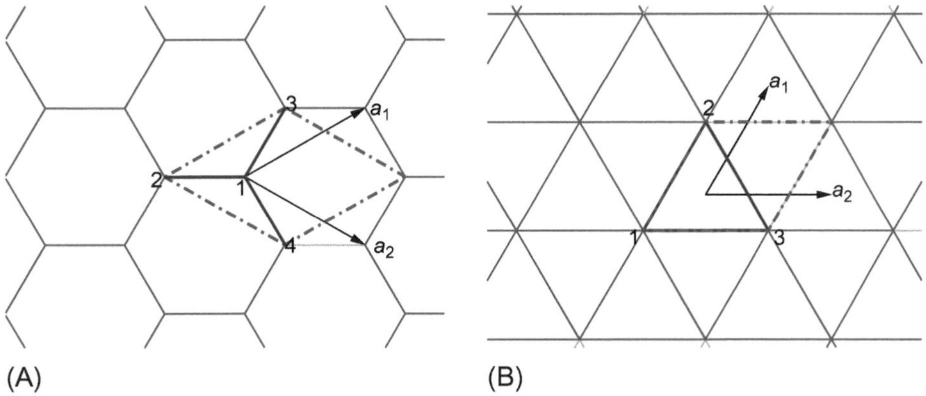

FIG. 2

Sample topologies. The unit cell elements are marked in *thick line*, a_1 and a_2 are the tessellation vectors; the RVE boundary is marked with *dashed lines*. (A) Hexagonal lattice and (B) triangulated lattice.

equilibrium. The above assumptions are equivalent to consider that either the size of the RVE negligible with respect to the size of the macroscopic domain, or the typical length of the lattice elements is negligible with respect to the local wavelength of the macroscopic strain field. In other words, we consider the RVE as a material point of our macroscopic continuum. Therefore, in this formulation, we completely decouple the length scale of the macroscopic and the microscopic domains.

To find the deformation work, we consider a FE model of the RVE. Let $\boldsymbol{s}$ the array of the nodal degrees of freedom of our RVE; by means of the FE model of the RVE, we can calculate the corresponding array of the nodal forces, $\boldsymbol{F}(\boldsymbol{s})$. Applying the principle of virtual work, the first variation of the total strain energy from a variation of the macroscopic strain is given by:

$$\mathrm{d}W = \int_{V_{\mathrm{RVE}}} P_{ij}\mathrm{d}G_{ij}\,\mathrm{dV} = \boldsymbol{F}^T\mathrm{d}\boldsymbol{s} \tag{2}$$

where P_{ij} and G_{ij} are, respectively, the components of the macroscopic 1PK stress tensor and of the macroscopic displacement gradient; $\mathrm{d}\boldsymbol{s}$ is the variation of the nodal displacements corresponding to $\mathrm{d}G_{ji}$. Assuming P_{ij} and G_{ij} are uniform over the RVE, we obtain the macroscopic stress tensor as

$$P_{ij} = \frac{1}{V_{\mathrm{RVE}}}\frac{\partial W}{\partial G_{ij}} = \frac{1}{V_{\mathrm{RVE}}}\mathbf{F}^T\frac{\partial \mathbf{s}}{\partial G_{ij}} \tag{3}$$

As we can observe, all the boundary nodes of the RVE are necessarily corresponding along the periodic vectors, while the internal nodes do not have any corresponding node in the RVE. For instance, in Fig. 2A, it results $r_3 = r_2 + a_1$ and $r_4 = r_2 + a_2$, and node 1, which is internal, had no corresponding node along any periodic direction of the RVE. Similarly in Fig. 2B, it results $r_2 = r_1 + a_1$, and $r_3 = r_1 + a_2$, and, in this case, we do not have any internal node. It follows that the position of all the nodes of the unit cell can be expressed through the position of some independent nodes and the periodic directions as follows:

$$\boldsymbol{r}_i = \boldsymbol{r}_{0j} + k\boldsymbol{a}_1 + h\boldsymbol{a}_2 \tag{4}$$

with $k,h \in \{-1,0,1\}$, and $k=h=0$ for the internal nodes. Let $\boldsymbol{\Delta a}_i$ be the change in the ith direction, and we can easily write the displacement of each node of the unit cell as

$$s_i = s_{0j} + k\boldsymbol{\Delta a}_1 + h\boldsymbol{\Delta a}_2 \tag{5}$$

Collecting all the DoFs of the UC in the array $\boldsymbol{s}$, we can write the following equation in a compact matrix form

$$\boldsymbol{s} = \boldsymbol{B}_0 \boldsymbol{s}_0 + \boldsymbol{B}_1 \boldsymbol{\Delta a} \tag{6}$$

where the entries of the matrices $\boldsymbol{B}_0$ and $\boldsymbol{B}_a$ are $1, 0$, or -1, and do not change during the deformation; $\boldsymbol{s}_0$ is the array that collects all the independent DoFs of the unit cell; and $\boldsymbol{\Delta a}$ collects the components of $\boldsymbol{\Delta a}_i$ and is defined as

$$\boldsymbol{\Delta a} = [\boldsymbol{\Delta a}_{11}, \boldsymbol{\Delta a}_{12}, \boldsymbol{\Delta a}_{21}, \boldsymbol{\Delta a}_{22}]^T \tag{7}$$

where the first index refers to the periodic direction and the second index refers to the Cartesian component. If $G_{ij} = \dfrac{\partial u_i}{\partial X_j}$ are the components of the macroscopic displacement gradient, the following holds:

$$\boldsymbol{\Delta a}_{ki} = G_{ij} a_{kj} \tag{8}$$

After collecting the components of the gradient in the array $\boldsymbol{G} = [G_{11}, G_{12}, G_{21}, G_{22}]^T$, we can express the change in the periodic directions as a function of the components of the displacement gradient as.

$$\boldsymbol{\Delta a} = \boldsymbol{B}_a \boldsymbol{G} \tag{9}$$

where the matrix $\boldsymbol{B}_a$ collects the components of the tessellation vectors rearranged as.

$$\boldsymbol{B}_a = \begin{bmatrix} a_{11} & a_{12} & 0 & 0 \\ 0 & 0 & a_{11} & a_{12} \\ a_{21} & a_{22} & 0 & 0 \\ 0 & 0 & a_{21} & a_{22} \end{bmatrix} \tag{10}$$

thus we can express all the DoFs of RVE directly as a function of the DoFs of the independent nodes and of the components of the displacement gradient as.

$$\boldsymbol{s} = \boldsymbol{B}_0 \boldsymbol{s}_0 + \boldsymbol{B}_G \boldsymbol{G} \tag{11}$$

where $\boldsymbol{B}_G = \boldsymbol{B}_1 \boldsymbol{B}_a$, which depends only on the topology of the RVE. Eq. (11) expresses the kinematics of the RVE, and relates the macroscopic displacement gradient $\boldsymbol{G}$ to the nodal displacements of the RVE. The deformed shape of the RVE can be found after the DoFs of the independent nodes, $\boldsymbol{s}_0$, have been determined, a step that can be done by the periodic equilibrium condition on the RVE.

From assumption (2), we can write the equation for the equilibrium of the RVE under the action of the surrounding elements, in terms of the nodal forces of the RVE. In particular, for the hexagonal lattice the following holds:

$$\begin{aligned} F_1 &= 0 \\ F_2 + F_3 + F_4 &= 0 \end{aligned} \tag{12}$$

Similarly, for the triangulated lattice, we can write the following periodic equilibrium equation:

$$F_1 + F_2 + F_3 = 0 \tag{13}$$

It is possible to show [14] that for an arbitrary lattice the periodic equilibrium equation can be written in terms of the $\boldsymbol{B}_0$ matrix as.

$$\boldsymbol{B}_0^T \boldsymbol{F}(\boldsymbol{s}) = 0 \tag{14}$$

and by means of Eq. (11) as follows:

$$\boldsymbol{B}_0^T \boldsymbol{F}(\boldsymbol{B}_0 \boldsymbol{s}_0 + \boldsymbol{B}_G \boldsymbol{G}) = 0 \tag{15}$$

The above equation expresses the equilibrium of the unit RVE under the action of the surrounding cells, for a uniform deformation field. Given $\boldsymbol{G}$, Eq. (15) can be solved by means of Newton Raphson Iterations.

Recalling the statement of the virtual work, we can obtain the following expression for the macroscopic stress tensor and its Jacobian:

$$\boldsymbol{P} = \boldsymbol{B}_G^T \boldsymbol{F}(\boldsymbol{s}) \tag{16}$$

$$\frac{\partial \boldsymbol{P}}{\partial \boldsymbol{G}} = \left(\boldsymbol{B}_0 \frac{\partial \overline{\boldsymbol{s}}}{\partial \boldsymbol{G}} + \boldsymbol{B}_G \right)^T \frac{\partial \boldsymbol{F}}{\partial \boldsymbol{s}} \left(\boldsymbol{B}_0 \frac{\partial \overline{\boldsymbol{s}}}{\partial \boldsymbol{G}} + \boldsymbol{B}_G \right) \tag{17}$$

Where $\frac{\partial \overline{\boldsymbol{s}}}{\partial \boldsymbol{G}}$ is given by the equation that follows, where $(\cdot)^+$ denotes the Moore-Penrose pseudo inverse

$$\frac{\partial \overline{\boldsymbol{s}}}{\partial \boldsymbol{G}} = -\left(\boldsymbol{B}_0^T \frac{\partial \boldsymbol{F}}{\partial \boldsymbol{s}} \boldsymbol{B}_0 \right)^+ \boldsymbol{B}_0^T \frac{\partial \boldsymbol{F}}{\partial \boldsymbol{s}} \boldsymbol{B}_G \tag{18}$$

The methodology described in this section enables the modeling of a component in lattice material by means of displacement-based FE analyses. For a given displacement gradient $\boldsymbol{G}$, at each integration point of the macroscopic structure, Eqs. (16) and (17) can be used to evaluate the 1PK tensor and its Jacobian for any given $\boldsymbol{G}$.

In the formulation described above, all boundary conditions are on the components of $\boldsymbol{G}$, which ensures the existence of a solution. Nonetheless, it is also possible to pose the inverse problem, which is to consider boundary conditions given on the $\boldsymbol{P}$ tensor and solve for $\boldsymbol{G}$. In the following sections, we consider mixed boundary conditions, where certain components of P and G are prescribed, and others are free. In this case, the solution still can be found by solving a nonlinear problem by means of the Newton-Rhapson method, which iterates on the not prescribed components of $\boldsymbol{G}$, and defines the residual on the components of $\boldsymbol{P}$, whose values are constrained. In this case special, proper care should apply, as bifurcation points might appear because of local buckling of the lattice elements.

4 MECHANICAL PROPERTIES OF SKIN PANELS IN LATTICE MATERIAL

4.1 ANALYSIS OF SELECTED LATTICE TOPOLOGIES

In this section, we focus on the selection of lattice topologies suitable for the design of the skin of a morphing wing. In particular, we consider a honeycomb topology, where the horizontal and the inclined sides can have different measures and the angle between the two is arbitrary (Fig. 3A), and the chevron lattice (Fig. 3B).

With reference to the lattices in Fig. 3, the kinematic and the equilibrium conditions can be formulated as shown in Table 1, where $\boldsymbol{a}_1$ and $\boldsymbol{a}_2$ are the periodic directions of the lattice.

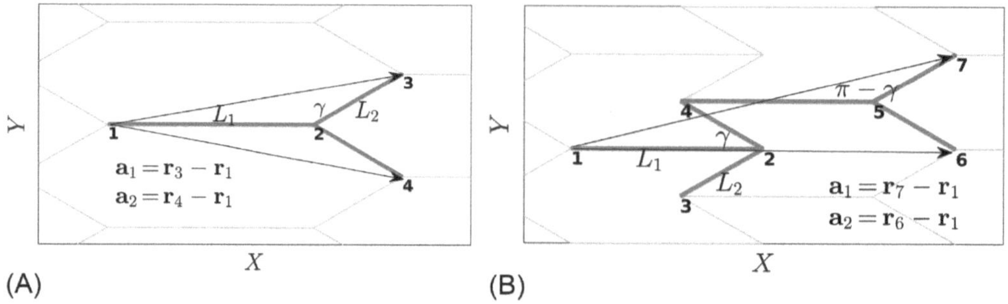

FIG. 3

Sample topologies: (A) hexagonal lattice and (B) chevron lattice. The thick lines represent unit cell elements; while a_1 and a_2 are the tessellation vectors; r_i is the coordinate vector of the node i; L_1 and L_2 are, respectively, the length of the horizontal and inclined members; γ is the angle between the horizontal and the inclined members.

Table 1 Compatibility and Periodic Equilibrium Conditions

	Hexagonal Lattice	Chevron Lattice	
Compatibility	$s_3 = s_1 + G\mathbf{a}_1$ $s_4 = s_1 + G\mathbf{a}_2$	$s_7 = s_1 + Ga_1$ $s_6 = s_1 + Ga_2$ $s_4 = s_3 + Ga_1 - Ga_2$	(19)
Periodic equilibrium	$F_2 = 0$ $F_1 + F_3 + F_4 = 0$	$F_2 = 0$ $F_5 = 0$ $F_1 + F_6 + F_7 = 0$ $F_3 + F_4 = 0$	(20)

G *is the displacement gradient,* $\mathbf{a}_i$ *are the periodic vectors defined in Fig. 3,* $\mathbf{s}_i$ *are the nodal displacements and* $\mathbf{F}_i$ *are the nodal forces.*

As previously observed, compatibility equations allow us to express the displacements of all the nodes of the RVE in terms of G and the displacement of a subset of the nodes of the RVE, which we call independent nodes. A possible choice of the independent nodes are nodes 1 and 2 for the hexagonal lattice (Fig. 3A), and 1, 2, 3, and 5 for the chevron (Fig. 3B). Because the number of equilibrium equations is always equal to the number of the independent nodes, the problem is closed, and given the components of the displacement gradient, we can determine the corresponding configuration of the RVE.

In this study, we restrict our attention to honeycomb, extruded planar topologies, as shown in Fig. 4. We observe that the size of the lattice in direction X_3, represents the thickness of the skin panel and therefore is a macroscopic parameter; on the other hand, the in-plane dimensions of the lattice, i.e., the length and the thickness of its elements, are microscopic parameters, specified at a length scale smaller than the macroscopic dimension of the panel. Let L_i be the typical length of the walls in the plane of the plate, t the wall thickness in the same plane, and z the out-of-plane height of the walls

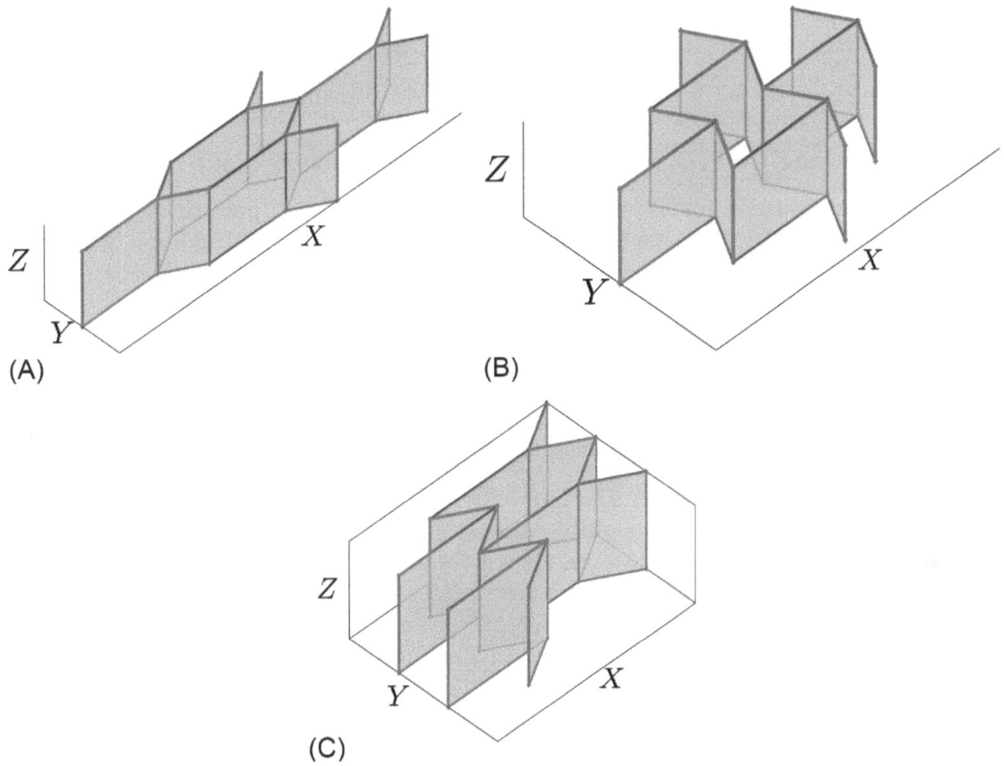

FIG. 4

Sample lattices: (A) classical hexagonal honeycomb, (B) auxetic hexagonal honeycomb, (C) chevron honeycomb. Auxetic lattices are obtained from the classic ones by setting $\gamma \in]\pi/2, \pi[$. Note that the two hexagonal configurations coexist in the chevron lattice.

corresponding to the macroscopic thickness of the plate. For these conditions, we assume the following relation holds: $t \ll L_i \ll z$. In this work, the lattice walls are modeled as three-dimensional plates. Thus for the macroscopic deformation modes that remain in the plane of the skin panel, the strain energy is essentially stored as pure bending of the walls, whereas for the deformation modes out of the plane $X_1 - X_2$, which correspond to the macroscopic bending of the plate, the strain energy is stored as membrane shear of the walls.

An essential parameter of a morphing skin panel is its bending stiffness, which is proportional to the thickness of the plate. In our case, the skin panel is structured rather than uniform; therefore as an indicator of its macroscopic bending stiffness, we select the macroscopic shear stiffness of the lattice in the out-of-plane directions $X_1 - X_2$; that is, we consider the components of the macroscopic material stiffness tensor, Eq. (17), such that $\frac{\partial P_{3i}}{\partial G_{3j}}$ with $i, j \in \{1, 2\}$.

To compare the performance of the lattices in Fig. 4, we prescribed their bending stiffness to be equal. This means that the macroscopic thickness of their skin in the X_3 direction, $\bar{z}$, is given and satisfies $\frac{1}{E_s}\frac{\partial P_{3i}}{\partial G_{3j}}\frac{\bar{z}}{L_1} = \chi^*_{oop} = 1 \times 10^{-6}$. In addition, to compare the structural efficiency of each design, we consider $P^*_{11} = \frac{P_{11}}{E_s}\bar{z}$ and $P^*_{22} = \frac{P_{22}}{E_s}\bar{z}$, (with E_s Young's modulus of the cell wall), respectively, as measures of the required actuation forces and the actuation work. L_1 is also prescribed. Hence, P^* is an actual measure of the actuation tension for panels of given shear stiffness. In addition, with the above definition of P^*, the actuation energy per unit surface of the skin panel is given by the following.

$$W = E_s\bar{z}\int P^*_{22}\mathrm{d}G_{22} \tag{21}$$

We now compare the mechanical responses of the topologies shown in Fig. 4 under uniaxial stretching.

The lattices under investigation are assumed with prescribed $L_1 = 10^{-3}$ m, $L_2/L_1 = 0.5$, $E_s = 70$ GPa, $\frac{L_1}{t} = 30$. Table 2 reports their stiffness in the initial configuration for alternative macroscopic constraint conditions. The first column shows the lattice stiffness in direction X_2 under uniaxial stress, i.e., the lattice is free to deform in the other directions upon deformation applied in direction X_2. The stiffness of the lattices is small, and they have similar values. The second column reports the stiffness under planar uniaxial strain, i.e., the lattice deformation is prevented in direction X_1 upon deformation applied in direction X_2, and the lattice is not constrained in direction X_3. In these conditions, the hexagonal lattices are from two to three orders of magnitude stiffer, whereas the stiffness of the chevron lattice is unchanged. In particular, the initial Poisson's ratio of the chevron lattice is zero, which corresponds to the decoupling of the loads in the directions X_1 and X_2. In contrast, the hexagonal lattices the Poisson's ratio have a finite value, which prevents deformation in direction X_1 and brings an increase in the lattice stiffness. The third column of the table reports the direct cross term for the plane $X_1 - X_2$ of the initial material stiffness matrix, i.e., the marginal stress increment in direction X_1 corresponding to the deformation in direction X_2. This term is an effective measure of the coupling between the directions X_1 and X_2 for a uniaxial stretch of the lattice. For hexagonal lattices, this term has magnitude of the same order, or larger, than that of the direct stiffness in direction X_2. On the other hand, it is zero for the chevron lattice, where the direction X_1 and X_2 are decoupled. The last two columns in Table 2 report

Table 2 Initial Stiffness Terms Normalized With Respect to the Young's Modulus, E_s, of the Solid Material

	$\frac{\partial P_{22}}{\partial G_{22}}\frac{1}{E_s}$ With $P_{11}=0$	$\frac{\partial P_{22}}{\partial G_{22}}\frac{1}{E_s}$ With $G_{11}=0$	$\frac{\partial P_{11}}{\partial G_{22}}\frac{1}{E_s}$ With $G_{11}=0$	$\frac{\partial P_{32}}{G_{32}}\frac{1}{E_s}$ With $G_{11}=0$	$\frac{\partial P_{31}}{\partial G_{31}}\frac{1}{E_s}$ With $G_{11}=0$
Classical hexagonal	2.30×10^{-6}	8.25×10^{-4}	8.17×10^{-2}	8.88×10^{-3}	1.48×10^{-2}
Auxetic hexagonal	2.28×10^{-5}	1.64×10^{-2}	-1.63×10^{-2}	2.94×10^{-2}	4.47×10^{-3}
Chevron	3.29×10^{-6}	3.29×10^{-6}	0	2.05×10^{-2}	6.41×10^{-3}

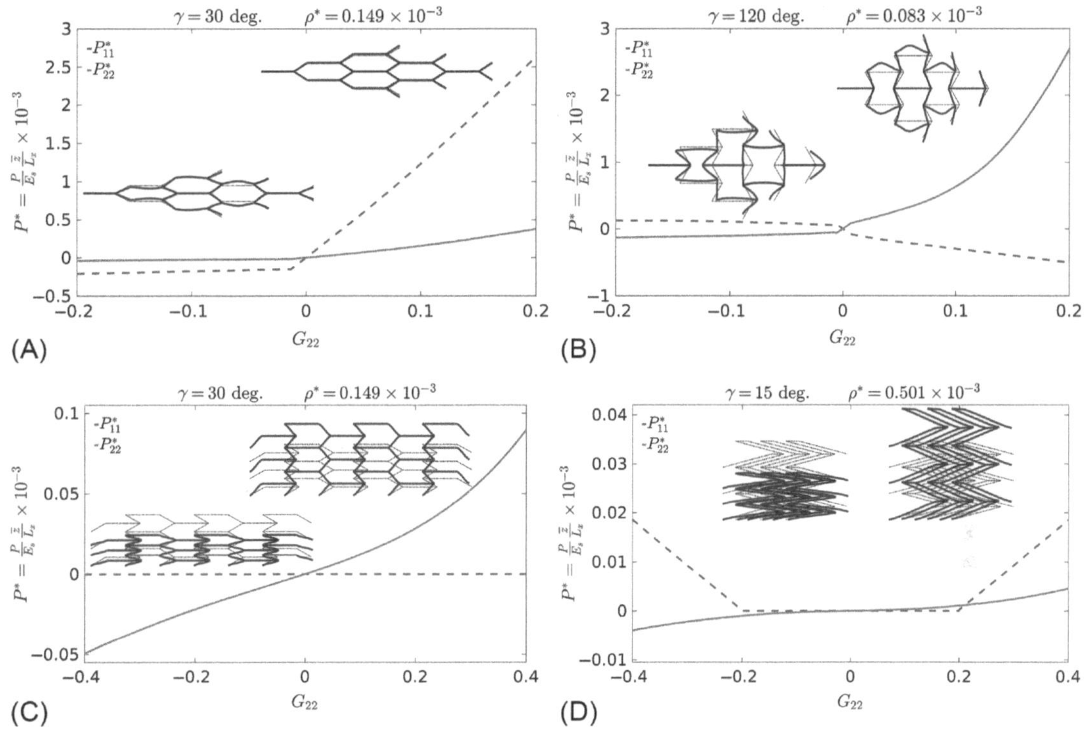

FIG. 5

Normalized stress-strain relations for the lattices under investigation. Normalized stress in direction 2, P^*_{22}, is shown in solid line, while normalized stress in direction 1, P^*_{11}, is shown in solid line. Each plot includes the deformed shape of the RVE in compression and in tension for the largest value of the deformation. For each topology, the value of the tension has been evaluated for configurations where the skin thickness, z, has been selected to yield a given macroscopic shear stiffness in direction 3, given by $\chi = \min\left(\frac{\partial P_{i3}}{\partial G_{j3}}, \frac{\partial P_{3i}}{\partial G_{3j}}\right)$ with i and $j \in \{1, 2\}$. This choice allows us to compare the effective actuation power, which is given by the area under the P^*_{22} curve, for skin panels with given bending stiffness.

the macroscopic shear stiffness of the lattice for out-of-plane deformation $X_1 - X_2$; the values represent the bending stiffness of the skin panel.

To compare the skin performance for given bending stiffness, we impose a reference thickness for all lattices, a condition that results in prescribing the out-of-plane bending stiffness. As shown in Fig. 5A and B, the hexagonal lattice, because of its nonzero Poisson's ratio, is very compliant under uniaxial stress, whereas its stiffness is quite high for uniaxial stretch. P_{11}, the reaction force in the directions orthogonal to the applied stretch, has the order of magnitude of the applied tension, or even larger for the case of the classical hexagon. The chevron lattice, however, is very compliant with reaction tensions essentially negligible even for very large deformation. Therefore, we can conclude that the chevron lattice is particularly promising in the design of a material for morphing wings.

4.2 THE DESIGN SPACE OF THE CHEVRON LATTICE

In this section, we analyze in detail the response of the chevron lattice for different configurations and explore its design property space. We assume L_1 is given and let $L_2/L_1 \in [0.25, 4]$, and consider two values of the initial angle between the members, namely $\gamma = 15$ degrees and $\gamma = 30$ degrees.

We first plot in Fig. 6A the out-of-plane stiffness of the plate, $\chi = \min\left(\frac{\partial P_{13}}{\partial G_{13}}, \frac{\partial P_{23}}{\partial G_{23}}\right)$. As we can observe for small value of L_2/L_1 the stiffness of the material is controlled mainly by the length of the horizontal member, thus it is constant. For larger values of L_2/L_1, however, as the length of the inclined member increases, the lattice becomes increasingly compliant for the out-of-plane shear in direction 2, and the overall shear stiffness of the plate drops. As a result, the reference thickness of the skin has to increase so as to ensure that the normalized macroscopic bending stiffness of the plate remains constant and equal to χ^*.

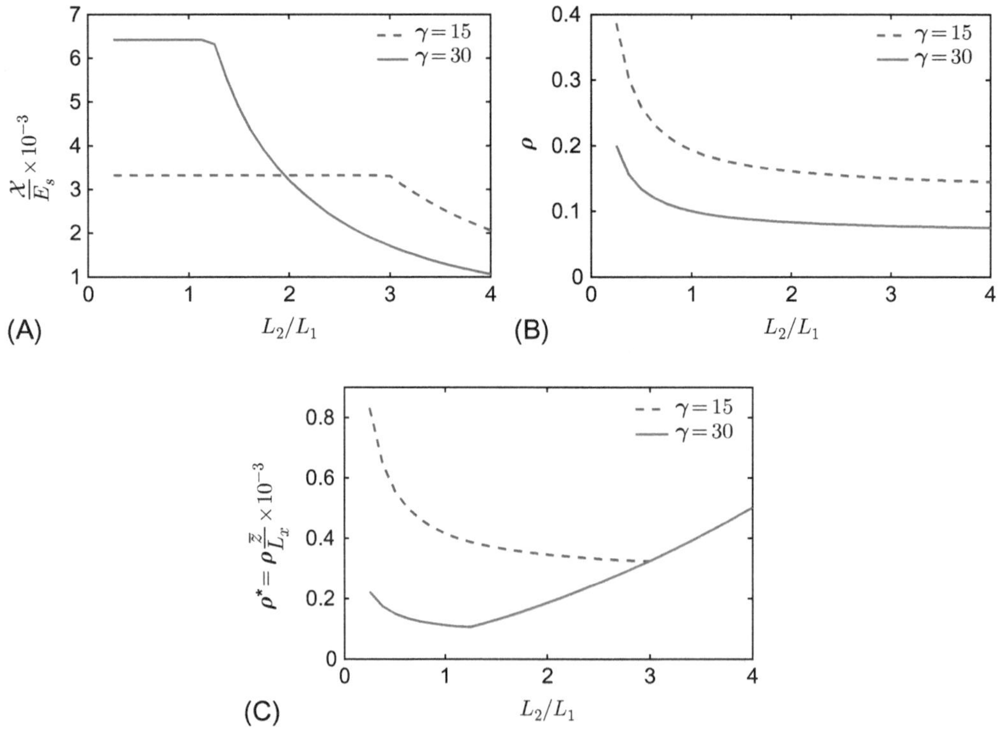

FIG. 6

Shear stiffness and normalized relative density for given parameters of the chevron lattice.

$\chi = \min\left(\frac{\partial P_{13}}{\partial G_{13}}, \frac{\partial P_{23}}{\partial G_{23}}\right)$ is the minimum out-of-plane shear stiffness of the material; $\rho = \frac{(L_1 + 2L_2)t}{2L_1 L_2 \sin\gamma}$ is the geometric relative density of the material, ρ^* is the effective relative density corrected to account for the material shear stiffness.

Fig. 6B shows the plot of the geometric relative density of the lattice material corresponding to the ratio between the volume occupied by the solid material and the volume of the RVE. Fig. 6C shows the effective relative density of the plate, defined as $\rho^* = \rho\frac{\overline{z}}{L_1}$, where ρ is the geometric relative density of the material. It is governed by the geometrical parameters of the unit cell only, and $\frac{\overline{z}}{L_1}$ is a correction term that accounts for the shear stiffness of the lattice material; it is defined such that the volume of the solid material per unit surface of the skin panel is given by $\rho^* L_1$. Even if the geometric relative density of the lattice material decreases when the aspect ratio of the unit cell increases (Fig. 6B), the effective relative density of the skin material (Fig. 6B) increases because it is necessary to increase the panel thickness to compensate for the reduction of the macroscopic shear stiffness (Fig. 6A). It is also interesting to observe that the combined variations of the shear stiffness and relative density of the material are such that after a certain value of L_2/L_1, they yield the same effective relative density, regardless of γ. In fact, the curves for $\gamma = 15$ an $\gamma = 30$ merge after $L_2/L_1 \asymp 3$.

Figs. 7A and 8A show the normalized actuation force per unit area necessary to obtain a 40% uniaxial deformation of the skin, both in extension and in compression. The chevron lattice becomes

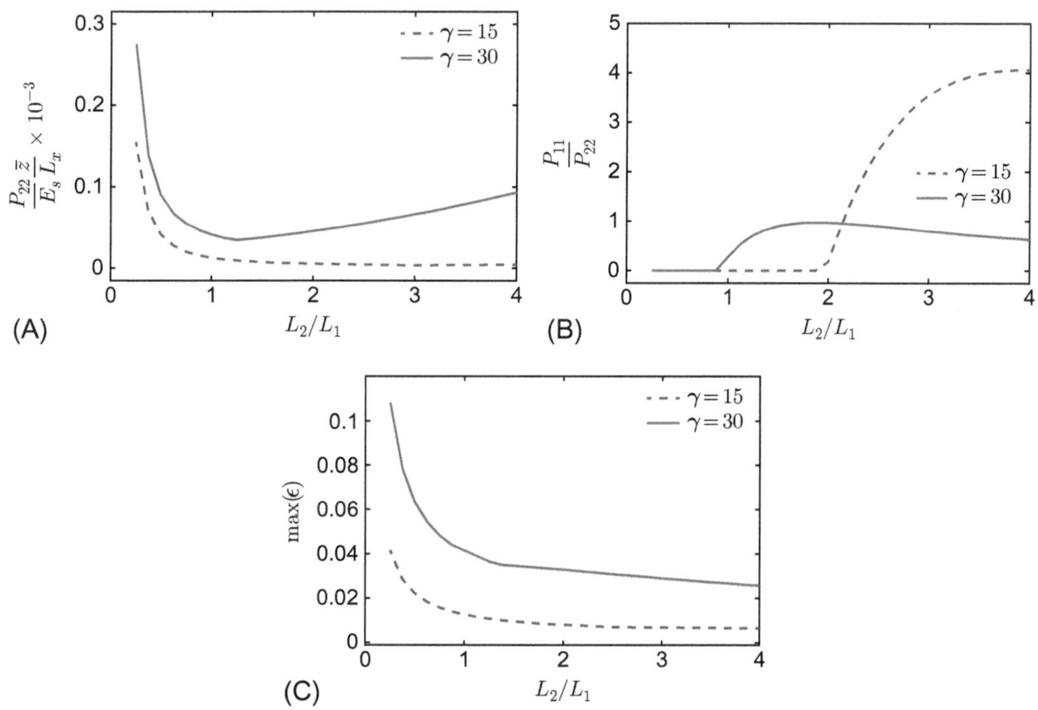

FIG. 7

Normalized actuation stress, actuation ratio, and maximum deformation of the solid material under tensile macroscopic applied strain. Curves obtained for different parameters of the chevron lattice for an actuation corresponding to 40% tension.

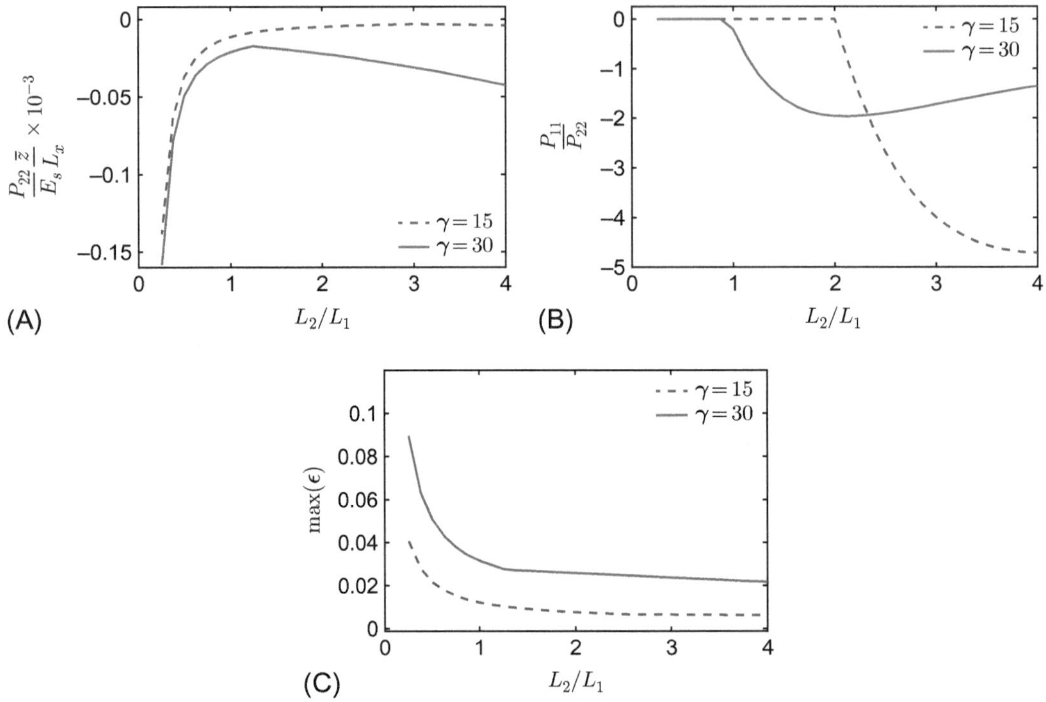

FIG. 8

Normalized actuation stress, actuation ratio, and maximum deformation of the solid material under compressive macroscopic applied strain. Curves obtained for given parameters of the chevron lattice for an actuation corresponding to 40% compression.

progressively more compliant as L_2/L_1 increases, because the bending stiffness of the inclined member decreases as its length, L_2, increases. Nevertheless, the net actuation force increases because it is necessary to increase the skin thickness to ensure that χ^* remains constant. This behavior is more evident for $\gamma = 30$ degrees, although an increment is also observed for $\gamma = 15$ degrees.

Figs. 7B and 8B show the ratio between the applied force, P_{22}, and the reaction force in direction 1, at 40% deformation in extension and in compression, respectively. For small values of L_2/L_1, the reaction forces are very small both for $\gamma = 30$ degrees and $\gamma = 15$ degrees. Yet, as the inclined member becomes slenderer, the equilibrium at the joints between the horizontal and the inclined members becomes unstable; a bifurcation point appears and induces a finite joint rotation, which in turn leads to a loss of lattice neutrality.

Figs. 7C and 8C show the maximum nominal deformation developed in the solid material of the lattice. This quantity specifies the allowable limit of macroscopic deformation for the skin, and it offers a means of comparison with the elastic limit of the solid material. The lattice is capable of undergoing macroscopic deformation as large as 40%, while maintaining the microscopic deformation of the material below 5%. This demonstrates that the Chevron lattice is very effective, the best among those considered in this study, in transforming in plane stretches into bending and large rotation of the lattice members, keeping actuation forces and solid material deformation quite low.

5 CONCLUSIONS

This chapter has focused on design challenges posed by the selection of a suitable material for the manufacturing of the skin of a morphing wing. A preliminary analysis of the requirements showed that no existing material is capable of simultaneously fulfilling all of the contrasting requirements of a morphing skin. A potentially viable solution has been identified in lattice materials. Lattice materials are characterized by an ordered periodic microstructures obtained by replicating a unit cell along independent tessellation vectors. By properly selecting the topology of the unit cell and the solid material of the lattice, it is possible to manufacture extremely stiff and strong materials, and to obtain unusual macroscopic properties.

In this chapter, we have focused in particular on the properties of the chevron lattice by exploring its design space. This work has shown that no single selection of the lattice parameters maximizes concurrently all the design objectives for a morphing wing skin. Rather, an optimal solution is to be found by establishing priorities on the relative importance of conflicting design objectives. For example, large values of L_2/L_1 ensure low actuation energy and small strain in the solid material (see Figs. 7A, C and 8A, C). On the other hand, they also produce a heavier skin and larger forces in the cross-direction. Furthermore, there is no ideal angle for the inclined members. Lower values of γ improve the compliance, but heavier lattices or stronger coupling between the actuation and the constraint directions are possible by controlling the value of L_2/L_1. To summarize, the final concept for the skin needs to take into account the actual properties of the solid material, the overall geometrical dimensions of the wing, and the strength of the structure surrounding the morphing flap.

REFERENCES

[1] S. Barbarino, O. Bilgen, R.M. Ajaj, M.I. Friswell, D.J. Inman, A review of morphing aircraft, J. Intell. Mater. Syst. Struct. 22 (9) (2011) 823–877.

[2] S. Barbarino, E.L. Saavedra Flores, R.M. Ajaj, I. Dayyani, M.I. Friswell, A review on shape memory alloys with applications to morphing aircraft, Smart Mater. Struct. 23 (6) (2014) 063001.

[3] J.C. Gomez, E. Garcia, Morphing unmanned aerial vehicles, Smart Mater. Struct. 20 (10) (2011) 103001.

[4] I. Kuder, A.F. Arrieta, W.E. Raither, P. Ermanni, Variable stiffness material and structural concepts for morphing applications, Prog. Aerosp. Sci. 63 (2013) 33–55.

[5] J.N. Kudva, Overview of the DARPA smart wing project, J. Intell. Mater. Syst. Struct. 15 (4) (2004) 261–267.

[6] E. Stanewsky, Adaptive wing and flow control technology, Prog. Aerosp. Sci. 37 (7) (2001) 583–667.

[7] I.-R. Dayyani, S. Ziaei-Rad, M.I. Friswell, The mechanical behavior of composite corrugated core coated with elastomer for morphing skins, J. Compos. Mater. 48 (13) (2014) 1623–1636.

[8] W. Raither, M. Heymanns, A. Bergamini, P. Ermanni, Morphing wing structure with controllable twist based on adaptive bending-twist coupling, Smart Mater. Struct. 22 (6) (2013).

[9] A. Spadoni, M. Ruzzene, Numerical and experimental analysis of the static compliance of chiral truss-core airfoils, J. Mech. Mater. Struct. 2 (5) (2007) 965–981.

[10] A. Spadoni, M. Ruzzene, Static aeroelastic response of chiral-core airfoils, J. Intell. Mater. Syst. Struct. 18 (10) (2007) 1067–1075.

[10a] R. Pecora, F. Amoroso, M. Magnifico, I. Dimino, K. Concilio, Kinematic rib-based structural system for innovative adaptive trailing edge, in: Proceedings of SPIE–The International Society for Optical Engineering, vol. 9801, 2016, Article number 980107.

[10b] I. Dimino, G. Diodati, A. Concilio, A. Volovick, L. Zivan, Distributed electromechanical actuation system design for a morphing trailing edge wing, in: Proc. SPIE 9801, Industrial and commercial applications of smart structures technologies 2016. SPIE Smart Structures/NDE, Las Vegas, Nevada, 2016, 980108. https://doi.org/10.1117/12.2219223.
[11] Y.J. Chen, F. Scarpa, I.R. Farrow, Y.J. Liu, J.S. Leng, Composite flexible skin with large negative Poisson's ratio range: numerical and experimental analysis, Smart Mater. Struct. 22 (4) (2013) 045005.
[12] K. Olympio, F. Gandhi, L. Asheghian, J. Kudva, Design of a flexible skin for a shear morphing wing, J. Intell. Mater. Syst. Struct. 21 (17) (2010) 1755–1770.
[13] A.Y.N. Sofla, S.A. Meguid, K.T. Tan, W.K. Yeo, Shape morphing of aircraft wing: Status and challenges, Mater. Des. 31 (3) (2010) 1284–1292.
[14] C. Thill, J. Etches, I. Bond, K. Potter, P. Weaver, Morphing skins, Aeronaut. J. 112 (1129) (2008) 117–139.
[15] N.M. Wereley, F. Gandhi, Flexible skins for morphing aircraft, J. Intell. Mater. Syst. Struct. 21 (17) (2010) 1697–1698.
[16] S. Hengsbach, A.D. Lantada, Direct laser writing of auxetic structures: Present capabilities and challenges, Smart Mater. Struct. 23 (8) (2014) 085033.
[17] M. Vaezi, H. Seitz, S. Yang, A review on 3D micro-additive manufacturing technologies, Int. J. Adv. Manuf. Technol. 67 (5–8) (2013) 1721–1754.
[18] V.S. Deshpande, M.F. Ashby, N.A. Fleck, Foam topology: Bending versus stretching dominated architectures, Acta Mater. 49 (6) (2001) 1035–1040.
[19] L.J. Gibson, M.F. Ashby, G.S. Schajer, C.I. Robertson, The mechanics of two-dimensional cellular materials, Proc. R. Soc. Lond. A Math. Phys. Eng. Sci. 382 (1782) (1982) 25–42.
[20] A.K. Noor, Continuum modeling for repetitive lattice structures, Appl. Mech. Rev. 41 (7) (1988) 285–296.
[21] A. Vigliotti, V.S. Deshpande, D. Pasini, Non linear constitutive models for lattice materials, J. Mech. Phys. Solids 64 (1) (2014) 44–60.
[22] A. Vigliotti, D. Pasini, Linear multiscale analysis and finite element validation of stretching and bending dominated lattice materials, Mech. Mater. 46 (2012) 57–68.

CHAPTER

9

COMPOSITE CORRUGATED LAMINATES FOR MORPHING APPLICATIONS

Alessandro Airoldi, Giuseppe Sala, Luca Angelo Di Landro, Paolo Bettini, Alessandro Gilardelli
Politecnico di Milano, Milano, Italy

CHAPTER OUTLINE

NOMENCLATURE

$[K]$	complete stiffness matrix of an orthotropic laminate
$\mathbf{N}$	vector of forces per unit width applied to a laminate
$\mathbf{M}$	vector of moments per unit width applied to a laminate
$\overline{\mathbf{A}}$	membrane stiffness matrix of an orhotropic laminate at the homogenized level
$\overline{\mathbf{D}}$	bending stiffness matrix of an orhotropic laminate at the homogenized level
$\overline{\mathbf{B}}$	membrane-bending coupling matrix of an orhotropic laminate at the homogenized level
$[F]$	complete compliance matrix of an orthotropic laminate
$\boldsymbol{\varepsilon}_0$	vector of mid-plane strains of an orthotropic lamante
$\boldsymbol{\kappa}$	vector of curvature of an orthotropic laminate
E	Young modulus

Morphing Wing Technologies. https://doi.org/10.1016/B978-0-08-100964-2.00009-5

l	half length of vertical arm in rounded and square-shaped corrugated geometries
r	radius of curvature and fillet in rounded and square-shaped corrugated geometries
t	thickness of composite laminate
c	semiperiod of a corrugated geometry
p	period of a corrugated geometry
h	total height of a corrugated geometry
ϑ	characteristic angle of trapezoidal corrugated geometry
s	total profile length in a semiperiod of corrugated geometry
A_{ij}	terms of the membrane stiffness matrix of a composite laminate
D_{ij}	terms of bending stiffness matrix of a composite laminate
I_1, I_2, I_3	geometrical parameters for the evaluation of axial stiffness of a corrugated laminate
E_{ii}, G_{ij}, ν_{ij}	Young moduli, shear stiffness moduli and Poisson's ratios of a composite ply
b	half length of horizontal arm in square-shaped corrugated geometry
k	transverse shear correction factor for composite laminate section
G_t	transverse shear stiffness of a composite laminate
t_{eq}	thickness of a flat laminate with the same weight of a corrugated geometry
k_s	in-plane shear stiffness efficiency index of a corrugated laminate
k_s	bending stiffness efficiency index of a corrugated laminate in nonmorphing direction
K_a	torsional stiffness of a thin-walled beam section
Ω	area enclosed by a thin-walled beam section
S	perimeter of a thin-walled beam section

1 INTRODUCTION

In the last two decades, many researches have been conducted in the aeronautical field to develop innovative structures, which are capable of smooth and progressive shape changes. This type of structures, which are currently referred to as morphing structures, modify their shape without experiencing damage or permanent deformations and can provide novel solutions for the generation of aerodynamic forces required to guide, control, and stabilize aerospace vehicles as well as to optimize them for different missions or mission segments. Seamless and gapless shape changes provided by morphing structures can perform with more efficiency and versatility the functions of rigid aerodynamic surfaces, which are conventionally adopted in the aerospace field. For such reasons, in recent years, researchers have devised and analyzed a large variety of morphing solutions, sometimes inspired by bio-mimesis or conceived by exploiting the properties of advanced materials and actuation concepts [1,2].

From a structural standpoint, a distinction can be proposed between morphing solutions that rely on internal mechanisms made of rigid parts covered by flexible skins, such as in Refs. [3–6] and solutions based on completely deformable structures, such as in Refs. [7–10]. However, in both types of architectures, a fundamental role is played by the skin, which must carry the aerodynamic pressures acting over a surface that can progressively change shape and transfer the resultant to the underlying structure. For such a reason the capability of undergoing large recoverable strains without failure is the basic requirement for a morphing skin system.

The study presented in [11] pointed out other fundamental requirements. Considering a numerical model of an internal morphing structure made of deformable beams, a sensitivity study was conducted by varying the axial and bending stiffness of the skin. A low axial compliance was found to be a basic requirement to reduce the actuation work, but bending stiffness cannot be reduced too much, to avoid the risk of skin bubbling and instability. The development of oscillations in the distribution of coefficient of pressure was observed when skin bending stiffness is excessively reduced [8].

The previous considerations focus on the behavior of skins along the main directions involved in the change of shape, which can be defined as morphing directions. Examples are chord direction in variable camber airfoils, including morphing leading and trailing edge, span direction for out-of-plane transformation, and generic in-plane direction for wing planform alteration.

Additional considerations can be introduced considering the properties in nonmorphing directions and the role of the skin in conventional aircraft stressed-skin constructions. This is related to a fundamental critical issue in the design of morphing structures, which must change shape but should also retain adequate load bearing capability and controlled stiffness properties to provide a valid structural contribution while minimizing weight penalties. Accordingly, a morphing skin should maximize the contributions to shear, torsional, and bending stiffness of the aerodynamic surface, if such properties are not directly involved in morphing performances. If such contribution is inadequate, other structural parts must be designed to provide the missing stiffness and load bearing capability, and significant weight penalties can be expected. A deformable skin with strongly anisotropic characteristics could be exploited to accomplish the aforementioned requirements.

Among the large number of engineering solutions proposed by researchers for morphing skins [12], corrugated composites are particularly promising from the structural point of view. They represent a particularly appealing concept, as they can easily be stratched and bent in the direction of the corrugated profile and exhibit high stiffness and strength properties in the normal direction of the corrugated profile. The production of the corrugated sheet by lamination of composite plies enhances design flexibility, as the properties in morphing direction and degree of structural anisotropy can be tuned by selecting the most convenient lay-ups. The use of composite corrugated laminates for morphing skins was proposed in Refs. [13,14]. Their potential was recognized quite early in literature reviews dedicated to morphing skins [12].

The interest in corrugated panels for morphing applications has renewed the research of efficient methods to predict the stiffness of such components. They can be considered as orthotropic plates characterized by a generalized stiffness matrix that links the applied force and moments per unit width to membrane strains and bending curvatures. The identification of such a matrix represents a homogenization of the corrugate geometry at a larger structural scale and has been proposed well before corrugated panels were suggested for morphing [15]. Although formulation for the axial and bending properties in morphing directions were already proposed in the seminal work [14], a complete framework to find the homogenized characteristic of a corrugated geometry made of circular segments was proposed in Ref. [16]. Properties in nonmorphing directions have been considered in Ref. [17], where an analytical formulation is provided for the calculation of the complete equivalent stiffness matrix of sinusoidal and trapezoidal laminates. The homogenization technique proposed in Ref. [17] can be adopted to obtain the complete stiffness matrix for a generic corrugated shape, although in the case of thick or asymmetric laminates a finite element approach could be required [18]. The use of homogenized stiffness matrices in numerical simulations to identify the optimal requirements for a morphing skin is exemplified in Refs. [8,19]. A detailed comparison between analytical predictions, finite element models, and experiments is presented in [20]. Recently, the properties of more complex corrugated geometries made of composite lamiantes have been analyzed to enhance design flexibility of the concept [21–23].

The detailed studies carried out in the last decade confirm the great design flexibility of corrugated laminates, but their adoption as a morphing skin in an aerodynamic surface presents some difficulties from the aerodynamic point of view, since the most interesting applications require that stream lines are transverse to the corrugation direction.

The problems caused by waviness of the surface and possible detrimental effects on aerodynamic performance were anticipated in [14], where it was proposed the use of an elastomer to fill the "valleys" of the corrugated shape. In the investigation carried out in Ref. [24] the need to limit the ratio between the depth of corrugation and aerodynamic chord was pointed out to reduce the increment of drag and the reduction of aerodynamic efficiency. For such a reason, an externally segmented skin was adopted in the wind tunnel test described in Ref. [25]. The need of an efficient solution for the integration of a smooth cover in the corrugated support is emphasized by the results presented in Ref. [26], which show that the decrement in lift-to-drag ratio could balance the advantages of smooth variation of profiles due to morphing structures and should be properly considered. A solution for the development of an efficient skin system has been proposed in Refs. [21,23,27], consisting of a corrugated support and an elastomeric cover providing a smooth aerodynamic surface.

Manufacturing aspects are also important to develop corrugated composite panels that can be applied to real-world structures. The examples in literature show that lamination technology can be used to produce corrugated composite laminates, though special techniques must be adopted for the lay-up or special molds should be used [20,25]. The research presented in [28] is focused on maximum elongation limits of corrugated composite laminates. It points out the importance of a proper manufacturing process and the use of a counter-mold to improve the quality of the laminates and increase mechanical performance.

This chapter is aimed at providing an exhaustive overview regarding corrugated composite laminates for morphing applications. In the following sections, all the issues presented in this introduction will be analyzed, focusing on some of the corrugated geometries that have been studied, which are presented in Section 2. Section 3 is dedicated to quantify both stiffness anisotropy of corrugated laminates and compliance in morphing directions. Section 4 shows a review of studies performed on structural strength and define structural efficiency indices to measure the structural contribution in the nonmorphing directions that can be obtained by corrugated laminates. In Section 5, the technological aspects related to the manufacturing of corrugated laminates will be discussed. Section 6 will discuss and propose a solution to address critical aerodynamic issues, which include the risk of skin bubbling and the detrimental effects of corrugation on the aerodynamic performances. The most relevant conclusions are finally summarized in Section 7.

2 TYPES OF CORRUGATED LAMINATES

Corrugated sheets have been studied for a long time, due to stiffening effects of corrugation that greatly enhances bending stiffness in one direction, as well as buckling loads, at a moderate weight cost. Applications range from roof structures, corrugated bulkheads in ship building [29], and floors in railway vehicles or bridges [30]. Historically, corrugated sheets were used in aircraft industries as primary structures, in the period between World War I and World War II [31]. Corrugated sheets are used as cores in sandwich structures, as in corrugated cardboard [32].

The appeal of corrugated sheets for morphing applications relies on the combination of such stiffening effects with high compliance and strain at failure in the corrugation direction. Therefore, morphing applications are especially interested in the significant anisotropy offered by corrugated sheets, which determine completely different responses in the morphing direction (defined *corrugation direction* and often *transverse direction*) and in the nonmorphing direction (often referred to as *longitudinal direction*).

Several types of corrugation have been proposed for morphing applications, with some typical shapes that are identified in Fig. 1. Rounded and trapezoidal shapes were the first shapes suggested for morphing aeronautical structures in Ref. [13,14]. The square shape is a special case of the trapezoidal shape, though the square-rounded shape considered in Ref. [23], shown in Fig. 1C takes explicitly into account the need for rounded fillets between the horizontal and vertical sections, a feature that arises from technological considerations. The sinusoidal shape, typically used for corrugated cardboard [15,32], was also considered by some authors for analytical studies regarding morphing, but was never practically implemented as a solution for morphing skins. Finally, some authors have considered circular shapes, which are made of circular arcs with variable angular amplitude [28,33,34], as shown in Fig. 1D.

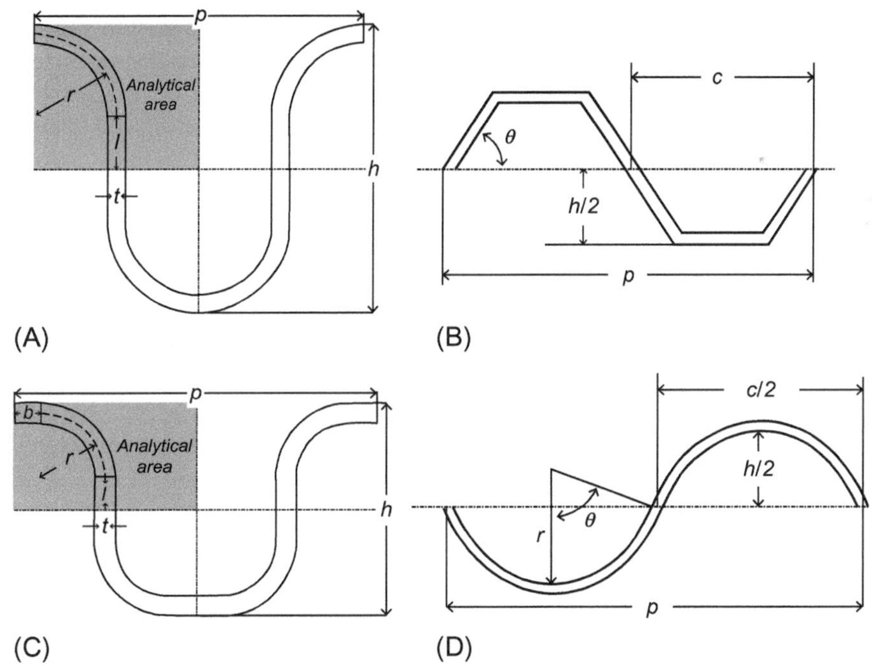

FIG. 1

Typical corrugated shapes for morphing skins: rounded (A), trapezoidal (B), square-shaped (C), and circular (D).

The corrugated laminate can be considered as a special orthotropic plate, having a total thickness represented by the height h, subjected to forces and moment per unit width that can be related to the mid-plane strains and to the curvatures of the mid-plane. The stiffness matrix $[\boldsymbol{K}]$ that describes the elastic linear response of such a plate can be given in the form of Eq. (1).

$$\begin{Bmatrix} \mathbf{N} \\ \mathbf{M} \end{Bmatrix} = \begin{bmatrix} \overline{\mathbf{A}} & \overline{\mathbf{B}} \\ \overline{\mathbf{B}}^T & \overline{\mathbf{D}} \end{bmatrix} \begin{Bmatrix} \boldsymbol{\varepsilon}_0 \\ \boldsymbol{\kappa} \end{Bmatrix} = [\mathbf{K}] \begin{Bmatrix} \boldsymbol{\varepsilon}_0 \\ \boldsymbol{\kappa} \end{Bmatrix} = [\mathbf{F}]^{-1} \begin{Bmatrix} \boldsymbol{\varepsilon}_0 \\ \boldsymbol{\kappa} \end{Bmatrix} \tag{1}$$

where $\mathbf{N}$, $\mathbf{M}$ are vectors representing the three forces and moment components per unit width applied to the plate, and $\boldsymbol{\varepsilon}_0$, $\boldsymbol{\kappa}$ are the three components of deformations and the three curvatures of plate mid-plane, respectively. In what follows, direction 1 is considered as the corrugation direction (morphing direction) and direction 2 as the longitudinal (nonmorphing) direction. The inverse of $[\boldsymbol{K}]$ is the compliance matrix $[\boldsymbol{F}]$, which provides the values of mid-plane strains and curvatures knowing the force and moments per unit width applied to the plate.

3 ANISOTROPY AND STIFFNESS PROPERTIES IN MORPHING DIRECTION

3.1 ANISOTROPY INDICES OF STIFFNESS PROPERTIES

The analytical expressions provided in Refs. [14,29] for the round and trapezoidal shapes can be used to evaluate the level of anisotropy that can be obtained by corrugated sheets, with simple variations of geometrical parameters.

For the round geometry, the stiffness in the corrugation direction, $\overline{A}_{11}$, and the one in the longitudinal direction, $\overline{A}_{22}$, can be obtained by the equivalent Young moduli of the homogenized corrugated laminate given in Ref. [14]. Stiffness terms can be obtained by multiplying the original expression for Young moduli by the height, h, of the orthotropic plate representing the corrugated laminate. Considering a homogeneous isotropic material with a Young modulus E and the nomenclature provided in Fig. 1, the following expressions are obtained:

$$\begin{aligned} \overline{A}_{11} &= \frac{Et^3}{12} \frac{r}{\left\{ \dfrac{l^3}{3} + r\left[\dfrac{\pi}{4}\left(2l^2 + r^2\right) + 2lr \right] \right\}} \\ \overline{A}_{22} &= \frac{2\pi r + 4l}{4r} Et \end{aligned} \tag{2}$$

The analytical expressions for the bending stiffness of a corrugate made of an isotropic material with a rounded geometry are provided in Eq. (3). They are obtained from the formulations given in Ref. [14], considering an isotropic material with a Young modulus E.

$$\begin{aligned} \overline{D}_{11} &= \frac{Et^3}{12} \frac{2r}{r\pi + 2l} \\ \overline{D}_{22} &= \frac{16l^3 + 24\pi l^2 r + 3\pi r\left(4r^2 + t^2\right) + 8l\left(12r^2 + t^2\right)}{48r} Et \end{aligned} \tag{3}$$

The anisotropy of membrane properties can be expressed as a ratio $\overline{A}_{22}/\overline{A}_{11}$, which turns out to be a function of the radius, r, of the vertical arm length, l and of thickness t. Moreover, the ratio is proportional to $1/t^2$. For technological reasons, radius r influences the values of the parameter l that can be

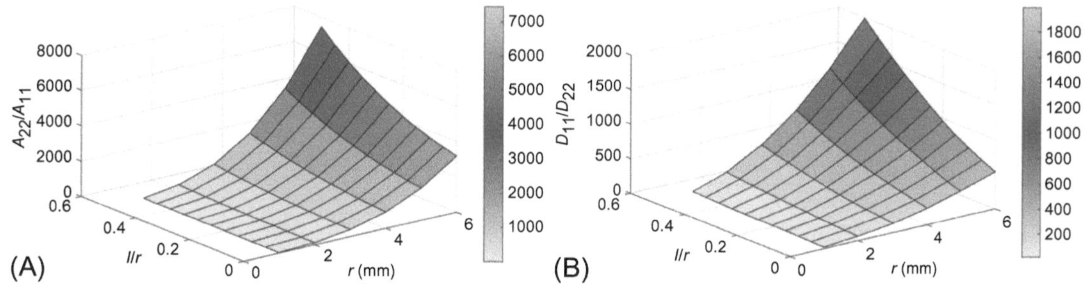

FIG. 2

Example of achievable ratios of membrane stiffness (A) and flexural stiffness (B) in the corrugation and transverse directions for rounded corrugation geometry.

adopted (see Fig. 1A). Therefore, an indication of the membrane anisotropy indices $\overline{A}_{22}/\overline{A}_{11}$ that can be achieved is provided by choosing a reasonable range for r, which is varied in the range 1 mm ÷ 6 mm, and assuming l values in the interval $0 \le l \le 0.5r$. An unitary laminate thickness, $t = 1$ mm, is fixed. The obtained results are reported in Fig. 2A.

Analogously, the ratio $\overline{D}_{22}/\overline{D}_{11}$ indicates the anisotropy of flexural properties. An example of the values that can be achieved for such ratio are provided in Fig. 2B, which refers to a laminate with a laminate thickness of 1 mm and the same range of parameters r and l considered for the $\overline{A}_{22}/\overline{A}_{11}$.

For the trapezoidal corrugation shown in Fig. 1B, the expressions in Eqs. (4), (5), taken from Ref. [29] can be used to evaluate the corrugation and longitudinal direction of the membrane and flexural stiffness, respectively.

$$\overline{A}_{11} = \frac{E\left(\frac{t}{h/2}\right)^2 t}{6\left[\frac{2}{c}\left(\frac{c - s\sin\theta}{1-\cos\theta}\right) + \frac{4}{3c}\frac{h/2}{\sin\theta}\right]}$$
$$\overline{A}_{22} = Et\frac{s}{c} \tag{4}$$

$$\overline{D}_{11} = \frac{c}{s}\frac{Et^3}{12}$$
$$\overline{D}_{22} = \frac{Et}{2c}\left[\frac{4(h/2)^3}{3\sin\theta} + 2(h/2)^2\left(c - \frac{h}{\tan\theta}\right)\right] \tag{5}$$

In the formulations of Eq. (4), the laminate thickness, t, the semiperiod, c, the total thickness h and the angle θ can be considered as four independent geometrical variables. Indeed, the total length s of the laminate, in a semiperiod c, can be expressed as a function of such quantities, as indicated in Eq. (6).

$$s = \frac{h}{\sin\theta} + c - \frac{h}{\tan\theta} \tag{6}$$

An indication of the values that can be achieved for the ratios $\overline{A}_{22}/\overline{A}_{11}$ and $\overline{D}_{22}/\overline{D}_{11}$ can be provided by fixing the semiperiod $c = 12$ mm, so to obtain a unit cell with a length identical to the cell of the

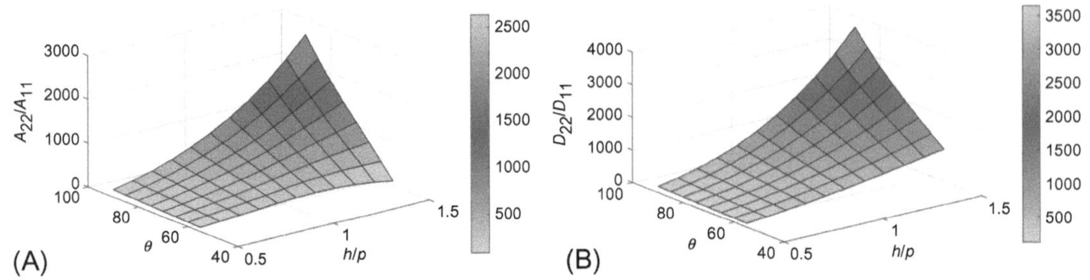

FIG. 3

Example of achievable ratios of membrane stiffness (A) and flexural stiffness (B) in the corrugation and transverse directions for trapezoidal corrugation geometry.

rounded geometry with a radius $r = 6$ mm. The angle θ and the total thickness, h, are varied in the range $55° \leq l \leq 90°$ and in the range $0.5p \leq h \leq 1.5p$. As in the case of rounded geometry, both ratios are proportional to $1/t^2$, The results of the sensitivity study are provided in Fig. 3 for a thickness of 1 mm.

The trends indicated by the plots in Figs. 2 and 3 are similar. In general, the indexes representing anisotropy increase with the total height of the laminate and can reach values of several thousand. For thin laminates, considering the dependency from the plate thickness, the ratios can achieve values higher than those indicated in plots. Moreover, if composite laminates are considered, the lamination sequence can be chosen to tailor the stiffness properties in the different directions, per specific requirements.

Indeed, the very thin composite corrugated laminates tested in [14] reached a $\overline{A}_{22}/\overline{A}_{11}$ of 4600 and a $\overline{D}_{22}/\overline{D}_{11}$ of 6000. In tensile tests an elongation of 1% in the longitudinal direction required the application of a force per unit width of 120 N mm^{-1} in the longitudinal direction, whereas, in the morphing direction, the application of a force per unit width of about 1 N mm^{-1} led to an elongation of about 40%.

The experiments reported in [13] were conducted on composite laminates made of different ply types with trapezoidal corrugations. Laminate thickness was varied between 0.11 and 0.62 mm. The obtained ratios $\overline{A}_{22}/\overline{A}_{11}$ varied from 186 to 10,891. Other specimens were tested in bending, obtaining ratios of flexural stiffness $\overline{D}_{22}/\overline{D}_{11}$ beyond 2000. Other results for trapezoidal corrugation are reported in [20], where composite laminate with a thickness of 0.3 were used, obtaining $\overline{A}_{22}/\overline{A}_{11}$ ratio up to 1800 and $\overline{D}_{22}/\overline{D}_{11}$ of about 1000. Such experimental results confirm that it is possible to manufacture composite corrugated laminates achieving the anisotropy indices predicted by the analytical formulations.

3.2 COMPLIANCE IN MORPHING DIRECTIONS OF DIFFERENT TYPES OF COMPOSITE CORRUGATED LAMINATES

The previous section provided some examples regarding the anisotropy of corrugated panels in morphing and nonmorphing directions. However, functional aspects of corrugated sheets for morphing applications depend on their axial and bending compliance. According to [11], morphing skins should exhibit high axial compliance, but also an adequate bending stiffness, to withstand aerodynamic pressures without excessive deformations. Although such consideration is valid when morphing performance requires axial extension, which is typical of camber variations in airfoils, an adequate bending compliance can be requested when morphing skin must accomplish significant changes of curvatures, as in the case of applications for leading edges and droop noses [35,36].

The estimation of compliance in morphing directions of a composite corrugated laminate can be carried out by comparing the axial and bending stiffness, $\overline{A}_{11}$ and $\overline{D}_{11}$, respectively, with the ones of a flat laminate having the same lay-up, which will be indicated as A_{11} and D_{11}. Hence, the ratios $(\overline{A}_{11}/A_{11})^{-1}$ and $(\overline{D}_{11}/D_{11})^{-1}$ indicate how the axial and bending compliance are magnified with respect to a conventional flat panel made of the same material. Such indices can be denoted as normalized axial and bending compliance, respectively, where normalization is referred to the compliance of a flat composite panel with the same lay-up of corrugated laminates.

For a trapezoidal panel, shown in Fig. 1B, the formulations provided in Ref. [17] are used. The axial and bending stiffness, turn out to be obtained through the following equations:

$$\overline{A}_{11} = \frac{2c}{\frac{I_1}{A_{11}} + \frac{I_2}{D_{11}}} \qquad \overline{D}_{11} = \frac{c}{s} D_{11} \tag{7}$$

where I_1 and I_2 are parameters depending on the geometry of the corrugated panel corrugate, with the following expressions:

$$I_1 = \frac{4(h/2)\cos\theta}{3\sin\theta} + 2c - \frac{4(h/2)}{\tan\theta} \qquad I_2 = \frac{4(h/2)^3}{3\sin\theta} + 2(h/2)^2\left(c - \frac{h}{\tan\theta}\right) \tag{8}$$

and s is the length of the corrugated profile, which can be evaluated through Eq. (6).

Since $\overline{A}_{11}$ depends on A_{11} as well as on D_{11}, the ratio $(\overline{A}_{11}/A_{11})^{-1}$ turns out to be a function of the specific lay-up used. The analysis can be limited to the case of a homogeneous lay-up $[0]_5$ of carbon/epoxy reinforced fabric plies, with the characteristics reported in Table 1.

In Fig. 4A, the normalized axial compliance for a trapezoidal corrugated laminate is plotted against the ratio of the total height to the semiperiod, h/c, keeping the semiperiod c fixed at 12 mm, similarly to the parametric analysis reported in Fig. 3. In this case, analysis was carried out considering a range $h/c = 0.5 \div 1.5$ and $\theta = 45° \div 90°$. Even adopting the smallest h/c ratio of 0.5, the axial compliance is increased by two order of magnitude with respect to the flat panel. Compliance increases with h/c ratio and with angle θ, until the maximum angle of 90°. Considering an $h/c = 1$, the normalized compliance

Table 1 Stiffness Properties of Carbon-Reinforced Fabric Plies Considered for Corrugated Laminates

$E_{11} = E_{22}$	MPa	56,550
ν_{21}	–	0.05
G_{12}	MPa	4040
$G_{13} = G_{23}$	MPa	3000
Ply thickness	Mm	0.100

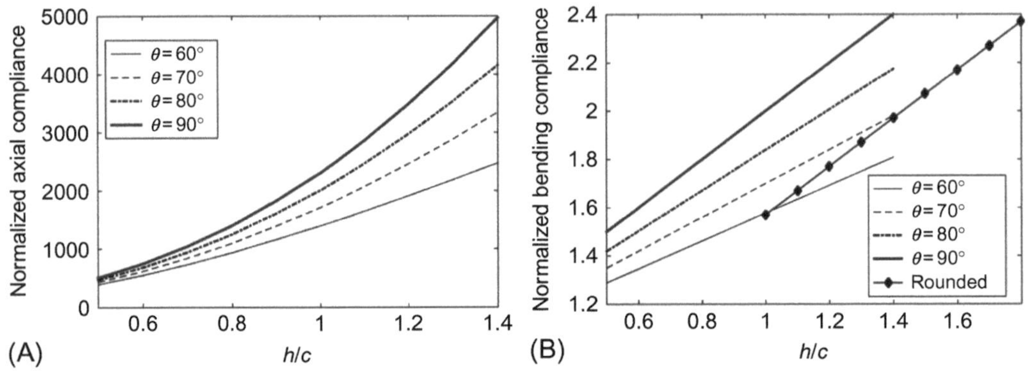

FIG. 4

Normalized axial compliance (A) for trapezoidal corrugate and normalized bending compliance for trapezoidal and rounded geometry (B).

ranges from 814, for $\theta = 45°$, to 2305 for $\theta = 90°$. Bending stiffness $\overline{D}_{11}$, depends only on laminate stiffness D_{11}, as it is shown in Eq. (7), so that the normalized bending compliance has the same value for any lay-up. In Fig. 4-B, the plot of the normalized bending compliance, for a fixed period of 24 mm, is plotted against the h/c ratio, for different angles. It can be observed that the corrugated laminate is always more compliant than the flat panel, as it can be immediately derived from Eq. (7). However, the values of normalized compliance are noticeably smaller than the ones referred to the axial case. Maximum values of 2.4 are obtained in the considered range of design variables. For $h/c = 1$, the values are bounded between $\sqrt{2}$, for $\theta = 45°$, and 2, for $\theta = 90°$.

For the rounded geometry, presented in Fig. 1A, both the seminal work [14] and a more recent paper [17] provide formulations for the prediction of properties in morphing direction. Formulations provided by Yokozeki [14] have been used for the study of anisotropy properties, though they refer to isotropic material (Eqs. 2, 3). According to [17] the axial stiffness $\overline{A}_{11}$ can still be represented by the general formula given by the first of Eq. (7). Hence, the inverse ratio between the stiffness of the corrugated and the plain laminate, $\left(\overline{A}_{11}/A_{11}\right)^{-1}$, still depends on the lay-up and a case study will be carried out for a $[0]_5$ stack sequence of plies with the mechanical properties indicated in Table 1. However, the expression of the parameters I_1 and I_2 are different and are given in Eq. (9):

$$\begin{aligned} I_1 &= \pi r \\ I_2 &= \frac{4l^3}{3} + 2\pi l^2 r + 8lr^2 + \pi R^3 \end{aligned} \tag{9}$$

where the meaning of all geometrical quantities is given in Eq. (1).

For the bending stiffness, $\overline{D}_{11}$, both [14,17] propose the formulation given in Eq. (10):

$$\overline{D}_{11} = \frac{2r}{s} D_{11} \tag{10}$$

where s is length of the corrugated profile in a semiperiod:

$$s = \pi r + 2l \tag{11}$$

For an isotropic material, such formulation is clearly equivalent to Eq. (3). Bending stiffness of the corrugated laminate depends only on the bending stiffness of the flat laminate, so results regarding the $\left(\overline{D}_{11}/D_{11}\right)^{-1}$ ratio are of general validity. Moreover, given the period $c = 2r$, the ratio of total height to the semiperiod, h/c, is determined by the choice of the vertical arm length, l:

$$\frac{h}{c} = \frac{2r + 2l}{2r} \Rightarrow l = r\left(\frac{h}{c} - 1\right) \tag{12}$$

If the result of Eq. (11) is introduced in Eq. (12), it can be deduced that the ratio $\left(\overline{D}_{11}/D_{11}\right)^{-1}$ will not depend on the radius, and consequently on the period, but only on the ratio h/c. Moreover, it can be observed that the ratio of total height to semiperiod for this geometry is equal or greater than the unit value, depending on the value of l.

The normalized axial and bending compliance, $\left(\overline{A}_{11}/A_{11}\right)^{-1}$ and $\left(\overline{D}_{11}/D_{11}\right)^{-1}$ have been investigated, by applying the formulations in Eqs. (7)–(10). The results in the range $r = 4$ mm $\div$ 10 mm and $h/c = 1 \div 2$ are reported in Fig. 5 for the axial compliance and have been included in Fig. 4B for the bending compliance, thus allowing for a direct comparison with trapezoidal corrugated laminates.

In Fig. 5, it can be observed that the axial compliance of the corrugated laminate for the rounded geometry strongly depends on the radius, even for a fixed h/c. A comparison between the rounded and the trapezoidal geometry can be carried out considering the radius of 6 mm, which corresponds to the same semiperiod of 12 mm used for the trapezoidal laminates. For $h/c = 1$, the rounded geometry leads to a normalized compliance of 1347, which is within the range obtained for the same h/c in the case of trapezoidal corrugation at different angles θ. The value obtained with the rounded geometry is like the one of a trapezoidal corrugation with the same period, same height and $\theta = 60°$.

Considering the normalized bending performances, values vary from 1.57 to 2.57, for h/c passing from 1 to 2, as indicated in Fig. 4B. At $h/c = 1$, the rounded geometry obtains the same normalized bending compliance of a trapezoidal geometry with $\theta = 60°$, but slope of the compliance versus h/c is higher for rounded geometry than for the trapezoidal case. However, in the range of variables where

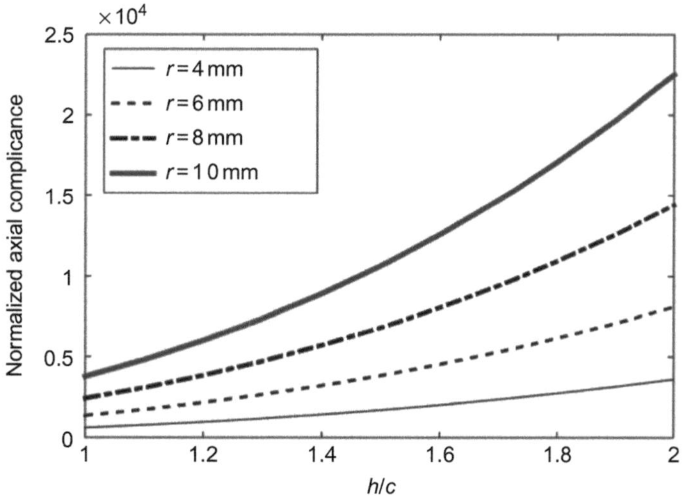

FIG. 5

Normalized axial compliance for rounded geometry.

the two sensitivity studies can be directly compared, the values of bending compliance are quite similar for the two geometries.

A third parametric analysis has been carried out considering the square-shaped geometry presented in Fig. 1C. For such geometry, the prediction taken from [17] are available, based on Eq. (7) with a proper expression for the geometrical parameters I_1 and I_2. However, an enhanced formulation is hereby proposed, based on the application of Timoshenko beam theory to a corrugated profile with a unit depth. Such theory considers the transverse shear stiffness of the laminate and could provide more accurate results for relatively thick profiles. If such effect is included, the formula for the evaluation of axial and bending stiffness can be expressed as in Eq. (13).

$$\begin{aligned}\overline{A}_{11} &= \frac{4(r+b)}{\left[\frac{I_1}{A_{11}}+\frac{I_2}{D_{11}}+\frac{I_3}{(kG_t t)}\right]}\\ \overline{D}_{11} &= \frac{r+b}{\frac{\pi}{2}r+l+b}D_{11}\end{aligned} \tag{13}$$

where

$$\begin{aligned}I_1 &= \pi r - 4b\\ I_2 &= 4l^2\left(b+\frac{l}{3}\right)+2rl(4b+\pi l)+r^2(4b+8l+\pi r)\\ I_3 &= \pi r + 4l\end{aligned} \tag{14}$$

The equivalent transverse shear stiffness G_t is used in Eq. (13), which turns out to be the out-of-plane shear stiffness $G_{13} = G_{23}$ reported in Table 1 for an homogeneous lay-up $[0]_5$, with a shear correction factor, k, that can be set to 5/6, considering a stripe with a unitary depth and a rectangular section. It can be observed that the second line of Eq. (13), referring to the bending stiffness, is the same formulation already reported in the first line of Eq. (7) and in Eq. (10). In these formulations, the multiplying factor required to obtain the bending stiffness of the corrugated laminate from the one of the flat laminate is equal to the ratio between the semiperiod and the semilength of the corrugated profile.

In a first sensitivity study, the fillet radius r, shown in Fig. 1C, has been kept fixed at 2 mm. A $[0]_5$ lay-up of fabric plies with properties reported in Table 1 has been considered and the length of the horizontal arm, b, has been varied so to obtain a range of period $c = 10$ mm $\div$ 16 mm.

As in the previous cases, the ratio h/c is fundamental to determine the compliance of the corrugated laminates. For the axial compliance, the period also plays a relevant role, as it is shown in Fig. 6A. Quantitatively, the square-shaped corrugation obtains very high compliance values that are like the ones achieved by the trapezoidal with $\theta = 90°$. For instance, the normalized compliance for $h/c = 1$ and $c = 12$ mm is 2030, whereas, for the trapezoidal case ($\theta = 90°$), the corresponding figure is 2305. The rounded geometry obtained significantly lower values (1347, for the corresponding values of design variables). Considerations regarding bending compliance are analogous, as it can be observed from Fig. 6B, where results obtained for the square-shaped geometry are compared with the study performed on rounded geometry. For this sensitivity study focused on square-shaped geometry, the period has a very limited influence.

It can be observed that a perfect trapezoidal composite corrugated laminate without fillets is not feasible from the technological standpoint. The squared-shaped corrugated laminates are actually a

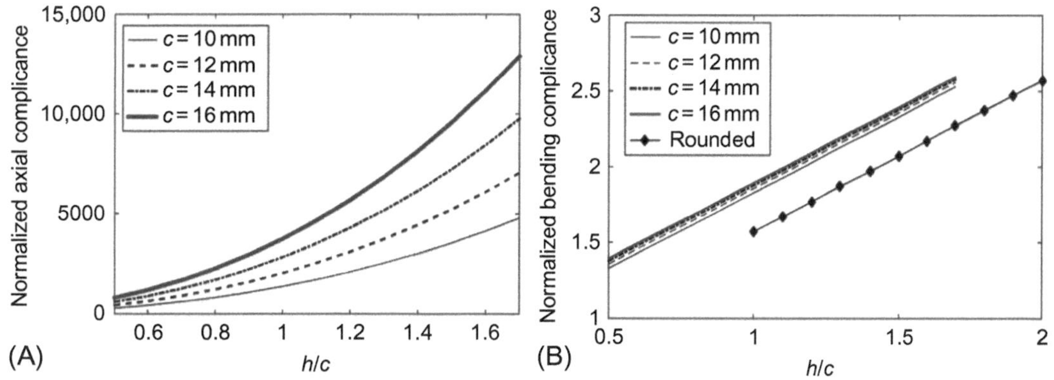

FIG. 6

Normalized axial (A) and bending (B) compliance for square-shaped corrugated geometry at fixed $r = 2$ mm.

technologically feasible version of a trapezoidal shape with $\theta = 90°$, with axial and bending compliances that decrease with the fillet radius r. As the fillet radius is increased, the compliance of the square-shaped corrugation becomes closer to the one of the rounded geometry. Such considerations have been quantitatively confirmed by performing a sensitivity analysis at a constant semiperiod $c = 12$ mm (Fig. 7).

In conclusion, it can be observed that trapezoidal corrugations with $\theta = 90°$ maximizes the compliance that can be obtained by considering both axial and bending responses, among all the geometries that have been investigated in this sensitivity study. On the contrary, the lowest values of compliance are reached by the rounded corrugation. The study verified that the square-shaped corrugation obtains values that tend to the ones of trapezoidal corrugation with $\theta = 90°$ as the radius of the fillet among vertical and horizontal arms is reduced. However, it should be remarked that square-shaped corrugation with fillet is indeed a feasible configuration from the technological point of view, whereas trapezoidal geometries, which are characterized by sharp corners, would require special technologies and would be characterized by poor strength performances.

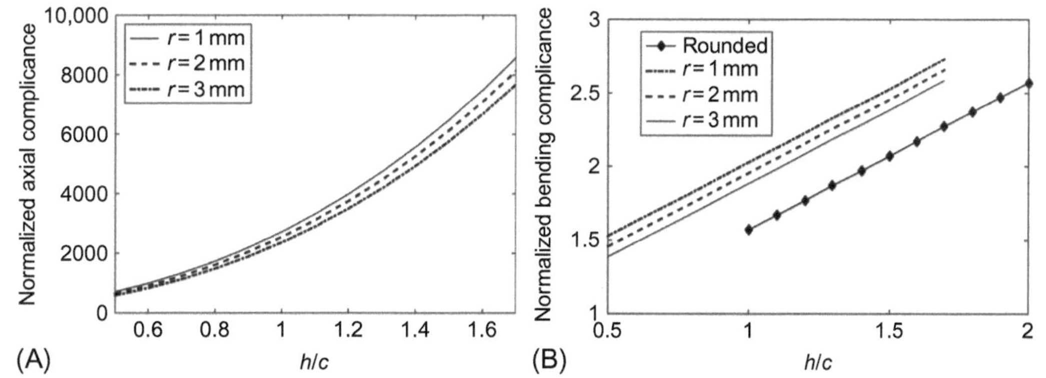

FIG. 7

Normalized axial (A) and bending (B) compliance for square-shaped corrugated geometry at fixed $c = 12$ mm.

4 STRENGTH AND STIFFNESS CONTRIBUTIONS IN NONMORPHING DIRECTIONS

4.1 FAILURE MODES OF COMPOSITE CORRUGATED LAMINATES AND STRAIN LIMITS

The capability of undergoing large recoverable strain is a fundamental requirement for a morphing skin. The deformability of composite corrugated laminates can reach significantly high levels, particularly for the maximum elongation in the morphing direction, evaluated under pure tensile conditions.

A parametric analytical study for the evaluation of strain limits is provided in [34], referring to circular shapes as the one exemplified in Fig. 1D. An analytical formulation, based on classical lamination theory, is used to evaluate the relation between the membrane strain, ε_0, and the curvature, $\boldsymbol{\kappa}$, of the corrugated plate at the homogenized level and the local strains in the plies of the composite laminate. A maximum strain criterion is used to calculate the maximum admissible strains and curvature in the homogenized corrugated plate and to compare them with the ones of a flat plate made of the same laminate. The parametric study is carried out by considering corrugated laminates made of glass-reinforced plies, with two cross-ply laminations sequences: [0/90/90] and [90/0/0/90].

Results show that tensile elongation at failure in the morphing direction noticeably increases with the ratio of corrugation amplitudes to the period, h/p. For $h/p = 100$, failure limit in tension turns out to be higher than 100 times the one of a flat panel with identical lamination sequence. Indeed, very high strain at failure in tension is a peculiar characteristic of corrugated laminates, as it was pointed out in Ref. [14], where failure strain higher than 40% were obtained. Results obtained in Ref. [20] for different combinations of geometrical parameters indicate failure strain between 15% and 40%. Fig. 8 refers to a test performed on a sample of a square-shape corrugated laminate, with a total length

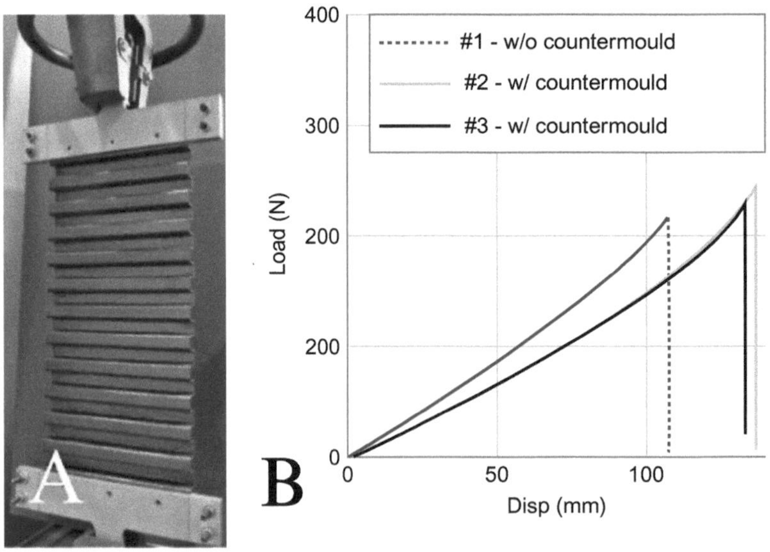

FIG. 8

Tensile tests (A) and force versus displacement response (B) for square-shaped corrugated laminates.

of 280 mm. The two curves presented in Fig. 4B report the responses of samples produced with different technological processes, which will be discussed in the next section. Failure occurs at the fillet, with a strain in the order of 30%.

According to [34] shear failure limits are also higher than the one of a flat panel and also depends on ratio h/p, but the variation of such index is much more limited with respect to the tensile failure in morphing direction. Indeed, the maximum value of the shear strain for $h/p = 2$ is two times the failure limit for the one of a flat panel. The behavior for bending, in morphing direction, is like the one referring to shear, with failure limits that varies between one and two times the values of a flat panel.

However, it should be remarked that the study reported in Ref. [34] is theoretical and considers plane stress states. Unfortunately, a critical issue for any curved composite laminate in bending conditions, including the curved parts in the various type of shapes proposed for corrugated skins, is represented by delamination. The interlaminar stress state induced by curvature in the bending of corrugated laminate is studied in Ref. [28], where geometries based on mutually tangent circular segments are considered. Charts are produced by applying an analytical model, which identifies two types of regions in the design space defined by geometrical parameters, including laminate thickness: regions where failure is due to in-plane stress states and regions where failure is induced by delaminations. For failures produced by in-plane stress states, results in Ref. [28] substantially confirm the study reported in Ref. [34], but in the regions dominated by interlaminar failure, failure limits in bending are typically lower than those of a flat panel. Thickness, depth of corrugation and period are the most important parameters that can lead to unfavourable combinations. Results reported in Ref. [13] also confirm that failure occurs at the corner of trapezoidal shapes, where stress states are influenced by three dimensional effects. As a consequence of this, it is apparent that an excessive reduction of fillet radius between differently inclined arms of the corrugated profile has a detrimental effect on strength performance and that trapezoidal configuration with sharp corners would not only be technologically difficult to obtain, but could be also characterized by a limited maximum achievable elongation and curvature.

4.2 EVALUATION OF STRUCTURAL STIFFNESS CONTRIBUTION IN NONMORPHING DIRECTIONS

Anisotropy of stiffness properties of corrugated shapes was analyzed in Section 3, showing that the stiffness parameters in the nonmorphing direction can be three order of magnitude higher than the ones in morphing directions. Such aspect is particularly appealing, since it indicates the possibility to design a skin that exhibits morphing characteristics and can provide at the same time a structural contribution in directions not involved in morphing functionalities. In particular, attention can be paid to the stiffness parameters that are fundamental for the structural role of skin in aerospace constructions: bending stiffness in direction perpendicular to corrugation direction (nonmorphing) and in-plane shear stiffness. Indeed, a panel made with a corrugated laminate can be assumed as an integrally stiffened panel. Corrugation profiles play the role of stringers applied to a flat conventional panel, thus increasing bending stiffness with advantages regarding the capability of transferring aerodynamic pressure without requiring closely spaced supports to limit excessive deformation, and preventing instability phenomena under the action of compressive stress applied to the panel. As far as in-plane shear stiffness is concerned, it is well known that in the stressed skin constructions, which represent a typical structural solution in most of the aircrafts, the external skin carries the shear stress and provides a fundamental contribution to torsional stiffness. Both functions rely on the in-plane shear stiffness of the skin panels.

Therefore, a fundamental issue in the design of a morphing skin for an aerodynamic surface is the evaluation of efficiency of structural contributions regarding bending stiffness in the direction normal to the morphing one and in-plane shear stiffness. Such contributions should be compared with those provided by a flat panel, and the comparison would be particularly meaningful if the structural stiffness is evaluated for both the morphing and the conventional flat panel having the same weight.

Based on the aforementioned considerations, a corrugated laminate made of a homogeneous lay-up $[0]_n$ is taken into account, with the plies having the properties listed in Table 1. A flat panel is also considered, made of plies with the same material properties of the corrugated one, with an equivalent thickness, t_{eq}, which is calculated to match the weight of the corrugated laminate. The expression shown in Eq. (15) provides the formula for the evaluation of the equivalent thickness for trapezoidal, rounded and square-shaped corrugation.

For trapezoidal:

$$t_{eq} = \frac{2t}{c}\left(\frac{h/2}{\sin\theta} + \frac{c}{2} - \frac{h/2}{\tan\theta}\right)$$

For rounded:

$$t_{eq} = \left(\frac{h}{c} - 1 + \frac{\pi}{2}\right)t \tag{15}$$

For square shaped:

$$t_{eq} = \frac{l + b + \frac{\pi R}{2}}{b + r}t$$

Based on the previous considerations, two indices can be defined to indicate the structural efficiency of the morphing skin in nonmorphing directions. The first one is a bending structural efficiency index, which represents the ratio between the bending stiffness $\overline{D}_{22}$ and the one of the flat panel of equivalent weight:

$$K_b = \overline{D}_{22} \Bigg/ \left(\frac{t_{eq}^3}{12}\frac{E_{22}}{1 - \nu_{12}\nu_{21}}\right) \tag{16}$$

where E_{22}, ν_{12}, and ν_{21} are the stiffness properties of the ply material.

The bending stiffness $\overline{D}_{22}$ can be evaluated, according to [17] from the general formulation:

$$\overline{D}_{22} = \frac{1}{2c}[I_2 A_{22} + I_1 D_{22}] \tag{17}$$

where I_1 and I_2 parameters have the expressions provided in Eq. (8), for the trapezoidal geometry, in Eq. (9), for the rounded shape, and in Eq. (14) for the square-shaped corrugation.

The second index is a shear structural efficiency index, which expresses the ratio between the in-plane shear stiffness of the corrugated laminate, $\overline{A}_{66}$, and the corresponding shear stiffness of a flat panel made of the same homogeneous ply material, with identical areal weight:

$$K_s = \overline{A}_{66} / (G_{12} t_{eq}) \tag{18}$$

where G_{12} is the shear stiffness provided in Table 1.

The general expression of the $\overline{A}_{66}$ term is [17]:

$$\overline{A}_{66} = \frac{c}{s} A_{66} \tag{19}$$

where s is the length of the corrugation profile in a semiperiod.

Eq. (19) indicates that the corrugated profile is characterized by a reduced shear stiffness and that the reduction factor is equivalent to the ratio between the total length of the profile and the projection of the profile on the plane of the homogenized laminate. This expression is related to the reduction of torsional stiffness of a beam with a thin-walled section, typical of aeronautical applications, as the one that is sketched in Fig. 9. In such simply connected section, where the panels define a single closed loop, the sectional torsional stiffness K_α, can be expressed according the following formula [37]:

$$K_\alpha = \frac{M_t}{d\alpha/dz} = \frac{4\Omega^2 G t}{S} \tag{20}$$

where a constant thickness of the panels, t, and an isotropic material with shear modulus, G, were assumed.

M_t is the applied torsional moment, α is the angle of torsion, and z is the beam axis. The quantities Ω and S, which are also indicated in Fig. 9, represent the area enclosed by the loop of the panels and the perimeter of the loop, respectively.

The sketch in Fig. 9 indicates that the area Ω of the section with the conventional noncorrugated panels can be considered almost identical to the area enclosed by the loop in the presence of corrugated panels, Ω_c. On the other hand, the perimeter of the loop with corrugated panel, S_c, is increased with respect to the perimeter of the conventional box, S. In the corrugated part, the incremental factor is the ratio of the length of the corrugation panel over the projection of the panel along the direction of the conventional panel, which is the inverse of the ratio that appears in Eq. (20). Hence, per Eq. (20), such ratio also provides the reduction of the contribution to torsional stiffness when a conventional panel is substituted by a morphing corrugated panel.

The response of a corrugated laminate under the action of an in-plane shear stress is provided considering a numerical analysis of such a case. In Fig. 10, the results of a Finite Element analysis on a representative unit of a square-shaped corrugate, shown at the upper left corner, are presented. Displacement boundary conditions are shown in the lower left corner. Kinematic constraints were used to obtain a pure shear deformation [21,23]. The color map refers to the shear stress in the wall of

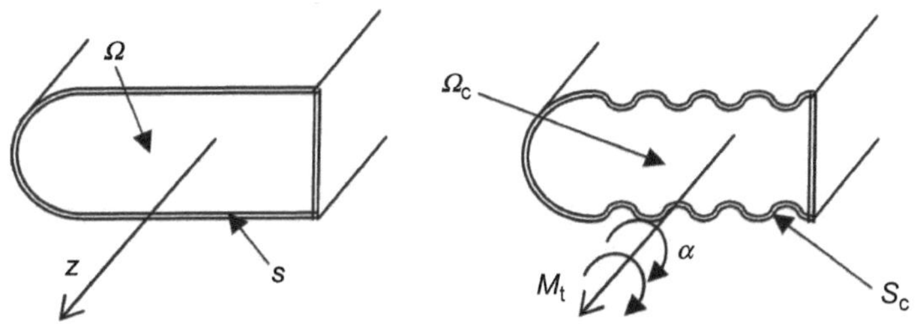

FIG. 9

Sketch of a thin-walled beam with plane and corrugated panels.

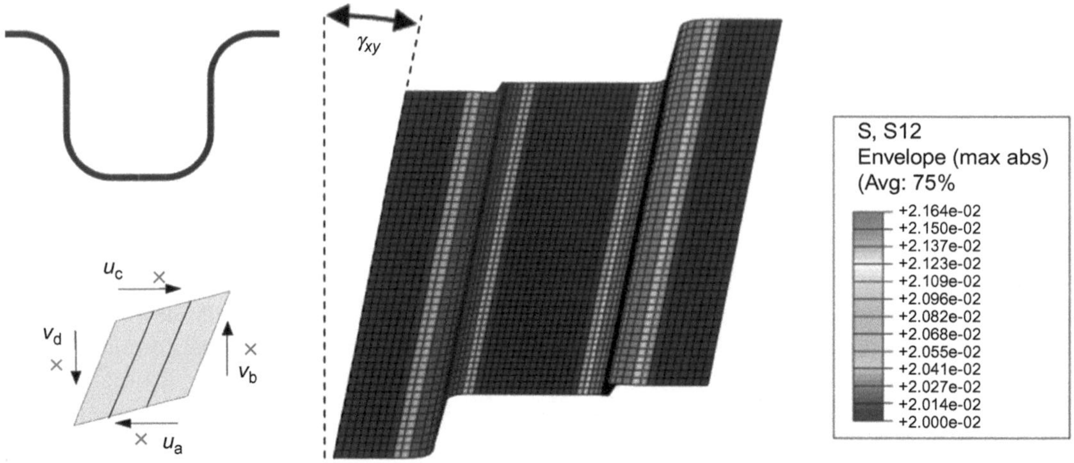

FIG. 10

In-plane shear deformation mechanism of a square-shaped laminate.

From S. Fournier, A. Airoldi, E. Borlandelli, G. Sala, Flexible composite supports for morphing skins, in: AIDAA Napoli XXII, Napoli, XXII Conference of Italian Association of Aeronautics and Astronautics, Naples, Italy, 9–12 September, 2013, pp. 1–12, ISBN: 9788890648427.

the laminate and it can be observed that stress levels are almost constant in the horizontal and vertical arms. Therefore, vertical arms undergo shear deformation, thus giving a contribution to the shear strain γ_{xy}. However, the stress oriented in the vertical direction cannot contribute to the shear stress resultant on the mid-plane of the laminate, so that the shear strain is increased with respect to the one of a flat panel at the same shear stress level.

The previous consideration highlights the importance of the shear structural efficiency index defined in Eq. (18), which should be maximized to reduce the weight penalty related to the adoption of a morphing solution based on a corrugated skin, if the torsional stiffness of the section must be maintained.

However, it should be noted that if a corrugated panel is free to elongate in the axial direction and, in particular, if its edges are free to undergo different elongations, the application of forces to the edges can lead to in-plane shape variations that do not depend only on the shear stiffness of the homogenized plate, but also on the very low axial stiffness of the corrugations. In such conditions, the corrugated panel cannot provide adequate stiffness to oppose to shape variations. Accordingly, the structural efficiency quantified by index K_s cannot be exploited to transmit forces and torques, since they tend to vary the in-plane shape of the panel under specific boundary conditions (depending on the constraints with internal structures) due to the possible effects of the extremely low axial stiffness at the level of the structural element scale.

The structural efficiency indices defined in Eqs. (16), (18), have been investigated for the trapezoidal, rounded and square-shaped corrugations considering the same design variable spaces used for the investigation of morphing performances in the previous section.

The results for the trapezoidal corrugation are shown in Fig. 11, considering a fixed semiperiod of 12 mm. It can be observed that the bending stiffness of the corrugated panel is more than two orders of magnitude higher than the one of a plane panel with the same weight. The index generally increases

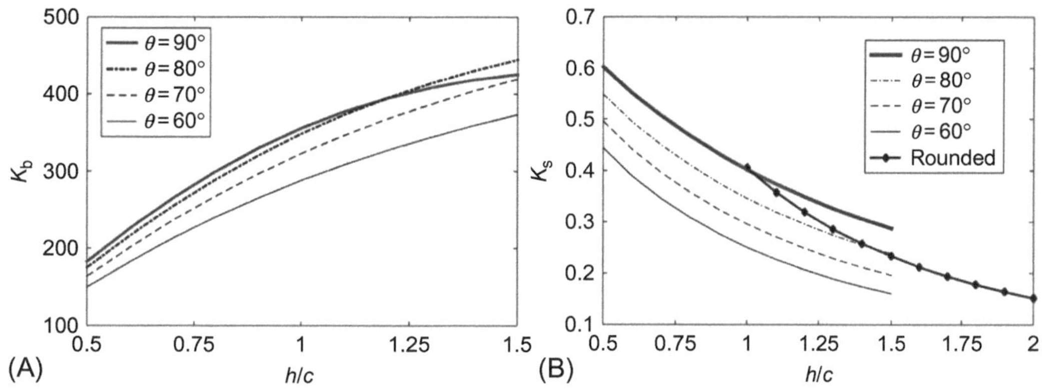

FIG. 11

Bending structural efficiency index for trapezoidal corrugation (A) and shear structural efficiency index for trapezoidal and rounded corrugation (B).

with the angle θ and the height to semiperiod ratio, h/c. Conversely, the shear efficiency index decreases with ratio h/c and the increment of θ, since when the in-plane shear stresses in the corrugated laminate are oriented in the vertical direction, shear deformation is produced without contributing to the shear stress resultant in the plane of the homogenized panel. Considering the results obtained in the previous section, a trade-off between morphing compliance and contribution to in-plane shear stiffness is apparent. For instance, if a highly compliant skin is required and h/c ratio is increased up to 1.5, where values of $(\overline{A}_{11}/A_{11})^{-1}$ can exceed 5000, the contribution to the shear stiffness turns out to be lower than 0.2 times the one of a plane panel having the same weight. However, if an h/c of about 0.8 is adopted, the panel will be 1000 times more compliant in the axial morphing direction (see Fig. 4A) than a flat panel of the same thickness and will retain an in-plane shear stiffness 0.5 times the one of a noncorrugated panel of the same weight. Indeed, the adoption of lay-up including plies oriented at ±45° can increase both axial compliance in morphing direction and the shear structural efficiency, thus showing the additional design flexibility that can be provided by composite corrugated laminates.

A sensitivity analysis of structural efficiency indices for the rounded corrugation (Fig. 1A), is shown in Fig. 12. In this case, radius r and consequently semiperiod c are varied ($c = 2r$). For the rounded geometry, the radius value determines the minimum value of the height, which can be changed by varying the length of the vertical arm L, so that different h/c can be obtained. For such a reason, the radius parameter has a dominant influence on the bending index K_b. The comparison with the trapezoidal corrugation can be carried out considering $r = 6$ mm ($c = 12$ mm). For such a choice, K_b varies from 350 to 432 for $h/c = 1.0 \div 1.5$ and turns out to be about 15%÷20% higher than the values obtained for trapezoidal corrugations.

Contrarily to K_b index, the shear structural efficiency index, Ks, does not vary with the radius value, but only depends on the h/c parameter. The index is plotted in Fig. 11B, together with the results referred to trapezoidal corrugation. It is worthwhile to note that the index reaches values like the ones of trapezoidal corrugation for $\theta = 60°$, if $h/c = 1$, but rapidly decreases when h/c is increased. Such behavior can be understood considering that the increment of h/c is obtained by increasing the length of

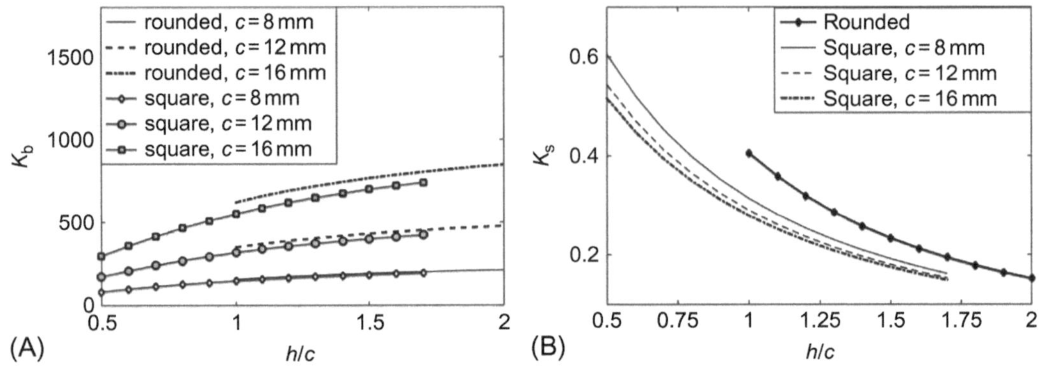

FIG. 12

Bending (A) and shear (B) structural efficiency index for rounded and square-shaped, at $r = 2$ mm, corrugation.

the vertical arm, which has no projection on the plane of the homogenized plane, with no contribution to the results of in-plane shear flow.

The value of K_b index for the square-shaped geometry with a fixed fillet radius $r = 2$ mm are also provided in Fig. 12, for the same values of the semiperiod considered for the rounded corrugation. It can be observed that in the region where the considered h/c ratios have the same values, square-shaped corrugation obtains slightly lower values of bending structural efficiency. Hence, although the square-shaped profile is characterized by high moment of inertia about neutral axis, its weight is higher than the one of the rounded corrugation, which eventually turns out to be more efficient.

The structural shear efficiency index for the square shaped geometry, at $r = 2$, is reported in Fig. 12 and compared with the results obtained for the rounded geometry. In the sensitivity study regarding the square-shaped profiles, a little effect of the semiperiod can be observed, whereas the rounded geometry is not sensitive to such a parameter, as it was already commented. Such trend is explained considering that, in the square-shaped configuration, the length of vertical arms, with no projection on the homogenized plate plane, reduces the structural efficiency, which turns out to be always lower than the one of the rounded corrugation and quite like the one of the trapezoidal corrugation for $\theta = 90°$. The rounded fillets between vertical and horizontal arms have a beneficial effect on this index, so that when the period is increased, the fillet radius being kept constant, the influence of the fillet is reduced and the index decreases.

The effect of the fillet radius in square-shaped geometries is analyzed in Fig. 13, where K_b and K_s indices are reported for a square-shaped corrugated laminate with a semiperiod fixed to $c = 12$ mm and different fillet radii. The increment of the radius reduces the bending structural efficiency index (Fig. 13A), whereas the shear structural efficiency increases and tends to the results of the rounded geometry, which can be considered a limit case, with $b = 0$ and $c = 2r$.

Overall, it can be observed that rounded geometry leads to the best values for considered structural efficiency indices, which have been defined to compare on a weight basis the different types of corrugation. Accordingly, it seems that rounded geometry, which produced the poorest performance in terms of axial and bending compliance, achieves the most interesting performance when load carryng capabilities in nonmorphing directions are taken into account.

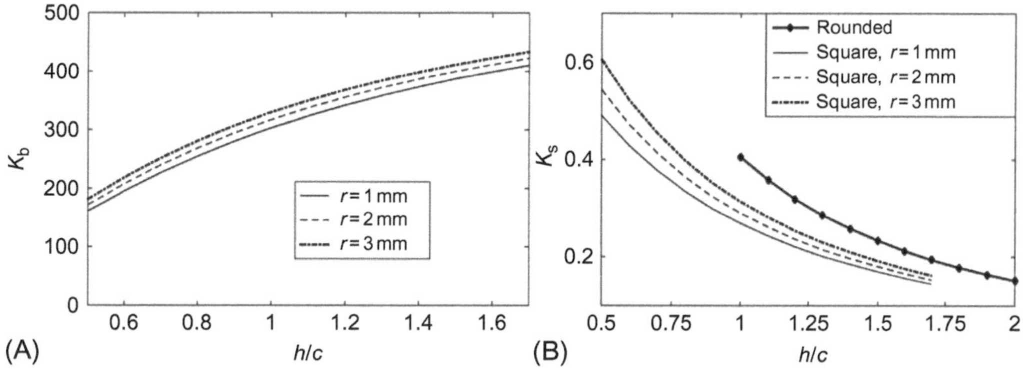

FIG. 13

Bending (A) and shear (B) structural efficiency index for square shaped corrugation at fixed semiperiod $c = 12$ mm and variable fillet radius.

Actually, differences are not particularly high considering the bending response in the nonmorphing direction, as in Fig. 12A. However, it is apparent that the rounded shape is particularly suited to maximize bending stiffness with a minimum weight cost. Differences become more significant considering in-plane shear structural efficiency. The value of θ angle in the trapezoidal shape must be decreased to obtain the same figure that characterizes the rounded corrugation, although for very high h/c, trapezoidal geometry exhibits some advantages. On the other hand, the fillet radius in the square-shaped corrugation must be significantly increased to obtain a shear structural efficiency closer to the one of the rounded geometry. However, it should be noted that the required adaptation of trapezoidal and square-shaped geometry reduces the axial and bending compliance that can be achieved. Hence, generally speaking, a trade-off must be carried out in the design of a corrugated morphing skin, between morphing performances (compliance and strain or curvature limits) and the contribution to the shear and torsional stiffness of the section where the morphing skin is introduced.

5 MANUFACTURING OF COMPOSITE CORRUGATED LAMINATES

The manufacturing of corrugated composite laminates involves several issues that may have a significant influence on the morphing performance and on overall strength characteristics of the morphing panel. As a general observation, it can be said that manufacturing difficulties depend on the shape of the corrugation, since the lamination of deeply corrugated shapes and the presence of fillets with small radii complicates the technological process.

The process used in Ref. [14], where composite corrugated laminates have first been proposed for the morphing application, is not described, though a proprietary process is mentioned for the production of the corrugated laminates made of carbon/epoxy plain weave fabric plies, which are used in this work.

The main problem in the lamination phase is the difficulty of achieving very small radius of curvature and to form relatively thick lay-ups of composite plies into the corrugated geometry. Forming can be achieved through a step-by-step lamination procedure, as described in Ref. [13,25] for the manufacturing of trapezoidal corrugations. In Ref. [13] the procedure was applied to three composite lay-ups made of aramidic, E-glass and carbon reinforced fabric plies with epoxy matrices. In Ref. [25] a similar procedure was applied to unidirectional E-glass reinforced epoxy prepregs to manufacture a skin demonstrator that was tested in a wind tunnel. In such processes, a manual lay-up technique was adopted and plies were separately introduced in the mold with the help of a heat gun, so to facilitate the prepreg forming. Then the assembly was kept in place during lamination by the application of metallic bars. Thereafter, a vacuum bag was applied, without application of a counter mold and an autoclave process was carried out. The manual lay-up of individual plies was also the methodology followed in Ref. [28] to produce circular corrugations, as the one sketched in Fig. 1D.

In [28] a metallic mold was used for curing in autoclave, but authors' findings indicate that the adoption of a metallic counter-mold turns out to significantly improve the quality of the laminate and the consequent strength properties.

A special procedure was presented for a trapezoidal corrugation [20], where curing was carried out at room temperature, by adopting a mold made of plywood and polyethylene foam.

The experience gained at Politecnico di Milano for the production of square shaped corrugated laminates [21,23,27] confirmed the difficulties in the lamination phase that led to adopt the individual manual lay-up procedure. The objective of the technological activity was the production of a square-shaped corrugated laminate, with a $[0]_n$ lay-up of carbon/epoxy fabric plies. Semiperiod was set to 12 mm, with $h/c = 1$ and a fillet radius of 2 mm.

The layers were individually laid on the mold by using a heat gun to facilitate forming, as in Ref. [24]. Metallic bars coated with PTFE adhesive layers were used to help forming the layers during the deposition. Final curing was carried out in autoclave, by adopting two different procedures. In the simplest process, a vacuum bag was directly applied to the mold, whereas, in the second method, an elastomeric counter mold was used. The counter mold was made of a nonsiliconic elastomer (Airpad), having a service temperature higher than 200°C. The elastomeric counter mold was cured at 176°C in autoclave by using the same mold prepared for the corrugated panel. The aluminium mold and the elastomeric counter mold are also presented in Fig. 14A. The quality of the laminates obtained with and without counter mold is presented in Fig. 14. In Fig. 14A, voids and surface defects are clearly

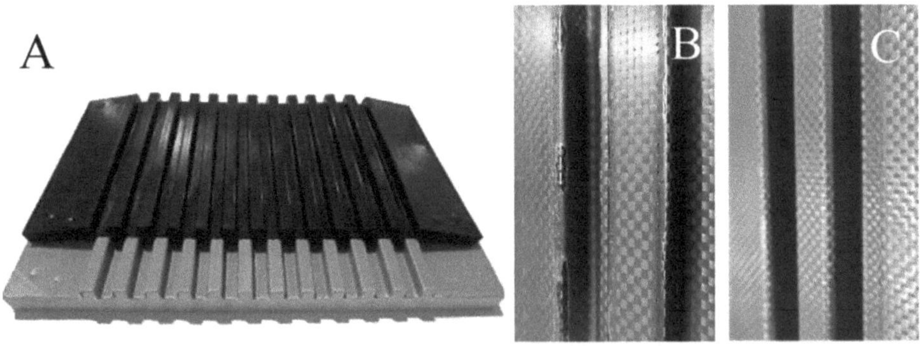

FIG. 14

Metallic mold and counter-mold for square-shaped corrugated panel production (A), detail of a corrugated laminate produced without (B) and with (C) the countermold [21,27].

visible in the square-shaped corrugated panel, particularly in the fillets. This panel, which was produced without counter mold, can be compared with the one presented in Fig. 14B, which was produced by using the elastomeric counter-mold and presents an appreciably better surface finishing.

Tensile tests were carried out on the produced specimens. The result of the tests are presented in Fig. 8B. It can be observed that the two tests performed on the panels manufactured with the counter mold are characterized by a lower stiffness, but show higher displacement at failure and maximum load than the coupon produced without counter mold. The difference in stiffness is due to a different compaction of the plies, which lead to a slightly lower thickness in the laminates produced with the counter mold, whereas the improvement in ultimate load and displacement give indication of the improved quality of the laminate. Accordingly, the results confirmed the difficulties in the production of composite corrugated laminates and the need of a step-by-step deposition process. Moreover, both visual inspection and experimental tests confirmed the observation reported in [28] regarding the advantages that can be achieved by using an elastomeric counter mold to improve the quality of the obtained laminates.

6 DEVELOPMENT OF AERODYNAMICALLY EFFICIENT MORPHING SKINS

6.1 AERODYNAMIC ISSUES IN THE APPLICATION OF COMPOSITE CORRUGATED LAMINATES

The most significant drawbacks in the adoption of corrugated laminates as morphing skins are associated to the potential aerodynamic penalties. The high bending stiffness in nonmorphing direction, which was analyzed in Section 4, prevents buckling in such a direction and limits the need of supporting the skin in such a direction. However, in the morphing direction, bending stiffness is very low. This can be an advantage for morphing performance, but leads to a possible risk of excessive deformation under the action of aerodynamic pressures, with undesired effects on the flow over the aerodynamic surface. The risk of skin bubbling, in the space interval between supports in the morphing direction, has been considered in the study performed in Ref. [11], and was numerically verified in the design of the morphing profile reported in Ref. [8]. The problem was addressed in Refs. [12,13,25] by bonding together two corrugated laminates, and inserting a foam core in the obtained closed cell. A rough sketch of the solution is presented in Fig. 15.

Actually, the solution presented in Refs. [12,13,25] defines a new type of corrugation, obtained by combining two simply corrugated laminates. The effects due to the inadequate bending stiffness in morphing directions can be prevented by properly selecting the geometry of the corrugated laminates and by varying the distances between the supports of the skin in the morphing direction. Hence, prevention of excessive bending deformation in a morphing direction considering the internal structure of the morphing surface is an issue that can be properly addressed. In the optimization carried out in Ref. [8], design variables included both parameters of the internal structure, which was based on a composite chiral honeycomb, and stiffness levels of the skin in the morphing direction. Such procedure made possible the identification of axial and bending stiffness requirements for the skin to fulfill morphing requirements without undergoing excessive bending deformation and local waviness under the action of aerodynamic pressures. Such requirements were fulfilled in Ref. [21,38], by selecting a square-shaped composite corrugated laminate with adequate geometrical properties and lay-up.

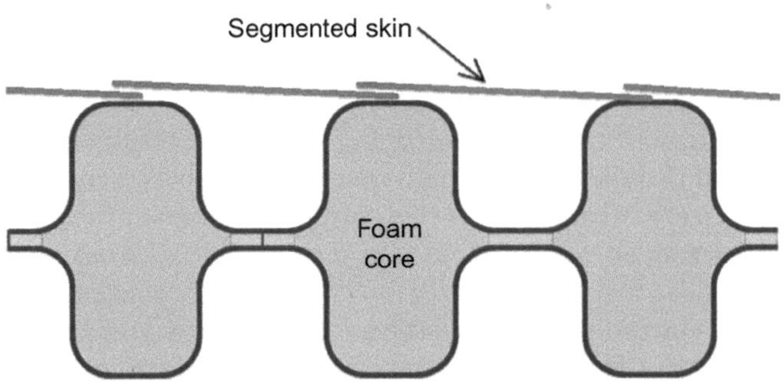

FIG. 15

Sketch of a possible solution to sustain aerodynamic pressures.

A more critical issue for the development of morphing skins based on corrugated laminates is represented by the direct effects of corrugation on aerodynamic performance. The results of numerical and experimental analyses show that airfoils equipped with corrugated trailing edges present significant increase of drag and a reduced slope of the lift curve versus the angle of attack with respect to conventional airfoils with smooth skins [24,26]. Such works also indicate that detrimental effects increase with the ratio of the height of corrugation to the chord of the aerodynamic profiles. Therefore, this ratio must be kept at very low levels (<<1%) to prevent a degradation of the performance, otherwise the aerodynamic effects of corrugation should be taken into consideration in the design of a morphing solution, to guarantee that potential advantages of morphing are not vanished by the reduction of aerodynamic performance due to corrugation.

Indeed, problems originated by waviness of the aerodynamic surface had already been considered in Ref. [14], where it was proposed to fill the "valleys" of the corrugation by using elastomeric material. However, such a solution can result in significant additional weight, particularly for deeply corrugated shapes. A different strategy was adopted in Ref. [25], where an externally segmented composite skin was bonded to the corrugated laminate to improve aerodynamic performances in wind tunnel tests. In such a solution, small strips of composite laminates, bonded at one end, cover the "valleys" of the corrugation (see Fig. 15). Such segmented skin can reduce the detrimental effects of corrugation to the aerodynamic performances, but, in real world applications, the presence of platelets bonded to surfaces could present some drawbacks concerning structural integrity, aerodynamic efficiency, vibrations and limited applicability in the case of large curvatures.

The solution presented in Refs. [21,23,27] involves the integration in a square-shaped corrugation of an elastomeric cover that is potentially able to guarantee, at a moderate weight cost, a reduction of two orders of magnitude of surface waviness with respect to the adoption of a corrugated surface. Such a solution will be presented in more detail in a following subsection.

6.2 PERFORMANCE INDEX BASED ON RATIO BETWEEN BENDING AND AXIAL COMPLIANCE

The need of preventing surface waviness or excessive distortions of aerodynamic profiles requires the adoption of skin with adequate bending stiffness, whereas axial stiffness must be selected mainly considering morphing requirements. In particular, in many applications, axial stiffness should be

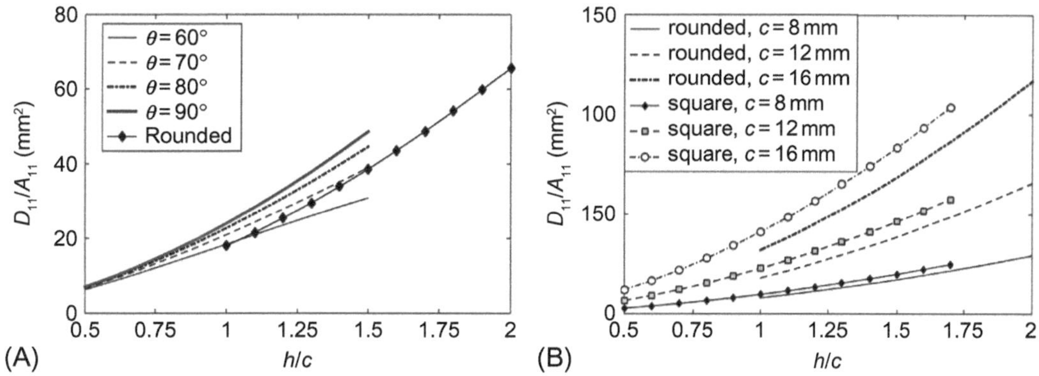

FIG. 16

Bending to axial stiffness ratio in morphing direction for trapezoidal shapes with $c = 12$ mm (A), generic rounded shapes and square-shaped corrugations with $r = 2$ mm (B).

minimized to facilitate shape variations and to reduce actuation forces in active morphing solutions. Design flexibility is one of the appealing properties of corrugated laminates, so that the ratio between bending and axial stiffness in morphing direction can be varied within a large range of values to find the optimal mix between axial and bending response.

The ratio between $\overline{D}_{11}/\overline{A}_{11}$ is plotted for trapezoidal shapes, at a fixed semiperiod $c = 12$ mm, in Fig. 16A and for rounded and square-shaped laminates in Fig. 16B; the fillet radius in the square-shaped case is fixed at $r = 2$ mm. It can be observed that even considering a fixed period, $\overline{D}_{11}/\overline{A}_{11}$ varies in the range 5 mm$^2 \div 30$ mm^2 for the trapezoidal shapes depending on the angle θ and the height to semiperiod ratio h/c. A much larger variability can be achieved by changing the semiperiod, as is shown in Fig. 16B. It is worth observing that the absolute values of the stiffness can be varied, in a composite corrugated laminate, by varying the lay-up and the material properties.

As far as the comparison between the different geometries is concerned, the trapezoidal shape with $\theta = 90°$ maximizes the ratio $\overline{D}_{11}/\overline{A}_{11}$ and, for the same semiperiod, square-shaped corrugated laminates with $r = 2$ mm obtain higher values than corresponding rounded shapes. If the fillet radius is increased, the $\overline{D}_{11}/\overline{A}_{11}$ ratio of square-shaped corrugations decreases and tend to become equal to those of the rounded shapes.

6.3 INTEGRATION OF AN ELASTOMERTIC COVER ON A SQUARE-SHAPED CORRUGATED LAMINATE

Detrimental aerodynamic effects due to corrugation appear whenever the direction of airflow is in the corrugation direction. Literature data confirm, both numerically and experimentally, that the amplitude of the corrugation, which is represented by the corrugation height h, must be limited to a very small fraction of the aerodynamic chord to avoid significant increment of drag and of drag-to-lift ratio.

Filling the valleys of the corrugation with an elastomer, as proposed in Ref. [14] can represent a solution, although, if period and amplitude of the corrugated shape are large, the consequent weight penalty can be unacceptable for aeronautical applications. An elastomeric cover could be adopted, but it should be adequately sustained in the valleys of the corrugation, particularly if the semiperiod is relatively large.

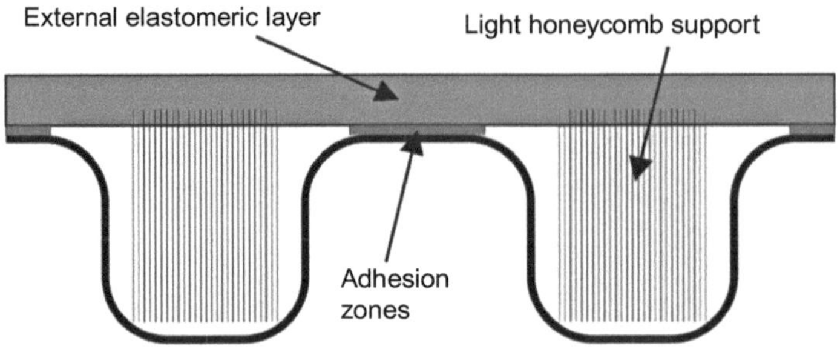

FIG. 17

Sketch of a square-shaped laminate with the integration of an elastomeric cover sustained by honeycomb stripes [23].

An efficient and lightweight solution presented in Refs. [21,23,27] is based on an elastomeric external cover that is sustained, in the "valleys" of the corrugation, by strips of conventional hexagonal-cell honeycomb, with cell axes directed perpendicular to the cover. In the solution, which is sketched in Fig. 17, the elastomeric cover is bonded to the horizontal arms of the corrugation. The honeycomb provides a support to the elastomer in the vertical direction, with a minimal interference on the overall deformation mechanism. Indeed, such a solution relies on the orthotropic behavior of the honeycomb, which presents very low stiffness properties in the plane of hexagonal cells.

A feasible technological process to produce such a skin system is summarized in Fig. 18. The composite corrugated panel is produced by using a vacuum bag technology and a specifically designed mold (Fig. 18A). Honeycomb inserts are cut and positioned in the valleys of the corrugate, which is supported by the original mold (Fig.18B). Thereafter, the mold is capsized and introduced in a

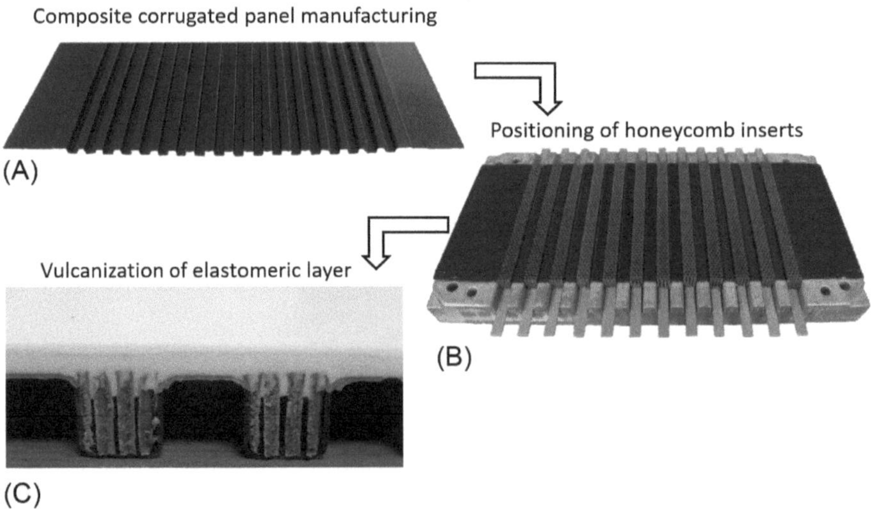

FIG. 18

Technological process for the integration of an elastomeric cover in a square shaped laminate [27].

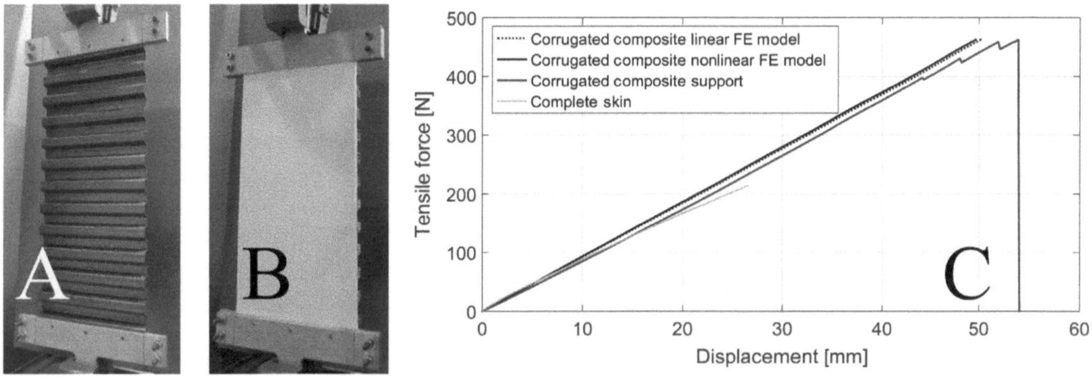

FIG. 19

Tensile test performed on corrugated laminate (A), skin system with integrated elastomeric cover (B) and comparison between experimental results and finite element predictions (C) [23].

rectangular filler that contains a liquid silicon rubber. During vulcanization, the rubber penetrates inside the cells of the honeycomb by capillarity and integrates the honeycomb in the elastomeric cover. At the same time, adhesion is obtained between the rubber layer and the horizontal sides of the corrugated support, thanks to the application of a special primer. The final result is shown in Fig. 18C.

Results of testing activities are summarized in Fig. 19, where the lay-out of a test on the corrugated laminate that support the elastomeric cover (Fig. 19A) and of a test on the complete skin system (Fig. 19B) are presented. Experimental results are reported in Fig. 19C, which also presents the comparison with the predictions of linear and nonlinear finite element analyses. Tests indicate that the presence of the elastomeric layer and of the supporting honeycomb stripes does not interfere with the axial compliance of the skin. The deformation of the corrugated support produces some waviness on the surface of the elastomer, particularly at high tensile stretching, as can be seen in Fig. 19C. However, the amplitude of such waviness is almost two orders of magnitude lower than that of the original corrugation.

The presented configuration therefore seems a promising solution for the development of aerodynamically efficient skins. The selection of adequate elastomeric layer and a proper set up of technological process can lead to an applicable lightweight smooth skin, potentially without detrimental effects on aerodynamic performances.

7 CONCLUSIONS

This chapter explored the mechanical properties of composite corrugated laminates for morphing skin systems. Different shapes of corrugation have been considered and different formulations have been adopted to quantify the stiffness properties of such types of panels. The anisotropy of properties has been analytically evaluated for different corrugated shapes with results that are confirmed by a review of the experiments performed in literature. The anisotropy is fundamental for the suitability of corrugated laminates in morphing applications; these applications are indeed characterized by the need of structural elements with tuneable and relatively low stiffness and high recoverably strain in selected directions, and load carrying capability and high stiffness in other directions. In this work, both axial and bending compliance in morphing direction of composite corrugated laminates have been thoroughly analyzed with sensitivity studies performed by varying main geometrical parameters of different corrugated shapes. A range of design variables was selected to consider technologically feasible configurations. According to the results, axial compliance can be up to three orders of magnitude higher than the one of a flat panel with the same lay-up. Moreover, the bending compliance of the considered corrugated laminates is 1.0÷2.5 times higher than that of a flat panel with the same lay-up.

The absolute value of the period and the height to semiperiod ratio have been found to be the most fundamental parameters that can be used to vary the mechanical response in morphing directions. Maximum compliance has been obtained by square-shaped corrugation with sharp corners, which can actually be considered a limit case of trapezoidal corrugations. The presence of rounded fillet, which is a necessary feature for the manufacturing of corrugated composite laminates, reduces the compliance levels, until the limiting case of a completely rounded geometry, which turns out to be the stiffer configuration for the same values of period and heights. Failure limits in morphing directions have also been discussed, showing that composite corrugated laminates can be stretched until levels exceeding the 40% of the original length. Moreover, corrugated panels can be bended at higher curvature than flat panels of identical lay-up, although experimental and analytical results presented in literature underline the risk of delamination-driven failures, particularly when corrugated profiles are thick and in the presence of fillets with small radii.

A study has been conducted considering the stiffness contribution in nonmorphing directions, by comparing flat and corrugated laminates on a weight basis. For bending response in the direction perpendicular to corrugation, corrugated laminates achieve stiffness levels up to several hundred times those of composite flat panels having the same lay-up and weight. Such quantification confirms the appealing characteristics that are conferred by the anisotropy of properties, which can be exploited, for instance, to reduce the number of the ribs required to support the morphing skin. The studies presented also indicate that the corrugated composite laminates can provide a significant contribution to shear and torsional stiffness of the section in aeronautical morphing beams. Actually, the efficiency in performing such a structural role, which is typical of the skin in aeronautical construction, is in conflict with the achievement of very high compliance in morphing directions. However, very interesting trade-offs can be obtained. From this standpoint, the rounded shape achieves the most interesting results, since it can provide 0.4 times the shear stiffness of a flat panel having identical weight, with no need of additional stringers to increase bending response in the transversal direction. The same configuration provides an axial compliance that is one thousand times higher than that of a conventional panel and a very high stretching strain before failure. Such characteristics confirm the advantages that can be obtained by the adoption of corrugated laminates. Moreover, all the results have been produced considering a homogeneous, unidirectional lay-up, but design flexibility of this structural concept can be largely enhanced considering the possibility of multidirectional lay-ups, which can be optimized to obtain the required mix of functional and structural properties.

This chapter has also discussed the main technological issues in the production of composite corrugated laminates, pointing out the need of an individual lamination of the plies and the advantages offered by the adoption of a counter-mold in the curing process.

Finally, the issues regarding the aerodynamic efficiency of the corrugated skin have been taken into consideration. The need of preventing skin bubbling has led to define an additional performance index, based on the ratio between bending and axial stiffness, which should be maximized in many morphing applications to prevent undesired local waviness of the aerodynamic surfaces. A sensitivity analysis on the main geometrical parameters indicate that very different values can be achieved for this index, so that the fulfillment of requirements for an adequate bending stiffness without excessively compromising the compliance in the morphing direction can be obtained. However, one of the most critical issues in the application of composite corrugated laminates is represented by the detrimental effects of corrugation on aerodynamic performances, which have been numerically and experimentally confirmed by studies reported in literature. The solution proposed for the integration of an elastomeric cover supported by honeycomb stripes in the valleys of the corrugations, indicates a possible and feasible technological path toward the solution of such critical problem and to the development of structurally and aerodynamically efficient morphing skins for future application in aviation.

REFERENCES

[1] S. Barbarino, O. Bilgen, R.M. Ajaj, M.I. Friswell, D.J. Inman, A review of morphing aircraft, J. Intell. Mater. Syst. Struct. 22 (2011) 823–827.
[2] A.Y.N. Sofla, S.M.A. Meguid, K.T. Tan, W.K. Yeo, Shape morphing of aircraft wing: status and challenges, Mater. Des. 31 (2010) 1284–1292.
[3] G. Amendola, M. Magnifico, R. Pecora, I. Dimino, Distributed actuation concepts for a morphing aileron device. Aeronaut. J. 120 (1231) (2016) 1365–1385, https://doi.org/10.1017/aer.2016.64.
[4] H.P. Monner, Realization of an optimized wing camber by using form-variable flap structures, Aerosp. Sci. Technol. 5 (7) (2001) 445–455.
[5] R. Pecora, A. Concilio, I. Dimino, F. Amoroso, M. Ciminello, Structural Design of an Adaptive Wing Trailing Edge for Enhanced Cruise Performance, 24th AIAA/AHS Adaptive Structures Conference, AIAA Sci-Tech, 2016, pp. 4–8.
[6] R. Pecora, F. Amoroso, M. Magnifico, I. Dimino, A. Concilio, KRISTINA: kinematic rib-based structural system for innovative adaptive trailing edge, in: Proc. SPIE—The International Society for Optical Engineering, Vol. 9801, 2016 Article number 980107.
[7] S. Barbarino, R. Pecora, L. Lecce, A. Concilio, S. Ameduri, Airfoil morphing architecture based on shape memory alloys, in: Proc. ASME Conference on Smart Materials, Adaptive Structures and Intelligent Systems, SMASIS2008, Vol. 1, (2008) 729–737.
[8] A. Airoldi, M. Crespi, G. Quaranta, G. Sala, Design of a morphing airfoil with composite chiral structure, J. Aircr. 49 (4) (2012) 1008–1019.
[9] L.F. Campanile, D. Sachau, The belt-rib concept: a structronic approach to variable camber, J. Intell. Mater. Syst. Struct. 11 (3) (2000) 215–224.
[10] J. Martin, J.J. Heyder-Bruckner, C. Remillat, F. Scarpa, K. Potter, M. Ruzzene, The hexachiral prismatic wingbox concept, Phys. Status Solidi B 245 (3) (2008) 570–577.
[11] F. Gandhi, P. Anusonti-Inthra, Skin design studies for variable camber morphing airfoils, Smart Mater.Struct. 17 (2008) 1–8.
[12] C. Thill, J. Etches, I. Bond, K. Potter, P. Weaver, Morphing skins, Aeronaut. J. 112 (2008) 117–139.
[13] C. Thill, J. Etches, I. Bond, K. Potter, P. Weaver, in: Corrugated composite structures for aircraft morphing skin applications, Proceedings of the 18th international conference of adaptive structures and technologies. Ottawa, Ont., Canada, October 3rd–5th, 2007.
[14] T. Yokozeki, S. Takeda, T. Ogasawara, T. Ishikawa, Mechanical properties of corrugated composites for candidate materials of flexible wing structures, Compos. Part A 37 (2006) 1578–1586.
[15] Briassoulis, Equivalent orthotropic properties of corrugated sheets, Comput. Struct. 23 (2) (1986) 129–138.
[16] G. Kress, M. Winkler, Corrugated laminate homogenization model, Compos. Struct. 92 (3) (2010) 795–810.
[17] Y. Xia, M.I. Friswell, E.I. Saavedra Flores, Equivalent model of corrugated panels, International Journal of Solid and Structures 49 (2012) 1453–1462.
[18] M. Winkler, G. Kress, Modelling of corrugated laminates, Compos. Struct. 109 (2014) 86–92.

[19] Airoldi A., Quaranta G., Beltramin A. and Sala G. Design of a morphing actuated aileron with chiral composite internal structures, Adv. Aircr. Spacecr. Sci., Vol. 1, N. 3, 2014, p. 329–349.
[20] P. Ghabezi, M. Golzar, Mechanical analysis of trapezoidal corrugated composite skins, Appl. Compos. Mater. 20 (2013) 341–353.
[21] A. Airoldi, S. Fournier, E. Borlandelli, P. Bettini, G. Sala, Design and Manufacturing of Skins Based on Composite Corrugated Laminates for Morphing Aerodynamic Surfaces, Smart Materials and Structures 26 (4) (2017) art. no. 045024.
[22] F. Previtali, G. Molinari, A.F. Arrieta, M. Guillaume, P. Ermanni, Design and experimental characterisation of a morphing wing with enhanced corrugated skin, J. Intell. Mater. Struct. 27 (2016) 278–292.
[23] Fournier S., Airoldi A., Borlandelli E., Sala G. Flexible composite supports for morphing skins, in: AIDAA Napoli XXII, Napoli, XXII Conference of Italian Association of Aeronautics and Astronautics, Naples, Italy, 9–12 2013, ISBN: 9788890648427, p. 1–12 Sept. 2013.
[24] C. Thill, et al., Aerodynamic study of corrugated skins for morphing wing applications, Aeronuat. J. 114 (2010) 237–244.
[25] C. Thill, J.A. Etches, I.P. Bond, K.D. Potter, P.M. Weaver, Composite corrugated structures for morphing wing skin applications, Smart Mater. Struct. 19 (2010) 1–10.
[26] Y. Xia, O. Bilgren, M.I. Friswell, The effects of corrugated skins on aerodynamic performances, J. Intell. Mater. Syst. Struct. 25 (7) (2014) 786–794.
[27] P. Panichelli, A. Gilardelli, A. Airoldi, G. Quaranta, G. Sala, Moprhing composite structures for adaptive high lift devices, Proceedings of 6th Internaltional Conference on Mechanics and Materials in Design, Ponta Delgada, Portugal, 2015, pp. 26–30.
[28] A. Schmitz, P. Horst, Bending deformation limits of corrugated unidirectionally reinforced composites, Compos. Struct. 107 (2014) 103–111.
[29] A. Samanta, M. Mukhopadhyay, Finite element static and dynamic analyses of folded plates, Eng. Struct. 21 (1999) 277–287.
[30] M. Winkler, G. Kress, Influence of corrugation geometry on the substitute stiffness matrix of corrugated laminates, Compos. Struct. 94 (2012) 2827–2833.
[31] A. Kay, Junkers Aircraft & Engines 1913–1945, Putnam Aeronautical Books, London, 2004, p. 134.
[32] Z. Aboura, N. Talbi, S. Allaoui, M.L. Benzeggagh, Elastic behavior of corrugated cardboard: experiments and modeling, Compos. Struct. 63 (2004) 53–62.
[33] G. Kress, M. Winkler, Corrugated laminate analysis: a generalized plane-strain problem, Compos. Struct. 93 (2011) 1493–1504.
[34] M. Winkler, G. Kress, Deformation limits for corrugated cross-ply laminates, Compos. Struct. 92 (2010) 1458–1468.
[35] O. Heintze, S. Steeger, A. Falken, J. Heckmann, in: P.C. Woelchen, M. Papadopoulos (Eds.), Enhanced adaptive droop nose: from computer model to miutli-functional integrated part, Smart Intelligent Aircraft Structure (SARISTU): Proceedings of the Final Project Conference, Springer Internation Publishing, 2016, pp. 97–111.
[36] M. Kintscher, J. Kirn, S. Storm, F. Peter, in: P.C. Woelchen, M. Papadopoulos (Eds.), Assessment of the SARISTU enhanced adaptive droop nose, Smart Intelligent Aircraft Structure (SARISTU): Proceedings of the Final Project Conference, Springer Internation Publishing, 2016, pp. 113–140.
[37] T. Megson, Aircraft Structures for Engineering Students, 5th, Butterworth-Heinemann, 2012.
[38] A. Airoldi, P. Bettini, P. Panichelli, M.F. Oktem, G. Sala, Chiral topolgies for composite morphing structures—Part I: development of a chiral rib for deformable airfoil, Phys. Status Solidi B 252 (7) (2015) 1435–1445.

SECTION 4

Systems Design

Editor:

Ruxandraa Botez
Ecole Technique Superieure (ETS), Montreal, QU, CANADA

List of Chapters

CHAPTER

ACTIVE METAL STRUCTURES

10

Rosario Pecora*, Salvatore Ameduri†, Francesco Rea*

University of Naples "Federico II", Naples, Italy The Italian Aerospace Research Centre, CIRA SCpA, Capua (CE), Italy†*

CHAPTER OUTLINE

NOMENCLATURE

1D	one-dimensional
2D	two-dimensional
3D	three-dimensional
A/C	aircraft
$A_{eq(f)/(r)}$	front/rear spar equivalent areas
A_{SMA}	SMA element cross-section area
b	number of blocks
B_i	ith block (of the rib)
$BM_{i(f)}/BM_{i(r)}$	amount of $BM_{i(U)}$ absorbed by (front/rear) spar caps
$BM_{i(U)}$	ultimate bending moment at S_i
c (/c^*)	wing chord (/trailing edge chord)
CAD	computer aided drawing
C_D	drag coefficient

Morphing Wing Technologies. https://doi.org/10.1016/B978-0-08-100964-2.00010-1

CFD	computational fluid dynamics
C_L	lift coefficient
DOF(s)	degree(s) of freedom
E_{SMA}	SMA element (equivalent) Young modulus
F	force
FE (/FEM)	finite elements (/finite elements method)
g	mechanism kernel function
$h_{i(f)}/h_{i(r)}$	height of (front/rear) spar web at S_i
K	active degrees of freedom of a kinematic chain
l	number of links
L	length
$l_{i(sk)}$	chord-wise length of the (upper/lower) skin panel at S_i
MRF	magneto-rheological fluids
P_i	perimeter of the box cell at S_i
q_i	cell's shear flow at S_i
$q_{i(fw)/(rw)/(sk)}$	shear flow along the front spar web/rear spar web/skin panels at S_i
S_i	ith cross section (of the trailing edge box)
$S_{i(U)}$	ultimate shear at S_i
SMA	shape memory alloy
SRS	(trailing edge) section reference system
TE	trailing edge
$t_{i(fw)/(rw)/(sk)}$	thickness of the front spar web/rear spar web/skin panels at S_i
$TM_{i(U)}$	ultimate torque at S_i
TRL	technology readiness level
Xsc_i	chord-wise position of the shear center at S_i
Γ	iteration function
Ω_i	area enclosed by the box cell at S_i
α_T	thermal expansion coefficient
ε	strain
σ_{ref}	allowable stress (in tension or compression)
σ_{tu}	failure stress in tension or compression
τ_{ref}	allowable stress (in shear)

1 INTRODUCTION

Two different design approaches for metallic morphing wing structures are illustrated in this chapter. The first involves the segmentation of the traditional load-bearing box into a set of structural cells interconnected by kinematic chains. The second is based on monolithic or quasimonolithic compliant architectures with suitably distributed stiffness to enable the required morphing.

The architectures of the first type are characterized by a pseudo-conventional arrangement recalling a typical multitab layout with parts in relative motion according to specific gear ratios. A suitably designed actuation ensures the shape change of the overall system while significantly contributing to withstand external loads in all operative conditions. The conventional shape-fixed structural arrangement then is replaced by a mechanized layout that implements morphing through the rigid-body motion of stiff subcomponents. Although characterized by the highest maturity and reliability level, mechanized solutions imply the use of a large number of parts, which in some cases can be prohibitive because of higher maintenance costs and weight increase. The number of parts is somehow linked to the "level of resolution" the morphed shape must attain. Let us consider, as an example, the finger of the hand. If the finger were made of one single bone, its internal structure would be simple, but just one shape would be possible. If more bones are interconnected, the structural complexity of the system increases, as does the number of possible shapes. The right balance between the structural complexity and morphing resolution is the main parameter to be taken in account when addressing the design of a mechanized solution; it naturally drives the potential benefits/drawbacks ratio of the overall device, and it is intrinsically linked to the specific end-application for which morphing is required.

Whenever the smoothness of the achieved morphed geometry represents a paramount requirement to be accomplished, the adoption of compliant architectures can be a valid solution to ensure high morphing resolution while lowering structural complexity.

Compliant mechanisms result from an optimization approach relying upon the strict cooperation of different subcomponents, collaborating to absorb and distribute loads (by minimizing the stress level) and enabling large deformations. Distributed actuators are usually adopted to elastically deform the structure to modify its shape. Unlike mechanized solutions, which morph through relative motion of parts, compliant mechanisms implement morphing through bending and/or torsion of specific structural elements. This strategy, while accurately reproducing a given set of morphed shapes with a quite simple overall layout, leads to penalties in terms of needed actuation power, fatigue life, and (sometimes detrimental) aeroelastic impacts at aircraft level. The right balance between the necessary compliance to accommodate morphing with low actuation power and the necessary stiffness to withstand aerodynamic loads while avoiding aeroelastic instability is the goal of a successful overall design.

Which architectural arrangement then is the best? The one based on kinematic chains or the one based on compliant mechanisms?

In order to help the reader in making (or changing) his own idea, this chapter presents different architectural concepts all conceived for the camber morphing of a metallic wing trailing edge.

By referring to similar end-applications, the adopted design strategies and obtained outcomes can be compared easily, thus better highlighting the advantages and weaknesses of each morphing solution.

2 MORPHING ORIENTED KINEMATIC CHAINS: WORKING PRINCIPLES AND DESIGN APPROACHES

Kinematic chains represent the most effective way to implement morphing while ensuring high levels of compliance of the morphing structure with the airworthiness requirements [1–7]. This can be seen as a direct consequence of conventional arrangements and standard materials and actuators.

A morphing solution based on kinematic chains is not so different from a classic multitab system that modifies the airfoil camber through differential deflection of consecutive segments; higher system complexity is necessary to better cope with more complicated shape transitions induced by morphing. The augmented complexity, however, is inherent only to embedded mechanics, actuation and controls, leading to very small impacts on system readiness for flight, which is, in turn, strongly sustained by the applicability of traditional design approaches widely approved by airworthiness authorities.

Generally speaking, the design methodology of mechanized morphing architectures consists of four main steps:

- Shape transition analysis;
- Mechanization of the kinematic chain;
- Preliminary structural design;
- Advanced structural design and layout assessment.

In order to better outline the process behind each design step, it is convenient to refer to a specific application scenario, being implicit that the methodology has a wider validity and is therefore not dependent on the particular case under investigation. In the light of this consideration, let's tailor our discussion to the implementation of wing trailing edge (TE) conformal camber morphing. In such a case, the shape transition is the same in all wing cross-sections and is usually described in terms of chord-wise extension of the morphable region (c^*), baseline, and morphed camber line geometry (dotted lines in Fig. 1). Airfoil thickness distribution is commonly considered unaffected by morphing.

These high-level geometrical input results from preliminary aerodynamic analyses aiming at the optimization of a specific performance at aircraft (A/C) level.

The most natural way to enable the transition of TE sections from the reference (baseline) shape to the target (morphed) one consists in the arrangement of a morphable layout for the ribs of the TE structure. The typical single-body rib can be assumed to be segmented in consecutive blocks hinged to each other along the airfoil camber line (points A, B, C Fig. 2).

The introduced segmentation turns the airfoil camber line into an articulated polygonal line that can assume different shapes depending on the relative rotation between adjacent blocks. The number of blocks and their chord-wise extension are commonly defined in order to satisfactorily match the camber line in both baseline and morphed configurations while lowering the structural complexity as much as possible. A large number of small blocks ensures more accurate reproduction of unmorphed/morphed camber lines; however, it can lead to considerable impacts on the feasibility of the structural arrangement as well as on its manufacturing and maintenance costs (high number of parts). Conversely, a small number of large blocks leads to simple structural layout but can be inadequate to enable morphing in compliance with high-fidelity shape reproduction requirements. The issue is the same as that which occurs when defining the mesh size in a given structural problem; refined meshes can produce

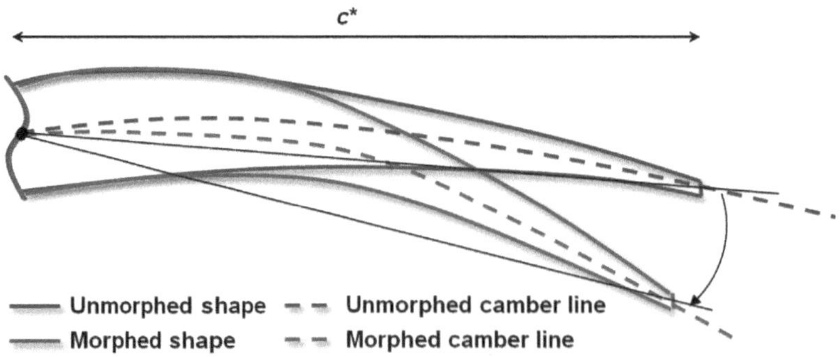

FIG. 1

Wing trailing edge morphing, conceptual scheme.

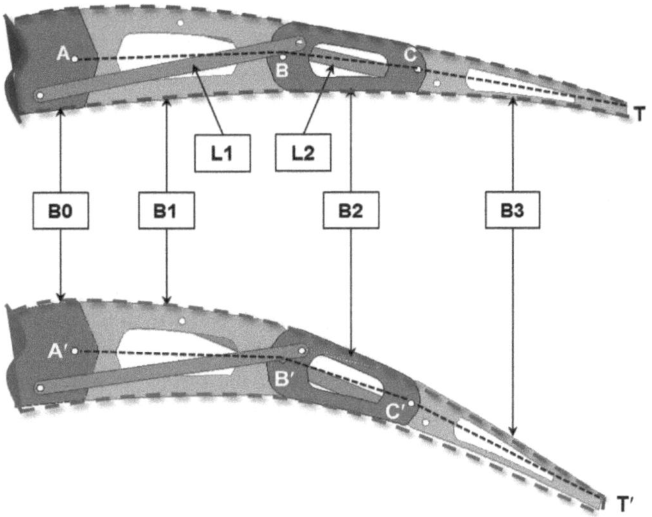

FIG. 2

The segmented rib concept.

accurate results but need more computational time, while coarse meshes speed up the analysis run but can prohibitively decrease the quality of the analysis outcomes.

The correct solution lies in properly balancing needs and requirements in order to come out with an optimal compromise that works well for the specific problem to be addressed.

In this case, the process leading to the optimal number and size of blocks is referred as *shape transition analysis*. An iterative analysis is done involving aerodynamic computations at each loop. A first trial is made on block number and length. In automated algorithms, this action can be easily performed referring to a configuration vector $\underline{C} = [c_1, c_2, \ldots, c_b]$ whose size, b, represents the number of blocks and whose generic component c_i stands for the chord-wise extension of the ith block.

For each component c_i a specific range R_i is defined coherently to structural constraints and feasibility considerations. A starting value C_0 is assumed for C in compliance with the above ranges.

Morphed airfoil shape is then rebuilt by considering the polygonal line defined by C_0, and aerodynamic analysis is carried out to evaluate the degradation (γ) of the target performance because of the adoption of the trial segmented solution. Such degradation can be measured as the difference between the C_L, C_D, or C_L/C_D of the target morphed shape and those related to the airfoil shape approximated by $\underline{C}_0$. A new set of components is then assumed for C, and the process is repeated until all ranges R_i are fully mapped. Optimal configuration vector is then obtained as the one minimizing the function $\gamma = \gamma(c_1, c_2, \ldots, c_b)$ over the variable ranges $R_1, R_2, \ldots, R_b$.

The shape transition analysis directly gives the position of camber line hinges and, indirectly, the amount of rotation required to each camber line segment to suitably fit the target morphed shape.

Let us now assume that the outcomes of the analysis are those sketched in Fig. 2, in terms of number of blocks and hinge positions along camber line in both unmorphed (A, B, C) and morphed (A′, B′, C′) configurations. The indirect results are represented by the rotations that the segments AB, BC, CT, have to make around A, B, C to turn the polygonal line ABCT into A′B′C′T′.

In order to implement these rotations, further linkages between nonadjacent blocks are clearly necessary.[1] The number of additional links (l) is defined on the basis of the active degrees of freedom (DOFs) to be assigned to the rib mechanism; if K denotes the active DOFs of the rib mechanism, it results:

$$K = b - (l + 1) \tag{1}$$

where b is the number of rib bocks.

Since the rib has to be at least intrinsically isostatic, K is equal to the number of actuators that should be installed to suppress (and control) the active DOFs of the mechanism.

In the example of Fig. 2, l is equal to 2, b is equal to 4 and the mechanism clearly has one DOF.

Linking rods L_1 and L_2 hinged to nonadjacent blocks force the camber line segments to rotate according to specific gear ratios. If the rotation of any of the blocks is prevented, no change in shape can be obtained; if one actuator moves any of the blocks, all the other blocks follow the movement according to the above-mentioned gear ratios.

The position of the linking rod hinges is determined by the specific rotation each segment has to perform to enable the transition from the baseline to the target morphed shape. Indirectly, the shape-transition analysis results in the definition of the linking elements hinges; this process is referred as *mechanization of the kinematic chain* and involves the solution of an *inverse kinematic* problem.

The positions of the hinges along the camber line (both in unmorphed and morphed configurations) are the input data of the problem, the positions of the links are the unknown variables to be determined.

The example case of Fig. 2 explains how the problem can be solved.

Because the position of each link is determined in a generic section reference system (SRS) by the coordinates of its tip hinges, the number of unknown variables is equal to eight (two links and two coordinates for each tip hinge). Referring to the equivalent representation of the rib mechanism shown in Fig. 3, and to its symbols, the mathematical formulation of the problem consists in finding the vector $\vec{v}$ $(= [x_{L_{1-1}}, z_{L_{1-1}}, x_{L_{1-2}}, z_{L_{1-2}}, x_{L_{2-1}}, z_{L_{2-1}}, x_{L_{2-2}}, z_{L_{2-2}}])$ defined over the domain $\mathcal{D}$ and satisfying the equation:

$$\vec{w}_m = g\left(\vec{w}, \vec{v}\right) \tag{2}$$

where $\mathcal{D}$ is the region delimited by the unmorphed airfoil; $\vec{w} = [x_A, z_A, x_B, z_B, x_C, z_C, x_T, z_T]$ is the vector defining the positions (in SRS) of the hinges A, B, C and of the airfoil tip point (T) in unmorphed configuration; $\vec{w}_m = [x_{A'}, z_{A'}, x_{B'}, z_{B'}, x_{C'}, z_{C'}, x_{T'}, z_{T'}]$ is the vector defining the positions (in SRS) of the hinges A, B, C and of the airfoil tip point in morphed configuration; $\vec{w}_d = [x_{A''}, z_{A''}, x_{B''}, z_{B''}, x_{C''}, z_{C''}, x_{T''}, z_{T''}]$ is the vector defining the positions (in the SRS) of the hinges A, B, C and of the airfoil tip point in downwards-morphed configuration; and g is the mechanism *kernel function*, governing morphing kinematics. Such function is generally nonlinear and noninvertible with respect to the variable $\vec{v}$.

The space of the solutions of Eq. (2) can be found through a methodology consisting of three main steps.

Step 1. Forced reduction of the unknown variables

[1]Linkages cannot be placed between adjacent blocks since they would prevent their relative rotation and therefore disable the shape transition.

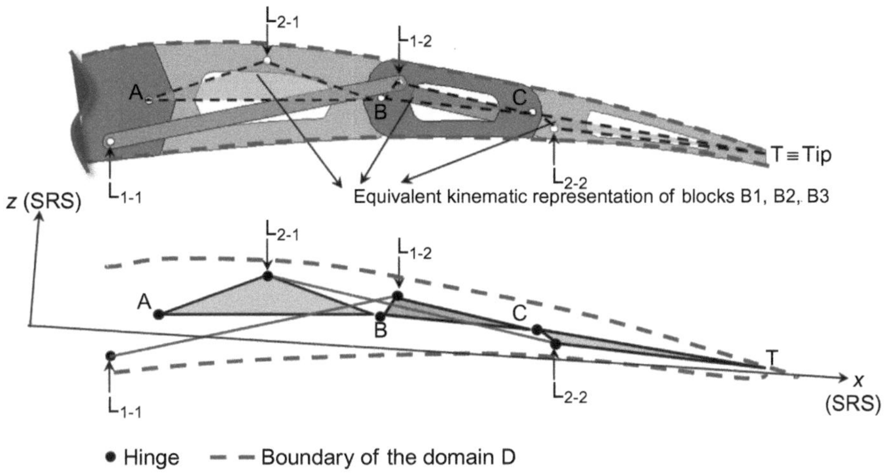

FIG. 3

Mechanization of the kinematic chain.

The positions of the hinges $L_{1\text{-}2}$ and $L_{2\text{-}2}$ are arbitrarily imposed in the domain $\mathcal{D}$. The unknown vector $\rightharpoonup v$ is then replaced by the following summation:

$$\vec{v} = \vec{v}_1^* + \vec{v}_2^* + \vec{t} \tag{3}$$

where $\vec{t} = [0, 0, x_{L_{1-2}}, z_{L_{1-2}}, 0, 0, x_{L_{2-2}}, z_{L_{2-2}}]$ is the vector defining the (imposed) positions of the hinges $L_{1\text{-}2}$ and $L_{2\text{-}2}$; $\vec{v}_1^* = [x_{L_{1-1}}, z_{L_{1-1}}, 0, 0, 0, 0, 0, 0]$ is the vector defining the unknown position of the hinges $L_{1\text{-}1}$; and $\vec{v}_2^* = [0, 0, 0, 0, x_{L_{2-1}}, z_{L_{2-1}}, 0, 0]$ is the vector defining the unknown position of the hinge $L_{2\text{-}1}$.

Step 2. Solution of the linearized morphing equations

In force of Eq. (3), the equation governing the morphing kinematics changes into:

$$\vec{w}_m = g\left(\vec{w}, \vec{t}\right) + g^*\left(\vec{v}_1^*\right) + g^*\left(\vec{v}_2^*\right) \tag{4}$$

where the kernel functions result decomposed into a nonlinear and noninvertible part depending on variables $\vec{w}$, $\vec{t}$ and into a generally linear and invertible part (marked with the apex "*") depending on variables $\vec{v}_1^*$ and $\vec{v}_2^*$. Since the terms, $\vec{w}_m$ and $g\left(\vec{w}, \vec{t}\right)$ are all known vectors defined in $\mathfrak{R}^8$, Eq. (4) can be solved algebraically to determine the unknowns $\vec{v}_1^*$ and $\vec{v}_2^*$.

Step 3. Iteration

A new value for $\vec{t}$ is imposed, and steps 1, 2 are repeated.

The iteration can be conceptually regarded as a function Γ operating on the variable $\vec{t}$ (coordinates of hinges $L_{1\text{-}2}$ and $L_{2\text{-}2}$) and providing as output the vectors $\vec{v}_1^*$ and $\vec{v}_2^*$ (coordinates of hinges $L_{1\text{-}1}$ and $L_{2\text{-}1}$):

$$\Gamma : \forall \vec{t} \in D \subset \mathfrak{R}^8 \rightarrow \left(\vec{v}_1^*, \vec{v}_2^*\right) = \Gamma\left(\vec{t}\right) \in D \subset \mathfrak{R}^8 \tag{5}$$

The function Γ (i.e., the iteration of the first two steps) allows for the evaluation of the locus described by the hinges $L_{1\text{-}1}$ and $L_{2\text{-}1}$ while the positions of the hinges $L_{1\text{-}2}$ and $L_{2\text{-}2}$ range in the domain of the airfoil section.

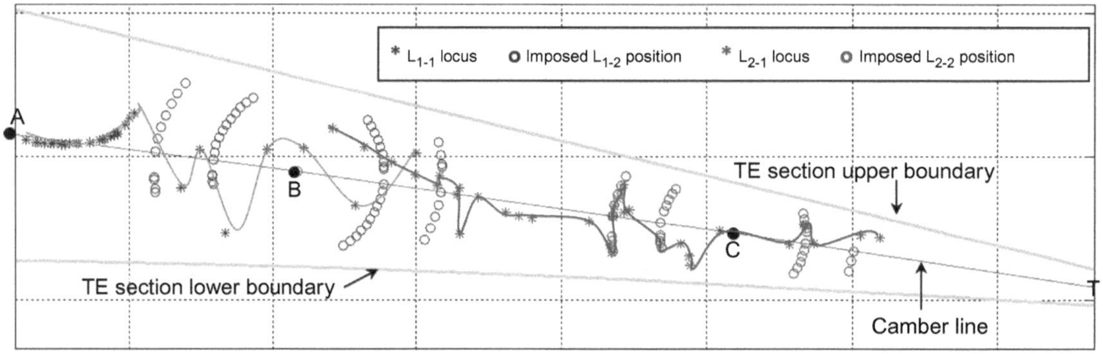

FIG. 4

Hinge locus.

A graphical representation of the hinges' locus obtained for the TE section of Fig. 2, has been reported (as explanatory example) in Fig. 4.

In order to find the best position of hinges $L_{1\text{-}1}$, $L_{1\text{-}2}$, $L_{2\text{-}1}$, $L_{2\text{-}2}$, practical considerations addressing the feasibility of the mechanism, are translated into an automatic selection algorithm implemented by suitable software tools. Two control parameters are defined for each hinge: the distance h from the camber line, the distance k from the closest airfoil boundary (either upper or lower).

The best positions of hinges $L_{1\text{-}1}$, $L_{1\text{-}2}$, $L_{2\text{-}1}$, $L_{2\text{-}2}$, are then evaluated by selecting among all vectors $\vec{v}_1^*$, $\vec{v}_2^*$, $\vec{t}$ satisfying the relation $\left(\vec{v}_1^*, \vec{v}_2^*\right) = \Gamma\left(\vec{t}\right)$, the ones whose components maximize the values of the control parameters h and k. In other words, the best position of the hinges is considered to be the one ensuring the:

- Highest clearance from the camber line, thus avoiding dangerous alignment between links L_1, L_2 and camber line hinges (potentially resulting in a bi-stable mechanism);
- Highest clearance from the airfoil boundary, thus avoiding interference problems between the physical hinges and the skin panels.

After the mechanism enabling rib morphing is defined in terms of the size of the rib blocks, and positions of hinges and links, the *preliminary design* of the structural system can be addressed. Elementary methods based on the theory of thin-walled section beams [8] can be used for this purpose.

The ribs' kinematic chain is transferred to the overall TE structure by means of a multibox arrangement. Each box of the structural arrangement is assumed to be characterized by a single-cell configuration delimited along the span by homologue blocks of consecutive ribs, and along the chord by longitudinal stiffening elements (spars or stringers). Upon the actuation of the ribs, all the boxes are put in movement, thus changing the external shape of the TE. If the shape change of each rib is prevented by actuators, the multibox structure is elastically stable under the action of external aerodynamic loads. A statically equivalent system—based on lumped elastic properties—is then considered for each box; this equivalent system is sketched in Fig. 5 with reference to a generic bay delimited by two consecutive TE ribs.

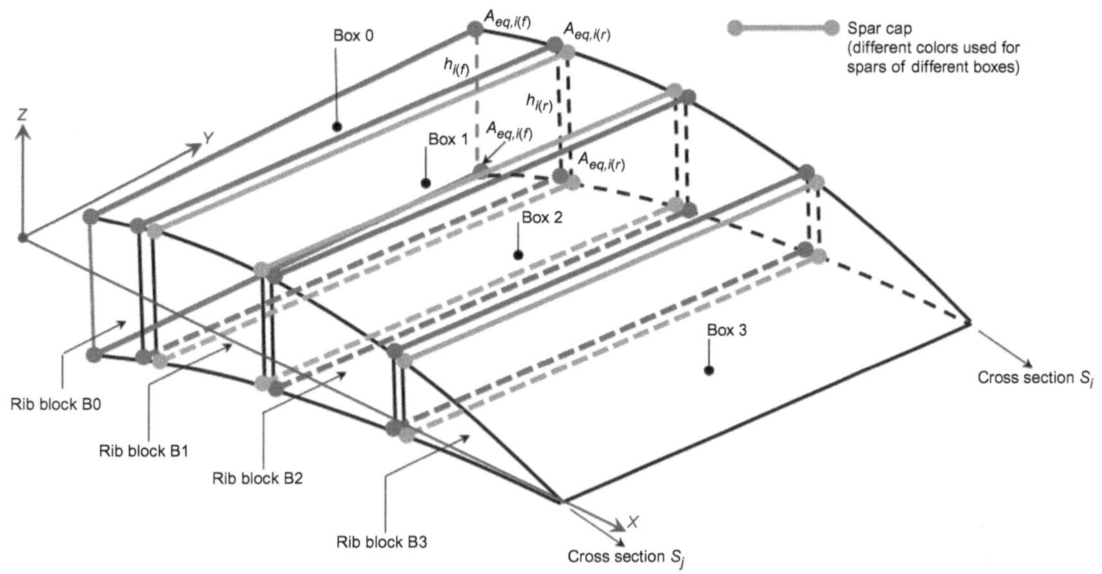

FIG. 5

Multibox arrangement, equivalent static scheme of a generic bay.

The structural sizing of each box is carried out under the following assumptions:

- No load transfer between adjacent boxes; each box of the multibox arrangement is treated separately;
- Most conservative condition in terms of multibox configuration (morphed/unmorphed) and pertaining aerodynamic loads. This condition is identified as the one producing the highest internal solicitations (bending moments, shears, torque) along all the cross-sections of the multibox arrangement.

For the cross-section of each TE box, the following approaches can be used to assess the dimensions of the load bearing components: spar webs, spar caps, skin panels, rib plates.

2.1 SPAR CAPS SECTION AREA AT GENERIC CROSS-SECTION

The ultimate bending moment $BM_{i(U)}$ at section S_i is supposed to be absorbed by front and rear spar caps according to Eqs. (6) and (7):

$$BM_{i(f)} = \frac{h_{i(f)}^2}{h_{i(f)}^2 + h_{i(r)}^2} BM_{i(U)} = \gamma \cdot BM_{i(U)} \tag{6}$$

$$BM_{i(r)} = \frac{h_{i(r)}^2}{h_{i(f)}^2 + h_{i(r)}^2} BM_{i(U)} = (1-\gamma) \cdot BM_{i(U)} \tag{7}$$

where $BM_{i(f)}$ is the amount of section ultimate bending moment absorbed by front spar caps; $h_{i(f)}$ is the height of the front spar; $BM_{i(r)}$ is the amount of section ultimate bending moment absorbed by rear spar caps; and $h_{i(r)}$ is the height of the rear spar.

Normal stress induced by bending moments in front and rear spar caps is then imposed equal to σ_{ref}, with σ_{ref} being equal to $f \cdot \sigma_{tu}$, with σ_{tu} ultimate tensile stress of the material and f a safety factor lower than 1. A good value for f is usually 0.7 in order to address a conservative design that includes all simplifications made for preliminary sizing purpose (while avoiding excessive weight increase).

Required equivalent areas for front and rear spar are obtained as:

$$A_{eq,i(f)} = \frac{BM_{i(f)}}{\sigma_{ref}} \cdot \frac{1}{h_{i(f)}} \tag{8}$$

$$A_{eq,i(r)} = \frac{BM_{i(r)}}{\sigma_{ref}} \cdot \frac{1}{h_{i(r)}} \tag{9}$$

Eqs. (8) and (9) always give a not-null value for the equivalent areas; for practical reasons related to the manufacturability of the structural components, a lower boundary A^* is established for cap areas. The equivalent areas are therefore considered equal to A^* in all cross-sections where Eqs. (8) and (9) provide values lower than A^*.

2.2 SPARS WEBS, SKIN PANELS, RIB PLATE THICKNESS AT GENERIC CROSS-SECTION

The chord-wise position of the ultimate shear force can be obtained as:

$$X_{S_{i(U)}} = -\frac{TM^*_{i(U)}}{S_{i(U)}} \tag{10}$$

where $TM^*_{i(U)}$ is the ultimate torque moment at section S_i and $S_{i(U)}$ the ultimate shear at station S_i.

The shear center is assumed to be chord-wise located at the middle of the section:

$$Xsc_i = \frac{X_{i(f)} + X_{i(r)}}{2} \tag{11}$$

where $X_{i(f)}$ and $X_{i(r)}$, respectively, indicate the chord-wise coordinates of front and rear spar webs.

According to the assumption about shear center position, the effective ultimate torque moment can be replaced by a dummy value given by:

$$\overline{TM}_{i(U)} = S_{i(U)} \cdot |X_{S_{i(U)}} - Xsc_i| \tag{12}$$

By invoking Bredts formula, cell's shear flow results equal to:

$$q_i = \frac{\overline{TM}_{i(U)}}{2\Omega_i} \tag{13}$$

Ω_i being the area enclosed by the cell.

The shear flows, arising in each element of the section, are then obtained by referring to the following equations:

$$q_{i(fw)} = q_i \frac{h_{i(f)}}{P_i} \tag{14}$$

$$q_{i(rw)} = q_i \frac{h_{i(r)}}{P_i} \tag{15}$$

$$q_{i(sk)} = q_i \frac{l_{i(sk)}}{P_i} \tag{16}$$

where $q_{i(fw)}$ is the shear flow in the web of the front spar; $q_{i(rw)}$ is the shear flow in the web of the rear spar; $q_{i(sk)}$ is the shear flow in the (upper/lower) skin panel; $h_{i(f)}$ is the height of the front spar; $h_{i(r)}$ is the height of the rear spar; $l_{i(sk)}$ is the length of the (upper/lower) skin panel; and P_i is the perimeter of the cell.

Shear stress induced by shear and torque moment along section elements are finally imposed to be equal to τ_{ref}, in which τ_{ref} is assumed equal to $\sigma_{\text{ref}}/\sqrt{3}$. Webs and skin panel thickness can be obtained by the equations:

$$t_{i(fw)} = q_{i(fw)}/\tau_{\text{ref}} \tag{17}$$

$$t_{i(rw)} = q_{i(rw)}/\tau_{\text{ref}} \tag{18}$$

$$t_{i(sk)} = q_{i(sk)}/\tau_{\text{ref}} \tag{19}$$

where $t_{i(fw)}$ is the thickness of the front spar web; $t_{i(rw)}$ is the thickness of the rear spar web; and $t_{i(sk)}$ is the thickness of the (upper/lower) skin panels.

In order to take in account manufacturability constraints, Eqs. (17)–(19) have to be combined with a lower boundary condition $t_{i(.)} \geq t^*$, where t^* is the minimum acceptable value for the thickness of the structural elements. In other terms, the thickness $t_{i(.)}$ is assumed equal to t^* if Eqs. (17)–(19) provide values lower than t^*. For preliminary sizing purposes, at each cross-section S_i, the thickness of the rib is generally assumed to be equal to the greatest between $t_{i(rw)}$ and $t_{i(sk)}$.

It is evident that this approach allows for an *isostress* preliminary sizing of the structural elements at each cross-section, making the dimensions of the structural elements suitably distributed along the span so that at each section they aren't oversized with reference to the internal forces they must withstand.

The preliminary sizing allows for the definition of a starting point for the main dimensions of the subcomponents and consequently for the elaboration of a detailed computer aided drawing (CAD) of the entire assembly.

The adequacy of the adopted arrangements is then investigated and assessed during the *advanced structural design* phase. Finite element analyses are carried out to verify and—if necessary—to update the structural layout while removing all the simplifications made in the preliminary design.

It is in fact in this phase that the *static interaction* between adjacent boxes is taken in account, and the links joining nonadjacent blocks are properly sized. Actuation chain, an integral part of the load-bearing structure, is investigated, together with its interfaces to the surrounding structure. Three analysis loops usually are necessary to finalize the design. In the first loop, main changes to the preliminary layout are determined, based on static analyses in the most severe combination of loads and configuration (morphed/unmorphed). In the second loop, the changes are proved to be adequate, including also other load distributions expected in service. In the third and final loop, dedicated analyses are carried out at interfaces and joint levels to suitably assess efficient structural solutions for the linkage of all the subcomponents of the system.

3 COMPLIANT MECHANISMS: WORKING PRINCIPLES AND DESIGN APPROACHES

A compliant mechanism is a structure, monolithic or quasimonolithic, able to withstand external loads (thus keeping its shape) and, at the same time, flexible enough to enable morphing actions.

This kind of structure poses different issues, including the health of the structure undergoing large deformations (making it prone to stress concentrations), the manufacture of a prevalently monolithic structure (strictly related to the choice of dedicated materials), the allocation of the different components (actuation, sensing) supposed to be highly distributed, coherently with the philosophy of a compliant mechanism [9–15].

The strategy adopted by designers when they size a compliant mechanism is characterized by unconventional approaches, essentially because classical architecture such as hinge connections, rigid and concentrated components, are not compatible with the compliant philosophy, and so unusual systems and subsystems are generally preferred. In this view, the following aspects play a critical role:

- *The topology chosen for the internal architecture*: In this sense, a dedicated optimization approach, basically heuristic (genetic), is usually adopted to set the different parameters (geometric boundary and locations of the structural subcomponents, their inertia features, the type of connection with other elements such as the skin). This approach, described in the next section, is driven by the necessity of fitting the target shape, having as constraints the structural integrity (prevention from overshooting allowable stress or any instability), the abatement of the weight, and of the energy necessary to actuate the system.
- *The skin*: Along with the monolithic approach suited to a compliant mechanism, the design of the skin is more often considered at the same time as the design of the internal mechanism. This occurs because the combined work of the internal architecture and the skin is considered a distinctive feature of a compliant system. For this reason, the topology definition often includes the identification and the assessment of skin parameters, such as thickness distribution or tissue typology (e.g., made of cells according to hierarchic approach).
- *The materials*: Working range is wider than usual (e.g., higher yield stress to withstand larger deformations, better manufacture properties to make the integration easier) and adaptability (i.e., possibility of modifying some of their properties to deal with potentially extreme working conditions) are two critical aspects to take into account during the selection of the materials. For example, the NiTiNol alloys has superelastic and shape memory properties that are exploited for particularly stressed parts (e.g., skins), and to actuate and, at the same time, to withstand (through the load-bearing properties) the internal architecture. Another example are the magneto-rheological fluids, whose tunable viscosity can be used to alter the dynamic response of the structure, acting both on the damping and on the stiffness.
- *The sensing*: As previously discussed, a compliant mechanism poses specific monitoring problems, focusing on both the actuation level (shape reconstruction) and on structural health. Different working parameters are added to the definition of the current status of the compliant mechanism:

kinematic (strain-displacement), mechanic (stress), thermodynamic (temperature) information can ensure an adequate description of the working condition of the device and can be handled adequately by a logic of control. Distributed sensors, from a hardware point of view, seem to better answer to the necessity of monitoring parameters that vary with continuity; furthermore, the monolithic nature of the compliant mechanism drives the choice toward sensors easy to be integrated and quickly inspected. Fiber optics, while exhibiting some features of the classical systems (such as cabling, continuous deployment, and so on), can assure a significant improvement of the channel numbers and wiring needs, as well as relying upon extreme multiplexing capability. Furthermore, light as an information carrier, allows the use of nimbler, nonshielded wires, preventing interference with on-board instrumentation, confining to laser sources and similar electronic equipment the unavoidable certification issues. Sensors based on graphene-sprayed technology are another interesting option both for the possibility of spraying (and so distributing) them into zones difficult to access and for the already tested opportunity of transferring the information by Wi-Fi (with obvious reduction of cabling issues). Finally, piezoelectric sensors, despite their intrinsic fragility, if suitably integrated within specific zones of the structure (not subjected to large deformations), constitute precious elements of a net dedicated to monitoring structural health.

- *The logic*: The main scopes of the logic of control are the achievement of the target shape/s, even in the most severe conditions, and the coordination of sensor signals for structural health monitoring. To ensure the achievement of a certain shape, the logic must contain a dedicated part able to perform the shape reconstruction. Note that the algorithm of reconstruction, differently from a kinematic mechanism (for which it is possible to identify some specific concentrated DOFs and extrapolate the entire configuration), has to take into account the distributed nature of the strain-displacement field and, consequently, manage a wider amount of data coming from several sensors, suitably distributed onto the structure. Also the structural health monitoring presents additional issues with respect to traditional, conventional structures; the large deformations and the use of unconventional materials as shape memory alloys (SMAs) require the adoption of specific detection logic strategies.

These aspects highlight the benefits and drawbacks of a compliant mechanism. The accurate estimation of them represents a fundamental part of the draft project of a morphing device. The designer, among others, is asked to decide whether to adopt the unconventional compliant mechanism philosophy.

A conventional design, mainly characterized by the sizing of single subsystems assumed to be rigid enough and rigidly moving one respect to each other, cannot be adopted. The compliant mechanism requires a global vision of the entire system whose final movement is the synergic result of the deformation of each single parts. The advantage of this is to potentially achieve the target shape relying upon distributed DOFs, with a final better aerodynamic finesse than a kinematic chain. This has one big disadvantage: The design schemes are necessarily unconventional, more complicated.

- *Unconventional*: It is generally difficult to find theoretical schemes allowing for a preliminary sizing of the mechanism because of the large deformations expected and because of the use of unconventional, nonlinear materials (for instance SMA).
- *More complicated*: An FE model suited for nonlinear static analyses is a lighter, easier model that can be adopted in the draft design phase. Despite the synergistic nature of the compliant device, each part of the model has to be modifiable, without upsetting and compromising its global

structure. This implies a modular approach during the models preparation. The designer must expend greater effort to envision the model from a global perspective, while at the same time considering the possibility of modifying different parts.

The algorithm used for the optimization will act on the single parts, modifying some features, according to genetic logic. The multiobjective approach (target shape approach, weight and cost minimization, abatement of the energy for the activation) represents itself an additional complication, needing an adequate number of runs to build a Pareto front.

The optimization has to involve not only the topology of the architecture, the skin features, and the amplification chain of the actuator system, but also to detail the connections among the different parts (e.g., the thickness distribution), with the aim of reducing as much as possible any stress concentration.

The materials chosen for the compliant mechanism present great potentials, with specific reference to superelastic alloys; their integration within skins might guarantee higher flexibility and at the same time an adequate stiffness, especially in critical zones such as leading edges. The drawbacks are:

- *The weight*: The density is generally more than twice that of aluminum alloys; weight penalties have to be adequately contained by an accurate sizing of the NiTiNol parts.
- *The integration*: The tenacity of the connection for highly stressed parts (junctions) and the finesse of the linkage on zones of aerodynamic interest (skin) represent issues the designer has to solve.

The main advantage of SMAs used as actuation systems is their compactness and the absence of mechanical parts moving with respect to each other. Drawbacks, including low efficiency and limited operative frequency range (because of thermal activation), limit the alloy percentage on the entire structure.

Similar considerations (involving the weight penalty and the integration strategy) can be done for other materials such as MRF.

The solutions for a compliant device must show an intrinsic lightness. They must be easily integrated, and the cabling must be limited because of the deep level of integration and distribution of the sensors required by a compliant mechanism.

This approach, however, might require the use of a new generation of sensors, whose technology readiness level (TRL) still must be enhanced for practical applications. The logic able to perform a shape reconstruction of a certain complexity is difficult to implement, for previously discussed reasons.

In conclusion, a compliant mechanism might offer advantages such as the possibility of fitting the best prescribed target geometries, an optimal stress distribution enhancing the fatigue life performance, and limited weight and power consumption. The price tag includes the adoption of new and unproven design strategies and technologies (materials, integration processes), all of which need further development. This situation currently keeps the TRL of a compliant mechanism generally lower than a conventional kinematic chain.

4 APPLICATIONS OF MORPHING ORIENTED KINEMATIC CHAINS

As practical example of the kinematic chain application to morphing structures, this section offers a case involving an innovative architecture for the wing flaps of large airplanes.

Within the framework of the JTI-Clean Sky (CS) project, and during the first phase of the Green Regional Aircraft Integrated Technological Demonstration (GRA-ITD), research was carried out on the design and technological demonstration of a novel architecture enabling the camber variation of a flap segment to be installed on next-generation, open-rotor Green Regional Aircraft (EASA CS-25 category). The driving idea was to replace a conventional double-slotted flap with a single-slotted morphing flap in order to improve A/C high-lift performances—in terms of maximum attainable lift coefficient and stall angle—and to reduce the noise emitted by the high-lift system.

Studies were limited to a portion of the flap element obtained by slicing the actual flap geometry (0.62 m chord) with two cutting planes distant 0.8 m along the wing span [16]. Target morphed shapes were evaluated on the base of 2D computational fluid dynamics (CFD) optimization analyses and were provided as input data for the design activities. A structural concept then was assessed in order to ensure the reversible transition from the nominal to the target shape of the flap segment. Segmented (finger-like) ribs driven by electromechanical actuators [17,18] were considered as a smart solution to enable the shape modification of a multibox architecture. The functionality of the concept was proved through an experimental ground test on a full-scale test article.

Further activities were addressed in order to increase the TRL of the validated technology within the second phase of the CS-GRA. Relying upon the already assessed concept, an innovative flap architecture was designed in order to enable two different morphing modes on the basis of the A/C flight condition/flap setting:

Morphing mode 1	Overall camber morphing to enhance high-lift performances during take-off and landing (flap deployed);
Morphing mode 2	Tab-like morphing mode. Upward and downward deflection of the flap tip during cruise (flap stowed) for load control at high speed.

A larger true-scale segment of the outer wing flap was selected for testing the new architecture in order to face the challenges posed by real-wing installation issues, especially with reference to the tapered geometrical layout and 3D aerodynamic load distributions.

The investigation domain covered the flap region spanning 3.6 m from the wing kink (Fig. 6) and was characterized by a taper ratio equal to 0.75 with a root chord equal to 1.2 m.

On the basis of specific aerodynamic analyses, the chord-wise extension of the flap tip—to be deflected according to morphing mode 2—was set equal to 10% of the local wing chord (dark-gray-colored portion in Fig. 6).

Maximum and minimum deflection required for a typical A/C mission were, respectively, set to ± 8 degree.

4.1 MORPHING CONCEPT OVERVIEW

The bimodal morphing capability was assumed to be implemented by active ribs playing the role of inner movable articulation of the flap structure. In order to define the positions and the nominal shapes of the morphing ribs along the flap span, eight reference planes were defined.

Reference planes (S_i, $i = 1{:}8$) were imposed to be perpendicular to the wing TE and equally spaced by a distance of 0.5 m (Fig. 7).

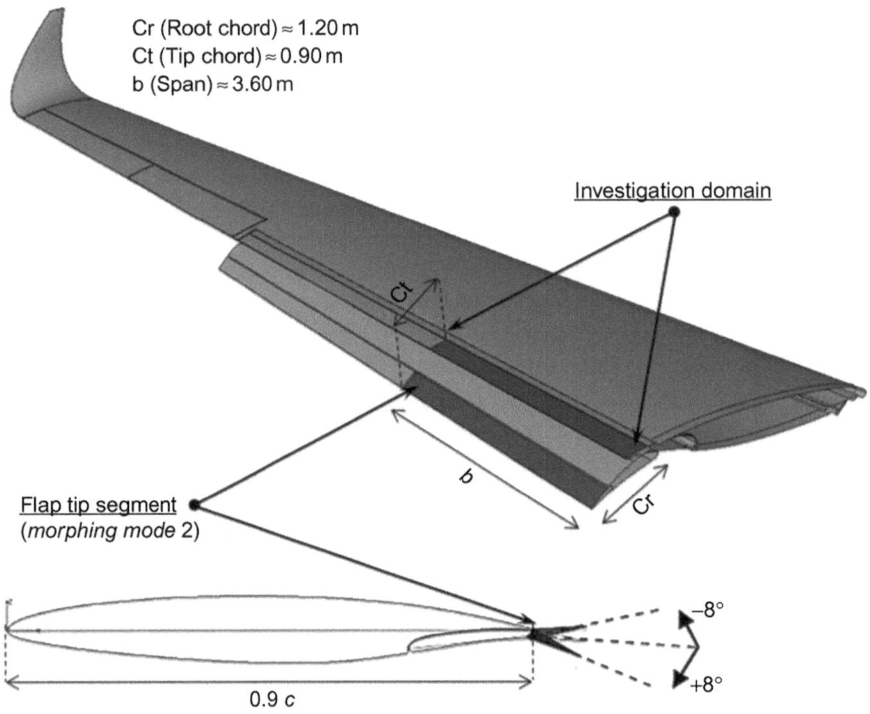

FIG. 6

Morphing flap, investigation domain [19].

They were numbered progressively according to the wing root-wing tip direction; the span-wise position of the first reference plane was set by imposing the coordinates of its intersection point (*P*) with wing TE.

The layout of the morphing rib mechanism was conservatively defined with reference to the last flap section (S8) because this section was characterized by the minimum available room for the installation of the actuation chain.

A parametric 3D-CAD, however, was generated in order to quickly scale the S8 layout according to the geometric features of the other flap sections.

The actuation strategy developed in the first phase of the CS-GRA project [16] was preserved in order to enable the overall camber morphing during take-off and landing phases (flap deployed, morphing mode 1).

The rib is composed by four consecutive blocks (B1, B2, B3, B4, Fig. 8) connected by three frictionless hinges (A, B, C) located along the airfoil camber line [19].

Blocks are allowed to rotate with respect to each other, thus making the airfoil camber line to morph. Blocks B1 and B3 are interconnected by means of a suitably shaped beam (L) having two hinges at the edges; internal leverage (M1) connects blocks B2 and B4 and is activated by the external rotary actuator (A1).

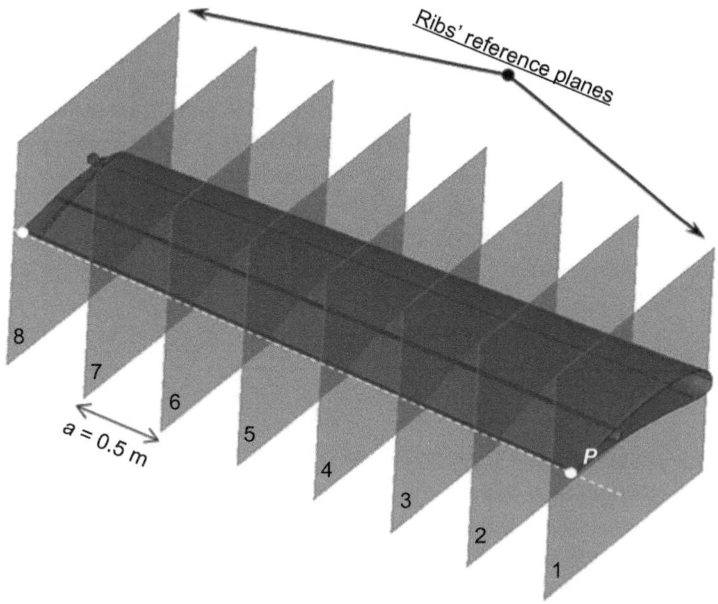

FIG. 7

Ribs' reference planes [19].

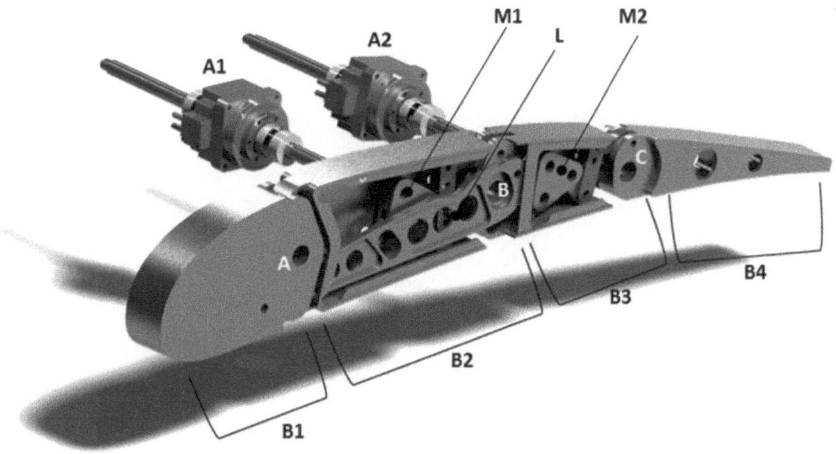

FIG. 8

Smart rib layout [19].

The rotation induced by A1 makes M1 to move, and therefore changes the relative position of block B2 with respect to block B4. At the same time, blocks B1 and B3 are forced to follow the movement being mutually interconnected to the remaining blocks. The position of the link L and of the pivots of the leverage M1 are selected in a way that, upon the rotation of A1 shaft, all the rib blocks rotate around hinges A, B, C according to specific angles compliant with the external morphed shapes to be achieved.

A secondary leverage (M2) links B4 to B3 and is driven by the rotation of the actuator A2. The secondary leverage amplifies the torque of the actuator and makes B4 to rotate around hinge C, thus implementing the tab-like morphing. During morphing mode 1, the secondary leverage is fixed because only actuator A1 is activated. On the other hand, actuator A1 is powered off during morphing mode 2.

The layout of the morphing rib was properly adapted to fit the geometry of the flap device at each spanwise section defined in Fig. 7.

The adaptation included only a slight reshaping of the upper and lower boundaries as well as the extension of the chord-wise length of blocks B1 and B4 while moving from section S8 to S1 (Fig. 9).

The chord-wise length of blocks B2 and B3 was kept equal for all flap sections, thus obtaining hinge lines perpendicular to rib reference planes.

Two C-shaped continuous spars (Al2024-T351) were positioned at 5% and 70% of the local airfoil chord in order to link all blocks B1 and B4 along the span.

These spars were suitably conceived to provide the greatest contribution in carrying external loads while ensuring adequate deformation levels to the entire assembly.

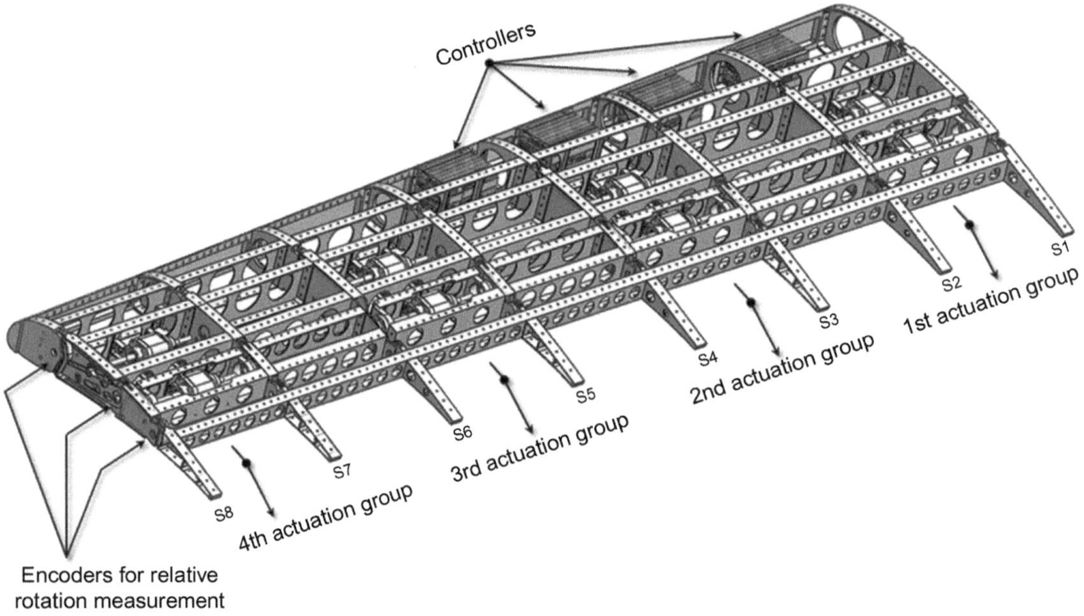

FIG. 9

Bimodal morphing flap layout: load-carrying structure and main equipment [19].

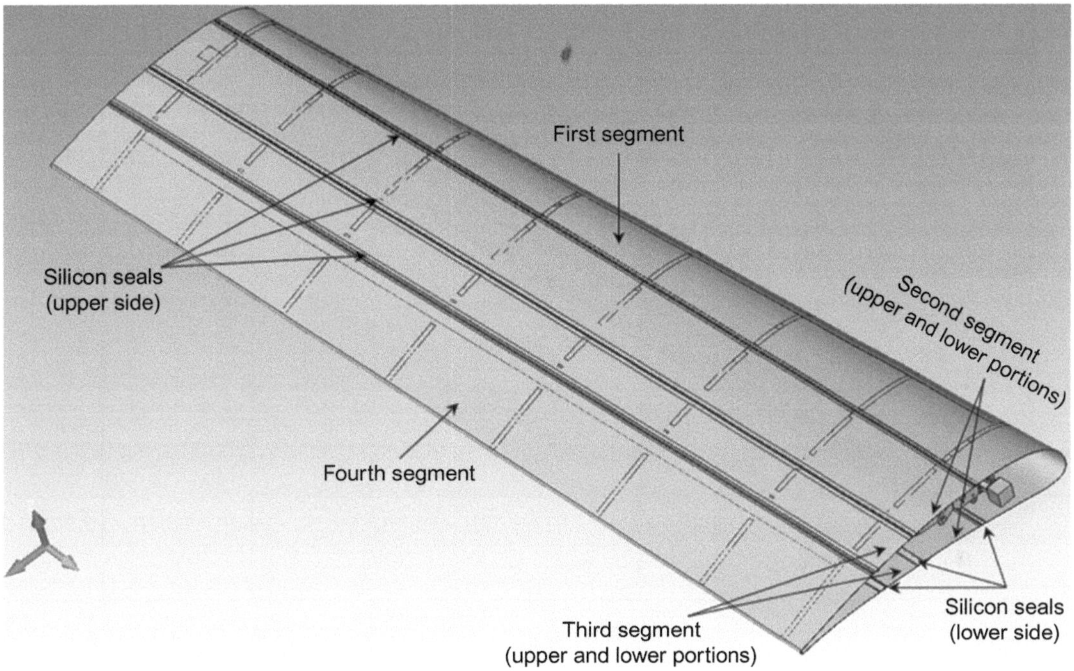

FIG. 10

Bimodal morphing flap layout: skin arrangement [19].

Segmented spars and stringers were adopted to generate a multibox arrangement elastically stable under bending and torsion. For these longitudinal elements, Al2024-T351 was selected.

Both ribs and longitudinal stiffening components were shaped properly in order to enable the installation of a segmented skin solution characterized (on both upper and lower side) by four elements in Al2024-T4 sliding on each other during morphing (Fig. 10). The shape of the silicon seals was defined by means of advanced kinematic analysis simulating flap morphing in both modes. These analyses also allowed for the verification of assembly tolerances while proving the absence of clashes between adjacent subcomponents in relative motion.

On the basis of the selected actuator type (Table 1), four actuation groups were considered adequate to move the entire device in operative conditions. They were duly coupled to HarmonicDrive gearboxes (Table 2), which acted as amplifiers of the actuation torque (and reducers of shaft rotation speed).

Four controllers were installed in the first bays of the leading edge in order to drive the actuation groups. Encoders for relative rotation measurement were placed around hinges A, B, C (Fig. 8) of the ribs at station S1, S4, S8. Their output was used in feedback to drive the actuators during morphing and to preserve the commanded flap shape in case a variation was induced by external perturbations (including deformations induced by aeroloads).

Table 1 Hollow Shaft Brushless Motor KBMS-14 Kollmorgen, Main Characteristics [19]

Specification/Actuator Model	KBMS-14	
Continuous stall torque	N m	2.11
	lb-in	18.67
Peak stall torque	N m	5
	lb-in	44.25
Maximum speed	RPM	8000
Peak current	Arms	10
Weight	kg	2.5
Number of poles	–	8

Table 2 HFUC-17-120-2UH-SP2983 Harmonic Drive AG, Main Characteristics [19]

Space Reducer Model		HFUC-17-2UH
Gear ratio		120
Maximum torque	N m	54
	lb-in	
Maximum speed	rpm	60
Transmission accuracy	arc-min	1.5
Maximum moment load	N m	
	in-lb	
Moment stiffness	N m/rad	16×10^3
	lb-in/rad	
Mass	kg	0.64
Ambient operating temperature	°C	0–60

4.2 STRUCTURAL ANALYSES

The thickness distribution of the load-carrying components along the span (for instance, the thickness of spar webs, skin, or rib flanges) was preliminarily evaluated on the base of the elementary methodology illustrated in paragraph 2.

Overall dimensions of structural components then were duly assessed by means of finite elements (FEs) analyses based on a refined and detailed structural model.

The blocks of each rib were meshed with solid element CTETRA10 [20] (Fig. 11A), while the inner link (L, Fig. 8) and the leverages (M1, M2; Fig. 8) were meshed with solid CHEXA8 (Fig. 11B and C).

CQUAD4 were conveniently adopted for longitudinal stiffeners, spars (Fig. 12) and for the skin.

Internal cylindrical hinges connecting adjacent rib blocks (Fig. 11D) and the items of the actuation chains were modeled with beam and rigid elements; the pin, used for fastening together adjacent items, was modeled with a beam element. Two rigid body elements were defined to connect the edges of the pin with the inner surface of the hole hosting the pin. Pin edges and grids on the hole surface were respectively considered as master and slave nodes of the rigid connection. Release of torsion rotation was applied at one edge of the pin in order to enable rotation around the hinge axis. Riveted and screwed joints were modeled similarly; no torsional release was considered in such cases.

In order to check the kinematic consistency of the FE model, actuator failures were simulated (by truncating the transmission line at shaft/rib interface) and a modal analysis was conducted with the flaps leading edge constrained in all DOFs. Two rigid modes reproducing the rib kinematic were successfully detected.

The first mode reproduced the tab-like motion (Fig. 13); the second mode captured the tab motion coupled with the camber morphing of the first three blocks (Fig. 14), which rotate according to the specific gear ratios imposed by the inner links and leverages. No other rigid modes were found.

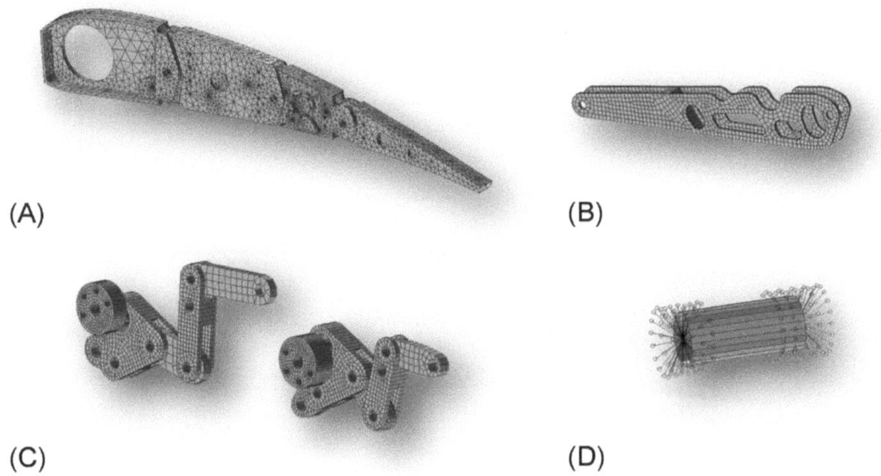

FIG. 11

FEM of some relevant structural items [19].

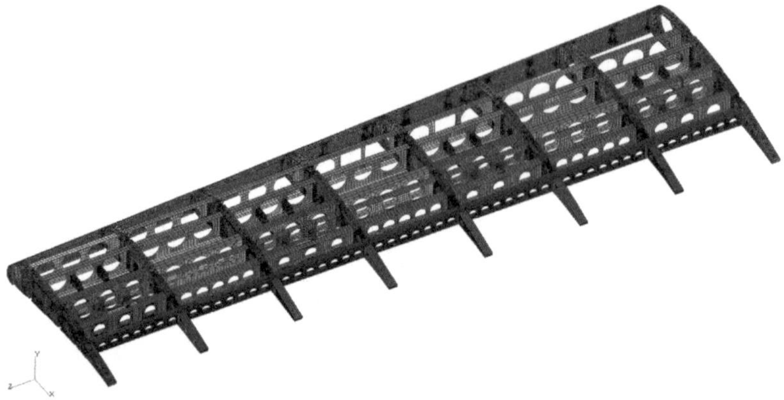

FIG. 12

FEM of the inner structure (isometric view) [19].

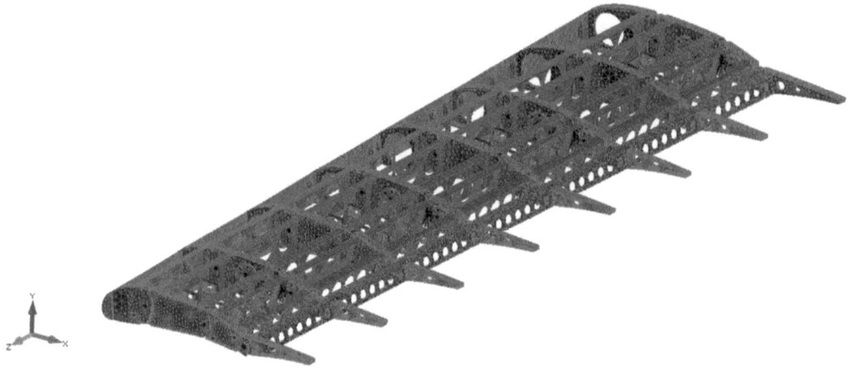

FIG. 13

Rigid mode reproducing tab-like morphing (actuation disconnected) [19].

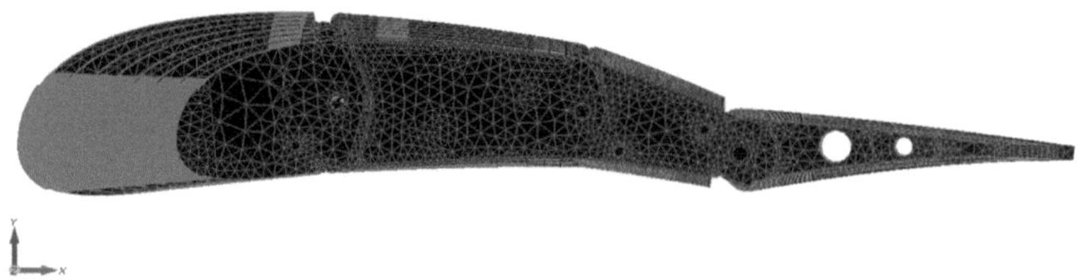

FIG. 14

Rigid mode reproducing the camber morphing coupled to tab-like morphing (actuation disconnected) [19].

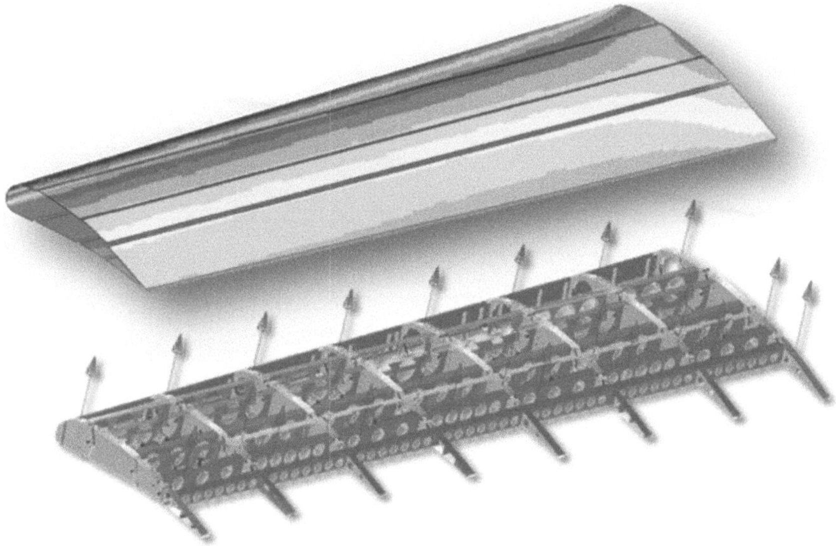

FIG. 15

Limit pressure distribution: (up: contour plot; low: equivalent lumped forces system) [19].

After the reliability of the model was proved in terms of smooth simulation of morphing kinematics, the actuators transmission line was reconnected to the ribs and the actuators shaft rotation was prevented in order to address static analysis.

Design pressure distributions were evaluated by CFD with reference to the operative condition of flap down at 35 degree, and a flap speed equal to 95 m/s at sea level.

The distributions were then converted into an equivalent set of lumped forces acting at the center of rib blocks B1, B3, and B4 (Fig. 15). The flap was constrained at the B2 blocks of the third and the sixth ribs (from the root section), thus simulating a rigid deployment track. Lumped forces pertaining to pressure distributions along B2 blocks were transferred to adjacent blocks while preserving overall shear, bending moment, and torque resultants.

Linear static analyses were carried out in MD-Nastran environment [20] at both limit and ultimate (1.5 times the limit) load conditions. For each condition, the reaction at the locked actuator shaft was calculated and verified to be less than the maximum actuation torque provided by the implemented system.

At limit load condition, the maximum displacement of 15 cm was found at the TE tip (Fig. 16); the torque required to restore the undeflected configuration was calculated by means of a dedicated linear analysis carried out on the flap-deformed shape with enforced motion of the actuator shafts. Also in this case, the obtained torque was compliant with actuation system performances.

In compliance with the applicable sections of airworthiness requirements, the absence of any local plasticization or elastic instability was verified up to limit loads, while the clearance from any failure was proved up to ultimate loads.

Very low stress values were found along the skin; its participation in loads adsorption resulted much lower than that occurring in case of conventional structural arrangements.

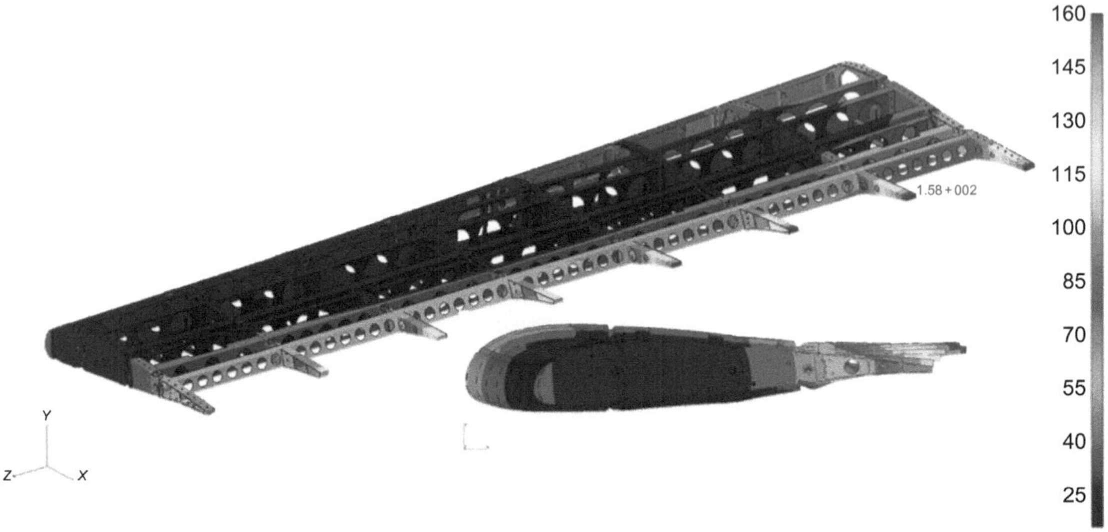

FIG. 16

Max displacements (mm) (limit load condition) [19].

Table 3 Maximum Von Mises Stress Arising in the Main Structural Elements [19]

Item	Max Von Mises Stress (Limit Load) (N/m^2)	Material Yield Stress (N/m^2)	Max Von Mises Stress (Ultimate Load)	Material Failure Stress (N/m^2)	Ref. Figure
Spar	276	330	414	450	Fig. 17A
Ribs	250		375		Fig. 17B
Rib leverages	266		399		Fig. 17C
Rib links	291		437		Fig. 17D

As expected, the highest stress values arose around the constrained regions simulating the links to the flap track.

The highest stress concentration was detected at the constrained blocks of ribs 3 and 6, as well as in their internal mechanisms. The spars adjacent to the constrained ribs were more stressed than the others. As evident from Table 3 and recalled figures, no element of the load-carrying structure was affected by stress values greater than material allowables.

5 APPLICATIONS OF THE COMPLIANT MECHANISM APPROACH

In this chapter two different compliant mechanisms are illustrated; the mechanisms have been designed in order to morph the trailing edge of a regional A/C wing [21–23].

The idea of the devices is to entirely substitute conventional flaps (characterized by rigid rotations and translations thanks to a dedicated kinematic chain) with morphing compliant mechanisms,

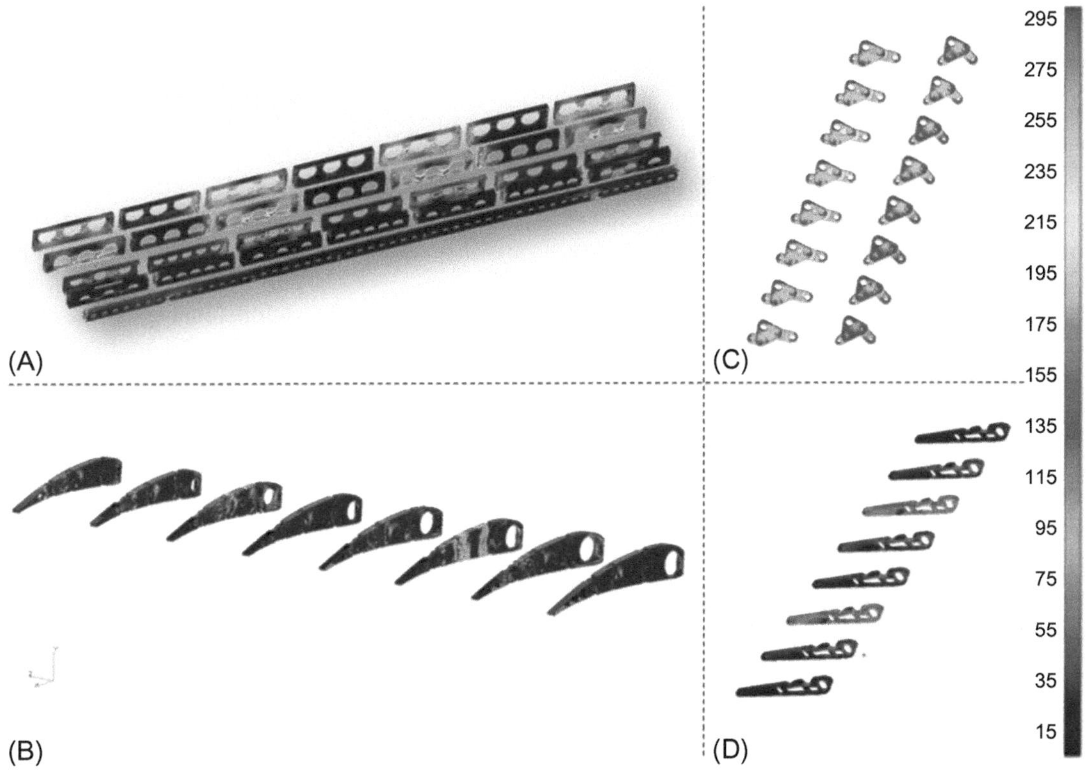

FIG. 17

Max von Mises stress (N/m^2) (limit load): (A) spars, (B) ribs, (C) leverages, (D) links [19].

producing large variations of the curvature along the airfoil region spanning from the 70% up to the 100% of the chord. The smoother morphed geometry would ensure the same high-lift performance of a traditional flap system and also achieve a remarkable abatement of the radiated noise and of the complexity of the deployable mechanism.

Both the devices were conceived to be monolithic or quasimonolithic: there are no lumped hinges to enable the camber variations; the morphing involves only the mean line of the airfoil, while the armadillo strategy was adopted for the skin (i.e., the skin was segmented and the relative moving of the different parts was admitted); components with a certain flexibility (arcs, beams) constitute the skeleton of the architectures and are mounted in serial way thus playing the role of distributed hinges; shape memory alloy actuators, integrated at the edges of these flexible elements, produce their deflections when activated, with a final global result of a camber increase.

First, we will discuss the working principle of each device by highlighting the main design parameters (e.g., arc and beam dimensions and their effect onto the global performance); second, we will produce dedicated simplified models with the specific aim of enabling the next phase: the optimization; suitable schemes of draft design were conceived depending upon the specific features of the two devices and their working ability; a dedicated strategy, based on the equivalent stiffness of the structure was used to model the SMA action; then the advanced design was faced through the FE approach used for modeling the details of the architecture; finally, a laboratory prototype of the arc-based architecture was manufactured and tested.

5.1 ARC-BASED FLAP, ACTUATED BY SMA ACTIVE ELEMENTS

The arc-based morphing flap, covering the 30% (72 cm) of the total chord and 30 cm spanned, is characterized by an unconventional internal architecture. The supporting structure is made of four arc elements per rib, linked in serial way as shown in Fig. 18. SMA ribbons connect the edges of each arc (see Fig. 19); they cooperate with the arc by absorbing upward bending and, when activated, they recover part of their original shape making the edges of the arcs to approach each other. This effect at arc level leads to a global increase of the curvature (as shown in Fig. 18, bottom). The external shape of the TE is ensured by a foam structure mounted at the arc edges and shaped to assure a smooth geometry even for maximum deflection.

The strategy adopted for sizing the structure is based on the equivalent rigidity concept. In practice, regardless of the complexity of the supporting structure (in this case the arc, made of titanium alloy), it is possible to identify the rigidity that a 1D actuator (in this case the SMA ribbon) perceives when integrated (preload condition) and activated. Through this model, both the structure and the SMA 1D element can be represented onto a force-displacement or a stress-strain plane, as illustrated in Fig. 20. Structure rigidity (namely K_a in the figure) is represented by a straight line (AB segment), rotated with respect the vertical axis to take into account of the opposite displacement of the edges of the SMA and the structure undergoing the actuation. Two other straight lines (AB* and A_1B_1 segments) represent the cases of a structure less rigid but equally preloaded (point A) and of a structure with the same rigidity but less preloaded (point A_1). On the same graph, curves for SMA load and unload are plotted for three different temperatures (the higher the temperature, the higher the curves).

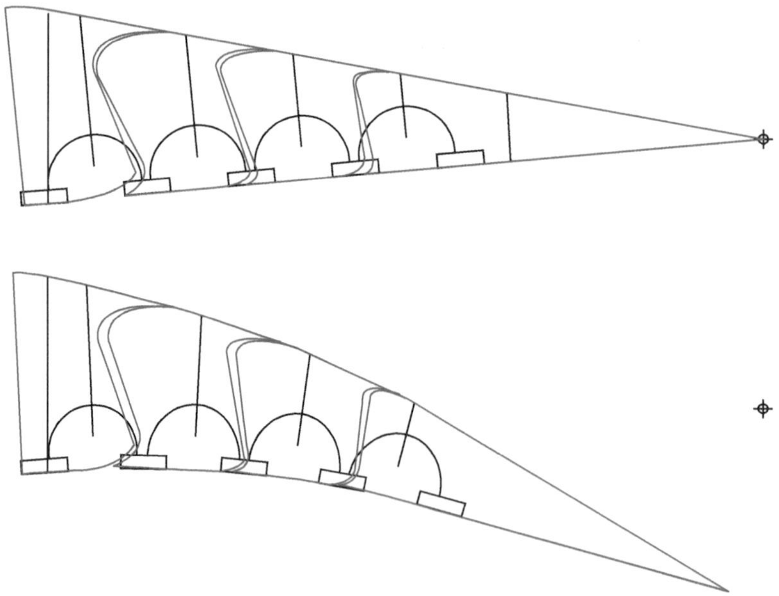

FIG. 18

Schematic of arc-based architecture: not actuated (top), actuated (bottom).

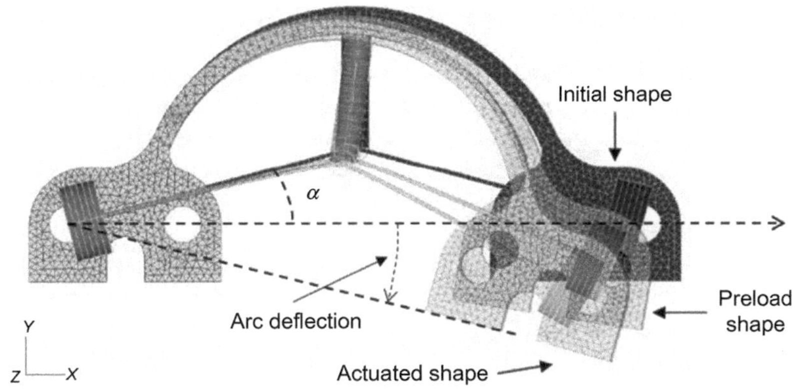

FIG. 19

Arc detailed FE model.

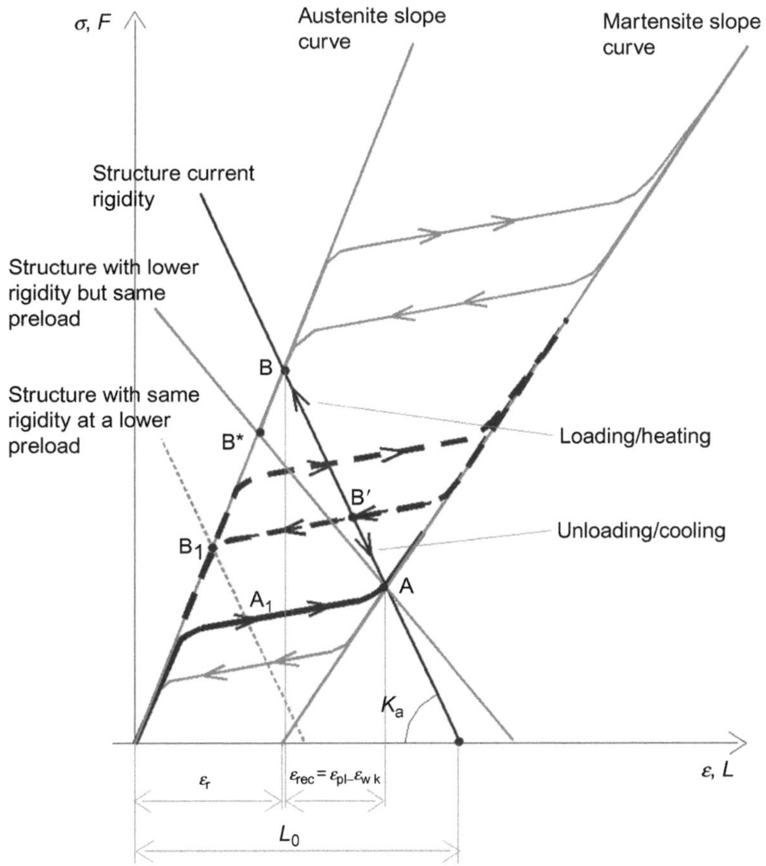

FIG. 20

SMA and structure material curves: design steps (Austenite ◇ Martensite path). Axes are stress-strain and force-displacement domains.

The integration process of the SMA, the preload generated, and the activation can be explained as:

- SMA integration onto the arc: The arc is compressed; after having fixed the SMA ribbon (whose material is characterized by a ε_r recoverable deformation at room temperature), the arc is left free to expand (see intermediate configuration illustrated in Fig. 19), so producing a certain stretching (ε_{pl}) into the ribbon (point A of Fig. 20, the intersection between the SMA curve and the structure straight line); a certain preload is generated and will be exploited to have the ribbon recovery when the SMA is not activated;
- SMA activation: When the SMA is heated, an upward translation of the SMA load and unload curves is produced; as a result, the intersection with the structures straight line moves up and leftward (see the intermediate point B′). This process continues up to full activation is reached (point B); the recovered strain (ε_{rec}) is now equal to the strain corresponding to point B (ε_{wk}) minus the preload deformation (ε_{pl}). Further temperature increases do not produce additional recovery of strain; this is essentially because of the specific geometry of the curves of the SMA, (see Austenite and Martensite slope lines) that are linear outside the transition.

Table 4 provides features of the SMA material used for the two compliant mechanisms presented in this chapter.

A simplified FE model (beam element-based) of the arc was built taking into account the characteristics shown in Table 5.

The apparent rigidity of the arc was computed by assigning a virtual negative thermal coefficient, α_T, to the SMA material and applying stepped thermal loads, T. The force, transmitted by the actuator to the structure, F_{SMA}, was then obtained through the classic relationship:

$$F_{SMA} = E_{SMA} A_{SMA} (\varepsilon - \alpha_T T) \tag{20}$$

where E_{SMA} is a virtual Young modulus assigned to the SMA material, A_{SMA} the SMA element cross-section, and ε the resulting deformation from the FE analysis. F_{SMA} is related to the component of the attained displacement onto the force line. In this way, the straight line was defined, representing the structure onto the plot of Fig. 20.

The main features of the titanium alloy arc were fixed, constrained by the available space in the TE and the structural integrity (maximum stress level in the arc and in the SMA lower than the allowable ones—800 MPa for the arc material and 450 MPa for the SMA ribbon), and assuming the target of at least 5 degree of deflection (here intended as the angle between the segment linking the edges of the arc

Table 4 Features of the SMA Material

SMA Ribbon	
Material	NiTiCu
Austenite/Martensite Young mod. (GPa)	40.0/10.0
Ultimate elongation (%)	12.0
Recoverable strain (%)	2.8
Martensite start/finish temperature (°C)	43.3/27.5
Austenite start/finish temperature (°C)	47.3/58.9
Stress of Martensite finish at room temperature (20°C) (MPa)	150

Table 5 Simplified FE Model Features	
Complete FE model	
Total number of nodes	940
Total number of elements	907
Arcs	
Material (Young, Poisson mod.)	127 GPa, 0.32 (titanium alloy)
Element type	CBEAM
Number of elements	240 (60 per arc)
SMA ribbons	
Material (Young, Poisson mod.)	Table 4
Element type	CBEAM
Number of elements	8 (1 per SMA ribbon)

and the horizontal line—see Fig. 19). The resulting arc has a mean radius of 55 mm, a thickness of 5 mm and a depth of 18 mm; it was integrated with a SMA ribbon 2 mm thick and 10 mm depth; an angle α of 13 degree between the SMA ribbon and the horizontal line was assumed, as shown in Fig. 19. A maximum deflection of 6.61 degree was predicted in this configuration, corresponding to a maximum stress level in the SMA and in the arc of 228 and 670 MPa, respectively. Further information is reported in the second column of Table 8. These results were verified through a more detailed FE model (shown in Fig. 19), reproducing the arc and the ribbon details, with specific focus on SMA-arc interface. This model was used to generate the FE model of the entire compliant architecture. The downward displacement of the TE was estimated with and without aerodynamic loads. The aerodynamic loads, assumed to vary linearly along the four foam blocks, were concentrated on the left edge of each arc: 150, 75, 80, 120 N, moving from the leading edge to the tip of the flap. These forces correspond to the most severe load absorbed by the flap. A maximum vertical displacement of 13.1 and 18.5 cm (corresponding to 10.3 and 14.4 degree of rotation, here intended as the slope of the line connecting the TE with the flap root) were estimated with and without loads.

After the assessment of the design phase, prototypes of the arcs and of the entire TE were manufactured and tested [21], proving the functionality of the concept in the presence of the most severe load condition (the maximum aerodynamic load acting on a fully deployed flap in landing phase) also in good agreement with the numerical outcomes.

A dedicated experimental campaign was then organized to validate the numerical predictions and to appreciate the effect of the combined action of four arcs connected in serial way, working in presence of external loads. Fig. 21 shows the prototypes of a single arc and the entire TE constituted of two ribs, each made of four arcs.

The setup used for both the single arc and the entire architecture was constituted by a DC power supply, a thermocouple reader, an acquisition system for strain gauges, a reference sheet, and a timer (Fig. 22).

Two strategies were taken into account to further enhance the performance of the concept. First, a parametric investigation of the effect of the angle α (sketched in Fig. 19) was performed, by highlighting the influence of this parameter on the arc performance. Then a genetic optimization involving all the main design parameters was implemented, finally leading to a more efficient architecture.

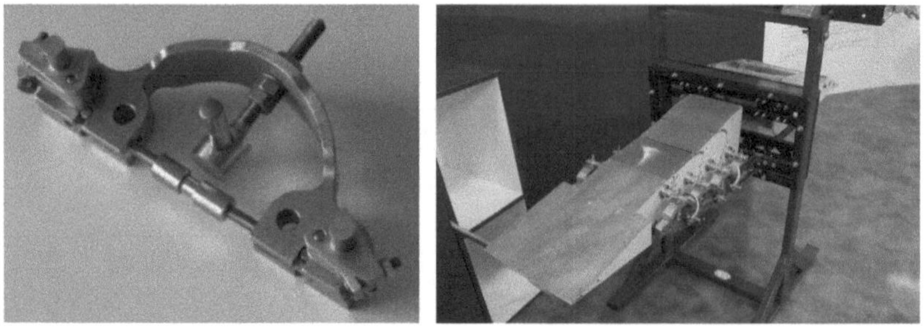

FIG. 21

Prototypes of the single arc (left) and of the trailing edge (right) [21].

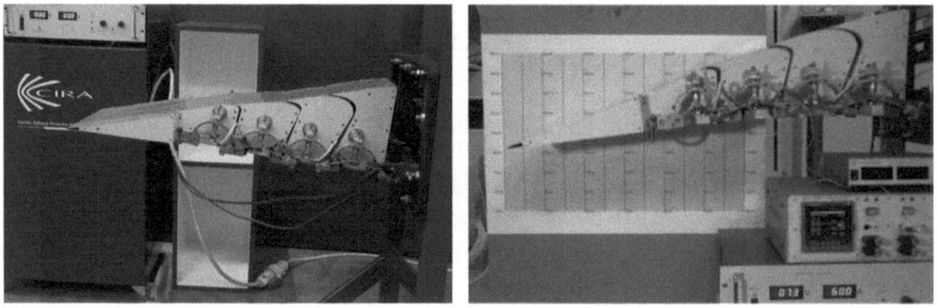

FIG. 22

Prototype in off (left) and on states (right) [21].

The parametric investigation was performed by assuming the same arc geometry and the SMA ribbon cross-section dimensions of the original arc device. As a result, the configuration shown in Table 6 was identified.

After having explored the parameter α, the problem of the combined effect of all the other parameters was investigated and a multiobjective optimization approach was adopted.

The scheme of Fig. 23 was adopted for the design task. At first, a buckling analysis was performed to estimate the minimum force causing instability, that is, the upper limit for the actuation.

The system working excursion was then defined, by identifying the two extreme configurations, moving from the preload condition (i.e., the prestressed cell state, able to ensure the complete recovery after operation, corresponding to point A of Fig. 20) to the full-deployed configuration (i.e., the cell maximum deflection, following a complete SMA activation, corresponding to point B of Fig. 20). These points were individuated by crossing the arc force-displacement characteristic elastic curve with the Martensite and Austenite linear characteristics of the SMA.

If the force, produced by the SMA device in the full-activated condition, was less than buckling limits, and the consequent stress level was within the material physical boundaries, then the arc ability in bearing the external loads while keeping the attained deformed shape, was investigated. According to an established discretization, the aerodynamic forces and moments were applied to the structural system, and its upward deflection was evaluated.

Table 6 Result of the Parameterization Study

Parameter	Value
α (degree)	21.0
Arc directive stiffness (N/m)	3.82E+5
SMA rigidity (N/m)	6.26E + 6
SMA recoverable ε	−3.18
SMA σ in operative condition (MPa)	270.0
Arc σ in operative condition (MPa)	567.0
Arc rotation (degree)	−7.21
Activation T (°C)	76.5
Energy required for arc activation (J)[a]	151.0

[a]*Energy was computed for a room temperature of 20°C and a specific heat of 322 J/kg K; the Austenite start temperature and the temperature-stress rate for the alloy considered are 47.3°C and 8 MPa/°C.*

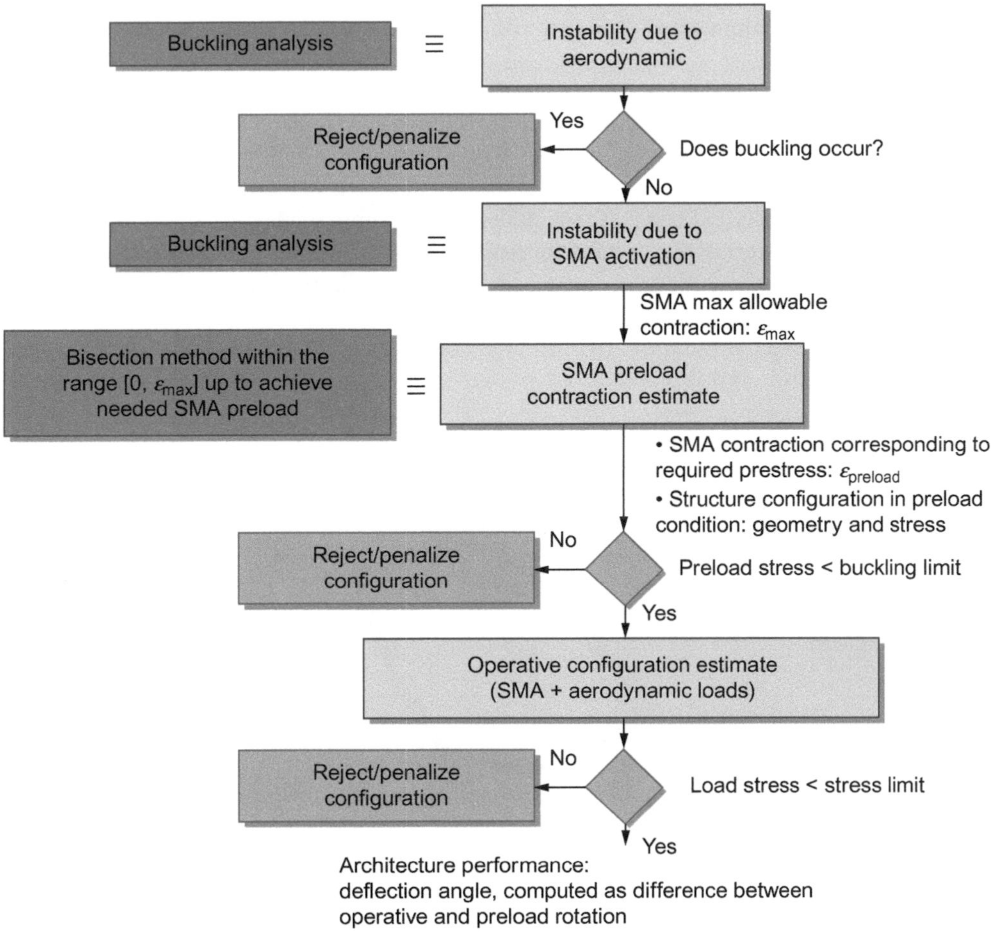

FIG. 23

Scheme adopted for the design and the optimization of a compliant structure.

Table 7 Design Parameters Considered for the Optimizations

Chromosome Id	Description	Range
1	Arc mean radius	4–10 cm
2	Radial thickness on the center of the arc	4–8 mm
3	Radial thickness on the edges of the arc	4–8 mm
4	Transversal thickness on the center of the arc	4–8 cm
5	Transversal thickness on the edges of the arc	4–8 cm
6	Inclination of the ribbon	0–40 degree
7	Thickness of the ribbon	0.5–2 mm
8	Width of the ribbon	10–15 mm
9	Material of the arc	Titanium, steel, aluminum

Table 7 shows how parameters were optimized and assumed as chromosomes for the genetic process.

Twenty initial populations (i.e., sets of arc configurations) were considered, each one consisting of 50 individuals (i.e., configurations). At the end of the optimizations, 20 configurations were identified, with the aim of maximizing the deflection angle. Coherently to the multiobjective approach, the best configurations were further evaluated on the basis of: the rotation achieved through a full actuation, the structural rigidity (the ability to keep its shape under aerodynamic loads), the stress level, the weight, the percentage of SMA material (relevant for the power supply), the cost of the material. Polar diagrams compared the different configurations in clusters of five; the resulting four configurations were finally compared in the diagram of Fig. 24.

The performance of configuration 14, the best compromise based on the different criteria, is compared to the previous ones in Table 8.

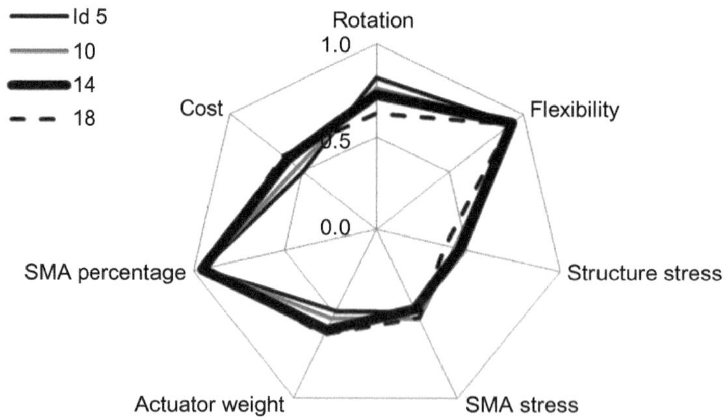

FIG. 24

Polar diagram to compare the best configurations.

Table 8 Comparison Between the Optimized and the "α" Parameterized Configurations

Parameter	Original Config.	Parameterized Config.	Optimized Config.	% Deviation (Parametrized vs. Optimized)
Arc rotation (degree)	6.6	7.2	9.89	37.5
SMA stress in operative condition (MPa)	228	270.0	236.9	−12.2
Arc stress in operative condition (MPa)	670	567.0	431.3	−24.0
Total weight (g)	505	410.0	228.0	−44.4
Activation T (°C)	75.8	76.5	72.9	−4.7
Energy required for arc activation (J)	142.7	151.0	192.5	27.2
Architecture rotation (degree)	12.3	15.0	19.9	32.7

As shown in the table, the target of increasing the transmitted rotation was reached, with an improvement of 37.5%. The stress level also improved, with a decrease of 12.2% and of 24% within the SMA and the structure, respectively. The weight decreased by 44.4%, which is particularly relevant for an aerospace application. The lower stress within the SMA also led to a lower temperature for the complete activation: 72.9°C versus 76.5°C. Finally, the greater mass of SMA caused a 27.2% increase of the energy for the complete activation.

5.2 X-CELL ARCHITECTURE FOR A SINGLE SLOTTED FLAP

The X-cell architecture was conceived to be integrated within the single slotted flap of a regional A/C. A dedicated conventional mechanism causes the deployment of the flap for specific flight conditions (e.g., take-off and landing), which generates hyperlift. The compliant architecture presented here is completely integrated within the flap to alter its curvature and produce additional hyperlift. It is accomplished without any external mechanism, but only through SMA actuators suitably integrated within the flap domain. The fully actuated morphed shape, as predicted through aerodynamic investigations, is illustrated in Fig. 25 and compared with a conventional, fully deployed configuration.

The architecture, shown in Fig. 26, is made of serially connected elastic metallic elements (cells). Each cell is formed by two clamped crossing plates (making an X-shape). This configuration guarantees a distributed bending flexibility, being the rotational pivot practically spread along the extension of the cell itself. The consequent stress distribution leads to an improved fatigue life and achievable deflections. Cell bottom edges were hinged to SMA elements, contributing to overall system rigidity and able to induce bending when activated.

An FE model of each cell was made, assuming a basic configuration in terms of elastic plates length, thickness distribution, and relative angle, SMA length and cross-section dimensions, and position of the SMA linear actuator with respect to the cell. The apparent rigidity of the cell was then computed by adopting the same procedure used for the arc (computation of the structure force-displacement curve through Eq. (20); estimation of the points A and B of preload and full activation in graph of Fig. 20);

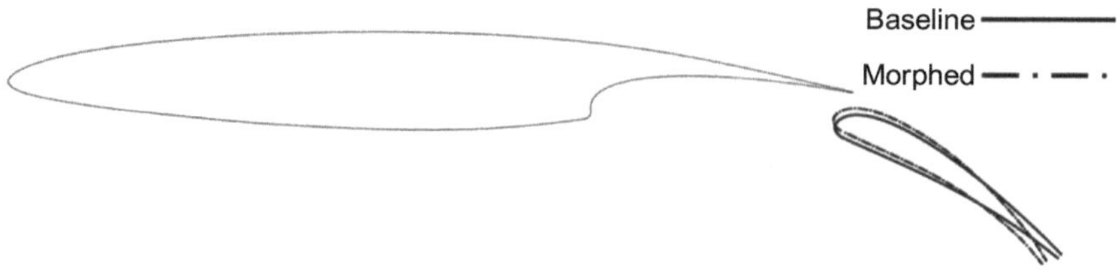

FIG. 25

Fully deployed, single slotted flap morphed *(dashed line)* and unmorphed *(solid line)*.

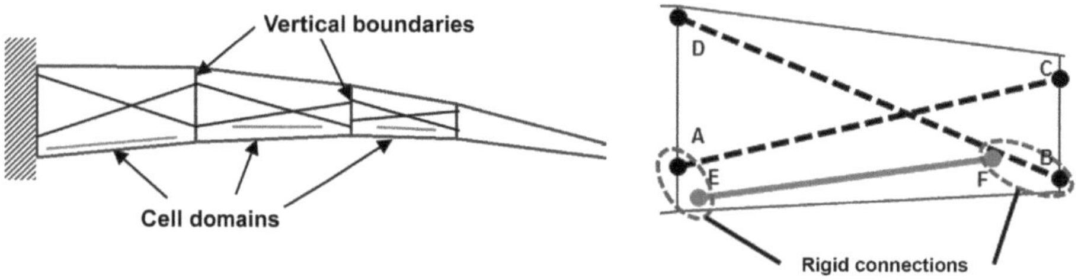

FIG. 26

Cell domains (left), schematic of a cell (right).

finally, the performance of the cell was estimated implementing within a genetic algorithm the procedure illustrated in the block diagram of Fig. 23. The performance of each cell was expressed in terms of:

- Vertical coordinates of the elastic plates edges (points A, B, C, and D in Fig. 26); abscissas were kept constant and assumed coincident with the vertical boundaries of the cells, Fig. 26 (maximum horizontal extension);
- Plates thickness and width, having assumed a linear distribution of these parameters over the length;
- SMA hinges location inside each cell (points E and F in Fig. 26) and SMA cross-section;
- Cell material, chosen among aluminum, 7075-T6, stainless steel, and titanium.

A summary of the design parameters of each cell and the corresponding variation ranges is provided in Table 9.

Twenty initial populations were generated randomly and given as input to the genetic algorithm with the aim of maximizing the cell deflection, within stability and material limits.

The optimal individuals were then compared with each other in clusters of five, selecting the best compromises in terms of achievable deflection, rigidity under the external aerodynamic loads, stress level within the cell and the SMA materials referred both to the instability limit and material structural allowable and cell weight.

The configurations this way selected were then further compared to each other; Fig. 27 shows this comparison for the first cell (closer to the root of the flap).

In Table 10, the results obtained for the three types of cells are summarized.

After the main features of the single cells were assessed, an FE model of an entire rib was analyzed. The rib had 241 elements of different types (beams for the cells and the SMA, plates for the rigid TE,

Table 9 Optimization Parameters Taken into Account for the X-Cells and Corresponding Variation Ranges

	Range—Allowed Values		
Design parameter	**Cell 1**	**Cell 2**	**Cell 3**
A, ordinate (mm)	−13.2 to 0.00	1.60–13.3	9.0–2.40
B, ordinate (mm)	1.60–12.6	8.70–15.7	5.40–13.4
C, ordinate (mm)	6.24–7.20	4.48–5.55	28.8–37.6
D, ordinate (mm)	5.70–7.30	5.48–7.20	40.1–54.1
E, abscissa (mm)	140.5–180.5	270.0–324.0	399.0–418.1
E, ordinate (mm)	−13.2 to 5.20	1.60–16.1	9.10–16.2
F, abscissa (mm)	23.3–27.0	352.0–399.0	464.5–485.6
F, ordinate (mm)	10.1–1.60	15.1–17.6	4.90–13.4
Plate thickness (mm)[a]	1.00–10.0	1.00–15.0	0.50–5.0
Plate width (mm)[a]	10.0–120.0	10.0–100.0	1.0–50.0
SMA cross section (mm^2)	10.0–300.0	10.0–240.0	10.0–180.0
Cell possible materials	Stainless steel; aluminum alloy; titanium alloy		
Aerodynamic force resultant (N)	855.9	552.6	287.7
Aerodynamic moment resultant (N m)	278.3	131.4	43.9

[a] *Initial and final values are searched for; a linear variation of the thickness and width is assumed, along the plate length.*

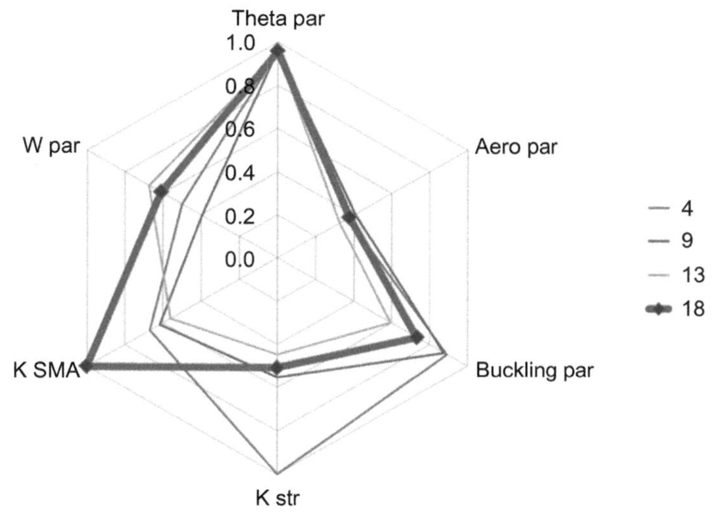

FIG. 27

Comparison among the best individuals of the 1st cell.

Table 10 Parameters Assessed for the Three Cells

		Cell		
		1st	**2nd**	**3rd**
Cell points	A, abscissa (mm)[a]	140.5	269.7	399.0
	A, ordinate (mm)	−9.6	1.6	21.1
	B, abscissa (mm)[a]	269.7	399.0	485.3
	B, ordinate (mm)	4.3	11.0	13.1
	C, abscissa (mm)[a]	269.7	399.0	485.3
	C, ordinate (mm)	64.1	44.8	35.7
	D, abscissa (mm)[a]	140.5	269.8	399.0
	D, ordinate (mm)	58.5	62.2	51.9
SMA points	E, abscissa (mm)	140.5	323.0	417.2
	E, ordinate (mm)	−6.8	15.2	9.1
	E, abscissa (mm)	239.2	375.1	464.5
	E, ordinate (mm)	8.4	15.4	5.0
Cell plates	Min/max thickness (mm)	1.0/2.0	1.1/1.5	0.6/1,3
	Min/max width (mm)	4.7/5.9	4.7/5.1	2.3/4.1
SMA features	SMA length (mm)	100.0	52.0	47.0
	SMA cross section (mm^2)	103.0	33.1	69.8
Cell performance	Material		Aluminum	
	Overall cell weight (g)	124.2	58.8	34.2
	Rotation—preload (degree)	8.7	4.3	6.2
	Rotation—SMA action (degree)	20.0	20.1	10.4
	Rotation—SMA & aerodynamic loads (degree)	18.8	17.5	10.2

[a]*Even though the abscissas of the edges of the cells (A,B, C, and D) were not object of optimization, for the sake of completeness they have been anyway reported.*

rigid links for the connections of the cells). In Fig. 28, the FE model, together with the clean shape of the flap (light blue) and the target morphed shape (red) is sketched. It is important to note that:

- The original configuration of the rib is globally rotated upward with respect to the clean configuration. This takes into account the fact that (as for the arcs) the cell bent to allow SMA ribbon integration and to generate the necessary preload through the elastic recovery of the cell itself;
- The target-morphed shape is reached through an intermediate activation, since the cell was designed and optimized to overdeflect with respect the assigned target shape.

Nonlinear static investigations allowed for the estimation of the behavior of the combined cells system under preload and when activated (Fig. 28, bottom) as well as to predict TE vertical displacements and related stress levels within the cell materials and inside the SMA for each of the considered conditions (clean, intermediate configuration). All these data are reported in Table 11, together with the safety factors and required temperatures for driving the SMAs actuators. The related necessary energy level

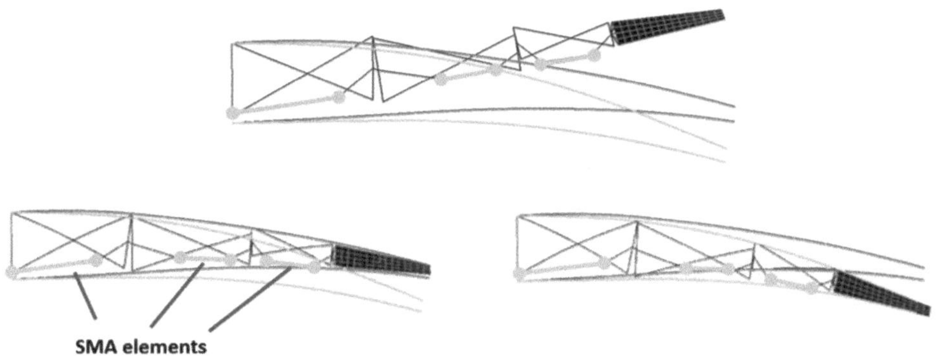

FIG. 28

Top: FE model of an entire rib with clean and actuated configurations; bottom left: clean configuration; bottom right: intermediate activated configuration.

Table 11 Single Rib Performance

Feature	Preload	Intermediate	Fully activated
Max vertical disp. (mm)[a]	92.2	137.6	336.4
Max stress within the cells (MPa)	61.6	90.2	249.9
Max stress within SMA (MPa)	150.0	224.7	823.0
Overall safety factor[b]	6.3	4.2	1.2
Activation temperatures (°C)[c]	–	85.4	128.4
Energy (kJ)[d]	–	4.1	6.4

[a]*With respect to the deformed configuration, following aerodynamics action.*
[b]*Min between SMA elements and cells safety factors.*
[c]*Max temperature among the three cells.*
[d]*Referred to a 25°C environmental temperature.*

is estimated by multiplying the SMA mass by its specific heat and the difference between the required and the environmental temperature (25°C).

After assessing the main parameters, the detailed design was dealt with. A CAD model of the flap is shown in Fig. 29, left, where the internal actuation architecture can be seen. A 3-bay flap 1 m span and having a chord of 230 mm, is considered. A detail of a cell is given on the right: the upward and downward plates (AC and BD segments, Fig. 26) are divided into two further equal elements, and distributed span-wise. This architecture guarantees the absence of any interference during motion.

As for the single rib FE model, to assure the proper SMA preload, each elastic cell is first bent downward and integrated with the SMA elements. The elastic reaction stretches the SMA segments, attaining the required preload at the desired reference configuration. It is worth noting that if the cells are mounted in this condition, an additional curvature is introduced to the clean flap configuration. To compensate this effect, each cell has to be mounted with an opposite pitch angle. Furthermore, the

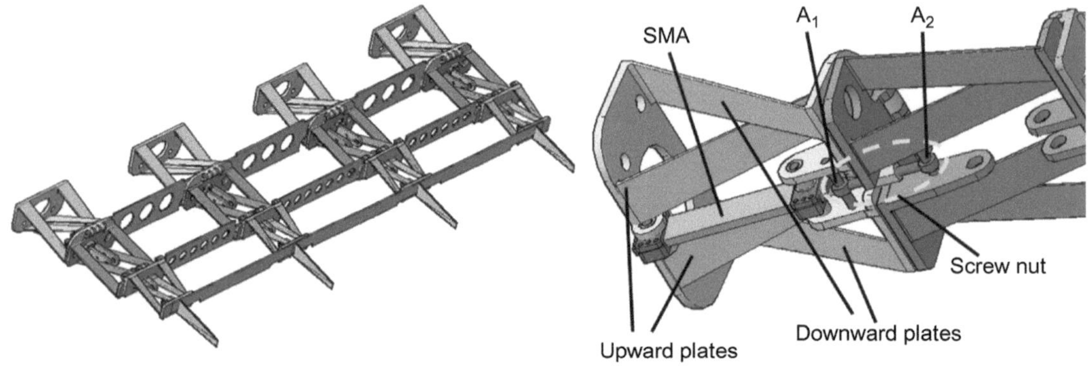

FIG. 29

Top: CAD of the internal architecture (left); detail of the connection between two cells (right).

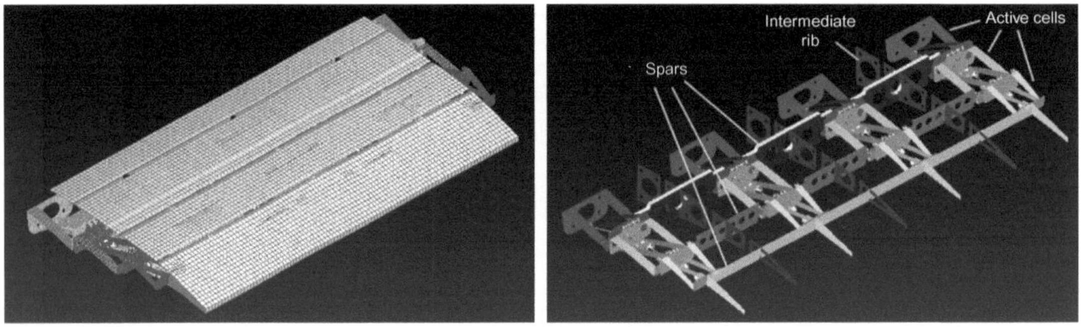

FIG. 30

FE model of the entire flap (left); internal components (right).

installation should be tunable, in order to compensate for possible assembly errors or variation in material characteristics. A dedicated mounting solution is thus devised and sketched in Fig. 29. The cells are connected to each other through hinges; screw-nut elements allow proper spacing of points A_1 and A_2, so to regulate their relative angle and ensure a suitable blockage around the pivots.

On the basis of the CADs, a detailed FE model of the entire flap was generated (Fig. 30).

Preloaded, intermediate, and fully activated configurations were simulated through nonlinear static analyses. The investigations provided information about the morphing ability and structural status (stress level) of the architecture. A brief summary of the results is reported in Table 12.

The aerodynamic loads, applied as a 2D distributed pressure on the skin, were concentrated (in terms of forces) on the downward connections of each cell; they are reported in Table 9. Their effect on the shape of the flap proved to be practically negligible when compared to the other forces acting on the structural system (related tip rotation and displacement resulted −0.39 degree and 4.2 mm, respectively). An even more negligible effect was observed for the preloaded configuration, because of the reduced magnitude of the aerodynamic loads at this attitude.

Table 12 3D FEM Results

	Aerodynamic Loads	Intermediate Config.	Fully Activated Config.
TE vertical displ. (mm)[a]	−4.2	92.7	227.8
Global rotation (degree)[b]	−0.39	8.5	15.2
Max stress within the structure (MPa)[c]	159.7	154.2	251.3

[a]*With respect to the preloaded condition.*
[b]*Slope of the straight line connecting the flap root to the TE position.*
[c]*For the stress within the SMA elements, refer to data of Table 11.*

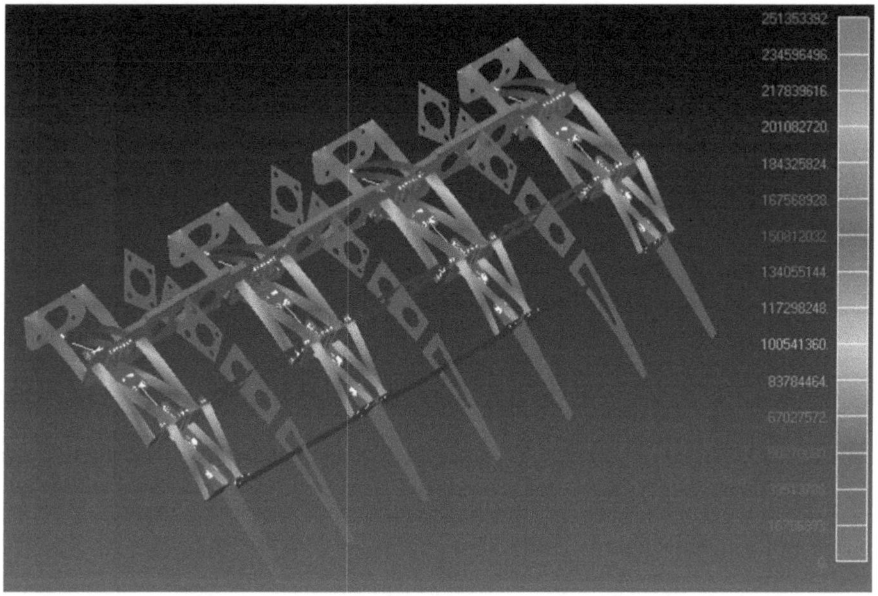

FIG. 31

Stress level on the elastic elements when actuated.

The arising stress field (max value estimated at 159.7 MPa) affected the skin in presence of the aerodynamic loads alone. When the SMA actuators caused the downward movement of the adaptive flap, the most affected components were cell plates. The max stress level within the structure was 251.3 MPa (see Fig. 31), well below the material stress limit (483 MPa), with a safety factor of about 2.

6 CONCLUSIONS

Morphing of metallic wing structures has fascinated generations of researchers; numerous and sometimes bizarre architectures have been proposed, tailored to specific end-applications and A/C type. Although different for layout, all of them can be categorized in two basic groups: mechanized architectures and compliant mechanisms.

In this chapter, the design philosophy behind each type of morphing structure has been presented, together with practical applications referred to as wing TE camber adaptation.

Mechanized architectures implement morphing through the rigid-body motion of stiff subcomponents interconnected by suitably designed kinematic chains and actuation leverages.

Each subcomponent of the kinematic chain is sized to provide its own contribution to the adsorption of external solicitations arising in operative conditions; actuators and actuation transmission lines are sized to enable the motion of the system and to preserve given shape configurations while counteracting aerodynamic loads with the minimum need of power.

Compliant mechanisms involve the deformation of structural elements to enable the required shape change; mechanical properties of the structure have to be properly distributed in order to ensure adequate morphing compliance and adequate stiffness to withstand external loads. In this case, the actuators play a significant role in the adsorption of external loads, but their power demand also depends on the amount of energy necessary to deform the compliant structure.

Conventional systems based on actuators and transmission chains are therefore much more effective for a kinematic chain than for a compliant mechanism at a parity of external operative loads. The adoption of conventional actuation systems for a compliant mechanism inevitably results in weight and power penalties. Compliant architectures, therefore, are more reasonably coupled with smart elements working as load-bearing components and morphing actuators; SMAs have been presented here as a good solution to physically implement such a combination of functions.

By taking these considerations into account and by carefully analyzing the architectural solutions proposed in this chapter, we see that kinematic chains are generally characterized by a higher level of maturity for in-flight applications. This is not only because of the adopted structural arrangements—somehow recalling and improving classical multitabs—but rather to the adoption of standard materials, equipment, design, and manufacturing processes.

It is indeed unquestionable that whereas innovation in aeronautics is pursued while lowering unconventionality, a smoother path to airworthiness approval is expectable.

On the other hand, just in force of their higher TRL, mechanized solutions need to comply with ever-demanding requirements in terms of production time/cost, inspectability, and maintainability. All these attributes push toward the minimization of parts and layout complexity.

Although compliant mechanisms are less complex than kinematic chains and generally capable of ensuring smoother shape morphing, they are, however, prone to other issues that are intimately related to their more unconventional nature.

For large-scale applications, the fatigue life of flexural hinges is remarkably lower than the one exhibited by bushings and bearings. The replacement of a damaged flexural hinge in a compliant mechanism, however, is likely to be a much more invasive operation than the conventional hinge repair in a kinematic chain. Because the stiffness of the overall architecture is suitably reduced to accommodate morphing with low actuation power, clearance from aeroelastic instability phenomena becomes a paramount requirement to be taken in account after the preliminary design phase. A compliant mechanism perfectly morphing under aerodynamic loads might indeed be responsible for detrimental impacts on A/C aeroelastic behavior or on its dynamic response to discrete or turbulent gust. Finally, the adoption of nonstandard materials or procedures for the design of compliant systems increases the level of uncertainty about structural performances of the resulting architectures, and, above all, on their reproducibility in a series production.

At the beginning of the chapter, we asked which architectural arrangement is the most promising? The one based on kinematic chains or the one based on compliant mechanisms?

As is often the case for the most complex questions, there is no absolute answer. The appeal of a given technology is indeed the function of the specific application as well as of the requirements to be satisfied to efficiently perform it.

Taking inspiration from nature, it would come easy to conclude that the optimum might be represented by a hybrid solution characterized by articulated and compliant mechanism, similar to the skeleton and muscles of a birds wing.

Nevertheless, we believe that easy answers to difficult problems are often wrong conclusions.

We prefer to observe that nature makes its own creatures to evolve according to the specific missions they need to accomplish during their lives. In this sense, A/C are not birds, and their evolutionary process has just begun.

REFERENCES

[1] E. Stanewsky, Adaptive wing and flow control technology, Prog. Aerosp. Sci. 37 (2001) 583–667.

[2] I. Chopra, Review of state of art of smart structures and integrated systems, AIAA J. 40 (11) (2002) 2145–2187.

[3] Bowman J., Sanders B., and Weisshaar T., Evaluating the impact of morphing technologies on aircraft performance, Proceedings of the 43rd AIAA/ASME/AHS/ASC Structures, Structural Dynamics and Materials Conference, Denver, Colorado, April 22–25, 2002.

[4] J. Spillman, The use of variable chamber to reduce drag, weight and costs of transport aircraft, Aeronaut. J. 96 (1992) 1–8.

[5] Bilgen O., Kochersberger K.B., Inman D.J., and Ohanian III O.J., Novel, bi-directional, variable camber airfoil via macro-fiber composite actuators, Proceedings of the 50th AIAA/ASME/ASCE/AHS/ASC Structures, Structural Dynamics and Materials Conference, Palm Springs, California, May 4–7, 2009.

[6] Wildschek A., Grünewald M., Maier R., Steigenberger J., Judas M., Deligiannidis N., and Aversa N., Multifunctional morphing trailing edge device for control of all-composite, all-electric flying wing aircraft Proceedings of the 26th Congress of International Council of the Aeronautical Sciences (ICAS), Anchorage, Alaska, September 14–19,2008.

[7] McGowan A.R., Horta L.G., Harrison J.S., and Raney D.L., Research activities within NASA's morphing program", Proceedings of the RTO AVT Specialists' Meeting on "Structural Aspects of Flexible Aircraft Control", Ottawa, Canada, October 1999.

[8] E.F. Bruhn, Analysis & Design of Flight Vehicle Structures, Tri-State Offset Company, Cincinnati, OH, 1969 (Chapter A19).

[9] S. Ameduri, A. Brindisi, B. Tiseo, A. Concilio, R. Pecora, Optimization and integration of shape memory alloy (SMA)-based elastic actuators within a morphing flap architecture, J. Intell. Mater. Syst. Struct. 23 (4) (2012) 381–396.

[10] S. Barbarino, R. Pecora, L. Lecce, A. Concilio, S. Ameduri, E. Calvi, A novel SMA-based concept for airfoil structural morphing, J. Mater. Eng. Perform. 18 (5) (2009) 696–705.

[11] Bye D.R. and McClure P.D., Design of a morphing vehicle, Proceedings of the 48th AIAA/ASME/ASCE/AHS/ASC Structures, Structural Dynamics and Materials Conference, Honolulu, Hawaii, April 23–26, 2007.

[12] Ivanco T.G., Scott R.C., Love M.H., Zink S. and Weisshaar T.A., Validation of the Lockheed Martin morphing concept with wind tunnel testing, Proceedings of the 48th AIAA/ASME/ASCE/AHS/ASC Structures, Structural Dynamics and Materials Conference, Honolulu, Hawaii, April 23–26, 2007.

[13] A. Hasse, L.F. Campanile, Design of compliant mechanisms with selective compliance, Smart Mater. Struct. 18 (2009) 1–10.

[14] Baker D., and Friswell M.I., The design of morphing aerofoils using compliant mechanisms, Proceedings of 19th International Conference on Adaptive Structures and Technologies, Ascona, Switzerland, October 6–9, 2008.

[15] A.V. Popov, T.L. Grigorie, R.M. Botez, Y. Mébarki, M. Mamou, Modeling and testing of a morphing wing in open-loop architecture, J. Aircr. 47 (3) (2010) 917–923.

[16] R. Pecora, F. Amoroso, G. Amendola, A. Concilio, Validation of a smart structural concept for wing flap camber morphing, Smart Struct. Syst. 14 (4) (2014) 659–678.

[17] R. Pecora, S. Barbarino, A. Concilio, L. Lecce, S. Russo, Design and functional test of a morphing high-lift device for a regional aircraft, J. Intell. Mater. Syst. Struct. 22 (2011) 1005–1023.

[18] R. Pecora, A. Concilio, I. Dimino, F. Amoroso, M. Ciminello, Structural design of an adaptive wing trailing edge for enhanced cruise performances, in: 24th AIAA/AHS Adaptive Structures Conference, AIAA Sci-Tech, San Diego, CA, January 4–8, 2016 (Paper ID: AIAA 2016-1317).

[19] R. Pecora, F. Amoroso, M. Magnifico, Toward the bi-modal camber morphing of large aircraft wing flaps: the CleanSky experience, in: Proceedings of SPIE—The International Society for Optical Engineering, vol. 9801, 2016.

[20] MSCMDNASTRAN®, Software Package, Ver. R3-2006, Reference Manual.

[21] S. Barbarino, R. Pecora, L. Lecce, A. Concilio, S. Ameduri, L. De Rosa, Airfoil structural morphing based on S.M.A. actuator series: numerical and experimental studies, J. Intell. Mater. Syst. Struct. 22 (2011) 987–1003.

[22] S. Barbarino, S. Ameduri, R. Pecora, Wing camber control architectures based on SMA: numerical investigation, Proc. SPIE 6423 (2007). 64231E-1–64231E-8.

[23] S. Ameduri, A. Concilio, R. Pecora, A SMA-based morphing flap: conceptual and advanced design, Smart Struct. Syst. 16 (3) (2015) 555–577.

CHAPTER

SENSOR SYSTEMS FOR SMART ARCHITECTURES

11

Monica Ciminello*, Rolf Evenblij†, Gianluca Amendola*

The Italian Aerospace Research Centre, CIRA SCpA, Capua (CE), Italy Technobis Fibre Technologies, Alkmaar, The Netherlands†*

CHAPTER OUTLINE

NOMENCLATURE

~	nearly
A	strain gauge circuit amplifier gain
AC	alternating current
ASPIC	application Specific Photonic Integrated Circuit
ATED	adaptive trailing edge device

Morphing Wing Technologies. https://doi.org/10.1016/B978-0-08-100964-2.00011-3

AWG	arrayed Waveguide Grating
$\vec{B}(s)$	binormal vector
b	piezoelectric width
CCD	charge-coupled device
C_p	capacity
d_{31}	piezoelectric coefficient
DC	direct current
ε	strain
EB	expanded beam
FBG	fiber Bragg grating
FORT	Fiber Optic Ribbon Tape
GRIN	graded index
IPA	isopropyl alcohol
l	resonance-determining dimension
$\vec{N}(s)$	normal vector
n	harmonic number
n_{eff}	effective index of refraction
R	resistance
ρ	mass density
$r(s)$	position vector
R_0	resistance at zero load and time
S	sensor sensitivity
s	curvilinear abscissa
$\vec{T}(s)$	tangent vector
T	temperature
TDM	time division multiplexing
Typ.	typically
V_o	output voltage
V_{ext}	excitation voltage
WDM	wavelength division multiplexing
WSN	Wireless Sensors Network
Y	effective elastic stiffness constant
α	thermal expansion
ξ	thermo-optic coefficients
$\rho\alpha$	photoelastic thermal expansion
Λ	grating period
$\kappa(s)$	curvature
λ_B	Bragg wavelength
$\tau(s)$	torsion vector

1 INTRODUCTION

Over the past decade most industrial sectors have been forced to reduce cost and become more competitive. World markets are expanding and this represents both an exciting challenge and a potential threat to industry. Many industrial sectors such as transportation are consequently under significant pressure to reduce costs and improve their effectiveness [1]. This climate is conductive to the introduction of new technologies leading to radical changes in the way structural systems are designed, built, and made operative. The end of the 20th century was witness to the merger of several technological disciplines that could eventually revolutionize engineering design philosophy and lead to the creation of intelligence within otherwise inanimate structures. In the view of improving structural system performance, morphing is a concept that is gaining more space in future design perspectives [2]. Research has the objective of realizing structures that can accommodate variations occurring during a typical operative working condition by changing their shape. This capability is often linked to large displacements, which in turn is linked to large strains. With current knowledge, such morphing structures can only be achieved by using dedicated kinematics, that can meet the incompatible requirement of systems. Kinematics can exhibit large displacements while preserving the capability of bearing external loads; functional advanced materials, rely on sensory data for knowledge acquisition and deliver them to the brain of the control network for decision making [3]. The first step to mimic nature's paradigm is to build structures with the ability to monitor their status and health. This new technology will endow a structure with senses with the ability to react to its environment and change its state. Furthermore, continuous monitoring of structural loads could lead to ongoing improvements in design that would avoid the occurrence of excessive strains, especially in high-stress regions. Structural sensor systems could substantially reduce life-cycle cost as it would provide the means of checking structures from birth to retirement [4]. Structurally integrated sensors form the key element of any smart structure. Integrating sensing systems is the first logical phase in the evolution of multifunctional materials. The resident sensing system within such an adaptive smart structure would provide the necessary information on its shape and load distribution and structural health monitoring [5]. Shape control of propellers, helicopter rotor blades, aircraft wings, and even wind power generators offer the potential for increased efficiency, improved maneuverability, reduced vibration, and less noise [6–8]. A longer life can be expected from reduced vibration as this also lessens structural fatigue. Distributed fiber optic, graphene-based sensors and wireless data transfer are destined to become some of the core technologies of the 21st century smart structures. The purpose of this chapter is to show that certain aspects of this exciting technology could have an important role to play in industrial sectors, such as the aeronautical one, which is concerned with smart structural monitoring.

2 STRAIN SENSORS

The purpose of a sensor is to respond to an input physical parameter and to convert it to a signal that is compatible with electronic circuits. In structural applications, the measurement of strain energy is mainly required. Strain is the deformation of a physical body under the action of applied forces. By measuring the strain produced in an elastic element by an unknown force, it is possible to retrieve information about the structural displacement, applied loads, and energy level. The sensor's output signal may be in the form of voltage, current, charge, or wavelength [9]. A short list of the most used sensors for structural applications is following discussed.

2.1 STRAIN GAUGE FOILS

A metallic strain gauge consists of a very fine wire or, more commonly, metallic foil arranged in a grid pattern [10]. The grid pattern maximizes the amount of metallic wire or foil subject to strain longitudinal direction. The cross-sectional area of the grid is minimized to reduce the effect of shear strain and Poisson strain. The grid is bonded to a thin backing, called the carrier, which is attached directly to the test specimen. Therefore, the strain experienced by the test specimen is transferred directly to the strain gauge, which responds with a linear change in electrical resistance. The strain gauge is a resistive elastic unit whose change in resistance is a function of the applied strain:

$$\frac{dR}{R} = S\varepsilon \tag{1}$$

where R is the resistance, ε is the strain, and S is the gauge factor of the material. Strain gauges are available commercially with nominal resistance values ranging from 30 to 3000 Ω, with 120, 350, and 1000 Ω being the most common values [10]. In most applications, the evaluation of strain requires accurate measurement of very small changes in resistance. For such small changes in resistance, strain gauges are almost always used as the resistive portion of the Wheatstone bridge [http://www.electrical4u.com/wheatstone-bridge-circuit-theory-and-principle/]. The general Wheatstone bridge consists of four resistive balanced arms with an excitation voltage, V_{ext}, that is applied across the bridge. The formula for voltage output (V_{o}) is:

$$V_{\text{o}} = \frac{V_{\text{ext}}AS}{4}\varepsilon \tag{2}$$

where V_{ext} is the excitation (input) voltage and A is the strain gauge circuit amplifier gain (external amplifier unit, typ. ~100); S generally results ~2.0.

2.2 PIEZOELECTRIC DEVICES

The piezoelectric effect can convert a variable strain into a variable AC electrical signal, while a steady-state condition produces no electrical response. A good design would be based on the effect of an applied deformation modulating the mechanical resonance of the piezoelectric crystal [11]. A basic idea behind the sensor's operation is that certain cuts of quartz crystal, when used as resonators in electronic oscillators, shift the resonant frequency when mechanically loaded. The equation describing the natural mechanical frequency spectrum of a piezoelectric oscillator is:

$$f_n = \frac{n}{2l}\sqrt{\frac{Y}{\rho}} \tag{3}$$

where n is the harmonic number, l is the resonance-determining dimension (e.g., the thickness of a relatively large thin plate or the length of a thin long rod), Y is the effective elastic stiffness constant (e.g., the shear stiffness constant in the thickness direction of a plate or Young's modulus in the case of a thin rod), and ρ is the mass density of the crystal material. A typical piezoelectric sheet can be treated as a parallel plate capacitor C_p. Considering only the effect of the strain along the 1-direction, and assuming the value of ε to be averaged over the gauge length, the equation relating strain and voltage (V_o) generated by the sensor is:

$$V_o = \frac{d_{31} Ylb}{C_p} \varepsilon \tag{4}$$

where d_{31} is the piezoelectric coefficient and b is the width.

2.3 GRAPHENE-BASED POLYMERS

In general, the functional properties of nanocomposites are sensibly affected by the characteristics of the nanofillers (both the structural and physical properties) and by their dispersion in the host matrix. The piezoresistive response of the fabricated strain sensors is characterized by measuring the variation of the electrical resistance as a function of the applied flexural strain under quasistatic monotonic and dynamic cyclic loading conditions. The sensitivity S of the sensor can be defined as the derivative of the electrical resistance with respect to the induced mechanical flexural strain (ε). The dimensionless expression [12] can be given by:

$$S(\varepsilon, t) = \frac{1}{R_0} \frac{\partial R(\varepsilon, t)}{\partial \varepsilon} \tag{5}$$

where R_0 is the resistance at zero load and time (before the test). The partial derivative underlines that the resistance can also be functional in other environmental (i.e., temperature, humidity, etc.) and operating parameters (i.e., source current, voltage, etc.) while the time variable t states that the material response can be generally affected by hysteresis and memory effects. Graphene-based polymeric composites are produced through the thermal expansion-liquid exfoliation route. After the treatment, thermally expanded graphite oxide is sonicated using processor tip horn ultrasonicator [13].

2.4 FIBER OPTICS

A fiber Bragg grating (FBG) is a local modulation of the refractive index in the optical fiber which can be created in the core by using a holographic method or a phase mask method. Although several grating types (uniform, chirped, apodized, etc.) exist [14], uniform FBGs will be the focus of this discussion. The FBG serves as a wavelength selective mirror: light travelling down the fiber is partially reflected but these reflections interfere destructively at most wavelengths. At one narrow range of wavelengths, constructive interference occurs and light is returned down to the fiber. Maximum reflectivity occurs at the so-called Bragg wavelength λ_B and depends on the effective index of refraction n_{eff} and on the period Λ of the grating per the well-known Bragg equation:

$$\lambda_B = 2 n_{eff} \Lambda \tag{6}$$

Both the effective refractive index and the Bragg period of the grating will be affected by any applied mechanical strain as well as by temperature. This situation is expressed in Eq. (7), where the first term gives the effect of strain and the second describes the effect of temperature:

$$\Delta\lambda_B = \lambda_B (1 - \rho_\alpha) \Delta\varepsilon + \lambda_B (\alpha + \xi) \Delta T \tag{7}$$

In Eq. (7), $\Delta\lambda_B$ is the change in Bragg wavelength, $\rho\alpha$, α and ξ are respectively the photoelastic, thermal expansion, and thermo-optic coefficients of the fiber, $\Delta\varepsilon$ is the change of strain and ΔT is the temperature change [14]. For a typical grating written in a silica fiber and with $\lambda_B \asymp 1550$ nm, sensitivities to strain and temperature are approximately 1.2 pm/με and 10 pm/°C respectively.

2.4.1 Associated Electronics

As described a FBG is a type of distributed Bragg Reflector embedded in a short segment of an optical fiber core that reflects specific wavelengths of light and transmits all the others. Therefore, an FBG can be used as an inline optical filter to block certain wavelengths, or as a wavelength-specific reflector. With the latter, as the fiber is stretched (or modulated in any other way causing an elasto-optic or thermo-optic effect) the reflected wavelength will shift accordingly and with a linear relation (Fig. 1).

Basically, this means that the strain transferred to the fiber at the location of the FBG can be determined by observing the wavelength change. The observation of this wavelength shift is generally referred as "interrogation." Thus, the interrogator is the system that is monitoring wavelength shifts. There are several ways of monitoring a wavelength shift based on the principles of spectrometry and interferometry. The interrogator sends broad-band light down to the optical fiber and accurately measures the wavelengths of the light reflected from the FBGs. In case of many FBGs printed into a single optical fiber, the interrogator must be able to determine which reflection comes from which FBG. There are principally two ways to accomplish this task:

- Wavelength division multiplexing (WDM)
- Time division multiplexing (TDM)

For TDM systems the individual FBGs are identified per the "time of flight" of a light pulse to travel from the interrogation unit, to the FBG and return. FBGs located at different distances along the optical

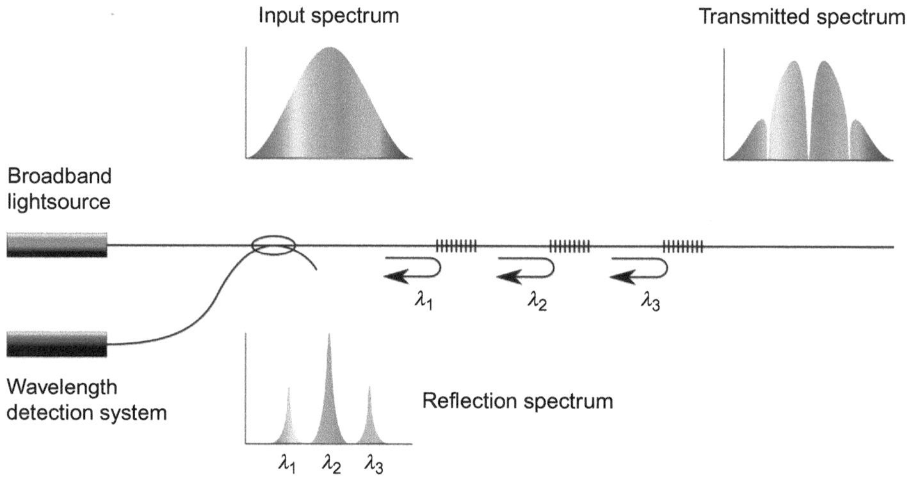

FIG. 1

Wavelength division multiplexing principle. The Deminsys FBG interrogator from a Technobis device is based on this principle.

fiber will therefore exhibit a different "time of flight." In WDM each FBG on a single optical fiber must be printed at a wavelength that is slightly different from the one of all its neighbors. Any operational strain or temperature change must not bring the FBG outside its allotted wavelength band so the interrogation system is clear from wrong identifications. Each of these approaches has its pros and cons. In reality, WDM systems are preferred for their advantages. WDM systems require uniquely identifiable FBG wavelengths. This constraint is used to limit the availability of sensors, but new FBG manufacturing basically allows every wavelength to be manufactured with significantly lower costs and moreover there is no limit to what can be achieved in terms of performance.

Based on the WDM principle there are many suitable approaches for analysis of the optical spectrum of the reflected FBG signals. For laboratory tasks, interferometers are often used. These systems are not so robust for very demanding environments like aerospace, industry, space, etc. For such applications other generic approaches are preferred:

- Edge filter interrogator;
- CCD based interrogator;
- Tunable laser based interrogator;
- Integrated photonics based interrogator.

The edge filter principle uses special optical filters with wavelength-dependent transparencies. In a differential configuration, the reflected light from the FBG is split into two equal beams and then fed through two different edge filters with opposite characteristics. The transparency of filter A increases with wavelength while the transparency of filter B decreases. A differential photo-diode behind the filters A and B generates an electrical signal proportional to the difference of light behind the two filters. A shift of the FBG wavelength therefore changes the amplitude of the output signal of the differential diodes. This s provides high bandwidth at low price. On the other hand, only one single FBG can be interrogated with an edge filter channel at a time. This enables only the value of the peak wavelength to be output but cannot provide information about a spectrum. Since it is an analog (level-depending) its accuracy and stability at lower and higher levels of scattered light can corrupt the results completely.

Another approach is represented by the charge-coupled device (CCD) based interrogator which is specifically suited for FBGs in the 850 nm range. A holographic grating reflects different wavelengths to different pixels of the CCD-line. The CCD technique can be implemented quickly (>20,000 M/s) and provides good accuracy and a good signal-to-noise ratio. It captures the whole spectrum and therefore can measure many FBG signals with different wavelengths simultaneously. Normally this approach is used in the 850-nm band because affordable CCDs are based on silicon technology and therefore work only in the near infrared light-band.

Within state of the art systems, the Deminsys series is the Technobis reference. This interrogator operates with four channels (fibers), each of which have eight sensors, 1 pm resolution, and a sampling speed of 20 kHz for all sensors simultaneously (Fig. 2).

The following image shows the concept of the spectrometry system in the Deminsys. As each FBG reflects a light signal with a different (center)wavelength, each signal is projected on the CCD at a different reflection angle (which is a typical behavior of a reflection grating). The projected signal on the CCD has a Gaussian shape form. Using a fast and accurate center of gravity (CoG) algorithm, any change in wavelength will be detected as Gaussian like CoG variation.

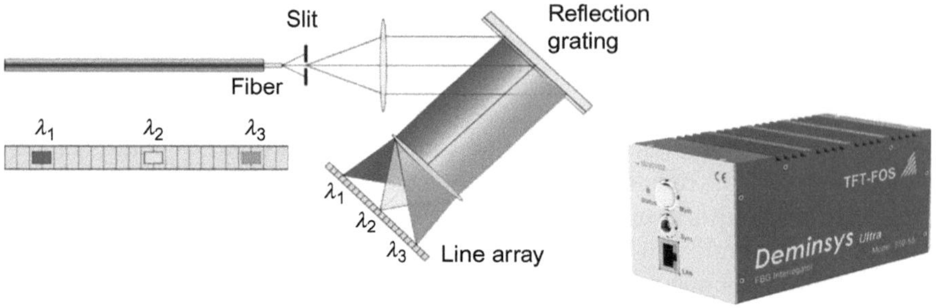

FIG. 2

Deminsys FBG interrogator system. Basically, the device is a spectrometer with the ability to detect wavelength shifts for 32 sensors simultaneously at 20 kHz.

For the C-band (1550 nm), interrogators with tunable lasers represent the best performing instruments. Other interrogator principles use broad-band light sources and therefore only a part of the light energy is related to the bandwidth of a FBG. Therefore, the reflected peak energy is low. A tunable laser concentrates all its energy in an extremely small bandwidth, and by sweeping over the whole bandwidth range, it scans the spectrum with high power and can provide very good signal-to-noise ratios.

Because of this high signal-to-noise ratio, FBGs with reflectivity down to $<1\%$ can still be measured with high precision. Because of the high energy budget the tuned laser output signal can be split into many output channels without losses in signal quality. Fig. 17 shows a four channel interrogator. Each channel can measure many Bragg sensors positioned on its fiber. Because one receiver per channel exists it is also possible to capture a full spectrum per channel. This is a big advantage because the available bandwidth results four times 80 nm = 320 nm. So FBGS with equal wavelengths can be measured if they are on different channels and the total number of connectable FBGS is four times higher in comparison to interrogator principles which work with only one receiving and signal conditioning unit.

In the last few years Next Generation Interrogator systems based on Integrated Photonics have been successfully developed by the company, Technobis. These systems also operate in the C-Band. Like electronics, Application Specific Photonic Integrated Circuits (ASPICs) allows many solutions, all based on a set of basic and advanced components and optical functionality. The interrogation principle of the ASPIC based interrogator series called Gator from Technobis and affiliated systems is spectrometry and additional interferometry for extreme resolutions.

The Gator ASPIC design consists of a modified spectrometer function called the Arrayed Waveguide Grating (AWG) and a photodiode at the output of each AWG channel (Fig. 3). The AWG is modified such that it operates as a spectrometer (patented technology), and in some cases extended with interferometry functionality, capable of performing extremely accurate resolution wavelength shift detections and at sampling speeds that are only defined by back-end electronics capabilities.

The development of ASPIC technology for aerospace applications has been mainly carried out in the framework of both national and international research programs. These programs combined specifically required performance aspects for different sensing applications: damage detection, impact detection, load monitoring, and shape sensing, including redundancy aspects and time division features.

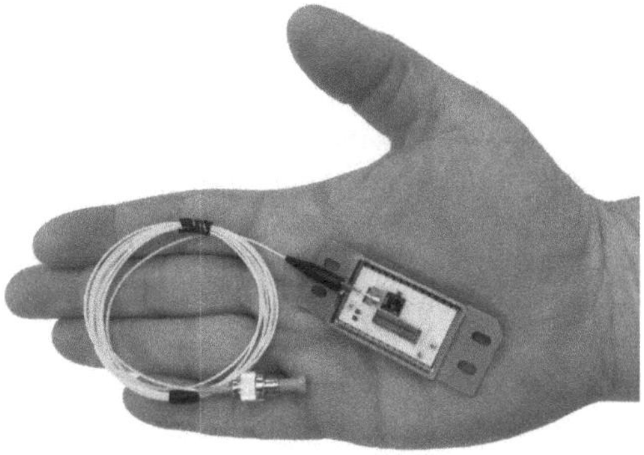

FIG. 3

The next generation FBG interrogation systems from Technobis, called Gator, based on integrated photonics.

2.4.2 Connectors

Most optical fiber connectors are based on the spring-load concept. That means that fiber faces are pressed together when the connectors are mated. The resulting glass-to-glass (or plastic-to-plastic) contact eliminates signal loss that would be caused by an air gap between the joined fibers. Every fiber connection experience two types of loss:

- Attenuation or insertion loss
- Reflection or return loss

Measurements of these parameters are defined in IEC standard 61753-1 [15]. The standard gives five grades for insertion loss from A (best) to D (worst), and M for multimode. The other parameter is return loss, with grades from 1 (best) to 5 (worst). Today >100 types of optical fiber connectors are available; SC and LC connectors are the most common types [16]. The main difference among types of connectors are dimensions and mechanical coupling methods. Generally, an organization will standardize on one kind of connector, depending on what equipment they commonly use. Different applications and environments require different performance aspects. In settings for harsh environments, protective enclosures are often used, and fall into two broad categories: hermetic (sealed) and free-breathing. Hermetic cases prevent entry of moisture and air but, lacking ventilation, can become hot if exposed to sunlight or other sources of heat. Free-breathing enclosures, on the other hand, allow ventilation, but can also admit moisture and airborne contaminants. Selection of the correct housing depends on the cable and connector type, the location, and environmental factors. Careful assembly is required to ensure good protection against the elements. Depending on user requirements, housings are tested under various environmental simulations, which could include physical shock and vibration, water spray, water immersion, dust, etc., to ensure the integrity of optical fiber connections and housing seals.

The aerospace environment addressed by this book suggests the use of high-performance fiber optic connectors. An appropriate connector selection is imperative to assure sufficient optical, environmental, and mechanical performance. The ability to choose the best interconnect system and subsequent connector depends upon knowing and understanding the optical performance expectations, operating conditions, mechanical and environmental constraints, reliability requirements, and maintenance plans. In general, optical fiber connectors for harsh and demanding environments can be organized into two categories: physical contact and expanded beam (EB).

The most common physical contact connectors are available in single terminus or multitermini varieties, circular or rectangular form factors, housing the fiber directly in singular or multicavity ferrules. Physical contact connectors usually offer exceptional optical performance, exemplary environmental performance and are available in dozens of form factors. In short, physical contact fiber optic connectors, when mated, provide the alignment and mechanism for two fiber optic elements (glass or plastic fiber) to touch and pass signal. Several alignment mechanisms are featured in most fiber optic connectors (Fig. 4).

The connector shell provides the mating mechanism and coarse connector alignment either through keyways or mating threads. The shell or housing for a multitermini connector usually holds and locates an insert and/or alignment sleeve holder. Finer alignment can be realized by locating pins contained in the insert. The insert typically captures the free floating individual terminus. Precise optical alignment between two mating termini is performed by alignment sleeves held captive within an alignment sleeve retainer or onto one of the mating termini.

Each terminus contains a ferrule that is usually ceramic (cubic zirconia). The optical fiber is epoxied into the ferrule and the end face polished. There is usually a spring on one or both termini to ensure adequate mating force between two termini. A cutaway section of an optical connector is shown in Fig. 5 and shows the connection between two fibers located in two mating termini.

FIG. 4

Fiber optic connector types.

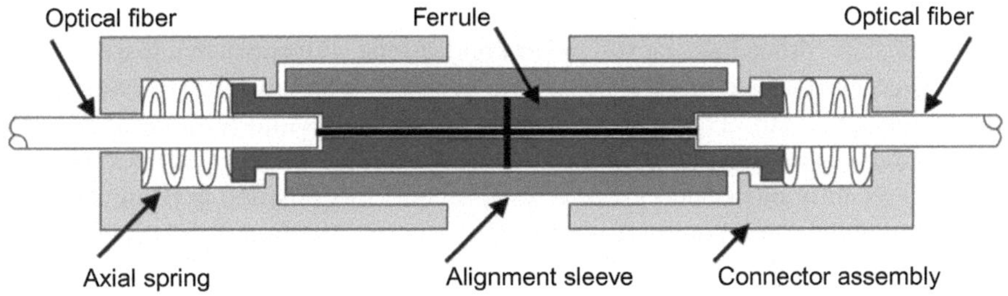

FIG. 5

Physical contact (FC).

Expanded beam [17] connectors typically use two lenses to expand, collimate, and then refocus the light from the transmitting fiber into the receiving fiber (Fig. 6). The lenses are generally either ball lenses or GRIN (graded index) rod lenses. The ball lens has a constant refractive index and is manufactured using commonly available glasses. The light from the source changes direction at the curved boundaries, and travel through the inside of the lens in straight lines. The GRIN lens has a cylindrical form factor, but the refractive index is not constant. Rather, the refractive index distribution varies radially with a parabolic profile, with the maximum index of refraction along the axis of the lens. Unlike the ball lens, the light curves through the inside of the GRIN lens. Light collimation is obtained by placing a point light source at the focal point of one of the two lenses.

The use of EB optic interfaces result in reduced signal loss due to contamination at the optic interface. The lens design also facilitates cleaning and because the light path travels over an air-gap, there is no physical contact. Lack of physical contact potentially eliminates the mechanical wear found in physical contact connectors, allowing more connector mating cycles. Fig. 6 shows a typical ball lens EB fiber optic connection system.

Like physical contact connectors, EB connectors typically use precision ceramic ferrules to align the fiber to the lens. As with PC connections, tight tolerance ceramic ferrules are used in conjunction with high quality termination and polishing procedures. The connector shell, or insert, which aligns the ferrule to the lens, becomes the critical subcomponent in enabling a low-loss EB connection. The achievement of the tight dimensions necessary for a quality EB insert is required in order keep

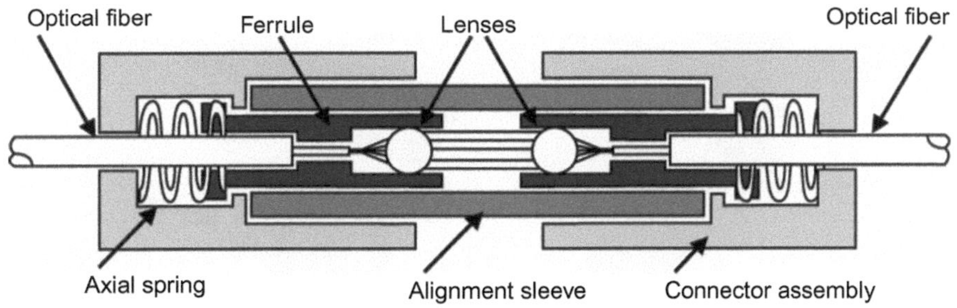

FIG. 6

Expanded beam (EB).

information losses low. The combination of precision subcomponents and ceramics with the proper termination and polishing procedures can yield EB connections with consistent losses values <1.0 dB.

In general, EB connectors are more expensive to produce, which has limited their use in telecom and data-com applications. However, EB technology is used in several military and commercial applications where frequent mating and unmating may expose the optical interfaces to contamination. Several companies recently introduced lower cost EB connectors targeted at commercial medical applications, where reliability and thousands of mating cycles are required. In harsh environment applications, standards have been developed utilizing both types of solutions (Fig. 7). End users will demand adherence to these standards. The engineer must know the difference between the two types of connectors. Successful connector designs can minimize the aforementioned weaknesses, but differences between PC and EB solutions remain.

2.4.3 Splicers

Fusion splicing [18] is the act of joining two optical fibers end-to-end using heat. The goal is to fuse the two fibers together in such a way that light passing through the fibers is not scattered or reflected by the splice, and so that the splice and the region surrounding it are almost as strong as the virgin fiber itself. The source of heat is usually an electric arc, but can also be a laser, or a gas flame, or a tungsten filament through which current is passed. The process of fusion splicing normally involves the adoption of localized heat to melt or fuse the ends of two optical fibers together. The splicing process begins by preparing each fiber end for fusion.

Stripping is the act of removing the protective polymer coating around the optical fiber in preparation for fusion splicing. The splicing process begins by preparing both fiber ends for fusion, which requires that all protective coating is removed or stripped from the ends of each fiber. Fiber optical stripping is usually carried out by a special stripping and preparation unit that uses hot sulfuric acid or a controlled flow of hot air to remove the coating. While like copper wire strippers, those for fiber

FIG. 7

Expanded beam (EB) multichannel connectors.

FIG. 8

Fiber splicing procedure.

optics are more precise: both require training and skill. Fiber optical stripping and preparation equipment used in fusion splicing is commercially available through a small number of specialized companies, which usually also design machines used for fiber optical recoating.

The customary method to clean bare fibers is based on the use of alcohol and wipes. However, high purity isopropyl alcohol (IPA) is hygroscopic: it attracts moisture to itself. This is problematic as IPA is either procured in presaturated wiper format or in (host) containers ranging for USA quart to gallon to drums. From the host container, the IPA is transferred to smaller, more usable containers. The hydroscopic nature of IPA is such that the highest quality at 99.9% is also the most hygroscopic. This means that moisture absorption into both the host container as well as the actual user's container begins with the time the original container is opened and continues as amounts are transferred and removed from both.

The fiber is then cleaved using the score-and-break method so that its end face is perfectly flat and perpendicular to the axis of the fiber. The quality of each fiber end is inspected using a microscope. In fusion splicing, splice loss is a direct function of the angles and quality of the two fiber-end faces. The closer to 90 degree the cleave angle is the lower optical loss the splice will yield.

Current fusion splicers are either core or cladding alignment. The two cleaved fibers are automatically aligned by the fusion splicer and then are fused together by using one of these methods (Fig. 8). Prior to removing the spliced fiber from the fusion splicer, a proof-test is performed to ensure that the splice is strong enough to survive handling, packaging, and extended use. The bare fiber area is protected either by recoating or with a splice protector. A splice protector is a heat shrinkable tube with a strength membrane.

3 SENSOR SYSTEMS FOR LARGE SCALE INTEGRATION

Future aircraft will possess adaptive capabilities, from supervised morphing to active deployable subcomponents, and self-sensing properties for auto-inspection. Deployable and adaptive structures are also attractive for space missions, envisaging a better matching of the different mission phases and an easier storage and release of some essential components, like antennas and solar cells [19]. All these functionalities are promising but challenging and a crucial role is played by integrated sensing networks (for real-time monitoring and control aims), that should be widely distributed all over the structure, implying a huge number of terminals and an impressive quantity of cables. Structural systems are in fact characterized by infinite degrees of freedom, in principle, excited over a wide frequency range. Furthermore, morphing and deployability also implies the occurrence of very large deformations.

When speaking of morphable structures, it is natural to think of deployable architectures (portable radars and antennas, military shields) or inflatable structures (airbags, seals), generally characterized by large displacements and rotations and demanding for highly flexible structural components. Information on both deployment status and structural integrity of the most critical components could be acquired through this type of sensor. Specific materials, able to recover large deformations, like foams, silicones and polymeric shape memory alloys, may also be monitored; the challenge of following and withstanding the large deformations of the structure is accompanied by the adhesion to the external surface and by extreme environmental conditions in space. Integration, cable routing, information management, and monitoring wide strain and temperature ranges, are critical tasks in this framework. Morphing or active deployable functionalities seem highly dependent on the availability of an extended sensor network on the basis of consolidated knowledge from several industrial research projects. The use of available sensors can originate serious limitations in the field of practical operability, like:

- limited strain values, noncompliant with super-deformable materials (fiber optics or even typical graphene-based sensors);
- system integrability;
- impossibility of a reliable bonding on silicones or foams;
- electrical wires integrability and related electromagnetic interferences;
- cable routing complexity.

Such limitations can be overcome relying upon the intrinsic nature of new generation sensor systems and related technologies:

- The possibility of laterally spraying sensors is easy and independent on geometric shapes (f.i. high curvatures) and surface extension; direct deposition is namely independent on the target material, overcoming typical problems related to the bonding on silicones or foams;
- The possibility of wireless communication and data transfer makes the installation and integration of sensors in complex systems (full aircraft) simple to manage as cable routing layout and substitution for maintenance does not involve anymore structure unpacking.
- The possibility of continuous monitoring at sub-cm resolution of the inner material structure status (composite and laminated) withstanding high thermal gradients is yet another nonconventional specification aspect.

In what follows, some technologies that can make these things possible are briefly described.

3.1 WIRELESS TECHNOLOGY

Wireless environments represent the next evolutionary development step in building, utilities, industrial, home, shipboard, and transportation systems automation. Regarding the physical placement of the devices, it is appropriate to recall that one of the advantages of wireless networks lies in the degree of simplicity with which the sensor nodes can be placed. A Wireless Sensors Network (WSN) consists of densely distributed cordless nodes that support embedded sensing, and signal processing [12,20]. Due to high density, the nodes are very close to each other and this results in achieving multihop network algorithms to reach the correct addressee of the information, but, at the same time, problems of mutual interference between separate sensors must be managed. The use of multihop strategies may allow for the use of low transmission power, improving the characteristics of the nodes in terms of power requirements; this latter turns out to be one of the most stringent requirements since the sensors use power sources that cannot be in general replaced, or at least cannot be replaced frequently. This is why an efficient implementation of wireless sensor networking must provide mechanisms that give the user the possibility to choose the best compromise between performance and battery life. In general, within the sensor field, WSNs employ contention oriented random access channel sharing and transmission techniques that are now incorporated in the IEEE 802 standards [21].

3.2 SPRAYED TECHNOLOGY

In recent years, thermal spray technologies have evolved from crude processes that were relatively difficult to control, into increasingly precise tools where the process is tailored to take into account the properties of both the deposited material and the required coatings. Cold spray is an all solid-state process, making it suitable for depositing oxygen-sensitive materials such as aluminum, copper and titanium or temperature-sensitive materials such as nanostructured and amorphous powders. Cold spray presents other characteristics that offer unique advantages compared to existing spray technology:

- the coating can exhibit wrought-like microstructures with near theoretical density values;
- the spray trace is small (typically 1–25 mm^2) and well defined, allowing for precise control on the area of deposition;
- the coating can be produced with compressive stresses, thus ultra thick (5–50 μm) coatings can be built-up without adhesion failure;
- coatings can be deposited on glass or polymers.

This process consists of applying coatings by exposing a metallic or dielectric substrate to a high velocity (300–1200 m/s) jet of small (1–50 μm) particles, accelerated by a supersonic jet of compressed gas at a temperature lower than the melting point of the material (Fig. 9) [22]. Low temperature allows for minimizing or eliminating some deleterious effects related to high-temperature processes such as oxidation, evaporation, melting, crystallization, residual stresses, deboning, and gas release.

Cold-sprayed coatings result in a very low number of defects and low oxygen content; therefore, they exhibit bulk-like properties for electrical or thermal conductivity that are not attainable by usual processes [23,24].

Cold spray technology can be used in strategic industries such as defense and aerospace as well as in various other general industries such as steel, utilities, paper, and so on; high performance materials such as super-alloys and nanostructured materials can benefit with such a process.

3.3 DISTRIBUTED TECHNOLOGY

Sensing system solutions, enabling truly continuous distributed measurement, may be implemented by using backscattered light along a fiber optic. With a spatial resolution of 5 mm or less, it is possible to know in real time the strain field throughout a structure and to monitor its evolution during this time. This unique technology enables strain or temperature measurements at all points along a simple optical fiber with gauge lengths as small as a few millimeters. One essential part of an optical sensor system making distributed measurements is a reflectometer to provide spatial resolution. For application cases involving long distances (in the order of tens of km), Raman and Brillouin systems [25] are generally

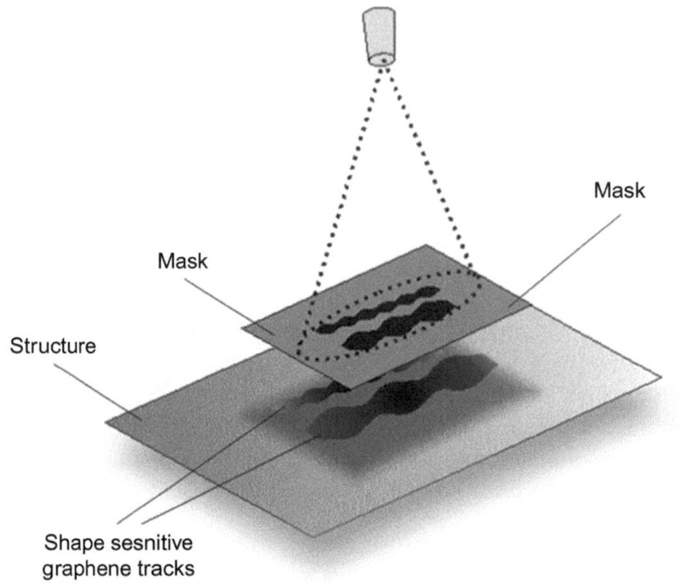

FIG. 9

Spray deposition technique for graphene based sensor.

used. They work with an optical time domain reflectometer, in which the monitoring unit transmits a short light pulse and uses the time of flight of the back scattered light to determine the location of the reflection. The Rayleigh-scattering based sensor requires resolution significantly higher than the one that can be obtained referring to a coherent optical frequency domain reflectometer [26]. On the other hand, quasidistributed monitoring systems can be realized using the draw tower grating technique [27], offering spliceless FBG chains embedded at any position along very long or customer-specific lengths of even ultrathin (50 μm) cladding diameter fibers. Thinner fibers are less invasive when embedded within laminate composite structures or textiles; and are more "elastic" in respect to bending properties for improved use in applications where small bending radii are required. In addition, using the Femto second laser technology, FBGs can as well be written into any fiber even with complete coating [28]. Due to this printing process, it is possible to write a huge number of different gratings into the same fiber. This process provides the highest stability gratings for temperature and strain sensing applications.

3.4 SOME INSTALLATION ISSUES

When structural sensors are taken into consideration, strain gauge featuring is immediately evident: simple construction with a small mass and volume; good static response; and electrical output for easy data processing. When using strain gauges, the full bridge circuit can avoid the continuous compensation of additional interference effects such as temperature variations. It also makes strain gauge sensors more suitable for long-term measurements of stability and linearity.

Moreover, the cost of the strain gauge itself is not ordinarily a primary concern for gauge selection, since the most relevant economic measure is represented by the total cost of the complete installation, while the gauge cost is just a small fraction of it.

On the other hand, the basis of the piezoelectric effect is that crystals under compressive loading generate an electric charge that is directly proportional to the applied force. This charge is then converted to a proportional output voltage with the aid of an amplifier. The high natural frequencies of piezoelectric sensors makes them ideal for dynamic applications. The sensitivity of a piezoelectric sensor does not usually depend on its size or on the volume of quartz, but on the material being used and its geometry. By using piezoelectrics, it is possible to implement a measured range larger than the one related to strain gauges, without loss of accuracy or resolution.

In terms of highly integrated structural monitoring, the extraordinary versatility of certain fiber optic sensors is unrivalled by conventional measurement technology (Fig. 10).

The extreme sensitivity of fiber optic strain sensors has been demonstrated with nanostrain resolution [29], while the bandwidth potential has been indicated by the megahertz frequency response [30]. Optical sensors permit measurements that are either impractical or uneconomic with conventional measurement technology, such as resistive foil strain gauges or piezoelectric. Fiber optics are extremely small in diameter, very light, resistant to corrosion and fatigue. They are dielectric by nature so they are safe to handle, unlikely to initiate fires or explosions, immune to electrical interference, and do not represent electrical pathways within a host structure. A single fiber optic may contain a string of structural sensors that can undertake distributed strain measurements and, as a matter of fact, replace the equivalent number of foil strain gauges or piezoelectric, each with several welding electrical leads.

In the monitoring example, piezoresistive graphene based sensors have attracted much research due to their potentials as next generation strain sensors. Piezoresistive sensors can detect very small strains with very high sensitivity, but a polymeric matrix would ensure the applicability to very large deformable systems if properly designed. A recent study [22] demonstrated the feasibility of a high sensitive piezoresistive sensor, based on graphene. The obtained experimental results demonstrated that the adopted manufacturing process permits achieving a sensor network with repeatable qualities and a high constant gauge factor (~200) in the investigated strain range 0.001%–0.02%. Furthermore, even if cable wires are needed, a spray coating system was developed to produce such films which makes this kind of sensor very simple to install. A graphene-based sensor would permit an easier installation (painting-like spray deposition) and in perspective, the implementation in a wireless network [22].

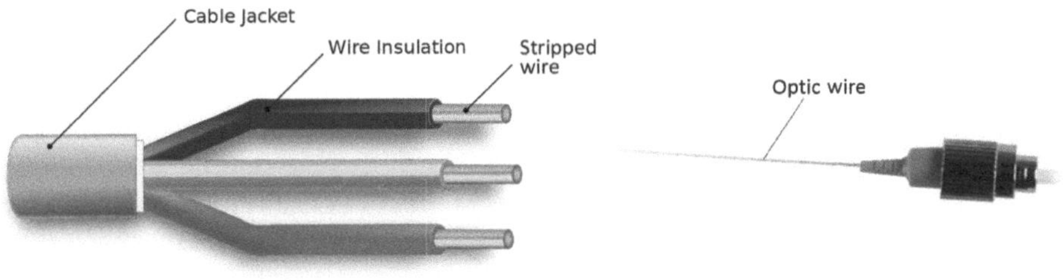

FIG. 10

Strain sensors cabling: strain gauges (left), fiber optic (right).

4 CASE STUDIES

Two meaningful applications are now described: SARISTU [31] and CRIAQ-MDO505 [32] projects.

SARISTU (Smart Intelligent Aircraft Structures) is a level 2, large-scale integrating project aimed at achieving reductions in aircraft weight and operational costs, as well as an improvement in the flight profile specific aerodynamic performance. The project focused on integration activities in three distinct technological areas: airfoil conformal morphing [33,34], self-sensing, and multifunctional structures using nano-reinforced resins. Coordinated by Airbus, the SARISTU Consortium brought together 64 partners from 16 countries. The total budget of the project was 51 M€, partially funded by the European Commission under FP7-AAT-2011-RTD-1 (Grant Agreement Number 284562).

The CRIAQ MDO505 project was launched to foster scientific cooperation between Italian and Canadian academies and research centers (CIRA, Unina, ETS, NRC), and leading industries (Alenia Aermacchi, Bombardier). The main project goal was to design a high technology readiness level Morphing Wing for a Bombardier-type regional aircraft [35]. The test article is composed of a morphing wing box and a camber morphing aileron. The combined effect of these two adaptive devices is devoted to improve aerodynamic performance by off-design drag reduction.

4.1 SHAPE RECONSTRUCTION OF A VARIABLE CAMBER WING TRAILING EDGE

In the SARISTU project a fiber optic sensing [36] based approach for chord wise shape reconstruction of an adaptive trailing edge device (ATED) (Fig. 11) is realized with the extrinsic implementation of FBG sensors.

Thanks to this implementation, the capability of closed loop control is provided for the morphing mechanism at a given set of the target shapes. The following paragraphs give an overall description about the development and implementation of the fiber optic sensing system as well as the applied methodology for shape reconstruction.

Based on a mechanical device, integrated with an optical fiber transducer, the shape information is translated into strain information and properly modulated. By specific data processing mathematics, these strain values are converted back into an approximation of the produced shape.

The shape sensor assembly consist of multiple rapid prototyped parts: four support structures and the sensor beam itself. The dimensions of the sensor beam are 300 × 10 × 3 mm; the sensor is characterized by 0.3 mm thick groove running along the upper and lower side of the beam, and the groove integrates the optical fiber. The red support structures are bonded to the aluminum parts of the skin structure by means of appropriate adhesives. The first Support Structure fixes the beam while the others allow the beam to slide through freely as the ATED morphs up and down (Fig. 12) [37].

The shape sensor assembly is based on the principle of noncontact shape monitoring. This means that the optical fiber is mechanically not directly connected to the object of measurement, the ATED in this case. The sensor assembly is mechanically designed such that the sensing beam is following the deflection freely through the guiding support structures as the ATED morphs through different shapes.

Referring to the interpolated strain distribution, it is then possible to recover the shape of the sensing strip (Fig. 13). The Frenet-Serret formulas [38] is used to achieve this task. Differential geometry principles establish that a "shape" can be defined as a group of scalar and vector quantities along a curvilinear abscissa (s): the curvature $\kappa(s)$, torsion $\tau(s)$, tangent vector $T(s)$, normal vector $N(s)$, binormal vector $B(s)$, and position vector $r(s)$. The Frenet-Serret formulas tell us that the vector

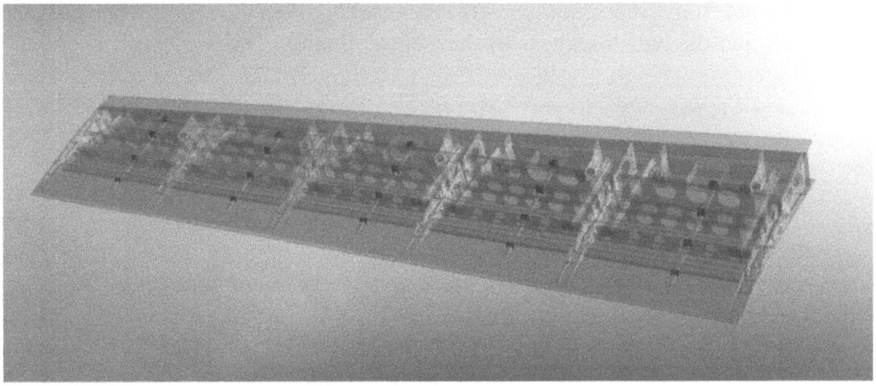

FIG. 11

Adaptive trailing edge device (ATED), SARISTU project.

FIG. 12

Fiber optic shape sensor assembly, SARISTU-AS02/04.

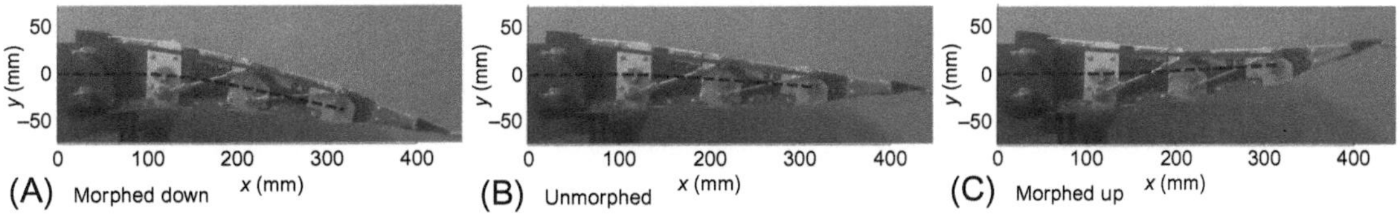

FIG. 13

Shape sensing experiments at Technobis facilities. Specifically, for this experiment a rudimentary morphing structure was constructed, based on the ATED concept. Different ATED configurations are provided: (A) morphed down, (B) unmorphed baseline, and C) morphed up.

quantities can be derived from the scalar ones. This means that given the relation between the strain at a certain location $\varepsilon(s)$, and the half-thickness of the Sensor Beam $y_{\max}$:

$$k(s) = \frac{\varepsilon(s)}{y_{\max}} \tag{8}$$

the Frenet-Serret formulas can be applied as follows:

$$\frac{d}{ds}\vec{T}(s) = k(s)\vec{N}(s)\frac{d}{ds}\vec{N}(s) = -k(s)\vec{T}(s) + \tau(s)\vec{B}(s)\frac{d}{ds}\vec{B}(s) = -\tau(s)\vec{N}(s) \tag{9}$$

With the initial values of $T(0)$, $N(0)$, and $B(0)$, the shape can be reconstructed by iteration where the position vector $r(s)$ of the sensor beam is computed by:

$$\vec{r}(s) = \int_0^s \vec{T}(s)\cdot ds + \vec{r}(0) \tag{10}$$

Preliminary experiments were performed with a 300 mm cantilever system integrating optical fibers and FBG sensors. In correspondence of 15 mm repeated deflection at the tip of the cantilever, the test results show a root mean square value of <0.5 mm for the position accuracy at the tip, i.e., an error margin of <0.2%. The next experiment regarded the sensor beam in an extended cantilever system like the ATED structure. Changing tip deflection from −3 to 3 mm (Fig. 14), the reconstruction

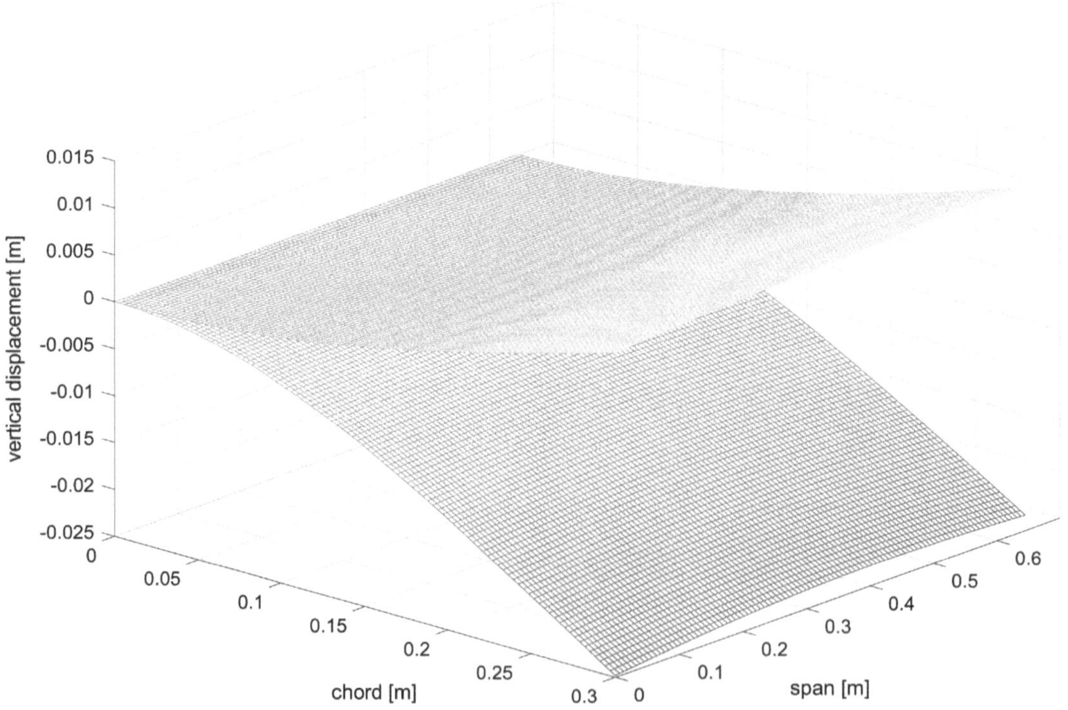

FIG. 14

Visualization of the shape measurements performed on the 2-bay ATED demonstrator at Aernnova facilities in Vitoria, Spain.

analysis demonstrated the average reconstruction error to be <4.3% in full compliance with design requirements.

In general sensing technology based on optical fibers is facing a bright future. The demonstrated studies and implementation models during and prior SARISTU progress, validated the feasibility for sensing applications. The mathematical model for the interpretation of strain and conversion into shape information has proven more than sufficient.

The ATED shape sensor assemblies have been implemented in the 2-bay ATED dummy demonstrator and finally in the 5-bay full scale demonstrator and tested at TsAGI wind tunnel facilities in Moscow [39].

4.2 DAMAGE AND LOAD MONITORING

To perform damage and load assessments on composites structures, fiber optic sensors were implemented through a reliable and repairable sensor system. For the integration of the FBG sensors, a concept called Fiber Optic Ribbon Tape (FORT) has been developed to integrate the sensors in a tape with a connector casing protection the fiber connectors during processing. The goal of the ribbon tape development [40] was to design a supportive carrier for the optical fiber, which can be cocured (initially secondary bonded) to the product to be monitored. In case the ribbon tape is damaged, it can easily be removed and replaced (Fig. 15).

The ribbon tape is a complex component from the optical fiber point of view. A tube-protected optical fiber enters the composite ribbon and then is assembled into the connector case (Fig. 16).

In the process of the development of the ribbon tape there was the need to assess the optical integrity of the optical fiber after embedding and locating points of excessive loss or even complete failures.

Two solutions for sensor installation have been discussed, cobonding during the autoclave process and secondary bonding using adhesive. Through extensive fatigue testing it is shown that FORTs can

FIG. 15

The repair principle (easily removing) of the ribbon tape is integrated in the new design of the connector casing. This is done by making an initiation point to peel-off the ribbon tape.

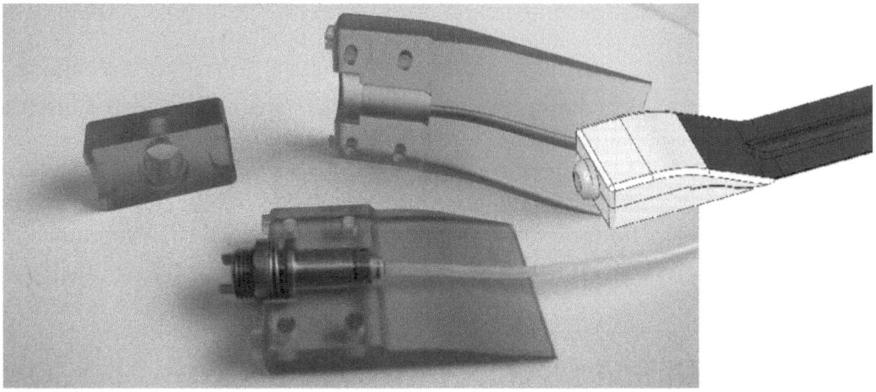

FIG. 16

Machined connector casing, an integral part of the ribbon tape sensor assembly.

FIG. 17

ATED ribbon tapes.

be reliably used as they provide accurate strain measurements over an extended operating spectrum of 106 duty cycles. In addition, the FORT measurements are more consistent as compared to the electrical strain gauges that detach from the surface of the structure after a certain number of cycles. Furthermore, FORTs offer all the advantages of FBG sensors plus they are readily replaceable in case of sensor failure or for maintenance issues. Also, fatigue testing at high and low temperatures provide acceptable responses of FORT strain measurement systems, demonstrating its proper working under combined mechanical and thermal loading (Fig. 17).

4.3 ROTATION ANGLE MONITORING

This paragraph describes the innovative research conducted on integrated sensors network specifically suited for adaptive structures. In light of these considerations, the proposed technology is an innovative Fiber optic based sensor transducer aiming at adaptively monitoring and reconstructing, afterwards, the actual shape of a morphing device. The design was conducted on a structural rib with numerical simulations conducted at progressively increasing levels of detail. A direct strain measurement is always difficult in a shape-changing structure due to the large deformations induced by morphing. The design of an innovative system which indirectly measures the strain is an appropriate solution for this issue. The transducer implements a relation between the target strain measurement and appropriate physical parameters such as rotation, deflection, and so on. Such devices are constituted of a thin simple supported postbuckled beam that is anchored to adjacent bodies of the rib by means of hinges [41]. The angular position of the FBG sensor along the beam neutral axes govern the ratio between the rib pivot rotation angle and the measured deformation, that was then arbitrarily set in a very large range of values per the beam local curvature. Fibers were simply bonded along the beam and the FBG position was arbitrarily defined in order to measure different strain amplitudes. Indeed, the supporting beam acts as a strain modulator for the FBG, whose range of measures was tailored on its position. The modulation property was set in the preliminary design phase per the working condition of the structure. The mathematical background beyond the architecture was derived by the postbuckling Timoshenko theory [42] relating design parameters (e.g., cross-section plate geometry, thickness, and material) to the achieved displacement/deformation ratio; then, the plate displacement field was related to a rotation angle of the structural parts around a pivot. The rib segmentation (blocks B1, B2, and B3) is shown in Fig. 18, where the FBG-PB transducer is integrated onto a finger-like rib mechanism.

A peculiarity of the proposed device is its ability to work in postbuckling configurations, with the advantages of an easy, plug and play installation: in fact, the supporting beam can manually be bent and plugged within its connection hinges; and a tuning effect of the sensitivity/range of measure, based on the FBG location onto the beam and of the initial postbuckling level. The optimal structural

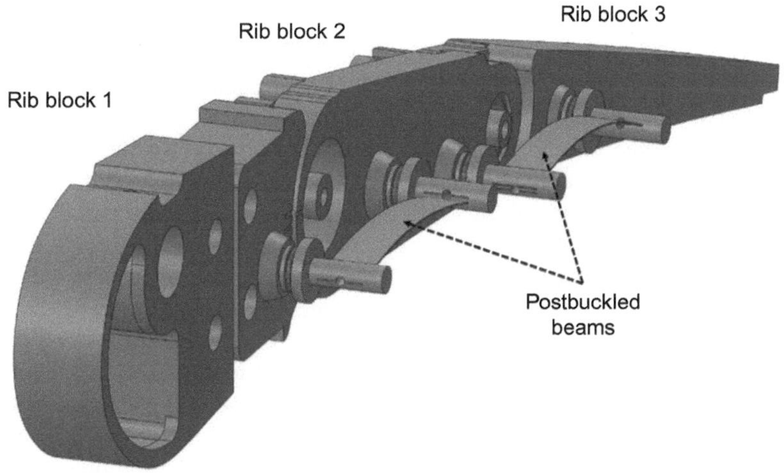

FIG. 18

Aileron rib integrated with the strain/angle transducer.

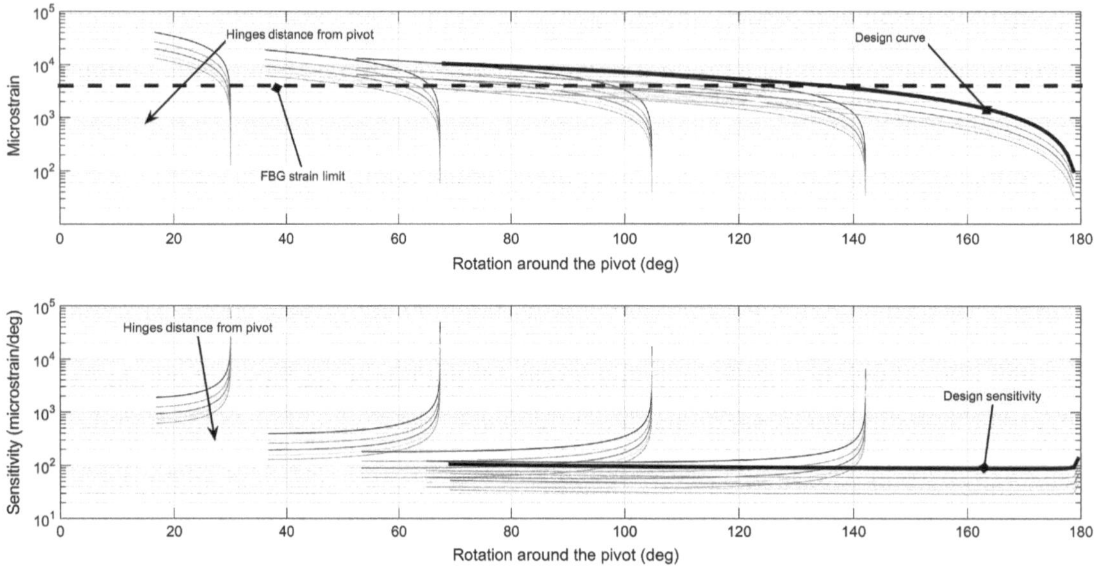

FIG. 19

Multiparameter diagrams for the FBG-PB sensor.

configuration of the sensor has been identified on the basis of several criteria such as sensitivity maximization, wide measurement range, and integration. A trade-off among these aspects was assessed. The bold lines in the following parametric graph (Fig. 19) represent the functionality curves that best fit the design specifications and structural constraints for the sensor. The curves are collected in different families, each one corresponding to a specific initial value of curvature; thus, e.g., the second family of curves moving from the left side corresponds to an initial angle of about 68 degree. If we increase the distance from the hinge at the central pivot, the curves shift downward as indicated by the arrow.

The measurement range (>18 degree) widely covers the global deflection assumed for the aileron (±7 degree), for wing shape optimization, before arriving to the FBG allowable limit which clearly occur at the intersection between the bold and dashed lines. In addition, a sensitivity of 100 μstrain/degree is obtained. The other families of curves are on the left, even though characterized by higher values of sensitivities (see the bottom part of the graph), were not considered because of the lower measurement range. The conceptual idea of the sensor was studied by means of numerical simulation and then validated through experimental tests. The main results are now discussed. The analyses were carried out on the basis of several finite element models with increasing levels of detail.

Firstly, the sensor supporting beam was considered isolated with pinned boundary conditions to simulate the effective mounting constraint; a displacement was imposed to the beam edge node and the obtained strains were compared with the Timoshenko approach, showing a satisfactory agreement. This trend is shown in Fig. 20, where a qualitative and schematic graphical representation is shown. The detailed graph has been discussed in Ref. [43]. To improve the level of accuracy of the nonlinear simulation, the rib deformations out of its plane were constrained since the aileron structure is stiffened by longitudinal spars and stringers preventing any rib displacement along aileron span. It is worthwhile to note that the strain obtained with this constrain condition is higher than the one occurring in case of unconstrained out-of-plane rib displacements. This is due to the ability of the unconstrained layout of

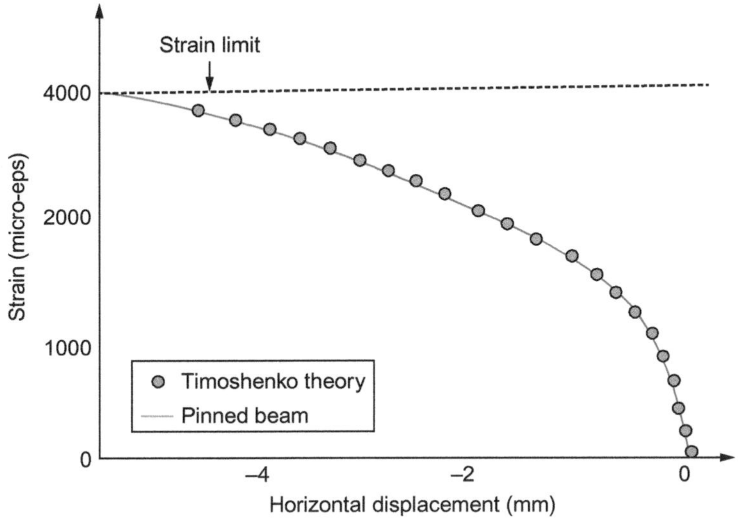

FIG. 20

Analytic and numeric model.

absorbing deformation and distributing it not only onto the sensor supporting plate, but also onto the rib structure. The experimental set-up consisted of two steel plates interconnected by means of bolts which allowed for relative rotation of the components (Fig. 21). The sensorized beam is a harmonic steel plate with 0.3 mm of thickness. To correlate measures, two sensors were mounted on the beam. A strain gauge and an FBG in colocated configuration were considered.

The signal acquired by the two sensors and the angle between the two plates are shown in the drawing of Fig. 22. The sensitivity of FO and strain gauge sensors is represented and a peak close to the un-morphed configuration (0 degree) is detected; this result is coherent with the buckling start of the sensor supporting plate, that amplifies displacements and deformations. As the deflection increases, the sensors (here intended as FBG or strain gauge integrated onto the plate) become less sensitive: a minimum of about 27.5 μstrain/degree for the strain gauge is achieved at the widest excursion of interest (15 degree).

Finally, a comparison of the strain/rotation curves obtained through the numerical models and experiments is shown in Fig. 23. Also in this case, it has been presented with a graphical representation of the curves while the authors refer the detailed graph in Ref. [43]. Rigid rib means that the effect of the span-wise stiffeners were considered, flexible rib labels was used otherwise.

The difference between the numerical curves is essentially due to the constraint condition of the sensor as previously explained. On the other hand, the difference between the strain gauge and the fiber optic measurements (a maximum of 90 μstrain), can be attributed to the nonperfect coincidence along the beam length of the two sensors. This deviation was estimated by means of the mathematical model. In fact, it was used to estimate the deformation at the beam center line in correspondence of a point shifted along the beam length of 5 mm, simulating noncoincidence between the installation points of the two sensors. Compared to conventional rotational sensors, which may namely be embedded into the morphing ribs kinematics, this concept allows a more distributed assessment of the wing camber variation by taking advantage of a supporting structure hosting FBG sensors. Further steps of this work

FIG. 21

Detail of the experimental setup.

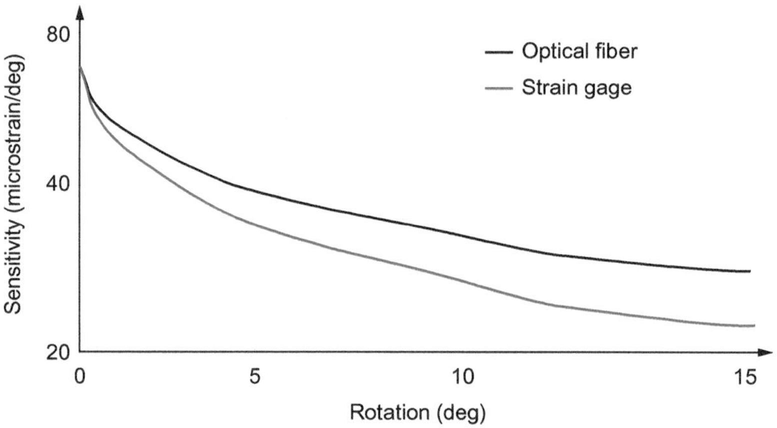

FIG. 22

Sensor sensitivity [43].

involve some sensor-related topics that are not covered in the presented research, such as sensor reliability and fly-ability. Fatigue life of the optimized beam material and fiber coating and adhesive interface will also be studied. This will allow the predicted morphing aileron conditions with the actual rib shapes, and, hence, to enable a more precise assessment of its enhanced aerodynamic performance and benefits.

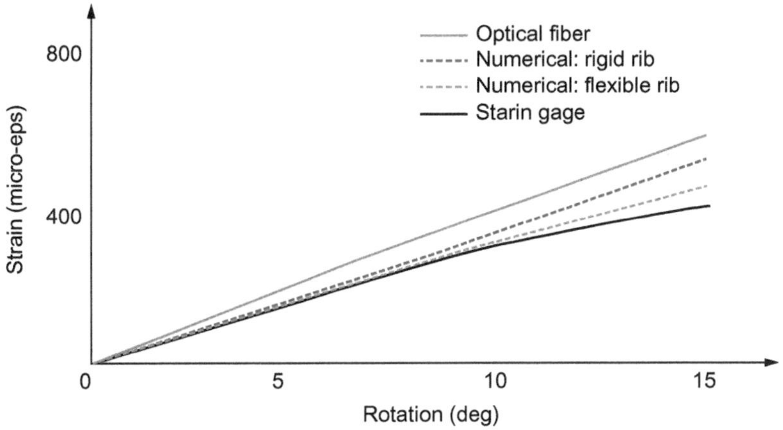

FIG. 23

Comparison between experimental measurements and numerical analysis.

5 CONCLUSIONS AND PERSPECTIVES

Large scale monitoring is the challenging issue to deal with. The basic components of structural sensing include sensor network, software, and hardware. The distributed capability of fiber optic sensors, can be considered the core technology of the 21st century smart structures. Nevertheless, wireless data transfer and spray deposition technologies can be considered a way to make large scale sensor smart integration possible. Indeed, to implement the sensing technology into structures, it becomes an integrated multifunctional structural design issue, not simply a sensor installation problem, which involves reliability and durability, quantification, validation, and certification. Traditional design methodologies may have to be modified and updated to accommodate the new technology from design to implementation. New design methodology suitable for the integrated multifunctional structures may also need to be developed. The material properties, dimension, configuration, and operational environment of structures may affect significantly how a structural sensing system shall be designed and implemented to achieve the objectives which involve optimal sensor/actuator placement, number of sensors and actuators, appropriate algorithms for maximizing the system performance, and hardware design, etc. Accordingly, key challenges in implementing the structural sensing technology shall include:

(1) *Reliability and durability*: The integrated sensing system must be reliable and durable. Not only sensors themselves, but the network interface between structures and the sensors must be able to sustain and remain functioning in the environments that the structure is expose to. Critical needs are durable/reliable sensors, advanced interface materials, and embedding/integrated manufacturing techniques.

(2) *Quantification*: The sensing capability must be quantifiable for its targeted level of application. Relating sensor measurements to physical changes in structures with a high capability of detection is still a major challenge.

(3) *Validation*: Adequate experimental and numerical methods are required to validate the system reliability and capability of detection. Traditional Non-Destructive Evaluation (NDE) testing techniques for laboratory applications may not be adequate for evaluating the integrated multifunctional structures in the field of large scale components. Improved or new test methods, techniques, and equipment may need to be developed.

(4) *Certification*: Existing certification procedures may not be adequate to accommodate the new design methodology and validation techniques for highly integrated structures. Therefore, existing procedures may need to be modified or new procedures might need to be developed.

Sensing is the most basic and fundamental element in the design of "intelligent" structures with multi-functional capabilities. With smart sensing, structures can sense, monitor, diagnose, and detect their health and physical conditions in any given environment and can respond effectively to optimize their performance, enhance safety, and minimize operation cost. A design with the structural sensing capability may result in major paradigm changes on how we will design structures in the future.

REFERENCES

[1] http://www.aviationtoday.com/am/repairstations/Understanding-MSG.
[2] http://www.economist.com/news/technology-quarterly/21651920-how-technology-changing-passenger-cabin-whatever-class-you-fly-flying.
[3] T.A. Weisshaar, Morphing aircraft technology – new shapes for aircraft design. In multifunctional structures/integration of sensors and antennas, Meeting Proceedings RTO-MP-AVT-141, Overview 1, Neuilly-sur-Seine France, 2006 RTO, pp. O1-1–O1-20, Available from: http://www.rto.nato.int/abstracts.asp.
[4] R. Measures, Structural Monitoring With Fibre Optic Technology, Academic Press, Elsevier, Toronto, Canada, 2001.
[5] C. Bockenheimer, H. Speckmann, Validation, verification and implementation of SHM at Airbus, in: Proceedings of the 9th International Workshop on Structural Health Monitoring (IWSHM 2013), 10–12 September, Stanford University, Stanford, CA, 2013.
[6] R. Pecora, F. Amoroso, M. Magnifico, Toward the bi-modal camber morphing of large aircraft wing flaps: the CleanSky experience, in: Proc. SPIE—The International Society for Optical Engineering, Volume 9801, 2016 Article number 980106.
[7] S. Barbarino, R. Pecora, L. Lecce, A. Concilio, S. Ameduri, Airfoil morphing architecture based on shape memory alloys, in: Proc. ASME Conference on Smart Materials, Adaptive Structures and Intelligent Systems, SMASIS2008, Volume 1, 2008, pp. 729–737.
[8] R. Pecora, F. Amoroso, M. Magnifico, I. Dimino, A. Concilio, KRISTINA: kinematic rib-based structural system for innovative adaptive trailing edge, in: Proc. SPIE—The International Society for Optical Engineering, Volume 9801, 2016 Article number 980107.
[9] J. Fraden, Handbook of Modern Sensors: Physics, Designs, and Applications, Springer-Verlag, New York, 2010,.
[10] http://elektron.pol.lublin.pl/elekp/ap_notes/NI_AN078_Strain_Gauge_Meas.pdf.
[11] http://www.piceramic.com/piezo-technology/fundamentals.html.
[12] A. Tamburrano, F. Sarasini, G. De Bellis, A. D'Aloia, M.S. Sarto, The piezoresistive effect in graphene-based polymeric composite, Nanotechnology 24 (2013), 465702.
[13] F. Bonaccorso, et al., Production and processing of graphene and 2D crystals, Mater. Today 15 (12) (2012) 564–589.

[14] S. Yin, P.B. Ruffin, F.T.S. Yu, Fibre Optic Sensors, second ed., CRC Press, London, United Kingdom, 2008.
[15] https://webstore.iec.ch/preview/info_iec61753-1%7Bed1.0%7Den.pdf.
[16] http://www.thefoa.org/tech/connID.htm.
[17] http://www.fibersystems.com/pdf/whitepapers/Expanded-Beam-White-Paper.pdf.
[18] http://www.thefoa.org/tech/ref/termination/fusion.html.
[19] U. Geppert, B. Biering, F. Lura, J. Block, M. Straubel, R. Reinhard, The 3-step DLR-ESA Gossamer road to solar sailing, in: Proc. Second International Symposium on Solar Sailing, ISSS, 2010 Brooklyn.
[20] J. Yick, B. Mukherjee, D. Ghosal, Wireless sensor network survey, Comput. Netw. 52 (12) (2008) 22.
[21] http://www.ieee802.org/.
[22] A. Rinaldi, M. Ciminello, et al., High sensitive grapheme-based strain sensors for SHM of composite laminates, SARISTU Final Conference, May 19–21, Moscow, Russia, 2015.
[23] R. Ghelichi, et al., Coating by the cold spray process: a state of the art, Frattura ed Integrità Strutturale, 8 (2009) 30–44.
[24] A. Papyrin, et al., Cold Spray Technology, Elsevier Science, London, United Kingdom, 2006.
[25] C.A. Galindez-Jamioy, J.M. López-Higuera, Brillouin distributed fibre sensors: an overview and applications, J. Sens. 2012 (2012) 3338–3356.
[26] B.J. Soller, D.K. Gifford, M.S. Wolfe, M.E. Froggatt, High resolution optical frequency domain reflectometry for characterization of components and assemblies, Opt. Express 13 (2) (2005) 666–674.
[27] http://www.fbgs.com/products/draw-tower-gratings/overview/.
[28] C.R. Liao, D.N. Wang, Review of femtosecond laser fabricated fibre Bragg gratings for high temperature sensing, Photon. Sens. 3 (2) (2013) 97–101.
[29] http://lunainc.com/installing-fiber-strain-sensors-2/.
[30] http://lup.lub.lu.se/luur/download?func=downloadFile&recordOId=609219&fileOId=610288.
[31] www.saristu.eu.
[32] www.criaq.aero.
[33] I. Dimino, A. Concilio, R. Pecora, Safety and reliability aspects of an adaptive trailing edge device (ATED), in: Proc. 24th AIAA/AHS Adaptive Structures Conference, Jan. 4–8, San Diego, United States, 2016.
[34] I. Dimino, G. Diodati, A. Concilio, A. Volovick, L. Zivan, Distributed electromechanical actuation system design for a morphing trailing edge wing, SPIE Smart Structures/NDE, Las Vegas, Nevada (USA) March 2016. in: Proc. SPIE 9801, Industrial and Commercial Applications of Smart Structures Technologies, 2016 980108 (April 16, 2016). https://doi.org/10.1117/12.2219223.
[35] G. Amendola, I. Dimino, A. Concilio, F. Amoroso, R. Pecora, Preliminary design of an adaptive aileron for the next generation regional aircraft, J. Theor. Appl. Mech. (Poland) 55 (1) (2017) 307–316.
[36] P. Wölcken, M. Papadopoulos (Eds.), Smart Intelligent Aircraft Structures (SARISTU), Proceedings of the Final Project Conference, Springer, Cham, Switzerland, 2016.
[37] R. Pecora, A. Concilio, I. Dimino, F. Amoroso, M. Ciminello, Structural design of an adaptive wing trailing edge for enhanced cruise performance, in: Proc. 24th AIAA/AHS Adaptive Structures Conference, San Diego, United States, Jan. 4–8 2016.
[38] F. Kong, R. Evenblij, Shape sensing of a morphing wing using FBG strain sensors: reconstruction based on the Frenet-Serret formula, Technobis Fibre Technologies, 2014.www.technobis.com.
[39] R. Evenblij, C. Koimtzoglou, M. Ciminello, Shape sensing of morphing structures, SARISTU Final Conference, May 19–21, Moscow, Russia, 2015.
[40] M.B. Bosboom, M.J. van Wijngaarden, et al., Ribbon tapes, shape sensors and hardware, SARISTU Final Conference, May 19–21, Moscow, Russia, KVE Composites, 2015.www.kve.nl.
[41] G. Amendola, M. Ciminello, I. Dimino, S. Ameduri, An FBG based sensor for morphing rib architectures: conceptual design and numerical modeling, 25th International Conference on Adaptive Structures and Technologies (ICAST 2014) will be Held on October 6–8, The Hague, The Netherlands, 2014.

[42] S.P. Timoshenko, J.M. Gere, Theory of elastic stability, McGraw-Hill Book Company, New York, Toronto, London, 1961.

[43] G. Amendola, M. Ciminello, I. Dimino, S. Amenduri, Real time monitoring of a variable camber aileron by original strain-angle transducer. J. Intell. Mater. Syst. Struct. (2015). https://doi.org/10.1177/1045389X15620036.

CHAPTER

12 CONTROL TECHNIQUES FOR A SMART ACTUATED MORPHING WING MODEL: DESIGN, NUMERICAL SIMULATION AND EXPERIMENTAL VALIDATION

Teodor L. Grigorie*,†, Ruxandra M. Botez*

École de Technologie Supérieure, ETS, Montreal, QC, Canada University of Craiova, Craiova, Romania†*

CHAPTER OUTLINE

NOMENCLATURE

α	angle of attack
c	wing chord
C_D	drag coefficient
CFD	computational fluid dynamics
C_L	lift coefficient
COA	centroid of area method
C_p	pressure coefficient
CRIAQ	Consortium for Research and Innovation in Aerospace in Quebec
Δe	error variation

Morphing Wing Technologies. https://doi.org/10.1016/B978-0-08-100964-2.00012-5

dY_i	stroke of the *i*th actuator
e	error
ETS	École de Technologie Supérieure
FFT	fast Fourier transform
FIS	fuzzy interference system
GUI	graphical user interface
I	current intensity
IAR-NRC	Institute for Aerospace Research at the National Research Council Canada
K_{ce}	derivative gain in PD control stage from PD-PI fuzzy logic controller
K_D	derivative gain
K_e	proportional gain in PD control stage from PD-PI fuzzy logic controller
K_I	integral gain
K_O	change in output gain
K_P	proportional gain
LAMSI	Laboratory of Memory Alloys and Intelligent Systems
LARCASE	Laboratory of Research in Avionics and Aero-Servo-Elasticity
LVDT	Linear Variable Differential Transformer
M	Mach number
mfi	*i*th membership function
N	negative
NSERC	Natural Sciences and Engineering Research Council of Canada
P	positive
PB	positive big
PD	proportional-derivative
PID	proportional-integral-derivative
PS	positive small
q	dynamic pressure
RMS	root mean square
SMA	shape memory alloys
U	output of the (fuzzy logic) controller
x_{tr}	chord-wise position of the transition point from laminar to turbulent flow
Z	zero

1 INTRODUCTION

The fast growth of aerial traffic in the last decade has spurred the development of new technologies in the design and manufacturing of modern aircraft. The main problem in the field is related to the green character of the newly developed aircraft, largely reflected by reduction in fuel consumption. In order to achieve this goal, research into the design of the new generation aircraft is taking three main directions: (1) development of full fly-by-wire aircraft, with lower weights and drags; (2) development of green and efficient propulsion systems; (3) drag reduction through improvement of aircraft aerodynamics (laminar flow control and development of new aircraft configurations) [1]. One of the actual green aircraft trends is related to the reduction of fuel consumption by reducing drag, which is directly related to airflow around the aerodynamic aircraft body design. The drag reduction concept is connected to the laminar flow and to the displacement of the transition point between laminar and turbulent flows toward the trailing edge [2].

A lot of theoretical and experimental studies have proved that the transition between laminar and turbulent flows is influenced by the shape of the wing airfoil. For a certain flight condition characterized by a given Mach number and a given Reynolds number, the airflow around a wing airfoil is laminar at the leading edge, but becomes turbulent at a certain point because of air viscosity. Turbulent flow is not desired because of its negative effect in terms of drag increase, which, over time, leads to high fuel consumption, and consequently, increased operating costs [3,4]. Many of these studies were related to the aircrafts' wing shape optimization [1,5–18], launching on the market a new concept suggestively called "morphing wing."

This new concept opened an important research direction in the aerospace field and united a multitude of multidisciplinary research teams, with partners from universities, R&D institutions, and industry. Moreover, it put together researchers and research teams from different countries united in one purpose, i.e., to expand the knowledge in the field and through it improve the quality of human life. As a consequence, a lot of morphing wing architectures were and are still imagined, designed, studied, and developed. One of these is our team project, a CRIAQ (Consortium for Research and Innovation in Aerospace in Quebec) project developed in Canada, related to "*Laminar Flow Improvement on an Aeroelastic Research Wing*." It aimed to realize a theoretical and experimental aerodynamic wind tunnel study of a rectangular wing equipped with a flexible skin, smart actuators and pressure sensors, and the ability to change its shape using an active controller. The goal was to move the transition zone from laminar to turbulent flow to the trailing edge, obtaining in this way a larger laminar flow region, and, as a consequence, a drag reduction.

This research was part of this major project initiated and funded by aerospace companies Bombardier Aerospace and Thales Avionics, as well as by CRIAQ and the Natural Sciences and Engineering Research Council of Canada (NSERC), and developed by several collaborating teams from the École de Technologie Supérieure (ETS) in Montreal, the Laboratory of Memory Alloys and Intelligent Systems (LAMSI), the Laboratory of Research in Avionics and Aero-Servo-Elasticity (LARCASE), École Polytechnique in Montréal, and the Institute for Aerospace Research at the National Research Council Canada (IAR-NRC) [2–4,19–37].

2 PROJECT BACKGROUND

The academia and industries collaboration in this project called for both aerodynamic modeling as well as conceptual demonstration of the morphing principle on real models placed in the wind tunnel. The following responsibilities were assigned to each team:

- The LAMSI team was charged with designing and manufacturing the actuators and flexible skin as well as the model internal structure.
- The École Polytechnique team had the responsibility of designing the optimized airfoils for each airflow condition using CFD codes in order to analyze the transition from laminar to turbulent flow.
- The NRC-IAR team had to organize and run the wind tunnel tests.
- The LARCASE team was responsible for the integration and validation of the control and monitoring systems of the morphing wing model [2].

The wing model that was used was rectangular (0.5 m × 0.9 m), with a laminar wing trailing edge airfoil (WTEA) as reference airfoil. To achieve the targets, several steps were completed [26]: (1) aerodynamic optimization study; (2) actuation system design, optimization, and experimental development; (3) flexible skin design and experimental execution; (4) design and experimental implementation of a pressure sensors system installed on the upper surface of the morphing wing to detect and visualize airflow characteristics; (5) design and experimental development of a control system for the morphing wing, in open loop and in closed loop architectures; (6) morphing wing system integration; and (7) experimental validation of the developed model in bench tests and wind tunnel tests.

The model was equipped with a flexible skin made of composite materials (layers of carbon and Kevlar fibers in a resin matrix) morphed by two actuation lines (Fig. 1 [4]). Each actuation line used shape memory alloys (SMA) wires as actuators. At the same time, 16 optical pressure sensors and 16 Kulite pressure sensors were placed on the flexible skin in different positions along the chord (Fig. 2 [2]). The sensors were positioned on two lines, in a V-configuration, at angles of 15° from the symmetry line (Fig. 3 [26]). The rigid lower structure was made from aluminum, and was designed to allow space for the actuation system and wiring [2–4,19–34].

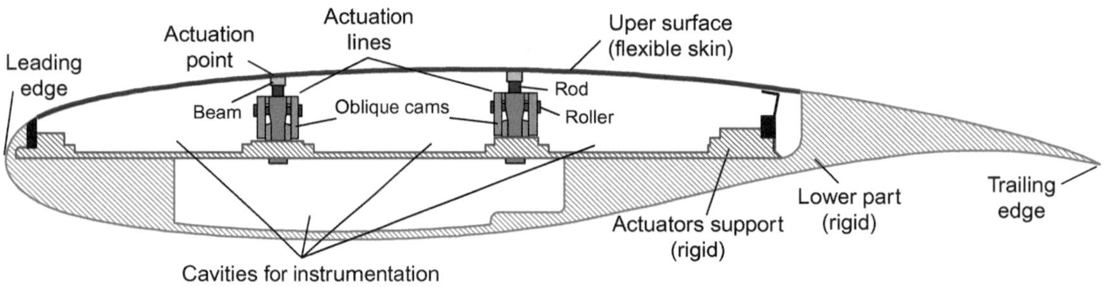

FIG. 1

Morphing wing model—cross-section.

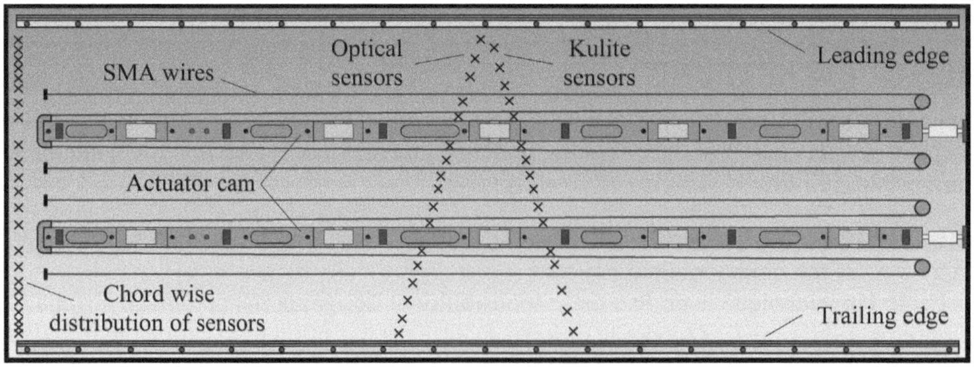

FIG. 2

Pressure sensors distribution on the flexible skin.

FIG. 3

Pressure sensors distribution on the morphing wing experimental model.

The flexible skin was manufactured in a four-ply laminate structure in a polymer matrix, with two unidirectional carbon fiber inner plies and two hybrid Kevlar/carbon fiber outer plies. The hybrid Kevlar/carbon fiber was used in the chord-wise direction, in which flexibility was needed for profile modification, whereas the low-modulus unidirectional carbon fiber was installed span-wise, where rigidity was preferred. The total thickness of the skin was 1.3 mm, the total Young modulus was 60 GPa, Poisson's ratios were 0.12 for carbon/Kevlar hybrid and 0.25 for unidirectional carbon [4,19,32].

In the first stage of the project, the determination of some optimized airfoils was made. The optimized airfoils were derived by Ecole Polytechnique team for 35 different flow conditions, expressed in terms of five Mach numbers (0.2, 0.225, 0.25, 0.275, and 0.3) and seven angles of attack (−1°, −0.5°, 0°, 0.5°, 1°, 1.5°, and 2°) [34]. The criteria used to locate these optimized airfoils was to move the transition point from laminar to turbulent flow closer to the trailing edge, promoting in this way large laminar regions on the wing surface, and thus reducing drag; the optimized airfoils were derived from the laminar WTEA-TE1 reference airfoil provided by the NRC-IAR team. The optimized airfoils were used as a starting point in the actuation system design (Fig. 4 [2]).

The design of the actuation system was performed by the LAMSI team in three steps: optimization of the number and positions of flexible skin actuation points, establishment of each actuation line's architecture, and analytical modeling of the smart materials actuators used in this application [2,19]. The actuation mechanism was designed to convert the horizontal movement along the wing span into vertical motion, perpendicular to the chord. In this way, two oblique cams with sliding rods, positioned span-wise, were included in the actuation system (Fig. 5 [4]). The oblique cams with sliding rods had a horizontal versus vertical ratio 3:1 (the cam factor was 1/3), thus giving a conversion of the horizontal stroke of x mm into a vertical stroke $z = x/3$.

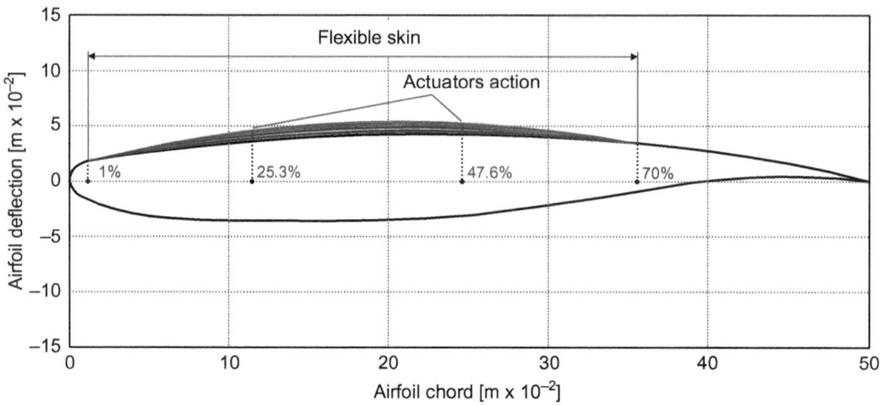

FIG. 4

Morphed airfoil shapes for different flow cases.

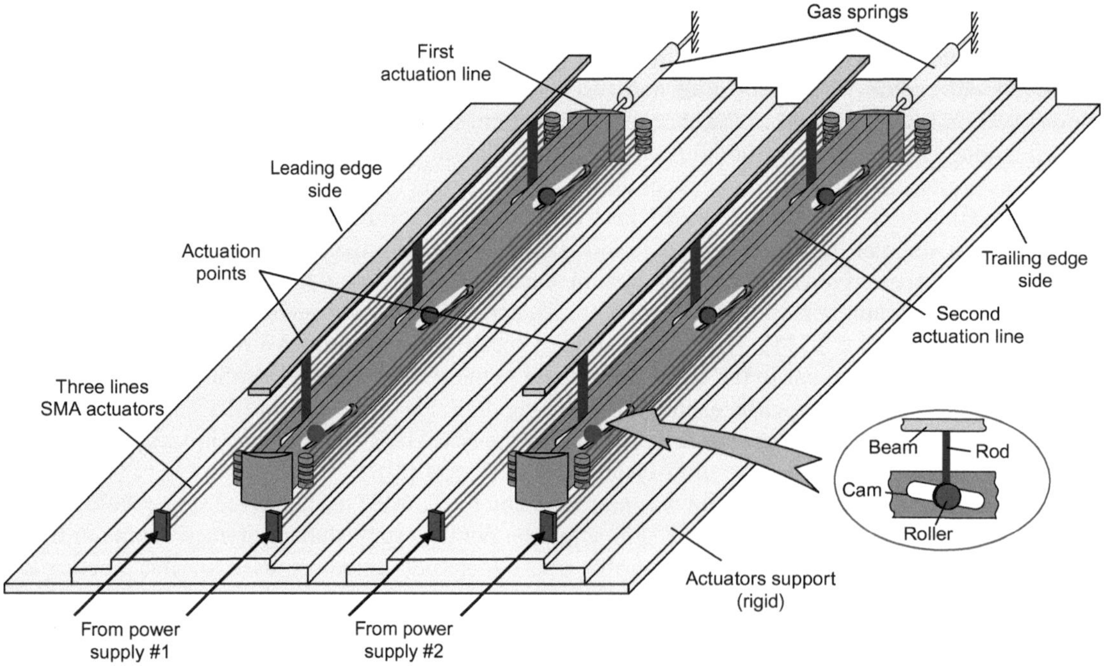

FIG. 5

Actuation system of the morphing wing.

For each actuator, the actuation distance was correlated with the desired airfoil shape for each of the 35 airflow conditions, and was given by the mechanical equilibrium between the nickel-titanium alloy SMA wires, which pulled the sliding rod in one direction, and the gas springs, which pulled the sliding rod in the opposite direction. Each sliding rod was actuated by means of three parallel SMA wires connected to an electrical current controllable power supply. From the optimized airfoils, an approximately 8 mm maximum vertical displacement was obtained for the rods, requiring a maximum horizontal displacement of 24 mm from the actuators.

The gas springs role was to counteract the pulling effect of aerodynamic forces acting in the wind tunnel over the flexible skin when the SMAs were inactive. The pulling action of the gas springs retracted the flexible skin in the unmorphed-reference airfoil configuration, while the pulling action of the SMA wires deployed the actuators in the load mode, i.e., morphed airfoil in the optimized airfoil configuration [4].

Two architectures were developed for the morphing wing control system: (1) an open loop architecture, which controlled the SMA actuators system electrically and (2) a closed loop architecture, which included the open loop architecture as an internal loop and controlled the transition point based on the information from the pressure sensors installed on the flexible skin. To design and validate the morphing wing control system, some numerical simulations, bench tests, and wind tunnel tests were performed. For the open loop architecture, several controllers were developed and tested [2–4,19–33]. Some experimentally validated open loop controllers are outlined in the next sections, as well as the architecture and validation results of the optimized closed loop control method developed throughout the morphing wing project.

3 GENERAL STRUCTURES OF THE OPEN LOOP AND CLOSED LOOP CONTROL ARCHITECTURES

To design the control system for the actuation lines, numerical simulations of the open loop morphing wing system were performed first using an SMA model. The inputs of the SMA model were the alloy initial temperature, the electrical current (heating), and the applied force, while its outputs were the actuator displacement and the alloy temperature. To use the SMA shape-changing characteristics, an initialization by an external force was required to pass initially through the transformation phase and then return to the initial phase at the end of the cooling phase. Subsequently, the open loop control system was experimentally validated in some bench tests and wind tunnel tests. Simultaneously, the transition point position detection and visualization were realized, and aerodynamic study was validated. This control strategy was defined as "open loop," because the controller had no feedback about the airflow state and executed the command from the information stored in a database built with the 35 calculated optimized airfoils (Fig. 6 [4,27]).

For each optimized airfoil, a pair of vertical deflections ($dY_{1\mathrm{opt}}$, $dY_{2\mathrm{opt}}$) characterized the differences between this airfoil and the reference airfoil at the two actuation points. In this way, the open loop controllers were designed to control the elongation of the SMA actuators until the real vertical displacements ($dY_{1\mathrm{real}}$, $dY_{2\mathrm{real}}$) in the two actuation points equaled the desired vertical displacements ($dY_{1\mathrm{opt}}$, $dY_{2\mathrm{opt}}$) of the optimized airfoil corresponding to a given flow condition [2–4,19–33].

The two SMA actuators had six wires each, supplied with power by the two AMREL SPS power supplies, controlled through analogue signal by a NI-DAQ USB 6229 data acquisition card. The NI-DAQ was connected to a laptop through an USB connection. The open loop-developed controllers were implemented in Simulink and provided the needed SMA current values through an analogue

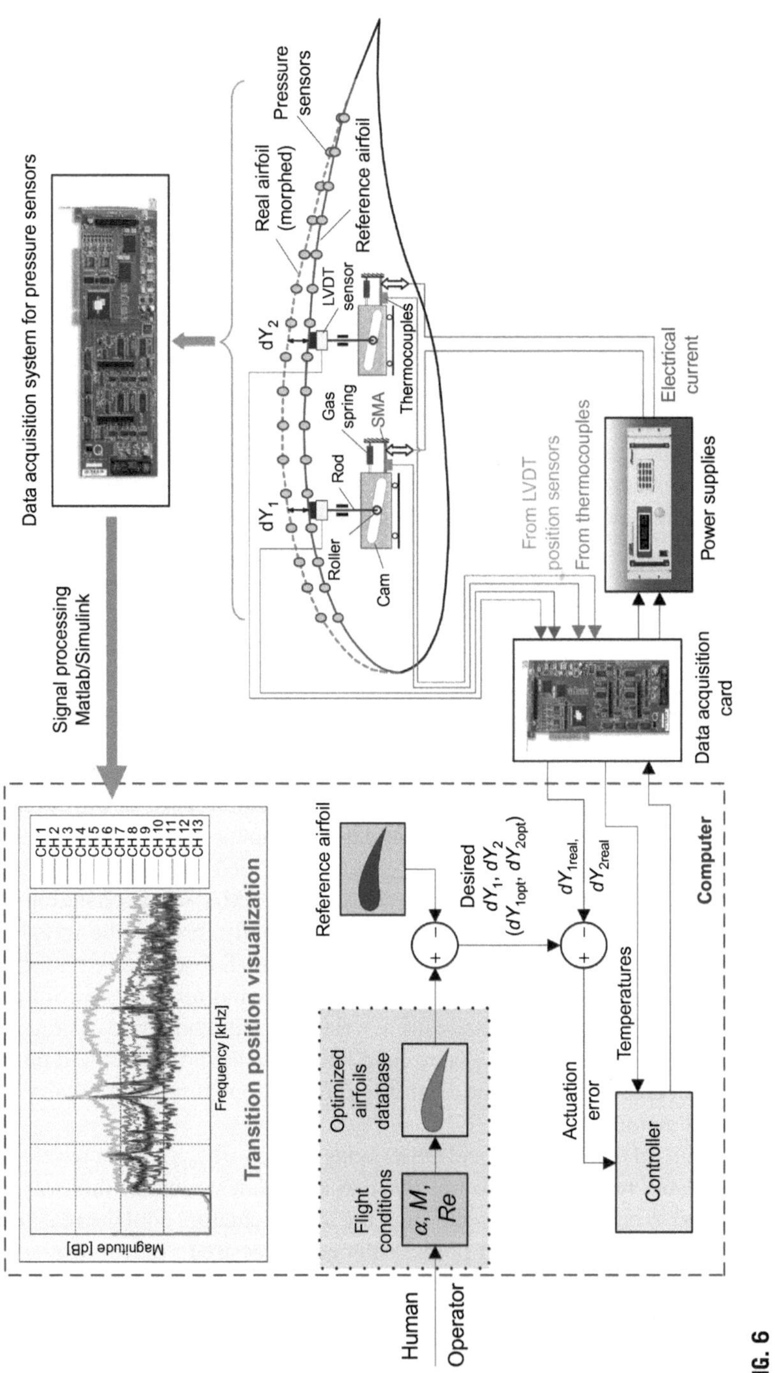

FIG. 6

The architecture of the open loop morphing wing model.

signal as shown in Fig. 6. In the scheme, the controller had feedback from the SMA lines behavior by using the information from two position sensors (linear variable differential transformer [LVDT] potentiometers). As a safety feature for the experimental model, the temperature of the SMA wires was monitored and limited by the control system. Therefore, the signals from the two LVDT potentiometers and the six signals from the thermocouples installed on each of the SMA wires' components were considered data acquisition card inputs, while the outputs consisted of four channels used to initialize and to control each power supply through their analogue/external control features by means of a DB-15 I/O connector [27].

The pressure data were acquired by using a NI-DAQ USB 6210 data acquisition card (Fig. 6); its input channels were connected directly to the IAR-NRC analogue data acquisition system. One extra channel was used for the wind-tunnel dynamic pressure (q) acquisition to calculate the pressure coefficients C_p from the values measured by the pressure sensors. The used sampling rate was 10 kS/s, in frames of 1024 points, for each channel; in this way, it was possible to perform (for all channels) a spectral decomposition of the boundary-layer pressure fluctuation FFT up to 5 kHz. The transition between laminar and turbulent flow was detected by means of each pressure signal's RMS values [2–4,20,23–27,29,31–33].

To allow the user to choose the optimized airfoils shape from database stored on the computer hard disk and give the controller the needed vertical displacements in order to obtain the desired optimized airfoil shape, a graphical user interface (GUI) was implemented in Matlab/Simulink. Fig. 7 [27] shows an example of GUI in which all the aerodynamic and morphing shape information are centralized. Along with the control buttons associated to the control software. The left part of Fig. 7 shows the wing in unmorphed configuration, while the right part shows the wing in morphed configuration [27].

Fig. 8 depicts the open loop control strategy, based on the optimized airfoils database. The input of the control loop consists of the vertical displacements of the actuators associated with the optimized airfoil, which corresponds to the airflow conditions in wind tunnel. This optimized airfoil is requested by the operator to be reproduced by the flexible skin, being selected from the computer database through the GUI listbox "Airfoil" and charged into the software by activating the button "Charge" (see Fig. 7). The software sends to the control loop the vertical coordinates, at the actuators lines level, required to reproduce the optimized airfoil (deviations dY_1 and dY_2 from reference WTEA airfoil). When operator selects the button "Optim," the controller adjusts the position of the actuators as required. In the control loop mechanism, the real position of the actuators is measured through the LVDT and compared with the desired dY_1 and dY_2 values. The horizontal displacement of the SMA oblique cam/actuator is converted in vertical displacement by division by 3 [2–4,20,23–29,31–33].

The second control architecture, "*optimized closed loop architecture*," was developed in order to generate real time optimized airfoils starting from the information received from the pressure sensors and targeting the morphing wing main goal: the improvement of the laminar flow over the wings upper surface (Fig. 9). The previously calculated optimized airfoils database was bypassed in this control strategy, and was used just to see if the closed loop real time optimizer had similar results for morphed airfoil in a flow case. To achieve the control, a mixed optimization method was used, between "the gradient ascent" or "hill climbing" method and the "simulated annealing" method [38]. Two variants were tested for the starting point on the optimization map control: (1) $dY_1 = 4$ mm, $dY_2 = 4$ mm (Fig. 9) and (2) dY_{1opt}, dY_{2opt} of the theoretically obtained optimized airfoil.

The reason for using a mixed method was that the cost function for such a complex problem (minimize the C_D, maximize the C_L/C_D, or maximize the transition point position x_{tr} for a morphing wing) was not defined analytically, and the implementation of "gradient ascent" method was not suitable. Moreover, because of the time cost (very long time response of the SMA actuators because of heating,

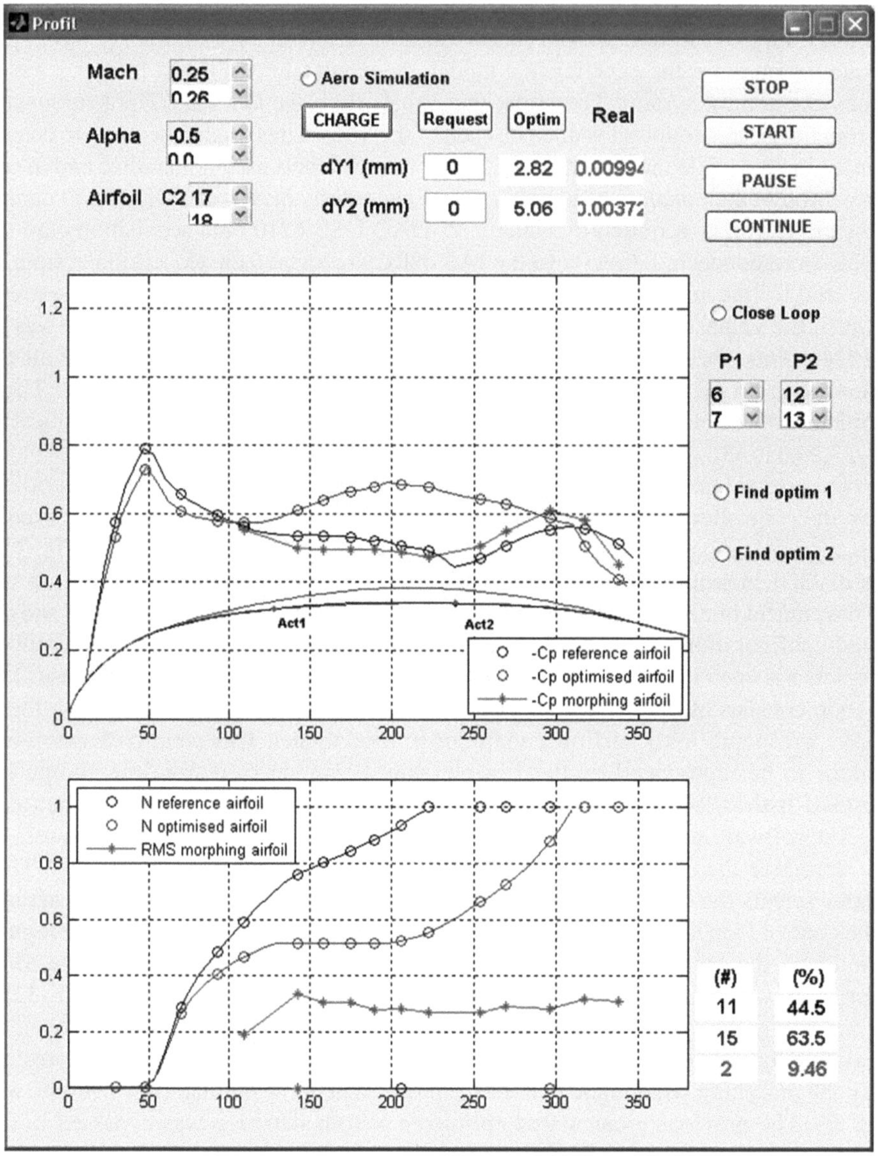

(A)

FIG. 7

Graphical user interface. (A) Unmorphed configuration.

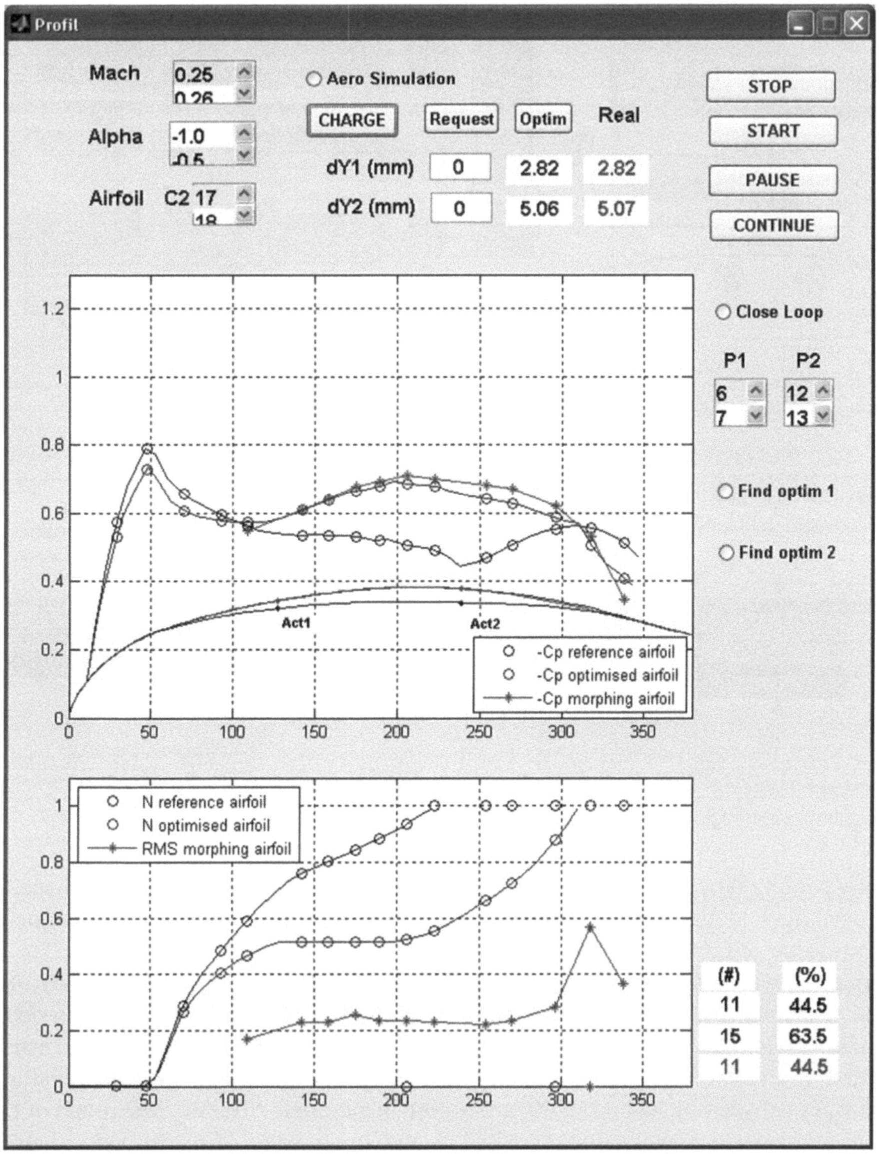

(B)

FIG. 7, CONT'D (B) Morphed configuration.

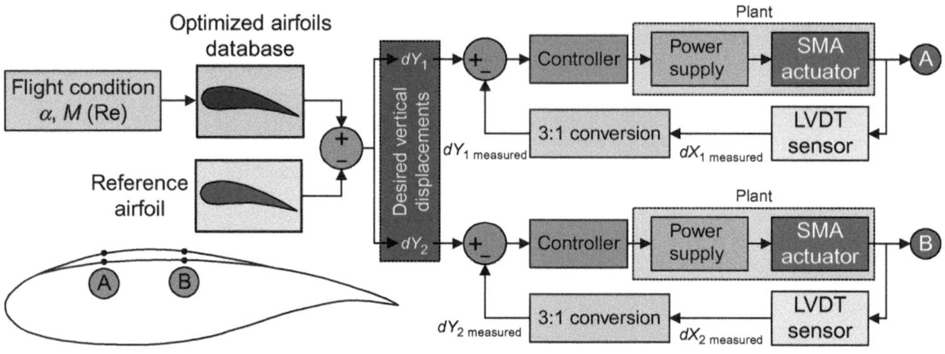

FIG. 8

Open loop control using optimized airfoils database and actuator position as feedback.

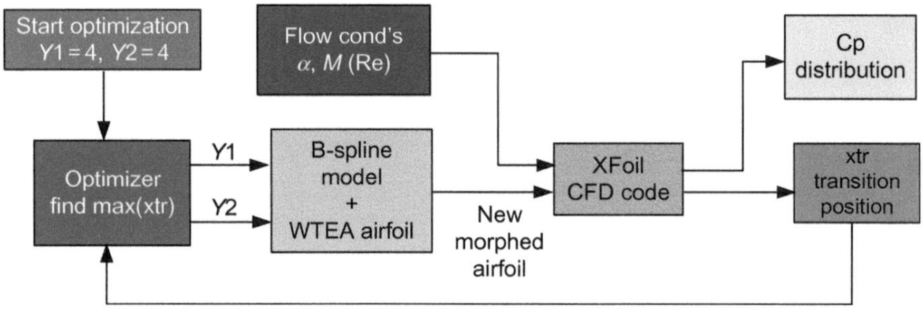

FIG. 9

Closed loop optimization logic schematic.

but especially cooling time), a purely probabilistic metaheuristic search algorithm was not suitable either. The mixed method was found to be the quickest and, at the same time, most accurate for finding the maximum value of transition point position x_{tr}.

The domain in which the optimizer searches the best configuration was defined by the bidimensional space of actuators strokes $\{dY_1 = [0, 8], dY_2 = [0, 8]\}$. The initial point of the optimizer was defined as $dY_1 = 4$ mm and $dY_2 = 4$ mm. Then the first round of evaluation points consists of eight points situated in a circle with the ray of 4 mm around the initial point. For each evaluation point, the x_{tr} value was evaluated by XFoil [39,40] and stored in memory. After the first round of evaluations, the optimizer decided which evaluating point had the maximum value of x_{tr}; this value then became the initial point for the next round of evaluations [27,31,33].

4 OPEN LOOP CONTROLLERS

Because of the strong nonlinear character of the adopted smart materials actuators, several variants for open loop controllers were developed using fuzzy logic techniques. Fuzzy logic is an interface between logic and human reasoning, providing an intuitive method for describing systems in human terms and automating the conversion of those system specifications into effective models [2,4,23,26–28]. It may be used to overcome the lack of a rigorous mathematical model, a priori developed for system, and, therefore, to compensate the loss of precision from linearization and uncertainties in the system's parameters, which negatively influences the quality of the resulting control. At the same time, advantages offered by fuzzy logic techniques (namely intuitive handling, simplicity, and flexibility) were fully exploited because of their proximity to human perception and reasoning. In the next subsections, four different variants of open loop controllers are considered: a fuzzy logic PD controller; two controllers based on fuzzy logic proportional-integral-derivative (PID) and on-off combination; and a self-tuning fuzzy logic controller (FLC) combined with an on-off one. All developed controllers were simulated numerically, and experimentally tested in the ETS laboratory (bench testing) and in the IAR-NRC wind tunnel.

4.1 FUZZY LOGIC PD CONTROLLER

The selected PD FLC structure has as its inputs the error (difference between the desired and measured vertical displacement) and the change in error (the derivative of the error), and as output the voltage controlling the power supply output current (Fig. 10 [4,41]). Widely accepted for capturing expert knowledge, a Mamdani-type controller was used, because of its simple structure of "min-max" operations [4,42].

The fuzzy controller internal mechanism was relatively simple. Based on the membership functions stored in the knowledge base, the fuzzifier converted the crisp inputs into linguistic variables. For our system, three membership functions were chosen for the inputs (N-negative, Z-zero, P-positive), while five membership functions were considered for the output (Z-zero, PS-positive small, PM-positive medium, PB-positive big, PVB-positive very big) (Fig. 11 and Table 1); a triangular shape was used, defined by a lower limit a, an upper limit b, and a value m $(a \leq m \leq b)$ [4]

$$\mu_A(x) = \begin{cases} 0, & \text{if } x \leq a \\ \dfrac{x-a}{m-a}, & \text{if } a < x \leq m \\ \dfrac{b-x}{b-m}, & \text{if } m < x < b \\ 0, & \text{if } x \geq b \end{cases} \tag{1}$$

$[-1, 1]$ interval was considered to be the universe of discourse for the two inputs, while $[0, 1]$ interval was used for the output.

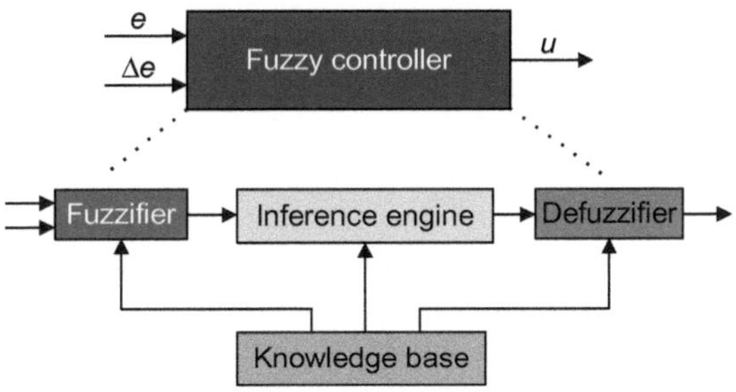

FIG. 10

Fuzzy controller architecture.

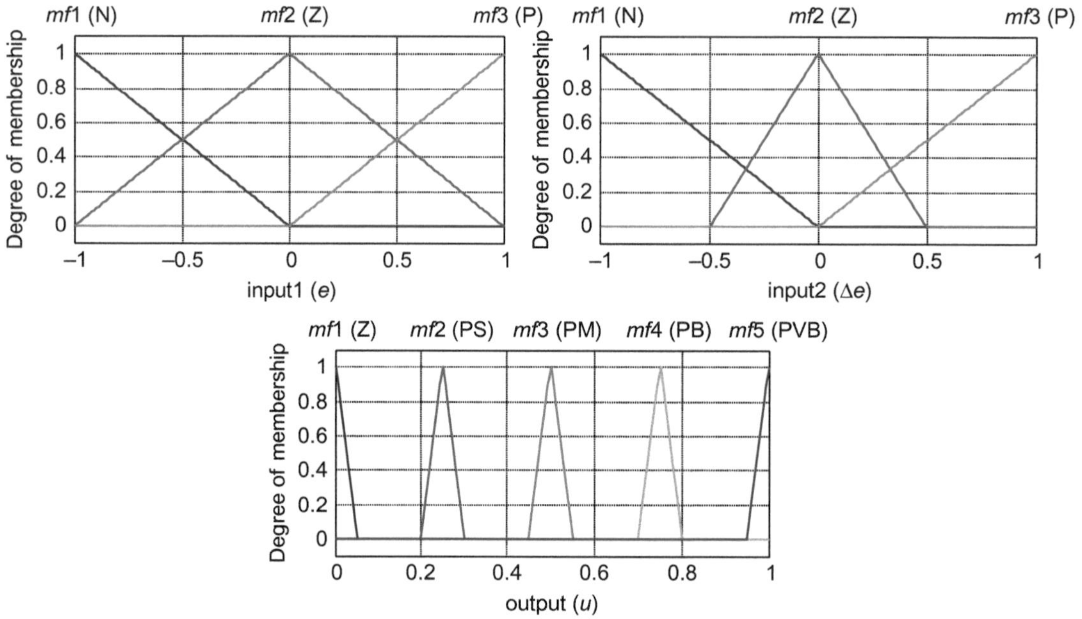

FIG. 11

Membership functions of the fuzzy logic PD controller.

Table 1 Parameters of the Input-Output Membership Functions (*mf*)

	Input 1 (*e*)			Input 2 (Δ*e*)			Output (*u*)				
***mf*/Parameter**	***mf*1**	***mf*2**	***mf*3**	***mf*1**	***mf*2**	***mf*3**	***mf*1**	***mf*2**	***mf*3**	***mf*4**	***mf*5**
a	−1	−1	0	−1	−0.5	0	0	0.2	0.45	0.7	0.95
m	−1	0	1	−1	0	1	0	0.25	0.5	0.75	1
b	0	1	1	0	0.5	1	0.05	0.3	0.55	0.8	1

Further, the inference engine converted the fuzzy inputs to the fuzzy output, based on the "if-then" type fuzzy rules in Table 2. The fuzzified inputs were applied to the antecedents of the fuzzy rules by using the fuzzy operator *and*; this produced a single number, representing the result of the antecedent evaluation. To obtain the output of each rule, the antecedent evaluation was applied to the membership function of the consequent and the clipping (alpha-cut) method was used; each consequent membership function was cut at the level of the antecedent truth. Unifying the outputs of all eight rules, the aggregation process was performed and a fuzzy set resulted for the output variable [4].

Because the output of the fuzzy system should be a crisp number, a defuzzification process was realized (Fig. 12) by using the centroid of area (COA) method [41]. The resulting control surface is shown in Fig. 13. Using the Matlab/Simulink simulation scheme illustrated in Fig. 14, the

Table 2 Inference Rules for the Fuzzy Logic PD Controller			
$e/\Delta e$	N	Z	P
N	Z	Z	Z
Z	PS	Z	–
P	PM	PVB	PB

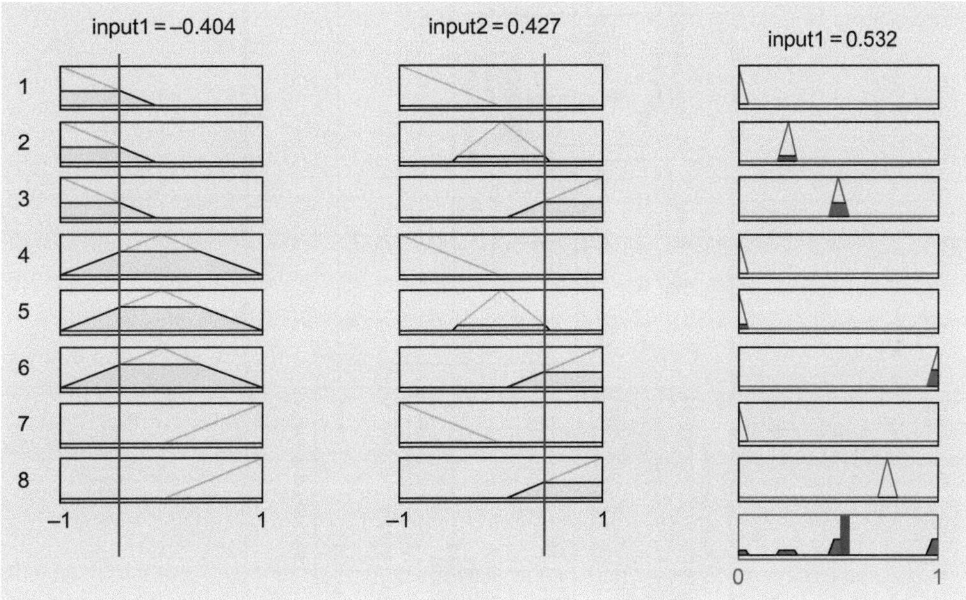

FIG. 12

Fuzzy system operating mechanism.

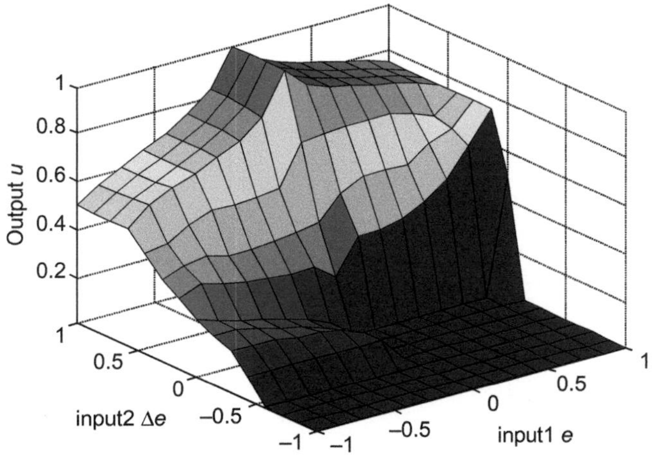

FIG. 13

Control surface.

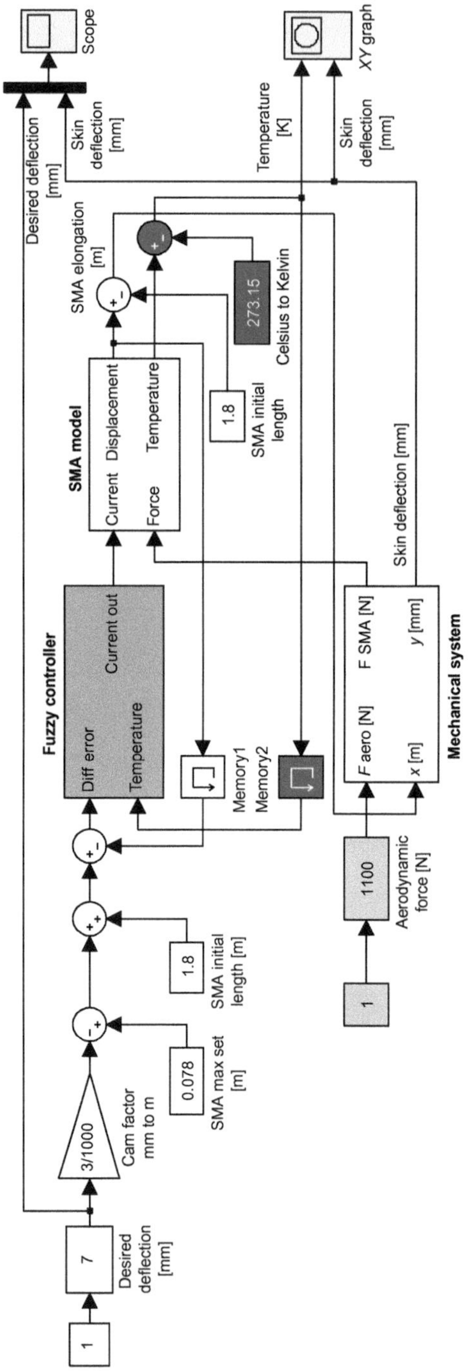

FIG. 14

Matlab/Simulink simulation scheme.

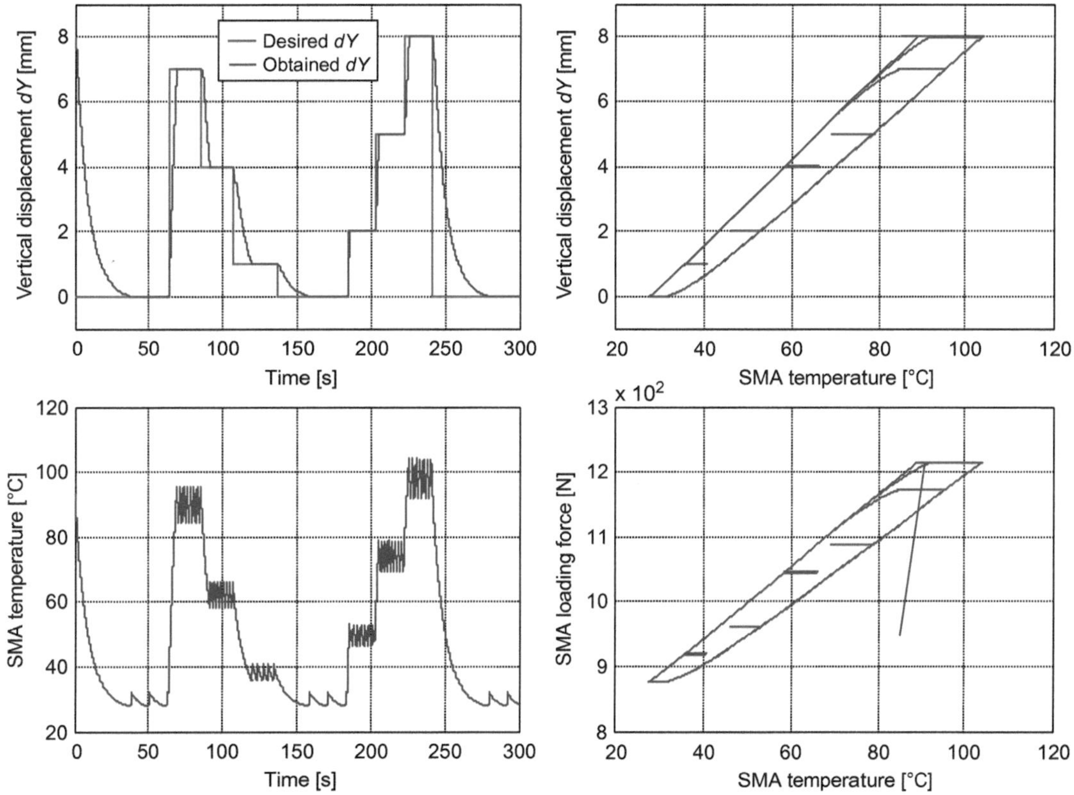

FIG. 15

Numerical simulation results.

characteristics in Fig. 15 were obtained, when a repeated step signal was used for the desired displacement. The main blocks in Fig. 14 have the next functions:

1. Fuzzy controller block models the controller and the conversion voltage/electrical current obtained from the power supplies.
2. SMA model is an S-function modeling the SMA nonlinear behavior (the model was built in the SMA and Intelligent Systems Laboratory at ETS, using Lickhatchev's model [43]).
3. Mechanical system models the morphing wing system's mechanical part behavior in terms of forces (aerodynamic force, skin elastic force, gas spring loading force) and skin vertical deflection [4].

Fig. 15 shows the response of the actuator relative to desired vertical displacement, the SMA actuator envelope (obtained vertical displacement vs. temperature), the SMA temperature in time, and the SMA loading force versus temperature. The relative pull of the obtained and desired displacements proved the good functioning of the controller; the system's response is a critically damped one, an easier latency being observed in the cooling phase of the SMA wires in comparison with their heating phase [4].

Based on the general architecture of the experimental model in Fig. 6, developed for the morphing wing open-loop control system, the bench testing and wind tunnel testing of the controller were performed. The bench testing also provided a means to observe the quality of the experimental reproducibility of the theoretical reference and optimized airfoils. When the controller was tested on the bench, two steps were performed: (1) independent testing of the two actuation lines, without the flexible skin on the model (Fig. 16 [23]) and (2) actuation lines simultaneous testing with the flexible skin on the experimental model. For both steps in the bench test, the actuators were asked to perform vertical displacements corresponding to the pairs (dY_1, dY_2) obtained for each of the 35 optimized airfoils.

Bench testing gave us the opportunity to make some adjustments in the mechanical part of the model regarding the gas springs' pretension force, which was finally set to the value of 1000 N; in lab conditions, the mechanical model was not loaded by the aerodynamic forces. Also, LVDT transducers were calibrated using a laser beam, which precisely measured the vertical displacements produced by each of the two actuation lines (Fig. 17). The same laser beam was used to evaluate the quality of the experimental reproducibility of the theoretical reference and optimized airfoils by scanning the center line of the wing model [4].

All of the bench test results were very good and proved the proper functioning of the designed control system. For example, in Fig. 18 [4], the results obtained for the independent control of the two actuation lines (step 1) at $M = 0.2$, $\alpha = -1°$ airflow case ($dY_1 = 3.86$ mm, $dY_2 = 4.02$ mm) are presented.

FIG. 16

Bench tests experimental model.

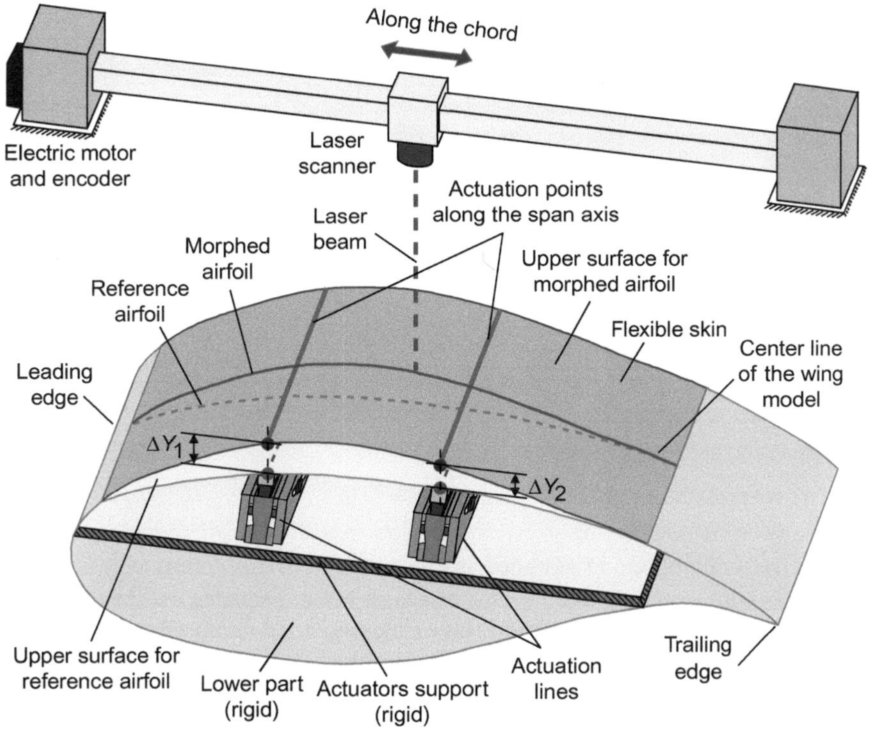

FIG. 17

Bench test morphing wing laser scan.

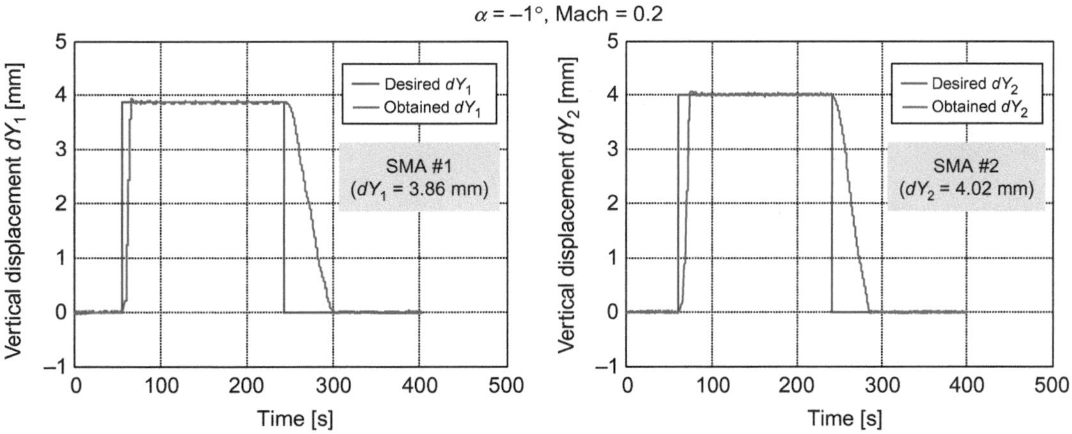

FIG. 18

Bench test control results for $M = 0.2$ and $\alpha = -1°$.

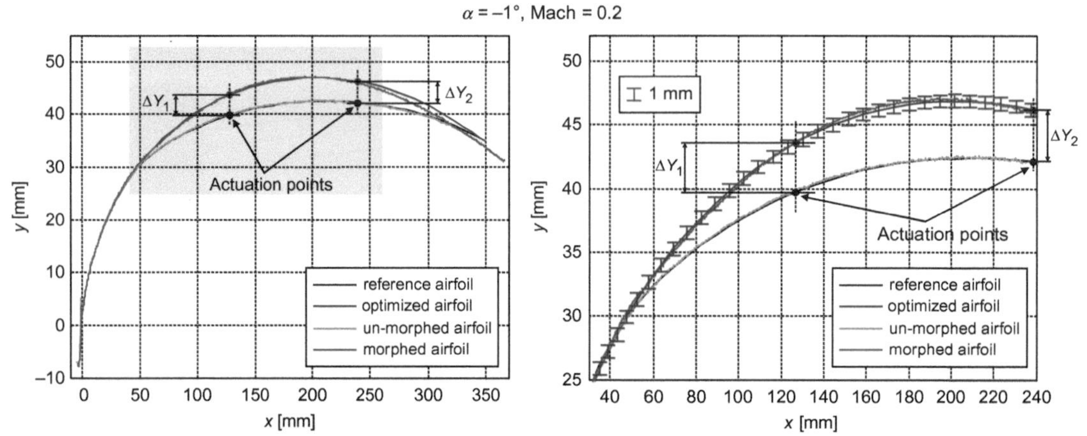

FIG. 19

Theoretical versus experimental airfoils for $M = 0.2$ and $\alpha = -1°$.

Fig. 19 shows the unmorphed and morphed airfoils obtained in that controlled case by experimental measurement of skin point coordinates by means of a laser beam. Depicted are the theoretical reference airfoil (black), the theoretical optimized airfoil (blue), the measured unmorphed airfoil (green), and the measured morphed airfoil (red). The left side of the figure shows the entire airfoil, while the right side presents a zoom of the airfoil between 30 and 240 mm. The global deviation obtained for the optimized airfoil (brown) is less than 0.3 mm, while in the actuation points, it is less than 0.05 mm. From the numerical values obtained for all 35 optimized airfoil cases, it was observed that the position control error because of the heating-cooling cycles, when the actuator maintains a desired position, was less than 0.05 mm [4].

After the experimental architecture was validated in bench tests, the morphing wing model was considered ready for wind tunnel tests, which were performed at the Institute for Aerospace Research at the National Research Council Canada (IAR-NRC) in Ottawa (Fig. 20 [2,23]). For open loop architecture validation, all 35 optimized airfoils were tested in corresponding flow cases. The configuration

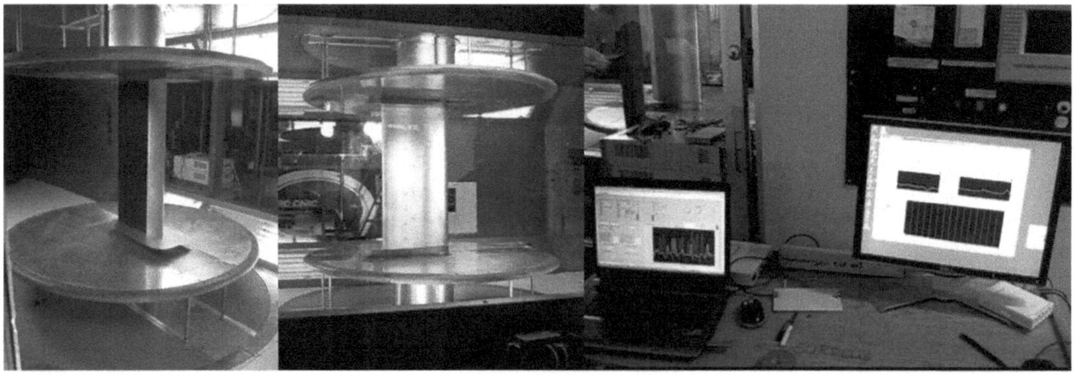

FIG. 20

Wind tunnel morphing wing model.

included an analysis of the wind flow characteristics, based on the information provided by the pressure sensors; the transition between laminar and turbulent flows was visualized in real-time [4].

For all tested cases, an improvement in the laminar regions on the wing surface was noticed, the transition point from laminar to turbulent flow being moved closer to the trailing edge. As an example, we report here the wind tunnel test results for $M = 0.3$, $\alpha = -1°$. airflow case ($dY_1 = 6.26$ mm, $dY_2 = 7.88$ mm). The FFT of the Kulite pressure sensors data (Fig. 21), and the pressure data RMSs, the C_p and N factor (for transition positioning) distributions (Fig. 22), for unmorphed and morphed airfoils have been shown [4].

The distributions were estimated by using the XFoil computational fluid dynamics; XFoil code is free licensed software in which the e^n transition criterion is used. In these graphs, the C_p and N values calculated by XFoil for various sensors are defined by circles [4].

The predicted transition position calculated by XFoil for the reference and optimized airfoils was found, respectively, to be the seventh position of the available sensor positions, corresponding to $0.44 \cdot c$, and the tenth position of the available sensor positions, corresponding to $0.59 \cdot c$. In the morphed-to-optimized airfoil case, the RMS plot displayed in Fig. 15 with star symbols, showed that the sensor with the maximum RMS, located at $0.63 \cdot c$, became the eleventh sensor plotted. The spectral decomposition of the pressure signals in Fig. 21 confirmed the Tollmien-Schlichting wave's occurrence in the eleventh sensor, visible in the highest power spectra (eleventh channel in the right side plots) in the frequency band of 3–5 kHz [4].

4.2 COMBINED ON-OFF AND PID FUZZY LOGIC CONTROLLER

Another developed open loop control structure was based on using a combination between an on-off controller and a PID FLC. The control strategy considered that the actuators would not be powered during the SMA cooling phase. Therefore, the fuzzy controller was chosen to work in tandem with

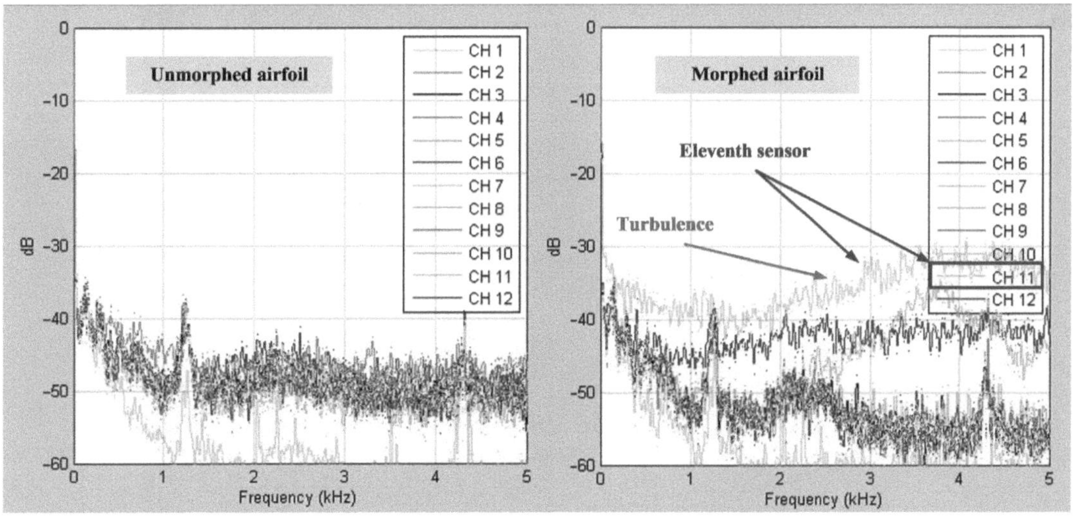

FIG. 21

FFT of the Kulite pressure sensors data.

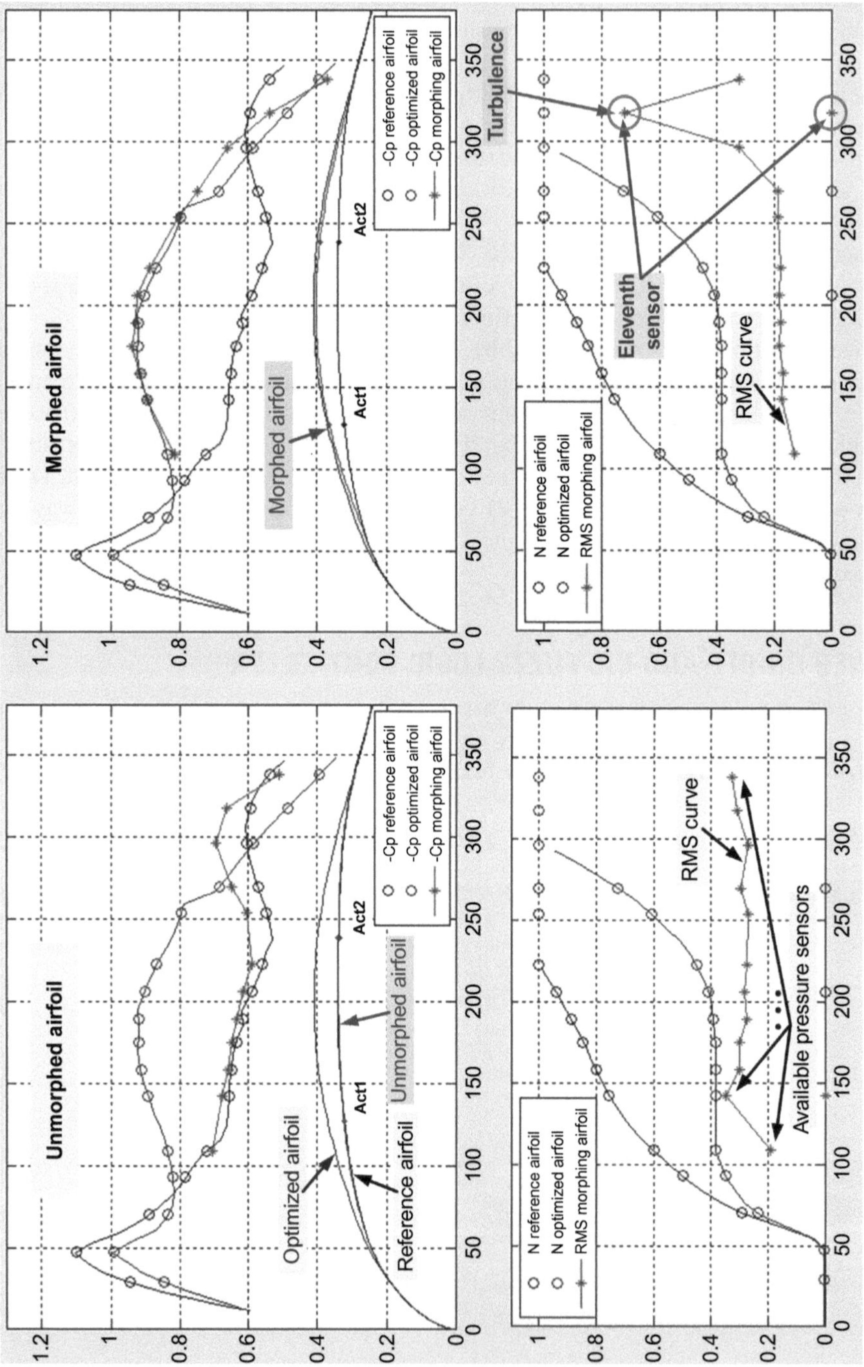

FIG. 22

Pressure data RMSs and the C_p distribution.

a bipositional controller (particularly an on-off one). The cooling phase can occur not only when controlling a long-term phase, when a switch between two values of the actuator displacements is commended, but also in a short-lived phase, which happens when the real value of the deformation exceeds its desired value and the actuator wires need to be cooled. As a consequence, the final controller should behave as a switch between the SMA cooling and heating phases, in which the output current is 0 A, or is controlled by the FLC. The resulted controller operational scheme was organized as in Fig. 23, being based on the fuzzy proportional-integral-derivative (PID) architecture shown in Fig. 24 [23].

The inputs of the controller are the error e and the change in error Δe, while its output is the electrical current i provided to the actuators. The elements in Fig. 24 are: K_I—integral gain, K_P—proportional gain, K_D—derivative gain, and K_O—change in output gain (which includes the conversion from output voltage of the data acquisition card to electrical current obtained at the output of the power supply) [23].

Considering [−0.1, 0.1] interval as universe of discourse for both of the controller inputs, three membership functions (*mf*) were chosen for each of the two inputs (*mf*1(e), *mf*2(e), *mf*3(e), respectively, *mf*1(Δe), *mf*2(Δe), *mf*3(Δe)). The inputs linguistic terms are N (negative), Z (zero), and P (positive), while for output are Z (zero), PS (positive small), and PB (positive big). The inputs membership function shapes are triangular ones as in Eq. (1). The membership function parameters are given in Table 3 [23].

To define the rules, a Sugeno fuzzy model proposed by Takagi, Sugeno, and Kang was selected [44]. Considering a zero-order Sugeno fuzzy model, the membership functions of the output were chosen as constants in the [0, 5] universe of discourse (Z = 0, PS = 2.5, PB = 5), and a set of seven inference rules were derived. With these considerations, the control surface resulted as shown in Fig. 25; the membership functions for each input and the rules are presented in Fig. 26. The fuzzy control surface was chosen in this way because of the combination with the on-off controller [23].

The Matlab-Simulink implementation of the open loop general control scheme is similar to that shown in Fig. 14, the fuzzy controller block modeling the controller presented in Fig. 23. Its detailed Simulink scheme is shown in Fig. 27 [23]. The block inputs are the control error and the SMA wire temperatures, while its output is the electrical current used to control the actuators. The first switch ensured the functioning in tandem of the fuzzy controller with the on-off controller selecting one of the two options shown in Fig. 23 (error is positive or not), while the second one protected the system by switching the electrical current value to 0 A when the SMA temperature was higher than the imposed limit. As a supplementary protection measure, a current saturation block was used to prevent the current from going higher than the physical limit supported by the SMA wires.

After a tuning operation, the controller was tested through numerical simulation to ensure that it worked properly. A series of step signals was applied to the input of the controlled actuator; the obtained characteristics are shown in Fig. 28. The graphical characteristics present the transient response of the actuator relative to the desired vertical displacement, the SMA actuator envelope (elongation versus temperature), and the SMA actuator loading force and temperature evolutions in time. As can be seen from the figure, the controlled system response was good, the transient regime being by the order of seconds; a small delay can be seized between the cooling and the heating phases. The maximum value of the loading force was 1214 N, corresponding to the maximum value of the desired vertical displacement, i.e., 8 mm, while the minimum value was 877 N, being associated with the 0 mm required displacement. For the SMA temperature, the maximum and the minimum values were,

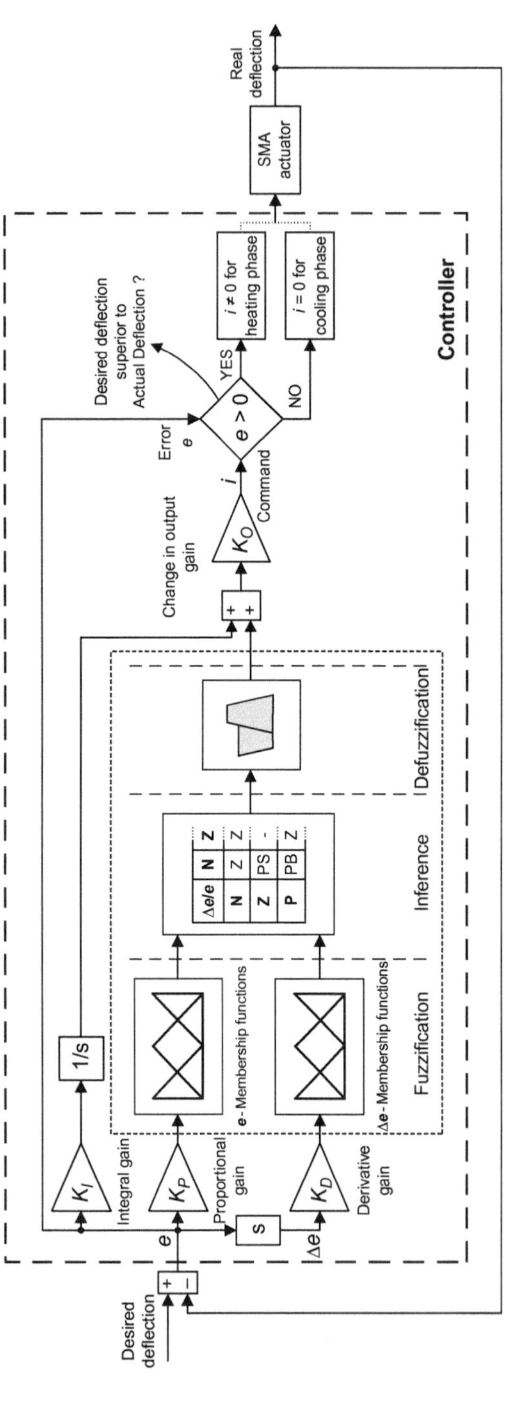

FIG. 23

Operational scheme of the controller.

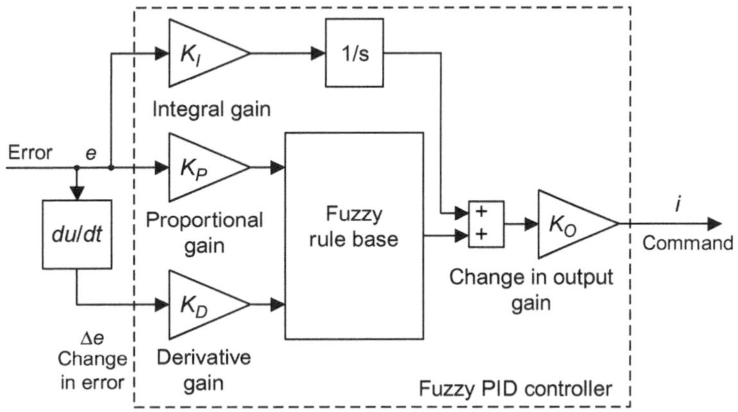

FIG. 24

Fuzzy PID controller architecture.

Table 3 Parameters of the Inputs' Membership Functions

mf	*a*	*m*	*b*
$mf1(e)$ and $mf1(\Delta e)$	−0.1	−0.1	0
$mf2(e)$ and $mf2(\Delta e)$	−0.1	0	0.1
$mf3(e)$ and $mf3(\Delta e)$	0	0.1	0.1

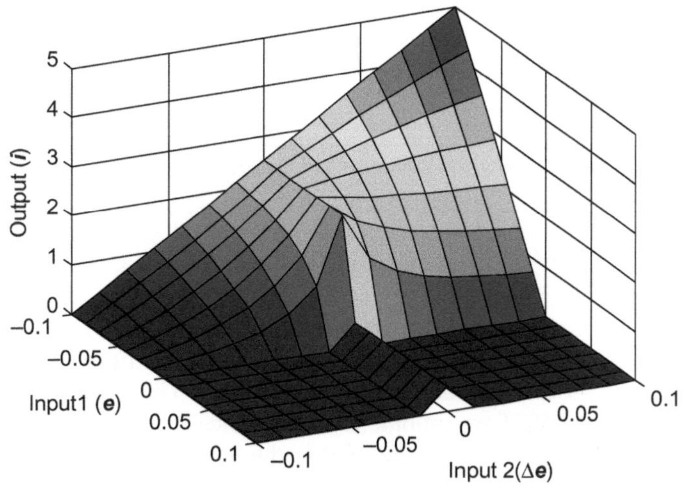

FIG. 25

Fuzzy control surface.

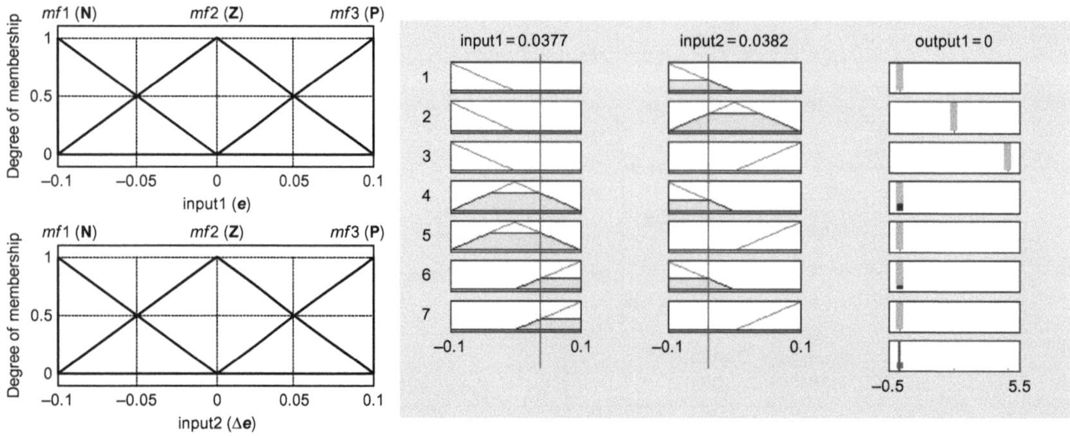

FIG. 26

Membership functions and rules.

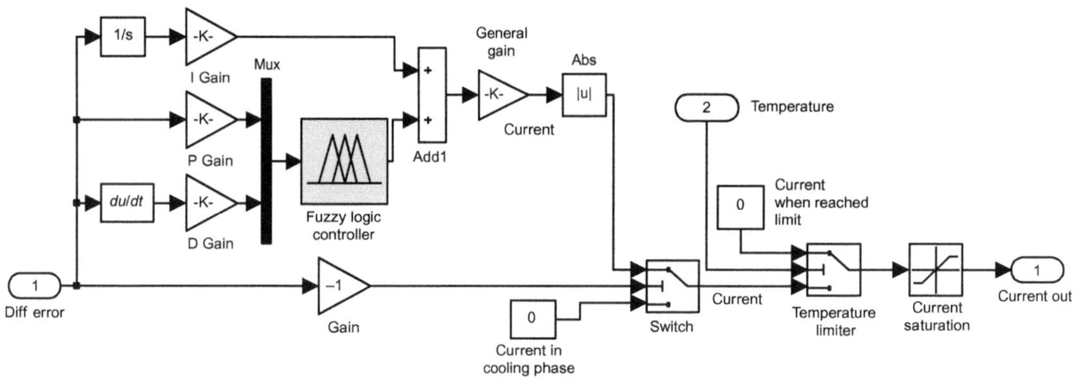

FIG. 27

Fuzzy controller block in Fig. 14 modeling the controller in Fig. 23.

respectively, 104.02°C and 27.8°C, being obtained for the desired displacements with the values of 8 and 0 mm [23].

For the bench test phase, a system response corresponding to the $M = 0.2$ and $\alpha = 2°$ flow condition ($dY_{1\text{opt}} = 5.99$ mm, $dY_{2\text{opt}} = 7.73$ mm) is shown in Fig. 29. From the figure, small oscillations of the obtained displacements are observed around their desired values. The amplitude of the oscillations in this phase was because of the LVDT potentiometers' mechanical link (they were not firmly fixed because the upper surface of the wing was absent) and to the SMA wires thermal inertia [23].

The controller also was tested and validated during wind tunnel tests carried out at the Institute for Aerospace Research at the National Research Council Canada. The controlled actuation system response for the flow condition $M = 0.3$ and $\alpha = 0°$ ($dY_{1\text{opt}} = 4.91$ mm, $dY_{2\text{opt}} = 7.34$ mm) is

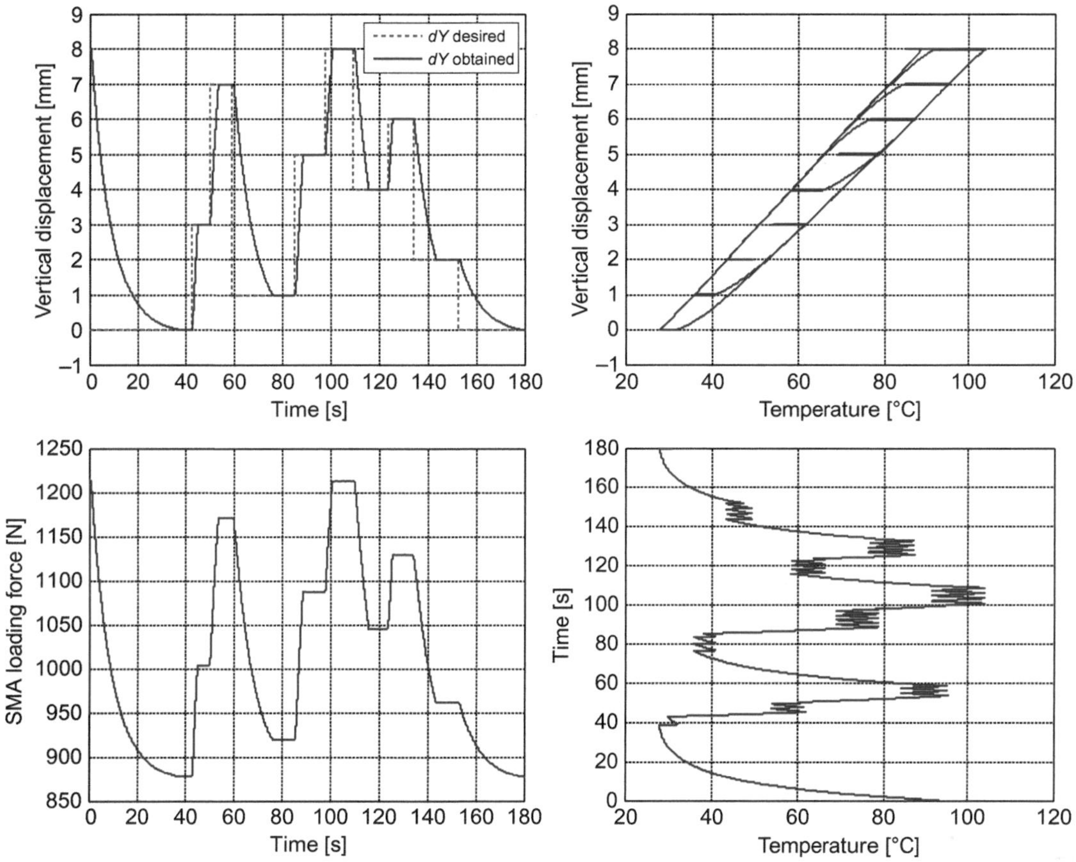

FIG. 28

Numerical simulation of the open loop morphing wing to the series of step signals.

presented in Fig. 30. As shown in the figure, the designed controller worked very well during the wind tunnel tests [23].

Fig. 31 displays the real-time results respectively for the morphed (optimized), and the unmorphed (reference) airfoils corresponding to the $M = 0.3$ and $\alpha = 0°$ flow condition. According to this figure, for the addressed flow condition and morphed airfoil, the transition is located near the pressure sensor 10 (see the peak of the RMS) [23].

4.3 COMBINED ON-OFF AND CASCADE PD-PI FUZZY LOGIC CONTROLLER

A third configuration developed for open loop controller used a combination between an on-off controller and a cascade PI-PD FLC. The controller operational scheme resulted as in Fig. 32, and was based on the cascade PI-PD FLC architecture in Fig. 33 [26]; it combined a fuzzy-PI and a fuzzy-PD controller, resulting in a fuzzy-PID controller [45,46]. For the control action adjustment, four

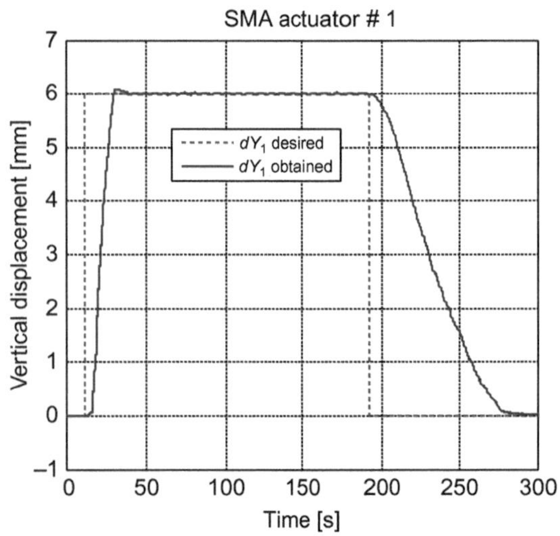

FIG. 29

Bench test ($M = 0.2$ and $\alpha = 2°$).

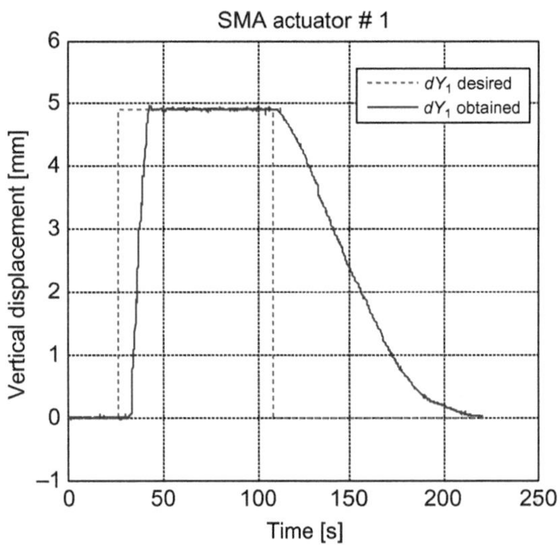

FIG. 30

Wind tunnel test ($M = 0.3$ and $\alpha = 0°$).

input-output gains were used: K_e, K_{ce}, K_I, and K_D. The controller inputs were the error e and the change in error Δe, while output was considered the command voltage of the power supplies.

The fuzzy control surface resulted as in Fig. 34. s—function, π—function and z—function were used, respectively, for the shapes of the input variables membership functions as shown in Fig. 35, while three constant values were chosen for the output membership functions: 0, 0.5, and 1.

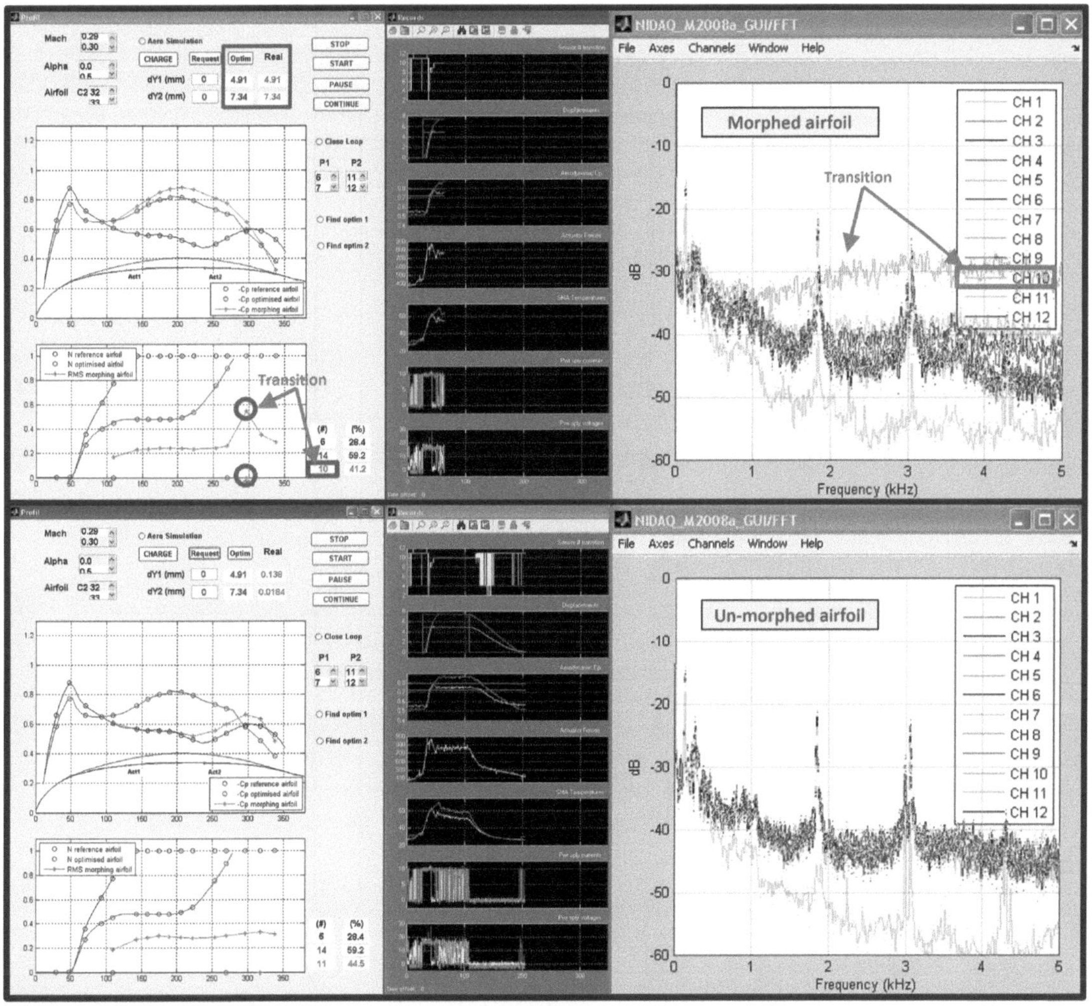

FIG. 31

Real time displayed results for $M = 0.3$, $\alpha = 0°$ flow condition.

The parameters characterizing the inputs' membership functions are shown in Table 4 (x_{left}—left breakpoint, x_{right}—right breakpoint, $[x_{m1}, x_{m2}]$—middle interval over that floating the π function peak). The rule-based inference chosen for each is presented in Fig. 36. To define the rules, a zero-order Sugeno fuzzy model was used [26].

The fuzzy controller was chosen to work in tandem with an on-off controller; the controller behaved as a switch between the SMA cooling and heating phases in which the output current is 0 A, or is controlled by the FLC [26].

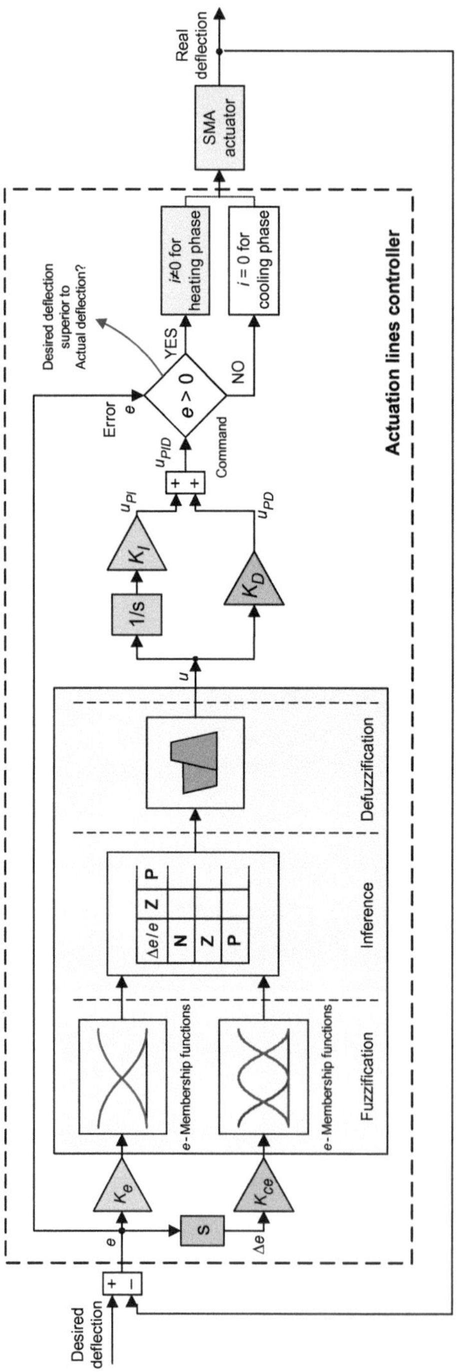

FIG. 32

Operational scheme of the controller.

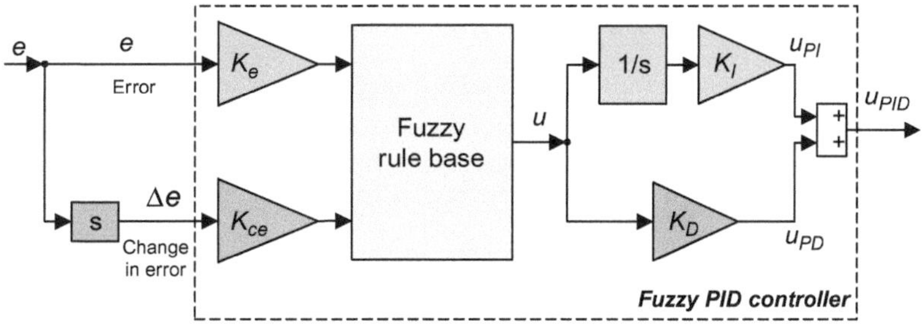

FIG. 33

Architecture of the fuzzy controller.

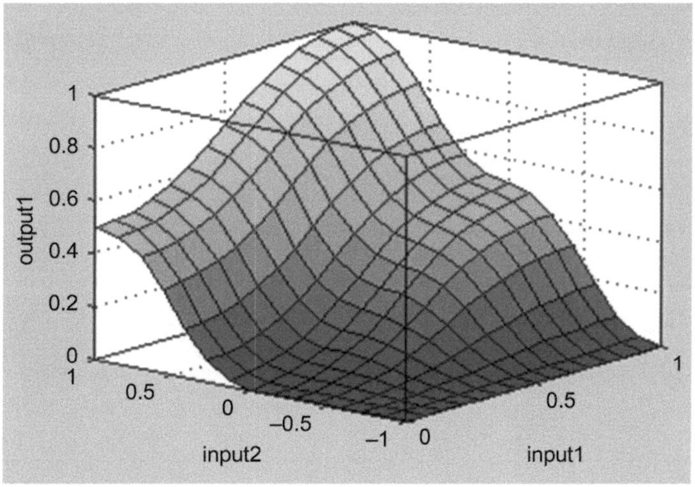

FIG. 34

The fuzzy control surface.

As in the previous control configurations, the controller was numerically simulated and experimentally tested in the ETS lab and in the IAR-NRC wind tunnel. The numerical simulations allowed the tuning of the controller in order to obtain the desired configuration, which was then implemented on the experimental model. The simulation of the tuned controller for a repeated step input led to the results presented in Fig. 37; for the wind tunnel test case at $M = 0.275$ and $\alpha = -0.5°$ ($dY_{1\text{opt}} = 2.62$ mm, $dY_{2\text{opt}} = 4.72$ mm), the obtained results are exposed in Fig. 38, while the bench testing results related to $M = 0.275$ and $\alpha = 2°$ flow condition ($dY_{1\text{opt}} = 6.70$ mm, $dY_{2\text{opt}} = 5.46$ mm), are shown in Fig. 39 [26].

During the current wind tunnel test, the pressure data signals obtained from the Kulite pressure sensors [47] were acquired using the IAR-NRC analogue data acquisition system; the sampling rate of each channel was 15 kHz; it allowed a pressure fluctuation FFT spectral decomposition of up to 7.5 kHz for all channels. The results shown in Fig. 38 for the wind tunnel test showed that, because of the gas springs pretension forces, the controller worked even if the required vertical displacements for the actuation lines were zero millimeters. Also, some noise parasitizing the LVDT sensors

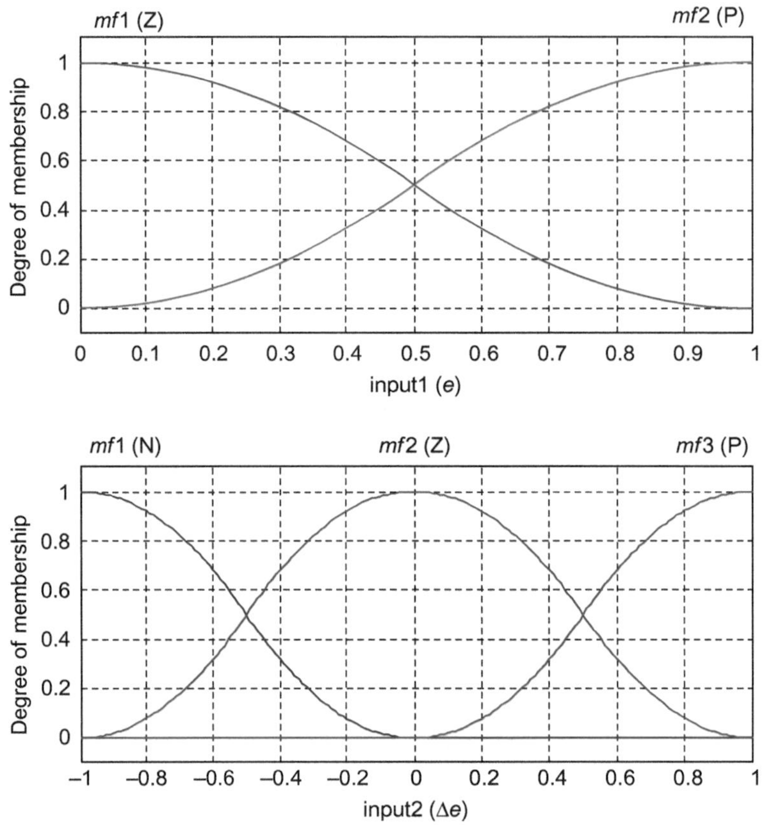

FIG. 35

Inputs' *mf*.

Table 4 Parameters of the Inputs' Membership Functions

Input	*mf*	*mf* Type	*mf* Parameters			
			x_{left}	x_{m1}	x_{m2}	x_{right}
e	*mf*1	z—function	0	–	–	1
	*mf*2	s—function	0	–	–	1
Δe	*mf*1	z—function	−1	–	–	0
	*mf*2	π—function	−1	0	0	1
	*mf*3	s—function	0	–	–	1

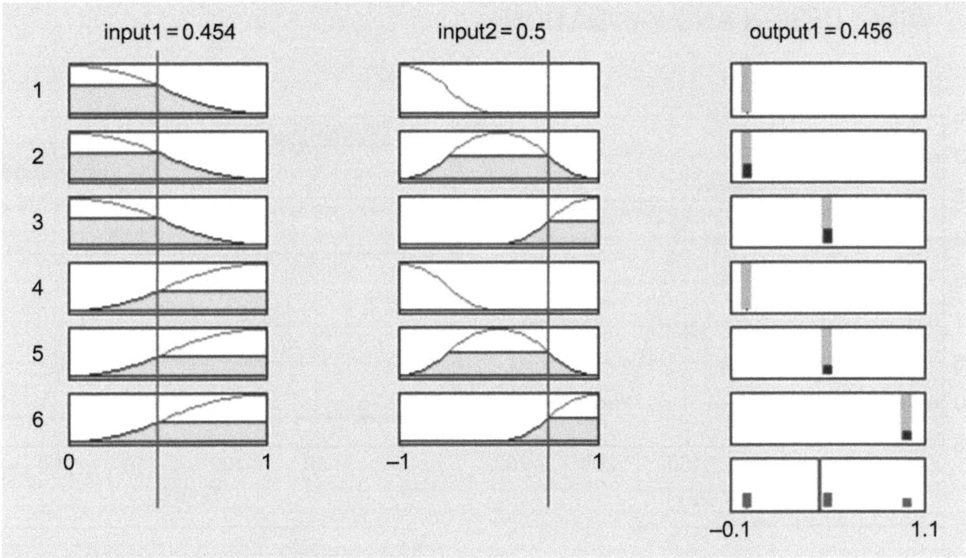

FIG. 36

Rule-based inference for fuzzy controller.

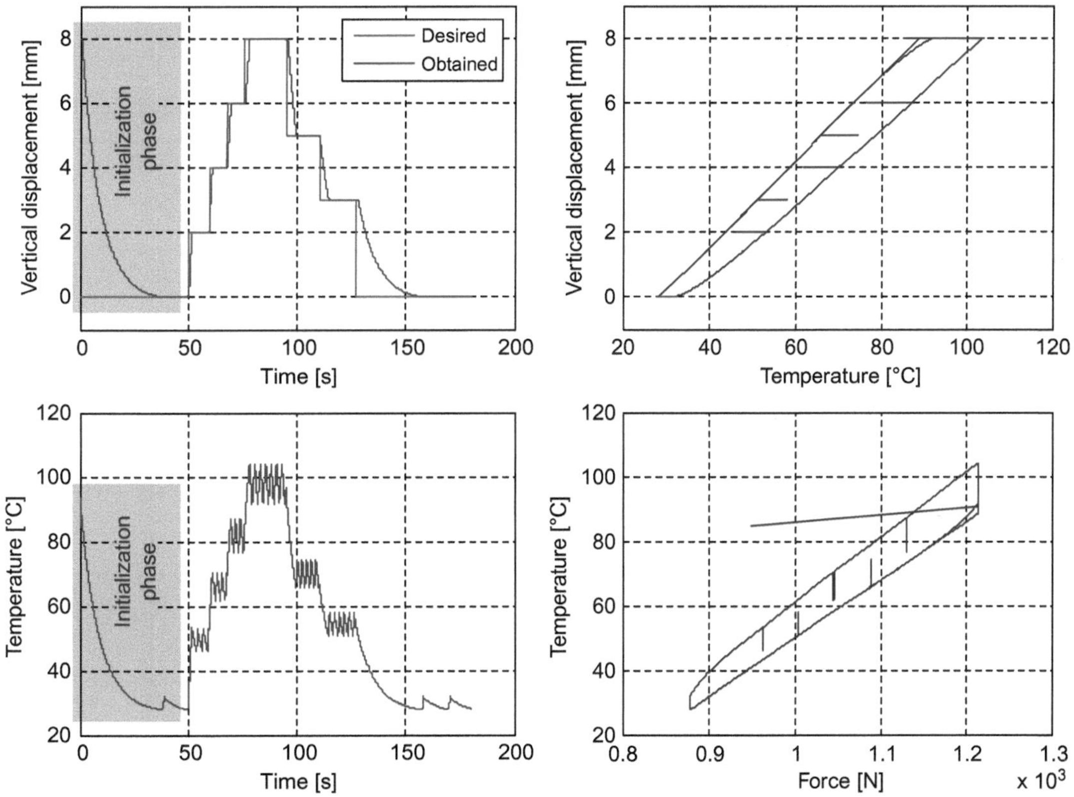

FIG. 37

Numerical simulation results.

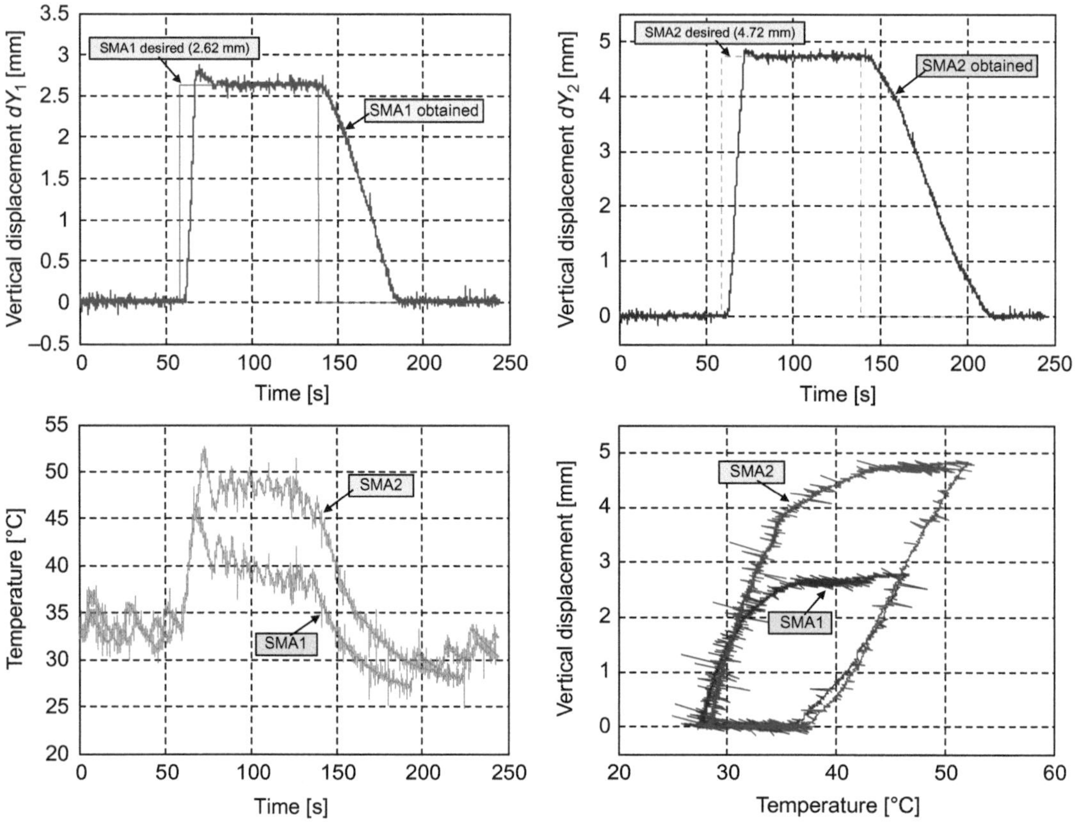

FIG. 38

Wind tunnel test ($M = 0.275$, $\alpha = -0.5°$).

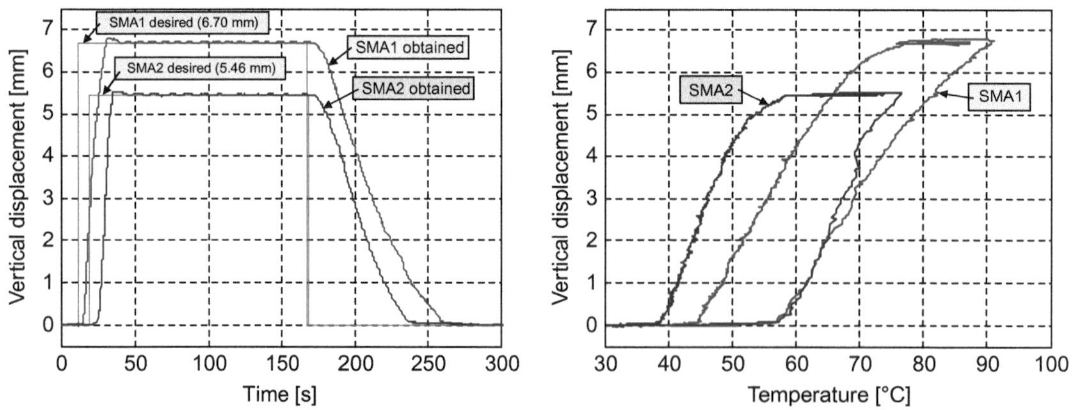

FIG. 39

Bench test results for $M = 0.275$ and $\alpha = 2°$ flow condition.

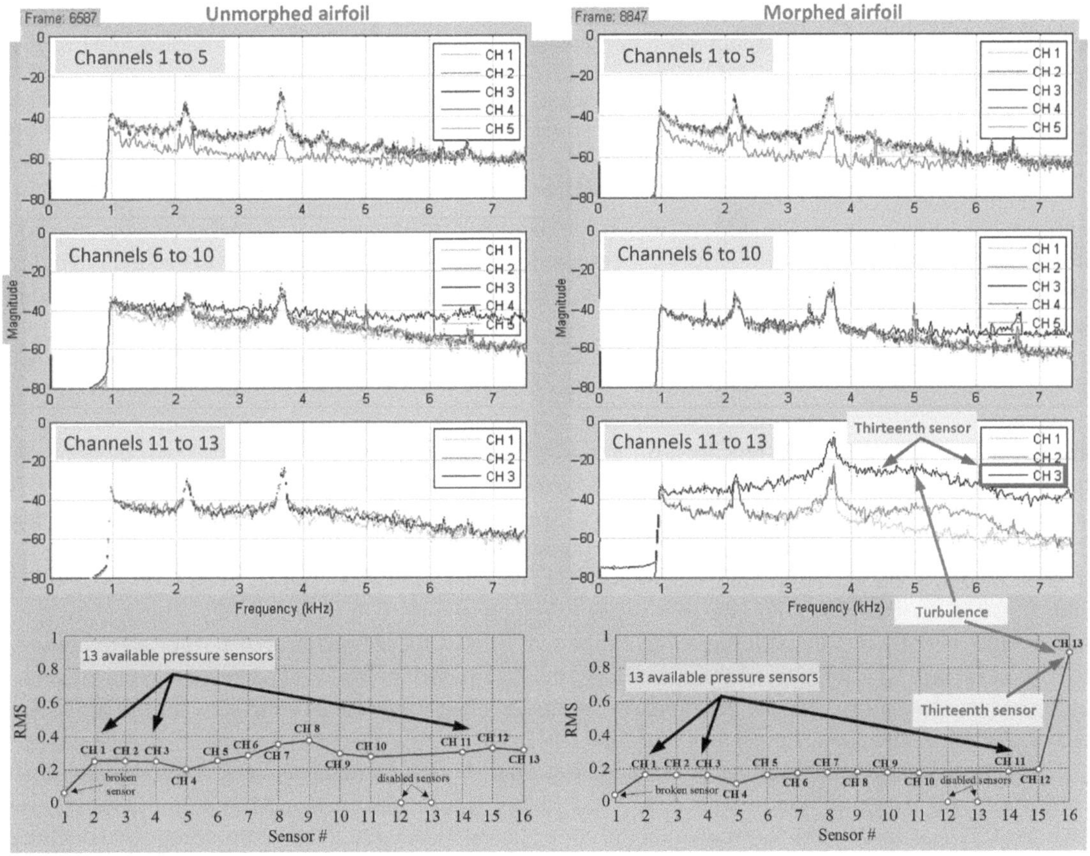

FIG. 40

Transition monitoring in wind tunnel test for $M = 0.275$ and $\alpha = -0.5°$ flow condition.

measurements appeared in this test because of the wind tunnel power sources and its instrumentation equipment. The transition monitoring proved that this noise level did not influence the transition point position significantly [26].

Fig. 40 depicts the results obtained in the wind tunnel for $M = 0.275$, $\alpha = -0.5°$ flow condition; the plots of the instantaneous RMS and spectrum for the pressure signal channels with unmorphed and morphed airfoil have been shown. From 16 Kulite pressure sensors initially mounted on the flexible skin, only 13 channels were available at this test (CH1 to CH13): sensor #1 was broken before the wind tunnel test, while sensors #12 and #13 were removed from plots because of bad dynamic signals that showed electrical failure. The left column of Fig. 40 presents the results for the reference (unmorphed) airfoil, and the right column displays the results for the optimized (morphed) airfoil. The spike of the RMS and the highest noise band on the spectral plots (CH 13 black spectra on the right low plot) for the morphed airfoil case suggested that the flow was already turned turbulent on the sensor of channel 13 (thirteenth available Kulite sensor), near the trailing edge; therefore, the transition point position was somewhere between CH12 and CH 13. For unmorphed airfoil, the transition was localized by the sensor on channel 8, with maximum RMS and the highest noise band on the spectral plots (CH 8 black spectra on the left middle plot) [26].

4.4 COMBINED ON-OFF AND SELF-TUNING FUZZY LOGIC CONTROLLER

This new control structure (Fig. 41) is fitted in the fuzzy controller block of the general open loop control structure implemented in Simulink and presented in Fig. 14. As can be easily observed, it is a hybrid controller architecture given by the combination of a bipositional controller (particularly an on-off one) and a FLC [2].

The upper FLC in the scheme from Fig. 41 is the main fuzzy controller, while the other one is used to adjust the coefficients. To obtain these blocks, two fuzzy inference systems (FIS) were developed and implemented: FIS1 for the main fuzzy controller, and FIS2 for the second fuzzy controller. The shapes chosen for inputs membership functions of both FIS were respectively s—functions, π—functions, z—functions, while the product fuzzy inference and the center average defuzzifier were applied (Sugeno). To define the rules, a zero-order Sugeno fuzzy model was chosen for both FISs. According to the values in Table 5, the membership functions for the FISs inputs were in the form depicted in Fig. 42 [2].

For the output membership functions of the FIS1 five constant values were chosen ($-1, -0.5, 0, 0.5,$ and 1), while for the FIS2 four constant values were chosen (0, 0.4, 0.8, and 1.2). Starting from the inputs' and output's membership functions of the two FISs, a set of nine inference rules were obtained for the FIS1, and 12 inference rules for the FIS2; the rule-based inference chosen for each is presented in Fig. 43, for both FIS. With the previous considerations, the fuzzy control surfaces for the FISs result characterized by the shapes presented in Fig. 44 [2].

The experimental testing of the developed controller, both in bench tests and in wind tunnel tests, confirmed its very good functioning. The wind tunnel test results presented in Fig. 45 for run test 42 (case C107, $M = 0.2, \alpha = 2°$) show that the controller worked even if the actuator required displacement was zero because of the gas springs pretension forces [2].

The associated instant plots of the RMSs and the spectrum for the pressure signals channels with unmorphed and morphed airfoils in this run test are displayed in Fig. 46. In this test the sensors #11 and #12 were removed from plots because of bad dynamic signals that showed electrical failure. In addition, sensor #5 showed a misalignment of the C_p value with its neighbors, which meant a leak or a pinched reference tube. The dynamic signal of sensor #5 was good, and for that reason the sensor value was kept on the plots. The left column of Fig. 46 presents the results for the reference airfoil (unmorphed), and the right column the results for the optimized (morphed) airfoil. The spike of the RMS suggested that on sensor #13, near the trailing edge, the flow was turbulent. The sampling rate of each channel for the pressure sensors data acquisition was 15 kHz, which allowed a pressure fluctuation FFT spectral decomposition of up to 7.5 kHz for all channels [2,32,33].

All the wind tunnel experimental results showed a decrease of the SMA wires operating temperatures vis-à-vis numerically simulated cases. An explanation can be found in the presence of the aerodynamic loads, with particular values for each test condition, which decreased the needed values of the actuation forces. The decrease of these temperatures was beneficial, taking into account the negative impact of a high thermal field on the other component of the system, especially on the flexible skin and on the pressure sensors [2].

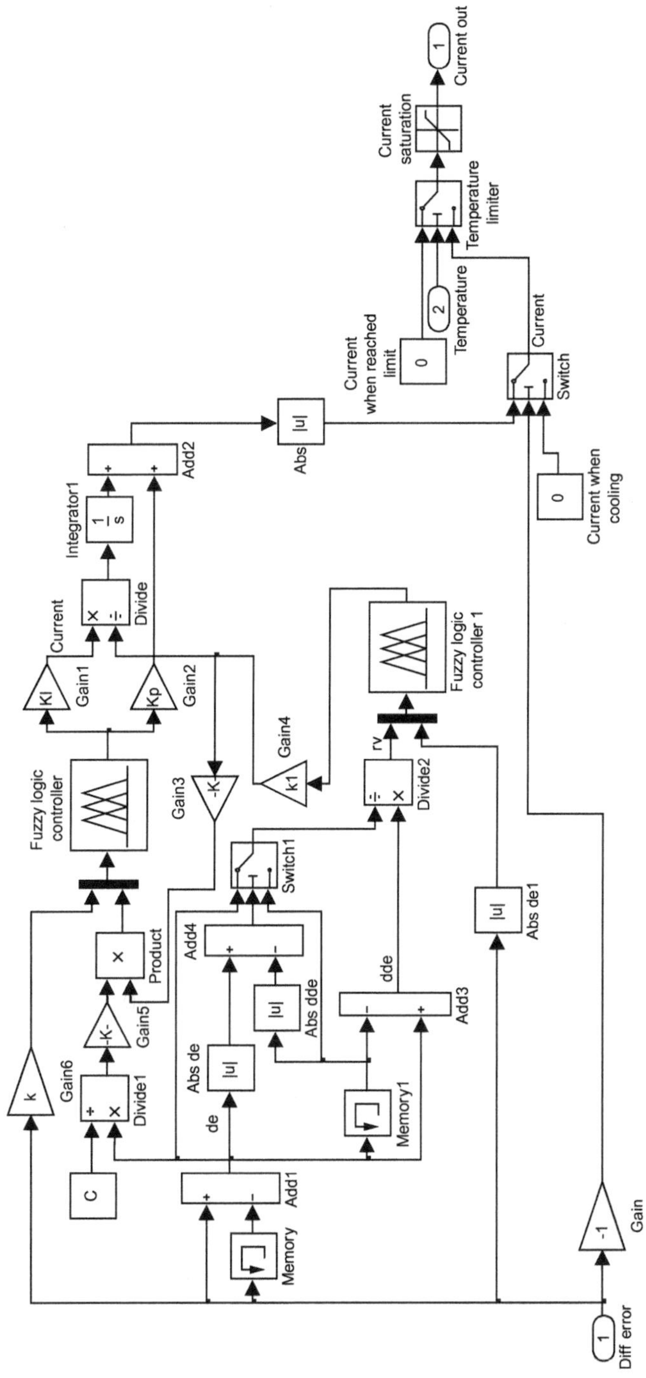

FIG. 41
The self-tuning fuzzy logic controller architecture.

Table 5 Parameters of the Inputs' Membership Functions

FIS	Input	*mf*	*mf* Type	*mf* Parameters			
				x_{left}	x_{m1}	x_{m2}	x_{right}
FIS1	1 and 2	*mf*1	*z*—function	−1	–	–	0
		*mf*2	*π*—function	−1	0	0	1
		*mf*3	*s*—function	0	–	–	1
FIS2	1	*mf*1	*z*—function	0	–	–	1
		*mf*2	*π*—function	0	1	1	2
		*mf*3	*s*—function	1	–	–	2
	2	*mf*1	*z*—function	0	–	–	0.4
		*mf*2	*π*—function	0	0.4	0.4	0.8
		*mf*3	*π*—function	0.4	0.4	0.8	1.2
		*mf*4	*π*—function	0.06	0.54	0.66	1.14

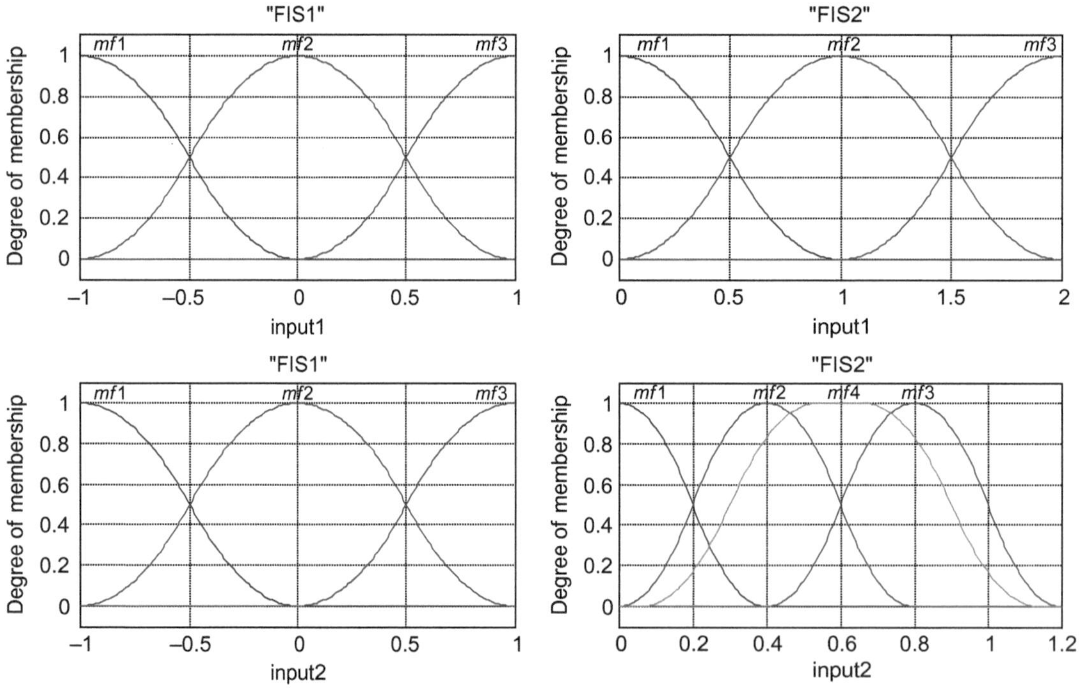

FIG. 42

Membership functions for the FISs inputs.

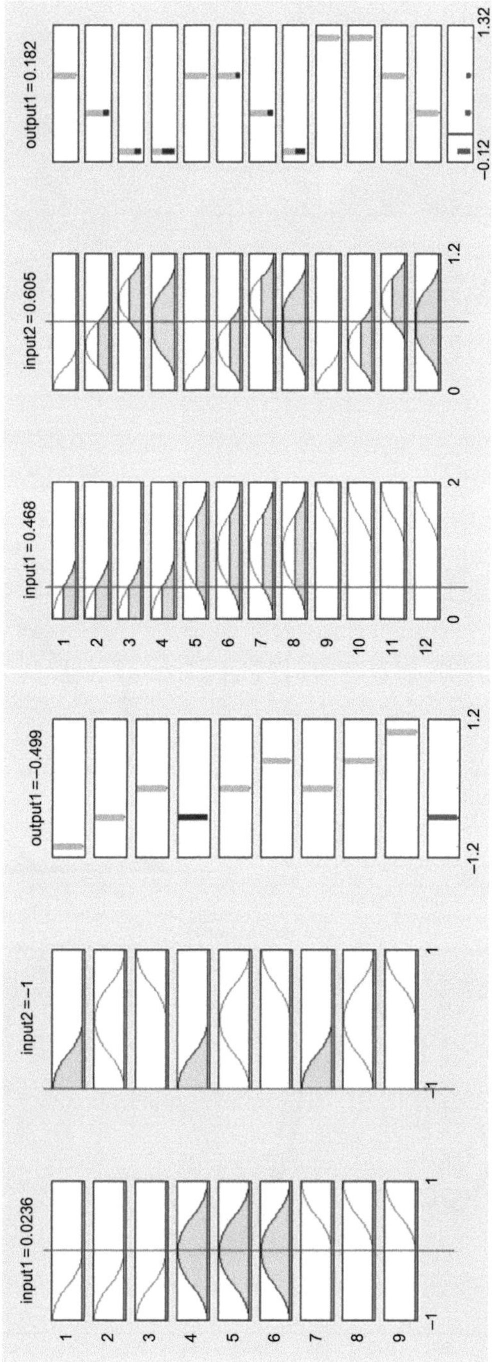

FIG. 43

Rule-based inference for the fuzzy inference systems.

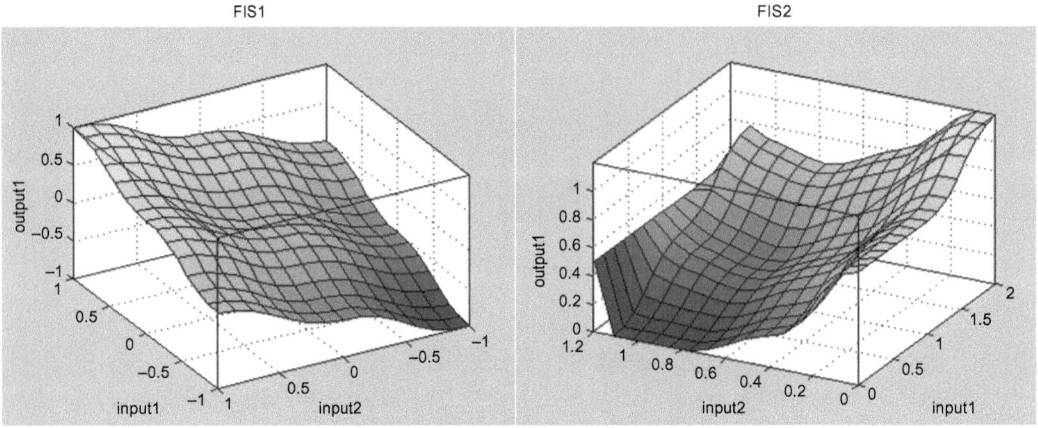

FIG. 44

The fuzzy control surfaces.

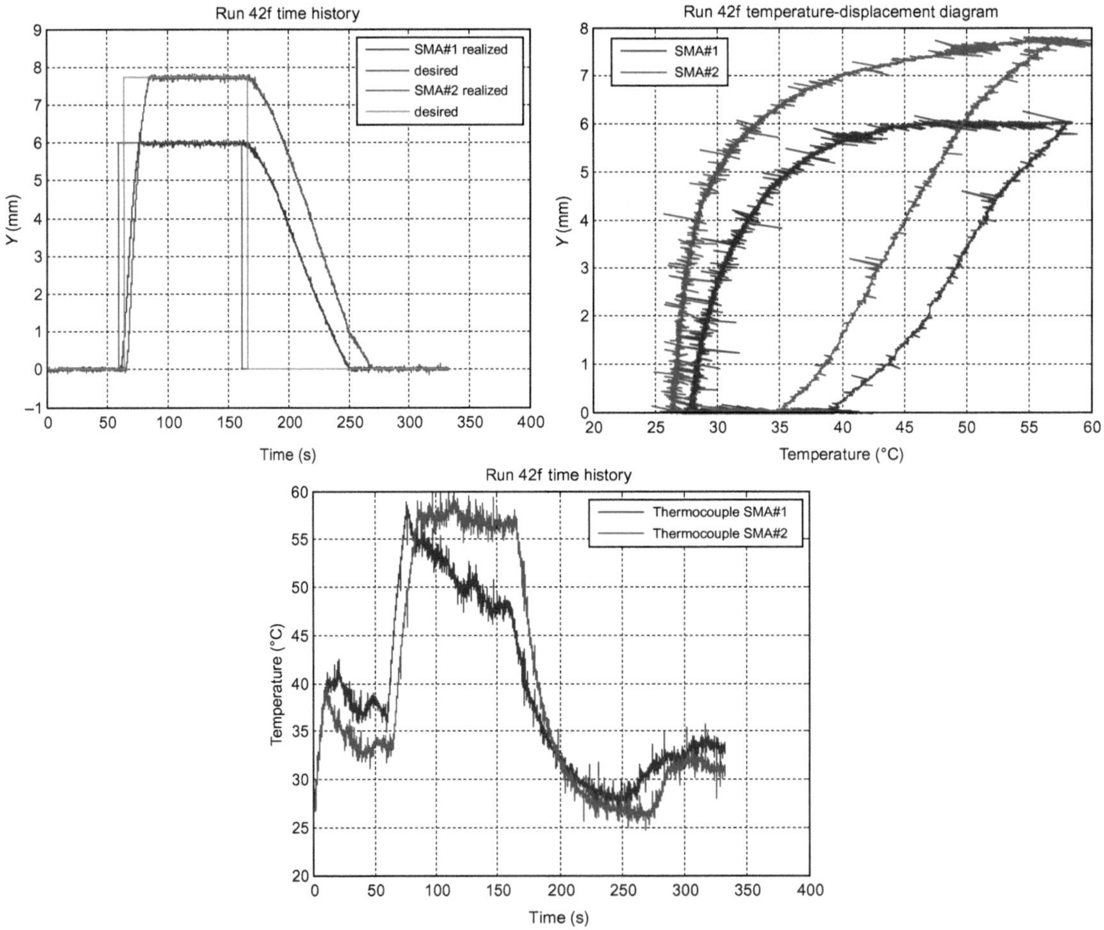

FIG. 45

Wind tunnel results for airfoil case C107: $M = 0.2$, $\alpha = 2°$.

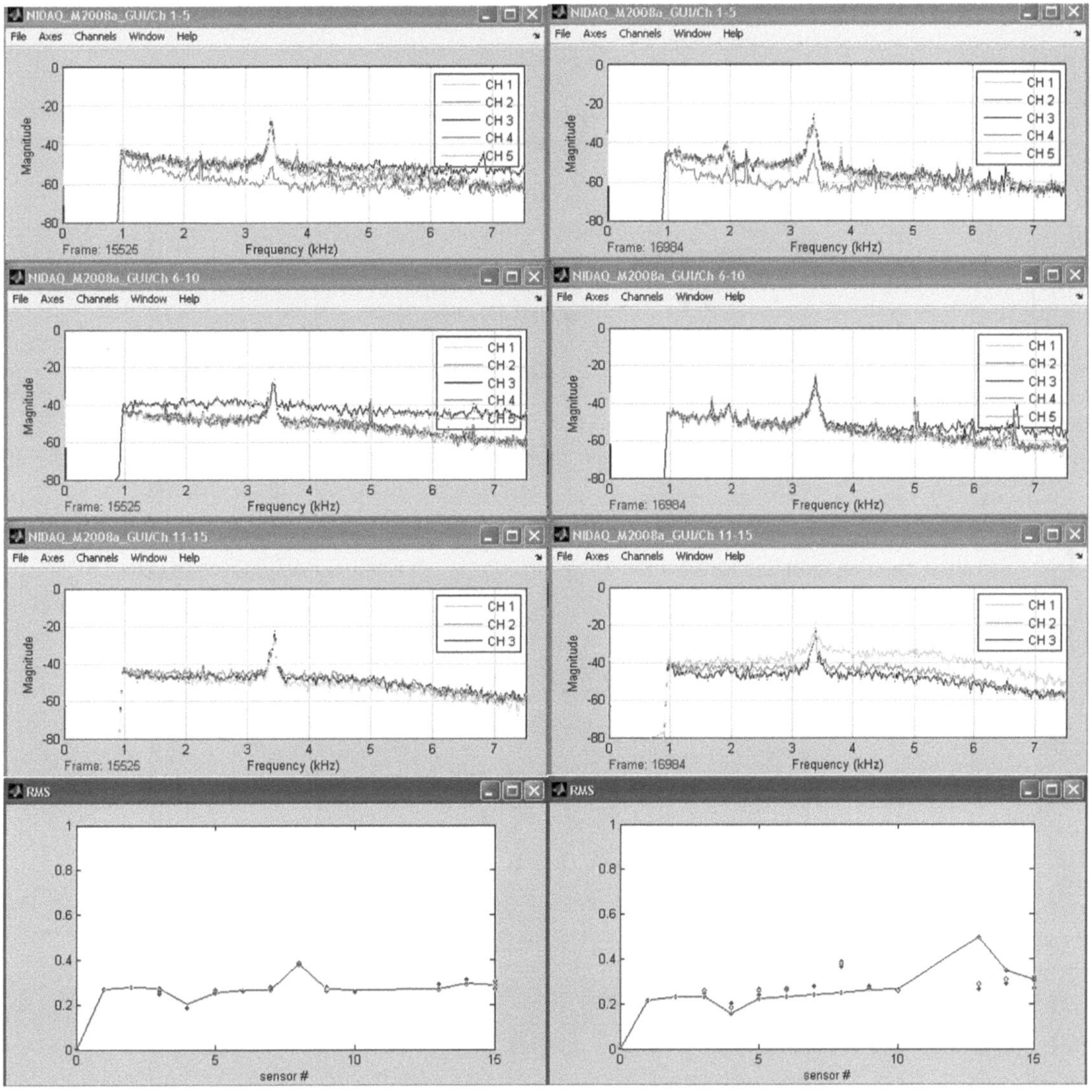

FIG. 46

Results obtained on the transition monitoring for the case $M = 0.2$, $\alpha = 2°$.

5 OPTIMIZED CLOSED LOOP CONTROL METHOD

Starting from the optimization algorithm developed for the *optimized closed loop architecture* (schematically presented in Fig. 9), some numerical simulations were performed in order to test it in a preliminary phase. The numerical simulation results were further used as comparison terms for real-time optimization during wind tunnel tests. Fig. 47A and B exposes the result of WTEA airfoil numerical

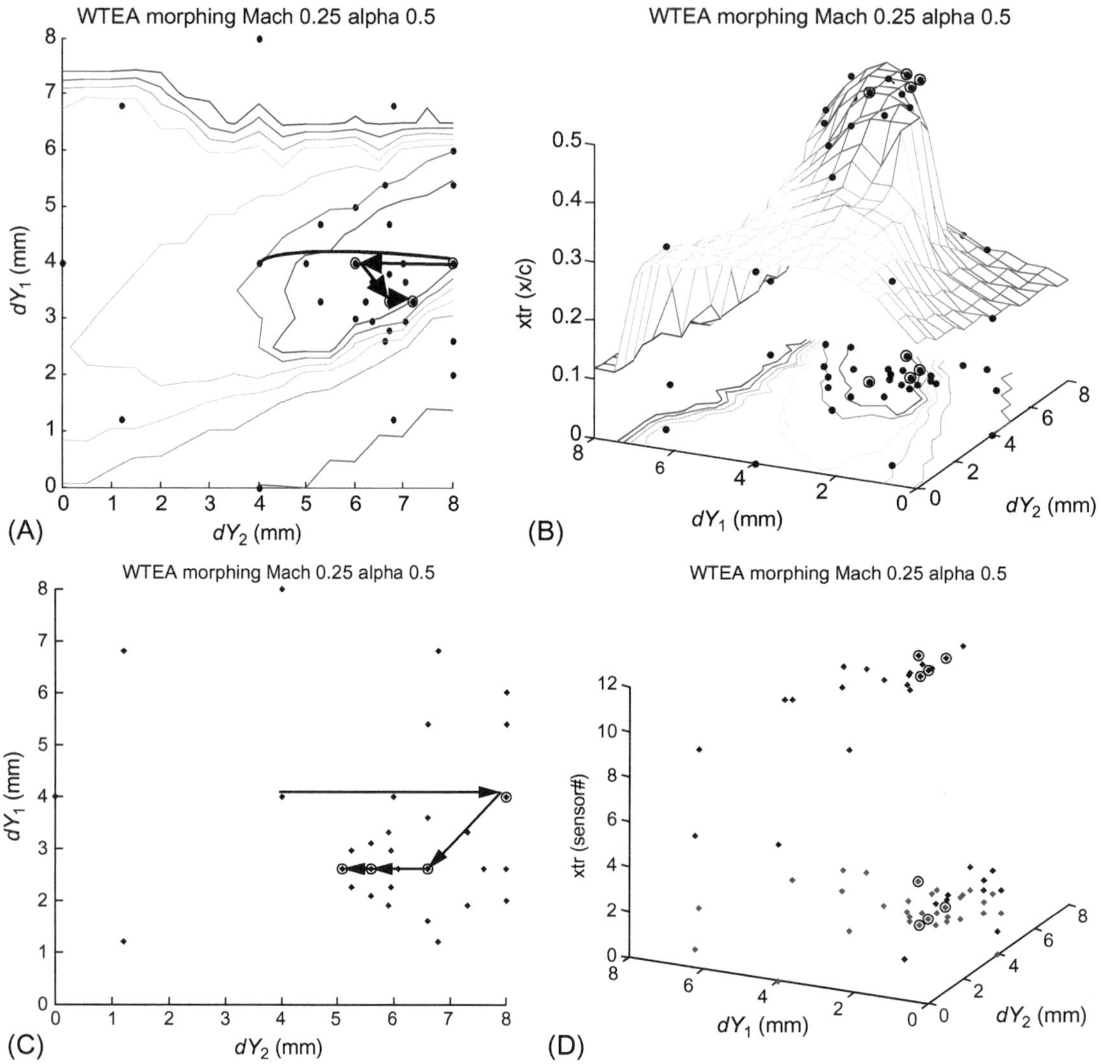

FIG. 47

Optimization during simulation using XFoil code (A and B) versus optimization in real time during wind tunnel tests (C and D) for the same airflow conditions $M = 0.25$, $\alpha = 0.5°$.

simulation optimization in the airflow case $M = 0.25$ and $\alpha = 0.5°$ and after four evaluation rounds: with rays of 4, 2, 1, and 0.5 mm. From Fig. 47B it can be concluded that the last round of evaluation was almost unnecessary because the maximum x_{tr} was found inside a plateau of maximums with very small differences between them. Before doing the optimization, a mapping of the search domain was performed, i.e., for each combination of dY_1 and dY_2 in the interval [0 mm, 8 mm] with a step of 1 mm, it was found the x_{tr} and the surface $x_{tr} = f(dY_1, dY_2)$ was built [27,28,32,33].

Fig. 47C and D shows the same optimization routine while running during the wind tunnel tests and in the same airflow conditions as the ones simulated, except that there was no map of the searched function. The result is slightly different because the airfoil shape of the real flexible skin under wind tunnel conditions is different from the airfoil shapes numerically defined through B-splines. Still, the result is similar in terms of actuator strokes dY_1 and dY_2 and position of the transition. Similarly, in Fig. 47D it can be observed a plateau of evaluation points that had the transition occurrence on the 11th pressure sensor [27,28,32,33].

Fig. 48 shows the results obtained by the optimization control in wind tunnel tests for C221 ($M = 0.225$, $\alpha = 1.5°$) and C225 ($M = 0.275$, $\alpha = 0°$) airflow cases.

Figs. 49 and 50 depict the pressure data RMSs for unmorphed (reference) and closed loop real-time optimized airfoils, and the FFTs of the Kulite pressure sensors data, for $M = 0.3$, $\alpha = 0.5°$ flow case. In the morphed-to-optimized airfoil case, the RMS plot displayed in Fig. 49 with star symbols, showed that the sensor with the maximum RMS has become the 10th plotted sensor. Starting point in

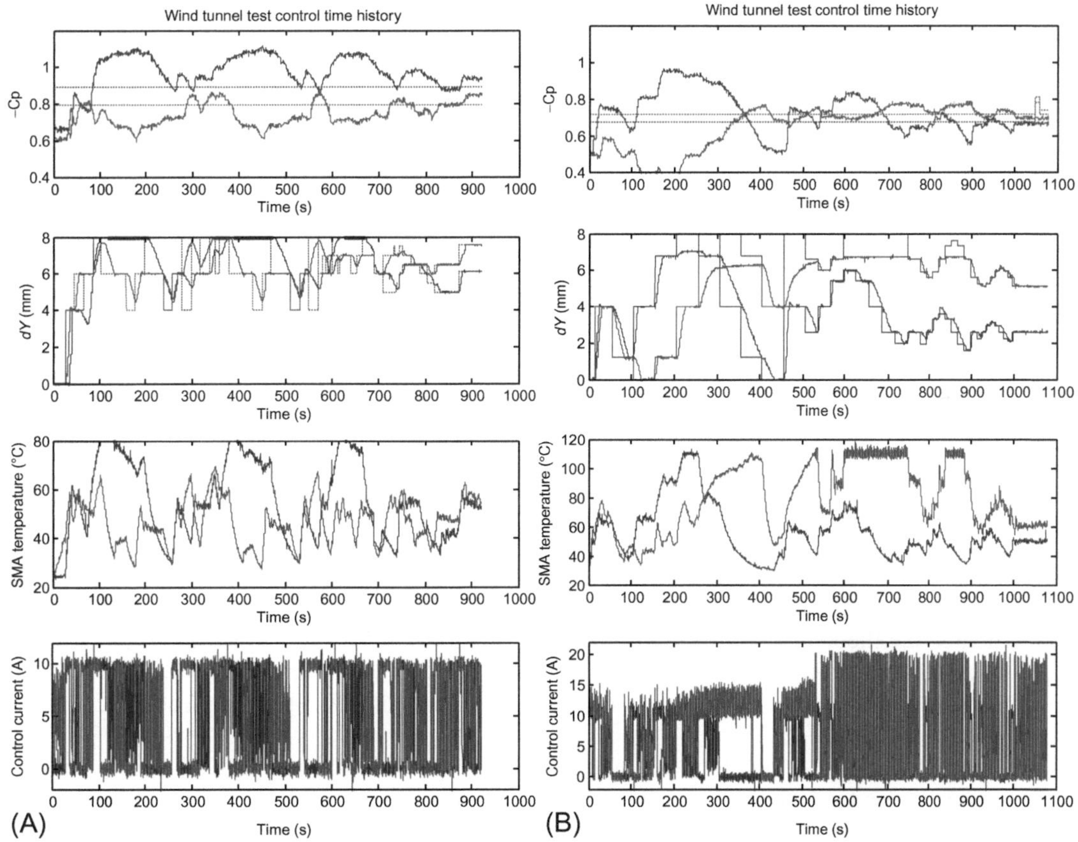

FIG. 48

Results for C221 and C225 airflow cases. (A) C221 ($M = 0.225$, $\alpha = 1.5°$). (B) C225 ($M = 0.275$, $\alpha = 0°$).

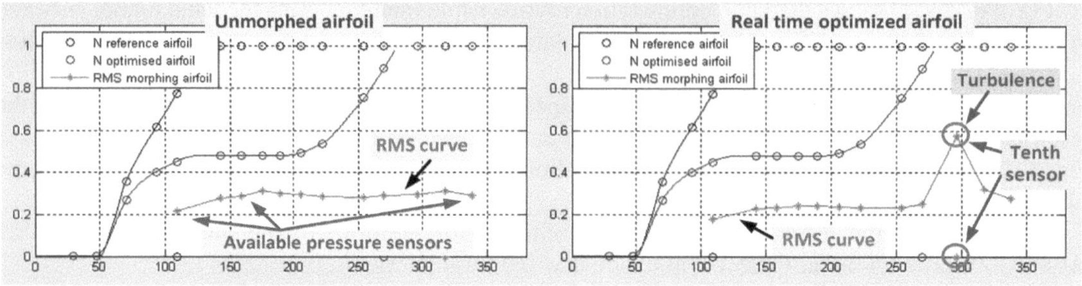

FIG. 49

The pressure data RMSs and the *N* factor distribution.

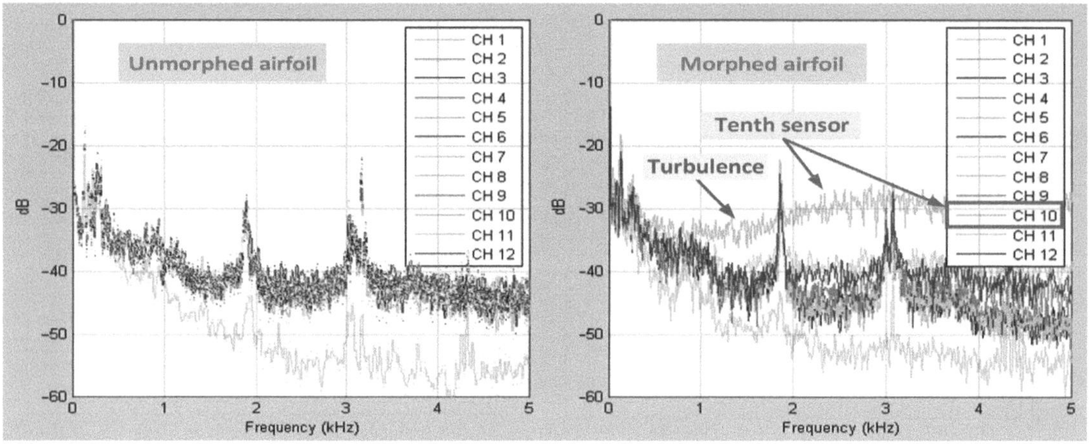

FIG. 50

Pressure signals FFT for reference and real time optimized airfoils ($M = 0.3$, $\alpha = 0.5°$).

optimization was the point with the coordinates dY_{1opt} and dY_{2opt}, characterizing the theoretically obtained optimized airfoil: $dY_{1opt} = 4.81$ mm, and $dY_{2opt} = 7.45$ mm. The experimentally obtained results validated the theoretical optimized airfoil, the optimization method implemented in the closed loop conducting to a morphed airfoil almost identical, $dY_{1opt\_cl} = 4.66$ mm, and $dY_{2opt\_cl} = 7.28$ mm. Also, the transition was detected on the same pressure sensor with the open loop case, i.e., the 10th sensor in the array [27,28,32,33].

The spectral decomposition of the pressure signals in Fig. 50 confirmed the Tollmien-Schlichting wave's occurrence in the 10th sensor, visible in the highest power spectra (12th channel in the right side plots) in the frequency band of 2–5 kHz; for this wind tunnel test the sampling rate of each channel for the pressure sensors data acquisition was 10 kHz, which allowed a pressure fluctuation FFT spectral decomposition of up to 5 kHz for all channels [27,28,32,33].

6 CONCLUSIONS

The objective of the presented research work was to develop and test various control systems for a morphing mechanism using smart materials, such as SMA, as actuators. These smart actuators modify the upper surface of a wing made of a flexible skin so the laminar to turbulent transition point moves close to the wing airfoil trailing edge. The final purpose of the research was to obtain a drag reduction as a function of flow condition by changing the wing shape.

Two architectures were developed for the used control system: an open loop and a closed loop. The strong nonlinearities of the SMA actuators' characteristics and the system requirements have led to various intelligent control structures for the inner loop of the control system, called the open loop architecture. The closed loop control architecture, called the *optimized* closed loop architecture, was developed in order to generate real-time optimized airfoils starting from the information received from the pressure sensors and targeting the morphing wing main goal: the improvement of the laminar flow over the wings upper surface.

All developed control structures were validated in wind tunnel tests in parallel with the transition point real-time position detection and visualization. The numerical and experimental results obtained with all developed control structures were very good, the controllers fully matching the requirements imposed for the research purpose achievement.

ACKNOWLEDGMENTS

The authors would like to thank the Consortium for Research and Innovation in Aerospace in Quebec (CRIAQ), Thales Avionics Inc., Bombardier Aerospace, and the National Sciences and Engineering Research Council (NSERC) for their financial support. We would also like to thank Dr. George Henri Simon for initiating the CRIAQ 7.1 project and Dr. Philippe Molaret from Thales Avionics and Dr. Eric Laurendeau from Bombardier Aerospace for their help and fruitful discussions.

REFERENCES

[1] Salgado, N., "Current and future aircraft technologies", Fuelling Aviation with Green Technology, ICAO HQ, Montreal, Canada, September, 9–10 2014.

[2] Grigorie, T.L., Popov, A.V., Botez, R.M., Mamou, M., and Mebarki, Y., "A new morphing wing mechanism using smart actuators controlled by a self-tuning fuzzy logic controller", 11th AIAA Aviation Technology, Integration, and Operations Conference, 20–22 Sept., SUA, 2011.

[3] T.L. Grigorie, A.V. Popov, R.M. Botez, M. Mamou, Y. Mébarki, Design and experimental validation of a combined PI and bi-positional Laws controller for delaying the transition from laminar flow to turbulent flow over a morphing wing, in: Informatics in Control, Automation and Robotics, Lecture Notes in Electrical Engineering (LNEE), vol. 89 (Part 2), Springer-Verlag, Berlin, Heidelberg, 2011, pp. 51–76.

[4] T.L. Grigorie, A.V. Popov, R.M. Botez, M. Mamou, Y. Mébarki, Design and experimental validation of a combined PI and bi-positional Laws controller for delaying the transition from laminar flow to turbulent flow over a morphing wing, in: Informatics in Control, Automation and Robotics, Lecture Notes in Electrical Engineering (LNEE), vol. 89 (Part 2), Springer-Verlag, Berlin, Heidelberg, 2011, pp. 51–76 Part 2.

[5] P. Gamboa, J. Vale, F.J.P. Lau, A. Suleman, Optimization of a morphing wing based on coupled aerodynamic and structural constraints, AIAA J. 47 (9) (2009) 2087–2104.
[6] S. Hiroharu, R. Kenichi, T. Asei, Laminar airfoil modification attaining optimum drag reduction by use of airfoil morphing, J. Aircr. 47 (4) (2010) 1126–1132.
[7] D. Munday, J.D. Jacob, Active control of separation on a wing with conformal camber, J. Aircr. 39 (1) (2002).
[8] A. Nadar, et al., Design and analysis of multi-section variable Camber wing, IJMER 1 (1) (2013) 122–128.
[9] Namgoong, H., Crossley, W.A., and Lyrintzis, A.S., "Morphing airfoil design for minimum aerodynamic drag and actuation energy including aerodynamic work", AIAA Paper 2006-2041, pp. 5407–5421, 2006.
[10] Raymond P. Le Beau, Jr., Anthony K., Nan-Jou P., and Jacob J., "Analysis of low speed flow over an adaptive airfoil with oscillating camber," 48th AIAA Aerospace Sciences Meeting Including the New Horizons Forum and Aerospace Exposition, AIAA Paper No. 2010-93, FL, 2010.
[11] B. Sanders, Aerodynamic and aeroelastic characteristics of wings with conformal control surfaces for morphing aircraft, J. Aircr. 40 (1) (2003) 94–99.
[12] S. J. Miller., "Adaptive Wing Structures for Aeroelastic Drag Reduction and Loads Alleviation", Ph.D. Dissertation, School of Mechanical, Aerospace and Civil Eng., Manchester, UK, 2010.
[13] C. Thill, J.D. Downsborough, J.S. Lai, I.P. Bond, D.P. Jones, Aerodynamic study of corrugated skins for morphing wing applications, J. R. Aeronaut. Soc. 114 (2010) 237–244.
[14] T.D. Usher, K.R. Ulibarri, G.S. Camargo, Piezoelectric microfiber composite actuators for morphing wings, ISRN Mater. Sci. (2013) 1–8.
[15] Zingg, D.W., Diosady, L., and Billing, L., "Adaptive airfoils for drag reduction at transonic speeds", 24th Applied Aerodynamics Conference, AIAA, Reston, VA, June, 5–8 2006.
[16] R. Pecora, F. Amoroso, M. Magnifico, Toward the bi-modal camber morphing of large aircraft wing flaps: the CleanSky experience, in: Proc. SPIE—The International Society for Optical Engineering, Volume 9801, 2016, Article No. 980106.
[17] I. Dimino, A. Concilio, R. Pecora, Safety and reliability aspects of an adaptive trailing edge device (ATED), in: Proc. 24th AIAA/AHS Adaptive Structures Conference, San Diego, CA, United States, Jan. 4–8, 2016.
[18] I. Dimino, A. Concilio, M. Schueller, A. Gratias, An adaptive control system for wing TE shape control, in: Proc. SPIE International Conference on Smart Structures 2013, San Diego, CA, USA, Mar. 10–14, 2013.
[19] Coutu, D., Brailovski, V., Terriault, P., and Fischer, C. "Experimental validation of the 3D numerical model for an adaptive laminar wing with flexible extrados", Proceedings of the 18th International Conference of Adaptive Structures and Technologies, Ottawa, Canada, October, 2007.
[20] L.T. Grigorie, R.M. Botez, Adaptive neuro-fuzzy inference system based controllers for smart material actuator modeling, J. Aerosp. Eng. 223 (G6) (2009) 655–668.
[21] T.L. Grigorie, R.M. Botez, A.V. Popov, Adaptive neuro-fuzzy controllers for an open loop morphing wing system, Proc. Inst. Mech. Eng. G J. Aerosp. Eng. 223 (7) (2009) 965–975.
[22] T.L. Grigorie, R.M. Botez, New adaptive controller method for SMA hysteresis modelling of a morphing wing, J. R. Aeronaut. Soc. 114 (1151) (2010).
[23] Grigorie, T. L., Popov, A.V., Botez, R.M., Mamou, M., and Mebarki, Y., "An intelligent controller based fuzzy logic techniques for a morphing wing actuation system using shape memory alloy", 52nd AIAA/ASME/ASCE/AHS/ASC Conference, Denver, USA, April, 4–7 2011.
[24] T.L. Grigorie, A.V. Popov, R.M. Botez, M. Mamou, Y. Mebarki, On-off and proportional-integral controller for a morphing wing. Part 1: actuation mechanism and control design, Proc. Inst. Mech. Eng. G J. Aerosp. Eng. 226 (2) (2012) 131–145.
[25] T.L. Grigorie, A.V. Popov, R.M. Botez, M. Mamou, Y. Mebarki, On-off and proportional-integral controller for a morphing wing. Part 2: control validation—numerical simulations and experimental tests, Proc. Inst. Mech. Eng. G J. Aerosp. Eng. 226 (2) (2012) 146–162.

[26] Grigorie, T.L., Popov, A.V., and Botez, R.M., Control of actuation system based smart material actuators in a morphing wing experimental model, AIAA Atmospheric Flight Mechanics Conference, Boston, MA, USA, August, 19–22 2013.
[27] Grigorie, T. L., Popov, A.V., and Botez, R.M., "Control strategies for an experimental morphing wing model", AIAA Atmospheric Flight Mechanics Conference, Atlanta, GA, June, 16–18, 2014.
[28] Grigorie, T.L., Botez, R.M., and Popov, A.V., Self-adaptive morphing wing model, smart actuated and controlled by using a multiloop controller based on a laminar flow real time optimizer, 24th AIAA/AHS Adaptive Structures Conference, San Diego, CA, USA, January, 4–8 2016.
[29] A.V. Popov, M. Labib, J. Fays, R.M. Botez, Closed loop control simulations on a morphing laminar airfoil using shape memory alloys actuators, J. Aircr. 45 (5) (2008) 1794–1803.
[30] A.V. Popov, R.M. Botez, T.L. Grigorie, Variations in optical sensor pressure measurements due to temperature in wind tunnel testing, J. Aircr. 46 (4) (2009) 1314–1318.
[31] A.V. Popov, T.L. Grigorie, R.M. Botez, Y. Mébarki, M. Mamou, Modeling and testing of a morphing wing in open-loop architecture, J Aircr 47 (3) (2010) 917–923.
[32] A.V. Popov, T.L. Grigorie, R.M. Botez, M. Mamou, Y. Mebarki, Closed-loop control validation of a morphing wing using wind tunnel tests, J. Aircr. 47 (4) (July–August 2010) 1309–1317.
[33] A.V. Popov, T.L. Grigorie, R.M. Botez, M. Mamou, Y. Mebarki, Real time morphing wing optimization validation using wind-tunnel tests, J. Aircr. 47 (4) (2010) 1346–1355.
[34] Sainmont, C., Paraschivoiu, I., and Coutu, D., "Multidisciplinary approach for the optimization of a laminar airfoil equipped with a morphing upper surface", AVT-168 NATO Symposium on the Morphing Vehicles, 20–23 April, 2009, NATO, Evora, Portugal, 2009.
[35] G. Amendola, M. Magnifico, R. Pecora, I. Dimino, Distributed actuation concepts for a morphing aileron device. Aeronaut. J. 120 (1231) (2016) 1365–1385, https://doi.org/10.1017/aer.2016.64.
[36] G. Amendola, M. Ciminello, I. Dimino, S. Ameduri, Real-time monitoring of a variable-camber aileron rib by original strain-angle transducer, J. Intell. Mater. Syst. Struct. 28 (8) (2017).
[37] G. Amendola, I. Dimino, A. Concilio, F. Amoroso, R. Pecora, Preliminary design of an adaptive aileron for the next generation regional aircraft, J. Theor. Appl. Mech. (Poland) 55 (1) (2017) 307–316.
[38] S. Luke, Essentials of metaheuristics, http://cs.gmu.edu/~sean/book/metaheuristics, 2009.
[39] Drela, M., "Implicit implementation of the full *en* transition criterion", 21st Applied Aerodynamics Conference, Orlando, FL, 23–26 June, pp. 1–8, 2003.
[40] M. Drela, M.B. Giles, Viscous-inviscid analysis of transonic and low Reynolds number airfoils, J. Aircr. 25 (10) (1987) 1347–1355.
[41] Z. Kovacic, S. Bogdan, Fuzzy Controller Design—Theory and Applications, Taylor and Francis Group, Boca Raton, FL, USA, 2006.
[42] G. Castellano, A.M. Fanelli, C. Mencar, Design of transparent fuzzy mamdani systems, in: Design and Application of Hybrid Intelligent Systems, IOS Print, Amsterdam, 2003, pp. 468–476.
[43] P. Terriault, F. Viens, V. Brailovski, Non-isothermal finite element modeling of a shape memory alloy actuator using ANSYS, Comput. Mater. Sci. 36 (4) (2006) 397–410.
[44] M. Mahfouf, D.A. Linkens, S. Kandiah, Fuzzy Takagi-Sugeno Kang Model Predictive Control for Process Engineering, IEE, Savoy Place, London, 1999, p. 4. WCPR OBL. UK.
[45] Erenoglu, I., Eksin, I., Yesil, E., and Guzelkaya, M. "An intelligent hybrid fuzzy PID controller", Proceedings of 20th European Conference on Modeling and Simulation, Bonn, Germany, 2006.
[46] H.X. Li, L. Zhang, K.Y. Cai, G. Chen, An improved robust fuzzy-PID controller with optimal fuzzy reasoning, IEEE Trans. Syst. Man Cybern. B Cybern. 35 (6) (2005) 1283–1294.
[47] Kulite Semiconductor Products, Inc., http://www.kulite.com/home.asp.

SECTION 5

Numerical Simulation

Editor:

Rosario Pecora
University of Naples "Federico II", Naples, Italy

List of Chapters

CHAPTER

INFLUENCE OF THE ELASTIC CONSTRAINT ON THE FUNCTIONALITY OF INTEGRATED MORPHING DEVICES

13

Salvatore Ameduri

The Italian Aerospace Research Centre, CIRA SCpA, Capua (CE), Italy

CHAPTER OUTLINE

NOMENCLATURE

A	three hinge arcs interface with the front spar web (Fig. 2A)
A/C	aircraft
ATED	adaptive trailing edge device
B	three hinge arcs interface with the front spar web (Fig. 2A)
C	three hinge arcs interface with the front spar web (Fig. 2A)
D	three hinge arcs interface with the front spar web (Fig. 2A)
DOF/DOFs	degree/degrees of freedom
f	subscript for front spar
FE	finite element
FEM	finite element model

Morphing Wing Technologies. https://doi.org/10.1016/B978-0-08-100964-2.00013-7

LE	leading edge
M	point in the middle of the wing span
r	subscript for rear spar
R	point at the root of the wing
T	point at the tip of the wing
TE	trailing edge
WL	winglet

1 INTRODUCTION

The effects of the constraint conditions play a fundamental role in the structural analysis.

Theoretically, the type of constraint influences the strategy adopted to solve the structural systems. In principle, isostatic systems can be directly solved, as the total number of equilibrium equations is equal to the unknown reactions. By contrast, hyperstatic systems require additional congruence conditions because of the higher number of constraints. Systems missing the minimum number of constraints give rise to instabilities. In addition to this classification, the intermediate and more practical configurations should be recalled, characterized by different typologies of yielding constraints: elastic linear or nonlinear supports, gaps or sliding connections, dissipative links.

From a practical point of view, the constraint typology affects the behavior of a structural system at several levels. Statically, the yielding (even though admitted) of the base of a bridge determines an additional stress field the designer must consider, as a function of the geologic profile of the ground. The transmission of forces and displacements assured by mechanical kinematic chains can be strongly penalized, as the actuation system discharges a part of the energy onto a yielding support; this is the case of an excavator or a tower crane, whose foundation soils must be adequately prepared, to guarantee a correct functionality and safety [1,2]. Dynamically, to draw another example from building sector cases, the rigidity of the foundation soil strongly affects the seismic response. As discussed in Ref. [3], soft foundations strongly penalize the seismic behavior. From the structural health point of view, the typology of the constraint plays a fundamental role, affecting both the time, the propagation, and the path of a fracture [4,5].

In the aerospace field, the elasticity of the constraint conditions is more directly related to the effectiveness of the aerodynamic surfaces. Conventional (flaps, slats, ailerons) and nonconventional (morphing) control devices are usually designed assuming support by a rigid structure. Designers attention is consequently devoted to other critical aspects, such as the kinematic chain, the attainment and maintenance of a target shape, the behavior of critical components like the skin, or the role of the device within the context of the wings global control architecture. However, the system is generally installed onto the wing structure, linked to the front (slat, morphing droop nose), the rear spar (flap, aileron, variable chamber trailing edge), and the tip rib (winglet flaps, variable chamber trailing edge). In any case, this connection necessarily occurs through structural, and then elastic, components. When actuated, therefore, a part of the action is inevitably transmitted to the interface of the wing. This results from one side in an induced deformation onto the wing structure and from the other side, in an alteration (and generally a penalization) of the movable-morphing performance of the single device. This induced deformation of the wing causes two drawbacks: (1) an additional stress concentrated at the interface and (2) an alteration of the movement of the other devices (due to the offsets induced at several levels and in the different zones of the wing).

The scope of the work at hand is to estimate the effects related to the actuation of the control surfaces due to the wing elasticity, regardless of aeroelastic effects already explored in other studies [6–8]. The work will focus on a finite element (FE) model representative of a complex morphing system integrated into a general wing. The outer wing box of a regional aircraft was assumed as numerical test specimen. A dedicated finite element model was built reproducing the adaptive system main details: layout and materials, interface connections, and actuation strategy of the movable surfaces. Three representative morphing technologies were considered: a droop-nose leading edge, a variable-camber trailing edge, and a spanwise double-flapped winglet. The droop nose motion, [10,11], was assured through an internal bending of the internal architecture, moved by two spanwise actuators. The rear hinges of the spanwise architecture are directly connected to the front spar; the skin is connected to the wing box at the top and bottom caps. The idea behind the morphing leading edge is to guarantee high lift performance equivalent to conventional systems (slats), without priming turbulence in cruise configuration. In contrast with conventional devices, in fact, the solid or quasi-solid nature of the droop nose leading edge guarantees a smoother geometry, without gaps and asperities inducing turbulence [12]. The achievement of the prescribed geometry at any flight condition and even in presence of an elastic support is thus of primary importance. The variation of the curvature of the trailing edge, [13,14], was modeled through linear actuators installed between the rear spar (connected to the top cap zone) and the upward rib rotating elements. The single degrees of freedom (DOF) feature of the architecture assures the transmission of motion throughout the system, up to the downward component. The scope of the device is to guarantee a curvature adaptivity to compensate for off-design conditions caused by weight variations during a typical aircraft (A/C) mission. The nature of the support also plays a fundamental role in this case, affecting the trailing edge geometry (local slope) and, thus, its contribution to the aerodynamic efficiency. The winglet consists of two flaps that were independently actuated through linear actuators, in turn installed between the flap hinges and the front spar of the winglet. The idea behind this device is to perform the spanwise load control and alleviation operations. In this sense, the effects due to the elasticity of the wing could be relevant for the induced variations on the attitude of the aerodynamic surfaces, whose effectiveness is crucial both for the correct setting of the load distribution and for certification issues, arising from the outboard location of the movable flaps [15,16].

The wing integrated model was at first investigated from a dynamic point of view, by implementing a normal mode analysis. The first mode shapes highlighted the main modalities of the wing to absorb elastic energy, then confirmed by the deformation field induced by the single morphing devices. The consequences of the actuation of each device was then investigated, by performing a nonlinear static analysis. The scope was to estimate the effects that were caused onto the entire wing as a function of the percentage actuation effort: the stress induced at the interface, the deformation (bending and torsion) transmitted onto the wing, and the cross-side effects on the other morphing systems (i.e., the variations of the deformed shape due to a certain device as a consequence of the cross-effects linked to the action of the other active elements). Synoptic tables were finally presented to provide a global view of the side effects (deviation of deflection and stress level) due to the percentage of activation of each device.

2 FEATURES OF THE FE MODELS

The results illustrated in this work were obtained through numerical analyses onto the FE model of the outer wing box of a regional aircraft, integrated with morphing devices. The model, sketched in Fig. 1A and B, is 4.66 m spanned and reproduces one-quarter of a wing; the internal layout is characterized by two main spars, representing the backbone of the entire wing part (from the inner section to the tip of the winglet). The main body of the model includes six ribs; the four intermediate ones connect the front to the rear spar, do not have any extension inside the leading edge and are segmented in the trailing edge domain, to enable the prescribed chamber-morphing variations. Both the spars keep a "C" shaped cross section up to the wing tip; at about three-quarters of their total span, they rise with a slope of 36 degrees, to form the winglet. The model is tapered by a factor of 0.25 and is swept at 18 degrees and 5 degrees on the leading and the trailing edge lines, respectively.

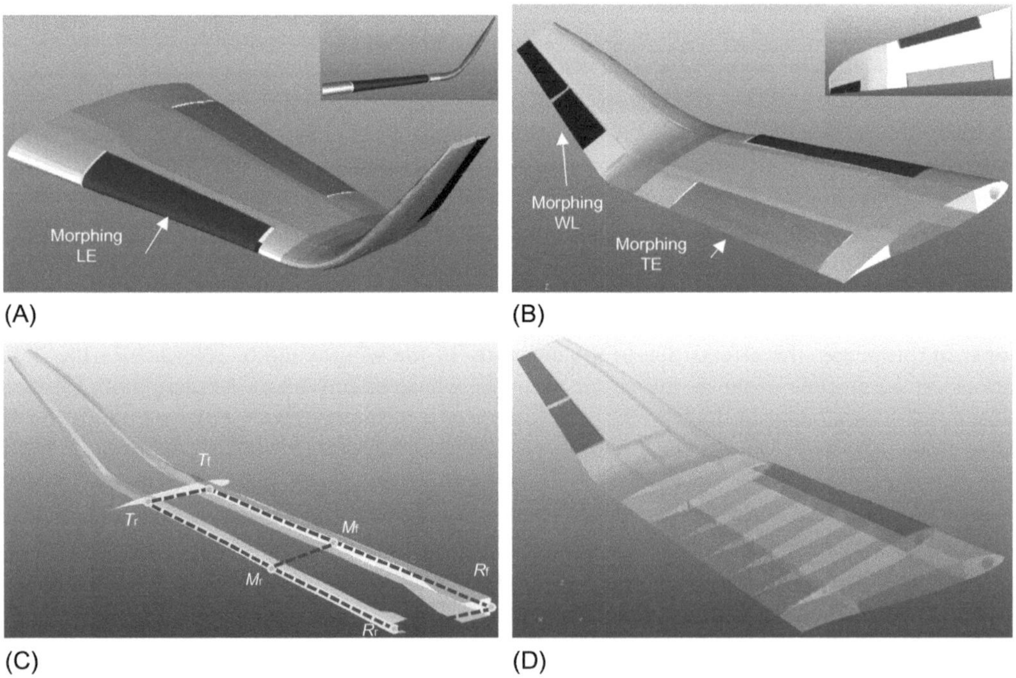

FIG. 1

Reference wing section—CAD representation: active systems details (A, B), their interfaces (C), layout sketch (D).

The following morphing devices were installed:

- a morphing droop nose, whose internal kinematic chain was derived from the architecture presented in [17];
- a variable chamber trailing edge, whose rib components are connected each other through the same architecture presented in [14]; for the actuation an alternative strategy was adopted); and
- an active winglet, integrated with two movable surfaces toward the span, with two independent linear actuators used to produce their rotations (this system was an original architecture, developed as an alternative to some literature example [15]).

The internal layout (see Fig. 1D) consists of: a front and a rear spar, representing the interfaces of the LE and of the TE (see details of the spars in Fig. 1C), five ribs connecting the spars along the entire wing span, a "tip" rib (the only one sketched in Fig. 1D) as connection element between the wing box and the winglet domain, and a couple of additional small ribs linked on the middle by stub spar, to stiffen the rib connection. In Fig. 1C also, some specific points and the ideal midplane (dashed lines) were sketched. In more detail, the capital letters, R, M, and T, state the spanwise position of the estimation points, root, mid, and tip, respectively; the subscripts, r and f identify parts of the rear and front spars. The effects due to the elastic nature of the wing and thus to the interaction of the different morphing devices, was expressed in terms of out-of-plane displacements of these points and rotation of the connecting segments. Both the front and the rear spars were "C" shaped and tapered at the root to house a solid stainless steel block; the latter, connected to a supporting structure, assures adequate clamping conditions during the wind tunnel tests. The thickness of the web of the front and rear spars was fixed to 7 mm to assure an effective flange-to-flange bolt connection to the leading-edge architecture and rib-to-flange connection to the ribs of the trailing edge. The caps of the spars were modeled through a composite layup element including both the spar cap in isotropic material and the skin laminate.

The morphing trailing edge extends from the 10 to the 58% of the model span and takes about the 30% of the chord, being structurally supported by the front spar. The rotating-movable part is not tapered: the rotation occurs in parallel to the trailing edge line and a tapered fixed structure connects the flap architecture to the rear spar. The morphing leading edge extends from the 15 and the 64% of the model span and takes about the 22% of the chord. It is tapered and the motion occurs in parallel to the front spar axis.

Table 1 collects the main features of the referred model.

The wing layout was realized through 2D FE elements, plates, and composite shells. The former were used for modeling ribs, spars, webs, and, generally, the skin (both standard and morphing); the latter were adopted for the overlapped zones generated by the connection among the spar caps and the riveted skin. The actuation of all the morphing concepts was modeled through thermal expanding beams, pinned at the edges of the designed transmission points. The internal architecture of the leading edge was linked to the web of the front spar through rigid elements. Gap elements were integrated between the internal leverage and the skin to adequately taking into account possible contact events in full activation conditions. The fixed rib elements of the morphing trailing edge were linked to the rear spar, merging the contact nodes, while the hinges among the rib pieces and the kinematic chain were modeled through rigid elements, linking all the translational DOFS and only the rib out-of-plane rotational DOFS. The movable flaps of the winglet were hinged to the main body through rigid elements pinning the rotational axis and the out-of-plane rotational DOFS. The kinematic chains used

Table 1 Main Features of The Outer Wing Box Model

Geometry	
Outer wing span [m][a]	4.66
Wing chords (root, winglet interface, winglet tip) [m]	2.01; 1.28; 0.49
Wing thickness (root, winglet interface, winglet tip) [m]	0.19; 0.12; 0.04
Sweep angle (LE, TE lines) [deg]	15.0; 9.0
Dihedral angle [deg]	3.65
Materials	
Front/rear spars	Stainless steel[b]
Ribs	7075-T6 Al[c]
Skin	Composite laminate[c]
FEM	
Number of elements	24,996
Number of nodes	21,776
Element types	Beam for linear actuators and stiffeners
	Plate for skin, ribs and spar webs
	Laminate shell for overlapped metallic layers (spar cap + skin)
	Rigid for LE kinematic chain connections
	Gaps for contact modeling among the LE skin and the internal architecture

[a] *Wing top view including the winglet.*
[b] *Young, Poisson mod., density [GPa, –, kg/m^3]: 210; 0.32; 7830.*
[c] *Young, Poisson mod., density [GPa, –, kg/m^3]: 71; 0.32; 2750.*

for all the devices are 1 DOF systems in which the actuator not only provides motion, but also cooperates with the remaining structure in absorbing the external loads.

2.1 LE MODELING STRATEGY

The leading edge concept integrated in the wing model is an evolution of concepts already presented in literature, [9–11]. The architecture was split in two independent portions, adequately tapered along the span. Fig. 2 depicts the main features of the device. In practice, per each portion, it is possible to distinguish two main structural subsystems: a spanwise structure, made of transversal actuators (red lines) and three arc-like structures (orange segments); a chordwise mechanism depicted in Fig. 2 (B—stick model) and (C—CAD drawing), in charge of transmitting deformation from the spanwise subsystem to the skin. Due to their geometrical features and the way they work within the global mechanism, stick elements were used for the simulation.

The upper and lower edges of the LE skin are connected to the top and bottom panels of the wing box. The internal architecture is hinged to the web of the front spar, that transmits directional forces to the rest of the structure) at the points A, B, C, and D, edges of the three hinged, Fig. 2A. Moreover, the

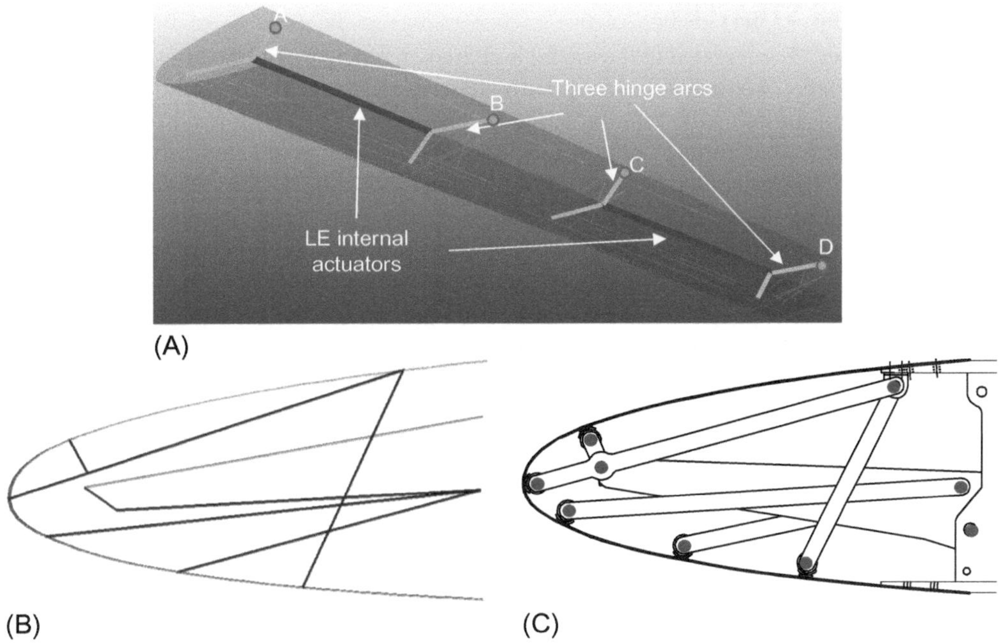

FIG. 2

Sketch of the morphing LE architecture (A); FEM stick model (B); implemented scheme (C).

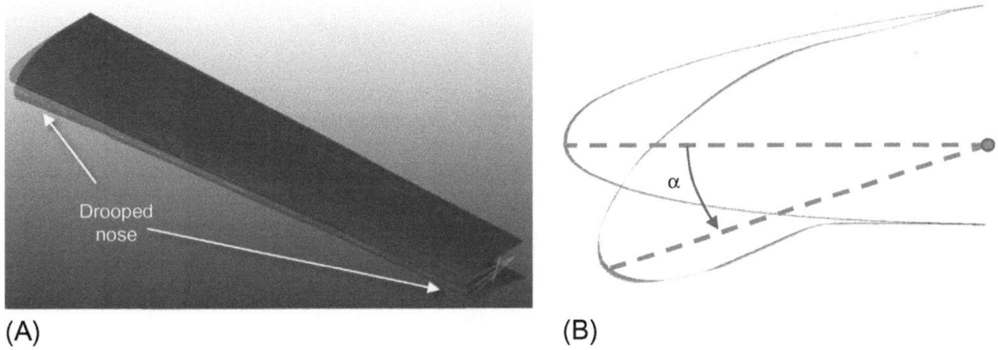

FIG. 3

Droop nose effect obtainable onto the tapered morphing LE: full skin (A) and inboard cross section (B) overlaps.

airfoil in-plane kinematic chain, Fig. 2B and C, was connected to the spar web through rigid links, enabling required rotational DOFs. The actuation was modeled through a thermal load acting onto the spanwise bar elements (namely LE internal actuators in Fig. 2). A pictorial representation of the movement (droop nose effect), produced by the implemented system, is given in Fig. 3, where the expected effect on the entire skin (A) and the inboard cross section (B) is illustrated.

2.2 TE MODELING STRATEGY

The variable chamber TE integrated in the model is characterized by partitioned ribs, [12]. The internal layout is illustrated in Fig. 4A; a detail of a representative rib and the actuation architecture (constituted of the expanding bars enforcing the rotation of the first rib component) is shown in Fig. 4C. The components are linked each other through hinges and alternatively through rod elements, this way forming a one-DOF system. The upstream parts are bolted through L-shaped flanges to the rear spar; this connection was modeled by thickening the contact zone with the spar. Plate elements were used both to model the core of the ribs and their thickened boundaries; provided the function of these latter parts, that is to contribute to the in-plane stability of the ribs and to guarantee an adequate support for the connection to the segmented skin, the 2D elements were placed normally to the rib, practically reproducing local foldings. The hinges were modeled through rigid connections constraining all the DOFS but the rib-in-plane rotation. To avoid undesired merge of the nearest nodes of two rib parts, an offset of the mid-plane at the junction was created (see details in Fig. 4B). The mechanical actuation was modeled through linear expanding elements hinged between the top cap of the rear spar and the first movable rib partition. Even though six actuators were integrated, one per rib, they were functionally grouped in two clusters (three inboard and three outboard), thus enabling possible independent actuation along the span. A 4-cm maximum vertical displacement of the TE line was considered in this

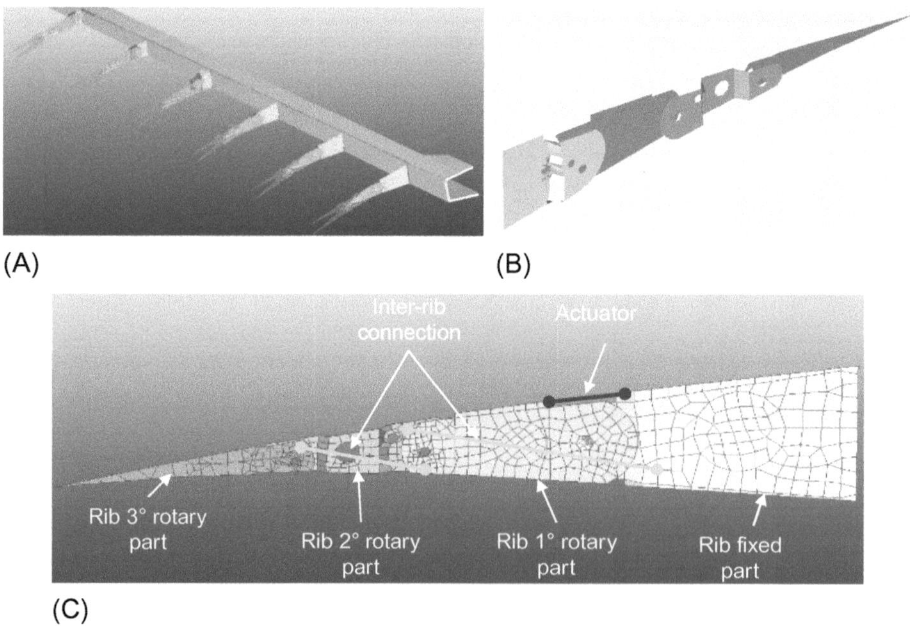

FIG. 4

Variable chamber TE: internal layout, clean and full actuated configurations (A), detail of a rib folding (B), rib partitions, connection rods and actuation (C).

work, corresponding to a global rotation of 5 degrees. This equivalent rotation was computed as the angle defined by the line joining the mean camber line at the rear spar with the trailing edge tip, with respect to the original, nondeformed chord.

2.3 WL MODELING STRATEGY

The winglet considered in this study was integrated with two movable surfaces along its span. Outboard and inboard independent rotations were enabled. A sketch of the layout is shown in Fig. 5. The actuation is obtained in a way similar to that proposed for the TE: two independent actuators, once more simulated through expanding bars, were hinged between the front spar and the movable surfaces of the WL, eccentrically deployed with respect to their pivot. Rigid elements were used to link the movable parts to the fixed structure; only the rotational DOF around the pivot axis was allowed. The inboard movable surface was connected to a rubber fairing, Fig. 5, [15] to guarantee an adequate aerodynamic smoothness even for large deflections. This arrangement is proved to strongly increase the rotational rigidity of the movable surfaces, with a consequent more relevant stress transmission at the attachment rear spar, where the surface is hinged, and on the front spar, on which the actuator insists; any transmission on the front spar has to be carefully taken into consideration for their impact on the other device supported by the spar, that is the LE. A $\pm$ 15-degree angular rotation range was considered.

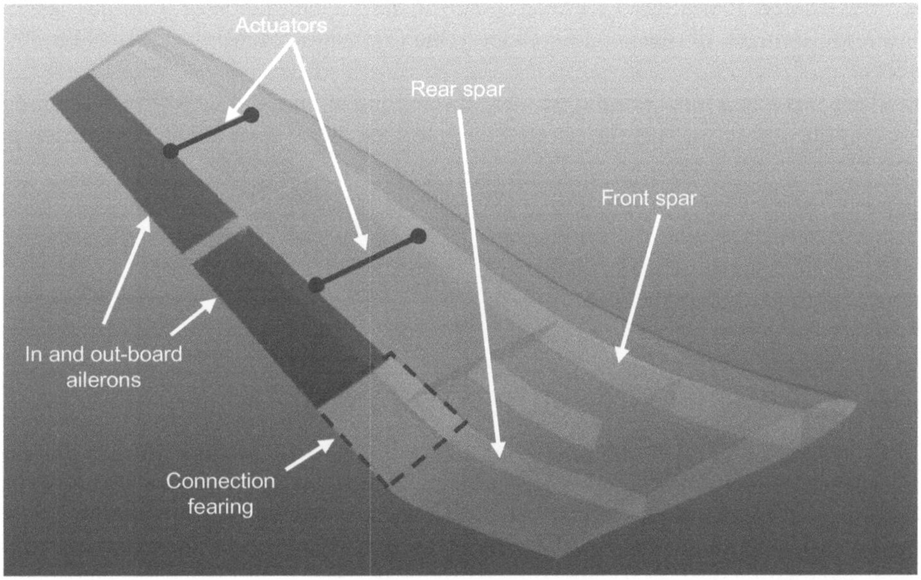

FIG. 5

Adaptive WL: internal layout, actuators - expanding bars, fairing connection to the inboard aileron (dashed line).

3 ISOLATED DEVICES BEHAVIOR

Before considering any combined effect, the three devices were at first singularly characterized, assuming rigid constraints. The target of this investigation was essentially to estimate the rigidity of each device to have an idea of the contribution to the overall rigidity of the entire outer wing model. Another aspect taken into account was the specific rigidity in the direction of the actuation (here intended as torque rigidity of the devices around the axis of rotation).

The LE, TE, and WL interfaces the web of the front spar and of the rear spar and the rib of connection to the wing, respectively, were constrained.

At first, a normal mode analysis was run per each device. Almost all the LE modal shapes prevalently involved the skin element (see Fig. 6A and B, 2nd and 3rd modes, respectively). Only the first mode involved the internal architecture, causing a transversal motion of the kinematic chain.

Then, the TE was investigated. As shown in Fig. 7 on the top, the first modal shape is characterized by a chordwise chamber bending; the other ones instead involve the spanwise direction.

Finally, the WL dynamic response was estimated. (Fig. 8) The first two modes involve only the outboard movable part, while the next ones have also some impact on the fairing and on the top skin.

The second type of investigation was aimed at estimating the rigidity of the devices referred to the direction of actuation, that is to say, out of the plane of the aerodynamic surface. To this scope, the rotation of the devices produced by unitary forces acting normally to the surfaces was computed through a static analysis. Fig. 9 shows the schemes of application of the forces and the results obtained. Due to the specific architecture of the LE (i.e., split into an inboard and outboard kinematic chains), the force was divided and distributed onto the inner edges of the three arc structures (Fig. 9A). Different strategies were adopted for the TE and the WL: a unique force was applied on the aft stringer of the ATED and on the TE segment of the WL (Fig. 9B and C, respectively).

The rigidities were computed as ratio between the applied moment (here intended as product of the unitary force by the arm respects to the constrained interface of the device) and the rotation produced. A summary of the results is reported in Table 2.

Both the LE and TE resulted much more rigid than the WL (five to six orders). The more compact geometries of the LE and of the TE, jointly to the stronger effect of the constraint (closer to the kinematic chains and directly supporting the architectures), contributed to this result.

4 GLOBAL STIFFNESS OF THE OUTER WING BOX

The investigations on the complete wing model started with the estimation of some global performance indexes. In this sense, a nonlinear static analysis was implemented to find representative values of the overall bending and torque rigidity. For that, a vertical concentrated force and a torque moment were applied on the interface rib between the WL and the wing box to measure the produced max vertical displacement and rotation (Fig. 7). The bending stiffness, defined as the ratio between the applied vertical force multiplied by the span and the produced rotation, and the torque rigidity, defined instead as the ratio between the applied torque moment and the resulting rib rotation, were evaluated as 1.18×10^7 and 6.53×10^2 Nm/rad (Fig. 10).

(A)

(B)

FIG. 6

Modal shapes of the LE rigidly supported. (A) 2nd mode: 52.2 Hz. (B) 3rd mode: 57.8 Hz.

After this first investigation, a normal mode analysis was carried out. The modal shapes of the wing were visualized, highlighting both its dominant dynamic behavior and the main contributions of the different morphing devices to the single shapes, Fig. 8. The first mode (11.7 Hz) was characterized by a spanwise bending behavior. It is worth to note the different contribution given by the two WL tabs: the deflections of the inboard one are less relevant, because of the fairing connection to the wing. This effect is more evident in the second modal shape (in which the torque component dominates). The contribution of the LE to both the modes is relevant and highlights a poor capability of the architecture in absorbing loads out of the wing plane. The curvature of the skin and its closed cross section favorably contribute to the bending and torque stability of the wing The third and the fourth modes, not reported in

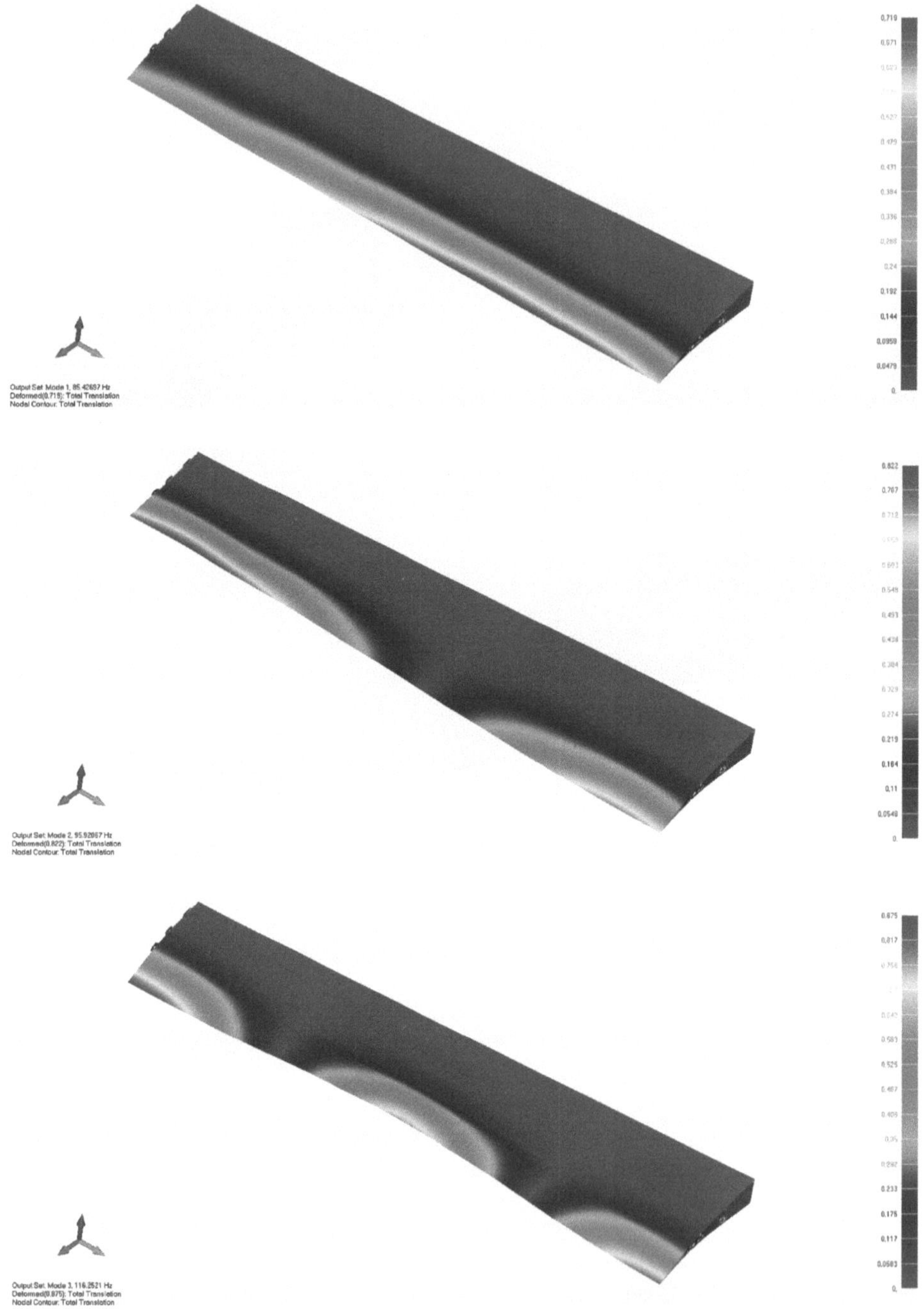

FIG. 7

Modal shapes of the TE rigidly supported. 1st mode: 85.4 Hz. 2nd mode: 95.9 Hz. 3rd mode: 116.2 Hz.

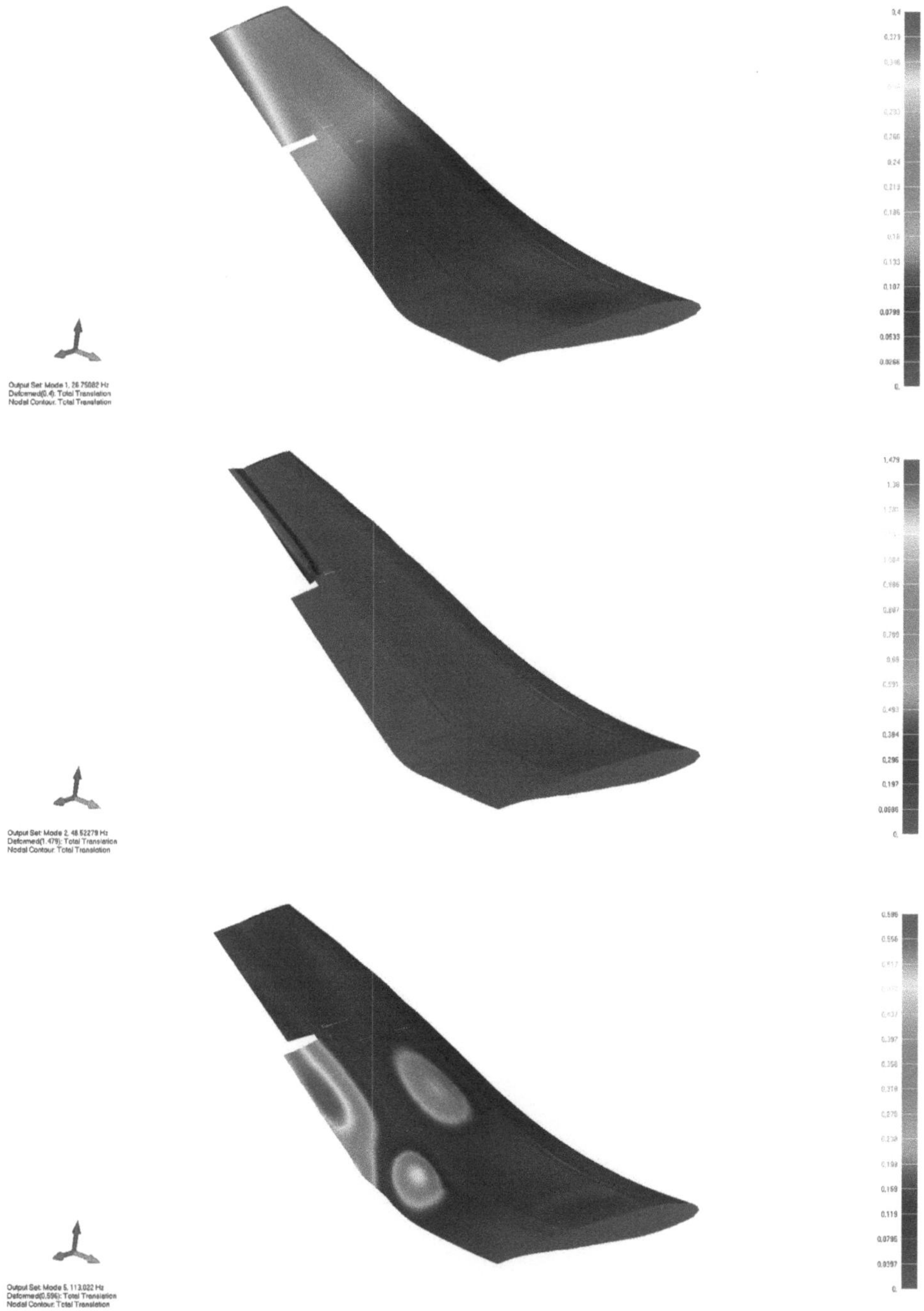

FIG. 8

Modal shapes of the WL rigidly supported at the connecting rib. 1st mode: 26.7 Hz. 2nd mode: 48.5 Hz. 5th mode: 113.0 Hz.

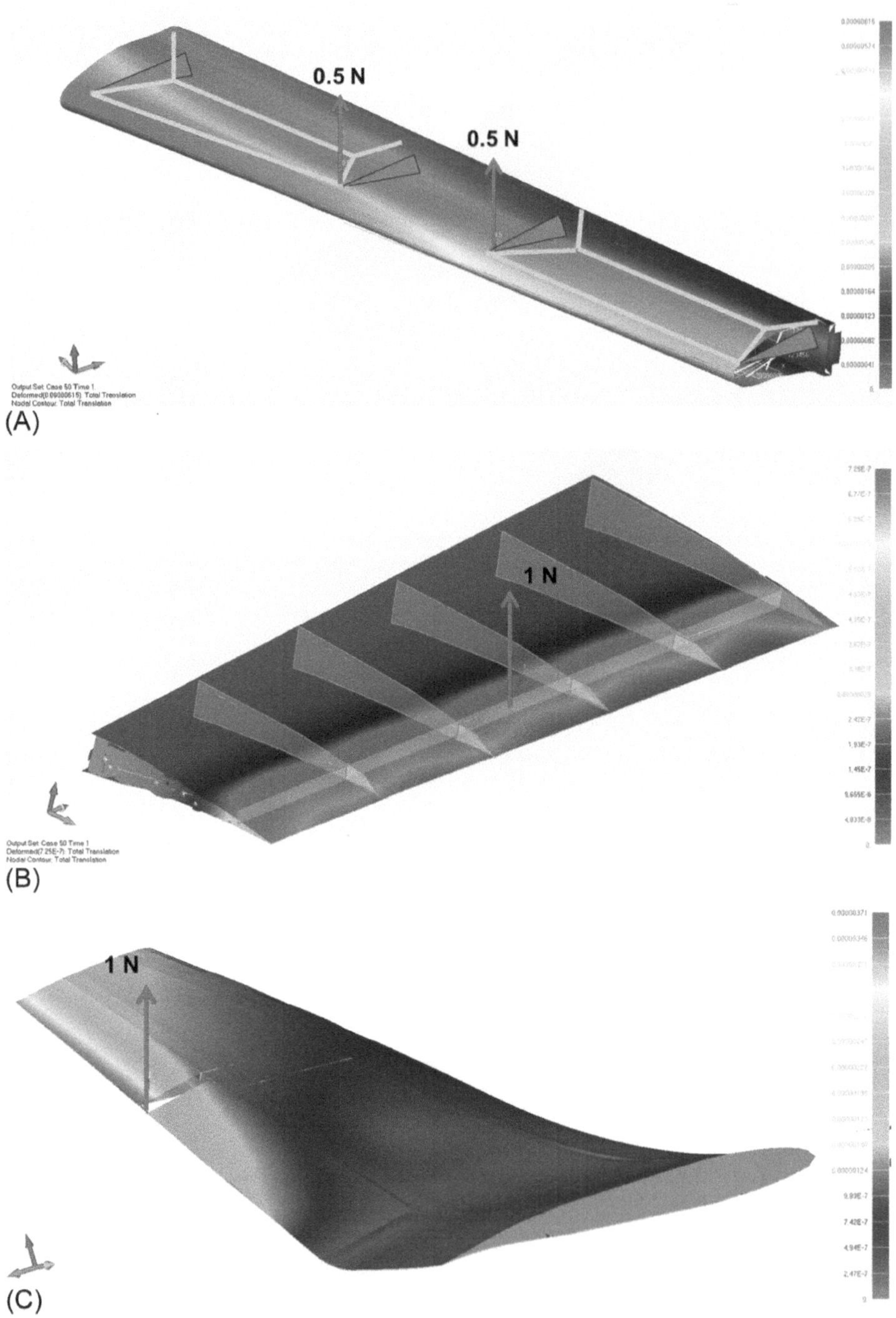

FIG. 9

Schemes of application of the forces to the morphing devices and numerical result: leading edge (A), trailing edge (B), and winglet (C). All applied force systems have a resultant of 1 N.

Table 2 Summary of the Out-Of-Plane Rigidities of the Morphing Devices

Morphing Device	Applied Force (N)	Arm (m)	Corresponding Torque (Nm)	Computed Rotation (rad)	Out of Plane Rigidity (Nm/rad)
LE	1	0.251	0.25	1.54E−05	1.63E+04
TE	1	0.366	0.37	1.38E−06	2.65E+05
WL	1	0.716	0.72	8.55E−01	8.37E−01

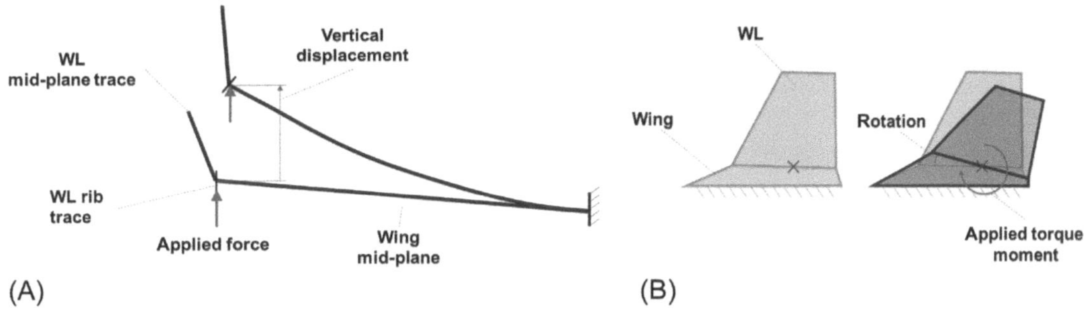

FIG. 10

Schemes adopted for the estimate of the bending (A) and torque (B) rigidity of the wing.

the figure, are yet characterized by the strong contribution of both the LE and the WL architectures. The first relevant contribution of the TE is evident only at the fifth modal shape (69.0 Hz), with a typical flapping motion. It is worth to note that due to the lower rigidity of the single devices, the WL and the LE give their contribution at the very early modes: first and second mode, LE and WL flapping (Fig. 11).

5 EFFECTS OF THE ACTUATION OF THE MORPHING DEVICES

The scope of the activities illustrated below is to estimate the effects produced by the target level of actuation of each single device. The results were expressed in terms of:

- devices deflection or vertical displacement versus the percentage level of actuation;
- bending and torque induced at some wing stations;
- transmitted stress, produced by the actuated device at the wing box interface.

The angles of deflection were herein defined as the slope of the straight lines linking the extreme of the devices (leading or trailing edges) with the airfoil chord intersection with the interface with wing box (see the dashed line in Fig. 3B, LE). The bending effect on the wing was expressed in terms of out-of-plane displacement of some points at root, middle and tip stations of the front and rear spar axes ($R_{f,r}$, $M_{f,r}$, and $T_{f,r}$) in Fig. 1. Finally, the torque effects were expressed by computing the slope of the straight lines connecting these points along the chord direction.

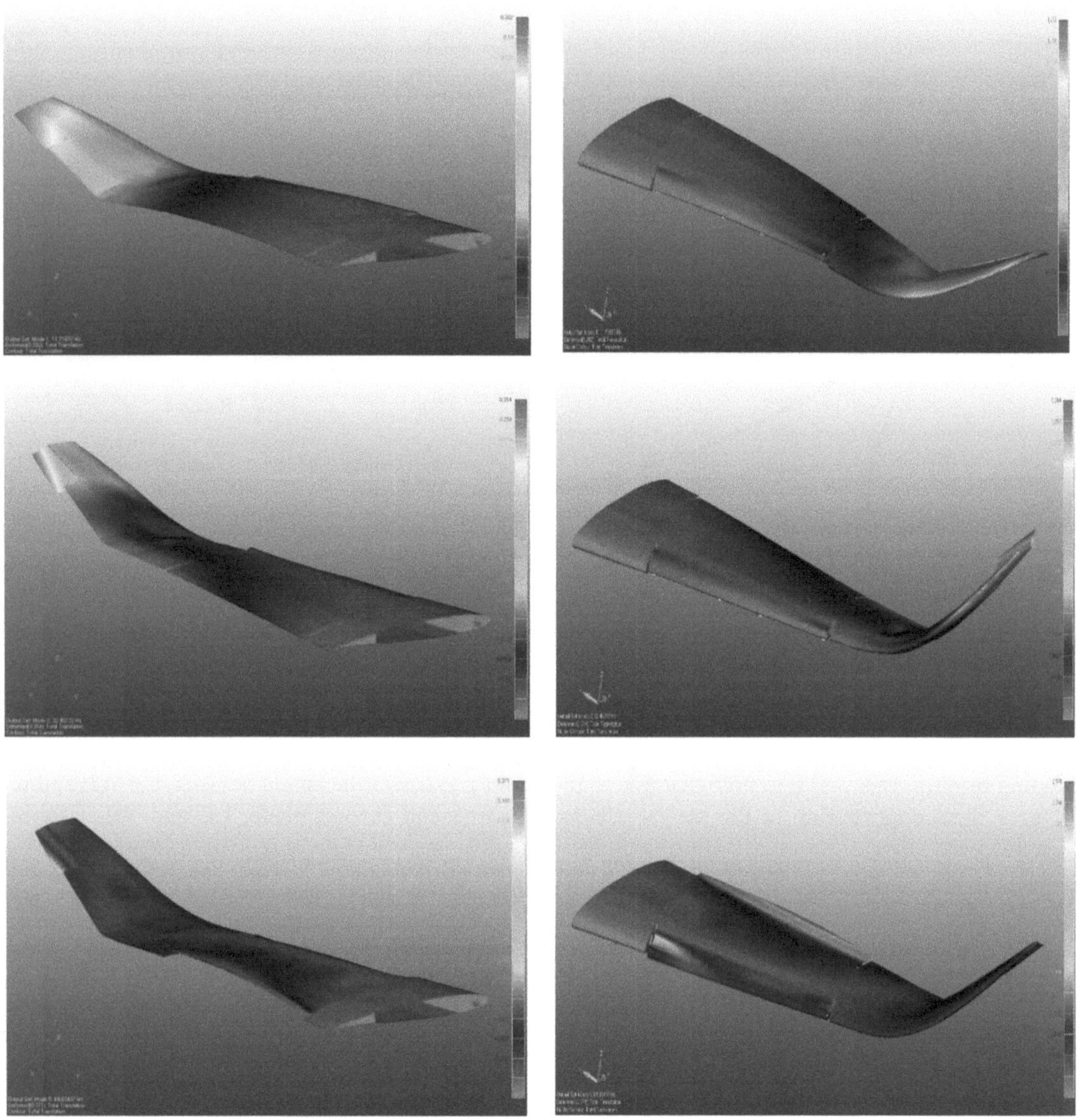

FIG. 11

First modal shapes of the outer wing model. 1st mode: 11.7 Hz. 2nd mode: 32.4 Hz. 5th mode: 69.0 Hz.

5.1 CROSS EFFECTS

A set of four graphs was dedicated to each device, the first three plots reporting the rotation induced on each device by the actuated one, the last plot showing the stress level at the interface between the wing and the actuated device. A dashed horizontal line in the last plot indicates the maximum allowable stress; any actuation over this level is not allowable. The same limit was taken back to the first three graphs, defining the prohibited red domain on the right side. All the results were presented as a function of the percentage of actuation of each single device. Again, it is worth noting that because the devices were designed under the assumption of rigid constraint, only a portion of the entire actuation level can be effectively exploited before running afoul of the stress limitation. Due to the intrinsic rigidity of the wing box, rotations of the actuated device are higher (at least of two orders) than the ones induced onto the other ones.

At first the effects of the actuation of the LE were estimated, Fig. 12. The results were expressed in terms of percentage of possible effort (actual vs. max capability of the selected actuator; for the chosen extensional devices, their full potential was equal to the maximum stroke). The basic hypothesis was that the morphing systems action does never lead to overcome the allowable stress level at the interface, here assumed equal to 200 MPa for the stainless steel constitutive material. As shown in the bottom graph, extensions no higher than 73% are possible. This constrains the inboard and outboard LE rotations to 4.2 degrees and 7.2 degrees, respectively. The induced rotations on the active TE and WL are three and two orders smaller, respectively. Analogous graphs were plotted for the activation of the TE, Fig. 13, and the WL, Fig. 14.

The maximum activation level of the TE actuator that matched the stress limit, resulted equal to 74%, again. This corresponds to around 7-degree maximum inboard rotation. The induced rotations on the other devices gave rise to results well different for the other two active elements. The LE was practically not affected, showing an induced maximum rotation of 0.35 degrees upward; on the contrary, the winglet showed value of 1.2 degrees, similar to the effect of a direct action. This larger effect is mainly due to the induced deformation around the wing axis, further amplified for the outboard cross sections.

Finally, the WL, due to its peripheral location, did not produce any appreciable effect on the other systems, several magnitudes smaller (10^{-6} to 10^{-7}).

It is finally worth noting that, due to its lowest rigidity, the WL is subjected to larger induced rotations than the other devices and this is particularly evident as the TE, the most rigid device is actuated. In this case (see the third plot of Fig. 14), the induced rotation arrives at 1.1 degrees at the stress limit against 0.027 degrees produced by the LE (see the third plot of Fig. 13). Once more, this result is in good agreement with the rigidity estimate reported in Table 2.

5.2 EFFECTS ON THE WING BOX

The effects of the morphing systems on the wing box structure are also worth noting. The devices activation proved, in fact, to alter somehow the wing geometry, with significant out-of-plane induced displacements. On the other hand, the wing box structure is the medium of interexchange of the actions among the single active devices and, thus, affects their performance.

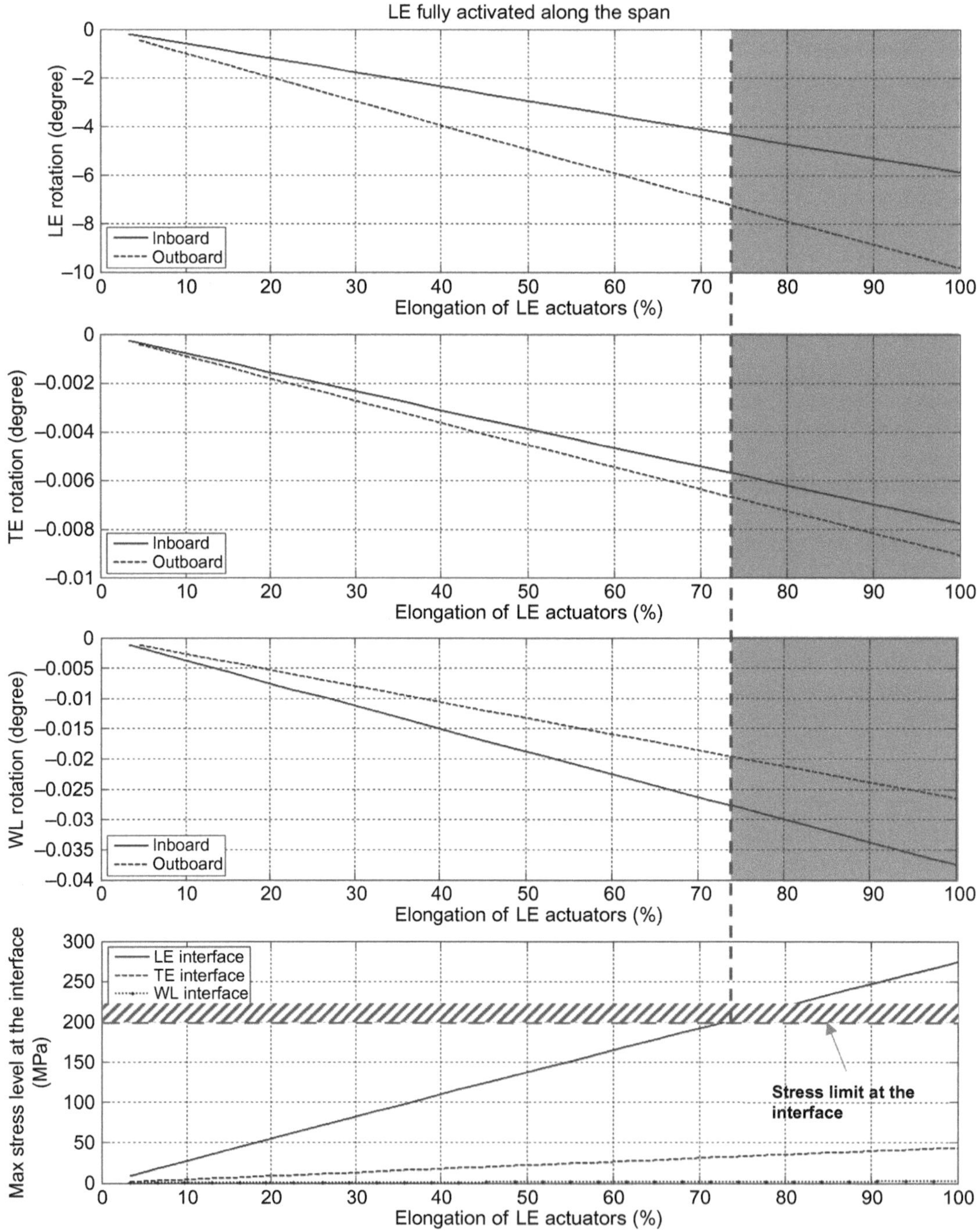

FIG. 12

Percentage of extension of the LE actuator versus induced rotation on the LE and on the other devices (first 3 graphs); stress transmitted at the interfaces (bottom graph).

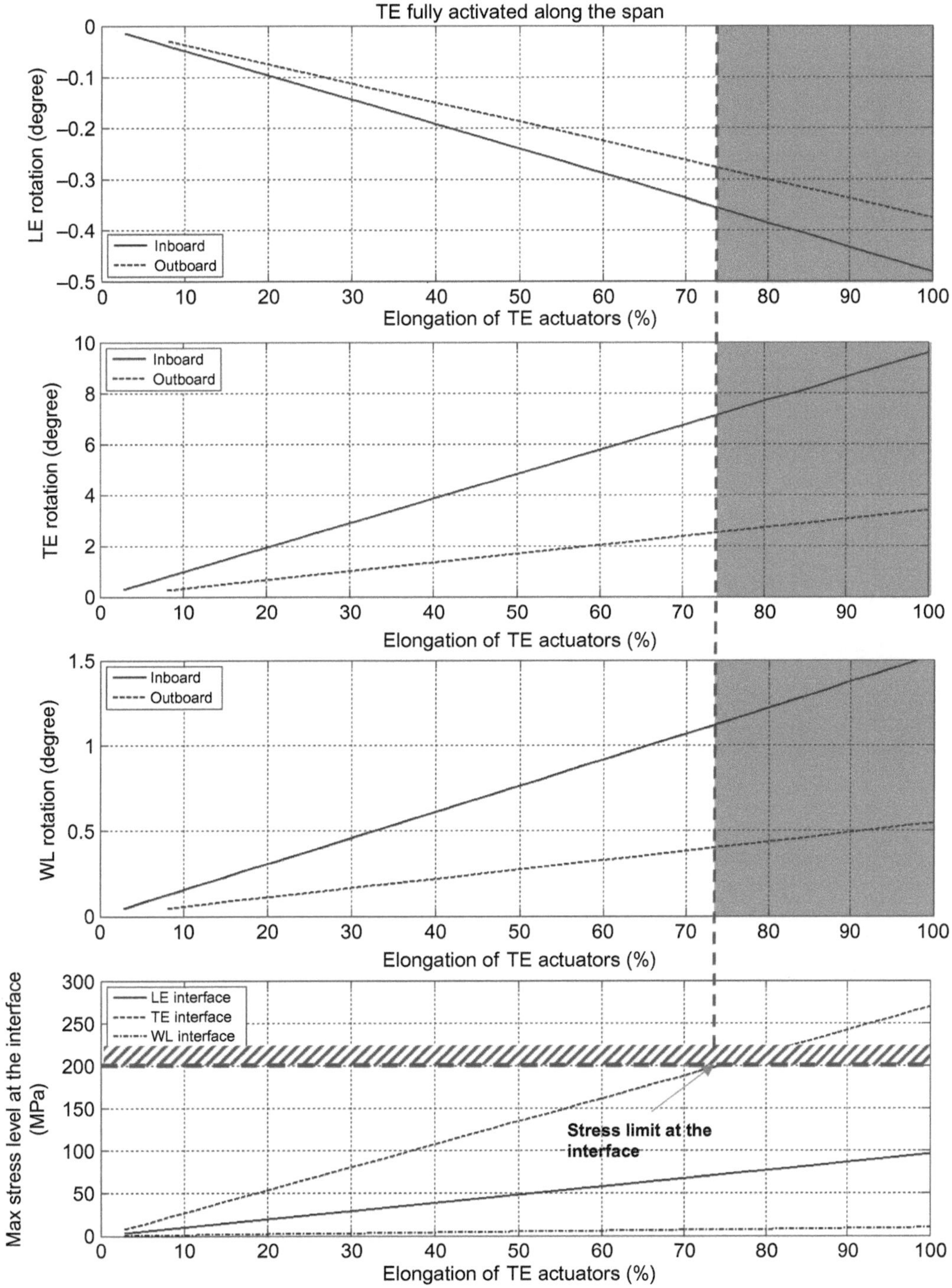

FIG. 13

Percentage of extension of the TE actuator versus induced rotation on the TE and on the other devices (first 3 graphs); stress transmitted at the interfaces (bottom graph).

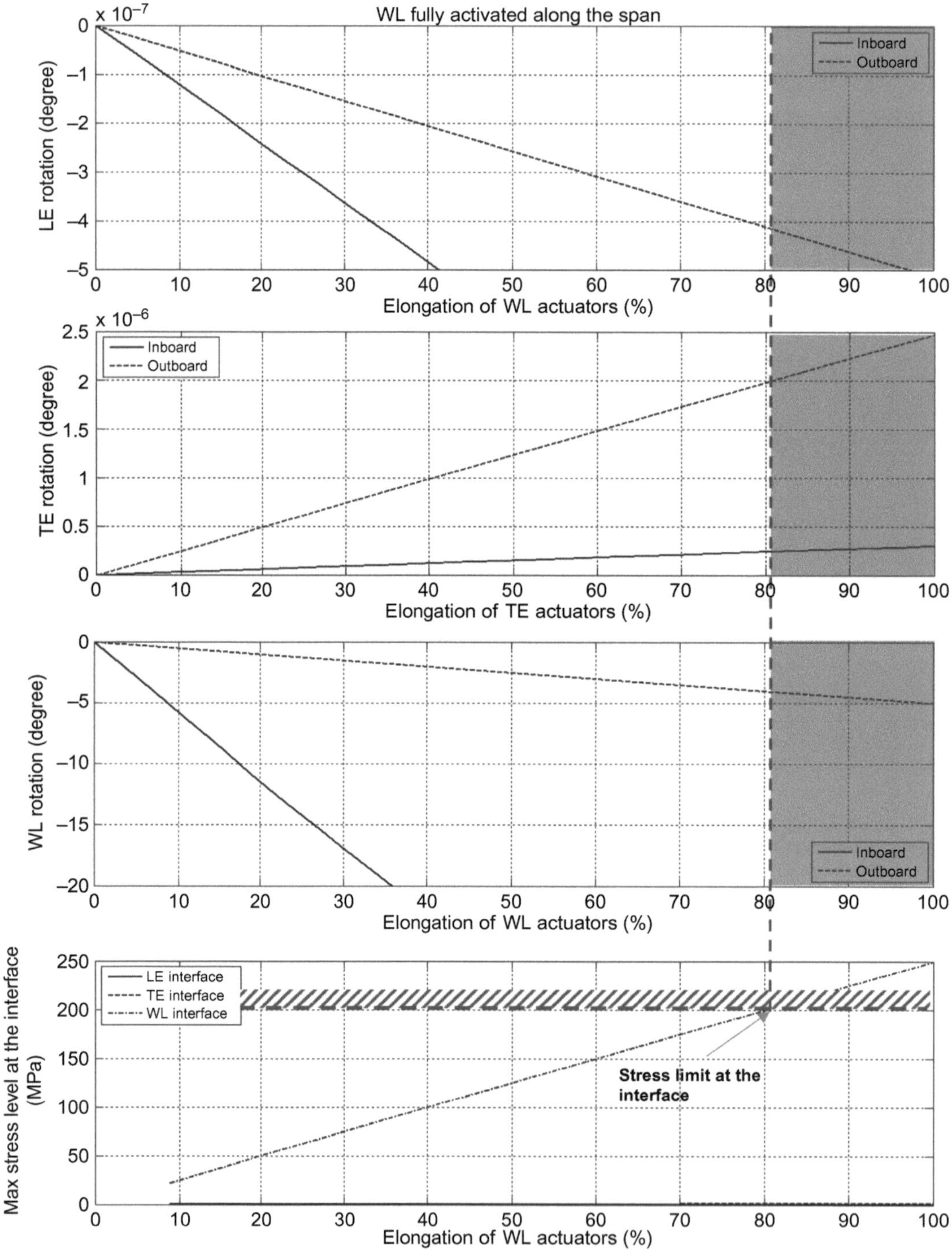

FIG. 14

Percentage of extension of the WL actuator versus induced rotation on the WL and on the other devices (first 3 graphs); stress transmitted at the interfaces (bottom graph).

The largest displacements, moving outward, were recorded for the rear spar because of its lower bending rigidity (if compared to the other structural elements) and the amplification effect generated due to its proximity to the wing tip. It is noteworthy that the TE caused a displacement of the rear spar remarkably larger than the front spar (Fig. 15, middle graph), if compared to the analogous effect given by the LE and the WL. Here, two factors concur: the different rigidity and the fact that the TE is directly connected to this structural element. The differential out-of-plane displacements of the front and rear spars are at the basis of the considerable rotations produced in this case with respect to what happens after the LE and WL actions. This is evident from the curves of the induced wing torque, plotted in Fig. 16, showing (middle graph) a maximum induced rotation of 1.4 degrees downward, 2 and 3 orders greater than the LE and TE, respectively. Coherently with the out-of-plane displacement plots and the cross-influence graphs shown in Fig. 15, the WL has the lowest impact onto the wing structure, 3 orders lower than the LE and the TE. Both displacement and rotation plots present an ostensible slope discontinuity on the middle point. This is due to the fact that only three points were taken into account (R, M, and T) to describe the evolution of the kinematic parameters toward the span. The higher slope of the curves outboard (right side of the plots of Figs. 15 and 16) can be explained through the stiffness decrease of the tapered wing.

Contour maps were finally plotted to give an idea of the overall stress distribution at the interface between the morphing devices and the wing (Fig. 17). At the front and rear spar interface, the stress spots are particularly evident due to the contact points of the actuators that pushes when expanding. Also, a further stress concentration is evident close to the junction between the WL and the outer wing box. This effect is caused by the fairing, representing a further constraint to the motion of the inboard WL tab. As a result, the rotary rigidity of the movable surface is higher with a consequent remarkable force transmission onto the rear spar, supporting the hinge, and onto the front spar, supporting the actuator.

Table 3 aims at giving a compact view of the ratios between the induced rotations and the % of actuation of each device. A graphical representation is reported in Fig. 18. The lateral axis reports which morphing system is activated, one by one. The frontal axis contains the same devices plus the wing box, to indicate the effect that each of them undergoes when a generic device of the lateral column is activated. The value of this estimated interaction is given on the vertical axis. The diagonal terms indicate the direct effect, while the off-diagonal values refer to the cross effect. Of course, the former are the most relevant, and only few of the others attain considerable levels. For the chosen activation architecture (extensional beam, that could represent a classical linear actuator for instance, combined with kinematic systems, even though arranged along very different schematics), the adaptive WL proved to have the greatest authority, followed by the active LE and TE. Both LE and WL exhibit a negligible effect both on the other devices and on the wing box. On the other hand, the TE proved to have a remarkable effect on all the three other structures (LE, WL, and wing box, cited in ascending order). In Table 4 the same ratios are reported, but normalized with respect to the "direct effect value." In this way, the effect of the trailing edge appears to be still more relevant with respect to the other, while the winglet confirms its almost "standing alone" behavior. The leading edge, instead, appears definitely as to contribute significantly to spread the strain field all across the reference structure. There is a fundamental agreement among these data and the rigidities estimated per each device (Table 2). To cite an example, the TE, the most rigid system, has also the highest impact on the wing structure, with an induced rotation of 2.75E−2 degrees per % of activation (absolute, Table 3) and 4.53E−01 (referred to the self-induced rotation, Table 4), 2 and 3 orders higher than the LE and the WL.

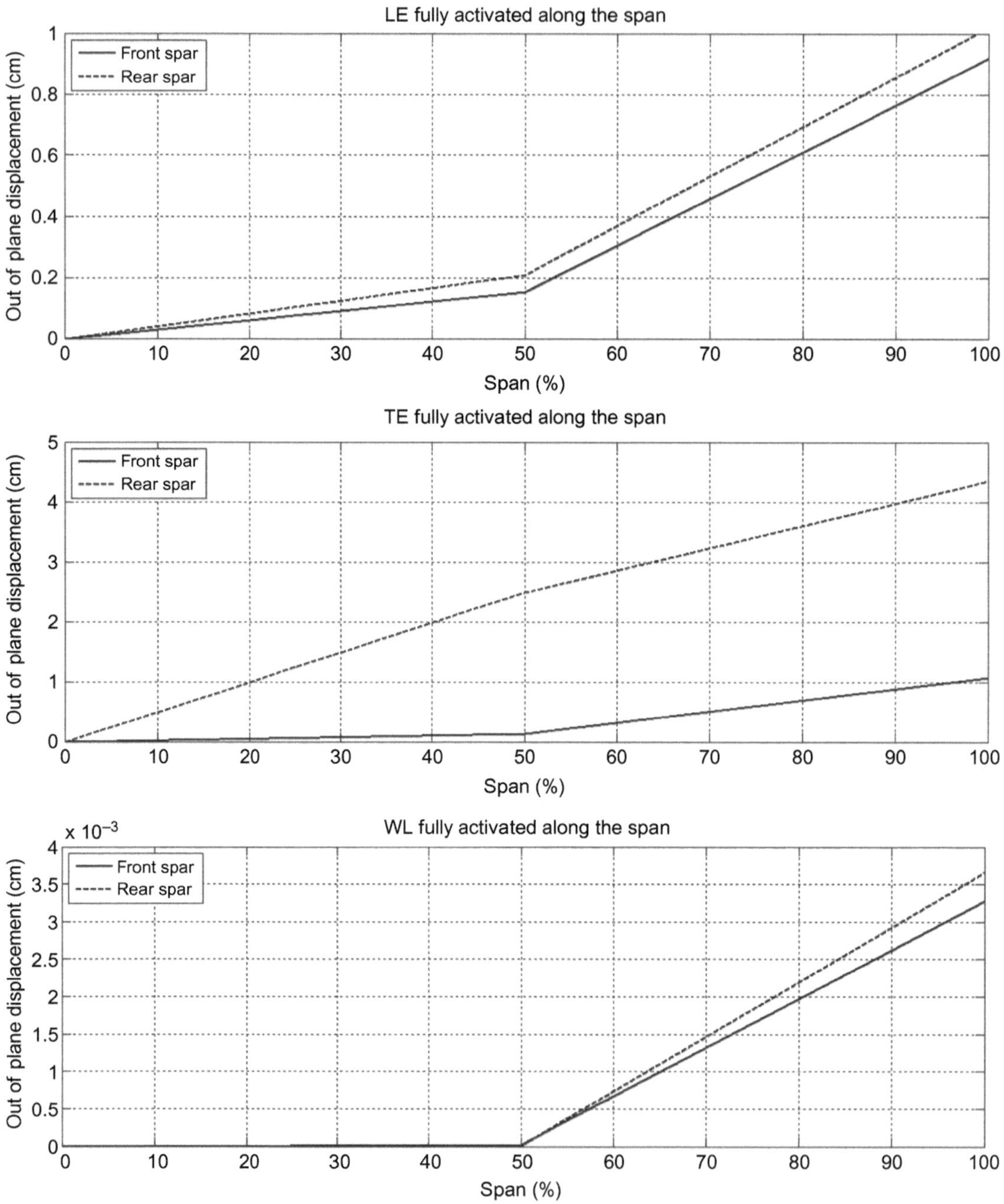

FIG. 15

Out-of-plane induced displacement onto the wing due to LE, TE, and WL activation (top, middle, and bottom, respectively).

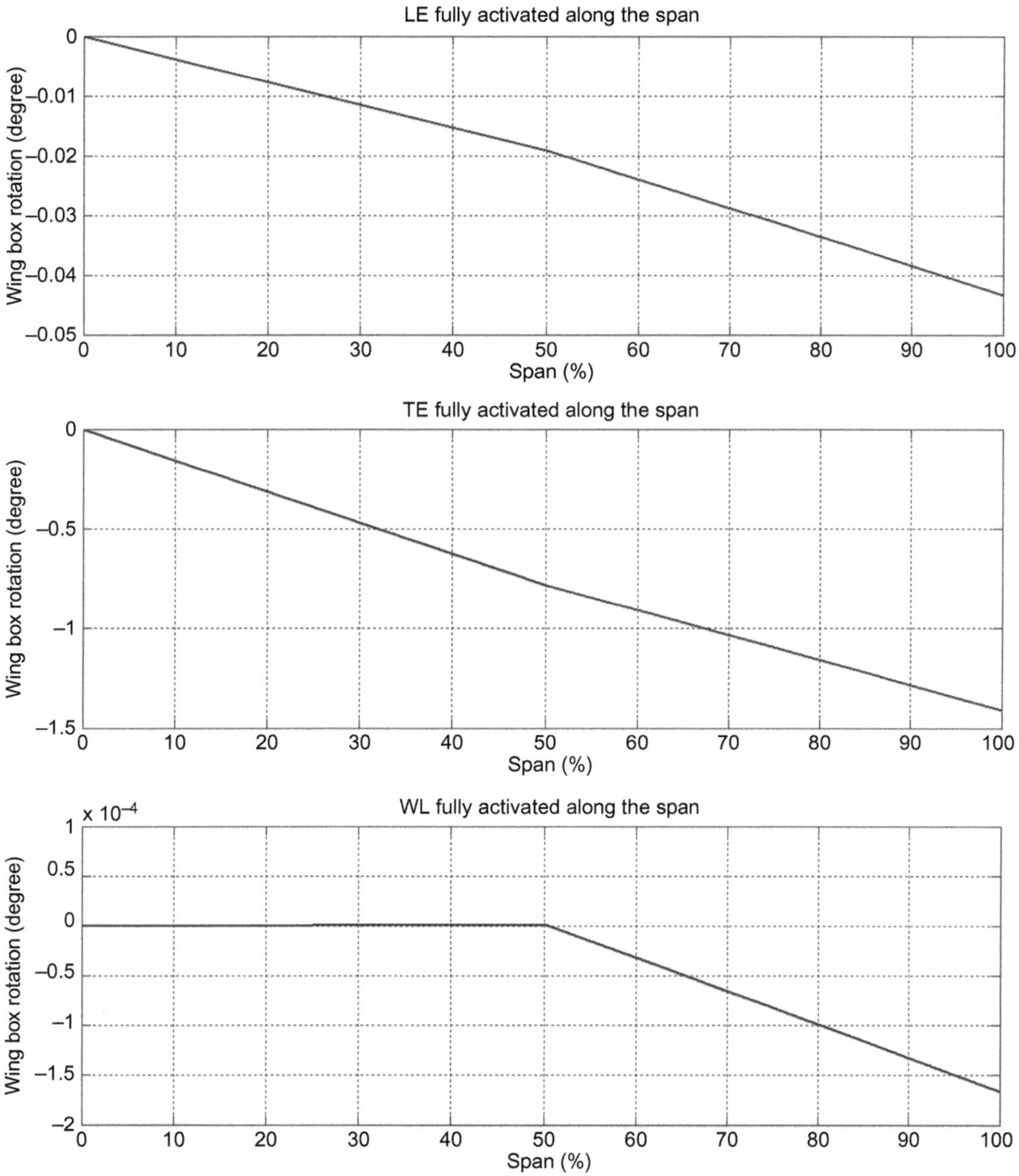

FIG. 16

Rotation induced onto the wing due to LE, TE, and WL activation (top, middle, and bottom, respectively).

FIG. 17

Stress distribution for a combined and full activation of all the morphing devices; max Von Mises stress: 329 MPa.

Table 3 Ratios Between the Induced Rotations and the % of Actuation Versus Actuated Device

		Actuated		
		LE	**TE**	**WL**
Cross effects	LE [deg/%]	1.50E−1	4.28E−3	8.33E−9
	TE [deg/%]	8.28E−5	6.07E−2	1.37E−8
	WL [deg/%]	3.28E−4	1.07E−2	2.50E−1
	Wing [deg/%]	5.89E−4	2.75E−2	3.45E−5

For the same reason, the TE has a higher effect on the WL, being less rigid: 1.07E−2 against 4.28E−3 deg./% for the LE. The WL, on the contrary, being less rigid, is the most subjected to cross-related effects and barely affects the motion of the other devices.

6 CONCLUSIONS AND FURTHER STEPS

The design of the movable parts of a wing is usually performed focusing on the mechanism enabling certain functions following aerodynamic specifications (geometry or shape envelope to achieve), in compliance with the target benefits in terms of performance (aerodynamic, maneuvers, stability) and alleviation of the structural stress field, with particular emphasis to the reduction of the root bending moment or the capability to alleviate the gust effects. This approach is based on the assumption that, whatever is the kinematic chain, it will be attached to the main structure by means of a rigid constraint. According to this strong hypothesis, all the energy provided by the actuation system is transmitted to the specific device, without any effectiveness losses (for instance, following undesired deformations of the support).

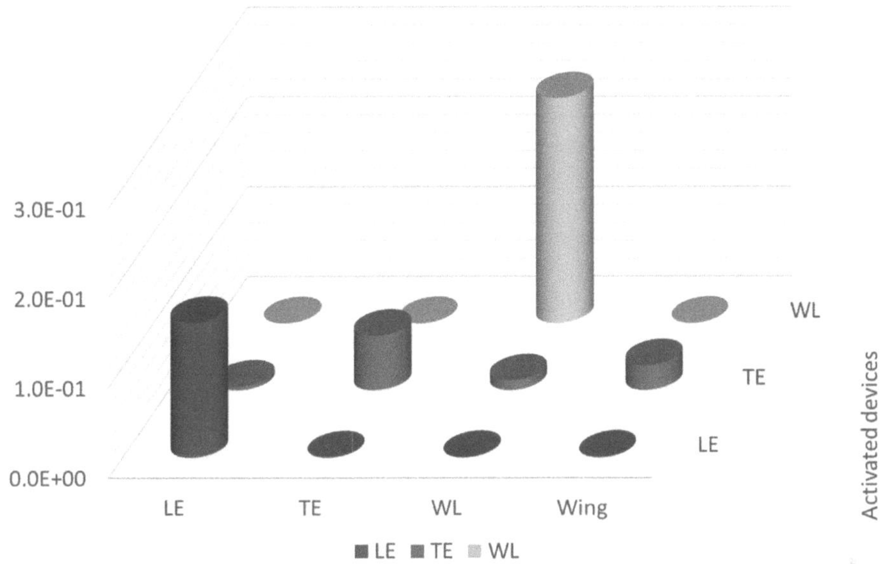

FIG. 18

Cross-related effects in terms of induced rotations per % of actuation (absolute values).

Table 4 Relative (Direct Effect Referred) Ratios Between the Induced Rotations and the % of Actuation Versus Actuated Device

		Actuated		
		LE	**TE**	**WL**
Cross effects	LE	1.00E+00	7.05E−02	3.33E−08
	TE	5.52E−04	1.00E+00	5.48E−08
	WL	2.19E−03	1.76E-01	1.00E+00
	Wing	3.93E−03	4.53E−01	1.38E−04

Although this approach makes easier the project task, allowing the designer concentrating exclusively on the morphing system, ignoring the induced deformation at the interface with the supporting wing structure may leads to incorrect approximations, whose impact on the ultimate effect of the morphing systems may be large.

This aspect is already relevant for conventional hinged active parts (like flaps and slats), but appears essential in case of morphing concepts, whose working logic is completely based on strain actuation. In this case, the nonconventional mechanical architectures that become an intrinsic part of the novel structural system and the lower global rigidity (because the referred architecture has augmented degrees of freedom) requires to adopt a novel design philosophy. In this case, for instance, the effects of the elasticity of the attachments shall be adequately taken into account, even during the concept generation.

The scope of this work is to investigate those effects, as the spread deformations after the activation of a certain morphing device. It should be clear that these considerations also apply to wings where a single adaptive element is considered, since it can modify the overall system geometry (the adaptive system and the reference system make in fact a large, single isolated object). In order to explore this phenomenon, three mechanical concepts were taken into account: an adaptive leading edge, an adaptive trailing edge, and an adaptive winglet, integrated onto a large wing section. The final objective was to estimate the cross effects of their actions. The three architectures did represent indeed a large spectrum of morphing devices, ranging from the classical mechanic (TE), to the compliant (LE), to the almost classical systems (WL).

In practice, this chapter tries to formulate answers to questions as: "What happens on the wing structure, as the leading edge is activated? Is there any appreciable bending and/or torque? If the trailing edge is activated, does the winglet work the same? Is the supporting constraint (the interface with the structure, i.e., the wing box) able to bear the acting loads?" and so forth. All these questions were answered through a dedicated FE modeling approach. A numerical representation of the entire wing section was realized, taking into account its capability to produce a controlled deformability. This theoretical tool allowed to describe the strong nonlinear phenomena that underlie the load transmission to the main structural elements, together with the working principles of the different morphing devices. Nonlinear analyses are mandatory because of the particular characteristics of the adopted devices that are called to produce large or very large displacement fields. The complexity of the numerical model was limited because of that and the necessity of carrying out a test campaign, sufficient to perform a parametric study.

As a first step, the devices were singularly investigated with the aim of estimating their own out-of-plane rigidity, in presence of a rigid support. This task was addressed by means of a normal mode and a static analysis, the former to highlight the energy absorption modality, the latter to quantify the torque rigidity of the devices around the axis of rotation.

Then, the global rigidity of the wing was estimated in terms of bending and torque stiffness with respect to selected and classical load configurations (concentrated tip forces and moments). Hints about the energy absorption modality of the structure were derived through a normal mode analysis. The achieved mode shapes highlighted in fact the individual involvement of each morphing device that cooperated with the wing to respond to specific types of solicitations. The cross effects analysis was finally faced. Nonlinear analysis runs were then carried out, by separately activating the different apparatuses and acquiring the complete morphed (or deformed) configurations against the activation level. Depending on what concept was activated, including its location, different levels of transmission were estimated. A global agreement was found out between this type of investigation and the previous one focused on the single devices: the higher the rigidity, the higher the effect on the wing structure and on the other morphing systems. In this sense, the adaptive TE, the most rigid one, proved to be the most effective in affecting the other structural parts and, therefore, their functionality most of all the wing and the functionality of the other devices. In this case, all the other smart components were biased, although in a different measure: the wing box, the winglet, and the leading edge, in that order. The WL is in fact eccentric and linked to the main structure through a stiff extremal rib. Therefore, it is strongly influenced by even small deformations that are actuated at the wing root, whether consistent or differential between the front and the rear spars.

The presented work highlighted the necessity of a different approach for the design of the morphing systems. This approach requires a suitable modeling of the adaptive elements and the supporting structures, together with the prediction of their coupled behavior, since the early design phases. Since the beginning, researchers should identify and classify a suitable flowchart incorporating the activities to be implemented. Both the reliability of the descriptive model of complete wing systems and the ability of the designer will be key to a simplified approach to the working modality of different adaptive devices. The dependability aspect is strictly related to higher level requirements, while the model reduction strategy can be driven by the experience, in conjunction with some specific support tools (for instance, based on neural networks prediction). Next, it is crucial to define the features of a highly integrated multidisciplinary approach in which the structural design aspects should be conveniently modulated and integrated with information coming from aerodynamics, aeroelasticity, materials characterization (for certain special parts, made of novel constituents), control logics (if any), and so on. Multidisciplinary design tools should be then selected and, in turn, integrated. The formulation of an adequate and exhaustive numerical test matrix is another crucial point. It should include functionality tests performed onto a dedicated and representative test rig, able to support the morphing devices and to reproduce the effect of the elastic nature of the actual interfaces.

REFERENCES

[1] R.A. MacCrimmon, Guide for the Design of Crane-Supporting Steel Structures, second ed., Canadian Institute of Steel Construction, Hatch, Niagara Falls, 2009.

[2] M. Grandi, I rischi nell'utilizzo delle attrezzature da lavoro, E-book WKI, 2015.

[3] E. Carrara, A. Rapolla, N. Roberti, in: Liguori (Ed.), I metodi geoelettrico e sismico per le indagini superficiali del sottosuolo, Liguori, Napoli, 2012, pp. 199–217.

[4] Y. Huang, W. Brocks, Quantification of constraint effects in elastic-plastic crack front field, J. Mech. Phys. Solids 46 (2) (1998) 219–241.

[5] S. Seitl, P. Hutar, T.E. Garcıa, A. Fernandez-Canteli, Experimental and numerical analysis of in- and out- of plane constraint effects on fracture parameters: Aluminium alloy 2024, Appl. Comput. Mech. 7 (2013) 53–64.

[6] Active aeroelastic aircraft structure (3AS), G4RD-CT-2002-00679, http://cordis.europa.eu/project/rcn/63059_it.html.

[7] Future Fast Aeroelastic Simulation Technologies (FFAST), 233665, http://cordis.europa.eu/result/rcn/55978_it.html.

[8] Novel Air Vehicles Configurations: From Fluttering Wings to Morphing Flight (NOVEMOR), 285395, http://cordis.europa.eu/project/rcn/93282_en.html.

[9] S. Ameduri, Device for modifying the airfoil geometry, European Patent EP 2727825 A1, 2014.

[10] G. Tomassetti, S. Ameduri, A. Carozza, Innovative streamline-flow preserving actuation strategies for wing airfoil nose, Int. J. Struct. Integr. 2 (4) (2011) 437–457.

[11] S. Ameduri, A SMA based morphing leading edge architecture, Adv. Mater. Res. 1016 (2014) 383–388. Trans. Tech. Publications, Switzerland. ISSN: 1662-8985, www.scientific.net/AMR.1016.383.

[12] M. Kintscher, J. Kim, S. Storm, F. Peter, in: P.C. Woelcken, M. Papadopoulos (Eds.), Assessment of the SARISTU enhanced adaptive droop nose, Smart Intelligent Aircraft Structures (SARISTU)—Proceedings of the Final Project Conference, Springer International Publishing, Cham, ISBN 978-3-319-22413-8, 2016.

[13] I. Dimino, D. Flauto, G. Diodati, A. Concilio, R. Pecora, Actuation system design for a morphing wing trailing edge, Recent Patents Mech. Eng. 7 (2014) 138–148.

[14] R. Pecora, M. Magnifico, F. Amoroso, L. Lecce, M. Bellucci, I. Dimino, A. Concilio, M. Ciminello, in: P.C. Woelcken, M. Papadopoulos (Eds.), Structural Design of an Adaptive Wing Trailing Edge for Large Aeroplanes, Smart Intelligent Aircraft Structures (SARISTU)—Proceedings of the Final Project Conference, Springer International Publishing, Cham, ISBN 978-3-319-22413-8, 2016.

[15] A. Wildscheck, S. Storm, M. Herring, D. Drezga, V. Korian, O. Roock, P.C. Woelcken, M. Papadopoulos (Eds.), Design, optimization, testing, verification and validation of the wingtip adaptive trailing edge, Smart Intelligent Aircraft Structures (SARISTU)—Proceedings of the Final Project Conference, Springer International Publishing, Cham, ISBN 978-3-319-22413-8, 2016.
[16] http://aviationweek.com/business-aviation/pilot-report-active-winglets.
[17] S. Ameduri, A SMA based morphing leading edge architecture, Adv. Mater. Res. 1016 (2014) 383–388. Trans. Tech. Publications, Switzerland. ISSN: 1662-8985, www.scientific.net/AMR.1016.383.

CHAPTER

APPLICATION OF THE EXTRA-MODES METHOD TO THE AEROELASTIC ANALYSIS OF MORPHING WING STRUCTURES

14

Modesto Pecora*, Rosario Pecora[†]

*The Italian Aerospace Research Centre, CIRA SCpA, Capua (CE), Italy**

University of Naples "Federico II", Naples, Italy[†]

CHAPTER OUTLINE

NOMENCLATURE

$\underline{\underline{\bullet}}$	matrix
$\underline{\bullet}$	vector
$\bullet_{\mathbf{BS}}$	matrix or vector related to BS
$\bullet_{\mathbf{SS}}$	matrix or vector related to SS
$\bullet_{\mathbf{BS\_ex}}$	matrix or vector related to overall system given by (BS + SS)
$\underline{0}$	null vector
@	at
$\underline{\underline{A}}^{-1}$	matrix of unsteady aerodynamic influence coefficients
A/C	aircraft
b	wing span
$\underline{\underline{B}}/\underline{\underline{\sigma}}_G$	physical/generalized damping matrix.
BS	basic system
cg	center of gravity

Morphing Wing Technologies. https://doi.org/10.1016/B978-0-08-100964-2.00014-9

c_r	reference chord
D	dissipative energy
DMIG	direct (stiffness/mass) matrix input at grids
$\underline{\underline{\Delta S}}$	diagonal matrix of boxes surface areas
$\underline{\Delta c_p}$	vector of lifting pressure arising on the boxes
$\underline{\underline{D}}$	substantial derivative matrix
δL	virtual work (of aerodynamic forces)
DLM	doublet lattice method
Eq.	equation
$\underline{exi}$	displacements induced by the *i*th extramode
$\underline{F}_{ext}$	Fourier transform of the external forces not depending on lifting surfaces oscillations
$\underline{h}_c$	displacements at box control points
$\underline{h}_i$	displacements at box integration points
$\underline{\underline{\Phi}}$	matrix of structural model normal modes
$\underline{\underline{H}}$	aeroelastic impedance matrix
j	imaginary unit
K_r	reduced frequency
$\underline{\underline{K}}/\underline{\underline{K}}_G$	physical/generalized stiffness matrix.
M	Mach number
$\underline{\underline{M}}$	mass matrix
$\underline{\underline{M}}_G$	generalized mass matrix
$\tilde{q}$	dynamic pressure
$\underline{q}$	Lagrangian coordinates vector
$\underline{q}'$	Fourier transform of the Lagrangian coordinates vector
$\underline{\underline{Q}}$	matrix of the complex generalized aerodynamics forces
$\underline{r}$	box position vector
$\underline{\underline{S}}$	displacements interpolation matrix
SS	substructure
T	kinetic energy
t	time variable
U	potential (elastic) energy
u_0	asymptotic flow speed (modulus)
V_A	maneuvering speed
V_D	dive speed
w	normal wash (modulus)
$\underline{z}$	displacements of the structural model grids
$\underline{\Delta F}$	vector of elementary forces on the boxes
ω	circular frequency of the oscillation

1 INTRODUCTION

During the preliminary structural design of a conventional wing, few basic rules may be followed to avoid poor aeroelastic performances of the elaborated layout [1–5]. These rules rely upon the best common practice and are generally the result of lessons learned on similar structural configurations.

Morphing wing structures currently represent a breakthrough and emerging technology, [6–8] and therefore no similitude criteria may be invoked to preliminary assess their aeroelastic behavior. On the other hand, due to the augmented elasticity and embedded servo-actuated mechanical solutions, they are more sensitive to aeroelastic issues and more likely to produce detrimental impacts at A/C level [9–12].

In the light of these considerations, the aeroelastic analysis of the conceived structural arrangements is strongly advised since the earliest stage of the design so that the maturation of un-flyable solutions is naturally avoided.

The *extra-modes method* [2] is herein detailed as a dramatically efficient tool to accomplish this paramount task. The morphing device is treated as a *substructure* and the A/C as the *basic system* on which the device is installed. Substructure's contribution to the aeroelastic response of the global system is expressed in terms of generalized parameters pertinent to additional and strategically defined modes capturing the substructure dynamics.

In order to provide a rigorous and clear description of the approach, the standard formulation of the aeroelastic equilibrium and stability equations in frequency domain and generalized coordinates, is firstly recalled. The reference equations' decomposition resulting from the partition of the overall aeroelastic system into *basic structure* and *substructure* is then addressed. Finally, two case-studies are reported [13,14] to practically show the application of the extra-modes method to the preliminary aeroelastic assessment of innovative morphing concepts.

2 AEROELASTIC EQUILIBRIUM EQUATION AND STABILITY

The unsteady aerodynamic effects due to the oscillations of the lifting surfaces in the airflow are commonly accounted for by means of influence coefficients conveniently evaluated through the doublet lattice method (DLM) [15]. A 3D aerodynamic mesh composed of flat panels is generally more than adequate to effectively simulate lifting and control surfaces behavior as well as the mutual interference among them. Panels are then subdivided into elementary boxes whose spatial density depends on the level of accuracy desired for the specific meshed item (Fig. 1).

The normal-wash ($\underline{w}$, Fig. 2) and the lifting pressure arising on oscillating boxes ($\underline{\Delta c_p}$) are related by the matrix relations:

$$\frac{\underline{w}}{u_0} = \underline{\underline{A}} \cdot \underline{\Delta c_p} \Rightarrow \underline{\Delta c_p} = \underline{\underline{A}}^{-1} \cdot \frac{\underline{w}}{u_0}$$

where u_0 is the modulus of the asymptotic airflow speed and $\underline{\underline{A}}^{-1}(\underline{r}, K_r, M)$ is the matrix of the unsteady aerodynamic influence coefficients, function only of the box position (vector $\underline{r}$), reduced frequency $\left(K_r = \frac{\omega c_r}{u_0}\right)$ and Mach number (M).

The associated elementary forces $\Delta \underline{F}$ are then given by:

$$\underline{\Delta F} = \tilde{q} \cdot \underline{\underline{\Delta S}} \cdot \underline{\Delta c_p} = \tilde{q} \cdot \underline{\underline{\Delta S}} \cdot \underline{\underline{A}}^{-1} \cdot \frac{\underline{w}}{u_0} \tag{1}$$

where $\underline{\underline{\Delta S}}$ is the diagonal matrix of boxes' surface areas and $\tilde{q}$ is the dynamic pressure.

If $\underline{h_c}$ represents the vector of the normal displacements at boxes' control points (3/4 of boxes' mean chords) and x the streamwise coordinate, then the normal wash may be written as:

$$\frac{\underline{w}}{u_0} = \left(\frac{d}{dx} + j\frac{K_r}{c_r}\right) \cdot \underline{h_c}(\underline{r}) = \underline{\underline{D}} \cdot \underline{h_c}(\underline{r}) \tag{2}$$

with $\underline{\underline{D}}$ standing for the substantial derivative matrix.

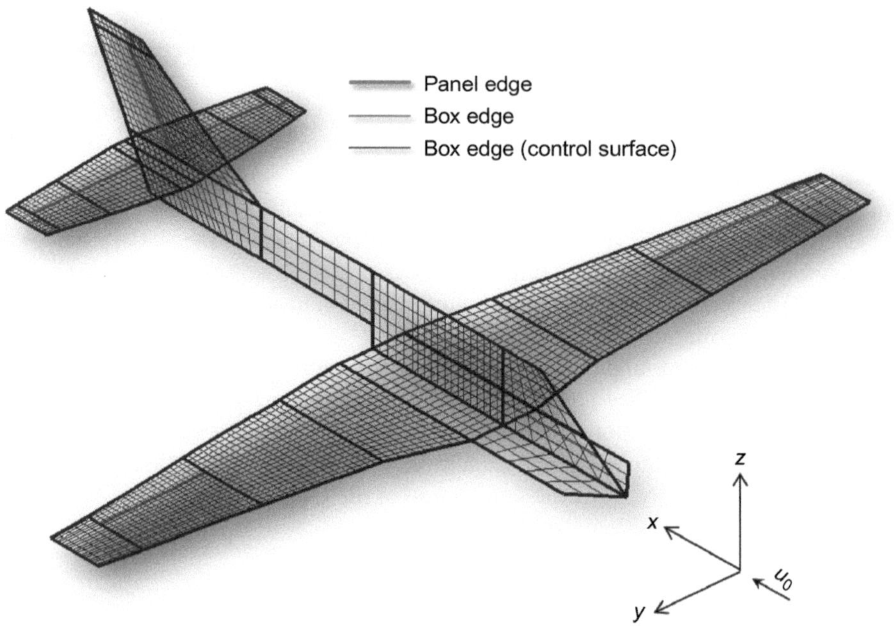

FIG. 1

Example of A/C aerodynamic model for DLM implementation.

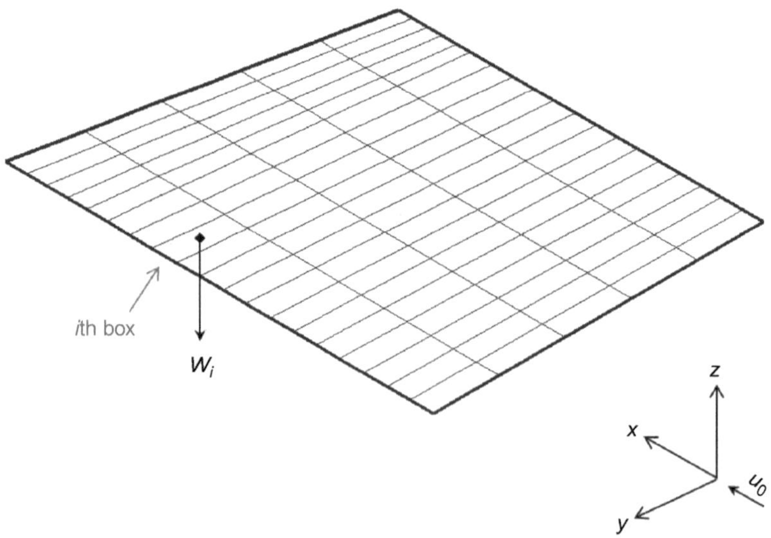

FIG. 2

Box downwash.

The virtual work of the elementary forces along the normal displacements is given by:

$$\delta L=\delta \underline{h}_{\mathrm{i}}^{\mathrm{T}}\cdot \underline{\Delta F}=\bar{q}\cdot \delta \underline{h}_{\mathrm{i}}^{\mathrm{T}}\cdot \underline{\underline{\Delta S}}\cdot \underline{\underline{A}}^{-1}\cdot \underline{\underline{D}}\cdot \underline{h}_{\mathrm{c}} \tag{3}$$

where $\underline{h}_{\mathrm{i}}$ is the vector of the normal displacements at boxes' integration points (¼ of boxes' mean chord).

$\underline{h}_{\mathrm{i}}$ and $\underline{h}_{\mathrm{c}}$ are related to the displacement of the structural model grids ($\underline{z}$) by interpolation matrices[1] ($\underline{\underline{S}}$) which are only function of the box locations ($\underline{r}$):

$$\underline{h}_{\mathrm{c}}=\underline{\underline{S}}_{\mathrm{c}}(\underline{r})\cdot \underline{z};\ \ \underline{h}_{\mathrm{i}}=\underline{\underline{S}}_{\mathrm{i}}(\underline{r})\cdot \underline{z} \tag{4}$$

Taking in account Eq. (4), the virtual work may be rewritten as:

$$\delta L=\tilde{q}\delta \underline{z}^{\mathrm{T}}\cdot \underline{\underline{S}}_{\mathrm{i}}\cdot \underline{\underline{\Delta S}}\cdot \underline{\underline{A}}^{-1}\underline{\underline{D}}\,\underline{\underline{S}}_{\mathrm{c}}\cdot \underline{z} \tag{5}$$

Referring to modal (or generalized) coordinates, it results:

$$\underline{z}=\underline{\underline{\Phi}}\underline{q}(t) \tag{6}$$

where $\underline{\underline{\Phi}}$ is the matrix of structural model's normal modes and $\underline{q}(t)$ is the vector of Lagrangian coordinates depending only on time variable (t).

The virtual work may be therefore rearranged in the space of modal coordinates as:

$$\delta L=\tilde{q}\left(\delta \underline{q}\right)^{\mathrm{T}}\cdot\left[\underline{\underline{\Phi}}^{\mathrm{T}}\cdot \underline{\underline{S}}_{\mathrm{i}}\cdot \underline{\underline{\Delta S}}\cdot \underline{\underline{A}}^{-1}\underline{\underline{D}}\,\underline{\underline{S}}_{\mathrm{c}}\cdot \underline{\underline{\Phi}}\right]\cdot \underline{q}=\tilde{q}\left(\delta \underline{q}\right)^{\mathrm{T}}\cdot \underline{\underline{Q}}\cdot \underline{q} \tag{7}$$

where $\underline{\underline{Q}}$ is the matrix of the complex generalized aerodynamics forces.

The kinetic (T), elastic (U) and structural dissipative (D) energies in physical and generalized coordinates are given by:

$$T=\frac{1}{2}\dot{\underline{z}}^{\mathrm{T}}\underline{\underline{M}}\,\dot{\underline{z}}=\frac{1}{2}\dot{\underline{q}}^{\mathrm{T}}\underline{\underline{M}}_{\mathrm{G}}\dot{\underline{q}}$$
$$U=\frac{1}{2}\underline{z}^{\mathrm{T}}\underline{\underline{K}}\,\underline{z}=\frac{1}{2}\underline{q}^{\mathrm{T}}\underline{\underline{K}}_{\mathrm{G}}\underline{q}$$
$$D=\frac{1}{2}\dot{\underline{z}}^{\mathrm{T}}\underline{\underline{B}}\,\dot{\underline{z}}=\frac{1}{2}\dot{\underline{q}}^{\mathrm{T}}\underline{\underline{\sigma}}_{\mathrm{G}}\dot{\underline{q}} \tag{8}$$

where $\underline{\underline{M}}/\underline{\underline{M}}_{\mathrm{G}}$ is the *physical/generalized mass matrix*; $\underline{\underline{K}}/\underline{\underline{K}}_{\mathrm{G}}$ is the physical/*generalized stiffness matrix;* $\underline{\underline{B}}/\underline{\underline{\sigma}}_{\mathrm{G}}$ is the *physical/generalized damping matrix.*

Recalling Lagrange equations:

$$\frac{d}{dt}\frac{\partial T}{\partial \dot{\underline{q}}}+\frac{\partial D}{\partial \dot{\underline{q}}}+\frac{\partial U}{\partial \underline{q}}=\frac{\delta L}{\delta \underline{q}} \tag{10}$$

and referring to the complex frequency domain, the aeroelastic equilibrium may be finally formulated as:

$$(j\omega)^2\underline{\underline{M}}_{\mathrm{G}}\underline{q}'+j\omega\underline{\underline{\sigma}}_{\mathrm{G}}\underline{q}'+\underline{\underline{K}}_{\mathrm{G}}\underline{q}'-\tilde{q}\underline{\underline{Q}}\underline{q}'=\underline{0}, \tag{11}$$

[1]Several methodologies can be applied for the evaluation of interpolation matrices; among these, spline functions are generally preferred due to the high reliability/computational time ratio [16,17].

where $\underline{q}'$ is the Fourier transform of $\underline{q}$.

Since the expression of δL was obtained with reference to the elementary forces generated only by lifting surfaces oscillations, it results that Eq. (11) does not take in account any external force independent by the motion. If we call the Fourier transform of these forces $\underline{F}_{ext}(j\omega)$, than Eq. (11) may be generalized in the following manner:

$$(j\omega)^2\underline{\underline{M}}_G\underline{q}' + j\omega\underline{\underline{g}}_G\underline{q}' + \underline{\underline{K}}_G\underline{q}' - \tilde{q}\underline{\underline{Q}}\underline{q}' = \underline{\underline{\Phi}}^T\underline{F}_{ext}(j\omega) = \underline{F}_{G,ext}(j\omega) \tag{12}$$

Recalling a notation typical of sytems science, Eq. (12) may be rewritten as:

$$\underline{\underline{H}}(j\omega,\tilde{q})\underline{q}'(j\omega) = \underline{F}_{G,ext}(j\omega) \tag{13}$$

where $\underline{\underline{H}}(j\omega,\tilde{q}) = \left[(j\omega)^2\underline{\underline{M}}_G + j\omega\underline{\underline{g}}_G + \underline{\underline{K}}_G - \tilde{q}\underline{\underline{Q}}\right]$ is commonly referred as the *aeroelastic impedance matrix*.

The solution of Eq. (13) in generalized coordinates and time domain can be formally obtained as:

$$\underline{\underline{H}}(j\omega,\tilde{q})\underline{q}'(j\omega) = \underline{0};\ \ \underline{\underline{H}}(j\omega,\tilde{q})\underline{q}'(j\omega) = \underline{F}_{G,ext}(j\omega)$$

The solution of the aeroelastic equilibrium equation in generalized coordinates and time domain may be formally obtained as:

$$\underline{q}(t) = \mathfrak{I}^{-1}\left[\underline{\underline{H}}^{-1}(j\omega,\tilde{q})\cdot\underline{F}_{G,ext}(j\omega)\right] = \frac{1}{2\pi}\int_{-\infty}^{+\infty}\underline{\underline{H}}^{-1}(j\omega,\tilde{q})\cdot\underline{F}_{G,ext}(j\omega)e^{j\omega t}d\omega \tag{14}$$

where $\mathfrak{I}^{-1}$ stands for the inverse Fourier transform operator.

Several numerical methods are available in literature to effectively evaluate the integral at the second member of Eq. (14); once $\underline{q}(t)$ is known the system response in physical coordinates may be easily found by invoking Eq. (6).

The inverse of $\underline{\underline{H}}(j\omega,\tilde{q})$ represents the *aeroelastic transfer matrix*; the aeroelastic system will be asymptotically stable if, for a given impulsive excitation $\underline{F}_{G,ext}(j\omega) = \underline{1}$, it is satisfied that:

$$\left|\underline{q}(t)\right| \leq 0 \text{ for } t \to +\infty,\ \text{ i.e.: } \int_{-\infty}^{+\infty}\left|\underline{\underline{H}}^{-1}(j\omega,\tilde{q})\right|d\omega \leq 0$$

It can be demonstrated that the above reported condition is verified if and only if the equation:

$$\det\left[\underline{\underline{H}}(j\omega,\tilde{q})\right] = 0 \tag{15}$$

has no roots with real part greater than zero.

The minimum value of $\tilde{q}$ for which Eq. (15) has at least one root with real part equal to zero is the *flutter* dynamic pressure.

3 EXTRA-MODES FORMULATION

Eqs. (13), (15) represent the reference mathematics for the assessment of the dynamic response and stability of a given aeroelastic system.

They both refer to generalized or modal coordinates in frequency domain; this means that they can be solved once the following steps have been accomplished:

a. modes evaluation;
b. modes interpolation on aerodynamic lattice; and
c. aerodynamic generalized forces evaluation.

These steps involve numerical procedures generally four times longer than those required for the computation of the system's time-response or flutter speeds.

As a consequence of this, if changes occur to an already investigated structural system, most of the computational time needed to reassess its aeroelastic behavior will be surely spent in the repetition of steps a, b, and c.

On the other hand, changes to a specific sub-component are quite likely to occur and sometimes are even imposed during the preliminary design phase to perform trade-off studies on different configurations.

In order to effectively deal with this issue, the reference system may be decomposed into a basic one (BS) and a substructure (SS) which is the only part of the systems supposed to undergo several design changes.

Steps a, b and c are then carried out just one time with reference to BS; the contribute of SS into Eqs. (13), (15) is then added through extra-matrices accounting for extra-modal parameters specifically related to SS. The formulation behind this process is referred as *extra-modes method* and represents one of the most powerful tool for sensitivity analyses, at the same time extremely suitable also for the aeroelastic survey of morphing structures.

In order to better outline the general approaches and equations, let us refer to a classical application case in which BS is a given aircraft and SS is constituted by the aileron and its control line.

Recalling Eq. (11), the aeroelastic equilibrium equation of the basic system in generalized coordinates and frequency domain may be written as:

$$\left[(j\omega)^2 \underline{\underline{M}}_{\mathrm{G,BS}} + j\omega \underline{\underline{\sigma}}_{\mathrm{G,BS}} + \underline{\underline{K}}_{\mathrm{G,BS}} - \tilde{q}\underline{\underline{Q}}_{\mathrm{BS}}\right] \underline{q}'_{\mathrm{BS}} = \underline{0} \tag{16}$$

where $\underline{\underline{M}}_{\mathrm{G,BS}}, \underline{\underline{\sigma}}_{\mathrm{G,BS}}, \underline{\underline{K}}_{\mathrm{G,BS}}, \underline{\underline{Q}}_{\mathrm{BS}}$ are respectively the generalized mass, damping, stiffness and aerodynamic forces matrices of BS while $\underline{q}'_{\mathrm{BS}}$ is the Fourier transform of the generalized coordinates selected for BS. By definition it results:

$$\underline{\underline{M}}_{\mathrm{G,BS}} = \underline{\underline{\Phi}}^{\mathrm{T}}_{\mathrm{BS}} \underline{\underline{M}} \underline{\underline{\Phi}}_{\mathrm{BS}},\ \underline{\underline{\sigma}}_{\mathrm{G,BS}} = \underline{\underline{\Phi}}^{\mathrm{T}}_{\mathrm{BS}} \underline{\underline{B}} \underline{\underline{\Phi}}_{\mathrm{BS}},\ \underline{\underline{K}}_{\mathrm{G,BS}} = \underline{\underline{\Phi}}^{\mathrm{T}}_{\mathrm{BS}} \underline{\underline{K}} \underline{\underline{\Phi}}_{\mathrm{BS}},\ \underline{z}_{\mathrm{BS}} = \underline{\underline{\Phi}}_{\mathrm{BS}} \underline{q}_{\mathrm{BS}} \tag{17}$$

where $\underline{\underline{M}}, \underline{\underline{B}}, \underline{\underline{K}}$ are the mass, damping and stiffness matrices of the entire system (BS + SS), $\underline{z}_{\mathrm{BS}}$ is the vector of BS displacements at structural grids and $\underline{\underline{\Phi}}_{\mathrm{BS}}$ is the matrix of BS modes.

Let us now introduce two extra-modes defined over SS structural grids:

- $\underline{ex1}$ ⇒ displacements vector of aileron structural grids due to aileron rotation around its hinge axis by an angle β and
- $\underline{ex2}$ ⇒ displacements vector of control bar grids due to a bar rotation γ coherent with β.

The generalized displacements of all the structural grids along the two extra-modes may be conveniently written as:

$$\underline{\underline{\Psi}}_{\mathrm{EL}} \begin{bmatrix} \underline{ex1} \\ \underline{ex2} \end{bmatrix} = \underline{\underline{\Psi}}_{\mathrm{EL}} \underline{q}'_{\mathrm{ex}} \tag{18}$$

where the matrix $\underline{\underline{\Psi}}_{\mathrm{EL}}$ is built so that only the displacements of the nodes belonging to SS (aileron and stick) are nonzero.

Recalling Eq. (17) and referring to Eq. (18), the generalized mass matrix of the overall system may be expressed in the following (expanded) way:

$$\begin{pmatrix} \underline{\underline{\Phi}}^{T}_{BS} \\ \underline{\underline{\Psi}}^{T}_{EL} \end{pmatrix} \underline{\underline{M}} \left(\underline{\underline{\Phi}}_{BS} \ \underline{\underline{\Psi}}_{EL} \right) = \begin{bmatrix} \underline{\underline{\Phi}}^{T}_{BS} \underline{\underline{M}}\, \underline{\underline{\Phi}}_{BS} & \underline{\underline{\Phi}}^{T}_{BS} \underline{\underline{M}}\, \underline{\underline{\Psi}}_{EL} \\ \underline{\underline{\Psi}}^{T}_{EL} \underline{\underline{M}}\, \underline{\underline{\Phi}}_{BS} & \underline{\underline{\Psi}}^{T}_{EL} \underline{\underline{M}}\, \underline{\underline{\Psi}}_{EL} \end{bmatrix} = \begin{bmatrix} \underline{\underline{M}}_{G,BS} & \underline{\underline{M}}_{BS,ex} \\ \underline{\underline{M}}_{ex,BS} & \underline{\underline{M}}_{ex,ex} \end{bmatrix} = \begin{bmatrix} \underline{\underline{M}}_{G,BS} & \underline{\underline{0}} \\ \underline{\underline{0}} & \underline{\underline{0}} \end{bmatrix} + \begin{bmatrix} \underline{\underline{0}} & \underline{\underline{M}}_{BS,ex} \\ \underline{\underline{M}}_{ex,BS} & \underline{\underline{M}}_{ex,ex} \end{bmatrix} \tag{19}$$

where $\underline{\underline{M}}_{ex,ex}$ is the generalized mass of SS and $\underline{\underline{M}}_{BS,ex}$, $\underline{\underline{M}}_{ex,BS}$ are the *coupling matrices* accounting for cross inertial effects between elastic modes and extra-modes.

In a similar manner, the overall contribution of the stiffness (/damping) matrix to SS ($\underline{\underline{K}}_{\beta,\gamma}/\underline{\underline{B}}_{\beta,\gamma}$ may be incorporated in the generalized expression of the overall system's stiffness (/damping) matrix:

$$\begin{pmatrix} \underline{\underline{\Phi}}^{T}_{BS} \\ \underline{\underline{\Psi}}^{T}_{EL} \end{pmatrix} \underline{\underline{K}} \left(\underline{\underline{\Phi}}_{BS} \ \underline{\underline{\Psi}}_{EL} \right) = \begin{bmatrix} \underline{\underline{K}}_{G,BS} & \underline{\underline{0}} \\ \underline{\underline{0}} & \underline{\underline{0}} \end{bmatrix} + \begin{bmatrix} \underline{\underline{0}} & \underline{\underline{\Phi}}_{BS} \underline{\underline{K}}_{\beta,\gamma} \underline{\underline{\Psi}}_{EL} \\ \underline{\underline{\Psi}}^{T}_{EL} \underline{\underline{K}}_{\beta,\gamma} \underline{\underline{\Phi}}_{BS} & \underline{\underline{\Psi}}^{T}_{EL} \underline{\underline{K}}_{\beta,\gamma} \underline{\underline{\Psi}}_{EL} \end{bmatrix} = \begin{bmatrix} \underline{\underline{K}}_{G,BS} & \underline{\underline{0}} \\ \underline{\underline{0}} & \underline{\underline{0}} \end{bmatrix} + \begin{bmatrix} \underline{\underline{0}} & \underline{\underline{K}}_{BS,ex} \\ \underline{\underline{K}}_{ex,BS} & \underline{\underline{K}}_{ex,ex} \end{bmatrix} \tag{20}$$

$$\begin{pmatrix} \underline{\underline{\Phi}}^{T}_{BS} \\ \underline{\underline{\Psi}}^{T}_{EL} \end{pmatrix} \underline{\underline{B}} \left(\underline{\underline{\Phi}}_{BS} \ \underline{\underline{\Psi}}_{EL} \right) = \begin{bmatrix} \underline{\underline{\sigma}}_{G,BS} & \underline{\underline{0}} \\ \underline{\underline{0}} & \underline{\underline{0}} \end{bmatrix} + \begin{bmatrix} \underline{\underline{0}} & \underline{\underline{\Phi}}_{BS} \underline{\underline{B}}_{\beta,\gamma} \underline{\underline{\Psi}}_{EL} \\ \underline{\underline{\Psi}}^{T}_{EL} \underline{\underline{B}}_{\beta,\gamma} \underline{\underline{\Phi}}_{BS} & \underline{\underline{\Psi}}^{T}_{EL} \underline{\underline{B}}_{\beta,\gamma} \underline{\underline{\Psi}}_{EL} \end{bmatrix} = \begin{bmatrix} \underline{\underline{B}}_{G,BS} & \underline{\underline{0}} \\ \underline{\underline{0}} & \underline{\underline{0}} \end{bmatrix} + \begin{bmatrix} \underline{\underline{0}} & \underline{\underline{B}}_{BS,ex} \\ \underline{\underline{B}}_{ex,BS} & \underline{\underline{B}}_{ex,ex} \end{bmatrix} \tag{21}$$

Finally, the following expression may be obtained for the generalized aerodynamic forces:

$$\begin{pmatrix} \underline{\underline{\Phi}}^{T}_{BS} \\ \underline{\underline{\Psi}}^{T}_{EL} \end{pmatrix} \underline{\underline{Q}} \left(\underline{\underline{\Phi}}_{BS} \ \underline{\underline{\Psi}}_{EL} \right) = \begin{bmatrix} \underline{\underline{Q}}_{BS} & \underline{\underline{0}} \\ \underline{\underline{0}} & \underline{\underline{0}} \end{bmatrix} + \begin{bmatrix} \underline{\underline{0}} & \underline{\underline{\Phi}}_{BS} \underline{\underline{Q}}\, \underline{\underline{\Psi}}_{EL} \\ \underline{\underline{\Psi}}^{T}_{EL} \underline{\underline{Q}}\, \underline{\underline{\Phi}}_{BS} & \underline{\underline{\Psi}}^{T}_{EL} \underline{\underline{Q}}\, \underline{\underline{\Psi}}_{EL} \end{bmatrix} = \begin{bmatrix} \underline{\underline{Q}}_{BS} & \underline{\underline{0}} \\ \underline{\underline{0}} & \underline{\underline{0}} \end{bmatrix} + \begin{bmatrix} \underline{\underline{0}} & \underline{\underline{Q}}_{BS,ex} \\ \underline{\underline{Q}}_{ex,BS} & \underline{\underline{Q}}_{ex,ex} \end{bmatrix} \tag{22}$$

For the overall system (BS + SS) the aeroelastic equilibrium equation in frequency domain takes the form:

$$\left[(j\omega^2) \underline{\underline{M}}_{G,BS\_ex} + (j\omega) \underline{\underline{\sigma}}_{G,BS\_ex} + \underline{\underline{K}}_{G,BS\_ex} - \tilde{q} \underline{\underline{Q}}_{BS\_ex} \right] \underline{q'}_{BS\_ex} = \underline{0} \tag{23}$$

where $\underline{\underline{M}}_{G,BS\_ex}$ $\underline{\underline{K}}_{G,BS\_ex}$, $\underline{\underline{\sigma}}_{G,BS\_ex}$, $\underline{\underline{Q}}_{BS\_ex}$ are the generalized mass, stiffness, damping, and aerodynamic forces matrices given by Eqs. (19)–(22) and:

$$\underline{q'}_{BS\_ex} = \begin{bmatrix} \underline{q'}_{BS} \\ \underline{ex1} \\ \underline{ex2} \end{bmatrix} = \begin{bmatrix} \underline{q'}_{BS} \\ \underline{q'}_{ex} \end{bmatrix} \tag{24}$$

Once the modes of the basic system are evaluated, they can be interpolated on aerodynamic lattice and the associated generalized forces may be computed. This set of operations leads to the definition of matrices $\underline{\underline{M}}_{G,BS}$, $\underline{\underline{K}}_{G,BS}$, $\underline{\underline{\sigma}}_{G,BS}$, $\underline{\underline{Q}}_{BS}$ and is not affected by any change in SS properties.

Extra-matrices related to the aileron + stick extramodes ($\underline{\underline{\bullet}}_{BS,ex}$,$\underline{\underline{\bullet}}_{ex,BS}$,$\underline{\underline{\bullet}}_{ex,ex}$) may be then evaluated by carrying out just simple matrix operations (basically products and transposition of matrices) and finally added to $\underline{\underline{M}}_{G,BS}$, $\underline{\underline{K}}_{G,BS}$, $\underline{\underline{\sigma}}_{G,BS}$, $\underline{\underline{Q}}_{BS}$ in order to solve the equilibrium equation in the form given by Eq. (23).

In force of the mathematics being involved, the operations leading to BS characteristic matrices need much more computational time than those required for extra-matrices definition.

If sensitivities studies need to be performed on SS (for instance, on control line stiffness or aileron mass-balance), the characteristics matrices of BS are computed one single time while the extra-matrices related to SS are calculated for each SS configuration to be investigated.

The extra-matrices are then added to BS matrices to rapidly end effectively assess the aeroelastic equilibrium of the overall system under a given configuration of SS.

4 AEROELASTIC ANALYSES OF MORPHING WINGS USING THE EXTRA-MODES METHOD

The assessment of morphing wings aeroelastic behavior is a key-factor for the definition of suitable arrangements in terms of structural architecture and embedded actuators.

The investigation for potential aeroleastic issues is extremely recommendable since the preliminary design phase; only in this way the risk to consolidate a malfunctioning or unsafe device is dramatically reduced.

Unfortunately, the preliminary design phase of a morphing system is generally characterized by relevant levels of uncertainty due to the large number of variables to be set in a multi-disciplinary manner. Sensitivity or trade-off studies represent then the most effective solution to explore relevant trends and, consequently, to find out optimal combinations of configuration parameters. To keep the *aeroelastic survey* in the trade-off analysis loops, a convenient method is needed to effectively elaborate several system's configurations in reasonable time and with adequate accuracy.

In the light of what was discussed in the previous paragraph, it is obvious that the extra-modes method offers all the right facilities to successfully achieve this target.

The morphing device may indeed be regarded as *the substructure SS* while the surrounding structure integrating the device *as the basic system BS*; *morphing modes* peculiar of SS unconventional nature may be added to the extra-mode vector and their impact on the aeroelastic behavior of the overall system may be smartly simulated.

In order to easily show how this simulation can take place, two interesting case studies have been reviewed in this section.

The first one deals with the assessment of morphing wing twist effectiveness in roll control; extra-modes are used here to solve the elastic aircraft roll maneuver in time domain with reference to an expanded version of Eq. (12).

The second one focuses on trade-off flutter analyses of a morphing wing trailing edge; it provides a clear example of extra-modes application to the iterated solution of the aeroelastic stability equation (Eq. 15).

4.1 EFFECTIVENESS OF WING TWIST MORPHING AS ROLL CONTROL STRATEGY

Referring to a high-aspect ratio wing of a given sailplane, it was asked to investigate the effectiveness of wing twist morphing in roll control [13].

The study was carried out mainly to demonstrate the greater maneuvering performances attainable with respect to a conventional solution based on ailerons deflection. In addition, the most efficient twist distribution to be implemented by a potential morphing architecture was individuated.

In order to address the elastic-aircraft roll dynamics, Eq. (12) was manipulated to account for the roll rigid mode and the time-dependent forces induced by the controls' action (namely by aileron deflection or wing twist morphing):

$$\left[(j\omega^2)\underline{\underline{M}}_G + (j\omega)\underline{\underline{\sigma}}_G + \underline{\underline{K}}_G\right]\underline{q'} = \underline{\underline{A}} \tag{25}$$

where

- $\underline{q}'$ is the Fourier transform of the generalized coordinate vector defined as follows:

$$\underline{q}' = \begin{bmatrix} \phi_x \\ \underline{d} \end{bmatrix} \tag{26}$$

ϕ_x is the rotation angle induced by the rigid roll mode and $\underline{d}$ is the vector of modal displacements induced by the k elastic modes used to represent the aircraft dynamic behavior;

- $\underline{\underline{M}}_G = \begin{bmatrix} I_{xx} & & & \\ & M_{G1} & & \\ & & \cdots & \\ & & & M_{Gk} \end{bmatrix}$ is the generalized mass matrix, with I_{xx} representing the A/C inertia about the roll axis and M_{Gi} the generalized mass of the ith elastic mode;
- $\underline{\underline{K}}_G = \begin{bmatrix} 0 & & & \\ & K_{G1} & & \\ & & \cdots & \\ & & & K_{Gk} \end{bmatrix}$ is the generalized stiffness matrix, with K_{Gi} representing the generalized stiffness of the ith elastic mode;
- $\underline{\underline{\sigma}}_G = \begin{bmatrix} 0 & & & \\ & 2M_{G1}\omega_1\zeta_1 & & \\ & & \cdots & \\ & & & 2M_{Gk}\omega_k\zeta_k \end{bmatrix}$ is the generalized damping matrix, with ω_i, ζ_i $(i = 1\ldots k)$ respectively representing the angular frequency and the modal damping of the ith elastic mode.
- $\underline{\underline{A}}$ is the matrix of generalized aerodynamic forces given by:

$$\underline{\underline{A}} = \tilde{q}\begin{bmatrix} Q_{RR} & \underline{Q}_{RE} \\ \underline{Q}_{ER} & \underline{\underline{Q}}_{EE} \end{bmatrix}\begin{bmatrix} \phi_x \\ \underline{d} \end{bmatrix} + \tilde{q}\begin{bmatrix} Q_{R\delta} \\ \underline{Q}_{E\delta} \end{bmatrix}\delta = \tilde{q}\underline{\underline{Q}}\begin{bmatrix} \phi_x \\ \underline{d} \end{bmatrix} + \tilde{q}\underline{Q}_{\delta}\delta \tag{27}$$

and Q_{RR} is the work done by the aerodynamic forces, due to the roll rigid-mode displacements, along the displacements induced by the roll rigid mode; $\underline{Q}_{RE}$ is the vector of the work done by the aerodynamic forces, due to the elastic-mode displacements, along the displacements induced by the roll rigid mode; $\underline{Q}_{ER}$ is the vector of the work done by the aerodynamic forces, due to the roll rigid-mode displacements, along the displacements induced by the elastic modes; $\underline{\underline{Q}}_{EE}$ is the matrix of the work done by the aerodynamic forces, due to the elastic-mode displacements, along the displacements induced by the elastic modes; $Q_{R\delta}$ is the work done by the aerodynamic forces, due to the control mode (extra-mode) displacements, along the displacements induced by the roll rigid mode; $\underline{Q}_{E\delta}$ is the vector of the work done by the aerodynamic forces, due to the control mode (*extra-mode*) displacements, along the displacements induced by the elastic modes; δ is a scalar value representative of the control action (e.g., aileron deflection).

By using a quadratic approximation of the generalized aerodynamic forces in the frequency domain, the following expressions are obtained:

$$\tilde{q}\underline{\underline{Q}} = \eta_0 \underline{\underline{Q}}_0 + \eta_1 \underline{\underline{Q}}_1 + \eta_2 \underline{\underline{Q}}_2$$
$$\tilde{q}\underline{Q}_\delta = \eta_0 \underline{Q}_{0\delta}$$

and by substituting into Eq. (27):

$$\underline{\underline{A}} = \eta_0 \underline{\underline{Q}}_0 + \eta_1 \underline{\underline{Q}}_1 + \eta_2 \underline{\underline{Q}}_2 + \eta_0 \underline{Q}_{0\delta} \tag{28}$$

where $\eta_0 = \tilde{q} = \frac{1}{2}\rho V^2$, $\eta_1 = \frac{1}{2}\rho V c_R$, $\eta_2 = \frac{1}{2}\rho {c_R}^2$ and c_R is the reference chord here assumed to be equal to the mean aerodynamic wing chord.

The matrices $\underline{\underline{Q}}_0, \underline{\underline{Q}}_1, \underline{\underline{Q}}_2$ and the vector $\underline{Q}_{0\delta}$ can be partitioned in the same way as the matrix $\underline{\underline{Q}}$ and the vector $\underline{Q}_\delta$:

$$\underline{\underline{Q}}_0 = \begin{bmatrix} Q_{0RR} & \underline{Q}_{0RE} \\ \underline{Q}_{0ER} & \underline{\underline{Q}}_{0EE} \end{bmatrix}, \ \underline{\underline{Q}}_1 = \begin{bmatrix} Q_{1RR} & \underline{Q}_{1RE} \\ \underline{Q}_{1ER} & \underline{\underline{Q}}_{1EE} \end{bmatrix}, \ \underline{\underline{Q}}_2 = \begin{bmatrix} Q_{2RR} & \underline{Q}_{2RE} \\ \underline{Q}_{2ER} & \underline{\underline{Q}}_{2EE} \end{bmatrix}, \ \underline{Q}_{0\delta} = \begin{bmatrix} Q_{0R\delta} \\ \underline{Q}_{0E\delta} \end{bmatrix} \tag{29}$$

Under the assumption of quasi-steady motion and invoking Eqs. (28), (29), Eq. (27) leads to the following set of equations in the time domain:

$$I_{xx}\ddot{\phi}_x = \eta_0 \underline{Q}_{0RE}\underline{d} + \eta_1 Q_{1RR}\dot{\phi}_x + \eta_0 Q_{0R\delta}\delta(t) \tag{30}$$

$$\underline{\underline{K}}_G \underline{d} = \eta_0 \underline{\underline{Q}}_{0EE}\underline{d} + \eta_1 \underline{Q}_{1ER}\dot{\phi}_x + \eta_0 \underline{Q}_{0E\delta}\delta(t) \tag{31}$$

By solving (Eq. 31) for $\underline{d}$ and by substituting the obtained solution into (Eq. 30), the following equation may be written:

$$I_{xx}\ddot{\phi}_x + C\dot{\phi}_x = C_\delta \delta(t) \tag{32}$$

where

$$C = -\left(\eta_0 \underline{Q}_{0RE}\underline{\underline{H}}\eta_1 \underline{Q}_{1ER} + \eta_1 Q_{1RR}\right), \ C_\delta = \eta_0 \underline{Q}_{0RE}\underline{\underline{H}}\eta_0 \underline{Q}_{0E\delta} + \eta_0 Q_{0R\delta}, \ \underline{\underline{H}} = \left(\underline{\underline{K}}_G - \eta_0 \underline{\underline{Q}}_{0EE}\right)^{-1}$$

Recalling the expression of the generalized aerodynamic forces given by Eq. (27) and focusing on the term $\underline{Q}_{E\delta}$, it's possible to observe that for this specific case-study, BS and SS are the same since the entire wing is interested by morphing. The extra-mode is here representative of the control action, it is defined along all the grids of the wing structural model and has been assumed to have impacts on generalized aerodynamic forces only. In other terms, both the aileron deflection mode and the wing morphing modes are supposed to not produce any alteration of wing stiffness, inertial and damping properties. Extra-modes shapes were selected coherently to the type of control to be simulated; the displacements induced by the extra-modes along the aerodynamic lattice of the wing have been depicted in Fig. 3.

These displacements are related to a unitary δ^* being δ^*:

- the aileron rotation around its hinge axis, for the conventional roll control strategy;
- the maximum twist of the wing (respect to the leading edge axis), for the morphing wing based control strategy; for reasons intimately connected to the structural feasibility of twist morphing, only monotonic twist distributions were assumed along the wing span with a

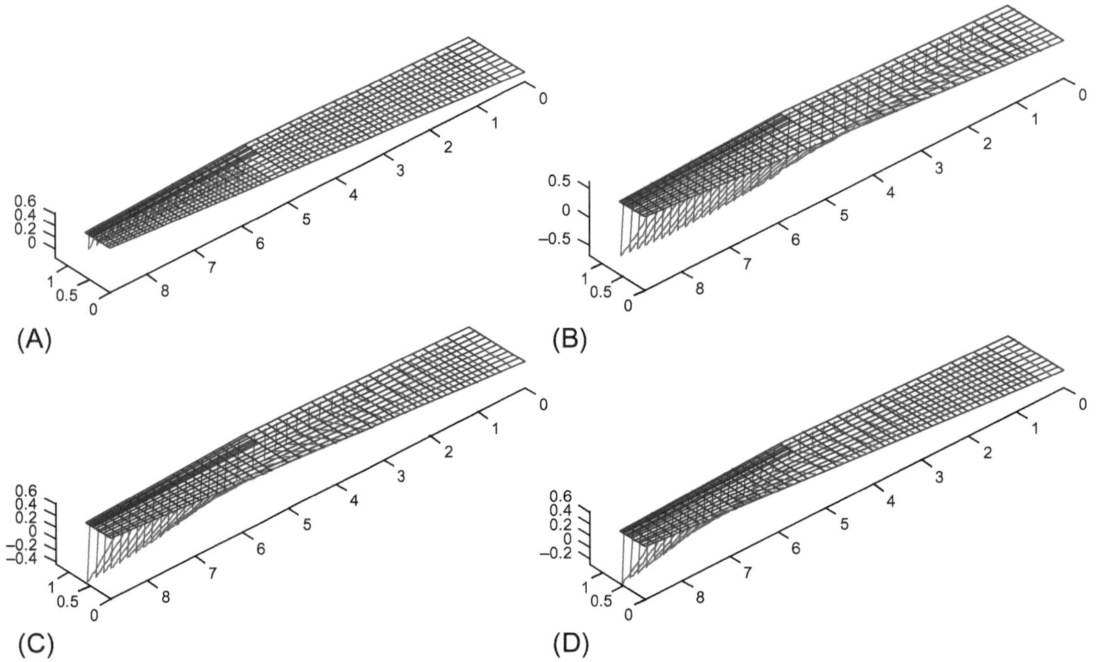

FIG. 3

Extra-mode displacements on wing aerodynamic lattice: (A) aileron rotation; (B) linear twist distribution; (C) quadratic twist distribution; (D) cubic twist distribution [13].

maximum absolute value at the wing tip. More in detail linear, quadratic and cubic distributions were imposed according to the equation:

$$\delta(y) = (2y/b)^h \delta^*$$

where y stands for the spanwise coordinate, b is the wing span and h is an integer equal to 1, 2, or 3 depending on the twist distribution to be implemented.

Eq. (32) was solved with reference to a modal representation of the entire A/C covering only the antisymmetric elastic modes in the range [0–80 Hz]; modes were numerically evaluated by Lanczos method [18] in free-free A/C conditions.

The control law $\delta(t)$ was assumed to be given by:

$$\delta(t) = \begin{cases} \delta^* & 0\,\text{s} \le t \le 2\,\text{s} \\ 0 & t > 2\,\text{s} \end{cases}$$

and the A/C asymptotic roll rate was evaluated in correspondence of unitary δ^* and two characteristic flight speeds at sea level, the maneuvering speed (V_A) and the dive speed (V_D).

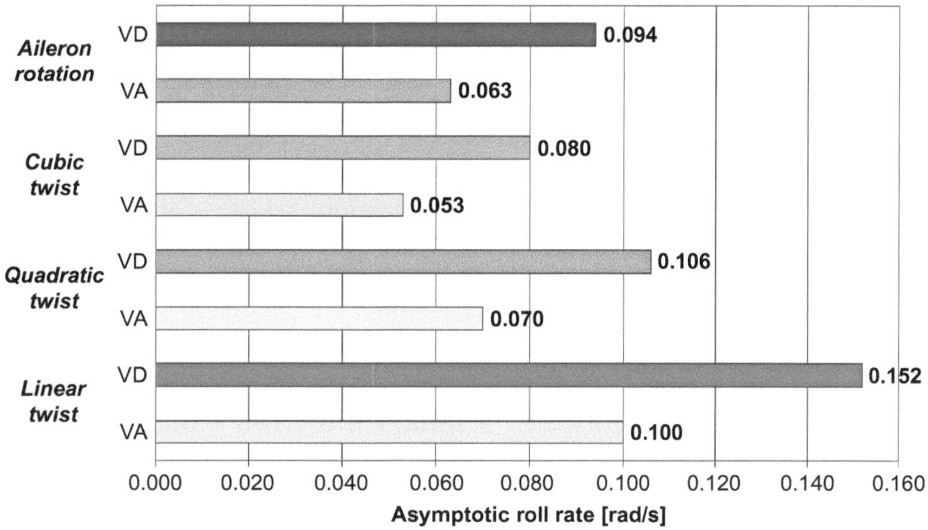

FIG. 4

Asymptotic roll rate comparison.

For each morphing extra-mode the equivalent δ^* required to achieve the same asymptotic roll rate induced by aileron deflections equal to 25°[2] at V_A and 8.3° (=25°/3) at V_D were also evaluated.

Asymptotic roll rates and equivalent δ^* were both used as indicators of control strategy effectiveness. Obtained results are recapped in Fig. 4 and Table 1.

It can be observed that roll control based on wing twist morphing is more effective than the roll control based on ailerons if the implemented twist distribution is linear or quadratic; moreover, linear

Table 1 Asymptotic Roll Rates and Equivalent δ^* to Get the Same Roll Rate Induced by Maximum Aileron Deflection

Flight Speed	Control Mode	Asymptotic Roll Rate (δ^* = 1°)	δ^* Useful to Obtain Same Asympt. Roll Rate Induced by Aileron Deflections Equal to 25° @ V_A/8.3° @ V_D
V_A	Linear wing twist	0.100 rad/s	15.8°
V_D		0.152 rad/s	5.1°
V_A	Quadratic wing twist	0.070 rad/s	22.6°
V_D		0.106 rad/s	7.4°
V_A	Cubic wing twist	0.053 rad/s	29.9°
V_D		0.080 rad/s	9.8°
V_A	Aileron deflection	0.063 rad/s	25°
V_D		0.094 rad/s	8.3°

[2]Maximum aileron deflection of the reference aircraft at V_A. The value at V_D was derived according to the general prescriptions of the applicable airworthiness requirements.

twist is much more effective than quadratic. These results can be justified by the fact that, for the strategies based on wing twist morphing, the necessary lift for roll control is generated by the variation of the local leading angles across the entire wing span, while in the conventional case, the variation is limited only to the wing portions upstream of the ailerons. The aerodynamic strips located outside the ailerons make a significant contribution in additional lift generation, thus reducing the local incidences required at the strips where the ailerons are located.

In the case of the cubic twist law, the imposed twist angle distribution is essentially localized in the outer part of the wing (see Fig. 3), thus leading to a smaller contribution from the aerodynamic strips located outside of the aileron area. In other words, the larger the portion of the wing span that is morphed, the lower the values of the twist angles necessary to obtain the same roll rates as those induced by aileron deflections. Moreover, the greater is the flight speed the bigger is the authority of the control.

4.2 TRADE-OFF FLUTTER ANALYSIS OF A MORPHING WING TRAILING EDGE

In the framework of SARISTU [19], a novel morphing structure was investigated to enable the controlled modification of the trailing edge shape for large civil aircraft applications [20]. Trailing edge camber adaptations were investigated to achieve significant benefits in aircraft fuel consumption whose reduction ranges from 3% to 5% depending on flight mission. Target morphed shapes—to be reproduced in flight—were determined through CFD-based optimizations analyses; the same applies to the overall dimensions of the trailing portion interested by morphing (chord/span, see Fig. 5). The mechanical concept adopted to produce camber variations, was based on segmented adaptive ribs consisting of *finger-like* segments covered by a highly flexible skin (Fig. 6). Each rib was assumed to be segmented into four consecutive blocks (B0, B1, B2, and B3) connected to each other by means of hinges located on the airfoil camber line. Block B0 is rigidly connected to the rest of the wing structure (namely to the rear spar of the wing), while all the other blocks are free to rotate around the hinges on the camber line, thus physically turning the camber line into an articulated chain of consecutive segments. Linking rod elements (L1, L2), hinged on nonadjacent blocks, force the camber line segments to rotate according to specific gear ratios compliant with the shapes to be achieved.

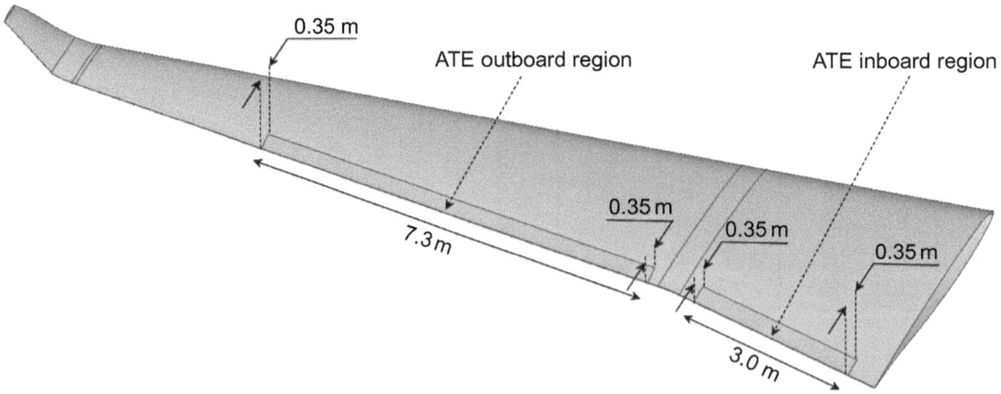

FIG. 5

Morphing trailing edge overall dimensions (meters) [22].

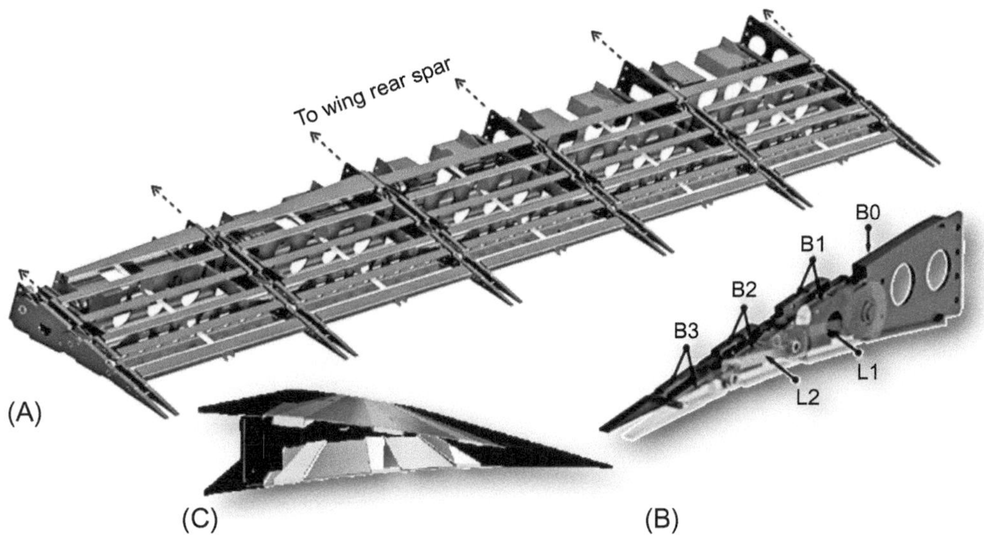

FIG. 6

(A) Morphing trailing edge architecture (5 bays only), (B) morphing rib, (C) compliant skin [22].

A single-degree-of-freedom mechanism is therefore implemented to enable rib morphing. If an actuator moves any of the blocks, all the other blocks follow the movement accordingly. On the other hand, if the rotation of any of the blocks is prevented by actuators, no change in shape can be obtained. Electromechanical actuators coupled with a suitable transmission chain were implemented to assure the controlled camber variation of the trailing edge as well as to preserve a given shape under the action of aerodynamic loads. Further details on the conceived architecture are given in Chapter 17 together with the methodologies implemented for the verification of the adopted structural arrangements (on this topic see also Chapter 15).

The layout therein investigated (and depicted in Fig. 6A with reference to the first 5 bays) resulted from an intense preliminary design phase in which the main dimensions and mechanical properties of the structural components were assessed.

These parameters were defined through iterative stress analysis based on elementary methods; at each iteration, the inertial and stiffness distributions attained for the overall architecture were checked from the aeroelastic stability standpoint. Trade-off flutter analyses were then performed at the very beginning of the preliminary design phase [14] to properly define safety ranges for trailing edge inertial and stiffness properties, as well as for control line harmonics (basically actuators' stiffness), assuring overall system compliance with EASA CS-25 airworthiness requirements, paragraph 25.369 (a) and (b)-(1) (absence of any aeroelastic instability up to 1.15 times the aircraft dive speed [21].

An equivalent beam representation characterized by two lumped masses per bay[3] was adopted for the wing of the reference aircraft (Fig. 7); two DMIG elements [22] were introduced to link right and

[3]One lumped mass for the leading edge and one lumped mass for the box (including the trailing portion if not related to the morphing regions of Fig. 5).

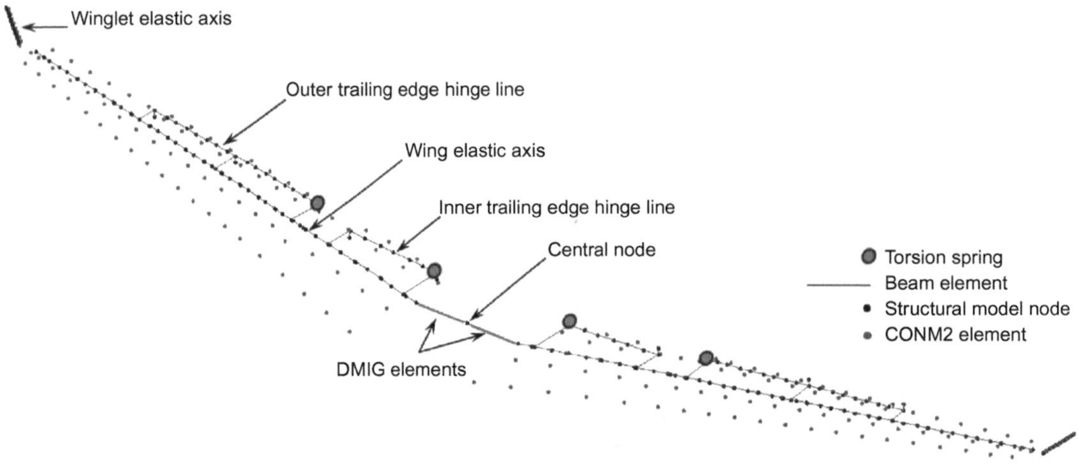

FIG. 7

Reference A/C structural model [14].

left wing to a central node located in correspondence of A/C centre of gravity. DMIG were obtained through the reduction of the wing-fuselage FEM to properly selected condensation grids. Fuselage and empennages were neglected since irrelevant for the assessment of flutter instabilities potentially involving the wing; for a proper evaluation of wing elastic modes in free-free condition, their inertial contribute was however considered by means of a lumped mass referred to the central node.

The morphing trailing edge was also represented by means of equivalent beams whose inertial and stiffness distributions were defined in correspondence of a baseline configuration and then scaled to simulate the increment/decrement of elasticity and weight.

The condensation of dynamic properties was carried out only with reference to the primary hinge line of the device, namely the one between block B0 and B1 of Fig. 6. A four lumped-masses representation was adopted for morphing trailing edge bays, each mass representing the inertial properties of a block of the multi-box arrangement of Fig. 6. At each bay, support grids were added in correspondence of the hinges between blocks 1–2 and 2–3, their movement was slaved to the rotation of the grid on the primary hinge through an extra-mode accounting for the gear ratios between the rib blocks. In other terms a *multitab-like* mode was imposed at the trailing edge to reproduce morphing induced displacements at each rib block; the frequency of this mode was set by acting on a torsional spring (accounting for actuators' stiffness) and connecting the primary hinge line to the wing elastic axis. In order to get a given harmonic frequency f, the torsional stiffness K_a was imposed equal to:

$$K_a = (2\pi f)^2 I$$

where I stands for the inertia of the movable trailing edge around its primary hinge line.

The condition $f = 0$ Hz was taken in account to simulate actuators failure; in such a case the movable trailing edge harmonic coincides with its fundamental (free rigid rotation around the hinge line).

The morphing trailing edge was then regarded as the substructure SS and the remaining part of the model as the basic system BS.

The aeroelastic stability equation was iteratively solved to evaluate flutter speed trends vs. morphing trailing edge inertial/stiffness distributions and morphing extra-mode frequency.

For such a purpose, the generalized parameters of the basic system were evaluated one single time and for each investigation case only the substructure's relevant matrices were updated and added to the overall system equations.

From a mathematical standpoint the addressed stability equation (see Eq. 15) was written in the form:

$$\det\left[\underline{\underline{H}}_{BS\_ex}(j\omega, \tilde{q})\right] = 0 \tag{33}$$

being (see Eq. 23):

$$\underline{\underline{H}}_{BS\_ex}(j\omega, \tilde{q}) = \left[(j\omega^2)\underline{\underline{M}}_{G,BS\_ex} + (j\omega)\underline{\underline{\sigma}}_{G,BS\_ex} + \underline{\underline{K}}_{G,BS\_ex} - \tilde{q}\underline{\underline{Q}}_{BS\_ex}\right]$$

Matrices $\underline{\underline{M}}_{G,BS\_ex}$ $\underline{\underline{K}}_{G,BS\_ex}$, $\underline{\underline{\sigma}}_{G,BS\_ex}$, were evaluated according to Eqs. (19)–(21) with $\underline{\underline{K}}_{\beta,\gamma}$ here intended as the stiffness matrix of the morphing trailing edge structure plus the torsional springs and $\underline{q}'_{ex} = \lfloor ex1 \rfloor$ as the vector of displacements induced by the morphing extra-mode.

Matrices of unsteady aerodynamic influence coefficients (and consequently $\underline{\underline{Q}}_{BS\_ex}$, see Eq. 22) were evaluated by means of DLM referring to an aerodynamic lattice representative of the wing only (Fig. 8). 8 macro-panels for each side of the wing were generated and further meshed into elementary boxes. Higher boxes density was considered for the movable trailing edge and to assure mesh uniformity- also for the not-movable trailing edge; four chordwise boxes were assumed for each strip of the trailing edge in correspondence of blocks B0–B3 locations. Elastic modes and extra-mode displacements at the center of aerodynamic boxes were obtained by means of linear spline functions attached to support-nodes of the structural model. In Fig. 9 the displacements induced by the morphing extra-mode along the outer trailing edge boxes have been reported.

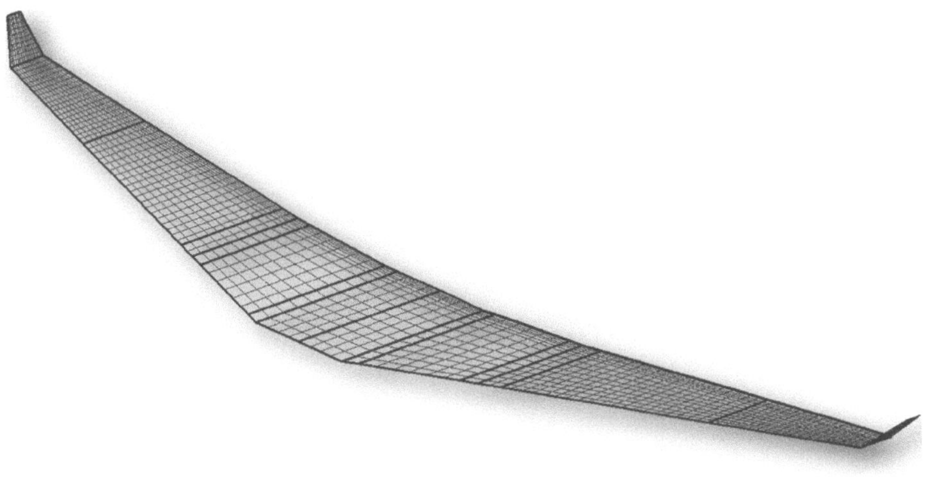

FIG. 8

Reference aircraft aerodynamic model (thicker lines represent panels' boundaries) [14].

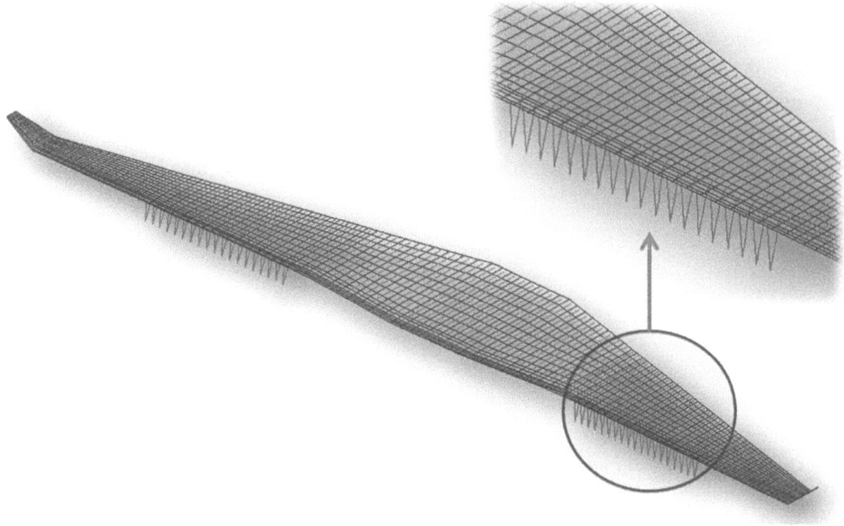

FIG. 9

Interpolation of morphing extra-mode along the outer trailing edge (circled are zoomed).

Flutter analyses were carried out by solving Eq. (33) under the following assumptions:

- PK-continuation method with rationalization of generalized aerodynamic forces for the evaluation of modal frequencies and damping trends versus flight speed.
- BS elastic modes association in the frequency range 0–60 Hz (elastic modes being pertinent to free-free aircraft, with only plunge and roll motions allowed).
- SS extra-mode and extra matrices mapping the configuration cases defined by any combination of the trade-off parameters recapped in Table 2.
- Modal damping (conservatively) equal to 0.01 for all modes (BS modes and extra-mode).
- Sea-level altitude, flight speed range 0–300 m/s ($\approx 1.5\ V_D$, being the dive speed $V_D \approx 200$ m/s for the reference A/C).

Table 2 Trade-Off Flutter Analyses Parameters

Movable Trailing Edge Inertial Distribution	Movable Trailing Edge Stiffness Distributions	Movable Trailing Edge Harmonic Frequency (Hz)
Baseline	Baseline	0
Baseline*1.3	Baseline*1.15	15
Baseline*1.6	Baseline*0.85	30
		45
		60

The results of the 45 analysis cases resulting from all the combinations of Table 2 parameters were arranged in the carpet plot of Fig. 10.

Only five of the forty-five investigation cases produced curves covered by the carpet plot; this is due to the fact that the remaining 40 cases resulted in flutter speeds greater than 300 m/s.

The red horizontal line at constant speed equal to 1.15 V_D, divides the carpet plot in two different regions:

- the region above the red line, defining the safe domain for flutter speed according to CS-25, paragraph 629 and
- the region below the red line, defining the unsafe domain for flutter speed occurrence.

If the inertial distribution of the movable trailing edge is kept equal to its nominal value and its stiffness distributions is reduced by the 15%, flutter occurs at flight speeds lower than $1.15 \cdot V_D$ whatever is the frequency of the trailing edge harmonic. The same happens if the inertial distribution is increased by the 30% and the stiffness distributions are equal to the nominal ones. These dangerous combinations of inertial and stiffness distributions are however quite unlikely to occur, since stiffness increase generally implies mass increase.

For all the other inertial/stiffness configurations covered by the carpet plot, a minimum frequency of 18 Hz for the trailing edge harmonic was found to be sufficient to avoid any flutter instability up to $1.15 \cdot V_D$. In case of actuators failure (harmonic frequency equal to 0 Hz) flutter always occurs at speeds within the certification envelope. Although the simultaneous failure of all actuators represents an event

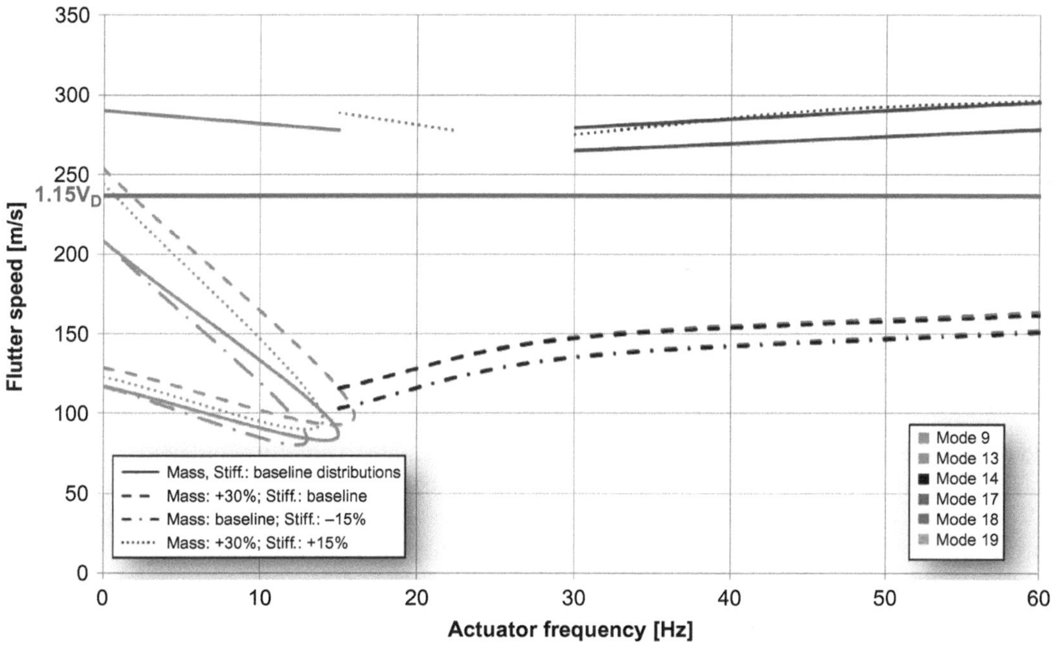

FIG. 10

Carpet plot of flutter speeds [14].

of extremely remote probability, it was however proved that, even in such a case, flutter might be avoided by a suitable mass-balancing of the movable trailing edge. The mass-balancing of the movable trailing edge was duly implemented by properly shaping the ribs blocks portions located forward of the hinge lines. Moreover, the requirement on the minimum allowable harmonic frequency was taken in account to adequately select morphing device's actuators.

5 CONCLUSIONS

When dealing with the design of morphing wings, conventional structural arrangements are commonly replaced by innovative solutions enabling shape changes through actively controlled elasticity or mechanical systems.

In both cases no simplified rules coming from consolidated experiences may be invoked to guarantee that a specific design will be characterized by stable and sustainable aeroelastic response to external loads expected in service.

In order to avoid the consolidation of fully inadequate configurations, aeroelastic issues must be considered since the very preliminary design stage of the morphing device.

The *extra-modes method* has been here presented as a powerful tool to address this task; thanks to a rational and robust formulation, it allows for an efficient aeroelastic assessment while properly dealing with the uncertainties and the continuous configuration changes typical of the earliest design loops.

The fundamental equations of the aeroelastic equilibrium and stability have been recalled and manipulated to account for a strategic partition of the modal database and associated generalized forces; the morphing system is treated as a *substructure* and its modes and generalized parameters are isolated from those related the rest of the A/C (which is in turn regarded as the *basic system*). In this way, two main goals are achieved:

- the computationally onerous procedure leading to the definition of basic system's parameters is not repeated each time a change to the substructure occurs;
- changes in the substructure may be efficiently taken in account by simple matrix operations thus also enabling fast sensitivity analyses on different substructure's configurations.

In order to show the great potentials of the method, two case studies were finally presented with reference to:

- the assessment of morphing wing twist effectiveness for A/C roll control and
- the multi-parametric flutter analysis of a wing equipped with a morphing trailing edge.

In both cases the method has proved to be a tremendous instrument to rapidly appreciate the aeroelastic behavior of a given design, irrespectively of the maturity level of the design itself.

Very embryonal arrangements were indeed investigated and crucial indications were anyway found to drive the design maturation process toward aeroelastically efficient solutions.

Although the extra-modes method has been here introduced for the preliminary aeroelastic assessment of morphing structures, it is worthwhile to point out that its convenience is not limited only to this framework.

The formulation has indeed a general validity and allows for a rational aeroelastic assessment of any substructure's update regardless of the design phase in which the update is applied.

With specific reference to morphing substructures, the method may follow the entire evolution of the addressed concepts through a progressive refinement of the extra-modes parameters up to their full correlation (or replacement) with experimental data coming from ground resonance test on physical prototypes.

REFERENCES

[1] R.L. Bisplinghoff, H. Ashley, R.L. Halfman, Aeroelasticity, Dover Publications Inc., New York, 1996.

[2] E.G. Broadbent, Flutter and Response Calculations in Practice, AGARD Manual on Aeroelasticity, vol. 3, NASA, 1963.

[3] E.H. Dowell, M. Ilgamov, Studies in Nonlinear Aeroelasticity, Springer-Verlag, 1968.

[4] H. Ashley, J. Dugundji, A.G. Rainey, in: Notebook for aeroelasticity, AIAA Professional Seminar Series, 1969.

[5] T.H.G. Megson, Aircraft Structures for Engineering Students, third ed., Butterworth-Heinemann, Burlington, MA, 2003.

[6] I. Chopra, Review of state of art of smart structures and integrated systems, AIAA J. 40 (11) (2002) 2145–2187.

[7] J. Browman, B. Sanders, T. Weisshaar, in: Evaluating the impact of morphing technologies on aircraft performance, Proceedings of the Forty-Third AIAA Conference on Structures, Structural Dynamics and Materials, AIAA Paper 2002-1631, April, 2002.

[8] E. Stanewsky, Adaptive wing and flow control technology, Progr. Aerosp. Sci. 37 (7) (2001) 583–667.

[9] A.R. McGowan, L.G. Horta, J.S. Harrison, D.L. Raney, in: Research activities within NASA's morphing program, Proceedings of the RTO AVT Specialists' Meeting on "Structural Aspects of Flexible Aircraft Control", Ottawa (Canada), October, 1999.

[10] H.P. Monner, D. Sachau, E. Brietbach, in: Design aspects of the elastic trailing edge for an adaptive wing, Proceedings of the RTO AVT Specialists' Meeting on "Structural Aspects of Flexible Aircraft Control", Ottawa (Canada), October, 1999.

[11] S. Courchesne, A.-V. Popov, R.M. Botez, New aeroelastic studies for a morphing wing, INCAS Bull. 4 (2) (2012) 19–28.

[12] G. Andersen, D. Cowan, D. Piatak, in: Aeroelastic modeling, analysis and testing of a morphing wing structure, 48th AIAA/ASME/ASCE/AHS/ASC Structures, Structural Dynamics, and Materials Conference, Structures, Structural Dynamics, and Materials and Co-located Conferences, 2007.

[13] R. Pecora, F. Amoroso, L. Lecce, Effectiveness of wing twist morphing in roll control, J. Aircr. 49 (6) (2012) 1666–1674.

[14] R. Pecora, M. Magnifico, F. Amoroso, E. Monaco, Multi-parametric flutter analysis of a morphing wing trailing edge, Aeronaut. J. 118 (1207) (2014) 1063–1078.

[15] W.P. Rodden, A Comparison of Methods Used in Interfering Lifting Surfaces Theory, AGARD Report R-643, March 1976.

[16] R.L. Harder, R.N. Desmarais, Interpolation using surface splines, J. Aircr. 9 (2) (1972) 189–191.

[17] R.V. Doggett Jr., R.L. Harder, Subsonic Flutter Analysis Addition to NASTRAN, (1973), NASA Report.

[18] C. Lanczos, An iteration method for the solution of the eigenvalue problem of linear differential and integral operators, J. Res. Natl. Bureau Stand. 45 (4) (1990).

[19] P.C. Woelcken, M. Papadopoulos (Eds.), Smart Intelligent Aircraft Structures (SARISTU)—Proceedings of the Final Conference, Springer-Verlag, 2016.

[20] R. Pecora, F. Amoroso, M. Magnifico, I. Dimino, A. Concilio, in: KRISTINA: kinematic rib-based structural system for innovative adaptive trailing edge, Proceedings of SPIE—The International Society for Optical Engineering, vol. 9801, 2016.

[21] European Aviation Safety Agency, Certification Specifications and Acceptable Means of Compliance for Large Aeroplanes—CS-25, Amendment 11, July 2011.
[22] MSCMDNASTRAN®, Software Package, Ver. R3-2006, Reference Manual.

CHAPTER

STRESS ANALYSIS OF A MORPHING SYSTEM 15

Marco Bellucci*, Maria Chiara Noviello†, Francesco Amoroso†, Rosario Pecora†, Ignazio Dimino‡, Antonio Concilio‡

MARE Engineering, Naples, Italy University of Naples "Federico II", Naples, Italy† The Italian Aerospace Research Centre, CIRA SCpA, Capua (CE), Italy‡*

CHAPTER OUTLINE

NOMENCLATURE

2D, 3D two-, three-dimensional
A cross-sectional area of the bolt

Morphing Wing Technologies. https://doi.org/10.1016/B978-0-08-100964-2.00015-0

Abr	smaller bearing area
As	minimum bearing area
At	net tension area; Ad = $(W - D) \times t$
ATE	adaptive trailing edge
ATED	adaptive trailing edge device
CAD	computer aided engineering
CFD	computational fluid dynamics
D	bolt, lug hole diameter
DMU	digital mock-up
DOF	degree of freedom
EIG	eigenvalue
FE	finite element
FEA	finite element analysis
FEM	finite element model or method
Fbru	ultimate bearing strength
Fbry	yield bearing strength
Fcy	compressive yield strength of the bushing material
Fsu	ultimate material shear stress
Ftu	ultimate tensile strength of the material
ftu	actual tension under UL
Fty	yield tensile strength of material
fty	actual tension under LL
Kbru	shear bearing efficiency factor for ultimate strength [1]
Kbry	shear bearing efficiency factor for yield strength [1]
Kt	stress concentration factor [1]
LL	limit Load
M4, M5	fasteners type (diameter)
MOS, MS	margin of safety
MSLL	MOS at limit load
MSUL	MOS at ultimate load
Pbru	product of Fbru $\times D \times t$; product of Kbru × Ftu × Abr
Pbry	product of Fbry $\times D \times t$; product of Kbry × Fty × Abr; product of 1.85 × Fcy × As
Pnetu	product of Kt × Ftu × At
Pnety	product of Kt × Fty × At
Ptrav	transverse load
R_a	axial load divided by the smaller value between Pnetu and Pbru
RBE	rigid body element
R_{tr}	transverse load divided by the inverse of Pbru
Sa	applied shear load
Sall	product of Fsu times A
SARISTU	Smart Intelligent Aircraft Structures
SM	safety margins
SMA	shape memory alloy
T	plate thickness
Ta	applied tension load
Tall	product of Ftu times A
UL	ultimate load

W	lug width
WB	wing box
ε_l	elastic limit—strain
ε_{MAX}	actual strain field, max value
σ_l	elastic limit—tension
σ_{MAX}	actual stress field, max value
°C	centigrade degree
f	frequency
Hz	hertz

1 INTRODUCTION

Conventional control surfaces, like flaps, slats, and ailerons produce discontinuities that affect aircraft aerodynamic efficiency. Their benefits are confined to specific aircraft flight segments, but are moderate with respect to those that could be obtained from an inherently deformable or locally adaptable wing capable to achieve dissimilar, multiobjective mission roles efficiently and effectively.

Morphing devices could provide adaptive capabilities to maximize aircraft performance in any flight operation. Such benefits may range from enhanced aerodynamic performance [2] up to considerable noise reduction [3], and better load control capabilities with benefits in structural weight and fuel saving. However, the higher level of complexity resulting from a more closely integrated multidisciplinary design makes it more challenging to meet standard aircraft airworthiness certification schemes without undergoing structural collapses.

The development of new materials and architectures, together with the ever-present need for a practical realization of innovative solutions, makes any proposed morphing airframe a demanding platform to be simulated. Among the others, some key aspects have to be considered since the preliminary design phase, such as:

- Simulation models of single parts (2D or 3D elements) to predict thin shell or thick solid components behavior, that are typically present on morphing structures.
- Accurate modeling of local geometrical singularities (holes, fillets, local thickness variations) in order to capture stress concentration and possible plasticization effects.
- Detailed modeling of internal connecting members: links, rods, lugs, joints, etc.
- Correct modeling of single degrees of freedom (DOF) mechanisms: hinges and rotation centers, kinematic chain of actuation systems, avoiding an internal over constraining of the system (for example, nonphysical bonding of some parts).

Nevertheless, such recommendations shall be considered only as general guidelines in order to generate an accurate and predictable numerical model of a morphing structure. All the other general FE modeling checks (convergence analysis check, global equilibrium check, correct evaluation of local stress peaks, and geometrical detail effect, etc.) must be taken into account to avoid any trivial mistake.

In this chapter, some general concepts related to the structural design and simulation of a morphing structure are given. A benchmark system, following referred to as adaptive trailing edge device

(ATE or ATED), is used as example to validate the proposed design approach. All the results, described and displayed in the following paragraphs, were reached in the framework of the European research project Smart Intelligent Aircraft Structures (SARISTU).

2 DESIGN OF A MORPHING STRUCTURE

A two-step design approach typically guides the design of structural-mechanical systems and related motion control. Such a methodology decomposes the interrelated topology and sizing problem into a simpler geometry optimization using engineering principles. However, this implies seeking the optimal solution for each mechanical subsystem individually and then refining the whole architecture in the course of the global system assembly and performance evaluation.

On the contrary, due to its multidisciplinary nature, the design process of a morphing system goes well beyond a pure design for stiffness with given loads. It requires a thorough combination of many design variables to optimize structural strength and strains distribution along with deformation shapes under operative loads by taking into account actuation kinematics, conformal skin, and the most relevant manufacturing constraints as well as light-weight targets, among others. In a combined FE-model (skin and mechanism), free design parameters, such as thickness distribution, fiber orientation, actuation mechanisms (rotational or translational) shall be then optimized in a multidisciplinary frame.

It is then obvious that all the structural design related aspects shall be covered by advanced FE analyses. Each part, such as plates, rods, actuators, skin, etc., incorporated into the self-standing one-piece device shall be flexible but strong enough to respond to varying operating conditions and external disturbances, without affecting system integrity, thus avoiding insurgence of failure or plastic deformations in the most stressed areas, global or local buckling, and so on.

Morphing systems design consists of several steps. In the herein proposed vision, the first one involves high-fidelity aerodynamic simulations aiming at identifying the optimal morphing aeroshapes in compliance with top-level high-lift requirements pertinent to the reference wing. The search for optimal solutions with respect to drag/lift-to-drag ratio (for instance) may take advantage of well-assessed CFD codes or be based on simple 2D aerodynamic simulations. The benefits at aircraft level in terms of aerodynamic performance, load control, and load reduction due to integrated use of morphing devices shall be then assessed through multifidelity analyses.

After that, the advanced design of a morphing structure involves the following steps:

- The generation of a parametric CAD model allowing design space investigation studies and interactions with pertinent aerodynamic and structural meshes.
- Creation of a FEM model based on the CAD model for detailed structural analysis. A condensed model can be also coupled with the aerodynamic model to form an aeroelastic framework that will allow for the aeroelastic assessment of the effects of flexibility.
- Detailed design and structural analyses including detailed drawings.
- Definition of a digital mock-up (DMU).
- Prediction and evaluation of assembly and integration issues. This task may be devoted to the detailed analysis of the mutual structural interaction between different morphing systems, which may be deployed simultaneously and whose performance may be affected by the wing box (WB) elasticity.

Starting from a conceptual design of the generic morphing device, the structural sizing of the morphing structure is an iterative process involving weight targets, load predictions and physical integration constraints. To this aim, the materials shall be carefully considered in close coordination with tooling, manufacturing, and assembly partners. The design of fittings, joints and tracks becomes fundamental once the major components, such as skins, spars and ribs are properly defined. The morphing skin is probably one of the most critical parts because it shall be able to guarantee:

- the continuity of the shape (in order to preserve the aerodynamic performance of the profile), and
- adequate flexibility and strength to withstand the external pressure loads while guaranteeing the desired geometry.

Following the proposed methodology, the adaptive trailing edge device (ATED) design started with the definition of the optimal morphing aeroshape, shown in Fig. 1. The aerodynamic computation of the system aeroshapes is out of the scope of this chapter. The morphing trailing edge was designed to reduce drag in off-design flight points by adapting chord-wise camber variations in cruise to compensate aircraft weight reduction due to fuel consumption. In principle, camber morphing may also enhance high-lift performance during take-off and landing and perform load control at high speed.

Such a desired chord-wise camber variation was implemented through a multifinger single DOF concept. The layout of an articulated (finger-like) rib structure was then assessed to physically realize the transition from the baseline airfoil configuration to the target aeroshapes envelope. After that, the actuation system was designed via a lever mechanisms driven by load-bearing actuators combing load carrying and command functions. Such a morphing mechanism was designed to reproduce a variation of $\pm 5°$ with respect to the neutral "unmorphed" configuration (see Fig. 1).

Four main plates characterize the rib structural concept, shown in Fig. 2: B0, B1, B2, and B3. B0 and B2 have the same middle plane; B1 and B3 are shifted with respect to them along the normal to their surface, always sharing a common middle plane. Each plate is connected to the adjacent one by a hinge, located on the rib camber line (points A, B, and C, respectively at 20%, 50%, and 70% of rib chord, Fig. 2). Block B0 is rigidly connected to the main wing structure, while all the other blocks are free to rotate around the hinges on the camber line that is then made as

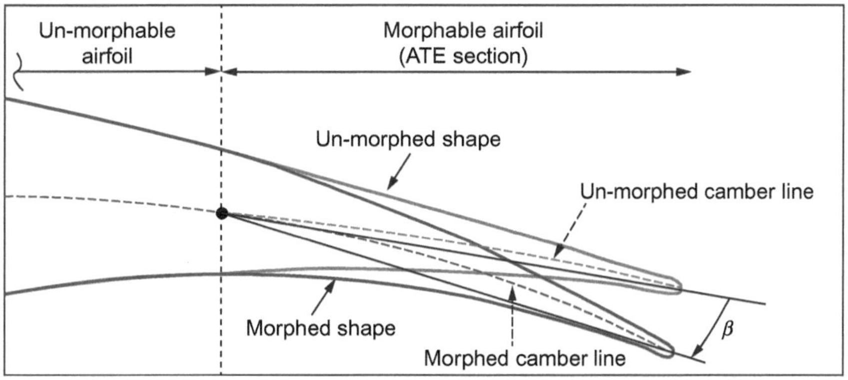

FIG. 1

Morphing shape definition.

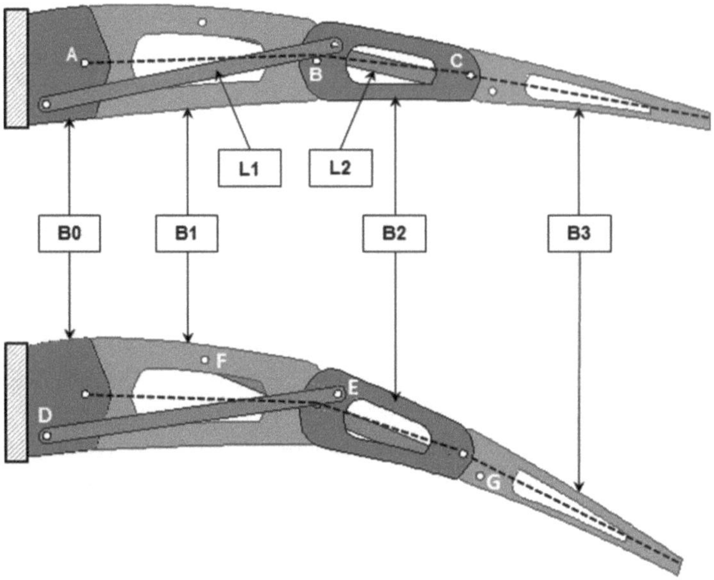

FIG. 2

Morphing rib architecture design.

an articulated chain of consecutive segments. Dedicated linking rods, hinged on nonadjacent blocks, force the camber line segments to rotate according to specific gear ratios. The position of the hinges can be obtained with an inverse kinematic problem approach fixing, for example, the chord wise extensions of each block, as done in [4].

Plate B0 is linked to plate B2 by means of a rod element hinged at points D and E, with hinges D and E being respectively located on B0 and B2. Plate B1 is linked to plate B3 through a second rod hinged at points F and G. Crossed links (DE and FG) positions are conceived in order to assure specific rotation ratios between adjacent plates and an overall plates movement to match the target morphed shape. More in detail, considering the plate B0 fixed on the flap strut, a 3° downward rotation of B1 around A makes all the other plates to move so that the positions of the hinges B and C move along the camber line of the target airfoil shapes.

Same applies if 11° downward rotation is imposed to plate B2 around B or 3° downward rotation is imposed to plate B3 around C. In other words, if a single plate is driven by a unique actuator, all the other plates are forced to move in compliance with the objective shape envelope. The resulting morphing rib solution is shown in Fig. 3. The spanwise stiffness of the structure was assured by a box arrangement characterized by spar connecting the plates of two consecutive ribs, Fig. 4.

The morphing system thus incorporated four different classes of main components: ribs, actuation system, spars and the morphing skin. An elastomer-based skin stiffened by aluminum profiles was the solution identified to achieve the desired shape change at cruise altitudes (temperature range up to −55°C). FEM simulations, shown in Fig. 5, and mechanical tests found optimal fatigue and aging properties of this new multimaterial device.

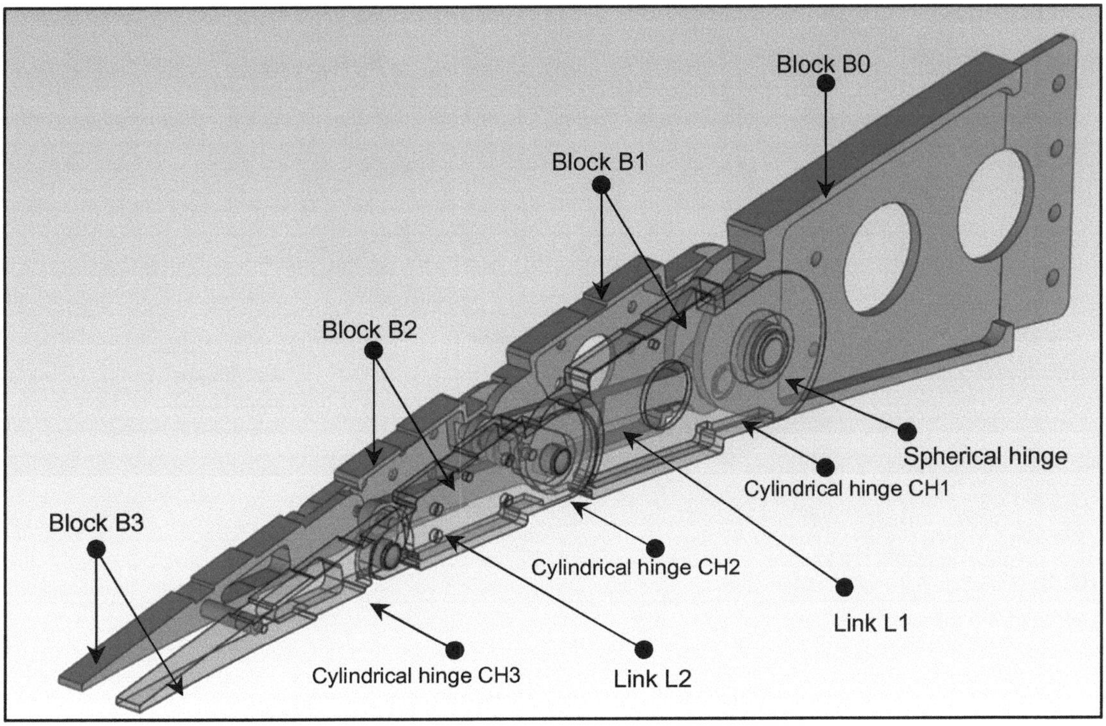

FIG. 3

Single morphable rib element.

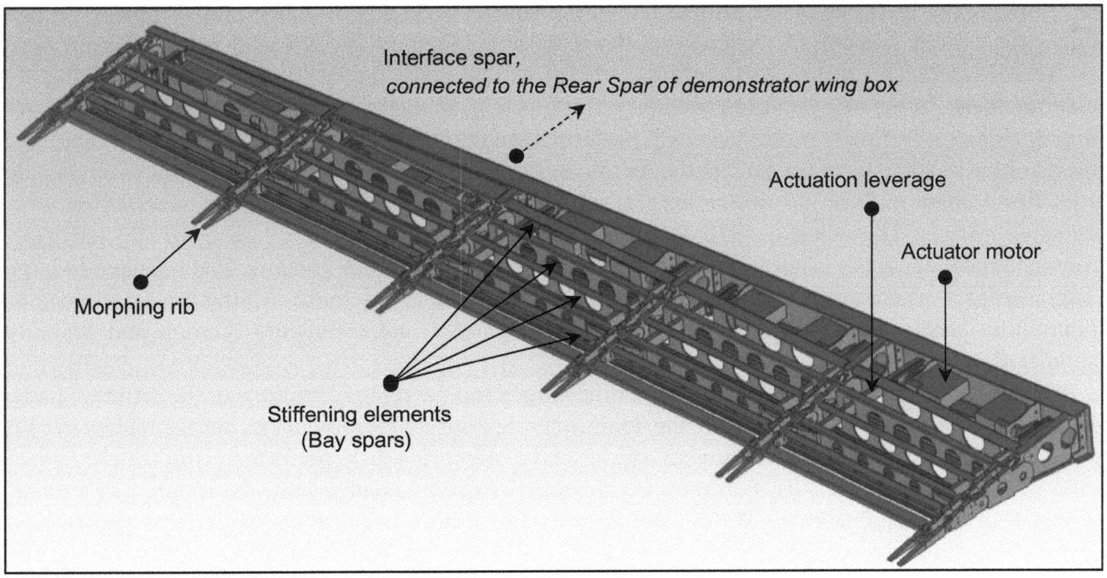

FIG. 4

Main Structure of the Adaptive Trailing Edge (5-bay system).

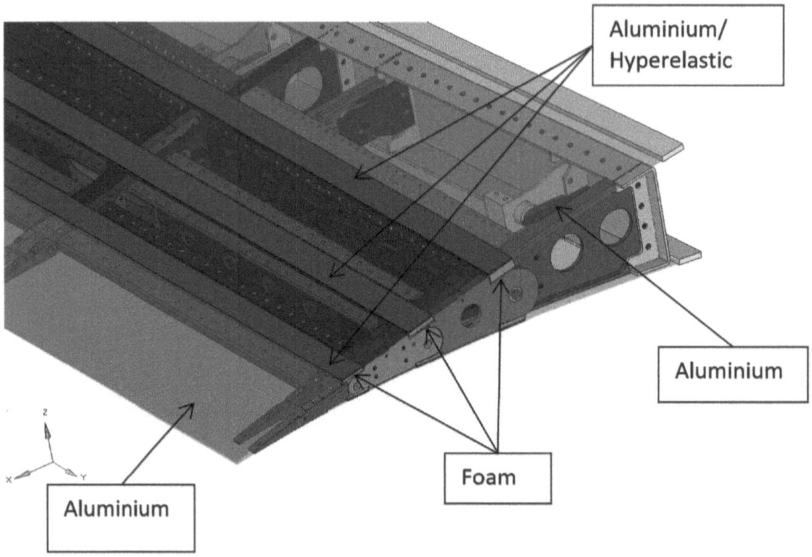

FIG. 5

Layout of the morphing skin.

3 FINITE ELEMENT MODELING OF MORPHING STRUCTURES

The FE model presented in this work takes into account all the components that contribute to make up the morphing trailing edge. The attention is particularly focused on the primary structure, the actuation system and the deformable skin panels. The primary structure is intended as the reference skeleton, with the goal of distributing the efforts among the several elements of which it is composed. It is a kinematic system, and only the insertion of the actuator architecture gives it load-bearing capabilities.

The electromechanical motors are the components that make the system capable of resisting the external loads. In the safe configuration, they work as a local clamp in the simulation; they are therefore imagined as infinitely rigid. A check is performed on the actual capability of such systems to afford the resulting incident force. In the end, the morphing skins are those components that allow macroscopic stretches without making the system lose its surface continuity, a property that is essential for aerodynamic surfaces. They are exposed to very large deformations (in the studied case, up to 5%) and therefore a nonlinear analysis is necessary. The morphing concept itself is something that is linked to large displacements, where small perturbation hypotheses cannot hold any more. All the three-components (aluminum panels, hyper-elastic sheets, and foam segments) are taken into account and properly simulated.

One of the major issues is the modeling of the hinges and the relative motions of the different parts. In fact, the configuration should ensure the noninterference during the movement, hardly evident in a FE simulation. This is the reason why several studies have been previously performed with the use of specific multibody codes [2,5–7]. Furthermore, the small radii, associated with the rotary connections, would lead to unreliable value of the local stresses and strains, because of the extreme geometrical layout. This limitation of the used model is discussed during the comment of the results. Going into the detail of the simulation process, various finite elements are taken into account in construction of the model, each with a different purpose: solid, shell, beam, and rigid elements:

- Parabolic tetra elements for ribs and rods modeling.
- Linear hexa elements for foam and actuator system modeling.
- Linear quadrilateral elements for skin and span wise spars modeling.

In the following, features of the numerical simulation of the above components are detailed along with some results.

3.1 RIB AND SPARS

Since the single block elements of morphable ribs are solid machined parts, solid elements seem to be a good solution to model them; the irregular geometry of the ribs lead to the use of tetrahedral elements. As well known in literature, tetrahedral 10 node parabolic elements should be used to assure accurate results [6,7]. A sketch showing the FEM model mesh of a morphable rib is shown in Fig. 6, including rib blocks (dark red) and links (dark) meshes. The rib blocks are made in Al 2024, while the links in Steel 17-4 pH.

The internal hinges (spherical and cylindrical, connecting the different blocks and the attachments between the links and the rib blocks) are modeled according to the usual scheme of rigid body element (RBE) connections [4,8]. Such a connection is an algebraic representation of a kinematic constraint, which is an explicit equation that couples more DOFs of a set of nodes. In a RBE connection, the unknown DOF of a certain node (named master node) is equal to a combination of the DOFs of other nodes, belonging to the same RBE set (slave nodes). In this way, the group has a single DOF for each

Color	Type	Description
Solid	Aluminum	Al2024
Solid	Steel	Steel 17-4 PH

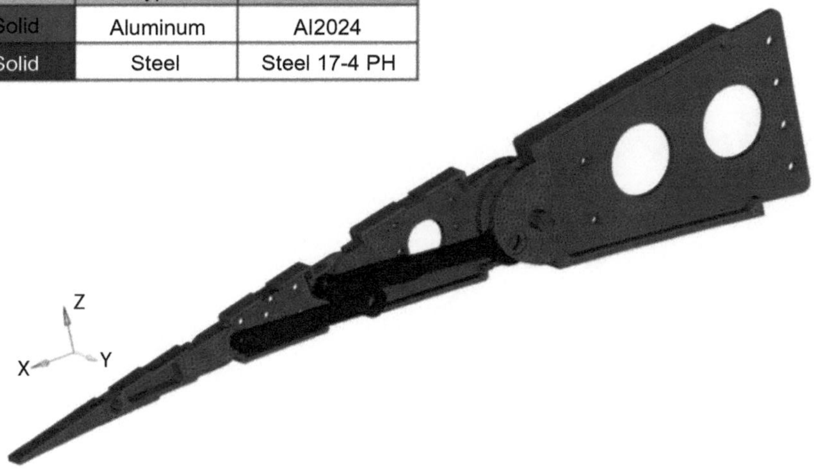

FIG. 6

Rib blocks and links, modeled with solid elements.

direction, making it a rigid body with only six translational and rotational motion capability and no internal deformability. RBE is generally used for two different applications:

- To simulate an extended rigid body.
- To simulate joints (pins, fasteners or screws, hinges).

In fact, by properly modulating the constraints among the nodes, concurring into the RBE set, different architectures may be performed, leaving the DOFs that better represent a particular joint. In other words, namely rigid or partially rigid connections may be realized, as in the case of the link between two parts or two blocks, via a couple of hinges. Indeed, several types of joints may be simulated by establishing a link between two sets of RBEs, connected with a rod element:

- Fasteners: all DOFs are coupled.
- Pins: all DOFs are coupled, except for a rotation around the pin axis.
- Multidimensional hinges: all translational DOFs are coupled while all rotational DOFs are uncoupled; in this case, the connecting element has a zero length; rotational DOFs may be reduced coming to a 2D hinge or reducing the system to a pin, as above.

An example of pin modeling is shown in Fig. 7. Therein, the independent node is seen as the ideal center of rotation. For any single hinge, a local coordinate system related to the rigid body is defined. The hinge is then modeled by coupling the DOFs of the master independent with the slave nodes, accordingly. For the case of a cylindrical hinge, as the one that is here discussed, all DOFs but the local cylindrical rotation are constrained. The mutual rotation between the connected parts is guaranteed by a BEAM, linking the two master nodes.

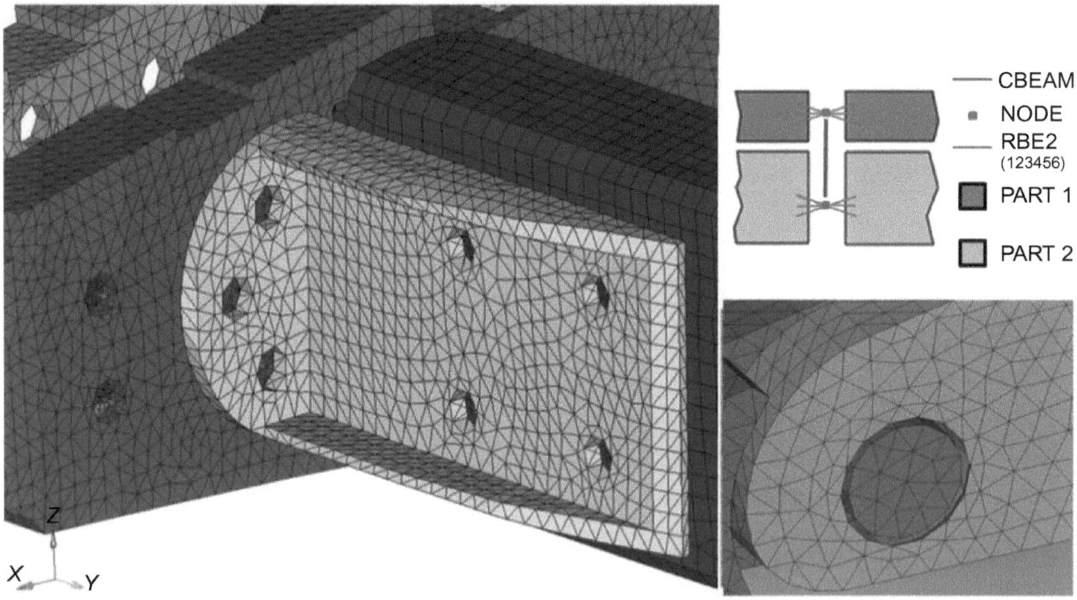

FIG. 7

Detail of a RBE connection that simulates a point to point link (Pin, rivet, bolt, etc.).

The FE mesh around the holes shall be very fine for two main reasons:

- in order to accurately capture local stress flow, and potential stress peaks due to the Kt concentration effect [1];
- to make an accurate evaluation of the connections Safety Margins (SM—bearings, lugs, joints, and fasteners), that are modeled with RBEs, connected to the central node of linking part.

Indeed, this step is not trivial because the local curvatures are usually very small and they would require a very dense mesh. This would in turn cause an explosion of the number of the DOFs and an excessive complication of the overall discretization. This is the reason why, usually, alternative methods are preferred. For instance, a comprehensive analysis is first run. The results show some critical points, where the stress or the strain field may attain high values. Where it is suspected this is due to insufficient mesh refinement, the single element may be extracted, together with the incident forces and modeled apart, with a very detailed grid. Achieved output is taken as the reference one, more representative of the reality.

The following images show some examples of the FE model of single parts. In detail, examples of the mesh around the holes are presented in Fig. 8, with reference to a generic block (where the voids are inserted for reducing the component weight) or the motor room (with function of lightening and for hosting the locking screws). In Fig. 9 instead, examples of the hinges modeling are reported. In the above picture, a classical cylindrical hinge is represented, connecting the rib block 1 to the rib block 2, while in the bottom sketch, a typical spherical hinge is shown, linking the rib block 2 to the rib block 3. Both the cases herein illustrated are relative to the use of solid elements (HEXA).

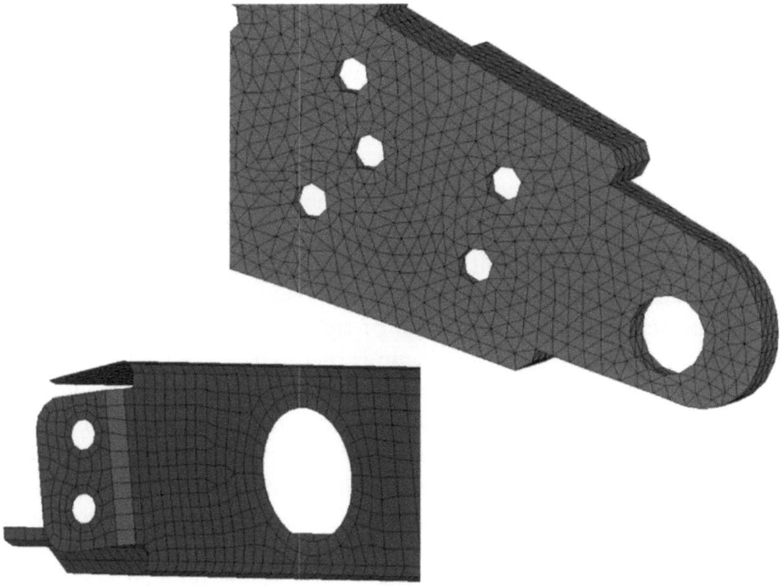

FIG. 8

Detail of FE modeling around the holes.

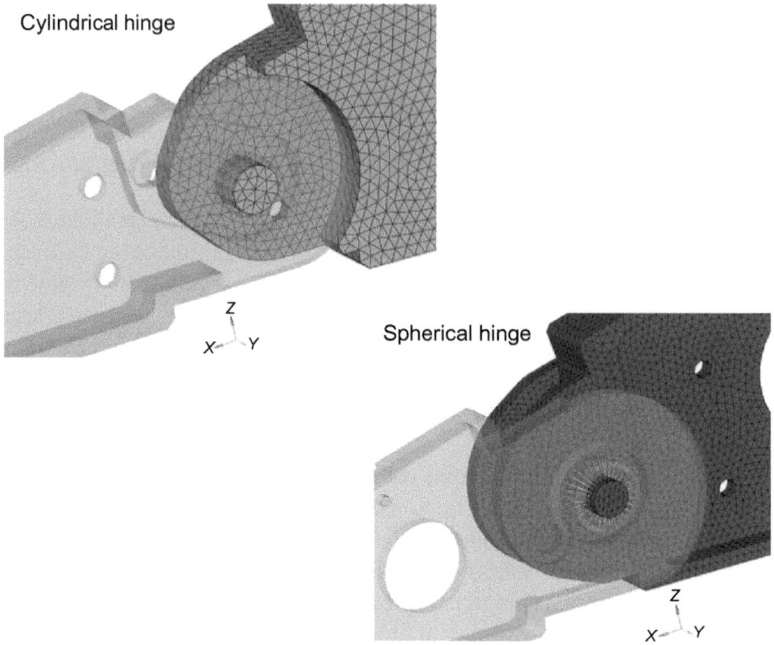

FIG. 9

Detail of FE models of internal hinges.

Finally, in Fig. 10, a detail of the FE model mesh of the spars is reported. These structural elements give the wing section a proper torsional stiffness. Because of their limited thickness and the easiness of the design and the connections, shell elements are preferred, limiting somehow the number of total DOFs.

3.2 FASTENERS

A generic fastener is modeled through a beam element whose extremities are rigidly linked to the connecting parts. The relative scheme is depicted in Fig. 11, right, which shows the detail of a joint between a spar (the longitudinal element on the right in the picture) and a rib block (the massive element on the left). The beam extremities are tied to a number of points, representing the regions that are in physical contact. These links are schematized as a number of rigid elements (RBEi), making up a "bundle." In the left side of Fig. 11, the realization schematic is reported.

In the following image, Fig. 12, some of the regions where spars and skin are fastened, are shown as an example, while in Table 1 an indication of the adopted diameters is reported. As it can be seen, for both the structural elements the items are "removable." This choice is selected in the design phase in order to have the possibility to continuously access the different sections, to facilitate maintenance and adjustments of the different components on the experimental prototype. In the same way, the very dense distribution of connection points is linked to the necessity of having a high safety margin, to minimize risks during the wind tunnel test campaign. Once again, the fact to deal with an advanced technology demonstrator drives this selection.

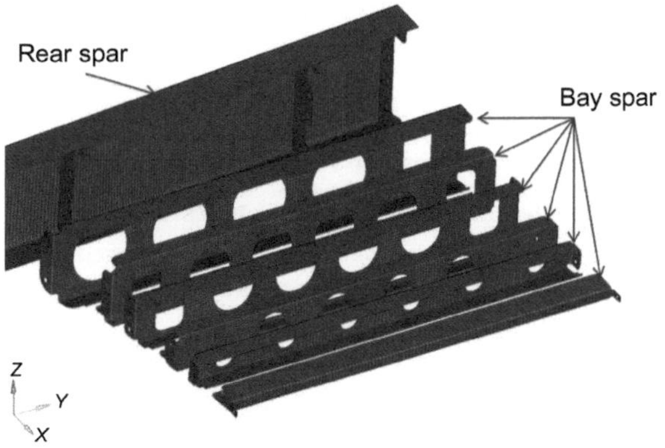

FIG. 10

Detail of the FE modeling of spars.

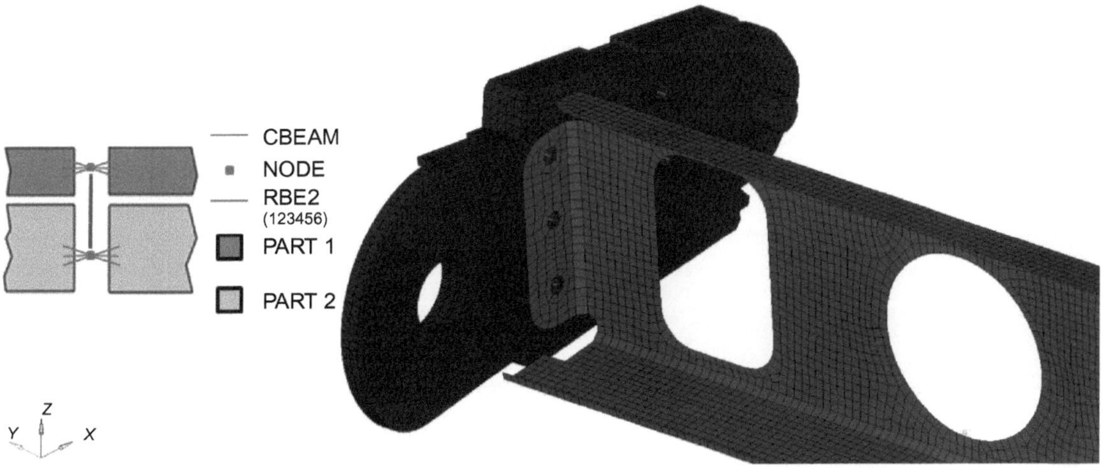

FIG. 11

Fasteners modeling scheme for rib-spar connection.

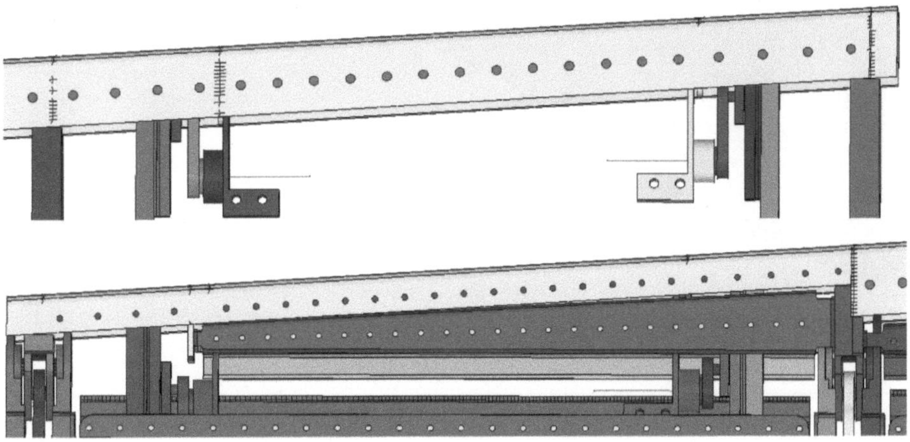

FIG. 12

A zoom on the fastened zones: 1st to 2nd bay (top), 5th bay (bottom).

Table 1 Fasteners Type, Connecting the Rear Spar and the Skin

Bays 1–4	Type	Bay 5	Type
M5	Removable	M4	Removable

3.3 SKIN

The skin of a morphing system [9] shall be compliant with its deformability. In synthesis, it does mean it must guarantee the continuity of the shape from a geometrical point of view, while ensuring the proper structural connections from a physical point of view. With reference to the morphing trailing edge and the architecture that have herein been proposed, the skin is made of aluminum segments permitting a rigid connection to the rib blocks and foam inserts allowing the necessary compliance of the ribs, by granting relative rotations between each couple of modules. The manufacture of the morphing skin is completed by a hyper-elastic sheet (silicone), having the basic function of protecting the underlying components, above all the foam [10]. In Fig. 13, the FE mesh of the complete morphing skin system is shown (aluminum, hyper-elastic, and foam portions).

Both the external cover (silicone) and the metal parts are modeled through plate elements (QUAD), while foams are schematized through solid elements (HEXA). Once again, this choice is a balance

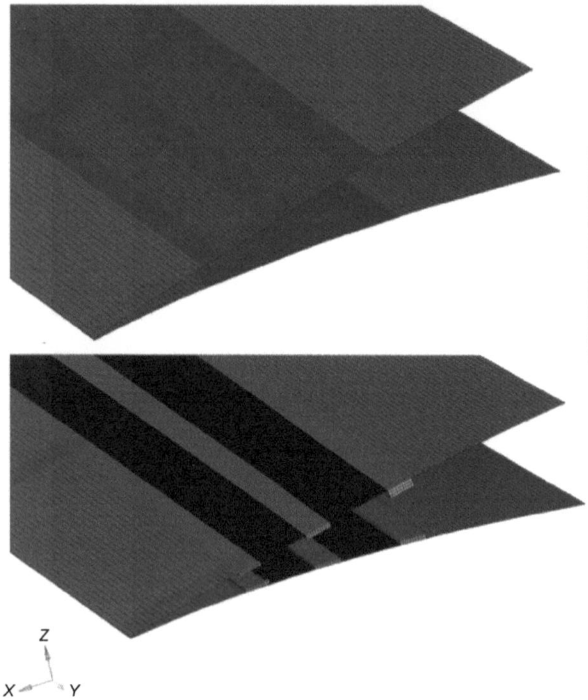

Color	Type	Description
Th 2.5mm	Aluminum	Al2024
Th 1.0mm	Hyperelastic	Elastometric skin
Th 1.5mm	Aluminum	Al2024
Solid	Foam	Hyperfoam

FIG. 13

Geometry of morphing skin: silicone and metal skin (top); deformable skin segment (bottom).

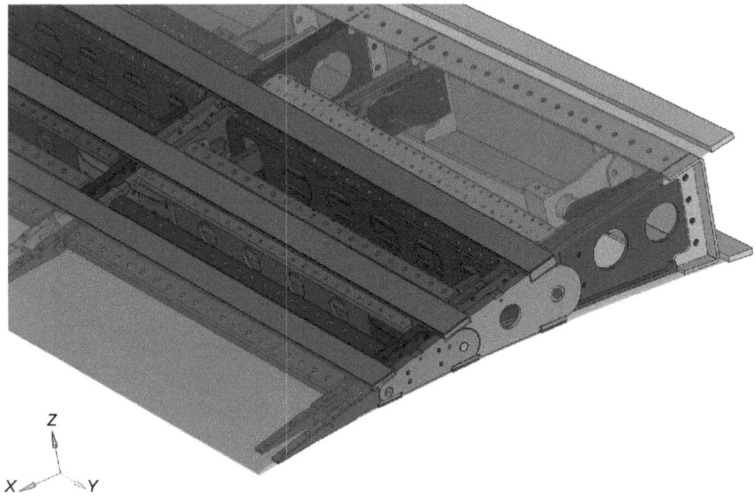

FIG. 14

Details of skin-structure fasteners.

among different necessities: the need to achieve a high-level discretization and the need to somehow contain the number of DOFs and the complexity of the model itself. The resulting scheme, already reported in Fig. 13, is complex enough, made a bit simpler because of the uniform width of the segments, spanwise. The metal part of the morphing skin (namely, nondeformable) is connected to the primary structure with fasteners, following the layout illustrated in Fig. 14 and according to the same schematic, previously shown in Fig. 11. The soft foam strips are instead bonded to both the silicone layer and the connected aluminum plates.

3.4 ACTUATION SYSTEM

The actuation system consists of the motor, a linear guide, in turn composed of a rail, a gliding element, and a carriage, all made of aluminum, and an actuation leverage that may be further split into a crank and a connecting plate, both of steel [5,11]. All these components are modeled with solid elements (HEXA) but the actuator shaft that is inserted in the crank through the connecting plate and is schematized with BEAMs, Fig. 15. A concentrated mass simulates the presence of the motor, positioned at its center of gravity, Fig. 16. After the schematization process, the global actuation system mass is adjusted to match the real one.

The basic assumption in the actuation system design is that the most penalizing cases occur when the system is blocked (clamped-like condition). This is what happens when the mechanism has reached a certain position and is stopped to maintain that configuration for the necessary time. In this circumstance, the relative displacements between the carriage and the gliding component is negligible (namely zero), so that the parts may be connected via rigid elements. This schematic gives rise to the most critical stress-strain distribution inside the system. In Fig. 17, the inner constraint system is reported.

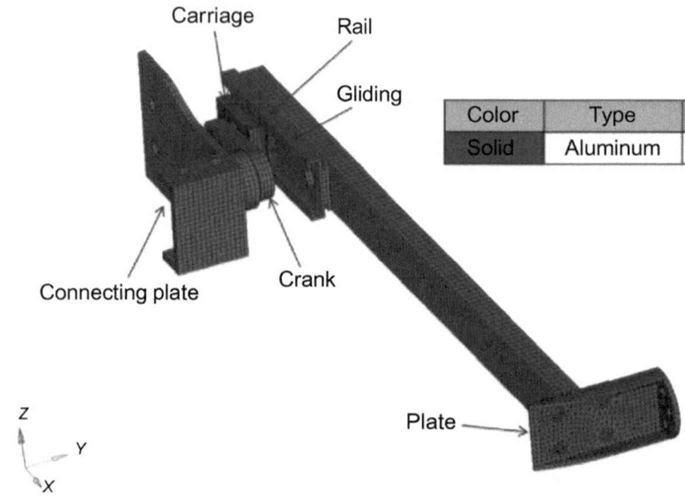

Color	Type	Description
Solid	Aluminum	Al2024

FIG. 15

FE representation of the actuator system.

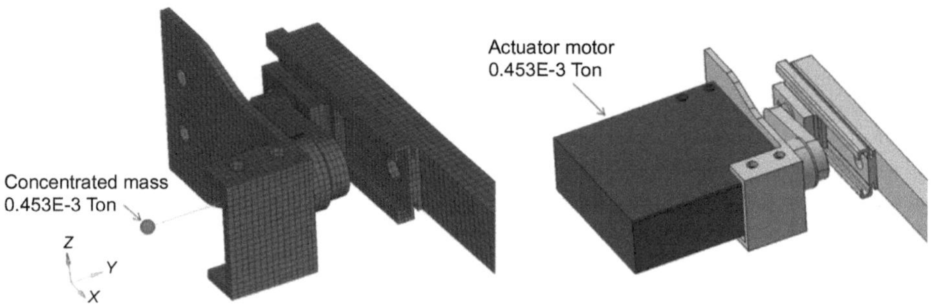

FIG. 16

Detail of the numerical representation of the actuator mass.

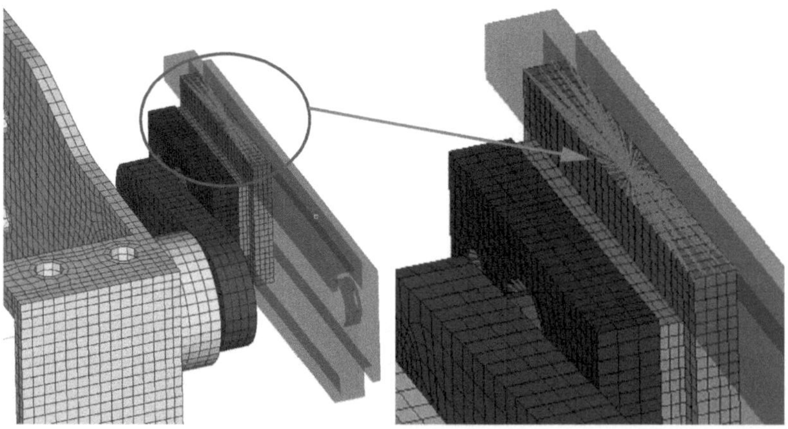

FIG. 17

Carriage-rail interface: detail of the inner constraint system.

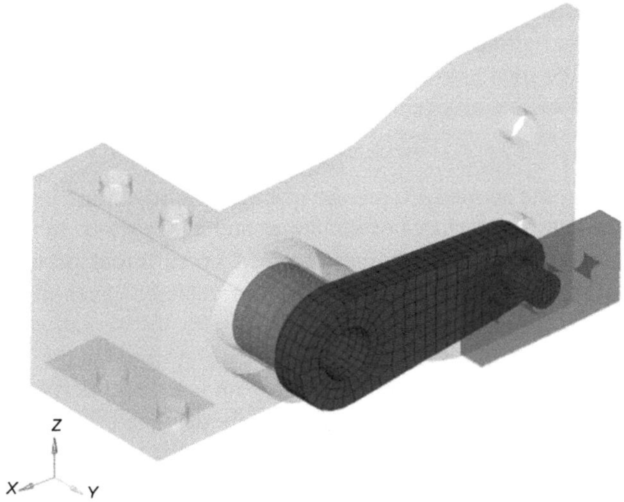

FIG. 18

Detail of the crank-pin model, with the actuator shaft.

The kinematic transmission system (transforming the motor rotary motion into a linear movement) is made of different parts centered on the crank, a squat beam aimed at transferring displacement and forces, Fig. 18. In detail, a pin connects the crank (the solid protrusion on the right side of the referred picture) to an interface plate (running along the linear guide light gray in the same figure) while a shaft links the crank to the inner actuator axis (light gray cylinder, on the left side of the reported schematic). This latter element is modeled with a classical beam linked to the crank with RBE elements, similarly to usual fasteners.

4 DESIGN LOADS AND CONSTRAINTS

Aeronautical structure design loads are distinguished in the:

- limit loads (LL), the most severe loads, actually occurring in service conditions and
- ultimate loads (UL), the loads that really drive the design process [12].

Requirements are different for the two cases. At the LL, the structure must not exhibit permanent strain in any part; this means that the internal stress field is completely contained into the elastic domain. At the UL, the structure must remain safe also in presence of plastic deformations; in other words, induced plastic deformations shall not determine any type of local or global structural collapse. The basic rule holds:

$$\mathrm{UL} = 1.5 \times LL \tag{1}$$

In practical terms, the structure is dimensioned at the UL and shall be verified at the LL. All the structural components are then sized to present internal tensions that are under the failure stresses. In this case, plastic behavior is tolerated. Nonlinear analysis shall be then conducted, in order to take into account the nonlinearity connected to the material behavior. The other part of nonlinearity, linked to geometrical appearance, is absorbed in the simulation technique. In fact, the occurring large displacements are not an effect of the external loads but the kinematics. Because the system shall be verified at each different configuration it may assume, the process simply takes into account different shapes. Any of them shall be then proved at LL, by confirming that the arisen stresses (and strains) remain under the threshold of linearity (elasticity domain):

$$\sigma_{MAX} \leq \sigma_l; \; \varepsilon_{MAX} \leq \varepsilon_l \tag{2}$$

In the investigated problem, the applied LL are the resultant aerodynamic loads and consist of a pressure distribution over the external surfaces of the upper and lower skins. In the analysis, the pressure field is simplified into a set of constant values, applied on different portions of the ATED. This "stepped" or "constant block" scheme is made of four different pressure levels per single bay (chord-wise), corresponding to a total of 20 different loads for the overall structure, Fig. 19.

In the analyzed configuration, the ATE is assumed to be connected to the WB by fasteners, joining the complementary parts of the respective skins. This layout is a direct derivation of the chosen architecture logic. Different partners at different sites separately designed the four parts of the SARISTU adaptive wing [13]. In order they could match each other during the assembly phase, great attention was in fact devoted to the interfaces. The ATE is connected to the WB through the rear spar that was decided to be split in two in order to ease the integration and the construction process. In fact, the two parts, independently attached on the ATE and the WB respectively, should have been merged during the final construction phase. However, several reasons concerning the used materials (different for the two items), and the installed sensor networks (widely distributed in each component of the WB, rear spar included), led the designers to imagine a different solution. In the end, the designers agreed that the two parts would have been connected through the upper and lower skins, also relying on the large thickness of the external panels of the WB.

For the assembly process purposes, limited connections between the two spars were however maintained. This unusual approach revealed to be safe, when the whole wing prototype was tested on ground. The internal loads between the ATE and the WB are then transferred by the ties, connecting the respective external coverings. Since the WB is much stiffer than the trailing edge, these constraints are modeled as perfect local clamps (all the DOF of each single grid, corresponding to the pins location were constrained, Fig. 20). The fasteners reaction forces are therefore the transmitted loads, with opposite sign.

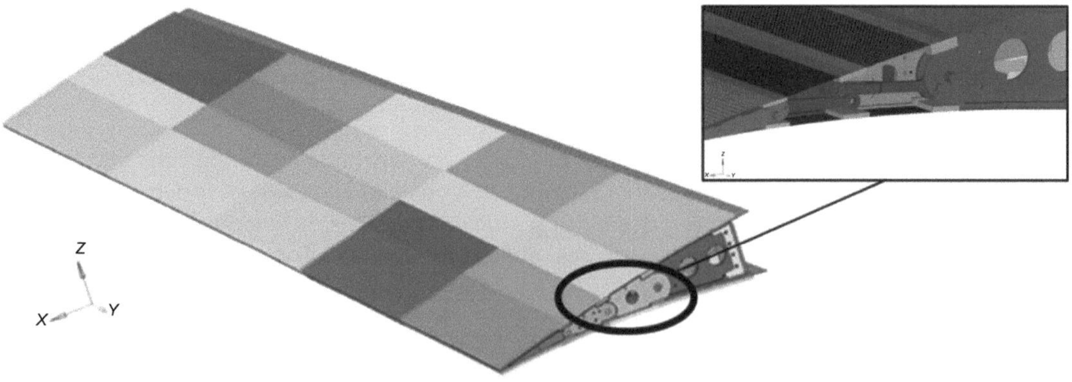

FIG. 19

Pressure distribution scheme (left) and local detail (right).

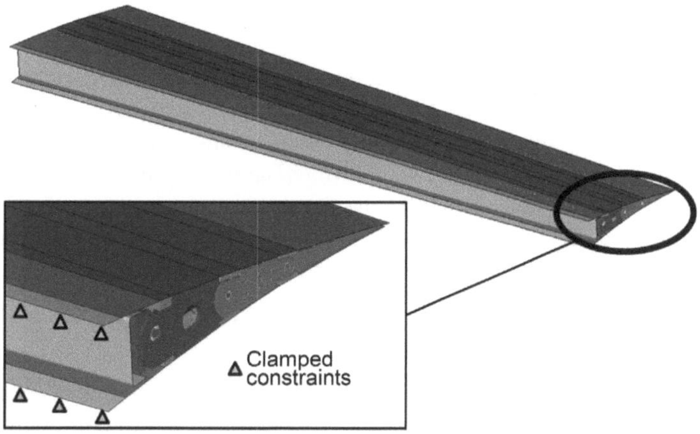

FIG. 20

Detail of the ATED constraints (upper and lower spar caps).

It should be also remarked that, in reality, these links are not rigid and therefore an elastic constraint should better represent the situation. However, the real stiffness of the other structural parts was unknown. Furthermore, the selected is a conservative approach, because the resulting constraint forces result to be larger than in the other instance. These two considerations definitely supported the adopted decision. The final experimental tests, carried out within the SARISTU project, confirmed those expectations.

5 STRUCTURAL DESIGN AND SIMULATIONS

The ATED structural analysis is carried out in accordance with the common practice in the aerospace structure design and agree with the traditional literature [1,12,14]. The main type of analysis, conducted on a generic structure for aerospace applications are:

- Static analysis at LL and UL, to ensure that the structure is correctly sized with respect to service and design loads.
- Buckling analysis, to verify that no elastic instability phenomena occur in the slender parts of the structure.
- Modal analysis, to avoid that natural frequencies and mode shapes could give rise to undesired aero-elastic phenomena.
- Stress analysis, to calculate the margin of safety (MOS) [15] for all the internal parts of the structure.

These same steps should be accomplished for a morphing structure. It is noteworthy that, as a natural consequence of the selected adaptive architecture (a classical kinematic chain), the design criteria cannot be different from the ones applied for a traditional system.

In the following, the results of these analyses are shown for the ATED structure. Concerning the static analyses, the computations were performed under the assumption of geometrical nonlinearity insurgence under the action of the LL and the UL (large elastic displacements are in fact assumed to occur, apart the morphing, quasirigid movement). Pressure loads were modelled as an equivalent set of lumped forces. Material nonlinearity is also taken into account. MOSs are calculated for the two cases, directly using the results coming from the FE analysis for the structural parts and some selected specialized design formulae for the internal connections (lugs, bushings, and fasteners). Buckling analysis is also performed, again deriving the relative MOS. Modal analysis completes this investigation, aimed at computing the natural modes and providing the aeroelastic specialists with the necessary data to highlight possible aerodynamic-structure local instabilities, whether static or dynamic.

5.1 STATIC ANALYSIS AT LIMIT AND ULTIMATE LOADS: LINEAR AND NONLINEAR ANALYSIS

Aerospace structures are quite complex since they must respect many and often conflicting requirements: they must be light but resistant, durable and, in a word, safe. Among the many consequences, this statement means that, from a structural design point of view, such a system must be designed in order to accurately distribute the mass in the region where it is necessary, since any unjustified weight increment may lead to inefficiency. This introduction and the previous statements lead to some basic conclusions:

- Aerospace structures are generally a combination of thin elements.
- Flexibility of the structural members must be carefully taken into account for two reasons:
  - Deformed structural shape can be very different from the original one (unloaded).
  - Unwanted static (buckling) and dynamic (aeroelastic) instabilities may arise.
- Evaluation of structural plasticity is a critical aspect.

Thanks to the use of computing and advanced methods, it is possible to develop numerical models that include all the effects mentioned above. The first and the second point lead to the already cited necessity to run nonlinear analyses. Also, the elastic stability must be checked. The third point means that the analysis at ultimate loads must be generally evaluated by adopting nonlinear material formulations (elastoplastic models), because it is accepted that some local plasticization may occur.

5.2 STRESS ANALYSIS

Result maps of a morphing structure are herein shown, referring to the practical example of the ATED. Because morphing structures may assume different macroscopic configurations, all of them should be independently verified, under the relative load profile. In fact, aerodynamics strongly depend upon the shape and the structural response itself is a function of the assumed shape. In Figs. 21 and 22, an example of global stress and displacement distribution related to the ATED structure is reported, in the un-morphed configuration (basic line, 0° deflection). Overall tension distribution allows evaluating the stress peaks and, then, the presence of possible critical points. It is worth to observe that, in this case, many peaks did appear at the junctions, i.e., where holes are drawn with their relative high-value curvatures. This kind of result is generally due to the adopted discretization that should be finer in such regions. Therefore, a local verification should occur in that case. The analysis of the reported pictures let conclude that the target structural system is generally oversized with respect to the actual loads and then, in this practical case, it is possible to attain some significant weight reduction through the implementation of a generic optimization process (in other words, some mass may be removed).

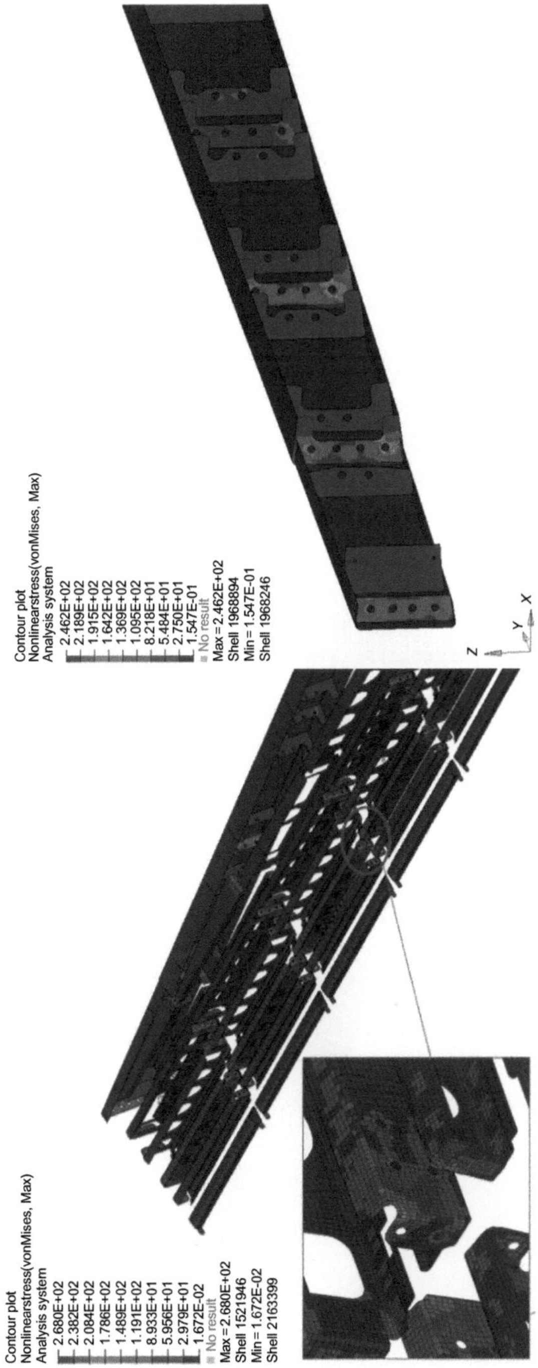

FIG. 21

LL: equivalent stress distribution on the inner spars (left) and on the rear spar (right).

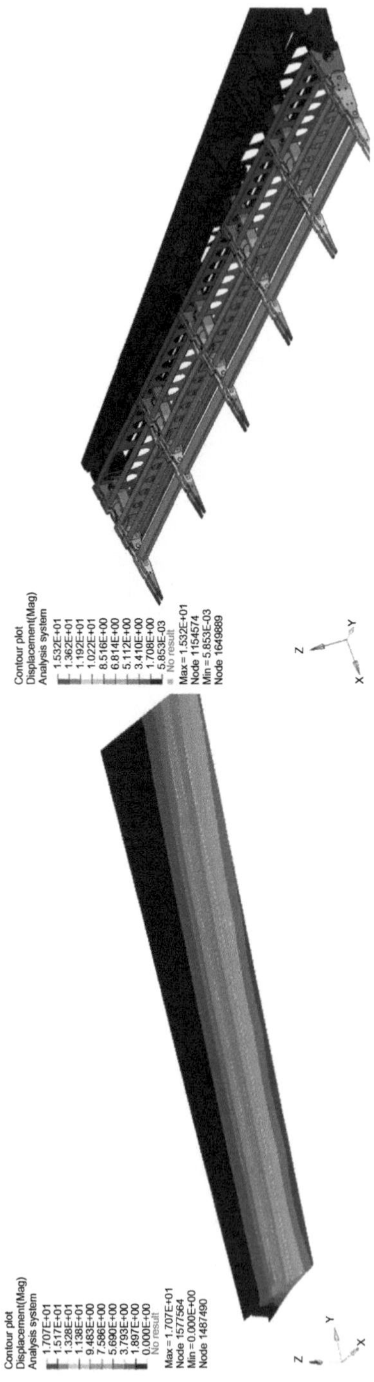

FIG. 22

Global ATE displacements distribution at LL condition.

5.3 BUCKLING ANALYSIS

Some examples of the results of the FE buckling analysis, carried out on the model of the ATED are herein shown. It is interesting to evaluate the effect separately for the different structural elements, namely the ribs and the panels. In fact, depending on the type of the occurring load, different instabilities may onset. The first buckling eigenvalue is −3.275. This means that the first critical load is three times larger than the occurring pressure (in terms of absolute value) and it should act in the opposite direction with respect to the real load. In this case, to speak about MOS may not have sense, unless the load is considered reversible (for instance, as the consequence of a certain backward flight). In that case, it would result that abs(MOS) is equal to 2.275 being MOS = abs(EIG) – 1. If the eigenvalue is equal to 1, MOS is zero.

The first buckling deformation mode is shown in Fig. 23, and involves the fourth rib, only (numbering from the wing root). It is also convenient to underline as this kind of analysis gives point information on the structural behavior without having the risk a sort of generalization takes place, with a relative (and relevant) increase of weight (and inefficiency). After the ribs, instabilities are found at the skin panels location. The first buckling mode involving those structural elements corresponds to the fifth mode, with an eigenvalue equal to −3.768. Such mode shape is shown in Fig. 24, where a very

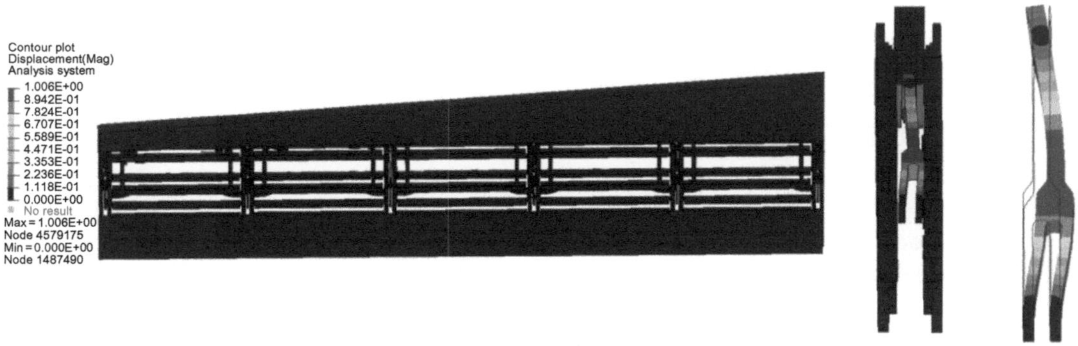

FIG. 23

First buckling deformation mode on the 5-bay FE model demonstrator.

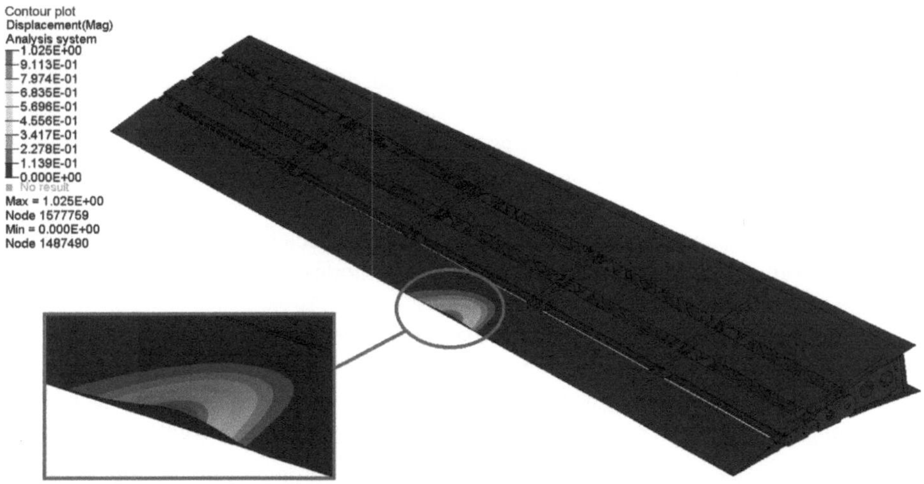

FIG. 24

First panel buckling mode.

Table 2 Summary of Eigenvalue Buckling Mode

Mode	Eigenvalue	MOS
1	−3275	NA
2	−3397	NA
3	−3428	NA
4	−3581	NA
5	−3768	NA
6	3803	2803
7	−3903	NA
8	4294	3294
9	4737	3737
10	4878	3878

limited phenomenon happens. In this case, then, it is expected that only a limited intervention on the structure would be necessary (if the design loads should ever increase by a factor of 3 and would be reversed! In the referred case, the structural system is definitely safe). Buckling analysis MOS are synthetically reported in Table 2.

5.4 MODAL ANALYSIS

The modal analysis is of primary importance to evaluate the system dynamics, particularly it give hints on the possibility some natural vibration modes could couple with themselves and with the external aerodynamic loads. This aspect is critical in aeronautical structures, mainly for the potential arising of aeroelastic phenomena, which can lead to catastrophic failures. Above all, according to the regulations, in the operational diagram (i.e., in all the conditions an aircraft can meet during its missions) a safe distance from the flutter velocity shall be ensured [16]. In Fig. 25 the results of the ATE modal analysis are shown, relative to the range up to 100 Hz. The principal conclusions that can be underlined are:

- Modal analysis shows that no local modes exists; in other words, all the shapes involve all the components.
- The first modal shape is the morphing deformation mode, confirming the peculiarity of the adaptive structural systems to include new DOF in the system.
- Also, the first mode occurs at relatively high frequencies, well beyond the "rigid mode zone" (namely, around 0 Hz); this is a clear signal that the selected configuration has a certain capability to resist the external loads.
- The modal density is reasonably low; this is a good index because many major instabilities, like flutter, derive from the coupling between two or more modes (flutter occurs when a torsional couples with a longitudinal bending eigenvector).
- The first torsional mode appears at 30 Hz, while the first bending mode takes place at 54 Hz; they are well identified and show a certain regularity.
- The last segment of the morphing trailing edge does appear the one that is characterized by the largest motion; this is an index of potential free mechanisms and highlights a weak point of the

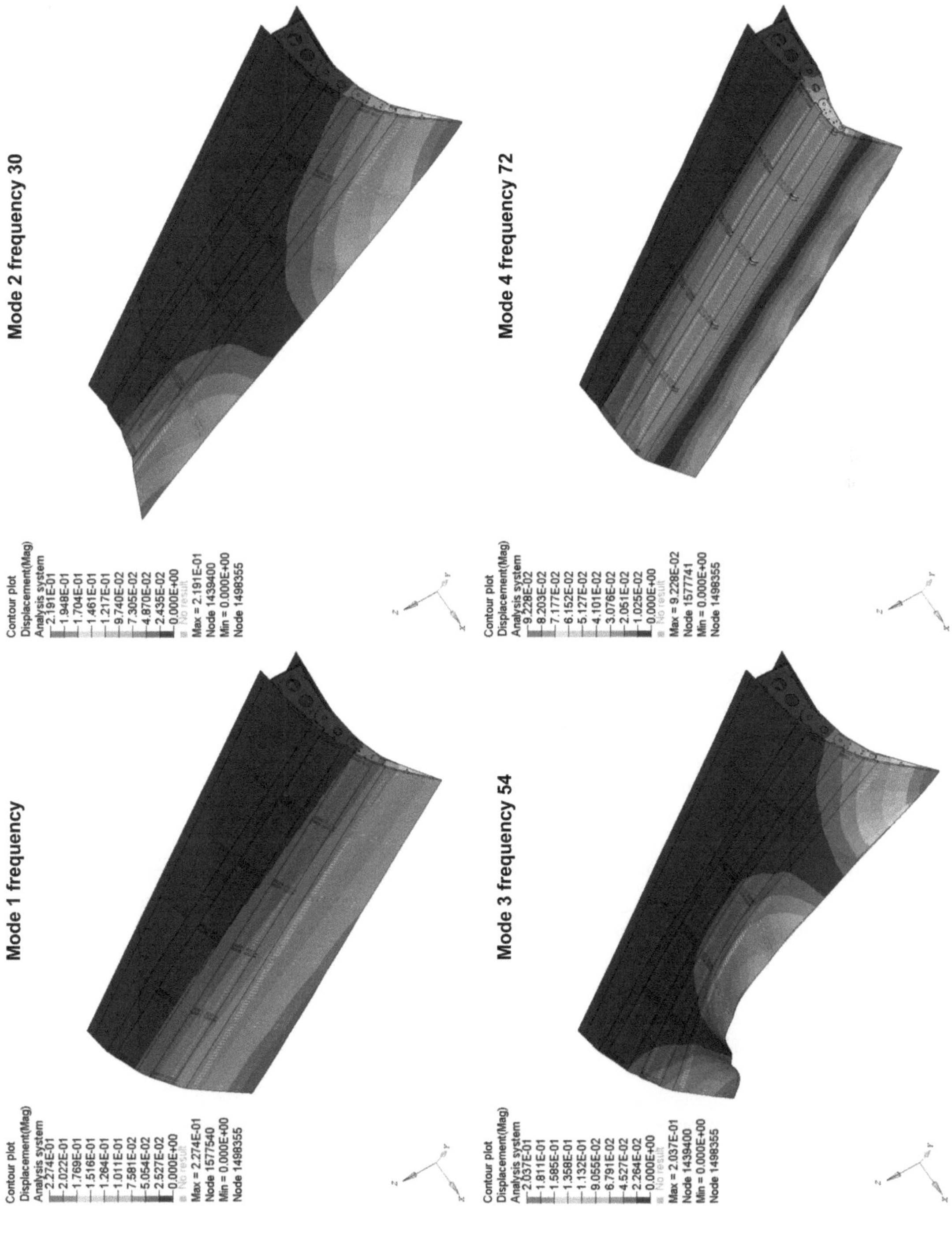

FIG. 25

Deformed shape of the first 4 natural modes of ATE.

designed kinematics; substantial improvements in the global structural dynamic response may arise by a modification of the local arrangement.

- In the current configuration, the same block gives significant contributions to the first and the fourth modal shapes in the considered range (chord-wise deformation), while a slight influence is revealed at the second and the third modes (torsion and spanwise deformation).

The main limitations that should be taken into account while discussing and analyzing these results are the following:

- The output of this study is strongly conditioned by the applied constraints (completely clamped at the spar caps extremities) and certain variations may be expected when the system is inserted on the wing section.
- The length of the device, though representative, is far from the real one, that could reasonably extend up to tens of meters for the referred wing; nevertheless, it should be considered that whatever the extension could be, the device could be segmented, in order to limit somehow the relative impact:

Keeping these limitations, we can state that modal analysis gives hints about the quality of the design and gives indications on possible structural modifications that can give rise to significant global improvements. Among the possible:

- The last is individuated as the most critical block; along with its kinematics, it is the element that gives rise to the largest displacement and, independently on the modes where it concurs with an absolute or relative contribution, its presence is always visible, for any considered mode.
- A rigid connection, as the one herein implemented (ideal clamped conditions at the main spar caps) definitely helps in separating modes and this hypothesis hold when real conditions apply; in other words, it arises that a stiff constraint facilitates to keep distinct the different morphing components behavior (it should be reminded that an actual morphing wing could host several adaptive devices, each interacting with the others).
- In that sense, the contribution of the skin should be evaluated; in fact, in the last segment of the morphing system its presence is very limited whether for its narrow extension in that region or the thickness of its deformable components (foams), very thin at the device tip.
- Interventions aimed at raising the torsional modes frequency should always consider the impacts on the bending modes, because the growth of one, without the corresponding increase of the other would lead to dangerous superpositions.

6 STRESS MARGINS OF SAFETY

6.1 SOLID PARTS

The global results obtained from FE static analysis are used to check the compliance with the structural requirements: the strength of each component is evaluated by comparing the Von Mises (VM) stress with the yield material strength. The technique that is used for the point extraction of this parameter provides the identification of the most stressed regions and, then, the elements that are more exposed to possible failures. MOS for such strength analysis are reported in Table 3.

Table 3 MOS Summary at LL and UL for Solid Parts

Part	Fty	fty	MSLL	Ftu	ftu	MSUL	Higher Stress Distribution
Rib	330	262.5	0.26	448	394.4	0.14	
Rod	1150	448.6	1.56	1227	672.9	0.82	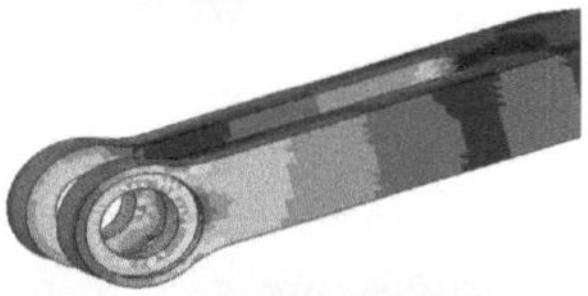
Spar	330	268	0.23	448	401.9	0.11	
Actuator	330	166	0.99	448	248.6	0.80	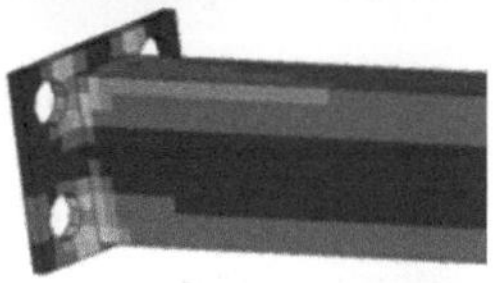
Crank	330	50.4	HIGH	448	55.4	HIGH	
Al skin	330	107	2.08	448	160.8	1.67	
Rear spar	330	246	0.34	448	369	0.21	
Foam				1.17	1.13	0.035	NOT REPORTED

Therein, several columns are reported: Part, indicates the structural component; Fty and fty represent the yield material and the actual tension under the LL, respectively; MSLL is the MOS at LL; in the same way, Ftu and ftu represent the ultimate material and the actual tension under the UL, in the order; MSUL is the MOS at UL. In the first column is depicted a representation of the most-solicited element of the reported part. As anticipated, it is possible to see how the higher stress levels are associated to the holes representation and, then, could be reasonably reduced by a proper mesh refinement.

6.2 INTERNAL CONNECTIONS

An important step in the structural system verification is the stress analysis of the connections (bushings, lugs, joints, and rivets). In the illustrated case, the study is performed through the information gained by the RBE (rigid beam) elements that, together with the corresponding network of connections, are used to model the different links. The internal resultant force that was acting on the independent node (master node) is extracted and, on those bases and the implementation of classical formulae, the MOS is calculated for each part. In synthesis, the process is structured as follows:

- Extraction of the forces that are acting at the connections from the FE analysis at both LL and UL.
- Use classical design formulae for calculating the MOS at LL and UL.

In Fig. 26, it is schematically shown how it is possible to extract the local internal forces for the generic connection system, moving from the FEM analysis. Operatively, the single forces associated to each RBE element are taken. The resultant of those forces is split, according to the rules of the vector computation, in two components: one, normal to the connection plane; the other, in the plane itself. In the order, the first represents the normal or tension stress, while the second is a measure of the plane or shear stress. Once these values are computed, the respective MOS may be calculated.

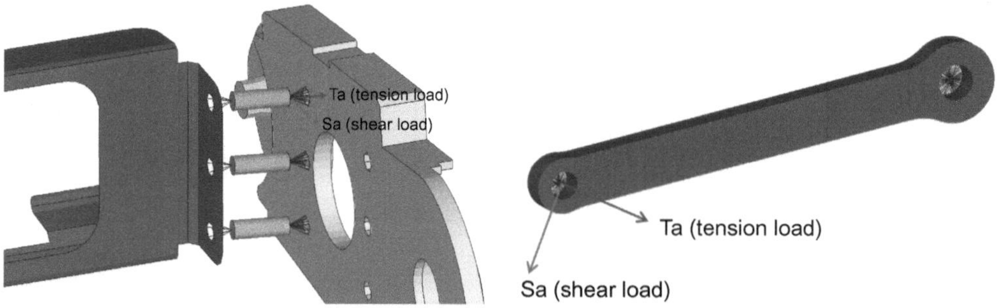

FIG. 26

Connection forces computation: connection schematic and force synthesis (left); resulting sketch (right).

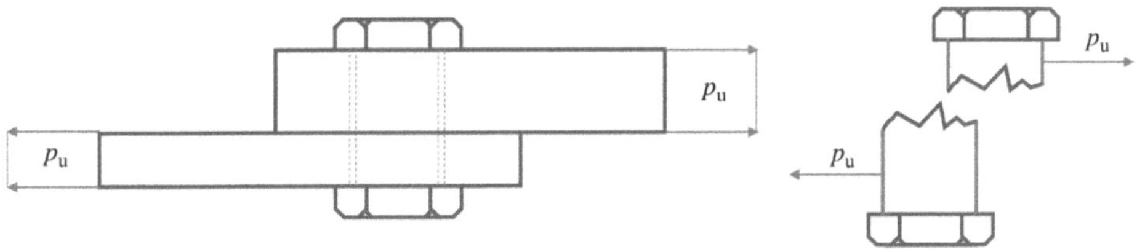

FIG. 27

Bolt failure: schematic.

6.2.1 Fasteners

The fasteners are verified at ultimate loads. The classical sketch representing the investigated problem is reported in Fig. 27. Usual technical formulas are used for this computation: they refer to a comparison between the max allowable ultimate or limit stress and the actual occurring force per unit of area. In general, a 1.5 safety factor is included. Fasteners MOS (or MS in the formulas) calculation includes [17–19]:

- Fastener shear check:

$$\text{MS shear} = \frac{\text{Sall}}{1.5 \times \text{Sa}} - 1 \tag{3}$$

where

- Sall = Fsu × A;
- Fsu = ultimate shear stress of the bolt material;
- A = cross-sectional area of the bolt;
- Sa = applied shear load;
- 1.5 = safety factor.

- Fastener tension check:

$$\text{MS tension} = \frac{\text{Tall}}{1.5 \times \text{Ta}} - 1 \tag{4}$$

where

- Tall = Ftu × A;
- Ftu = ultimate tensile stress of the bolt material;
- A = cross-sectional area of the bolt;
- Ta = applied tension load;
- 1.5 = safety factor.

- Fastener shear + tension check:

$$\frac{(1.5 \times \text{Sa})^3}{\text{Sall}^3} + \frac{(1.5 \times \text{Ta})^2}{\text{Tall}^2} < 1 \tag{5}$$

The exponentials of this last equation follow empirical evidence. However, some explanations can be found by some considerations, starting from the introduction of an effective shear to allowable tensile load ratio, $K =$ Allowable Shear/Allowable Tension, Sall/Tall [19]. For a threaded fastener, the ratio of the body over the threads strength is given by the ratio of the respective areas. For a 0.375–24 UNJ bolt, it is equal to 1.161 [20–22]. Generally, the material allowable shear to tensile stress may be assumed equal to 0.5; factor K is then, for this particular case equal to $0.5 \times 1.161 = 0.581$. It may be verified that, if the theoretical failure curve is plotted for the max normal and shear stress for such a value, Eq. (5) approximately follows that envelope [19]. Such a relation, then, produces results very similar to what predicted by the normal and shear stress theories.

6.2.2 Lugs and Bushings

Lugs and bushings are verified at both LL and UL. MOS are calculated by using classical design criteria [1], referring to a comparison between the maximum allowable ultimate or limit stress and the actual force per unit of area. The reference schematic is proposed in Fig. 28. In general, a 1.5 safety factor is included. Lug MOS calculations include separate equations for bearings net section and separate items for shear, transverse load and oblique loads. Finally, the equation concerning the actual insisting force is taken into account for bushings. In the equations, the suffix y stands for "yield" (LL) while the suffix u stands for "ultimate" (UL).

- Lug Analysis—bearing check:

The following equations hold:

$$\text{MS bearing y} = \frac{\text{Pbry}}{\text{Sa}} - 1 \tag{6}$$

$$\text{MS bearing u} = \frac{\text{Pbru}}{1.5 \times \text{Sa}} - 1 \tag{7}$$

where

- Pbry = Fbry × D × t;
- Pbru = Fbru × D × t;
- Fbry = yield bearing strength;
- Fbru = ultimate bearing strength;
- D = bolt diameter;
- t = plate thickness;
- Sa = applied shear load.

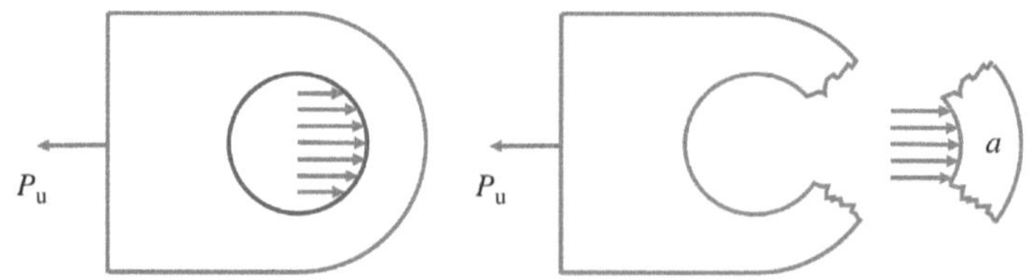

FIG. 28

Lug failure (bearing): schematic.

- Lug Analysis—net section check:

The following equations hold:

$$\text{MS net y} = \frac{\text{Pnety}}{\text{Ta}} - 1 \tag{8}$$

$$\text{MS net u} = \frac{\text{Pnetu}}{1.5 \times \text{Ta}} - 1 \tag{9}$$

where

- Pnety = Kt × Fty × At;
- Pnetu = Kt × Ftu × At;
- Kt = stress concentration factor [1];
- Fty = tensile yield strength of material;
- Ftu = ultimate tensile strength of material;
- At = net tension area $(W - D) \times t$;
- W = lug width;
- D = lug hole diameter;
- Ta = axial load.

In the following figure, Fig. 29, a qualitative representation of the data reported in bibliography [1] are shown, together with a geometrical sketch representative of the lug bearing behavior, and the indication of some reference parameters.

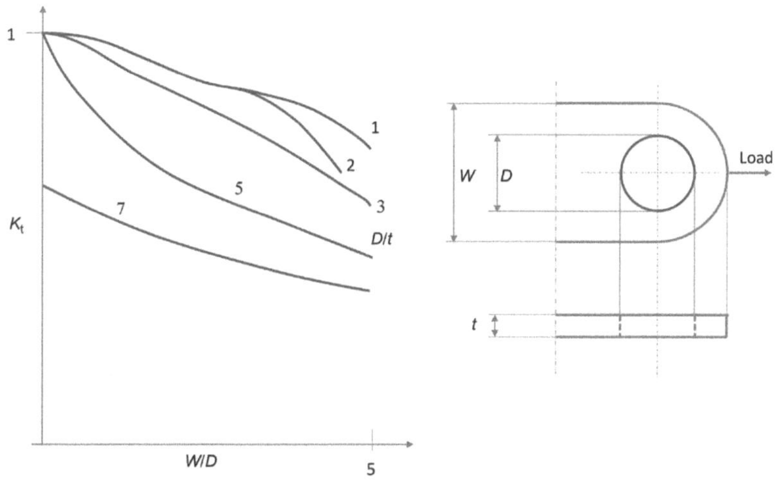

FIG. 29

Lug failure: Kt evaluation [10] and some parameters identification.

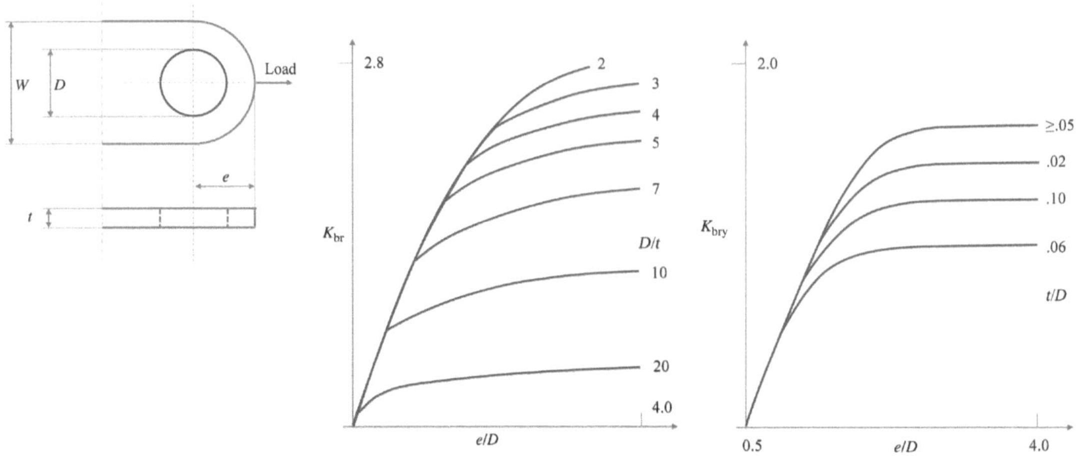

FIG. 30

Lug failure: Kbr, Kbry evaluation [10] and definition of some key parameters.

- Lug analysis—shear out-bearing check

The following equations hold:

$$\text{MS brsh y} = \frac{\text{Pbry}}{Ta} - 1 \tag{10}$$

$$\text{MS brsh u} = \frac{\text{Pbru}}{1.5 \times Ta} - 1 \tag{11}$$

where

- Pbry = Kbry × Fty × Abr;
- Pbru = Kbru × Ftu × Abr;
- Kbry = shear bearing efficiency factor for yield strength [1];
- Kbru = shear bearing efficiency factor for ultimate strength [1];
- Fty = yield tensile strength of the material;
- Ftu = ultimate tensile strength of the material;
- Abr = smaller bearing area;
- Ta = axial load.

In Fig. 30, a qualitative representation of design features reported in bibliography [1] is shown, with the rough indication of suitable values of the parameters Kbry and Kbru. In the picture, such parameters depend on other geometrical factors that do not appear in the formulae above drafted but are reported in the sketch. In detail, the key value does appear to be "e," defined as the distance of the center of the lug bearing from its shortest extremity. The bundle of curves depends on the ratio D/t (or its inverse) that is the relative importance between the lug bearing hole diameter and its thickness.

- Lug analysis—transverse load check

The following equations hold:

$$\text{MS brtl y} = \frac{\text{Pbry}}{\text{Ptrav}} - 1 \tag{12}$$

$$\text{MS brtl u} = \frac{\text{Pbru}}{1.5 \times \text{Ptrav}} - 1 \tag{13}$$

where

- Pbry = Kbry × Fty × Abr;
- Pbru = Kbru × Ftu × Abr;
- Kbry = shear bearing efficiency factor for yield strength;
- Kbru = shear bearing efficiency factor for ultimate strength;
- Ptrav = transverse load;
- Fty = yield tensile strength of material;
- Ftu = ultimate tensile strength of material;
- Abr = smaller bearing area.

In Fig. 31, a qualitative representation of a bibliography schematic is reported [1], with the rough indication of suitable values of the parameters Kty and Ktu. As in the former case, such parameters depend on a particular ratio, involving geometrical factors that were not recalled in the formulas above. In this case, the key factor is a ratio between two nominal areas, Aav and Abr, defined as follows. The first one is defined

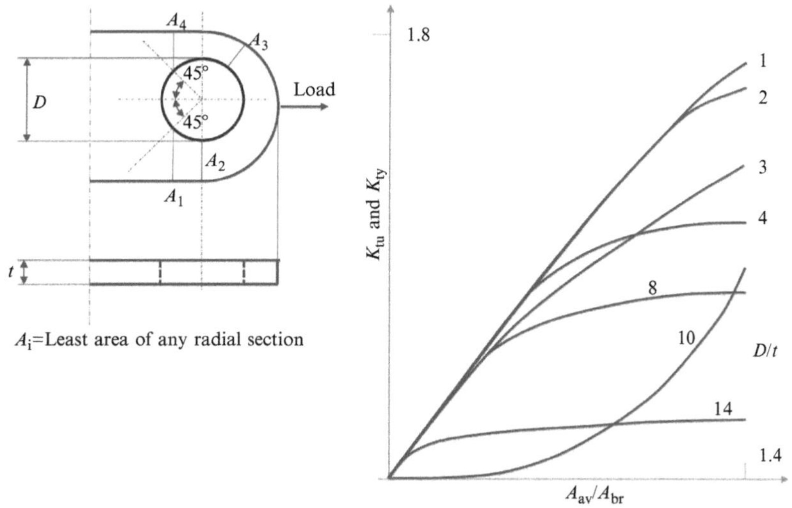

FIG. 31

Lug failure: Kty, Ktu evaluation [10] and definition of some key parameters.

as the product between the lug bearing hole diameter and its thickness (Aav $= D \times t$). The other one represents a more or less complex dependence of geometrical data that somehow describe the position of the hole within the mechanical element. In Fig.31, a sketch of a generic bearing allows to graphically define those factors (Ai). The expression for Abr is reported in the formula below, while the bundle of curves depends on the ratio D/t:

$$\mathrm{Abr} = \frac{6}{\frac{3}{A_1} + \frac{1}{A_2} + \frac{1}{A_3} + \frac{1}{A_4}} \tag{14}$$

- Lug analysis—oblique load check:

The following equation holds:

$$\mathrm{MS\ ol} = \frac{1}{\left(R_{\mathrm{a}}^{1.6} + R_{tr}^{1.6}\right)^{0.625}} - 1 \tag{15}$$

where

- R_{a} = axial load divided by the smaller value between Pnetu and Pbru;
- R_{tr} = transverse load divided by the inverse of Pbru.

- Lug analysis—bushing:

The following equation holds:

$$\mathrm{MS\ bush\ y} = \frac{\mathrm{Pbry}}{Sa} - 1 \tag{16}$$

where

- Pbry = 1.85 × Fcy × As;
- Fcy = compressive yield strength of the bushing material;
- As = minimum bearing area; and
- Sa = applied shear load.

A representation of a generic bushing is reported in Fig. 32, with a schematic of its failure:

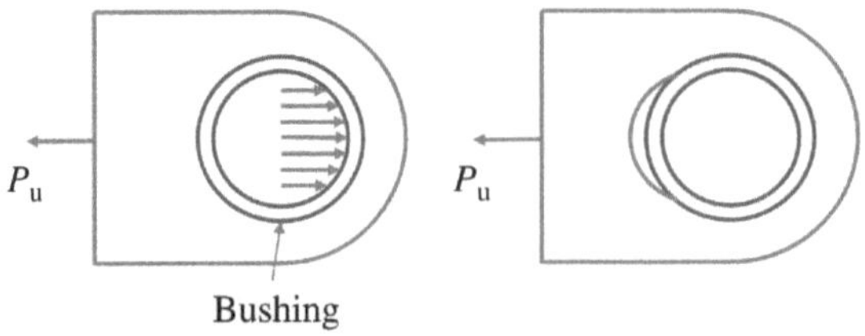

FIG. 32

Lug failure: representation of bushing collapse.

6.2.3 MOS Summary

MOS for each typical fastener, bearing, and bushing of the investigated structure (ATED) are reported in Tables 4–6:

It can be clearly seen as the MS are well beyond the classical values of 1.0–1.5 with some limited exceptions. This brings immediately to some considerations:

- The structure may be lighter.
- The excess of stress strain field, occurring at the holes may derive from the adopted mesh (modeling) and should be refined by an appropriate numerical design.

Concerning the first point, it confirms what was seen on the stress analysis on the different parts of the system. In fact, the structure proves to be oversized, together with its joints distribution, whose density is generally high. The second point is an extrapolation of what was already evidenced. In fact, because

Table 4 Summary Table of MOS for Fasteners (HIGH Means ≥3; italics stands for 0 ≤ MOS ≤ 0.5)

Fasteners Check			
Joints Location	**MS Shear**	**MS Tens**	**MS Combined**
Skin to wing box	HIGH	HIGH	OK
Skin to spars	*0.24*	HIGH	OK
Rib to spar	HIGH	HIGH	OK
Rib to rear spar	HIGH	HIGH	OK
Spar to actuator plate	HIGH	HIGH	OK
Rib to Rod1	HIGH	HIGH	OK
Rib to Rod2	2.41	HIGH	OK
	2.95	HIGH	OK

Table 5 Summary Table of MOS Calculations for Bearings (HIGH Means ≥3; italics stands for 0 ≤ MOS ≤ 0.5)

Bearing Check			
Joints Location	**Parts**	**MS LL**	**MS UL**
Skin to wing box	Skin	HIGH	HIGH
Skin to spars	Skin	*0.48*	*0.39*
	Spars	0.97	0.85
Rib to spar	Rib	HIGH	1.41
	Spars	2.98	2.75
Rib to rear spar	Rib	HIGH	2.13
	Rear Spar	HIGH	HIGH
Spar to actuator plate	Spar	HIGH	HIGH
	Actuator Plate	HIGH	HIGH
Rib to Rod1	Rib	HIGH	HIGH
Rib to Rod2	Rib	0.70	0.60
	Rib	*0.36*	*0.28*

Table 6 Summary Table of MOS Calculations for Bushings (HIGH Means ≥3; italics stands for 0 ≤ MOS ≤ 0.5)

Bushing Check	
Joints Location	**MS**
Rib to Rod1	HIGH
Rib to Rod2	1.40
	0.40
Rib to rib	HIGH

the connectors are placed at the holes, the same points where the high stresses values occurred from the analysis, they should reveal limited safety margins, coherently. The fact that the resulting values are instead so large, generally, means the structure is not so solicited.

Where the safety margins are reduced, coming to values around 0.3–0.4, some intervention shall be needed in the sense of a design refinement. The most relevant is the case of the actuation rod N.2, linking the second to the third rib block. The modal analysis already indicated this as a critical element, showing an excessive displacement in the eigenvector, symptom of a poor local stiffness. This is herein confirmed. The only other parts concern the connection between skin to spars (skin versus), that seems to require different and more solid links and the use of thicker plates.

7 CONCLUSIONS

Using as a reference an ATED, this text dealt with the main aspects related to the FE modeling and the structural analysis procedures, applicable to a morphing structure for aeronautic applications. General concepts were pointed out regarding the processes of numerical simulations, the analyses type, the MOS calculations, and the typical stress analysis procedures, with reference to both local and global aspects.

Peculiarities of the investigated system were its kinematic architecture (finger-like), the inclusion of a polymer-based morphing skin, and a distributed servo-actuation system, made of commercial motors and a suited mechanics. It is relevant to underscore that the selected layout was perfectly compatible to aeronautical standards and permitted to access well-consolidated practices. With the major exception of the elastomeric skins, all the other components are commercial items or made of typical COTS. The illustrated approach may be directly applied to other kind of morphing structures, built with the same philosophy. In the case of compliant systems, some adaptations could be necessary.

Skipping the structural skeleton elements, made of segmented ribs and spars, hinge connections and other kinematic parts, the core of the morphing device is the actuation system that has to meet two targets: morphing authority and load bearing capacity. It is generally associated to a lightweight, compact, and robust kinematic chain with anti-jamming features, using the servo rotary motors to allow the system to assume a prescribed bundle of shapes. In detail, an actuation arm, driven by the internal architecture, transmits the motion to the ribs block system, that in turn drags the whole architecture. The movements are commanded and supervised by a reliable control system; it verifies that the achieved deformation corresponds to the basic requirements within a certain predefined level of accuracy. Because the motor is simply seen as a point mass with no structural function (with the main exception of the shaft), while the control system does not modify the structural response, these two elements were not considered in this study. In fact, also in the case of a feedback solution, the slow response required (well under 10 Hz) prevents the control action from modifying the structural transfer function.

Summarizing, the design is similar to a classical aeronautical structure process, with specific reference to kinematic chains, already existing on a typical aircraft (as in the case of hyper-lift surface systems, like flaps or ailerons). Indeed, the aspect that is more distant from a typical aerodynamic structure, concerns the verification of the system under the external loads action. In fact, because a morphing surface is the target of the investigation and, by definition, may assume different configurations, it should be tested in a suitable range of geometries, each with its own pressure distribution. In other words, a single becomes a bundle of structures, each characterized by a specific layout. This concept significantly complicates the verification step, taking also into account that is impossible to handle the aerodynamic results for each possible configuration. If this occurrence is extrapolated to craft hosting many morphing systems, with the entire range of cross effects, it is well understood how such an operation can be very delicate and give rise to massive computations. The actual impossibility of running the verification process for the infinite possible layouts (continuous morphing), imposes that the simulations take place for a limited set of shapes, only.

The classical analysis include static, buckling and modal investigations. The first and the second ones are run into the usual mode and there is no particular comment about, unless a remark on the importance of correctly analyzing the structure at the connection points (articulations), where the movement is transferred from an element to another. Modal analysis is instead particular because it shall give information, essential for further aeroleastic studies. Because the systems have augmented ways of deforming, denser modal characteristics are expected, with essential output to the design process. Particularly important is the modal shapes examination for at least two reasons: to individuate the largest displacement that can give evidence of some kinematics not well supported by a sufficient rigidity (therefore giving rise to mechanisms); to highlight the real possibility of interaction between two or more eigenvectors, at the basis of the insurgence of aeroelastic instabilities. It is worth to remark that, because the structure can assume different configurations, different modal analyses should be performed. In fact, the eigensolution may change as the system modifies its shape.

REFERENCES

[1] E.F. Bruhn, Analysis and Design of Flight Vehicle Structures, S.R. Jacobs, Carmel, IN, ISBN 978-09-615-2340-4, 1973. 650 pp.
[2] R.M. Botez, P. Molaret, E. Laurendeau, in: Laminar flow control on a research wing project presentation covering three year period, Canadian Aeronautics and Space Institute Annual General Meeting, Montreal (Canada), January 2007, 2007.
[3] G. Scarselli, F. Marulo, A. Paonessa, in: Sensitivity investigation of aircraft engine noise to operational parameters, Proceedings of the 16th AIAA/CEAS Aeroacoustics Conference (31st AIAA Aeroacoustics Conference), Stockholm (Sweden), June, 2010.
[4] Predictive Engineering, Inc., in: Small Connection Elements (RBE2, RBE3 and CBUSH) and How Amazingly Useful They are for FEA Modelling, Engineering Mechanics White Paper, 2013.
[5] I. Dimino, D. Flauto, G. Diodati, A. Concilio, R. Pecora, Actuation system design for a morphing wing trailing edge, J. Recent Patents Mech. Eng. 7 (2014) 138–148.

[6] J. Carl, D. Müller-Hoeppe, M. Meadows, in: Comparison of tetrahedral and brick elements for linear elastic analysis, Term Project AFEM Spring 2006, ASEN 5367, Department of Aerospace Engineering Sciences, University of Colorado, Boulder, .
[7] E. Wang, T. Nelson, R. Rauch, in: Back to elements—tetrahedra vs. hexahedra, ANSYS 2004 International Conference, Pittsburgh, PA, May 24th–26th, 2004.
[8] MSC Software, RBEs and MPCs in MSC Nastran—A rip-roarin' review of rigid elements, (2011).
[9] Thill C., Etches J., Bond I., Potter K., Weaver P., "Morphing skins", Aeronaut. J., Vol. 112, N. 1129, pp. 117–139, 2008.
[10] O. Schorsch, A. Lühring, C. Nagel, in: P.C. Wölcken, M. Papadopoulos (Eds.), Elastomer-based skin for seamless morphing of adaptive wings, Smart Intelligent Aircraft Structures (SARISTU)—Proceedings of the Final Project Conference, Springer International Publishing, Switzerland, ISBN 978-3-319-22412-1, 2016.
[11] I. Dimino, G. Diodati, A. Concilio, A. Volovick, L. Zivan, in: Distributed electromechanical actuation system design for a morphing trailing edge wing, SPIE Smart Structures/NDE, Las Vegas (NA-USA), 20–24 March, 2016.
[12] E. Torenbeek, Advanced Aircraft Design: Conceptual Design, Technology and Optimization of Subsonic Civil Airplane, John Wiley & Sons, Chichester, ISBN 978-1-118-56811-8, 2013. 440 pp.
[13] P.C. Woelcken, M. Papadopoulos (Eds.), Smart Intelligent Aircraft Structures (SARISTU), Proceedings of the Final Project Conference, Springer International Publishing, Cham, ISBN 978-3-319-22413-8, 2016.
[14] J. Roskam, Airplane Design Part I–VIII, 8 vols, DAR Corporation, Lawrence, KS, ISBN 978-9-993-24960-3, 1989.
[15] D. Peery, Aircraft Structures, Dover Publications, Mineola, NY, ISBN 978-9-993-24960-3, 2011. 575 pp. (originally published in 1950 by McGraw-Hill).
[16] European Aviation Safety Agency, Certification Specifications for Large Aeroplanes—CS-25, European Aviation Safety Agency, Cologne, 2007. p. 617.
[17] J. Chambers, Preloaded Joint Analysis Methodology for Space Flight Systems, NASA/TM–1995–106943, Lewis Research Center, Cleveland, OH, .
[18] NASA, Criteria for Preloaded Bolts—Rev. A (replaces NSTS 08307 Baseline, 1989), Johnson Space Center, NASA/NSTS–1998–08307, Johnson Research Centre, Houston, TX, .
[19] B.E. Steeve, R.J. Wingate, Aerospace Threaded Fastener Strength in Combined Shear and Tension Loading, NASA/TM-2012-217454, Marshall Space Flight Center, Huntsville, AL, .
[20] Aerospace Industries Association of America, "Bolt, Shear, Hexagon Head, 108 ksi Shear, Long Thread", NAS 1953-1970, Aerospace Industries Association of America, January 1986 (former version), January 2015.
[21] SAE Committee E-25 General Standards For Aerospace And Propulsion Systems, Areas for Calculating Stress or Load Values for Externally and Internally Threaded Fasteners, SAE AS5054, SAE International, Warrendale, PA, .
[22] SAE Committee E-25 General Standards For Aerospace And Propulsion Systems, Screw Threads—UNJ Profile, Inch, Controlled Radius Root with Increased Minor Diameter, SAE AS8879, SAE International, Warrendale, PA, 2004.

FURTHER READINGS

[1] A. Gratias, M. Schueller, M. Lipowski, I. Dimino, in: Actuator net design for morphing wing structure, Smart System Integration Conference, Amsterdam (NL), 13–14 March, 2013.
[2] I. Dimino, G. Diodati, A. Concilio, A. Volovick, L. Zivan, in: Distributed electromechanical actuation system design for a morphing trailing edge wing, SPIE Smart Structures/NDE, Las Vegas, NA, 20–24 March, 2016.

SECTION 6

Morphing Wing Systems

Editor:

Antonio Concilio
The Italian Aerospace Research Centre, CIRA SCpA, Capua (CE), Italy

List of Chapters

CHAPTER

16 MORPHING OF THE LEADING EDGE

Salvatore Ameduri
The Italian Aerospace Research Centre, CIRA SCpA, Capua (CE), Italy

CHAPTER OUTLINE

NOMENCLATURE

CBEAM	beam element
CQUAD	plate element
$\boldsymbol{F}$	actuation force necessary to balance the external loads and the elastic reaction of the structure
$\boldsymbol{f}$	cost function (scalar)
$\boldsymbol{K}$	elastic constant, representing the elastic reaction of the structure
$\boldsymbol{L}$	arc rod length
$\boldsymbol{p}$	position vector of the leading edge external profile
$\boldsymbol{R}$	resultant of the external loads acting on the leading edge
$\boldsymbol{r}$	transmission factor
$\boldsymbol{s}$	curvilinear abscissa
target	refers to parameters to be targeted by the optimizer

Morphing Wing Technologies. https://doi.org/10.1016/B978-0-08-100964-2.00016-2

TRL	technology readiness level
x	displacement of the actuator shaft (spanwise direction)
y	displacement of the three-arc point in contact with the structure
α_0	initial angle of the three-arc structures
$\Delta\alpha$	the three-arc structures angle variation

1 SUMMARY

In the present paper, the development and realization of a true-scale droop nose laboratory demonstrator is illustrated. After an overview on the most recent works and programs on the same topic, the working principle adopted for the device is presented, highlighting the main design parameters. The conceptual and advanced design is then examined; a dedicated optimization approach was implemented to achieve as closely as possible the prescribed morphed shape (the droop nose configuration) while complying with structural constraints (which are particularly severe for the skin component). After having assessed of the advanced design, the manufacture and integration task were finalized. A dedicated setup was then assembled to prove the ability of the device to effectively produce the droop nose effect. The work ends with a critical review of the achieved numerical and experimental results and with recommendations for the next steps of the research.

The chapter is organized as it follows:

Introduction
Conceptual Approach to the Morphing of the Leading Edge
Working Principle of the Architecture Selected to Produce the Drop Nose Effect
Architecture Design
- Identification of the Kinematic Chain in the Rib Plane
- Topologic Optimization of the In-Plane Rib Architecture
- Spanwise Architecture and Actuation Design
- Modelling and Working Simulation of the Complete Architecture

Prototyping
Experimental Campaign
- The Setup
- Experimental Results
- Numerical—Experimental Comparison

Conclusions and Further Steps

2 INTRODUCTION

A typical mission of an aircraft, whether civilian or military, is characterized by different phases (e.g., take-off and landing, low and high speed cruise, battle maneuvers) whose duration depends generally on the class of vehicle [1]. Conventional design approaches cannot assure the satisfaction of the requirements foreseen for all the flight phases, as they often conflict and imply strong practical penalties on each others. On the contrary, the traditional design privileges some specific flight regimes, the most extended during the operative life (e.g., cruise), with a consequent penalization of the other operational phases [2,3].

To guarantee a certain level of adaptivity and, thus, to compensate for the above-mentioned drawbacks on other flight conditions, dedicated movable parts (slats, flaps, ailerons) have been introduced. Despite their ability to alter wing performance, the design solutions adopted to integrate these devices lead to penalties in weight (the kinematic chains and actuation systems are heavy and can constitute up to 50% of the wings total weight), dimensions (the impossibility of integrating these architectures into the wing body detrimentally impacts aerodynamic efficiency), and stress concentration (connecting these architectures to the wing structure creates stress accumulation zones which must be periodically monitored). Gaps in aerodynamic surfaces may penalize zones of the wing (e.g., the leading edge) particularly critical for preserving laminar flow.

The advent of new "smart" materials enables exploitation of novel, unconventional design strategies, producing adaptivity through smooth, distributed, yet significant geometry variations.

Interest in this approach is evidenced by the large number of studies seeking to demonstrate and quantify advantages inherent to applying a morphing approach to specific zones of the wing [1,4,5]. Both conventional and novel actuators/materials, like the shape memory alloys (SMAs), are being investigated as critical components for actuating morphing architecture and assuring the absorption of loads [4,6].

In this research scenario, there were many critical projects, including "Active Aeroelastic Aircraft Structures (3AS)," a European project [1,7,8] pursuing the development of an adaptive stiffness structure, enabling geometric variations only under the action of the external loads. This philosophy resulted in a variable-stiffness all-movable vertical tail, based on magnetorheologic technology, exhibiting adaptivity in both subsonic and supersonic regimes [9]. Experiments proved that loads which generally cause undesired deviations from the original surface geometry can be usefully exploited to enable the adaptation to different flight regimes.

Another wing hot spot, the leading edge, is object of widespread investigation due to its strong effect upon the aerodynamics of the entire wing.

The joint USAF-NASA flight test program demonstrated the advantages caused by camber variations in aircraft flight envelope extension. Later other research bodies, such as NASAs Morphing Program [10] and DARPAs SMART Wing Project [11], investigated the capacity of smart materials to alter aerodynamic surfaces; the first program investigated leading and trailing edge active zones, integrated with piezo and SMA components, while the second program investigated low-frequency morphing architectures.

Statkus [12] conceived and developed an approach based upon two groups of actions influencing the leading edges camber, producing horizontal and vertical displacements and inducing skin rotation to mitigate the stress field. Another actuation of the droop nose effect was conceived by Zimmer [13]. Geissler et al. [14] proposed a piezo-actuated leading edge rotating around an internal hinge. Another architecture (applicable also to rotary wings), proposed by Kota et al. [15], applicable also to rotary wings, maintained its optimal shape throughout the azimuth, thus enhancing blade performance in terms of stall delay.

Within the "Smart High Lift Devices for Next Generation Wings" SADE, (2008–12) European project [16,17], strategies to alter leading and trailing edges shapes were developed and demonstrated at several levels. Alternatives to conventional hyper lift generators (in which characterized gaps worsen the laminar flow regime) were investigated, including a quasimonolithic morphing leading edge which enabled the droop nose effect. Operating jointly with a slotted, variable-camber flap, this concept performed differently from traditional slats (conceived as a separate part of the wing); the droop nose approach not only assured an adequate high lift performance under critical conditions (e.g., take-off and landing), but also helped preserve laminar flow due to the absence of gaps or surface asperities.

Among the leading edge morphing concepts, two deserve mention due to their relevance in terms of achieved technology readiness level (TRL): the architecture proposed by DLR (a skin composite leading edge tested in a wind tunnel at the TsAGI plant), and the architecture developed by CIRA [18], even (which produced the droop nose effect on a metallic skin and was tested in laboratory at the CIRA plant).

Within SARISTU, DLR developed a novel droop nose architecture which was tested in a wind tunnel at TsAGI plants. The functionality of the droop nose concept was proven, integrated into a full-size wing model, under airflow limited to 0.3 M. The device was designed to reduce drag and radiated noise by optimizing the laminar flow in a range of angles of attacks [19]. It was originally dedicated to drag optimization during other crucial phases of the flight envelope (climb, descent, cruise); however, tests were limited to verifying its capacity to generate high lift. Functional integration into the wings leading edge played a fundamental role in the effort [19] with researchers demonstrating maintenance of the architectures de-/antiicing, bird strike protection, erosion protection, and lightning strike protection capabilities. Such design constraints led to a structural system consisting of a multi-layered skin composite cooperating with a lightweight internal kinematic chain. Additional internal protection (as opposed to an additional external layer, which would have made the skin excessively rigid) helped dissipate impact energy of bird strikes, preventing damage [19]. Different concepts were numerically simulated and validated during an extensive bird strike test campaign. In the end, the bird-splitter approach won. The idea underlying the design of the structural morphing system was to enforce droop nose deformation by limiting membrane stress at the maximum extent [19]. This was accomplished by defining a proper stack sequence and making the plies work mainly in bending. The combined effect of enforced deformation by the inner actuators along the chord, and by the aerodynamic loads along the span, defined the general layout. The inner, structural, kinematic chain was made of metallic elements. In detail, two main components could be identified: a lever hinged to the front spar (providing a hub for other elements) and struts hinged to the lever and the skin (in charge of transmitting the motion between those members). The latter was linked to the struts through omega-shaped stiffeners, contributing strongly to the spanwise rigidity of the whole architecture.

In this chapter, attention is paid to the aforementioned morphing leading edge, developed by CIRA within SADE.

At first, the conceptual approach to the problem is illustrated from a structural and aerodynamic point of view. Secondly, the most promising strategies to achieve the droop nose effect are critically discussed, highlighting the pros and cons. Thirdly, the selected architecture is illustrated with specific emphasis on the innovation contents. Then the modelling problem is discussed, highlighting the influence of the main design parameters on system functionality. The model is then implemented into a genetic algorithm to assess the main parameters influencing accurate achievement of the desired morphed shape. After having established the advanced design, the executive drawings were finalized and a full-size laboratory prototype manufactured. Finally, a dedicated experimental campaign was organized to test the functionality of the concept for even the most severe load conditions.

3 CONCEPTUAL APPROACH TO THE MORPHING OF THE LEADING EDGE

A system which enables morphing of a wing part must meet specific accuracy requirements with regards to the stiffness and controllability of the target shape. These requirements are much more stringent if the zone in question is the leading edge, as it plays a fundamental role in the aerodynamics of the entire airfoil and is subject to the greatest pressure. Thus, during the preliminary draft phase of a morphing leading edge, it is important to consider some specific criteria [20,21].

- It must assure hyper lift performance, equivalent to or higher than conventional slats; from an aerodynamic point of view, this means that the deformed shape has to fit a target.
- It must preserve as much as possible the laminar flow condition in cruise regime, thus delaying the transition along the chord; this can be achieved through a dedicated droop nose geometry without any gap or curvature discontinuities.
- It must withstand the greatest peak pressure of the entire chord; conventional architectures deal with this problem by relying upon the front spar contribution (adequately close to the peak) to transmit the load (shear flux and bending) to the ribs and panels of the leading edge. This requires an adequate rigidity in the upward structure. Such an architectural approach cannot be adopted for a droop nose morphing leading edge because the rigidity would strongly penalize actuation and, in case of a high-authority actuator, the necessary deformations would dramatically damage the structure. To solve this paradox, load bearing elements must be considered.
- It must adequately contribute to reduce the load, balancing the elastic reaction of the structure to aerodynamic actions (especially as they increase along the leading edge as it deflects), while reducing weight and costs. A nonlinear kinematic chain, whose load reduction ability increases with the external load, can solve this problem.

The droop nose deflection tends to increase the tensile stress on the top of the skin, while the bottom part is prone to buckle. This phenomenon is emphasized in Fig. 1.

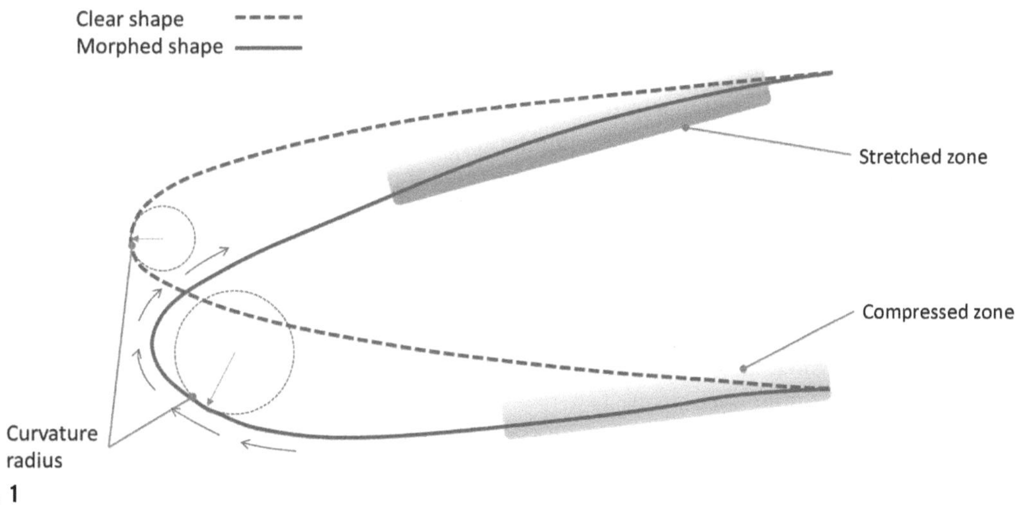

FIG. 1

Most loaded zones during the droop nose actuation: the skin upper zone tends to stretch while the lower zones tends to buckle.

The architecture of the morphing actuator has to induce a global clockwise rotation (see the arrows in Fig. 1) in order to alleviate the tensile stress on the upper surface and compression on the bottom, while simultaneously maintaining the target shape (generally characterized by a larger curvature radius close to the leading edge).

Under these intense, high amplitude aerodynamic loads, designing the actuator is challenging. It must maintain wing shape, yet be strong enough to bend the wing skin, while withstanding dynamic aerodynamic stresses. It is crucial to amplify the authority of the actuator yet minimize its dimensions and weight.

All these considerations lead to an articulated design in which each component has a prevalent function, but must cooperate with the others to absorb loads, alleviate stress, and manipulate the skin to the target shape.

The architecture described in the next section was conceived according to these requirements.

4 WORKING PRINCIPLE OF THE ARCHITECTURE SELECTED TO PRODUCE THE DROP NOSE EFFECT

The proposed architecture reproduces the leading edge zone of a wing of a transport regional aircraft. The main components are

- A kinematic chain assuring enough structural rigidity and transmitting deformation; it is made of two architectures: one distributed along the span, and the other within the rib plane.
  - A transmission chain suitably connected to the skin transmitting the deformation generated by the actuation system.
  - Structural elements cooperating with the skin in maintaining the external shape.
- A supporting structural element acting as wing spar.
- An electrical linear actuator.
- A metallic skin.

A lateral view (in plane rib) of the structural scheme is illustrated in Fig. 2. The actuator is mounted in span-wise direction. This both exploits spanwise space (generally wider than the rib plane) and implements a suitable kinematic chain reducing the forces necessary to produce deformation.

Along the span, the architecture is characterized by a three-hinges arc. As shown in Fig. 3, the actuator is mounted in a spanwise direction and connected through three-hinges arcs to the in-plane rib chains.

The actuator expansion (Fig. 3b) leads to a displacement toward the span direction of its edges (A and B points), with a consequent spacing of points C, D, E, and F; this causes the in-plane rib chains to bend down.

This architecture (three-hinges arcs) is advantageous because it reduces the force required by the actuator to produce the morphing effect. This reduction factor is not constant, but increases with the actuator extension. This is remarkably favorable, because the aerodynamic loads increase as the leading edge is bent down.

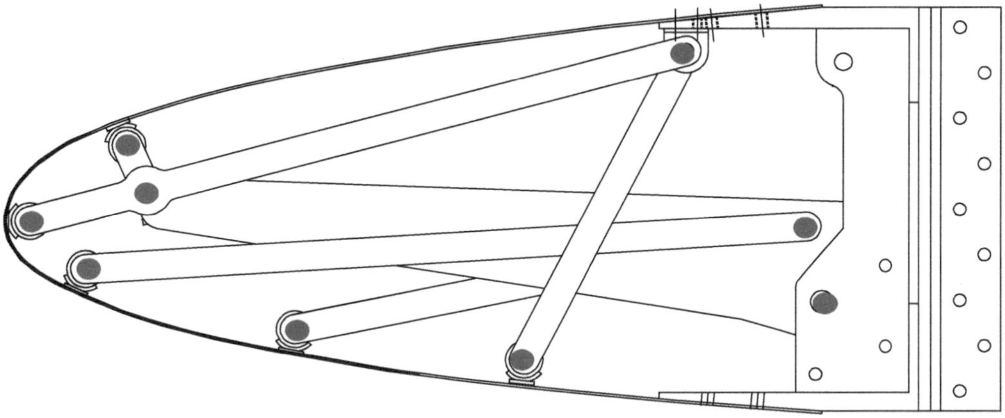

FIG. 2

Device lateral view; connecting hinges (*filled circles*).

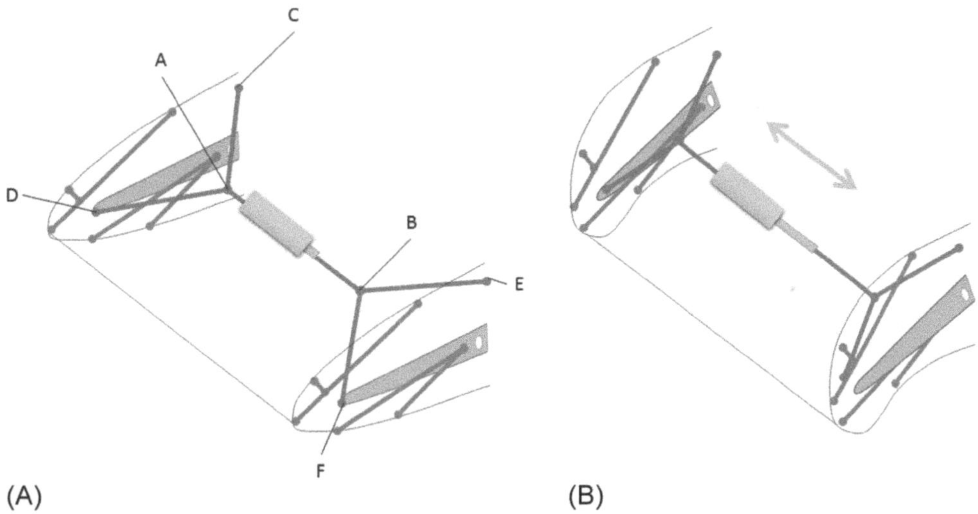

FIG. 3

Spanwise architecture, not activated (A) and activated (B).

5 ARCHITECTURE DESIGN

The design of the device was carried out as follows:

- Identification of a kinematic chain in the rib plane, assuring the achievement of the deformed droop nose configuration, as dictated by the aerodynamic specifications.

- Topologic optimization of the architecture identified in the previous task; the position of the hinges was defined through a genetic optimization, aimed at minimizing the deviation between the achieved deformation and the target.
- Design of the kinematic chain connecting the in-plane rib architecture with the actuator disposed along the wing span; this chain is made of a couple of three-hinged arcs, each one connected to one end of the actuator. The angle between the trusses of the arc represents a fundamental design parameter, as it determines the transmission factor law.

5.1 IDENTIFICATION OF THE KINEMATIC CHAIN IN THE RIB PLANE

The architecture was dictated by the typology of deformation desired. As already mentioned, the architecture has to guarantee:

- Enough rigidity to maintain the prescribed shape under the most severe load conditions.
- A global increase of the curvature of the airfoil (droop nose effect).
- A local increase of the radius onto the leading-edge zone.
- Structural integrity of all the components, particularly the skin.

To this purpose, the architecture illustrated in Fig. 2 was identified. The main lever (see gray component in Fig. 3) hinged to the front spar rotates down, transmitting the movement to the skin through connecting rods.

The preservation of the airfoil shape is assured by means of structural elements that connect and constrain different points of the skin; these elements enforce the skin motion to minimize the stress level.

The architecture functionality was simulated through the FE approach. A 2D model was initially realized, reproducing a strip of the leading edge, including the in plane rib architecture (see Fig. 4A).

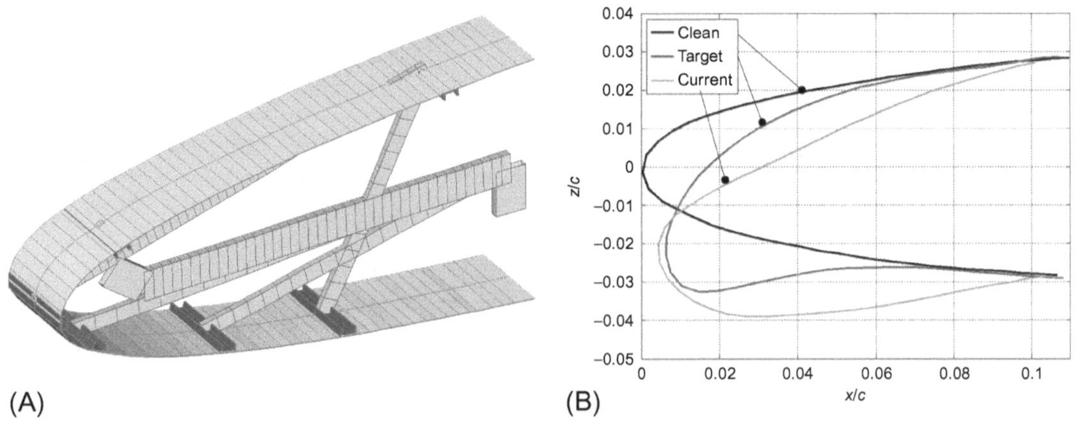

FIG. 4

2D FE model of a leading edge strip, integrated with the in plane rib architecture (A); shape comparison (B—starting with the top): base line, target shape [17], typical no optimized shape.

Table 1 Main Features of the 2D Leading Edge Strip Model

Number of nodes	414
Number of elements	338
Element type	CQUAD for the skin [22]
	CBEAM for the kinematic chain [22]
Materials	Skin: 7075-T6 aluminum alloy
	Architecture: 7075-T6 aluminum alloy
Skin thickness (mm)	0.75
Dimensions of the main lever cross section (mm)	40 × 10

A static nonlinear analysis was performed to investigate the ability of the kinematic chain to effectively produce the elastic deflection of the leading edge. The actuation was obtained through a lumped moment applied on the hinge connecting the main lever to the front spar. Finally, the effect of the aerodynamic loads was simulated by applying a variable pressure on the skin plate elements. The main features of the model are summarized in Table 1. A comparison among the base line, the target shape and a nonoptimized deformed geometry is provided in Fig. 4B. The FE model was conceived to meet two opposite requirements: accuracy in simulation (for the prediction of the displacement and stress field within the skin) and low computational time. This latter aspect resulted to be fundamental for the next design task, optimization.

5.2 TOPOLOGIC OPTIMIZATION OF THE IN-PLANE RIB ARCHITECTURE

As described, the internal mechanism was ideated according to practical engineering considerations. However, the preliminary internal configuration was not able to assure the optimized morphed shape. An advanced design process was thus necessary to assess the most important design parameters.

Due to the large number of optimization parameters (i.e., the in-plane coordinates of the 9 hinges highlighted in Fig. 2), and being the problem highly not linear (for the large expected displacements) a heuristic approach was preferred.

To avoid premature convergences, 20 random populations each one constituted by 50 individuals (characterized by a set of location of hinges) were processed for a number of 100 iterations. The inverse of the norm of the differences between the point position vectors of the target ($\vec{p}_{\text{target}}$) and of the current, ($\vec{p}$) shapes was chosen as cost function (f):

$$f = \left(\sum \left\| \vec{p}\,(s) - \vec{p}_{\text{target}}(s) \right\| \right)^{-1} \tag{1}$$

being s the curvilinear abscissa. Points of the same abscissa, s, were compared. In Fig. 5 the convergence diagram obtained for the configuration fitting the best the target shape is illustrated. No more than 20 iterations were necessary to reach the final configuration. The corresponding optimized shape is shown in Fig. 6A and compared with the baseline and the target one. The configuration proved to comply also with the stress limit requirements (achieved stress: 347 MPa vs. a material yield stress limit of 503 MPa), as shown in Fig. 6B.

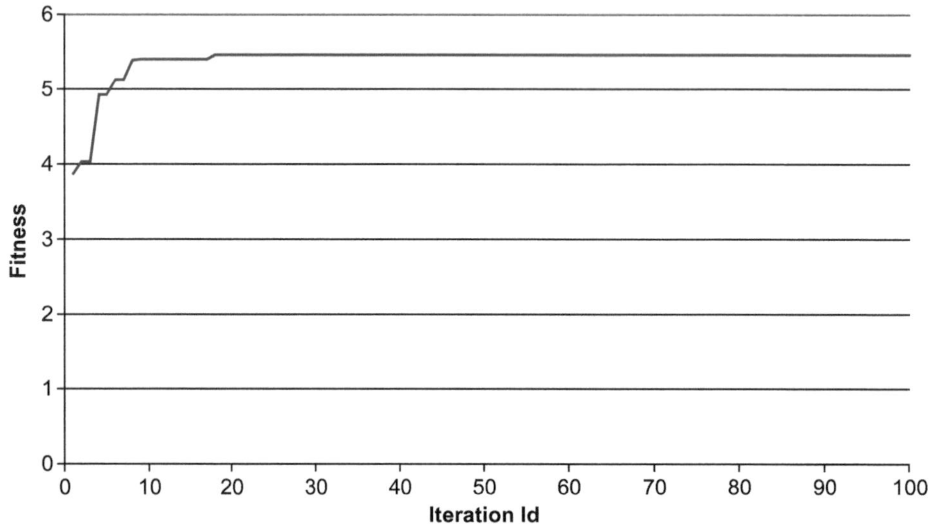

FIG. 5

Convergence curve of the optimal configuration.

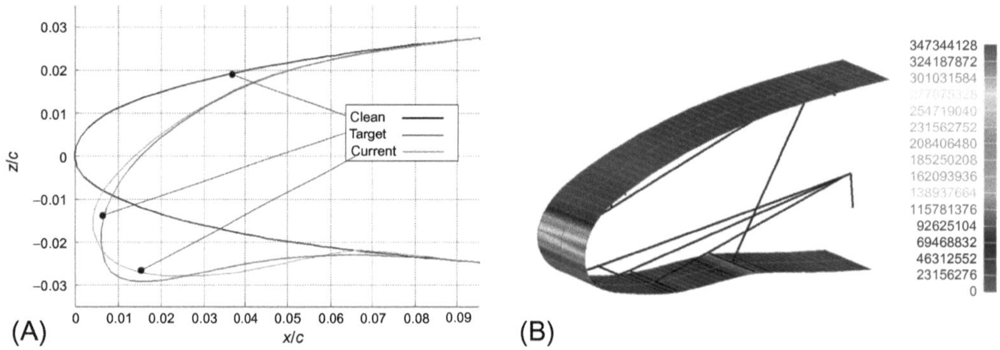

FIG. 6

Starting with the top, baseline, target and optimized shape comparison (A); stress field distribution within the skin [MPa] (B).

5.3 SPANWISE ARCHITECTURE AND ACTUATION DESIGN

After assessing the topology of the in plane rib architecture, the actuation chain in the span direction was defined.

The spanwise system can be schematized as shown in Fig. 7, representing the top view of the architecture (A) and a detail of the three hinges arc acting on the rib (B). The horizontal line represents the linear actuator whose expansion generates the forces F, pushing externally the central pivots of the arcs (x displacement, shown in Fig. 7B). The other two hinges of the arcs are connected to the front spar and to specific points of the rib in plane architecture (see points E and F in Fig. 3a); these constraints are respectively represented by the pinned points on the top of the scheme and the vertical elastic

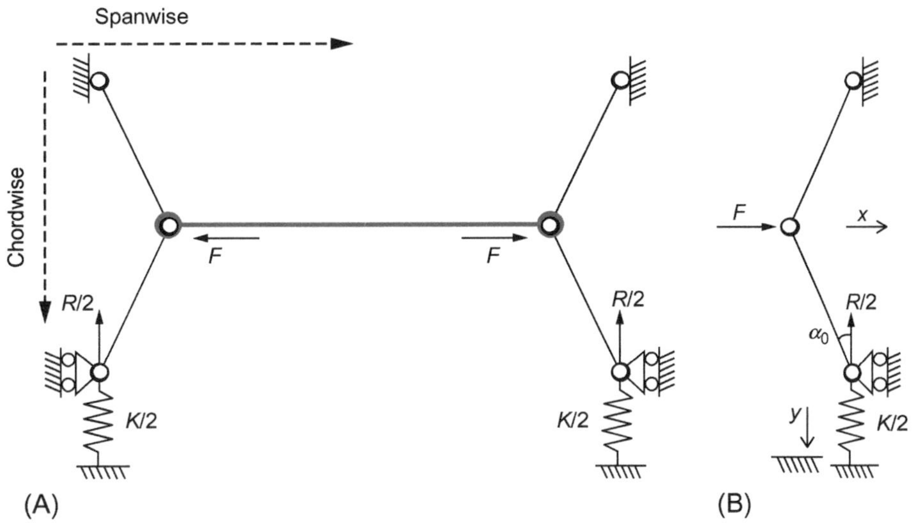

FIG. 7

Structural scheme of the spanwise reduction system (A); detail of the three hinge arc, acting on the rib (B).

constraints on the bottom. As evident, the only allowed motion for these hinges is along the chord direction (y); moreover, the elastic reaction of the structure and the transmitted action due to external aerodynamic loads were, respectively, simulated through springs with an equivalent rigidity, K, and by means of a force R. By applying the energy balance to the scheme (B)

$$F \cdot x - \frac{R}{2} \cdot y = \frac{1}{2} \cdot \left(\frac{K}{2}\right) \cdot y^2 \tag{2}$$

and assuming the following relations among the displacements x and y and the length L of the arc rods and the initial inclination α_0

$$x = L \cdot [\sin(\alpha_0) - \sin(\alpha_0 - \Delta\alpha)] \tag{3}$$

$$y = 2L \cdot [\cos(\alpha_0 - \Delta\alpha) - \cos(\alpha_0)] \tag{4}$$

it is possible to find the ratio r, namely transmission factor, between the force F applied by the actuator and the force R due to the external loads

$$r = \frac{F}{R} = \frac{1}{2} \cdot \left(\frac{y}{x}\right) + \frac{1}{2} \cdot \left(\frac{K}{2R}\right)\left(\frac{y^2}{x}\right) \tag{5}$$

As evident, this ratio is not constant, but depends on the actuation level (i.e., by the y displacement).

The initial angle, α_0, proved to be an important design parameter. In Fig. 8, the F/R ratio was plotted vs. the actuator displacement (x) nondimensionalized with respect to the length of the arc arm, vs. α_0 values ranging within 10 and 45 degrees. No elastic reaction was considered ($K = 0$) and the external action R was assumed constant. High values of α_0, while ensuring greater exploitability of the actuator excursion, determine high and thus unfavorable force transmission ratios at the very beginning of the motion.

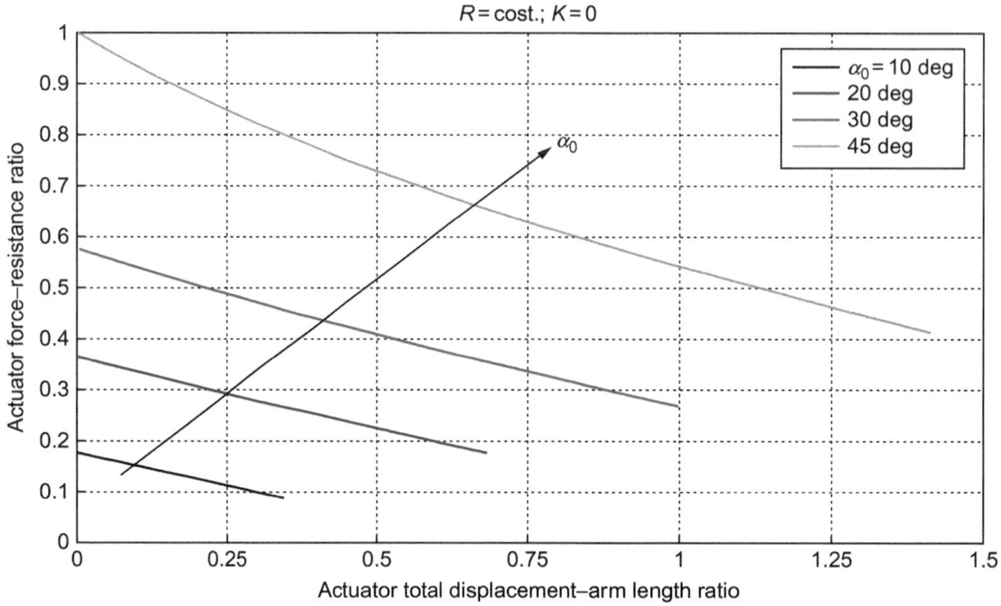

FIG. 8

Transmission ratio (F/R) vs. normalized actuator displacement vs. α_0.

Another important design aspect is represented by the structural rigidity, affecting the slope of the F/R curves and, therefore, the rapidity at which the transmission factor decreases. In Fig. 9A F/R curves were plotted for KL/R ratios of 0, 1, and 6.5 (the last one typical of the architecture). The slope of the curves for $KL/R = 6.5$ is initially positive, with an increase of the amplification factor that, for $\alpha_0 = 45$ degrees, becomes >1 and thus unfavorable. For this reason, a value of 20 degrees was assumed for the prototype; in this case, the F/R ratio is not >0.4, and the full exploitation of the actuator's stroke is ensured.

Finally, as the the aerodynamic load was influenced by the airfoil geometric configuration, R was assumed linearly variable from 25% to 100% of its maximum value, corresponding respectively to the undeformed and to the full activated/morphed configurations. In Fig. 9B the F/R curves were plotted for the maximum value of R and considering it variable. As evident, low initial values for R determine a slope increase, justified by the growth of the second addend at the right hand member of Eq. (2). It is noteworthy that, even in this case, for $\alpha_0 = 20$ degrees, the transmission factor is still lower than 0.5.

5.4 MODELLING AND WORKING SIMULATION OF THE COMPLETE ARCHITECTURE

After assessing the main aspects of the in-plane rib architecture and actuation kinematic chain, the modelling of the complete prototype followed. A FE approach was utilized, reproducing a 1 m-span leading edge, confined by two morphable ribs. Both the in-plane and the spanwise kinematic chains were represented through beam elements. Finally, the working actuator was simulated through an expanding bar. The external view of the model and the internal architecture are shown in Fig. 10.

A nonlinear static analysis was implemented to verify the devices ability to withstanding the applied load without any structural damage; the most severe aerodynamic load condition and the full actuated

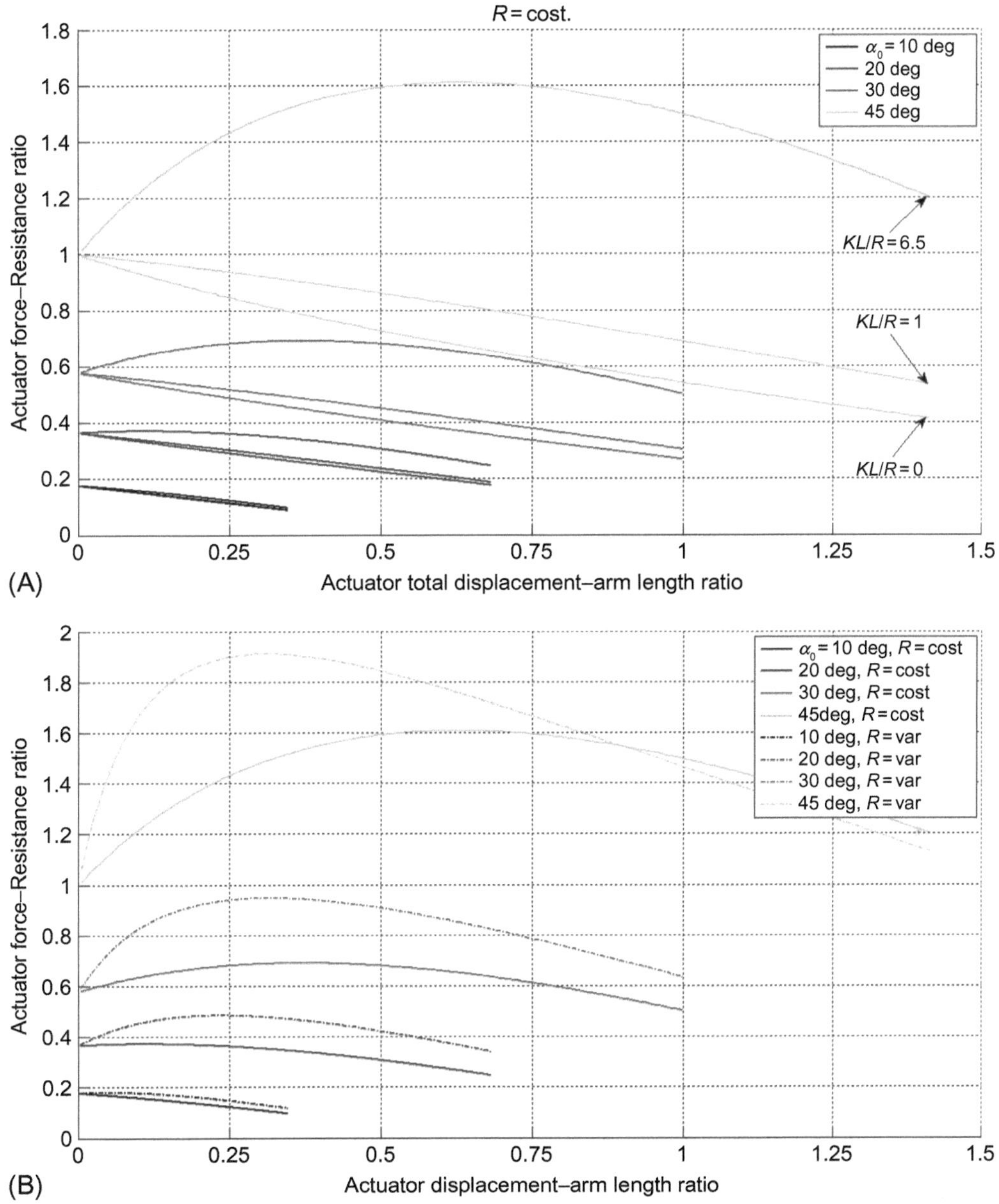

FIG. 9

Transmission ratio (F/R) vs. normalized actuator displacement vs. α_0, vs. KL/R ratio (A); transmission ratio (F/R) vs. normalized actuator displacement vs. α_0, $R = \text{cost}$ and R linearly variable (B).

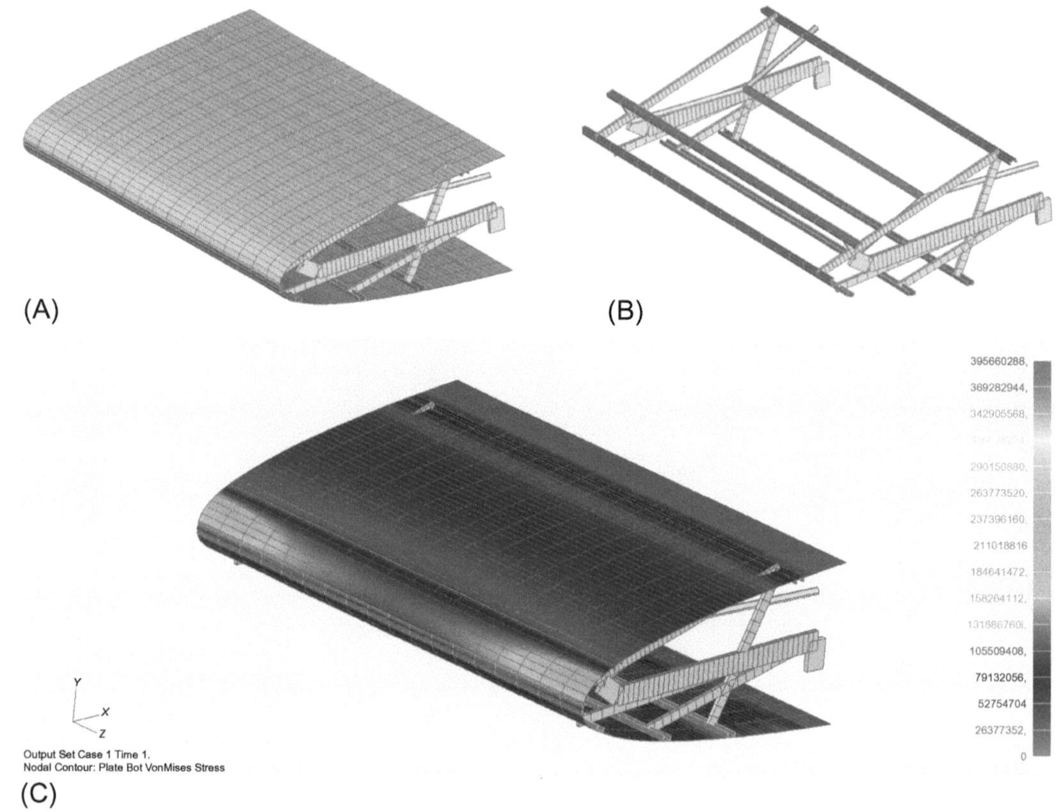

FIG. 10

3D FE model of the morphing leading edge (A); internal kinematic chain (B); stress analysis (C).

configuration were considered to this scope (see Fig. 10C). The achieved stress level, 395 MPa, did not exceed the allowable maximum.

Different behaviors in terms of force amplitude and trend were observed in the presence and absence of aerodynamic loads (Fig. 11, respectively the top and bottom curves). The necessary actuation force, in the presence of aerodynamic loads, is quite higher (three orders of magnitude) within the whole displacement normalization range. Without aerodynamic loads, according to the transmission chain model, the actuation force dramatically increases at the beginning (up to a normalized displacement of 0.1); this is due to the transmission factor, initially unfavorable. Then, as the excursion of the actuator and structural reaction rise together, the transmission factor becomes favorable and practically stable. The aerodynamic loads proved to have a strong influence on the actuation force, altering the slope of the curve as described in the previous paragraph. The effect of the aerodynamic loads is initially prevalent upon the structure elastic reaction; this, together with the unfavorable initial transmission factor, justifies the high initial force value. As the deformation level of the structure arises, its effect prevails upon the aerodynamic loads and the improvement of the transmission factor leads to a slow decrease of the actuation force.

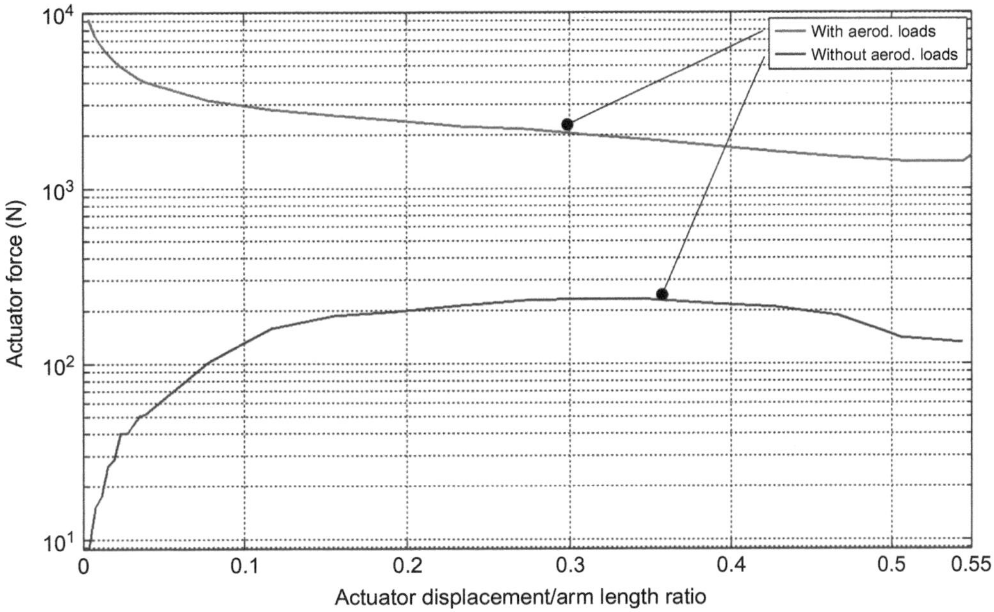

FIG. 11

Starting with the top, actuation force vs. normalized displacement with and without aerodynamic loads.

6 PROTOTYPING

On the basis of the numerical outcomes, the executive drawings of the prototype were elaborated (Fig. 12), and a laboratory specimen was manufactured.

The main aspects of the prototype are depicted in the following: Fig. 13 (diametric view), Fig. 14 (linear actuator integration details). The actuator properties (HIWIN LAS3-1) are summarized in Table 2.

7 EXPERIMENTAL CAMPAIGN

A dedicated experimental campaign was carried out with the aim of characterizing the demonstrator and estimating eventual deviations from the numerical predictions.

7.1 THE SETUP

The setup (Fig. 15) was composed of the following components:

- Full-size prototype.
- Power supply (1.5 ÷ 24 VDC @ 1.0 A): to supply prototype actuator and displacement sensors.
- Actuator switch: to command actuator expansion and retraction.

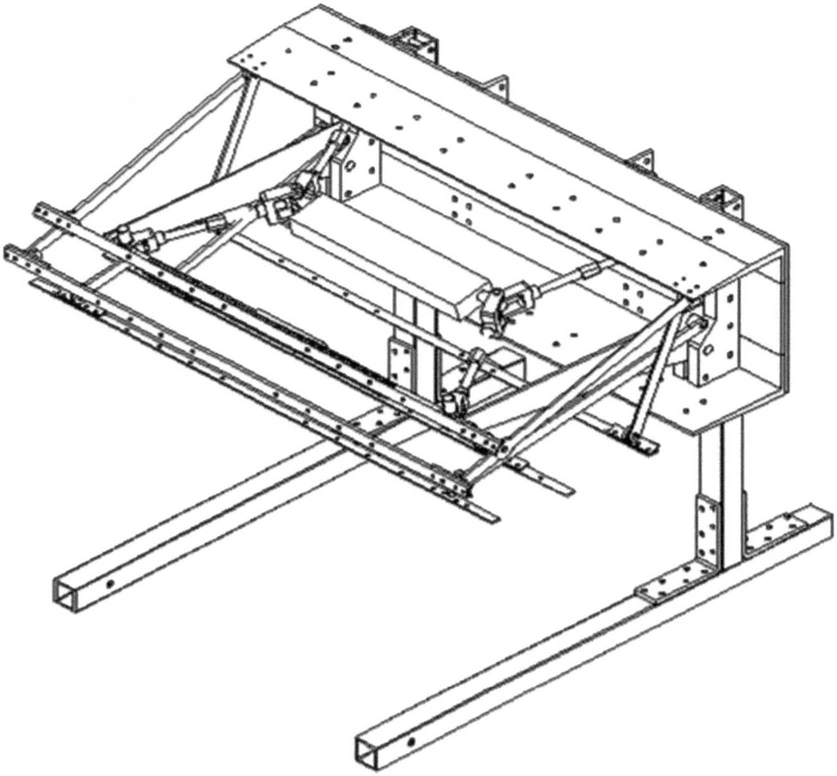

FIG. 12

Real size demonstrator executive drawings.

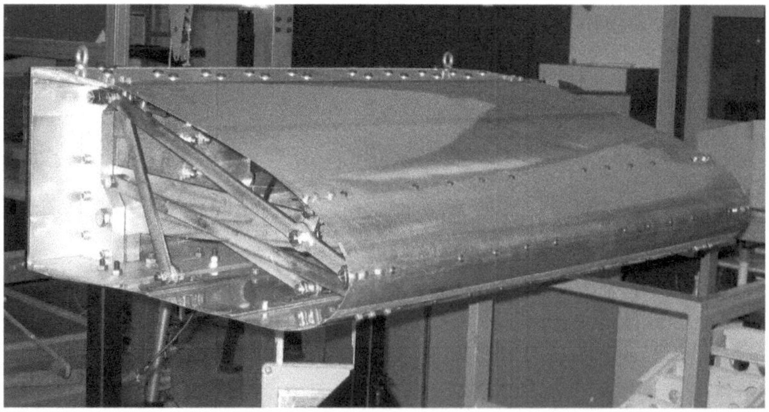

FIG. 13

Lab demonstrator diametric view.

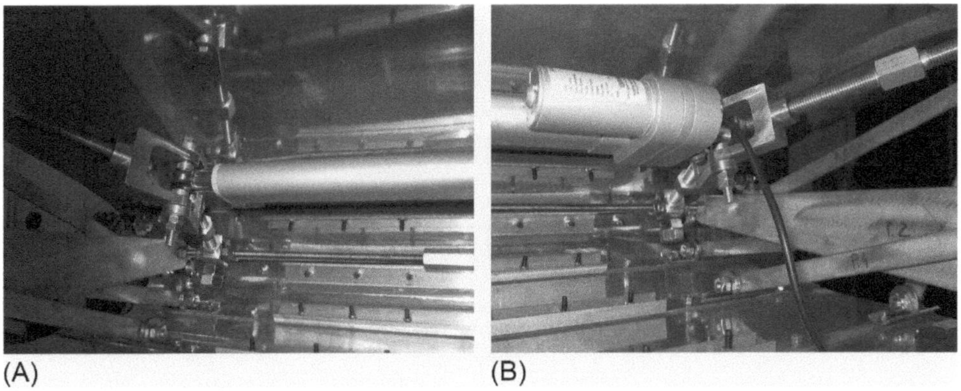

FIG. 14

Linear actuator integration details left (A) and right (B) triple-hinge connections.

Table 2 Actuator Characteristic	
Max push force (N)	1200
Max pull force (N)	1200
Max current (A) @ 24 VDC	2.5
Speed (mm/s)	8–12

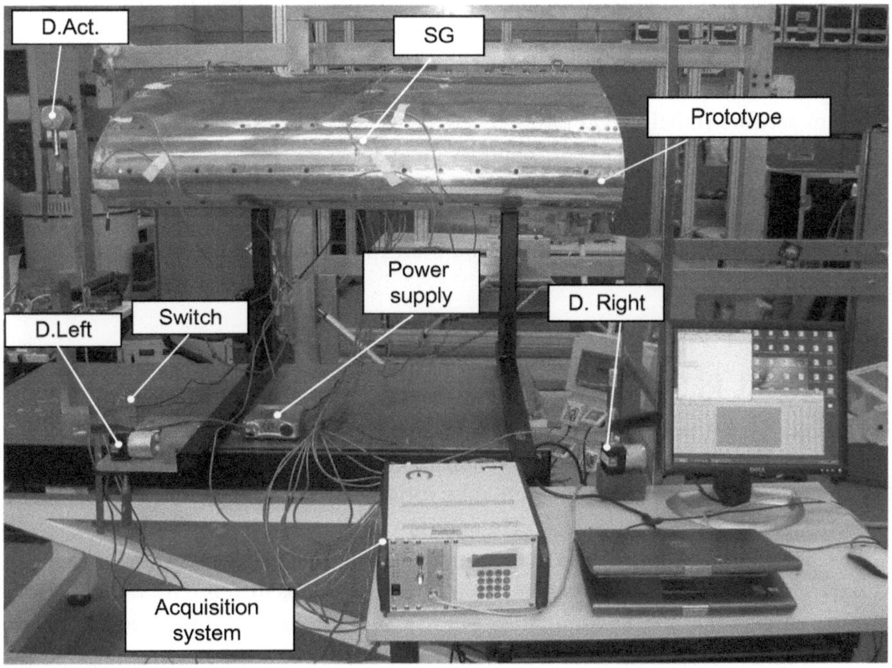

FIG. 15

Experimental setup.

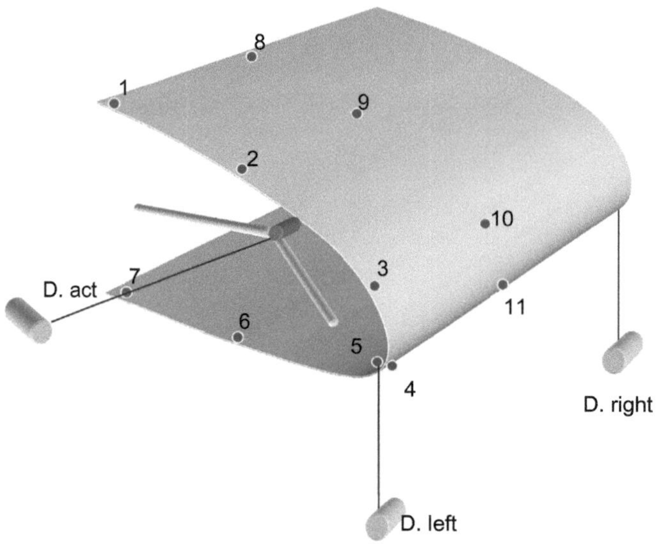

FIG. 16

Location of the sensors.

- Acquisition system: IMC Cronos PL8, to collect signals produced by each sensor (displacement sensors and strain gauges).
- Displacement (potentiometer) sensors: 3 draw-wire displacement sensors series WDS-500-P60-SR-U.
- Strain gauges: 11 sensors EP-08-125BT-120.

The displacement sensors (labelled "D.left," "D.right," and "D.act" in Fig. 15) were used to measure skin vertical displacement of "Point experimentally measured" shown in Fig. 16, appreciating eventual asymmetries for the actuation; the last one, the "D.act," was used to measure the linear actuator displacement. A map of the locations of the strain gauges (named "SG" in Fig. 15) is provided in the scheme of Fig. 16. Some details of the sensors and their installation on the demonstrator are also provided (Figs. 17 and 18).

7.2 EXPERIMENTAL RESULTS

Different on and off cycles were monitored through displacement and strain sensors. The linear actuator was activated up to reach the maximum extension compatible with the architecture, i.e., perfect alignment of the transmission arms hinged on actuator edges. This condition is illustrated in Fig. 24, depicting structure deformation for a complete activation.

In Figs. 19–21, some activation cycles are illustrated in terms of displacements and most remarkable strains acquired over time. The displacements on the two edges differed for a maximum of 6.4 mm on a span of 1 m, corresponding to an inclination of the leading-edge line of 0.37 degrees with respect to the horizontal line. The highest strain was measured very close to the leading edge stagnation line

FIG. 17

Details of strain gages location.

FIG. 18

Actuator displacement sensor.

(1360 με); here the asymmetry of compressive and tensile measurements proved the presence of normal strain (297 με) in addition to pure bending strain.

All the graphs show a good repeatability level; the forth cycle is noteworthy because of two spikes evident both for displacement and strain acquisitions. They were caused by the overcoming of the aligned hinge condition (i.e., actuator hinge deadlock), with a consequent sudden equilibrium change.

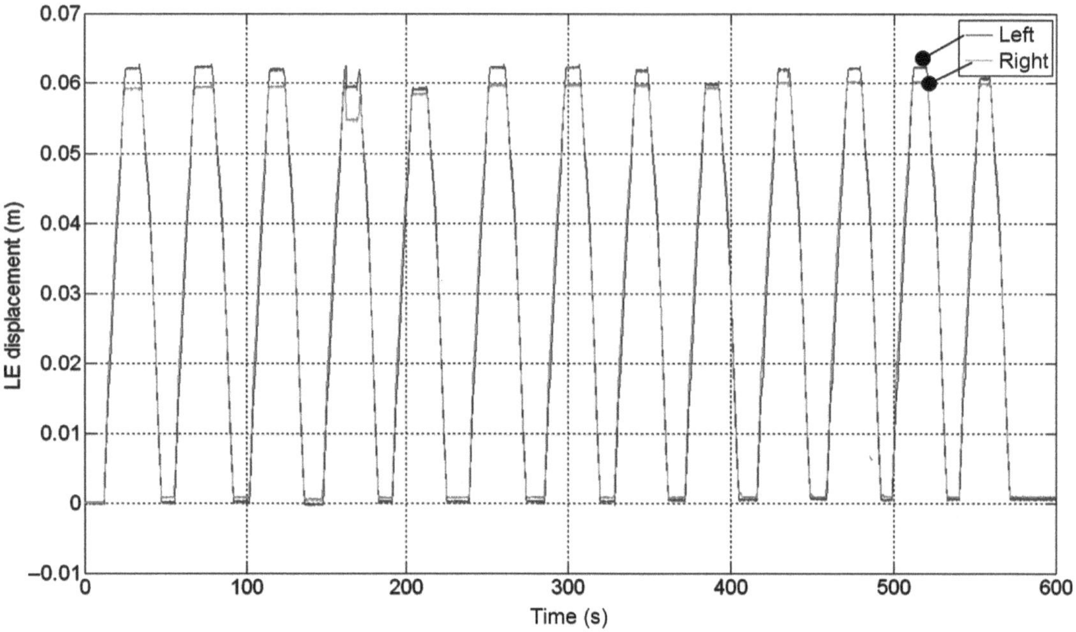

FIG. 19

Left and right leading edge points displacement vs. time.

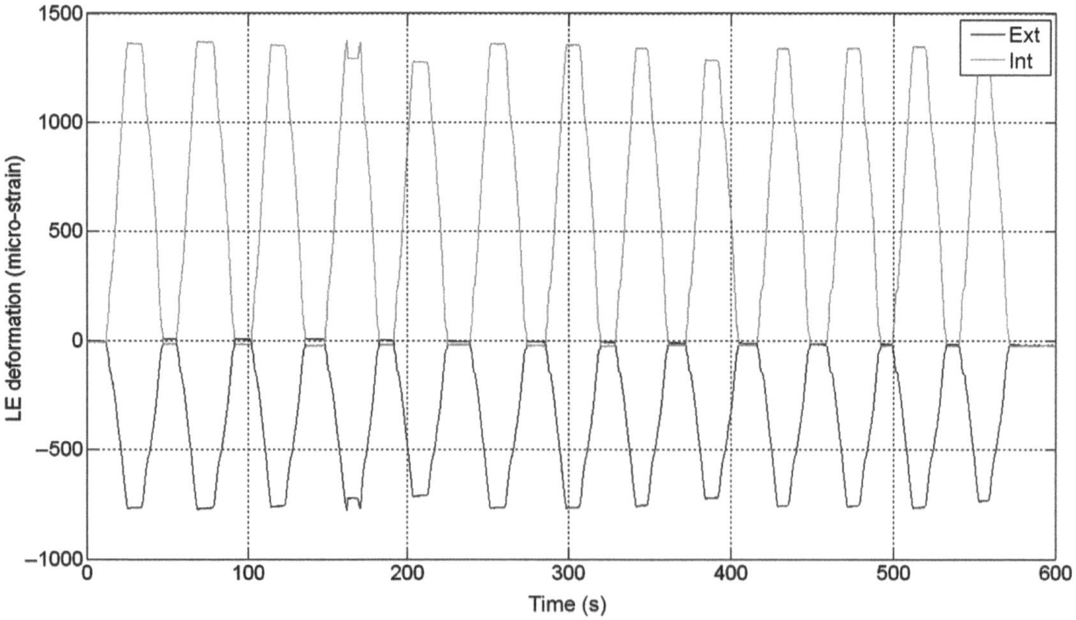

FIG. 20

Leading edge external (curve on the bottom)—internal deformation (curve on the top) vs. time.

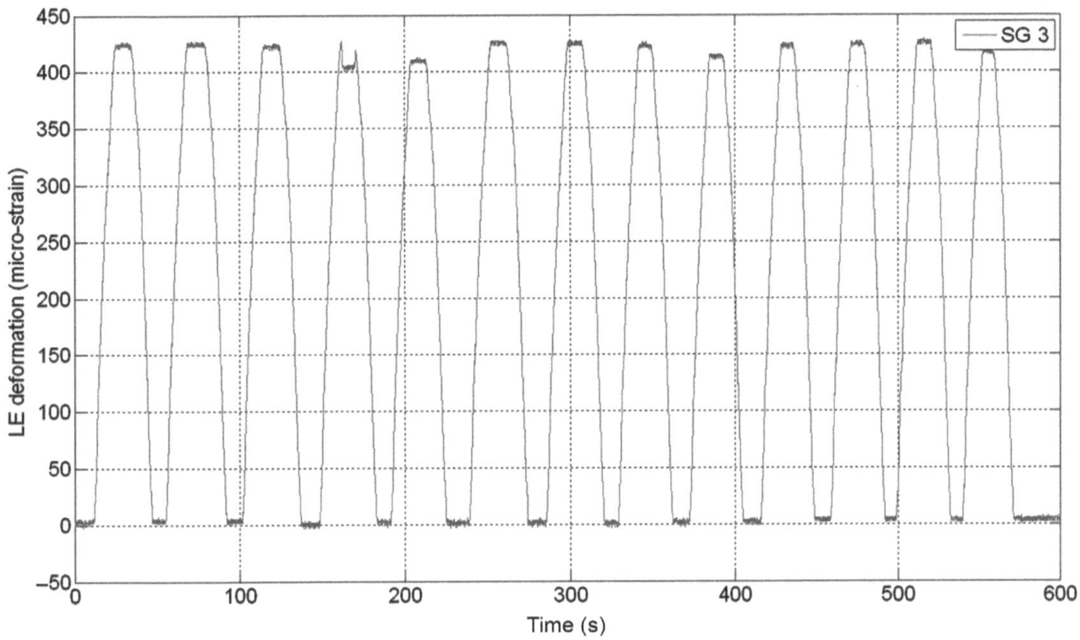

FIG. 21

Leading edge deformation on the upper zone (SG3) vs. time.

In practice since the alignment condition corresponds to the largest spacing among the points C and F, and C and D, shown in Fig. 3, any crossing of this condition leads to a decrease of the actuation level The first peak was caused by the overcoming during the expansion, the second one by the overcoming during the retraction. These peaks are important since they give an estimate of the theoretically maximum exploitable extension of the actuator.

7.3 NUMERICAL—EXPERIMENTAL COMPARISON

A comparison between numerical predictions and experimental data was carried out.

The data compared refer to the displacement of the "Point experimentally measured" and the strain close to the leading-edge zone (strain gauges 3, 4, and 5 with reference to Fig. 16). Both numerical and experimental displacements (D right displacement [see schematic in Fig. 16] vs. actuator displacement) were compared in Fig. 22; three curves were plotted in more detail: numerical with and without aerodynamic loads and experimental without external loads. Note that the slope of the curve becomes zero, approaching the max excursion allowable for the actuator. This behavior is consistent with what was observed in the previous section regarding the effects of passing the alignment hinge condition (Figs. 23 and 24).

The strain distribution (dashed and solid curves for external and internal skin sides) versus the curvilinear abscissa is illustrated together to the local measurements of aforementioned strain gauges (*circles*).

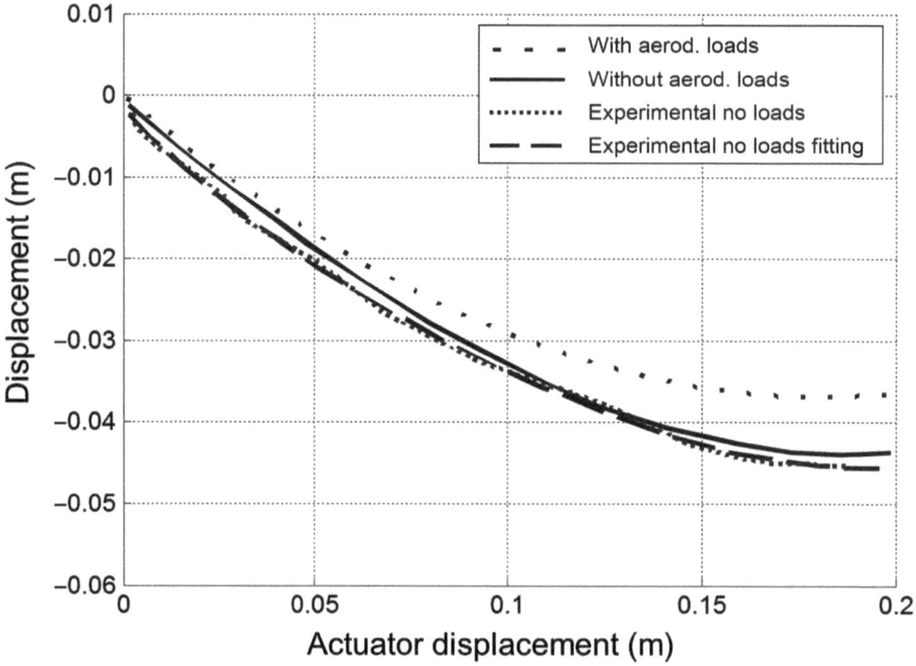

FIG. 22

Displacement comparison: vertical vs. actuator displacement.

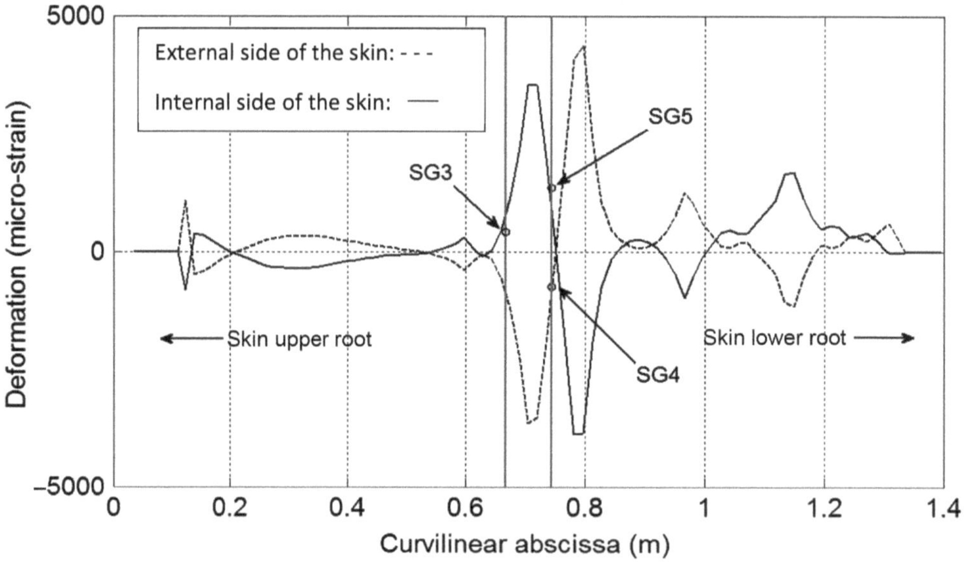

FIG. 23

Strain comparison: numeric (external, internal skin sides, *dashed, solid*), experimental *black circles*.

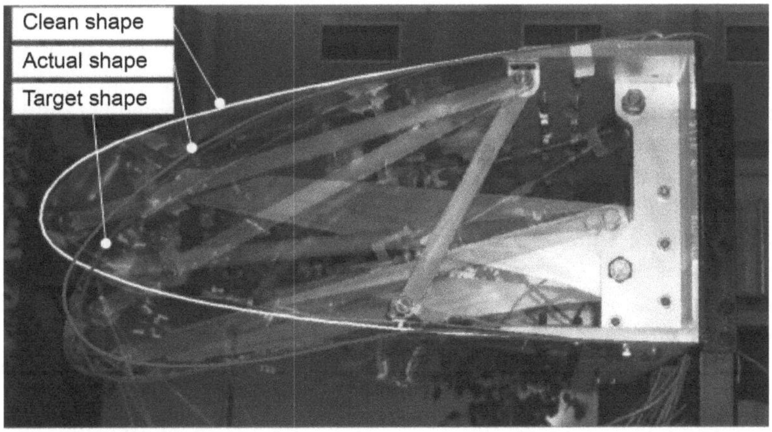

FIG. 24

Unmorphed and morphed configurations, compared with numerical data (starting with the top, clean geometry, actual deformation and target deformation).

Finally in Fig. 24, starting with the top unmorphed and morphed configurations are compared together to the baseline, the actual, and the target deformation, numerically predicted.

8 CONCLUSIONS AND FURTHER STEPS

This chapter focused on the possibility of changing leading edge wing zones geometry by producing large deflections without gaps and discontinuities. The resulting morphed geometry, called "droop nose," will assure an equivalent hyperlift generation unattainable via traditional systems (slats); at the same time, the absence of gaps and superficial discontinuities will help preserve the laminar flow in cruise condition.

The aerodynamic delicacy of the leading edge zone, and the presence there of the highest level of load, are only some of the challenges posed by designing droop nose architecture.

Next, the author provided an overview of the main studies on the leading edge morphing and on the programs concerning this topic. The most significant achievements on the topic were cited and discussed, with specific focus on two studies: the droop nose system developed and tested in a wind tunnel facility by DLR during the SARISTU Project, and the droop nose architecture developed by CIRA in the SADE Project.

The next section then offered a critical discussion of the different steps necessary to develop the morphing leading edge conceived by CIRA. First, the design approach was described with specific focus on strategies to actuate the deformation and achieve the desired target shape, without compromising the structural integrity. The original system for reducing the necessary actuation force was described. Then the advanced design was dealt with, including all the details necessary to produce a laboratory demonstrator, which was subsequently assembled and tested, effectively proving the functionality of the concept.

The idea herein described paves the way to further research aimed at enhancing the TRL of the concept. Research should examine the most critical aspects of the current device, beginning with the skin. Here, a metallic skin was tested and proven aerodynamically suitable for deformation. However, more versatile materials (e.g., composites) should be investigated not only for deformation, but for their ability of withstanding other stresses (e.g., lightning, bird strike, and hail impact). More research is needed to investigate the skins interface with the wing box structure, especially the leading edge skins connection to the wing box, and the structural connection of the architecture to the front spar. The wing box structures elasticity is crucial to the correct working of the droop nose system, as well as the stability of the leading edges skin, which can deform or wrinkle, priming turbulence. Future research should also focus on actuation. As mentioned, reducing the actuation force necessary to both withstand aerodynamic forces and generate sufficiently powerful actuation is crucial. The architecture selected for generating actuation forces will be nonlinear, with an adequate force reduction ratio, guaranteeing force balance in even the most severe conditions without an overly large actuation device. Finally, there must be further research into control and sensing. Due to the aerodynamic sensitivity of the leading edge zone, assuring achievement of the target shape is fundamental. To this purpose, a system capable of sensing shape reconstruction, and a dedicated logic of control thereof, are absolutely necessary. Last but not least, there must be investigation of the fatigue issues due to the cyclic deformations of the skin during droop nose actuation; these must be counteracted through choice of the materials and design strategy. Future research must provide more detail on advanced concepts like corrugated skins. Dedicated layouts able to address typical issues for the wing leading edge zone (e.g., bird impact, icing, etc.) must be identified and incorporated into the design, manufacture, and validation processes.

REFERENCES

[1] S. Ameduri, A. Concilio, C. Costantin, A. Gianvito, R. Sorrentino, in: Design of an MRF variable stiffness device to produce torsion stiffness variations of aerodynamic control surfaces, Proceedings of the Actuator ‘04, 9th International Conference on New Actuators, Bremen, 14–16 June, 2004, pp. 730–734.

[2] F. Bauer, P. Garabedian, D. Korn, A. Jameson, Supercritical Wing Sections I, II, III: Lecture Notes in Economics and Mathematical Systems, Springer, New York, NY, 1977.

[3] K.L. Bonnema, S.B. Smith, in: AFTI/F-111 mission adaptive wing flight research program, Proceedings of the Fourth AIAA Flight Test Conference, San Diego, CA, 18–20 May, 1988, pp. 155–161.

[4] I. Chopra, in: Review of state of art of smart structures and integrated systems, Paper Presented at the 42nd AIAA/ASME/ASCE/AHS/ASC Structures, Structural Dynamics, and Materials Conference, Seattle, WA, 2002 (AIAA Journal 40(11)).

[5] J. Spillman, The use of variable chamber to reduce drag, weight and costs of transport aircraft, Aeronaut. J. 96 (1992) 1–8.

[6] A. Concilio, S. Ameduri, Influence of structural architecture on linear shape memory alloy actuator performance and morphing system layout optimization, J. Intell. Mater. Syst. Struct. 25 (16) (2013) 2037–2051. November 2014.

[7] S. Kuzmina, G. Amyriants, J. Schweiger, J. Cooper, M. Amprikidis, O. Sensberg, in: Review and outlook on active and passive aeroelastic design concept for future aircraft, Proceedings of the 23rd ICAS 2002 Congress, Toronto, 8–13 September, 2002, pp. 432.1–432.10.

[8] J. Schweiger, F. Weiss, T. Kullrich, in: Active aeroelastic design of a vertical tail for afighter aircraft, Paper Presented at the 8th AIAA/USAF/NASA/ISSMO Symposium on Multidisciplinary Analysis and Optimisation, AIAA-2000-4828, Long Beach, CA, 6–8 September, 2000.
[9] S. Ameduri, A. Concilio, A. Gianvito, Design of an MR based on device for the adaptive stiffness control of tail shafts, J. Intell. Mater. Syst. Struct. 20 (2009) 837–848.
[10] A.-M.R. McGowan, L.G. Horta, J.S. Harrison, D.L. Raney, in: Research activities within NASA's morphing program, Paper Presented at the RTO AVT Specialist' Meeting on IJSI 2,4 456Structural Aspects of Flexible Aircraft Control, Ottawa, 18–20 October, 1999. published in RTO MP-36.
[11] J.N. Kudva, Overview of the DARPA smart wing project, J. Intell. Mater. Syst. Struct. 15 (April) (2004) 261–267.
[12] F.D. Statkus, Continuous skin variable camber airfoil edge actuating mechanism, United States Patent No. 4,351,502, The Boeing Company, Seattle, WA, 1980.
[13] Zimmer, H. (1979), "Quertriebsko¨rper mit vera¨nderbarer Profillierung, insbesondere Flugzeugtragflu¨gel," German Patent No. DE 2907912-A1, German Patent and Trademark Office, Munich.
[14] W. Geissler, H. Sobieczky, M. Trenker, in: New rotor airfoil design procedure for unsteady flow control, Proceedings of the 26th European Rotorcraft Forum, Paper No. 31, The Hague, 26–29 September, 2000.
[15] S. Kota, G. Ervin, R. Osborn, R.A. Ormiston, in: Design and fabrication of adaptive leading edge rotor blade, Proceedings of the American Helicopter Society 64th Annual Forum, Montreal, 29 April-1 May, 2008.
[16] K. Schröder, M. Pleißner, SADE—flexible high-lift structures—aerodynamic requirements, in: in: AIRBUS Internal Report ME0834651, EDGACWD Department, Bremen, 2008, pp. 1–15. 26 November.
[17] SADE, Annex I—Description of Work, 27 March, 2009, pp. 5–15.
[18] Ameduri, S., Device for modifying the airfoil geometry, European Patent EP 2727825 A1, 2014.
[19] P.C. Wolcken, M. Papadopoulos, in: Smart intelligent aircraft structures (SARISTU), Proceedings of the Final Project Conference, Springer International Publishing AG, Cham, ISBN 978-3-319-22412-1, 2015. pp. 41, 45, 74, 117–121. https://doi.org/10.1007/978-3-319-22413-8.
[20] G. Tomassetti, S. Ameduri, A. Carozza, Innovative streamline-flow preserving actuation strategies for wing airfoil nose, Int. J. Struct. Integr. 2 (4) (2011) 437–457.
[21] S. Ameduri, A SMA based morphing leading edge architecture, Adv. Mater. Res. 1662-89851016 (2014) 383–388. www.scientific.net/AMR.1016.383.
[22] MSC/Nastran, Version 70.5, Quick Reference Guide, The Macneal-Schwendler Corporation, University of California, 1998.

CHAPTER

AN ADAPTIVE TRAILING EDGE

17

Antonio Concilio*, Ignazio Dimino*, Monica Ciminello*, Rosario Pecora†, Francesco Amoroso†, Marco Magnifico†

*The Italian Aerospace Research Centre, CIRA SCpA, Capua (CE), Italy**

University of Naples "Federico II", Naples, Italy†

CHAPTER OUTLINE

Morphing Wing Technologies. https://doi.org/10.1016/B978-0-08-100964-2.00017-4

Nomenclature

A/C	aircraft
ATED	adaptive trailing edge device
BL	distance between actuation force and arm hinge
$BM_{i(f)}$	bending moment absorbed by front spar caps
$BM_{i(r)}$	bending moment absorbed by rear spar caps
$BM_{i(U)}$	active (ultimate) bending moment
BR	distance between actuation force and motor hinge
CAD	computer aided design
CFD	computational fluid dynamics
CIRA	The Italian Aerospace Research Centre
COTS	commercial-off-the-shelf
C_p	pressure coefficient
CS-25	certification specifications for large airplanes
D	drag
DMU	digital mock-up
DOF	degree of freedom
EASA	European Aviation Safety Agency
EMA	electro-mechanical actuator
FBG	fiber bragg grating
FC	failure condition
FE	finite element
FEM	finite element method
FH	flight hour
FHA	fault hazard assessment
FO	fiber optic
HAZ	hazardous
$h_{i(f)}$	height of the front spar
$h_{i(f,\ r,\ sk)}$	height of front, rear spar; length of up/low skin panel
$h_{i(r)}$	height of the rear spar
L	distance between the motor and the actuation hinges
L	landing
L	lift
m	meter
M	Mach number
MA	mechanical advantage
MAJ	major
MIN	minor
P_i	perimeter of the box cell
PK	an aeroelastic method (pole vs. reduced frequency)
$q_{i(fw,\ rw,\ sk)}$	web shear flow at front, rear, upper/lower skin panel
R	length of the motor arm
RBM	root bending moment
S	second
$S_{i(U)}$	ultimate shear at section S_i;
SM_L	safety margin—limit load

SM_{UL}	safety margin—ultimate load
SRF	safety, reliability and failure
TE	trailing edge
$t_{i(fw,\ rw,\ sk)}$	front, rear spar web, up/low skin panels thickness
$TM_{i(U)}$	ultimate torque moment at si, referred to Y-axis
TO	take-off
V_{c}	cruise velocity
V_{d}	dive velocity
V_{g}	an aeroelastic method (velocity vs artificial damping)
V_{max}	max velocity
A	angle between the ref. plane and the actuation arm
Γ	angle between the actuation and the motor arm
$\sigma_{\mathrm{max,L}}$	Max Von Mises stress—limit load
$\sigma_{\mathrm{max,UL}}$	Max Von Mises stress—ultimate load
1D, 2D, 3D	one, two, three-dimension

1 INTRODUCTION

Modifying a wing trailing edge is perhaps the most trivial way to implement camber morphing [1,2]. On the other hand, historically, the use of flaps and ailerons were the first answer to attain different profiles for take-off and landing with respect to cruise. In this case, different steady configurations are adopted to face the varied flight conditions (low vs. high speed, low vs. high altitude, and so on). However, external situations can change a lot during a single flight and even flight by flight. A commercial aircraft can move for a short or long range; it can be empty or full-loaded. Center of gravity itself may move considerably with important effects on the aircraft trim. Fuel consumption alone determinates weight variations that can be estimated around 25% of the total weight for typical commercial aircraft. Because the design point is generally one for standard structures, it means the plane is not projected for others, optimally. This in turn implies the performance will not stay optimal for the entire mission [1].

Early works on the subject may be found in the 1980s, with specific reference to the implementation of variable-shape wings on large commercial aircraft [3,4]. In those papers, in terms of purely aerodynamic studies, the benefits of having a variable characteristic wing are dealt with. It is relevant to mention that, in the cited works, both adaptive trailing and leading edges are treated in terms of plain elements (single-hinged). Indeed, studies on the opportunity of installing morphable wings on-board of civil and military aircraft are almost contemporary to the Wright Brothers [5], while documented visions about the implementation of these still futuristic devices may be found at the end of the 19th century! [6–9] At those times, technology limited any suitable development of those systems. Perhaps, the most challenging difficulty in realizing morphing wing components is to develop architectures that withstand external aerodynamic loads without suffering appreciable deformations from the equilibrium (design) line, yet permitting the insurgence of high-level strains to match the target flight condition.

In principle, a wing may be deformed in different ways: for instance, by modifying its camber, its span, or its twist [10]. Remaining in 2D configurations, the curvature may in turn be adjusted in several ways. However, per a generic profile, the three regions in which it can usually be divided are not at all equal. From a structural point of view, the nose is usually very rigid and complex. It must host the de-icing systems, be sufficiently stiff to satisfy bird-impact requirements, and ensure a high-quality geometry as aerodynamic flow is highly affected by its shape. The box, ranging between the two main spars, is hard to be modified from a practical point of view because it usually accommodates the tank, with strong implications for safety and allowable room. The trailing edge should guarantee a certain shape rather than having a real structural role; moreover, the flow therein is usually detached and geometry imperfections barely affect the quality of aerodynamic behavior. In other words, it is the easiest region to implement shape variation devices.

Speaking of morphing trailing edges, some specific characteristics should be ensured [10,11]. Skins should be highly deformable in-plane while they must be stiff in the normal direction; furthermore, that material should have a Poisson module next to zero to avoid transversal stretching [12]. The underlying structural skeleton requires a compromise between high load-carrying capacity and adequate mobility. This means that the system should be able to migrate from one geometry to another, being able to freeze at a certain state with minimal deviations (elastic displacements) from the nominal outline [13,14]. This target necessitates innovative structural and actuation solutions [15,16]. Also, the motors should be completely embedded in order the profile is not macroscopically modified. In principle, their number should be large to reduce their size and ensure suitable degrees of freedom [17,18]. An accurate detection system should be implemented to keep optimal shape irrespectively of the external and internal conditions. It should be based on a distributed sensor network that can provide dense enough information to reconstruct its deformed aspect [19].

From these few words the basic elements of a morphing trailing edge system are easily individuated: a deformable structural system, distributed actuation and sensor systems, a governing control system, and a sophisticated structural skin. The level of complexity increases because of its augmented functionality, moving from a namely passive to an active architecture. In turn, this at least implies manufacture, weight, and maintenance penalties [20]. The first item is strictly related to enhanced mobility. In fact, because the system is highly deformable, it would need dedicated jig and tools to be assembled in order to avoid the insurgence of unwanted prestresses and prestrain [21]. The introduction of additive degrees of freedom causes weight penalties, mainly due to the installation of kinematics and other devices to move and block the system from one state to another. Finally, the maintenance process is trivially affected by the introduction of a new system in the aircraft [22]. While manufacture has an impact on nonrecurrent costs, the latter two become part of the adaptive trailing edge life. Most of the these comments hold for other adaptive architectures.

In this chapter, the main steps concerning the design of an adaptive trailing edge are discussed, with special focus on each of its relevant architectural elements. In detail, the skin, structural skeleton, actuator, sensor, and control systems are dealt with. It is discussed the necessity of defining precise specifications of the occurring load in order to come to a proper design. In fact, despite being smart, the reference structure is designed to bear load and to assume specific configurations under continuous action of external forces. Some attention is devoted to aspects that are essential to come to a finalized product: namely, aeroelastic and safety analyses. The former assumes an absolute relevance because the system has augmented degrees of freedom with respect to a standard layout and then, a more complex dynamic response and higher risk of matching instabilities. The latter is required to envisage a future certification process for this kind of devices that requires the development of a dedicated path.

The research leading to these results gratefully received funding from the European Union Seventh Framework Programme (FP7/2007–13) under Grant Agreement no. 284562.

2 THE CONCEPT

Civil commercial aircraft missions are quite standard but different configurations may be displayed as a function of real carried weight and other environmental parameters. Actual lift coefficient can furthermore span over several tenths of a unit as fuel burns and the weight reduces by around 25%. The best pattern will change, as it strongly depends on the flight conditions that continuously move from certain stochastic initial settings. The work herein discussed, addresses the assessment of morphing wing smart systems for medium-range passenger airplanes (around 3 h cruise flight). Synthetically, lift-over-drag (*L*/*D*) ratio is the referenced parameter to meet these performance improvements, kept to its optimal value while weight and angle of attack change. Camber wing adjustments are realized through chord-wise trailing edge (TE) adaptations to compensate for occurring state variations, aimed at guaranteeing instantaneous optimal overall geometry, among a predefined quasicontinuous domain.

Shape variation is attained through a multifinger architecture driven by load-bearing actuator systems. The working mode is illustrated in Fig. 1, where the enveloping configuration (max top and bottom deflections) is shown. The control system drives the actuator action after receiving information from a widely distributed strain sensor network. An adaptive, highly deformable skin absorbs part of the external load and insures a smooth profile. The system keeps its structural properties while actuated, then allows the preservation of a specific shape regardless of the action of the operational load and internal insurgent stresses. Forced static and dynamic responses are considered. Such systems allow for achieving upgrades that can be estimated in terms of needed fuel reduction or, equivalently, range increase, to around 3% or more.

The adopted system uses nominally constant span-wise camber variations: in this case, the distributed actuators are told to work "synchronously." An added action variability along the wing extension could lead to some design weight decrease by alleviating the root bending moment (RBM) or, in the same way, reducing the stress field at the wing attachment to the fuselage. This potentiality could lead to interesting developments in future activities. Nevertheless, the system as it already is allows for a differential actuation between consecutive motors of 0.5 degrees corresponding to something like 1.2 degrees per meter. The prototype is designed at $M = 0.7$ and in conformity with the standard

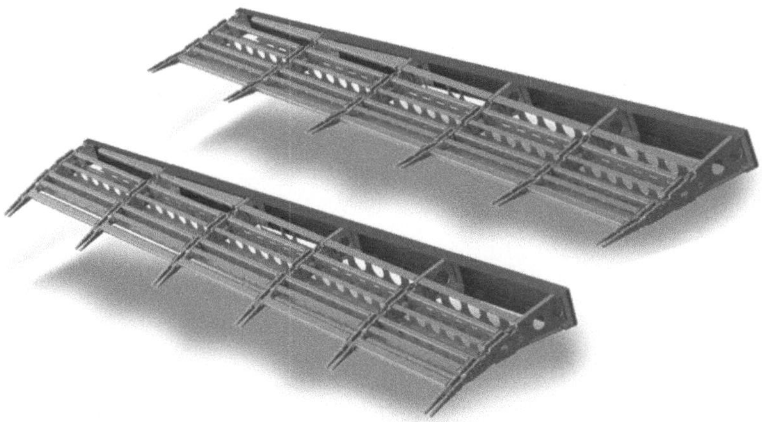

FIG. 1

Mechanical concept.

aeronautical regulations for the reference class of aircraft (CS-25). It is manufactured to be integrated within a 5.5 m span full wing section (combining other morphing devices—an adaptive winglet and an adaptive leading edge), prepared for wind tunnel tests ($M = 0.3$) [22]. The segment referred to the outboard region of the wing, so to allow the proper installation of the proposed systems and the investigation of the edge effects.

2.1 LAYOUT

2.1.1 The structural architecture

The transition of the trailing edge from the baseline to the target shape is ensured by giving the ribs adaptive properties. Any of these elements is segmented into four consecutive blocks, connected to each other by hinges that are located on the airfoil camber line. The first segment is rigidly connected to the main wing structure, while the other ones are assigned with a rotational degree of freedom (DOF) around the respective hinges. The camber line is then turned into a kinematic chain. The three degrees of freedom of the ribs are reduced to one by stiff links that further connect alternate (nonadjacent) blocks, forcing certain specific gear ratios among different elements, compliant with the target shapes envelope.

Each rib results in a 1-DOF mechanism: as one of its movable elements is fixed, the whole rib freezes into a particular configuration. Consistently, the motion of a generic block determines the movement of the complete system. A 3-segment polygonal line then approximates the trailing edge camber, while the thickness distribution is approximately unchanged. Such adaptive ribs are connected to each other by transversal spars joining each couple of analogous blocks. The system made of parallel spars, parallel blocks, and bottom and top skin coverage does resemble a closed "box," and the whole structure resembles a "multibox" arrangement, Fig. 2.

It is relevant to remark that, per the previous statements and particularly the fact that the structural system is a "mechanism," the motors have the unique role to guarantee a certain shape is reached and held. The structural concept has been finalized after an iterative design process consisting of three main loops; the executive layout has been obtained by progressive updates of a sketched configuration, on the basis of feasibility assessment and stress analyses outcomes. Management of the digital mock-up (DMU) is crucial in individuating possible interferences and other manufacture and integration issues. Structural design considers both operational loads (dive velocity, V_d) and aeroelastic limits.

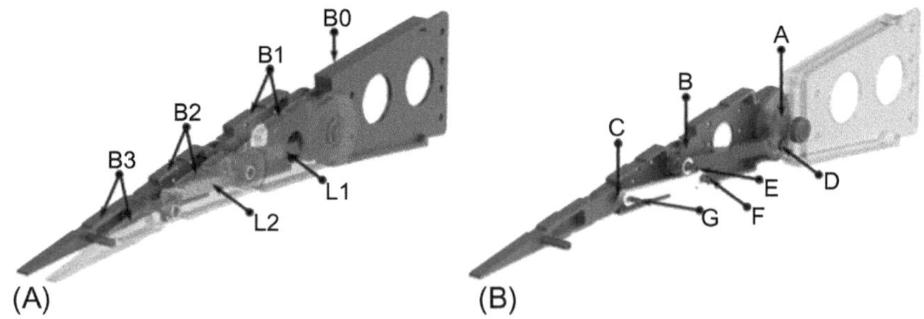

FIG. 2

The morphing rib. (A) Indication of the structural blocks (B0, B1, B2, and B3) and the mechanical links (L1 and L2). (B) Indication of the hinges connecting the structural blocks (A, B, and C in the order) and defining the attachment of the mechanical links to the structural blocks (D and E for L1; F and G for L2).

2.1.2 The skin system

Skins for morphing structural parts have the role of ensuring a smooth transition between movable and fixed parts, necessary for preserving air flow quality and, then, avoid negative impacts on overall performance. The adopted solution uses a combination of soft and stiff segments, distributed along the chord, to guarantee necessary deformations (at the hinges location) and a sufficient resistance to air pressure gradients, respectively. Because deformability is confined to specific regions, strain levels are very high (around 5%) and this is a clear criticality of the selected architecture, above all for the cyclic nature of the referred application. Upper and lower skin sides are riveted to the transversal spars, running between active rib components, and to the ribs themselves.

In detail, soft segments are made of elastomer foams, while hard segments consist of aluminium profiles. Material system smoothness is ensured by an overall thin silicone layer. The morphing skin is able to follow the adaptive trailing edge movement, driven by the active ribs, while maintaining the conformity in the deflected configuration. The elastomer foam geometry is optimized after several finite element method (FEM) calculations. Operational strain levels are fixed while air pressure actions forced shape deviations that are limited to 1 mm. This value shows no relevant effects on aerodynamics, mainly for the position of the investigated area, where flow is already turbulent or nonattached, Fig. 3.

Adhesive joints are used to guarantee a permanent connection between foams and structural materials. Fracture surface analysis are used to evaluate interface stability. Mechanical tests prove the reliability of the proposed solution: they include static and fatigue loads derived by general requirements and consider temperature-aging aspects in the range between −55°C to +80°C. In that interval, the properties of used materials are confirmed to be stable. In the end, a complete morphing skin panel 2.2 m long is manufactured and delivered, demonstrating the feasibility of the established process [23].

2.1.3 The actuation system

The movement of the adaptive trailing edge is ensured by a devoted and embedded actuation system, based on electromechanical actuators (EMA, working in a static or quasistatic mode). A CAD representation, together with some details about the installation inside the wing body, is shown in Fig. 4 [16]. An SRF (safety, reliability, and failure) approach is implemented to evaluate the systems airworthiness level when deployed onto the trailing edge of a low-critical aircraft control surface, by pursuing:

- Simplicity: "direct-drive" actuation, without gearboxes.
- Redundancy: duplication of actuation per morphing rib.
- Fault tolerance: continued operation after single-actuator fault.

FIG. 3

The adaptive skin layout.

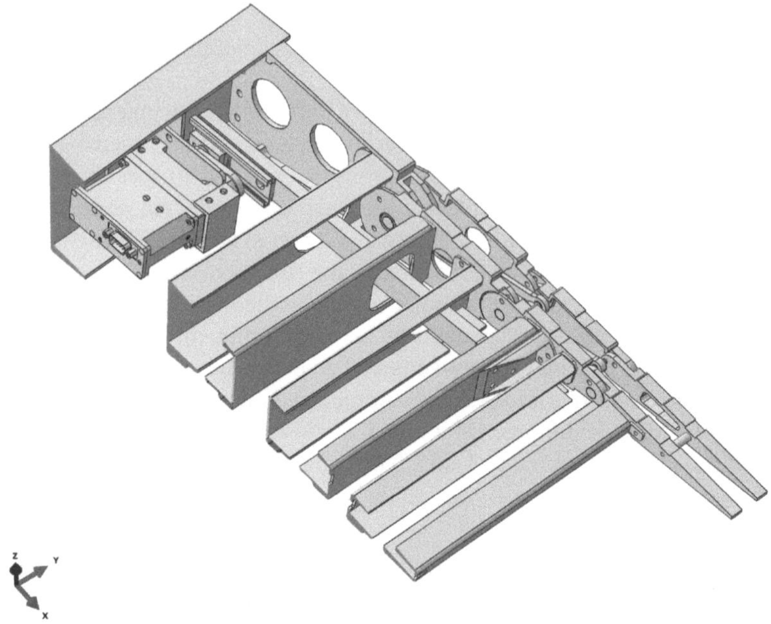

FIG. 4

The actuation system: installation details.

The integration of full-electric motors reduces classical drawbacks of hydraulic systems (like weight and room allowances) and minimizes the overall complexity of the actuation architecture while yielding maintenance benefits. Lack of supply buses, improved controllability, enhanced efficiency, removal of fluid losses, and flammable fluids bring further advantages. A general limit of EMA is the possibility of jamming failures that can lead to critical aircraft fault conditions. For that, the employment of certified, flyable systems is imperative. Operative loads and available space force the use of several actuators, whose size should be kept to a minimum to fit the reference geometry. These conditions lead to the use of a couple of actuators per rib. The resulting unshafted layout allows further weight savings. The overall actuator cabling system is defined after the structural design is completed, assessing the connections to the control system hardware, signal conditioning electronics, and routing to each actuator.

2.1.4 The sensor system

A distributed sensor network is necessary to have feedback on the achieved shape. Considerations concerning the cabling and other installation issues lead to the prefererance of fiber optic (FO) sensors for this task, instrumented with fiber bragg gratings (FBG) for strain evaluation. Devices are deployed chord-wise, at the middle-bay sections, and span-wise, at the tip and main spar, to form a set of closed cells. In this way, the actions over the morphing skin and impacts on the assembly chain are reduced to a minimum. Two main data sets are extracted, on the bases of which the complete shape is reconstructed:

- Span-wise bending.
- Chord-wise bending and torsion.

The chord-wise sensor is a cantilevered sliding beam with integrated FOs, aimed at providing data on the deflected shape. It is constrained to the skin profile, modulating the expected large strains so that standard gauges may be used, and act as an independent structural system for camber measurement. This original solution is a product of the joint work among structural (University of Naples "Federico II", CIRA), manufacturing (Aernnova) and sensor (Inasco, Technobis) technology partners [24]. Longitudinal deformations are much less relevant. Nevertheless, they are important to "close the box." In this case, the FO is installed on the metal tip and the main rear spar through ribbon tapes, a very thin and flexible glass fiber reinforced patch. Strain data detected by this sensor net are read and stored by a system made of the following components:

- A Deminsys interrogator (produced by Technobis).
- A laptop for signal monitoring.
- An optical switch for multiplexing signals.
- Power supply.

The quasistatic nature of the referred application (refresh should occur every 9 min for a 3 h period of flight) and the relatively wide instrumentation bandwidth (20 kHz) permit using an optical switch to reduce the number of active channels, simplifying the overall architecture. The complete system (made of sensible devices and their associated electronics, including the data-bus) is designed per the max number of allowable FBG per channel, accessible timeframe, and available machines. A sketch of the complete layout is shown in Fig. 5 [25], where the transversal and longitudinal sensor segments are evident.

2.1.5 The control logic

The control architecture uses both inherent rotation sensors on each actuator and a FBG-based distributed sensor system to synchronize the different element actions and to monitor localized failures. Distributed actuation can also enable the motion of individual ribs, enforcing, for example, twist angles, leading in turn to further features as load alleviation. Relative actuator movement with regards to transition times and velocity discrepancies between each rib segment are considered to synchronize each actuator. Control system components are selected as a last step. The connection between control

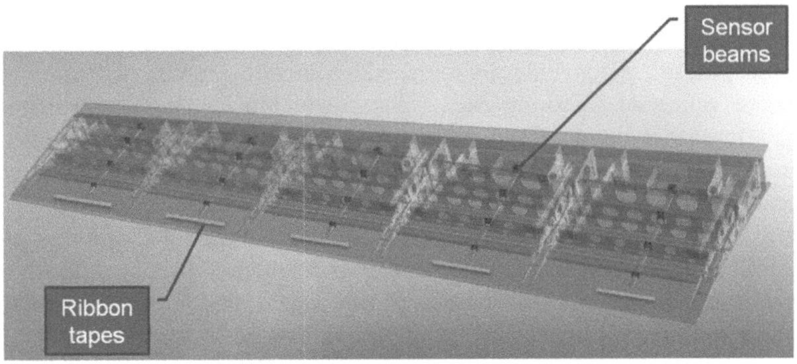

FIG. 5

The sensor system layout.

system (dSPACE embedded system) and actuator controller is specifically designed and verified together with the implemented equipment.

Different kinds of logic are implemented. At first, a classical feed-forward solution drives the single bays through a pc (also enabling differential rotations of the different ribs, within established structural limitations) or a joystick command (operating the various ribs synchronously). In this case, no feedback is used to adjust the achieved shape, but encodes data that ensures the achieved rotations are the driven ones. Subsequently, an automatic feed-back system is implemented, where the shapes are tuned to the desired target through a comparison between the achieved and the predicted strain vector or the corresponding deformed shapes, reached after a numerical integration process. In the frame of real operations, the control system may also refer to direct system data like the flight configuration or, more easily and direct, the fuel valve opening. In Fig. 6 [19], a detailed block diagram is sketched, illustrating the way the developed control system works.

3 DESIGN

3.1 DESIGN LOADS

The conceived articulated mechanism, made of actuators and suitable kinematics, is designed to resist significant aerodynamic loads, corresponding to a velocity around $M = 0.7$. Those are much higher than internal (due to skin stiffness) and inertial (generally negligible) forces. Among different load cases, defined through high-fidelity CFD computations, two different conditions are selected. First, cruise operational loads are computed, as the device should mainly operate in that situation. Second, limit loads are considered, related to dive speed, representing the most severe case. As a further sizing consideration, actuator loads are expected to increase moving towards the wing tip, because of decreasing dimension of the wing chord and constant value of the adaptive trailing edge chord. Finally, wind tunnel test loads at $M = 0.3$ on the reference full-scale wing demonstrator are also computed to predict actuator power consumption and validate actuator failure monitoring tools.

A three-dimensional nonviscid vortex lattice method is used to calculate pressure coefficient distribution along the wing in correspondence of a trimmed symmetric maneuver, with limit positive load factors and maximum take-off weight, Fig. 7. This is proven to be the most severe condition for the investigated system. C_p diagrams (upper and lower) are interpolated along the adaptive trailing edge rib planes; at each plane, those are conveniently replaced by linear piecewise distributions, characterized by constant average values of C_p along each rib segment block. The obtained numbers are then multiplied by the reference dynamic pressure and then by a further factor of 1.5 to get ultimate loads. Integration over the area finally leads to limit and ultimate bending moment, shear and torque distributions along each cross section.

Using a simplified block model of the morphing device, integrating rigid parts, rigid hinges, and elastic links, the torque acting on each actuator device is calculated. In this case, for a more accurate evaluation, skin stiffness contribution is considered as concentrated spring elements. The computed moments then allow defining the reference specifications for the motors. The Bental RSA-06 is selected for its limited weight (<0.5 kg) and size (a box of $10 \times 3 \times 5$ cm), compared to other commercial candidates.

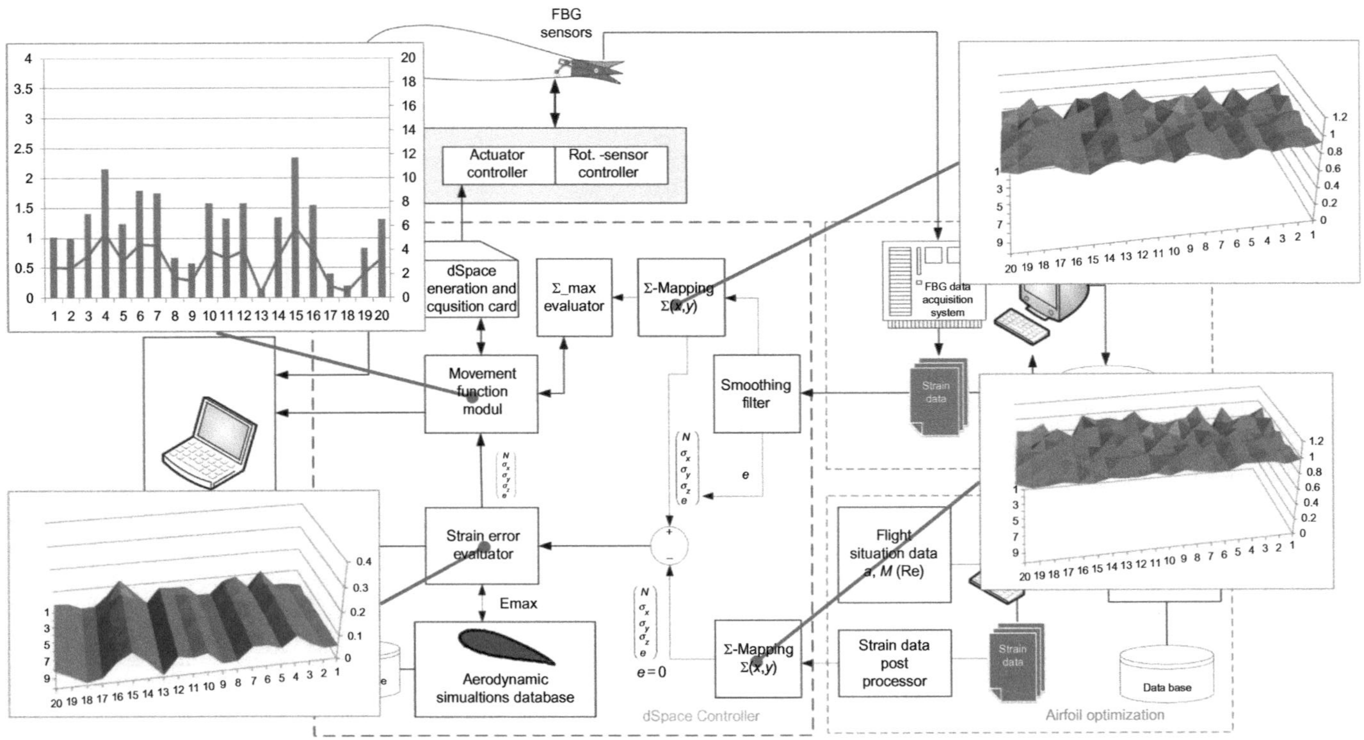

FIG. 6

The control system logic.

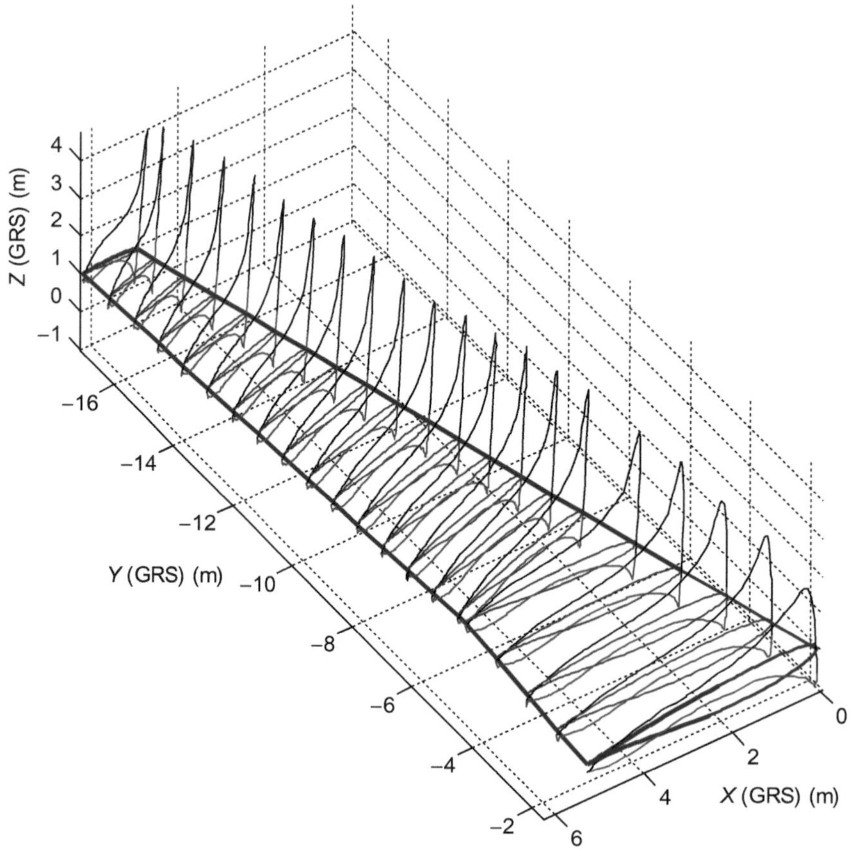

FIG. 7

Representation of the aerodynamic load (C_p).

3.2 STRUCTURAL SIZING

A thin walled section beam theory is used to address the sizing of the structure. To this aim, the architecture is idealized as a combination of single-cell boxes, rigidly connected to each other, chord-wise. The following figure illustrates a conceptual sketch, representative of the arrangement adopted for a generic bay, intended as the portion of the system which is delimited by two consecutive ribs and limited by the rear wing box spar and the tip, Fig. 8. Classic approaches are used to dimension load bearing components (spar webs, spar caps, skin panels, rib plates).

- Spar caps section area at generic section S_i.

  The front and rear spar caps are assumed to absorb the ultimate bending moment BMi according to the formula:

$$BM_{i(f,r)} = \frac{h^2_{i(f,r)}}{h^2_{i(f)} + h^2_{i(r)}} BM_{i(U)}$$

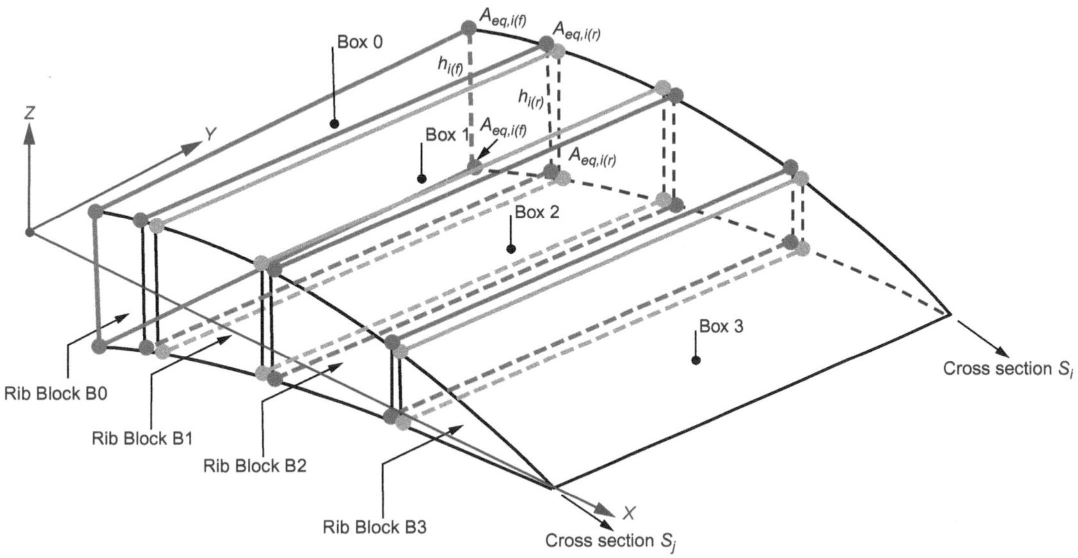

FIG. 8

Schematic of the single mechanical bay.

$BM_{i(f)}$	Bending moment absorbed by front spar caps
$h_{i(f)}$	Height of the front spar
$BM_{i(r)}$	Bending moment absorbed by rear spar caps
$h_{i(r)}$	Height of the rear spar
$BM_{i(U)}$	Active (ultimate) bending moment

Normal stresses (σ_{ref}) induced by bending moments at the front and rear spar caps are imposed equal to 70% of the ultimate material tensile stress. Required equivalent areas for front and rear spar are then obtained as:

$$A_{eq,i(f,r)} = \frac{BM_{i(f,r)}}{\sigma_{ref} h_{i(f,r)}}$$

- Spar webs, skin panels, rib plate thickness at generic section S_i.

  The ultimate shear force position along the abscissa is computed as:

$$X_{S_{i(U)}} = \frac{TM_{i(U)}}{S_{i(U)}}$$

$TM_{i(U)}$	Ultimate torque moment at section S_i, referred to Y-axis
$S_{i(U)}$	Ultimate shear at section S_i

The shear center location along the chord (x-axis) is assumed to be equal to:

$$X_{SC_i} = \frac{X_{i(f)} + X_{i(r)}}{2}$$

being $X_{i(f)}$ and $X_{i(r)}$ the positions along the x axis of the front and rear spars.

Consequently, the effective ultimate torque moment is replaced by a dummy value given by:

$$\overline{\mathrm{TM}}_{i(U)} = \mathrm{S}_{i(U)} \left| X_{S_{i(U)}} - X_{SC_i} \right|$$

and the close-cell shear flows are calculated as:

$$q_i = \frac{\overline{\mathrm{TM}}_{i(U)}}{2\Omega_i}$$

(Bredt's formula) where Ω_i is the area enclosed by the box cell.

The shear flow in each element of the section is then:

$$q_{i(fw,rw,sk)} = q_i \frac{h_{i(f,r,sk)}}{P_i}$$

$q_{i(fw,\ rw,\ sk)}$	Shear flow in the web of the front, rear, upper/lower skin panel
$h_{i(f,\ r,\ sk)}$	Height of the front, rear spar; length of the upper/lower skin panel
P_i	Perimeter of the box cell

Shear stress induced by shear and torque moment along the section are assumed equal to $\tau_{\mathrm{ref}}, \sigma_{\mathrm{ref}}/\sqrt{3}$, respectively. Web and skin panel thicknesses are finally obtained as:

$$t_{i(fw,rw,sk)} = \frac{q_{i(fw,sw,sk)}}{\tau_{\mathrm{ref}}}$$

$t_{i(fw,\ rw,\ sk)}$	Thickness of the front, rear spar web, upper/lower skin panels

The thickness of the rib plate is chosen as the max of the computed thicknesses. For practical reasons, web and skin panel thicknesses are selected equal or larger than 2 and 1.5 mm, respectively. The described approach allows a constant-stress sizing of the structural elements at each cross section. In this way, the structural element dimensions at each section do not result largely oversized with respect to the actual internal solicitations. Light aluminium aeronautical alloys (2000 and 7000 series) are preferred for all components of the primary structure.

3.3 ACTUATOR SELECTION

To ensure a clean airflow, the target architecture is thought as a completely embedded mechanism. The actuation torque for holding and moving the morphing device must be reduced to limit actuator size and guarantee that it fits the allowable room. This modulation can be achieved through a devoted Fairbairn mechanism that diminishes imposed displacement while amplifying forces. Because each rib is an

SDOF system, a single actuation chain may be considered. In detail, a crank mechanism is implemented, dimensioned with respect to the actual moments that are computed with respect to the "virtual hinge" position (that is, the point around which the system rotation occurs). It is defined for three different outlines: morphed up (max top deformation), morphed down (max bottom deformation), and unmorphed. Its path is then extrapolated for intermediate positions.

The structural actuator system (that concurs to bear external and internal loads) is completed with an arm, connected to the motor shaft and to a sliding guide. This guide moves in turn another arm, rigidly connected to the second rib block that transmits a torque to the system at the attachment point. A simplified scheme is shown in Fig. 9. The thickness and length of the referred trailing edge constrain the possible force arm values. The load is distributed along the architecture by transversal spars. The ensemble of these spars, the rib blocks and the skin, is a closed configuration able to resist high forces. Having defined the mechanical advantage (MA) of the mechanism as the ratio between the loading moment and the driving moment, the relation between the actuator and the rib block rotation angles (γ and α, respectively) may be computed as (meaning of the symbols is shown in Fig. 9) [19]:

$$MA = \frac{B_L}{B_R}$$

$$B_R = R\cos\gamma; B_L = L\cos\alpha + B_R$$

$$MA = \frac{L\cos\alpha}{R\cos\gamma} + 1$$

$$R\sin\gamma = L\sin\alpha \rightarrow \gamma = \arcsin\left(\frac{L}{R}\sin\alpha\right)$$

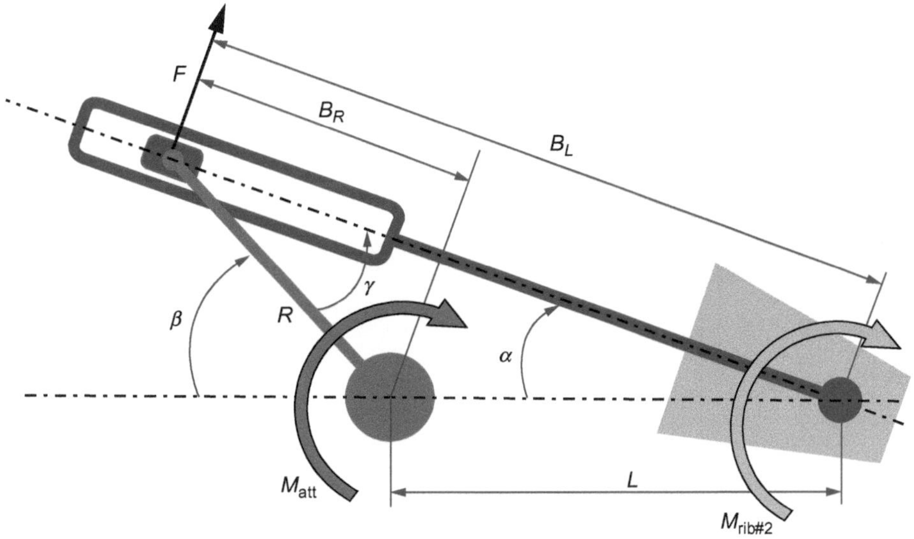

FIG. 9

The actuation system schematic.

From the above formulas it is understood that the second rib block rotation, α, grows slower than γ and therefore the ratio $MA = L \cos \alpha / R \cos \gamma$ shall increase as that angle increases. This is much more evident as the ratio between the arm length L (distance between the second rib block hinge) and the actuation crank radius R is higher. Similar formulas may be derived for the actuator rotation angle, β, versus the rib block rotation angle. The higher the L/R ratio, the higher the actuator rotation angle. This affects the max mechanical advantage because standard commercial servo-actuators have rotation limits. This architecture is selected among many others, essentially for:

- The high value of MA, compared to the limited room available.
- The system simplicity and reliability.

Starting from these results, the motor specs are defined, Table 1. A certain number of flight-certified servo-actuators are then screened in order to derive the most suitable ones for the envisaged application. In the end, the Bental RSA-06 is selected for its reduced weight (<0.5 kg) and limited size. Its rotation is confined within the range ±45 degrees. The worst load condition among the analyzed configurations, corresponding to the dive velocity, generates a stress field that is well below the material limits. In the same way, the linear guide makes use of COTS, very compact self-lubricated components. DryLin linear bearings operate on glide pads unlike the common recirculating ball bearing systems. These bearings do not depend on the travel length and hence do not pose any condition on the minimum stroke length. Certain geometries could produce the insurgence of high frictions between carriage and guide that can hinder movement. This condition does not occur for the selected configuration. Among the main advantages of this COTS components, the following may be cited:

- Short installation height (between 6 and 12 mm).
- Limited weight.
- Several options available (for instance, pretensioned).
- Maintenance-free, self-lubricating.
- Corrosion-resistant.
- Low wear with low coefficient of friction.
- Widely used.

Table 1 Servo-Actuator Specifications

Parameter	Value	Units
Type of actuator	Piezoelectric or electromechanical (stepped motor)	
Max torque	6 (dynamic) or 15 (static)	Nm
Displacement	±45 (pk to pk)	degrees
Resolution	0.55 ÷ 1.1 (max actuator backlash); 0.1–0.05 (ATE device resolution)	degrees
Dimensions	100 × 50 × 200 (W × H × L)	mm
Weight	<1	kg
Number of actuators	10	–
Actuator speed	>10	degrees/s
Actuation signal max latency	<10	ms
Nominal voltage	12 or 24	V
Power consumption	<100	W

3.4 RESULTS

3.4.1 Finite element analysis

After having set the dimensions of the most relevant structural components, the updated CAD of the structural layout is used to address detailed stress verifications based on finite element (FE) models and analyses. A detailed schematization of all the subassemblies is necessary, including ribs, spars, skin, and the actuation mechanical system, as well as spherical and cylindrical joints, rivets, and screws. To ensure the model compatibility with the main structure discretization, a customary approach is adopted; mesh size and general properties are defined for getting detailed and reliable distributions of stress and strains (especially around holes and, generally, discontinuities) while limiting the required computational time.

Because of the possible insurgence of large displacements, nonlinear analysis is required to evaluate deformation and stresses. Such computations are performed under limit loads, with the actuators "locked." Furthermore, all grids belonging to the upper and lower skins and joint to the wing rear spar, are fully constrained to the structural skeleton. The overall force field is simulated through a pressure distribution (dynamic pressure of 26,000 N/m^2, dive speed, sea level), acting on the upper (suction) and lower skin (compression), in compliance to the C_p trends, evaluated through inviscid 3D vortex lattice method. Conservatively, inertial alleviation is not taken in account. The ultimate load condition is finally achieved and evaluated by multiplying such limit loads by the reserve factor of 1.5.

Reaction forces at fastening holes are used to verify both fasteners and lugs. Fasteners are checked with respect to either their failure in tension or shear and tension combined to shear. Lugs are checked with respect to failure by bearing, net section crack, shear-out, transverse, and oblique load. Positive safety margins are obtained in correspondence of all performed checks. The absence of structural instabilities in the considered load range is proven through a classical linear buckling analysis. The first nonnegative buckling eigenvalue is found equal to 3.80. In Fig. 10 [13], the elastic displacements contour, following the action of the limit loads, is shown. A max upward displacement equal to just 15 mm

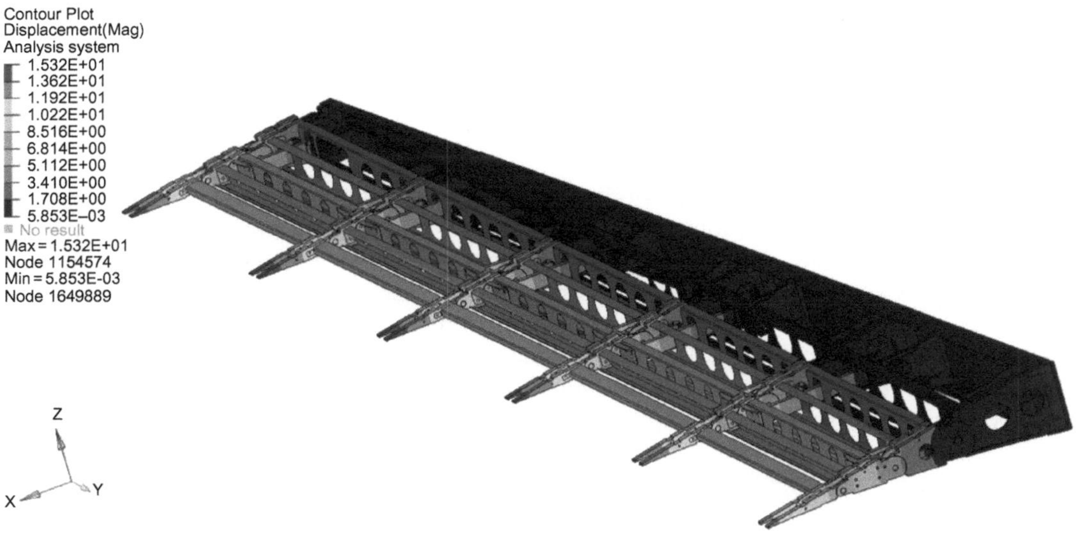

FIG. 10

FEM representation—structural response to static loads.

ensues, localized at the tip region. The structural arrangement then results stiff enough to withstand aerodynamic loads without undergoing significant shape change due to elastic deformations. Such deviations may be fully recoverable through a proper setting (zero shift) of the installed motors. As per EASA CS25 part C, the absence of any permanent deformation (failure) up to limit (ultimate) load condition is proven. In Table. 2, some examples of that analysis are shown.

3.4.2 Shape control

Shape control is implemented via open and closed loop architectures. In the former, the controller ensures the actuator shaft rotations, previously associated to reach specific deflection trailing edge angles. There is no feed-back on the achieved shapes on the basis of which the system tries to obtain a best match. In the latter, the controller monitors shape information through elaboration coming from the FBG-network, distributed all over the structure, and the rotation levels that are output from the motors encoder. The comparison between the expected and achieved displacement vectors (made of strains and angles information) is elaborated to adjust the current system position. In both cases, the controller is implemented real-time in a Digital Signal Processing (DSP) board.

In the experimental campaigns, the mechanism is driven to achieve deflections in the range between ± 5 degrees, measured as the angle made from the straight line joining the tip and the root of the adaptive trailing edge and the reference chord. A double comparison is carried out: experimental measures are compared with the CAD expectations and the FE numerical predictions. The error index is assumed as the mean square of the errors, evaluated at each of the reference points. This same error measure is the one that the devoted control system uses to drive the structure: its logic assumes that the optimal position is achieved as the error attains its minimum (it is stationary around that point). Examples of the attained results are shown in Fig. 11 [19]. Control system stability is checked by verifying its behavior after impulsive excitations (hammer hits, occurring before, during, and after morphing deployment).

Another multidimensional index is implemented, using the actual shapes of the adaptive trailing edge, reached after actuation. Both loaded and unloaded conditions are investigated. In detail, the strain maps are off-line processed to somehow reconstruct the attained morphed shape. This vector is compared with numerical results. The resulting deviation is then elaborated as a new map and then condensed as another distributed fault index. Also in this case, the results are satisfactory.

3.4.3 Damping levels

Even though the system is designed to work statically, dynamic information is essential for facing aeroelastic instability problems, for instance. The structure is a metal kinematic architecture so that its behavior is well known. The presence of foam skin may produce important and unfamiliar contributions. For these and other reasons, four different dummy test articles are designed, manufactured and tested, conceived as a quasirigid, span-wise limited reproductions of the final prototype. The structural skeleton, common to the specimens, is made of steel to reduce any structural elastic coupling effect with the morphing skin during static and dynamic tests. The different test architectures are respectively named as [22]:

- Dummy 1—Skin over dummy—including the morphing skin.
- Dummy 2—Actuators over dummy—including the servo-rotary actuator systems.
- Dummy 3—Control system over dummy—including the sensor and control systems.
- Dummy 4—Control skinned dummy system—as mentioned, plus the morphing skin.

Table 2 Max Von Mises Stresses and Safety Margins at Limit and Ultimate Loads

	Limit Load		Ultimate Load		Stress Map: the max stress values are found within the hinges (Ribs and Rods) or at the attachment points (Spar)
Part	$\sigma_{max,L}$	SM_L	$\sigma_{max,UL}$	SM_{UL}	
Rib	262.5	0.26	394.4	0.14	
Rod	448.6	1.56	672.9	0.82	
Spar	268	0.23	401.9	0.11	

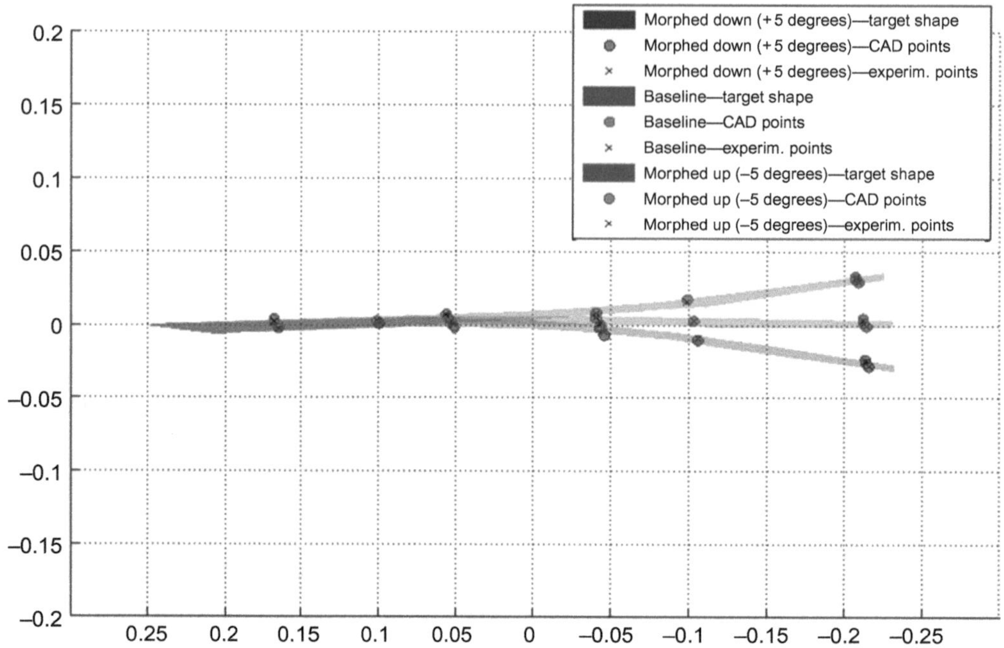

FIG. 11

Actual vs. reference shape.

In the following discussion, Dummy 2 results are not analyzed because they can be seen from the specific point of view of the evaluation of the system capability to dissipate energy, as a repetition of what Dummy 3 attained. In order to estimate the damping effectiveness of the morphing skin, absolutely beneficial for the flutter behavior of the investigated object, modal analyses are carried out on the 1st, the 2nd, and the 4th dummy prototypes. Results are synthetically reported in Table 3.

Dummy 3 and 4 are tested on the same test rig in the same location, while Dummy 1 is tested in a different lab. The way the articles are manufactured, allows accessing diverse dissipation contributions: it is expected that the first would give hints on the pure skin dynamics; the second on the structure response (both skeleton and actuator system); the third on the whole integrated system behavior. Though the boundary conditions are not the same for all, some general conclusions can be derived. Before of all, different damping contributions do not seem additive and the phenomenon therefore shows nonlinear characteristics. The high values of damping seem to indicate that the structural kinematic system (other from the actuation kinematic system) has a large capability of blowing energy, also linked to its complexity (large number of mechanical degrees of freedom). This aspect is common to the three realizations and does not affect global results; therefore, it could be stated that the experiments moved from a certain "damping threshold."

Within the aforementioned similarity limits, both skin and actuator systems seem to give significant increments of damping levels. In both the first two cases, results approach two percentage units against the classical unit for typical structural configurations. While skin provides a clear contribution in terms of viscous damping, the kinematic system brings a further complexity that is directly connected to the introduction of added devices (and then, parts and joints). The most relevant phenomenon is associated

Table 3 Dummy 1, 3, and 4; Modal Parameters

Boundary Condition	Mode Type	Frequency (Hz)	Damping (%)
Dummy 1			
Free-Free	–	–	–
	Torsion	59.0	0.8
	Rib antiphase bending	80.0	0.3
	Rib in-phase bending	125.6	1.8
Clamped	Kinematic Mode	35.0	3.0
	Torsion	65.0	2.9
	Rib antiphase bending	145.0	3.5
	Rib in-phase bending	214.0	2.3
Dummy 3			
Free-Free	Kinematic	18.0	–
	Torsion	32.2	1.6
	Transversal bending	59.0	1.5
	Tip oscillation	80.0	2.7
Clamped	Longitudinal bending	18.9	–
	Torsion	34.2	4.5
	Torsion, 2	48.1	1.3
	Torsion, 3	54.2	1.5
Dummy 4			
Free-Free	Long. bending	–	–
	Torsion	49.4	9.8
	Torsion, 2	95.3	4.6
	Torsion, 3	104.6	3.7
Clamped	Long. bending	–	–
	Torsion	43.5	7.4
	Torsion, 2	59.8	3.8
	Torsion, 3	65.1	3.5

with the inclusion of the skin over the active structure. It should be noted that this is the most massive configuration and some effects may also be linked to this aspect. In any case, the major increase occurs in the passage from the actuated to the complete system, with damping values moving from two/three to almost ten percentage units. This can be a clear advantage for real applications because it ensures higher aeroelastic stability, well over predicted limits (regulations conservatively consider 1% damping).

3.4.4 Flutter analysis

Flutter analyses considers the complete adaptive trailing edge, installed overall the target wing section, to assure wind tunnel test safety. Moving from the investigated device model, the following steps are run. The FE model, previously used for static analyses, is condensed into a dynamic equivalent one, characterized by reduced DOF. Dynamic condensation at certain grids is based on the generation of direct input

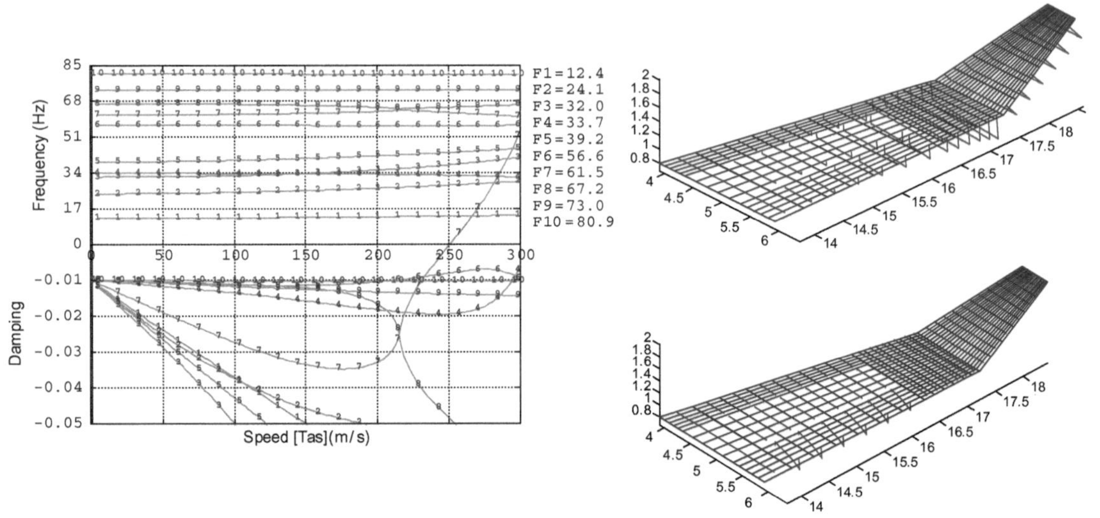

FIG. 12

Flutter V_g plot (left) and some relevant modes (right): bending and trailing edge elastic deflection.

matrices, able to capture both inertial and stiffness properties of the device. Reduction grids are selected to get an optimal mode shape resolution. Afterwards, the condensed dynamic model is assembled onto the FEM of the complete test article, allowing the computation of the global system modes while also accounting for the flexibility of the test rig, joining the test article to the wind tunnel floor.

A doublet lattice method is adopted for evaluating the unsteady aerodynamic coefficients; for such a purpose, a suitable 3D panelling is implemented. Surface splines are used to interpolate modal displacements along the centers of the aerodynamic panels. Generalized aerodynamic forces are evaluated regarding the airflow conditions that are expected during operation. A PK method is adopted to investigate the occurrence of flutter in the speed range up to three times V_{max}, the max working speed (100 m/s) of the reference wind tunnel. Results demonstrates that no flutter can be expected up to 240 m/s. The first instability is detected at 250 m/s, due to the coupling between the wing bending and adaptive trailing edge elastic deflection. In Fig. 12, a short summary of this result is shown, with specific reference to the modal shapes and the damping vs. flow velocity diagram.

4 SAFETY AND RELIABILITY ASPECTS

4.1 GENERALITIES

Although electromechanical (EMA) are generally slower, weaker, and less reliable than hydraulic actuators for actuation of aircraft aerodynamic systems, digital servo controllers have made them a viable solution for secondary surface control. Undoubtedly, the possibility of avoiding complex and heavy feeding of oil circuitry and the opportunity of integrating these devices into an all-electric architecture, are clear points of strength. Installed hydraulic motors driving primary control surfaces may be used on current planes in parallel to mitigate the hazard, associated to the installation of these “innovative” elements. A light-weight and compact actuation system is realized on the adaptive trailing edge, based on a lever mechanism driven by EMA, whose details have been presented and discussed in the previous paragraphs.

4.2 DISTRIBUTED ACTUATION

The implemented actuation arrangement is unshafted. In other words, each kinematic system is independently moved by a couple of actuators that are directly attached on it. Because each driving device is sized for the torque that acts on a single rib, it is much smaller than an equivalent conventional hydraulic actuator working on the complete control surface. A major advantage of such an architecture is the possibility to move individual ribs in a synchronous or independent way, to achieve some twist angles (in the reference layout, this possibility is limited to ± 0.5 degrees per bay). Nevertheless, such a configuration has a potential to fullfil airworthiness requirements, essentially because of its large redundancy and the fact that it works on a minor-critical surface like the trailing edge tip.

4.3 THE ATED FUNCTION

Starting from the aircraft, the morphing functions may be mapped and linked. This approach permits assessing functional failures and associated effects. In detail, the main adaptive trailing edge function can be classified as a "continuous and quasi-static wing trailing edge shape optimization control," included in the standard sections "2.—Plan, Generate and Control A/C movement, Sub 2.2.5.1—Control Lift and Drag." [26]. This function tries to combine together the morphing targets for a "Drag minimization" and a "Lift adaptation." Along this view, the *L/D* ratio seems a good parameter to track performance variations.

4.4 FAULT HAZARD ASSESSMENT

The fault hazard assessment (FHA) is a qualitative stage for evaluating the actual applicability of any device on-board a generic aircraft. With specific reference to the herein presented adaptive trailing edge, the following considerations hold:

- A partial damage of an adaptive rib would require the concurrent failure of both actuators, the single motor able to drive the other, in case of its breakdown.
- The total damage of the adaptive trailing edge would require concurrent failure of all the adaptive ribs.
- Symmetric or nonsymmetric jam is not catastrophic, because the system is not essential (like spoilers, ailerons, and flap, for instance) but a facilitator for the flight.
- The control system may implement filters that identify anomalies in the sensor network and, consequently, may shut the entire system down.
- In case of failure (different from jamming) the actuator can be locked or reverted to "damping mode" (switched off), both safe for the flight.

Table 4 Fault Hazard Assessment of the ATED

Failure Scenario	Failure Condition	Title	A/C Flight Phases	Safety Requirement	
				Severity	Objective
Loss of wing shape opt. function	*FC01* Loss of ATED control	Loss of ATED function	*All*	MIN-MAJ	1E-3/FH 1E-5/FH
Erroneous provision of wing shape opt. function	*FC02* Erratic ATED control	Erratic ATED function	*All s*	MAJ	1E-5/FH
Inadvertent provision of wing shape opt. function	*FC03* Uncommanded ATED control	Uncommanded ATED function	*All*	MIN-MAJ	1E-3/FH 1E-5/FH
Partial loss wing shape ctrl capabil. (symmetric)	*FC04* Symmetric partial loss of ATED function	Symm. partial loss of ATED function	*All*	MIN-MAJ	1E-3/FH 1E-5/FH
Asymm. partial loss of wing shape ctrl. capabil.	*FC05* Asymm. partial loss of ATED function	Asymm. partial loss of ATED function	*All s*	MAJ	1E-5/FH
Partial loss of wing shape ctrl capabil. (asymmetric) associated with one engine loss at land.	*FC06* Partial loss of ATED (asymm.) combined with one engine loss at land	Loss of ATED control on one wing (asymm.) w one engine loss at land	*Landing*	MAJ-HAZ	1E-5/FH 1E-7/FH
Partial loss of wing shape ctrl capab. asymm. associated w strong cross-wind at TO/L	*FC07* Partial loss of ATED (asymm.) comb. w strong cross wind at TO/L	Loss of ATED control on one wing (asymm.) w strong cross-wind at TO/L	*Landing*	MAJ-HAZ	1E-5/FH 1E-7/FH

Before the phases of design, development and operation, FHA analysis identifies seven failure scenarios that allows addressing safety requirements at a subsystem level, expressed as fault rate per hour, Table 4. Malfunctions are classified as per the potential causes, the relative severity and the related viable mitigation plans.

4.5 FUNCTIONAL HAZARD ASSESSMENT

FHA involves the production of several so-called "fault trees" that permit evaluating global system failure rates by considering each failure mode contribution to its functionality and reliability. Several indexes, such as actuator position monitor, motor shaft angle sensors, and absorbed power are considered to reveal breakdowns. The construction of the Fault Tree is based on relationships between events and causes, represented by means of logical AND/OR gates, which resolve the system complexity and highlight its potential crashes. Probability of such seizures result from the analysis and integration of failure probability of each basic component. For instance, jamming is a typical failure condition (FC) for electromechanical actuators. Typically, it can lead to aircraft failures that are ranked between hazardous and catastrophic for primary control surfaces. For the ATED application, the asymmetric failure that can be caused by an asymmetrical jamming is classified as "major." This means that the associated impact on wing loads may be reduced by acting on conventional aircraft control surfaces. However, although minimal impacts are expected on A/C controllability, the pilot workload may increase to trim the aircraft, both longitudinally and laterally. Block fuel margins should also be considered to compensate the consequent reduced aircraft efficiency.

5 DISCUSSION: IMPLEMENTATION ON REAL AIRCRAFT

5.1 SYSTEM DEVELOPMENT

The technology herein discussed is developed targeting a real aircraft. Its specifications are derived from realistic load computations and assembly conditions. Its configuration is however defined with respect to an isolated clean wing, without command and control surfaces. The whole aircraft should instead be considered to properly assess the deriving polars and the consequent *L/D* estimations. Otherwise, the presence of the fuselage, horizontal tail-plane, and other structural parts and forms, are neglected. Dramatic variations are not expected from the point of view of the design, because these additional inputs cannot lead to drastic variations of the absolute value or the distribution of the resultant loads. The same applies to algorithm logic or sensor installation. From the point of view of the integration into a real wing, furthermore, an overall simplification should be even derived, because of the larger available room available around the wing root.

Some further considerations should be discussed regarding the control algorithm. The experimental system, both in its feed-back and feed-forward versions, is in fact based on a slow control logic that compared measured strains with target ones, derived from target shapes. The different geometries are directly associated to achieved deflections and fuel resulting performance is extrapolated from the nominal performance of the engines. This approach can be improved by implementing at least three further steps. The real shape may be reconstructed from available information and compared with the reference shape, either analytical or numerical. The actual performance may be estimated by inserting the actual morphed geometries into an aerodynamic code that would allow predicting real benefits and provide information to adjust the control effort. Finally, because fuel consumption is addressed, control actions should directly be linked to the instantaneous fuel consumption.

Along with these considerations, the passage from a mock-up to a real aircraft should at least consider these upgrades:

- The design should refer to a movable surface (flap) instead of a block-wing, that would also mean a different approach for the system installation, including cable routing, for instance.
- CFD computation of the real *L/D*, following the reconstruction of the achieved shapes, and a consequent better estimate of the attainable benefits.
- Modulation of the load, computed on a whole aircraft system.
- *Re*-design of the installation, in principle easier for larger available space.
- Use of the instantaneous fuel consumption rate as a reference variable for real-time controllers.

5.2 OPERATIONAL ASPECTS

The proposed morphing device is a combination of classical and new technologies. There are well-assessed components (sensor, control, actuator, and structural systems) combined with innovative products, like the skin. While classical components are not generating complex issues on maintenance and reliability (classical kinematic chain models or sensor networks), the skin should undergo a long verification period to fully characterize its behavior, above all with respect to fatigue and aging. Following these considerations, a reliable maintenance plan should be defined to properly assess the fly-ability of these devices.

Another issue concerns the use of a large or small number of motors. Without going into detail of controllability, it is clear that the deployment of many devices discloses many possibilities, like implementation of differential actuation for twisting purposes, while a reduced set confines its function to a uniform camber adaptation. Another point is that redundancy is guaranteed by the presence of several motors: it increases the global survivability of the system even if some motors break, but leads to shorter maintenance periods. FHA is the conclusive step of this path. Summing up all these terms, the final cost of the device is derived, a good parameter to evaluate the applicability to real aircraft. So, from the operational point of view, at least these main objectives shall be achieved:

- Maturation of morphing skin technology.
- Preliminary aging and fatigue characterization of the overall system.
- Definition of suitable maintenance plans.
- Detailed computation of costs and benefits.

5.3 AEROELASTIC ISSUES

As a point, aeroelastic studies are expected to have relevance on the general design. In fact, the deformability of the system says that it has more degrees of freedom than a standard one and a higher modal density is expected, in turn arising a complex coupling among different modes. Furthermore, because the camber deformation devices are distributed, numerical reduction processes to get suitable models for the relative analyses, should be properly adjusted. In the end, this requires an update of the assessed, almost standardized, processes and codes.

6 CONCLUSIONS AND FUTURE DEVELOPMENTS

This chapter deals with the development and the implementation of a morphing trailing edge for cruise performance enhancement. The adaptive system modifies the global wing curvature to compensate weight variation of the aircraft during cruise, because of fuel consumption. In the same way, it can adapt the wing camber as a function of real take-off weight, depending on the passenger number and luggage (or the boarded goods for cargo). The complex device is made of integrated actuators and sensors and is driven by a control architecture, using well-assessed technology: the proper assembly of standard commercial components allows achieving an innovative, smart architecture. This choice permits bypassing many complications connected to the realization and, in perspective, the certification of a novel element, dealing with reliable and well-known parts.

Nevertheless, issues to be solved remain, because the wing acquires new capabilities. These problems are mainly related to spec assessment, architecture definition, installation aspects, and implementation strategy. The specifications should consider a complete aircraft, so to predict the overall effect of the proposed device on the airplane aerodynamic polar. The morphing system layout should also derive from the global geometry, so to assess better engineering deployment along the wing span and along its chord. In the operative case, the adaptive trailing edge should be installed along the flap zone, then integrating a complex kinematic system onto a movable surface. This is a clear complication of the original arrangement. On the other hand, this configuration offers more available room to host the additional component than the wing tip zone. Indeed, studies to verify the possibility of inserting such systems even in the aileron segment are currently being performed by many teams, worldwide (2016). Finally, instead of extracting the control reference parameters by complex extrapolations performed on the reconstructed geometry, adjustments should instead refer to data directly linked to the final objective as, in this case, the fuel consumption rate.

Some further aspects can be improved in the implemented architecture without extensive changes in the approach. First, the actuator can be integrated within the structural system design for realizing a unique active structural load-bearing architecture, instead of merging two separate components that are independently developed. As a second point, it should be recalled that the sensor system is constructed as an extended network, made of FO FBG (Fiber Bragg's Gratings). This choice drastically reduces the number of wires. However, real implementation shows that cabling remains an issue. This aspect can be improved by the use of wireless systems. Of course, complications arise also in this case, for instance linked to the need of defining a suitable inner architecture, compliant with an adequate signal transmission path to the feeding sensors and compatible with the existing systems. Finally, the control device should consider the whole status of the reference aircraft and provide information on its condition; therefore, it should be included in the overall avionics.

ACKNOWLEDGMENTS

The authors want to thank O. Schorsch, C. Nagel, A. Lühring and M. Wolf at Fraunhofer IFAM for the morphing skin design; M. Schueller and A. Gratias for the outstanding support for the control system design of ATED; and G. Diodati at CIRA and D. Flauto at the University of Palermo for their valuable support for the design of the ATED actuation system.

REFERENCES

[1] J. Browman, B. Sanders, T. Weisshaar, Evaluating the impact of morphing technologies on aircraft performance, Proceedings of the Forty-Third AIAA Conference on Structures, Structural Dynamics and Materials, AIAA Paper 2002-1631, April, 2002.

[2] S. Vasista, L. Tong, K.C. Wong, Realization of morphing wings: a multidisciplinary challenge, J. Aircraft 49 (2012) 11–28.

[3] H. Hilibig, H. Wagner, in: Variable wing camber control for civil transport aircraft, ICAS Proceedings, ICAS-84-5.2.1, Toulouse, 1984, pp. 107–112..

[4] R.F. Brissenden, A.R. Heath, D.W. Conner, M.L. Spearman (technical monitors), the Boeing Preliminary Design Department, Assessment of variable camber for application to transport aircraft, NASA Contractor

Report CR-158930, National Aeronautics and Space Administration Langley Research Center, Hampton, VA, October, 1980.

[5] H.S. Wolko (Ed.), The Wright Flyer: An Engineering Perspective, The Smithsonian Institution Press for the National Air and Space Museum, Washington, DC, ISBN 0-87474-979-4, 1987.

[6] L. Kennet (Ed.), Military aviation, reprint of the Clement Ader's work, "L'Aviation Militaire", originally published by Berger-Levrault, Paris (F), 1909, Air University Press, Maxwell Air Force Base, AL, 2003, ISBN 978-1-58566-118-3.

[7] C. Ader, Appareille Ailè pour la Navigation Aerienne, France Patent N.205155, Priority Date, Apr. 19th, 1890; Publication Date, Aug. 11th, 1890.

[8] A.A. Holle, Plane and the like for aeroplanes, Germany Patent N.DE366693C, Priority Date, Oct. 15th, 1915; Publication Date, Jan. 10th, 1923.

[9] A.A. Holle, Plane and the like for aeroplanes, United States Patent N.1225711, Priority Date, Dec. 22nd, 1916; Publication Date, May 8th, 1917.

[10] A.R. McGowan, L.G. Horta, J.S. Harrison, D.L. Raney, in: Research activities within NASA's morphing program, Proceedings of the RTO AVT Specialists' Meeting on Structural Aspects of Flexible Aircraft Control, Ottawa, October, 1999.

[11] D.A. Perkins, J.L. Reed, E. Havens, Morphing wing structures for loitering air vehicles, Proceedings of the Forty-Fifth AIAA Conference on Structures, Structural Dynamics and Materials, AIAA Paper 2004-1888, April, 2004.

[12] Thill C., Etches J., Bond I., Potter K., Weaver P., "Morphing skins", Aeronaut. J., Vol. 112, N. 1129, pp. 117–139, 2008.

[13] R. Pecora, M. Magnifico, F. Amoroso, L. Lecce, M. Bellucci, I. Dimino, A. Concilio, M. Ciminello, Structural design of an adaptive wing trailing edge for large aeroplanes, in: P.C. Wölcken, M. Papadopoulos (Eds.), Smart Intelligent Aircraft Structures (SARISTU), Proceedings of the Final Project Conference, Springer International Publishing, Cham, Switzerland, ISBN 978-3-319-22412-1, 2016.

[14] A.P. Milojević, N.D. Pavlović, Development of a new adaptive shape morphing compliant structure with embedded actuators, J. Intell. Mater. Syst. Struct. 27 (10) (2016) 1306–1328 DOI: 10.1177/1045389X15590270.

[15] K. Zhao, J.P. Schmiedeler, A.P. Murray, Design of planar, shape-changing rigid-body mechanisms for morphing aircraft wings, J. Mech. Robot. 4 (4) (2012) 041007.

[16] I. Dimino, D. Flauto, G. Diodati, A. Concilio, R. Pecora, Actuation system design for a morphing wing trailing edge, J. Recent Pat. Mech. Eng. 7 (2014) 138–148.

[17] I. Dimino, M. Ciminello, A. Concilio, A. Gratias, M. Schueller, R. Pecora, in: Control system design for a morphing wing trailing edge, 7th ECCOMAS Conference on SMART Structures and Materials, SMART 2015, Azores, 3–6 June, 2015.

[18] Ciminello M., Concilio A., Ameduri S., Flauto D., Mennella F., "Hinge rotation of a morphing rib using FBG strain sensors", J. Smart Struct. Sys. Vol. 15, N.6, pp. 1393–1410, 2015.

[19] I. Dimino, M. Ciminello, A. Concilio, R. Pecora, F. Amoroso, M. Magnifico, M. Schueller, A. Gratias, A. Volovick, L. Zivan, Distributed actuation and control of a morphing wing trailing edge, in: P. C. Wölcken, M. Papadopoulos (Eds.), Smart Intelligent Aircraft Structures (SARISTU), Proceedings of the Final Project Conference, Springer International Publishing, Cham, Switzerland, ISBN 978-3-319-22412-1, 2016.

[20] Y. Essa, F.M. de la Escalera Cutillas, I. Dimino, M. Ciminello, A. Concilio, Manufacturing and testing of smart morphing SARISTU trailing edge, in: P.C. Woelcken, M. Papadopoulos (Eds.), Smart Intelligent Aircraft Structures (SARISTU), Proceedings of the Final Project Conference, Springer International Publishing, Cham, ISBN 978-3-319-22413-8, 2016.

[21] M.H. Tooley, D. Wyatt, Aircraft electrical and electronic systems: principles, operation and maintenance, in: Aircraft Engineering Principles and Practice, Butterworth-Heinemann, Burlington, MA, 2009. 978-0-750-68695-2.

[22] P.C. Woelcken, M. Papadopoulos (Eds.), Smart Intelligent Aircraft Structures (SARISTU), Proceedings of the Final Project Conference, Springer International Publishing, Cham, ISBN 978-3-319-22413-8, 2016.

[23] O. Schorsch, A. Lühring, C. Nagel, Elastomer-based skin for seamless morphing of adaptive wings, in: P. C. Wölcken, M. Papadopoulos (Eds.), Smart Intelligent Aircraft Structures (SARISTU)—Proceedings of the Final Project Conference, Springer International Publishing, Switzerland, 2016. ISBN 978-3-319-22412-1.

[24] M. Bosboom, M. van Wijngaarden, R. Evenblij, P. Bettini, T. Loutas, V. Kostopoulos, D. Habas, M. Tur, N. Gorbatov, A. Bergman, U.B. Simon, I. Kressel, C. Koimtzoglou, M. Ciminello, A. Weisser, C. Paget, Ribbon tapes, shape sensors, and hardware, in: P.C. Wölcken, M. Papadopoulos (Eds.), Smart Intelligent Aircraft Structures (SARISTU)—Proceedings of the Final Project Conference, Springer International Publishing, Switzerland, 2016. ISBN 978-3-319-22412-1.

[25] R. Evenblij, F. Kong, C. Koimtzoglou, M. Ciminello, I. Dimino, A. Concilio, Shape sensing for morphing structures using fiber bragg grating technology, in: P.C. Woelcken, M. Papadopoulos (Eds.), Smart Intelligent Aircraft Structures (SARISTU)—Proceedings of the Final Project Conference, Springer International Publishing, Cham, ISBN 978-3-319-22413-8, 2016.

[26] I. Dimino, A. Concilio, R. Pecora, Safety and reliability aspects of an adaptive trailing edge device (ATED), Proceedings of the 24th AIAA/AHS Adaptive Structures Conference, AIAA SciTech Forum (AIAA 2016-1318), https://doi.org/10.2514/6.2016-1318.

FURTHER READING

[1] J. Spillman, The use of variable camber to reduce drag, weight and costs of transport aircraft, Aeronaut. J. 96 (1992) 1–8.

[2] H.P. Monner, D. Sachau, E. Brietbach, in: Design aspects of the elastic trailing edge for an adaptive wing, Proceedings of the RTO AVT Specialists' Meeting on Structural Aspects of Flexible Aircraft Control, Ottawa, October, 1999.

[3] R. Pecora, F. Amoroso, M. Magnifico, I. Dimino, A. Concilio, KRISTINA: kinematic rib-based structural system for innovative adaptive trailing edge, Proceedings of SPIE—The International Society for Optical Engineering, vol. 9801, 2016.

[4] Barbarino S., Pecora R., Lecce L., Concilio A., Ameduri S., Calvi E. A., "A novel SMA-based concept for airfoil structural morphing", J. Mater. Eng. Perform. Vol. 8, N.5, pp. 696–705, 2009.

CHAPTER

MORPHING AILERON

18

Gianluca Amendola*, Ignazio Dimino*, Antonio Concilio*, Rosario Pecora†, Francesco Amoroso†, Maurizio Arena†

*The Italian Aerospace Research Centre, CIRA SCpA, Capua (CE), Italy**

University of Naples "Federico II", Naples, Italy†

CHAPTER OUTLINE

NOMENCLATURE

α	wing angle of attack
A/C	aircraft
AOA	angle of attack
β	actuator shaft rotation
CBEAM	beam element [1]
CD	drag coefficient
CL	lift coefficient
CNC	computer numerical control
COTS	commercial of the shelf
Cp	pressure coefficient
CTETRA	Tetrahedral element [1]

Morphing Wing Technologies. https://doi.org/10.1016/B978-0-08-100964-2.00018-6

d.o.f	degree of freedom
δ	rigid aileron rotation
FE	finite element
φ	actuation beam deflection angle
GVT	ground vibration tests
γ	morphing deflection angle
I.C.	instantaneous center of rotation
L	distance from actuator shaft to kinematic rotation center
MA	mechanical advantage
M_{att}	actuator torque
MPC	master point constraint [1]
$M_{rib\#3}$	applied external moment on the rib plate 3
N	number of actuators
PWM	pulse width modulation
R	crank length
RBE2	rigid body element type 2 [1]
R_i	reaction moment oriented as i axis
T/A	test article
T_i	reaction force oriented as i axis
TRL	technology readiness level
ϑ	morphing aileron angle in wind tunnel
UAV	unmanned aerial vehicle
VM	von Mises
VLM	vortex lattice method

1 INTRODUCTION

The term "morphing" is often used to describe any unconventional method of vehicle shape modifications [2]. There are numerous military aircraft that demonstrate shape changes such as the variable-sweep wings on the F-14. On the other hand, in civil transport, the use of flaps, slats, and spoilers constitute an attempt to realize an adaptive wing in order to address multiple flight condition such as take-off and landing. In order to design flying structures with an advanced level of adaptability, many engineers took inspiration from nature [3]. The wings of all flying creatures are, in fact, multi-taskers, which means that they adapt simultaneously to the environment [4]. On the other hand, conventional wings work efficiently around their design point while they are poorly efficient in other flight phases since their shape is not largely modificable to equally accommodate the requirements demanded by multiple (off-design) flight conditions [5] This indicates that aircraft wings are only seldom optimal with performance penalties in off-design conditions [6,7]. The most prominent way to realize wing morphing is to develop an adaptive trailing edge that can modulate the local airfoil camber —for instance—to follow the aircraft weight variations [8–10]. The design challenge is then to design according to loads-carrying and large deformation capability. Apparently this seems a paradox in force of the conflicting design requirements. Compliant and rigid-body structures both represent different approaches to solve the paradox [11–14]. In the European scenario, in the last decades, important research projects were launched, supported by industrial partners. The main objective was to improve the technology readiness level (TRL) of morphing sysstems by means of full-scale wind tunnel demonstrators in order to reach the goal of flight tests in 2020. CIRA (the Italian Aerospace Research Centre) and other European partners have been deeply involved in projects such as Clean Sky [15] and SARISTU [16], which are the most relevant of the last decades. Their aim is to develop breakthrough technologies to significantly increase the performances of airplanes and air transport, fostering less noisy and more fuel-efficient aircraft [17,18]. In the wake of the very promising results obtained in the framework of the mentioned projects, it was considered extremely interesting to investigate on the applicability of morphing enabling technologies also to the aileron region. This is a very delicate zone, where aeroelastic phenomena may be very important following the reduced local structural stiffness and the complex aerodynamics. Furthermore, the aileron is a safety-critical primary control surface, and its failure is catastrophic for the entire aircraft. On the other side, this region exhibits higher performance for small deflections. The morphing aileron system is aimed at working in cruise to modify a limited chord segment to accommodate the aircraft weight variations following fuel consumption. However, during classical maneuvers, this morphing part remains rigid, and the aileron works in the usual manner [19]. The system is therefore made of two independent actuation systems: one devoted to roll maneuvers and the other devoted to the implementation of morphing. The former is a simple linear actuator that allows the aileron rigid rotation around its main hinge axis; the latter is a self-contained kinematic distributed along the aileron span and driven by rotary actuators. Load control constitutes another framework suited for this technology; in fact, with small deflections, both positive and negative, it is possible to redistribute the lift along the span in order to reduce the root bending moment. The morphing aileron has been developed in the framework of the CRIAQ (Consortium for Research and Innovation in Aerospace in Quebec) MD0-505 [20] project which promoted the cooperation between Canadian and Italian partners from the aeronautical research and industrial fields. The project goal was to test a full-scale wingtip model in a relevant environment. The test article (T/A) was designed in order to validate wing morphing capability from both aerodynamic and structural standpoints. The wing model could modify the wing upper skin (aerodynamic bump) with the aim to delay transition from laminar to turbulent flows, while the morphing aileron was activated to improve lift over drag ratio through trailing edge camber morphing. The present chapter is aimed then at describing all the phases relevant to the morphing aileron design and validation, starting from the theoretical approaches, to going through the definition and the elaboration of suitable numerical models, until the experimental campaign. Finally, an overview of the wind tunnel test set up is also provided, with a description of the main results in terms of pressure distributions and experimental aerodynamic data.

2 CONCEPTUAL APPROACH

A variable camber device is designed according to shapes and stiffness requirements. These requirements are very severe if the aileron region is the area under investigation. This zone plays a fundamental role for the aircraft roll control while is subjected to the external loads. Thus, during the preliminary design phase it is important to consider some specific critical aspects:

- The aileron constitutes a primary control surface which is safety critical. Its failure is a catastrophic event for the aircraft.
- The morphing capability is added to the conventional aileron, which remains free to rotate around its hinge axis.

- The aileron is a crucial item from the aeroelastic standpoint. Morphing will introduce normal modes and additional degrees of freedom to the conventional ones, potentially driving flutter instability.
- The wingtip region is characterized by very reduced space, leading to a difficult integration of actuator and kinematic.

These considerations led to the adoption of a rigid body mechanized structure in which each component cooperates to distribute the stress due to external loads and to move the architecture from the baseline shape to the target morphed shapes.

3 WORKING PRINCIPLE AND T/A ARCHITECTURE

The morphing aileron architecture was designed while referring to a typical transport regional aircraft application. The morphing aileron is mainly composed of:

- Five segmented ribs connected by means of rotational hinges positioned on the camber line, defining a kinematic chain assuring enough structural robustness and transmitting deformation. The rib is driven by a dedicated actuation system that, in conjunction with the skin, maintains the external shape.
- Spanwise stiffening elements such as spars and stringers in a multi-box arrangement.
- Three servo-rotary actuators.
- A segmented skin ("armadillo-like" configuration) with silicon gap fillers to avoid discontinuities between adjacent parts.

The morphing aileron design started from the definition of the aerodynamic target shapes that have to be matched. The morphing aileron hinge axis is located at 70% of the chord as in a real regional aircraft. Aerodynamic requirements and structural constraints drive the aileron outline, as shown in Fig. 1 where the baseline and the morphed-up and morphed-down shapes are depicted. Shapes must satisfy the following geometrical specifications:

- morphed configuration has been obtained assuming a parabolic shape for the airfoil camber line equivalent to a plain flap rigid rotation comprised between $-7°$ and $+7°$,[1]
- continuous monotonic curvature of the morphed camber lines (no changes in the sign of the slope of aileron camber), and
- no elongations in the morphing trailing edge chord.

The main structural components are ribs, spars and skin. The rib was considered segmented along the chord in a "finger-like [21]" layout properly tailored to enable camber morphing upon actuation. More in detail, each rib was assumed to be composed by three consecutive blocks as shown in Fig. 2, respectively called B0, B1, and B2. The rib blocks were connected to each other by means of rigid hinges positioned on the airfoil camber line, indicated with A and B. In order to design a 1 d.o.f. mechanism, nonconsecutive rib plates (B0 and B2) were connected by means of a link L that forced the rib segments to rotate according to a specific gear ratio function of the target shapes to be realized.

[1]Airfoil morphing is obtained by turning its camber line into a parabolic arch connecting the hing point at $x/c = 0.7$ to the tip point placed in the same position it would have if the aileron rigidly rotates by an angle φ around the hinge point. This angle is referred in the text as "equivalent deflection" or "morphing angle." In the specific case under investigation the range of φ was set to $[-7°:7°]$ where the negative sign stands for upwards deflection.

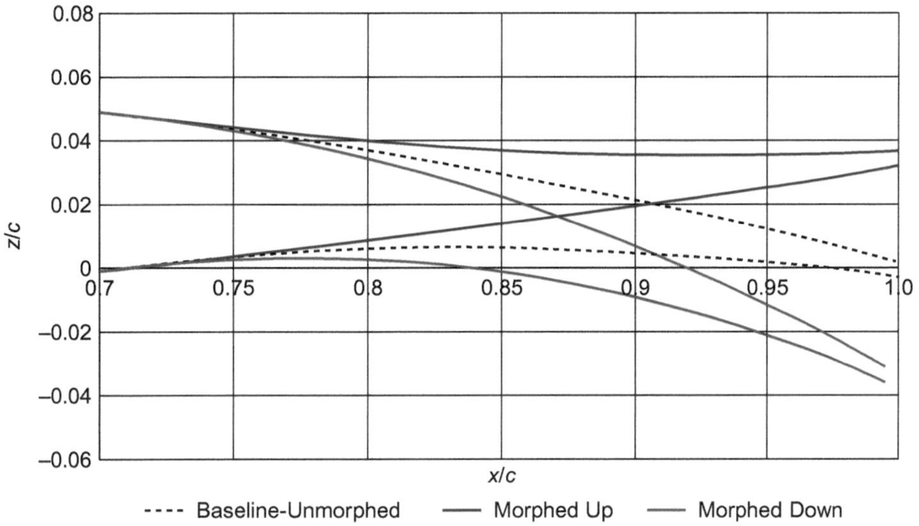

FIG. 1

Morphing aileron aerodynamic shapes.

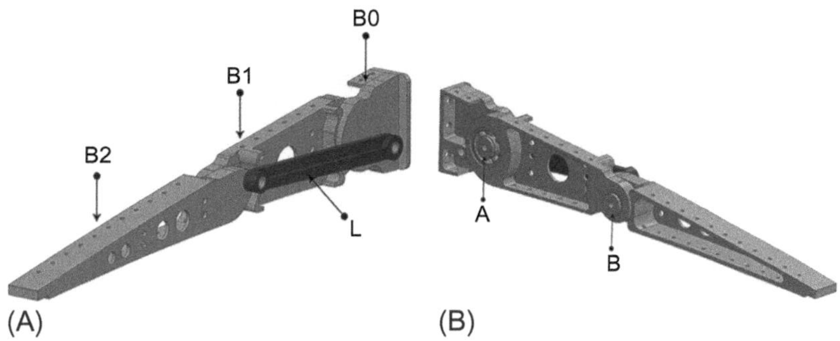

FIG. 2

Morphing rib architecture: (A) blocks and links and (B) hinges [22].

The rib can be then actuated at one block, and all the others follow the movement accordingly. The mechanism has been designed referring to a three-segment polygonal line in order to approximate the camber of the airfoil and to morph it into the desired configuration while the thickness distribution remains unchanged. The described rib architecture ensures the following advantages:

- It is easy to manufacture because it is made of components that can be easily manufactered by numerical control machining.
- All the hinges are realized with commercial items (bushing and bearings).
- The structure is easy to assemble.
- It allows reliable morphing.

The ribs' kinematic was transferred to the overall aileron structure by means of a *multi-box* arrangement as shown in Fig. 3.

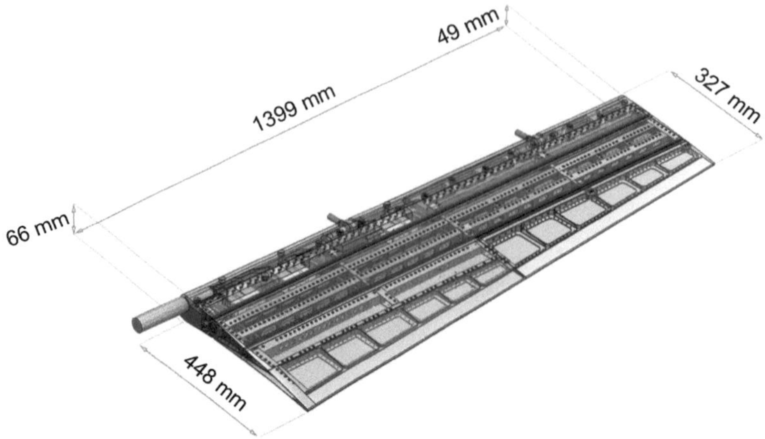

FIG. 3

Morphing aileron structure: multi-box arrangement [18].

Each box of the structural arrangement is characterized by a single-cell configuration delimited along the span by homologous blocks of consecutive ribs, and along the chord by longitudinal stiffening elements (spars and/or stringers). Upon the actuation of the ribs, all the boxes are put in movement, thus changing the external shape of the aileron; if the shape change of each rib is prevented by locking the actuation chain, the multi-box structure is elastically stable under the action of external aerodynamic loads. A four-bay (five-rib) layout was considered for an overall (true-scale) span of 1.5 m; AL2024-T351 alloy was used for spars, stringers and rib plates, while 17-4pH steel was used for ribs' links. Off-the-shelf airworthy components were properly selected for the bearing and bushings at the hinges and coupled to torsional springs to recover potential free-plays. A multi-module skin was considered in conformity to the multi-box segmentation; three aluminium-alloy panels were then adopted, each panel sliding over the consecutive one in an "armadillo-like" configuration. Airflow leakage at the skin segments interfaces was prevented through low-friction silicone seals. The wind tunnel T/A consisted of the morphing wing box developed by ETS (École de Technologie Supérieure Montreal) and the variable camber aileron designed by UNINA (University of Naples "Federico II") and CIRA (Fig. 4A); the system was dimensionally representative of a true-scale outer wing segment belonging to a typical regional A/C produced by Bombardier Aerospace. The wing box and the aileron were both equipped with internal and independent actuators enabling morphing capabilities, namely, generation of skin bumps for the wing box and camber variation for the aileron. Aileron rotation around its hinge axis was assured by a linear actuator connected to a stiff cylindrical support holding the entire T/A (Fig. 4B) and in turn constrained to the wind tunnel by means of bolts. Root and tip shafts (circles in Fig. 4) were positioned into interface holes made at wing root and wing tip sections. The linear actuator and the cylindrical support were both placed out of the test chamber and therefore not subjected to the aerodynamic flow.

Furthermore, the morphing aileron was also hinged to the main wing structure by means of two inner tie rods which were installed on the web of the wing rear spar (Fig. 5).

The aileron structure was sized according to the aerodynamic loads related to the most severe operative condition expected in the wind tunnel. The vortex lattice method (VLM) method was used to

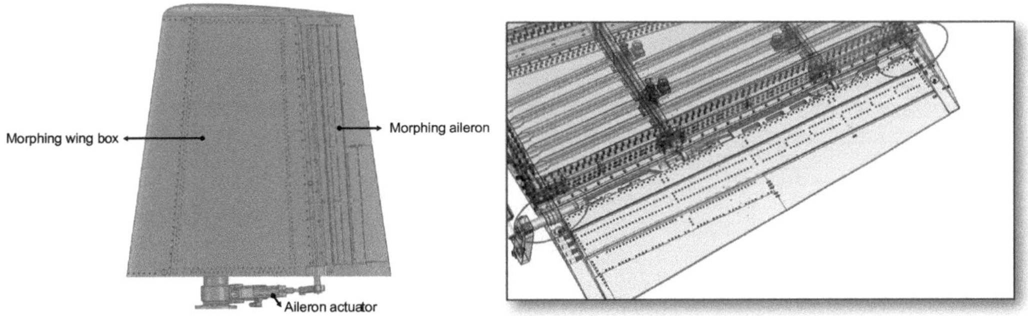

(A) (B)

FIG. 4

T/A Assembly (A) and shaft installation (B).

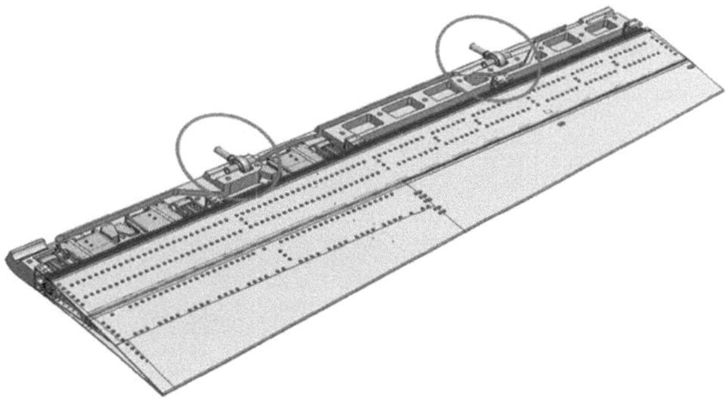

FIG. 5

Hinges that connect the morphing aileron to the model wing box.

evaluate experimental setting in terms of wing angle of attack, speed, and aileron equivalent deflection. The most severe conditions occurred in correspondence of 7° aileron equivalent morphing down deflection and at dynamic pressure $q = 4425\ \text{N/m}^2$. Spanwise pressure distributions along each segment of the aileron is represented in Fig. 6.

4 ACTUATION SYSTEM DESIGN

The actuation system of the morphing aileron had to be compact due to very restricted dimensions and stiff enough in order to withstand external loads, during morphing and while preserving a given aerodynamic shape. When the aileron is subjected to load, the aerodynamic pressure is transferred by the skin to the underlying ribs structure and then sustained by the actuators thanks to a kinematic chain. Loads and volumetric requirements then drive the design of the actuation system. Precisely, as shown in Fig. 7, only the three near-root ribs can be actuated directly, while the last two, located near the tip, are slaved, and their movement is passive. In the first design loop, a rotary actuator was selected. The Bental Industries RSA-06 model was chosen due to its light weight, compactness, and high actuation load capability [23].

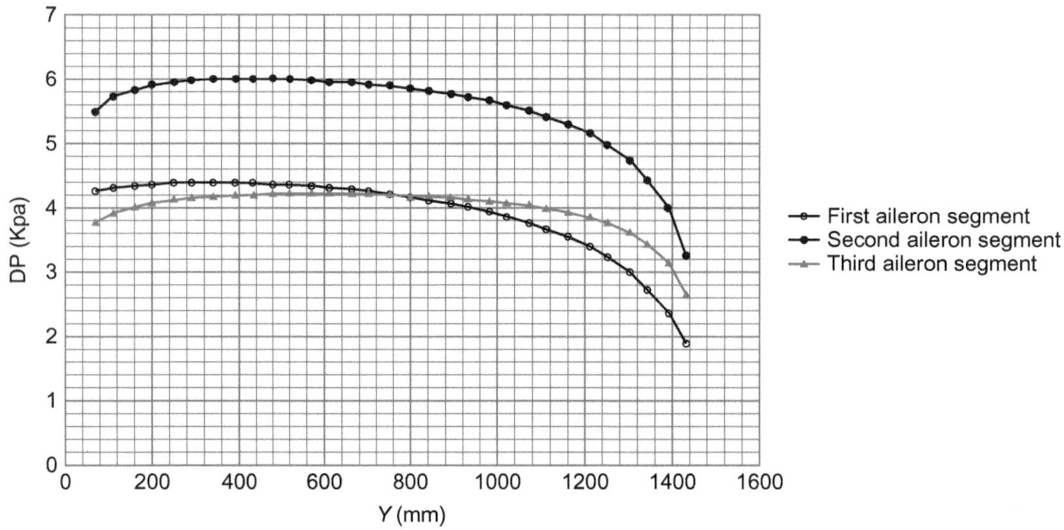

FIG. 6

Pressure distribution along aileron span [18].

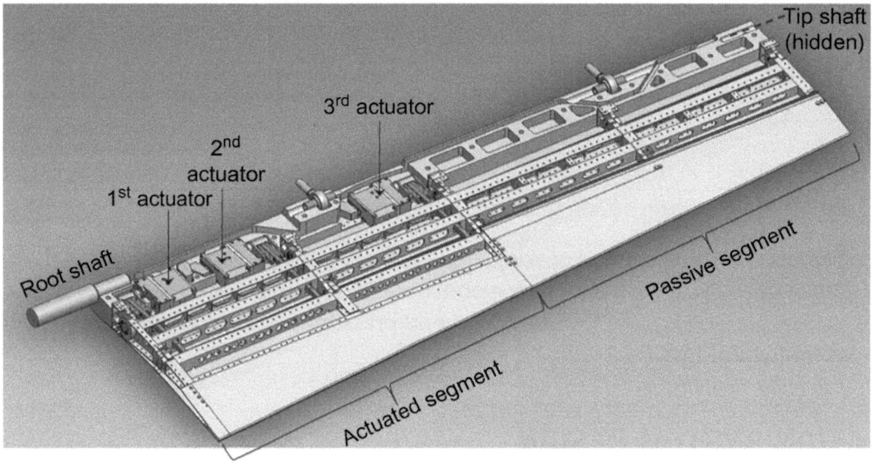

FIG. 7

Internal view of the aileron with actuated and passive segment highlight [21].

The analytical scheme of the actuation kinematic is represented in Fig. 8, and it is based on the oscillating glyph mechanism [24,25]. The actuator shaft is located in correspondence of the point O, while point V represents the instantaneous rotation center (I.C.) of the third chordwise segment of the rib.

The actuator torque M_{att} is transmitted to the aileron structure by means of a beam, indicated as B_L directly connected to the rib segment. The beam deflection φ is related to the crank rotation β which is transformed in a linear oscillating movement of a sliding component having a stroke B_R. Many devices can be used as sliding elements. Some of them are, for example, linear guides with recirculating balls,

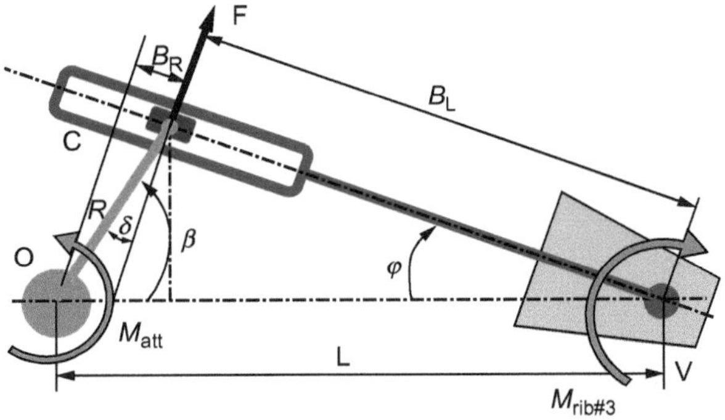

FIG. 8

Oscillating glyph connected to the third rib segment of the morphing aileron [26].

cam followers, cylindrical bearings, and also roller-based guides. The component selection was carried out on the base of its structural allowable solicitations. The force F is generated by the contact between the selected linear guide and the rail in which it slides. By connecting the actuator shaft to the crank hinge O and the beam to the third rib segment (B2), the actuation torque is transmitted first to the crank and then to the rib, which rotates around the hinge V in order to counterbalance the external moment $M_{\text{rib\#2}}$. The entire mechanism was designed with a certain mechanical advantage (MA); it is a torque amplification factor that depends on the main geometrical characteristics according to the following equation:

$$\text{MA} = \frac{\text{LOAD}}{\text{DRIVER}} = \frac{M_{\text{rib\#2}}}{M_{\text{att}}} = \frac{FB_{\text{L}}}{FB_{\text{R}}} = \frac{B_{\text{L}}}{B_{\text{R}}} \tag{1}$$

Greater B_{L} corresponds to greater mechanical advantage for the fixed value of the crank length R. It is then important to estimate accurately the position of point V, and then the actuation MA curves as a function of the aileron deflections. Fig. 9 shows the circular trajectories of the rib block B2 during the morphing. Each point, E, F, and G, rotates around the instantaneous rotation center V.

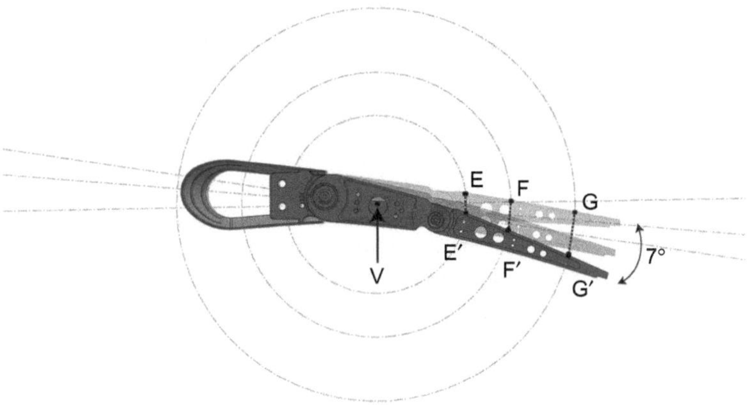

FIG. 9

Circular trajectories of sample points (E, F, and G) during morphing [21].

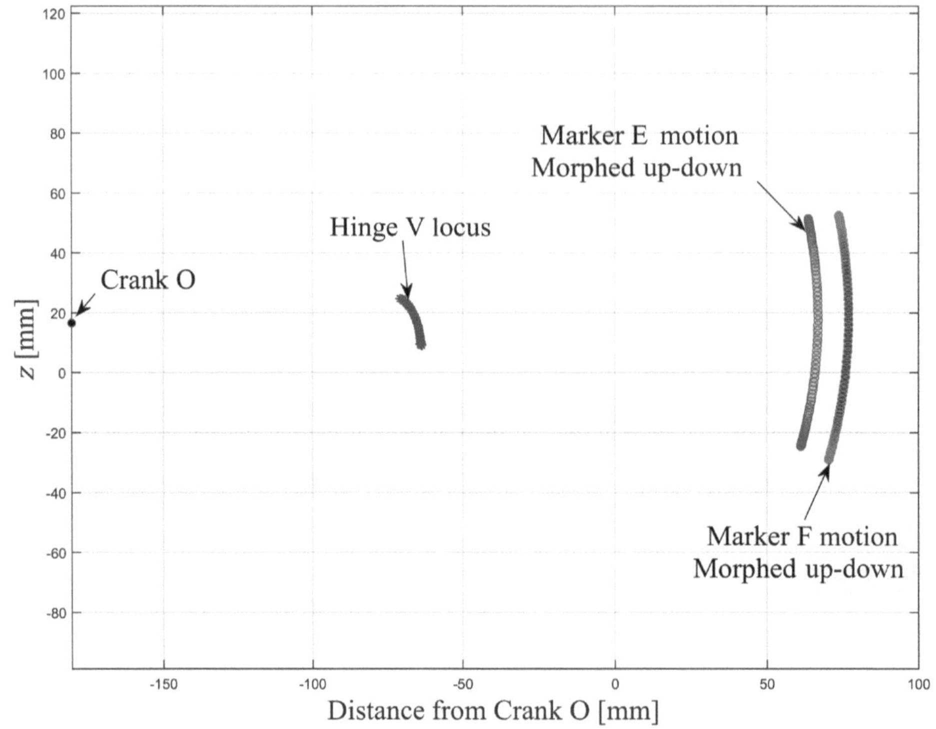

FIG. 10

Locus of the mechanism I.C. during the morphing excursion from morphed up to morphed down.

The position of V is constantly changing as B2 rotates. During the entire morphing excursion (from $-7°$ to $+7°$ equivalent rotation), it describes a locus of points. The position of the markers E and F in Fig. 9 during the morphing movement is reported in Fig. 10.

According to such considerations, MA has been calculated for different positions of the virtual hinge V, specifically for the positions corresponding to the equivalent rotations of $-7°$ and at $+7°$ (extreme morphed up and morphed down configurations). In morphed-down configuration, the hinge V is situated farther with respect to the crank position O, rather than in morphed-up configuration, thus changing the MA curve. The curves then translate upward as hinge V moves away from O, as shown in Fig. 11, resulting in a higher amplification factor.

The actuator torque with respect to the external moment and the V locus have been then computed and reported in Fig. 12. The torque as function of the morphing angle (φ) for both morphed up and down configurations is represented. It has been estimated by dividing the aerodynamic moment by MA. It is noticeable that the highest load occurs in correspondence of aileron downward deflection, thus making it the most severe condition for actuator sizing. The maximum actuator torque is however below 13 Nm, which is the allowable value prescribed by the producer [23].

The actuation beam represents the most important part of the mechanism because it is subjected to the simultaneous action of force F and the external moment $M_{\text{rib\#2}}$, producing bending stress to be taken into account in the sizing procedure and numerical verification. Another important kinematic equation that can be derived from the analytical scheme is the following:

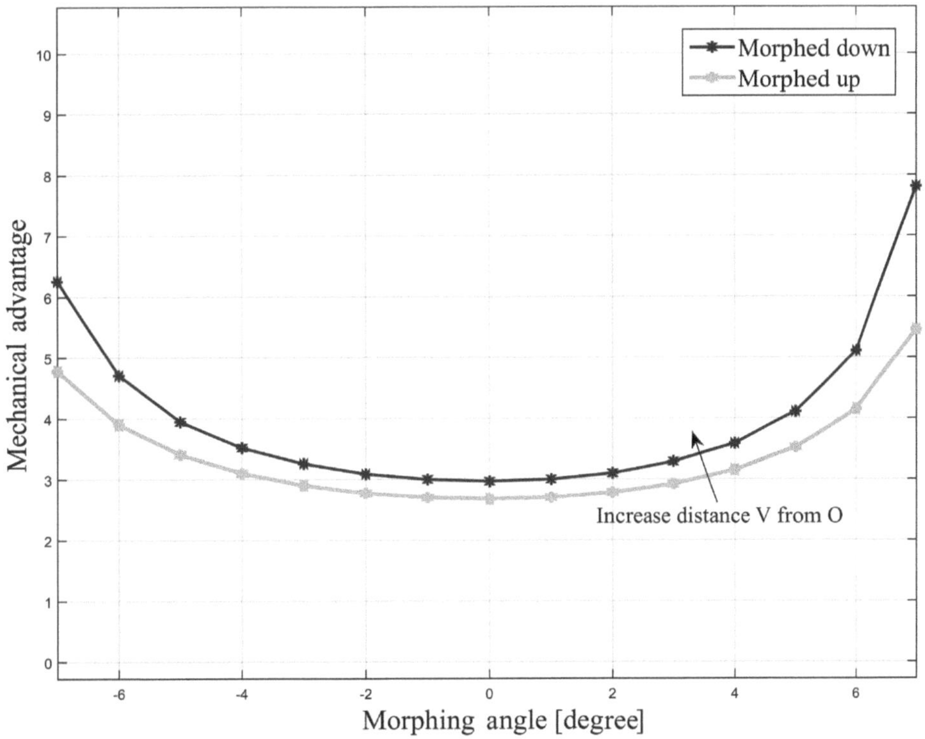

FIG. 11

Curves of the MA versus V location.

$$\cot\varphi = \frac{L}{R \cdot \sin\beta}\cot\beta \tag{2}$$

Eq. (2) allows for calculating the actuator shaft rotation (β) needed to achieve a given morphing angle (φ). It allows the designer to estimate the actuator rotation range to obtain the morphing deflection between $-7°$ to $+7°$. The actuation system was indeed selected on the basis of a trade-off analysis among all the design parameters such as *MA*, volumetric constraints, actuator rotation β, crank dimension, and stress. They are reassumed in Table 1 where the following commercial of the shelf (COTS) architectures have been investigated:

- linear guides with rollers;
- cam follower;
- recirculating ball guide;
- nonrecirculating ball guide;
- high rigidity linear guide.

A selection of the different systems has been carried out on the bases of their performances in terms of max allowable excursion (ME) and other parameters. The trade-off led researchers to choose the High Rigidity linear guide configuration. This arrangement permitted to achieve a suitable aileron deflection range, combined with a good capability to resist external loads (maximum force MF) and a relevant

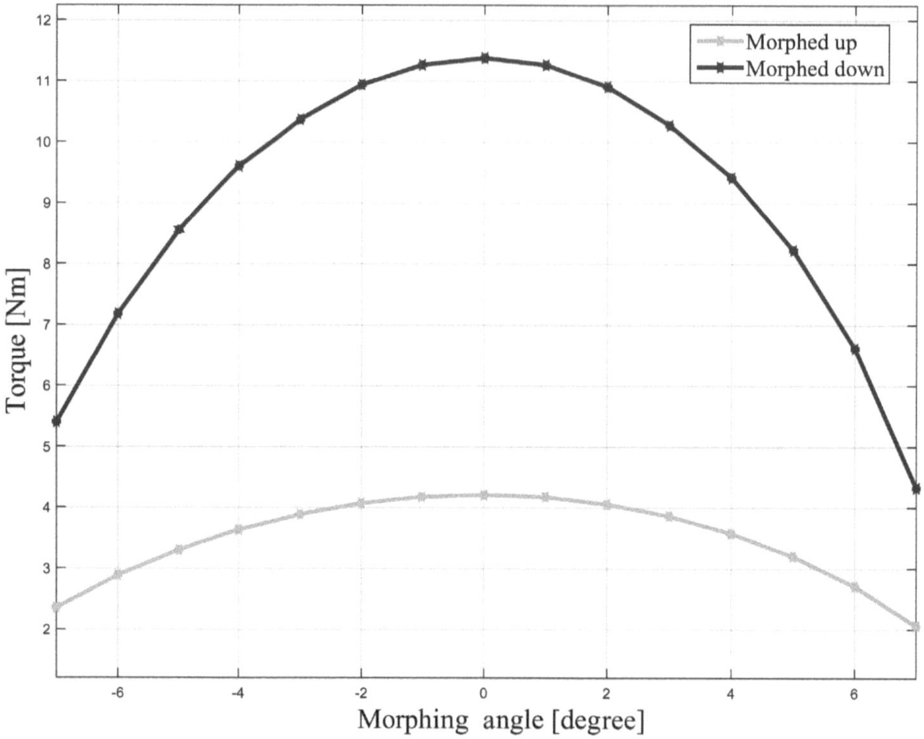

FIG. 12

Curves of the actuator torque variation between morphed up and morphed down.

Table 1 Aileron Actuation System Trade-Off Table

Configuration	ME [deg]		*L* [mm]	*R* [mm]	MA 7°	AR [deg]	Stroke (mm)	MF [N]	*N*
	MU	MD							
Linear guide with rollers	7	−7	92.83	30	2.3	15.2	2.2	820	5
Cam follower	5	−4	117.91	30	7.3	49.7	16	785	5
Recirculating ball guide	6.5	−4	121	30	8.5	52.7	17.65	311	3
Nonrecirculating ball guide	6	−4	119.2	35	4.2	37.3	12.68	233	3
High rigidity linear guide	6	−4	119.2	35	4.2	37.3	12.68	1091	3

actuator absolute rotation (AR). Therefore, the actual architecture gave an equivalent aileron deflection of −4 to +6°. in morphed-up (MU) and morphed-down (MD) modes. It can be noticed that gaining the highest MA can have a negative effect on the actuator rotation (greater than the allowable) and on the aileron morphing deflection (limited by structural interference). Furthermore, the actuation system was also analyzed by means of finite element (FE) models. Each linear guide can resist normal (upward/

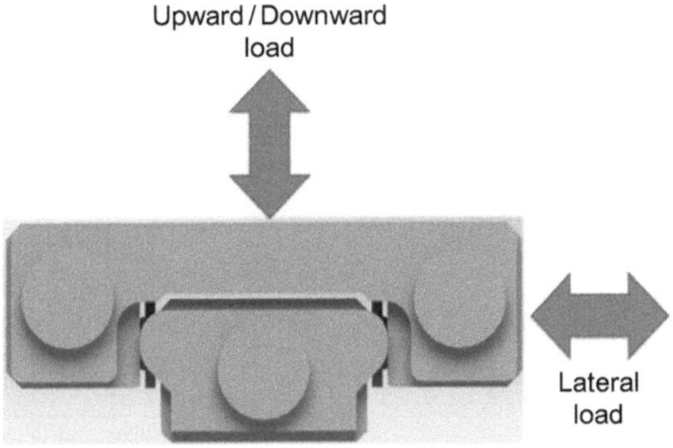

FIG. 13

Lateral and normal direction on the linear guide [22].

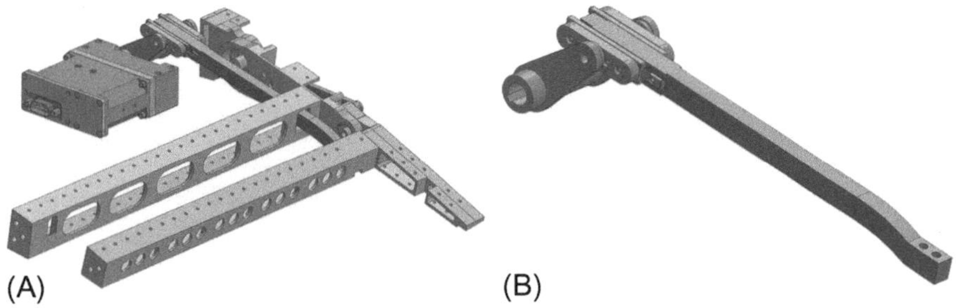

FIG. 14

Actuation system final architecture with high rigidity linear guide [21].

downward) and transversal (lateral) forces, lying in the generic plane that is perpendicular to the sliding reference direction, Fig. 13.

The allowable forces are reported in technical data sheet (prepared by the producer of the hardware) and refer to the loads inducing stresses so relevant to compromise the correct working principle of the components. This device exhibits substantial benefits, mostly resulting in an increased allowable static force equal to 1091 N [27]. In addition, in order to mitigate the maximum counterbalancing load acting on the guide to equilibrate the aerodynamic moment, a fork-shaped crank coupled with a double sided linear guide was preferred (as shown in Fig. 14A and B).

5 NUMERICAL SIMULATIONS

The numerical simulation has been conducted at ever increasing level of detail, starting from the isolated kinematic to the 3D aileron structure integrated with the actuation mechanism. A preliminary FE model was realized in order to investigate on the actuation kinematic without the other structural parts which contribute to withstand the external loads. The model simplification then consisted of concentrated loads applied on rigid elements, in such a manner that the elasticity of the remaining elements

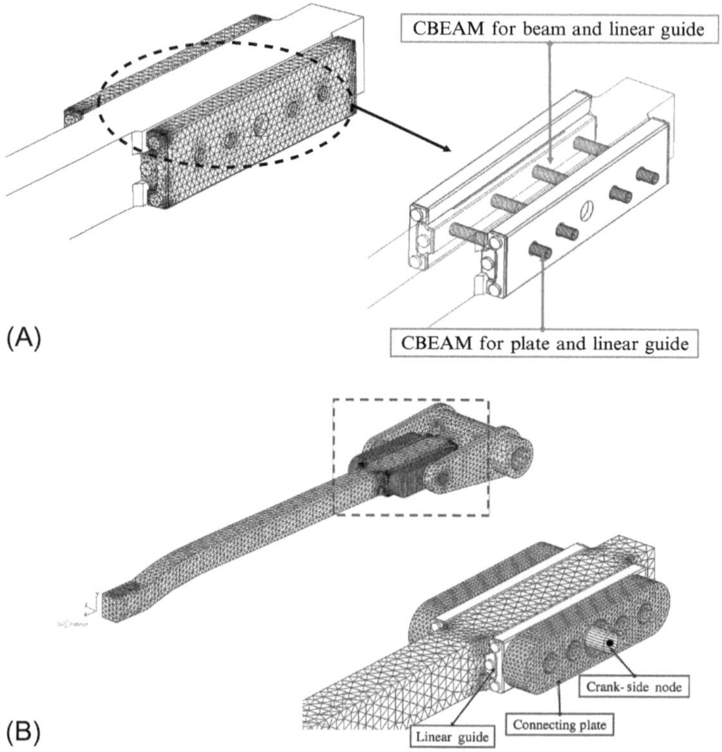

FIG. 15

Connection pins between linear guides components (A) and fem detail (B) [21].

was not considered. All the components of the actuation system were connected to each other by means of several pins which were simulated using CBEAM elements [1], Fig. 15.

The FE model of the entire aileron was then generated. FE model data have been recapped in Table 2.

The most relevant components of the primary structure were modelled with solid CTETRA [1] elements, namely, ribs, actuation kinematic chains, spars, and skin. The aileron leading edge was not

Table 2 FE Model Characteristics

FE Model General Data	
Number of elements	2.138E+6
Number of nodes	1.393E+6
Estimated DOFs	3.638E+6
Total estimated mass [kg]	21.00
Moment of inertia about aileron hinge-line [kg m^2]	0.403

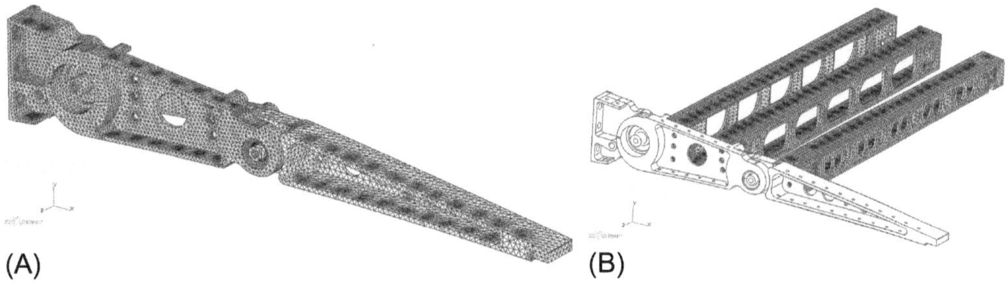

FIG. 16

(A) Aileron rib solid mesh (CTETRA) and (B) spar solid mesh (CTETRA) [21].

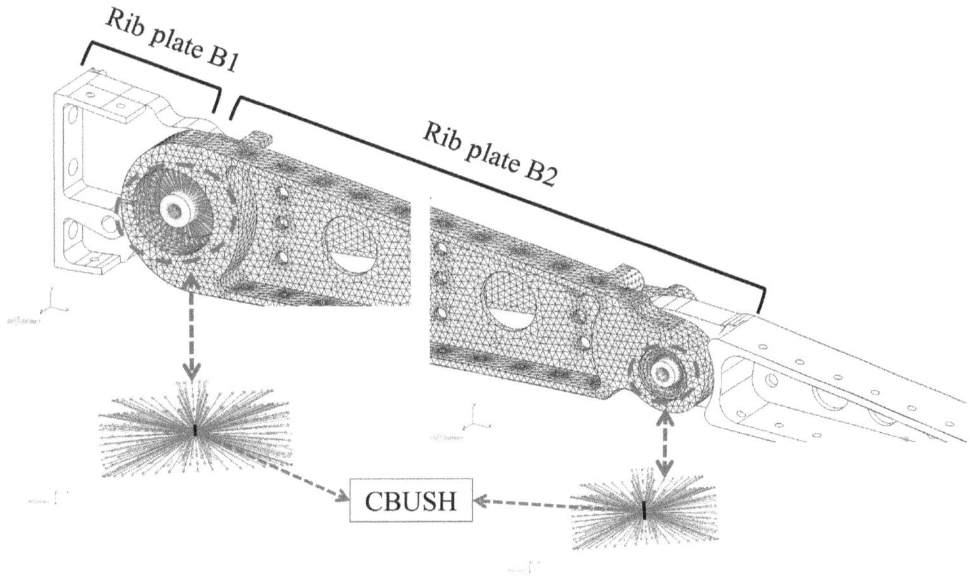

FIG. 17

Detail of the model of internal hinges [21].

modelled for stress analysis purposes; it was considered only to properly evaluate the interface loads transmitted by the aileron to the wing box. In Fig. 16, mesh details of rib and spars are shown.

Hinges were modelled referring to the usual scheme of rigid body connections. At each of the two sides of the hinge housing, a master node was placed at the center of the circular hole, and nodes on the edge of the circular hole were slaved to it through MPC-RBE2 connection. Master nodes belonging to the two sides of the hinge housing were finally joined through a CBUSH element showing low torsional stiffness about the rotation axis. In Fig. 17, the models of the hinges connecting rib block 0 with 1 and rib block 1 with 2 are shown.

The modelling approach used to connect ribs and spars together has been sketched in the Fig. 18. For each fastener (or screw) hole, a master node was generated at the center and connected to all the nodes on

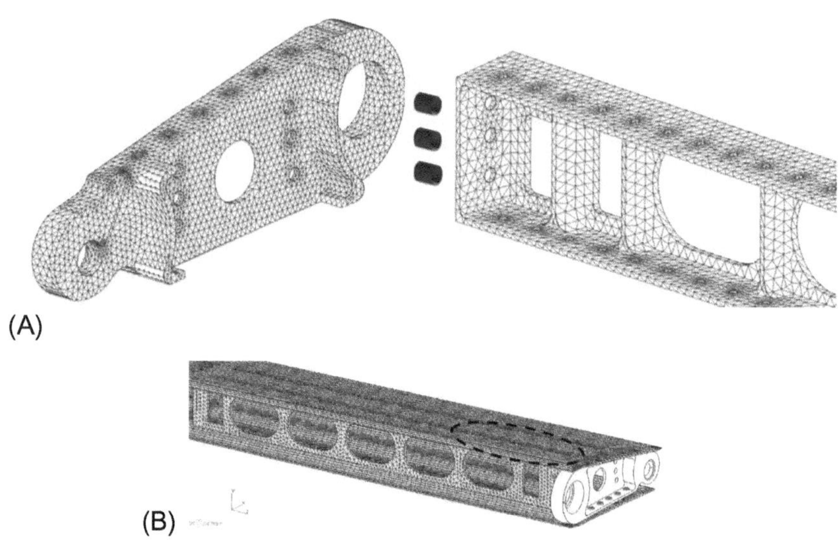

FIG. 18

Fasteners modelling scheme for rib-spar connection (A) and skin (B) [21].

the edge through an RBE2 element. The fastener (or screw) was then modelled with a beam element joining the master nodes at the center of each hole (Fig. 18A). The same procedure was used for modelling the fastening connection between spars and skin and between rib segments and skin (Fig. 18B).

After modelling the structural parts, the loads (estimated by VLM [28]) were applied as distributed pressure on the aileron lower skin. In Fig. 19, the pressure load is represented by small arrows on the aileron upper and lower skin, Fig. 19.

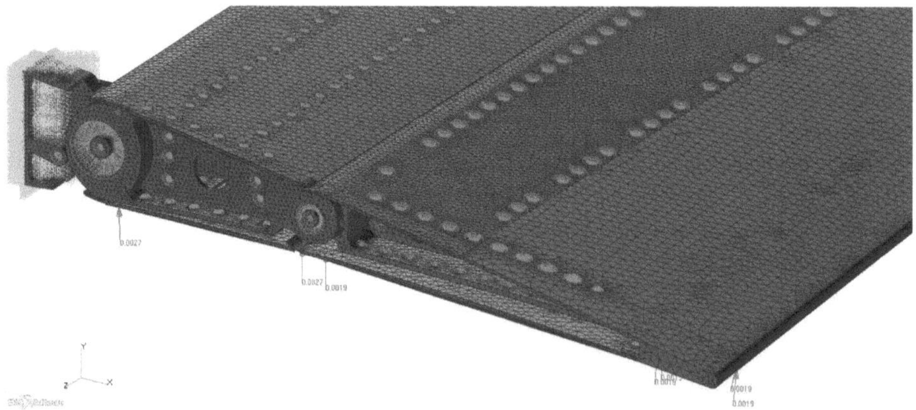

FIG. 19

Distributed pressure applied to the structure.

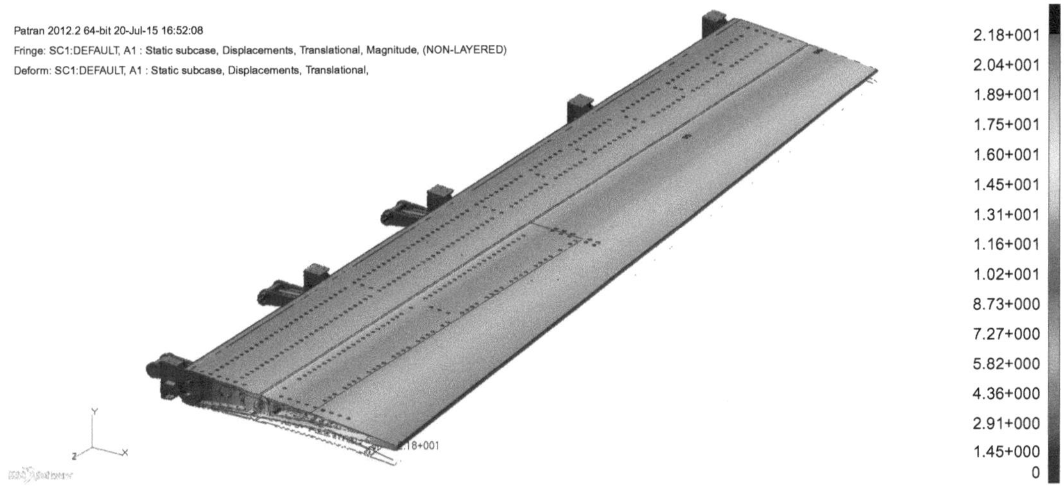

FIG. 20

Global aileron displacement distribution at LL condition [21].

The first movable block of the morphing aileron was considered constrained in correspondence of its joints to the leading edge with all displacements and rotations suppressed. In addition, other two different boundary conditions were generated. The first one consisted in vertical displacement equal to zero, and it was imposed to the nodes on the linear guides. In such a manner, the total vertical force acting on the sliding elements was evaluated as reaction load. On the other hand, the second constraint condition was imposed to the nodes in the fork-shaped crank to simulate the locking of the actuator shaft and consisted in null displacements and rotations. Static analysis results have been here reported with reference to the limit load (LL) condition only. In Fig. 20, the global magnitude of the displacements exhibited by the aileron at limit load condition is shown. The maximum value (21.8 mm) is located at the trailing edge in correspondence of the root section.

The maximum stress on the ribs occurred at the third block of the rib 2, and resulted equal to 231 MPa, Fig. 21. It was localized in a very small area around the hinge connecting block 2 with block 1.

The elements with stress levels higher than 150 MPa are also shown in Fig. 22. This result indicates that a rapid stress reduction occurs while moving away from the hinge axis.

For what regards the rib links, Fig. 23, the highest VM stress arises in correspondence to the element connected to the third block of rib 4; however the stress value (257 MPa) resulted well below the yield stress of the link material (steel, ~1000 MPa).

The VM stress distribution on the spars is shown in Fig. 24. The elements are not particularly stressed, with the exception of localized areas around the hole between the spar and the rib located at aileron tip. The value is close to the yield stress of AL 2024 but in a very small region.

The stress values along the skin resulted uniformly distributed with some (small) peaks around the fasteners, Figs. 25 and 26.

In the real operative conditions, the linear guide, being free to move, is not subjected to stress in the direction of motion. Forces are transmitted in the vertical (with respect to the guide axis) and, partially, normal (with respect to the guide plane) directions. For the current application, the actuator system was

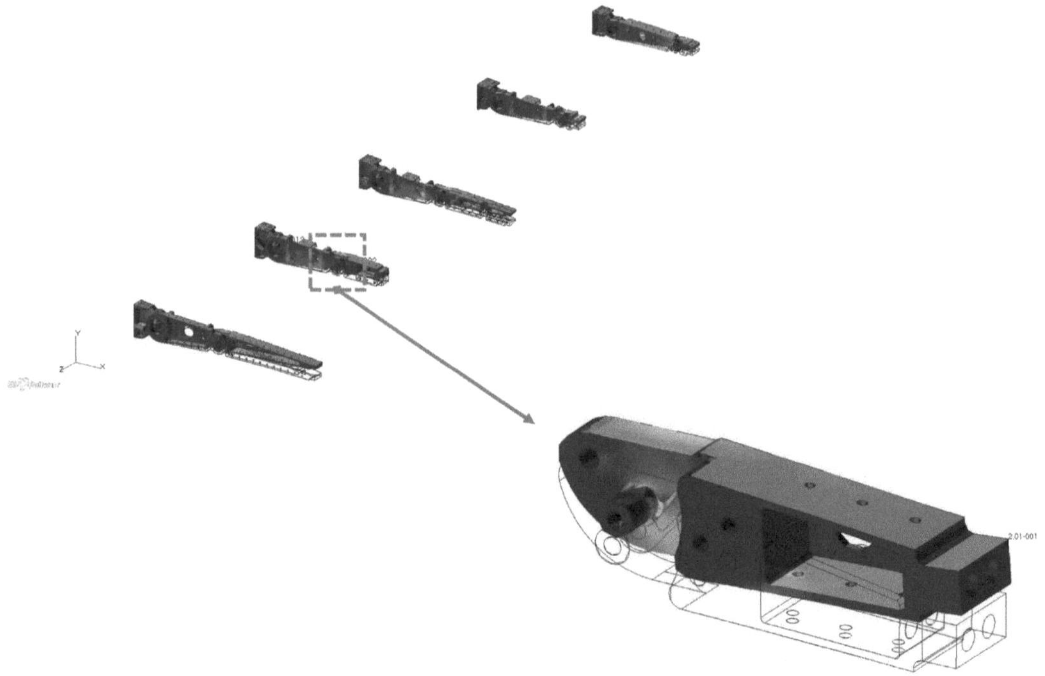

FIG. 21

Global von Mises (VM) stress distribution on ribs at LL condition [21].

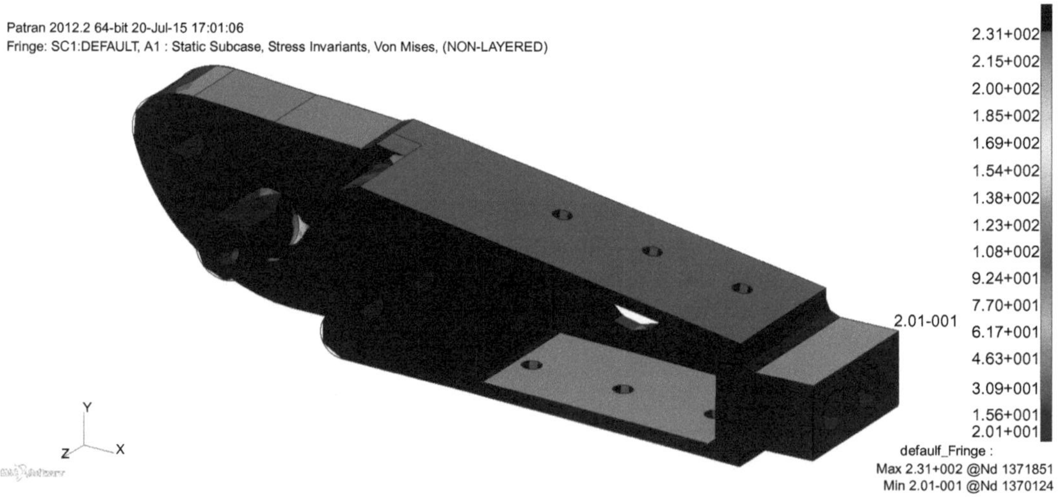

FIG. 22

Element stress distribution above threshold of 150 MPa [21].

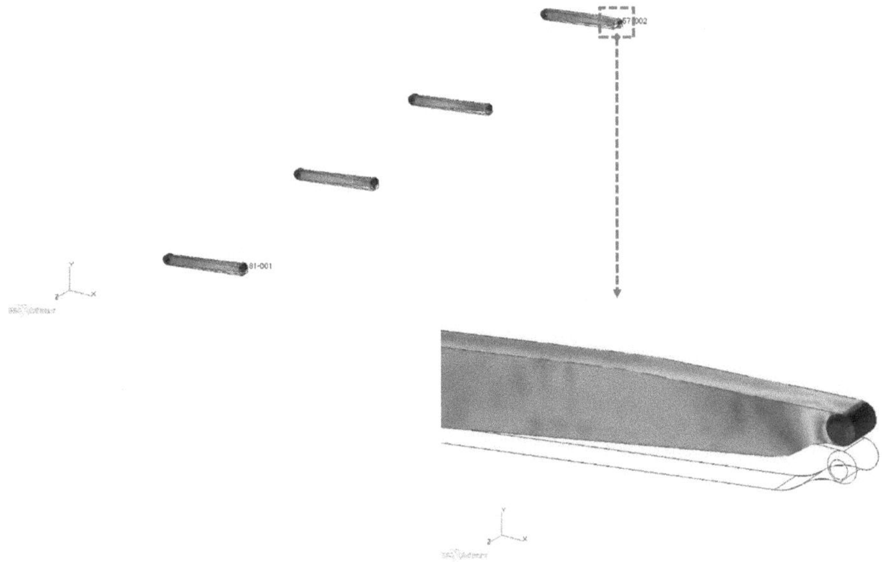

FIG. 23

Global VM stress distribution on links at LL condition [21].

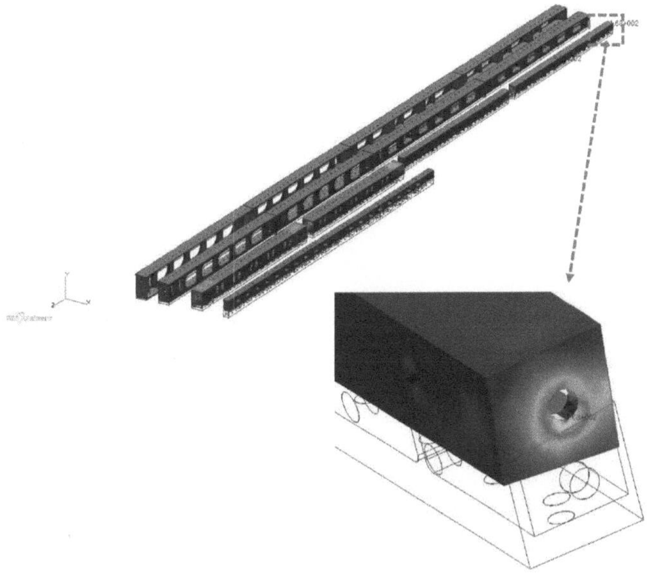

FIG. 24

Global VM stress distribution on spars at LL condition [21].

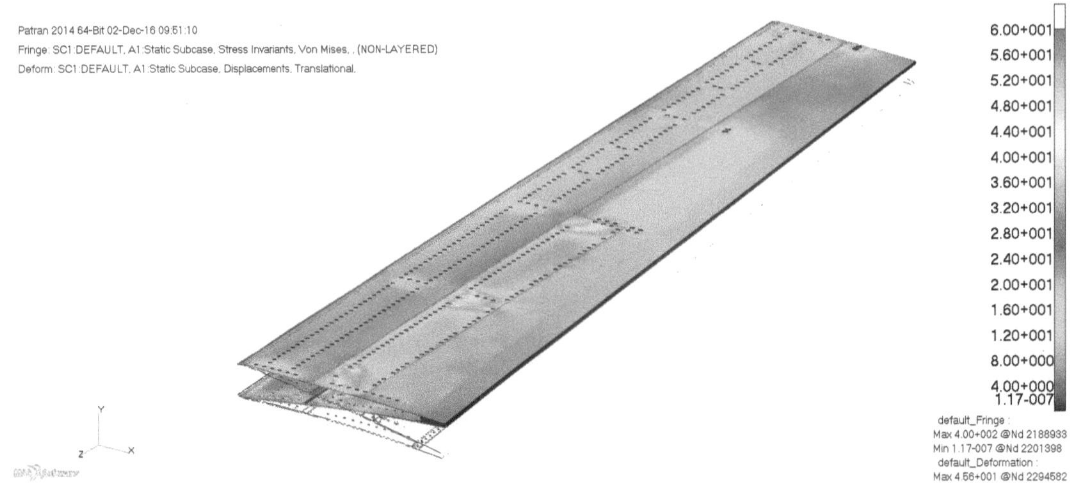

FIG. 25

Global VM stress distribution on skin at LL condition.

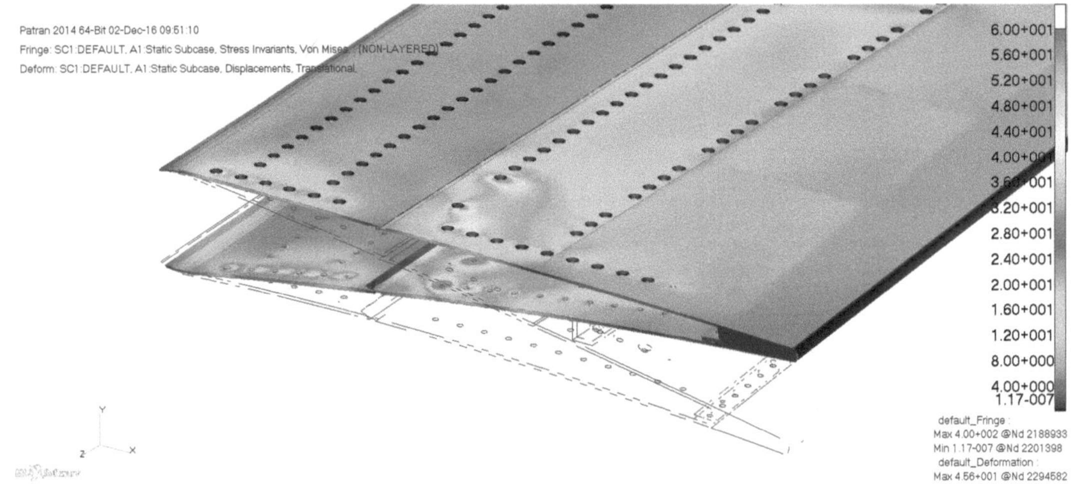

FIG. 26

Stress detail on the skin stress distribution at LL condition.

sized, referring to the jamming condition, considered as the most critical. This effect was simulated by means of a perfect bonding between the rail and slider. In fact, as visible in Fig. 27, the larger extent of the constraints (additional clamps) is expected to lead to higher stresses, both locally (in the contact region) and distributed (over all the linear guide). When the linear guide is locked, the actuation beam is simultaneously loaded with the external aerodynamic moment (respect to point V of Fig. 8), the vertical static force due to the jamming, both producing a pure bending with a higher stress level than the one related to the condition relative to the free sliding motion. In Fig. 28A is reported the stress

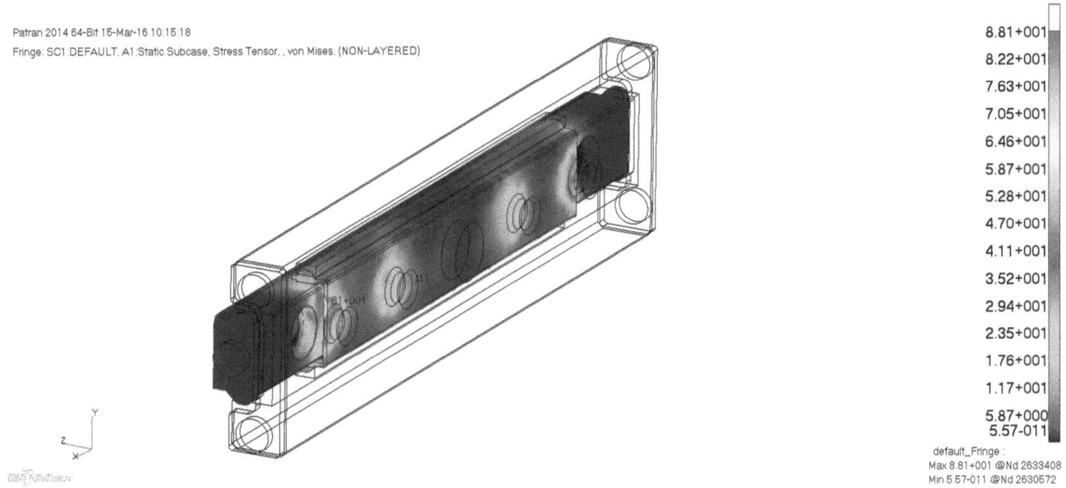

FIG. 27

Linear guides VM stress distribution.

distribution along the actuation beam, while in Fig. 28B is reported a stress contour with the elements characterized by solicitation levels higher than 320 MPa. Actuation levers showed the typical stress distribution of the beams in bending; stress with peaks greater than 320 MPa is localized close to un-chamfered notches.

5.1 INTERFACE LOAD

In order to evaluate the reaction loads at wing box interface, the FE model was completed with the mesh of the aileron leading edge as well as of all those elements assuring its connection to the wing box. The aileron hinge axis was virtually reproduced by adopting specific constraint conditions implemented with reference to the coordinates system[2] depicted in Fig. 29; tie-rods were fully modelled, and their bases were rigidly connected to master nodes, in turn constrained in all DOFs (grids number 4820028, 4820046 in Fig. 29 and Table 3). Hinges at root and tip ribs were modelled by using the same approach herein described.

Linear static analysis was then carried out and limit reaction loads were finally evaluated. The results are summarized in Table 3.

In order to evaluate the actuator torque required to equilibrate aerodynamic loads at limit condition (Mach $= 0.25, \alpha = 0°, \delta = 0°, \gamma = 7°$), the reaction loads acting on the linear guides were estimated for each actuation chain. The estimated lateral forces and the corresponding torque are reported in Table 4. The first column identifies the FE model nodes where the reaction loads were calculated. The vertical forces are reported in the second column and the torque in the third one. The torque was obtained by multiplying the force by its distance from the hinge axis of the rib block, the result was then divided by the mechanical advantage. The reference nodes which have been used to estimate the vertical forces are depicted in Fig. 30.

[2] *Y* axis along aileron hinge axis, *XZ* plane parallel to root rib plane.

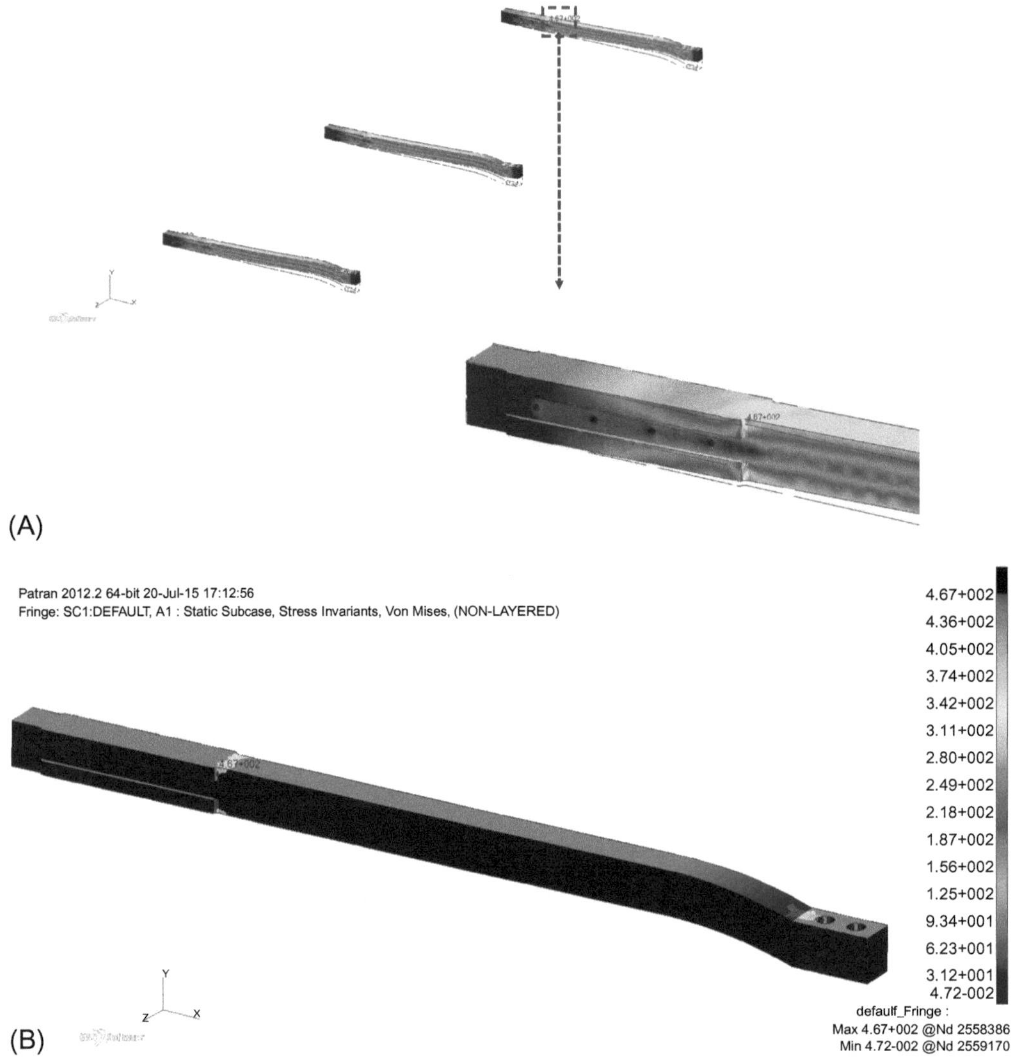

FIG. 28

Beam element VM stress distribution at LL (A) and stress above threshold values of 320 MPa (B) [21].

Each reaction force resulted well below the allowable static load of the linear carriage (1091 N) and also the actuator torque resulted below the design peak torque of 13 Nm.

6 PROTOTYPING

On the base of the numerical outcomes, the executive drawings of the prototype were elaborated for its manufacture. The main structural parts such as rib, spar, actuation beam, and skin were manufactured with a numerical control machine. The main phases of the fabrication process are shown in Figs. 31–33.

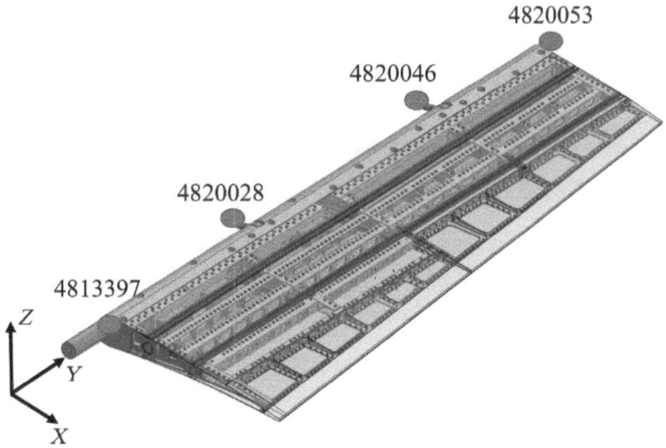

FIG. 29

FE: evaluation of reaction loads at the wing box interface (the 4-point attachment has been indicated through the identification number of the linking nodes).

Table 3 Reaction Loads at Wing Box Interface (Limit Condition)

GRID	T1 [N]	T2 [N]	T3 [N]	R1 [Nm]	R2 [Nm]	R3 [Nm]
4813397	−6.9792E+2	0.0	−5.1779E+2	3.6005	4.3742E+1	−1.2532
4820028	1.6819E+2	0.0	−4.5130E+2	5.7469	2.5147E+1	1.4515E+1
4820046	1.7739E+2	0.0	−3.0889E+2	−1.3898	3.0719E+2	5.7412E−1
4820053	−7.9091E+2	0.0	−7.8816E+1	3.1102	0.0	−2.8114

Table 4 Actuator Torque Evaluation

Node ID	Lateral Force [N]	Actuator Torque [Nm]
2634247	297.14	12.15
2634246	260.47	
2634245	282.72	12.17
2634244	275.58	
2634243	277.70	12.88
2634242	313.41	

Firstly, ribs are represented in Fig. 31, while the actuation kinematic chains are reported in Fig. 32, also with the integration of the actuation system inside the aileron main structure.

The assembled aileron prototype (after painting) is shown in Fig. 33.

After the manufacturing, the aileron was tested by means of a dedicated experimental campaigns. The experimental characterization was aimed at:

- Verifing the control logic algorithm by means of actuators bench tests. In these tests, the actuators working conditions were simulated both in a fast and a slow movement by means of a “step” or a “sine” function.

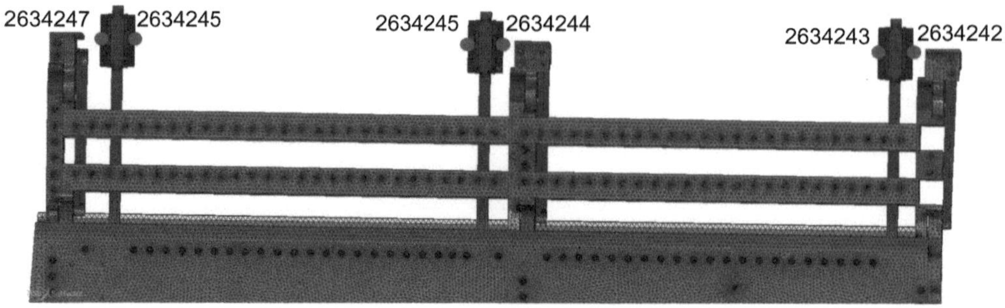

FIG. 30

Linear guide nodes.

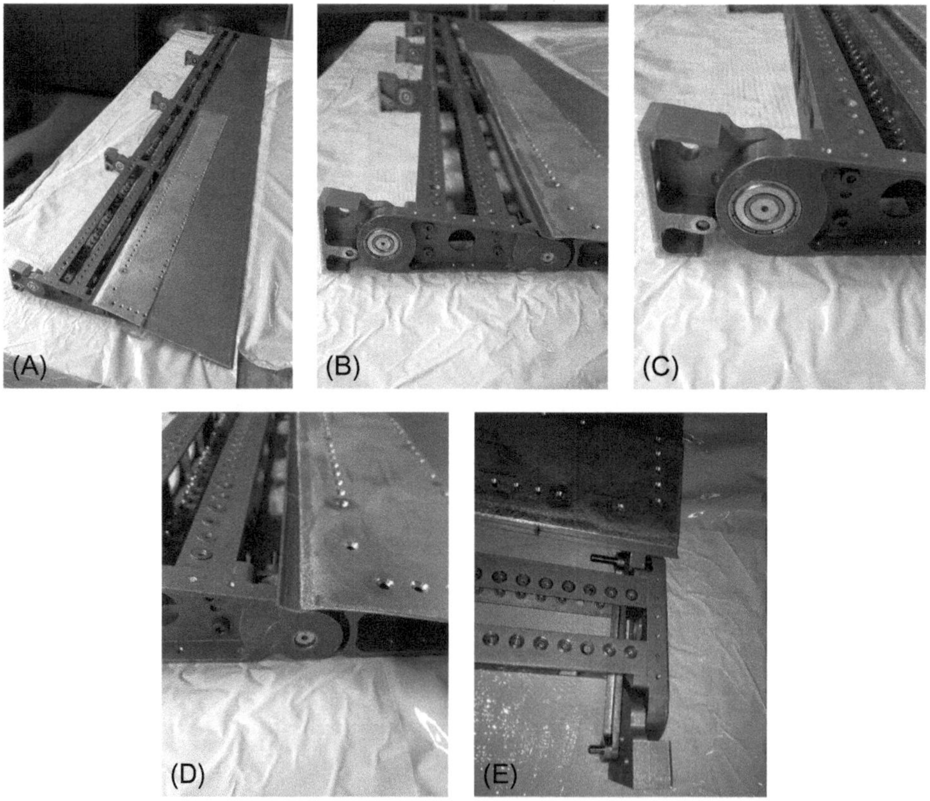

FIG. 31

Aileron Manufacturing with detail on hinges and rib, (A)–(E) [33].

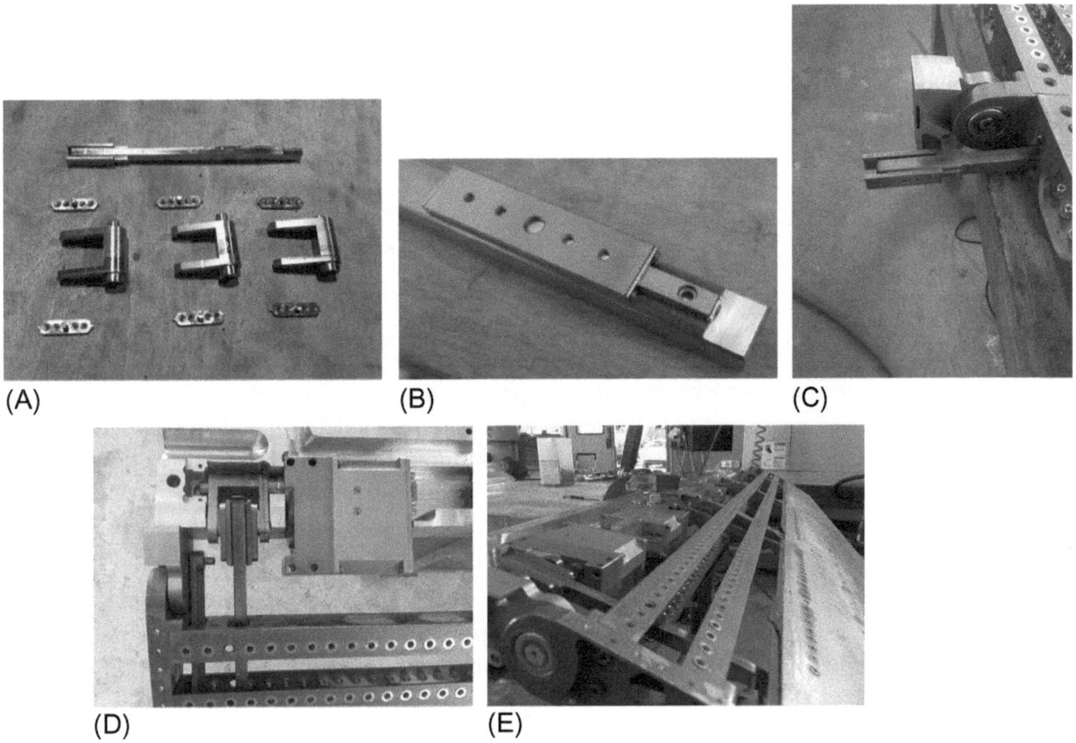

FIG. 32

Detail on aileron actuation system, (A)–(E) [29].

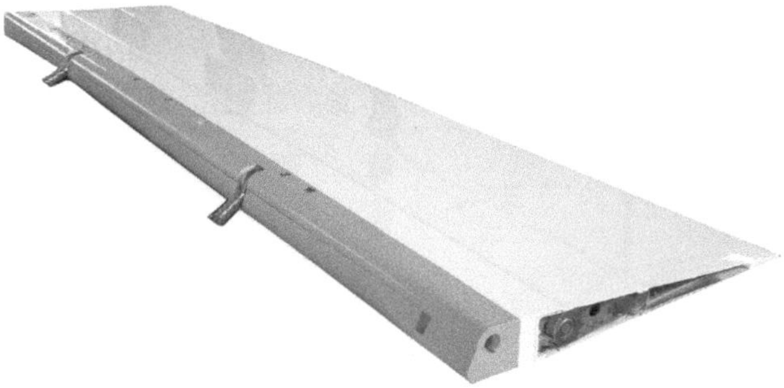

FIG. 33

Photo of the actual aileron prototype [30].

- Demonstrating the morphing capability of the conceived structural layout.
- Characterizing the dynamic behavior of the morphing structure through the identification of the most significant normal modes.
- Validating the implemented numerical model.

7 EXPERIMENTAL TESTS AND MAIN OUTCOME

7.1 GVT AND NUMERICAL CORRELATION

The experimental setup dedicated to the aileron ground vibration tests (GVT) is depicted in Fig. 34, while in Fig. 35, the aileron in free-free condition is reported. The following instrumentation was used:

- power supply (1.5 ÷ 24 VDC @ 1.0 A) to move the actuators;
- DSP controller;
- tri-axial accelerometers;
- excitation shaker;
- LMS-SCADAS® tool for signal processing and modal analysis.

In Fig. 36, the measurement points on the aileron upper skin are represented.

The aileron GVT was conducted to validate the numerical FE model of the aileron and then to provide input for aeroelastic stability analyses based on experimental modes association in the frequency range 0–100 Hz. Two normal modes were detected in the frequency range of interest from:

- torsional mode and
- morphing mode.

In Fig. 37, the numerical and experimental modal shapes have been compared, while the graph of Fig. 38 shows a satisfactory level of correlation between numerical and experimental frequencies.

FIG. 34

Setup for aileron experimental campaign.

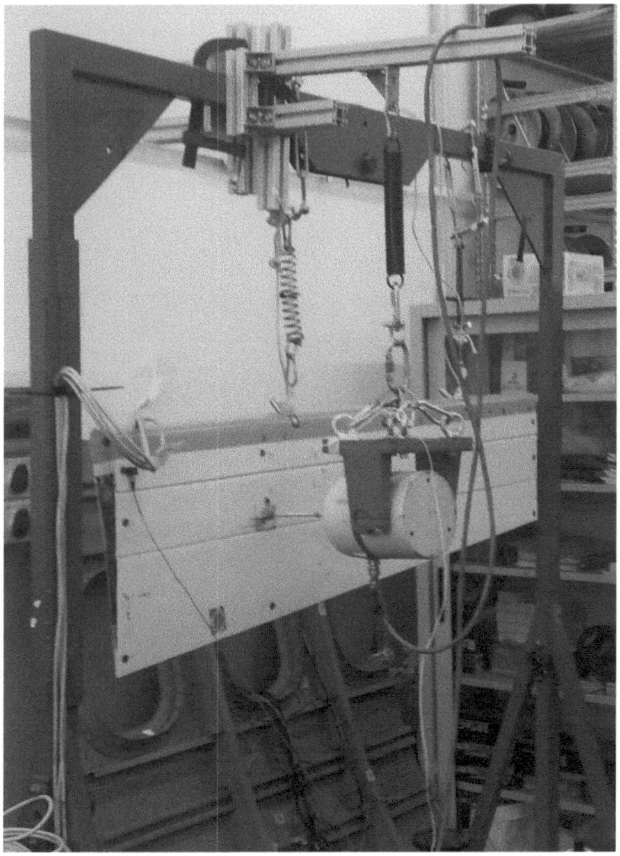

FIG. 35

Aileron GVT experimental setup.

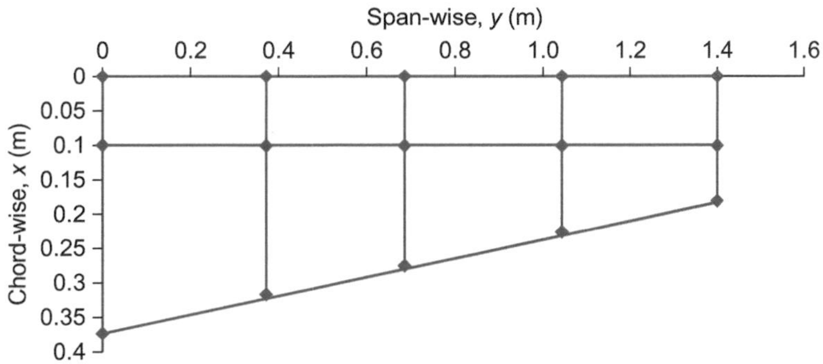

FIG. 36

Aileron measurement points [30].

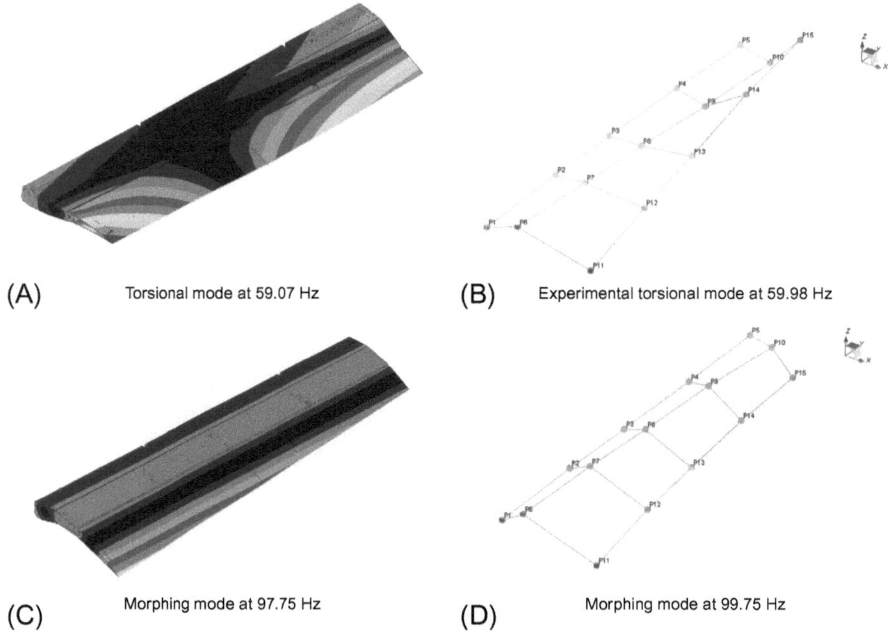

FIG. 37

Numerical and experimental normal modes comparison [34].

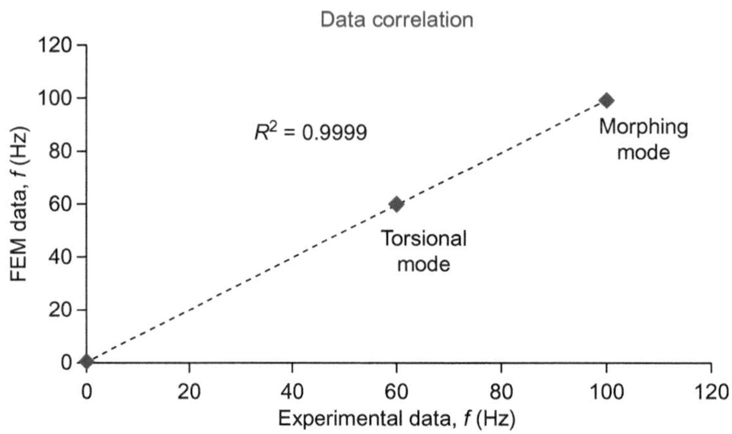

FIG. 38

Normal modes correlation [34].

7.2 FUNCTIONALITY TEST

The functionality test consisted in characterizing the aileron kinematic in terms of actuator rotation and rib deflection. This is analytically expressed by Eq. (2). For each morphing angle, within the effective range [−4;+6] degrees, a given actuator shaft rotation was imposed and the tip displacement was measured as reported in Fig. 39. The picture represents the wing prototype installed within the wind tunnel

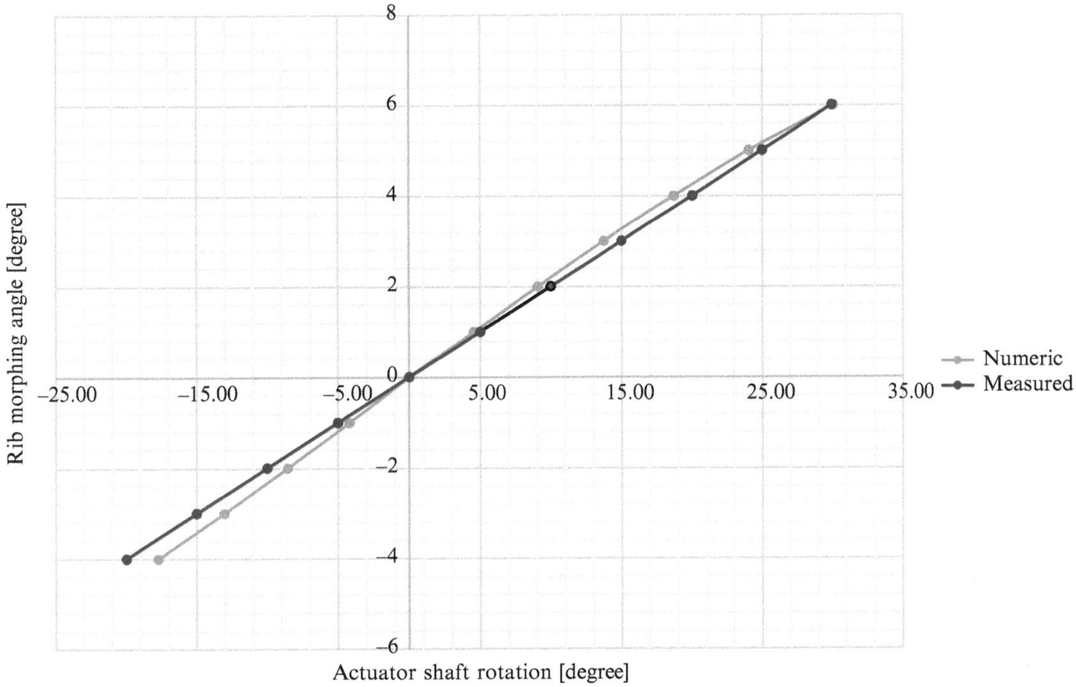

FIG. 39

Correlation between numerical and experimental: actuator rotation versus morphing deflection.

test chamber in an upper view. Along that direction, the wing section is clearly visible, as it was a 2D layout. Some captions help to distinguish the top wing fairing from the aileron. The sequence shows the movement of the aileron from two to six degrees. In each position, the tip displacement was measured and is graphically reported as a function of the actuator rotation. A good match with the numerical results is shown. The very small deviations observed are due to the fact that the implemented measurement method was based on images elaboration.

7.3 EXPERIMENTAL SHAPES

A comparison between the numerical (target) and experimental (real) aileron shapes was addressed. In Fig. 40, the reference design shapes are shown. Such geometries were then approximated with a set of discrete points in order to have a proper comparison with the measured shapes. The aileron is represented in both morphed down and morphed up configurations. In lab, the upper aileron surface was constellated with a set of circular retro-reflective markers, allowing direct image acquisition. Measures were performed by NRC (the Canadian National Research Council), via a high precision photogrammetry procedure using 3D-tracking cameras [31]. Design (CAD) and measured shapes were overlapped, Fig. 41. In that representation, the circular markers represent the position of the reference points as it was measured experimentally, while the upper skin target geometry is reported as a best fitting surface between the design positions of the same reference points. Deviations between the two shapes were computed as the least square difference between numerical and experimental marker values.

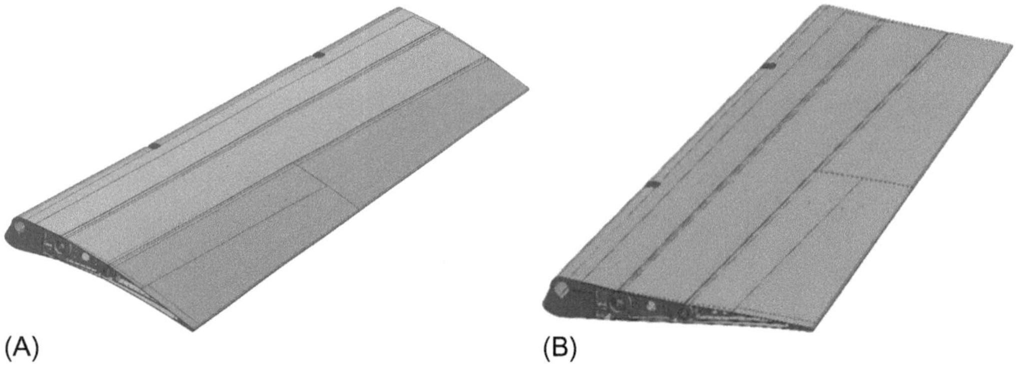

FIG. 40

Aileron CAD morphed shapes: downward (A), upward (B) [30].

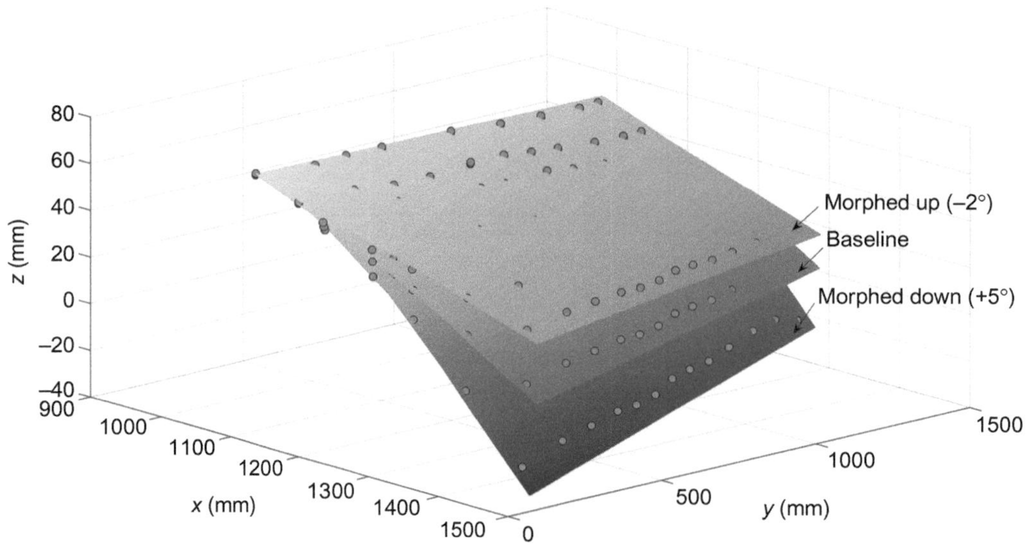

FIG. 41

Comparison between numerical and experimental aileron shapes.

It can be noticed that there a good level of matching was found except in some points for the morphed-up configuration where small deviations were observed. This is particularly due to the stiffness of the last rib, which resulted lower than expected. This is in turn linked to the implemented architectural design (very sensible to tolerances) and the large distance of the actuators (largely contributing to the overall rigidity of the system) from the monitored region.

Table 5 Displacement Error Between Numerical and Experimental Morphed Aileron Shapes

Configuration	Max ϵ [mm]	Locations
Morphed-up	4.29	Tip Rib TE
Baseline	4.32	Tip Rib TE
Morphed-down	5.65	Tip Rib TE

The deviation between the shapes (error) is reported in Table 5 for the three aileron configurations.

8 WIND TUNNEL TESTS

The wind tunnel tests were performed at the National Research Council Canada [32] in the 2 m × 3 m section, closed circuit, subsonic wind tunnel. The facility exhibits a maximum Mach number of 0.33. The wing model equipped with the morphing aileron at the trailing edge installed in the test chamber is shown in Fig. 42.

The upper surface flexible skin of the wing demonstrator was equipped with 32 Kulite piezoelectric-type transducers for pressure measurement, and their data were processed to determine the laminar-to-turbulent flow transition location. In addition to the Kulite sensors, 60 static pressure taps were installed on the wing leading edge, lower surface, and aileron, thus providing complete experimental pressure distribution around the wing cross section located at 40% of the wing model span. Infrared (IR) thermography camera shots were performed to capture the transition region over the entire wing model surface [33]. The IR thermography visualization allowed for the identification of the transition region between laminar and turbulent regimes, based on the analysis of the model surface temperature [34]. A wind tunnel balance was used for measuring the aerodynamic forces and moments. In order to check for the onset of aeroelastic instability phenomena during the wing tunnel test, three accelerometers were installed. The three accelerometers were positioned on the wing box, aileron, and wind tunnel balance, respectively, as shown in Fig. 43.

The wind tunnel test sessions covered 49 flight cases with different combinations of wing angle of attach (±3°), speed (from Mach 0.15 to 0.20) and aileron deflection (−4, +6°). For all these flight cases, the objectives of the measurement were the delay of the flow transition from the laminar to turbulent states and the validation of the morphing aileron technology from both the functionality and the aerodynamic performances standpoints Fig. 44 shows the comparison between the Cp distributions numerically evaluated and experimentally measured for two generic test cases. It can be observed that the two pressure distributions show a very good correlation level proving that for each test case the aero-shape actually implemented by the aileron was satisfactorily matching the design one.

Obtained aerodynamic results have been summarized in terms of the following curves: lift versus angle of attack (CL-α), Fig. 45A); drag polars (CD-CL), Fig. 45B). The first one shows a typical linear

FIG. 42

Wing model and morphing aileron.

trend. The curve slope (CLα) remains unchanged, the camber increase (from baseline to 6°) shifts the curves upwards. The aerodynamic polars cross each other in correspondence of a pivot point for high CL and move to the right side of the Cartesian plane for low CL, enveloping an optimal curve maximizing aircraft aerodynamic efficiency throughout the cruise [29].

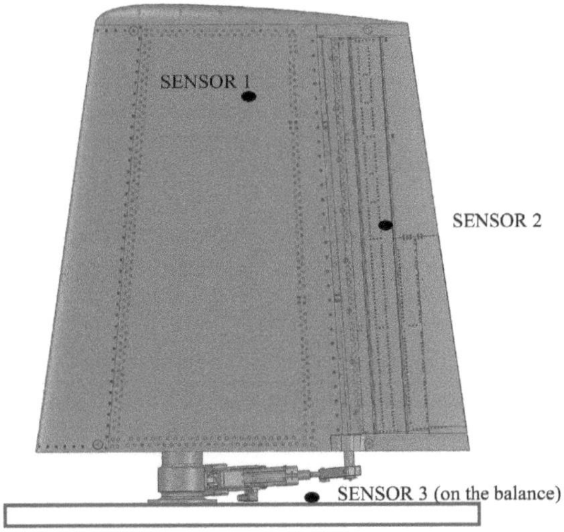

FIG. 43

Wing accelerometers position and orientation [35].

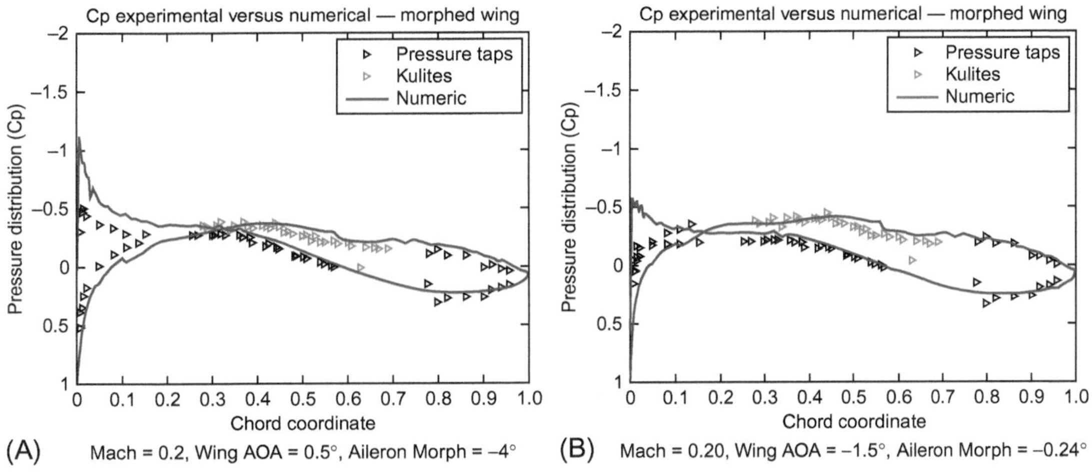

FIG. 44

Numerical versus experimental pressure distribution [35].

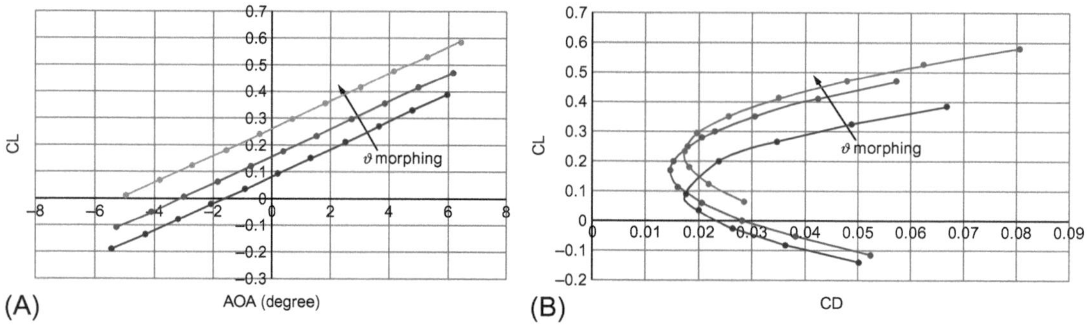

FIG. 45

Lift coefficient versus angle of attack curve (A), drag polars with increasing morphing angles (B) [33].

9 CONCLUSIONS

In the present chapter, an interdisciplinary project done in the framework of an international cooperation between Canada and Italy was presented and discussed. The project was aimed at designing and tesing an innovative morphing wing tip prototype under constraints similar to those imposed for real aircraft end-application. The demonstrator was equipped with two morphing systems: the upper wing box skin and the variable camber aileron. The composite flexible upper surface was capable of sustaining deformations induced by four electrical actuators (bump generators) positioned inside the wing box structure. The innovative morphing aileron was conceived as a technological extension of a mechanical concept already assessed for a smart wing trailing edge. It was designed for a symmetrical deflection during cruise in order to compensate A/C weight variation due to fuel burned. In such a manner, it is aimed to increase aerodynamic efficiency (reduce drag) in off-design points. Additionally, the deflection of a morphing aileron it is expected to redistribute the spanwise wing distribution in order to reduce wing root bending moment. Furthermore, the conventional aileron roll-control functionality is preserved. The novel aileron ensures then an augmented functionality with respect to the conventional one. Being the aileron a safety critical surface [36], the structural design of a reliable morphing solution is rarely addressed in literature. Such an original work provides thus evidence and arguments that contribute to the knowledge of morphing systems. The camber variation is performed thanks to a "finger-like" rib architecture and a kinematic chain composed of compact electromechanical actuators and a leverage with double-sided guides and a fork-shaped crank. The mechanisms and the primary structure were designed and verified—through advanced numerical models—with reference to relevant operative loads conditions. The laboratory demonstrator was then manufactured, assembled and tested, proving the effective functionality of the concept. Finally, wind tunnel tests assessing the aerodynamic impacts of such innovative architecture have been tackled. During wind tunnel tests, the wing box upper surface and the aileron camber were actively modified with the main purpose of delaying the flow transition from the laminar to the turbulent regime. The aerodynamic curves of Lift-AOA and drag polars have been reported for different morphing deflection. The trend shows promising results to be further validated with future research projects. The idea herein described leads the way to further researches aimed at enhancing the TRL of the device. To this scope some key-aspect deserve further and more detailed investigation; among these, the following seem to be the most relevant:

- an embedded sensing network for enhanced control of the achieved aero-shapes;
- aerodynamic benefits comparison between conventional and morphing aileron on the base of a larger amount of test cases enveloping a wider set of operative conditions;
- enhanced design with Topology optimization;
- segmented skin aerodynamics comparison with a tailored complaint skin technology; and
- high speed simulations and tests.

REFERENCES

[1] MSC-MD/NASTRAN®, Software Package, Ver. R3–2006, "Reference Manual".
[2] G. Reich, B. Sanders, Introduction to morphing aircraft research. J. Aircraft 44 (4) (2007) 1059, https://doi.org/10.2514/1.28287.
[3] M. Abdulrahim, in: Flight performance characteristics of a biologically—inspired morphing aircraft, Presented at 43rd AIAA Aerospace Science Meeting and Exhibit. 10–13 January 2005, Reno, NV, 2005.
[4] J. Valasek, Morphing Aerospace Vehicles and Structures, A John Wiley & Sons, Ltd., Hoboken, 2012.
[5] S. Barbarino, O. Bilgen, R.M. Ajaj, M.I. Friswell, A review of morphing aircraft, J. Intell. Mater. Syst. Struct. 22 (9) (2011) 823–877.
[6] G. Amendola, I. Dimino, F. Amoroso, R. Pecora, A. Concilio, Preliminary design of an adaptive aileron for next generation regional aircraft, J. Theor. Appl. Mech. 55 (1) (2017) 307–316. Warsaw, 10.15632/jtam-pl.55.1.307.
[7] M. Arena, M. Noviello, F. Rea, F. Amoroso, R. Pecora, G. Amendola, in: Modal stability assessment for a morphing aileron subjected to actuation system failures: numerical analysis supported by test evidence, Proceending of ICMAE 2016, 7th International Conference on Mechanical and Aerospace Engineering, 2016, https://doi.org/10.1109/ICMAE.2016.7549580.
[8] H.P. Monner, D. Sachau, E. Breitbach, in: Design aspects of the elastic trailing edge for an adaptive wing, Presented at RTO AVT Specialists' Meeting on "Structural Aspects of Flexible Aircraft Control", Held in Ottawa, Canada, 18–20 October and Published in RTO MP-36, 1999.
[9] J. Bowman, M.B. Sanders, T. Weisshaar, in: Evaluating the impact of morphing technologies on aircraft performance, 43rd AIAA/ASME/ASCE/AHS/ASC Structures, Structural Dynamics, and Materials Conference 22–25 April 2002, Denver, CO, 2002.
[10] H. Hilbig, H. Wagner, in: Variable wing camber control for civil transport aircraft, ICAS Proceedings, ICAS-84-5.2.1, Toulouse, 1984, pp. 107–112.
[11] S. Kota, R. Osborn, G. Ervin, D. Maric, in: Mission adaptive compliant wing—design, fabrication and flight test, Presented at the RTO Applied Vehicle Technology Panel (AVT) Symposium held in Evora, Portugal on 20–24 April, 2009.
[12] R. Pecora, F. Amoroso, G. Amendola, A. Concilio, Validation of a smart structural concept for wing-flap camber morphing, Smart Struct. Syst. 14 (4) (October 2014) 659–679, https://doi.org/10.12989/sss.2014.14.4.659.
[13] S. Barbarino, R. Pecora, L. Lecce, A. Concilio, S. Ameduri, Airfoil morphing architecture based on shape memory alloys, in: Proc. ASME Conference on Smart Materials, Adaptive Structures and Intelligent Systems, SMASIS2008, Volume 1, 2008, pp. 729–737.
[14] R. Pecora, F. Amoroso, M. Magnifico, I. Dimino, A. Concilio, KRISTINA: kinematic rib-based structural system for innovative adaptive trailing edge, Proc. SPIE—The International Society for Optical Engineering, Volume 9801, 2016, Article No. 980107.
[15] www.cleansky.eu.
[16] www.saristu.eu.
[17] R. Pecora, F. Amoroso, M. Magnifico, in: Toward the bi-modal camber morphing of large aircraft wing flaps: The Clean Sky experience, Proceedings of SPIE Smart Structures/NDE Conference 2016, Las Vegas (NV), USA, March, 20–24, 2016.

[18] P.C. Wölcken, M. Papadopoulos, in: Smart Intelligent Aircraft Structures (SARISTU), Proceedings of the Final Project Conference, Springer, 2015. ISBN 978-3-319-22413-8.
[19] G. Amendola, I. Dimino, M. Magnifico, R. Pecora, Distributed actuation concepts for a morphing aileron device, Aeronaut. J. 120 (2016) 1365–1385, https://doi.org/10.1017/aer.2016.64.
[20] http://www.criaq.aero/ (Web site of the CRIAQ consortium).
[21] R. Pecora, F. Amoroso, M. Magnifico, I. Dimino, A. Concilio, KRISTINA: kinematic rib-based structural system for innovative adaptive trailing edge, in: Proc. SPIE—The International Society for Optical Engineering, Volume 9801, 2016, Article No. 980107.
[22] G. Amendola, I. Dimino, A. Concilio, M. Magnifico, R. Pecora, in: Numerical design of an adaptive aileron, Proceeding SPIE 9803, Sensors and Smart Structures Technologies for Civil, Mechanical, and Aerospace Systems 2016, 98032A (April 20, 2016), Las Vegas, 2016, https://doi.org/10.1117/12.2219167.
[23] Bental Actuators data sheet (http://www.bental.co.il).
[24] I. Dimino, G. Diodati, A. Concilio, A. Volovick, L. Zivan, Distributed electromechanical actuation system design for a morphing trailing edge wing, SPIE Smart Structures/NDE, Las Vegas, Nevada (USA) March 2016. in: Proc. SPIE 9801, Industrial and Commercial Applications of Smart Structures Technologies, 2016, 980108 (April 16, 2016). https://doi.org/10.1117/12.2219223.
[25] I. Dimino, A. Concilio, M. Schueller, A. Gratias, An adaptive control system for wing TE shape control, in: Proc. SPIE International Conference on Smart Structures 2013, San Diego, California, USA, March 10–14, 2013.
[26] R. Pecora, A. Concilio, I. Dimino, F. Amoroso, M. Ciminello, in: Structural design of an adaptive wing trailing edge for enhanced cruise performance, 24th AIAA/AHS Adaptive Structures Conference, AIAA Sci-Tech, 4–8 January, 2016.
[27] http://ikont.com/ (Website of the IKO Linear Guides Producer).
[28] Falkner. V.M., The Accuracy of Calculations Based on Vortex Lattice Theory, Rep. No. 9621, British A.R.C., 1946.
[29] G. Amendola, M. Ciminello, I. Dimino, S. Ameduri, Real-time monitoring of a variable-camber aileron rib by original strain-angle transducer, J. Intell. Mater. Syst. Struct 28 (8) (2017).
[30] G. Amendola, I. Dimino, F. Amoroso, R. Pecora, in: Experimental characterization of an adaptive aileron: lab tests and FE correlation, Proceeding SPIE 9803, Sensors and Smart Structures Technologies for Civil, Mechanical, and Aerospace Systems 2016, 98034P (April 20, 2016), Las Vegas, 2016, https://doi.org/10.1117/12.2219187.
[31] Y. Le Sant, B. Deleglise, Y. Mebarki, in: An automatic image alignment method applied to pressure sensitive paint measurements, International Congress on Instrumentation in Aerospace Simulation Facilities, 1997. ICIASF '97 Record, https://doi.org/10.1109/ICIASF.1997.644658.
[32] R.M. Botez, A. Koreanschi, O.S. Gabor, Y. Mebarki, M. Mamou, in: Concilio "Innovative wing tip equipped with morphing upper surface and morphing aileron for greener aviation"., 3AF CEAS green aviation Conference, 11–13 October 2016, 2016 Brussels.
[33] Y. Mebarki, M. Mamou, M. Genest, Infrared Measurements of the Transition Detection on the CRIAQ Project Morphing Wing Model, 2009. NRC LTR AL-2009-0075.
[34] A. Koreanschi, O. Sugar-Gabor, R.M. Botez, in: New numerical study of boundary layer behavior on a morphing wing-with-aileron system, 32nd AIAA Applied Aerodynamics Conference, Atlanta, GA, 2014.
[35] R.M. Botez, A. Koreanschi, O.S. Gabor, Y. Mebarki, M. Mamou, Y. Tondji, F. Amoroso, R. Pecora, L. Lecce, G. Amendola, I. Dimino, Concilio in "Numerical and experimental testing of a morphing upper surface wing equipped with conventional and morphing ailerons" in: 55th AIAA Aerospace Sciences Meeting (9–13 January 2017 Grapevine, TX), 2017 eISBN: 978-1-62410-447-3.
[36] I. Dimino, A. Concilio, R. Pecora, Safety and reliability aspects of an adaptive trailing edge device (ATED). in: Proc. 24th AIAA/AHS Adaptive Structures Conference, January 4–8, 2016 San Diego, United States. https://doi.org/10.2514/6.2017-1425.

SECTION 7

Full Scale Realization, Safety, and Reliability

Editor:

Ignazio Dimino
The Italian Aerospace Research Centre, CIRA SCpA, Capua (CE), Italy

List of Chapters

CHAPTER

MORPHING TECHNOLOGY FOR ADVANCED FUTURE COMMERCIAL AIRCRAFTS

19

Miguel Á. Castillo Acero*, Federico Martín de la Escalera†, Yasser Essa†
Aernnova Group, Madrid, Spain Aernnova Engineering Division S.A., Madrid, Spain†*

CHAPTER OUTLINE

NOMENCLATURE

3AS	active aeroelastic aircraft structure
3D	three-dimension
AS02	application scenario N.2
ATED	adaptive trailing edge device
CFD	computational fluid dynamics
CIRA	Centro Italiano Ricerche Aerospaziali (I) (Italian Aerospace Research Centre)

Morphing Wing Technologies. https://doi.org/10.1016/B978-0-08-100964-2.00019-8

CS 25	certification specification 25 (EASA)
CURVED	new concepts for aircraft rear end architecture
DATCOM	USAF Stability and Control Data Compendium
DR	Dutch roll
EADN	enhanced adaptive droop nose
EASA	European Aviation Safety Agency
EU	European Union
FE	finite element
FEM	finite element method
FhG	The Fraunhofer Institute
FP5	European Research Framework Programme N.5
FP7	European Research Framework Programme N.7
FTP	flight test platform
GVT	ground vibration test
HST	DuraForm HST composite
HTP	horizontal tail plane
IFAM	Institut für Fertigungstechnik und Angewandte Materialforschung (D) (Institute for Manufacturing Technology and Advanced Materials)
NLF	natural laminar flow
PA12	polyamide material, Type 12
SARISTU	smart intelligent aircraft structures
SDS	selectively deformable structures
SFWA	smart fixed wing aircraft
ЦАГИ (TsAGI)	Центральный аэрогидродинамический институт (RUS) (Central Aerohydrodynamic Institute)
UAV	unmanned aerial vehicle
UniNA	University of Napoli, Italia
VMC	minimum lateral control speed
VMCA	minimum lateral control speed in flight
VMCG	minimum lateral control speed on ground
VTP	vertical tail plane
WATE	winglet adaptive trailing edge
WT	wind tunnel
C_L	lift coefficient
C_n	yaw moment coefficient
C_y	lateral force coefficient
g	acceleration of gravity
q	dynamic pressure $= 0.5 * \rho * V^2$
S_{ref}	wing surface (reference)
V	velocity
Y_{ext}	lateral force, acting on the rudder
P	air density
Φ	roll angle
β	sideslip (yaw) angle
δ_a	angle of attack
δ_r	rudder angle
avail	available

1 INTRODUCTION

The capability of aircraft movable surfaces to change geometry during flight (morphing) has attracted many designers and researchers over the decades [1]. A morphing aircraft structure can be defined as one that is able to adapt its external shape, even considerably, to match the mission requirements during flight. The concept of morphing is not new, but its implementation has usually been linked to penalties in weight, complexity, or cost; only in certain conditions were these outweighed by system-level advantages. The aeronautical design trends toward eco-friendly and highly efficient green aircraft, and the recent technology attainments make such tradeoffs more acceptable, driving demand for advanced compliant or kinematic morphing designs with fewer limitations and more benefits. In particular, recent developments in the field of smart materials [2] may help to overcome classical drawbacks (like space allowance and weight requirements) and improve the benefits of even traditional design solutions. The simultaneous, contemporary evolution of electromagnetic actuators [3–5], may also lead to important advantages.

The challenge is to design a system that is able to withstand the applied loads but is capable of changing its shape as well [6,7]. A multidisciplinary approach is necessary for merging this ability with integrated structural actuation systems, starting from early design development. This need substantially increases the overall configuration complexity from the very beginning. There is a multiplicity of alternatives to the benchmark configuration, which expands the domain of the dynamic behavior and aeroelastic coupled response analyses. Currently, such a technology is still far from an actual industrial, consolidated level of development; for this reason, manufacturers are reluctant to apply this innovative concept to real aircraft. However, recent increases in demand for specific know-how in satellite and unmanned aerial vehicle (UAV) construction have spurred much research toward structural morphing, resulting in an impressive leap forward [8–10].

Design implications aside, morphing wings give rise to a number of new manufacture, assembly, and verification challenges for structural systems. The number of components drastically increases as traditional frozen architectures are supplanted by kinematic chains or compliant devices. The increased number of parts that make up this kind of mechanical systems, results in an increased number of difficulties, as well as an uprise in the required manufacturing effort and design attention, especially concerning the aspects related to tolerances and assembly. As a result, production and design scientists are expected to cooperate even more than normal due to the massive growth in schematic complexity. Furthermore, because the structure is movable, special assembly rigs must be designed to preserve the kinematic characteristics and avoid local and permanent deformations. Specialized jigs and tools must be created to execute certain particular operations on these newly-complicated components [11].

These same considerations hold true for final integration onto the reference wing (which could host more than one morphing component) but on an even larger scale. Finally, the verification process is lengthened and complicated, because each of the multiple configurations the structure can assume has to be verified and subjected to different load distributions. There is yet another major difference with respect to the traditional, rigid structures: the massive presence of electronic elements and related cabling. This causes further design (e.g., allocation of limited structural space) and assembly difficulties (deciding when in the process to install cables, etc.). In this case, manufacture and verification are not directly impacted [12].

Going into detail of the manufacture steps, some further aspects shall be better pointed out. The increased number of components, together with their limited dimensions, leads to the necessity of the definition of new standards and new procedures, to avoid the explosion of unique elements to be produced with an embarrassing increase of development, operational and maintenance costs. Tolerances themselves should be assessed to new benchmarks, respecting the innovative configurations that are going to be faced and the specific dimensions of the different parts, without neglecting the circumstance that the various elements move with respect to each other. Moreover, it is believed that the realization of morphing subsystems should require specialized and separated competences within the industry, because of their peculiarities, somewhat a halfway between a structure and a machine.

In this chapter, significant examples of the manufacture process of morphing wing components are illustrated. In detail, the attention will be focused on four projects where Aernnova used their own background to develop innovative capabilities for the construction of adaptive structural modules. The European project SARISTU [13,14] is the example where those concepts were developed at an extreme level, targeting the production of a full-scale wing section for operative wind tunnel (WT) tests. In active aeroelastic aircraft structure (3AS) [15] and CURVED [16], another European and a national Spanish project, respectively, the attention was instead focused on the development of structured materials with innovative characteristics, suitable for the integration in the proposed innovative concepts. Finally, these experiences have been exploited for the assessment of a breakthrough adaptive rudder [17,18].

SARISTU project was a big step into the future assessment of morphing devices. It included many brilliant concepts, largely based on the use of COTS (commercial off the shelf) elements and practical enough to be applied to commercial aircrafts [19–21]. SARISTU full-scale wing prototype underwent successful WT tests, proving the possibility to overcome weight penalties by applying multidisciplinary design solutions and clever and integrated structural actuation architectures. As a result, the system was able to achieve a high control authority with respect to a conventional wing with limited mass increments. Among the morphing devices that were implemented on that experimental model, the manufacturing activities concerning the adaptive trailing edge device (ATED) preparation are herein presented. The testing process started on a reduced system, a two-bay demonstrator; results showed high levels of controllability and a good aeroelastic behavior. Also, the displacement field exhibited correct morphing behavior and no shape discontinuities. These attainments were the bases for realizing the final five-bay test article.

Low-speed WT tests on the full-scale SARISTU wing confirmed that the implemented design was able to produce improved roll moments. Profile shape morphing demonstrated the expected aerodynamic efficiency increase with a negligible effect on zero-lift drag. The design achieved the objective to keep mass penalty to a minimum and respecting the room constraints. The future developments of morphing technology is essential to increase aircraft versatility in view of a large-scale application on commercial aircraft.

3AS was perhaps the first major activity within the European Union (EU), where the adaptive structural concepts were exploited. In that frame, the need for manufacturing new particular materials was evident, in order to face to the new challenges a morphing structure would have presented. Among all, the necessity of high-level strains, following the large displacement fields naturally associated with any shape variation. The CURVED project moved from those attainments to make a step forward to the development of a new class of structured materials, approaching bio-inspired mechanisms. Both those experiences are currently finding the way for more challenging applications in the development of a morphing rudder, where an innovative structural morphing machine can have a compulsory role for the proper attainments of the prescribed objectives.

2 ATED MANUFACTURING

SARISTU (smart intelligent aircraft structures, 2011–15) was a multidisciplinary thinking project since the very preliminary configuration approach, and was based on a tight team working with reference to both the challenging design and manufacture phases. One of the main activities of Aernnova in this project was related to the development of the Adaptive (Morphing) Trailing Edge Device (ATED), within the AS02 (application scenario N.2) "Adaptive Structural Tailoring of Trailing Edge for Enhanced Aircraft Performance," together with the project partners UniNA, CIRA, Technobis, Mare Engineering, and IFAM (Fraunhofer Institute).

2.1 THE MORPHING SYSTEM

In detail, Aernnova was fully responsible for the manufacturing and assembly of the two-bay and five-bay demonstrators, whose internal structures are reported in Figs. 1 and 2, respectively.

The system was aimed at providing a wing trailing edge with deforming capabilities along the chord, so to give rise to a variable camber wing section. For this reason, the component was made of segmented parts, assembled through distributed hinges and rigid, stiffening connections. This configuration led to a remarkable number of different parts and a noticeable increase of the manufacturing and assembly phases. The ATED was finally integrated within a full-scale WT demonstrator, hosting two other morphing devices: an adaptive droop nose (EADN) and an adaptive winglet (WATE), Fig. 3.

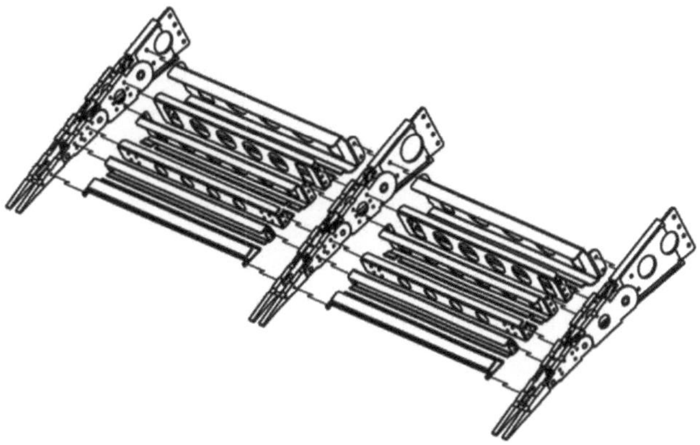

FIG. 1

SARISTU two-bay demonstrator. Exploded view.

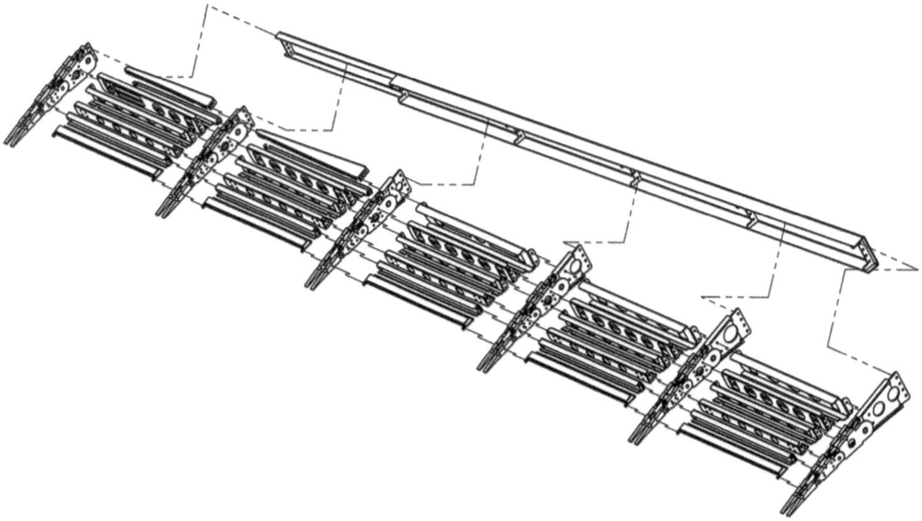

FIG. 2

SARISTU five-bay demonstrator [20]. Exploded view.

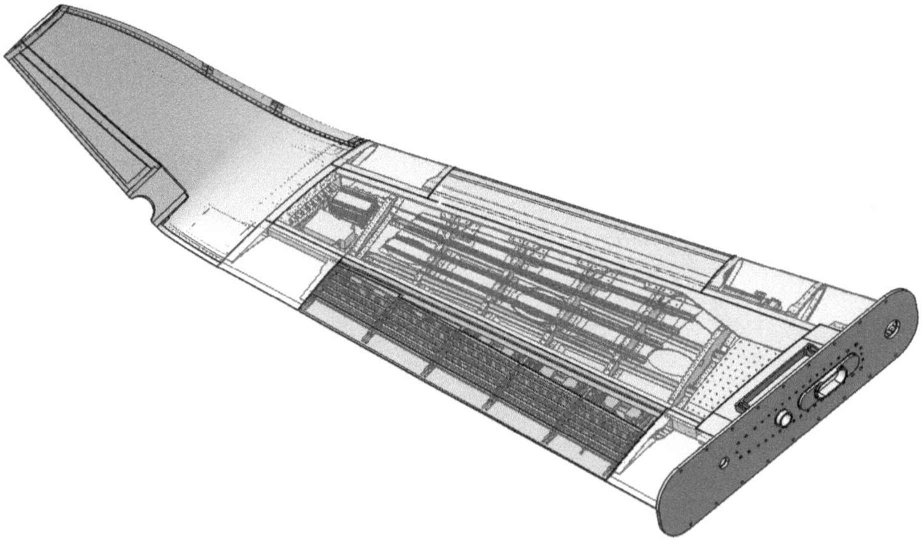

FIG. 3

Full SARISTU wind tunnel demonstrator.

2.2 MANUFACTURING

The SARISTU project gave essential experience from the adaptive structures manufacturing and assembly point of view. In particular, those activities allowed highlighting some very peculiar aspects of the processes needed to accomplish the realization of such components. Starting from the ATED two-bay, an intermediate demonstrator, Fig. 4, essential hints were obtained for the correct development of the final five-bay prototype.

Moving from the designers' drawings, the manufacturing process started from their critical review and the assessment of the optimal materials selection for machining and bending processes. In detail, the ATED had an internal skeletal structure with the role of ensuring the correct transmission of the movement from the actuator system and guaranteeing the structural integrity under the loads action. To accommodate the cable routing (from both the sensor and the actuator systems) and to attain a suitable weight saving, specific holes were included in the layout. The resulting architecture was made of a great number of elements, different from each other because of the object geometry, strongly tapered. The manufactured components included:

- Segmented ribs, aimed at allowing the rotational movements, necessary to give rise to the desired structural morphing.
- Longitudinal spar-x's, supporting the metallic sections of the morphing skin and contributing to the torsional rigidity of the ATED.
- A rear spar (the main spar), having the double role of carrying the actuator motors and assuring the interface with the wing box; in other words, this element sustained all the forces coming through the ATED, either internal or external.

The elements of the rotary actuator system transmission chain were manufactured and assembled to give rise to a system able to transform the produced rotations into longitudinal displacement of the actuation bar, in turn converted by the whole mechanism architecture into a vertical displacement. The resulting device is shown in Fig. 5. In the operative configuration, the motor was hosted within specific rooms, built along the main spar.

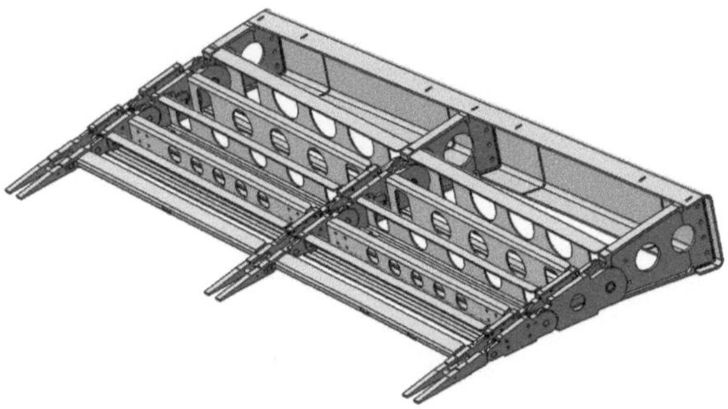

FIG. 4

SARISTU 2-bay demonstrator: internal structure display.

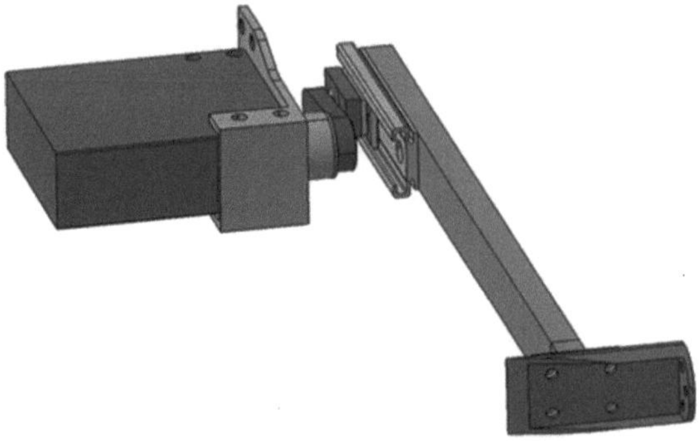

FIG. 5

Actuator chain: CAD representation.

The extremal part of the ATED accommodated a rigid cap, aimed at ensuring a smooth airfoil surface. Such a tip skin is reported in Fig. 6, where some steps can be seen, designed to have an easy and strong connection with the other surfaces.

The part of the trailing edge, running from the rear spar to the first movable segments of the adaptive device was called "dead box" (DB) to indicate it is a rigid zone, with no movement occurring inside. It hosts the core elements of the kinematics, that is the motors and the related transmission components. The DB skin was one of the most critical element of the proposed prototype. In fact, its strongly shaped geometry obliged the designers to draw a careful layout that could take into account the constraints of the sketch, the needs to ensure a strong connection with the spars and the skin and, finally, the necessity to avoid interferences with the mechanisms below. Furthermore, it had to secure a proper interface between the Wing Box and the Trailing Edge structure. In fact, according to the selected configuration, the loads between the two structural regions were transferred through the top and the bottom skins. The DB skin was made of upper and lower panels, as shown in Fig. 7. In that picture, a detail is shown, concerning the configuration at the first two ribs (toward the wing root). In this case, the structural elements that ensure the structural continuity were the rear spar flange, the first spar-x flange, and the rib flange.

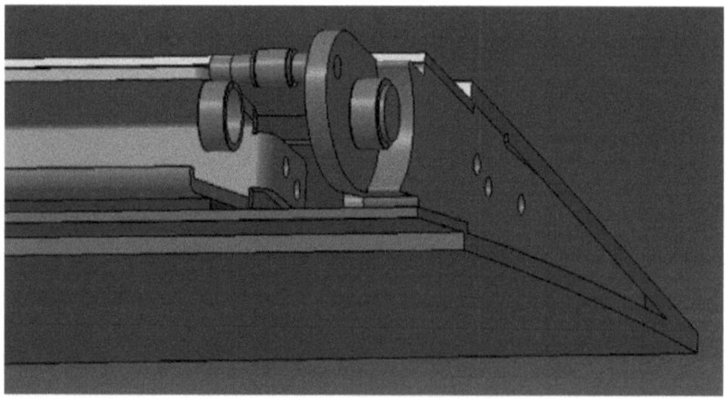

FIG. 6

Tip skin.

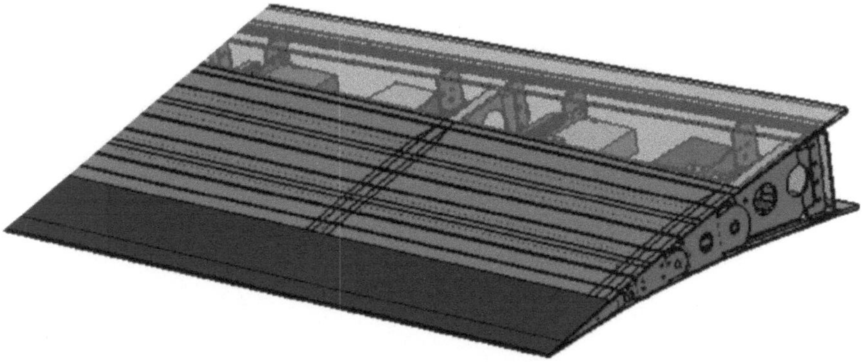

FIG. 7

Skins connecting ATED and torsion box (wing box).

The morphing skins were located along the upper and lower surfaces of the trailing edge. They were made of aluminum C-sections joined through flexible foam and a silicon cover, Fig. 8. The morphing skin design and its realization was performed by FhG IFAM [22].

2.3 ASSEMBLY

For the assembly process, an appropriate and original tooling device was necessary for correctly positioning the components and properly handling the drilling and riveting processes at the established points. This system was originally designed, developed, and built by Aernnova. Among its main features was the capability to ensure a perpendicular surface to the digging tool, compulsory to have a good match between the skins and the internal structure. Moreover, because of the silicone protective layer placed over the morphing skin, it was essential to have a certain easiness of operation over such a nonconventional and delicate surface. Otherwise, the risk of producing distortions and breakages would have been high. Furthermore, it allowed having a robust support where deploying the morphing

FIG. 8

Morphing skin.

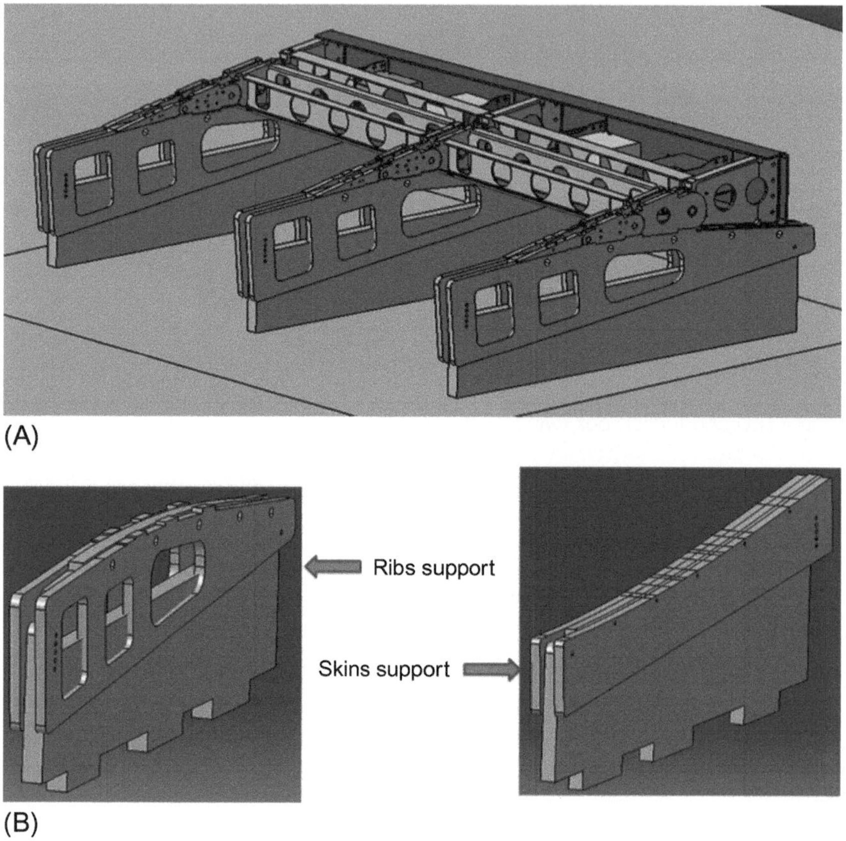

FIG. 9

Assembly tool for rib and skin system: general vision (top) and support details (bottom).

elements, without giving rise to particular strain or stress fields due to its mobility (a free kinematic chain, at this stage of the operations). In Figs. 9 and 10, it is evident how the assessed system was able to assume different positions, simply sliding onto the shaped supports. The tooling considered the natural wing twist across the spanwise direction in order to avoid any residual stress after the assembly process was concluded. The skeleton was then assembled in various subsequent steps starting from the ribs and going to the motors, the spars, and, finally, the actuator bars. For the high number of elements and the complex geometry, particular attention was paid to define the correct operations sequence to minimize possible installation interferences.

As stated, the drilling process on the morphing skin was very demanding. In fact, standard operations could damage the silicone coating. Milling was found to be a brilliant alternative solution, Fig. 11. The assessed tools allowed safe performance of this important task. The actions were different for the upper and the lower surfaces, because the former was riveted with counter-sink blind bolts, while the latter was fitted with rivet nuts and counter-sink bolts. It is important to emphasize that the complete milling operation was conducted before the sensor system installation and the cable routing deployment, as it was necessary to have clean surfaces along those other processes.

FIG. 10

Full vision of the assembly tool: supports and drilling facility (white portal on the left).

FIG. 11

Morphing skin milling.

The sensor system was made of two parts [23]. The spanwise system was arranged with ribbon tape sensors installed in the inner section of the upper tip skin. A ribbon tape sensor is made of fiber optics embedded into a composite layer (carbon, fiberglass, or other). It was bonded to the reference structural component to measure the spanwise elastic deflection of the ATED. The chord-wise system instead monitored the shape variation of the morphing system and recorded parameters related to the vertical

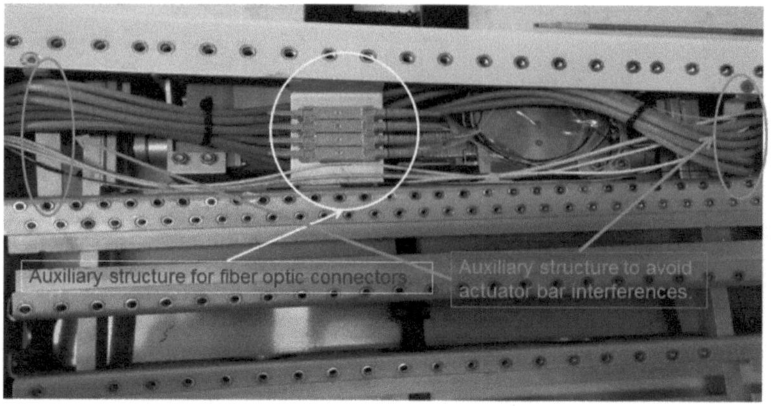

FIG. 12

Actuator cable routing.

kinematic displacements. In both cases, the movement info was gathered after an elaboration of strain data. Cable routing was carefully done in order to avoid interactions with the main structure and possible kinematic conflicts that could have arisen during morphing with the other system elements. Both electrical and optical cables were sized and deployed to pass through the preconditioned spar and rib holes. Cables were finally bounded together into one or more bundles and then constrained to some architectural path key-points through support fixtures, Fig. 12.

As a final step of the system integration, the upper and the lower morphing skins were riveted and screwed to the ATED internal structure, respectively. Figs. 13 and 14 show the AS02 two- and five-bay demonstrator at their final stage. In particular, in Fig. 13, some bigger holes on both the sides and along the middle line of the test article (chord-wise direction) are evident. They were created to host some eyebolts to transfer, safely, the external load to the inner ribs during the functionality and the GVTs (ground vibration tests) experimental campaigns. These features do not appear on the five-bay final

FIG. 13

Assembled final stage, two-bay demonstrator.

FIG. 14

Internal structure of the five-bay demonstrator.

ATED demonstrator, Fig. 14, because in that case the structure should have been mounted over the complete wing section and directly tested in WT, after a limited functionality verification and some general laboratory tests on the whole system (modal analysis, for instance).

The assemblage of the complete morphing wing took place at the Alenia Aermacchi facilities, in Pomigliano D'Arco (I). The 5.5 m span full scale segment included several morphing demonstrators, targeting different objectives: a morphing leading edge (Enhanced Droop Nose, EADN [19]), to augment take-off and landing performances; an adaptive trailing edge (herein discussed); an adaptive winglet (winglet adaptive trailing edge, WATE [21]), to increase the aircraft capability under gust actions and shape the load function along the wing. Each of them hosted in turn different embedded actuator and sensor systems, responding to different specs.

The complete wing was tested in the T-104 WT at TsAGI, Moscow (RUS), Fig. 15, up to a flow speed of about 120 m/s, proving the functionality of the different systems and their potentiality to reduce fuel consumption and to enable natural laminar flow (NLF). They were designed, sized and manufactured with respect to a reference wing for a 130 pax jet regional aircraft, 3000 nm range, 0.75 M cruise and 10,668 m flight level [24,25].

2.4 TEST CAMPAIGN

Soon after the realization of the two-bay demonstrator, and as an integral part of the five-bay complete ATED prototype manufacture, a test campaign was organized to get information about the test article behavior. In particular, it was aimed at getting indications about its aeroelastic performance through targeted vibration tests. In fact, these data could have been essential to confirm or change the design philosophy prior of the final demonstrator manufacture and, definitely, the WT tests.

The experiments referred to two different boundary conditions: clamped along the main spar, to simulate the "real" operative constraints (attached to the wing box), or completely free, to avoid any influence of the external elements (like the test rig). Aerodynamic loads were simulated by means of a "static load tree," Figs. 16 and 17. Loads were applied to both static and vibration excitations in order to

FIG. 15

T-104 wind tunnel facility at TsAGI (RUS): SARISTU demonstrator, installed on the WT support (right).

Courtesy of TsAGI.

evaluate possible effects of the occurring deformation field over the structural response. In this case, large strains arose, intrinsic to the implementation of morphing. Three basic referred shapes were considered: the baseline and the max excursions, both up and down (trailing edge deflected negatively or positively). However, several other morphed shapes were tested in order to have a complete overview of the system answer. Also, the control system was proved in both open (feedforward) and closed (feedback) loop configuration. In the latter case, in fact, the system transfer function stability could have been affected by the control presence.

Basically, a feed-forward control has the aim to impose a certain shape to the structure. It should guarantee that a specific configuration is maintained under impulsive or constant forces. A feedback control system, instead, incessantly modulates its action as a function of the information, gathered by the sensor network. Instabilities can happen or, simply, the control may show itself unable to preserve a certain geometry (for instance, leading the structure to another configuration). In this way, it is understood that the overall system transfer function may change. Therefore, the dynamic characterization of the augmented architecture is necessary in order to understand for instance the aeroelastic behavior, whose characteristics (including the instability region, consequence of certain interactions between the structure and the airflow) may exhibit important diversities with respect to a traditional configuration. Its original performance could then drastically change. After these preliminary investigations, conducted under static or shock forces, a GVT campaign did take place, under the same aforementioned conditions. A modal analysis was performed for any of the considered geometries, computing eigenfrequencies, complex eigenvalues, and eigenvectors. Test results were recorded via three-dimension (3D) displacement cameras and a dedicated vibration acquisition system.

FIG. 16

Aerodynamic load application system on the 2-bay demonstrator.

FIG. 17

Specimen free-free conditions and vibration testing.

Structure displacements showed the attainment of the target design shapes with minor geometrical deviations. The test article afforded well the action of the max design loads while the actuation system showed good authority in attaining the objective geometries. Modal results revealed the structure had a good dynamic response in terms of aeroelastic foresight. A representative example of the modal shapes is reported in Fig. 18 [11,26].

Some further tests were devoted to verify the response of the morphing skins to an internal gradient, simulating the difference between the aerodynamic (external) and the static (internal) pressure, in flight. This action was reproduced through an inflated balloon, placed into the structural body, Fig. 19. It was proved that the skin suffered very limited vertical excursions under that particular action

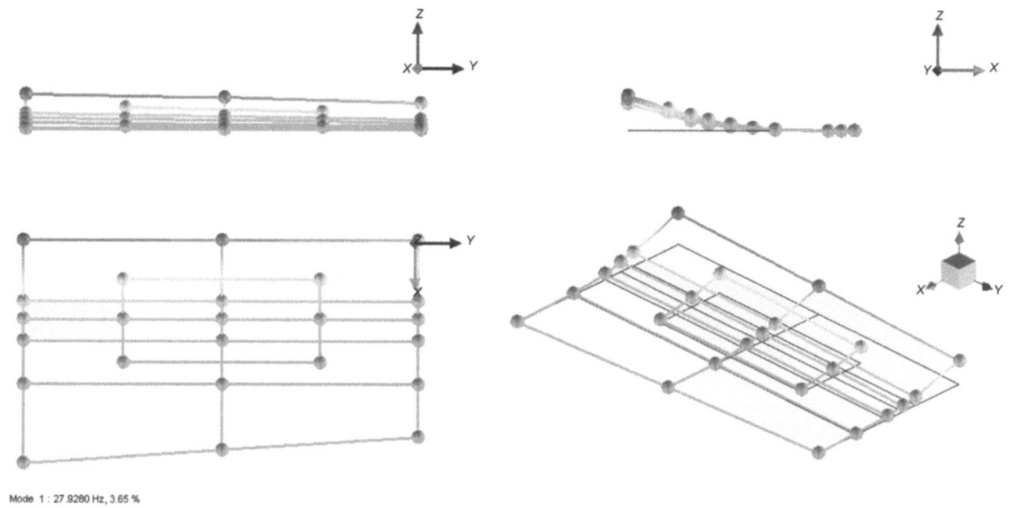

FIG. 18

Example of 2-bay demonstrator modal shape: 1st eigenvector.

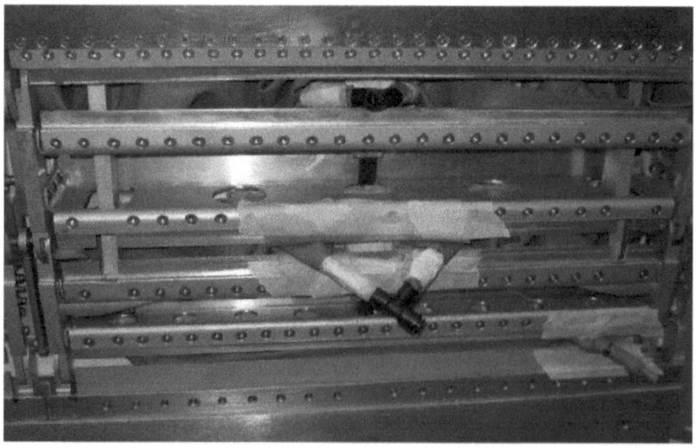

FIG. 19

Test set-up for imposing an inner pressure to the ATED skins.

(less than 1 mm height), i.e., predicted displacements were expected to give rise to minor aerodynamic effects, according to computations performed by the same research group. Structurally, tested balloons exploded without causing any permanent deformations.

2.5 CONCLUSIONS

SARISTU project proved the economic advantages of implementing morphing structures in-flight. These systems, in fact, ensured an appreciable increase in terms of the overall aircraft efficiency, although the increases of weight and complexity were not negligible. In particular, the manufacture of hundreds of pieces is not sustainable from an industrial point of view and should be strongly limited before such engineering solutions can be actually implemented. The simplification of the proposed devices may be then considered a main driver for future studies. Analogously, the use of lighter and better workable materials, specifically suited for these applications, can be another remarkable field of investigation.

3 OTHER EXPERIENCES

Aernnova has an extensive background of multidisciplinary experience related to the design of morphing structures. It started with an FP5 European project, continued within internal developments inside the company, and finally arrived to the abovementioned SARISTU and a project funded by the Spanish Government. Among other topics, these studies were also devoted to outstanding material candidates for the development of adaptive architectures.

3.1 3AS PROJECT

In the framework of the EC (European Commission) co-financed Fifth Framework project "Active Aero-elastic Aircraft Structures"—3AS [27], Grant No. GRD 1-2001-40,122, Aernnova was implicated in the development and analysis of "selectively deformable structures" known as SDS, load carrying cells of variable geometry, Fig. 20. The attractiveness of this kind of structures is in its unique

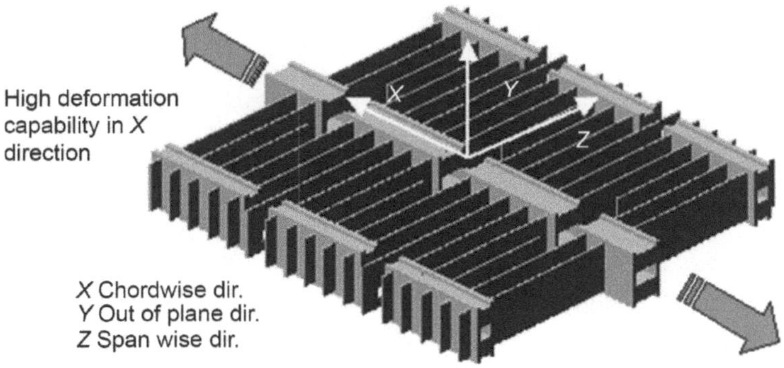

FIG. 20

SDS unit cell.

properties, allowing in-plane deformation with little transverse deformation, because of a Poisson ratio very close to zero [28]. Part of the work was focused on the feasibility and the applicability of these structures to the aircraft industry, with the objective of improving the aerodynamics and, possibly, the structural and aero-elastic characteristics. A lot of work was then performed on these topics, till arriving to a real-scale demonstrator, Fig. 21. Full scale validation tests were carried out, Fig. 22, confirming the predicted performance of the proposed inserts.

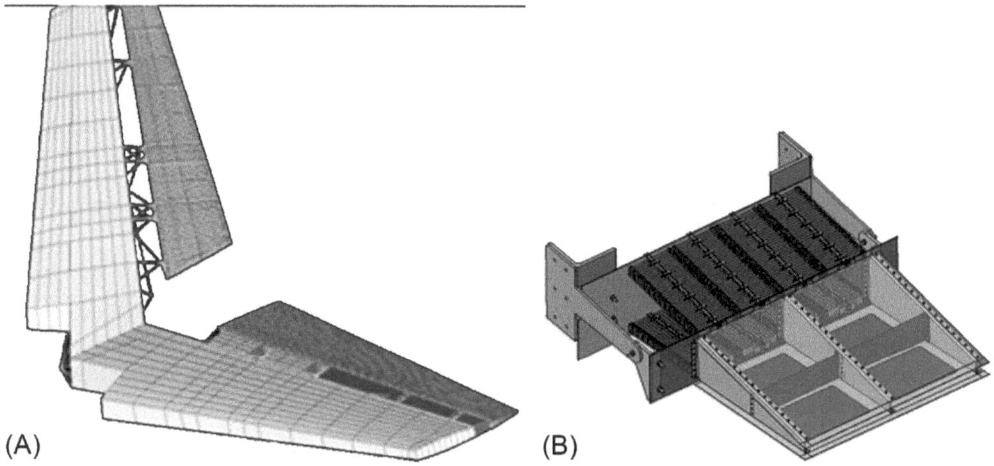

FIG. 21

A sketch of the full-scale specimen representing the stabilizer-elevator union with SDS inserts.

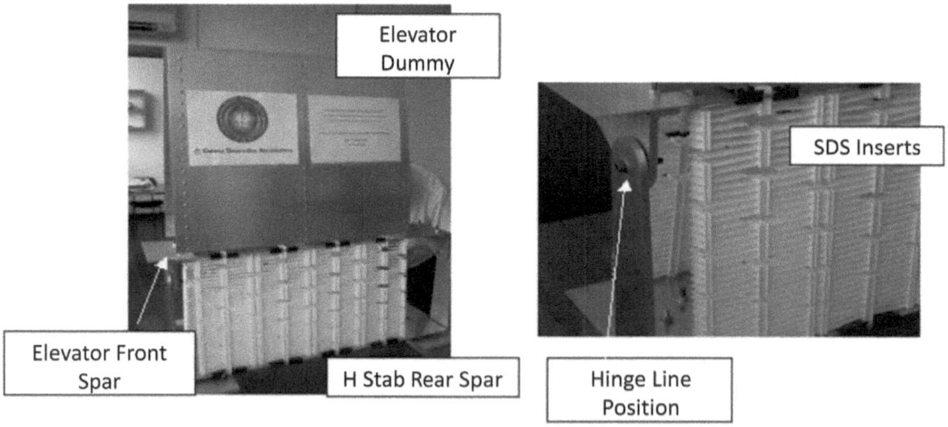

FIG. 22

The test rig to investigate the SDS inserts performance.

3.2 CURVED PROJECT

During the CURVED project [16], funded by the Spanish Government, substantial work in the design and development of morphing structures for a UAV trailing edge was performed. The internal design focused on the use of a bio-inspired system as a shape changing mechanism. It recalled the functionality of a vertebral apparatus, starting from a previous idea [29]. The architecture allowed alteration of the structural shape through the use of cylindrical bodies, Fig. 23. This concept permitted the development of a novel trailing edge architecture. Each cylinder was intersected with a very stiff web, terminating with two external pins. By connecting the subsequent solid pins, the various vertebrae were linked each other. This arrangement was the core of the system. The connections had the additional function of

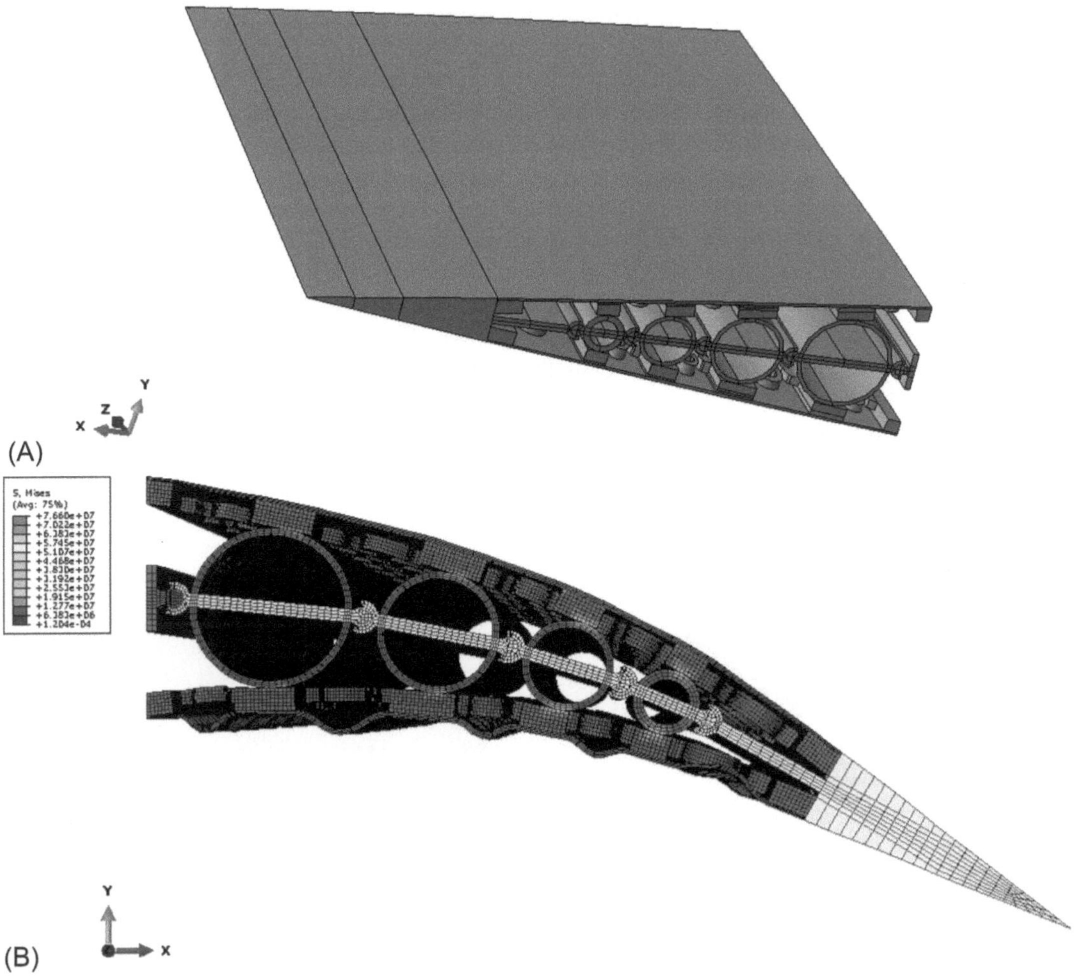

(A)

(B)

FIG. 23

Bio-inspired mechanism and achieved displacement.

allowing a rotational movement at fixed axes (the pin axes), inducing structural displacements for modifying the shape of the reference device. The vertebrae fitted the trailing edge geometry and departed from the torsion (wing) box up to the free end (wing tip). The upper and lower parts of the vertebrae were joined to the upper and lower skins and acted like the skeleton support. In order that this layout could be actually implemented, morphable skins were studied as suitable alternatives to classic rigid panel concepts. As resulted in several researches, the development of these elements is a crucial step for the attainment of deformable structural systems. Among the different solutions that were investigated, the "Double-C" outline, Fig. 24, permitted achieving a sound directional capability. They were made of series of stringers and other minor elements working as springs. As a result, the skin acquired flexible properties in the morphing direction (extension) and rigid properties in the normal direction (local bending) with very low transversal deformation.

The design of the morphing skin structures was the output of an investigation aimed at finding the optimal arrangement among several lattice architecture possibilities, Fig. 25. Their definition was based on numerical models and the mechanical characterization of the unit cell. Abaqus Finite Element (FE) solver was used in this process. After the individuation of the best elementary individual, a lot of work was performed to optimize and characterize the resulting architecture, Fig. 26.

In the same project, Aernnova carried out an extensive research to select a suitable process and a suitable material for the deformable skin. The research led to the selection of the SLS (Selective Laser Sintering) manufacturing technique and the PA12 polyamide material. This latter choice was mainly linked to its good ductile and toughness properties combined with a low commercial cost and low weight characteristics (extremely important for morphing architectures that naturally tends to be more complex and heavy with respect to classical systems). Among the PA12 drawbacks, remarkable aging effects should be cited, due to its hygroscopic nature: humidity impoverishes its properties, making it brittle. In this sense, experiments were performed by saturating manufactured specimens with water particles and then conducting static tests to verify tensile and fracture outcomes. Different material versions, achieved by incorporating hard particles, were considered for improving its properties and

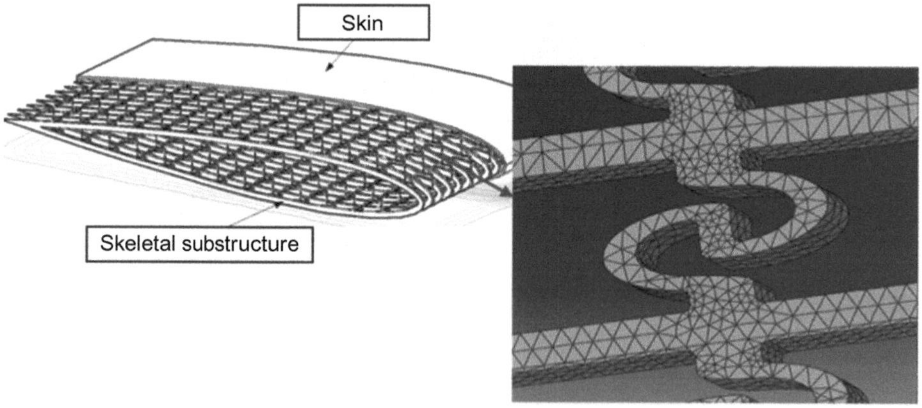

FIG. 24

Double C-shaped coating panels.

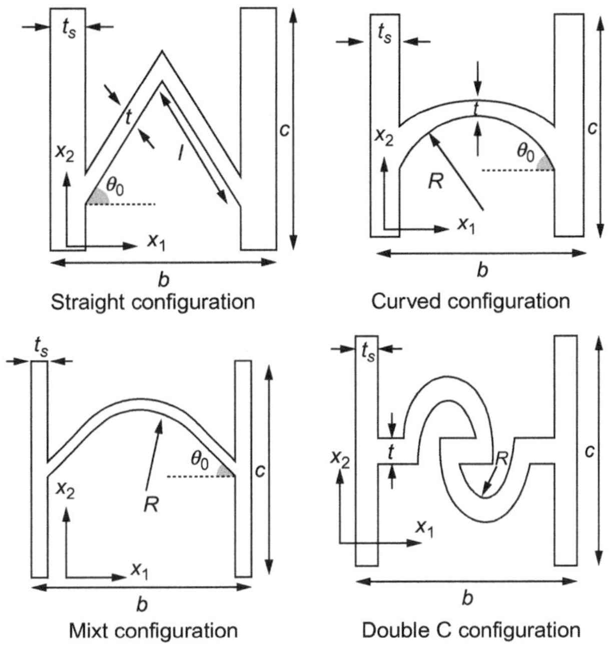

FIG. 25

Unit cell of the investigated skin lattice structures.

compensating the moisture disadvantages. The characterization process was simple. First, representative items were manufactured and submerged in hot water (90°C) at constant temperature for accelerating the saturation process. Other reference specimens were kept dry at the same temperature (dry condition testing). In the picture, representative of the described investigation process, two other materials behavior is also reported for comparison, namely GFN (Glass Filled Nylon, containing glass beads) and HST (DuraForm HST composite, containing ceramic fibers) Fig. 27.

After this process, a mechanical test campaign was performed to characterize the different types of PA12, before realized. Standard norms and regulations were followed at this step, involving static tensile, compression, fracture and fatigue analysis, Fig. 28. Interesting results were achieved. For instance, the yield strength was found to generally improve under dry conditions; furthermore, varieties of the target material, altered with spheres and fiber compounds, showed no general deviations in the structural characteristics with aging, while the original one presented corrupted toughness and tensile properties.

The project also investigated the real applicability of the developed concepts on commercial aircraft or UAV, for instance examining the effects of the application of such material on the aerodynamic behavior. A dedicated computational fluid dynamics (CFD) analysis was then carried out as it is crucial to understand the changes produced by this kind of skins. The investigated benchmark model showed the occurrence of some potential flow instability, Fig. 29. The final configuration of the morphing skin may be appreciated in Fig. 30, where some manufactured samples are shown.

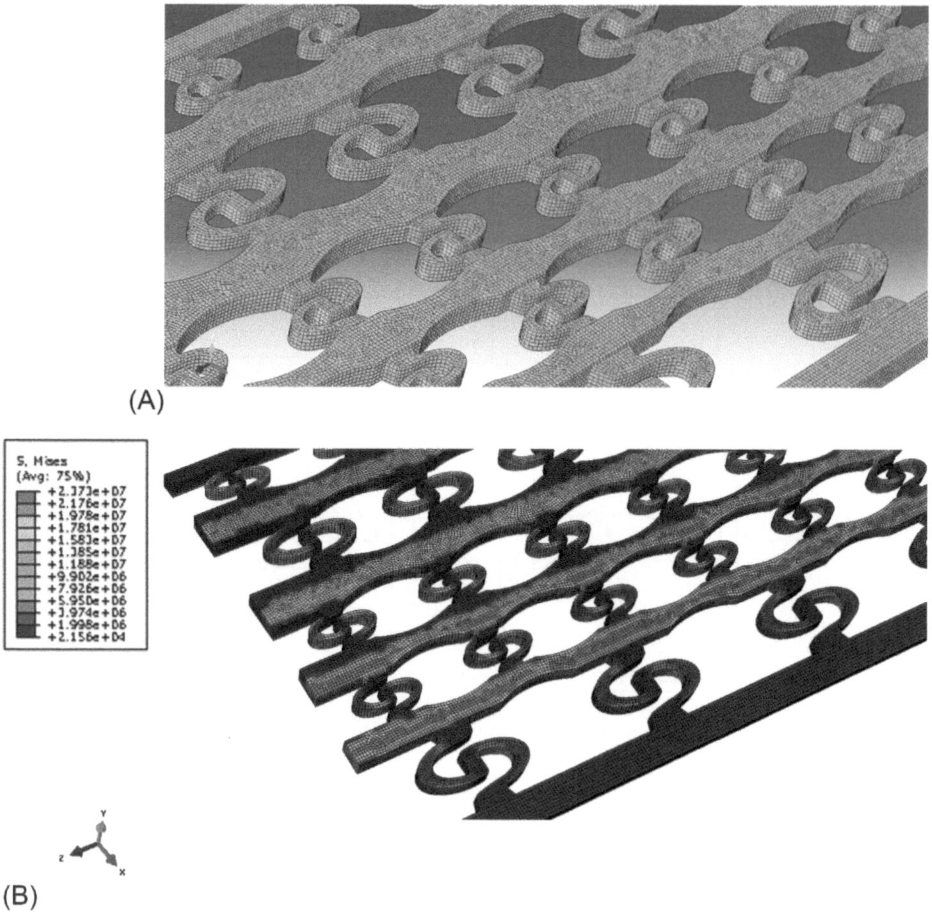

FIG. 26

Double C-shaped coating panel: Abaqus FEM analysis.

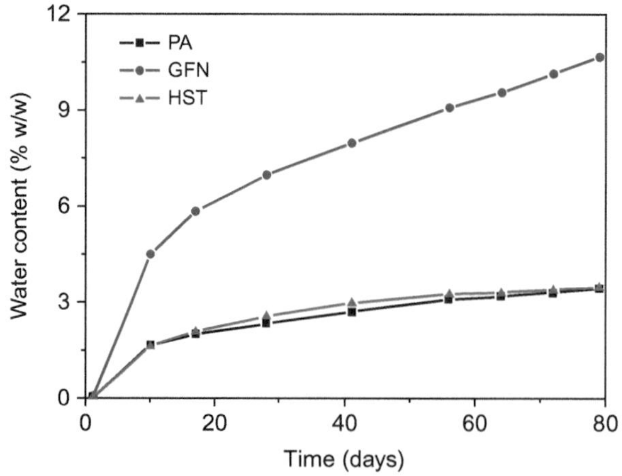

FIG. 27

Water saturation process.

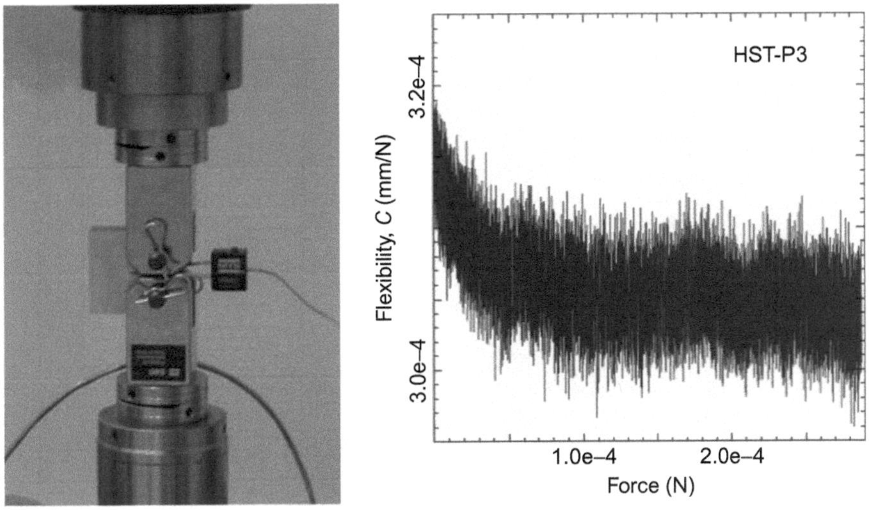

FIG. 28

Fatigue testing and cyclic load application.

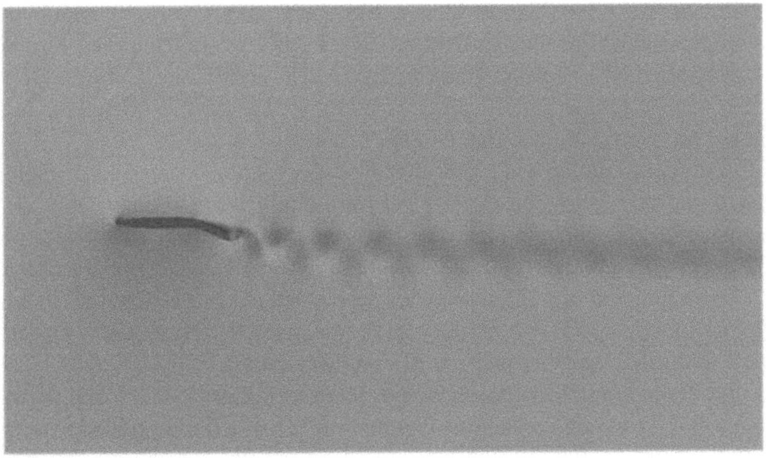

FIG. 29

Computed effect of the morphing skin presence on a selected airfoil (CFD analysis).

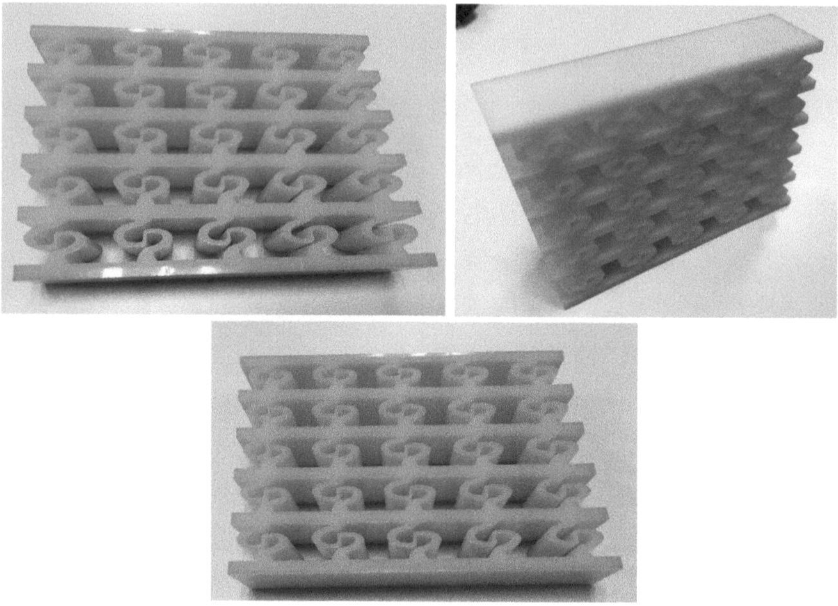

FIG. 30

PA12 double-C shaped panels, as manufactured via SLS.

4 FUTURE STUDIES—THE MORPHING RUDDER

Current developments on EU Clean Sky and Clean Sky 2 (2016) are going to enable the incorporation of different technologies up to flight demonstration [30]. In the case of Clean Sky FP7 smart fixed wing aircraft (SFWA), it is planned to demonstrate in-flight NLF and linked technologies on the external wing sections of an A340 flight test platform (FTP). One of the further objectives is to verify hybrid laminar flow (HLF) on a vertical tail plane (VTP).

As concluded in previous studies [17], the smoother aerodynamic pressure distributions originated by morphing control surfaces may constitute a relevant enabler for laminar flow systems. Adaptive wing dynamics variation and performance improvements were object of several researches in the recent past [31,32]. A possible future application of these innovative devices regards the rudder of commercial aircraft. The implementation of those concepts proved to bring a potential increment of 15% on lateral force. Similarly, advantages for boundary layer transition from laminar to turbulent (delay) were previously verified with SMA (shape memory alloys) actuators [33,34].

A significant number of studies in compliant mechanisms for aircraft [35,36], wind energy [37,38] and other UAVs [39–41] are available. At this time however, to the best knowledge of the authors, there are no previous publications on static and dynamic stability effects of morphing lateral control surfaces. The most similar recent studies were performed on the flight control of aircraft morphing tailless configurations from different perspectives, as aeroelastic improvements [42] or the use of distributed shape-change effectors arrays [43,44]. On the other side, the development of new UAV architectures gave rise to a relevant number of publications on lateral stability analysis and control [45].

4.1 SYNTHESIS

Aernnova developed an application case study on this topic, with reference to a two-jet wing for a 100-passenger plane. This layout is the preferred choice for the high fuel efficiency and the maintenance costs reduction with respect, for instance, to a four-engine motor system. The research aimed at increasing the rudder aerodynamic efficiency by the introduction of a morphing solution that could enable its size reduction. This objective was not limited to potential weight and drag diminutions but explored further operational benefits on static and dynamic stability. This activity took advantage of the previous studies concerning the development of suitable morphing materials, previously illustrated.

The proposed system could morph chord-wise, while maintaining straight lines spanwise. The deflection angle was defined as the space region between the nondeformed chord and the line joining the hinge with the morphed aft trailing edge tip, Fig. 31. For a given deflection angle, the adaptive fin section presented a curvature that is the base for the increased aerodynamic efficiency. The development of an airworthy structural solution was faced in previous works, [18,46] Fig. 32. CFD simulations were carried out by using the OpenFoam [47] code to better understand the global effects of replacing a conventional with a deformable vertical command plane, under the action of operational aerodynamic loads. Under the most critical conditions, both the system stiffness and strength were verified. Structural analysis was performed through the Abaqus FE solver, Fig. 33.

In the analyzed case, the most demanding excitations were related to lateral gusts during approach and landing, and the engine out during take-off. The latter was indeed the most challenging requisite, mainly due to the contemporary trend to increase engine thrust for ensuring higher transport capacities. That situation was therefore taken as the design condition. Studies like the one herein presented are required prior to flight clearance of the aircraft demonstrators where these devices should be installed. Those investigations partially covered the airworthiness requirements of a morphing rudder from the static stability point of view. The aircraft reference configuration referred to 1 g stationary load in cruise. Following, the results concerning the augmented lateral directional static stability of a transport aircraft are reported, as the rudder was reconfigured into an adaptive device to provide a constant 15% larger lateral force during deployment.

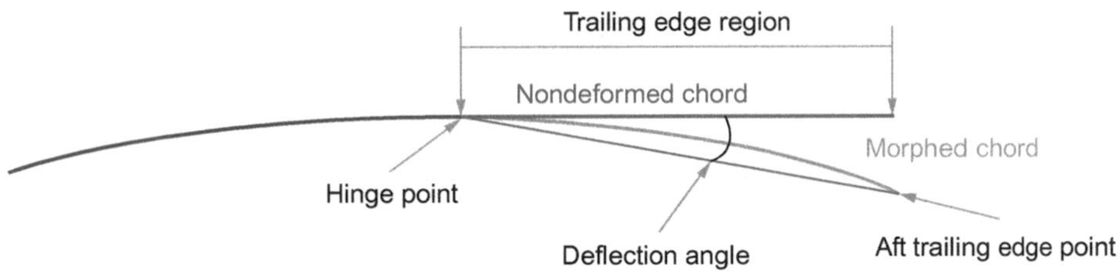

FIG. 31

Definition of the deflection angle of a morphing element.

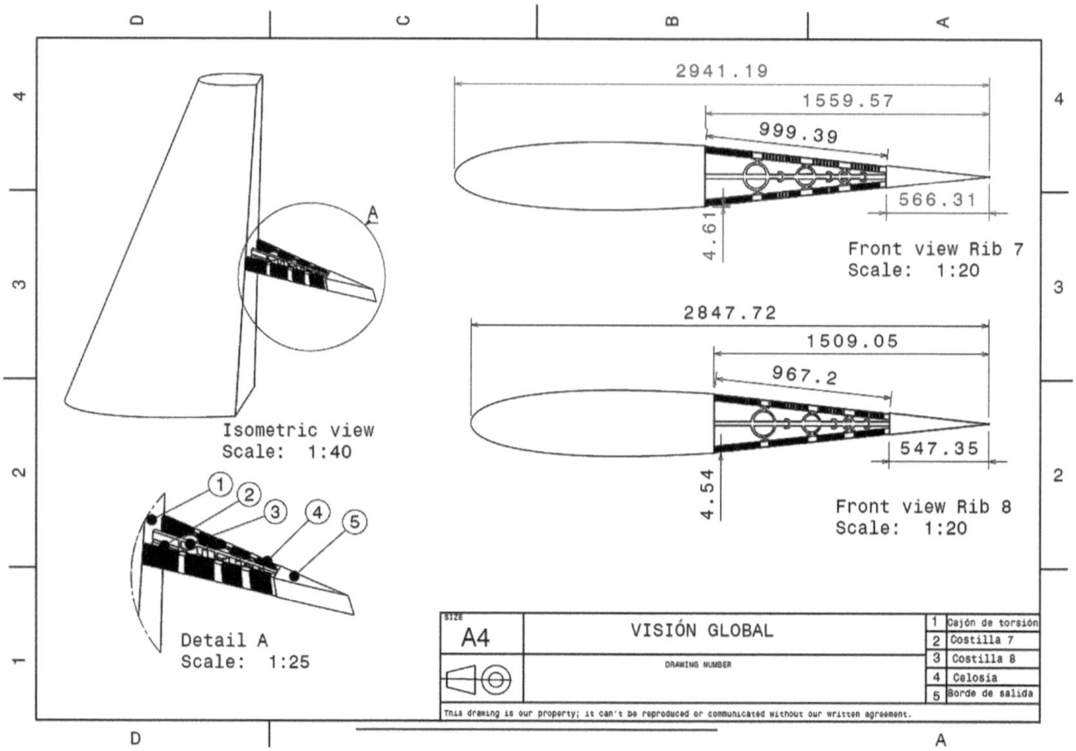

FIG. 32

Morphing rudder structural architecture.

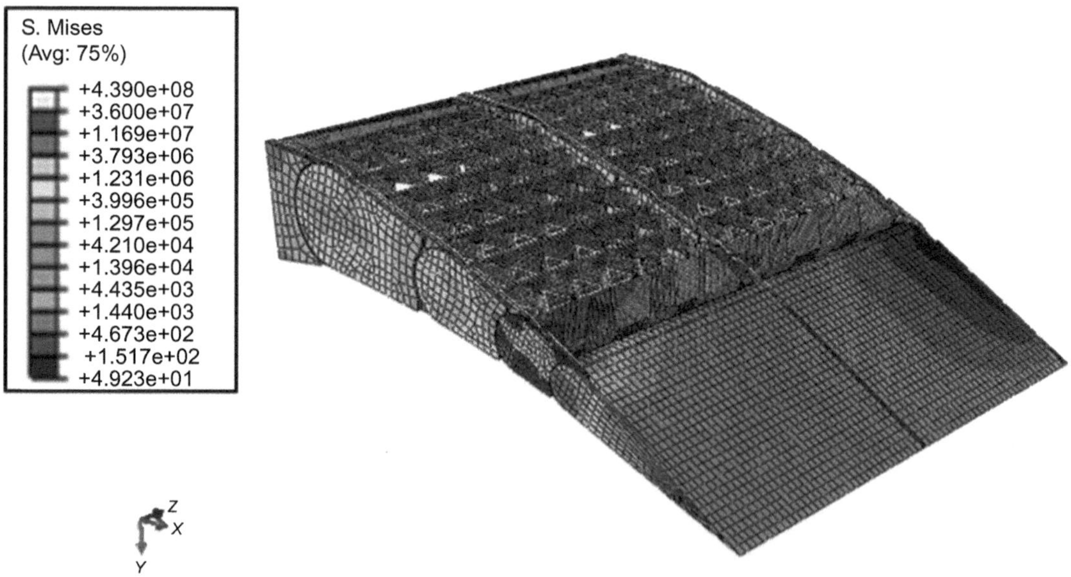

FIG. 33

Abaqus static mechanical testing for the proposed adaptive rudder.

4.2 MANUFACTURING CHALLENGES

There were expected difficulties in manufacturing and assembling a morphing rudder, fully compliant with the airworthiness requirement. The selection of the actuators was critical. They had to provide enough force and displacement while having a low weight and the capacity to fit within the space available in the trailing edge torsion box. The selection of the actuators was strongly dependent on the orthotropic stiffness of the morphing panels described in the previous chapter, also weight constrained. As usual the actuator power had to be large enough to deform adequately the structural system chord-wise while resisting the external aerodynamic loads.

Another challenge was represented by the manufacturing of highly orthotropic panels by a certifiable aerospace method, as already discussed before. The final integration of the proposed solution should finally comply with functional, weight, and performance requirements; for that reason, the assembly of the complex resultant mechanism should have been very accurate and should have allowed easy maintenance service.

4.3 LATERAL DIRECTIONAL STABILITY ANALYSIS

Lateral-directional static stability had to be investigated in balanced and unbalanced flight conditions. Under equilibrated and balanced situations, the three characteristics to be examined in steady flight, are:

- Static directional stability: variation of directional control forces and positions of the rudder as slip angle changes.
- Static lateral stability: variation of forces and positions of ailerons when the slip angle modifies.
- Lateral force characteristics: variation of roll angle when yaw angle changes.

For nonequilibrated (or dynamic) conditions, it is required to study the characteristics of three lateral directional eigenmodes: the Rolling mode, the Spiral mode and the Dutch Roll (DR). It results that the rudder modification into a variable-shape system does not affect the aircraft dynamic characteristics but the system control matrixes, respectively called [A] and [B] [47,48]. The associated variation modifies in turn the dynamic margins of stability and their effects.

The aircraft key parameters are included in Table 1. This set of data was partially obtained from reference [49]. In that table, HTP stands for horizontal tail plane.

4.3.1 Static stability analysis

As a first step, the aerodynamic influence coefficients, or simply aerodynamic coefficients, were derived, due to the lack of published lateral stability derivatives data for the referred type of aircraft. Indeed, there are many references in literature to obtain those parameters with CFD [35] or finite element methods (FEMs) [36]. Herein, the used approach is well documented [50], and the achieved information was compared with published data as DATCOM [51]. According to the reference configuration (engine out during take-off), an 85 m/s velocity was considered, at sea level (dry air density 1.237 kg/m^3).

That kind of flight situation is characterized by a rudder deflection that generates aircraft lateral force and heading moment that compensates thrust asymmetry. The consequent slippage angle requires asymmetrical deflection of the ailerons to balance the force system and achieve momentum equilibrium. In the studied case, the maximum rudder deflection was set to 30 degrees, for both conventional

Table 1 Reference Aircraft Key Parameters

Reference Aircraft Parameters	Value
Wing dihedral angle (degree)	5.0
Wing reference height (m)	1.3
Wing reference surface (m^2)	72.7
HTP reference height (m)	5.3
VTP chord length at root (m)	4.7
VTP chord length at tip (m)	1.4
VTP sweep angle at ¼ chord (degree)	35.0
VTP span (m)	3.5
VTP reference height (m)	3.1
VTP reference surface (m^2)	9.8
Wing span (m)	25.1
Wing sweep angle at 1/2 chord (degree)	23.5
Wing—VTP aerodynamic lift coefficient	1.1
HTP span (m)	9.8
HTP sweep angle at 1/2 chord (degree)	30.2
Fuselage diameter (m)	3.2
Fuselage length (m)	23.9

and morphing rudder and the aileron excursion to 28 degrees. The aircraft slip angle was then derived from the equation that compute the lateral force, Y_{ext}:

$$C_{y_{\delta_a}}\delta_a + C_{y_{\delta_r}}\delta_r + C_{y_\beta}\beta + C_L sen\,\Phi = -\frac{Y_{\text{ext}}}{qS_{\text{ref}}} \tag{1}$$

As the external lateral forces are null (equilibrated flight), and both $C_{y_{\delta a}}$ and $C_{y_{\delta a}}$ may be considered negligible for the purpose of this analysis, it follows:

$$\beta = -\frac{C_{y_{\delta_r}}\delta_r + C_L sen\,\Phi}{C_{y_\beta}} \tag{2}$$

For the reference airplane, in order to provide enough clearance between engines and wing tips with the ground, the limitation of the roll angle Φ was set to 5 degrees.

Standing these considerations, the output slippage angles, β, for the original and the morphed configurations, were obtained by means of an original Matlab code [51]. A procedure reported in literature [52] was implemented; in this phase, available data were derived from the already cited aircraft (Table 1). The results for conventional rudder are included in Table 2.

The complete attainments are included in the authors' article [52], recently published. With respect to the original configuration, the adaptive rudder was found to have potentiality to lead to a slip angle increase of about 10%. Taking to account that:

$$C_{n_{avail}} = C_{n_{\delta_a}}\delta_a + C_{n_{\delta_r}}\delta_r + C_{n_\beta}\beta \tag{3}$$

Table 2 Conventional Rudder: Lateral-Directional Aerodynamic Coefficients (rad^{-1})

Coefficient	Value
$c_{y\beta}$	−1.27
$c_{l\beta}$	−0.30
$c_{n\beta}$	0.26
$c_{y\delta a}$	0
$c_{l\delta a}$	0.14
$c_{n\delta a}$	−0.02
$c_{y\delta r}$	0.36
$c_{l\delta r}$	0.02
$c_{n\delta r}$	−0.20

As for both systems, $C_{n_{\delta a}}=0.0179$ and $C_{n_\beta}=0.595$, then the value of $C_{n_{avail}}$ was derived [52]. The proposed rudder showed an associated increase in the maximum available yaw moment between 5% and 10%. This result implied potential reductions of the minimum lateral control speed VMC, both on ground (VMCG) and in flight (VMCA), as defined by certification authorities as airworthiness requirements, paragraph CS 25.149 [53]. This was a significant improvement to flight qualities: decreasing the minimum control speed gives in fact the aircraft a greater margin of controllability with respect to the rate of commands control loss (stick shaking).

This issue is object of further investigations, because of the potential architecture benefits in the process of large transport aircraft design. Among the different applications of this technology, there is one that specifically concerns the novel variants of very large aircraft. Currently, there is in fact a tendency to enlarge available planes size, to increase their capacity: for instance, A350 XWB 1000 in relation to the "900" and Boeing 777 × in relation to the "777-300" versions. The possibility of achieving a reduction in the minimum control speed is critical to avoid the undesirable situation of stick shaker warnings, typically associated with augmented payload adaptations of already certified airplanes.

4.3.2 Dynamic stability analysis

Regarding the effects of the proposed morphing rudder on lateral directional dynamic stability, further studies were performed in order to assess certification requirements. These studies are currently in the latest analytical validation steps prior to publication (2016), but it is possible to anticipate that the effects on system frequencies and gains resulted acceptable from both handling qualities and controllability points of view. In other words, the study permitted stating that the incorporation of such a device on a large transport aircraft was practically feasible. The attainments considered uncertainties on the aerodynamic coefficients values that made necessary to investigate two critical scenarios: Spiral and DR.

The configuration with free flight controls showed negligible effects on Spiral case in terms of general gain margins and no effects at all in slip and roll control coefficients. Analogous results were attained for the DR case, except for the slip phase margin where a significant change of twice the original value came out, even though with a negligible frequency deviation.

The results with fixed flight controls showed changes around 25% in sideslip amplitude average for both Spiral and DR. The frequency response barely increased for both the cases. The roll mode frequency was affected for a percentage around 10% and the yaw divergence happened from 10% to 20% sooner than in the standard layout.

However, it has to be considered that the yaw damper controller used measurements of the roll angle and the yaw rate for actuating ailerons and rudder so to attain a specific roll and DR damping, according to an established optimal law that was designed for the reference aircraft arrangement. In the case of a morphing rudder replacing the traditional vertical plane, this new layout should be taken into account and, therefore, it is expected that the software can change. As a consequence, and prior to start an actual flight test campaign, a review of the implemented algorithms is required.

As a continuation of these studies, it would be desirable to analyze with more detail different aircraft condition scenarios (other than Spiral or Dutch Roll) and their effects on the parameter uncertainties. Then, these investigations might be further extended to other aircraft weight and balance combinations and alternative flight envelopes. The discussed results can then be considered as a basic step before further deeper analyses.

5 CONCLUSIONS

The desire of the airlines to have more efficient aircraft is driving current research into incorporating morphing technologies to increase airplane performance and improve economics. It is expected that such systems will produce promising benefits in the near future, increasing their application domains to several aircraft components like the VTP. A major drawback to the use of morphing systems is the characteristic weight and complexity penalty of specific subcomponents. Skins and mechanical design of such structures are key challenges in the development of shape control devices.

In this segment, Aernnova has acquired a long experience in different research projects, such as SARISTU, CURVED, 3AS, and the frame related to SFWA, within the JTI (Joint Technology Initiative) Clean Sky program (adaptive rudder). Tooling design and manufacturing were vital for the adequate final assembly and the precise drilling of the system, as demonstrated for the realization of the SARISTU two-bay and five-bay prototypes. The selection of the actuator system, the choice of highly orthotropic coating panels (the system skin), and weight and volume constraints are critical challenges in the proper development of a morphing system. During the reported investigations, various materials were tested under different conditions, and morphing functionality tests were carried out.

As of this writing (2016) and with reference to the morphing rudder development for a real large aircraft, the complete characterization of the highly orthotropic panels, developed through additive manufacturing processes, awaits complete mechanical analysis and testing. Study of fatigue and damage behavior, as well as of durability of the design solution configuration in relevant environments (not only under static conditions but also with cyclic loads in actual temperature and humidity conditions) is still needed. The available knowledge about the mechanical properties of these architectures, gave good hints about their potentiality to fulfill certification requirements. Although different alternatives were proposed, further work is necessary to discover even more competitive materials and create lighter devices, while remaining compliant with relevant standards.

The following conclusions were drawn with respect to the lateral directional stability analysis, carried out in the critical condition of one engine out during take-off:

- The variation of the derivative coefficients of directional stability for slippage and aileron deflection of lateral force, balance and yaw moments, was negligible.
- The lateral directional stability force coefficients increased as the aerodynamic lateral force, due to the augmented efficiency of the proposed device.
- The higher effectiveness of the proposed device induced an increase of the slippage angle.
- The available yawing moment arose, and consequently the minimum lateral control speed decreased.

These findings cannot be generalized since they resulted from comparing a single unconventional rudder configuration with a reference one, on the same aircraft with the same nominal layout. Furthermore, the absolute validity cannot be inferred due to the lack of robust aerodynamic data coefficients resulting from flight tests; in the reported example, they were numerically estimated.

From the dynamic stability point of view, aerodynamic studies showed a modification on the wake frequency with respect to the conventional arrangement (i.e., modification on the Strouhal number for a given Reynolds number). This occurrence directly affects at least one of the main aerodynamic excitation frequencies for the rudder aeroelastic studies. This consequence should be considered together with the rudder and fix empennage structure modifications that, in turn, influence bending and torsion stiffness. Therefore, the aeroelastic performance and behavior of the morphing rudder also requires detailed analyses and dedicated studies.

REFERENCES

[1] T.A. Weisshaar, Morphing aircraft systems: historical perspectives and future challenges, J. Aircr. 50 (2) (2013) 337–353.

[2] M. Schwartz, Smart Materials, 554 p,CRC Press, Boca Raton, FL, ISBN 97-8142-004-3-730, 2008.

[3] M. Torabzadeh-Tari, Analysis of electro-mechanical actuator systems in more electric aircraft applications (M. thesis), Department of Electrical Engineering, Royal Institute of Technology (S), 2005, ISBN 91-7178-099-8.

[4] B.K.S. Woods, M.I. Friswell, N.M. Wereley, Advanced kinematic tailoring for morphing aircraft actuation, AIAA J. 52 (2014) 788–798 (Special section on Multidisciplinary design optimization).

[5] I. Dimino, G. Diodati, A. Concilio, A. Volovick, L. Zivan, in: Distributed electromechanical actuation system design for a morphing trailing edge wing, SPIE Smart Structures and NDE Conference 2016, Las Vegas, NA, USA, March 20th–24th, 2016.

[6] G. Amendola, I. Dimino, A. Concilio, R. Pecora, F. Amoroso, Actuation system design for a morphing aileron, J. Appl. Mech. Mater. 798 (2015) 582–588.

[7] R.E. Slack, Electro-mechanical actuation of a morphing wing concept (M. thesis), Department of Aerospace Engineering, University of Bristol, UK, 2005, OCLC 931575530.

[8] M.I. Friswell, D.J. Inman, in: Morphing concepts for UAVs, 21st Bristol UAV Systems Conference, Bristol, UK, April 3rd-5th, 2006.

[9] V.P. Galantai, Design and analysis of morphing wing for unmanned aerial vehicles (M. thesis), Department of Mechanical and Industrial Engineering, University of Toronto, Canada, 2012.

[10] H. Baier, L. Datashvili, Active and morphing aerospace structures—a synthesis between advanced materials, structures and mechanisms, Int. J. Aeronaut. Space Sci. 12 (3) (2011) 225–240.

[11] Y. Essa, F.M. de la Escalera Cutillas, I. Dimino, M. Ciminello, A. Concilio, Manufacturing and testing of smart morphing SARISTU trailing edge, in: in: P.C. Woelcken, M. Papadopoulos (Eds.), Smart Intelligent Aircraft Structures (SARISTU), Proceedings of the Final Project Conference, Springer International Publishing, Cham, ISBN 978-3-319-22413-8, 2016, pp. 199–215.
[12] Concilio A. (editor), "Adaptive trailing edge device (ATED): exploitation of morphing—booklet of technologies", SARISTU Workshop, Chemnitz, December 11, 2014. http://www.cira.it/it/comunicazione/events/file/Booklet%20ATED.pdf.
[13] www.saristu.eu.
[14] www.saristu.eu/project/.
[15] cordis.europa.eu/project/rcn/63059_it.html.
[16] CURVED Partnership, New concepts for aircraft rear end architecture (CURvED), led by Airbus and financed by the Spanish Government-PID, Development of Industrial Technology Centre (CDTI), 2012.
[17] M.A. Castillo-Acero, C. Cuerno-Rejado, M.A. Gomez-Tierno, in: Aerodynamic modelling for a morphing rudder, Royal Aeronautical Society Applied Aerodynamics Conference, Bristol, UK, July 22-24, 2014.
[18] M.A. Castillo-Acero, C. Cuerno-Rejado, M.A. Gomez-Tierno, Morphing structure for a rudder, Aeronaut. J. 120 (1230) (2016) 1291–1314.
[19] M. Kintscher, J. Kim, S. Storm, F. Peter, in: P.C. Woelcken, M. Papadopoulos (Eds.), Assessment of the SARISTU enhanced adaptive droop nose, Smart Intelligent Aircraft Structures (SARISTU)—Proceedings of the Final Project Conference, Springer International Publishing, Cham, ISBN 978-3-319-22413-8, 2016, pp. 113–140.
[20] R. Pecora, M. Magnifico, F. Amoroso, L. Lecce, M. Bellucci, I. Dimino, A. Concilio, M. Ciminello, in: P. C. Woelcken, M. Papadopoulos (Eds.), Structural design of an adaptive wing trailing edge for large aeroplanes, Smart Intelligent Aircraft Structures (SARISTU)—Proceedings of the Final Project Conference, Springer International Publishing, Cham, ISBN 978-3-319-22412-1, 2016, pp. 159–170.
[21] A. Wildscheck, S. Storm, M. Herring, D. Drezga, V. Korian, O. Roock, in: P.C. Woelcken, M. Papadopoulos (Eds.), Design, optimization, testing, verification and validation of the wingtip adaptive trailing edge, Smart Intelligent Aircraft Structures (SARISTU)—Proceedings of the Final Project Conference, Springer International Publishing, Cham, ISBN 978-3-319-22413-8, 2016, pp. 219–255.
[22] O. Schorsch, A. Lühring, C. Nagel, in: P.C. Woelcken, M. Papadopoulos (Eds.), Elastomer-based skin for seamless morphing of adaptive wings, Smart Intelligent Aircraft Structures (SARISTU)—Proceedings of the Final Project Conference, Springer International Publishing, Cham, ISBN 978-3-319-22412-1, 2016, pp. 275–291.
[23] R. Evenblij, F. Kong, C. Koimtzoglou, M. Ciminello, I. Dimino, A. Concilio, in: P.C. Woelcken, M. Papadopoulos (Eds.), Shape sensing for morphing structures using fiber bragg grating technology, Smart Intelligent Aircraft Structures (SARISTU)—Proceedings of the Final Project Conference, Springer International Publishing, Cham, ISBN 978-3-319-22413-8, 2016, pp. 471–491.
[24] G.M. Carossa, S. Ricci, A. De Gaspari, C. Liauzun, A. Dumont, M. Steinbuch, in: P.C. Woelcken, M. Papadopoulos (Eds.), Adaptive trailing edge: specifications, aerodynamics, and exploitation, Smart Intelligent Aircraft Structures (SARISTU)—Proceedings of the Final Project Conference, Springer International Publishing, Cham, ISBN 978-3-319-22413-8, 2016, pp. 143–158.
[25] F. Peter, E. Stumpf, G.M. Carossa, M. Kintscher, I. Dimino, A. Concilio, R. Pecora, A. Wildschek, in: P. C. Woelcken, M. Papadopoulos (Eds.), Morphing value assessment on overall aircraft level, Smart Intelligent Aircraft Structures (SARISTU)—Proceedings of the Final Project Conference, Springer International Publishing, Cham, ISBN 978-3-319-22413-8, 2016, pp. 859–871.
[26] I. Dimino, M. Ciminello, A. Concilio, R. Pecora, F. Amoroso, M. Magnifico, M. Schueller, A. Gratias, A. Volovick, L. Zivan, in: P.C. Woelcken, M. Papadopoulos (Eds.), Distributed actuation and control of a morphing wing trailing edge, Smart Intelligent Aircraft Structures (SARISTU)—Proceedings of the Final Project Conference, Springer International Publishing, Cham, ISBN 978-3-319-22412-1, 2016, pp. 171–186.
[27] 3AS Partnership, Active aeroelastic aircraft structure (3AS), ref. G4RD-CT-2002-00679, FP5, 2002–2005.

[28] f. Martín de la escalera, n. caballero, in: Development of engineered morphing material for aircraft performance improvement, International Forum of Aeroelasticity and Structural Dynamics, IFASD 2005, Muenchen, June 28th-July 1st, 2005.
[29] D. Elzey, A. Sofla, H. Wadley, in: A bio-inspired, high-authority actuator for shape morphing structures, SPIE Smart Structures and NDE Conference 2003, San Diego, CA, USA, March 2nd-6th, 2003.
[30] M. Albert, in: Translation of innovative technologies into simplified conceptual aircraft models for technology evaluation in cleansky, 3AF International Conference on Greener Aviation: Clean Sky Breakthroughs and Worldwide Status, Bruxelles, March 12th-14th, 2014.
[31] H.E. Yuanyuan, G.U.O. Shijun, Modeling and experiment of a morphing wing integrated with a trailing edge control actuation system, Chin. J. Mech. Eng. 25 (2) (2012) 248–254.
[32] L. Falcão, A. Gomes, A. Suleman, in: Multidisciplinary design optimisation of a morphing wingtip, 2nd International Conference on Engineering Optimization, Lisbon, September 6th-9th, 2010.
[33] T.L. Grigorie, A.V. Popov, R.M. Botez, M. Mamou, Y. Mebarki, On-off and proportional-integral controller for a morphing wing. Part 1: actuation mechanism and control design, Inst. Mech. Eng. G J. Aerosp. Eng. 226 (2) (2011) 131–145. Publisher: Sage.
[34] T.L. Grigorie, A.V. Popov, R.M. Botez, M. Mamou, Y. Mebarki, On-off and proportional-integral controller for a morphing wing. Part 2: control validation—numerical simulations and experimental tests, Inst. Mech. Eng. G J. Aerosp. Eng. 226 (2) (2011) 146–162.
[35] G. Waller, in: CFD prediction of stability derivatives of a turboprop aircraft using a Cartesian grid based Euler code, 23rd International Congress of Aeronautical Science, ICAS 2002, Toronto, Canada, September 8th-13th, 2002.
[36] D.A. Babcock, Aircraft stability derivative estimation from FEM (PhD thesis), Oklahoma State University, 2004.
[37] X. Lachenal, S. Daynes, P.M. Weaver, Review of morphing concepts and materials for wind turbine blade applications, Wind Energy 16 (2) (2012) 283–307.
[38] S. Daynes, P.M. Weaver, A morphing trailing edge device for a wind turbine, J. Intell. Mater. Syst. Struct. 23 (6) (2012) 691–701.
[39] J.C. Gomez, E. Garcia, Morphing unmanned aerial vehicles, Smart Mater. Struct. 20 (10) (2011) 16, 103001.
[40] V.P. Galantai, A.Y.N. Sofla, S.A. Meguid, K.T. Tan, W.K. Yeo, Bio-inspired wing morphing for unmanned aerial vehicles using intelligent materials, Int. J. Mech. Mater. Des. 8 (1) (2012) 71–79.
[41] M. Abdulrahim, H. Garcia, G.F. Ivey, R. Lind, Flight testing a micro air vehicle using morphing for aeroservoelastic control, J. Aircr. 42 (1) (2005) 1–17.
[42] M. Trapani, S. Guo, in: Development of a rudderless aeroelastic fin technology, 26th International Congress of the Aeronautical Sciences, ICAS 2008, Anchorage, AK, USA September 14th-19th, 2008.
[43] G.T.G. Tao, S.C.S. Chen, J.F.J. Fei, S.M. Joshi, in: An adaptive actuator failure compensation scheme for controlling a morphing aircraft model, 42nd IEEE International Conference on Decision and Control (IEEE Cat. No.03CH37475), Maui, HI, USA, December 9th-12th, 2003.
[44] D.L. Raney, R.C. Montgomery, L.L. Green, M.A. Park, in: Flight control using distributed shape-change effector arrays exhibit, 41st AIAA/ASME/ASCE/AHS/ASC Structures, Structural Dynamics, and Materials Conference & Exhibit, Atlanta, GA, USA, April 3rd-6th, 2000.
[45] T. Kamuran, E.M. Jafarov, in: Lateral dynamic modeling of hezarfen unmanned aerial vehicle (UAV) and H-loop shaping robust control system design, 10th WSEAS International Conference on Systems, Athens, July 10th-12th, 2006.
[46] M.A. Castillo-Acero, C. Cuerno-Rejado, M.A. Gomez-Tierno, in: Highly orthotropic panels structural stability, farrar and bloch waves theory, 3rd International Conference on Buckling and Post-Buckling Behaviour of Composite Laminated Shell Structures, Braunschweig, March 25th-27th, 2015.
[47] Z. Tukovi, H. Jasak, A moving mesh finite volume interface tracking method for surface tension dominated interfacial fluid flow, Comput. Fluids 55 (2012) 70.

[48] D. McLean, Automatic Flight Control Systems, 593 p, Prentice-Hall, New York, NY, ISBN 97-8013-054-0-089, 1990.
[49] J. Roskam, Airplane Flight Dynamics and Automatic Flight Controls, Part I, 959 p, DAR Corporation, Lawrence, KS, ISBN 97-8188-488-5-174, 2003.
[50] J.A.A. Lyrio, P. Paglione, in: Wing and airfoil optimized design of transport aircraft, 11th Brazilian Congress of Thermal Sciences and Engineering, ENCIT 2006, Curitiba, December 5th-8th, 2006.
[51] J. Roskam, Airplane Flight Dynamics and Automatic Flight Controls, Part I, 959 p, DAR Corporation, Lawrence, KS, ISBN 97-8188-488-5-181, 2003.
[52] USAF, Digital Datcom, Public Domain Computer Programs for the Aeronautical Engineer, 2013. Available from: http://www.pdas.com/datcom.html.
[53] J. Roskam, Airplane Design Part VIII, 368 p, DAR Corporation, Lawrence, KS, ISBN 97-8188-488-5-556, 2002.

FURTHER READING

[1] Mathworks Inc., MATLAB, Version 8.3.0.532 (Ra), 2014, The MathWorks, Natick, MA, .
[2] M.A. Castillo-Acero, C. Cuerno-Rejado, M.A. Gomez-Tierno, in: Morphed vertical tailplane assessment for certification requirements, 5th CEAS Air & Space Conference, Delft, September 7th-11th, 2015.
[3] EASA (European Aviation Safety Agency), Certification Specifications and AMC for Large Aeroplanes, Cologne (D), CS-25, 2009.

CHAPTER

MORPHING WING INTEGRATION 20

Generoso Iannuzzo, Salvatore Russo, Alfonso Apicella, Lorenzo Rossi, Sergio Esposito
Leonardo S.p.A., Pomigliano D'Arco (NA), Italy

CHAPTER OUTLINE

NOMENCLATURE

3D	three-dimension
ATED	adaptive trailing edge device
CAD	computer aided engineering
CFRP	carbon fiber reinforced polymer
CH	coordination hole
COA	conditions of assembly
deg	degree
DMU	Digital Mock-Up
EADN	enhanced adaptive droop nose

Morphing Wing Technologies. https://doi.org/10.1016/B978-0-08-100964-2.00020-4

F1, 2, 3, 4	fairing N.1, 2, 3, 4
FBG	fiber Bragg gratings
FO	fiber optic
In/rev	inch per revolution
KH	key hole
LE	leading edge
LH	left hand
PF	power feed
PFD	positive feed drill
PH	pilot hole
RH	right hand
rpm	round per minute
SARISTU	Smart Aircraft Intelligent Structures
TE	trailing edge
ЦАГИ	Центральный аэрогидродинамический институт (RUS)
TSAGI	Central Aerohydrodynamic Institute
TH	tooling hole
WATE	wingtip active trailing edge
WT	wind tunnel

1 INTRODUCTION

The assembly of an aircraft structure is a high precision and critical activity: it must be performed within a very small range of tolerances due to the general great complexity, number of parts and dimensions; the final construction must ensure the absence of defects and must be compliant with the design specifications and the regulation requirements [1].

Since the last aircraft generation, wing structure comprises an extremely large number of parts: ribs spars, stringers, panels, and numerous secondary structures like clips and doublers. Consequently, putting these elements together is difficult and requires huge, bulky jigs which must be exactingly built to maximize tolerance chain reduction. Early in aircraft manufacture, applied methods were extremely complex and lengthy. For example, the assembly of conventional high-lift devices (at the leading and the trailing edge, LE and TE, respectively) on traditional structures required large, high-precision tooling and several additional parts (such as leverages and hydraulic actuators) which had to be compliant with the same tolerance requirements as the whole structure [2].

Lately, several novel concepts in wing box design, developed for new aircraft, provided higher integration levels for parts. Skins and ribs became integrated with stringers and clips, respectively, monolithic spars showed a variable thickness, and so on. Advanced composite materials ensured this extraordinary capability, intrinsically, and led also to significant weight savings. In spite of several difficulties connected with the manufacture of such components, the assembly of these new generation structures was simplified if compared to traditional ones in several ways [3]. First of all, the reduced number of elements (a consequence of the a priori incorporation of some of them) permitted the elimination of assembly tolerances among those formerly constituent parts. The high technology processes involved guaranteed high precision and very small tolerance deviations with respect to the nominal dimensions, allowing the use of less-complex and less-expensive jigs and tools which were simpler to use if compared to traditional instruments.

Currently, the development of morphing aerodynamic surfaces (aircraft components which can drastically change their shape while flying) presents a new challenge [4]. These new deformable structural systems can withstand large external load strains and displacements while maintaining their established form, selected within a bundle of possible shapes (the well-known "smart structures paradox"). Installation of these movable elements could lead to many kinds of manufacturing problems, including emergence of significant residual strains at the beginning of the development and the construction of the target architecture. The modern assembly line will have to take such characteristics into account, and must devise ingenious redesigns to accommodate these innovative devices.

This text will explore that new challenge and propose some criteria to overcome its associated difficulties while complying with common aeronautical manufacture rules and regulations. Many of the solutions devised and executed within the European project SARISTU (Smart Aircraft Intelligent Structures) [5] are applicable within this context. That research activity sought to maximize aircraft performance by developing new adaptive high-lift arrangements for the LE, TE, and winglet, with the ultimate objective of reducing the overall weight and fuel consumption. While the main wing box was developed through innovative integration techniques, related to the realization of a full CFRP (carbon fiber reinforced polymer) structure, dedicated jigs and methods were implemented to install the proposed morphing devices.

In this work, all the process for the integration and assembly of the wing section demonstrator are described, with particular attention to the instruments and procedures that were applied.

2 DEMONSTRATOR COMPONENTS

An adaptive wing may be made of several elements, depending on the objectives of the system. In fact, a generic morphable wing does not exist, at least not at the level of current technology. Its different components are designed and constructed for a specific target, thus a generic shape-changing system cannot be thought of as an absolute solution to all wing needs. Consequently, the number and the kind of devices that could be present on an aerodynamic surface are virtually infinite. However, it is reasonable to imagine that problems common to any kind of morphing architecture can be individuated and described, as the essential characteristics of different kinds of smart appliances are more or less standard. In particular, the following features are expected to be always (or often) found in an adaptive element:

- an electrical network, for power feeding and data acquisition and transmission;
- a fiber optic (FO) network, if FO sensors are used;
- an actuator system;
- a sensor system; and
- an adaptive skin (depending on the implemented architecture).

Generally, the control system is located outside the smart structural module. It should be noted that the connections with the main structural body occur at the component boundaries; for mechanical systems, these are mostly the same as for traditional structural elements. The main difference is the massive presence of cables and the unconventional deformability of the components. It is the required room

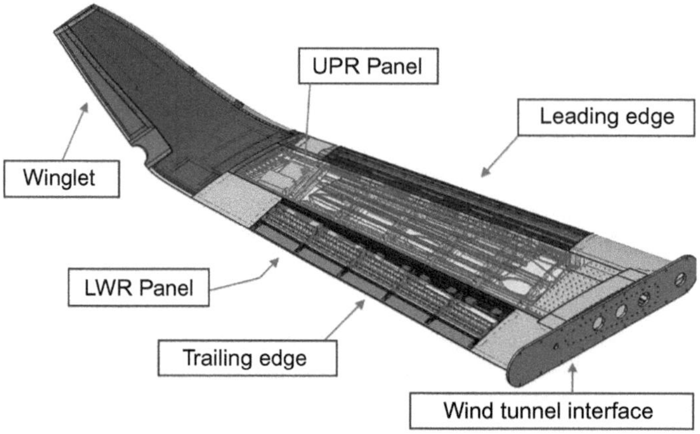

FIG. 1

Overall CAD representation of the adaptive wing section.

allowance for cabling and the need for special supports and tools for the adaptive parts that make the assembly and installation processes unusual.

In the analyzed case, the SARISTU smart wing was made of the following main subsystems that will be detailed herein:

- wing box primary structure;
- adaptive leading edge (EADN—enhanced adaptive droop nose [6]);
- adaptive trailing edge (ATED—adaptive trailing edge device [7]); and
- adaptive winglet (WATE—winglet adaptive trailing edge [8]).

The wing box was the main supporting structure and was fully instrumented with a FO system to detect its structural health state. The EADN, conceived and developed by an international team led by DLR [6], was aimed at improving the take-off and landing performance of the aircraft and used a single motor, connected to a number of mechanical actuators via a shaft beam. The ATED, conceived and developed by an international team led by CIRA [7], altered the wing camber during cruise, then modifying the wing shape to compensate the weight reduction related to the fuel consumption; its movements were driven by a distributed actuator system and the actual achieved form was discerned by a widely distributed and integrated sensor network. The WATE, conceived and developed by an international team led by EADS-IW [8], balanced the gust effects and modulated the load distribution over the wing, achieving both load alleviation (statics) and control (dynamics) capability; it was characterized by a single actuator system moving a tab-like surface.

A pictorial representation of the wing structure is presented in Fig. 1, while its schematic is shown in Fig. 2, where the main components are indicated with respect to their conceptual position and their attachment to the main body (interfaces). In the same picture, the main (extremal) ribs, used for connecting the test article to the main attachment at the wind tunnel (WT) facility (envisaged for running the final tests) and to the adaptive winglet device, are highlighted.

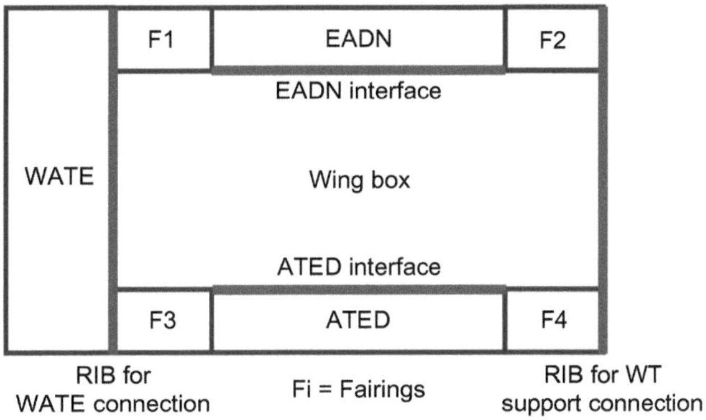

FIG. 2

Adaptive wing schematic.

This chapter deals with the integration of these smart components (designed and realized by different international teams, as above mentioned) on the main wing body to give rise to the adaptive wing section, 5.5 m span, of the SARISTU project.

Because it would not have been possible to let the various systems directly interface each other, some "respect" or "dead zones" were foreseen, filled with special fairings (rigid components), to preserve the geometrical continuity of the demonstrator. The reported scheme (Fig. 2) is not scaled and is meant to serve only as a structural reference. It should be noted that, while the connection between the WATE and the WT support (this latter specific element is not herein treated) occurs via traditional ribs, the ATED and the EADN were linked by joining their top and bottom skins with the wing box front and rear spars, respectively. The main spars of the adaptive devices and the wing box were not in fact connected to each other, apart from some small pins, used as a reference during construction. This unconventional choice was necessary to protect the parts of the wing box sensor system (FO cabling) running over the front and rear spars. This is one of the peculiarities directly related to the special nature of an adaptive wing.

2.1 WING BOX PRIMARY STRUCTURE

The wing box primary structure was oversized, as the target wing section was a prototype devised to undergo WT testing (in which the safety coefficients need to be very large). Any mishap could have caused the complete destruction of the WT or its complex operative system. However, there were many reasons to keep the design factors high. Because such a demonstrator must be unique, great care was devoted to manufacturing pieces which were resistant to any potential accidents during assembly, transportation, and installation. It was also necessary to support the adaptive components, and that support had to be stiff enough to prevent both structural static and dynamic coupling, while preventing deformation of the structure under aerodynamic loads. Static de-coupling is useful prevent morphing actions from propagating through the structure and involving other smart components. Dynamic de-coupling minimizes the possibility of aeroelastic instabilities.

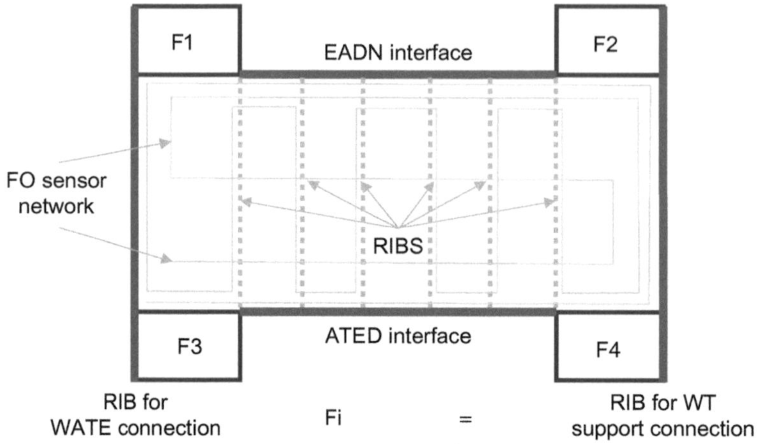

FIG. 3

Wing box schematic (top and bottom panels are not referenced, for the sake of clarity).

The above considerations directly applied to the panels and the ribs, including the WATE interface, while the rib designed to be linked to the WT support was sized according to specs coming from the WT operator (TsAGI). In summary, the main structural constituents of the wing box were, Fig. 3:

- front and rear spars;
- root and tip ribs;
- six central CRFP ribs, each connected to the spars through a pair of struts located forward, and an angle rib rearward; and
- top and bottom panels.

Among the handcrafting details worthy of mention, the upper and lower skins were respectively fixed by screws (allowing removal for needed interventions and modifications) and by the usual rivets. In this latter case, however, detachable cages allowed visual inspection and some limited access to the inner structural body. The main geometrical (dimensions) and physical (material) characteristics of the wing box are summarized in Table 1 as magnitudes, while a representative picture is reported in Fig. 4:

2.2 LEADING EDGE

The adaptive leading edge consisted of a massive and highly deformable fiberglass skin connected to stiff ribs which produced an established form and maintained it under the action of external forces. Such ribs worked as actuators, their movement driven by a shaft propagating the command of an external motor. In the final configuration, this motor was hosted inside one of the fore fairings, following the scheme of Fig. 5.

The device was mounted on the wing box through a connection between the top and bottom panels and its front spar caps plus another line of joints between the web of the two half-spars (one associated with the wing box, and another, part of the LE). As already illustrated in Fig. 5, the spars line region was the only connection between the active device and the main structure. The other boundaries (first and

Table 1 Geometrical and Physical Reference Characteristics of the Main Components of the Wing Box

	Width (mm)	Length (mm)	Thickness (mm)	Material
Overall dimensions	2070	5200	280	–
Panels (mean values)	880	3000	10	CFRP
Ribs (mean values)	200	800	5	CFRP
Fairings (mean values)	600	600	4	CFRP
Rib (WT connect.)	300	105	600	Steel
Rib (WATE connect.)	130	1200	100	Aluminum
Auxiliary spar (ATED connect.)	120	3000	60 flange	Steel
Auxiliary spar (EADN connect.)	200	3000	120 flange	Steel

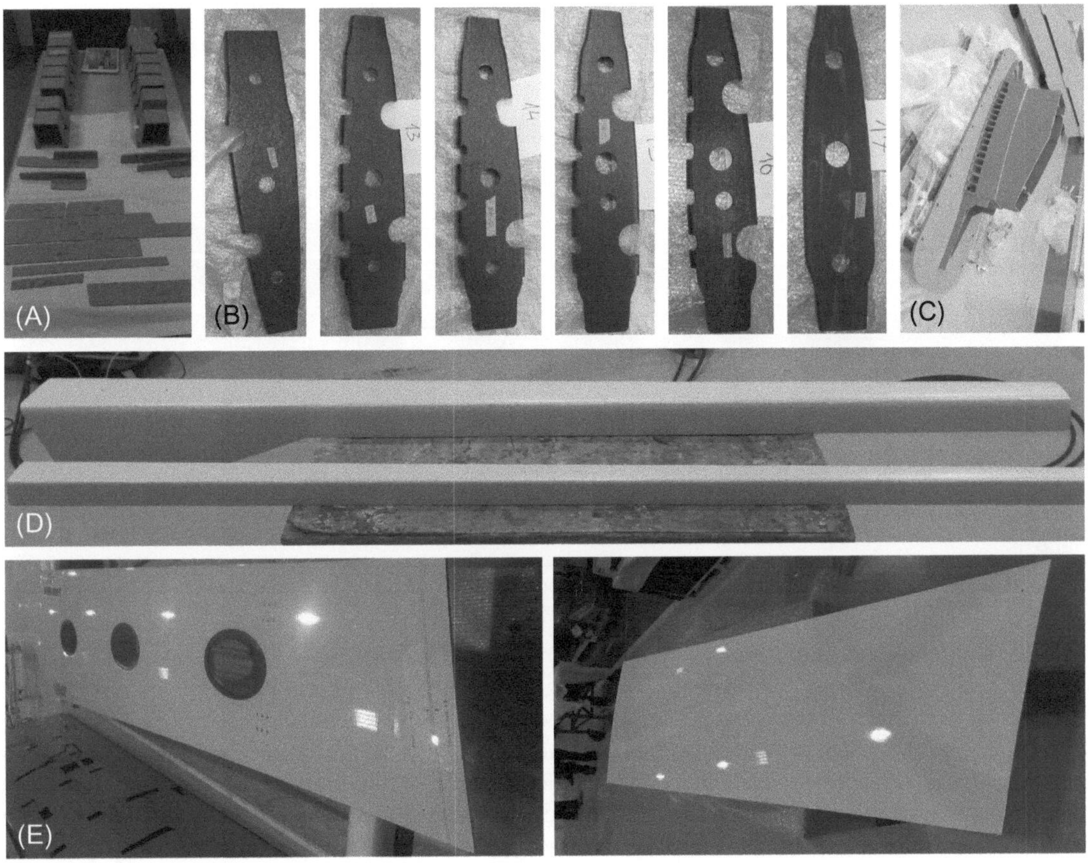

FIG. 4

Wing box main components: (A) forward ribs, coupling the wing box to the WATE; (B) the central ribs; (C) the root rib assembled; (D) the front spar and the rear spar; (E) the skins: lower skin (left), with the inspection doors, and upper skin (right).

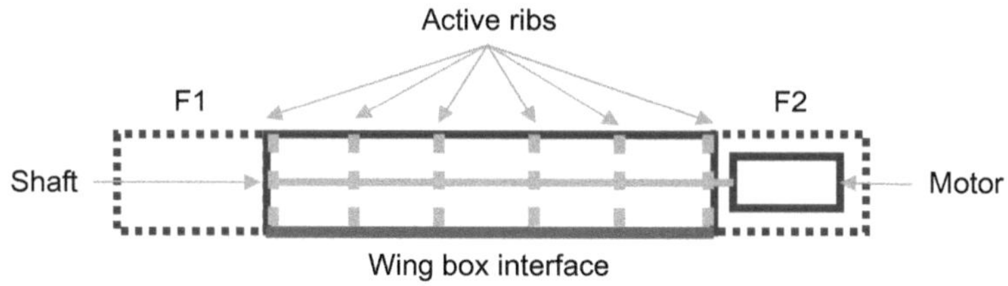

FIG. 5

EADN schematic.

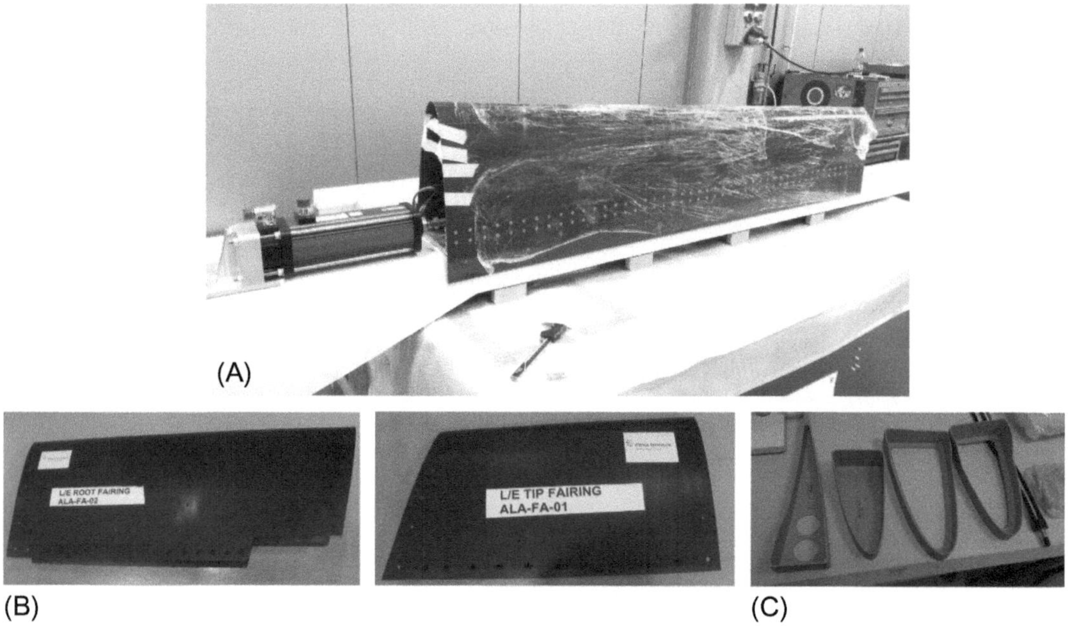

FIG. 6

Adaptive leading edge components: (A) morphing droop nose; (B) root and tip fairings; and (C) cap ribs.

last ribs) were then free to move without any constraints. This choice avoided any kind of interference between the morphing part and the rest of the wing. In order to guarantee a geometric continuity, two fairings were designed, manufactured, and installed on the prototype (black caps in Fig. 6, top). The active LE system, in fact, could not be extended along the entire span for several reasons. At the interface with the WT support, sufficient room would be required to permit the cables passage without affecting the morphing movement and other classical working operations in WT. The root fairing also served as protection for the motor and its associated electronics. Instead, towards the wingtip, an appropriate room should have been reserved for avoiding possible overlaps with the WATE and guaranteeing a degree of accessibility. In case of difficulty, the presence of a dummy volume would have caused no problem at all.

The Adaptive LE allowed deflection up to ±4 degrees ensuring, in perspective, a laminar flow zone extension along the surface of the wing, with many benefits in terms of drag reduction during take-off, landing, and cruise, resulting in reduction of fuel consumption. The absence of gaps along its curvature did strongly limit the noise emissions of this kind of device. Fig. 6 shows the main EADN components, before their assembly. The fairings aimed at giving the system a geometrical continuity. It is noteworthy that there was no physical connection between these fixed parts and the adaptive leading edge. The motor, its shaft, and the skin formed a unique working block. The motor was installed on the main spar, after the other two elements had been mounted. The ribs that are shown at the bottom of the picture are relative to the fairings, while the morphing ones are embedded inside the fiberglass deformable skin. In the same picture, the solid rib is the extremal rib (towards the wingtip), while the perforated ribs are located at the nominal interface with the EADN, where the shaft is supposed to pass through.

2.3 TRAILING EDGE

The adaptive trailing edge was a large finger-like architecture extending all along its span. Made of a metallic structure, it was covered with a highly deformable skin designed to withstand strains while preserving geometric continuity. To enable its movement, the ribs were divided in parts and linked to the driving motors.

Active ribs, transmission kinematics, and electrical actuators made up a compact kinematic structural system able to allow expected motion while bearing external and internal loads. In contrast with the EADN, in the ATED, the single segmented rib was driven by a dedicated couple of motors with the exception of the lateral ones, moved by a single actuator. The resulting system was then made of 10 embedded active systems and six mechanical chord-wise stiffeners. This choice led to several differences from the former architecture, among which the following may be noted:

- the mass of cables (more relevant for this case);
- the compactness (this architecture could be completely integrated in the morphing structure, so that it did not necessitate additional external covers);
- the costs (with reference to both the direct and the maintenance expense);
- the reliability (a conflict indeed: on the one hand, the whole system is redundant therefore safer than the former; on the other hand, the likelihood that a single actuator out of ten fails is higher); and
- the versatility (as contrasted with a single-motor configuration, a multiple-motor configuration can be structured to work in a "differential" way, enabling attainment of manifold shapes, such as a twist).

It should be noted that the choice of a single motor would have led to a very different layout, since it could not have been located within the reference structural system for lack of room to accommodate its increased dimensions. The schematic of the adaptive trailing edge is reported in Fig. 7.

Regarding the installation on the wing box, the same considerations and modalities (described for the adaptive leading edge above) applied. In this case, however, the device was connected to the main structure via the top and bottom panels only, to preserve the functionality of the health monitoring system segment that was placed along the aft spar of the wing box. There were two main reasons for this difference. Firstly, lower loads acted on that zone. Secondly, the rear spar was smaller than the front spar. This limited space made any kind of drilling on its web difficult without damaging the FO network therein deployed.

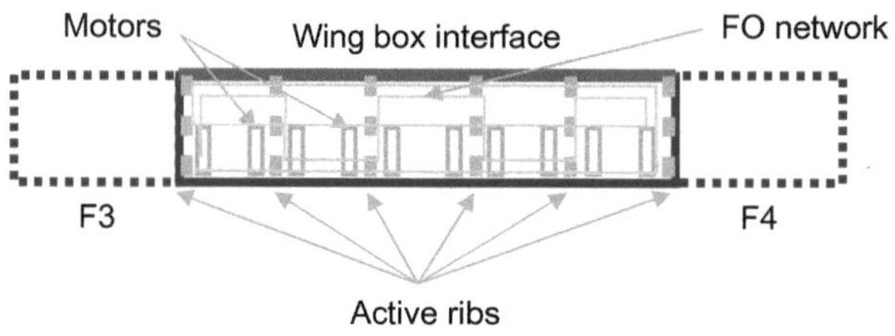

FIG. 7

ATED schematic.

As in the case of the EADN, the connection occurred only on that edge, leaving the other boundaries free to move. In the same way as before, two fairings were installed on the prototype to preserve the geometrical continuity. In this case, however, no motor was located under the fairings; instead, a significant space was left for the cables (which were more massive than in the former case). In contrast with the EADN, the ATED could not be extended all along its span for the wing taper, making it extremely difficult to arrange the embedded system layout in a space which became narrower and narrower.

The adaptive trailing edge was designed to attain symmetrical and continuous deflections in the range ±5 degrees to compensate for the weight reduction resulting from fuel consumption, while still seeking optimal trim configuration. Differential mode would have sought span-wise load modulation to alleviate the root bending moment and preserve the wing structure from the maximum impinging forces. This was not implemented in the prototype. It would be worthwhile to investigate slotted configurations which could enable larger deflections; noise footprint penalties would be irrelevant because of the operational altitude (cruise).

Fig. 8 shows the ATED appearance and the related geometrical fairings before their assembly on the complete wing demonstrator section. As for the previous case, there was no physical connection between these parts and the morphing device. As it can be seen, the adaptive trailing edge was a unique block, and so there was no chance to install any of its parts separately from one another. This aspect may be seen an advantage from many points of view (no need for alignment of the parts, no need to follow a special assembly logic shared with the manufacturer and inserted into the global process, and so on).

On the other side, it may be critical to move such a massive and deformable component prior to its installation on the wing box. Special jigs and tools would still be required for the assembly of such novel elements. In the second picture of Fig. 8, it is possible to glimpse the special tools assessed for the ATED parts integration before its release [9]. In the final sketch, the complete device is shown just before installation on the wing box; the morphing skin, made of alternate strips of metal and hyperelastic polymer [10], is clearly visible.

2.4 WINGLET

The adaptive winglet consisted of a classical winglet part with a movable tab, capable of quasistatic and dynamic deflections. The tab mechanical system, the motor, and the associated electronics were totally embedded within the body. A main difference with the other systems is the presence of an elastomeric

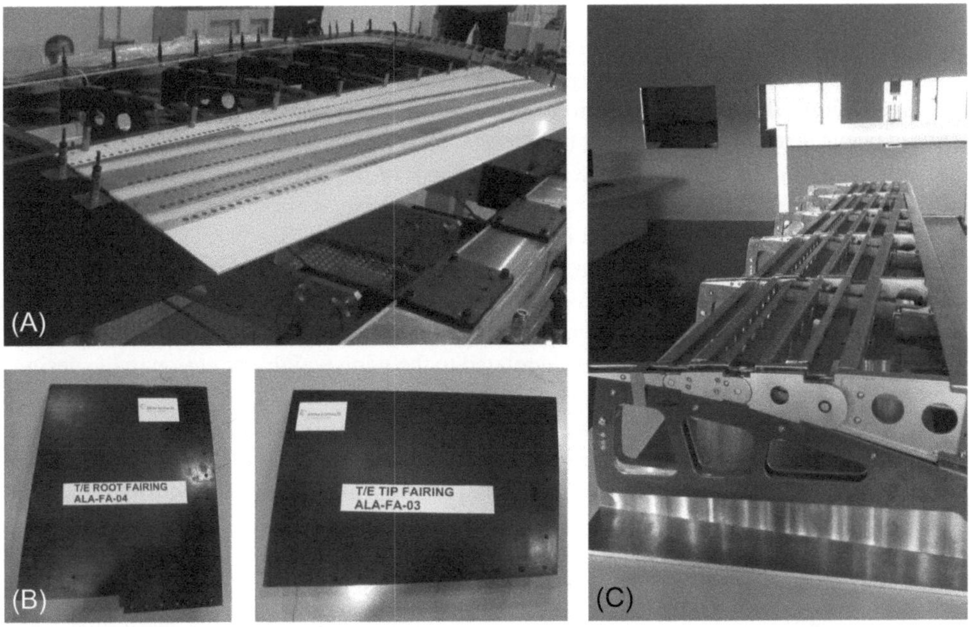

FIG. 8

The ATED components: (A) the trailing edge, while being installed on the demonstrator; (B) the fairings for the wing root (left) and the tip (right); and (C) the skeleton of the morphing component.

link between the tab and the remainder of the system, guaranteeing geometrical continuity along its movement. In summary, the morphing device consisted of a single-hinged aerodynamic surface (resembling a plain, classical flap), a single motor, and a single shaft, along with a highly-deformable piece that gave the local TE shape some uniformity during actuation. For this reason, the system was named WATE. Its schematic is reported in Fig. 9.

The issues related to the installation of the EADN on the wing box were definitely different for this device. First of all, there was no fairing, so connection occurred directly on the wing box. There was no structural health monitoring system in that zone, so the rib was manufactured from steel and joined classically. The attachment was rigid and very similar to that of a usual winglet, so there was no

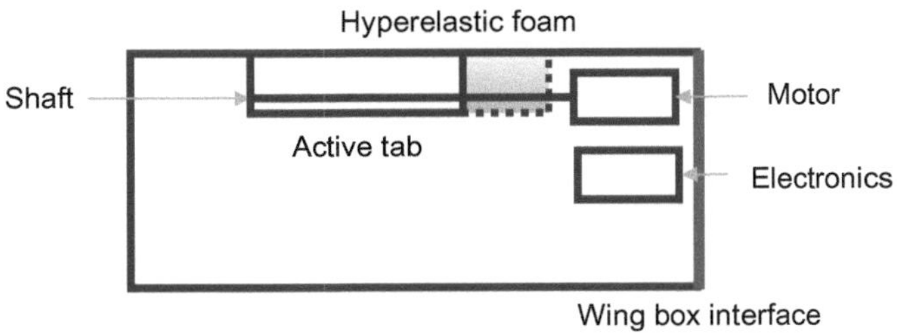

FIG. 9

WATE schematic.

interference between movable parts. As for the other systems, special attention was devoted to the cables which exited the WATE and entered the wing box through the external fairings and had to run all over the main structure to the other fairings.

In real applications, this would be a serious problem. Special room must be reserved for this passage which should occur through the wing box (where the tankers are located), through the LE (site of the morphing, antiicing, and bird impact protection systems), or the TE (where only the morphing kinematics are present, but where room is scarce). In this latter case, it should be noted that in actual operations, the ATED is a part of the flap. It is also debatable whether the WATE electronics should be effectively located at the winglet location, inducing a remarkable mass effect that could lead to an undesired aeroelastic behavior. The adaptive winglet was designed to achieve load redistribution (static deflection) and gust effects attenuation (dynamic movement). Its excursions were in the range between −20 degrees and 15 degrees. Its effects may be related to the wing weight reduction in the design phase, as the structure would be exposed to lower bending moments and, generally, to reduced gust effects. In addition, the winglet could also be used as a directional control surface, alleviating the rudder efforts.

A special mention could be given to the elastic polymeric element which linked the small flap to the main structure [11]. It is one of the first applications of the kind and could be used for existing on-board devices (flaps, slats, and so on) to increase their aerodynamic effectiveness (by reducing the extremity losses) and noise radiation (by reducing the gaps and the associated sound generation). For the aeroelastic impacts connected to the WATE presence (a heavy mass, added at the wing tip), this is a critical device. There are many definite benefits associated with this development, but increasing complexity can dramatically affect safety, maintenance, and manufacturing (especially as this development necessitates the realization of novel, specialized tools for assembly). Fig. 10 shows some interesting

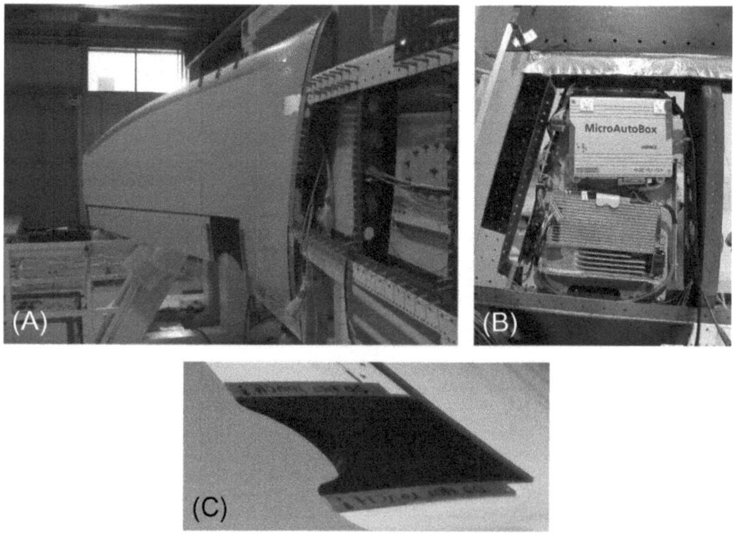

FIG. 10

The main components of WATE: (A) the winglet during the installation process on the outer wing; (B) the monitoring and actuation system, located in the wing box at the winglet root; and (C) a detail of the polymeric element linking the active trailing edge with the winglet body.

particulars of the WATE, from pictures taken during its installation on the wing box. At the right top, it is possible to see the system as it appears after the assembly. The small black area is the aforementioned highly deformable rubber, while the bottom rectangle is the active tab, limited by a dark line (physical division with the rest of the winglet). The same details are visible in the zoom at the bottom left part of the picture. Finally, on the right side, the electric driver and the control system appear in their slots within the last bay of the wing box, at the attachment with the WATE body.

3 CONDITIONS OF ASSEMBLY

A proper assembly process must provide the different components with either predrilled holes (both full-size holes or near full-size holes) or suitably defined drilling masks. Holes which permit the correct display of the components on the target jig or the fabrication tooling are respectively referred as "key holes" (KHs) or "tooling holes" (THs). Holes that allow mating of parts between themselves and the drilling mask are called "pilot holes" (PHs) or "coordination holes" (CHs). A general description of the implemented assembly strategy is herein reported.

What is usually required on the structural parts in terms of holes deployment is illustrated in Figs. 11 and 12. Only holes that are defined in this phase should be prepared, with no exception. The necessary remaining holes are then drilled or simply brought to their final diameter by means of drilling masks. Specifically, only the CH diameters are prepared to match their final size. The "assembly sequence" is the strategy, usually defined by the manufacturer, which outlines the order of different installations. Generally, several stages of realization are scheduled. Tolerances are associated with all the parts, whether composite (CFRP or fiberglass) or metallic (aluminum or steel). Throughout this process, the type of shimming to be used during the construction is also defined.

Coupling elements would necessarily have some gaps to fill; this is done by using liquid (resin) or solid (patch) shims. Shimming should not violate the designs aerodynamic or geometrical constraints.

● CH "Coordination Hole"
■ PH "Pilot Hole"
◆ KH "Key Hole"
▲ TH "Tooling Hole"

Final Dia < .190
Final Dia .250 – .3125
Final Dia > .375

Hole type	Diameter [inch]	DIA tolerance ∅	Position tolerance ⊕
PH Pilot hole	0.1285	–0.000"/+0.005"	0.014"
TH Tooling Hole	0.187	–0.000"/+0.005"	0.008"
KH Key Hole	0.250	–0.000"/+0.008"	0.010"
CH Coordination hole	0.0984	–0.000"/+0.003"	0.010"
	0.1285	–0.000"/+0.003"	0.010"
	0.250	–0.000"/+0.003"	0.010"
	0.3125	–0.000"/+0.003"	0.010"

FIG. 11

Representative characteristics of the reference holes (assembly operations).

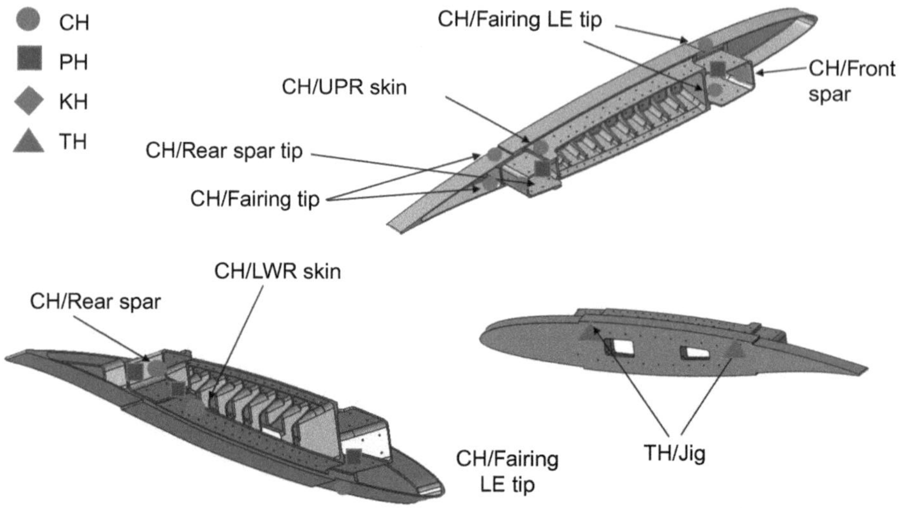

FIG. 12

Example of deployment of the reference holes on some characteristic elements.

The procedure is regulated by proprietary industrial conditions of assembly (COA) documents which mandate, among other things, that the design drawings include a list of all parts used for assembly and all pieces to be fastened.

From efforts to guarantee the best results while minimizing defects in the adaptive wing system, several innovative assembly technologies and state-of-the-art strategies emerged. Novel jigs and tools were designed, as was a user-friendly operative procedure which illustrated each step sequentially by directly referring to 3D numerical models. The new jig incorporated locators to facilitate proper placement of the parts during their assembly, based on the digital mock-up (DMU) nominal values. Drilling operations used customized masks. Specific fasteners were selected due to the galvanic interaction between different materials.

To allow flexibility in assembly, a rotating jig was designed and manufactured. It permitted operations in horizontal and vertical positions and was equipped with saddle profiles (provided with vacuum cups) in order to ensure maximum precision in meeting aerodynamic requirements. Few coordinated and THs were arranged on the parts, harmonized with the jig. User-friendly operative procedures containing a detailed visual description of each assembly step were developed. Whenever necessary, perfect coupling between the elements was ensured via strippable solid or liquid shims. Innovative drilling technologies were adopted to obtain very low-tolerance holes (without burrs/scratches/delamination) and short drilling times [14,15]:

- Positive high-speed feed drill motors (3000 rpm) with concentric collets to make one-shot holes in composite stacks.
- Positive feed drill (PFD) motors (320 round per minute, rpm) combined with an innovative "vibratory system" for achieving satisfactory big holes (up to 7/16 in) on mixed stack-ups at the root rib location (high strength steel + CFRP) and at the tip rib and rib post locations (aluminum 7075 + CFRP).
- Click bond nut plates for upper panel screwing.

At the end of the assembly phase, the cabling was completed, and the functionality tests were performed under laboratory conditions with the support of partners responsible for the morphing components. Strain gauges and optic fibers were used to detect the deformation field.

4 JIG

The jig modeling and construction were the first key steps in the creation of the SARISTU outer wing experimental demonstrator. The new wing box design made it possible to reduce the number of jigs, as compared to traditional wings, whose numerous parts required multiple stations to complete total assembly of the structure. In the new configuration, it was possible to perform multiple necessary operations on the same station. Furthermore, due to their single block nature, the adaptive components did not require the traditional jig reference points to align hinges, instead utilizing the aerodynamic profile of the whole structure.

In this application, a unique jig station where all the tasks could be performed was designed, ensuring precise, tolerance-sensitive positioning of all structural elements to be assembled. A user-friendly support system facilitated the operators comfort and access to work areas, while preventing any accidental handling and placement damage. Based on these requirements and other considerations, a rotating jig was created, composed of two distinct parts: a robust structure and a rotating frame. All the parts were indexed and equipped with THs, terminals, and pins, guaranteeing correct location of all the different elements to be assembled.

A system consisting of five levelling rods with vacuum suction pads was equipped to support and to fix skin positions in compliance with the airfoil design. It also allowed easy, correct drilling and assembly of other structural elements and permitted precise alignment of the trailing and the LE. Two dedicated mobile supports kept the winglet shape and sustained its weight during the installation procedure, ensuring precise fitting. The jig movement was carried out by means of five removable saddles. In order to ensure maximum accessibility and safety to the assemblers, the jig was equipped with a servo-mechanical actuation system, enabling rotations around the longitudinal axis and operation in both vertical and horizontal orientations. The jig is shown in Figs. 13 and 14. The first picture shows the machine as appears in the installation facility, before any component is mounted on it. The second picture details its different components, their location, and their meaning. It is remarkable that the different morphing devices were delivered as single-blocks, allowing each of them to be installed as a unique piece. Their weight had to be accounted for in the installation process, as did the need to guarantee nondeformability of the morphing components to avoid residual strains in both the devices and the structural system hosting them.

5 EQUIPMENT AND TOOLING

More than 90% of the assembly of all the parts comprising the wing-box primary structure was performed in a one-shot drilling process, representing the state-of-the-art in manual assembly [9–13] using semiautomated drills called power feed (PF) or positive feed drill (PFD). PFDs cut into the structure using a gear-powered drill bit rotation. This required a drilling mask and a Lubbering system (see Fig. 15) equipped with a concentric collet, allowing countersink on the aerodynamic surfaces in the

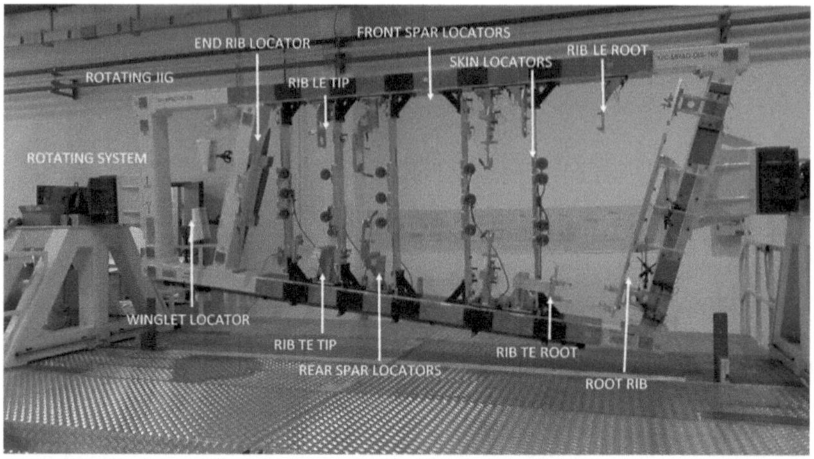

FIG. 13

The jig for the adaptive wing section assembly.

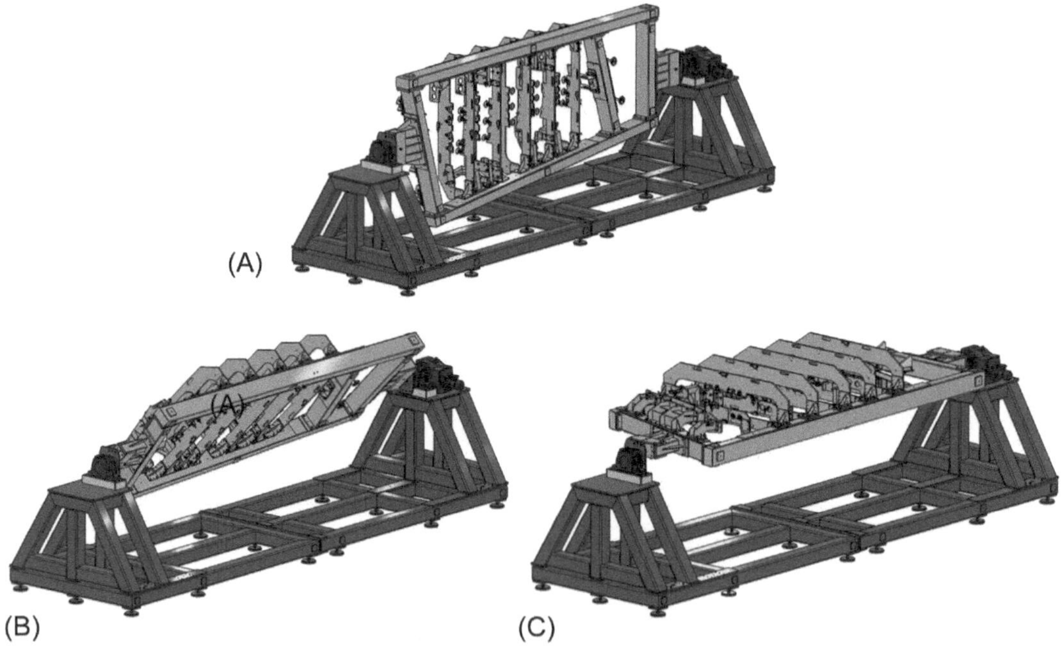

FIG. 14

The jig system (schematic): standing position (top); rotation stages (bottom).

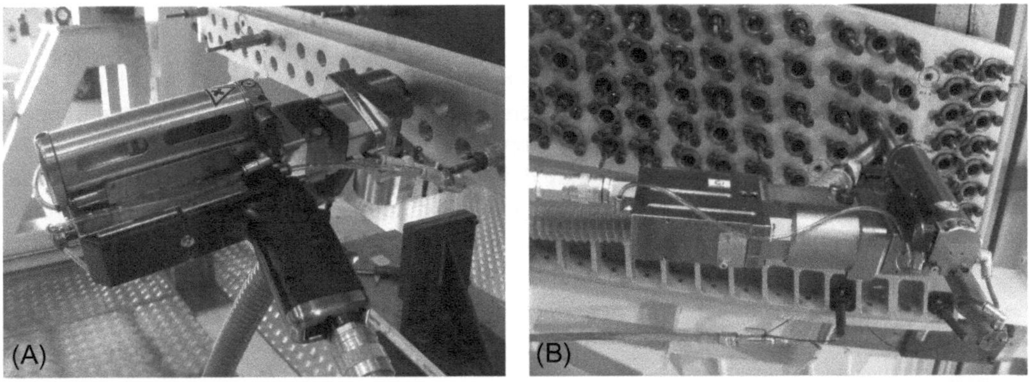

FIG. 15

Lubbering PFD (A) root rib with steel root rib and slow speed drill and (B) spar skin drill with high speed drilling and countersink with concentric collet machine.

same shot. The drill mask, positioned on the components via PHs, ensured the holes were correctly centered, sized, and finished. The PFD process parameters were:

- 3100 rpm and 0.05 inch per revolution (in/rev) for the coupling holes between CFRP elements and
- 550 rpm and 0.05 in/rev for the coupling holes between elements of different materials (CFRP vs. steel for the root and tip ribs; CFRP vs. aluminum for the middle ribs).

Tip jig components, especially the heavier and bulkier ones, were handled with an overhead crane. A specific tool mounting system was produced, Fig. 16.

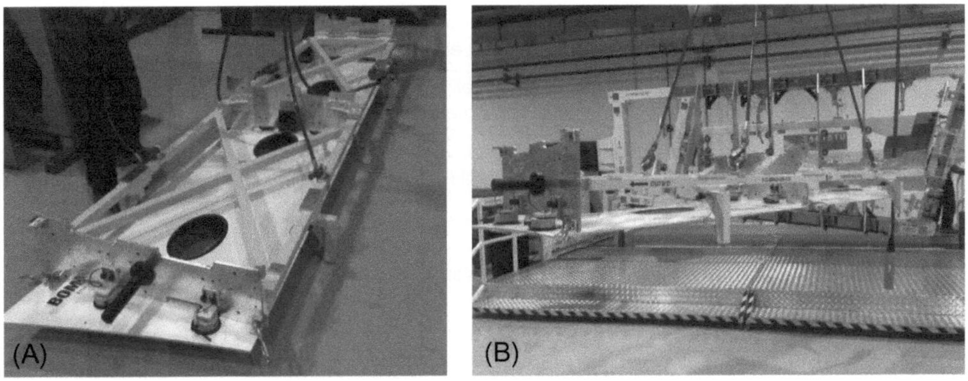

FIG. 16

Skin handling through slings system connected to overhead crane.

6 DEMONSTRATOR ASSEMBLY

Having described the peculiar characteristics of each element that made up the demonstrator, and having defined the technologies and processes that were used, as well as the implemented equipment and hardware, the step-by-step procedure that was adopted for the assembly execution is herein defined. The driving objective of the assembly procedure definition was to minimize risk of failure and ensure compliance with the delivery time. The operations flow is reported below (see Fig. 17), taking into account the different constituent elements and main substructures, i.e., the wing box, LE, TE, winglet, and embedded sensor system. It can be summarized in three main stages:

- The assembly of the wing box, subdivided into the following steps:
  - The positioning of the root rib, the end rib, and the spars, in order to generate the external perimeter of the structure (main frame).
  - The positioning and drilling of the ribs, in order to create the structure reinforcement.
  - The positioning and drilling of the skins.
- The positioning and drilling of the adaptive leading edge, TE, and winglet.
- The installation of the FOs system, the extensometers, and the wiring.

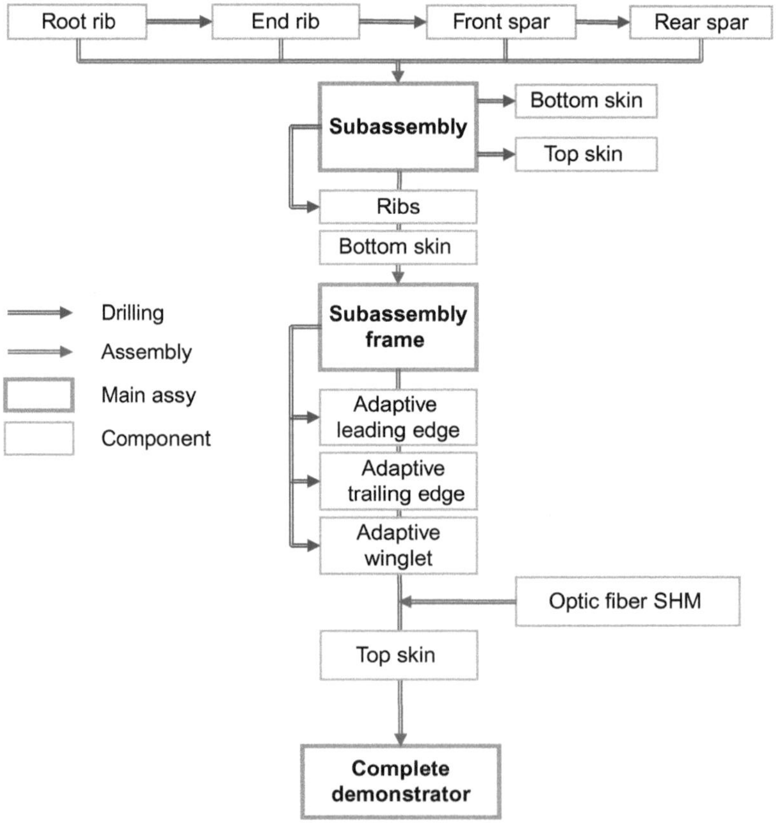

FIG. 17

Logic diagram of the assembly process.

6.1 THE ASSEMBLY OF THE WING BOX

The assembly of the demonstrator wing box primary structure (perimeter frame) was achieved with a sequence of 195 macro operations, grouped into four blocks, classified as positioning, drilling, testing, and, when necessary, application of shim junctions. The first block of operations consisted of placing the front and rear spars jointly with the root and end ribs via the jig THs and coordinated holes on the parts. This guaranteed correct alignment of the different components and proper definition of the reference substructure. The second block of operations consisted of the placement and drilling of the internal ribs to reinforce the structure. As in the previous step, this was performed through the use of coordinated holes conveniently located on the ribs, the post ribs (spar slots, installed for facilitating the ribs placement), and the spars. In Fig. 18, the results of these two preliminary phases are reported. The resulting frame is illustrated on the left. On the right, a photograph shows a detail of the frame placed on the jig, with the ribs already mounted. The premounted slots, displayed for simplifying the ribs installation, are clearly visible along the front spar on both the pictures.

In the third step, the lower skin was installed. In this case, in addition to CHs, profile saddles assured achievement of the designed aerodynamic shape (the main requirement of the demonstrator) and avoided prestresses and strains. Careful design management of tolerances between the spars and skin achieved the desired surface, which incorporated a gap to compensate for the stresses possible during the manufacturing and assembly process.

The resulting gaps between these elements were then filled with a shim that could have been either solid (fiberglass, aluminum, or capton, like the one in Fig. 19), or liquid (resin). The rotating rig permitted this by keeping the frame in the horizontal position and placing the skin underneath. A skin handling tool permitted correct alignment with the target substructure. Once the gaps verification was performed and assessed, the ultimate arrangement and definitive drilling of the rib post (numbered from 12 to 16, right hand to left hand) were executed, Fig. 20.

The drilling of the skin, spars, and ribs was performed using PF drills. Fig. 21 shows the resulting coupling holes distribution. At the end of this step, the lower skin was removed, and the parts were

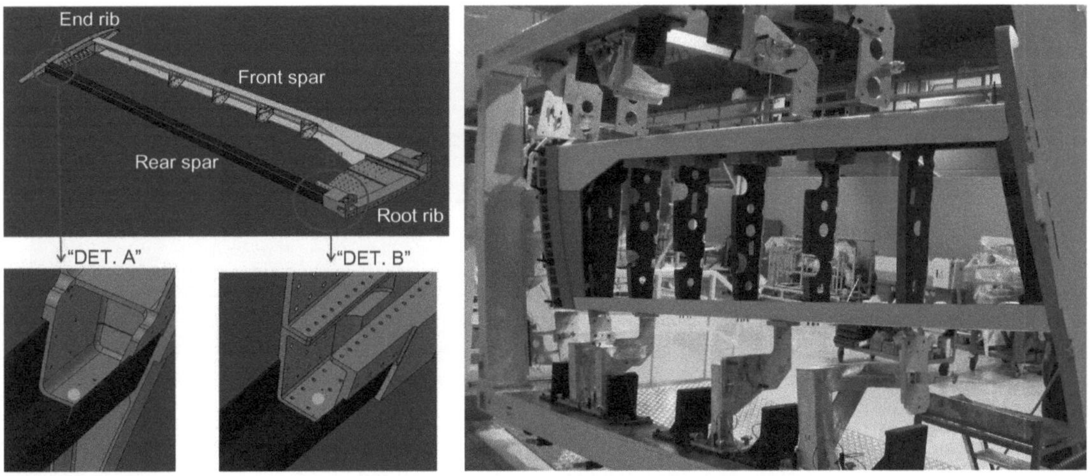

FIG. 18

Demonstrator structural frame and installation details (left); ribs mounting (right).

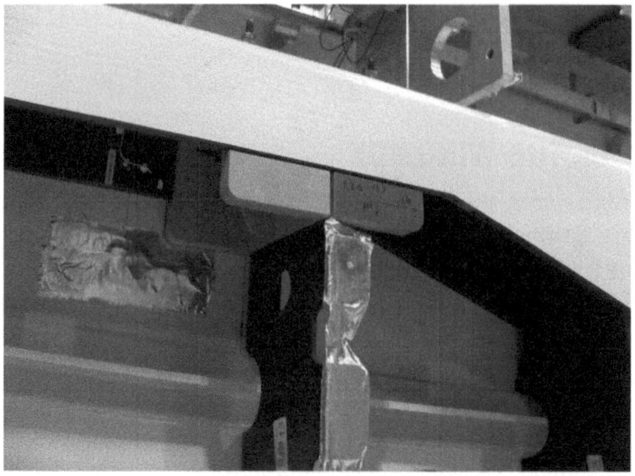

FIG. 19

Application of solid shim to a rib post to fill the gap present in the coupling with the lower skin.

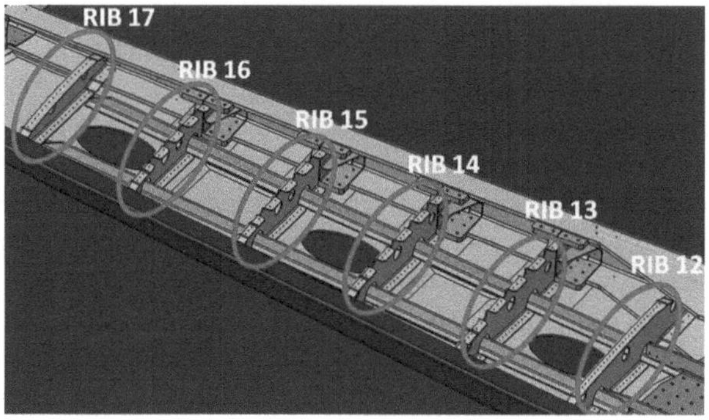

FIG. 20

CAD representation of the wing section demonstrator: detail on the ribs and rib posts location.

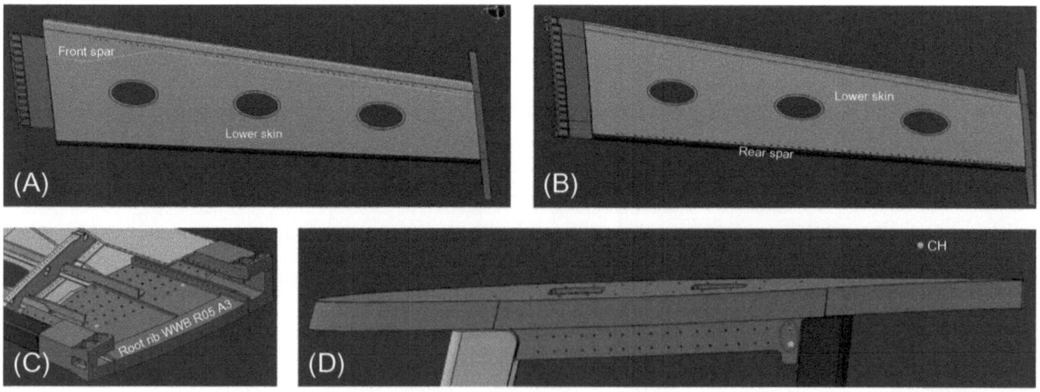

FIG. 21

CAD representation of the wing section demonstrator (some CH are indicated): coupling holes between the lower skin and: (A) the front spar; (B) the rear spar; (C) the root rib; and (D) the tip rib.

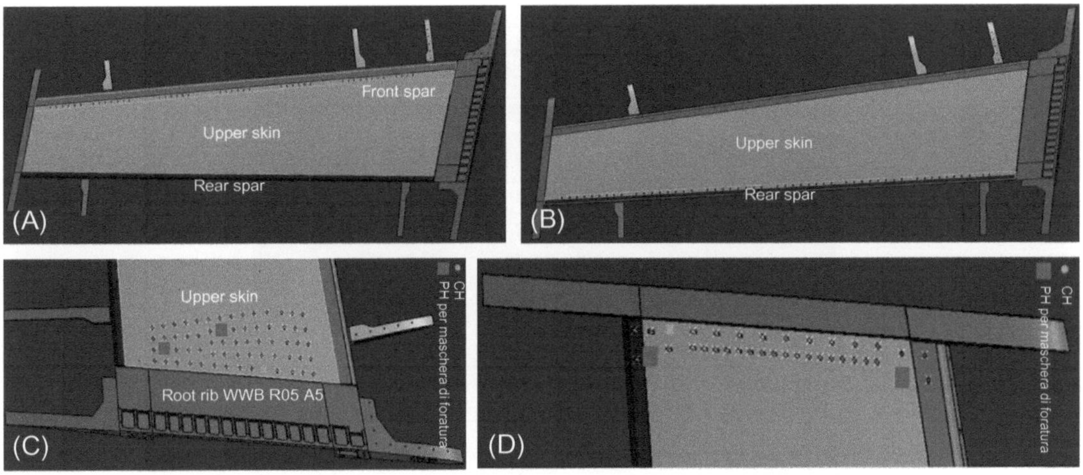

FIG. 22

CAD representation of the wing section demonstrator: coupling holes between the upper skin and: (A) the front spar; (B) the rear spar; (C) the root rib; and (D) the tip rib.

cleaned of chips and burrs. The fourth segment of operations started with the positioning and drilling of the upper skin. Once again, the saddles were sited, the jig was rotated to the opposite horizontal position, and the skin handling tool was revolved and operated. Once more, after the gaps were filled, all the parts were drilled at their final diameters, Fig. 22. Then, the upper skin was removed, while the lower skin was repositioned and definitively fastened. Assembly of the primary structure was completed by connecting all the other parts to the main frame, with the exception of the upper skin (which would have been screwed at the end of the assembly process).

6.2 MORPHING SYSTEMS INSTALLATION: THE LEADING EDGE

The second block of assembly operations consisted of mounting the morphing parts. The adaptive leading edge was the first to be installed, Fig. 23. The front spar was predrilled on the web where the actuator should have been located and in correspondence to the post ribs. In order to facilitate construction of the adaptive elements and their further integration onto the wing section, the structural front and rear spars were composed of main and auxiliary parts (part of the wing box structure and the morphing systems, LE and TE respectively). This solution allowed partners who developed the morphing parts to assemble these components and perform preliminary functionality and characterization tests autonomously. For the LE, parts were joined using the aerodynamic surface (with the profile saddles) and the predrilled post rib locations as reference.

The adaptive leading edge placement started by joining the respective spars, matching the predrilled holes, in a temporary fastening. The saddles were then positioned on the jig, and the skin of the morphing part was aligned with the skin of the wing box. Where necessary, design gaps were filled with shims. Then, using the drilling mask, all the holes between the LE and the flange of the spar were

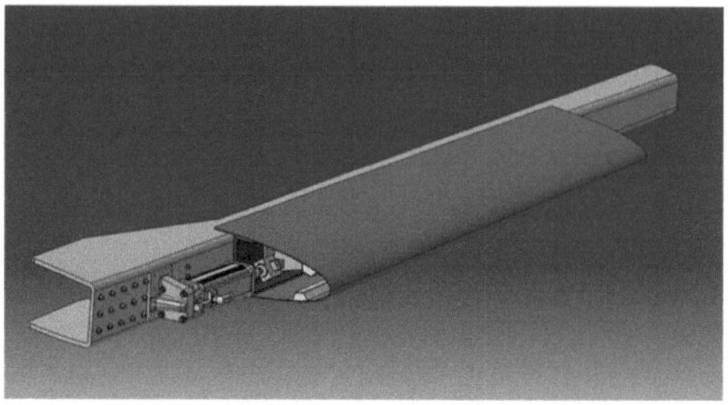

FIG. 23

LE assembly—CAD representation.

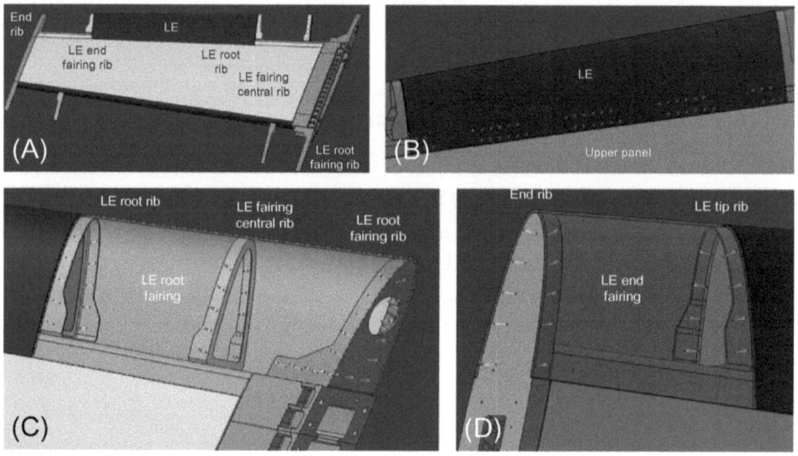

FIG. 24

CAD representation of the wing section demonstrator: (A) leading edge area layout; (B) CH between LE and front spar; (C) CH between the root fairing and the ribs; and (D) CH between the end fairing and the ribs.

drilled. After installation of the LE, the actuator and its shaft were fixed on the LE spar, and auxiliary ribs necessary for the fairings were positioned and drilled on the spar. At this point, fairings were positioned on the respective ribs, shimmed and drilled, Fig. 24. Alignment with the aerodynamic surfaces at the boundaries was verified with shaped saddles.

6.3 MORPHING SYSTEMS INSTALLATION: THE TRAILING EDGE

The basic steps of the assembly of the TE were developed in a way similar to that of the LE assembly. The TE as already described is terminated with an auxiliary spar, but unlike the LE, no connections on its web were required by design. Only skin and wing box rear spar connections were necessary.

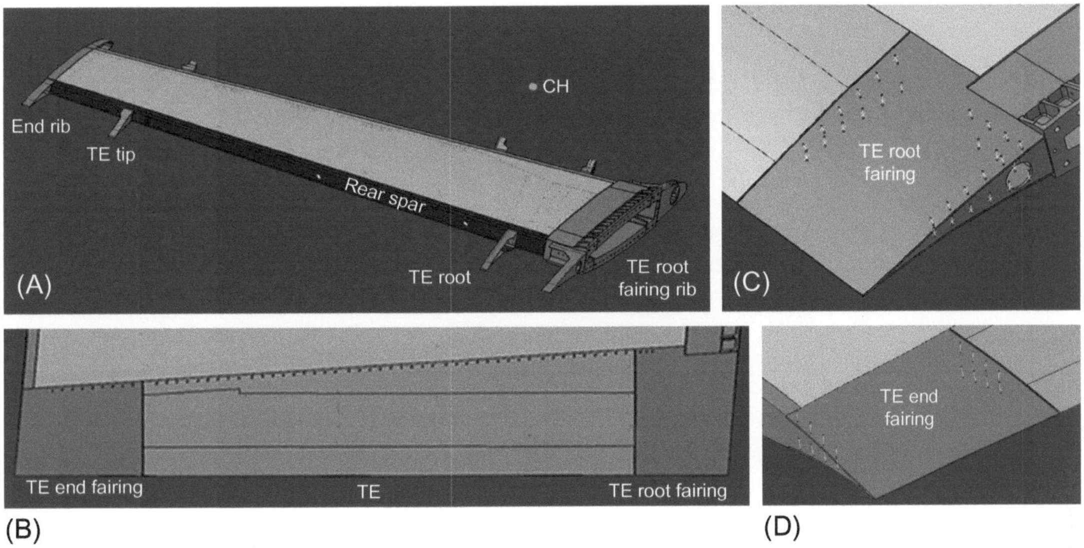

FIG. 25

CAD representation of the wing section demonstrator: (A) trailing edge area layout; (B) CH between the trailing edge and the rear spar; (C) CH between the root fairing and the ribs; (D) CH between the end fairing and the ribs.

The positioning was consequently done only by aerodynamic surfaces requirements. Shimming to guarantee this specification was performed, and drilling by a semiautomated drill machine was finally conducted to connect the TE. On the rear spar were two fairings, sustained by auxiliary ribs positioned at the flanges of the spar and drilled together. Fairings were aligned by aerodynamic surfaces and shimmed, Fig. 25.

6.4 MORPHING SYSTEMS INSTALLATION: THE WINGLET

The morphing winglet was realized by FACC, a leading manufacturer of these systems. The installation replicated the standards of the aeronautic industry. The end rib of the wing was mated with the root rib of the winglet through coordinated holes, drilled prior to installation (using a drilling mask by FACC for its side and by Leonardo for the wing rib). The jig was horizontally oriented to facilitate the winglet-positioning process. The weight of the component necessitated handling by highly specialized technicians. Special support on the jig allowed alignment between the holes on the outer face of the wing tip rib with those on the winglet outer section, Fig. 26. No connections were necessary on the skins, and alignment was guaranteed by the design tolerances of each part. Installation of all necessary electronic components proved to be a major difficulty in assembly of this morphing part, as they were located in the last bay of the main wing box. All instrumentation was located on an auxiliary support, fixed on the spars.

FIG. 26

The adaptive winglet, mounted on its saddle and installed on the wing box.

7 FBG SENSOR NETWORK

The installation of the fiber Bragg gratings (FBG) network occurred during assembly of the wing box [15–18]. The system used ribbon-tape technology [19]. Developed by a scientist seeking to protect optical fiber from the external environment [20], ribbon-tape technology offers long-term preservation of these fibers under typical aeronautic scenarios, while avoiding the need to directly incorporate them into structural components (and the attendant complex issues which would result from that incorporation). This also avoids direct handling of the fragile FOs and circumvents drawbacks associated with their limited bonding surface (usually, diameters well under 1 mm). Ribbon tape technology provides the single fiber with an added surface, normally made of fiberglass, protecting it and making connections easier. This also simplifies later detachment of ill-functioning or broken sensors, as the fiber is easily stripped away with the covering tape. Removed fibers are not reusable, but this method makes it easier to install new elements.

Among the different components that were present in the wing box, the main spars (both front and rear) proved the most interesting to monitor and also the most challenging locations for insertion of the sensor networks. Panels had to be manufactured by other firms (which was inconvenient) and then moved to the Leonardo plants to be mounted on the main structure. This meant exposing the network to risks during transportation. Ribs were basically meant to preserve the aerodynamic shape and were not expected to exhibit a relevant strain field. WT support ribs were very stiff and had a complex shape, like the WATE support. Moreover, the installation and deployment paths were complicated for many reasons (these elements are the first to be installed; they are very stiff; the cabling path to the external would have been difficult, and so on).

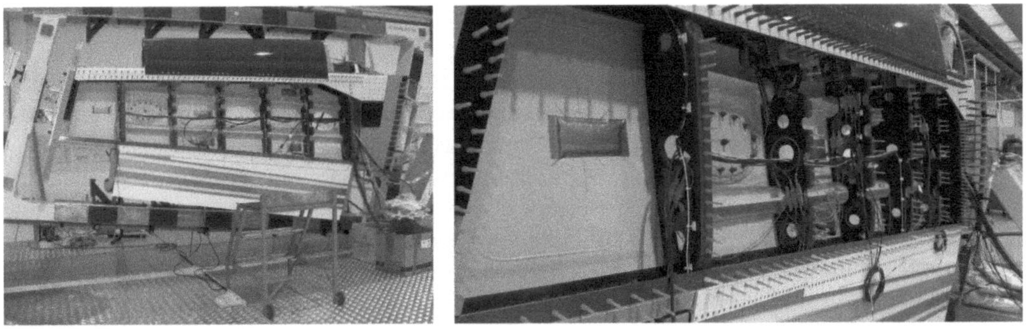

FIG. 27

FO monitor system installation: very first steps (left) and the final results (right).

In spite of that, many problems arose, even on the spars. Along the fiber deployment, the presence of the ribs and the post ribs had to be taken into account (it was not possible to configure installation before the structural system was mounted), as did further possible interferences with other structural elements, like secondary stiffeners, various kinds of junctions, the top and bottom panels, etc. After many considerations, it was decided to deploy the fiber network over the internal spar web, for both the elements. This was the safest, though not the most accessible, area. Its great advantage was it did not need to have a surface specifically designed for hosting the ribbons.

In detail, a group of six ribbons was considered for each bay of the wing box (a bay is the region between two consecutive ribs), up to a total of six bays. The displayed network took into account the presence of the ribs, the post ribs, the junction lines, and any other possible element displayed in the structure. In addition, the associated wiring also considered the auxiliary strain gauges system, to be used as a reference with respect to the information attained by the FO network. For the rear spar, the same architecture was achieved, but the number of active elements was reduced to four per bay because of the smaller size of this structural component.

The deployment of the FO system required full accessibility on the installation area, for proper bonding. This application had to be carried out after all drilling, but before the final fastening. The fibers were bonded through the use of specific polymeric-based adhesives. This process was supported by a specialists team, made of scientists from Technobis and the Polytechnic of Milano [15,18].

In Fig. 27, two different moments of the FO system integration are shown. In the picture on the left, a preliminary phase is reported with the cabling incomplete and many wires deployed on the various structural elements. In the picture on the right, the deployment was completed. In spite of the limited system that was implemented (only the main spars were affected), the mass of cables is impressive and may be considered an indicator of how a full-monitored structure could look. The green bundles are the FO connectors. The strategy was clear. The different wires were brought from the front to the rear spar and attached to the rib walls. All the wires were then brought out of the structure through suited holes and drilled in the last structural and the WT attachment ribs.

8 CONCLUSIONS

The assembly process of a morphing wing was herein illustrated. The study case was a prototype designed and developed within the SARISTU program. It consisted of a main wing box, equipped with a FO system for monitoring its structural health state, plus morphing elements independently built by three different international teams. The project strategy was to assign each device with specific and limited interface constraints so that each element could be designed autonomously, minimizing the issue related to their integration. Specifically, the interfaces between these components and the wing box were limited to the main spars (LE and TE) or the interface rib (winglet). This streamlined the final assembly step, significantly simplifying the procedure.

Special care had to be devoted to preserving the aerodynamic shape of these subsystems. Their weight and size necessitated special support saddles to facilitate the connection process and avoid unwanted residual strains on the parts. Deploying KHs or THs as a reference for the jig assured precision matching of characteristic elements. The assembly then followed a well-structured sequence of drilling and fastening. The main drawback in separately constructing the different main components was the need to adjust the interface area via shimming; in this large international consortium, however, it was mandatory to delegate roles to specific subteams. Still, even in an industrial setting, it could be advantageous to build the different morphing parts separately and then assemble them as was done in this project, since each of the adaptive components is a complex system unto itself with many structural and electronic parts and connections.

In addition to the saddles, it is important to address the design and creation of a unique, customized jig that had to take into account all the peculiarities of this novel project. The interface elements, in conjunction with the saddles, allowed the correct and controlled positioning of all the elements. Coordination of these precision connections was especially challenging due to the remarkable amount of wiring involved (because of the presence of numerous electromagnetic actuators and, in some cases, widely distributed sensor networks).

This system was a sort of hybrid of a structure and a machine, full of both wiring and cabling. A diffused sensor net was deployed within the wing box to monitor, not its deformed shape, but the strain field and its impact on the health of the wing box. The assembly stage had to accommodate this novel element, planning at the very early phases of the design how to deploy the cables and determine the feeding sources and associated electronics (except for the winglet, in which it was embedded).

In spite of all the attention dedicated to installing the morphing systems, at the end of this process, it was still necessary to adjust their position in order to reach a new nominal zero. It would be advisable, then, to take this into account during the design process, perhaps deliberately oversizing the mobility of the morphing devices so they can match their own reference shape without any impact on expected performance.

ACKNOWLEDGMENTS

The research leading to these results has gratefully received funding from the European Union Seventh Framework Programme (FP7/2007-2013) under Grant Agreement n° 284562, SARISTU.

REFERENCES

[1] J.P. Fielding, Introduction to Aircraft Design, Cambridge University Press, Cambridge, ISBN 978-0-521-65722-8, 2010. p. 259.

[2] F. Viadero, M. Ceccarelli (Eds.), New Trends in Mechanism and Machine Science: Theory and Applications in Engineering, Springer Science & Business Media, Dordrecht, ISBN 978-9-400-74902-3, 2012. p. 809.

[3] S. Mazumdar, Composites Manufacturing: Materials, Product, and Process Engineering, CRC Press, Boca Raton, FL, ISBN 978-1-42004-198-9, 2001. p. 416.

[4] J. Valasek (Ed.), Morphing Aerospace Vehicles and Structures, John Wiley and Sons, Chichester, ISBN 978-1-119-94030-2, 2012. p. 212.

[5] P.C. Woelcken, M. Papadopoulos (Eds.), Smart Intelligent Aircraft Structures (SARISTU), Proceedings of the Final Project Conference, Springer International Publishing, Cham, ISBN 978-3-319-22413-8, 2016.

[6] M. Kintscher, J. Kim, S. Storm, F. Peter, Assessment of the SARISTU enhanced adaptive droop nose, in: P. C. Woelcken, M. Papadopoulos (Eds.), Smart Intelligent Aircraft Structures (SARISTU), Proceedings of the Final Project Conference, Springer International Publishing, Cham, ISBN 978-3-319-22413-8, 2016.

[7] R. Pecora, F. Amoroso, M. Magnifico, I. Dimino, A. Concilio, KRISTINA: kinematic rib-based structural system for innovative adaptive trailing edge. in: Proc. SPIE Smart Structures and NDE Conference 2016, Las Vegas, NA, USA, 20–24 March 2016. https://doi.org/10.1117/12.2218516.

[8] A. Wildscheck, S. Storm, M. Herring, D. Drezga, V. Korian, O. Roock, Design, optimization, testing, verification and validation of the wingtip adaptive trailing edge, in: P.C. Woelcken, M. Papadopoulos (Eds.), Smart Intelligent Aircraft Structures (SARISTU), Proceedings of the Final Project Conference, Springer International Publishing, Cham, ISBN 978-3-319-22413-8, 2016.

[9] Y. Essa, F.M. de la Escalera Cutillas, I. Dimino, M. Ciminello, A. Concilio, Manufacturing and testing of smart morphing SARISTU trailing edge, in: P.C. Woelcken, M. Papadopoulos (Eds.), Smart Intelligent Aircraft Structures (SARISTU), Proceedings of the Final Project Conference, Springer International Publishing, Cham, ISBN 978-3-319-22413-8, 2016.

[10] O. Schorsch, A. Lühring, C. Nagel, Elastomer-based skin for seamless morphing of adaptive wings, in: P.C. Wölcken, M. Papadopoulos (Eds.), Smart Intelligent Aircraft Structures (SARISTU), Proceedings of the Final Project Conference, Springer International Publishing, Switzerland, ISBN 978-3-319-22412-1, 2016.

[11] C. Nagel, A. Fiedler, O. Schorsch, A. Lühring, Seamless morphing concepts for smart aircraft wing tip, in: P.C. Wölcken, M. Papadopoulos (Eds.), Smart Intelligent Aircraft Structures (SARISTU), Proceedings of the Final Project Conference, Springer International Publishing, Switzerland, ISBN 978-3-319-22412-1, 2016.

[12] A. Marino, P. Cirillo, D. Gerbasio, C. Natale, P. Chiacchio, in: A general low-cost and flexible architecture for robotized drilling in aircraft assembly lines, 2016 International Symposium on Power Electronics, Electrical Drives, Automation and Motion (SPEEDAM), Anacapri (I), 22nd–24th June, 2016, https://doi.org/10.1109/SPEEDAM.2016.7525936.

[13] I. Erdema, P. Helgossonb, H. Kihlmanb, in: development of automated flexible tooling as enabler in wing box assembly, 6th CIRP Conference on Assembly Technologies and Systems (CATS), Chalmers (S), May 15th–18th, 2016,https://doi.org/10.1016/j.procir.2016.02.065.

[14] FAA, Aviation Maintenance Technician Handbook—Airframe Volume 1: FAA Airframe Handbook V1—Volume 1 of Aviation Maintenance Technician Handbook—Airframe, AMT Exams, Oklahoma City, OK, 2012. p. 588.

[15] F. Kreith, The CRC Handbook of Mechanical Engineering, second ed., CRC Press, Boca raton, FL, 1998. ISBN 978-0-84939-418-8. p. 2624.

[16] D. Drezga, V. Korian, O. Roock, B. Lopez, A. Fiedler, S. Storm, V. Snop, Winglet design, manufacturing, and testing, load monitoring by means of optical fibres and strain gages, in: P.C. Wölcken, M. Papadopoulos (Eds.), Smart Intelligent Aircraft Structures (SARISTU), Proceedings of the Final Project Conference, Springer International Publishing, Switzerland, ISBN 978-3-319-22412-1, 2016.

[17] M. Bosboom, M. van Wijngaarden, R. Evenblij, P. Bettini, T. Loutas, V. Kostopoulos, D. Habas, M. Tur, N. Gorbatov, A. Bergman, U.B. Simon, I. Kressel, C. Koimtzoglou, M. Ciminello, A. Weisser, C. Paget, Ribbon tapes, shape sensors and hardware, in: P.C. Wölcken, M. Papadopoulos (Eds.), Smart Intelligent Aircraft Structures (SARISTU), Proceedings of the Final Project Conference, Springer International Publishing, Switzerland, ISBN 978-3-319-22412-1, 2016.

[18] A. Airoldi, G. Sala, R. Evenblij, C. Koimtzoglou, T. Loutas, G.M. Carossa, P. Mastromauro, T. Kanakis, Load monitoring by means of optical fibres and strain gages, in: P.C. Wölcken, M. Papadopoulos (Eds.), Smart Intelligent Aircraft Structures (SARISTU), Proceedings of the Final Project Conference, Springer International Publishing, Switzerland, ISBN 978-3-319-22412-1, 2016.

[19] P. Charlaftis, T. Loutas, C. Koimtzoglou, V. Kostopoulos, Structural health monitoring of aerospace components with optical fibre ribbon-tapes, 6th Symposium for NDT in Aerospace, AeroNDT 2014, Madrid (E), November 12th–14th, 2014.

[20] Z. Zhou, J. Ou, in: Development of FBG sensors for structural health monitoring in civil infrastructures, North American Euro-Pacific Workshop "Sensing Issues in Civil Structural Health Monitoring", Waikiki Beach, HI, Nov. 11th–13th, 2004.

CHAPTER

MORPHING DEVICES: SAFETY, RELIABILITY, AND CERTIFICATION PROSPECTS

21

Maurizio Verrastro*, Ignazio Dimino†

Leonardo S.p.A., Caselle (TO), Italy The Italian Aerospace Research Centre, CIRA SCpA, Capua (CE), Italy†*

CHAPTER OUTLINE

NOMENCLATURE

A/C	aircraft
AIAA	American Institute of Aeronautics and Astronautics
ARP	aerospace recommended practices
ATED	adaptive trailing edge device
CCA	common cause analyses

Morphing Wing Technologies. https://doi.org/10.1016/B978-0-08-100964-2.00021-6

CMR	certification maintenance requirement
CS	certification specifications
DAL	design assurance level
DO	(RTCA) document
EADN	enhanced adaptive droop nose
EASA	European Aviation Safety Agency
EUROCAE	European Organisation for Civil Aviation Equipment
ED	EUROCAE document
FC	failure condition
FDAL	functional design assurance level
FH	flight hour
FHA	functional hazard assessment
FMES	failure mode effect summary
FTA	fault tree analysis
FT	fault tree
HW	hardware
IEEE	Institute of Electrical and Electronics Engineers
PSSA	preliminary system safety assessment
RTCA	Radio Technical Commission for Aeronautics
SAE	Society of Automotive Engineers
SARISTU	Smart Intelligent Aircraft Structures
SSA	system safety assessment
SW	software
VC	cruise velocity
WATE	winglet active trailing edge

1 INTRODUCTION

Despite the considerable interest and the growing advances in morphing wing technologies, morphing devices continue to be perceived as highly difficult to certify by the aviation industry. Although several morphing wing concepts and demonstrators already were tested in representative environment [1–5] and, in some cases, even flight tested [6,7], the certification roadmap of these advanced mechanical systems still suffers from certain safety-related gaps in the attempt to comply with current certification requirements.

At this time, the authors are not aware of the use of morphing devices in any commercial aircraft. Two fundamental questions still remain to be addressed: What changes or additions to the traditional certification processes would be necessary to demonstrate that morphing devices comply with the certification standards? And is there any room for some nontraditional means of compliance? Answering to these questions not only would allow for safe use of such highly integrated systems that perform complex and interrelated aircraft-level functions, but also could result in a successful progress in the certification prospects of morphing wing structures.

In this chapter, some gaps in the design approach of morphing devices are discussed in order to provide sufficient consistency with current standards in aviation. The use of acceptable safety-related design methodologies, combined with more automated manufacturing, assembly, and integration processes, appear to be the most adequate means to certify these mechanical systems within the context of industry standards. Additionally, beyond structural factors or other considerations of airworthiness to prevent catastrophic failures, the novel aircraft functions enabled by morphing systems imposes a thorough examination of the associated risks that can have an impact on aircraft flight capabilities and crew workload. As with all the other aircraft control surfaces and systems, the real challenge is to show whether the safety targets are met, until the single system is assessed and the interaction between conventional and adaptive systems performing different functions are collectively evaluated at aircraft (A/C) system level. More reliance on verification by analysis, simulation, and limited proofs of concepts from the early design phase could lead to partial certification credit while preserving the essential features of safety.

Government Certification agencies such as the Federal Aeronautical Administration (FAA) and the European Aviation Safety Agency (EASA) require that aircraft manufacturers follow precise design and certification policies to manage and mitigate systems failure risks throughout the operational life of an airplane. Historical evidence indicates that the probability of a serious accident because of operational and airframe-related causes is approximately one per million hours of flight, and only about 10% of the total are attributed to failure conditions (FCs) caused by A/C systems [8]. The process of identifying risks and quantifying or qualifying the degree of risk they pose to individuals and resources is usually referred to as safety assessment [9]. A probability approach is typically used to manage these assessments. A logical and acceptable inverse relationship must exist between the average probability per flight hour (FH) and the severity of FC effects, as shown in Fig. 1. Catastrophic failures must be

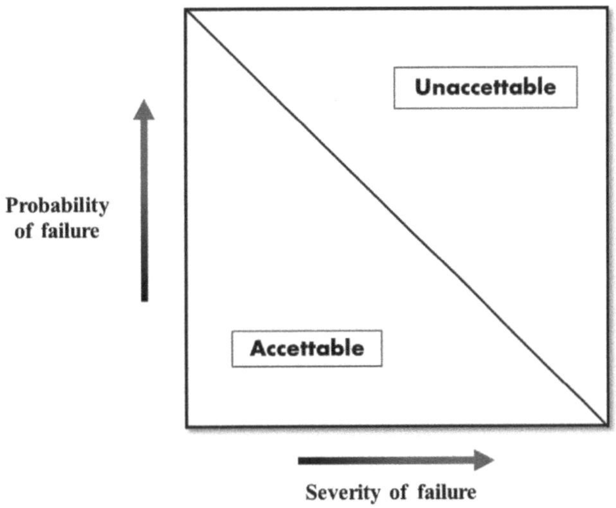

FIG. 1

Relationship between probability and severity of failure condition effects.

extremely improbable and must not result from a single failure [8]. The upper limit for the average probability per FH for catastrophic FCs shall be 1×10^{-9}, which establishes an approximate probability value for the term "extremely improbable." On the other hand, FCs having less severe effects could be relatively more likely to occur, with upper average probability limits equal to 1×10^{-7} and 1×10^{-5} per FH for hazardous and major, respectively.

According to the EASA CS-25 regulations applicable to large commercial aircraft, safety assessments consist of three major phases, i.e., functional hazard assessment (FHA), preliminary system safety assessment (PSSA), and system safety assessment (SSA), which consider the interactions among the aircraft systems, between software and hardware (HW) components, and all the interfaces, including pilot and crew. In this frame, systems functions are examined qualitatively to identify potential design, maintenance or crew faults, and external environment risks. The severity of these hazards is determined and placed in specific classes, indicative of the maximum tolerable probability of occurrence. After that, the link between faults at subsystems level and their end-effects are evaluated in the FHA, by taking into account the actual system constraints throughout the system lifetime. Each hazard is quantitatively examined in a top-down fashion, from the events to their causes, until failures of the basic components are classified. On the basis of the probabilities assigned to failure events of the basic components, the probability of occurrence of the top event is then calculated. Such a quantitative analysis usually is achieved with the fault tree (FT) technique.

Starting from a generic overview of A/C system functions, this chapter follows a standard safety analysis to define both qualitative and quantitative requirements to support the certification practices of morphing devices with respect to safety and reliability targets. Two main documents combined with the certification specifications (CS)-25 EASA regulations [8] are considered as guidelines: Society of Automotive Engineer (SAE) ARP 4754a [9] and SAE ARP 4761 [10]. A complete aircraft FHA, including morphing systems and functions, is outside the scope of this study. Integrated use of three morphing devices—droop nose, morphing trailing edge, and adaptive winglet—is assessed through the imposition of certain constraints after the design phase of the morphing architecture and permitted adaptation to meet specific system safety properties. Because of the concepts novelty, literature references about a safety-driven design of morphing devices are hard to find, for both new-generation aircraft and retrofit applications. Furthermore, the identification and classification of the potential failures of morphing devices might be a nontrivial exercise. Finally, not all the necessary analysis and modeling tools are available in the literature and further work might be required to bring such an approach to practical applications. Because of these points, a conservative approach is used to assign the functional systems failure severities and safety targets by developing and sharing a high-level mapping between a functional view of morphing devices and their A/C safety-related standards.

2 SYSTEM LEVEL APPROACHES TO THE CERTIFICATION OF MORPHING WING DEVICES

Depending on the time-scale of deployment, three different functions might be associated with a morphing device:

- *Morphing on the order of minutes*: Lift (and drag) control during long mission segments (mainly cruise) to compensate aircraft weight reduction because of fuel consumption.

- *Morphing on the order of seconds:* Active lift distribution control to minimize drag during short off-design mission segments (mainly climbing and turning operations).
- *Morphing in less than a second:* Wing load alleviation by reducing gust-induced root bending moment (RBM) peaks on aircraft wing.

For some recent-generation aircraft, such as Boeing 787 and Airbus A350, some novel aircraft functions, such as differential flap setting, already are performed by innovative flap actuation system concepts. Distributed actuation enables decentralized load control along the wing span, which is particularly suited for active lift distribution control for induced drag reduction. More tailored control systems and inherent positioning sensors contribute to guarantee this functionality.

A standard system safety assessment of aircraft systems, whose the key phases are shown in Fig. 2, must consider both the crew workload and operation implications of potential failures by establishing key capabilities and related limitations.

As part of the certification process, after requirement analysis has been completed, the top-level requirements assigned through the aircraft level FHA need to flow down to next level (the morphing wing system FHA), and subsequently to lower levels, such us morphing devices FHA. Without a specific A/C application, a generic set of aircraft level functions can be chosen arbitrarily to define high-level links between enhanced morphing system functions and standard aircraft functions by identifying, at the same time, the criticalities associated with the FCs. The analyses are intended at subsystem level, and performed with a typical bottom-up approach.

The SSA invokes a quantitative analysis of the FTs generated for potential failures with hazardous or catastrophic consequences. The reliability data of the system components are necessary to calculate

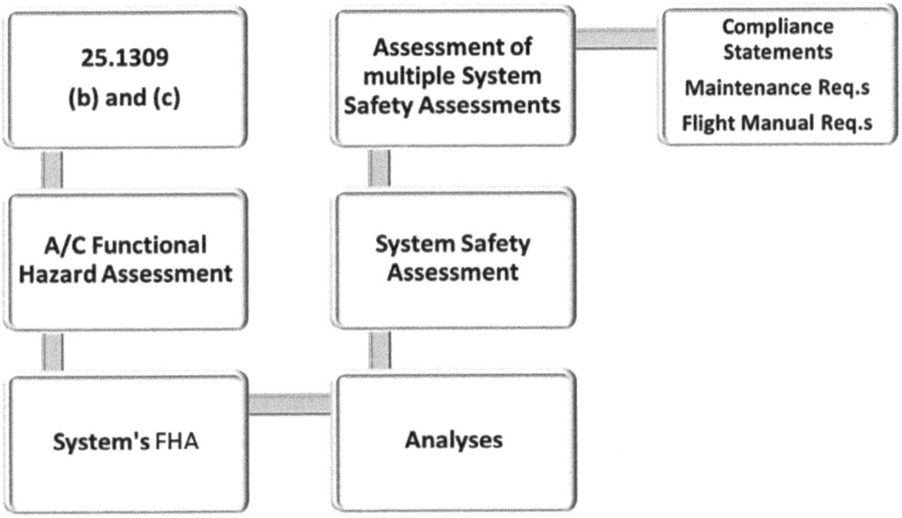

FIG. 2

Safety process overview diagram.

basic events probability figures which are typically requested to the components suppliers (commonly required by contract and by equipment specification) prior to the system integration and assembly phases.

We will elaborate on the safety process assessment approach by focusing on three morphing systems:

- Enhanced adaptive droop nose (EADN).
- Adaptive trailing edge device (ATED).
- Winglet active trailing edge (WATE).

A/C functions enabled by these systems are identified individually as explicitly requested by safety-related standards. It is assumed that for a safe integration of such morphing devices in a morphing wing, they are rigidly connected to a wing box without losing any deployment performance because of structural interactions. After that, their impact on A/C operation is evaluated by assessing their individual and combined A/C level functions and hazards.

2.1 ADAPTIVE DROOP NOSE

Aircraft use high-lift systems that create gaps when moving from high-lift to cruise flight position or vice versa [11]. Gaps create a discontinuous surface that interrupts the laminar flow and causes it to detach from the surface. In order to maintain a great region of laminar flow on the wing, the surface must be as continuous as possible in the desired aerodynamic shape with tight geometrical tolerances.

The adaptive droop nose is a movable morphing leading edge that can adapt its shape in order to reduce the drag during take-off and landing by acting as a high-lift device. This device can create a laminar flow by drastically enhancing aircraft aerodynamic performance in high-lift conditions, especially when combined with a morphing flap. Less drag directly reduces fuel consumption and affects the aircrafts weight by up to 3%. The concept of this smart leading edge device was developed to integrate functions such as de-/anti-icing, bird strike protection, erosion protection, and lightning strike protection. The structure-kinematics system was characterized by a low-complexity, lightweight kinematics, good fatigue performance [12].

2.2 ADAPTIVE TRAILING EDGE DEVICE

The main purpose of ATED is to reduce drag and increase efficiency (*L*/*D* ratio) in off-design flight points by adapting wing shape and lift distribution through a static deflection of the device [13] – [15]. ATED function can be referred to as continuous and quasistatic wing TE shape optimization control. While weight and angle of attack vary throughout the flight, ATED allows the aircraft trim configuration to remain optimal in terms of efficiency or minimal drag. The wing trailing edge shape optimization is aimed at compensating for a reduction in aircraft weight because of fuel burning. Because span-wise variations can be attained, weight savings potentially can be achieved in the design phase through RBM reduction. Smart Intelligent Aircraft Structures (SARISTU) demonstrated that aircraft fuel consumption can be lowered by at least 3% for a three-hour flight.

2.3 MORPHING WINGLET

Aircraft winglets are a proven way to reduce drag, save fuel, cut carbon dioxide and nitrogen oxide emissions, and reduce community noise. Blended winglets are offered as standard equipment on new aircraft designs and also are available as retrofit installations on existing commercial airplanes. The addition of winglets also increases aircraft range.

Winglets, however, introduce significant loads into the main wing structure that can diminish the expected benefits. These additional loads result in a heavier design, new wingtip interfaces, and an overall re-engineering of the wing box to allow for the winglet surface integration. The idea of an adaptive winglet has been successfully investigated in several theoretical studies and small-scale experiments. Adaptive winglets, where the geometry can be adjusted to the changing flow conditions, has the potential to improve the aerodynamic performance during climb and high-speed off-design conditions by providing adapted wing lift distribution throughout the A/C flight envelope. Additionally, they can significantly reduce aerodynamic loads at critical flight points (active load alleviation) that have variable trailing edge control.

Major aircraft manufacturers such as Airbus, Boeing, and McDonnell Douglas have received several patents that focus on changing the winglet shape to achieve minimal drag at multiple flight points [16,17]. Among the many prototypes of morphing winglets found in the literature, the adaptive winglet with active trailing edge, developed in the framework of the SARISTU project, is one of the latest and most advanced examples [18]. A full-scale carbon fiber reinforced plastics (CFRP) adaptive winglet device, including skin, stringers and four ribs, was designed, manufactured, and tested in a wind tunnel, and it showed very promising results. An active flap actuated by an electromechanical actuator and attached to the winglet's rear spar by a fail-safe connection (five single hinges) was commanded through a pure feed-forward control with no adaptation. In addition, a morphing skin covered the region between the fixed and movable parts, ensuring a smooth morphing shape. This design, however, led to additional actuation power to deform the morphing material under operative loads. Furthermore, a C-shape cut-out was necessary to avoid excessive membrane deformation at the winglet trailing edge, significantly reducing the expected aerodynamic benefits.

An adaptive winglet is definitively a "safety critical" structure. Past investigations on such a mechanical system have demonstrated that loss of the adaptive winglet control can be classified as catastrophic for aircraft [18]. Thus, the probability of its occurrence must be below the threshold value of $<10^{-9}$ per FH for safety reasons, as written in paragraph CS 25.1309.

The design of a morphing winglet design must follow a standard safety-critical system design approach, starting from a functional safety analysis. A failure hazard assessment is then needed to derive the design prerequisites for the system architecture and for the control system. After such qualitative safety classification is made for each functional failure, the overall system can be iteratively designed using fault tree analysis (FTA) on the basis of empirical values and experience for subsystem failure rates.

The safety classification and relevant safety figures are also a design driver for systems related to structural load alleviation/control functions. In fact, the recommended safety factor used for structural sizing purposes increases with the probability of being in FC, as shown in Fig. 3.

Although a FC related to degraded performance of an adaptive winglet can be classified as minor because of the minor safety repercussions on the aircraft occupants, a FT is always recommended for such systems in order to be able to compute the ultimate load for jam in the worst-case load positions.

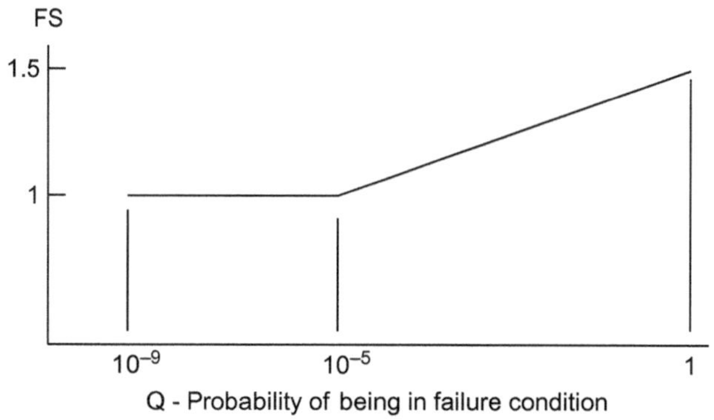

FIG. 3

Computation of the safety factor [8].

2.4 DEFINING THE SYSTEM LEVEL FUNCTIONS OF MORPHING DEVICES

A very generic overview of A/C level functions is proposed in Ref. [19]. This reference collects the results of a working group of specialists at several organizations such as AIAA, SAE, IEEE, and many manufacturers in aerospace (e.g., Boeing, Rockwell, Honeywell, Airbus). These top-level functions define the high-level requirements for the intended aircraft, Fig. 4. Their details are not investigated in this work but focus is given to the general process flow of current certification practices, starting from such a system functions assignment. As a result, the scope here is to give an overview and not a full descriptive narrative.

The list of a morphing wing device functions, which contribute to enhance standard A/C-level functions, is given in Table 1.

A high-level description of the morphing device system is required to understand the functional breakdown. Overall, these systems aim at reducing wing drag, controlling wing lift distribution, and reducing wing loads (including vibrations and fatigue loads). In addition, the wing structural

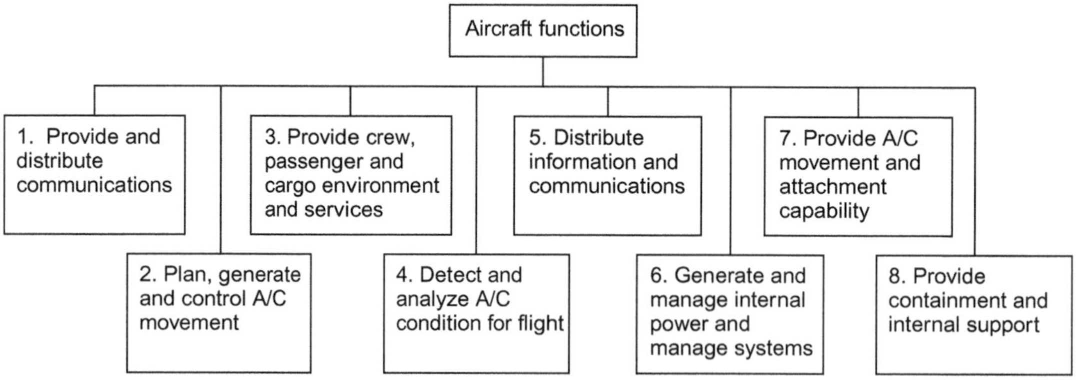

FIG. 4

Aircraft top-level functions.

Table 1 Morphing Wing System Functional Breakdown

Morphing System Functions
Drag minimization function
Lift adaptation function
Turbulence/gust load alleviation
Maneuvers load alleviation
Vibration and fatigue control

integrity also can be taken into account because of the structural load alleviation, protection, and control function, as seen in A/C function 8: Provide containment and internal support. An exploded view of the aircraft level functions is given in the Table 2. High-level functions that might be potentially involved by the use of an adaptive trailing edge device can be linked to the A/C function: Plan, generate, and control A/C movement.

The functional safety analysis of a morphing wing concept and integrating different and independently actuated morphing devices, must be performed at two different levels: the morphing wing level and the single morphing device level. In order to integrate such results, a clear mapping of aircraft level functions, morphing system functions, and physical devices becomes mandatory. These links create the basis for a clear matching between the top-down morphing wing FHA and the bottom-up morphing wing subsystems FHAs.

Table 2 Aircraft Functions Affected by Morphing Wing Concept

Aircraft Functions
2. Plan, generate and control A/C movement
2.2 Generate and control aircraft movement
2.2.5 Control A/C aerodynamics configuration
2.2.5.1 Control lift and drag
2.2.6 Protect aerodynamic control
2.2.6.1 To provide protection against turbulence effects
2.2.6.2 To provide protection against stall load
2.2.7 Provide aerodynamic control forces
2.2.8 Support supplemental flight control
2.2.8.1 To provide overload protection and A/C load protection
2.2.8.2 To provide protection against maneuvers effects
2.2.13 Generate lift
2.2.14 Provide aerodynamic stability
8. Provide containment and internal support
8.1 Provide containment
8.1.2 Provide structural integrity and loads distribution
8.1.2.1 To provide fatigue protection

Table 3 Functional Link Between Aircraft Level and Morphing System Functions

Morphing System Functions	Aircraft Level Functions	
Drag minimization function	2. Plan, generate and control A/C movement	2.2.5.1 Control lift and drag 2.2.7 Provide aerodynamic control forces
Lift adaptation function	2. Plan, generate and control A/C movement	2.2.5.1 Control lift and drag 2.2.7 Provide aerodynamic control forces 2.2.13 Generate lift 2.2.14 Provide aerodynamic stability
Turbulence/gust load alleviation	2. Plan, generate and control A/C movement	2.2.6.1 To provide protection against turbulence effects
Maneuvers load alleviation	2. Plan, generate and control A/C movement	2.2.8.2 To provide protection against maneuvers effects
Vibration and fatigue control	8. Provide containment and internal support	8.1.2.1 To provide fatigue protection
A/C load protection	2. Plan, generate and control A/C movement	2.2.6.2 To provide protection against stall load 2.2.8.1 To provide overload protection and A/C load protection

This work proposes a dual-level functional link, as reported in Tables 3 and 4. By combining these tables, it is possible to draft the functional, logical and architectural relationships among aircraft level functions, morphing wing system, functions and morphing device functions.

2.5 DUAL LEVEL SAFETY

The adaptation capability of the morphing wing investigated in this work is limited to a few predetermined states reached by the combination of three different morphing devices: a droop nose, a morphing trailing edge, and an adaptive winglet. The consequence is that the FHA analyses coming from the single devices or subsystems must be integrated and harmonized at both the wing level and the aircraft level, respectively.

The FHA harmonization/integration process, depicted in Fig. 5, is an iterative process in which the results from the two level analyses are compared and properly combined. This approach allows for the integration of the results coming from the single morphing devices in order to prove that functional and safety properties are satisfied under operational conditions. For instance, the safety classification of a generic failure reached by a single morphing device cannot be categorized with higher severity than the associated failure managed at wing level, because other morphing devices or control surfaces might be deployed to compensate for the loss of the corresponding wing level morphing function.

After the morphing wing FHA is drafted, then the lower level FTs can be prepared using the identified FCs as top events for every morphing device. After they are completed, the higher level (wing level) FTs can be finalized by combining the lower level results, with the addition of the proper conditioning factors and external events. Such a recommended approach is likely to be applicable to a wider variety of morphing systems having different adaptive surfaces or functions.

Table 4 Wing Morphing Subsystem Functional Matching

Morphing System Functions	Involved Subsystem	Way to Operate	FHA Functions
Drag minimization function	EADN	Continuous/ quasistatic operation	Wing shape optimization function "Drag minimization" and "lift adaptation" functions are connected together (i.e., a modification of the aerodynamic profile in order to increase lift also causes a drag coefficient change). The EADN will be used also as high-lift device, during take-off and landing phases
	ATED	Continuous/ quasistatic operation	
	WATE	Continuous/ quasistatic operation	
Lift adaptation function	EADN	Continuous/ quasistatic operation	
	ATED	Continuous/ quasistatic operation	
	WATE	Continuous/ quasistatic operation	
Turbulence/ gust load alleviation	WATE	Dynamic operation	Turbulence/gust load alleviation function
Maneuvers load alleviation	WATE	Dynamic operation	Maneuvers load alleviation function
Vibration and fatigue control	WATE	Fast-dynamic operation	Wing vibration and fatigue control function
A/C load protection	WATE	Dynamic operation	Wing loads protection function

3 FUNCTIONAL HAZARD ASSESSMENT

The FHAs are typically carried out at both aircraft and system levels. Such documents contain the following information managed for each function (aircraft or system level accordingly):

- Identified FCs.
- Effects of FCs.
- Classification of each FC based on the identified effects.
- A statement summarizing the assumptions used for classifying each FC (e.g., adverse operational or environmental conditions and phase of flight).

A discrete scale allows the categorization of the severity of the effects of a FC according to the CS-25 regulation criteria. The classification levels are defined as: Catastrophic, hazardous, major, major,

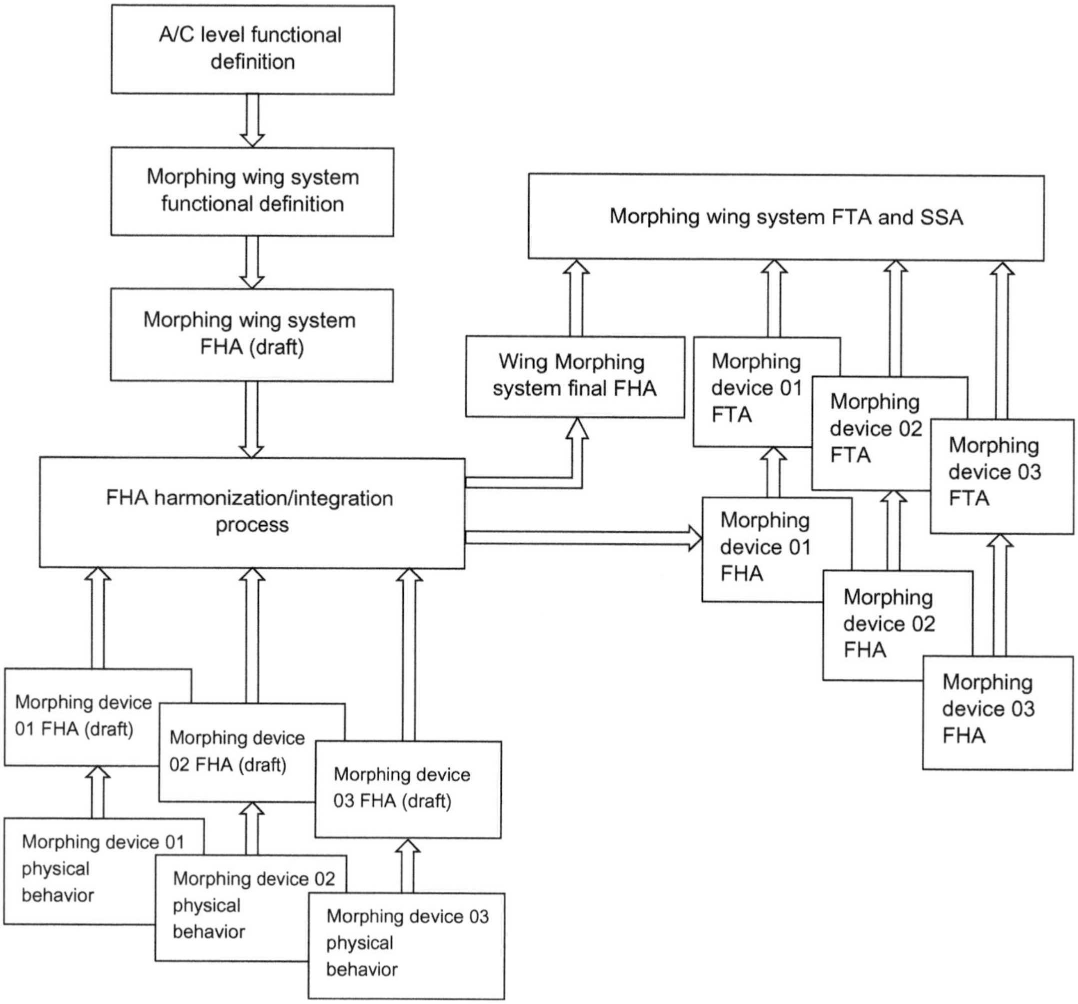

FIG. 5

Dual level safety process overview.

minor, or no safety effect, depending on the related impact on aircraft operation and crew workload, as follows [8]:

No safety effect: FCs that have no effect on safety and do not affect the operational capability of the airplane or increase crew workload. As a safety target, these failures have no probability requirement.

Minor: FCs that do not significantly reduce airplane safety, and that involve crew actions that are well within their capabilities. Minor FCs include, for example, a slight reduction in safety margins or functional capabilities, a slight increase in crew workload, such as routine flight plan changes, or some physical discomfort to passengers or cabin crew. When using quantitative analyses, these failures are commonly accepted as probable, i.e., FCs having an average probability per FH less than the order of 1×10^{-5}.

Major: FCs that reduce the capability of the airplane or the ability of the crew to cope with adverse operating conditions. For example, a significant reduction in safety margins or functional capabilities, a significant increase in crew workload or in conditions impairing crew efficiency, or discomfort to the flight crew, or physical distress to passengers or cabin crew, possibly including injuries. When using quantitative analyses, these failures are commonly accepted as remote, i.e., FCs having an average probability per FH greater than of the order of 1×10^{-5}, but less than of the order of 1×10^{-7}.

Hazardous: FCs that reduce the capability of the airplane or the ability of the crew to cope with adverse operating conditions resulting in a large reduction in safety margins or functional capabilities, an excessive workload to the flight crew or a serious or fatal injury to a relatively small number of the occupants other than the flight crew. When using quantitative analyses, these failures are commonly accepted as extremely remote, i.e., FCs having an average probability per FH greater than the order of 1×10^{-7} or less than of the order of 1×10^{-9}.

Catastrophic: FCs that result in multiple fatalities, usually with the loss of the airplane. When using quantitative analyses, these failures are commonly accepted as extremely improbable, i.e., FCs having an average probability per FH greater than of the order of 1×10^{-9}.

The goal of this step is to clearly identify the circumstances and severity of each FC along with the rationale for its classification. For every identified morphing wing function, the following FCs can be considered:

- Loss of function (total or partial).
- Erroneous provision of function.
- Inadvertent provision of function.

An example of subsystem level FHA developed in the framework of SARISTU project for the morphing trailing edge device is shown in Table 5 [20]. The main information from the FHA includes

- The failure mode identification number.
- The failure mode description.
- The flight phase in which the failure mode can occur.
- The severity classification referred to CS-25 norms and the resulting probability figure required to comply with CS-25 safety regulations.
- The FC details.
- A description of the A/C level effects.
- The detection method (if detection is possible).
- The flight crew reaction after failure detection (if detection is possible).
- The possible requirements coming from safety considerations (e.g., redundancy requirements, inspections, etc.).
- The external events involved in the hazard (if applicable).
- The justification for safety categorization following CS-25 regulations.

Both qualitative and quantitative requirements result from the FCs safety classification. The typical example is an electromechanical actuator (EMA) jamming. This failure mode leads to a FC that can be identified with a functional loss. In detail, if the actuator allows the EMA-based morphing device to be configured to reduce drag, its jamming leads to a drag minimization function loss. As a result,

Table 5 Fault Hazard Assessment of the ATED [20]

Failure Scenarios	FC	Title	A/C Flight Phase(s)	Safety Requirement	
				Severity	Objective
Loss of wing shape optimization function	*FC01* Loss of ATED control	Loss of ATED function	All flight phases	MIN-MAJ	1E-3/FH 1E-5/FH
Erroneous provision of wing shape optimization function	*FC02* Erratic ATED control	Erratic ATED function	All flight phases	MAJ	1E-5/FH
Inadvertent provision of wing shape optimization function	*FC03* Uncommanded ATED control	Uncommanded ATED function	All flight phases	MIN-MAJ	1E-3/FH 1E-5/FH
Partial loss wing shape control capability (symmetric)	*FC04* Symmetric partial loss of ATED function	Symmetric partial loss of ATED function	All flight phases	MIN-MAJ	1E-3/FH 1E-5/FH
Asymmetric partial loss of wing shape control capability	*FC05* Asymmetric partial loss of ATED function	Asymmetric partial loss of ATED function	All flight phases	MAJ	1E-5/FH
Partial loss of wing shape control capability (asymmetric) combined with one engine loss at landing	*FC06* Partial loss of ATED (asymmetric) combined with one engine loss at landing	Loss of ATED control on one wing (asymmetric) combined with one engine loss at landing	*Landing*	MAJ-HAZ	1E-5/FH 1E-7/FH
Partial loss of wing shape control capability (asymmetric) combined with strong cross-wind at take-off or landing	*FC07* Partial loss of ATED (Asymmetric) combined with strong cross wind at landing	Loss of ATED control on one wing (asymmetric) combined with strong cross-wind at landing	*Landing*	MAJ-HAZ	1E-5/FH 1E-7/FH

the qualitative requirement becomes a driver for the system architecture definition and affects both the actuation kinematics and the deployment logics. In case of a safety critical systems, the morphing device also must be designed with a proper integrity level and redundancies. Monitoring of safety critical functions is a safety target for system designers: morphing wing surfaces that cause forced oscillations or free floating, potentially resulting in wing structural collapse, must be designed with a fail-safe approach (e.g., mass balancing preventing free float induced vibrations) or with an independent control/monitor architecture that can prevent forced oscillations.

The ATED fault hazard assessment is also applicable at morphing wing system level for the identification of the FCs. They are listed in Table 6. The main criticalities associated with the simultaneous use of different morphing devices could include:

- The droop nose device can cause a sudden stall, possible catastrophic scenario in case of take-off and landing phases (no time for the pilot to perform recovery actions).

Table 6 Fault Hazard Assessment of a Morphing Wing Incorporating a Droop Nose, a Morphing Trailing Edge Device, and an Adaptive Winglet

				Safety Requirement			Traceability			
	ID Failure Mode	Title	A/C Flight Phase(s)	Severity	Objective	Failure Cause/ Involved Subsystem	WATE	ATE	EADN	External Event or Condition
1	2.2.5.1-01 Drag minimization/ lift adaptation	Loss of wing shape optimization function	All flight phases	MAJ	1E-5/FH	*FC 2.2.5.1-01*: Total loss of SARISTU wing shape control capability (including jamming) – Loss of WATE quasistatic operation – Loss of ATE control – Loss of EADN control	1.c (FC2); 1.e (FC6)	1.1–1.4	Row 2	
1.1	2.2.5.1-02 Drag minimization/ lift adaptation	Erroneous provision of wing shape optimization function	All flight phases	MAJ	1E-5/FH	*See FC 2.2.5.1-01*: Erratic provision of SARISTU wing shape control – Erratic WATE quasistatic operation or – Erratic ATE control or – Erratic EADN control	3.a-d	3.1–3.4	Row 10–12	
1.2	2.2.5.1-03 Drag minimization/ lift adaptation	Inadvertent provision of wing shape optimization function	All flight phases	MAJ	1E-5/FH	*See FC 2.2.5.1-01*: Inadvertent provision of SARISTU wing shape control function – Inadvertent WATE quasistatic operation	2.a, 2.b, 2.d (FC4, FC6, FC2)	2.2, 2.4	Row 8	

Continued

Table 6 Fault Hazard Assessment of a Morphing Wing Incorporating a Droop Nose, a Morphing Trailing Edge Device, and an Adaptive Winglet—cont'd

	ID Failure Mode	Title	A/C Flight Phase(s)	Safety Requirement		Failure Cause/ Involved Subsystem	Traceability			
				Severity	Objective		WATE	ATE	EADN	External Event or Condition
						– Uncommanded ATE control – Uncommanded EADN control				
2	2.2.5.1-04 Drag minimization/ lift adaptation	Partial loss wing shape control capability (symmetric)	All flight phases	MIN-MAJ	1E-3/FH 1E-5/FH	*FC 2.2.5.1-04*: Partial loss of SARISTU wing shape control function – Loss of WATE quasistatic operation or – Loss of ATE control or – Loss of EADN control	1.c (FC2); 1.e (FC6)	1.3, 1.4	Row 13–15, 34–39	
1.3	2.2.5.1-05 2.2.7-01 Drag minimization/ lift adaptation	Asymmetric partial loss of wing shape control capability	All flight phases	MAJ	1E-5/FH	*See FC 2.2.5.1-01*: Loss of SARISTU wing shape control on one wing only combined with symmetry check failure – Loss of WATE quasistatic operation on one wing or – Loss of ATE control on one wing or – Loss of EADN on one wing control	1.d (FC3, FC4)	2.1, 2.3	Row 31–33	

3	2.2.7-02 Drag minimization/ lift adaptation	Partial loss of wing shape control capability (asymmetric) combined with one engine loss at take-off or landing	Take-off and landing	HAZ	1E-7/FH	*FC 2.2.7-02*: Loss of SARISTU wing shape control on one wing only (with symmetry check failure) combined with one engine failure – Loss of WATE quasistatic operation on one wing or – Loss of ATE control on one wing or – Loss of EADN on one wing control N.B. For the reference A/C the classification is HAZ due to the engine collocation (on the fuselage) which causes a limited yaw moment in case of one engine failure	1.d (FC4)	2.3.1	Row 31–33	One engine loss
4	2.2.7-03 Drag minimization/ lift adaptation	Partial loss of wing shape control capability (asymmetric) combined with strong crosswind at take-off or landing	Take-off and landing	CAT	1E-9/FH	*FC 2.2.7-03*: Loss of SARISTU wing shape control on one wing only (with symmetry check failure) combined with strong crosswind – Loss of WATE quasistatic operation on one wing or	1.d (FC4)	2.1	Row 31–33	Strong crosswind

Continued

Table 6 Fault Hazard Assessment of a Morphing Wing Incorporating a Droop Nose, a Morphing Trailing Edge Device, and an Adaptive Winglet—cont'd

				Safety Requirement			Traceability			
	ID Failure Mode	**Title**	**A/C Flight Phase(s)**	**Severity**	**Objective**	**Failure Cause/ Involved Subsystem**	**WATE**	**ATE**	**EADN**	**External Event or Condition**
						– Loss of ATE control on one wing or – Loss of EADN on one wing control				
5	2.2.13-01 Lift adaptation	Loss of A/C control due to symmetric or asymmetric loss of EADN control as high lift device	Take-off and landing	CAT	1E-9/FH	*FC 2.2.13-01*: Loss of SARISTU wing shape control (high lift device function) – Loss of EADN control as high lift device			Row 1, 3, 4–6, 31–33	
5.1	2.2.13-05 Lift adaptation	Partial loss of EADN control as high lift device combined with icing conditions	Take-off and landing	CAT	1E-9/FH	*See FC 2.2.13-01*: Partial loss of SARISTU wing shape control (high lift device function) combined with icing conditions – Loss of EADN performance as high lift device and icing			Row 13, 15, 34, 36, 37, 39	Icing conditions
5.2	2.2.13-06 Lift adaptation	Erroneous provision of EADN control as high lift device combined with icing conditions	Take-off and landing	CAT	1E-9/FH	*See FC 2.2.13-01*: Erroneous provision of SARISTU wing shape control (high lift device function) combined with icing conditions – Erroneous EADN provision as high lift device AND icing			Row 10, 12	Icing conditions

6	2.2.13-02 Lift adaptation	Partial loss of EADN control as high lift device	Take-off and landing	MIN	1E-3/FH	*FC 2.2.13-02*: Partial loss of SARISTU wing shape control (high lift device function) – Loss of EADN control efficiency/ reduction of performance as high lift device			Row 13, 15, 34, 36, 37, 39	
61	*2.2.13-03* Lift adaptation	Erroneous provision of EADN control as high lift device	Take-off and landing	MAJ	1E-5/FH	*FC 2.2.13-03*: Erroneous provision of SARISTU wing shape control (high lift device function) – EADN erroneus provision as high lift device			Row 10, 12	
7	2.2.13-04 Lift adaptation	Unexpected provision of EADN control as high lift device	Take-off and landing	CAT	1E-9/FH	*FC 2.2.13-04*: Provision of SARISTU wing shape control (high lift device function) when not required – EADN high lift function (unwanted activation)			Row 7, 9	
N	2.2.14-01 Drag minimization/ lift adaptation	Loss (complete or partial), erroneus or inadvertent provision of wing shape optimization function	All flight phases	NSE	N/A	Total/partial loss, erroneus or inadvertent provision of SARISTU wing shape control function – WATE generic failure	1.c to 1.g	1.1–3.4	Row 13 to 15, 34 to 39	

Continued

Table 6 Fault Hazard Assessment of a Morphing Wing Incorporating a Droop Nose, a Morphing Trailing Edge Device, and an Adaptive Winglet—cont'd

	ID Failure Mode	Title	A/C Flight Phase(s)	Safety Requirement		Failure Cause/ Involved Subsystem	Traceability			
				Severity	Objective		WATE	ATE	EADN	External Event or Condition
						– EADN generic failure (low excursion only) – ATE generic failure				
8	2.2.6.1-01 Turbulence/ gusts load alleviation	Total loss of turbulence/gust load alleviation capability	All flight phases	NSE	N/A	*FC 2.2.6.1-01*: Loss of SARISTU turbulence/gust load alleviation: – Loss of WATE dynamic/ occasional gust load alleviation function	1.e, 1.g (FC5)	1.4		
9	2.2.6.1-02 Turbulence/ gusts load alleviations	Total loss of turbulence/gust load alleviation capability in case of gusts/ turbulences	All flight phases	MAJ-HAZ	1E-5/FH 1E-7/FH	*FC 2.2.6.1-02*: Loss of SARISTU turbulence/gust load alleviation combined with turbulence/gusts – Loss of WATE dynamic/ occasional gust load alleviation function	1.d, 1.f (FC5)	1.4		Gusts/ turbulences
10	2.2.6.1-03 Turbulence/ gusts load alleviation	Undetected total loss of turbulence/gust load alleviation capability in	All flight phases	HAZ	1E-7/FH	*FC 2.2.6.1-03*: Loss of SARISTU turbulence/gust load alleviation combined	1.d, 1.f (FC5)	1.3		Gusts/ turbulences

		case of gusts/ turbulences				with turbulence/gusts and SHM/BIT failure – Loss of WATE dynamic/ occasional gust load alleviation function				
8.1	2.2.6.1-04 Turbulence/ gusts load alleviation	Partial loss of turbulence/gust load alleviation capability	All flight phases	NSE	N/A	*See FC 2.2.6.1-01*: Partial loss of SARISTU turbulence/gust load alleviation – Loss of WATE dynamic/ occasional gust load alleviation function efficiency	1.e, 1.g (FC5)	1.4		
9.1	2.2.6.1-05 Turbulence/ gusts load alleviation	Partial loss of turbulence/gust load alleviation capability in case of gusts/ turbulences	All flight phases	MAJ	1E-5/FH	*See FC 2.2.6.1-02*: Partial loss of SARISTU turbulence/gust load alleviation combined with turbulence/gusts – Loss of WATE dynamic/ occasional gust load alleviation function efficiency	1.e, 1.g (FC5)	1.4		Gusts/ turbulences
10.1	2.2.6.1-06 Turbulence/ gusts load alleviation	Undetected partial loss of turbulence/gust load alleviation capability in case of gusts/ turbulences	All flight phases	HAZ	1E-7/FH	*See FC 2.2.6.1-03*: Partial loss of SARISTU turbulence/gust load alleviation combined with turbulence/gusts and SHM/BIT failure – Loss of WATE dynamic/ occasional gust load alleviation function efficiency	1.d, 1.f (FC5)	1.3		Gusts/ turbulences

Continued

Table 6 Fault Hazard Assessment of a Morphing Wing Incorporating a Droop Nose, a Morphing Trailing Edge Device, and an Adaptive Winglet—cont'd

				Safety Requirement			Traceability			
	ID Failure Mode	Title	A/C Flight Phase(s)	Severity	Objective	Failure Cause/ Involved Subsystem	WATE	ATE	EADN	External Event or Condition
11	2.2.6.1-07 Turbulence/ gusts load alleviation	Inadvertent/ undesired provision of turbulence/gust load alleviation function	All flight phases	MAJ-HAZ	1E-5/FH 1E-7/FH	*FC 2.2.6.1-07*: Inadvertent/ undesired provision of SARISTU turbulence/gust load alleviation – WATE dynamic/ occasional function provided when not needed	2.a, 2.b, 2.d (FC5, FC7)	2.1–2.4		
12	2.2.6.1-08 Turbulence/ gusts load alleviation	Erroneus provision of turbulence/gust load alleviation function	All flight phases	CAT	1E-9/FH	*FC 2.2.6.1-08*: Erroneus provision of SARISTU turbulence/gust load alleviation (during turbulences/gusts) – WATE dynamic/ occasional function provided with erroneus parameters	3.a-d (FC5)	3.1–3.4		
13	2.2.8.2-01 Maneuvers load alleviation	Total loss of maneuvers load alleviation capability	All flight phases	MAJ	1E-5/FH	*FC 2.2.8.1-01*: Loss of SARISTU maneuvers load alleviation – Loss of WATE dynamic/ occasional function in case of A/C maneuvers	1.e, 1.g (FC5)	1.4		Maneuver load

14	2.2.8.2-02 Maneuvers load alleviation	Undetected total loss of maneuvers load alleviation capability	All flight phases	HAZ	1E-7/FH	*FC 2.2.8.2-02*: Loss of SARISTU maneuver load alleviation with loss of SHM/BIT function – Loss of WATE dynamic/ occasional function in case of A/C maneuvers	1.d, 1.f (FC5)	1.3		Maneuver load
15	2.2.8.2-03 Maneuvers load alleviation	Partial loss of maneuvers load alleviation capability	All flight phases	MIN-MAJ	1E-3/FH 1E-5/FH	*FC 2.2.8.2-03*: Partial of SARISTU maneuver load alleviation – Loss of WATE dynamic/ occasional function efficiency in case of A/C maneuvers	1.e, 1.g (FC5)	1.4		Maneuver load
16	FC 2.2.8.2-04 Maneuvers load alleviation	Undetected partial loss of maneuvers load alleviation capability	All flight phases	MAJ	1E-5/FH	*FC 2.2.8.2-04*: Partial of SARISTU maneuver load alleviation with loss of SHM/BIT function – Loss of WATE dynamic/ occasional function efficiency in case of A/C maneuvers	1.d, 1.f (FC5)	1.3		Maneuver load
17	2.2.8.2-05 Maneuvers load alleviation	Inadvertent/ undesired provision of maneuvers load alleviation function	All flight phases	MIN-MAJ	1E-3/FH 1E-5/FH	*FC 2.2.8.2-05*: Inadvertent/ undesired provision of SARISTU maneuver load alleviation – WATE dynamic/ occasional maneuvers load alleviation function provided when not required	2.a, 2. b, 2.d (FC5, FC7)	2.1–2.4		Maneuver load

Continued

Table 6 Fault Hazard Assessment of a Morphing Wing Incorporating a Droop Nose, a Morphing Trailing Edge Device, and an Adaptive Winglet—cont'd

	ID Failure Mode	Title	A/C Flight Phase(s)	Safety Requirement		Failure Cause/ Involved Subsystem	Traceability			
				Severity	Objective		WATE	ATE	EADN	External Event or Condition
18	2.2.8.2-06 Maneuvers load alleviation	Erroneus provision of maneuvers load alleviation function	All flight phases	CAT	1E-9/FH	*FC 2.2.8.2-06*: Erroneus provision of SARISTU maneuver load alleviation – WATE dynamic/ occasional maneuvers load alleviation function parameters error	3.a-d (FC5)	3.1–3.4		Maneuver load
19	8.1.2.1-01 Wing vibration and fatigue control	Total loss of vibration and fatigue control function	All flight phases	MIN-MAJ	1E-3/FH 1E-5/FH	*FC 8.1.2.1-01*: Loss of SARISTU vibration and fatigue control function – Loss of WATE fast-dynamic/ continuous function to control fatigue loads	1.e, 1.g (FC5)	1.4		
20	8.1.2.1-02 Wing vibration and fatigue control	Undetected total loss of vibration and fatigue control function	All flight phases	MAJ	1E-5/FH	*FC 8.1.2.1-02*: Loss of SARISTU vibration and fatigue control function combined with SHM and/or BIT failure – Loss of WATE fast-dynamic/ continuous function to control fatigue loads	1.d, 1.f (FC5)	1.3		

21	8.1.2.1-03 Wing vibration and fatigue control	Partial loss of vibration and fatigue control function	All flight phases	MIN	1E-3/FH	*FC 8.1.2.1-03* (or see FC8.1.2.1-01 depending on its classification): Loss of SARISTU vibration and fatigue control function efficiency – Loss of WATE fast-dynamic/ continuous function efficiency to control fatigue loads	1.e, 1.g (FC5)	1.4		
20.1	8.1.2.1-04 Wing vibration and fatigue control	Undetected partial loss of vibration and fatigue control function	All flight phases	MAJ	1E-5/FH	*See FC 8.1.2.1-02*: Loss of SARISTU vibration and fatigue control function efficiency with the loss of SHM/BIT monitoring – Loss of WATE fast-dynamic/ continuous function efficiency to control fatigue loads	1.d, 1.f (FC5)	1.3		
22	8.1.2.1-05 Wing vibration and fatigue control	Erroneus provision of vibration and fatigue control function	All flight phases	CAT	1E-9/FH	*FC 8.1.2.1-05*: SARISTU vibration and fatigue control function erroneus provision – Erroneus parameters providing WATE fast-dynamic/ continuous function to control fatigue loads	3.a-d (FC5)	3.1–3.4		

Continued

Table 6 Fault Hazard Assessment of a Morphing Wing Incorporating a Droop Nose, a Morphing Trailing Edge Device, and an Adaptive Winglet—cont'd

				Safety Requirement			Traceability			
	ID Failure Mode	Title	A/C Flight Phase(s)	Severity	Objective	Failure Cause/ Involved Subsystem	WATE	ATE	EADN	External Event or Condition
23	2.2.8.1-01 2.2.6.2-01 Wing load protection	Total loss of wing loads/ overloads protection function	All flight phases	NSE	N/A	Loss of SARISTU wing overload/stall load protection – Loss of WATE dynamic/ occasional load protection	1.e, 1.g (FC5)	1.4		
24	2.2.8.1-02 Wing load protection	Total loss of wing loads/ overloads protection function in case of wing overload	All flight phases	HAZ	1E-7/FH	*FC 2.2.8.1-02*: Loss of SARISTU wing overload/stall load protection in conjunction with overload condition – Loss of WATE dynamic/ occasional load protection when needed	1.e, 1.g (FC5)	1.4		Overload condition
25	2.2.6.2-02 Wing load protection	Total loss of wing loads/ overloads protection function in case of high stall load	Take-off and landing	CAT	1E-9/FH	*FC 2.2.6.2-02*: Loss of SARISTU wing overload/stall load protection in conjunction with overload condition – Loss of WATE dynamic/ occasional load protection when needed	1.e, 1.g (FC5)	1.4		Stall load
23.1	2.2.8.1-03 2.2.6.2-03 Wing load protection	Partial loss of wing loads/ overloads protection function	All flight phases	NSE	N/A	Loss of SARISTU wing overload/stall load protection efficiency	1.e, 1.g (FC5)	1.4		

						– Loss of WATE dynamic/ occasional load protection efficiency				
26	2.2.8.1-04 Wing load protection	Partial loss of wing loads/ overloads protection function in case of wing overload	All flight phases	MAJ	1E-5/FH	*FC 2.2.8.1-04*: Loss of SARISTU wing overload/stall load protection efficiency – Loss of WATE dynamic/ occasional load protection efficiency	1.e, 1.g (FC5)	1.4		Overload condition
27	2.2.6.2-04 Wing load protection	Partial loss of wing loads/ overloads protection function in case of stall load	All flight phases	MAJ-HAZ	1E-5/FH 1E-7/FH	*FC 2.2.6.2-04*: Loss of SARISTU wing overload/stall load protection efficiency – Loss of WATE dynamic/ occasional load protection efficiency	1.e, 1.g (FC5)	1.4		Stall load
28	2.2.8.1-05 Wing load protection	Undetected partial loss of wing loads/ overloads protection function in case of wing overload	All flight phases	HAZ	1E-7/FH	*FC 2.2.8.1-05*: Loss of SARISTU wing overload/stall load protection efficiency and/or SHM/BIT: – Loss of WATE dynamic/ occasional load protection efficiency	1.d, 1.f (FC5)	1.3		Overload condition
28.1	2.2.6.2-05 Wing load protection	Undetected partial loss of wing loads/ overloads protection	All flight phases	HAZ	1E-7/FH	*See FC 2.2.8.1-05 or FC 2.2.6.2-04 if considered HAZ*: Loss of SARISTU wing overload/stall	1.d, 1.f (FC5)	1.3		Stall load

Continued

Table 6 Fault Hazard Assessment of a Morphing Wing Incorporating a Droop Nose, a Morphing Trailing Edge Device, and an Adaptive Winglet—cont'd

	ID Failure Mode	Title	A/C Flight Phase(s)	Safety Requirement		Failure Cause/ Involved Subsystem	Traceability			
				Severity	Objective		WATE	ATE	EADN	External Event or Condition
		function in case of stall load				load protection efficiency and/or SHM/BIT – Loss of WATE dynamic/ occasional load protection efficiency				
29	2.2.8.1-06 2.2.6.2-06 Wing load protection	Inadvertent/ undesired provision of overload/stall load protection function	All flight phases	HAZ	1E-7/FH	*FC 2.2.8.1-06*: Inadvertent/ undesired provision of SARISTU overload/stall load protection – WATE dynamic/ occasional function provided when not needed	2.a, 2. b, 2.d (FC5, FC7)	2.1–2.4		Overload or stall load
30	2.2.8.1-07 2.2.6.2-07 Wing load protection	Erroneus provision of overload/stall load protection function	All flight phases	CAT	1E-9/FH	*FC 2.2.8.1-07*: Erroneus provision of SARISTU overload protection function (in case of overload) – WATE dynamic/ occasional function provided with erroneus parameters	3.a-d (FC5)	3.1–3.4		Overload or stall load
31	8.1.2-01 Provide structural integrity and loads distribution	Loss of structural integrity due to SARISTU device free float	All flight phases	CAT	1E-9/FH	*FC 8.1.2-01*: Mobile surfaces free float (mechanical disconnection, total loss of system dumping)—WATE only TBC	1.b, 2.c (FC1)	1.1, 1.2		

- Possible induced oscillations for fast dynamic actuated devices.
- Possible morphing devices asymmetric configurations causing asymmetric drag, critical or catastrophic in combination with further external asymmetry causes (e.g., single engine loss, crosswind). Also in this case, the worst scenario can occur during take-off or landing.

The morphing droop nose device is a single subsystem that can cause a catastrophic failure because of the possible detachment of aerodynamic flow after an erroneous deployment. For this device, no common cause can lead to this scenario. A single functional design assurance level (FDAL) A or a dual FDAL B design is necessary to match the safety requirements [9] (see SAE ARP 4754a for details).

Furthermore, because of the dynamic deployment, an independent control/monitor architecture must be implemented into the adaptive winglet to prevent induced oscillations. A symmetric deployment check (with the possibility to stop the morphing devices if an asymmetry is detected) is also fundamental to prevent failure case scenarios and to comply with the quantitative safety targets to make the selected architectures certifiable and airworthy.

Regarding the possibility to achieve wing load/vibrations alleviation/control functions with fast-actuated morphing devices, the CS-25 regulations impose that the additional load caused by the failed morphing device shall be considered in the design load. The consequence is that every related FC (from the structural standpoint) is classified no more than minor in the frame of morphing wing FHA. Finally, it has to be noted that although software functions should be considered in the FHA and the proper DAL shall be assigned, these topics are not discussed because they are outside the scope of this chapter. These can be found within the referenced documents [9].

4 DUAL-LEVEL APPROACH FOR THE FTA OF A MORPHING WING

The System Safety Assessment is a document that reports all the identified hazards for the system under analysis and shows the compliance with the safety requirements. FTs are required for each FCs classified as HAZ or CAT by the FHA.

The FTA is a modeling technique able to represent the failure combination that can cause a hazard. The quantitative data for the FTs are usually taken from the reliability data and analyses provided by the device manufacturer. In the FTA, the FCs are usually linked with logic gates. For preliminary and simple FT, *and* and *or* logic gates are usually sufficient to represent the most common conditions. FTs typically are built for a standard flight, which is defined by the aircraft manufacturer on the basis of the aircraft type. Nevertheless, the safety requirements are expressed on the basis of a single FH.

With reference to the proposed morphing wing system, a dual-level approach was followed. This involves a bottom-up process for subsystems and devices and a top-down process for the whole wing. The following example shows how to implement the dual-level approach in the FTA. Fig. 6 represents a subsystem FC developed with a FT technique. Basic events represent HW (mechanical or electric/electronic) failures provided by supplier or by reliability books or FMES entry.

Concerning the ATED device, a number of FTs were produced to evaluate the impact of actuation kinematics on the device system morphing authority, reliability, and functionality up to the top-level aircraft functions. An example of FTA corresponding to the failure case "asymmetric partial loss of ATED function," whose FHA severity was identified as major, is shown in Fig. 7. The probability

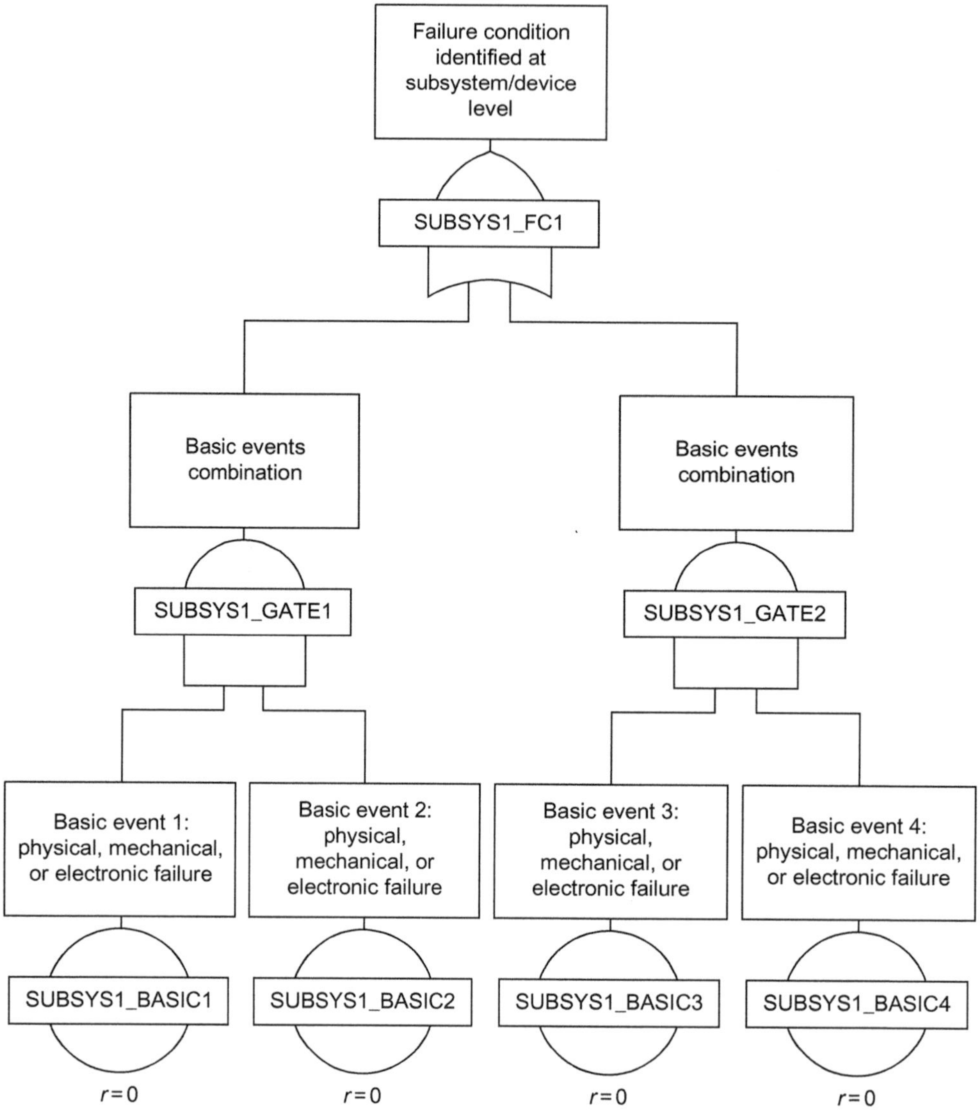

FIG. 6

Example of subsystem/device level FT.

of failure of the top event was then analyzed along with its compliancy with the expected severity with the purpose of making ATED a sufficiently reliable device for aircraft application.

Several indexes, such as actuator position control, motor shaft angle sensors, and actuator absorbed power, can help reveal ATED failures. The construction of the FT is based on relationships between events and causes, represented by means of logical and *or* gates, which solve the system complexity and

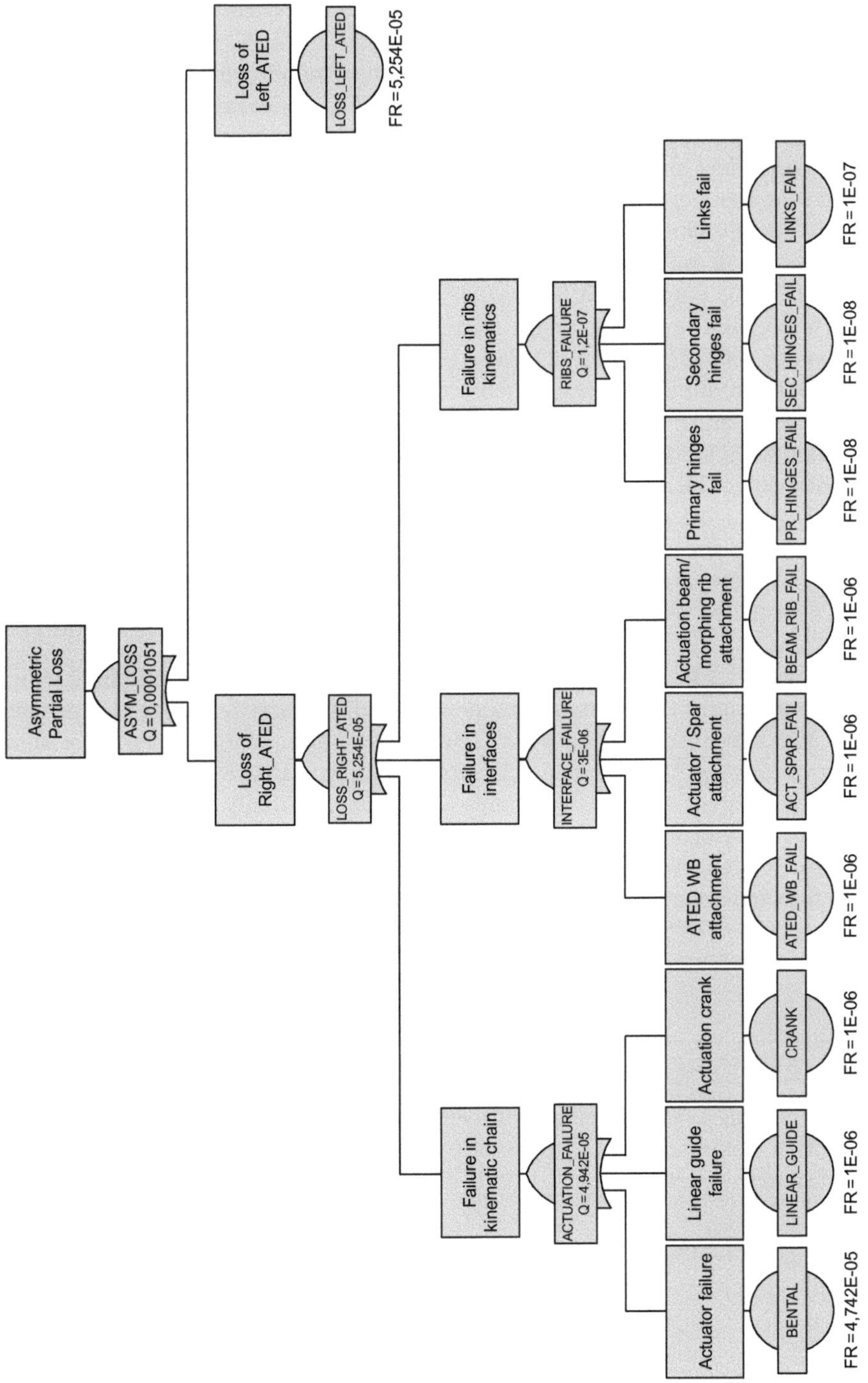

FIG. 7

Fault tree example with failure rates per hour for asymmetric partial loss of ATED function [20].

its potential failures. To this aim, the failure probability of each basic component of the morphing kinematics must be collected.

The main drawback associated with the use of electromechanical actuators is mechanical failure or jamming, which can typically lead to hazardous/catastrophic A/C FCs. Conversely, according to the ATED FHA, asymmetric ATED failure caused, for example, by actuators jamming, is classified as major. This means that the associated impact on wing loads can be reduced by acting on conventional aircraft control surfaces on the basis of the detected ATED shapes. Nevertheless, although minimal impacts are expected on A/C controllability, pilot workload increases in order to trim the aircraft longitudinally and laterally. In addition, safety margins on aircraft block fuel must be considered to compensate the potential loss of ATED functions in flight.

Although the failure rate of the top event is not fully compliant with the safety requirement (Failure rate for major events: 1.0E-5/FH), the investigated architecture does not incorporate any standard device for detecting and minimizing asymmetric conditions onboard aircraft. The easiest option can include, for instance, an ATED position sensor and a brake system that can be activated automatically to stop further ATED movement after the position sensors detect any asymmetric situation. Their use would allow the ATED meeting the safety requirements starting from a failure rate of 0.1 per FH. For morphing systems performing load (static or dynamic) alleviation such as the adaptive winglet, a dual command and monitoring lane with its own control unit (ECU) is mandatory to guarantee an adequate redundancy. In addition, an acceptable number of linear variable displacement transducers (LVDTs) mounted to the actuator ball screw and angular sensors are needed for operational reliability.

An example of higher level FTA is shown in Fig. 8. The potential failures at the morphing wing level are a combination of subsystems (single morphing devices) failure cases with proper exposure factors or external events. If a failure mode does not have immediate and detectable effect, the related events are called "latent" or "dormant." Commercial FT software can be used to evaluate these events through mathematical models. Latent or dormant failures can be detected only with specific functional tests or maintenance actions. Proper inspection intervals also can be scheduled to satisfy safety requirements. The currently available technology is able to minimize the certification maintenance requirement (CMR) with the implementation of self-test ability, in particular for complex electronic devices.

5 COMMON CAUSE ANALYSES

As already reported, the safety analysis is not limited to the verification of quantitative requirements for the FCs identified in the FHA. Common cause analyses (CCA) are also necessary to verify the independence of the systems failures, in terms of

- particular risk analysis
- common mode analysis
- zonal safety analysis.

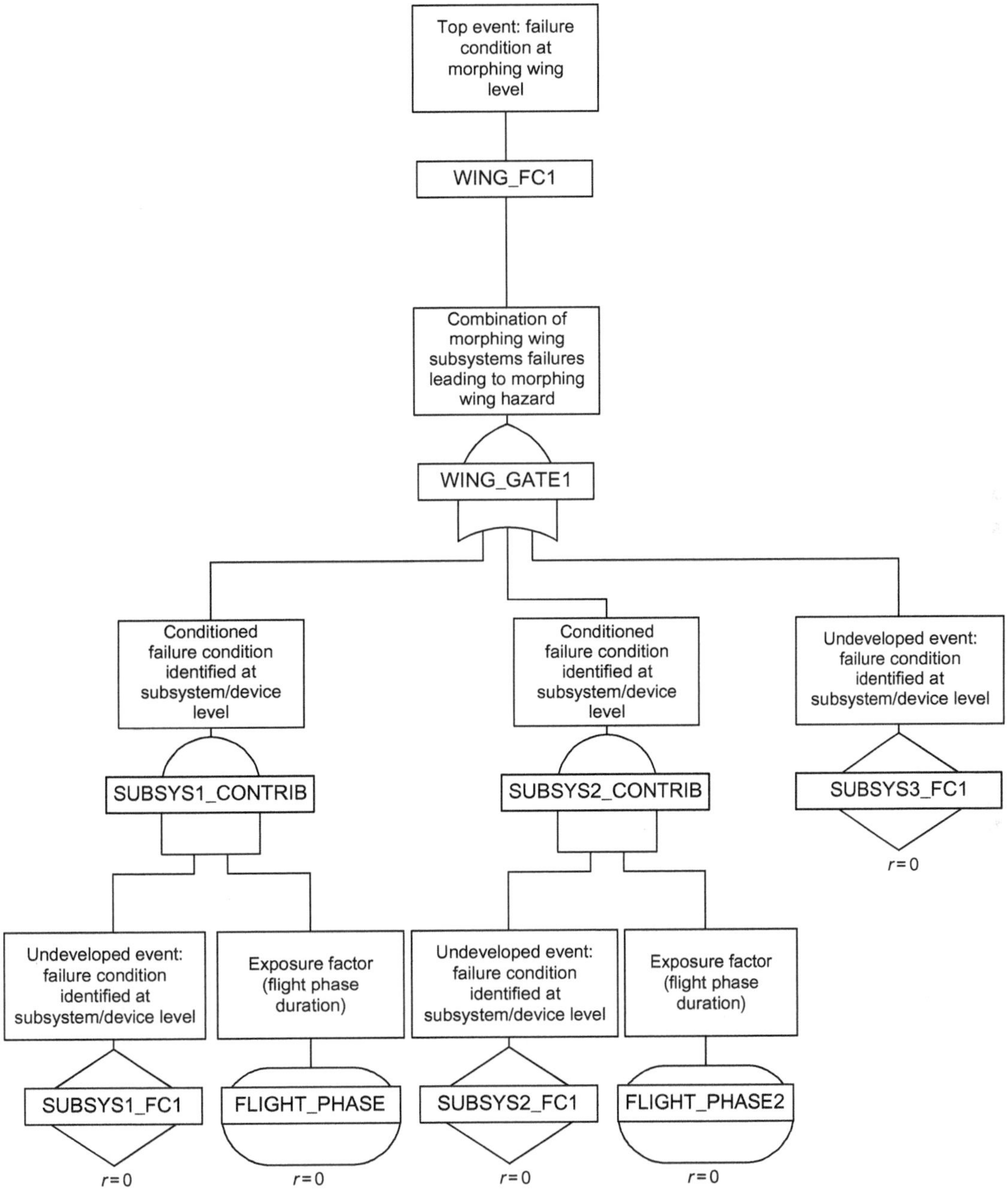

FIG. 8

Example of morphing wing level FT.

5.1 PARTICULAR RISK ANALYSIS

Particular risk analysis and zonal safety analysis are related mainly to the physical installation and design of the systems. Without the actual application and associated constraints, these analyses can be drafted only by supporting some assumptions about the general morphing wing concept. For the EADN device, a bird strike is one of the most important risks to be considered. According to the CS-25, the EADN design shall be designed to assure capability of continued safe flight and landing of the airplane after impact with a 4-lb bird when the velocity of the airplane (relative to the bird along the airplanes flight path) is equal to cruise velocity (VC) at sea level or 0.85 VC at 2438 m (8000 ft), whichever is the most critical. Compliance can be shown by analysis only when based on tests carried out on sufficiently representative structures of similar design.

5.2 COMMON MODE ANALYSIS

For catastrophic FCs, the quantitative requirement (failure probability <1E-9/FH) is not sufficient to make an aircraft system airworthy. An additional requirement is that no single failure shall lead to the top hazard. This condition can be checked by verifying the FT structure. A FT branch shall be combined with another one by means of an *and* gate before reaching the top event to be compliant with this requirement. This check is usually performed by the tool with the "minimal cut set" extraction.

In case of a software function executed by a dedicated HW leading to a catastrophic FC, a dissimilarity is required. For instance, the same function must be executed by two different sets of HW, with two different sets of software developed by independent teams. A HW/SW dissimilarity is not required for the wind tunnel test demonstrator, but the proposed architecture for the safety critical subsystems must take into account the possibility to be upgraded in case of real scenario application.

5.3 ZONAL SAFETY ANALYSIS

A common mode failure also can be caused by the interaction of nearly installed equipment without adequate protection/isolation. Thus it might be important to evaluate the possible zonal hazards in case of real scenario application after the early design stage. General zonal hazards can be because of

- Hot point or electrical cables/connectors installed in proximity of pipes, tanks, or accumulators containing flammable fluids (mainly fuel or hydraulic oil).
- Equipment with explosion risk or with high kinetic energy components (mainly rotating parts such as turbines) that can result in high energy debris release. In this case, every nearly installed component can be damaged without a proper protection.
- Equipment that can result in an uncontained fire (engine combustor, electrical equipment). The possible scenario is a propagation of damage to surrounding equipment or tubes, cables, etc.
- Equipment filled with aggressive fluids (such as acid). A fluid leakage can cause damage or contamination to surrounding components.

Proper protection of cable and pipe routings must be considered in order to minimize zonal safety hazards. Some previously exposed hazards seem to derive from the combination of at least two independent failures. For example, in case of electrical cables/connectors positioned near fuel or hydraulic pipes, the fire/explosion event can be envisaged only if an electrical arc/spark is generated (failure of cable or connector) in combination with a flammable fuel leakage (pipe rupture).

6 CONCLUSIONS

The design of morphing devices must not compromise the aircrafts general performance and intrinsic functions. This chapter covers the major safety and reliability aspects associated with the design and integration of morphing systems into an aircraft wing. The summary recommendation is that a three-step strategy is needed for a safe use and successful certification of morphing devices. This strategy consists of individual and combined safety assessments, followed by system design activities to comply with the corresponding safety requirements, and design assurance using the latest A/C safety standards. For future applications, theoretical and detailed simulations must support these steps along with model-based design techniques. Furthermore, the verification process of the system level functional and safety target can require different testing plans and tools to assess the associated risks. Although the chapter has focused on some particular morphing devices and types, such a flow and recommendations are likely to be applicable to a wider variety of morphing concepts.

REFERENCES

[1] www.saristu.eu (Accessed 17 December 2016).

[2] G. Amendola, I. Dimino, M. Magnifico, R. Pecora, Distributed actuation concepts for a morphing aileron device. Aeronaut. J. 120 (1231) (2016) 1365–1385, https://doi.org/10.1017/aer.2016.64.

[3] G. Amendola, I. Dimino, F. Amoroso, R. Pecora, Experimental characterization of an adaptive aileron: lab tests and FE correlation, SPIE Smart Structures/NDE, Las Vegas, NV, March 2016. Proc. SPIE 9803, Sensors and Smart Structures Technologies for Civil, Mechanical, and Aerospace Systems 2016, 98034P (2016); https://doi.org/10.1117/12.2219187.

[4] I. Dimino, A. Concilio, and R. Pecora, Primary structural components characterization of an adaptive trailing edge device (ATED), 24th AIAA/AHS Adaptive Structures Conference, AIAA SciTech, 4–8 January, 2016.

[5] Pecora, R., Concilio, A., Dimino, I., Amoroso, F., Ciminello, M., Structural design of an adaptive wing trailing edge for enhanced cruise performance, 24th AIAA/AHS Adaptive Structures Conference, AIAA SciTech, 4–8 January, 2016.

[6] NASA, US AFRL, "Adaptive Compliant Trailing Edge Flight Experiment", RC Soaring Digest, Vol. 31, N.14, pp. 85–86, 2014.

[7] NASA Website, http://www.nasa.gov/press/2014/november/nasa-tests-revolutionary-shape-changing-aircraft-flap-for-the-first-time, Release 14-308, November 7, 2014.

[8] European Aviation Safety Agency, Certification Specifications and Acceptable Means of Compliance for Large Aeroplanes. CS-25. Amendment 17, 2015.

[9] SAE ARP4754A, Guidelines for Development of Civil Aircraft and Systems, SAE Aerospace Recommended Practice, SAE International, 2010.

[10] SAE ARP4761, Guidelines and Methods for Conducting the Safety Assessment Process on Civil Airborne Systems and Equipment, SAE International, 1996.

[11] O. Heintze, S. Steeger, A. Falken, J. Heckmann, in: Enhanced adaptive droop nose—from computer model to multi-functional integrated part, Smart Intelligent Aircraft Structures (SARISTU). ISBN 978-3-319-22413-8, 2015, pp. 97–111, https://doi.org/10.1007/978-3-319-22413-8_5.

[12] M. Kintscher, J. Kirn, S. Storm and F. Peter, Assessment of the SARISTU enhanced adaptive droop nose, in Smart Intelligent Aircraft Structures (SARISTU), 2015, pp. 113–140, isbn 978-3-319-22413-8.

[13] R. Pecora, F. Amoroso, M. Magnifico, I. Dimino, A. Concilio, KRISTINA: kinematic rib based structural system for innovative adaptive trailing edge, SPIE Smart Structures/NDE, Las Vegas, NV, March 2016. Proc. SPIE 9801, Industrial and Commercial Applications of Smart Structures Technologies 2016, 980107, April 16 (2016); https://doi.org/10.1117/12.2218516.

[14] I. Dimino, G. Diodati, A. Concilio, A. Volovick, L. Zivan, Distributed Electromechanical Actuation System Design for a Morphing Trailing Edge Wing, SPIE Smart Structures/NDE, Las Vegas, Nevada (USA) March 2016, in: Proc. SPIE 9801, Industrial and Commercial Applications of Smart Structures Technologies 2016, 980108, April 16, 2016.

[15] I. Dimino, A. Concilio, M. Schueller, A. Gratias, An Adaptive Control System for Wing TE Shape Control, in: SPIE International Conference on Smart Structures 2013, San Diego, California, USA, 10-14 March 2013.

[16] Allen, J.B., Articulating winglets, US5988563 A, 1999.

[17] Irving, J., Davies, R., Wing tip device, US7275722 B2, 2007.

[18] A. Wildschek, S. Storm, M. Herring, D. Drezga, V. Korian, O. Roock, Design, optimization, testing, verification, and validation of the wingtip active trailing edge, Smart Intelligent Aircraft Structures (SARISTU), pp. 219–255.

[19] Working group document, Framework for the Application of Systems Engineering in the Commercial Aircraft Domain. Version 1.2a, July 28. 2000.

[20] I. Dimino, A. Concilio and R. Pecora, Safety and reliability aspects of an adaptive trailing edge device (ATED), 24th AIAA/AHS Adaptive Structures Conference, AIAA SciTech, 4–8 January, 2016.

CHAPTER

ON THE EXPERIMENTAL CHARACTERIZATION OF MORPHING STRUCTURES

22

Ignazio Dimino*, Gianluca Amendola*, Rosario Pecora†, Antonio Concilio*, André Gratias‡, Martin Schueller‡

The Italian Aerospace Research Centre, CIRA SCpA, Capua (CE), Italy University of Naples "Federico II", Naples, Italy† Fraunhofer ENAS, Chemnitz, Germany‡*

CHAPTER OUTLINE

NOMENCLATURE

ATE	adaptive trailing edge
ATED	adaptive trailing edge device
CAD	computer aided engineering
D	drag
D1,2,3,4	dummy number 1, 2, 3, 4
DSP	digital signal processing
E	error

Morphing Wing Technologies. https://doi.org/10.1016/B978-0-08-100964-2.00022-8

EMA	electro-mechanical actuator
f	frequency
FB	feedback (control system)
FBG	fiber bragg grating
FF	feed-forward (control system)
Hz	hertz
I/O	input/output
L	lift
L	load (experiments)
L^*	load value, producing a tip displacement compatible with the target deformed shape
L/D	lift over drag ratio
LE	leading edge
MCS	morphing control system
PC	personal computer
Re	Reynold's number
Rho	air density
S	actual shape
S_{opt}	reference shape
T/A	test article
V	flow speed
V^*	flow speed, operative conditions
W	weight
α	angle of attack
α^*	angle of attack, operative conditions

1 INTRODUCTION

In comparison with conventional aircraft control surfaces, such as flaps, slats, and ailerons, morphing devices may perform more complex and interrelated A/C level functions by combining lift adaptation, drag reduction, load alleviation, and more effective pitch and roll control. Thus, the interaction between conventional and adaptive systems to perform augmented functions may ensure higher and more comprehensive overall benefits at aircraft level. However, because of the novelty associated with the long-term integration of such systems, very little literature is available on the experimental verification and validation of morphing devices to demonstrate compliance and reliability with the current certification standards. In addition, the identification and classification of the potential failures detected during the experimental campaigns along with the associated risks may be a nontrivial exercise even on-ground.

The design complexity of morphing devices involves integration practices combining a wide range of scientific disciplines, such as, among others, topology optimization, materials and mechanical engineering, signal processing, and digital signal processing (DSP) programming. The use of both numerical and physical models is vital since the early design stages for the conceptual design of the components. The early identification of the actual and potential upscaling constraints is a matter of high priority during the system development as well as a continuous check of the industrial applicability of the developed concepts, the parallel prediction of the structural long-term stability, the installation constraints, and the potential safety and maintenance needs.

Since the early phase of a morphing system design process, it is highly recommended to plan and schedule the assembly of physical dummies, aimed at testing the developed concepts individually and assessing both the numerical and experimental uncertainties prior to entering the advanced development phase. This involves ranging from component-level demonstrators up to full-scale subassemblies following a standard "pyramidal" approach.

Dedicated laboratory experiments may be found in the literature to validate finger-like architectures implemented either in a morphing aileron [1] or in a morphing trailing edge device [2,3], showing good agreement with the corresponding numerical expectations. The targeted experimental campaigns aimed at demonstrating both the adaptive capability and the structural robustness under operative loads along with dynamic behavior for further aeroelastic analyses. Further experimental characterization was also proposed to validate the distributed electromechanically-based actuation systems driving those architectures [4,5]. Another component whose structural characteristics may be tested apart, is the morphing skin [6]. The specific needs it must accomplish require constant elasticity and strength properties at high and low temperatures ranging from −55°C to 80°C, which must be validated under static, quasistatic, and fatigue loading.

In consideration of these considerations, the primary challenge in development is to mature numerical and experimental tools, which may help to

- Assess both the individual and the integrated performance of the subsystems through dedicated equipment and tools.
- Predict the mutual interaction and, above all, the harmony of operation between different morphing systems and their impact on aircraft aerodynamics.
- Develop tailored structural interfaces starting from the reference wing box.
- Adapt tolerances and manufacturing processes for minimum cost.
- Manage design variations, components assembly, and integration to the test aircraft.

This chapter details a component and system verification procedure for both the assessment and experimental validation of a morphing device. The test objectives are defined at both component and system level, with practical notions on a specific application combining a compliant structural system and a multibox self-contained actuation mechanism. Tests rely on the use of special equipment and instrumentation to characterize the system performance from both the structural and aerodynamic standpoints and show if the design meets all given specifications. This allows determining the global system compliance with respect to a verification matrix, from the lowest level of the system hierarchy up to top level. Such a bottom up approach may also facilitate the identification of noncompliances and the definition of mitigation plans.

Starting from individual concepts up to the global experimental analysis of the integrated systems, the verification approach requires a modular test rig, several physical models and measurement tools. The structural design is validated through dedicated tests aiming at demonstrating both the morphing performance and the structural robustness under operative loads. The dynamic behavior is also investigated and validated by a direct correlation with the theoretical data, mainly concerning modal parameters and flutter trends.

Although such a general approach is applicable to any morphing system, a morphing trailing edge device is taken as a reference case study.

2 TESTING PRACTICES FOR MORPHING SYSTEMS

In general, numerical simulations are required to study the behavior of systems whose mathematical models are too complex to provide analytical solutions, as in most nonlinear systems. Such mathematical models can be thus either numerical, accounting for all features of the actual system, such as nonlinearities, temperature effects, etc., or alternatively, they can be experimental, previously-recorded, and stored in a database. It is often the case, when implementing an actual morphing device, that experimentally identified models are required for a correct operation. They have the advantage of being more precise and can be re-evaluated during the system lifetime if any change occurs in the environment or in the system itself.

With recent advances in computer technology for data acquisition, signal processing, and analysis, experimental models can accurately be identified from the measured responses and excitations using system identification techniques. Moreover, if the structural response does not change significantly over the timescale of operation, the system identification can also be performed off-line.

Due to their intrinsic nature, the experimental characterization of morphing devices implies different levels of investigation, which may be summarized in

- unit testing
- integration testing
- validation testing

Unit testing focuses on individual design items in isolation. The goal of unit testing is to isolate each part of the morphing device and characterize its single performance and properties. Integration testing is the phase in which individual morphing devices are combined and tested as a group. The purpose of integration testing is to verify performance and reliability expected on major items and their compliance with the requirements. All the components are assembled to interact correctly, and this is done after testing the individual systems. Finally, validation testing is the phase in which the assembled systems are tested to meet the intended system function. Every step is followed by functionality checks aimed at assessing the system integrity before and after system installation.

2.1 MORPHING TRAILING EDGE DEVICE

Throughout the aircrafts flying range, particularly in off-design flight segments such as take-off, landing, climbing, descent, and high speed cruise, it is highly desirable to increase aircraft aerodynamic performance by a differential control of the lift distribution over the wing span.

Recent European design studies [7,8] on morphing devices have largely proved the potential of novel aircraft structural systems, aiming at adaptively modifying the wing structural shape to reduce the induced drag penalty associated with off-design flight conditions. Wing camber variation through adaptive wing trailing edges exhibited the greatest potential because of the highly associated *L/D* ratio enhancements [9].

In the following paragraphs, the experimental steps necessary to validate the design of morphing devices are addressed for a case study. Such a process is tested on an adaptive trailing edge (ATE)

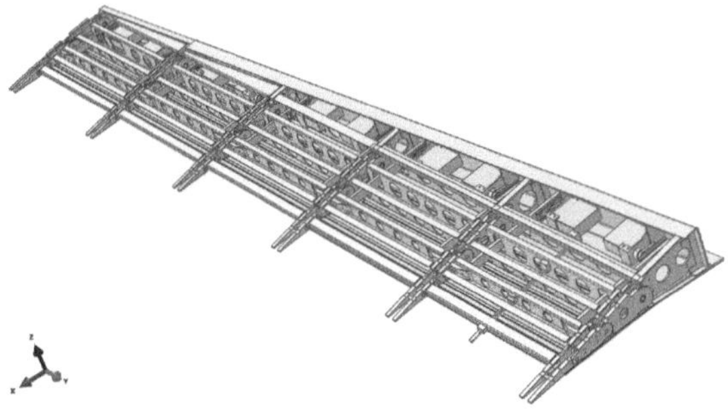

FIG. 1

Morphing trailing edge device.

demonstrator, shown in Fig. 1, consisting of a five-bay adaptive structure with modular architecture. The main system components are

- Multifinger morphing ribs designed to achieve target shapes during cruise.
- Morphing skin absorbing external loads and ensuring a smooth profile in operative conditions.
- Load-bearing servo-rotary actuators synchronously or differentially actuated for camber or twist variation.
- Control system based on a PWM-based closed loop motor servo controller providing actuator movement based on feedforward and feedback signals.
- Sensor system measuring chordwise and spanwise strain distribution to predict actual adaptive trailing edge device (ATED) shapes.

Such an adaptive structure is proposed to ensure a controlled and fully reversible transition from a baseline shape to a set of different configurations, each capable of withstanding the associated external load. To this aim, a distributed electromechanical actuation system, fully integrated within the control surface and suitable for morphing trailing edge devices, is designed with related tailored mechanical transmission, control, and sensor systems. Such a morphing device shows potential to enhance large commercial aircraft cruise performance by compensating the aircraft weight reduction during a long-range mission due to fuel consumption. These significant changes in flight conditions are compensated by varying the wing camber during the mission to obtain a near optimum wing camber preserving aerodynamically efficient flight. As a result, it allows the trimmed configuration to remain optimal in terms of efficiency (*L*/*D* ratio) or drag (*D*).

The testing procedure is a pyramid consisting of several tests planned at coupon, subsystem, and system component levels. The general concept of the unit testing approach is shown in Fig. 2.

For each component, ranging from the morphing skin coupon to an increasing number of assembled morphing units, the test strategies and test sequences are followed in detail to assess the single unit functionalities individually. The unit testing ends with the complete system (including subsystems and components) performance verification with respect to the design requirements.

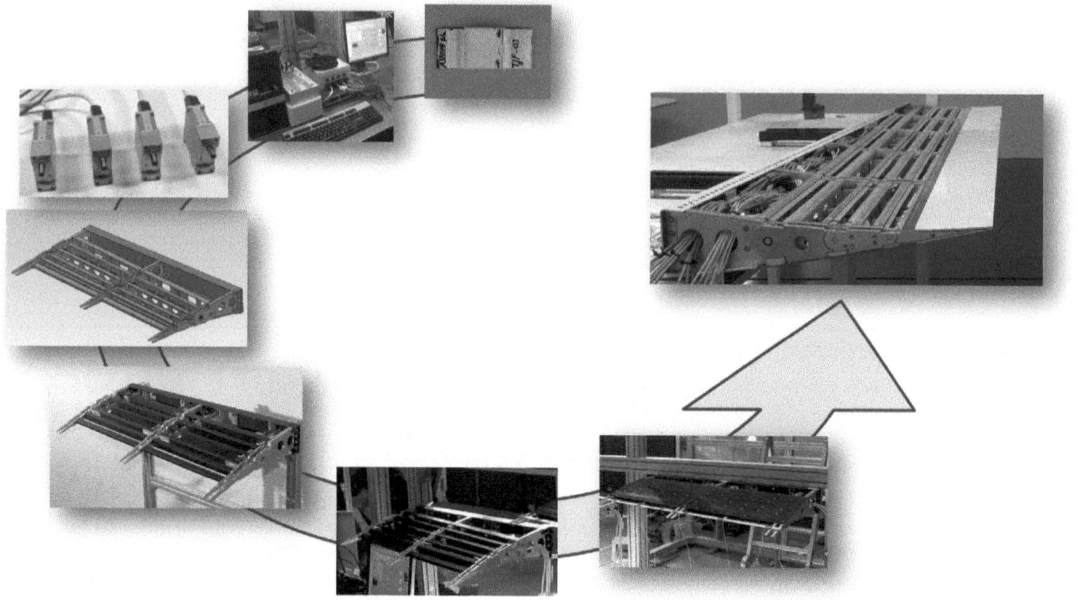

FIG. 2

Unit testing involving an increasing number of subcomponents.

3 UNIT TESTS: FROM COMPONENT TO MORPHING SYSTEM VERIFICATION

The iterative nature of the verification process of morphing structures makes it necessary for the need of testing assemblies with an increasing level of complexity while tracking the test assumptions and requirements. Each test configuration, aimed at validating some specific requirements while raising new questions to be addressed successively, gives the basis for discriminating the structural effects and performance of each subcomponent from possible combined properties. For this purpose, for instance, skin material properties may be derived from coupon level testing. In addition, isolated actuator system performance as well as control system accuracy and stability may be tested against simplified mock-ups.

To characterize the individual systems making the morphing structure, each subcomponent is assembled in representative mock-ups (dummy structures) including an increasing number of subcomponents and tested individually. The unit test involves five different test architectures and resulting demonstrators composed by an increasing number of ATED subcomponents. Four different dummy test articles and architectures are then manufactured and tested for component and system verification purposes. These test articles (dummies) are namely rigid reproductions of the morphing structural system made in steel in order to reduce as much as possible any structural elastic coupling effects between the structural skeleton and the morphing skin during static and dynamic tests.

The four different test architectures and objectives are shown in Fig. 3 and Table 1 respectively. They are referred to as:

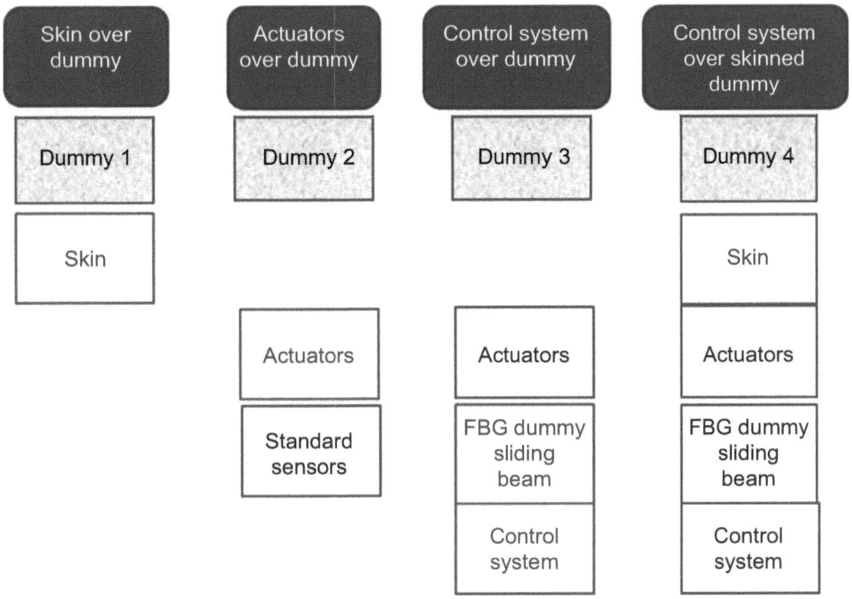

FIG. 3

Subcomponents of the dummy structures tested individually.

Table 1 Unit Test Items and Objectives

ID	Investigated Configuration	Declared Objectives
D1	Skin over dummy	Skin elastic behavior
D2	Actuators and structure	Coupled actuator-structural system elastic behavior
D3	Control system over dummy	Control system capabilities
D4	Control system over skinned dummy	Effects of skin characteristics over controllability
De	Complete system	Complete system verification

- Skin over dummy—composed of the dummy skeleton and the morphing skin.
- Actuators over dummy—composed of the dummy skeleton, the servo-rotary actuators, and actuation kinematics.
- Control system over dummy—composed of the dummy skeleton, the sensor system, the actuation system, and the control system.
- Control system over skinned dummy—composed of the dummy skeleton, the sensor system, the actuation system, the control system, and the morphing skin.

The test objectives of the complete morphing system are detailed in Fig. 4. The complete system verification involves the aluminum primary structure skeleton, the morphing skin, four servo-rotary actuators, the actuation system, the control system, and the sensor system.

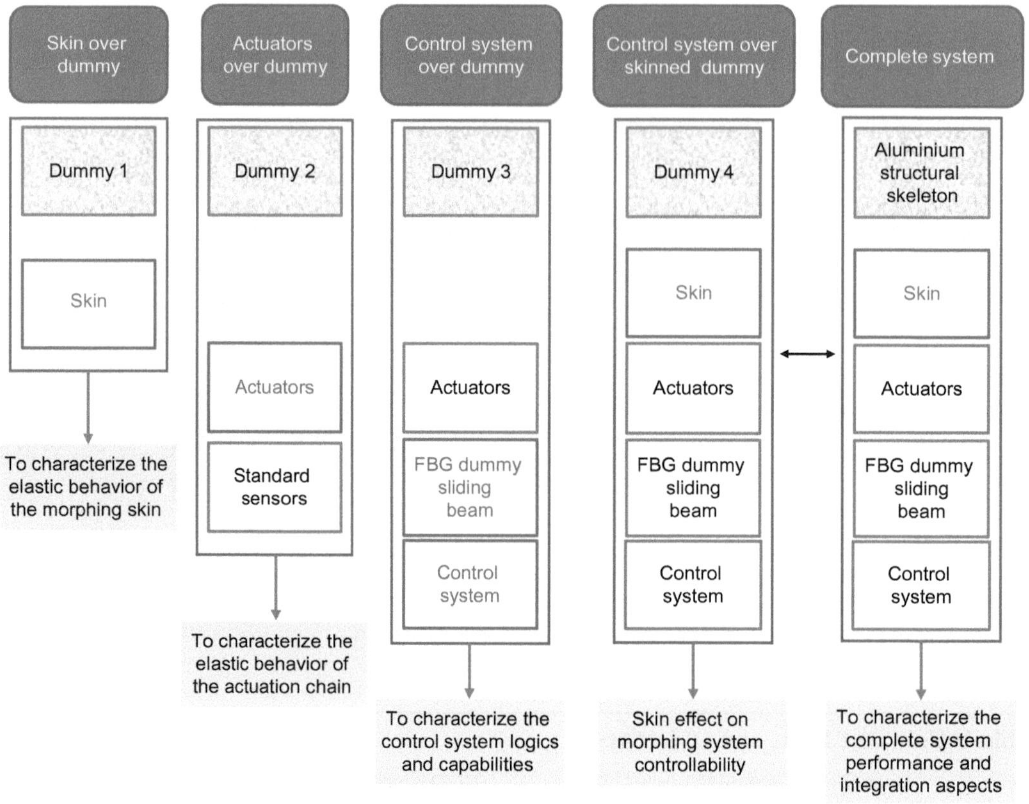

FIG. 4

Test logics on dummy structures and complete morphing system.

3.1 SKIN OVER DUMMY

The test article D1 is forced to morph downward through the application of three lumped loads (L) at the morphing device tip. L is progressively increased up to the value L^* at which the tip reaches the position compliant with the target morphed configuration. At each load increment, the following parameters are measured through a scanner-laser system:

- Displacements at acquisition points set 1.
- Displacements at acquisition points set 2.

As shown in Fig. 5, Displacements of set 1 points are used to monitor the actual shape reached by the inner structure. Displacements of set 2 points are used to measure deviations of skin hyperelastic strips from the target surface (morphing-induced bubbles).

L is progressively decreased from L^* to 0 and measurements of previous step are repeated for each load step. Experimental data are processed in order to define exhibited skin stiffness (force vs.

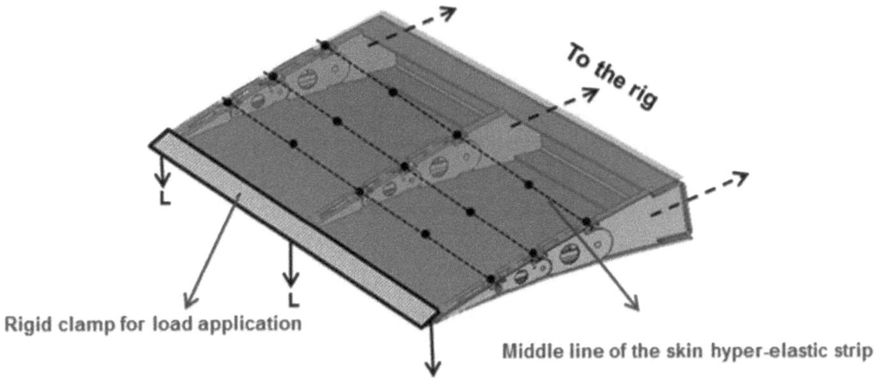

FIG. 5

Skin elasticity and morphing-induced deformation tests.

displacements curves, linear behavior) and hyperelastic strip performances in terms of target aerodynamic surface reproduction.

- Acquisition point (set 1).
- Acquisition point (set 2). (Lower skin acquisition points have been removed by the figure for clarity; they are homologue of the acquisition points located on the upper skin.)

3.1.1 Pressure-induced skin deformation

A dummy leading edge is installed on the same test article used for the previous test. Morphing is disabled through pin-locks in the ribs. The test article is connected to the wind tunnel rig and tested. Pressure sensors (kulites) are installed on the upper skin, as shown in Fig. 6. Angle of attack (α) and flow speed (V) are adjusted in order to get pressure levels fairly representative of operative conditions ($\alpha = \alpha^*$, $V = V^*$ in such condition). α is imposed equal to α^* and V is progressively increased up to V^*. Displacements at

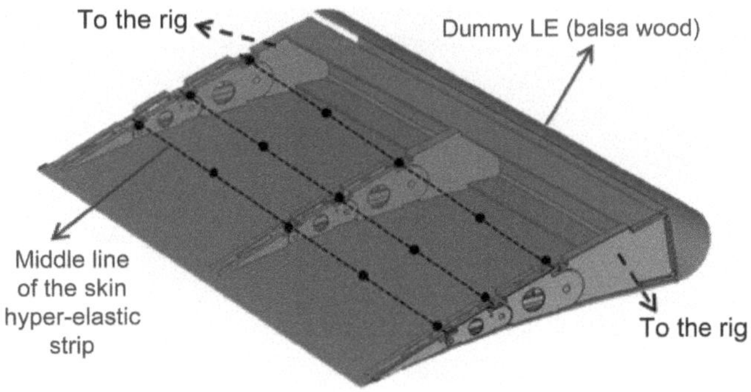

FIG. 6

Pressure-induced skin deformation.

acquisition points located along hyperelastic strips are measured through a scanner-laser system at each speed increment.

V is progressively decreased from V^* to 0 and measurements of the previous step are repeated for each speed decrement. Experimental data are processed in order to measure deviations of the upper skin hyperelastic strips from the target surface (pressure-induced bubbles).

- Displacement acquisition points

3.1.2 Skin normal modes

The same test article and rig of the previous test is used. Morphing is disabled through pin-locks in the ribs. T/A in unmorphed configuration and clamped to the rig. The structure is excited through hammer impact at the excitation points, as shown in Fig. 7. Accelerometers time-responses (along the normal to skin surface) are acquired in correspondence of 70 reference points over the T/A (35 on the upper skin +35 on the lower skin). The frequency response functions (FRFs) are synthetized by averaging over five measurements per acquisition point. Experimental data are processed in order to evaluate modal parameters (shapes, damping and frequency) in the range 0–100 Hz. The same procedure is repeated for the morphed down and morphed up configurations.

- Excitation point.
- Acquisition point (lower skin acquisition points have been removed by the figure for clarity; they are homologue of the acquisition points located on the upper skin).

3.2 ACTUATORS OVER DUMMY

The Test article D2, shown in Fig. 8, consist of skeleton and actuator systems, including actuation kinematics. The structural skeleton is a dummy one, whereas the actuator system is made according to its design characteristics.

The objectives of the tests are

- to identify actuation features (capabilities verification) in loaded and unloaded conditions
- to identify deformed shapes under external loads, actuation, and both

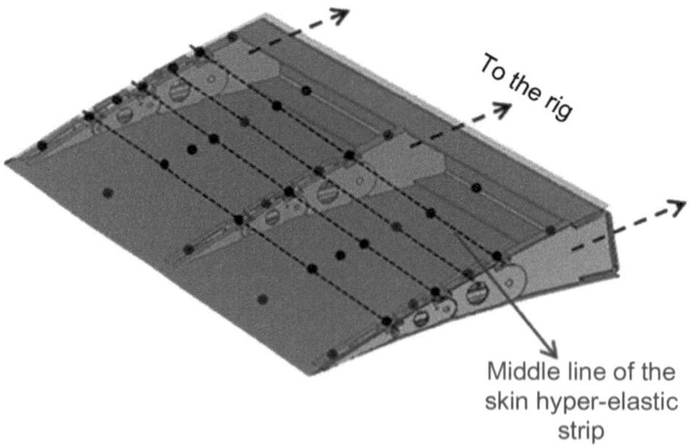

FIG. 7

Skin normal modes testing.

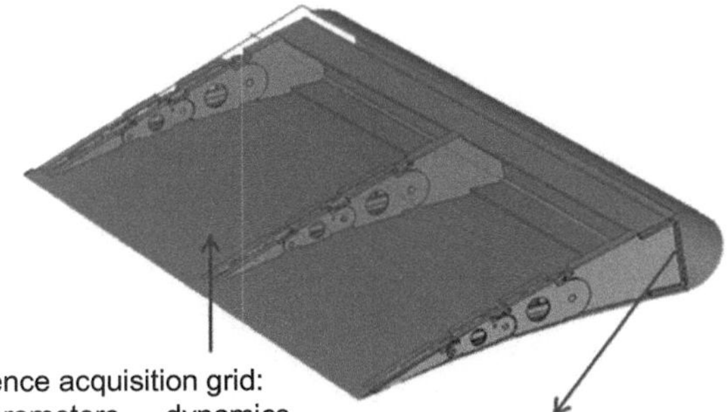

FIG. 8

Actuators over dummy.

- to characterize the dynamic properties (including eigenvectors and modal damping)
- to correlate numerical-experimental results

Static measurements include aerodynamic loads (static) and actuator action in order to evaluate static capabilities of the actuators in attaining the desired deformations. Aerodynamic loads are simulated by distributed weights over the structure. The structure is constrained to a given position after actuator action, or the actuators are blocked at their neutral position. Free-free conditions are used to detect dynamic properties of the system alone, without any contribution from the test rig and connections (f.i., damping). Dynamic measurements also include aerodynamic load and actuator action. Aerodynamic loads are simulated by point weights, per a multiple mass-spring configuration (very low eigenfrequencies).

3.3 CONTROL SYSTEM OVER DUMMY

3.3.1 Static and dynamic tests

The Test article D3 consists of a dummy skeleton and the control system, including actuation kinematics. Control and actuator systems are integrated as the final design.

The tests objectives are

- to test the control system (morphing platform) including integrated hardware and control logics, with and without aerodynamic loads
- to identify deformed shapes under external loads, actuation and both
- to characterize the dynamic properties (including eigenvectors and modal damping) of the combined control-actuator system
- to correlate numerical-experimental models including actuators and control logic

Both feed-forward and feedback control architectures are then tested, as sketched in Fig. 9. In both cases, the actuation mechanism is driven to enforce the structure to a desired shape. In the former, the control logic is driven by a target actuator rotation predicted by numerical models. In the latter,

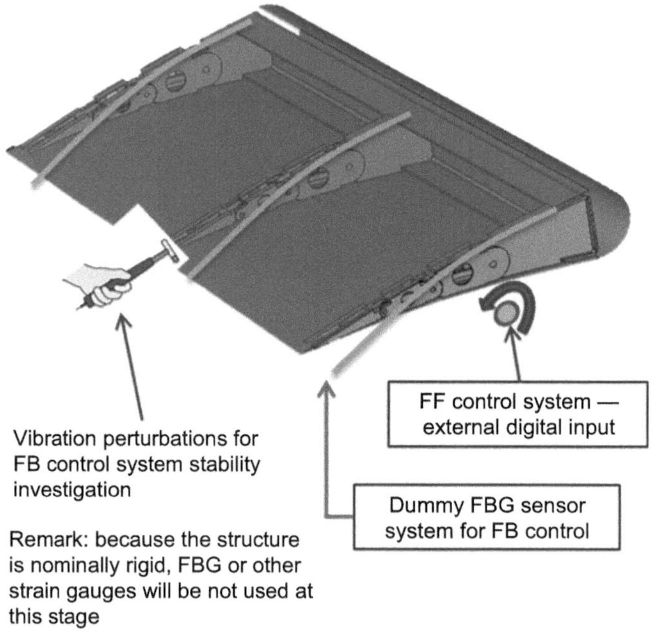

FIG. 9

Control system over dummy.

the driving error signal is extracted from the comparison between the actual and desired displacement (strain) vector. Actuation time, synchronous actuation, slew rate, resolution, and stability are experimentally assessed. Static measurements also include aerodynamic loads simulated by distributed weights over the structure. Dynamic measurements investigate the dynamic stability of the FB control system when subjected to small perturbations (vibrations). Finally, modal analysis is carried out by clamping the structure to a test rig. The control system is on-line (FB implemented) in order to characterize its effects on modal parameters.

3.4 CONTROL SYSTEM OVER SKINNED DUMMY

The Test article D4 consists of the dummy structural skeleton, the adaptive skin, the actuation system including kinematics and actuators, a sensing system, and the control system. The sensor network is representative of the spanwise and chordwise fiber bragg grating (FBG)-based sensor systems, needed for the actual shape reconstruction, Fig. 10. The objectives of the tests are:

- to test the control system platform including skin behavior (static-elasticity and dynamic-damping and elasticity-contribution) with and without aerodynamic loads
- to identify deformed shapes under external loads, actuation, and both
- to characterize the dynamic properties (including eigenvectors and modal damping) of the combined control-actuator system
- to correlate numerical-experimental models including actuators, sensors, and control logics

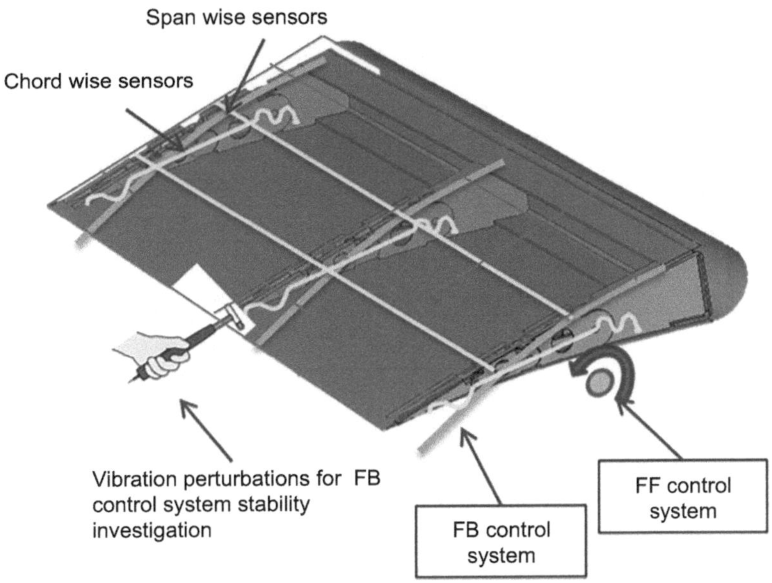

FIG. 10

Control system over skinned dummy.

Both feed-forward and feedback control logics are tested. The actuation mechanism is driven to enforce the structure to achieve a desired shape monitored through a strain sensor system.

In the FF architecture, the control logic drives the actuators to a target rotation, predicted by numerical models. In the FB architecture, after having imposed the target shape, the driving error signal is extracted from the comparison between the actual and desired shape, identified from the original shape vector. Actuation time, synchronous actuation effects, slew rate, resolution, and stability will be experimentally assessed.

The target strain vector is derived from the target morphing trailing edge shapes (aerodynamic inputs detailed in Fig. 18) through numerical predictions as well as experimental assessments. Data are correlated before starting the implementation of the control system.

Static measurements include aerodynamic loads simulated by distributed weights over the structure. Dynamic measurements also investigate the dynamic stability of the FB control system when subjected to small perturbations-vibrations.

3.5 COMPLETE SYSTEM

The structural system D5 is the complete morphing trailing edge demonstrator, made of an aluminum structural skeleton, the adaptive skin, the actuator system including kinematics, the sensor system, and the control system, shown in Fig. 11. The overall system consists of two bays for a total length of about 1 m.

FIG. 11

Complete ATED system [6].

The main test objectives are

- to detect the static and dynamic properties of the ATE as a whole
- to identify ATE shape in morphed/unmorphed configurations, loaded/unloaded conditions
- to verify control system functionality, performance, robustness
- to correlate the experimental with the numerical results

The key experimental tests are rapidly described per the approach previously detailed for dummy structures.

3.5.1 Modal analysis—Free-free boundary conditions

Free-free conditions are used to detect dynamic properties of the system alone, without any contribution from the test rig and connections (f.i., damping). Different configurations are tested (−5, 0, +5 degrees). Control system is off-line. Actuators are fed.

3.5.2 Static tests—FF control—Loaded/unloaded conditions

Static measurements include aerodynamic loads (static) and actuator forces in order to evaluate static capabilities of the actuators in attaining the desired deformations. Aerodynamic loads are simulated by distributed weights over its structure, as shown in Fig. 12. The structure is constrained to a certain position after the actuators action by the only effect of the actuator internal friction (operative condition), or the actuators are blocked at their neutral position (safety lock). The control system acts as a feed-forward system (commanded by a joystick). The actuator system operates both under and with no external load. The deformed shapes are measured under external loads (neutral position), actuation and both; control system functionality (feed-forward).

3.5.3 Static tests—FB control—Loaded/unloaded conditions

The same experiments are performed when the control system is activated and commanded to attain defined shapes at different times. Tests shall occur after the system is nominally stabilized (feedback control). The structure may experience smooth variations under test. Perturbations may be artificially induced and displacement and strain maps are compared before and after reaching the stabilized condition.

FIG. 12

Static tests on the adaptive trailing edge two-bay unit.

3.5.4 Static tests—Skin bubbling effects

Bubbling effects are simulated by including pressurized balloons in the structural body, under the elastomeric skin.

3.5.5 Modal analysis—Clamped conditions—FF control

Constrained conditions are realized by clamping the demonstrator to a test rig, simulating the wing half spar, as shown in Fig. 13. Dynamic measurements include aerodynamic loads and actuator action (blocked into their final configuration or connected to the activated control system) to evaluate dynamic properties of the system and assess control system stability under small dynamic shape variations. Aerodynamic loads are simulated by point weights, according to a multiple mass-spring configuration (very low eigenfrequencies). The structure is driven to different deformed configurations, moving from the neutral position (0 degree deflection). Modal analysis is carried out with the structure loaded and unloaded. Control system is off-line as the experiment takes place (feed-forward control law). Actuator system is blocked, following its own internal friction or a safety block.

3.5.6 Modal analysis—Clamped conditions—FB control

In the same configuration, the structure is again driven to different deformed configurations, moving from a neutral position (0 degree deflection). Modal analysis is carried out with the structure loaded and unloaded. Control system is on-line as the experiment takes place (feedback implemented). In this way, whether control system static or dynamic stability will be evaluated (ability of reaching a certain position and the ability of holding that position after the occurrence of small perturbations-vibrations). In this new configuration, the actuator system is free to move, according to the control system input.

FIG. 13

Modal analysis on the adaptive trailing edge two-bay unit.

4 SYSTEM INTEGRATION TEST BENCH FOR MORPHING SYSTEMS

The simultaneous need for monitoring target morphed shapes, actuation forces, and flight controls along with the counter-effects of aerodynamic loads under aircraft operating conditions, would suggest the use of ground-based engineering tools for the physical integration of morphing systems. Integration testing occurs after unit testing and before final system testing to verify morphing system functionalities in representative assemblies where all the components are installed and tested.

The integration testing phase requires a dedicated test bench, which is usually referred to as "Iron Bird." Such a test bench becomes even more crucial for morphing systems as the integration and validation tasks involve assembling, optimizing, and testing both systems individually and their combined effects. Such a facility and its related equipment is essential to validate control surfaces, actuation, control, and sensor systems on ground prior to the actual integration to the aircraft. The operational conditions are replicated, in terms of actual loads all the commanded actuators shall withstand, representative inertia, and stiffness of mechanical interfaces.

Fig. 14 shows a potential layout suitable for a system integration test bench for morphing systems. It includes different morphing devices installed on an aero-elastically reasonable aircraft wing box as well as the basic equipment needed to carry out "hardware in the loop simulations." Such a concept becomes essential to demonstrate advanced control technologies in a modular multilevel design that provides the robustness and the flexibility of a real aircraft integration. Manufacturing, assembly and integration issues including electrical and flight control may be extensively addressed in relation to the actual configuration of the aircraft. In addition, such a tool may confirm the characteristics of all system

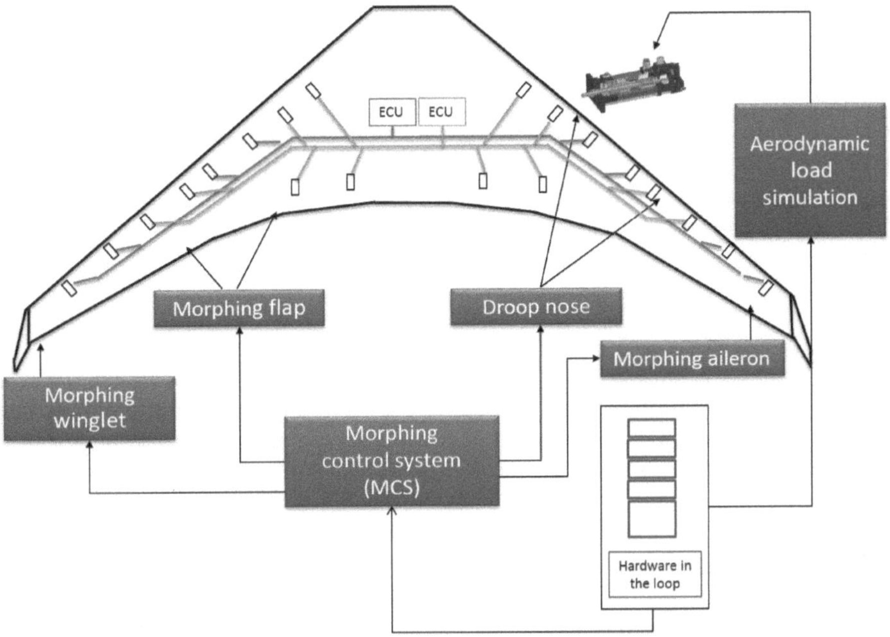

FIG. 14

Schematics of an "Iron Bird" tool for adaptive structures integration.

components or discover an incompatibility that may require modifications during early development stages, by accelerating the transition to test in a relevant environment. Finally, failures and mitigation actions can be studied in full detail and recorded for analysis.

The "Iron Bird" for testing morphing wing architectures enables test engineers to evaluate the real-time capabilities of morphing devices with the purpose of

- Demonstrating maturity, reliability, and integrated performance of morphing devices, that otherwise could only be achieved with more expensive, costly, and less safe methods such as wind tunnel tests or flight tests.
- Optimizing morphing wing architecture by testing both compliant and rigid-body mechanism-based morphing concepts and their related actuation, sensor, and control systems by monitoring aircraft weight and cost savings.
- Testing different actuation technologies (electromechanical actuators (EMAs), smart actuators) in controlled conditions.
- Investigating aircraft safety-related aspects by simulating system failures, such as jamming, runaways, one engine loss, strong cross-wind, aeroelastic effects to validate fault tree analyses and hazard assessments.
- Including operational loads that apply hinge moment forces to the aircraft morphing surfaces, representative of the aerodynamics forces applied during the simulated flight test and driven by the flight simulation model.

- Detailing cable routing and pathways.
- Validating the electrical consumption of each actuation system, in stationary and dynamic conditions, and the required command to A/C surface in each test case.

Once the different morphing prototypes are designed, manufactured, assembled, and integrated in the integration platform, they are "loaded" through a computer generated aerodynamic model of the aircraft simulating flight loads. Any changes and fine-tuning during such a development phase may be tested and validated using this valuable tool.

5 FULL-SCALE TESTING

The great interest in flying morphing technologies is driven by the promise of dramatic performance benefits and possible long-term cost savings in next generation aircraft. Significant design challenges require advances and multidisciplinary research efforts on a primary contributor to aircraft performance, outside of the propulsion system, which is the aircraft configuration. Given that a morphing wing is related to a configuration change in aircraft wing shape, the exploitation of these changes has led to the need for shape reconstruction tools and lift distribution monitoring and control capabilities through spatially distributed morphing using compact actuators.

A full-scale morphing wing can be developed and tested in different ways, depending on the timescale of the integrated concepts. If morphing devices operate in less than a second, they can reasonably assist in maneuvering the aircraft. As a result, the control system architecture and related simulations would require considering the A/C model as well as related flight control actions. If morphing occurs on the order of a few seconds, the adaptive device can be suited for active lift distribution control to maximize *L*/*D* during different off-design mission segments, such as climbing and turning flights. If wing shape changes occur on the order of minutes, the aircraft can truly take advantage of the benefits of morphing only during long mission segments, such as cruise, by compensating, for instance, aircraft weight reduction due to fuel consumption, thus enhancing aircraft aerodynamic performance and aerodynamic efficiency in such limited off-design conditions.

In all these cases, a specific control architecture must be designed and physically implemented to facilitate these types of experimental investigations.

5.1 SHAPE CONTROL OF ADAPTIVE WINGS

Aircraft flaps or slats are the common way to modify aircraft wing geometry during take-off and landing in order to provide additional lift at low speed. However, they have limited efficiency due to the geometric discontinuities, require high installation efforts and offer no functional flexibility in cruise. For some last-generation aircraft as B787 and A350, additional functions like differential flap setting are ensured by innovative flap actuation system concepts. Tailored control systems and inherent positioning sensors ensure this functionality. Additionally, distributed actuation enables decentralized control along the wing span, which is particularly suited for active lift distribution control for induced drag reduction. A sketch of distributed actuation and control applied to a morphing wing is given in Fig. 15.

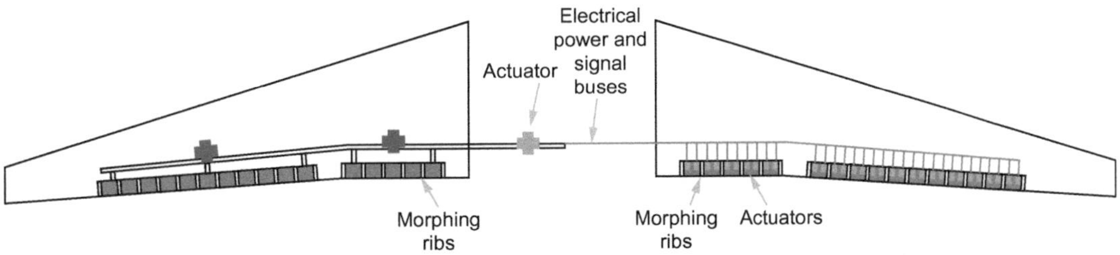

FIG. 15

Distributed actuation of a morphing trailing edge device: (Left) conventional actuation system; (right) distributed two-motors per bay approach [11].

Electro-hydraulic actuators with conventional feed-forward control logics are currently used solutions to control high lift movable devices and system mechanical components in aircraft because of their relative simplicity. However, they may be unsuitable for adaptive structural systems requiring local response modifications in flight, with a significant impact on the overall aerodynamic performance of morphing wings. Conversely, electromechanical actuation is increasingly receiving considerable attention in aeronautics, competing with conventional hydraulic systems, especially for their energy efficient use in secondary control surfaces. However, there are still some concerns for hazardous aircraft operations. The associated reliability and safety requirements are very stringent and involve specific needs in terms of failsafe protection in the event of emergency shut-down, diagnostics, and maintenance, which may hardly be met by state-of-the-art electromechanical based actuation concepts.

Symmetric flap actuation on both wings is paramount for safe flight and is traditionally ensured by coupling all flap surface actuators to a torque shaft system. It extends along the rear spar of the wing and is driven by a centralized actuator located near the movable flap surfaces, as displayed in Fig. 15 (left). Many of these elements are no longer necessary in a distributed actuation concept for a morphing wing trailing edge. A distribution of electromechanical motors, capable of withstanding the operative loads up to V dive conditions and namely enabling individual positioning of each morphing bay, may be used to replace the central actuator. Fig. 15 (right) shows the actuator layout, potentially driving the morphing ribs individually.

A major potential advantage of distributed actuation with respect to a shaft mechanism is the ability to move individual ribs either synchronously or independently to different angles (twist) [12]. Varying the angles b/w inboard and outboard wings as well as b/w different stations may lead to further aerodynamic benefits. This specific architecture uses both inherent rotation sensors on each actuator and a FBG-based distributed sensor system to synchronize the different element action and to monitor localized failures. Each actuator is rated for the torque of a single adaptive rib and is hence smaller than the equivalent conventional hydraulic actuation driving the whole control surface. As the number of motors drastically increases, such a configuration has potential to fulfill airworthiness requirements for a safety-critical surface and could be a viable solution for a less critical device such as an ATE.

The objective of a morphing control platform is to control the wing shapes by introducing actuation forces to match as accurately as possible each of the desired wing shapes exhibiting optimal aerodynamic performance. At each iteration, the actual shapes are compared with the optimal ones to

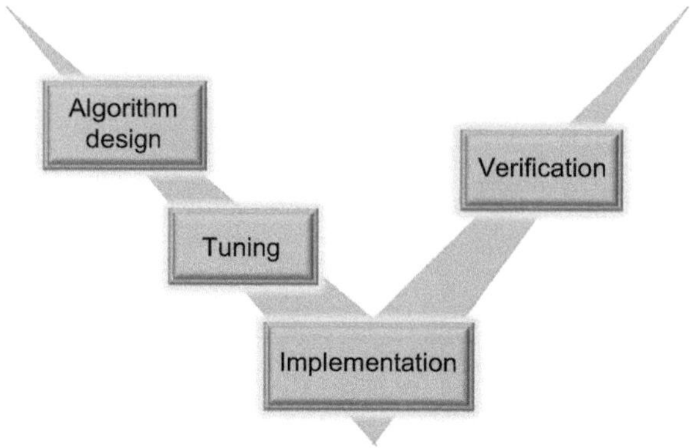

FIG. 16

Controller design work-flow diagram.

minimize the error function in approximating them. The process concludes when such a difference is lower than a threshold value.

Fig. 16 illustrates the controller design work-flow suitable for a morphing wing controller design application. It comes up from an adaptation of the well-known "V diagram," widely employed for the development of model-based control systems using the techniques of rapid prototyping and hardware-in-the-loop testing. The most important aspects of the process are

- Design of the control system: Use of MATLAB/Simulink to design and select the system coefficients of the controller.
- Development of a Simulink model: Use of Simulink blocks to graphically model the physical system.
- Run simulations in nonreal time: check the behavior of the model before creating a real-time application. Check the stability of the model.
- Hardware in the loop simulations: the controller design is tested with hardware in the loop. This step provides a validation feedback of communication protocol and parameter tuning in an actual dynamic environment.
- Create a real-time application: real-time workshop code generation (C code) from simulink model. The C/C++ compiler compiles C code to an executable code downloaded to the controller DSP board running in real-time.
- Run the application in real time.
- Analyze and visualize data.

As control performance depends on the accuracy of the identified model, a mathematical model adequately capturing the systems dynamical motion is needed for a successful controller design. This model can also be experimentally captured following both the unit and the integration testing outcomes. Before controlling the system in mathematical terms, it is fundamental to understand how the system behaves without control. To this aim, numerical simulations and experimental data of the morphing system and the rotary actuator are necessary. This provides the opportunity to prove out processes prior to field implementation enhancing the degree of confidence of the assumptions made in the controller design.

In a preliminary step, the plants of the morphing system can be numerically identified by assuming that

- The morphing structure is composed by rigid blocks connected by revolution joints. The rotary mechanism is reduced to a single degree of system freedom.
- The skin elasticity is simulated by springs applied between the rib segments. The stiffness of the spring is calculated by considering the geometrical and physical characteristics of the skin.

A representative second order system is then defined such that the steady-state response of the morphing system is approximately that of the original system. Such a model can provide a good approximation of the step response of the more complex system, thereby allowing its use in the controller design.

Three steps are necessary for an actuator movement function generation of a system consisting of several actuators. From the difference of the strain vectors achieved at different actuator locations, the maximum relative movement function must first be extracted. This function gives an idea of the cost function to be minimized through the control. Second, by knowing the maximum velocity of the actuator, which is dependent upon the physical constraints of the actuator system itself, the minimum transition time, which is the time to move the structure from one absolute position to the desired absolute position, is calculated. This function also depends upon the main acceleration parameter restrictions like the acceleration function (trapezium, sinusoidal or jerk-free ramp, etc.). Finally, the movement functions for any actuator is calculated by fixing the transition time corresponding to the maximum relative movement.

Two control logics may finally be implemented: open loop and closed loop. If the controller executes the command on the basis of the off-line predictions of the crank angle needed to morph the device, then the control strategy is defined as open loop (feed-forward), since the controller gives no feedback on the actual shape. If the controller processes the strain information given by the sensors over the structure as well as the actual actuator rotation measured by the actuator encoder so that the control actions can be real-time adjusted, then the control strategy can be defined as closed loop (feedback control). Both logics may be implemented in a morphing platform running, for instance, in a dSPACE system. The dSPACE system is high performance digital control system directly interfaced with MATLAB/SIMULINK running on a PC. The control logics are developed in SIMULINK block diagram, then converted to real time C running in real time in the DSP.

5.2 WING SHAPE CONTROLLER STRATEGIES AND EXPERIMENTAL VERIFICATION

Several methods for the experimental characterization and the actual control of adaptive wing shapes can be found in literature [13–16]. As a reference application, the controller design workflow is described considering its use for a novel variable wing camber trailing edge control. The designed adaptive control system uses individual rotary actuators driven by PWM signals with the aim of forcing the structure to achieve the desired shape necessary to enhance aircraft performance. No roll, yaw, or pitch control is considered. The actual trailing edge shape is monitored through FBG fiber optic sensors. They are used to monitor the strain distribution over the structure and characterize airfoil shape sensitivity to both the operational and actuation loads. The monitored shape information is fed back to the shape control system to achieve the desired shape while ensuring aerodynamically efficient surface.

As shown in Fig. 17, the process starts from a given set of desired wing shapes generated with the objective of minimizing drag while maintaining optimal lift coefficient. They result from an accurate aerodynamic analysis involving a number of flight conditions and lead to optimized shapes relying on complex numerical routines involving both theory and interpolation methods. First, these optimal ATE shapes are stored in a database. Then, the structural response to the actuation loads is numerically predicted by FEM simulations.

The process developed for the morphing wing trailing edge model control is illustrated in Fig. 18. The optimal airfoils are prescribed at a single flow condition (Mach = 0.75 in cruise) and for different combinations of angle of attack. For such conditions, theoretical smooth trailing edge shapes are generated by using polynomial functions and then aerodynamically studied to calculate the optimal performance. A continuous wing camber variation corresponding to a rigid TE deflection from −5 to 5 degrees is considered. The optimized airfoil database (*in the bottom right-hand box*) contains information about the strain vector corresponding to the target TE shapes on specific points of interest. Off-line aerodynamic simulations may be required to estimate the impact of the difference between the actual and the desired shape on the achieved aerodynamic performance and reduced fuel consumption. However, this target remains out of the scope of the controller whose performance will be estimated only by structural parameters describing the capability of achieving the target shape. The shape reconstruction module (*in the top right-hand box*) includes the software and hardware needs for measuring strain data through FBG sensors. Such data will be on-line processed by the controller (*in the left-hand box*) to calculate the error signal and hence the control action driving the actuators. Moreover, the actual angular position reached by the actuator will be provided by a rotary encoder and compared to the expected rotation value. The controller compares the information received from sensors with the information stored in a database in the computer memory.

Three steps are necessary for the definition of the actuator movement function, shown in Fig. 18, of a system consisting of several actuators [17]. From the difference of the strain vectors achieved at the different actuators locations, the maximum relative movement function has to be extracted first. This functions gives an idea of the cost function to be minimized through the control. Second, by knowing the maximum velocity of the actuator which is dependent upon the physical constraints of the actuator system itself, the minimum transition time, which is the time to move the structure from one absolute position to the desired absolute position, is calculated. This function also depends upon the main acceleration parameter restrictions like the acceleration function (trapezium, sinusoidal, or jerk-free ramp, etc.). Finally, the movement functions for any actuator is calculated by fixing the transition time corresponding to the maximum relative movement.

In the tuning phase, the actual shapes measured by videogrammetry, may be off-line processed and compared with the computer aided engineering (CAD) model shape expectations. The experimental results were compared with the CAD model expectations, as sketched in Fig. 19. The actual displacements of the morphed ATED structure were firstly measured and then compared with the numerical predictions. The measurement points are shown in Fig. 20. The error was evaluated by the sum of square errors between the experimental and simulated data. The position of the ATED device in terms of morphing angle was obtained by minimizing this error function. The results of the closed loop control obtained with applied external loading are reported in Fig. 21. The range of ATED morphing,

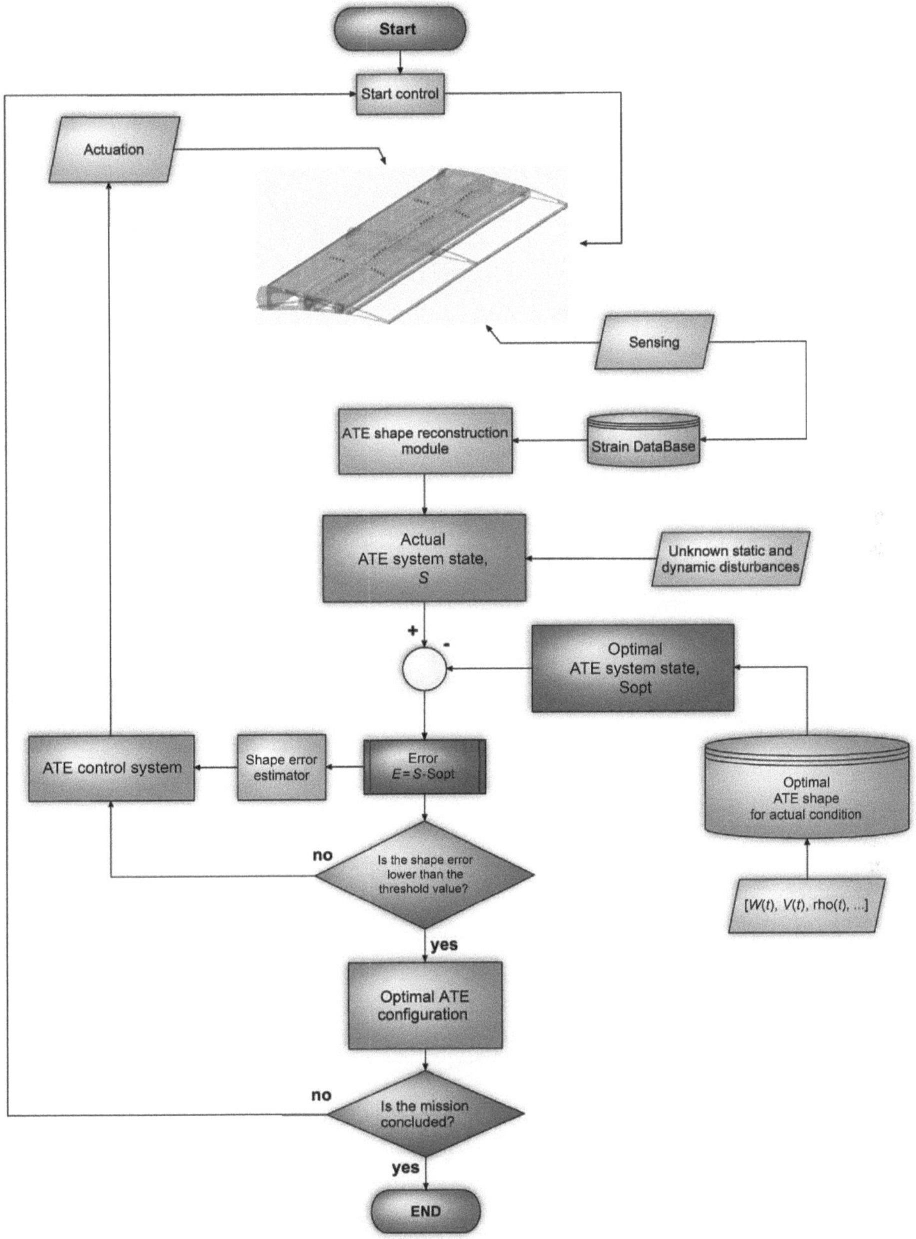

FIG. 17

Flowchart for shape control of morphing aircraft wings [10].

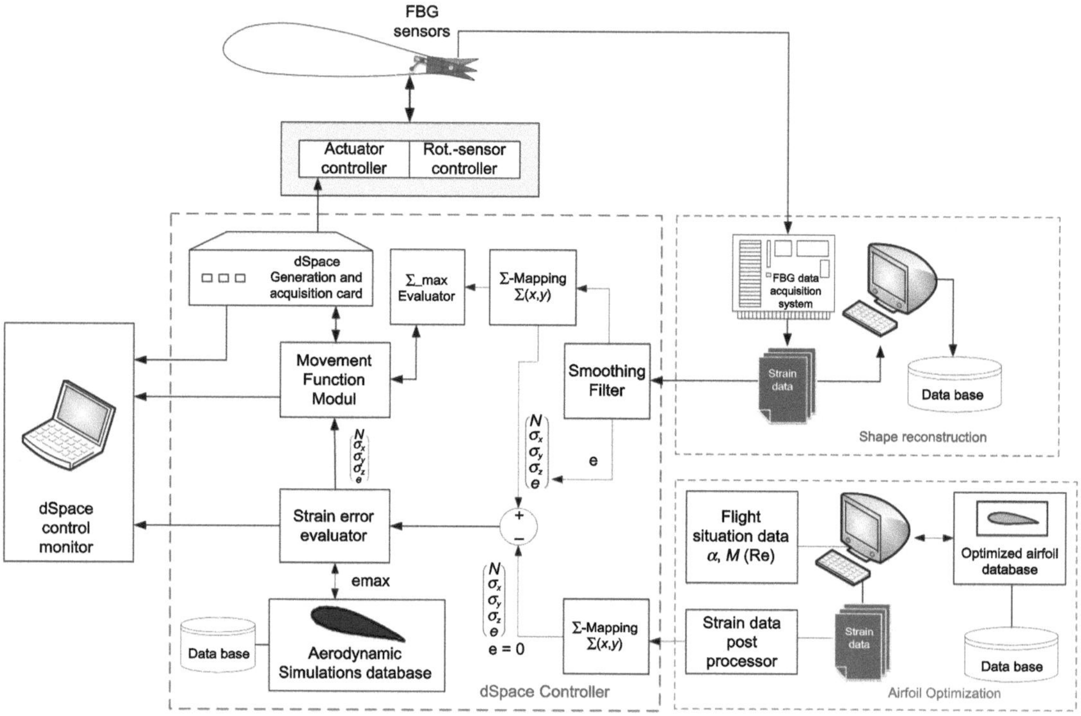

FIG. 18

Adaptive trailing edge control system platform [10].

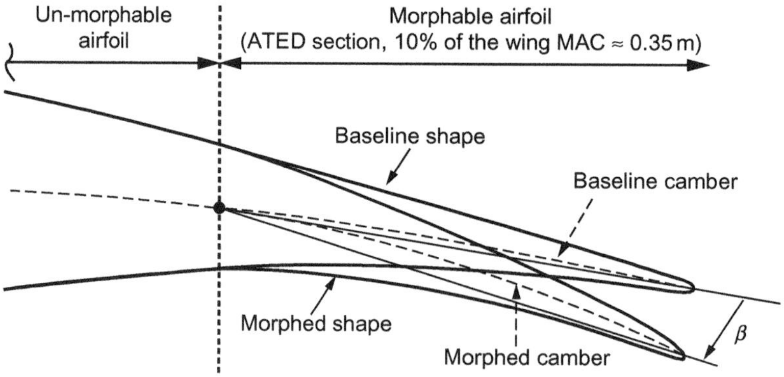

FIG. 19

Baseline and morphed airfoils.

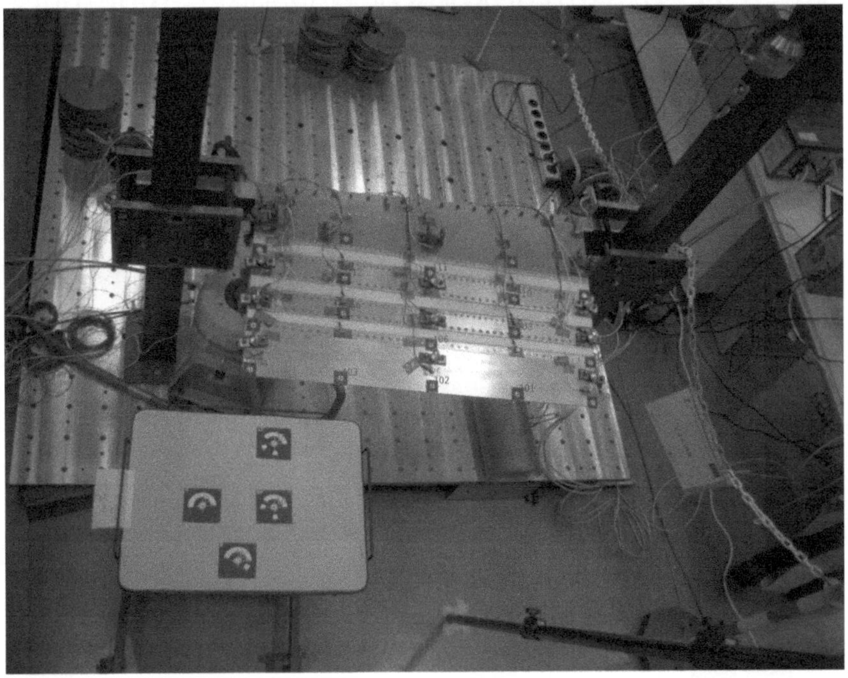

FIG. 20

Location of measurement points used for shape reconstruction.

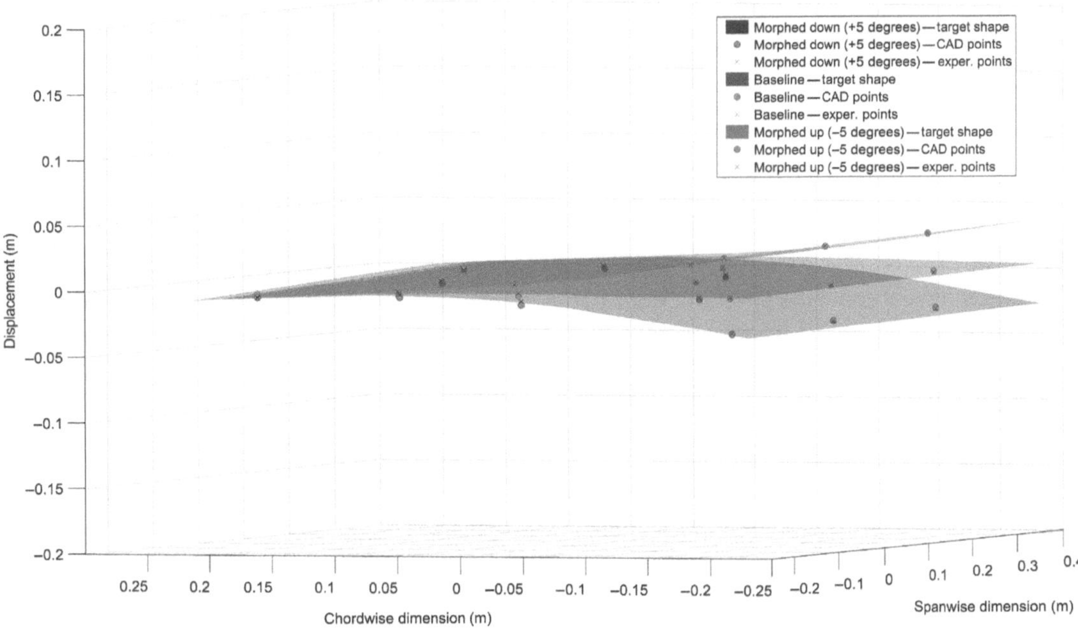

FIG. 21

Comparison between actual and expected shapes.

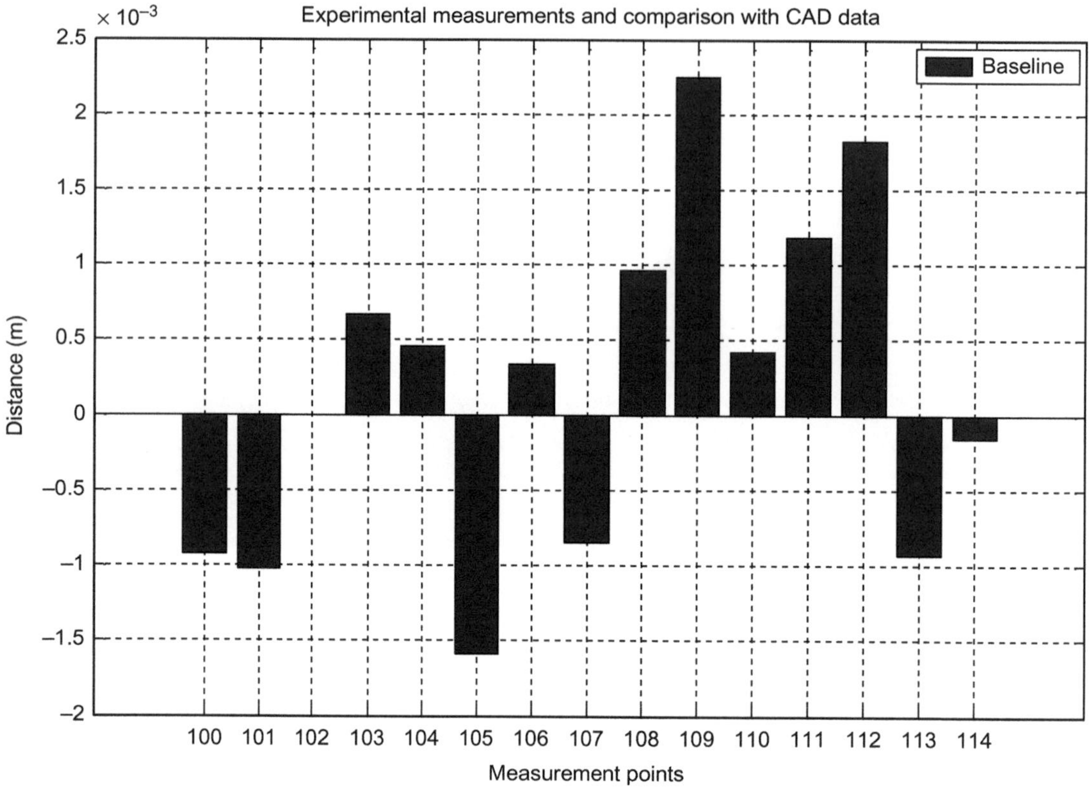

FIG. 22

Shape deviation in baseline configuration.

assessed by the minimum of the error function, was estimated in the range [−4.91, 5.21 degrees]. The deviations with respect to the CAD shape are given in Figs. 22–24 for the baseline and the full morphing deployment (+5, −5 degrees) respectively. Control system performance, such as actuation time, slew rate, resolution, and stability were also in this way validated.

Shape recovery capability of the feedback control architecture was evaluated along with controller stability and robustness. Such tests were performed by impacting the trailing edge tip with a hammer after commanding the morphing deployment, Fig. 25. Strain map distribution sensed by the sensor system was also off-line processed to reconstruct an ATED shape in morphing conditions. The differences between the actual and target shape were evaluated by the respective correlations with the strain levels and were found satisfactory.

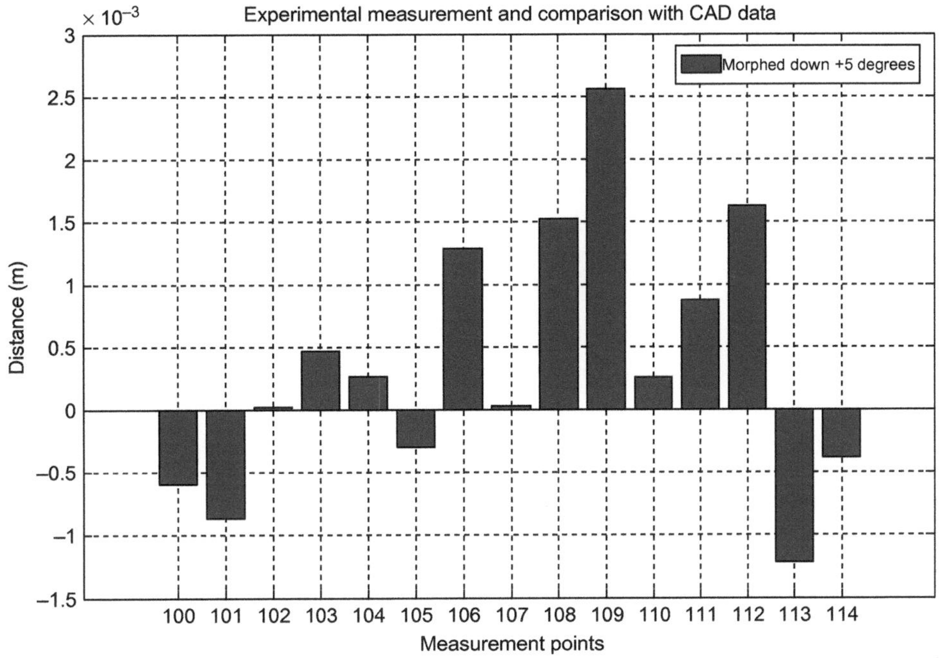

FIG. 23

Shape deviation in morphed down configuration.

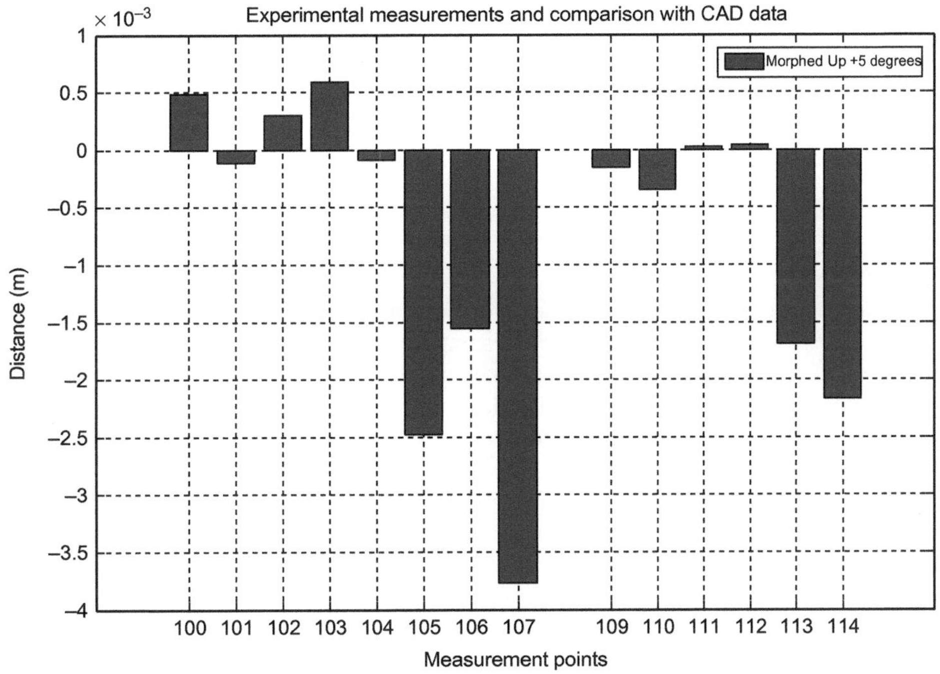

FIG. 24

Shape deviation in morphed up configuration.

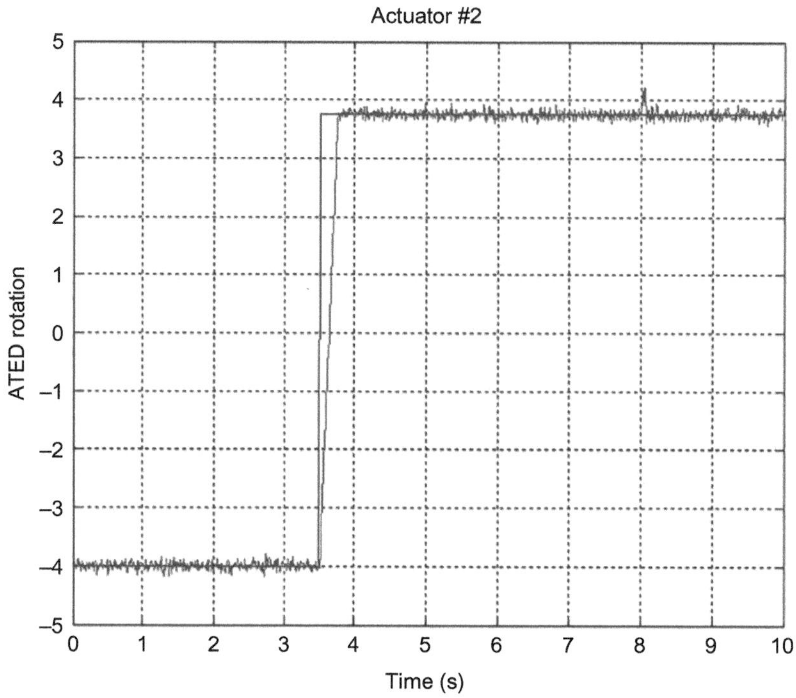

FIG. 25

Controller robustness tests through hammer excitation.

6 CONCLUSIONS

Although advanced experimental techniques have been successfully applied to the recognition of the structural properties of systems, morphing devices require tailored verification and qualification plans. Within this context, some testing practices for characterizing the structural properties and capabilities of morphing devices are presented and discussed. Component and system verification processes have been successfully applied to a benchmark structure, i.e., the morphing trailing edge device and the prospects of using such guidelines are analyzed and compared to standard experimental approaches.

First, isolated tests are illustrated to characterize the structural capabilities and material properties of the system subcomponents. After that, four different dummy structures have been manufactured and tested in order to verify system components in relevant scenarios presenting an increasing level of system complexity. The approach includes static and dynamic experiments with and without simulated aerodynamic loads and under the action of feed-forward and feed-back control logics. In the latter, the control system stability under small dynamic impulsive excitation has also been considered. Finally, a test plan involving the assembled demonstrator is detailed.

The use of a ground-based engineering tool for the physical integration of morphing systems is proposed to evaluate morphing system functionalities in representative assemblies where all the components are installed and tested. Besides the assessment of the static and dynamic properties of the system as a whole, morphed shapes, actuation forces, and flight controls may be investigated in detail along with the counter-effects of aerodynamic loads under operating conditions. The description of morphed shapes correlations between experimental and numerical results concludes the chapter.

REFERENCES

[1] G. Amendola, I. Dimino, F. Amoroso, R. Pecora, Experimental characterization of an adaptive aileron: lab tests and FE correlation, SPIE smart structures/NDE, Las Vegas, NV, March 2016, Proc. SPIE 9803, Sensors and Smart Structures Technologies for Civil, Mechanical, and Aerospace Systems, 2016, 98034P, April 20, 2016; https://doi.org/10.1117/12.2219187.

[2] I. Dimino, A. Concilio, and R. Pecora, Primary structural components characterization of an adaptive trailing edge device (ATED), 24th AIAA/AHS Adaptive Structures Conference, AIAA Sci Tech, 4-8 January 2016.

[3] R. Pecora, F. Amoroso, M. Magnifico, I. Dimino, A. Concilio, KRISTINA: kinematic rib based structural system for innovative adaptive trailing edge, SPIE Smart Structures/NDE, Las Vegas, NV, March 2016, Proc. SPIE 9801, Industrial and Commercial Applications of Smart Structures Technologies 2016, 980107, April 16 2016; https://doi.org/10.1117/12.2218516.

[4] G. Amendola, I. Dimino, M. Magnifico, R. Pecora, Distributed actuation concepts for a morphing aileron device, Aeronaut. J. 120 (1231) (2016) 1365–1385, https://doi.org/10.1017/aer.2016.64.

[5] I. Dimino, D. Flauto, G. Diodati, A. Concilio, R. Pecora, Actuation system design for a morphing wing trailing edge, Recent Pat. Mech. Eng. 7 (2014) 138–148.

[6] O. Schorsch, A. Lühring, C. Nagel, R. Pecora, I. Dimino, Polymer based morphing skin for adaptive wings, Proceedings of the 7th ECCOMAS Conference on Smart Structures and Materials—SMART 2015, Azores, 3-6 June. 2015.

[7] Green Regional Aircraft (GRA). Retrieved from http://www.cleansky.eu/green-regional-aircraft-gra. Accessed on June 2017.

[8] AS02 (Structural Tailoring of Wing) Trailing Edge Device. Retrieved from http://www.saristu.eu/project/activities/as-02-structural-tailoring-of-wing-trailing-edge-device/. Accessed on June 2017.

[9] G. Diodati, A. Concilio, S. Ricci, A. De Gaspari, F. Huvelin, A. Dumont, J. L. Godard, Estimated performance of an adaptive trailing-edge device aimed at reducing fuel consumption on a medium-size aircraft, Proc. SPIE 8690, Industrial and Commercial Applications of Smart Structures Technologies 2013, 86900E, 2013. https://doi.org/10.1117/12.2013685.

[10] R. Pecora, A. Concilio, I. Dimino, F. Amoroso, M. Ciminello, Structural design of an adaptive wing trailing edge for enhanced cruise performance, in: 24th AIAA/AHS Adaptive Structures Conference, Jan. 4–8 2016. San Diego, United States.

[11] I. Dimino, G. Diodati, A. Concilio, A. Volovick, L. Zivan, Distributed Electromechanical Actuation System Design for a Morphing Trailing Edge Wing, SPIE Smart Structures/NDE, Las Vegas, Nevada (USA) March 2016, in: Proc. SPIE 9801, Industrial and Commercial Applications of Smart Structures Technologies 2016, 980108, (April 16, 2016).

[12] I. Dimino, G. Amendola, F. Amoroso, R. Pecora, A. Concilio, Morphing technologies: adaptive ailerons, in: R.K. Agarwal (Ed.), Recent Progress in Some Aircraft Technologies, InTech, ISBN 978-953-51-2482-5, 2016. Print ISBN 978-953-51-2481-8. DOI: 10.5772/63645. Available from: https://www.intechopen.com/books/recent-progress-in-some-aircraft-technologies/morphing-technologies-adaptive-ailerons.

[13] F. Austin, M.J. Rossi, W. Van Nostrand, G. Knowles, A. Jameson, Static shape control for adaptive wings, AIAA J. 32 (9) (1994) 1895–1901.
[14] D. Baker, M.I. Friswell, The design of morphing aerofoils using compliant mechanisms, in: Proceedings of the 19th International Conference on Adaptive Structures and Technologies, October 6–9, Ascona, Switzerland, 2008.
[15] A.V. Popov, L.T. Grigorie, R.M. Botez, M. Mamou, Y. Mébarki, Closed-loop control validation of a morphing wing using wind tunnel tests, J. Aircr. 47 (4) (2010) 1309–1317.
[16] A.V. Popov, T.L. Grigorie, R.M. Botez, Y. Mébarki, M. Mamou, Modeling and testing of a morphing wing in open-loop architecture, J. Aircr. 47 (3) (2010) 917–923.
[17] I. Dimino, A. Concilio, M. Schueller, A. Gratias, An Adaptive Control System for Wing TE Shape Control, in: SPIE International Conference on Smart Structures 2013, San Diego, California, USA, 10–14 March 2013.

CHAPTER

WIND TUNNEL TESTING OF ADAPTIVE WING STRUCTURES 23

Svetlana Kuzmina, Fanil Ishmuratov, Mikhail Zichenkov, Vasily Chedrik, Gennady A. Amiryants, Vladimir Kulesh, Victor Malyutin, Alexander Chedrik, Viktor Timokhin, Sergey Shalaev, Alexander Chevagin, Roman Efimov, Innokentiy Kursakov, Ksenia Kuruliuk, Alexander Lysenkov, Vladimir Malenko, Mikhail Pronin, Andrey Saprykin

Federal State Unitary Enterprise Central Aerohydrodynamic Institute Named After Professor N.E. Zhukovsky (TSAGI), Zhukovsky, Russia

CHAPTER OUTLINE

Morphing Wing Technologies. https://doi.org/10.1016/B978-0-08-100964-2.00023-X

NOMENCLATURE

3AS	active aeroelastic aircraft structures
3-D	three-dimension
AWTC	aeroelastic wing tip controls
AMVT	all-movable vertical tail
ARGON	multidisciplinary software package for aircraft structural analysis and design
CAD	computer aided design
C_{MAC}	mean aerodynamic chord
DSM	dynamically scaled model
EURAM	European Research Aeroelastic Model
FE	finite element
FP5	(European Research) Framework Programme N.5
FRF	frequency response function
FS	full scale
G	AMVT rotational stiffness
GVT	ground vibration test
M	Mach number
MAC	mean aerodynamic chord
Mb	root bending moment
m_x^δ	roll effectiveness of the aileron
NASTRAN	NASA Structural Analysis (a structural FE code)
Nw	load factor at the wing tip
Ox	reference axis–stream direction
Oy	reference axis–normal to the chord
Oz	reference axis–span-wise direction
PC	personal computer
SADE	smart high lift devices for next generation wing
SDS	selectively deformable structures
SLE	smart leading edge
TA	tip aileron
ЦАГИ	Центральный аэрогидродинамический институт (RUS)
TSAGI	Central Aerohydrodynamic Institute
UWA	under wing aileron
V	velocity
VT	vertical tail
x	coordinate
X_F	Airplane Aerodynamic Center
y	coordinate
$\Delta x0$	geometry deviation, x direct, from non-deformed, non-flow to deformed under flow state
$\Delta y0$	geometry deviation, y direct, from non-deformed, non-flow to deformed under flow state
Hz	hertz
m	meter
N	newton
rad	radian
s	second

1 INTRODUCTION

Analytical-experimental advanced research for improving aircraft performance by properly tailoring the aeroelastic response of the lifting surfaces were performed in TSAGI within the framework of the active aeroelastic aircraft structures (3AS), smart high lift devices for next generation wings (SADE), and smart intelligent aircraft structures (SARISTU) European projects.

3AS intended to improve flight performance, economy, and efficiency of aircrafts by developing the concepts of active and passive aeroelasticity. These concepts allow considerable reduction in aerodynamic drag, structural weight, and operating costs. The European Research Aeroelastic Model (EURAM) was developed as an experimental platform for the demonstration of these new technical approaches. The wide, multifunctional possibilities and reliability of the EURAM were demonstrated during the 3AS project studies. The structure of the model proved that it was able to delay divergence, flutter onset, and withstand large gust loads, which were achieved without high risk of model failure. A comprehensive experimental database has been collected. It was shown that the use of nonconventional wing tip ailerons allowed for a reduction in structural weight due to their positive impact on the aeroelastic response of the wing.

The EURAM demonstrator, a four-engine, wide-body transport aeroelastic wind tunnel model of 1:10 length scale, span = 5.7 m, and mass = 200 kg, was designed and fabricated to test three specific concepts: aeroelastic wing tip controls, *all-movable vertical tail*, and selectively deformable structures.

Two aspects of the concept of active aeroelastic wing were studied:

- use of wing elasticity for increasing roll control characteristics, and decreasing the induced drag with the aid of controls located upstream of the wing elastic axis;
- use of rotational elasticity of the axis of an all-movable vertical tail of reduced size for the improvement of lateral stability and controllability; in this case, the vertical tail fixing is performed by using adaptive rotational stiffness versus flight speed.

The complete EURAM model and its parts were tested in TSAGIs T-103 and T-104 wind tunnels.

Development and evaluation of the potential of morphing airframe technologies based on selectively deformable structures, as one of the promising concepts for smart structures, was also investigated. It is important to be aware of the fact that for composite structures stiffness, elasticity, and strength are not scalar, but tensored properties. Therefore, it is possible to combine the required strength and stiffness with selective deformability as a perspective opportunity in the design of adaptive wings. In this case, a force applied to this material should result in a deflection along the direction of the acting load with only a minimal deformation in the transverse direction. This effect is achieved by designing the internal material structure as a special cellular network. The individual cell can be described essentially as a moment-resisting frame. Deformations in this type of composite structure may be resolved into two components: traction–contraction, and flexural deflections.

The main task of the SADE project was the investigation of the next generation smart, high-lift devices directed toward the lift-to-drag ratio, increasing fuel efficiency, and decreasing noise and emissions at all flight regimes, especially during takeoff and landing regimes. The SADE large-scaled demonstrator was represented by a straight wing with a 5 m span and a 3 m chord, and was tested on the TSAGI wind tunnel T-101. Three sections of the smart leading edge were attached to the front spar. To measure the deformations of the smart leading edge's upper and lower surfaces, a noncontact, videogrammetric method was used.

The SARISTU project was focused on the development of morphing technology and intelligent aircraft structures. Within the framework of the project the full-scale wing demonstrator and morphing devices were developed and manufactured by partners in SARISTU Application Scenario's AS01, 02, and 03. The wing had slitless control surfaces, with, namely, an enhanced adaptive drop nose, an adaptive trailing edge, and a winglet active trailing edge. The objective of the wind tunnel test was to validate the integrated morphing concepts functionality in realistic loading and within a full-scale environment. As one of the project partners, TSAGI was responsible for the organization and execution of tests on the full-scale wing demonstrator in the TSAGI's T-104 wind tunnel.

1.1 GENERAL TEST PROCEDURE FOR THE MORPHING ITEM

The three different classes of concepts have been investigated and demonstrated:

- Deformation of the structure by means of conventional or new aerodynamic control surfaces;
- Deformation of complete aerodynamic surfaces by attachment/actuation devices with adaptive stiffness to adjust for the desired aeroelastic effectiveness; and
- Direct deformation of the structure by integrated/embedded active and smart materials, or variable position/stiffness structural elements.

TSAGI specific experience coming from the application to European projects:

- 3AS
- SADE
- SARISTU

2 3AS

The EURAM was developed as an experimental platform for the demonstration of the new technical approaches and concepts developed as part of the European 3AS FP5 Project [1–3]. 3AS aimed to improve the flight performance, economy, and efficiency of aircraft through the development of active and passive aeroelastic concepts. These concepts allow considerable reduction in aerodynamic drag, structural weight, and operating costs. The EURAM demonstrator, a four-engine wide body transport aeroelastic wind tunnel model of 1:10 length scale, span = 5.7 m, and mass = 200 kg, was designed and fabricated to test three specific concepts:

- Aeroelastic wing tip controls (AWTC)—new control surfaces at the wing tip, forward of the elastic axis, that adjust the flexible wing deformation to the optimum shape for minimum induced drag;
- All-movable vertical tail (AMVT)—replacement of the existing vertical tail by a smaller, all-movable surface with variable rotational attachment stiffness. The rotational axis is located behind the elastic axis, and the torsional stiffness can be adjusted to the fight condition; and
- Selectively deformable structures (SDS)—modification of the inner aileron using a new kind of structure that allows large deformations with small internal forces, thereby creating a more continuous deformed shape.

The complete model was tested in TSAGI's T-103 and T-104 wind tunnels (Fig. 1). Components of the EURAM (semi-wings with ordinary and new control surfaces, and ordinary and all-movable vertical tails) were tested separately in different wind tunnels by partners of the 3AS Project. Finite Element

FIG. 1

The complete EURAM in TSAGI wind tunnel T-104.

(FE) models of the complete EURAM airplane were created for multidisciplinary optimization and analysis. In addition to the FE models (in NASTRAN format) the TSAGI software, ARGON [4], was used for computing strength, static and dynamic aeroelasticity, and aeroservoelasticity characteristics. The ARGON code was used to both predict the results as well as update analytical models. Updating the mathematical models of the elastic structure was carried out at different stages on the basis of stiffness measurements, the ground vibration test (GVT), and wind tunnel tests. Such an integrated approach validated the provided results.

2.1 REQUIREMENTS FOR THE EURAM AND EXPERIMENTAL FACILITIES

The following main requirements for the demonstrator and tests were formulated according to the 3AS project goals:

- Modular structure with the possibility to disassemble wings and vertical tails for separate tests in different wind tunnels;
- Large-scale and compartment-beam structure with easy changes between different configurations and control surfaces;
- Use of actuators for ailerons and non-traditional surface dynamic control;
- Several different attachment types incorporated into the fuselage for adjusting the AMVT rotational stiffness as well as the basic vertical tail with rudder;
- Wind tunnel tests of the complete model on a balance and under free-free conditions;
- Measurement of static aeroelasticity, flutter, and aeroservoelastic characteristics for complete model as well as for separate subcomponents.

2.2 MODEL DESIGN AND MANUFACTURE

The T-Flex 3-D CAD system was used for the design of new components of the model and for the preparation of the necessary drawings. As examples of this, the central parts of the fuselage, the tail structure, and the forward aileron with pylon and integrated actuator/sensor are shown in Fig. 2.

New control surfaces were employed, which gained a positive effect from the wings elastic deformation. The EURAM fuselage, wings, and tails were constructed using aluminum beams with rectangular and H-shaped cross-sections. Plywood was used as a main material for the compartment structures. The forward aileron and the pylon were realized by adopting carbon-fiber or plywood structures.

2.3 LABORATORY TESTS

Laboratory measurements of stiffness, mass, and modal characteristics of the EURAM components, and a ground vibration test of the complete model configurations, were carried out. The qualification of the model parameters and preparation data for creation of adequate mathematical models and their corrections was also performed. An example of the correlation between experimental and theoretical wing beam stiffness along the span is illustrated in Fig. 3.

The ground vibration test of the complete EURAM was conducted in the frequency range from 1 to 10 Hz. Natural symmetric/asymmetric elastic modes for five structural configurations were identified inside this frequency range. All the natural frequencies of the control surfaces were above 20 Hz.

Wind tunnel (WT) tests were performed on different EURAM components: AMVT, left semi-wing, forward aileron, and inner aileron with SDS, as shown in Figs. 4 and 5.

For the measurement of static aeroelasticity characteristics of the complete EURAM, the model was attached to a vertical strut with six-component strain-gage balances at the model's center of gravity (Fig. 6, left). The models suspension system enabled the angle of attack and sideslip angle to be changed. Deflection angles of ailerons, tip and under wing forward ailerons, basic rudder, and the all-movable vertical tail could also be mechanically changed and fixed.

For flutter tests in the T-104 tunnel, the complete EURAM was supported by a universal "two-points" cable low frequency suspension system designed in TSAGI especially for the investigation of dynamically scaled model flutter characteristics (Fig. 6, right). The suspension system provides five degrees of freedom for the model as a rigid body within a frequency range up to 1 Hz. Choosing these suspension system-structural parameters and geometry provides enough static and dynamic stability to the model so that it behaves as a rigid body for all configurations. Signals by strain gauges and

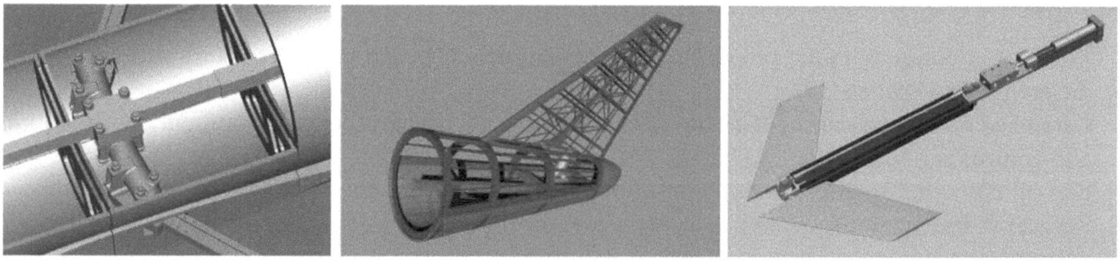

FIG. 2

Fuselage central part (left), vertical tail structure (middle), and forward aileron (right).

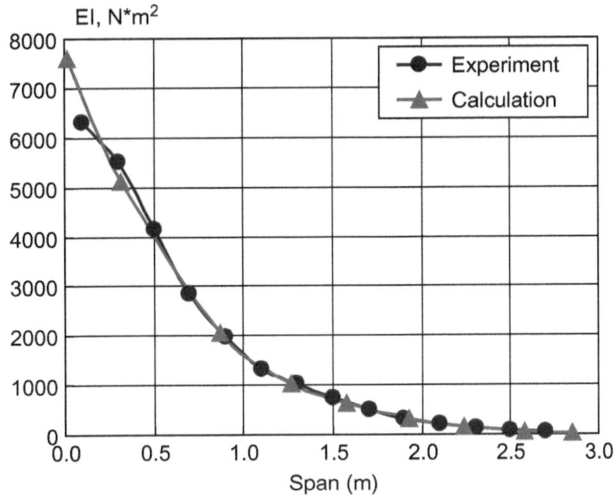

FIG. 3

Comparison between experimental and computed wing stiffness.

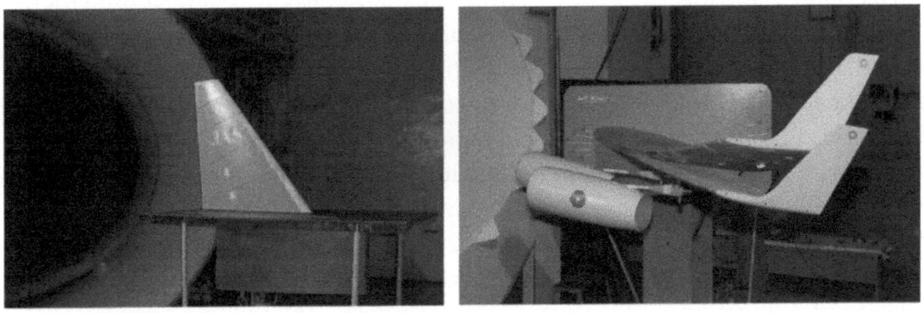

FIG. 4

AMVT model (left) and static deformation of half-wing (right).

FIG. 5

Forward aileron (left) and inner aileron with flexible SDS (right).

FIG. 6

Static aeroelasticity test (left) and flutter test (right) on the complete EURAM in WT T-104.

accelerometers, as well as video and visual information, were used for the analysis of model flutter characteristics. The analysis of measured data showed that necessary flutter margins were ensured for all test configurations of the EURAM. Flutter characteristics of the model with the new attached wing tip surfaces somewhat degraded at speeds >34 m/s; however, adequate safety margins were still maintained.

For the purposes of aeroservoelastic investigations, the complete EURAM was installed in the T-104 wind tunnel on the same cable low frequency, "two-points" suspension system as used for the flutter tests. The model was equipped with miniature hydraulic actuators with a maximum force of about 300 N. "Sensorex" inductive type displacement sensors with amplifiers were used both in the actuator feedback loop and for the measurement of the actuator rod displacements, ensuring adequate deflection of both ordinary and forward ailerons. A special analog unit provided the necessary feedback loop for the ailerons actuator. Two PCs with analog-digital and digital-analog transfer blocks provided the acquisition and control of sensor signals, excitation signals, and control law digital filters for the closed loop experiments. The sampling frequency for the open loop case was 250 Hz, and 1000 Hz for the closed loop. A cascade of two wings, controlled by the PC-driven hydraulic actuator, and located at the beginning of the test section allowed simulation of a single gust, of different harmonic gusts, and random turbulence (Fig. 7).

Two aspects of the active aeroelasticity concept were studied:

- use of wing elasticity for increasing the roll control characteristics and decreasing the induced drag with aid of controls located forward of the wing stiffness axis;
- use of rotational elasticity of the reduced size all-movable vertical tail for the improvement of lateral stability and controllability.

The computational beam model was designed using the ARGON software package, leading to the manufacture of the beam-compartment dynamically scaled model (DSM). The developed model was based on the method of prescribed deformation mode shapes (Ritz polynomial method) where the structural parts of the DSM were modeled with bending or torsion beams and lumped masses (Fig. 8). The structural parts were joined by rigid springs. The compartments of the DSM were not modeled and the deformation of the lifting surfaces was considered to be smooth.

Finite element (FE) models of structural parts and full DSMs were then also developed in NASTRAN. Compartments of the lifting surfaces were modeled using rigid plate elements to improve the visualization of the displacements (Fig. 9). The displacement field for the analysis of the aerodynamic forces was defined by one-dimensional splines generated on the nodal displacements of the spars.

FIG. 7

Gust response test on the EURAM.

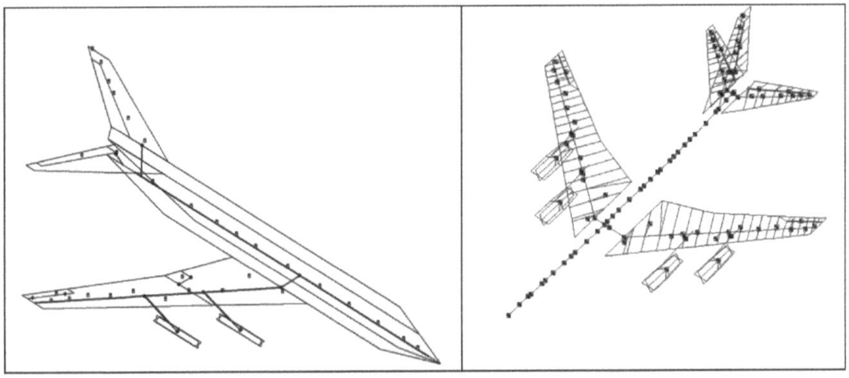

FIG. 8

Computation model in ARGON (left) and FE model of EURAM with DSM aileron (right).

It was necessary to validate the effectiveness of the concepts considered in this project on a full-scale computational airplane model developed on the basis of its DSM. The geometrical sizes of the mathematical model to the full-scale airplane were defined by multiplication of the DSM sizes by the length scale coefficient. The structural layout was chosen on the basis of known structural layouts of existing wide body airplanes and from experience. The traditional approach of modeling using two-dimensional shell elements and one-dimensional beam element was adopted for the development of the full-scale airplane model (Fig. 9, left). The aerodynamic model coupled to all structural models is shown in Fig. 9, right. Unsteady aerodynamic forces for the dynamic aeroelasticity problems were analyzed using the doublet-lattice method.

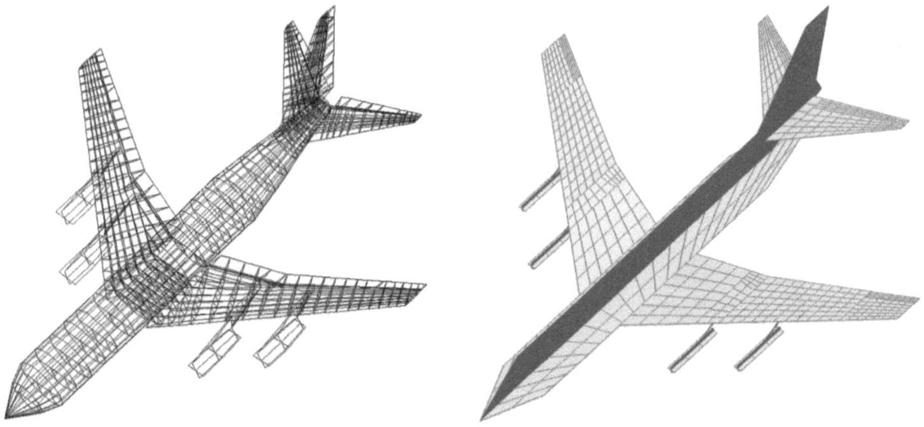

FIG. 9

FE model of full scale EURAM (left) and aerodynamic model (right).

The mathematical model of the full-scale airplane has a reasonable lift-to-drag ratio, which is related to induced drag only at cruise flight regime. Supercritical airfoils, with a thickness-to-chord ratio of 14.4% at the wing root and 9% at the wing tip, were used along with jig twist angles of 3° at the wing root and $-1°$ at the wing tip. Fig. 10 shows the comparison of the lift-to-drag ratios for different configurations. The configuration with tip aileron (TA) has a slightly greater value of lift-to-drag ratio, mainly due to its larger effective aspect ratio.

Most of the considered concepts in this work that aim to use aeroelastic deflections in a positive way are mainly related to static aeroelastic characteristics. Therefore, considerable attention was paid to the static aeroelasticity studies. The influence of structural elasticity on the location of the airplane aerodynamic center, X_F, for the actual airplane scale of dynamic pressures is shown in Fig. 11. The

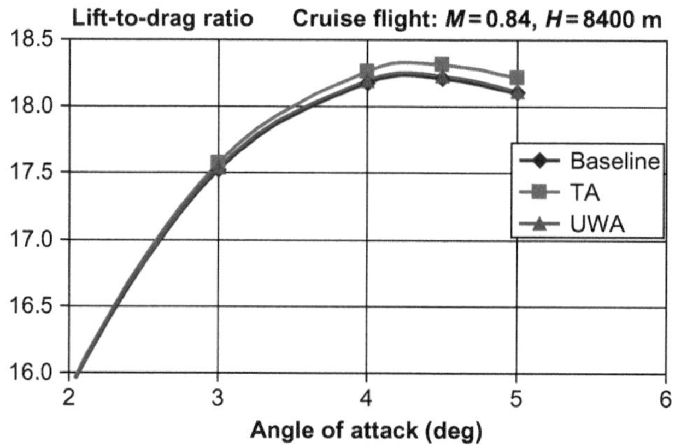

FIG. 10

Comparison of lift-to-drag ratio of full aircraft for different configurations: TA and UWA.

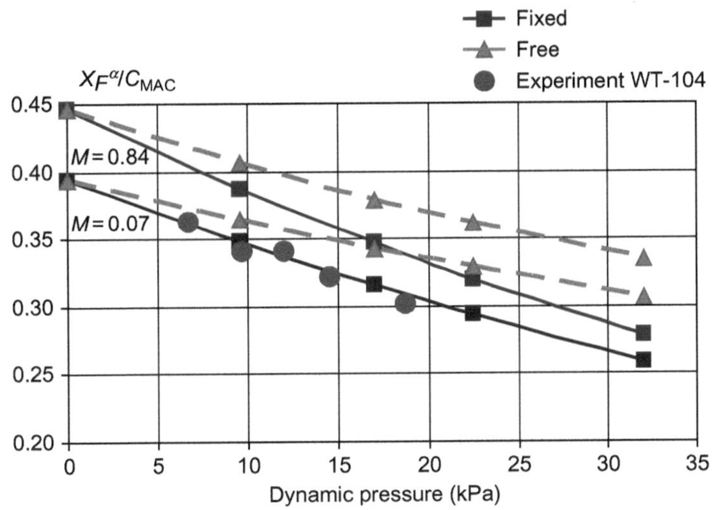

FIG. 11

Influence of structural elasticity on the aerodynamic center location.

analytical and experimental results are in good agreement for fixed structures in incompressible airflow ($M = 0.07$). The characteristics of free-free structures are different from those of a fixed structure. Also, it is necessary to take into account the influence of the Mach number on full-scale airplane behavior. Fig. 11 shows that, for an in-cruise flight $M = 0.84$, the location of the aerodynamic center of the full scale airplane in free flight is significantly different from the location found in the wind tunnel tests. Note the good correlation between analytical and experimental results.

2.4 AEROELASTIC WING TIP CONTROLS CONCEPT

The objective of the concept is to investigate new types of control surfaces that generate an active elastic deformation of the wing leading to achieve desired aerodynamic characteristics. Fig. 12 shows the comparison for the roll effectiveness, m_x^{δ}, of the wing tip aileron. For the fixed structure, the analytical results are in agreement with the experimental ones (recalculated for the scales of the actual airplane). Unlike a regular aileron, the effectiveness of the wing tip aileron has practically not decreased. On a free structure, the inertial forces, induced by roll angular acceleration, twist the wing into "useful" angles. Therefore, the effectiveness of the wing tip aileron on a full-scale (FS) airplane increases. However, inertial forces should be taken into account for the FS structure because the DSM is not similar to an FS airplane from the inertial standpoint.

One of the key aspects of the active aeroelasticity concept is the active control system. Frequency response functions (FRFs) for load factors at various structural sections and wing root loads are computed to evaluate the possibilities of using AWTC for an active control system. The results of this analysis were found to be in satisfactory agreement with the experimental data. For example, in Figs. 13 and 14 the diagrams of FRFs related to the load factor at the wing tip (Nw) and to the wing root bending moment (Mb) are shown with reference to the impact of symmetrical harmonic deflection of regular

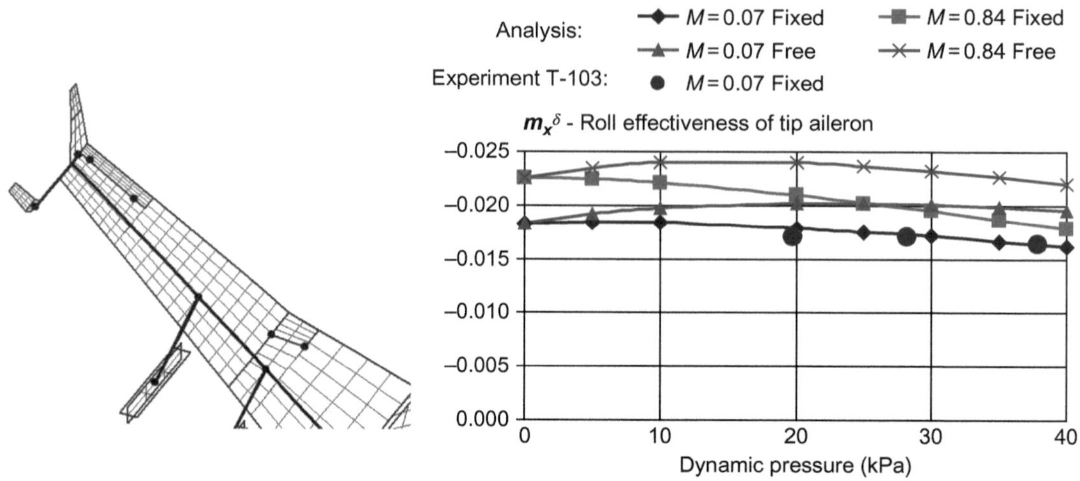

FIG. 12

Scheme of wing tip aileron (left), and roll effectiveness (right).

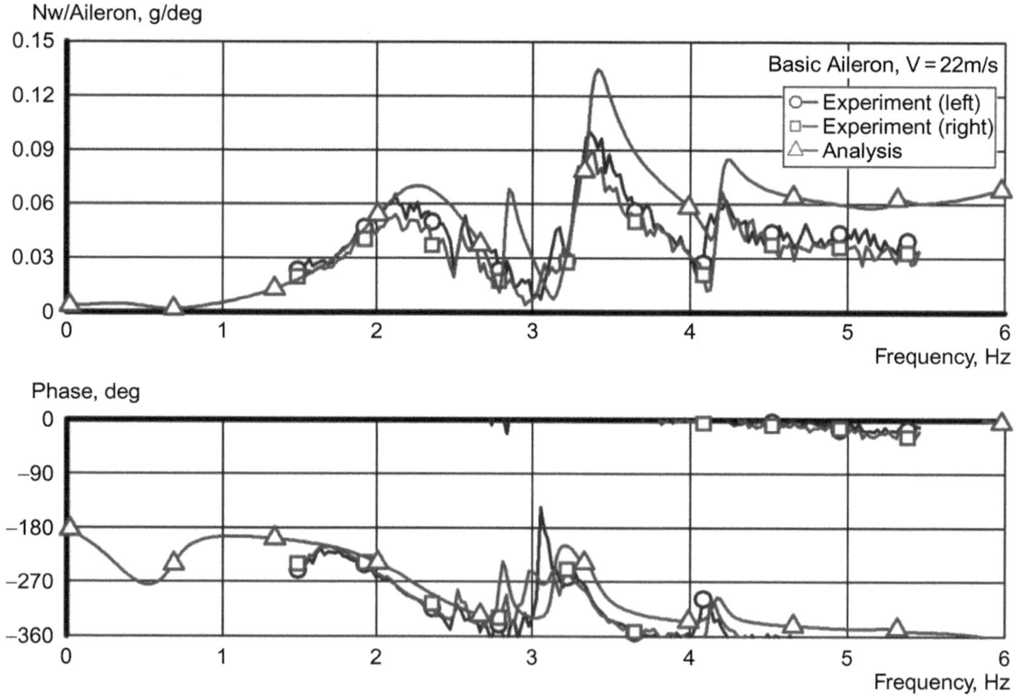

FIG. 13

FRF for load factor at the wing tip due to regular aileron oscillation.

ailerons, at airflow speed $V = 22$ m/s. These figures show that analysis characteristics well agree with the experimental data both in amplitude and phase. Some of the mismatch in amplitude can be explained by the well-known peculiarity of linear panel aerodynamics that somehow amplifies the lifting properties. Also, some difference in structural damping may have to be considered. It is interesting to compare the dynamic effectiveness of different wing control surfaces: regular (basic) aileron, tip aileron (TA), and aileron on a pylon under wing (UWA). Their effectiveness for the gust load alleviation system in time domain was studied in Ref. [5]. Here we consider the effectiveness of the AWTC on wing root bending moment in frequency domain for different airflow speeds (Fig. 15). The dynamic effectiveness of the basic aileron remains greater at airspeed $V = 22$ m/s, but at $V = 30$ m/s the AWTC has a considerably higher effectiveness in the frequency range of the first natural elastic mode. The regular aileron has a higher effectiveness in the frequency range of higher elastic modes.

2.5 ALL-MOVABLE VERTICAL TAIL CONCEPT

The objective of the concept was to develop and validate novel approaches for increasing the effectiveness of the vertical tail surface through the use of an adaptive attachment. All-movable vertical tails are a known design feature [6]. They were already used in the early days of supersonic flight. For example, they were used on famous airplanes such as the XB-70 and SR-71. Today, the upper part

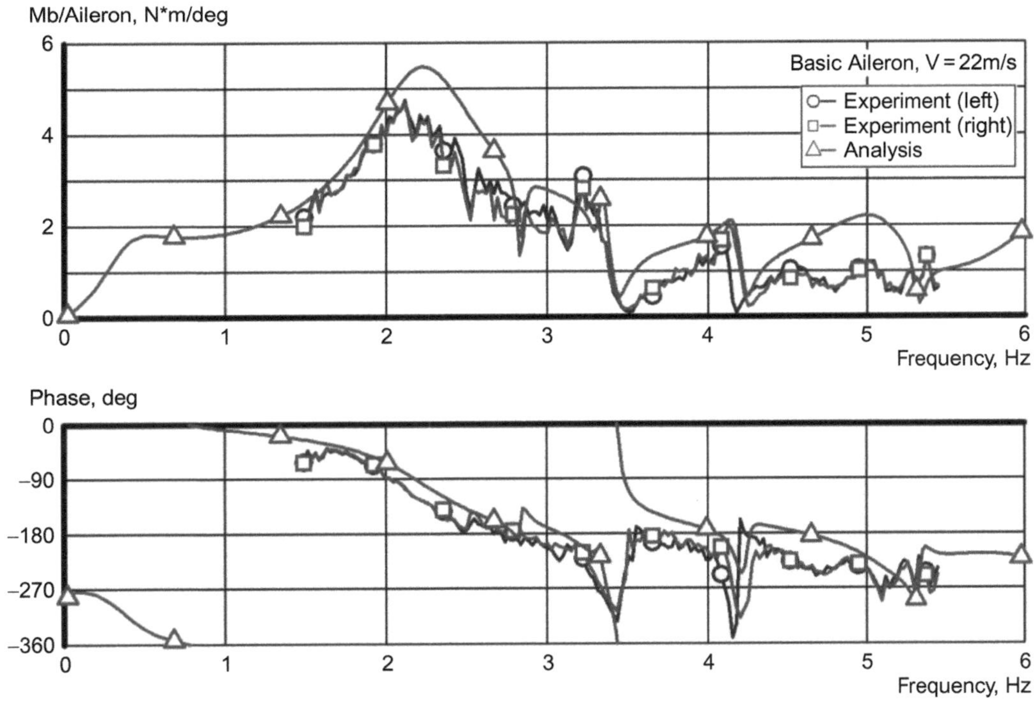

FIG. 14

FRF for wing root bending moment due to regular aileron oscillation.

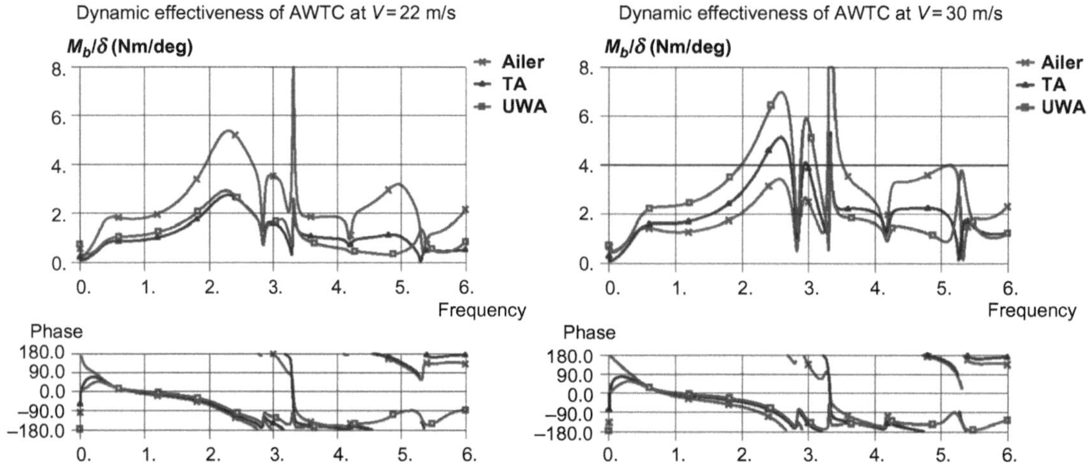

FIG. 15

Dynamic effectiveness of AWTC at $V = 22$ m/s (left) and $V = 30$ m/s (right).

of the vertical tail (VT) of the F-117 is all- movable. An advantage of the adaptive attachment stiffness concept for all-movable tail surfaces is that the size of the tail surface can be reduced by a factor corresponding to the chosen aeroelastic effectiveness increase. The same design failure criteria with respect to flutter can be applied as on conventional designs.

Finite element (FE) models of the traditional vertical tail with rudder, new all movable fin, and of the complete EURAM airplane were created for multidisciplinary optimization and analysis. In addition to the FE models (in NASTRAN format), the domestic TSAGI software, ARGON, was used for the design and optimization of the shape and attachment stiffness of the AMVT. The main design requirements were defined as follows.

- area of the AMVT is equal to 65% of the area of a basic VT
- the same position of 25% MAC point for basic VT and AMVT in *X* direction
- the same aspect ratio
- the same profiles

Minimizing the AMVT structural weight and accordingly decreasing its stiffness properties, have been restricted by flutter safety requirements in the considered range of rotational stiffness. A structural optimization procedure was performed, taking into account aeroelasticity constraints, by using ARGON code in order to determine the geometry and stiffness parameters of the AMVT [7]. The photos of the two vertical tails are shown in Fig. 16. Geometric parameters are compared in Table 1.

The aerodynamic side force was computed for an elastic VT, which was connected to the fixed point through rotational springs of different stiffness. The side force increases due to rotational stiffness for axis positions >30% C_{MAC} and slightly decreases due to the VT's own elasticity at high speeds (Fig. 17). Fig. 18 shows the required stiffness for a side force efficiency equal to 1.5.

FIG. 16

Comparisons between the two EURAM vertical tails.

Table 1 Vertical Tail Geometric Parameters

	VT	AMVT
Root chord (m)	0.900	0.74
Span (m)	1.00	0.8
χ_0 (degs)	46.86	38
Tip chord (m)	0.250	0.19
C_{MAC}	0.636	0.52
X (25% MAC)	0.59	0.59

Comparative flutter characteristics of the isolated AMVT and the complete model with the AMVT are shown in Fig. 19. Anti-symmetrical oscillations of the complete DSM change their behavior in the airflow in the presence of the adaptive attached AMVT. Dependence of divergence and flutter speeds on the AMVT rotational stiffness, G, are presented for a rotational axis position of 40% MAC. Two additional flutter modes have appeared in the case of the complete airplane model. Low frequency flutter mode—AMVT rotation + rigid body yaw—appears for the complete EURAM

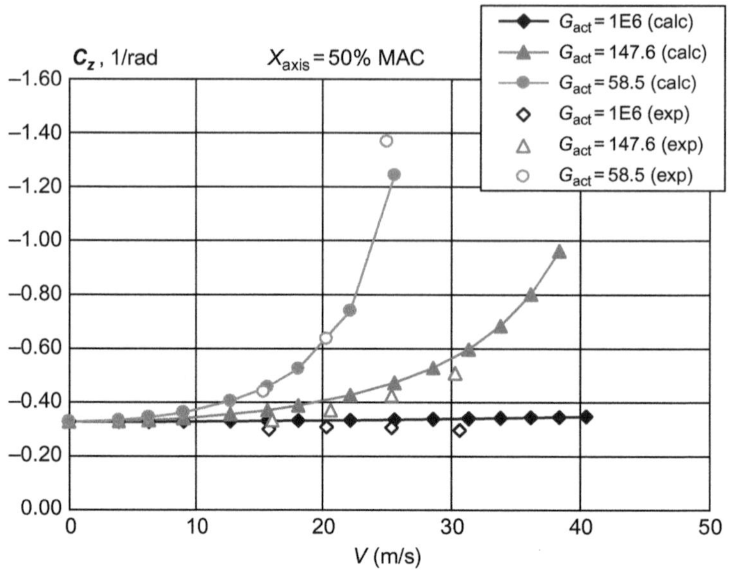

FIG. 17

Effectiveness of AMVT in side force for different values of rotational stiffness.

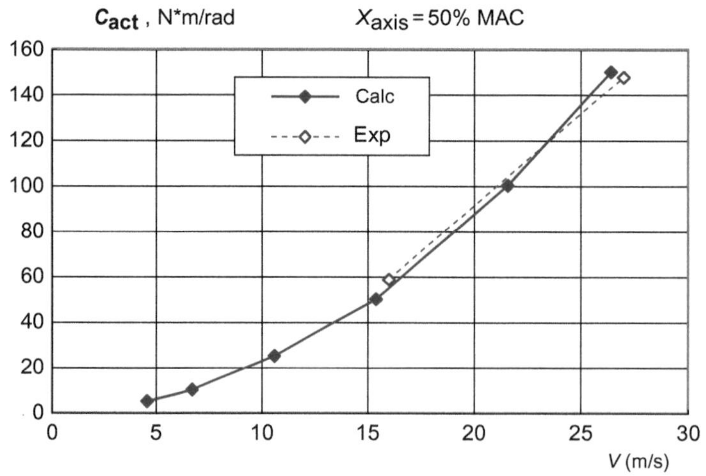

FIG. 18

Experimental and calculated results of requirements for the rotational stiffness to provide the side force efficiency of 1.5.

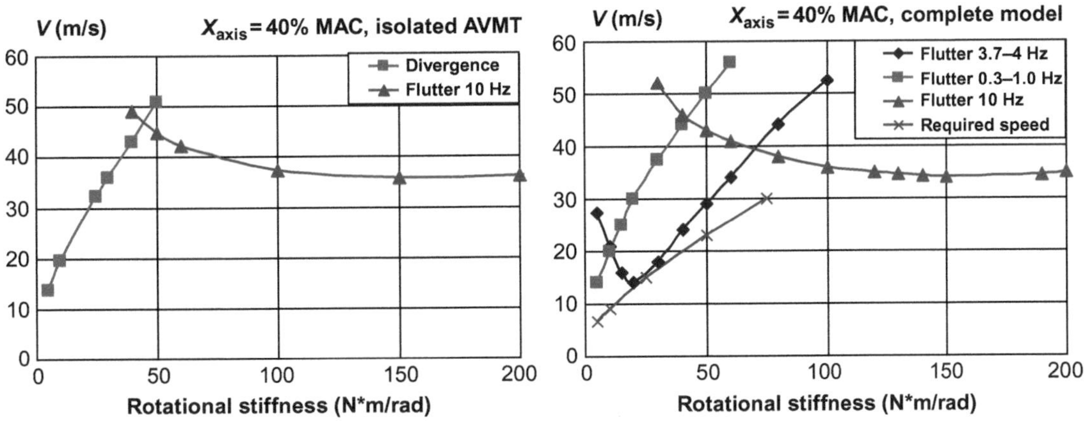

FIG. 19

Flutter and divergence of isolated AMVT (left) and of complete AMVT (right).

DSM (Flutter 0.3–1.0 Hz) instead of divergence for the isolated AMVT at low rotational stiffness. Due to the interaction of the rotational oscillations of the AMVT with the second antisymmetrical wing bending mode and fuselage bending, a new flutter mode (Flutter 3.7–4.0 Hz, Fig. 21) appeared. Fig. 19 (right) also illustrates the required flow speed versus rotational stiffness at which side stability and controllability is increased by 1.5 times. It can be seen that the flutter margin is not sufficient for stiffness in the range of 20–30 Nm/rad. Augmentation of the flutter margin can be reached by using active damping in the AMVT actuator. Comparisons of the numerical and experimental results are represented in Fig. 20.

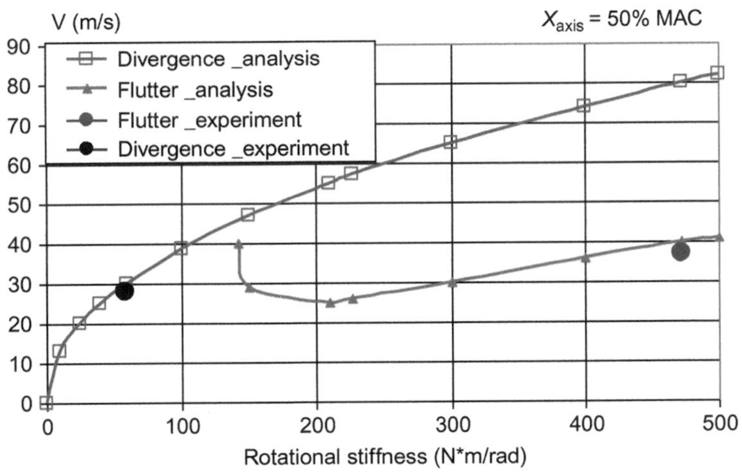

FIG. 20

AMVT divergence and flutter speeds versus rotational stiffness.

2.6 SELECTIVE DEFORMABLE STRUCTURE CONCEPT

The objective of the new passive structural design concept is to develop and investigate a slotless connection using smart elements of a selective deformable structure [8,9] that allows large continuous deformations of a load-carrying structure. A flexible connection of the aileron leading edge with the wing box was designed for the dynamically scaled model: EURAM. In addition to the ordinary (regular) inner aileron, the adaptive aileron was designed (Fig. 21). The static aeroelasticity properties of the adaptive aileron were studied experimentally and theoretically. The main attention was given to the behavior of the wing, where the adaptive aileron was connected.

Wind tunnel tests of the EURAM wing model with the adaptive and ordinary inner aileron were performed in the TSAGI subsonic wind tunnel T-103. The adaptive aileron was fabricated using SDS technology. Fig. 22 shows the wing with WT T-103 equipment (external strain gage balance of the wing).

Calculated and measured values of rolling moment due to the deflection of the ordinary inner and outer ailerons are presented in Fig. 23 (left). The main advantage of using an adaptive aileron structure is the increase of control effectiveness. The lift coefficient derivative with respect to control deflection for the wing compartment with an adaptive aileron is about 20% higher than that of a wing with a regular aileron; this leads to a better lift to drag ratio, which is about 15% higher. Fig. 23 (right) shows comparisons between an experimental lift to drag ratio of the wing with regular and adaptive inner ailerons.

The research leading to these results has received funding from the fifth framework program of the European Community for research, technological development, and demonstration activities under project ID G4RD-CT-2002-00679.

FIG. 21

Adaptive aileron.

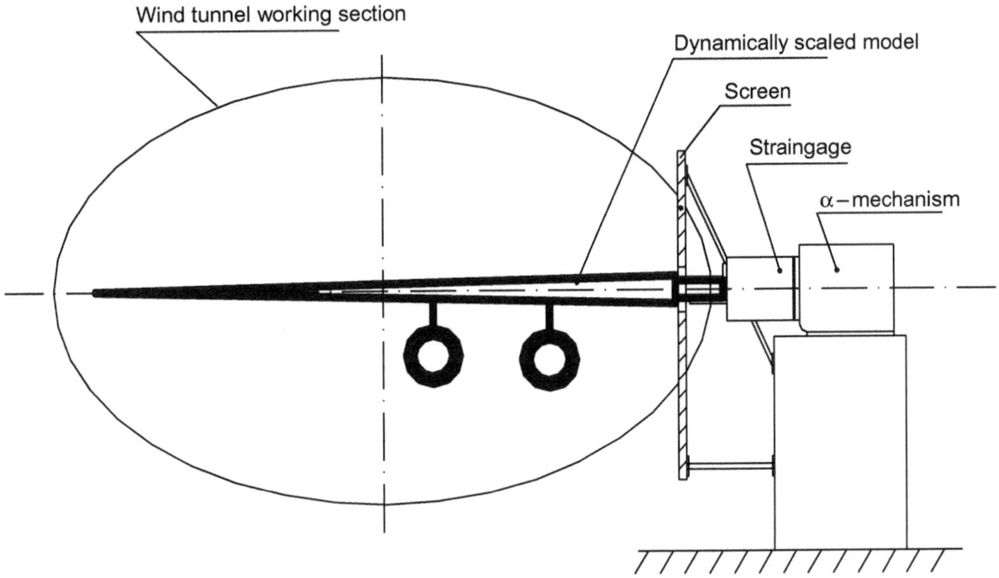

FIG. 22

Schematic of the EURAM wing in T-103.

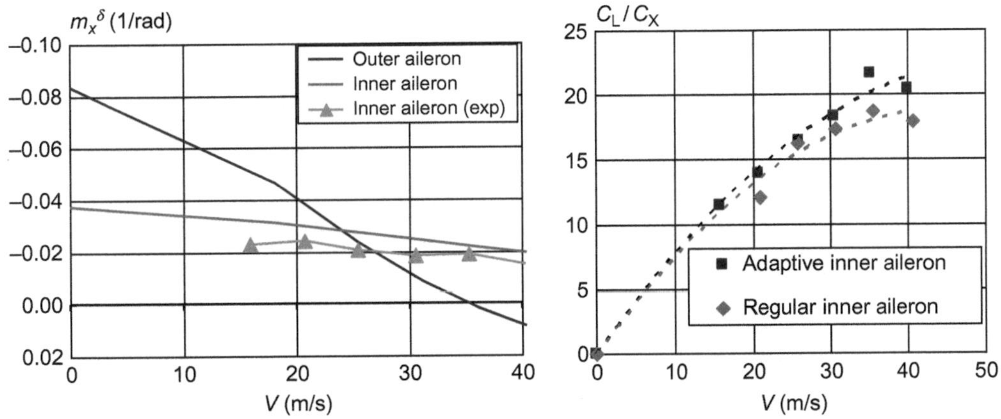

FIG. 23

Rolling moment coefficient of EURAM wing (left) and lift to drag ratio of the wing compartment (right).

3 SADE

The main task of the SADE project was the investigation of the next generation smart, high-lift devices to enhance fuel efficiency while decreasing noise and pollutant emissions at all flight regimes, especially during takeoff and landing. Some concepts of adaptive control were considered in the project [10–14]. A significant step in the investigation of the most promising concept was represented by the manufacturing and testing of a large-scaled demonstrator in the TSAGI big size, low-speed T-101 wind tunnel [15].

To measure the deformations of smart leading edge (SLE) upper and lower surfaces, the noncontact videogrammetric method and the double-channel measuring system have been developed. Digital cameras were located inside the left endplate above and under the wing. An additional controller based on compact computer in the left endplate was used to transfer distance image to the image acquisition system.

3.1 WING DEMONSTRATOR

The demonstrator represents a two dimensional wing with a 5 m span and a 3 m chord mounted on a T-101 remote-rotate base. It was installed horizontally in the work section on the upper structure of the wind tunnel and was limited by end plates. The wing box includes two metal spars connected to upper and lower milled metal panels. Three sections of the smart leading edge (SLE) were attached to the front spar. Each section included a flexible composite skin that was deflected by a drive mechanism during wind tunnel tests. The skin of the smart leading edge was fabricated using glass fiber reinforced plastic. A single-section flap was mounted on the rear spar. The deflecting angle of the flap was set up by means of rigid replaceable coupling plates. The setup of the demonstrator in the test-section of the wind tunnel T-101 is presented in Fig. 24.

The photo of the experimental setup is shown in Fig. 25. The constituent parts of the demonstrator are: a flexible adaptive nose (leading edge), the wing body (central power part of the wing), and the wing trailing-edge flap. Two flow shutters were mounted at the ends of the wing-demonstrator. The demonstrator was installed on the racks of the aerodynamic balance, which were placed above the control cabin. The nose part had a width of 650 mm along the chord, and consisted of three separate sections along the model span. The first one had a length of 2 m in the middle of the wing and the others spanned 1.5 m from each tip of the wing model.

3.2 VIDEOGRAMMETRY METHOD OF DEFORMATION MEASURING

The method of digital videogrammetry has been developed and used as an effective solution for noncontact measurements of geometric shape parameters and deformation of a model within a wind tunnel flow [15,16]. Implementation of the method involves placing a special optical system in a wind tunnel or a test bench, which is usually fixed and rigidly connected to the coordinate system of a tunnel or a bench. However, there is a problem regarding the measurement of deformation of large-scale models in large wind tunnels and in field conditions. The optical system in this case cannot be considered as stationary. Furthermore, the area of measurement of a large-scale model usually cannot be covered by a single optical system— two or more channels of the measuring system are required.

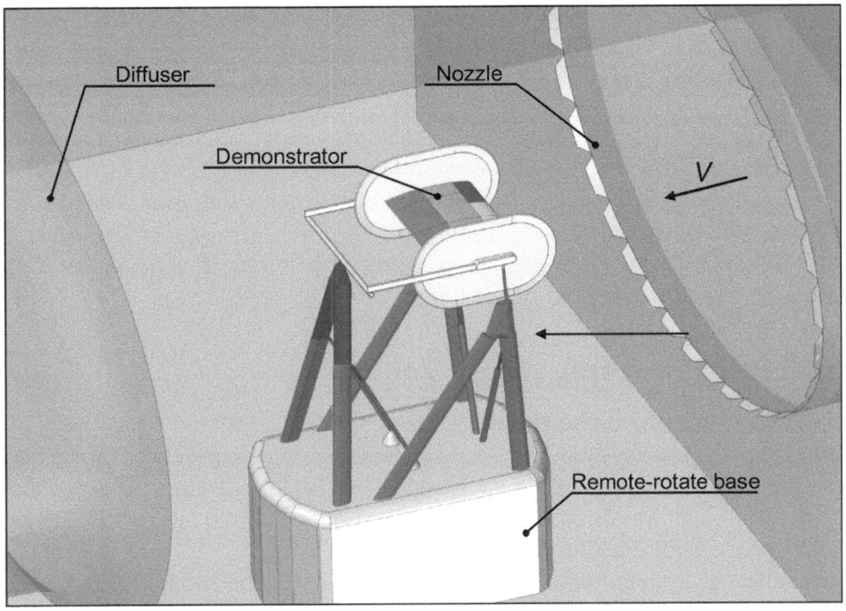

FIG. 24

Setup of the demonstrator in the working section of the wind tunnel T-101.

FIG. 25

Experimental facility in working section of wind tunnel T-101.

The aim of this work was to develop and apply the videogrammetry method for measuring the deformation of a large-scale wing model in TSAGI's wind tunnel T-101. This goal was achieved by accomplishing the following tasks:

- generation of a special multichannel optical system, providing simultaneous registration of two or more coherent images of different parts of the model surface;
- development of methods and means of mutual linking (group calibration) of several videogrammetry channels; and
- development of methods linking coordinate systems of different, in general, mutually movable, cameras with the coordinate system of the model.

3.3 TEST OBJECT AND EXPERIMENTAL FACILITY

The wing demonstrator scheme is shown in Fig. 26.

The origin of the coordinate system was placed at the point of the wing chord on the front wing body cut along the vertical plane of symmetry of the model. The *Ox* axis was directed along the chord in the stream direction, the *Oy* axis was directed upwards and perpendicular to the chord, and the *Oz* axis was placed along the span toward the direction of the cameras.

Three interconnected systems of markers were applied on the model surface (Fig. 27). Two measuring systems of markers were located on the top and bottom surfaces of the nose part within 10 sections, with values of z coordinate 1300, 1050, 950, 475, 0, 475, 950, 1050, 1600, and 2480 mm. Each section contained 14 markers. The first marker was located on the wing box at point $x = 25$ mm and the following ones along the profile arc to the nose with 50 mm step. The markers had an elliptical shape. Their sizes (along minor axis from 3 to 9 mm) and orientation were estimated so that markers in the images would have the form of circle with a diameter of about 5 pixels. The reference system of markers included two lines: the first line consisted of markers that were already

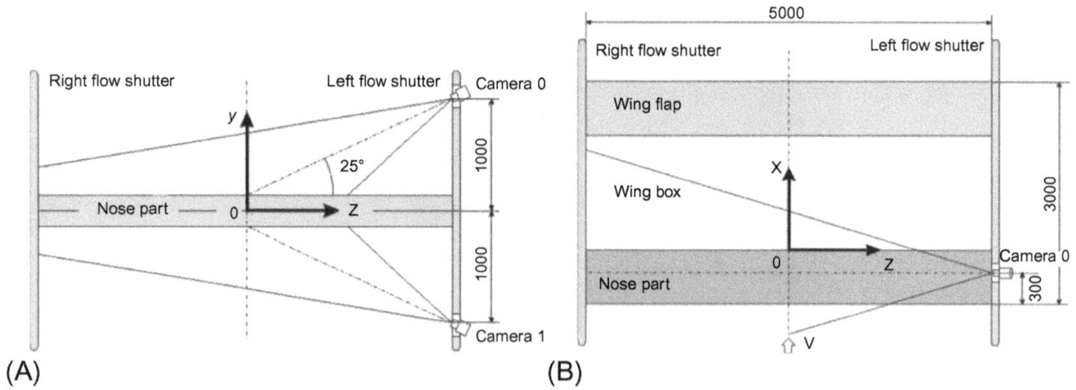

FIG. 26

Overall layout of demonstrator and camera placement. (A) Front view. (B) Top view.

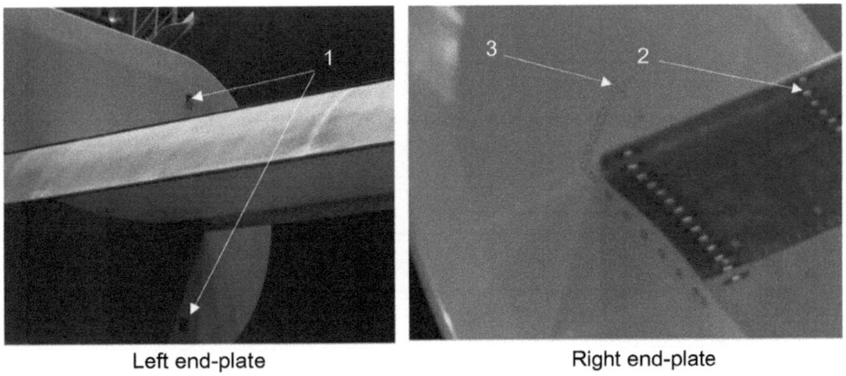

FIG. 27

Location of the cameras and marker system (1—location of the cameras, 2—measuring markers, 3—supporting markers).

on the top and bottom of the wing box at points with x coordinate 25 mm; the other line included a group of additional markers placed on the surface opposite (right) the flow shutter, which was visible by the two cameras. The coordinates of all markers were measured with a manual measuring instrument.

A regular lighting system of the test section of the wind tunnel served as a source of continuous illumination for the wing surface.

The main targets of the wind tunnel tests were to take measurements of the pressure distribution for three wing cross-sections, SLE skin deformations, and SLE structural strain under aerodynamic loading.

3.4 MEASURING PROCESS AND DATA HANDLING

The measurement of the wing model deformation was carried out on the SLE in a deflected position (takeoff and landing) and undeflected position (cruising regime). In each case, the measurements of the x, y coordinates for all the markers at the angles of attack of -10, -5, 0, 5, 10, 15, 19, and 22 degrees were carried out. The coordinates of undeformed model markers were preliminarily measured for all angles of attack without flow in the wind tunnel, and then the measurement was repeated at the same angles at flow speeds of 30, 40, and 50 m/s.

Image processing was performed using a set of standard and special programs. During processing, the following steps were followed: measurement of coordinates of marker projection centers on each image; calculation of coordinates of corresponding points on the surface; evaluation of deformation parameters by subtracting the coordinates of points on the surface in a non-flow state from the coordinates in the current with flow state. The results of the measurements of the SLE profile shape in deflected and undeflected states without flow are shown in Fig. 28.

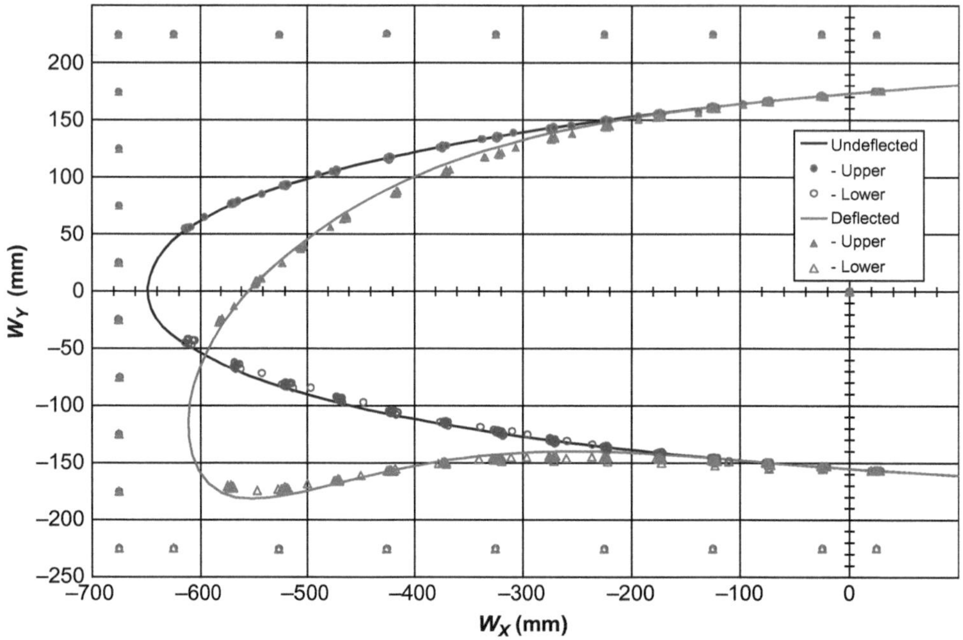

FIG. 28

Results of measurements of SLE shape in deflected and undeflected states.

The results of the measurements of the surface deformation, defined as the deviation $\Delta x0$ and $\Delta y0$ of the corresponding coordinates in the with-flow state from the undeformed non-flow state at each angle of attack, were presented in the form of two-dimensional matrices on a rectangular grid with the purpose of three-dimensional imaging capability of deformation fields. An example of a visualization of the deflection fields along the *Oy* axis of the upper and lower surfaces in the most deformed state of the test mode (with the undeflected SLE at angle of attack of 22 degrees and flow speed of 50 m/s) is presented in Fig. 29.

If compared to the model size, only small displacements due to aerodynamic loading were found. Longitudinal wavy variations of vertical deviations of the surface, which correlate with the configuration of the structure supporting the skin, were observed in all test regimes. The formation of ledges were also observed at the adjacent section joints of the wings leading edge, reaching maximum values of 1 mm on the lower surface on the right joint, and 1.5 mm on the left one.

- The research leading to these results has received funding from the European Union's Seventh Framework Programme for research, technological development, and demonstration under grant agreement no 213442.

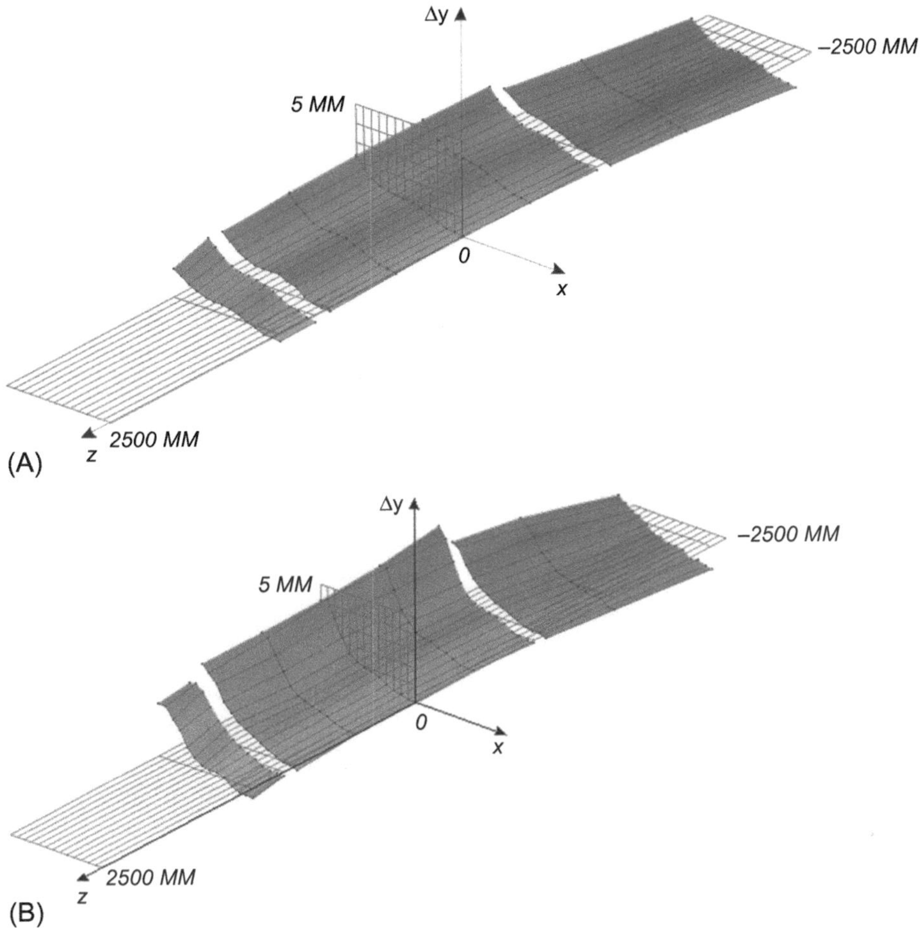

FIG. 29

Three-dimensional presentation of deformation fields of top (A) and bottom (B) sections of the SLE surfaces.

4 SARISTU

The SARISTU project was focused on the development of morphing technology and intelligent aircraft structures. The research received funding from the European Union's Seventh Framework Programme for research, technological development, and demonstration under grant agreement number 284562. Within the framework of this project, the full-scale wing demonstrator equipped with morphing devices was developed and manufactured by partners working Application Scenario's AS01, 02, and 03 [16]. The actual wing size mockup is presented in Fig. 30. The wing had slitless control surfaces, with, namely, an enhanced adaptive drop nose (EADN [23]), an adaptive trailing edge (ATE [24]), and a winglet active trailing edge (WATE [23]).

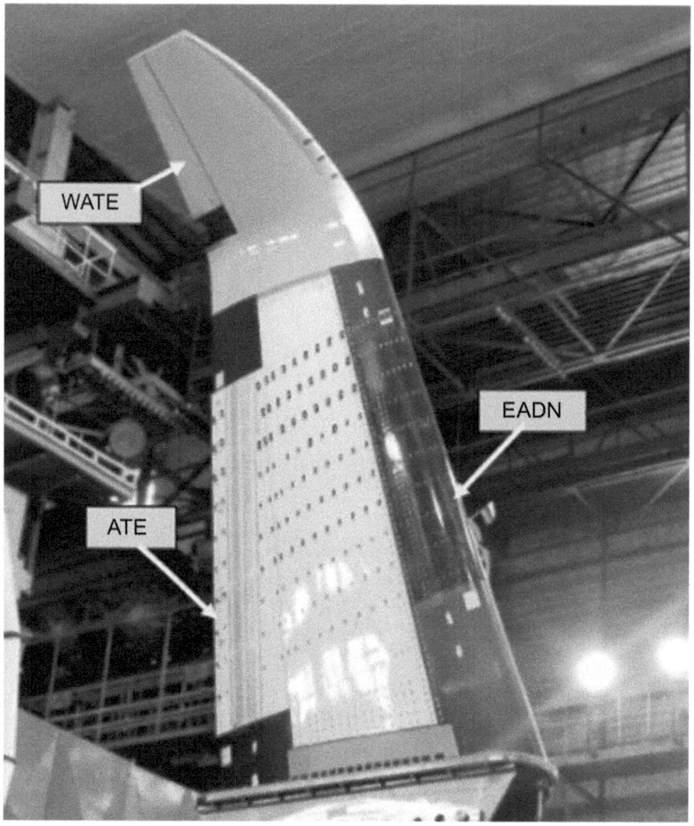

FIG. 30

The wing demonstrator.

The tests of the wing demo were carried out in the large, low-speed T-104 wind tunnel (WT), TSAGI, in Russia (Fig. 31). The main WT characteristics are listed below:

- WT nozzle diameter: 7 m
- open test section
- test section length: 13 m
- flow speed: 10–100 m/s
- total pressure P_t and temperature T_t correspond to ambient atmosphere
- Reynolds number per 1 m: up to $8 \cdot 10^6$

4.1 OBJECTIVES OF THE WIND TUNNEL TEST

The objective of the wind tunnel test was to validate the integrated morphing concepts functionality in a realistic loading, full-scale environment. In this case, the wind tunnel test at low speed was considered to be the best compromise between the costs of a full-scale, high-speed wind tunnel and safety, while

FIG. 31

Large low-speed T-104 WT.

also reducing the demonstrator structural loads, which reduced structural and aeroelastic issues and risks. In any case, the maximum speed level for this test had to be close to a real design functional flight envelope of the EADN. So, for this demonstrator, the functional test had to be very close to real flight conditions.

For ATE and WATE validation, these tests constituted a first step toward a higher technology readiness level (TRL) and were useful to collect information and to gain experience in perspective of future high-speed experimental campaigns not planned in SARISTU.

The wind tunnel tests objectives have been recapped below:

- Actuation, sensors and mechanical devices functional test. Morphing devices integrated into a "real" wing section, had to be actuated under loads during the wind tunnel test, simulating a realistic functional setting at different levels of real aero loading, which are variable with dynamic pressure, incidence, sideslip, and device shape.
- The actuation power and actuation loads initial predictions had to be realistically verified on the real mechanical system under real aerodynamic loadings. Devices where considered both fixed and actuated under loads simulating the possible morphed operative conditions.
- The morphing-achieved shapes quality had to be verified against actual aerodynamic loading conditions. This was extremely important for the EADN, as the wings leading edge is a critical region for leading lift level, wing stall, and for the general aerodynamic performance of the wing.
- Structural loads alleviation effectiveness (WATE) had to be proven in an open loop, simulating WATE tab dynamic deflections, with both low and high rates; and the internal loads within the structural elements (spars, skin, etc.) had to be measured in order to verify the finite element method (FEM) predictions.
- Structural loads had to be monitored in real time and, as per the previous bullet, effects of both external and internal loads had to be measured and loads modeling validated.

- The aeroelastic characteristics of the morphing devices had to be verified in the wing demo by a ground vibration test and by flutter limits expansion during wind tunnel testing. Moreover, a real time flutter monitoring system had to be active during all wind tunnel production tests. This to improve the confidence of the morphing mechanical and structural solutions with respect to aeroelastic instabilities.

The aims of the tests for TSAGI were

- to design and manufacture an appropriate test rig for a full-scale wing demo WT test;
- to prepare WT instrumentation and to provide synchronization with control and measuring systems of the partners involved;
- to carry out ground vibration tests;
- to carry out preliminary balance test at the air speed up to 50–60 m/s at acceptable (for WT balance) aerodynamics loads (lift $P \leq 20$ kN) in order to perform the preliminary wing box strain-gauges (SG) calibration, requested before the production WT runs;
- to carry out WT production runs without WT balance by using calibrated SG sections only for aerodynamic load control at acceptable lift $P \leq 80$ kN (for the test rig).
- to validate wing demo, control systems, and instrumentation functionality under WT test conditions;
- to measure real aerodynamic loads acting on the wing demo; and
- to measure the wing demo deformations.

The wing demo was adapted for the study of the morphing devices. The wing relevant size was limited by a wing span length of 4.695 m. The wing size was mainly chosen by taking into account the following three basic requirements:

- compliance of the demo structure and internal elements with the real wing properties;
- compliance with TSAGI's requirements for the models strength tests in the WT; and
- compliance with the model's requirements for positioning in the WT open test section under the conditions of maximum possible usage of available equipment and instrumentation.

The tests first-priority was to check and demonstrate the wing structures functional viability, but not the accuracy of the wing demos aerodynamic features. Thus, it was decided to increase the wings dimensions as much as possible to make it go beyond the requirements set for typical tests in the WT within the open test section. In this case, due to the oversized test benchs interaction with the WT flow, the reduction of accuracy reduction for the measurements of the aerodynamic features was evaluated by calculations.

The test rig for the wing demo tests was developed and manufactured by TSAGI. Fig. 32 shows the test rig 3-D model. The test rig was mounted on the WT cabin, Fig. 33. The wing angles of attack, AoA, was changed by rotating the WT cabin. The wing sweep angle, AoS, was changed within the range $-5° \div +10°$, with a step of 5°, by using the rig interface rotating system.

The WT test campaign consisted of three stages: the first stage was the preliminary GVT and flutter expansion test; during the second stage wing aerodynamic loads were measured for the calibration of wing SG; and, finally, in the third stage production runs took place for the validation of wing systems functionality. The following paragraph provides a general description of the methodology that was implemented by TSAGIs for the preparation of the production stage of the test. A summary of the most relevant results is also reported.

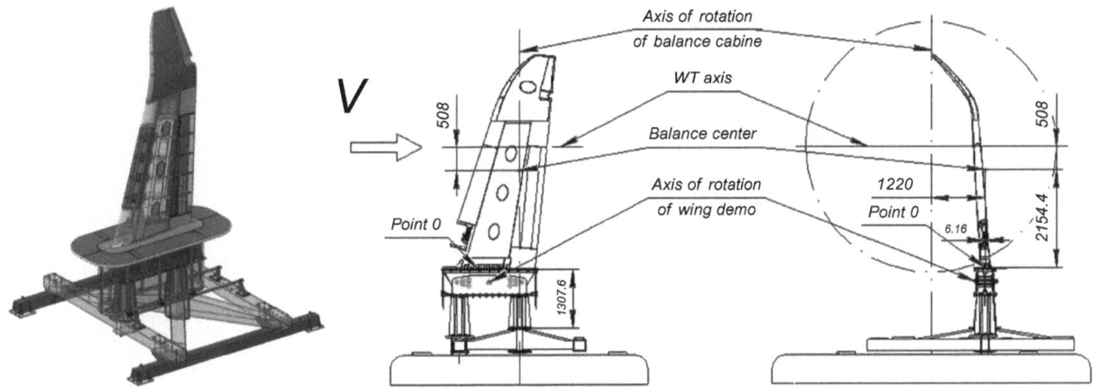

FIG. 32

Test rig.

FIG. 33

The test rig on the T-104 WT cabin.

4.2 GROUND VIBRATION TEST AND FLUTTER EXPANSION TEST

The modal parameters of the wing demonstrator were obtained during GVT. GVT was performed in two stages:

- First stage: demonstrator placed in T-101 WT area;
- Second stage: demonstrator installed in T-104 WT balance cabin.

The scheme and photo of the test setup are presented in Figs. 34 and 35. GVT were performed with the EADN and ATE surfaces set in their neutral position using the activation by the actuator. The WATE tab, instead, was locked by a mechanical link and, therefore, it was not possible to measure its proper resonance frequencies relevant to harmonic oscillations around its own hinge axis.

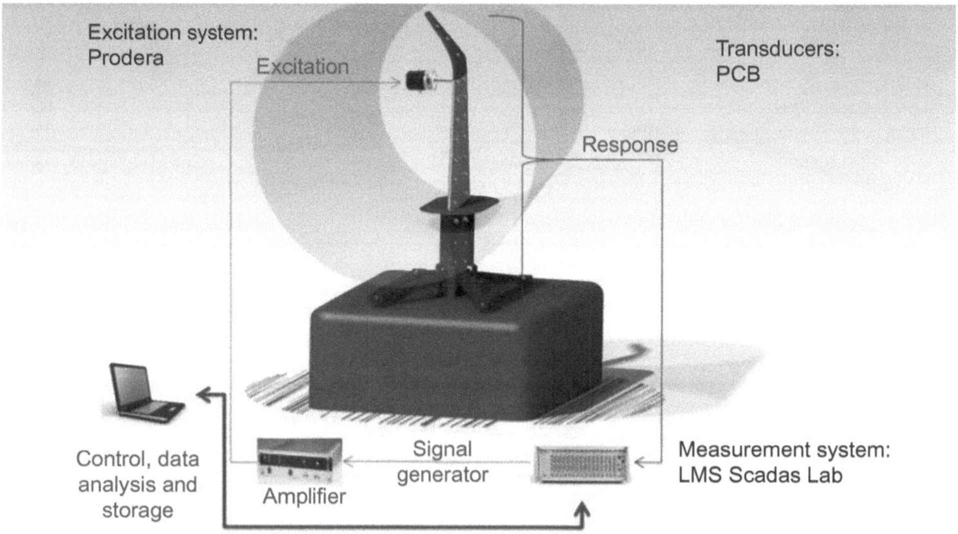

FIG. 34

Scheme of GVT measurements.

FIG. 35

Test rig configuration for GVT at T-104 WT.

GVT was performed with equipment provided by TSAGI. An LMS SCADAS Lab System was used for the generation of the drive signal and for data acquisition. For broadband excitation and analysis, a white noise signal was used. The modal frequencies of the test article were determined by the analysis of its response to a swept sine excitation from 5 to 80 Hz. The excitation force was given at a constant amplitude by using a single exciter during the whole sweep. The individual modes of vibration were investigated with appropriate excitation points. Modal parameters of each mode were investigated separately.

The wing demonstrator was excited with the PRODERA modal shaker that was suspended by means of low frequency springs. Seven different excitation points (Fig. 36) and 66 accelerometers were used.

Based on the obtained results, the final modal and flutter analysis was performed. Details of the GVT results can be found in the SARISTU technical documents [16].

Flutter expansion was carried out after the GVT test. The test article, fitted with the test rig used for GVT and for subsequent productive WT tests, was installed on the T-104 WT cabin. The first flutter WT test was carried out up to the cleared advisory airspeed of 50 m/s. Further expansion was performed starting from 50 m/s up to a Vmax of 102 m/s. The most important flutter accelerometers were taken under control in real time: the WATE, TAB, ATE, and one of the wing tips were selected. The relevant PSD (g^2/Hz), calculated in real time, was shown on the monitor. Since potential flutter criticality wasn't measured during flutter expansion nor was it foreseen by the analysis of experimental data, it was concluded that, in regards to the flutter aspects, all productive WT activities were allowed up to a Vmax of 102 m/s.

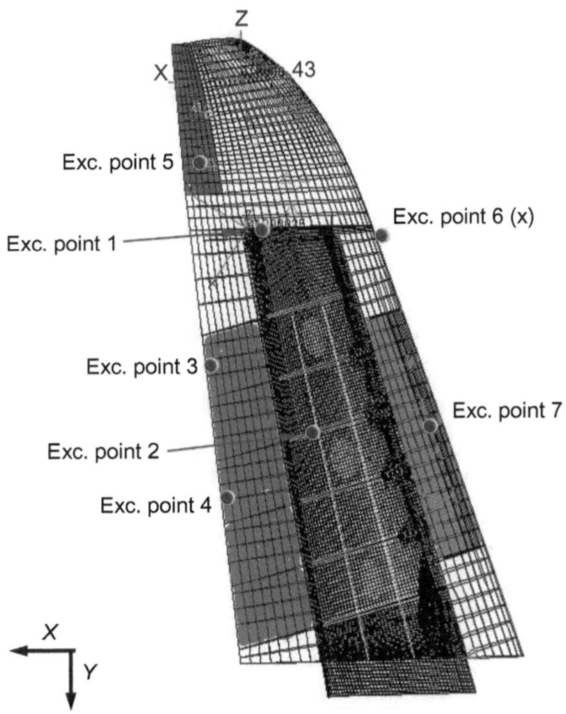

FIG. 36

Scheme of excitation points for GVT.

4.3 LOAD MEASUREMENTS

The main target of the preliminary stage of the test consisted in the measurements of the aerodynamic loads acting on the wing by the WT balance for the in-built strain gage calibration. During the calibration tests the measurements of drag, lift, and pitching moment were carried out. The aerodynamic load tests were executed within the velocity range of $V = 30 \div 60$ m/s at angles of incidence AoA $= -5° \div 15°$ and AoS $= -5° \div +10°$. The measurements of the aerodynamic loads were performed by the aerodynamic balances AB-104 M in the WT cabin; see Fig. 37.

Before the wing load tests, TSAGI performed the preliminary test for the evaluation of the aerodynamic forces (L_{ss}, D_{ss}) and moment ($M_{y\ ss}$) acting on the support system without the wing. The balance tests were executed at the flow velocity $V = 50$ m/s. During data processing, these loads were subtracted from the balances total readings L_{Σ}, D_{Σ}, and $M_{y\Sigma}$:

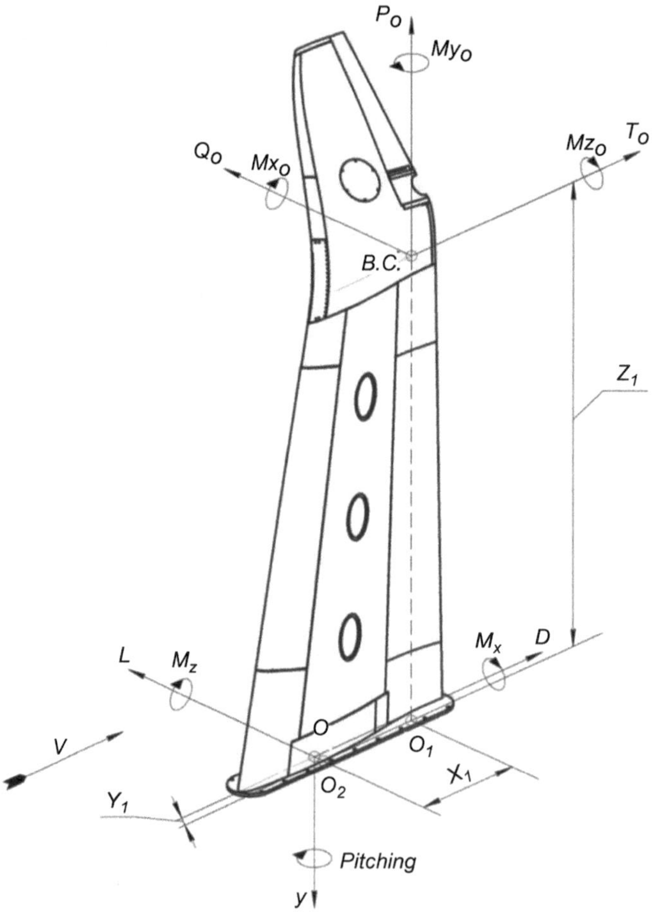

FIG. 37

Load measurement scheme and coordinate system.

$$L = L_{\Sigma} - L_{ss};$$
$$D = D_{\Sigma} - D_{ss};$$
$$M_y = M_{y\Sigma} - M_{yss}$$

The dimensionless aerodynamic coefficients reported below were used to validate the numerical models developed within the project:

$$c_{xa} = \frac{D}{\frac{\rho V^2}{2} \cdot S}, c_{ya} = \frac{L}{\frac{\rho V^2}{2} \cdot S}, \text{ and } m_{ya} = \frac{M_{ya}}{\frac{\rho V^2}{2} \cdot S \cdot b_a}$$

The coefficients were referred to the wing surface $S = 6.378\ \text{m}^2$ and the mean aerodynamic chord $b_a = S/l = 1.42$ m.

The graphs in Figs. 38 and 39 show that the wing baseline configuration has the typical aerodynamic features, namely the linear dependency of the lift coefficient from the angle of attack $C_{ya} = f$ (AoA). The maximal lift coefficient value at AoA = 12.5° resulted equal to 0.8. When changing the wing airfoil by deflecting the leading edge, trailing edge, and winglet tab, the $C_{ya} = f(\alpha)$ inclination line slightly increases and the lift force coefficient rises up to $C_{ya} \sim 0.9$.

During the loads calibration test campaign, 25 test runs were performed at flow speeds of 50, 55, and 60 m/s. The following configurations were tested:

- wing baseline configuration without morphing device deflections;
- leading edge deflected;
- trailing edge deflected (deflection angles +5° and −5°);
- winglet tab deflected (deflection angles +10° and −10°).

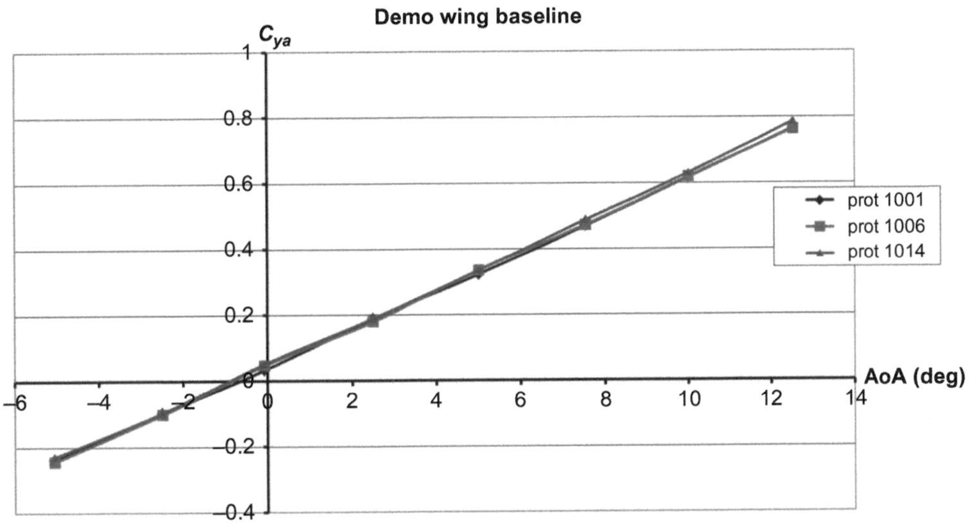

FIG. 38

Lift coefficient.

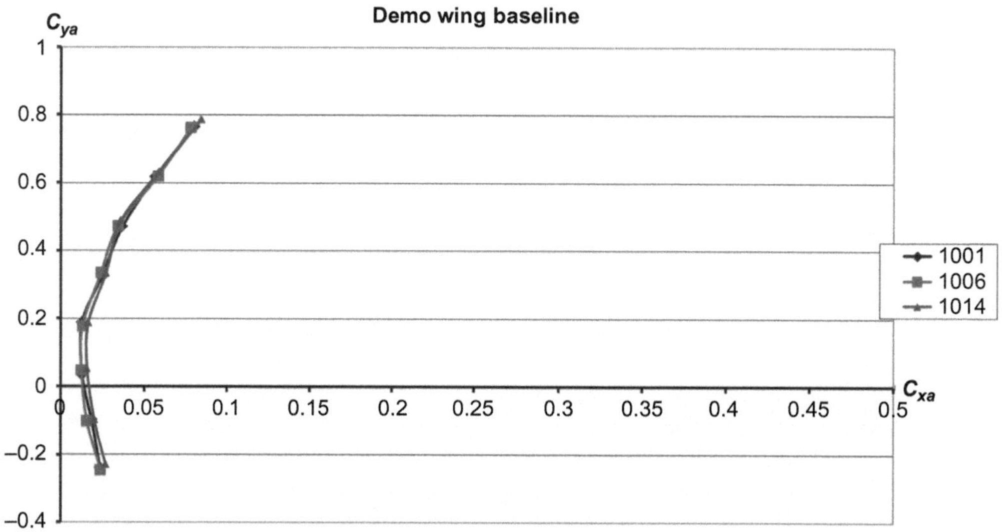

FIG. 39

Lift to drag coefficient.

The main aim of TSAGI's activity was to analyze the measured loads accuracy and the reliability of the WT tests results. The WTs raw data resulted in lower than expected loads. Therefore, TSAGI had to perform:

- balanced, iterative calibration after the load tests to check for the correct functioning of the acquisition system;
- detailed measurements of the WT flow and the test rig and the wing demo; and
- wing demo aerodynamic simulation with the test rig under the conditions of T-104 WT tests.

It was necessary to estimate the influence of the large size wing and rig on the following factors:

- flow velocity and flow downwash in the WT test section;
- actual AoA value at the test rig presence in the WT;
- possible change of the flow core boundaries;
- necessity of applying the corrections (and its possible value) to the obtained aerodynamic coefficients.

The WT flow measurements with the rig and wing were executed at a velocity $V = 50$ m/s, AoA $= 0$ and 10 degrees after the productive runs completion. The experiments were made using TSAGI's standard methodology. The six-point air pressure sensor was used. The sensor was moved by a sliding trolley in three directions X, Y, and Z. The initial point coordinates were in the center of the WT nozzle plane, Fig. 40. The dynamic pressures and flow downwash angles in the vertical and horizontal planes in front of the wing and nearby the flow boundaries (WT shear layer) were defined.

The analysis of the obtained data and their comparison with the empty WT showed that:

- The presence of the test rig and the wing demo slightly decreases the flow core dimensions compared to the empty WT. But a more significant transformation can be expected in the lower part of the WT flow due to the large rig support frame.

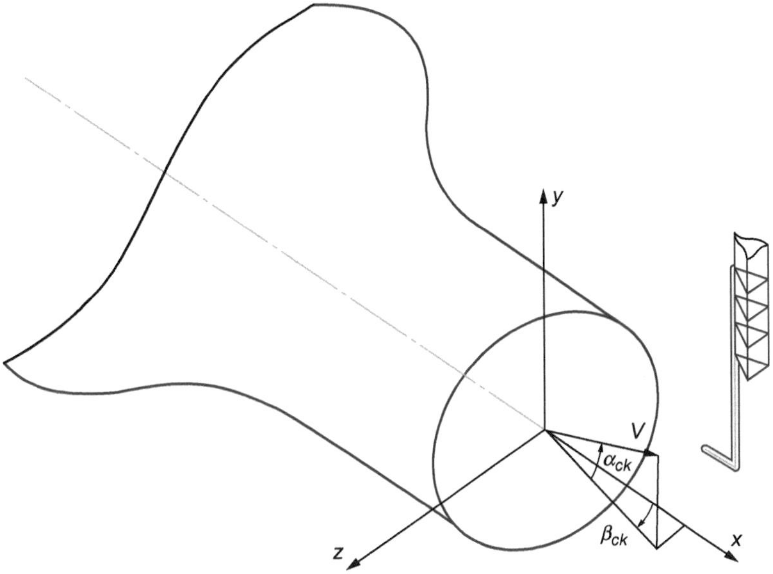

FIG. 40

Wind tunnel flow field measurements scheme.

- The actual reduced dynamic pressure coefficient $\mu = q/q_{\infty}$ is lower than the one registered in case of empty WT. This coefficient goes down from $\mu = 1$ (for empty WT) up to $\mu = 0.97 \div 0.98$ in front of the wing center and its upper parts at $X = 3.5$–4.5 m, Fig. 41. The difference between the actual coefficient μ from its reference value $\mu = 1.0$ shows some flow deceleration due to the large size of the wing and test rig if compared to the WT flow core.

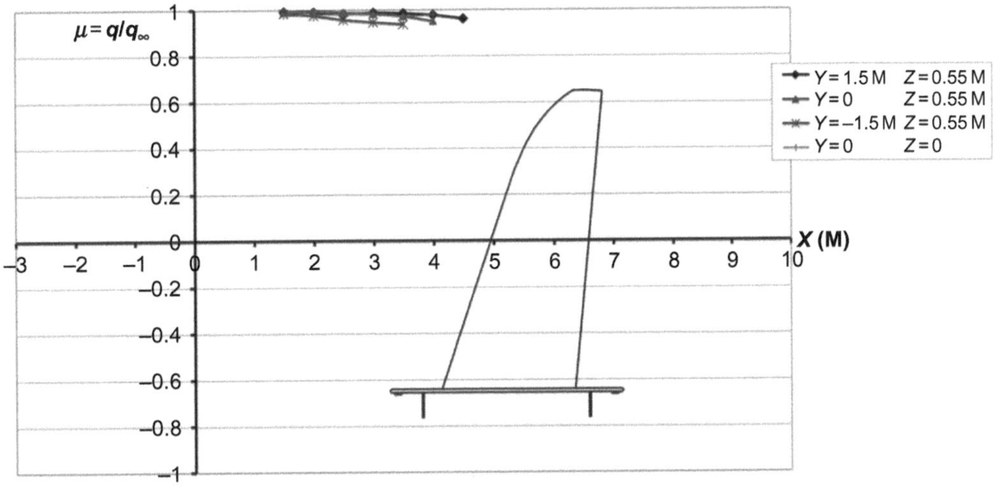

FIG. 41

Distribution of dynamic pressure coefficient $\mu = q/q_{\infty}$ at $V = 50$ m/s at AoA = 10 degrees.

- When the wing demo is installed in the WT, the actual flow angularity angle on the horizontal plane tends to increase in comparison with the empty WT flow core angle, even at a distance longer than the chord of the wing (up to 1.5 degrees at AoA = 10 degrees). The flow angularity growth means that the wings actual AoA decreases in correspondence of this value ($\triangle$AoA $\leq$ 1.5 degrees depending on AoA); more detail about this have been reported in the following.

The obtained data showed the necessity of angularity corrections to the aerodynamic moments and force coefficients, presented in Figs. 38 and 39, not taking into account the actual dynamic pressure, q, and angularity deviations in the WT in the presence of a big test model.

4.4 CALCULATIONS OF WING DEMO AERODYNAMICS IN T-104 WT

The methodology and results of the simulation of the wing with the test rig in a T-104 WT environment are presented in this paragraph. All computations were performed by means of an in-house code EWT-TSAGI [17,18]. This software is used by TSAGI to generate the mathematical models of wind tunnels (e.g., ETW and Cologne) for detailed in-tunnel simulation.

The investigation was carried out mainly:

- to make a visualization of the WT test and to present the detailed structure of the flow around an experimental facility;
- to estimate the aerodynamic loads and to compare them with the test data;
- to provide corrections for test data; and
- to estimate possible reasons for the obtained difference between the actual and predicted loads.

Calculations were carried out in the framework of the Navier-Stokes system of equations, and were closed by the SST turbulence model. A second order numerical scheme TVD (GKR) in the form of an explicit scheme with implicit smoother was used in the solver in Ref. [19]. Structural multiblock computational meshes were used in the solver. The mesh was rebuilt automatically for different geometry configurations.

The geometry corresponded to the real wing demo. The computations were performed in two formulations: with WT (Fig. 42) and without it. In the case of in-tunnel computations, the basic elements of the test rig (WT nozzle and diffusor, WT cabin, rig peniche, and additional WT cabin stand) were taken into account. The computational domain boundaries in general corresponded to the WT building walls. A right-handed system was selected with the X-axis directed downstream, the Y-axis directed along the wingspan, and the Z-axis oriented upwards.

The in-tunnel simulation was performed with flow parameters as follows: Mach number $M = 0.147$ ($V = 50$ m/s); and incidence and yaw angles AoA = 0, AoS = 0.

Let's consider the variation of relative dynamic pressure $\mu = q/q_{\infty} = (P_t - p)/q_{\infty}$ at different Z-coordinates (Fig. 43), with P_t, local total pressure, and p, local static pressure. It can be seen that the WT jet boundaries in both the computations and experiments agree with a satisfactory level of accuracy. Mach number variation is evaluated by $\Delta M \sim 0.005$. Flow deceleration in the jet center occurs within both computation and experimental results.

The obtained results indicate that the additional left stand on the WT cabin, which is located near the model, as well as the position of the model itself with respect to the jet axis have a weak effect on wing

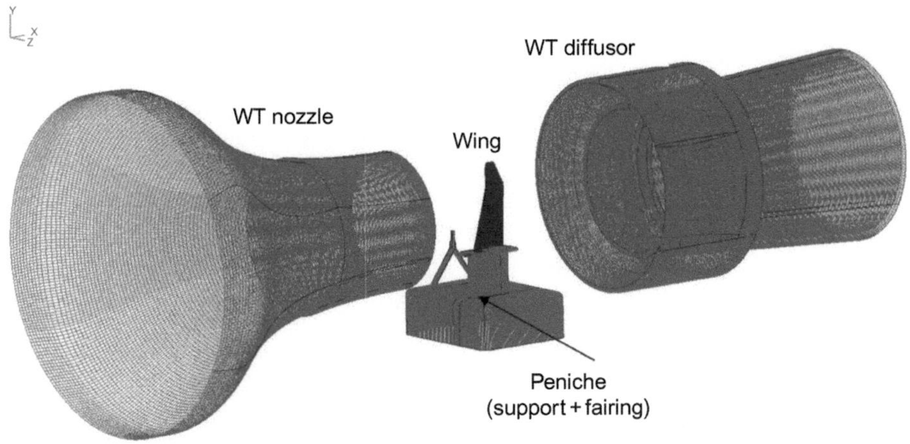

FIG. 42

Computational domain for in-tunnel calculations.

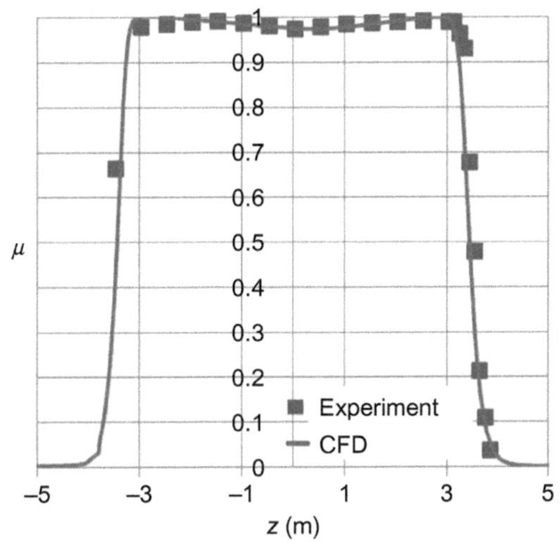

FIG. 43

Distribution of the relative dynamic pressure parameter μ along $X = 3.5$ m, $Y = 0$, wing installed, $V = 50$ m/s, AoA = 0.

characteristics (Fig. 44). Computations also indicate a small vortex flow near the peniche (Fig. 45). The estimation shows that it has no effect on wing characteristics.

The comparison between free-flight wing computations and the experiment was performed at flow parameters as follows: $M = 0.147$ ($V = 50$ m/s); $P_t = 102866$ Pa; $T_t = 289.4$ K; AoA—various; and AoS = 0. As shown in Fig. 46, there is a discrepancy between the computed free-flight conditions (TSAGI CFD) and the experimental (without corrections) values of the lift coefficient Cy and Cy^{α} (curve slope angle).

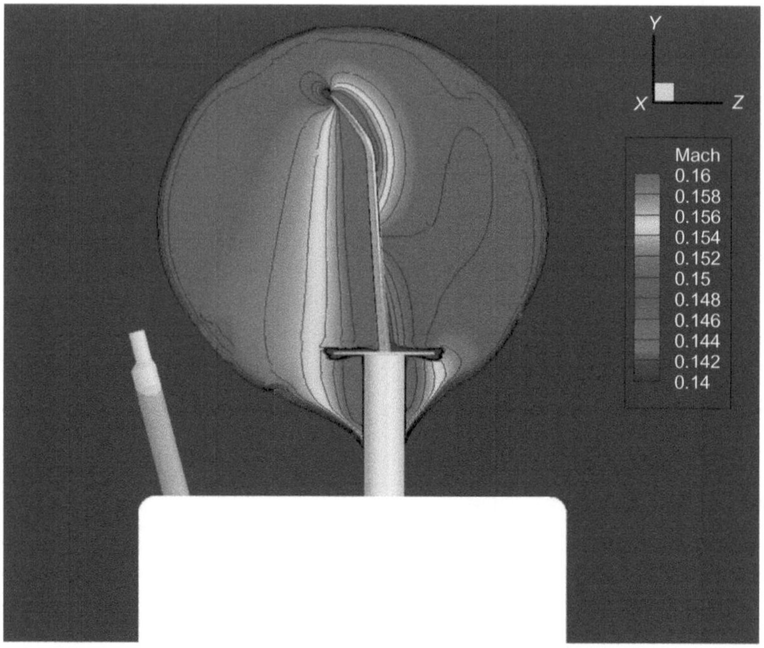

FIG. 44

Flow field of Mach number at section $X = \text{const}$, $V = 50$ m/s, AoA = 0.

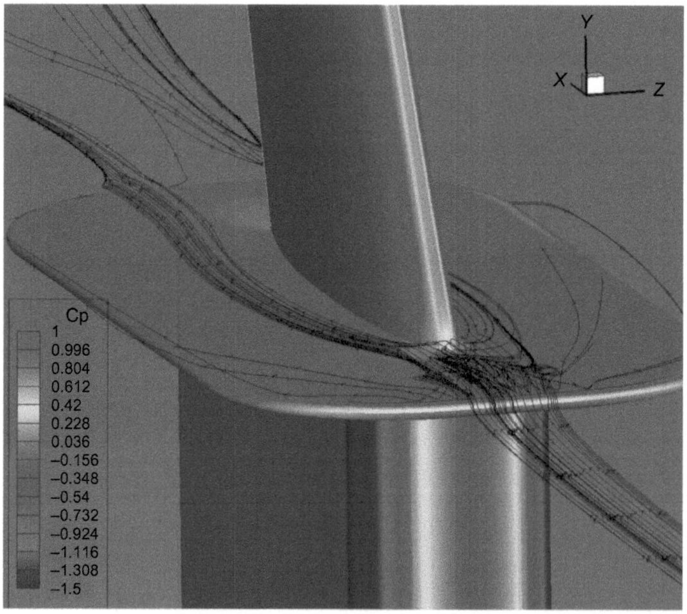

FIG. 45

Flow visualization near the peniche.

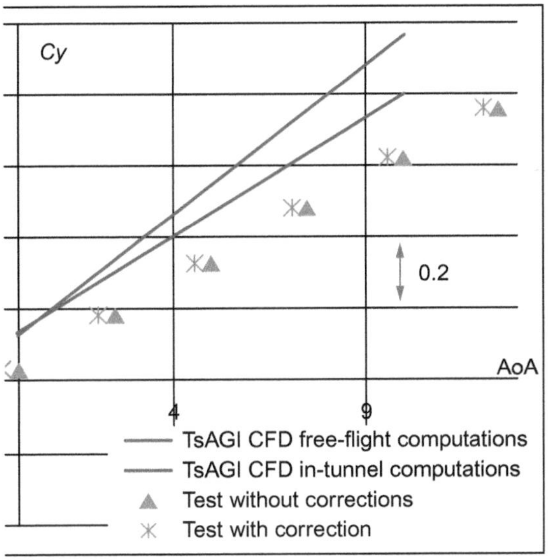

FIG. 46

Comparison of the computed and WT test data.

The main reason for this mismatch is the flow angularity due to interference with the WT core flow boundaries. According to the results of the calculation, flow boundaries interference leads to substantial transformation of flow direction near the wing (Fig. 47) and deviation of the Cy and Cy^{α} coefficients (Fig. 46, TSAGI CFD).

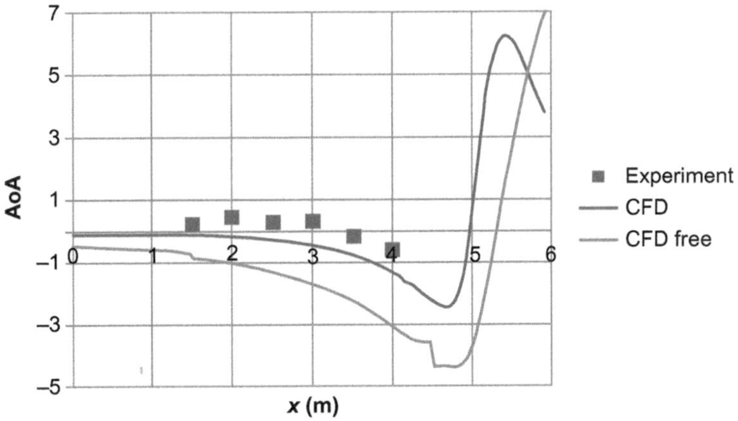

FIG. 47

Comparison of the local flow angularity on the horizontal plane for free-flight and for WT conditions at $Y = 0$, $Z = 0.55$ m for a wing installed at AoA = 10 degrees.

It is known [20–22] that the incidence angle correction caused by flow-boundary interference in the WT could be estimated as

$$\Delta \text{AoA} = \delta_\alpha C_y S/F \tag{1}$$

where S is the wing reference area, F is the WT nozzle cross section area, and δ_α is the interference coefficient. Usually values of δ_α for circle flow boundaries in a T-104 WT may vary from −0.125 to −0.25. For the incidence angle, AoA = 10 degrees, the previous values of the interference coefficients resulted in a decreasing AoA (0.7–1.5 degrees). This is valid in a case of moderate blockage of the WT. But in our case, the test rig and the wing model were oversized and the incidence angle correction obtained by Eq. (1) was underestimated. For obtaining the appropriate corrections, CFD computations were used. If it is assumed that the relation (1) is correct, the values of δ_α can be obtained by comparing the Cy and the corresponding AoA for in-tunnel and free-flight computations (Fig. 47). A new δ_α is chosen to make the slope of the CFD curve of Fig. 47 to match the experimental one. This leads to a new $\delta_\alpha = -0.35$ degrees. For example, for the incidence angle, AoA = 10 degrees, the obtained δ_α results in an angularity correction of ΔAoA = 2.1 degrees.

4.5 DEFORMATIONS MEASUREMENTS OF THE WING WITH ELASTIC CONTROLS IN WT T-104 FLOW

Practically important data about wing morphing parts deformation were obtained in the WT test. The videogrammetry system for simultaneous noncontact measurements of the deformation of elastic elements of wing mechanization was developed by TSAGI. The system had four optical channels. Channels 1 and 2 were intended to measure deformations of the upper and lower surfaces of the wing, comprised of enhanched adaptive drop nose (EADN), adpative leading edge (ATE), and wing box (WB). Channels 3 and 4 were intended to measure deformations of the morphing part of the winglet. The method of stereometry was applied by using two cameras for measuring the deformations of the wing.

Grids of markers were applied on the upper and lower surfaces of the wing demonstrator. The grid of markers was conventionally divided into three areas: EADN, ATE, and WB. Each cross-section of the EADN had 10 markers at approximately equal distances between each one. In total, there were 290 markers on each wing surface. Fig. 48 shows the marker points on all wing cross-sections on the Oxy coordinate plane.

A comparison of deformed (ATE deflection) and nondeformed state has been reported as example in Fig. 49 with reference to the wing cross-section located at 1.260 m from the wing root.

Estimates of instrumental measurement error (standard deviations) for channels 1 and 2 had given values of about 0.1–0.2 mm and 0.2–0.3 mm along the x and y coordinates, respectively. For the stereo system (channels 3 and 4) instrumental error was estimated to be, at most, 0.1 mm.

5 CONCLUSIONS

- Approaches for the experimental characterization of large morphing structures has been presented with reference to three main researches carried out by TSAGI. Three different classes of structures were addressed: structures with deformation levels controlled by specifically conceived movable surfaces, structures with adaptive stiffness and actively controlled shape-changing structures. For

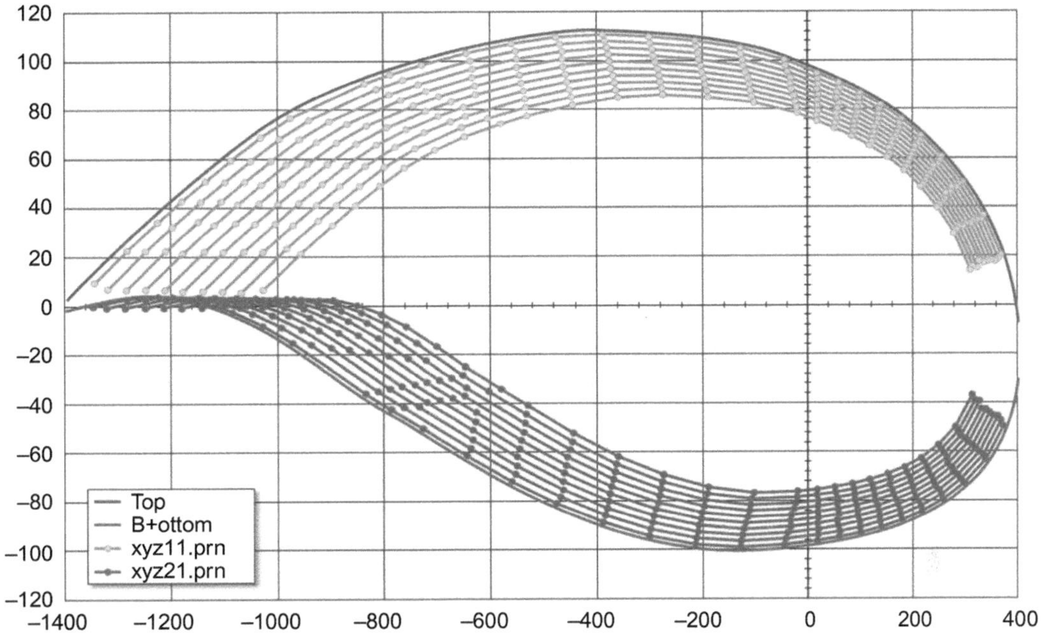

FIG. 48

Scheme of the layout of markers on the airfoils within cross-sections.

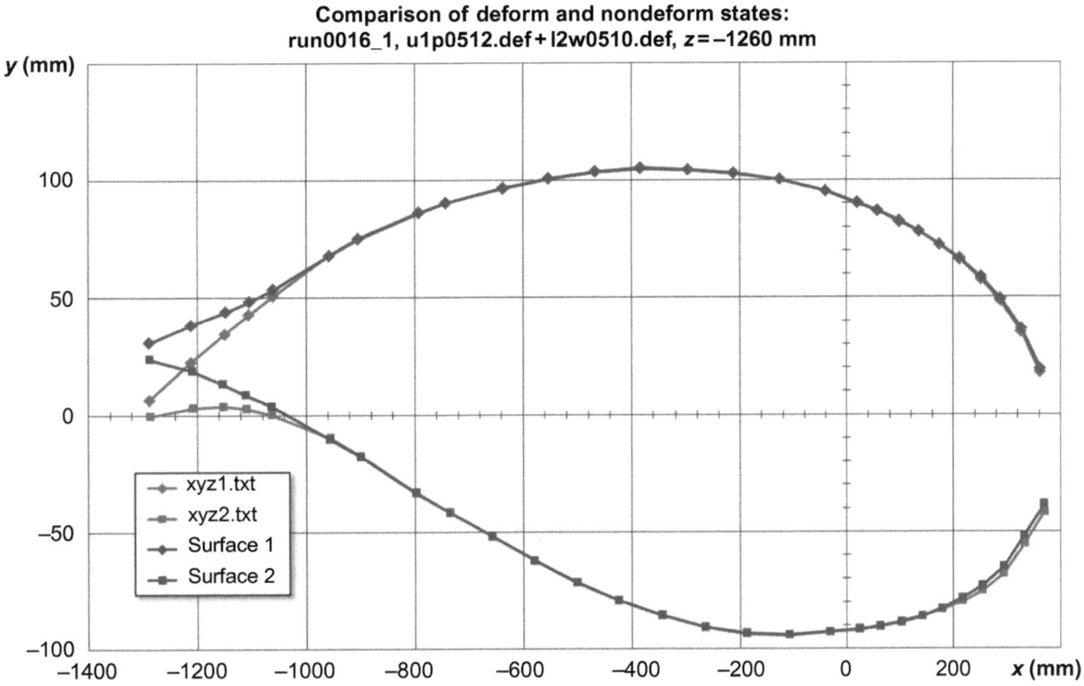

FIG. 49

Comparison of a deformed and nondeformed state of a wings surface during ATE deflection.

each of the above-mentioned classes a thorough experimental campaign in the wind tunnel was faced and the main steps of the performed activities have been described while pointing out the main issues encountered during the pretest, test, and data analysis phases. The most recent activity, carried out within the SARISTU project, represents a clear and emblematic example of the inherent complexity of morphing structures experimentation. In this framework the following main tasks were accomplished:

- Ground vibration tests of the wing demonstrator was performed. All of the required data (frequencies, modal shapes, generalized parameters, and linearity checks) were obtained. The results were used for the flutter clearance analysis needed to safely execute the WT test. On the basis of the obtained data, the modal database of the wing structure was built up.
- Wing demo balance measurements in the T-104WT were completed for in-built strain gage calibration for flow speeds $V = 30–60$ m/s. During the balance tests, the measurements of wing drag, lift, and pitching moment were carried out.
- Since WT raw aerodynamic loads resulted lower than expected, a detailed numerical simulation was carried out to find out the reason of this mismatch. According to the WT flow, detailed measurements, and numerical investigations, a large portion of the test rig and the essential interference of the wing demo with the WT core flow boundaries were discovered to cause a decrease in the actual angle of attack in comparison with free-flight conditions. The decrease of the actual angle of attack explained the low loads obtained in the WT.
- A propietary electronic wind tunnel approach was used for the calculation of the interference angularity correction needed for the recalculation of the actual angle of attack. For an incidence angle of 10 degrees, the angularity correction resulted up to $\triangle$AoA = 2.1 degrees. These corrections were proved to be in agreement with the obtained WT test data.
- A videogrammetry system for simultaneous noncontact measurements of the deformation of morphing elements of the wing demo was developed. Important, practical research results were obtained for further follow-up analyses.

ACKNOWLEDGMENTS

Portions of this chapter were previously published in ASD Journal (2011), Vol. 2, No. 2, IFASD-2015-148, 2015 and ICAS 2016 CD-ROM PROCEEDINGS, 2016.

REFERENCES

[1] Kuzmina S., Amiryants G., Schweiger J., Cooper J., Amprikidis M., Sensburg O., Review and outlook on active and passive aeroelastic design concepts for future aircraft, 23rd Congress ICAS, Toronto, 2002.

[2] S. Kuzmina, F. Ishmuratov, M. Zichenkov, V. Chedrik, Integrated numerical and experimental investigations of the active/passive aeroelastic concepts on the European research Aeroelastic model EuRAM, ASDJ 2 (2) (2011) 31–51.

[3] Simpson J., Suleman A., Cooper J., Review of European Research Project “Active Aeroelastic Aircraft Structures”, IFASD-2005, Munich, 2005.

[4] Ishmuratov F., Chedrik V., Argon code: structural aeroelastic analysis and optimization, IFASD-2003, Amsterdam, 2003.

[5] B. Moulin, M. Karpel, Gust loads alleviation using special control surfaces, J. Aircraft 44 (1) (2007) 17–25.

[6] Schweiger, J.; Weiss, F., Kullrich, T., Active aeroelastic design of a vertical tail for a fighter aircraft, 8th AIAA/NASA/ISSMO Symposium on Multidisciplinary Analysis and Optimization, CP-4848, Long Beach, CA, 2000.

[7] Kuzmina S., Ishmuratov F., Zichenkov M., Kuzmin V., Schweiger J., Integrated numerical and experimental investigations of the active adaptive all-movable vertical tail concept, IFASD-2005, Munich, 2005.

[8] Amiryants G., Selectively deformable structures; new concept in engineering, ICAS-1998, Melbourne, 1998.

[9] Kawiecki G., Amiryants G., Adaptive selectively deformable structures analysis, IFASD-03, Amsterdam, 2003.

[10] H.P. Monner, J. Riemenschneider, in: Background and recent results of the European project 'Smart High Lift Devices for Next Generation Wings', 1st EASN Association Workshop on Aerostructures, 7–8 October 2010, Paris, France, Aerodynamics - Sensors and Systems, vol. 3, 2004, pp. 22–27.

[11] M. Kintscher, O. Heintze, H. P. Monner, Structural design of a smart leading edge device for seamless and gapless high lift systems, 1st EASN Association Workshop on Aerostructures, 7–8 October 2010, Paris, France, 2010.

[12] T. Kühn, Aerodynamic optimization of a two-dimensional two-element high lift airfoil with a smart droop nose device, 1st EASN Association Workshop on Aerostructures, 7–8 October 2010, Paris, France, 2010.

[13] J. Kirn, T. Lorkowski, H. Baier, Development of flexible matrix composites (FMC) for fluidic actuators in morphing systems, 1st EASN Association Workshop on Aerostructures, 7–8 October 2010, Paris, France, 2010.

[14] Amiryants G., Ishmuratov F., Malyutin V., Timokhin V., Selectively deformable structures for design of adaptive wing smart elements as part of active aeroelastic concept, 1st EASN Association Workshop on Aerostructures, 7–8 October 2010, Paris, France. 2010.

[15] Amiryants G., Kulesh V., Malyutin V., Smotrov A., Chedrik A., Structural dynamic investigation of adaptive wing demonstrator within sade project, IFASD-2015-148, St. Petersburg, 2015.

[16] www.saristu.eu.

[17] V.Y. Neyland, S.M. Bosnyakov, S.A. Glazkov, A.I. Ivanov, S.V. Matyash, S.V. Mikhailov, V.V. Vlasenko, Conception of electronic wind tunnel and first results of its implementation, Prog. Aerosp. Sci. 37 (2) (2001) 121–145.

[18] S. Bosnyakov, I. Kursakov, A. Lysenkov, S. Matyash, S. Mikhailov, V. Vlasenko, J. Quest, Computational tools for supporting the testing of civil aircraft configurations in wind tunnels, Prog. Aerosp. Sci. 44 (2008) 67–120.

[19] E.V. Kazhan, Stability improvement of Godunov-Kolgan-Rodionov TVD scheme by a local implicit smoother, TsAGI Sci. J. 43 (6) (2012) 787–812.

[20] H. Glauert, Interference on wings, bodies and airscrews, ARC R & M 1566 (1933).

[21] Application guide for T-104 wind tunnel (In Russian). TsAGI, Russia, 1980, pp. 1–121.

[22] M. Pindzola, C.F. Lo, Boundary interference at subsonic speeds in wind tunnels with ventilated walls, AEDC. No. TR-69-47 (1969) 129.

[23] P.C. Wolcken, M. Papadopoulos, Smart Intelligent Aircraft Structures (SARISTU), in: Proc. Final Conference, Springer, 2015. ISBN 978-3-319-22412-1.

[24] R. Pecora, A. Concilio, I. Dimino, F. Amoroso, M. Ciminello, Structural design of an adaptive wing trailing edge for enhanced cruise performance, in: Proc. 24th AIAA/AHS Adaptive Structures Conference, San Diego, United States, 2016.

SECTION 8

Smart Helicopters

Editor:

Fabio Semperlotti
Purdue University, Purdue, IL, USA

List of Chapters

CHAPTER

ROTARY WINGS MORPHING TECHNOLOGIES: STATE OF THE ART AND PERSPECTIVES

24

Matthew L. Wilbur*, Mihir P. Mistry†, Peter F. Lorber‡, Robert Blackwell, Jr.‡, Silvestro Barbarino‡, Thomas H. Lawrence‡, Uwe T.P. Arnold§

US Army Research Laboratory, Adelphi, MD, United States The Boeing Company, Chicago, IL, United States† Sikorsky, A Lockheed Martin Company, Stratford, CT, United States‡ ZF Luftfahrttechnik GmbH, Calden, Germany§*

CHAPTER OUTLINE

NOMENCLATURE

ABC	advancing blade concept
ADASYS	adaptive dynamische systeme
AFC	active fiber composites
AMR	advanced material rotor
AMRDEC	Aviation and Missile Research Development and Engineering Center (US)
ARL	US Army Research Laboratory (US)
ATIC	Advanced Technology Institute of Commuter Helicopter
ATR	active twist rotor

Morphing Wing Technologies. https://doi.org/10.1016/B978-0-08-100964-2.00024-1

BVI	blade vortex interaction
CTEF	continuous trailing edge flap
DAL	design assurance level
DAP	directionally attached piezos
DC	direct current
DLR	Deutsche luft- und raumfahrt (D); German Aeronautic and Aerospace
EADS	European Aeronautic Defense and Space Company (NL)
GTech	Georgia Institute of Technology (US)
HART-III	Higher Harmonic Control Aeroacoustics Rotor Test-III
HHC	higher harmonic control
HP	horsepower
IBC	individual blade control
IDEPFC	interdigitated electrode piezo fiber composites
IFT	in-flight tuning
JAXA	Japan Aerospace Exploration Agency (J)
LEEMA	leading edge electromagnetic airfoil
MDHS	McDonnell Douglas Helicopter Systems (US)
MFC	macro fiber composite
MIT	Massachusetts Institute of Technology (US)
MTOW	max take-off weight
NASA	National Aeronautics and Space Administration (US)
ONERA	Office National d'Etudes et de Recherches Aérospatiale (F); French National Aerospace Research Centre
ONR	Office of Naval Research (US)
PSU	Pennsylvania State University (US)
Rev	revolution
RPM	rotations per minute
RRB	reconfigurable rotor blade
RSRA	NASA rotor systems research aircraft
RT&B	rotor track and balancing
SABT	smart active blade tip
SMA	shape memory alloys
SMART	smart material-actuated rotor technology
SPR	smart pitch rods
TDT	transonic dynamics tunnel
TRAC	telescoping rotor aircraft
TRL	technology readiness level
TU-Delft	Technische Universiteit Delft (NL); University of Technology of Delft
TWISCA	twistable section closed by actuation
UAV	unmanned aerial vehicle
UMD	University of Maryland (US)
UMich	University of Michigan (US)
VDR	variable diameter rotor
VDTR	variable diameter tilt rotor
VGART	variable geometry advanced rotor technology
VGR	variable geometry rotor
VTOL	vertical take-off and landing
ZFL	ZF Luftfahrttechnik GmbH (D)

1 INTRODUCTION

The roles helicopters play in aviation has increased since the availability of the world's first mass-produced helicopter in 1942, the Sikorsky R-4. The advantages helicopters lend to both the civilian and military sectors are irreplaceable, especially for search and rescue operations, where their capability to hover for periods of time is unmatched by other types of aircraft. Furthermore, the capability to takeoff and land without the need of a runway allows for the transportation of goods, people, and services to remote and inaccessible places.

Engineers have faced a plethora of issues to obtain stable flight of rotary-wing aircraft because of fairly complex operating aerodynamic environments as compared to fixed-wing aircraft. Even in the simple condition of hover, the rotor blades experience radially varying flow velocities, which are compensated by the design of blades with a built-in twist. However, in forward flight, the aerodynamic environment changes with azimuth as well. Classical control designs introduce azimuthal (periodic) changes in blade pitch via a swashplate, which is capable of introducing only once-per-revolution inputs. Although well proven, this approach rotates the blade as a whole about its radial axis and is far from optimal because it cannot completely compensate for periodic and higher harmonic excitations. The design selection of the fixed blade planform represents a compromise because the optimal configuration also varies with flight condition.

Aside from the performance issues that plague rotary wing aircraft, the rotor blades also generate a significant amount of vibration and noise. These problems are caused by a number of factors, including the increased asymmetrical loading of the rotor disk and blade tip shocks that occur in forward flight with increasing airspeed. The noise emitted by the rotor blades is compounded by the interaction of the blades with the tip vortices of the preceding blades (BVI noise). Therefore, rotors in tangential flow, as opposed to propellers operating in axial flow, inherently produce vibrations, noise, periodic loads, etc., because of the asymmetry of the flow condition. These problems, while not necessarily imperative to performance improvements, are essential to be solved in order to obtain a "jet smooth" ride.

Two possible ways to improve adaptability to the complex aerodynamic environment include either the quasi-static (as a function of flight condition) or dynamic (as a function of azimuth) rotor control inputs. The first method relates to the direct change of the blade planform to better suit the flight condition. Examples of this are variations of the twist, chord, camber, and span of the blades. The second method, instead, allows for the introduction of the N − 1, N, N + 1/rev inputs into the rotor system, which falls under the category of higher harmonic control (HHC). The HHC approach can be achieved by either controlling the motion of the swashplate in the fixed frame at more than once per revolution, or by using individual blade control (IBC), in which each blade is controlled individually in the rotating frame. The latter can be done by either varying the blade pitch input via a variable length pitch link or directly via dynamic blade morphing.

2 OVERVIEW OF ROTOR MORPHING TECHNOLOGIES

This section presents the state of the art of morphing technologies applied to rotor blades. The term "morphing" is used here in the broader meaning of the word, to encompass any technology that improves performance (e.g., envelope expansion, control load manipulation, etc.) or reduces vibrations or noise. As such, this section summarizes a broad spectrum of investigations, from numerical concepts to flight tests, including any concepts tested on a whirl stand or in a wind tunnel.

The different technologies will be discussed and organized based on the rotor blade parameter being adapted:

- Trailing Edge Flaps
  - Piezo ceramic based actuation
  - Non piezo actuation
- Active and Variable Twist
  - High frequency actuation
    - Surface mounted actuators
    - Internally mounted actuators
  - Quasi-static actuation
- Variable Span
- Emerging Rotor Morphing Technologies
  - Variable Chord
  - Variable Camber
  - Variable Nose Droop
  - Variable Rotor Speed

2.1 TRAILING EDGE FLAPS

Rotor blade morphing is not a new concept. The first example of morphing technology in the broad sense dates to 1947, when the prototype Kaman K-125 first flew using servo flaps. The servo flaps, located at the 3/4 radius position, twist the blade and hence morph its shape. They also provided in-flight blade tracking capability. However, it was not until the 1960s when this area of research garnered significant interest from government labs, private industry, and academia.

Flaps have been used in production helicopters since 1947, when Charles Kaman, founder of Kaman Helicopters, used them for primary control of the K-125. The servo flap design is composed of a flap fully exposed and attached to the trailing edge of the main rotor blade. The exposed servo flap enables very low control moments because it is aerodynamically balanced. A distinct disadvantage is the drag produced by the bracketry, control rod, and servo flap itself. Modern design focuses on plain flaps embedded in the trailing edge, like an aileron. These flap systems have been shown to not only reduce the noise and vibration [1–3] but also to improve performance [4]. It needs to be noted that, to date, high frequency flap motion has not produced simultaneous performance improvement and vibration reduction. A major problem in designing active trailing edge devices is to be able to produce the required forces to overcome the aerodynamic moments of the flap while being able to operate at high frequencies (N-1, N and N+1/rev) within the confines of the blade.

Table 1 summarizes the state of the art investigations on trailing edge and Gurney flaps [5–67].

2.1.1 *Piezo ceramic-based trailing edge flap actuation*

Piezoelectric actuators are appealing for on-blade actuation because of their simplicity. An overview of the piezo effect is presented by Chopra [5]. Most investigations about trailing edge flaps have used either piezo-bender or piezo-stack devices, all of which use the reverse piezo effect for actuation.

As the name implies, piezo-benders actuate when a voltage is applied to the piezo-ceramic, causing in-plane strain that leads to a vertical tip deflection if differentially applied. The actuation of the flap in

Table 1 Summary of Investigations on Trailing Edge/Gurney Flaps

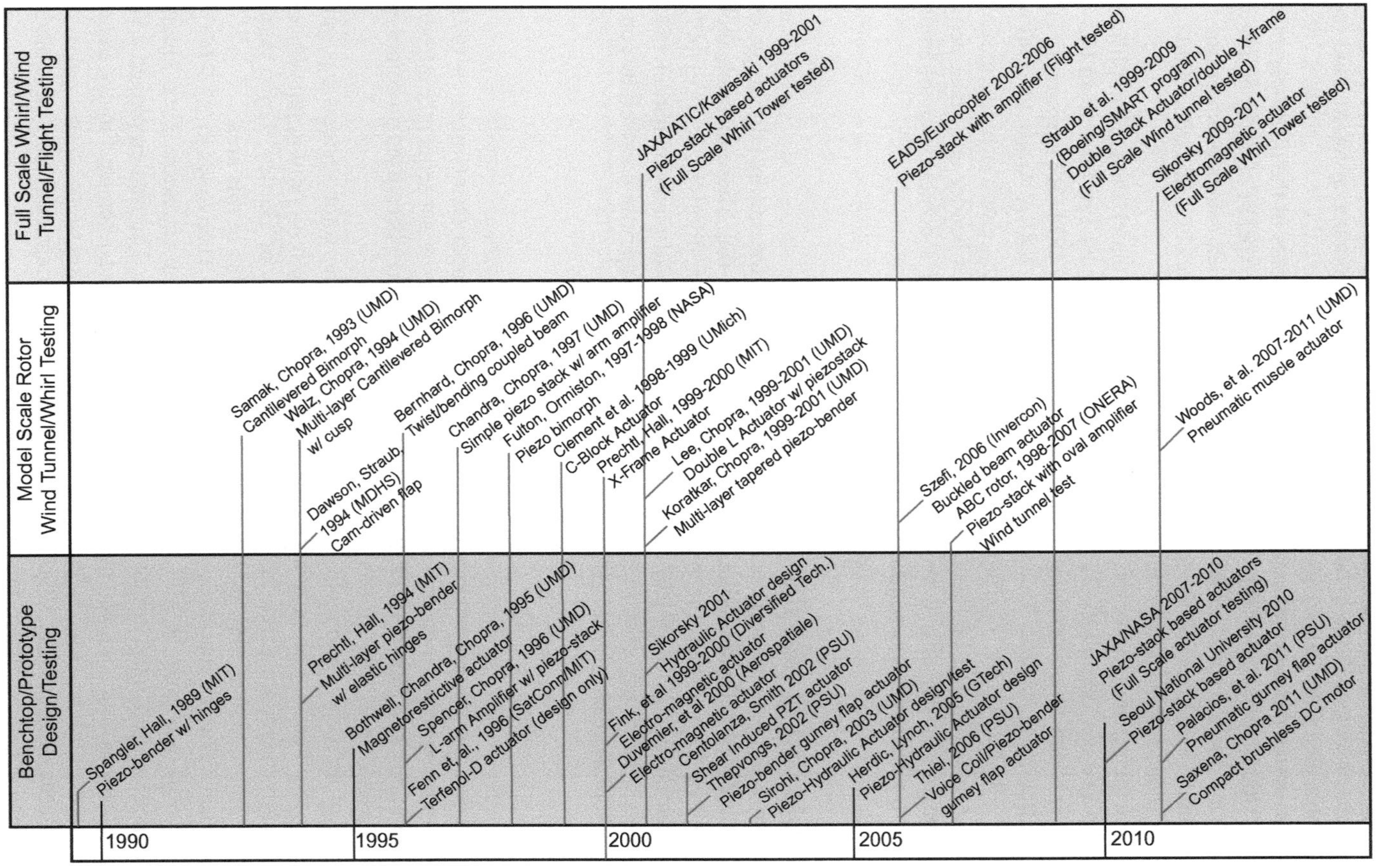

this arrangement is caused by the bending of the piezo-bender, which is translated to a pitching of the flap by means of linkages or hinges. Piezo-benders investigations range from bench-top to wind-tunnel test, mostly on small or scaled rotor blades [6–16].

Piezo-stack devices, unlike piezo-benders, use extensional displacements of the piezo-ceramics as the mode of actuation. This enables higher force capabilities, albeit at the expense of lower amplitudes. Therefore, in order to meet the force and stroke actuation requirements, piezo-stack actuators are used in cohort with a mechanical amplification device. The addition of these mechanisms often hinders solution effectiveness or reduces their performance.

Two solutions worth mentioning, however, have been tested in wind tunnel: the L-L amplification piezo-stack actuator by Lee & Chopra [19–21] and the X-frame actuator by Prechtl & Hall [22–24]. Both concepts were part of the Smart material-actuated rotor technology (SMART) program carried out by Straub et al. at Boeing [1,25–29]. This program conducted in the NASA Ames 40 × 80 wind tunnel marked the first full-scale active flap rotor testing in the United States. Additional discussions of investigations using piezo-stacks can be found in Refs. [17–29]. Besides bender and stack actuator designs, other piezo-based designs also were explored that considered a combination of multiple piezo-stacks or more exotic mechanical solutions [2,3,30–46].

2.1.2 Nonpiezo-based trailing edge flap actuation

Although there is a significant body of work using piezo-based actuators for trailing edge flaps, select research groups also have explored alternative methods of actuation [47–62]. These methods include the use of conventional motor and cam mechanisms [47,48], magneto-restrictive actuators [49,50], electro-magnetic actuators [51,52,56–58], hydraulic systems [53], hybrid-piezo actuation [54,55], and pneumatic actuators [59–62]. Sikorsky successfully tested an active flap rotor powered by conventional, albeit high performance, electromechanical actuators in 2009.

2.1.3 Gurney flap

Gurney flaps were invented by well-known race car driver and team manager Dan Gurney in 1971 for use on his team's cars to increase traction for better handling. The device was subsequently introduced to the aeronautics community by Liebeck in 1976 [63]. Since then, they have been used on commercial aircraft and rotorcraft as a simple high lift device.

Interest in the use of actively deployable gurney flaps for use in the rotor systems has gained momentum. A Gurney flap needs to be only about 2% of the chord length to yield approximately a 30% airfoil Cl_{max} improvement. Research has shown the effectiveness of Gurney flap systems in rotorcraft performance improvement [64], however, research is still sparse in this area.

Earlier work on actuated Gurney flap designs for rotor systems was by Thepvongs in 2002 [65] and was modeled after the work was done by MIT on piezo-bender active flap actuation. Then Thiel and Lesieutre [66] developed two separate Gurney flap bench-top prototypes. The first employed a voice coil connected to a mechanical amplifier, and the second used a tapered piezo-bender. Finally, Palacios et al. [67], in conjunction with Invercon Technologies, developed and wind tunnel-tested a pneumatically actuated active Gurney flap system. This concept used a pair of pneumatically inflatable bladders embedded in the blade to actuate a hinged Gurney flap.

2.2 ACTIVE AND VARIABLE TWIST

Active flap systems attenuate rotor system forces by two means: direct lift increment because of flap deflection and the increase in pitching moment that leads to the elastically twisting of the rotor blade. Another way to produce this type of control of the rotor forces is via direct active variations of the blade twist. This variation of blade twist can be both dynamic (N − 1, N, N + 1/rev frequencies) or quasi-static (change per flight condition). While dynamic variation of twist is intended to reduce the noise and vibration of the rotor system [68–70], quasi-static twist changes can be used both for performance improvement and for vibration reduction [71–75].

As active twist technologies change the planform of the entire rotor blade, a major challenge is to develop a system that can overcome the aerodynamic and dynamic forces experienced by the blade while maintaining structural integrity. Blade weight typically suffers because the actuation materials do not carry much of the design load.

Table 2 summarizes state-of-the-art investigations on variable twist for rotor systems [68–132].

2.2.1 High frequency active twist rotor designs

These types of systems can be further categorized into surface mounted actuation systems and internal actuation systems. The basic premise of surface mounted active twist technologies is to introduce a shear in the blade that would result in cross-section twist. To achieve that effect, longitudinal strain produced at ±45 degrees from the spanwise axis of the blade needs to be produced using piezo-ceramics. Considering piezo-ceramic sheets produce strain along both planar directions when an applied field about thickness is introduced, the first problem that needed to be solved was introducing directional strain. The directionally attached piezos (DAP) concept developed by Barrett and Chopra [76] in the early 1990s was the first documented effort of a piezo actuated active twist rotor blade concept and the starting point for many of the designs that followed [77–80]. It was found feasible to introduce the majority of strain generated by the piezos along a prescribed direction by using elements with high-aspect ratios attached to the substrate via a narrow bank of adhesive. Testing of these concepts showed a significant impact of active twist technology on vibratory rotor loads.

In parallel to the efforts on the DAPs, researchers at MIT were developing the next generation of piezo actuators, the interdigitated electrode piezo fiber composites (IDEPFC), commonly referred to as active fiber composites (AFC). These are composed of circular cross-section piezo fibers sandwiched between two layers of interdigitated electrode circuits [81].

This scheme allows for improved actuation strain capability because the interdigitated electrodes allow for even distribution of the electric field and the piezo fibers operate in the more efficient d33 mode. These were used in a variety of research projects [68–70,82–87], most notably on the NASA/Army/MIT active twist rotor (ATR) and Boeing advanced material rotor (AMR) that will be discussed in more detail in the coming sections. The goals of each of the projects were different, with the NASA/Army/MIT project focused on developing a sound understanding of the capabilities of an active twist rotor using a generic simple blade planform. The Boeing company project focused more on development and implementation issues of rotors with technology from a practical and operational standpoint.

The labor associated with the fabrication process of AFCs and their performance degradation because of imperfections led to the development of macro fiber composite (MFC) actuators. MFC actuators were developed by NASA Langley Research Center [88], but the technology eventually was licensed to Smart Material Corporation, and these actuators are now available commercially.

Table 2 Summary of Investigations on Variable Twist Rotor Systems

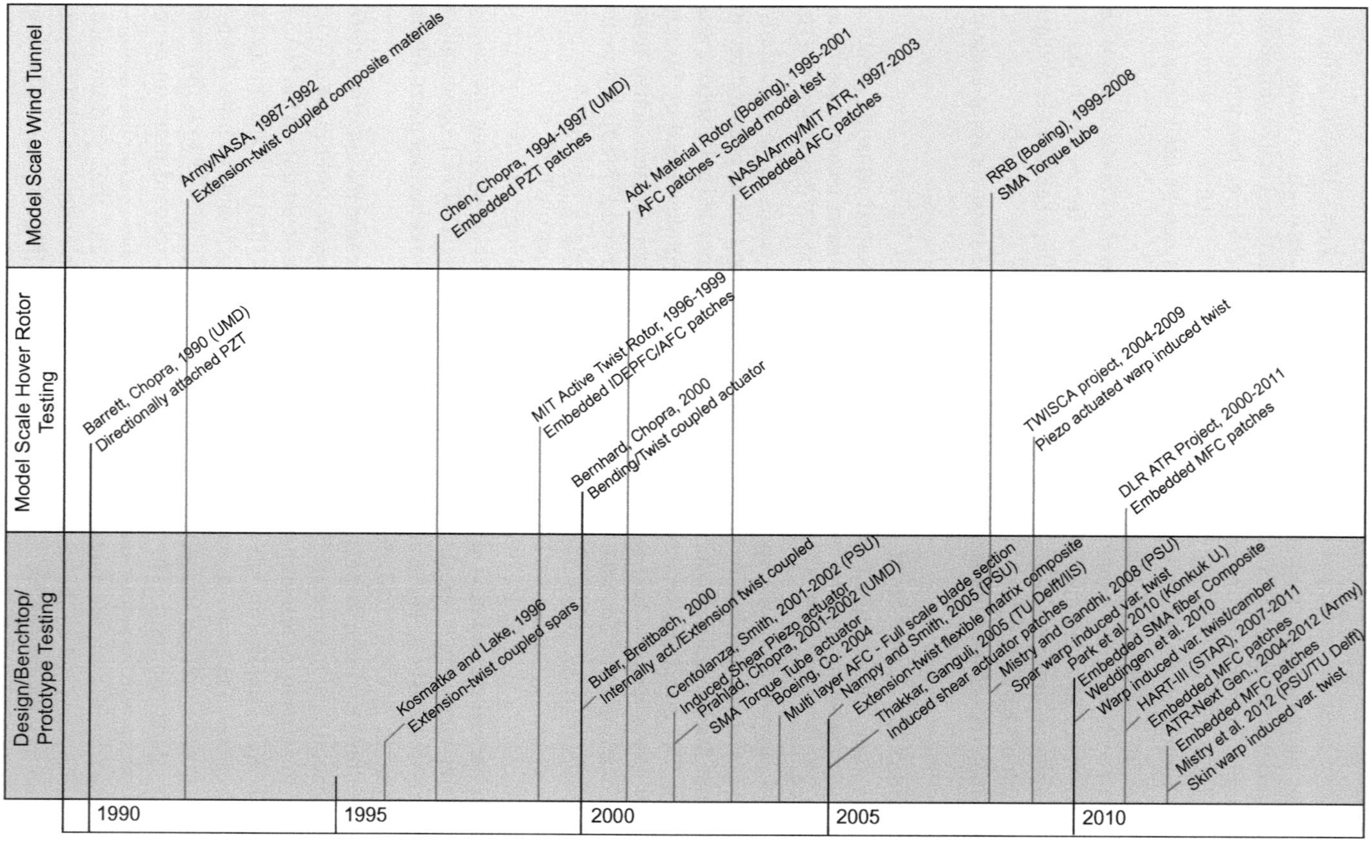

The key difference is that the fibers have a rectangular cross-section produced from piezo sheets sliced into thin strips. This process leads to more uniform cross-sections and fiber spacing. NASA Langley, DLR and others have used such actuators [71,89–103].

Researchers also have explored other actuation designs housed within the blade. Bernhard and Chopra [30,31,104,105] introduced a bending-twist coupled beam running along the blade span, referred to as smart active blade tip (SABT). DLR considered using extension-twist coupled composites and actuators [89,91]. ONERA pursued the TWIstable section closed by actuation (TWISCA) concept [106,107], which exploited the property by which open-section beams under torsional deformation undergo cross-sectional warping. The main drawback is that torsional stiffness is lower than classical rotor blades, and actuator failure could lead to flutter instabilities. Finally, Centolanza and Smith [108] analytically evaluated an induced shear piezo tube as an active twist actuator.

2.2.2 Quasi-static variable twist rotor designs

Variable twist rotors (quasi-static variation of twist; less than 1/rev) are not as well explored as the active twist rotor concept. The primary reason is the high amplitude, up to 30 degrees in some cases, required for these types of systems. Note these technologies do not only have to produce large twist variation, but also to do so while maintaining blade stiffness to withstand applied inertial or dynamic loads.

Starting in the late 1980s and through the early 1990s, NASA, in conjunction with the US Army, worked on development of extension-twist coupled composite blades for the quasi-static variation of blade twist for tilt-rotor applications [109–111]. The basic premise was to use RPM change and therefore centrifugal field change to passively alter blade twist using tailored extension-twist coupled composites. The primary benefit is eliminating an actuation device that reduces the manufacturing and operating complexity. Kosmatka and Lake [112] and Nampy and Smith [113] also investigated centrifugally actuated variable twist rotors. At this time, however, most manned aircraft have a fixed rotor speed design. Tilt-rotor aircraft and the optimum speed rotor (OSR) concept developed by Karem Aircraft are the exceptions. The latter is now embedded within the A-160 Hummingbird [114] in development with the Boeing Company. As such, such a centrifugal driven system would not be suitable as a retrofit for existing aircraft. This trend might change in the future as the interest of the rotorcraft community toward variable rotor speed concepts is increasing [115–121].

Active device systems also have been investigated. The ONR-funded reconfigurable rotor blade (RRB) program for tilt-rotor blade applications used two shape memory alloy (SMA) tubes to elastically twist the blade as the torque tube underwent rigid body rotation [122–126]. The blade/actuator system was wind tunnel-tested in the late 2000s on a 1/8th scale, three-bladed model. To date, this is the only actuated variable twist rotor wind-tunnel test of this kind. Other concepts based on SMA actuation also have been investigated [127–129].

Alternative methods to quasi-statically changing blade twist included warp-twist coupling of open section beams and warping of the blade skin [130–132].

2.3 VARIABLE SPAN

Variable span has been the focus of early investigations because of its potential for aircraft performance improvements. Linden et al. [133] provided a broad summary of variable span rotor technologies investigated prior to 1960s.

Table 3 Summary of Investigations on Variable Span Rotor Systems

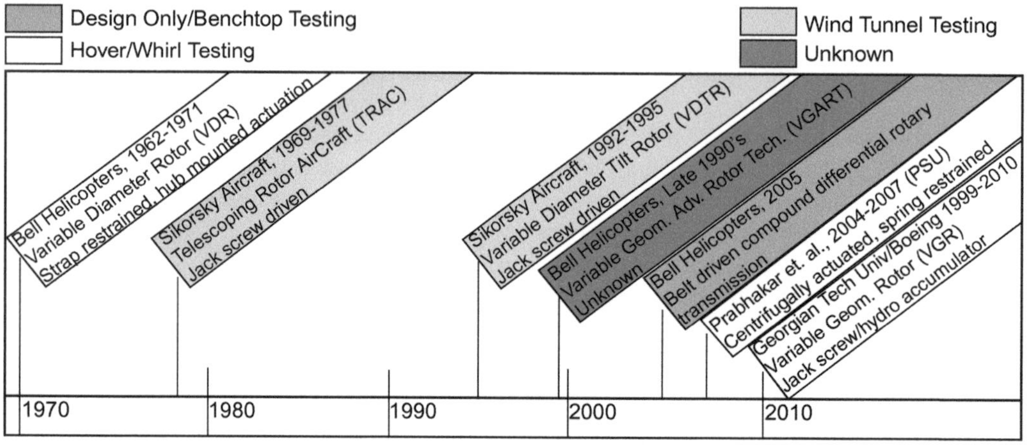

Table 3 summarizes the state of the art investigations on variable span for rotor systems after the 1960s [134–145].

During the early 1960s, Young et al. at Haig-K Helicopters developed a variable diameter rotor (VDR) system for conventional edgewise rotors [133]. The system, which was extensively hover tested, used telescoping blades whose motion was controlled via an assembly of cable/straps. During the early 1970s, full power tests of the latest revision of the project, the VDR-4 test article, were undertaken. This prototype design incorporated a locking mechanism to hold the blades in place during operation and later included an innovative mechanism method to move the telescoping sections. Following this testing of the VDR-4 prototype, the project was concluded and no further work was done until the mid 2000s when Fenny [134] at Bell Helicopters introduced a compound differential rotary mechanism for the cable system, which allowed for increased system reliability, a major problem with the original VDR project mechanism.

Sikorsky also investigated variable span in the Telescoping Rotor AirCraft (TRAC) project using a jackscrew mechanism to operate telescoping blade sections [135]. Unlike the Bell concept, this design called for an inboard torque tube on which an aerodynamic blade section was allowed to slide. The design allowed for the jackscrew to be actuated via the main rotor power system itself. A scaled model of this concept was extensively wind tunnel-tested and was shown to provide performance improvements [136–139]. This design was later modified for a tilt-rotor aircraft configuration and renamed the variable diameter tilt rotor (VDTR) project in the early 1990s [140,141]. As part of this program, significant wind tunnel and hover testing was conducted on a scaled prototype. The major difference for the scale rotor tests of this project was inclusion of a separate motor housed in the rotor hub to actuate the jackscrew mechanism.

Bell Helicopters again worked on variable rotor systems, during the late 1990s, as part of the variable geometry advanced rotor technology (VGART) program in conjunction with the Army. The program was closed-door collaboration, and as such, no detailed description of the actuation mechanism is available.

Simultaneously during the late 1990s, The Boeing Company worked on the development of a variable geometry rotor (VGR) using a jackscrew mechanism in conjunction with a hydro-pneumatic accumulator [142,143]. The concept also included a mechanism to automatically vary the blade built-in twist as the span was varied. This was done via helically curved grooves on the main span of the blade along which the blade section attach points would slide, causing a combined rotary and span-wise motion of the section.

More recently, Prabhakar, Steiner and Gandhi, worked on the development of a passive, centrifugally actuated variable span rotor system [144,145]. The authors envisioned a system wherein the rotor speed would be changed to control the blade span-wise location.

A recent study by Mistry [146] evaluated the effects of variable span for a UH-60A-type aircraft at different gross weights and altitudes. The results of the analysis showed that variable span in the range of +17% to −16% of baseline radius could reduce the power requirement of the helicopter by up to 20% for high altitude and high gross weight conditions.

2.4 EMERGING ROTOR MORPHING TECHNOLOGIES

Since the 2000s, researchers have explored three relatively new types of rotor morphing technologies: Variable camber systems, variable chord systems, and variable nose droop systems. Table 4 summarizes the state-of-the-art investigations on the emerging rotor morphing technologies [4,46,147–170].

2.4.1 Variable chord rotor systems

Variable chord systems have the potential for significant improvement in aircraft performance [4,46,147–154] and gained some traction in the early 2000s. The active tab mechanism project, undertaken by the Japan Aerospace Exploration Agency (JAXA), focused on the development and testing of

Table 4 Summary of Investigations on Emerging Rotor Morphing Technologies

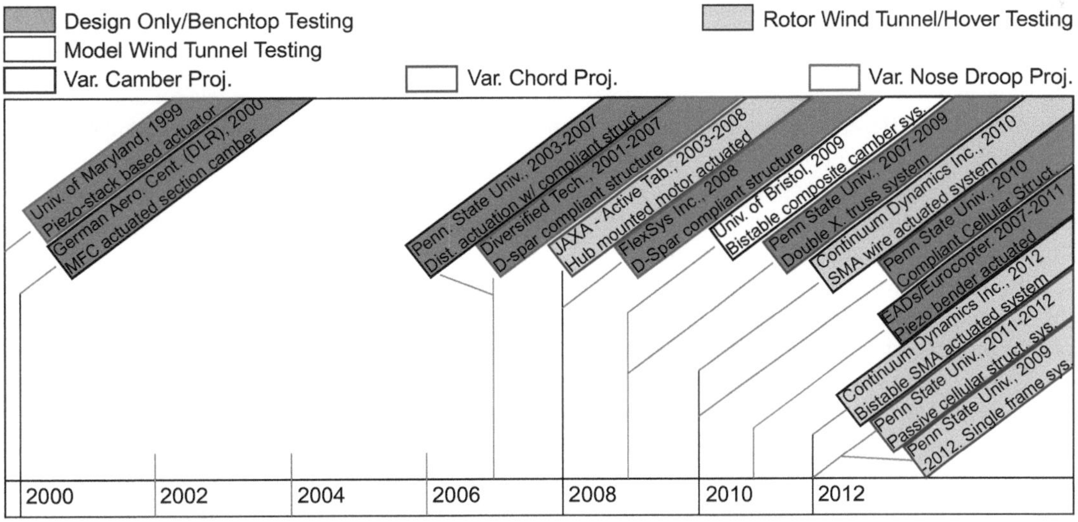

a high frequency (N – 1, N, N + 1/rev) actuated rotor tab for noise reduction. The program included model scale wind-tunnel test. The tab was actuated by a motor housed within the hub.

Khoshlahjeh and Gandhi [152] investigated the benefits of quasi-static variable chord. This led several researchers at Pennsylvania State University to develop multiple variable chord rotor systems for quasi-static actuation using a variety of actuation mechanisms. These include passive, centrifugally actuated discrete rigid plates [153], compliant cellular structures [151], electric and pneumatic actuators to drive an X-truss mechanism [4,147,150], and SMA-based amplification mechanisms [154,155]. The work by Moser et al. [153] demonstrated the first passively actuated (rotor speed variation driven) variable chord concept on a rotor test stand.

2.4.2 Variable camber rotor systems

Active flap systems generally are intended to produce an effective change in the camber of the system, however, larger chord flaps can lead to increases in rotor profile power because of the discretized nature of these systems. An alternative is to directly change the camber of blade sections to produce the necessary changes in the aerodynamic characteristics of the blade section. These types of systems are subject to the additional challenge of providing a smooth aerodynamic surface.

DLR proposed a concept (no prototype built) to induce camber of rotor blade sections by using MFCs [156]. Researchers at Pennsylvania State University worked on the development of an active camber system based on a compliant truss assembly [157,158]. The authors also considered the design of deformable skins for the use of such active camber systems [159]. Daynes et al. [160] developed an active camber system that uses a bi-stable composite skin actuated via an electromechanical actuator placed within the D-spar of the blade. Researchers at Continuum Dynamics [161] developed a shape memory alloy (SMA) wire (with a restoring spring) based actuation scheme to realize variable blade camber. This system was hover tested on a scale rotor test stand. They also developed another unique bi-stable structure that allowed for the variation of blade camber [162]. The design benefit was its simple manufacturing process and the ease with which it could be applied to existing blades. The European Aeronautic Defense and Space (EADS) and Eurocopter designed an active camber concept using piezo-bender actuated via specifically design ceramics that are encased in a compliant material affixed to the trailing edge of a rotor blade section [163–165].

In 2012, Mistry [146] evaluated the performance improvements from a variable camber for a BO-105 type rotor in hover. The design space considered included three different thrust levels for a range of altitudes and seven different camber distribution schemes. The analysis showed that variable camber was capable of reducing power up to 18% for high thrust levels at high altitudes. Furthermore, it was found that a linearly distributed camber configuration, wherein the maximum camber was at the root, showed the best power reduction. Also presented was a variable camber design that exploited the warp-twist relationship of open-section beams. A unique actuation structure was developed within a CH-46E blade section. This prototype was shown to be capable of producing up to 18 degrees of distributed camber with a relatively low input warping of up to 0.18 in. A finite element analysis indicated the requirement of a shear-deformable core for proper camber deformation in the presence of centrifugal and aerodynamic loads.

2.4.3 Variable nose droop rotor systems

Nose droop systems can be used effectively to alleviate issues related to blade section dynamic stall [166]. Rotor blades generally are composed of a D-spar or C-spar at the leading edge, which contributes most of the blade stiffness. Nose droop systems pose a significant design challenge, as they are in the same location as the significant blade structure.

Shaner and Chopra worked on the development of a piezo-stack based actuator for a leading edge nose droop system [167]. However, in spite of these design changes, the actuator did not perform at expected levels when bench-top tested. Fink et al. [168] at Diversified Technologies, in conjunction with the Aeromechanics Division, Aeroflightdynamics Directorate (AMRDEC), developed the leading edge electromagnetic airfoil (LEEMA) actuator for a variable nose droop rotor system. The system design was optimized to provide a smooth nose droop actuated via a linear electromagnetic actuator housed in the leading of the blade section. A bench-top prototype based on the actuator system was built and tested to show its effectiveness in producing desired nose droop.

FlexSys, in collaboration with AMRDEC, developed a conformable leading that was designed with compliant structure optimization techniques [169]. This system was designed to convert the rotary motion of a motor into the effective rotation of the blades leading edge. The relative simplicity of manufacturing because of the reduced number of moving parts makes this design very viable, however, it was not as strong as the conventional spar structure. Other issues regarding anti-ice and erosion were not addressed. A bench-top prototype of the concept was built and tested to show the actuation characteristics of the system.

2.4.4 Variable rotor speed systems

Traditionally rotating wing aircraft are designed to operate at a fixed rotor speed (RPM). This design feature was chosen for numerous reasons, the most prominent of which is weight, as this type of system would not require a variable ratio transmission. The area of variable speed (variable RPM) rotors, however, has recently gained momentum, with the introduction of the A160 Hummingbird unmanned aerial vehicle and its optimum speed rotor system [114].

Researchers at Pennsylvania State University examined the performance improvements with variation in rotor RPM only over the entire flight envelope of a UH-60A type aircraft model [115,116,170,171]. This study showed the plausibility of using only a limited rotor RPM variation to achieve a maximum power reduction of up to 14% for specified cases. A recent study by Mistry [146] expanded on this work by combining $\pm 11\%$ variation of the rotor speed with a variation between $+17\%$ and -16% of baseline blade span for a range of gross weights and altitudes. Results showed that, for high altitude and high gross weight conditions, it was possible to reduce the overall power required by the aircraft by up to 30%.

Bowen-Davies and Chopra [118] conducted an analysis on a similar type of aircraft to understand both the performance benefits and the vibratory characteristics of a variable RPM rotor system. Simultaneously, Berry and Chopra [119] conducted a model scale wind tunnel experiment to understand the behavior of a variable RPM system. Furthermore, an experimental study by Datta, Yeo, and Norman considered the effects of a slowed rotor configuration at high advance ratios [120]. A slowed rotor experiment was conducted on a full scale UH-60A rotor system at the NASA Ames 40×80 wind tunnel.

3 CRITICAL REVIEW OF SOME SIGNIFICANT EFFORTS

This section discusses in more detail some of the rotor morphing technologies that have had a greater impact in this field, accomplished significant performance improvements or deserve further investigation, have achieved a higher technology readiness level (TRL), have the potential for future use in production helicopters, or are at the cutting edge of current investigations.

3.1 ACTIVE TRAILING AND LEADING EDGE DEVICES

At the University of Maryland, between 1999 and 2001, Lee and Chopra worked on various designs for stack actuators and corresponding displacement amplification mechanisms [19,20]. The focus of this research was the development of a full-scale active trailing edge flap actuator. Three different prototypes of an L-L amplification device were developed. The first design was whirl-tested at the University of Maryland vacuum chamber, where it was shown to work in a centrifugal field. Several issues relating to the mechanical assembly of the device were discovered, however. A second device was built with special consideration given to losses from the elastic hinge and the stack actuator contact surface. This second generation included a curved contact surface for the piezo-stack and a specifically designed elastic hinge. This device underwent both vacuum spin tests and open jet wind-tunnel test. Wind-tunnel test observations showed the system is incapable of providing the required restoring force for a wide range of operating frequencies [21]. Based on the observation from these experiments, a third-generation bidirectional piezo-stack actuation device was proposed. This added another piezo-stack device, which allowed for a power stroke in both directions of flap actuation. This design eliminated the need for an external restoring spring and allowed for better actuation authority for a wider frequency bandwidth [21].

Starting in the late 1990s, significant efforts have been undertaken at the Massachusetts Institute of Technology by Prechtl and Hall [22,23] on the development of piezo-stack based full scale active flap actuators. This body of work was focused on a double-stack actuator X-frame actuator, and included rotor hover tests. The actuator was designed to produce a displacement perpendicular to the orientation of the stack actuation. This concept was further developed by Hall et al. [24] to include two X-frame devices to provide higher force stroke capability.

The Boeing Company, in the late 1990s, started working on the development of an active trailing edge flap system for a full-scale helicopter rotor system. The goal of the project was the full-scale flap actuator development for whirl, wind tunnel, and eventually flight tests. Several different actuation designs were considered [25,26], with two concepts eventually selected for bench-top prototype development and testing. The two concepts were based on the double stack L-L concept developed by Lee and Chopra [27,28] and the double X-frame actuator by Hall et al. [24] The double X-frame was finally selected for use in the smart material-actuated rotor technology (SMART) program which involved the hover [29] and wind tunnel [1] testing of the MD-900 light utility helicopter rotor system in the NASA Ames 40 × 80 wind tunnel. This program marked the first full-scale active flap rotor testing in the United States.

By the late 1990s, the idea of individual blade control (IBC) using active trailing edge flap had caught the attention of international researchers. The Japanese Aerospace Exploration Agency (JAXA), in conjunction with the Advanced Technology Institute of Commuter Helicopter (ATIC) and Kawasaki Heavy Industries designed, built and tested a full-scale active flap actuator for the ATIC rotor system [37,38]. This concept used two parallel piezo-stack actuators to produce the rotation of a bell crank, which in turn actuated the flap. This project culminated in the first full-scale whirl tower testing of an active flap rotor system. Alternative IBC technologies will be discussed more in detail in a separate section.

The European Aeronautic Defense and Space (EADS) Company, in conjunction with Eurocopter, between 2002 and 2006, developed, full-scale whirl [42] and flight tested [2,3] a piezo-stack based actuator for the BK-117 rotor system named Adaptive Dynamische Systeme (ADASYS). This design used the Poisson effect of an oval structure to amplify the strain of the piezo-stack, and two such actuators are used to produce a power stroke in both flap directions. This program also involved significant design work for the optimization of the elliptical amplifying structure and the flap module design for rotor blade installation. This program marks the first time a full-scale active flap system was flight tested: a significant step closer to commercial mass-production. The flight test of the actuation mechanism was a success, with experimental measurements showing significant reduction in both and noise and vibration emanating of the main rotor.

JAXA, in conjunction with NASA, is working on the development of a full scale piezo-stack based active flap actuator device for whirl testing at the Kawasaki Industries hover test stand [44–46]. The actuator design for this program is similar to the double piezo-stack used for the full ATIC rotor system [38].

In the mid 1990s, Dawson and Straub [47] at McDonnell Douglas developed a cam-driven tip-mounted rotor flap system. This program involved the design, build, and test of the first Mach-scaled active flap rotor system. The program was intended as a test bed for active flap systems to experimentally determine their effectiveness for reducing noise and vibration, and as such, involved extensive wind-tunnel test [48]. The actuation system used a fixed cam mounted at the hub along which the cam follower for the flap would ride. Different cam schedules were used to measure the effects of flaps on the noise and vibration of the rotor system.

During the late 2000s, Sikorsky worked on an active flap rotor based on the S-434 rotor system. The active flaps were conceived primarily as higher harmonic control devices to reduce rotor noise and vibration and improve aerodynamic performance. The first analytical and experimental studies were performed to characterize the aerodynamic effectiveness of the flaps for generating blade pitching moments and lift over a frequency range from zero to six times the blade rotation frequency. As described by Lorber [172], a 20% chord flap was found to be effective over the Mach number and reduced frequency ranges of interest. Operating the flap when the blade was undergoing deep dynamic stall had a detrimental effect. Following analysis of the flap on rotor dynamics, a high authority electromechanical actuator was developed to function in the high centrifugal load environment of a rotor blade. This device is composed of an electric motor connected to a roller screw enclosed in an oil reservoir [56]. A demonstration rotor was built [173] and underwent full-scale whirl [57] and wind-tunnel test [58] at the NASA Ames 40 × 80 wind tunnel. This is the second full-scale wind-tunnel test of an active flap rotor system in the United States. Noise and vibration benefits were demonstrated in the wind tunnel [173], but performance improvements were modest. This actuation system also was capable of producing large amplitudes at low frequencies, as would be required for use for primary flight control, referred to as a "swashplateless rotor."

Around the same time, Sikorsky also investigated active leading edge slats. These were intended to perform a similar function for the helicopter blade that a passive slat performs for a fixed wing aircraft: increase airfoil maximum lift capability at reduced airspeed. The helicopter application is much more challenging because of the operating environment. The slat needs to retract or relocate itself on the advancing blade side to avoid excessive drag at high relative Mach numbers experienced on the advancing side of the rotor. An actuate-retract frequency of 5 Hz would be typical at full scale. Initial studies demonstrated aerodynamic feasibility on an airfoil section, and then a fixed slat was mounted on a scale model rotor in a wind tunnel [174]. This experiment succeeded in increasing the rotor stall boundary to higher thrust and forward speed, but confirmed that an active system was required to preserve cruise efficiency. Follow-on activities investigated several actuation concepts, but none has yet met all operational requirements.

3.2 INDIVIDUAL BLADE CONTROL

3.2.1 IBC concept

Although comparably broad in its literal meaning, the term individual blade control (IBC) initially was crafted to describe a mechanism that would allow change of the pitch of all blades individually (i.e., independently) by a restricted angle small enough not to interfere with the primary control. These systems were designed to be installed as add-ons to the primary control system. In the typical embodiment, an active element (actuator) replaces the conventional constant-length pitch rod that links the rotating part of the swashplate to the blade pitch horn. Within the limits of the provided authority and bandwidth, such systems can introduce arbitrary blade pitch motions.

Based on the physics of the rotor aerodynamics, the most useful inputs turn out to be sinusoidal feathering motions at frequencies of integer multiples of the rotor RPM (n/rev). As one could expect, the optimum inputs are similar for each blade, however, phase-shifted corresponding to their azimuthal spacing. Such control schemes are equivalent to what is also known as higher harmonic control (HHC). At the time when the HHC concept, which could be viewed as the predecessor of IBC, was first investigated experimentally, the mechanical solution, however, was conceptually different. High bandwidth actuators had been placed below the swashplate in the non-rotating frame. By shifting and tilting the swashplate at the blade passage frequency (N/rev), it is possible to introduce higher harmonic blade pitch motions. In this case, however, the frequencies available at the blade root are restricted by the kinematics of the swashplate to the blade passage frequency N/rev and the neighboring multiples (N − 1)/rev and (N + 1)/rev. In addition, the inputs are exactly the same for each blade (despite the mentioned phase shift). Therefore, this type of HHC realization is much more restricted in its application, since it can neither provide all interesting frequencies (e.g., the valuable 2/rev for rotors of more than three blades, see below), nor use individual inputs to compensate any blade dissimilarities [175,176]. Nevertheless, an early HHC research program conducted by Hughes Helicopters' in the 1980s led to the successful vibration flight tests of a modified OH-6A helicopter at Yuma Proving Ground. The program demonstrated vibration reductions of approximately 80% in most flight conditions [177,178]. A similar approach was used for flight tests performed with an SA 349 Gazelle helicopter by Aerospatiale [179,180]. The descriptions of a further HHC flight test activity can be found in Ref. [181], whereas wind tunnel campaigns are described in Refs. [182–188], including the very comprehensive HART wind-tunnel test [189,190].

Sometimes the term IBC is also used for other blade morphing concepts as long as the blades are controlled independently from each other. Such technologies such as trailing edge (servo) flaps, variable nose droop, active twist, or variable camber are described in a separate section. Because of the complexity that comes with fitting a second independent control system, designers have started to think about combining the functionalities of the primary control (high authority at low bandwidth) and IBC (low authority at high bandwidth) within one system. It turns out that primary control systems often need to be oversized with respect to their performance for redundancy reasons alone. Therefore, such a system might bring sufficient power margin to inherently cover any additional IBC requirements unless a particular failure forces it into a degraded mode that demands all resources for primary control. Especially more recently, since the interest has shifted from hydraulic to electrical control systems, the combination of both elements is worth being investigated. Thus, even primary control systems, which by their novel architecture enable blade-individual actuation, are sometimes called IBC. This could be swashplateless designs that use powerful rotatory actuators to rotate the blades independently at their root.

As described before, it is the highly complex flow condition in the rotor plane, which is likewise created and endured by the blades, that makes the active spatial and temporal adaptation of the lifting surfaces so desirable. Whereas the invention of the conventional cyclic control (1/rev) was a necessary first step to allow the controlled and stable flight of a helicopter, it is obvious that for smooth and efficient flight much more complex control schemes are required. IBC, in the sense of independent higher frequency blade root control, can be viewed as the next big step toward the ideal blade, which can be morphed dynamically to optimize the profile section properties and local angle of attack at any point in time and space. By giving up the freedom to change the blade twist dynamically and by accepting a fixed pre-optimized profile section shape, the technical task is curtailed to a manageable challenge while realizing already a considerable part of the achievable effect. This was and still is the motivation to explore the benefits of IBC within the practical limits with respect to authority and bandwidth.

3.2.2 Demonstrated benefits

Several wind tunnel and flight test experiments have provided valuable insight into the capabilities and limitations of IBC [191–208]. In some cases, special focus was put on the extended capabilities of IBC compared to HHC. But since the design of IBC systems always includes all the HHC capabilities, harmonic inputs at the HHC frequencies are always included within the IBC investigations. A broad range of different blade pitch time histories has been investigated to address the full spectrum of classical deficiencies of a rotor in tangential flow, namely

- High vibrations (N/rev)
- High noise radiation
- Limited efficiency at higher forward speed (e.g., thrust per HP or lift over drag)
- High control forces, exponentially increasing with forward speed.

Beside these main objectives, other applications that, in particular, make use of the independence of the inputs at each blade include:

- Reduction of vibrations (1/rev) that are created by unbalances from inherent blade dissimilarities or external damage (in-flight tuning, IFT)
- Simplification of the rotor track & balance (RT&B) procedure
- Suppression of ground resonance.

Most operational IBC systems have been based, up to now, on hydraulic actuation. Electrical systems have been built so far only for the IFT application. The realized authorities range from 0.2 to more than 5 degrees blade pitch angle. Where discrete frequencies have been used explicitly, they have covered the range from 2/rev to (N+2)/rev. Interestingly enough, the analytical predictions often were considerably off from actual flight test results. Simulations of the complex rotor flow conditions today still lack the fidelity to reliably capture the higher harmonic effects especially. Table 5 lists the different goals and the corresponding control input parameters. The last application has been studied by several researchers through analysis and simulation but is difficult to validate in ground/flight test [191].

Table 5 Summary of IBC Methods and Effectiveness

Application	Most Effective Frequencies	Required Blade Pitch Authority	Remark
Vibration reduction (N/rev)	(N − 2)/rev–(N + 1)/rev	±0.1–1.5 degrees	60%–90% reduction depending on number of combined harmonics and number of monitored locations
Noise reduction	2/rev, 3/rev	±0.5–3 degrees	Primarily blade vortex interaction (BVI) noise, −3–9 dB in descending flight
Reduction of power required (main rotor torque) and/or increase of lift over drag ratio	2/rev	±1–2 degrees	Effect kicks in at medium forward speed and is strongest at high forward speed and high rotor loading
Reduction of control system loads	2/rev, 3/rev	±0.5–2 degrees	Suppression of higher harmonic contents through corresponding IBC frequencies, optimum 2/rev input phase angle coincides with optimum setting for power reduction
Reduction of mass and/or aerodynamic unbalances (1/rev vibrations)	Blade-individual quasi-static offset (0/rev)	0.1–0.2 degrees	Flat track not enforced as with traditional RT&B, but split tip paths used to compensate mass dissimilarities
Simplified RT&B	Blade-individual quasi-static offset (0/rev)	0.1–0.4 degrees	
Ground resonance suppression	Respective lead-lag frequency	±0.5 degrees	Blade-individual feedback of blade lead-lag motion states (position, rate); time domain control

Whereas most researchers have focused on one goal at a time, in some cases, the effect on different parameters was studied simultaneously. Few combinations are actually relevant for practical operation of an aircraft. For example, ground resonance suppression and high speed rotor efficiency do not need to be balanced nor do control load reduction and BVI noise suppression in a landing approach, because they fall into different flight regimes. In many cases, impressive effects have been demonstrated with just one single input frequency (single harmonic), but better results often are achieved when mixing multiple harmonics. Especially when different goals are pursued simultaneously, the combination of several frequencies can help avoid undesirable negative impacts on noncontrolled parameters. In general, the more blades the rotor has, the more harmonic frequencies are available to address the different goals and the better are the chances that multiple benefits can be achieved at the same time.

In many of the early IBC experiments, the input parameters were systematically varied and the change of the target parameter observed. Based on such data, sensitivities, transfer functions, and optimum input settings could be computed. These tests are referred to as "open loop." Later, the designers came up with increasingly sophisticated control algorithms that allow closed loop control. Especially in the case of multiharmonic inputs, the manual selection of suitable input parameters

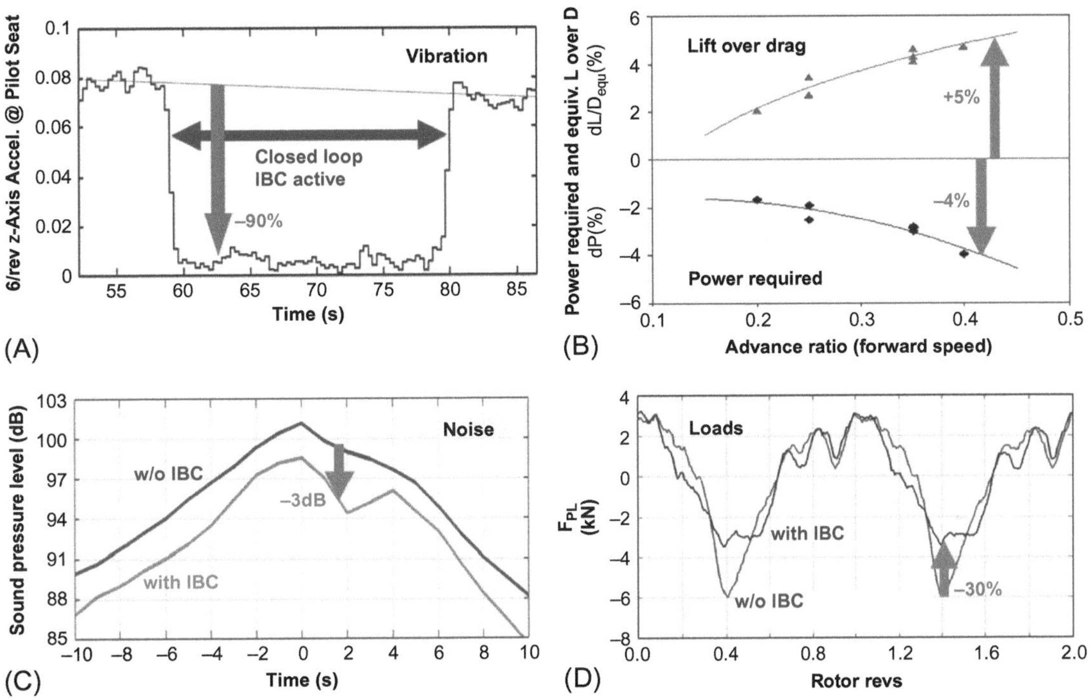

FIG. 1

Examples of test results showing the IBC effect on (A) vibration, (B) rotor power required, (C) radiated noise, and (D) pitch link loads (from CH-53G flight and UH-60 wind-tunnel test, respectively).

becomes increasingly difficult. Moreover, the optimum inputs have been shown to be highly dependent on flight conditions, in many cases dictating the application of a fast, convergent, and stable closed loop controller. This approach also allows the automatic but customizable weighting of the different goals. Fig. 1 shows selected examples of experimental data recorded during wind tunnel or flight tests:

- Fig. 1A Vibration reduction through 5/rev closed loop inputs of not more than 0.15 degrees blade pitch angle (measured in cockpit during 70 kts level flight of CH-53G testbed)
- Fig. 1B Improvement of equivalent L over D and rotor power required by optimized 2/rev inputs (measured in the wind tunnel at rotor balance for full-scale UH-60 rotor)
- Fig. 1C Noise reduction through optimized 0.67 degrees 2/rev inputs (measured by ground microphone during landing approach fly-over of CH-53G testbed)
- Fig. 1D Pitch link load reduction through closed loop 2/rev blade pitch inputs (measured in 130 kts level flight of CH-53G testbed)

Whereas systems designed exclusively for IFT application can be limited to those authorities that are harmless even in the worst failure cases, powerful IBC systems potentially can be dangerous if erroneously computed blade pitch commands are sent to the actuators. Therefore, special provisions need to be implemented that reliably prevent such scenarios. The realized systems usually use some kind of mechanical lock-out system that forces the active pitch rods into the default position in the case of any detected malfunction. An alternative option would be to use a corresponding level of redundancy for the mechanical, hydraulic, and/or electrical hardware and to develop the software to a higher design assurance level (DAL).

3.2.3 Examples of IBC projects

The potential benefits of HHC/IBC have been thoroughly investigated in theory by analytical calculations and numerical simulations. The first studies were carried out long before corresponding practical tests were started, and corresponding research activities continue [209–219]. Comprehensive overviews are presented in Refs. [220, 221]. Meanwhile, multiple whirl stand, wind tunnel, and flight test experiments have been carried out. The first successful flight tests with a real IBC system using blade-individual hydraulic actuators in the rotating frame was carried out on a BO-105 testbed of MBB/Eurocopter, which was intermittently operated between the mid-1980s and 2000 [176,196,202,203].

Within two test campaigns in 2001 and 2009, a high authority, high bandwidth IBC system was investigated in the large NFAC wind tunnel. The system had again been designed by ZF Luftfahrttechnik to be fitted onto a full-scale UH-60 rotor. It used four hydraulic actuators each featuring a pressurized gas-driven safety system to allow the immediate lock-out of the actuator piston in a neutral position in case of unexpected reactions. The first entry was focused on vibration and noise reduction [198]. In a second campaign, the focus was put on rotor performance improvement, control system load reduction, and automatic rotor track and balance/in-flight tuning [199,200].

In 1995, the German DoD contracted ZF Luftfahrttechnik to design, build, integrate, and flight test an IBC system on a CH-53G testbed of the German Army. The IBC system was powered by a separate hydraulic system. The actuators had a maximum authority of ± 1.3 degrees and were fitted with a spring-loaded lock-out system for fail-safe operation. The applications addressed during the two subsequent flight test campaigns were vibration, power required, noise, and load reduction [204]. In the second campaign, vibration measurements based on 12 accelerometers in the fuselage and strain gauges at the IBC actuators were used as real time inputs to an adaptive closed loop controller [205,206].

Initiated by the positive results with respect to the automatic optimization of the rotor track and balance condition achieved during the wind-tunnel test campaign, the German DoD commissioned the development of a reduced authority IBC system for the Army's CH-53G helicopters. For this application, referred to as in-flight tuning, very small authorities and a low bandwidth are sufficient. Therefore pure electrical actuation was considered to be the most efficient solution. The system uses six so-called smart pitch rods (SPR^{TM}), which are safely restricted to a range of 0.2 degrees of blade pitch angle and require 10 s to travel from minimum to maximum length position. The SPR^{TM} actuators contain all control and power electronics necessary to realize the desired pitch link length adjustment. By architecture, the IFT system is redundant and self-reconfiguring, which provides functional robustness in the case of local failures. The 50+ FH campaign successfully showed the optimization of the rotor track and balance condition throughout all flight conditions compared to what could be achieved by the fixed setting of passive pitch links. The experiments also demonstrated the largely simplified recommissioning of an aircraft after blade exchange or control system maintenance actions. The system layout and the test results are described in Refs. [207, 208, 222].

3.2.4 Maturity, benefits, and drawbacks

Whereas none of the full-authority HHC and IBC systems has yet seen fleet-wide application, the IFT variant has been converted into a production system. In general, IBC systems can be easily retrofitted to most rotorcraft types, usually even without redesign of the adjacent control system components. In contrast to on-blade concepts such as active twist or trailing edge flaps, the blade root actuation IBC uses the existing blades without any modification. The applied technologies upon which such systems are based are all state-of-the-art, and the maturity would allow immediate introduction with a production aircraft.

Past trade studies have indicated that hydraulic solutions are well-suited for higher authority systems because of the high power density of the corresponding linear actuators. However, electrical actuation has improved considerably in recent years, and even slightly heavier actuators might be tolerable in many cases. The transfer of hydraulic pressure into the rotating system is often considered undesirable, although the test campaigns have actually never suffered from problems in this field.

Installing an additional system always comes with some weight and power consumption penalty. The weight of a production system tends to be in the range of 1% of the aircraft MTOW, however, strongly depending on the used technology and the provided performance in terms of authority and bandwidth. In many cases, the application of IBC will allow removal of existing passive means of vibration reduction (pendulum absorbers in the rotating frame, gearbox insulation, compensating spring-restrained masses in the fuselage), which then easily compensates the added system weight. With respect to the power consumption, the net requirement again depends on the used technology (hydraulic, electric) and the achievable power reduction. In cruise speed, the power saved by IBC will always overcompensate the system power consumption; in hover, a few kW are needed to operate the system. If the technical layout allowed for recuperative operation, as in the case of certain electrical system designs, power consumption actually would be very small for many suitable IBC inputs, because the mechanical power in terms of actuator load times actuator linear speed yields negative values for a certain azimuth range of a rotor revolution. Within one of the wind tunnel campaigns, it was discovered that at the 2/rev IBC setting optimized to reduce the rotor power required, the averaged mechanical power needed to operate the actuators was negative, which means that power was actually fed back from the aerodynamically loaded blades into the actuators. By design, however, the hydraulic system used for this test did not allow recovering that power.

Although the addressed improvements are immediately obvious to the rotorcraft designer, the primary goal is to yield operational and financial benefits from the aircraft owner's perspective. Table 6 relates the intended IBC effects with the direct or indirect benefits.

3.3 ACTIVE TWIST

In the late 1970s, several researchers had considered improving helicopter rotor aerodynamic performance and reducing rotor system loads and fuselage vibration by incorporating design features that would change blade twist as a function of flight condition and/or blade azimuthal position. The US Army [223–226] sponsored wind tunnel testing of blade designs that showed how combinations of built-in twist, blade camber, and tip sweep, in conjunction with reduced blade torsional stiffness could be used to twist blades to improve hover performance while decreasing blade and control system loads in forward flight (when compared to a conventional rotor). Those tests, however, demonstrated blades that incorporate reduced torsional stiffness to promote live twist can increase the fuselage vibration

Table 6 Summary of IBC Effects and Benefits

Application	Operational Benefit	Financial Benefit
Vibration reduction (N/rev)	Less crew fatigue, reduced health thread, longer allowable exposure	Reduced component wear; less unscheduled maintenance; longer component life
Noise reduction	Fewer missions barred because of legal noise limits	Less restricted operation in noise-controlled neighborhoods; lower landing fees
Reduction of power required (main rotor torque) and/or increase of lift over drag ratio	Extended range; longer endurance; higher max. speed	Reduced fuel consumption; higher transport capacity
Reduction of control system loads	Extended operational limits (high speed, high G maneuvers)	Longer component life
Reduction of mass and/or aerodynamic unbalances (1/rev vibrations)	For vibration reduction effects see above; reconfiguration after battle damage provides emergency flight performance	For vibration reduction effects see above; reconfiguration after battle damage can prevent loss of aircraft
Simplified RT&B	Simplified blade exchange, control system rigging; faster aircraft return to service, less (specialized) personnel required	Saved RT&B flight hours; less restrictive spare blade selection reduces logistic effort
Ground resonance suppression		Spared maintenance of lag dampers; higher useful load because of saved lag damper weight

associated with blade-to-blade manufacturing differences. In turn, it might require tighter tolerances on blade characteristics during manufacturing (such as torsional stiffness and airfoil contour) or an active vibration control system to deal with fuselage vibration at frequencies other than N/rev.

Several research organizations began pursuing active twist rotor (ATR) technologies during the mid-1990s as high-performance planar piezoelectric actuators were being developed. The active fiber composite (AFC) actuator, developed by MIT, was the primary actuator of choice at that time. Since then, macrofiber composite (MFC) actuators, developed jointly by NASA Langley Research Center and the US Army Research Laboratory (ARL), have become the actuator of choice because of improvements in control authority, actuator robustness, manufacturability, unit cost, and availability. Regardless of the specific actuator used, the ATR concept has the following distinguishing features: (1) actuator coverage over the bulk of the blade radius, often arranged in multiple layers, (2) actuator orientation relative the blade span of ± 45 degrees to induce twisting motions of the blade, and (3) no mechanical blade components subject to wear or maintenance requirements.

In 2000, the first demonstration of an ATR in forward flight was conducted at the NASA Langley Transonic Dynamics Tunnel (TDT) under a joint NASA/Army/MIT project [69,227]. The four-bladed, articulated rotor used a generic 55-in. radius, rectangular blade with a NACA-0012 airfoil section, -10 degrees of built-in twist, and solidity of 0.0982. Two layers of AFC actuators were embedded within the upper and lower airfoil surfaces over the outer 70% of the blade radius. The primary purpose of the wind-tunnel test was to demonstrate ATR fixed-system vibration reduction under open-loop control, therefore, 3P, 4P, and 5P active-twist actuations were used throughout the test. Sound pressure level acoustic data also were acquired [70]. A second test in the TDT, conducted in 2002, demonstrated that closed-loop control laws were able to adequately reduce fixed-system vibrations to levels at or below those demonstrated during the initial open-loop-control test [84,228]. A limited amounted of rotor performance improvement data was collected using closed-loop 2P actuation [228]. The test program demonstrated the ATR is capable of offering reductions in individual fixed-system harmonic vibratory loads of up to 100%, combined fixed-system load reductions of up to 70%, blade-vortex interaction noise reductions of up to 2.8 dB, and rotor performance improvement of up to 2% [69,70,84,227,228].

Despite the relative success of the wind-tunnel test demonstrations of ATR technology and the promise of additional capability with more modern actuators and improved rotor design, no attempt to manufacture a full-scale rotor system—either for prototype testing, commercial, or military purposes—has been planned. While ATR technology suffers from the same concerns as most other active rotor concepts demonstrated to date—survivability, potential increased weight, and transfer of electrical power from the fixed to the rotating systems—the primary reason for disinterest in developing a full-scale rotor system appears, at least within the commercial rotorcraft industry, to be a lack of evidence that ATR can provide a significant improvement in rotor performance, which would result in increased aircraft efficiency, speed, payload, and range. Additionally, the cost of embedding piezoelectric actuators across most of the blade radius would be expensive, and it is unclear whether the payoff is sufficiently adequate to justify full-scale development and increased rotor blade production costs.

The advanced material rotor project, under the direction of the Boeing Company, involved the development of Mach scaled CH-47 Chinook blades with active twist capabilities [86]. This project also used AFCs, but unlike the NASA/Army/MIT effort, the blades had a more complex planform. The blades were hover and wind-tunnel tested at the Boeing VTOL wind tunnel. As part of the overall project, the researchers also considered active flap actuators, and interestingly, reached the determination that an AFC-driven active twist rotor system was more cost effective. In particular, the production and weight costs of the active twist rotor system were estimated to be lower because of its relatively simple blade integration procedure as compared to an active flap [86]. Although limited experimental data is available on these tests, the scaled CH-47 blades were shown to be capable of producing tip twists comparable to the NASA/Army/MIT active twist rotor. Following their initial investigations based on a scaled CH-47 rotor, researchers at Boeing shifted their focus to a full-scale CH-47 AFC actuated active twist rotor system [87]. As part of this effort, a section of a full-scale blade was built with AFC actuators incorporated into the spar. Because of the size and stiffness of the blade, several layers of the AFC actuators were embedded to increase the actuation authority. While twist was demonstrated on the bench-top using AFC actuation, the authority was lower than desired because of manufacturing issues with certain layers of actuators. Imperfections in fiber spacing and diameter led to nonuniformity in the field applied by the inter-digitated electrodes and the strain induced in the piezo fibers. The labor associated with the fiber placement and the fabrication process, and the performance degradation because of imperfections represents serious limitations of the AFC actuator technology. These efforts highlight the difficulty of the mass production of such an active twist rotor blade.

Following the conclusion of the ATR program, Army researchers began studying methods to attack the weaknesses in the ATR technology that led to the lack of full-scale development of the concept. While little can be done to eliminate the need to transfer power to the rotating system to drive an active rotor system, concepts have been developed to address actuator cost, weight, and survivability issues. Chief among these concepts is the continuous trailing edge flap (CTEF), in which a portion of the trailing-edge of the blade near the tip is outfitted with MFC-class actuators that distort the trailing-edge, thereby effecting a change in the local blade aerodynamic characteristics [229]. Such a concept places the actuators only in the region of highest dynamic pressure where blade distortion will provide the greatest control authority, thereby achieving goals similar to those of the ATR with far fewer actuators. The CTEF concept is beneficial as compared to discrete flap configurations because there are no moving parts and no discontinuity of the aerodynamic surface, because the rotor blade skin remains continuous across the passive and active blade sections. Much research and development remains for the CTEF concept to make its way onto a commercial or military aircraft. The technology, however, does offer several advantages over the predecessor ATR in that it might readily be developed for simple improvements such as automatic rotor blade tracking at a relatively low cost, while offering significant leaps in the technology by offering a solution for swashplateless rotor primary flight control [230]. Developing a CTEF system readily replaceable if it fails or [is] damaged also appears to be feasible. In contrast, a damaged ATR blade likely would have to be replaced at a significantly higher cost.

While the tip twist performance has improved significantly over the years with higher authority MFC actuators and properly designed blades, the main challenge of replaceability in the event of actuator failure remains. Because of the high degree of integration, actuator failure at this time eventually would require replacement of the entire blade. While this could be an option for small-scale rotors and UAVs, it is not cost-effective for full-sized aircraft. If the question of actuator replaceability can be adequately resolved, the actuation authority is clearly sufficient for vibration reduction (as an example, the NASA/Army/MIT rotor was capable of reducing up to 95% of the vibration levels with only 1.5 degrees tip twist).

In parallel to the AFC actuated active twist rotor, other researchers at Boeing focused on the development of a quasi-statically variable twist rotor under the auspices of the ONR-funded reconfigurable rotor blade (RRB) program [122–126]. The goal was to apply the technology to tilt-rotor blades. The actuation system consisted of two shape memory alloy (SMA) tubes operating in a complementary manner. When heated, these SMA tubes undergo large strain and twist and can provide a large output torque. The SMA actuator was housed within the root of the tilt-rotor blade with the output and connected to a torsionally stiff torque tube. The other end of this passive torque tube (which spans the remainder of the blade) was attached to the blade tip to elastically twist the blade as the torque tube underwent rigid body rotation. The project resulted in the development of technologies for efficient heat transfer to the SMA tubes, and the SMA actuator underwent several generations of design iterations to maximize efficiency before incorporation into the prototype blade. The blade/actuator system was tested in the wind tunnel the late 2000s on a 1/8th scale model three-bladed rotor (similar to tilt-rotors). Being the first, and to date only, wind-tunnel test of an actuated variable twist rotor, this work represents a milestone. However, with the baseline blade structure undergoing only minor redesign to incorporate the actuation system it was much too torsionally stiff, and the SMA actuators were able to produce only around 2 degrees of static twist [125,126]. Even if the SMA actuators could produce higher torque, the strains in the blade would increase. This work highlights the importance of suitably reducing the torsion stiffness and ensuring that the modified blade can withstand the strains associated with large quasi-static twist deformations.

3.4 VARIABLE SPAN

One much-studied concept for helicopters and tilt rotors has been the variable diameter rotor. For helicopters, this is normally applied to hybrid or compound designs that feature an aircraft with wings and conventional tail surfaces and fitted with a rotor on top and some form of forward propulsion.

In hover and at low speed, the rotor has a large diameter, which reduces both the downwash velocity and the power required. At higher speeds, the rotor diameter is reduced as the wing assumes more lift. This reduces the rotor drag significantly and also can serve as the basis for stopped and stowed rotors.

When applied to tilt-rotor aircraft, the benefits are similar, but for different reasons. In hover and low speed, larger diameter prop-rotors reduce downwash velocity and power required as they do for helicopters because of increased disc area. The rotors retract as it progresses into forward flight in order to both clear the fuselage and to raise blade angles of attack to increase section lift to drag ratio. In forward flight, the prop-rotor diameter is reduced to better optimize it for high-speed flight where it acts like a propeller and needs to produce much less thrust. Therefore, the compromise of designing prop-rotor for both hover and high speed is much reduced and results in a more efficient aircraft.

This concept was successfully tested as a reduced tip speed wind-tunnel model [141,231]. Follow-up work developed full-scale actuation and blade retention concepts, and evaluated aircraft performance for various applications [232].

Reducing the diameter of the blade can be approached in many ways, but none has yet led to flight of a manned aircraft. The approaches, discussed in detail by Linden [133], are:

- Telescoping blades with one or more sections
- Vertically folding blades
- Very thin, flexible blades
- Pneumatic retraction.

The first case is the most obvious method, and designs have been wind tunnel-tested for both conventional rotors and tilt rotors [140,233]. While the aerodynamic benefits have been shown clearly, the design of mechanisms has not progressed sufficiently to allow manned flight test. The retraction system has to resist the great centrifugal force on the outboard segment(s), the lift loads, and the high vibratory environment, while guaranteeing that the retractable portions of all blades are at exactly the same radial positions. In addition, the air vehicle has to be able to land safely with the rotor(s) at either extreme position.

Another proposed approach is to hinge the blade at a suitable outboard point and have it fold up vertically and over to lie on top of the inboard portion of the blade. While this is a simpler approach mechanically, it has never been applied to any manned aircraft.

The third case is completely different from the other first two. In this case, a thin flexible blade is wound on a drum, like a tape measure. In this case tip weights are used to generate a high centrifugal force to keep the blade stiff. However, torsional stiffness is a problem, and the aerodynamic efficiency is poor.

The last approach is to make the outboard portion of the blade out of flexible material and use high-pressure air to push it out of the inboard section and inflate it to an airfoil cross-section. While patents on such a design exist, no actual tests are known to have been made.

3.5 SLOWED/STOPPED ROTOR

In the search for high-speed VTOL aircraft, a hybrid or compound design is frequently selected. As noted earlier, this is a helicopter with wings and auxiliary propulsion added. However at high speeds of 250–300 knots, the drag of the main rotor continues to be high, even if reduced in diameter, and is detrimental to performance. The maximum forward speed attainable for a helicopter is limited by the need to keep the tip of the advancing blade from experiencing transonic flow.

An area in which morphing technology might play a role is for an aircraft that transitions from the helicopter mode for low-speed flight to a reduced rotor speed or a stopped rotor configuration for high-speed flight. Retractable rotors that could be stowed in the fuselage also have been envisioned.

Hughes aircraft proposed a rotor with a very large equilateral shaped hub and stub blades extending out of each vertex. At low speed, the rotor rotates in the normal manner with the blades providing both lift and control via collective and cyclic pitch changes. At higher speeds, with the wings producing more lift, the rotor is slowed and then stopped with one vertex forward. In this condition, the hub now acts like a triangular wing. While an interesting concept, no manned aircraft has even been built. Boeing has recreated this concept using a large circular hub with four blades extending from it. It worked in the same manner except that as the rotor slowed, the blades also retracted, so that at high speed the hub became a large circular wing on top of the airframe. Conventional wings, tail surfaces, and forward propulsion were retained.

One approach considered by NASA and the US Army in the 1980s was X-Wing. in which a four-bladed rotor was to be stopped at moderate forward speed with two blades swept forward at a 45-degree angle and two blades aft at a 45-degree angle [234]. This used a rigid four bladed rotor that employed circulation controlled airfoils to produce lift by blowing air out of a radial slot on the blade. For X-Wing, slots were located on both the leading edge and trailing edge of the blade so lift could be produced if the blade was in either forward flow or reverse flow. The rotor would rotate normally at low speed, and then, at about 200 knots, the rotor would slow and stop with the rotor in an X configuration. In this way, an additional wing was not required. In order to avoid divergence of the forward swept blades, however, they had to be attached to the hub rigidly, and flatwise bending stiffness had to be very high. Two consequences of high stiffness, according to simulations, were high blade loads and potentially high fuselage vibration in the rotary wing mode. If active or passive morphing features that avoided divergence could be incorporated into a blade of more conventional stiffness, rotary wing blade load and vibration issues could be mitigated, thereby enabling a concept such as this to succeed.

Extensive development was undertaken, starting with Lockheed performing analyses and testing a small demonstrator in the NASA Ames 40 × 80 wind tunnel. Sikorsky Aircraft later took over the program and made extensive modifications to the NASA rotor systems research aircraft (RSRA) to install a large X-Wing rotor for testing. The program ran over budget, so it was never flown with the X-Wing rotor operational.

This concept was later modified by Boeing into the Canard Rotor Wing (CRW) design and two UAVs were built and tested. In this case the rotor was two bladed and driven by tip jets. The blades were conventional and used normal collective and cyclic pitch. A forward canard and large horizontal tail provided lift and control while the rotor was being stopped. Unfortunately, both UAVs would crash due to failures not related to CRW concept.

4 CONCLUSIONS

To date, the only morphing technology that has made it to a production helicopter is servo flaps, first used in the Kaman K-125, the inaugural flight of which was Jan. 15, 1947, and more recently used on the K-Max (Fig. 2).

A plethora of innovative methods exist to produce the active twist of rotor blades. With the exception of the AFC actuator design, however, none of the programs discussed has been developed for

FIG. 2

Kaman K-Max: servo flap close-up [235].

full-scale active twist rotor systems. Despite the significant amount of work done in this area, significant room remains for developing innovative solutions for full-scale active twist rotor designs. That being said, the designs in the current state, as mentioned earlier, are mature enough to be used for UAVs.

IBC is being investigated in several projects on full-scale helicopters and is possibly the closest to production. The benefits of higher harmonic control have been investigated and tested by multiple authors. The use of IBC technology as primary control of the blade root actuation does not exclude its combination with on-blade morphing elements, which tend to be mounted farther outboard and primarily affect blade twist. By using a coordinated combination of higher harmonic blade root and active flap motions, a more sophisticated and spatially distributed approximation of the optimum profile section orientation could be realized.

Variable span has been one of the earliest technologies to be investigated, but it also is one of the most difficult to implement, with no success story to date. Complexity, actuation requirements, and safety concerns have hindered its implementation on full-scale helicopters with no known additional research currently being done in this direction.

REFERENCES

[1] F. Straub, V. Anand, T. Birchette, B. Lau, in: SMART rotor development and wind tunnel test, 35th European Rotorcraft Forum, Hamburg, Germany, 2009.

[2] O. Dietrich, B. Enenkl, D. Roth, in: Trailing edge flap for active rotor control aeroelastic characteristics of the ADASYS rotor system, 62nd American Helicopter Society Forum, Phoenix, AZ, 2006.

[3] D. Roth, B. Enenkl, O. Dietrich, in: Active rotor control by flaps for vibration reduction-full scale demonstrator and first flight results, 32nd European Rotorcraft Forum, Maastricht, The Netherlands, September 2006.

[4] O. Leon, Reducing helicopter main rotor power requirements using multiple trailing edge flaps and extendable chord sections, Master's thesis, The Pennsylvania State University, 2009.

[5] I. Chopra, Review of state of art of smart structures and integrated systems, AIAA J. 40 (11) (November 2002) 2145–2187.

[6] R. Spangler, Piezoelectric actuators for helicopter rotor control, Master's thesis, Massachusetts Institute of Technology, February 1989.

[7] E. Prechtl, Development of a piezoelectric servo-flap actuator for helicopter rotor control, Master's thesis, Massachusetts Institute of Technology, May 1994.

[8] S. Hall, E. Prechtl, Development of a piezoelectric servoflap for helicopter rotor control, Smart Mater. Struct. 5 (1996) 26–34.

[9] D. Samak, I. Chopra, in: A feasibility study to build a smart rotor: trailing edge flap actuation, SPIE Symposium on Smart Structures and Intelligent Systems, vol. 1917, 1993.

[10] C. Walz, I. Chopra, in: Design and testing of a helicopter rotor model with smart trailing edge flaps, 35th Structures, Structural Dynamics and Materials Conference, Adaptive Structures Forum, Hilton Head, SC, April 1994.

[11] M. Fulton, R. Ormiston, in: Hover testing of a small-scale rotor with on-blade elevons, 53rd American Helicopter Society Forum, Virginia Beach, VA, April 29–May 1, 1997.

[12] O. Ben-Zeev, I. Chopra, Advances in the development of an intelligent helicopter rotor employing smart trailing-edge flaps, Smart Mater. Struct. 5 (1996) 11–25.

[13] M. Fulton, R. Ormiston, in: Small-scale rotor experiments with on-blade elevons to reduce blade vibratory loads in forward flight, 54th American Helicopter Society Forum, Washington, DC, May 20–22, 1998.

[14] N. Koratkar, I. Chopra, in: Analysis and testing of Mach scaled rotor model with piezoelectric bender actuated trailing-edge flaps for helicopter vibration control, 40th AIAA/ASME/ASCE/AHS/ASC Structures, Structural Dynamics, and Materials Conference and Exhibit, St. Louis, MO, April 12–15, 1999.

[15] N. Koratkar, Smart helicopter rotor with piezoelectric bender actuated trailing-edge flaps, Ph.D. thesis, University of Maryland, 2000.

[16] N. Koratkar, I. Chopra, Wind tunnel testing of a Mach-scaled rotor model with trailing-edge flaps, Smart Mater. Struct. 10 (2001) 1–14.

[17] B. Spencer, I. Chopra, in: Design and testing of a helicopter trailing edge flap with piezoelectric stack actuators, SPIE Symposium on Smart Structures and Materials, vol. 2717, February 1996.

[18] R. Chandra, I. Chopra, in: Actuation of trailing edge flap in wing model using piezostack device, 38th Structures, Structural Dynamics, and Materials Conference and Adaptive Structures Forum, Kissimmee, FL, April 7–10, 1997.

[19] T. Lee, I. Chopra, Design of a piezostack-driven trailing-edge flap actuator for helicopter rotors, Smart Mater. Struct. 10 (2001) 15–24.

[20] T. Lee, I. Chopra, in: Design and static testing of a trailing-edge flap actuator with piezostaccks for a rotor blade, SPIE Symposium on Smart Structures and Integrated Systems, San Diego, CA, vol. 3329, March 1999.

[21] T. Lee, High displacement piezoelectric trailing-edge flap mechanism for helicopter rotors, Ph.D. thesis, University of Maryland, 1999.

[22] S. Hall, E. Prechtl, in: Preliminary testing of a Mach-Scaled Active Rotor Blade with a Trailing Edge Servo-Flap, SPIE Symposium on Smart Structures and Materials, 1999.
[23] E. Prechtl, Design and implementation of a piezoelectric servo-flap actuation system for helicopter rotor individual blade control, Ph.D. thesis, Massachusetts Institute of Technology, February 2000.
[24] S. Hall, T. Tzianetopoulou, F. Straub, H. Ngo, in: Design and testing of a double X-frame piezoelectric actuator, SPIE Symposium on Smart Strucutres and Materials, 2000.
[25] F. Straub, D. Merkley, in: Design of a smart material actuator for rotor control, SPIE North American Conference on Smart Structures and Materials, San Diego, CA, February 26–March 3, 1995.
[26] F. Straub, M. Ealey, L. Schetky, in: Application of smart materials to helicopter rotor active control, SPIE Symposium on Smart Structures and Materials, San Diego, CA, March 3–6, 1997.
[27] F. Straub, H. Ngo, V. Anand, D. Domzalski, in: Development of a piezoelectric actuator for trailing edge flap control of full scale rotor blades, SPIE Symposium on Smart Structures and Materials, Newport Beach, CA, 3668-104, 1999.
[28] F. Straub, D. Kennedy, D. Domzalski, A. Hassan, H. Ngo, V. Anand, T. Birchette, Smart material-actuated rotor technology - SMART, J. Intell. Mater. Syst. Struct. 15 (April 2004) 249–260.
[29] F. Straub, D. Kennedy, A. Stemple, V. Anand, T. Birchette, in: Development and whirl tower test of the SMART active flap rotor, SPIE International Symposium on Smart Structures and Materials, San Diego, CA, March 14–18, 2004.
[30] A. Bernhard, I. Chopra, in: Trailing edge flap activated by a piezo-induced bending-torsion coupled beam, 52nd American Helicopter Society Forum, Washington, DC, May 1996.
[31] A. Bernhard, Smart helicopter rotor with active blade tips, Ph.D. thesis, University of Maryland, 2000.
[32] J. Clement, D. Brei, A. Moskalik, R. Barrett, in: Bench-top characterization of an active rotor blade flap system incorporating C-block actuators, 39th AIAA/ASME/ASCE/AHS/ASC Structures, Structural Dynamics, and Materials Conference, Long Beach, CA, April 20–23, 1998.
[33] J. Clement, D. Brei, R. Barrett, in: Wind tunnel testing of a high authority airspeed insensitive rotor blade flap, 40th AIAA/ASME/ASCE/AHS/ASC Structures, Structural Dynamics, and Materials Conference, St. Louis, MO, April 12–15, 1999.
[34] L. Centolanza, E. Smith, B. Munsky, Induced-shear piezo-electric actuators for rotor blade trailing edge flaps, Smart Mater. Struct. 11 (2002) 24–35.
[35] L. Centolanza, Induced shear piezoelectric actuators for smart rotor blades, Ph.D. thesis, The Pennsylvania State University, 2001.
[36] J. Szefi, E. Mockensturm, E. Smith, W.K.P. Rehrig, L. Centolanza, in: Development of a novel high authority piezo-electric actuator for rotor blades with trailing edge flaps, 62nd American Helicopter Society Forum, Pheonix, AZ, May 9–11, 2006.
[37] T. Hongu, M. Sato, E. Yamakawa, in: Elementary studies of active flap control with smart material actuators, 12th European Rotorcraft Forum, September 14–16, 1999.
[38] Y. Hasegawa, N. Katayama, N. Kobiki, E. Nakasato, E. Yamakawa, H. Okawa, in: Experimental and analytical results of whirl tower test of ATIC full scale rotor system, 57th American Helicopter Society Forum, Washington, DC, May 9–11, 2001.
[39] H. Mainz, B. van der Wall, P. Leconte, in: ABC rotor blades: design, manufacturing and testing, 31st European Rotorcraft Forum, Florence, Italy, 2005.
[40] P. Leconte, H. Mercier des Rochettes, in: Experimental assessment of an active flap device, 58th American Helicopter Society Forum, Montreal, Quebec, Canada, June 11–13, 2002.
[41] Y. Delrieux, A. Le Pape, P. Leconte, P. Crozier, B. Gimonet, H. Mercier des Rochettes, in: Wind-tunnel assessment of the concept of active flaps on a helicopter rotor model, 63rd American Helicopter Society Forum, Virginia Beach, VA, May 1–3, 2007.
[42] B. Enenkl, V. Kloppel, D. Preissler, P. Janker, in: Full scale rotor with piezoelectric actuated blade flaps, 28th European Rotorcraft Forum, Bristol, UK, 17–19 September, 2002.

[43] J. Lee, J. Kwak, K. Son, S. Shin, in: Development of an SNU intelligent rotor blade with an active trailing-edge flap, 66th American Helicopter Society Forum, Pheonix, AZ, May 11–13, 2010.
[44] T. Fukami, N. Kobiki, T. Komura, and S. Saito (2007) "Design and performance evaluation of full scale on-board active flap system," in 63rd American Helicopter Society Forum, Virginia Beach, VA, May 1–3.
[45] K. Noboru, S. Saito, in: Performance evaluation of full scale on-board active flap system in transonic wind tunnel, 64th American Helicopter Society Forum, Montreal, Quebec, Canada, April 29–May, 2008.
[46] S. Saito, N. Kobiki, Y. Tanabe, W. Johnson, G. Yamauchi, L. Young, in: Overview of the novel intelligent JAXA active rotor program, Heli Japan 2010, AHS International Meeting on Advanced Rotorcraft Technology and Safety Operations, Ohmiya, Japan, November 13, 2010.
[47] S. Dawson, F. Straub, in: Design, validation and test of a model rotor with tip mounted active flaps, 50th American Helicopter Society Forum, Washington, DC, May 11–13, 1994.
[48] J. Milgram, I. Chopra, F. Straub, Rotors with trailing edge flaps: analysis and comparision with experimental data, J. Am. Helicopter Soc. 43 (4) (October 1998) 319–332.
[49] C. Bothwell, R. Chandra, I. Chopra, Torsional actuation with extension-torsion composite coupling and a magnetostrictive actuator, AIAA J. 33 (4) (April 1995) 723–729.
[50] R. Fenn, J. Downer, D. Bushko, V. Gondhalekar, N. Ham, Terfenol-D driven flap for helicopter vibration reduction, Smart Mater. Struct. 5 (1996) 49–57.
[51] D. Fink, T. Hawkey, M. Gaudreau, in: An electro-magnetic actuator for individual blade control, 56th American Helicopter Society Forum, Virginia Beach, VA, May 2–4, 2000.
[52] M. Duvernier, L. Reithler, J. Guerrero, R. Rossi, in: Active control system for a rotor blade trailing-edge flap, Proceedings of SPIE, vol. 3985, 2000.
[53] A. Bernhard, J. O'Neil, F. Kohlhepp, W. Welsh, P. Lorber, in: Active Rotor Control (ARC) of a mach-scale trailing edge flap rotor, 57th American Helicopter Society Forum, Washington, DC, May 9–11, 2001.
[54] J. Sirohi, I. Chopra, Design and development of a high pumping frequency piezoelectric-hydraulic hybrid actuator, J. Intell. Mater. Syst. Struct. 14 (March 2003) 135–147.
[55] S. Herdic, Development of piezo-hydraulic actuation systems technology for use on a helicopter trailing edge flap, Master's thesis, Georgia Institute of Technology, December 2005.
[56] Z. Chaudhry, B. Wake, A. Bagai, P. Lorber, A. Collins, in: Active rotor development for primary and secondary flight control, 65th American Helicopter Society Forum, Grapevine, TX, May 27–29, 2009.
[57] P. Lorber, J. O'Neil, B. Isabella, B. Wake, U. Jonsson, F. Sun, in: Whirl test of a large scale high authority active flap rotor, 66th American Helicopter Society Forum, Phoenix, AZ, May 11–13, 2010.
[58] P. Lorber, J. O'Neil, B. Hein, B. Isabella, K. Andrews, M. Brigley, J. Wong, P. LeMasurier, B. Wake, in: Whirl and wind tunnel testing of the sikorsky active flap demonstration rotor, 67th American Helicopter Society Forum, Virginia Beach, VA, May 3–5, 2011.
[59] B. Woods, E. Bubert, C. Kothera, J. Sirohi, N. Wereley, in: Experimental testing of pneumatic artifical muscles for trailing edge flap actuation, 48th AIAA/ASME/ASCE/AHS/ASC Structures, Structural Dynamics, and Materials Conference, Honolulu, HI, April 23–26, 2007.
[60] Woods, B., E. Bubert, C. Kothera, and N. Wereley (7–10 April, 2008) "Design and testing of a biologically inspired pneunatic trailing edge flap system," in 49th AIAA/ASME/ASCE/AHS/ASC Structures, Structural Dynamics, and Materials Conference, Schaumburg, IL.
[61] B. Woods, N. Wereley, C. Kothera, in: Wind tunnel testing of a helicopter rotor trailing edge flap actuated via pneumatic artifical muscles, ASME 2010 Conference on Smart Materials, Adaptive Structures and Intelligent Systems, Philadelphia, PA, September 28–October 1, 2010.
[62] B. Woods, C. Kothera, N. Wereley, in: Whirl testing of a pneumatic artifical muscle driven helicopter trailing edge flap, 66th American Helicopter Society Forum, Pheonix, AZ, May 11–13, 2010.
[63] R.H. Liebeck, Design of subsonic airfoils for high lift, J. Aircr. 15 (1978) 547–561.

[64] E. Bae, F. Gandhi, M. Maughmer, in: Optimally scheduled deployments of miniature trailing-edge effectors for rotorcraft power reduction, 65th American Helicopter Society Forum, Grapevine, TX, May 27–29, 2009.
[65] S. Thepvongs, Design and testing of an active Gurney flap actuator for helicopter rotor blade control, Bachelors thesis, Tech. rep, The Pennsylvania State University, 2002.
[66] M. Thiel, Actuation of an active gurney flap for rotorcraft applications, Master's thesis, The Pennsylvania State University, 2006.
[67] J. Palacios, A. Overmeyer, E. Smith, M. Maughmer, M. Kinzel, J. Szefi, in: Experimental measurement of unsteady lift and pitching moments for MiTEs on a S903 airfoil, 67th American Helicopter Society Forum, Virginia Beach, VA, May 3–5, 2011.
[68] M.L. Wilbur, W.T. Yeager Jr., W.K. Wilkie, C.E.S. Cesnik, S. Shin, in: Hover testing of the NASA/ARMY/MIT active twist rotor prototype blade, 56th Annual Forum of the American Helicopter Society, Virginia Beach, VA, May 2–4, 2000.
[69] Wilbur, M. L., Mirick, P.H., Yeager, W. T. Jr., Langston, C. W., Cesnik, C. E. S., and Shin, S., "Vibratory loads reduction testing of the NASA/ARMY/MIT active twist rotor," J. Am. Helicopter Soc., Vol. 47, (2), April 2002, pp. 123–133.
[70] M.L. Wilbur, W.T. Yeager Jr., M.K. Sekula, in: Further examination of the vibratory loads reduction results from the NASA/Army/MIT active twist rotor test, 58th Annual Forum of the American Helicopter Society, Montreal, Quebec, Canada, June 2002.
[71] M. Sekula, M. Wilbur, W. Yeager, in: A parametric study of the structural design for an advanced active twist rotor, 61st American Helicopter Society Forum, Grapevine, TX, June 1–3, 2005.
[72] M. Floros, W. Johnson, M. Scully, in: Advanced rotor aerodynamics concepts with application to large rotorcraft, American Helicopter Society Aerodynamics, Acoustics, and Test and Evaluation Technical Specialists Meeting, San Francisco, CA, January 23–25, 2002.
[73] C. Keys, F. Tarzanin, F. McHugh, in: Effect of twist on helicopter performance and vibratory loads, 13th European Rotorcraft Forum, Arles, France, September 8–11, 1987.
[74] W. Bousman, in: Airfoil design and rotorcraft performance, 58th American Helicopter Society Forum, Montreal, Quebec, Canada, June 11–13, 2002.
[75] G. Leishman, Principles of Helicopter Aerodynamics, Cambridge University Press, Cambridge, United Kingdom, 2006.
[76] R. Barrett, Intelligent rotor blade and structures development using directionally attached piezoelectric crystals, Master's thesis, University of Maryland, 1990.
[77] P. Chen, I. Chopra, Induced strain actuation of composite beams and rotor blades with embedded piezoceramic elements, Smart Mater. Struct. 5 (1996) 35–48.
[78] J. Jang, I. Chopra, Ground and air resonance of an advanced bearingless rotor in hover, J. Am. Helicopter Soc. 33 (3) (July 1988).
[79] P. Chen, Development of a smart rotor with induced-strain actuation of blade twist, Ph.D. thesis, University of Maryland, 1996.
[80] P. Chen, I. Chopra, Wind tunnel test of a smart rotor model with individual blade twist control, J. Intell. Mater. Syst. Struct. 8 (1997) 414–425.
[81] A. Bent, N. Hagood, Piezoelectric fiber composites with interdigitated electrodes, J. Intell. Mater. Syst. Struct. 8 (1997) 903–919.
[82] R. Derham, N. Hagood, in: Rotor design using smart materials to actively twist blades, 52nd Annual Forum American Helicopter Society, Washington, DC, 1996.
[83] J. Rodgers, Development of an integral twist-actuated rotor blade for individual blade control, Ph.D. thesis, Massachusetts Institute of Technology, February 1999.
[84] S. Shin, Integral twist actuation of helicopter rotor blades for vibration reduction, Ph.D. thesis, Massachusetts Institute of Technology, 2001.

[85] S. Shin, C. Cesnik, S. Hall, in: Closed-loop control test of the NASA/Army/MIT active twist rotor for vibration reduction, 59th American Helicopter Society Forum, Phoenix, AZ, May 6–8, 2003.
[86] R. Derham, D. Weems, M.B. Mathew, R. Bussom, in: The design evolution of an active materials rotor, 57th American Helicopter Society Forum, Washington, DC, May 9–11, 2001.
[87] D. Weems, D. Anderson, M.B. Mathew, R. Bussom, in: A large-scale active-twist rotor, 60th American Helicopter Society Forum, Baltimore, MD, June 7–10, 2004.
[88] W.K. Wilkie, R.G. Bryant, J.W. High, R.L. Fox, R.F. Hellbaum, in: Low-cost piezocomposite actuator for structural control applications, Smart Structures and Materials 2000: Industrial and Commercial Applications of Smart Structures Technologies, Proceedings of SPIE, vol. 3991, 2000.
[89] A. Buter, E. Breitbach, Adaptive blade twist – calculations and experimental results, Aerosp. Sci. Technol. 4 (2000) 309–319.
[90] P. Wierach, J. Riemenschneider, S. Opitz, F. Hoffmann, in: Experimental investigation of an active twist model rotor blade under centrifugal loads, 33rd European Rotorcraft Forum, Kazan, Russia, 2007.
[91] H.P. Monner, S. Opitz, J. Riemenschneider, P. Wierach, in: Evolution of active twist rotor designs at DLR, 49th AIAA/ASME/ASCE/AHS/ASC Structures, Structural Dynamics, and Materials Conference, Schaumburg, IL, April 7–10, 2008.
[92] F. Hoffmann, S. Opitz, J. Riemenschneider, in: Validation of active twist modeling based on whirl tower tests, 65th Annual American Helicopter Society Forum, Grapevine, TX, May 27–29, 2009.
[93] J. Riemenschneider, S. Opitz, P. Wierach, H. Mercier des Rochettes, L. Buchaniek, D. Joly, in: Structural design and testing of active twist blades—A comparison, 35th European Rotorcraft Forum, Hamburg, Germany, September 22–25, 2009.
[94] S. Opitz, J. Riemenschneider, F. Hoffmann, O. Schneider, in: Measurement of the dynamic tip twist angle of an active twist model scale rotor blade, 36th European Rotorcraft Forum, Paris, France, 2010.
[95] S. Opitz, J. Riemenschneider, M. Schulz, in: Prerequisites for the closed loop control of an active twist rotor blade, 67th American Helicopter Society Forum, May 3–5, 2011.
[96] Opitz, S. (June 2011) in Personal Correspondence, DLR Braunschweig.
[97] H.P. Monner, J. Riemenschneider, S. Opitz, M. Schulz, in: Development of active twist rotors at the German Aerospace Center (DLR), 52nd AIAA/ASME/ASCE/AHS/ASC Structures, Structural Dynamics and Materials Conference, Denver, CO, April 4–7, 2011.
[98] M.K. Sekula, M. Wilbur, W. Yeager, in: Aerodynamic design study of an advanced active twist rotor, American Helicopter Society 4th Decennial Specialists Conference on Aeromechanics, San Francisco, CA, January 21–23, 2004.
[99] M. Wilbur, M. Sekula, in: The effect of tip geometry on Active-twist rotor response, 61st American Helicopter Society Forum, Grapevine, TX, June 1–3, 2005.
[100] R. Thornburgh, A. Kreshock, M. Wilbur, in: Structural optimization of active-twist rotor blades, 67th American Helicopter Society Forum, Virginia Beach, VA, May 3–5, 2011. this is the Apache ATR paper.
[101] C. Cesnik, J. Mok, J. Morillo, A. Parikh, in: Design optimization of active twist rotor blades, 30th European Rotorcraft Forum, Marseille, France, September 2004.
[102] A. Kreshock, R. Thornburgh, M. Wilbur, C.W. Langston, in: Experimental bench testing of an active-twist rotor blade, 53rd AIAA/ASME/ASCE/AHS/ASC Structures, Structural Dynamics and Materials Conference, Honolulu, HI, April 23–26, 2012.
[103] D. Thakkar, R. Ganguli, Induced shear actuation of helicopter rotor blade for active twist control, J. Thin-Walled Struct. 45 (2007) 111–121.
[104] A. Bernhard, I. Chopra, Hover test of mach-scale active twist rotor using piezo-bendingtorsion actuators, J. Aircr. 39 (4) (2002) 678–688.
[105] A. Bernhard, I. Chopra, Analysis of a bending-torsion coupled actuator for a smart rotor with active blade tips, Smart Mater. Struct. 10 (2001) 35–52.

[106] H. Mercier des Rochettes, D. Joly, L. Buchaniek, P. Leconte, in: A new concept of active twist rotor blade applied to main rotor of helicopter, RTO Applied Vehicle Technology Panel (AVT) Symposium, Evora, Portugal, April 20–24, 2009.
[107] J. Riemenschneider, S. Keye, P. Wierach, H. des Rochetter, in: Review of the common DLR/ONERA project Active Twist Blade (ATB), 30th European Rotorcraft Forum, Marseille, France, September 14–16, 2004, pp. 22.1–22.9.
[108] L. Centolanza, E. Smith, in: Induced-shear piezoelectric actuators for active twist rotor blades, 43rd AIAA/ASME/ASCE/AHS/ASC Structures, Structural Dynamic and Materials Conference, Denver, CO, 22–25 April, 2002.
[109] Nixon, M. (April 1988) Improvements to Tilt Rotor Performance Through Passive Blade Twist Control, Tech. Rep., NASA TM 100583, USAAVSCOM TM 88-B-010.
[110] R. Lake, M. Nixon, M. Wilbur, J. Singleton, in: A demonstration of passive blade twist control using extension-twist coupling, 33rd AIAA/ASME/ASCE/AHS/ASC Structures, Structural Dynamics and Materials Conference, Dallas, TX, April 13–15, 1992.
[111] R. Lake, M. Nixon, M. Wilbur, J. Singleton, P. Mirick, in: Demonstration of an elastically coupled twist control concept for tilt rotor blade application, 33rd AIAA/ASME/ASCE/AHS/ASC Structures, Structural Dynamics, and Materials Conference, Dallas, TX, April 13–15, 1992.
[112] J. Kosmatka, R. Lake, in: Extension-twist behavior of initially twisted composite spars for tilt-rotor applications, 37th AIAA/ASME/ASCE/AHS/ASC Structures, Structural Dynamics and Materials Conference and Exhibit, Salt Lake City, UT, April 15–17, 1996.
[113] S. Nampy, Structural behavior and design of flexible matrix composite box beams with extension twist coupling, Master's thesis, The Pennsylvania State University, August 2005.
[114] J. DiOttavio, D. Friedmann, in: Operational benefits of an optimal, widely variable speed rotor, 66th American Helicopter Society Forum, Phoenix, AZ, May 11–13, 2010.
[115] W. Guo, Flight control design for rotorcraft with variable rotor speed, Ph.D. thesis, The Pennsylvania State University, December 2009.
[116] J. Steiner, F. Gandhi, Y. Yoshizaki, in: An investigation of variable rotor RPM on performance and trim, American Helicopter Society 64th Annual Forum, Montreal, Quebec, Canada, April 29–May 1, 2008.
[117] M. Mistry, F. Gandhi, in: Performance improvement with variable rotor span and RPM, 66th Annual AHS Forum, Phoenix, AZ, May 11–13, 2010.
[118] G. Bowen-Davies, I. Chopra, in: Aeromechanics of a variable-speed rotor, 67th American Helicopter Society Forum, Virginia Beach, VA, May 3–5, 2011.
[119] B. Berry, I. Chopra, in: Wind tunnel testing for performance and vibratory loads of a variable-speed mach-scale rotor, 67th American Helicopter Society Forum, Virginia Beach, VA, May 3–5, 2011.
[120] A. Datta, H. Yeo, T. Norman, in: Experimental investigation and fundamental understanding of a slowed UH-60A rotor at high advance ratios, 67th American Helicopter Society Forum, Virginia Beach, VA, May 3–5, 2011.
[121] B. Berry, I. Chopra, in: Performance and vibratory load measurements of a slowed-rotor at high advance ratios, 68th Annual American Helicopter Society Forum, Fort Worth, TX, May 1–3, 2012.
[122] D.J. Clingman, D. Jacot, in: Shape memory alloy consortium and demonstration, Smart Structures and Materials Conference (SPIE), Newport Beach, CA, March 7, 2000.
[123] R. Ruggeri, D. Jacot, D. Clingman, in: Shape memory actuator systems and the use of thermoelectric modules, Smart Structures and Materials Conference (SPIE), San Diego, CA, 18 March, 2002.
[124] N. Caldwell, E. Gutmark, in: Performance predictions of a blade twist actuator system, 45th AIAA Aerospace Sciences Meeting and Exhibit (2007), Reno, NV, January 8–11, 2007.
[125] D. Arbogast, R. Ruggeri, R. Bussom, in: Development of a -scale NiTinol actuator for reconfigurable structures, SPIE, Industrial and Commercial Applications of Smart Structures Technologies, vol. 6930, 2008.

[126] R. Ruggeri, D. Arbogast, R. Bussom, in: Wind tunnel testing of a lightweight -scale actuator utilizing shape memory alloy, 49th AIAA/ASME/ASCE/AHS/ASC Structures, Structural Dynamics, and Materials Conference, Schaumburg, IL, April 7–10, 2008.

[127] H. Prahlad, I. Chopra, in: Design of a variable twist tiltrotor blade using shape memory alloys (SMA) actuators, Smart Structures and Materials Conference (SPIE), Newport Beach, CA, 5 March, 2001.

[128] H. Prahlad, Development of Shape Memory Alloy (SMA) Torsional Actuators for Variable Twist Tilt Rotor (VTTR) blades, Ph.D. thesis, University of Maryland, 2002.

[129] J. Park, S. Kim, S.N. Jung, in: Application of shape memory alloy hybrid composites for variable-twist proprotors, 66th American Helicopter Society Forum, Phoenix, AZ, May 11–13, 2010.

[130] M. Mistry, F. Gandhi, in: Twist control of an I-beam through vlasov bimoment actuation, 49th AIAA/ASME/ASCE/AHS/ASC Structures, Structural Dynamics, and Materials Conference, Schaumburg, IL, April 7–10, 2008.

[131] M. Mistry, Induced warp systems to obtain active twist of rotor blades, Master's thesis, The Pennsylvania State University, 2008.

[132] Y. Van Weddingen, O. Bauchau, S. Kottapalli, S. Ozbay, Y. Mehrotra, in: Application of out-of-plane warping to control rotor blade twist, American Helicopter Society Aeromechanics Specialists Conference, San Francisco, CA, January 20–22, 2010.

[133] A.W. Linden, D. Beck, M. D'Onofrio, R. Flemming, E. Hibyan, R. Johnston, et al., Variable Diameter Rotor Study, Tech. rep, Air Force Flight Dynamics Laboratory, Air Force Systems Command, Wright-Patterson Air Force Base, Ohio, January 1972. Report No. AFFDL-TR-71-170.

[134] C. Fenny, in: Mechanism for varying the diameter of rotors using compound differential rotary transmissions, 61st American Helicopter Society Forum, Grapevine, TX, June 13, 2005.

[135] R.M. Segel, E.A. Fradenburgh, in: Development of the TRAC variable diameter rotor concept, AIAA/AHS VTOL Research, Design, and Operations Meeting, February 17–19, 1969.

[136] E.A. Fradenburgh, R.J. Murrill, E. Kiely, Dynamic Model Wind Tunnel Tests of a Variable Diameter, Telescoping-Blade Rotor System (TRAC Rotor), Tech. rep., Sikorsky Aircraft, USAAMRDL-TR-73-32, Eustis Directorate, U.S. Army AirMobility R&D Laboratory, Fort Eustis, VA, July 1973.

[137] E.A. Fradenburgh, in: Application of a variable diameter rotor system to advanced VTOL aircraft, 31st American Helicopter Society Forum, Washington, DC, May 13–15, 1975.

[138] E.A. Fradenburgh, L. Hager, N. Kefford, Evaluation of the TRAC Variable-Diameter Rotor: Preliminary Design of a Full-Scale Rotor and Parametric Mission Analysis Comparision, Tech. rep., Sikorsky Aircraft, USAAMRDL-TR-75-54, Eustis Directorate, U.S. Army Air Mobility R&D Laboratory, Fort Eustis, VA, February 1976.

[139] H. Frint, Design Selection Tests for TRAC Retraction Mechanism, Tech. rep., Sikorsky Aircraft, USAAMRDL-TR-76-43, Eustis Directorate, U.S. Army Air Mobility R&D Laboratory, Fort Eustis, VA, January 1977.

[140] E.A. Fradenburgh, D.G. Matuska, in: Advancing tiltrotor state-of-the-art with variable diameter rotors, 48th American Helicopter Society Forum, Washington, DC, June 3–5, 1992.

[141] K. Studebaker, D.G. Matuska, in: Variable diameter tiltrotor wind tunnel test results, 49th American Helicopter Society Forum, St. Louis, MO, May 19–21, 1993.

[142] R.S. Turmanidze, S.N. Khutsishvili, L. Dadone, in: Design and experimental investigation of variable-geometry rotor concepts, Adaptive Structures and Materials Symposium, International Mechanical Engineering Congress and Exposition, New York, NY, November 11–16, 2001.

[143] R.S. Turmanidze, L. Dadone, in: New design of a variable geometry prop-rotor with cable and hydraulic-system actuation, 66th American Helicopter Society Forum, Phoenix, AZ, May 11–13, 2010.

[144] T. Prabhakar, F. Gandhi, J. Steiner, D. McLaughlin, in: A centrifugal force actuated variable span morphing helicopter rotor, American Helicopter Society 63rd Annual Forum, Virginia Beach, VA, May 1–3, 2007.

[145] T. Prabhakar, A centrifugal force actuated morphing rotor, Master's thesis, The Pennsylvania State University, 2007.
[146] M. Mistry, Quasi-static rotor morphing concepts for rotorcraft performance improvements, Ph.D. Thesis, The Pennsylvania State University, 2012.
[147] O. Leon, E. Hayden, F. Gandhi, in: Rotorcraft operating envelope expansion using extendable chord sections, American Helicopter Society 65th Annual Forum, Grapevine, TX, May 27–29, 2009.
[148] M. Khoshlajeh, F. Gandhi, in: Extendable chord rotors for helicopter envelope expansion and performance improvement, 67th American Helicopter Society Forum, Virginia Beach, VA, May 3–5, 2011.
[149] K. Noboru, N. Kondo, S. Saito, T. Akasaka, Y. Tanabe, in: An experimental study of on-blade active tab for helicopter noise reduction, 30th European Rotorcraft Forum, Marseilles, France, September 14–16, 2004.
[150] E. Hayden, Design and testing of a helicopter rotor blade chord extension system, Master's thesis, The Pennsylvania State University, August 2012.
[151] Barbarino, S., Gandhi, F., Webster, S. "Design of extendable chord sections for morphing helicopter rotor blades, Special issue "Morphing"," J. Intell. Mater. Syst. Struct., vol. 22, no. 9, pp. 891–905 (June 2011), doi:10.1177/1045389X11414077.
[152] M. Khoshlahjeh, F. Gandhi, Extendable chord rotors for helicopter envelope expansion and performance improvement, J. Am. Helicopter Soc. 59 (2014), 012007. https://doi.org/10.4050/JAHS.59.012007.
[153] Moser, P.R., Barbarino, S., Gandhi, F., "Helicopter rotor blade chord extension morphing using a centrifugally actuated bi-stable von-mises truss," J. Aircr., vol.51, no.5, pp.1422–1431, doi: 10.2514/1.C032299.
[154] S. Barbarino, F.S. Gandhi, R. Visdeloup, in: A bi-stable von-mises truss for morphing applications actuated using shape memory alloys, Proceedings of the ASME 2013 Conference on Smart Materials, Adaptive Structures and Intelligent Systems (SMASIS2013), September 16–18, Snowbird, UT, paper SMASIS2013-3062, 2013.
[155] S. Barbarino, F.S. Gandhi, in: A cellular frame for morphing applications based on Bi-stable von-mises trusses actuated using shape memory alloys, Proceedings of the AIAA Science and Technology Forum and Exposition (SciTech), January 13–17, 2014, National Harbor, MD, 2014. ID: 1736921.
[156] A. Buter, C. Ehlert, U.D. Sachau, E. Breitbach, in: Adaptive rotor blade concepts - direct twist and camber variation, RTO AVT Symposium on Active Control Technology for Enhanced Performance Operational Capabilities of Military Aircraft, Land Vehicles and Sea Vehicles, Braunschweig, Germany, May 8–11, 2000.
[157] P. Anusonti-Inthra, R. Sarjeant, M. Frecker, F. Gandhi, Design of a conformable rotor airfoil using distributed piezoelectric actuators, AIAA J. 43 (8) (August 2005) 1684–1695.
[158] A. Nissly, P. Anusonti-Inthra, F. Gandhi, M. Frecker, in: Design optimization of a controllable camber rotor airfoil, 61st American Helicopter Society Forum, Grapevine, TX, June 1–3, 2005.
[159] F. Gandhi, P. Anusonti-Inthra, Skin design studies for variable camber morphing airfoils, Smart Mater. Struct. 17 (2008) 8.
[160] S. Daynes, S. Nall, P. Weaver, K. Potter, P. Margaris, P. Mellor, in: On a bistable flap for an airfoil, 50th AIAA/ASME/ASCE/AHS/ASC Structures, Structural Dynamics, and Materials Conference, Palm Springs, CA, May 4–7, 2009.
[161] T. Quackenbush, D. Wachspress, R. McKilllip, C. Solomon, in: Tiltrotor/tiltwing performance enhancement via active on-blade flaps, 66th Annual American Helicopter Society Forum, Phoenix, AZ, May 11–13, 2010.
[162] R. McKilllip, T. Quackenbush, in: The flight adaptive blade for optimum rotor response (FABFORR) concept, American Helicopter Society Future Vertical Lift Aircraft Design Conference, January 18–20, 2012.
[163] C. Maucher, B. Grohmann, P. Janker, A. Altmikus, F. Jensen, H. Baier, in: Actuator design for the active trailing edge of a helicopter rotor blade, 33rd European Rotorcraft Forum, Kazan, Russia, September 2007.
[164] B. Grohmann, C. Maucher, P. Janker, P. Wierach, in: Embedded piezoceramic actuators for smart helicopter rotor blades, 49th AIAA/ASME/ASCE/AHS/ASC Structures, Structural Dynamics, and Materials Conference, Schaumburg, IL, April 7–10, 2008.

[165] B. Grohmann, F. Muller, E. Ahci, R. Pfaller, M. Bauer, C. Maucher, O. Dietrich, S. Storm, P. Janker, in: Design, evaluation and test of active trailing edge, 67th American Helicopter Society Forum, Virginia Beach, VA, May 3–5, 2011.
[166] P. Martin, K. McAlister, M. Chandrasekhara, W. Geissler, in: Dynamic stall measurements and computations for a VR-12 airfoil with a variable droop leading edge, 59th American Helicopter Society Forum, Phoenix, AZ, May 6–8, 2003.
[167] M. Shaner, I. Chopra, in: Design and testing of a piezostack actuated leading-edge flap, SPIE symposium on Smart Structures and Materials, vol. 3668, March 1999.
[168] D. Fink, N. Reinhardt, R. Severance, R. Phillips, M. Gaudreau, R. Ormiston, in: Deformable leading edge electromagnetic airfoil, 63rd American Helicopter Society Forum, Virginia Beach, VA, May 1–3, 2007.
[169] S. Kota, G. Ervin, R. Osborn, R. Ormiston, in: Design and fabrication of an adaptive leading edge rotor blade, 64th American Helicopter Society Forum, Montreal, Quebec, Canada, April 29–May 1, 2008.
[170] W. Guo, J. Horn, in: Helicopter flight control with variable rotor speed and torque limiting, 65th Annual American Helicopter Society Forum, Grapevine, TX, May 27–29, 2009.
[171] J. Steiner, An investigation of performance benefits and trim requirements of a variable speed helicopter rotor, Master's thesis, The Pennsylvania State University, 2008.
[172] P. Lorber, in: Active flap full-scale oscillating blade section experiment, Bridgeport CT: AHS NE Region Technical Specialists' Meeting: Improving Rotorcraft Acceptance Through Active Controls Technology, October, 2000.
[173] B. Sim, et al., in: Non-harmonic active flap motions for low frequency in-plane rotor noise reduction, American Helicopter Society 68th Annual Forum, Fort Worth, TX, May 1–3, 2012.
[174] P. Lorber, A. Bagai, B. Wake, in: Design and evaluation of slatted airfoils for improved rotor performance, American Helicopter Society 62d Annual Forum, Phoeniz, AZ, May 9–11, 2006.
[175] R. Kube, K.-J. Schultz, in: Vibration and BVI noise reduction by active rotor control: HHC compared to IBC, 22th European Rotorcraft Forum, Brighton, UK, 1996.
[176] M. Müller, U.T.P. Amold, D. Morbitzer, in: On the importance and efficiency of 2/rev IBC for noise, vibration and pitch link load reduction, 25th European Rotorcraft Forum, Rome, Italy, 1999. paper no. G4.
[177] E.R. Wood, R.W. Powers, C.H. Cline, C.E. Hammond, On developing and flight testing a higher harmonic control system, J. Am. Helicopter Soc. 30 (1) (1985) 3–20.
[178] F.K. Straub, E.V. Byrns, in: Application of higher harmonic blade feathering on the OH-6A helicopter for vibration reduction, NASA CR 4031, December 1986.
[179] M. Polychroniadis, M. Achache, in: Higher harmonic control: flight tests of an experimental system on SA349 research Gazelle, 42nd Annual Forum of the American Helicopter Society, Washington, DC, 1986, pp. 777–789.
[180] M. Achache, M. Polychroniadis, Development of an experimental systems for active control of vibrations on helicopters—Development methodology for an airborne system, Vertica 11 (1/2) (1987) 123–138.
[181] W. Miao, S.B.R. Kottapalli, H.M. Freye, in: Flight demonstration of higher harmonic control (HHC) on S-76, 42nd Annual Forum of the AHS, Washington, DC, June 2–4, 1986.
[182] C.E. Hammond, Wind tunnel results showing rotor vibratory loads reduction using higher harmonic blade pitch, Journal of the AHS 28 (1) (1983) 10–15.
[183] G. Lehmann, The effect of higher harmonic control (HHC) on a four-bladed hingeless model rotor, Vertica 9 (3) (1985) 273–284.
[184] J. Shaw, N. Albion, E.J. Hanker, R.S. Teal, Higher harmonic control: wind tunnel demonstration of fully effective vibratory hub forces suppression, J. Am. Helicopter Soc. 34 (1) (1989) 14–25.
[185] T.F. Brooks, E.R. Booth Jr., The effects of higher harmonic control on blade-vortex interaction noise and vibration, Journal of the AHS 38 (3) (1993) 45–55.
[186] W.R. Splettstoesser, K.-J. Schultz, R. Kube, T.F. Brooks, R. Booth, G. Niesl, O. Streby, A higher harmonic control test in the DNW to reduce Impulsive BVI noise, J. Am. Helicopter Soc. 39 (4) (1994) 3–13.

[187] W.R. Splettstoesser, K.-J. Schultz, R. Kube, T.F. Brooks, E.R. Booth Jr., G. Niesl, O. Streby, A higher harmonic control test in the DNW to reduce impulsive BVI noise, Journal of the AHS 39 (4) (1994) 3–13.
[188] A. Murashige, N. Kobiki, A. Tsuchihashi, K. Inagaki, T. Tsujiuchi, Y. Hasegawa, H. Nakamura, Y. Yamamoto, E. Yamakawa, in: Second ATIC aeroacoustic model rotor test at DNW, 26th European Rotorcraft Forum, The Hague, Netherlands, September 26–29, 2000.
[189] G. Yu, P. Heller, P. Mercker, in: HHC aeroacoustic rotor test at DNW—the joint German/French/US HART Project, 20th European Rotorcraft Forum, Amsterdam, Netherlands, 1994.
[190] W.R. Splettstößer, et al., in: Key results from a higher harmonic aeroacoustic rotor test (HART) in the German-Dutch wind tunnel, 21st European Rotorcraft Forum, St. Petersburg, Russia, 1995.
[191] N.D. Ham, B.L. Behal, R.M. McKillip Jr., in: A simple system for helicopter individual-blade-control and its application to lag damping augmentation, 8th European Rotorcraft Forum, Aix-en-Provence, 1982.
[192] S.A. Jacklin, A. Blaas, S.M. Swanson, D. Teves, in: Second test of a helicopter individual blade control system in the NASA Ames 40-by-80 feet wind tunnel, 2nd AHS International Aeromechanics Specialists Conference, 1995.
[193] S.M. Swanson, S.A. Jacklin, A. Blaas, C. Niesl, R. Kube, in: Acoustic results from a full-scale wind tunnel test evaluating individual blade control, 51't Annual Forum of the American Helicopter Society, Fort Worth, TX, 1995.
[194] S.A. Jacklin, A. Blaas, D. Teves, R. Kube, in: Reduction of helicopter BVI noise, vibration, and power consumption through individual blade control, 51't Annual Forum of the American Helicopter Society, Fort Worth, TX, USA, 1995.
[195] S. Jacklin, S. Swanson, A. Blaas, D. Teves, G. Niesl, R. Kube, in: A full-scale wind tunnel investigation of a helicopter individual blade control (IBC) system, NASA TM, NASA Ames Research Center, Rotorcraft Aeromechanics Branch, CA, 1996.
[196] D. Roth, O. Dietrich, M. Bebesel, R. Pongratz, R. Kube, H. Munser, in: Individual blade root control demonstation—Recent activities, 27th European Rotorcraft Forum, Moscow, Russia, 2001.
[197] Lorber, P, et al. 2001. "Active rotor experiments at mach scale using root pitch IBC," American Helicopter Society 57th Annual Forum, Washington, DC, May 9–11, 2001.
[198] A. Haber, S.A. Jacklin, G. deSimone, in: Development, manufacturing, and component testing of an individual blade control system for a UH-60 helicopter rotor, Presented at the American Helicopter Society Aerodynamics, Acoustics, and Test and Evaluation Technical Specialists Meetings, San Francisco, CA, January 23–25, 2002.
[199] T. Norman, C. Theodore, P. Shinoda, D. Fuerst, U. Arnold, S. Makinen, P. Lorber, J. O'Neill, in: Full-scale wind tunnel test of a UH-60 individual blade control system for performance improvement and vibration, loads, and noise control, American Helicopter Society 65th Annual Forum, Grapevine, TX, May 27–29, 2009.
[200] H. Yeo, E. Romander, T. Norman, in: Investigation of rotor performance and loads of a UH-60A individual blade control system, American Helicopter Society 66th Annual Forum, Phoenix, AZ, May 11–13, 2010.
[201] N.D. Ham, T.R. Quackenbush, in: A simple system for helicopter Individual-Blade-Control and its application to stall flutter suppression, 7th European Rotorcraft and Powered Lift Aircraft Forum, Garmisch-Partenkirchen, 1981.
[202] D. Schimke, U.T.P. Amold, R. Kube, in: Individual blade root control demonstration—evaluation of recent flight tests, 54th Annual Forum of the American Helicopter Society, Washington, DC, 1998.
[203] D. Morbitzer, U.T.P. Arnold, M. Miiller, in: Vibration and noise reduction through individual blade control, 24th European Rotorcraft Forum, Marseilles, France, 1998.
[204] U. Arnold, in: Recent IBC flight test results from the CH-53G helicopter, 29th European rotorcraft Forum, Friedrichshafen, Germany, September 16–18, 2003.
[205] D. Fuerst, U.T.P. Arnold, in: Closed loop IBC system and flight test results on the CH-53 helicopter, American Helicopter Society 60th Annual Forum, Baltimore, MD, June 7–10, 2004.

[206] U.T.P. Arnold, D. Fuerst, Closed loop IBC results from CH-53G flight tests, Aerosp. Sci. Technol. 6 (5) (2005) 421–435. Special Issue Rotorcraft Research.
[207] U.T.P. Arnold, D. Fuerst, S. Hartmann, A. Hausberg, in: Flight testing of an in-flight tuning system on a CH-53G helicopter, American Helicopter Society 70th Annual Forum, Montréal, Quebec, Canada, May 2014.
[208] U. Arnold, D. Fürst, in: In-flight tuning system for the CH-53G helicopter, 41st European Rotorcraft Forum, Munich, 2015.
[209] Steward, W.Second harmonic control on the helicopter rotor, ARC reports and memoranda, Nr. 2997, 1952.
[210] P.R. Payne, Higher harmonic rotor control: the possibilities of third and higher harmonic feathering for delaying the stall limit in helicopters, Aircr. Eng. 30 (8) (1958) 222–226.
[211] P.J. Arcidiacono, Theoretical performance of helicopters having second and higher harmonic feathering control, J. Am. Helicopter Soc. 6 (8) (1961) 8–19.
[212] R.K. Wernicke, J.M. Drees, in: Second harmonic control, 19th Annual Forum of the AHS, Washington, DC, May 1–3, 1963.
[213] J.G. Yen, Higher harmonic control for helicopters with two-bladed and four-bladed rotors, J. Aircr. 18 (12) (1981) 1064–1069.
[214] W. Johnson, in: Self-tuning regulators for multicyclic control of helicopter vibration, NASA TP 1996, March 1982.
[215] K. Nguyen, I. Chopra, Application of higher harmonic control to rotors operating at high speed and thrust, J. Am. Helicopter Soc. 35 (3) (1990) 78–89.
[216] E.R. Wood, et al., in: On the unsteady aerodynamics of higher harmonic control, European Rotorcraft Forum, Cernobbio, Italy, 1993.
[217] Hall, S. R. and Werely, N. M., “Performance of higher harmonic control algorithms for helicopter vibration reduction,” J. Guid. Control. Dyn., Vol. 16, No. 4, October 1993, pp. 193–797.
[218] R. Kube, in: Effects of blade elasticity on open and closed loop higher harmonic control of a hingeless helicopter rotor, DLR-FB 91-26, Braunschweig, Germany, 1997.
[219] H. Yeo, in: Assessment of active controls for rotor performance enhancement, American Helicopter Society 62nd Annual Forum, Phoenix, AZ, May 9–11, 2006.
[220] Ham, N D., “Helicopter individual-blade-control research at MIT 1977-1985,”s.l.: Vertica, Volume 11, Issue 1/2, Pages: 109–122, 1986.
[221] C. Kessler, in: Active rotor control for helicopters: motivation and survey on higher harmonic control, 36th European Rotorcraft Forum, Paris, 2010.
[222] D. Fuerst, U.T.P. Arnold, D. Graham, in: In-flight tuning: concept, realization and application, Prsentation at the Airworthiness, CBM, and HUMS Specialists’ Meeting of the American Helicopter Society, Huntsville, AL, February 2013.
[223] R.H. Blackwell, D.J. Merkley, in: The aeroelastically conformable concept, 35th Annual Forum of the American Helicopter Society, May 1978.
[224] R.H. Blackwell, K.C. Frederickson, Wind Tunnel Evaluation of Aeroelastically Conformable Rotors, Applied Technology Laboratories, Ft. Eustis, VA, January 1981. USAAVRADCOM-TR-80-D-32.
[225] G.S. Doman, F.J. Tarzanin Jr., J. Shaw Jr., Investigation of Aeroelastically Adaptive Rotors, Army Air Mobility Research and Development Laboratory, Eustis Directorate, Ft. Eustis, VA, May 1977. USAAMRDL-TR-77-3.
[226] R.H. Blackwell, Investigation of the Compliant Rotor Concept, Army Air Mobility Research and Development Laboratory, Eustis Directorate, Ft. Eustis, VA, June 1977. USAAMRDL-TR-77-7.
[227] Booth, E.R. Jr., and Wilbur, M.L., “Acoustic aspects of active-twist rotor control,” J. Am. Helicopter Soc., Vol. 49, (1), January 2004, pp.3–10.
[228] A.P.F. Bernhard, J. Wong, in: Sikorsky active rotor control evaluation of the NASA/Army/MIT active twist rotor, American Helicopter Society 59th Annual Forum Proceedings, Phonix, AZ, May 2003.

[229] J. Shen, R.P. Thornburgh, Y. Liu, A.R. Kreshock, M.L. Wilbur, in: Design and optimization of an airfoil with active continuous trailing-edge flap, American Helicopter Society 69th Annual Forum Proceedings, Phonix, AZ, May 21–23, 2013.

[230] R.P. Thornburgh, A.R. Kreshock, M.L. Wilbur, M.K. Sekula, J. Shen, in: Continuous trailing-edge flaps for primary flight control of a helicopter main rotor, American Helicopter Society 70th Annual Forum Proceedings, Montreal, Quebec, Canada, May 20–22, 2014.

[231] D. Matuska, P. Lorber, A. Dale, in: Wind tunnel test of a variable diameter Tiltrotor model, NASA Report CR 177629, 1994.

[232] P. Gemanowski, B. Stille, M. Strauss, in: Technology assessment for large vertical-lift transport Tiltrotors, NASA Report CR-2010-216384, 2010.

[233] Brender, Scott, "The attributes of a variable diameter rotor system applied to civil Tiltrotor aircraft," NASA-CR-203092.

[234] http://www.sikorskyarchives.com/X-WING.php (accessed 22.03.2017).

[235] YSSYguy at the English language Wikipedia, November 8, 2006, https://cs.wikipedia.org/wiki/Soubor:N267KA.JPG (accessed 22.03.2017).

CHAPTER

AERODYNAMIC ANALYSES OF TILTROTOR MORPHING BLADES

25

Antonio Pagano
The Italian Aerospace Research Centre, CIRA SCpA, Capua (CE), Italy

CHAPTER OUTLINE

Morphing Wing Technologies. https://doi.org/10.1016/B978-0-08-100964-2.00025-3

NOMENCLATURE

BEMT blade element momentum theory
CFD computational fluid dynamics
CSD computational structure dynamics
DOE design of experiments
IADP innovative aircraft demonstrator platforms
LED leading edge deformations
MDA multidisciplinary design and analysis
MDO multidisciplinary design and optimization
MOO multiobjective optimization
NLPQL nonlinear programming by quadratic Lagrangian
NSEA nondominated sorting evolutionary algorithm
RPM round per minute
RSM response surface methodology
SETE static extended trailing edge
TED trailing edge deformations
B Prandtl's tip loss factor
c blade local chord
C_{lmax} maximum lift coefficient
C_P power coefficient
C_T thrust coefficient
D drag
F.M. figure of merit
J propeller advance ratio
L lift
M_{dd} drag divergence Mach
M_{WT} wind tunnel freestream Mach number
$M_{\Omega R}$ Mach number at blade tip
N_b Number of blades
P rotor power
R rotor radius
t airfoil thickness
T rotor thrust
u_i tangential induced velocity
V_∞ aircraft advance velocity
w_i axial induced velocity
y spanwise coordinate
β propeller pitch
γ sectional sweep angle
δ sectional dihedral angles
η propeller efficiency
θ rotor blade pitch angle
μ rotor advance ratio
ρ density
σ rotor solidity
φ inflow angle
Ψ local rotation angle for deformation
ω rotor angular velocity

1 INTRODUCTION

Aircraft equipped with convertible rotors, mainly tiltrotors and tilt wings, are about to enter the scenario of civil transport system. The AgustaWestland AW609, a 6–9 passenger civil tiltrotor, is close to certification, and UAV prototypes are demonstrating their potential in association with distributed electric propulsion, as exhibited by NASA's Greased Lightning tilt wing. In Europe, tiltrotors are increasingly discussed for the research challenges they pose. The Clean Sky 2 joint technology initiative under the European Union's Horizon 2020 Framework Programme for research and innovation includes the development of a tiltrotor demonstrator within the Fast Rotorcraft IADP (innovative aircraft demonstrator platforms).

These aircraft are able to take off and hover like helicopters and to cruise as turboprops by tilting the rotor. First rotor designs struggled against the conflicting requirements of the hover and cruise operating conditions. Rotor solidity, blade planform, blade twist (and its distribution), and tip speed and airfoils (and their distribution) were design variables investigated to find a compromise between greater propulsive efficiency in cruise and good hover performance.

Tiltrotor blades are subject to a significant number of flow conditions and passive blade shape optimization tends to favor blade designs with longer rotor radii and larger blade areas, as a consequence of a high loads demand inherent to helicopter mode. This impacts high-speed level flight, where rotors should be smaller to gain more efficiency, since it is in that flight regime that the tiltrotor is expected to spend much of its mission profile.

Morphing technology, with its high potential to adapt lifting shapes to meet specific aerodynamic needs, may surpass the limitations associated with the geometric compromises. In the context of morphing technologies, devices altering spanwise and chordwise blade shapes are investigated to equip helicopter rotors in particular. When considering tiltrotors, the adoption of devices allowing retractable and telescopic rotors or morphing blade sections (acting on camber or chord length) must be carefully evaluated because of the inherent complexity of the propulsion system.

In this regard, a concept of the adoption of morphing blades on next generation tiltrotors is proposed in this chapter. It is based on the experience matured by CIRA in the field of aerodynamic design and blade shape aerodynamic optimization of passive/active tiltrotor blades.

2 AIM AND STRUCTURE OF THE CHAPTER

Despite the high complexity of rotor tilting, some current blade-morphing technologies are investigated from the aerodynamic point of view in the following sections of this chapter to assess the potential benefit of shape modifying blades and their impact on rotor performance under examined flight conditions. Several research issues are addressed and the relative context is presented by outlining recent efforts.

By adopting a realistic tiltrotor blade geometry on meaningful aerodynamic conditions, new blade designs incorporating morphing devices are obtained; this text outlines the necessary numerical procedures and optimization processes (which make use of several software tools).

Though comprehensive description of the simulation and optimization environment including individual analysis components is outside the scope of this work, a brief outline of them is given. Blade morphing techniques are described and applied and their subsequent optimization provides new designs. The discussion on the potential benefit of these new designs is articulated on the aerodynamic performance comparison of morphing blades against passive-based solutions.

The nature of these investigations is exploratory, and many complexities (aeroelasticity implications, device mechanisms, absorbed power, for example) are not considered. The main focus is on rotor aerodynamic efficiency and performance gains.

3 RESEARCH CONTEXT

This document explores the adoption of morphing blades on tiltrotors with the aim of improving the rotor aerodynamic performance. Rotors for such aircrafts are designed to simultaneously address the peculiarities of both axial flights (hover, vertical ascent/descent, and cruise) and edge flights (flyover). Many design efforts concentrated in Europe concern the tiltrotor concept ERICA [1], which was conceived on a reduced rotor diameter so that the tiltrotor can also take off and land as a propeller-driven aircraft. DART [2], ADYN [3], and NICETRIP [4,5] are past European Commission-funded projects dealing with rotor design.

Indeed, aerodynamic blade shape optimization is one of the most widely investigated topics spanning over many application fields: wind turbines, marine propellers, turbomachinery, propeller-driven aircraft, helicopters, and tiltrotors. Without entering into the peculiarities of each of the studies, those on turboprop propellers and helicopter rotors are of great interest because convertible rotors (proprotors) operate in both working conditions. CSD-CFD (computational structure dynamics-computational fluid dynamics) is used to optimize helicopter rotor blade both in hover and forward flight [6–9]. The latest investigations on propeller blade optimization concern the involvement of BEMT (blade element momentum theory) [10,11], comprehensive rotor code [12], high-order CFD [13], and CSD-CFD coupling [14]. The wide design space and the cost of evaluations based on high-fidelity numerical simulations have thus far limited investigations on tiltrotor optimization to trade-off studies [3,15] or to the use of simple analysis methods [16] or comprehensive models [17]. Proprotor blade optimization is shifting towards blade morphing technologies, which theoretically may overcome the barriers of geometric compromises. This kind of investigation assumes an ever-increasing relevance since the rotorcraft community (which is constantly striving for rotor performance improvements) is looking for new breakthroughs and groundbreaking solutions.

Technologies allowing variable span rotors, variable angular velocity rotors, and airfoil-based morphing [18,19], applied both separately and simultaneously, have become increasingly popular and are being explored from mono/multidisciplinary approaches [16]. For these advanced configurations, airfoil selection, shape modification, and position may play a critical role; in fact, individual airfoils may undergo different flow characteristics depending on the actual peripheral speed to which they are exposed. The sectional peripheral speed may vary either when the distance of the blade section from the hub center is altered (elongable/retractable rotor) [20,21] or when the angular velocity changes (variable speed rotor) [22]. The combination of the two technologies [23] adds even more working conditions for airfoils so that the task of their selection, shape modification and placement is very complicated.

By means of Navier-Stokes CFD analyses, Romander [24] investigated the influence of airfoil thickness on the rotor performance in airplane mode by alternatively thinning only the inboard cross section (influencing the blade shape up to 50% of the span) or the whole blade. He found that scaling down the aerofoil thickness over the entire blade gave more benefits than scaling only the root section. Rotor comprehensive calculations by Acree [25] confirmed that a thinner inboard section was beneficial for cruise propulsive efficiency, allowing for reduced performance gain in hover as well. They also emphasized that the addition of airfoil-related parameters to the design variables made blade shape optimization almost impractical because of the significant size of the design space. Stahlhut and Leishman [26] optimized the thickness-to-chord ratio spanwise distribution by representing the properties (C_{lmax} and M_{dd}) of next generation aerofoils as functions of t/c by using an improved BEMT model. They observed that a baseline tiltrotor blade with a 0.12 spanwise constant value of t/c performed better when equipped with thicker inboard sections (maximizing hover performance) and thinner outboard sections (maximizing propeller efficiency).

The airfoil shape modification is very attractive when airfoil reshaping takes place without gaps or surface discontinuities. In fact, the lack of gaps between fixed and moveable parts of airfoils improves the morphing devices efficiency, as discussed by Yeo and Chopra [27] after comparing deformable trailing edges and discrete trailing edge flaps. Concepts on continuous variable geometry airfoils are emerging and many of them are successfully progressed to the prototype status. Many of these concepts appeared in the recent literature [27–32], showing, with respect to the un-morphed geometry, a lift increase which may be used to respond to the high thrust demand of tiltrotor blade sections in hover. Thus, their incorporation within a tiltrotor blade, especially if the correspondent blade morphing is considered from the early stages of design, may lead to significant aerodynamic performance improvements, since the compromise between hover and cruise requirements is no longer so stringent.

The use of devices for modifying the blade sectional geometry on rotorcraft configurations is largely debated. For example, in Ref. [33], the pros and cons of trailing edge extensions, trailing edge flaps and Gurney flaps are reviewed; leading edge and trailing edge deflections are studied in Refs. [34, 35] for helicopter performance enhancement. CIRA contributed to the subjects of aerodynamic shape optimizations of passive and active tiltrotor blades with several publications [36–38].

4 OUTLINE OF METHODS AND NUMERICAL TOOLS

This section describes the software tools necessary for a shape optimization procedure including morphing. A full description of the simulation library is outside the scope of this work. Many analysis tasks require collaboration between disciplinary codes and construction of numerical processes, for which software integration environments are of incomparable help. This section briefly illustrates features of the integration and optimization environment in which the present numerical processes were constructed. The performed aerodynamic analysis predicting rotor performance and trim consists of a certain multidisciplinary design and analysis (MDA) procedure which may involve CFD simulations. Further, since shape optimization results in modifications of the initial geometry, blade parameterization, and surface and volume grid generation must be involved in the numerical process.

4.1 INTEGRATION AND OPTIMIZATION ENVIRONMENT

The process integration and optimization tool OPTIMUS from Noesis Solutions [39] is invaluable for integrating arbitrary analysis codes, automating the process execution, controlling data exchange, splitting the process over a heterogeneous computational environment where analysis codes run on different computer platforms, and postprocessing results. The key functionalities of optimization methods are fully exploited in order to address the search of global optima. DOE (design of experiments), RSM (response surface methodology), Gradient/Genetic based algorithms are available for the exploration of the design space, the approximation of models, the design optimization, respectively. Optimus versions ranging from 10.10 to 10.15 are the versions of this software package used for all of the applications herein shown.

4.2 MDA PROCEDURES AND OPTIMIZATION PROCESSES

Only two MDA procedures are briefly described. The first one, based on a BEMT performance code, is applied to quickly screen the design space; the second one, built on a coupling procedure involving CFD, is invoked for more detailed analyses. The numerical procedure based on BEMT provides both the rotor trim and the performance evaluation.

4.2.1 MDA

CIRA MDA procedures are automatic numerical procedures where software tools are linked together. Typically, the software tools are analysis codes selected from a simulation library structured on a disciplinary basis (e.g., comprehensive rotor codes, CFD codes, grid generators, and aeroacoustic codes, all of them generally offered in multiple versions implementing different mathematical models ranging from simple to very sophisticated ones) and complementary software components (such as code interfaces).

The application of these MDA procedures is documented in several publications [40–43] where different optimization problems are addressed: aerodynamic shape optimization for helicopter and turboprop blades aiming at performance maximization, noise reduction, SMA device integration and characterization. Herein, two simplest MDA procedures are illustrated in Fig. 1 where the aerodynamic analysis of proprotors in trimmed conditions is performed within iterative loops which differ for the CFD involvement.

4.2.2 Optimization

Optimization may be based upon the exploration of the design space and/or upon the search for local/global optima. The exploration of the design space is done via DOE evaluations whose data are also useful for obtaining surrogate models. Multiobjective optimization (MOO) is used to search for optima in the case of two concurrent objectives, like performance in hover and in cruise. Analysis is accomplished by using either response surface methods (RSM) or direct simulations. The optimization may involve both gradient-based and evolutionary algorithms.

Fig. 2 offers a snapshot of the OPTIMUS working area where the optimization process based on BEMT procedure is arranged. With reference to the figure, blue arrows represent the data flow, orange icons are the numerical tools, green cylinders indicate the output data and red polygons stand for the objectives. The workflow, once the actual blade surface is generated on the basis of the design variables (light blue oval) that are set by the optimizer, is split into two streams in order to predict the rotor performance in helicopter and airplane mode.

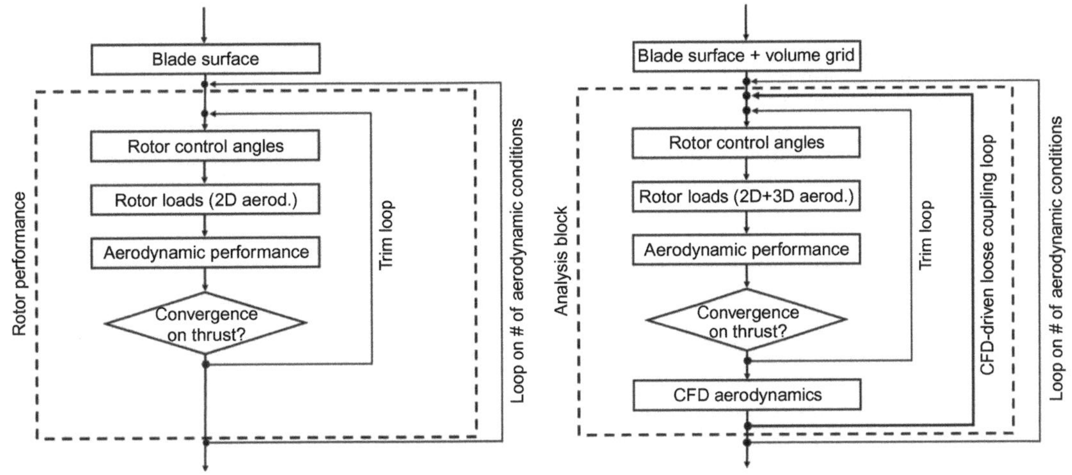

FIG. 1

Flowchart of BEMT procedure (left) and BEMT-CFD coupling procedure (right).

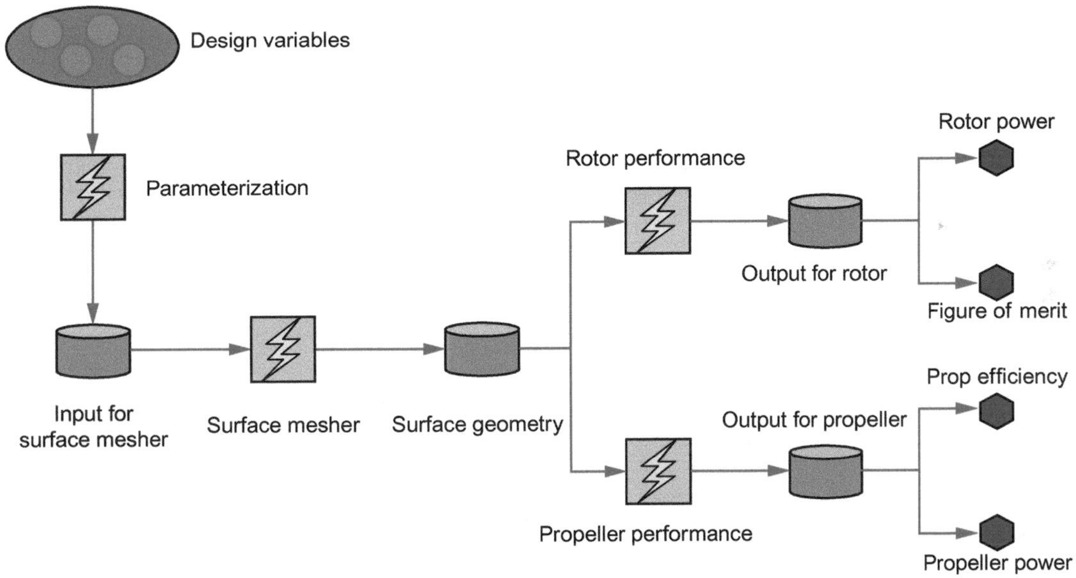

FIG. 2

Workflow of a CFD-driven optimization process.

4.2.3 Two level optimization

For the specific problems addressed by this work, it is convenient to have the possibility of performing an optimization on top of another optimization. As it is shown above, the workflow is split into two streams depending on the tiltrotor operative condition. Some design variables may affect both operative conditions, but some others may be peculiar to each condition. Furthermore, analysis may be peculiar to a specific task as well and based on different models, each of them with its own settings. Optimus in Optimus is a tool allowing a compound optimization, articulated into an upper level and a lower level. The design parameters affecting both the two tiltrotor operative conditions act on the upper level and the parameters which are peculiar to an individual operative condition act on two inner levels. This functionality is particularly powerful in the design and optimization of morphing blades. For example, if the aim of the optimization is to find the optimum planform of a blade with the capacity to alter the twist distribution when a tiltrotor converts from hover to cruise, the upper level optimization acts on design variables associated with the planform, whereas the two inner levels consist of twist optimization for different aerodynamic conditions, but for the same planform inherited by the upper level. The two level optimization process Optimus in Optimus visually appears as in Fig. 3 and it practically implies the tasks depicted in Fig. 4.

If the blade is able to extend out radially by means of a rigid translation, chord and twist distributions can be optimized in the upper level. Then, the inner level associated with the helicopter mode affects the search for the optimum angular velocity and the radius extension. The inner level associated with the airplane mode is only based on the optimum angular velocity search.

Within the hypothetical context of a full morphing blade, it is possible to arrange a two-level optimization process where the upper level deals with the airfoils selection, radial distribution, and morphing technology, and the inner levels find the optimum planform shapes both for helicopter and airplane mode acting independently on design variables such as chord length, geometric twist, vertical and horizontal leading edge offset, angular velocity, and radius extension.

4.3 BEMT ANALYSIS

The in-house CIRA BEMT code provides both the rotor trim and the performance evaluation. The BEMT code uses simplified aerodynamics which, for steady conditions, makes use of the 2D aerodynamic coefficients tables and, for unsteady conditions, a Beddoes-Lieshman type state-space formulation [44]. Several approximations are used for the stall treatment and the flow three-dimensionality.

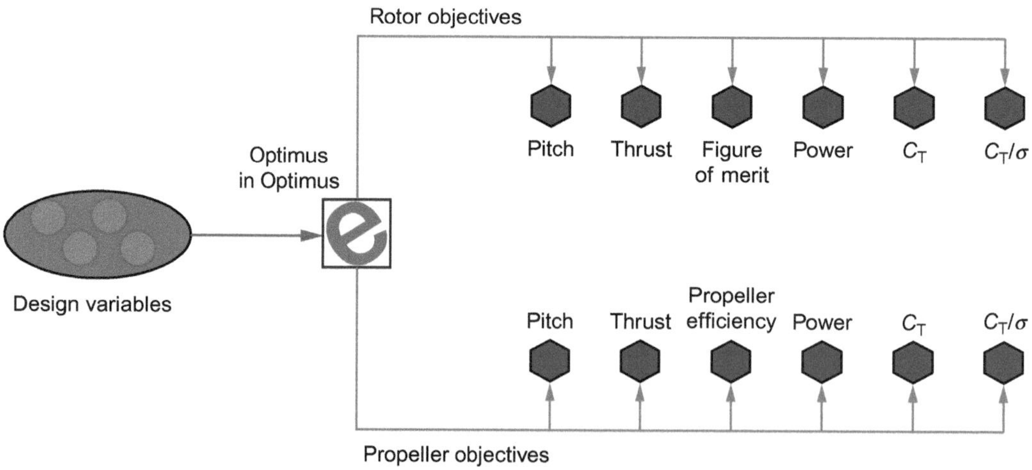

FIG. 3

Workflow for Optimus in Optimus process.

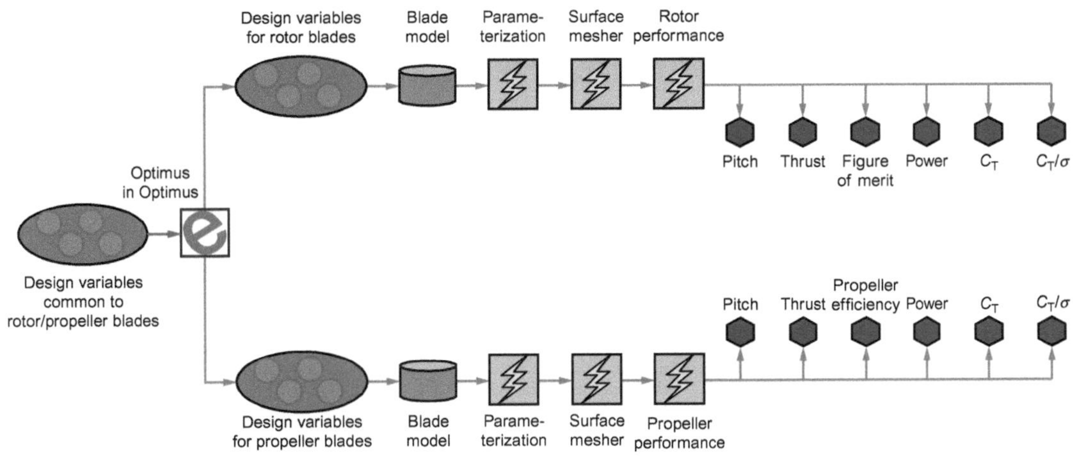

FIG. 4

Explanation of the Optimus in Optimus functionality.

Recently the BEMT code has been updated to remove some limitations [26], mainly related to small angles approximation, and to further improve the model with the inclusion of the swirl velocity. This document section includes a brief description of the new mathematical model and its validation.

4.3.1 Description

A blade airfoil at a radial station y of a rotor having radius R, axially advancing at a velocity V_∞ and rotating at a peripheral velocity ωy, is exposed to a relative velocity

$$V = \sqrt{(V_\infty + w_i)^2 + (\omega y - u_i)^2}, \tag{1}$$

where w_i and u_i are, respectively, the axial and tangential induced velocities. The inflow angle between the plane of rotation and V is

$$\varphi = \varphi_0 + \alpha_i = tg\frac{V_\infty}{\omega y} + \alpha_i. \tag{2}$$

By looking at Fig. 5, the forces along the reference system axes can be computed as

$$\begin{cases} dT = dL\cos\varphi - dD\sin\varphi \\ dF_x = dL\sin\varphi + dD\cos\varphi \end{cases}, \tag{3}$$

where the elementary lift and drag come from the blade element theory applied to a blade segment whose spanwise and chordwise dimensions are dy and c_y

$$\begin{cases} dL = \frac{1}{2}\rho V^2 c_y dy\, C_l \\ dD = \frac{1}{2}\rho V^2 c_y dy\, C_d \end{cases}. \tag{4}$$

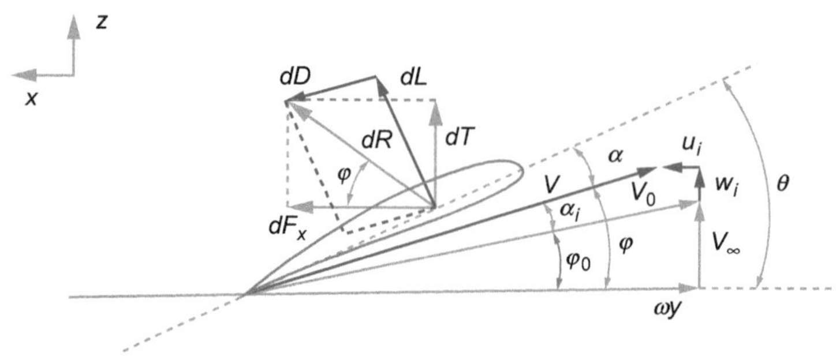

FIG. 5

Airfoil under peripheral and free stream velocity.

The thrust and torque extracted from a rotor annulus, according to the momentum theory, are:

$$\begin{cases} dT = 4(1-B)\pi\rho y dy|V_\infty + w_i|w_i \\ dF_x = 4(1-B)\pi\rho y dy|V_\infty + w_i|u_i \end{cases} \tag{5}$$

where B is the Prandtl's tip loss factor (N_b is the blade number):

$$B = \frac{2}{\pi}\cos^{-1}\left[\exp\left(\frac{N_b\left(\frac{y}{R}-1\right)}{2\frac{y}{R}\sin\varphi}\right)\right]. \tag{6}$$

By equating the elementary forces coming from the blade element and momentum theories, it is possible to obtain the following transcendental equations:

$$\begin{cases} \frac{V_\infty}{V} = \sin\varphi - \frac{1}{8B}\frac{N_b c_y}{\pi y}\frac{1}{\sin|\varphi|}[C_l\cos\varphi - C_d\sin\varphi] \\ \frac{\omega y}{V} = \cos\varphi + \frac{1}{8B}\frac{N_b c_y}{\pi y}\frac{1}{\sin|\varphi|}[C_l\sin\varphi + C_d\cos\varphi] \end{cases}. \tag{7}$$

When the right terms are called, respectively, $F(\varphi)$ and $G(\varphi)$, the transcendental equations can be rewritten as

$$\begin{cases} \frac{V_\infty}{V} = F(\varphi) \\ \frac{\omega y}{V} = G(\varphi) \end{cases}. \tag{8}$$

These two equations are subsequently combined into a single transcendental equation by extracting V from each of them and then subtracting:

$$V - V = 0 = F(\varphi) - \frac{V_\infty}{\omega y}G(\varphi). \tag{9}$$

Its complete form is:

$$\left(\sin\varphi-\frac{V_\infty}{\omega y}\cos\varphi\right)\sin\varphi-\sin|\varphi|\frac{1}{8B}\frac{N_b c_y}{\pi y}\left[C_l\left(\cos\varphi+\frac{V_\infty}{\omega y}\sin\varphi\right)-C_d\left(\sin\varphi-\frac{V_\infty}{\omega y}\cos\varphi\right)\right]=0. \quad (10)$$

This equation is solved by means of the *regula falsi* method [45]. Once φ is known, V is calculated and the axial induced velocity and the swirl velocity are computed as:

$$\begin{cases} w_i=V\sin\varphi-V_\infty \\ u_i=\omega y-V\cos\varphi \end{cases}. \quad (11)$$

4.3.2 Validation

With reference to tiltrotor configurations, Fig. 6 includes the rotor performance comparisons between numerical predictions and experimental data of the ADYN test campaign [46]. The rotational tip velocity is $M_{\Omega R}=0.504$ in hover and $M_{\Omega R}=0.491$ in cruise; the wind tunnel flow velocity is $M_{WT}=0$ in hover and $M_{WT}=0.30$ in cruise. The results shown in this figure comes from the postdictive phase when the code was trained with the help of experimental and CFD data. Recently, in the framework of the European Commission funded project ESPOSA [47], an experimental analysis of the effect of nonuniform inflow conditions on the performance of propellers was performed by conducting two test campaigns: the first one at DNW-LST wind tunnel on a four-bladed propeller and the second one at the TUD-OJF facility on an eight-bladed propeller.

Both propellers were assembled by adopting the same blade but they were installed into different wind tunnel models. The propellers were basically tested at a fixed pitch, under two free stream velocities (30 and 60 m/s), for several angular velocities spanning from 3000 to 11,000 rpm, and with flow unsteadiness coming either from a nonzero angle between the propeller axis and the free stream or from the presence of a wing downstream the propeller. The velocity components behind the propellers were also measured, and the comparison with CIRA BEMT code is illustrated in Fig. 7. The calculations on the two model propellers reveal a fairly good agreement with experiments in terms of thrust and torque (see also Figs. 8 and 9). At lower advance ratios the CIRA BEMT code predicts, especially for the four-bladed propeller, high sectional effective angles of attack which imply the airfoils work well beyond the stall angle.

4.4 CFD DRIVEN ANALYSIS

The BEMT code uses simplified aerodynamics, mostly based on 2D aerodynamic coefficients tables and several approximations such as for stall treatment. In many cases, especially in the presence of strong nonlinearities and three-dimensional effects, this kind of simplification significantly affects the performance prediction suggesting the use of more sophisticated aerodynamics. This is accomplished by coupling the BEMT code with a CFD analysis, so that the BEMT code provides the trim conditions and performance on the basis of the CFD aerodynamics. The coupling procedure herein applied involves the CIRA BEMT code [48], and the CFD code which is based on the Viscous-Inviscid Interaction between a boundary layer module and a full potential aerodynamic model [49]. The choice of a CFD tool based on potential theory is motivated by the fact that this model is a very cost-effective

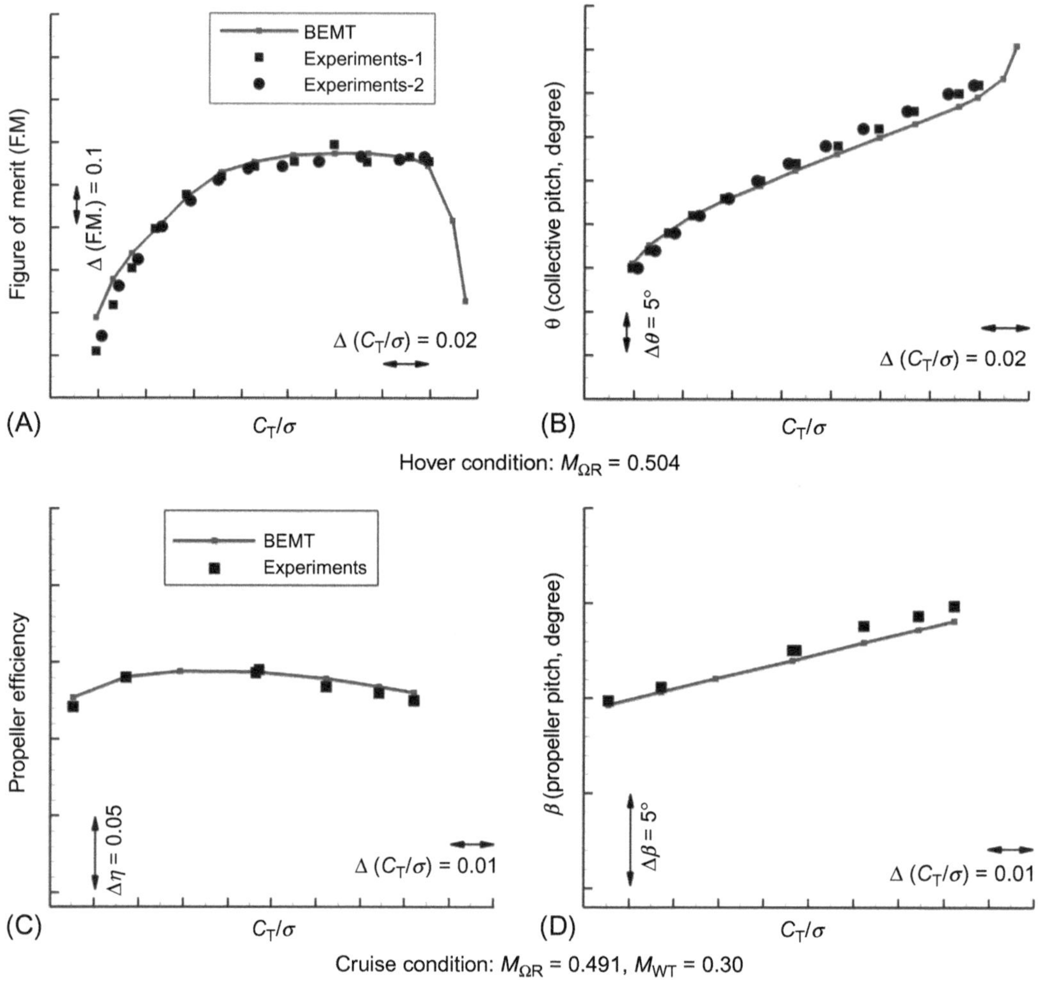

FIG. 6

Validation of the BEMT code on the ADYN rotor.

tool when the number of configurations to be analyzed, in addition to extended trim loops, is expected to be large and a good compromise on results credibility and computational resources is necessary.

4.5 BLADE PARAMETERIZATION

A critical step in the automated design optimization process is the selection of an efficient way to parametrically describe the geometry. The general aim is to reduce the number of design variables while retaining the ability to capture a global range of designs.

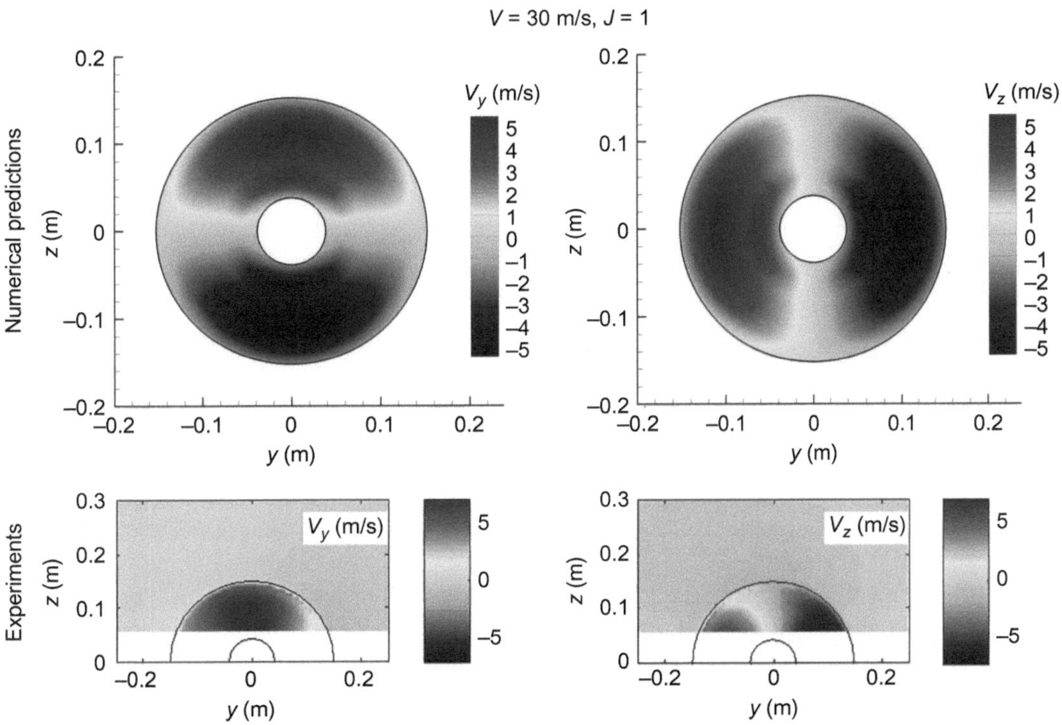

FIG. 7

Comparison of the velocity components; numerical results are obtained in the propeller disk, experiments refer to data at a distance of 56 mm from the disk.

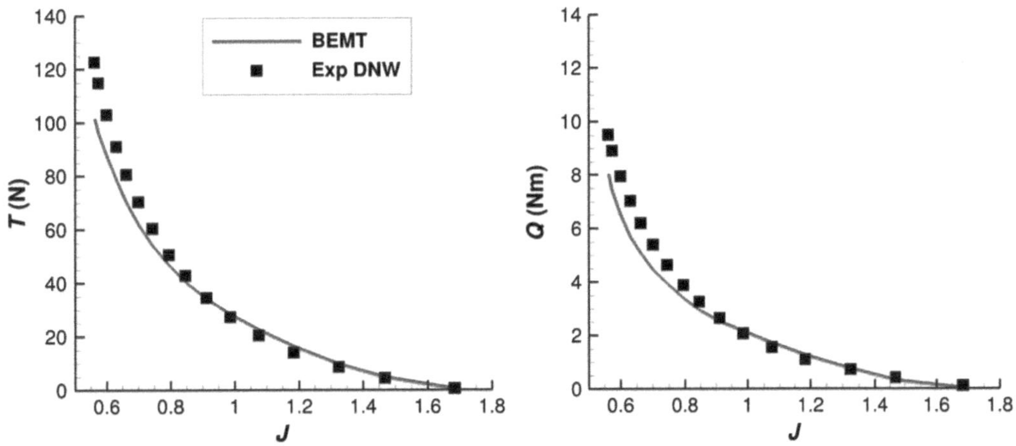

FIG. 8

Four-bladed propeller thrust (left) and torque (right) versus advance ratio.

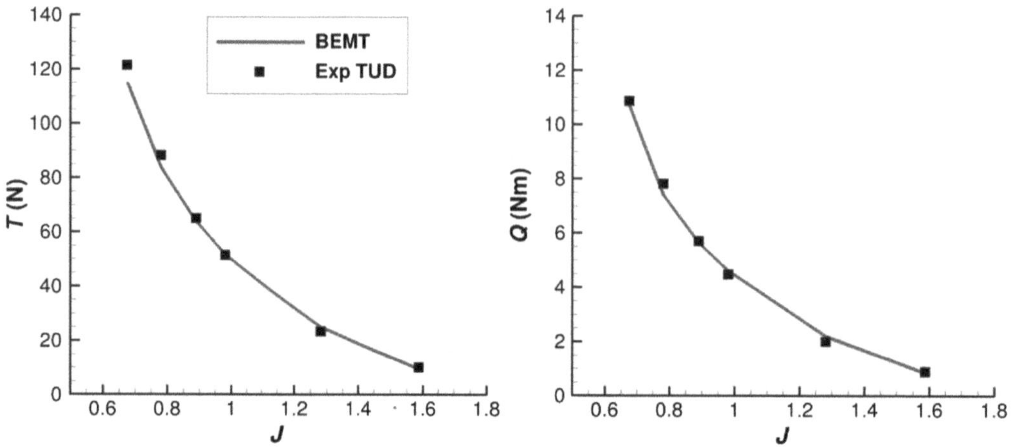

FIG. 9

Eight-bladed propeller thrust (left) and torque (right) versus advance ratio.

The parameterization module here adopted reads the initial geometry and computes for each section the updated sectional constructive parameters necessary to generate the new surface grid according to the current set of the design variables. As specified later, the blade geometry is modified by means of given control sections. If three control sections are used, the global number of design variables for a full parameterization is $6 \times 3+8 = 26$ (respectively, chord, twist, horizontal and vertical leading edge offset, sectional sweep and dihedral angles for three sections, the position of an intermediate section, an option to select the interpolating function for modifying the geometry, a variable for modifying the blade segments length which the blade is subdivided into, a parameter for selecting the set of airfoils being devoted to equip the blade [including those with morphing devices] the RPM value, the rotor radius, the percentage of the actual radius by which the blade may be elongated, and a variable for choosing the elongable/retractable strategy).

4.5.1 Planform

A helicopter rotor blade is herein generated by positioning, rotating, and scaling 2D airfoils (in nondimensional coordinates) along the blade span. The shape of each individual airfoil (including thickness) is studied separately because the aerodynamic and structural requirements are different depending on the airfoil spanwise position. When a set of airfoils is available, the designer act on some spanwise stations where he previously defined the blade constructive parameters such as chord length, geometric twist, and vertical and horizontal leading edge offset (with respect to a given reference axis) to give the blade the desired planform including sweep and dihedral angles.

The approach followed here to parameterize the blade planform geometry uses the traditional approach to design the blade, thus, the planform shape is separated by the sectional shape. This means that the blade shape is modified by means of constructive parameters affecting the blade planform whereas the sectional shape is modified by selecting the appropriate set of airfoils and distributing them along the blade span.

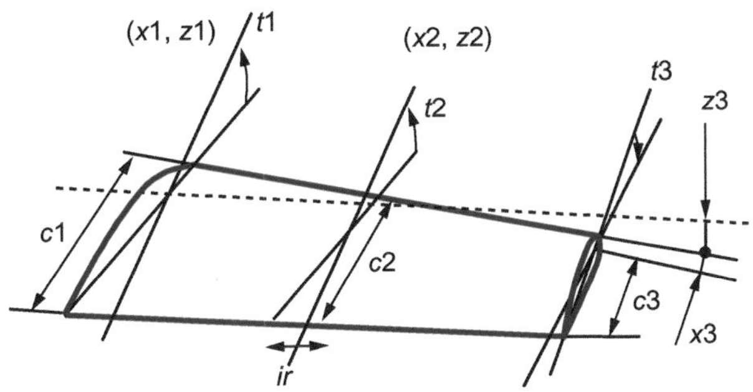

FIG. 10

Design variables for blade planform parameterization.

In order to perturb the design surface in a continuous way, six constructive parameters are identified for each spanwise station (see Fig. 10): chord length (c), geometric twist (t), vertical and horizontal leading edge offset (x,z), sectional sweep (γ), and dihedral angles (δ). Under the hypothesis of parallel planar sections, these parameters reduce to the first four. For simple blade shapes (e.g., rectangular shape), the constructive parameters at the inner and outer stations can be used. Nevertheless, the number of spanwise stations is expected to increase especially for complex blade shapes and highly non linear parameter variations. Of course, the number of sections needs to be limited anyway, otherwise the number of design variables becomes larger and larger. For this reason, the adopted parametric model is based on three or, at most, four sections.

Generally, the first and the last section correspond to the sections limiting the geometry to be optimized. On the contrary, the intermediate sections are chosen by the user. Indeed, the user chooses the position of the intermediate section and a software tool calculates the constructive parameters by interpolating on the closest spanwise stations. When more than four spanwise stations are necessary to appropriately characterize the blade, the constructive parameters associated to the sections in excess are not considered as design variables but they are modified according to predefined interpolation functions which distribute the deltas of the surrounding design variables.

4.5.2 Blade length morphing

The influence of rotor radius is investigated by considering three options as depicted in Fig. 11. In the first case (opt = 1), the geometry is radially stretched or squeezed by linearly distributing the initial planform parameters on the actual radius length. In the second case (opt = 2), the aerodynamic part of the blade is rigidly shifted outboard. In the third case (opt = 3), a portion of the blade tip radially extends outboard (as in the telescopic blade concept). The latter case implies limitations on the internal section and on the twist distribution.

Four design variables are used: the blade radius, the rotor angular velocity (expressed in terms of RPM), the option for implementing the three strategies described above, and the percentage of the actual radius by which the blade may be elongated (active for opt = 3). Rotor radius and angular velocity are linked together because of the definition of blade tip speed which has a significant impact on the aerodynamic conditions of the tests.

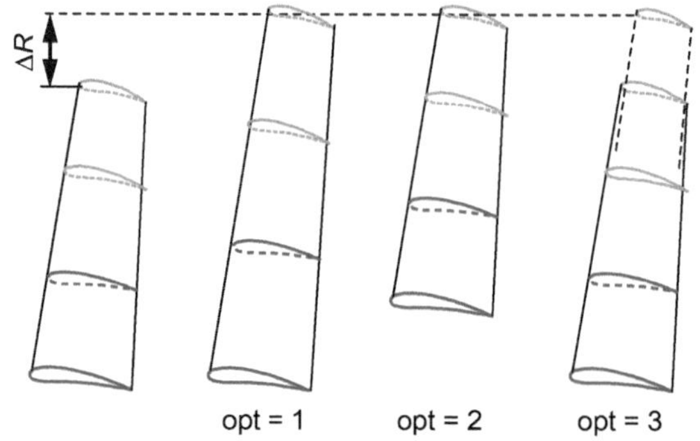

FIG. 11

Options for varying the rotor radius.

4.5.3 Airfoil geometry morphing

Because the 2D shape of the airfoils calls for their aerodynamic characterization (BEMT analysis is based on look up tables), morphing airfoils and the associated degree of deformation are seen as all the other airfoils mounted on the blade. The parameters associated to the selection of airfoils are discussed in Section 4.6. In this document section the techniques that are used for continuously deforming airfoil geometries are described; they concern with leading edge deformations (LED), trailing edge deformations (TED), and static extended trailing edge (SETE [18,29]) geometries.

To smoothly induce a nose droop or a continuous trailing edge flap, the airfoil part to be deformed is split in two so that one is rigidly rotated and the other accommodates the deformation of the nose or the trailing edge. With reference to Fig. 12, the rotation Ψ_{LED} affects the points of the segment comprised between x_{rig} and x_{rot} according to the following law:

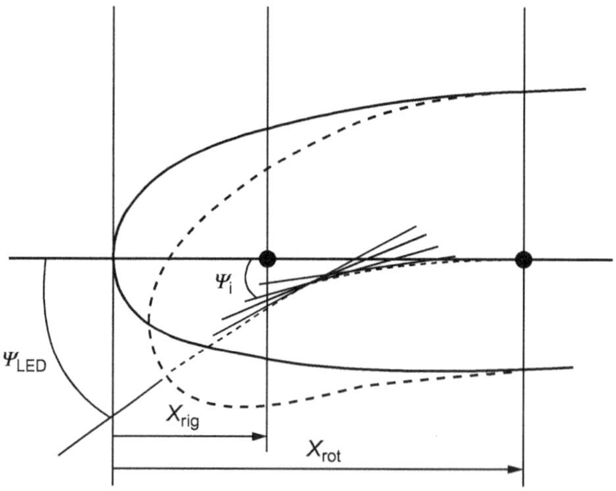

FIG. 12

Leading edge deformation.

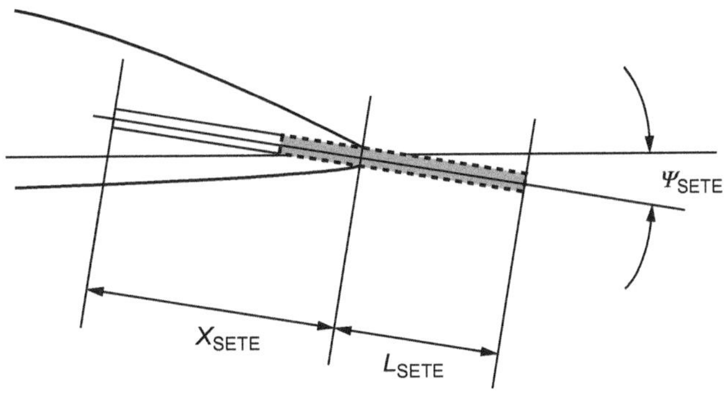

FIG. 13

Airfoil deformation for SETE.

$$\Psi_i = \Psi_{\text{LED}} \left(\frac{x_{\text{rot}} - x}{x_{\text{rot}} - x_{\text{rig}}} \right)^{2/3} \tag{12}$$

This strategy allows preservation of the shape of both the leading and trailing edges. The points x_{rig} and x_{rot} can be set by the user. For the present application they correspond, respectively, to 10% and 25% of the chord.

Concerning the device based on SETE, a flat plate of length x_{SETE} and thickness equal to the trailing edge thickness can extend out of the trailing edge according to a user-defined angle mainly depending on the available internal volume. Both the chordwise plate elongation ($\Delta_{\text{SETE}} = L_{\text{SETE}}/x_{\text{SETE}}$) and the associated angle (Ψ_{SETE}) can be set as illustrated in Fig. 13.

A value of 20% of the chord is used for x_{SETE}. When the trailing edge includes a tab, the plate comes out as a natural elongation of the tab. The user has in this case the possibility of modifying the tab angle. The airfoil deformation is performed as an off-line activity, and presently requires a visual inspection before proceeding with the aerodynamic characterization to ensure that there are no anomalous surface irregularities or volume violations.

4.6 AIRFOIL SELECTION

As stated in the surface modeler description, the blade geometry is divided into two separate input files: the constructive parameters file with the planform data (including twist), and the blade model with the airfoil shape. Thus, a blade with a given planform could be equipped with different airfoil shapes and spanwise distributions by simply replacing the blade model. A distinct software module dealing with airfoil selection and distribution accommodates the treatment of either morphing and nonmorphing sections. Basically, the blade model is an ASCII file containing a set of the available airfoils (that is, the name of the files with their nondimensional geometry and the associated aerodynamic look-up table), an initial estimate of the spanwise position and the number of morphing states if any.

The airfoil selection module initially divides the blade into constant-length radial elements of two kinds, those with an equal airfoil geometry and those whose geometry is linearly interpolated with the two different airfoils at the blade edges. The module produces a list of all the blade models with different combinations of airfoils and segments as output. The implemented strategy is based on some driving factors hereafter described:

- one airfoil results into one blade segment corresponding to the blade span;
- two or more different airfoils generate several blade segments;
- at this stage, a blade segment is only characterized by its length and the airfoil shapes;
- if two consecutive airfoils have different geometries, three blade segments are introduced so that two segments have a constant airfoil and the third in the middle allows for the transition to an airfoil geometry to the subsequent one;
- the more different airfoils (and morphing states) are used the larger the number of possible blade models;
- the user has the possibility of excluding those segments which have two different morphing states at their external sections;
- the airfoil chord-to-ratio thickness cannot increase as its spanwise position progresses towards the blade tip; and
- the length of blade segments is modified by the blade parameterization module.

In the simple case of a blade having just two different airfoil geometries (respectively, at the blade root and tip), this module generates six different blades as depicted in Fig. 14 and listed in Table 1. Each blade segment is characterized by a progressive number (starting from the inner one) and by the parameter *len*, which is used to stretch or squeeze its radial length. Fig. 15 illustrates the effect of the parameter *len* on the radial length of the blade segment 2. A negative value of *len* is used to block the blade segmant length to its initial value.

4.7 SURFACE GRID GENERATION

Fundamentally, using the information contained therein, the blade geometry is constructed into two separate input files: the first one with blade constructive parameters (the planform data including twist) and the other with the blade model (airfoil shapes and their radial distribution). The blade surface

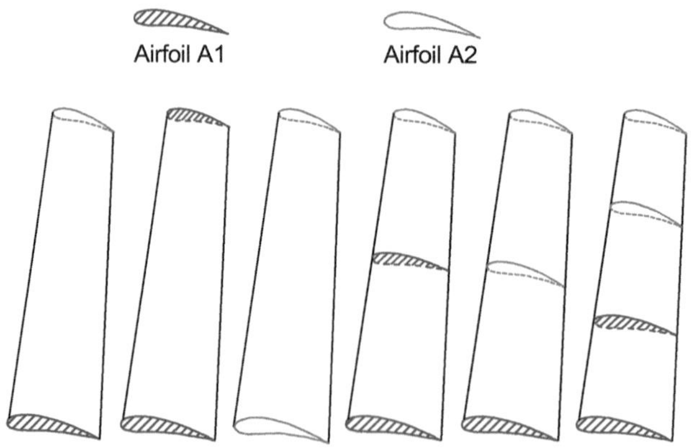

FIG. 14

Possible blade segmentations with two different airfoils.

Table 1 Different Blade Models From Airfoils A1 and A2.

n	Segment1	Segment2	Segment3
1	A1–A2	–	–
2	A1–A1	–	–
3	A2–A2	–	–
4	A1–A1	A1–A2	–
5	A1–A2	A2–A2	–
6	A1–A1	A1–A2	A2–A2

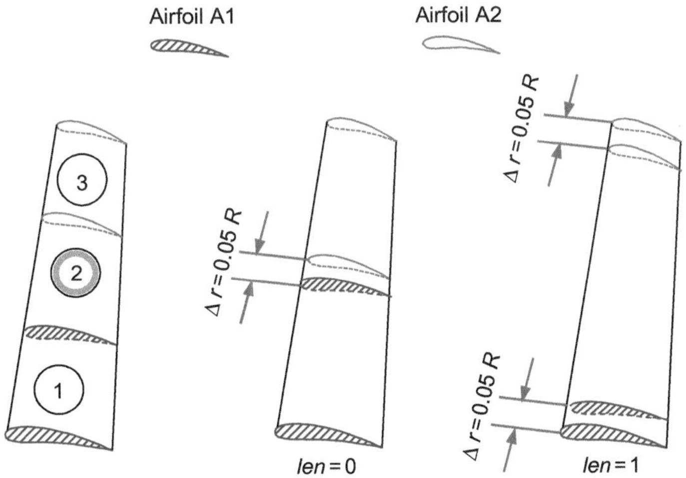

FIG. 15

Control of the blade segment length. $\Delta r = 0.05\ R$ is the minimum blade segment length.

generator first reads the blade model (with airfoil shapes and radial distribution) and constructive parameters, and then depending on the actual radial position, each airfoil geometry, is scaled according to the given chord, rotated according to the specified built-in twist, etc., for all of the constructive parameters. Thus, it is possible to obtain different blade surfaces by using the same constructive parameters (that is, basically the planform) and mounting different airfoils (which may differ in shape and radial position). As an example of the flexibility of the surface generator, Fig. 16 illustrates a blade surface after the modifications induced on a blade segment by a multiple morphing device.

4.8 VOLUME GRID GENERATION

When CFD is involved in the numerical process, the volume grid generation follows the surface grid generation. The perturbations produced by the optimizer on the body surface must be transferred to and propagated into the volume grid. Two basic techniques are available to propagate the surface grid-point

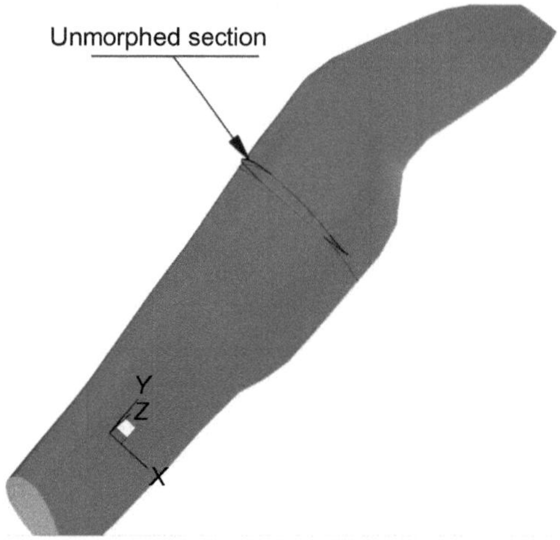

FIG. 16

Surface accommodating a blade segment based on an airfoil whose shape is modified by means of the simultaneous activation of LED, TED and SETE morphing devices.

movements into the field: grid deformation and grid regeneration. For structured single-block grids (that is, without the complexity coming from block topology creation typical of multiblock structured grids), the deformation of an existing grid relies mostly on transfinite interpolation (TFI) with exponential blending functions. This technique implies that the initial volume grid is not regenerated, and a file with the blade surface deformations must be read by the CFD solver which internally adapts the volume grid. The grid deformation technique, even if available in the software package here considered, is not applied since the choice of blending functions has a considerable influence on the quality of the volume grid when there is a large variation of the design variables. The grid regeneration technique mostly depends on the automation of the grid generation process. In the present work, an algebraic module of the grid generator system is used [50]. Since it allows for an automatic and extremely fast generation of high quality C—H grids (particular volume grids, having a C-shaped transversal section, repeated along the orthogonal direction), the strategy based on regeneration is preferred.

5 BACKGROUND

Within the 5th FP (Framework Programme) of the European Commission, some interrelated CTP (Critical Technology Project) research projects were launched. The CTP project ADYN was dedicated to rotor blade aero-acoustic optimization [3] of the ERICA [1] tiltrotor concept. That optimization was largely based on trade-off and parametric studies with the aim of optimizing the blade so that the rotor resulted in less noise during descent flight, maximizing the hover and cruise rotor performance. While the rotor radius was not changed, the parameters which researchers played on were planform and twist distribution. The airfoil set on the aerodynamic part of the blade was replaced by a new set of airfoils. Fig. 17 shows the baseline and the optimized rotor geometries, also referred as TILTAERO and ADYN

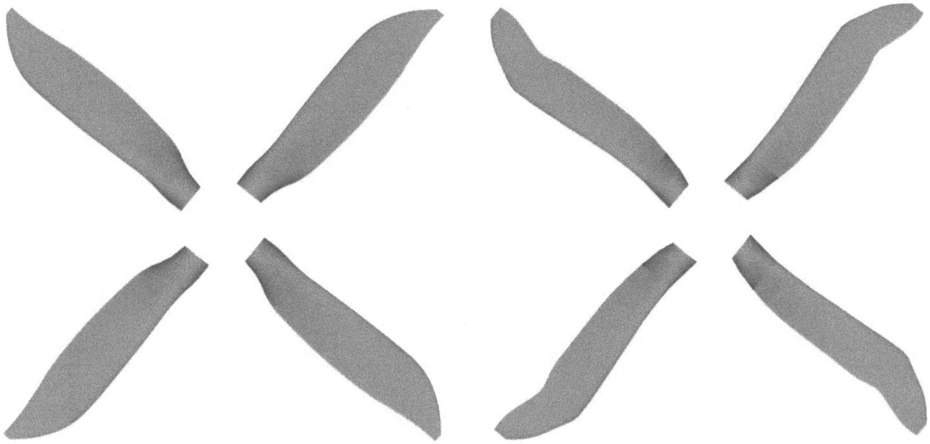

FIG. 17

TILTAERO (left) and ADYN (right) rotor planform geometries.

rotors. The two rotors were experimentally studied by means of an articulated wind tunnel campaign. The TILTAERO model rotor has a diameter of about 3 m and its four blades present a significant variation of the built-in twist, a maximum chord near the $r/R = 0.75$ spanwise station and a negative back sweep angle. The ADYN rotor has the same diameter and the blades are characterized by a double sweep angle, with anhedral angle at the tip and a more complex twist and chord span distribution.

The participation in that research initiative led CIRA to plan some activities on aerodynamic shape optimization of tiltrotor blades. The objective of those activities was twofold: the assessment of an efficient simulation environment [36] allowing code integration, and the maturation of expertise in aerodynamic tiltrotor blade shape optimization through numerical investigations where planform parameters (including built-in twist distribution) were involved. The assembly of code networks for rotor performance calculations and the optimization strategy deserved great attention to make the whole optimization task affordable. In particular, a flexible integration environment was chosen so that the most convenient analysis tools could be easily linked into code networks and the balance between the resulting increased simulation complexity and the computational effort could be controlled. After its assessment, a number of processes were captured, successfully validated, and made ready for subsequent investigations. Those processes differed in the design variables number, type of optimization strategy (DOE techniques, gradient/evolutionary algorithms), and analysis task (RSM-based, BEMT, BEMT-CFD).

The potential of that environment was discussed after performing an optimization exercise consisting on the aerodynamic shape optimization of the TILTAERO rotor geometry. After those investigations, a number of designs were available for synthesis. Table 2 collects data in terms of analysis type, number of analyses (for the two operative conditions), design technique, and optimization algorithm. RSM versus direct simulation, gradient based algorithm versus evolutionary based algorithm, or BEMT analysis versus BEMT-CFD analysis are some of the comparisons that were investigated for direct and cross correlations.

A visual representation of the most promising designs is included in Fig. 18. Many of the proposed designs indicate that, with respect to the baseline blade, the optimized blade tends to have: (a) at the

Table 2 Summary of Designs and Their Characteristics

Design #	Analysis Type	Analysis No.	Design Technique	Algorithm
Baseline	BEMT-CFD	1 × 2	–	–
1	BEMT	1000 × 2	DOE	Random
2	ART-HEL	1000 × 2	DOE	Random
3	RSM	–	MOO	NLPQL
4	RSM	–	MOO	NSEA
5	BEMT	121 × 2	MOO	NLPQL
6	BEMT	932 × 2	MOO	NSEA
7	BEMT-CFD	1254 × 2	MOO	NSEA

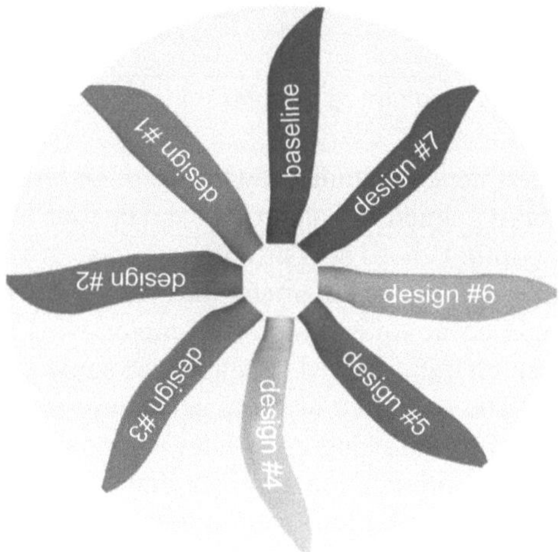

FIG. 18

Wheel representation of design planforms.

root, lower values of twist and chord; (b) at the intermediate section, smaller chords and both backward and forward sweep; (c) at the blade tip, negative values for twist, and small values for chord and backward sweep.

6 CASE STUDY

6.1 DESCRIPTION OF ACTIVITIES

The paper refers to the broad context of 3D aerodynamic shape optimization of rotary wing propulsion systems where morphing devices are explored. To evaluate the benefits of such devices, the newly optimized designs are compared to passive blades. In fact, the mechanical complexity together with structural penalties, weight increase, and actuation power must be offset by significant performance gains. For this reason, the shape of the baseline blade is first optimized without the intervention of any morphing device. In a subsequent phase, the devices acting on rotor radius variation and on airfoils are factored into the optimization task. The morphing devices, when involved, are herein assumed to be activated only for hovering.

6.2 BASELINE GEOMETRY

The baseline geometry is the ADYN blade, chosen as a reference because the optimization outcome is not expected to be verified by experimental tests. The idea of re-optimizing a known geometry also allows for exploitation of past efforts and comparison to available data.

Fig. 19 illustrates the ADYN blade where the aerodynamic part, from $r/R = 0.25$ to the tip, is outlined. The blade model is based on the distribution of five airfoils over seven segments as depicted in Fig. 20 where given airfoil shapes are designated with different labels and transitional airfoil shapes are shaded.

6.3 OPTIMIZATION OBJECTIVES AND STRATEGY

The rotor blade is optimized so that the aerodynamic performance is maximized at two nominal rotor load conditions. In hover flight, at a rotational tip velocity (in terms of Mach number) of $M_{\Omega R} = 0.63$, the thrust coefficient is $C_T = 0.021$; in level flight, at a rotational tip velocity of $M_{\Omega R} = 0.532$, and at an advance flight velocity of $M_{WT} = 0.58$ (350 Kn at 7500 m), the thrust coefficient is $C_T = 0.0157$.

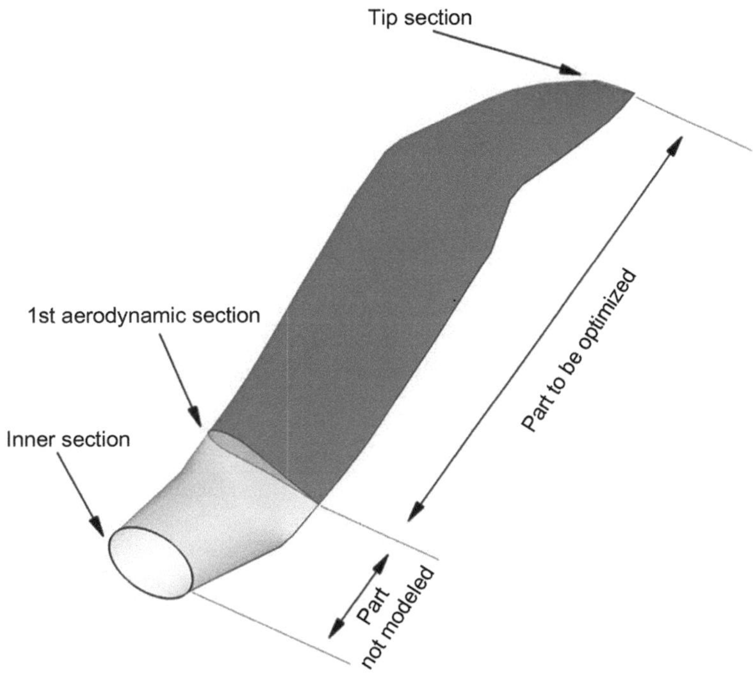

FIG. 19

3D view of the ADYN blade.

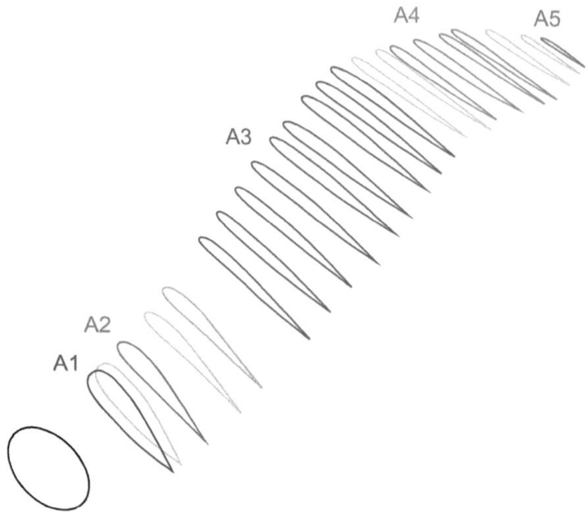

FIG. 20

Airfoil and segment distribution of the baseline blade.

The exploration of the design space is mainly done by means of DOE evaluations. MOO is accomplished by using direct simulations. The optimization exercises involve both gradient-based and evolutionary algorithms. Pareto fronts are available from MOOs and, in absence of any decision-making criterion, designs are extracted by giving priorities to the optima with equal gains on both objectives, and by discarding extreme (highly nonconventional) geometries after a visual inspection.

Furthermore, considering that a significant number of simulations are needed both for the exploration of the domain space and the optimization task, the blade is assumed to be rigid which, combined with the simulation of axial flight conditions (hover and cruise at zero incidence), simplifies the rotor trim phase. All of the rotor aerodynamic analyses are performed by trimming the collective pitch to obtain the required nominal thrust.

The length of the blade segments and the airfoil geometries allocated in there, the twist distribution, the rotor radius, the angular rotation are the main design variables investigated. The blade geometry can be modified in the radial range $r/R = [0.25, 1.0]$ where r and R are, respectively, the local and nominal radius.

7 UN-MORPHED BLADES

Besides the planform optimization, the influence of airfoil geometries and their distribution along the blade span are of the utmost importance, and the role they play in increasing the rotor performance may be greater than that of the planform shape. For this study, two sets of airfoils are used, each of them comprising five geometries. The performance characteristics of these two sets of airfoils (called simply set A and set B) are included in Tables 3 and 4.

Fig. 21 shows the impact of two airfoil sets on the TILTAERO blade performance while keeping all the other parameters constant, including the spanwise airfoil distribution. Even if the very significant

Table 3 Airfoil Performance of Set A

Airfoil	t/c	C_{lmax}	C_d @ $C_l = 0$	M_{dd} @ $C_l = 0$
A1	0.35	1.14 ($M = 0.3$)	0.016 ($M = 0.3$)	0.60
A2	0.20	1.84 ($M = 0.3$)	0.00825 ($M = 0.4$)	0.70
A3	0.12	1.52 ($M = 0.3$)	0.00795 ($M = 0.6$)	0.79
A4	0.09	1.35 ($M = 0.3$)	0.00741 ($M = 0.6$)	0.83
A5	0.07	1.24 ($M = 0.3$)	0.00727 ($M = 0.6$)	0.86

Table 4 Airfoil Performance of Set B

Airfoil	t/c	C_{lmax}	C_d @ $C_l = 0$	M_{dd} @ $C_l = 0$
B1	0.38	1.21 ($M = 0.3$)	0.038 ($M = 0.3$)	0.60
B2	0.20	1.32 ($M = 0.4$)	0.012 ($M = 0.4$)	0.75
B3	0.12	1.50 ($M = 0.4$)	0.0085 ($M = 0.6$)	0.80
B4	0.09	1.24 ($M = 0.4$)	0.0079 ($M = 0.6$)	0.85
B5	0.07	1.21 ($M = 0.4$)	0.00718 ($M = 0.7$)	0.90

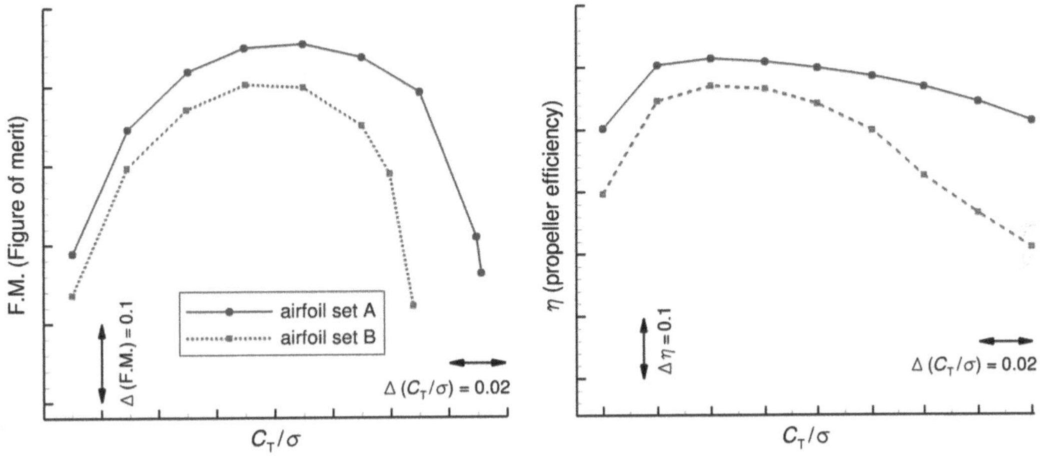

FIG. 21

Influence of two sets of airfoils on the tiltrotor performance. *Solid line* = set A; *dashed line* = set B.

improvement is caused by a fortuitous combination of effects, the exercise clearly demonstrates that further investigations are necessary. The following cases are hereafter discussed: (a) the airfoil number and geometry (coming from set A and set B) are optimized on the baseline planform; (b) the airfoil position, number, and geometry are optimized on the baseline planform; (c) the airfoil position, number, and geometry are optimized on the baseline planform for different values (see opt = 1 in Fig. 11) of rotor radius and angular velocity.

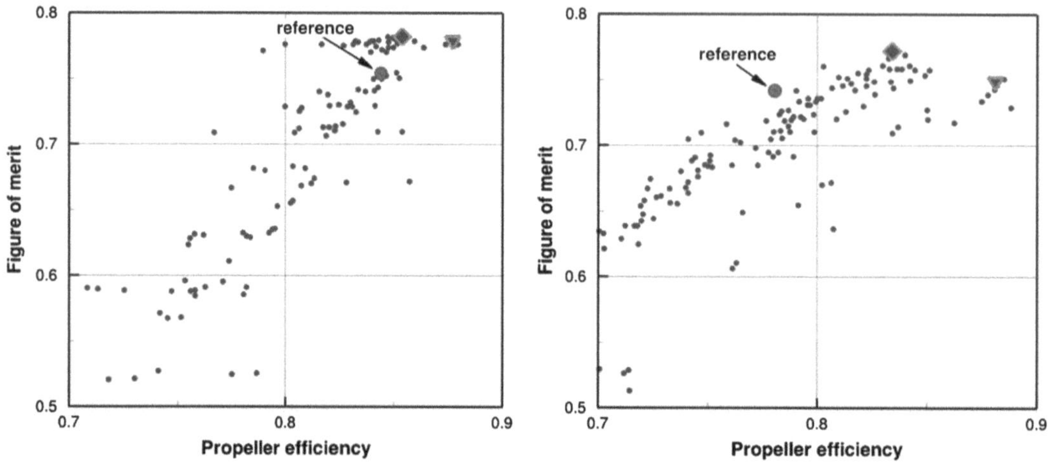

FIG. 22

Blade designs adopting airfoil set A (left) and set B (right) in the objective function space (*reference symbol* represents the blade with nonoptimized airfoil positions).

The subcase (a) can be synthesized as follows: starting from the existing ADYN blade planform to find the optimum distribution of the airfoils, the blades are constructed by using the same planform data and, alternately, the two sets of airfoils. Because of an equal number of distinct airfoils for the two sets, the same number of blades with different airfoils is generated. The two-level optimization process is arranged so that the optimization on the airfoil distribution is performed in the outer level and a single evaluation at the nominal conditions is conducted in the inner levels. Fig. 22 contains the results for the airfoil set A and B.

The performance gains, with respect to the blades where airfoil position was not optimized, are more relevant for set B. Indeed, airfoils set A was used when the ADYN blade was designed. In both figures the designs showing the highest figure of merit (FM) and the highest propeller efficiency (η) with limited penalty on FM are emphasized with the diamond and triangle symbols respectively; later they will be called, accordingly, diamond and triangle designs. The corresponding airfoil distributions can be extracted from Fig. 23 where the dashed and dash-dotted lines refer to set A and set B. The relevant difference in aerodynamic performance between the first two airfoils of each set (A1 vs. B1 and A2 vs. B2) motivates the inner airfoil distributions.

Anyway, the new distributions allow for substantial rotor improvements with respect to the performance of the baseline blade. Furthermore, the gap between the performance of the blades with the two sets of airfoils is limited for both diamond and triangle designs. Set A of airfoils allows for better rotor performance in helicopter mode; on the contrary, set B is more promising for rotors in airplane mode.

In subcase (b) the design space is extended so that the length of the blade segments is considered. Results in Fig. 24 are reasonably close to those of Fig. 22. The spanwise distribution of the airfoils is depicted in Fig. 25 where it can be observed that high propulsive efficiency for airfoils set B (triangle design) is obtained thinning almost the whole blade from $t/c = 0.2$ to $t/c = 0.07$.

The first two subcases try to accommodate the airfoil distribution to the frozen planform and in particular to the twist law. The investigations are based on the assumption that there is a

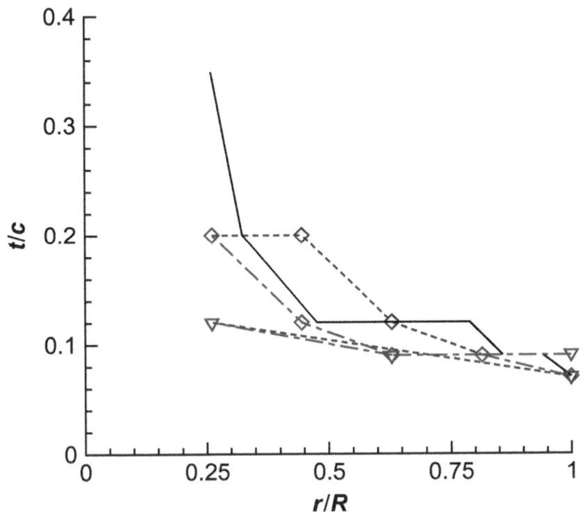

FIG. 23

Airfoil thickness distribution for subcase (a). *Solid line* = ADYN blade. *Dashed line* = blade with set A. *Dash-dotted line* = blade with set B. Symbols refer to diamond and triangle designs.

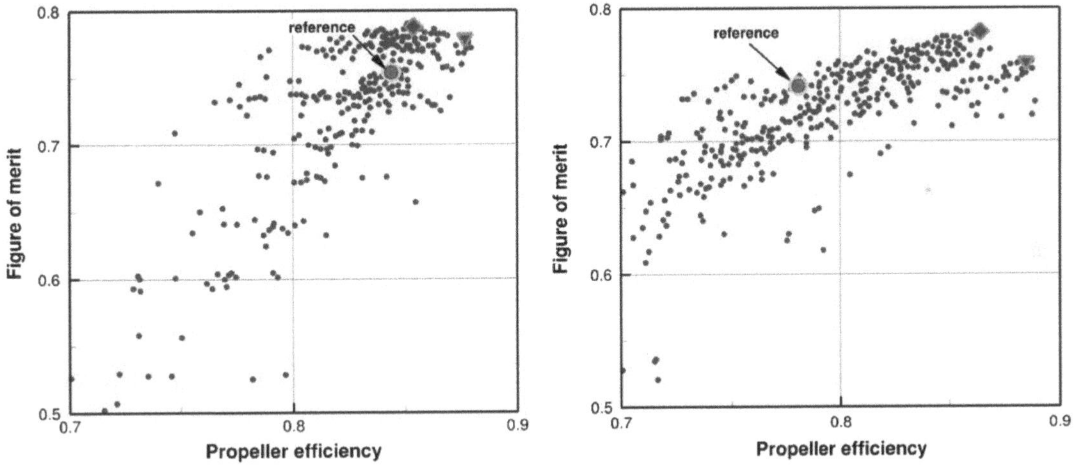

FIG. 24

Blade designs adopting airfoil set A (left) and set B (right) in the objective function space (*reference symbol* represents the blade with nonoptimized airfoil position and segment length).

correspondence between the linear variation of the airfoil geometries and the interpolated aerodynamic tables characteristics.

In order to explore the influence of the rotor radius, the twist law, and the rotor angular velocity on the objective functions, the subcase (c) is performed: by maintaining constant the airfoil distribution coming from the diamond designs of subcase (b), the two-level process manages a DOE including the variation of the rotor radius and the twist law (which are in common for both the investigated tiltrotor

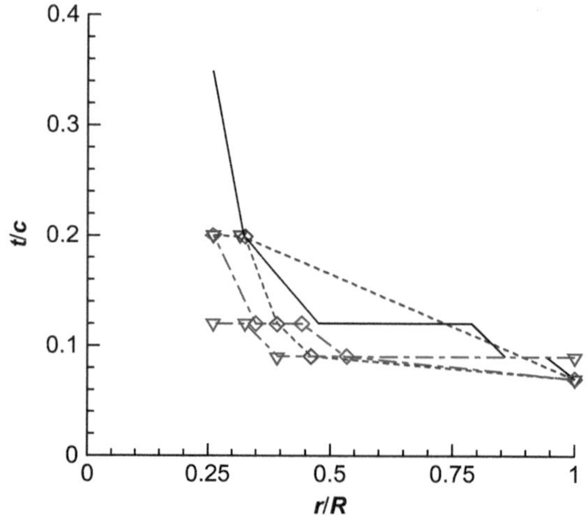

FIG. 25

Airfoil thickness distribution for subcase (b). *Solid line* = baseline blade. *Dashed line* = blade with set A. *Dash-dotted line* = blade with set B. Symbols refer to diamond and triangle designs.

operating conditions, upper level) and the optimization of the angular velocity with a gradient algorithm (inner level). This process allows exploration of those designs with the same optimum airfoil distribution and rotor radius length for hover and cruise, but with an optimum rotor angular velocity which is distinct for hover and cruise operative conditions.

Fig. 26 emphasizes that new improvements are still possible. As a consequence, the two-level process of subcase (c) is rerun after replacing the DOE with a MOO where an evolutionary algorithm (NSEA+) is applied on the upper level of the global optimization process (see Fig. 27).

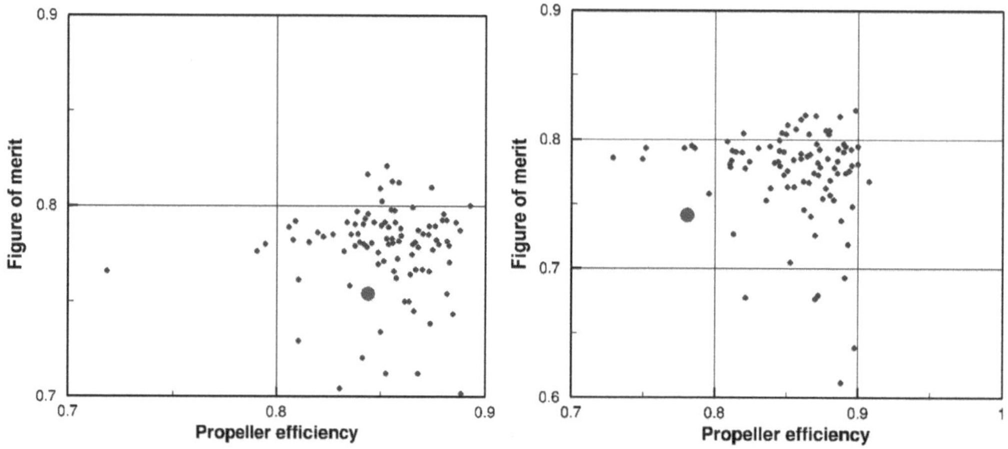

FIG. 26

DOE results relative to the diamond designs of Fig. 24, set A (left) and set B (right). Rotor radius and twist law are simultaneously optimized at optimum angular velocity.

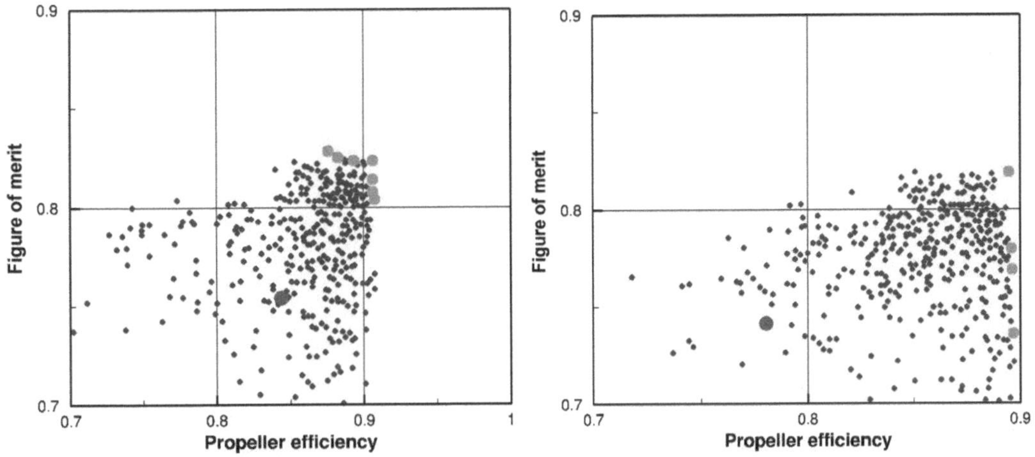

FIG. 27

MOO results relative to the diamond designs of Fig. 24, set A (left) and set B (right). Rotor radius and twist law are simultaneously optimized at optimum angular velocity. *Emphasized bullets* delimit the Pareto front.

Table 5 Summary of the Results of Subcase (c)

	DOE Set A	MOO Set A	DOE Set B	MOO Set B
FM	0.821	0.824	0.801	0.819
η	0.898	0.907	0.893	0.896
ΔR	−5.0%	−4.9%	−5.2%	−4.4%
ΔT_{w1}	+1.6° ($r/R = 0.7$)	+3.7° ($r/R = 0.7$)	+3.12° ($r/R = 0.8$)	+3.8° ($r/R = 0.6$)
ΔT_{w2}	+3.3° ($r/R = 1.0$)	+2.75° ($r/R = 1.0$)	+1.05° ($r/R = 1.0$)	+2.9° ($r/R = 1.0$)
ΔRPM_H	+0.7%	−0.5%	−0.1%	−0.7%
ΔRPM_A	−23%	−20%	−16%	−21%
$\Delta RPM_{A/H}$	−22%	−19%	−17%	−22%

Table 5 summarizes the results of subcase (c) with respect to the baseline blade in terms of percentage increment/decrement of the rotor radius (ΔR), increment of the built-in twist angle at two radial stations (ΔT_{w1}, ΔT_{w2}), the optimum angular velocity in helicopter (ΔRPM_H) and airplane (ΔRPM_A) mode and the ratio between these two velocities ($\Delta RPM_{A/H}$).

There is a general tendency to reduce the radius and the airplane mode angular velocity and increase the twist angles near the blade tip with respect to the baseline values. The ratio of angular velocities in airplane/helicopter mode is greater than the baseline rotor one. Fig. 28 compares the aerodynamic performance of the baseline blade and the MOO set A blade of Table 5.

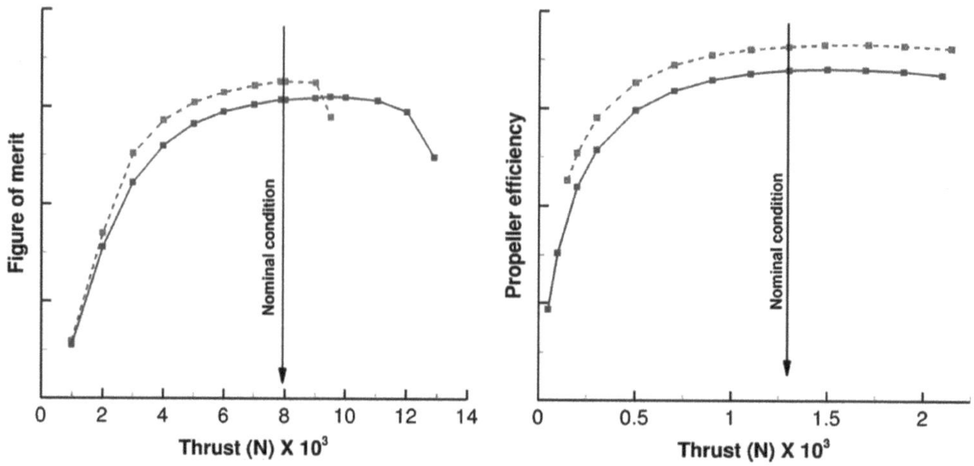

FIG. 28

Effects of the airfoil selection and distribution on rotor aerodynamic performance. *Dashed line* = MOO set A blade of Table 5; *Solid line* = baseline blade.

8 MORPHING BLADES

8.1 BLADE SPAN MORPHING AND VARIABLE SPEED ROTOR

Among the concepts developed over the past years, the concept based on the rigid radial movement of the whole aerodynamic part of the blade is investigated in this section. A two-level process is articulated as follows: the upper level concerns the optimization of airfoil distribution, length of the blade segments and twist law; the inner level is split into two further optimizations, the first one being relative to the angular velocity and the radius extension in the helicopter mode, and the second one relative to the angular velocity in the airplane mode.

Originally, in the upper level the rotor radius was also included as a design variable. This choice proved to be inconvenient since the upper level dominated on the inner level, and the best designs had almost zero radius variation and a blade configuration very close to the one of subcase (c) of the previous paragraph. The other alternative that is herein implemented, consisting of letting the two inner optimizations find the optimum rotor length for hover and cruise conditions, suffers from the following drawback: the blade segments are proportional to the rotor radius and the selected airfoils would not be in the same spanwise positions if the radius is different.

Since subcase (c) indicates that the optimum radius is slightly shorter than the nominal one, for the current subcase (d) the radius is fixed to 85% of the nominal value, and tip extension up to 40% *R are explored*. The spanwise chord distribution also varies. Fig. 29, which is relative to airfoil set A, shows encouraging results, but the rotor performance is not very far from that obtained for an un-morphed blade. The radial position of the tip is close to the tip of the optimum un-morphed blade, subcase (c) when extended to meet the helicopter mode requirements. Fig. 30 shows that the performance gains are more significant for the airplane mode, as expected.

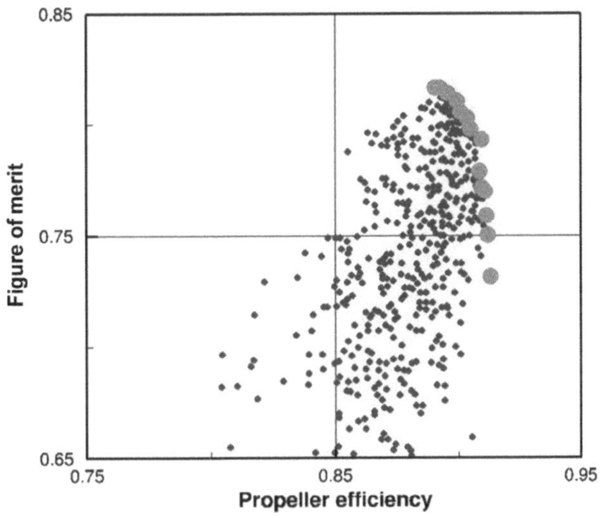

FIG. 29

Optimization relative to the radial extension of the blade.

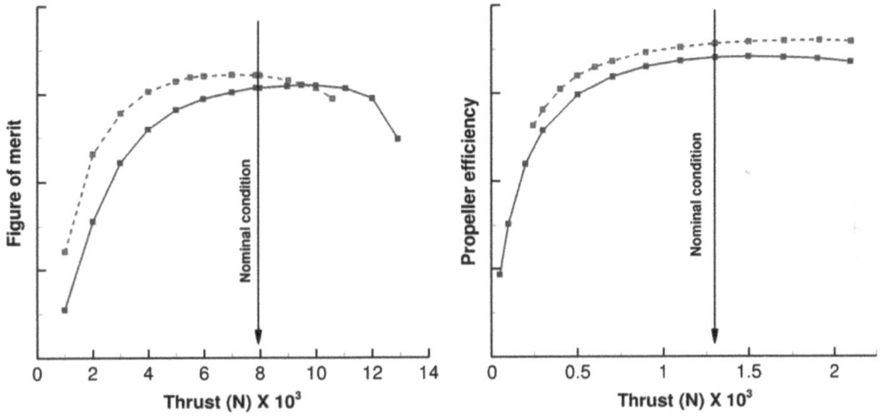

FIG. 30

Effects of the airfoil selection and distribution on rotor aerodynamic performance. *Dashed line* = blade radially extended; *solid line* = baseline blade.

8.2 BLADE SECTION MORPHING

8.2.1 Geometry morphing states

Among the airfoils specified in Table 3, the airfoil with 12% thickness-to-chord ratio is selected for the application of the morphing techniques described in paragraph 4.5.3. The applied deformations, which are summarized in Table 6, originate from literature review. In particular, the airfoil nose is drooped by

Table 6 Morphing Deformations Applied on Airfoil A3.

Id.	Ψ_{LED}	Ψ_{TED}	Δ_{SETE}
A3_00	0	0.0	0.0
A3_01	10	–	–
A3_02	15	–	–
A3_03	20	–	–
A3_04	–	2.5	–
A3_05	–	5.0	–
A3_06	–	10.0	–
A3_07	–	–	0.60
A3_08	–	–	0.75
A3_09	–	–	0.90
A3_10	10	–	0.60
A3_11	15	–	0.75
A3_12	20	–	0.90
A3_13	10	−2.0	0.60

10°, 15°, and 20°; three different rotations are set for the trailing edge (2.5°, 5°, and 10°); the trailing edge plate is extended 60%, 75%, or 90% of its length which is based on 20% of the local chord. The airfoil geometries resulting from the separate application of the morphing devices are depicted in Fig. 31. All of the morphing states listed in Table 6 (thus, including the clean airfoil) are numerically characterized and the relative look-up tables are computed by using a Euler/Boundary layer code. Fig. 32 compares the airfoil performance for the A3_10–A3_12 morphed shapes of Table 6. That table also includes multiple deformations coming from the combination of different devices.

The performance of the baseline rotor is recomputed because of the look-up tables now homogenously obtained by the same CFD code and at the appropriate Reynolds numbers. Fig. 33 gives evidence that the hover performance rapidly degrades after the thrust nominal value.

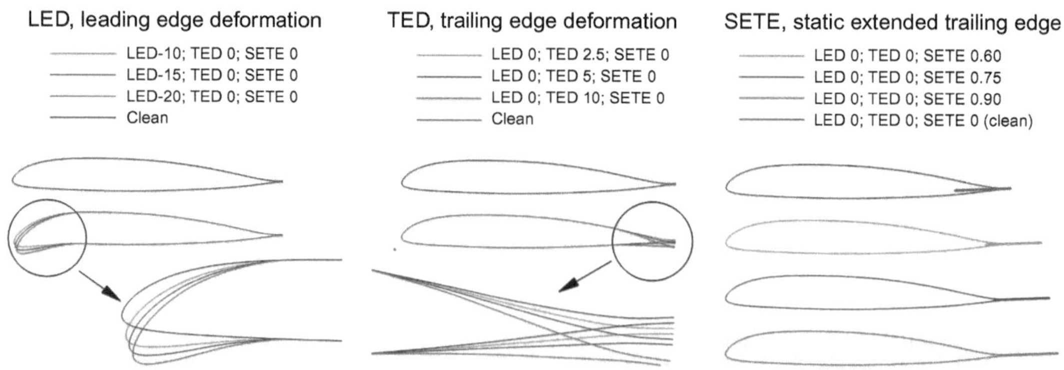

FIG. 31

Continuous morphing airfoil shapes.

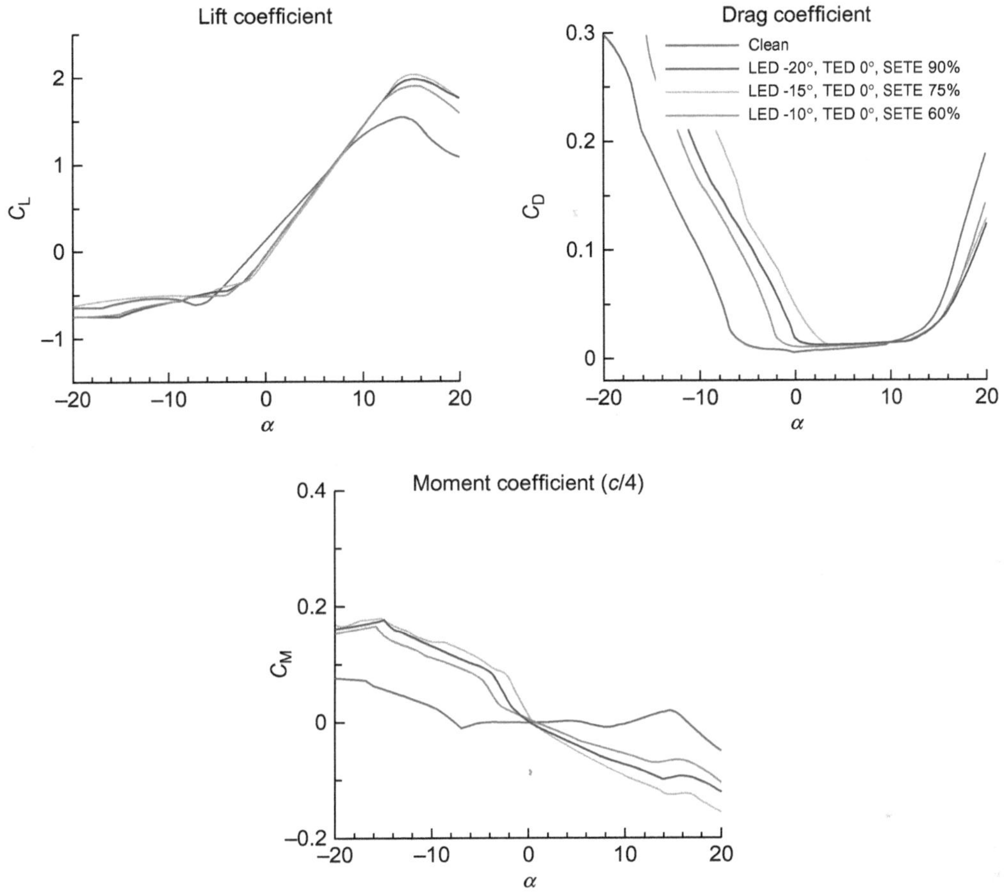

FIG. 32

Aerodynamic characterization of a section with multiple morphing devices at Mach = 0.4.

The first investigation which is performed aims at seeing whether the activation of morphing devices can ensure safer margins. The airfoil selection module is run starting from the airfoils equipping the baseline blade (five, in total), one of them with 13 morphing states. The module gives as output a list of 1649 possible blade models where the blade segments are not yet optimized. The process of Fig. 34 is launched with the goal of exploring all of the available designs at nominal conditions except for the hover thrust which is increased by 20% with respect to the nominal value. The analysis of Fig. 35 allows to see that the designs fulfilling the thrust increase in hover are both equipped with un-morphed and morphed sections. If some designs succeed by just positioning the airfoils differently (as already observed in Ref. 37), the number of those implementing morphing airfoils is significantly higher.

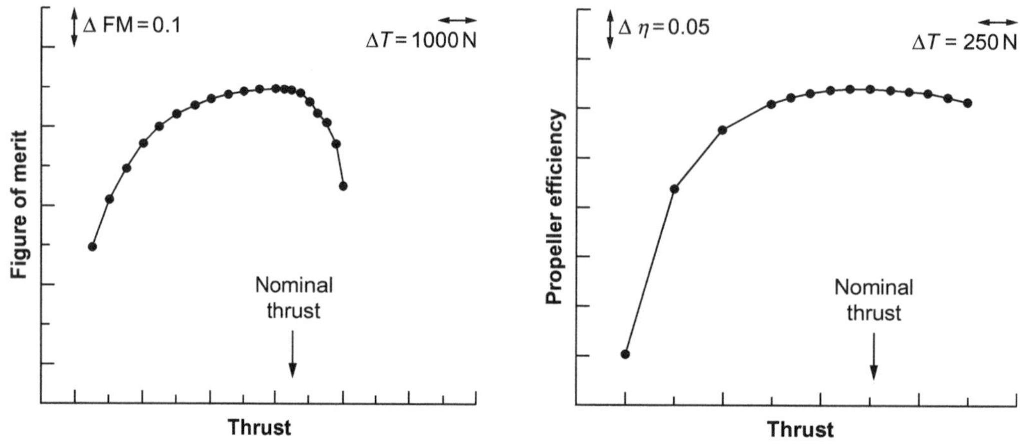

FIG. 33

Baseline rotor performance in hover (left) and cruise (right).

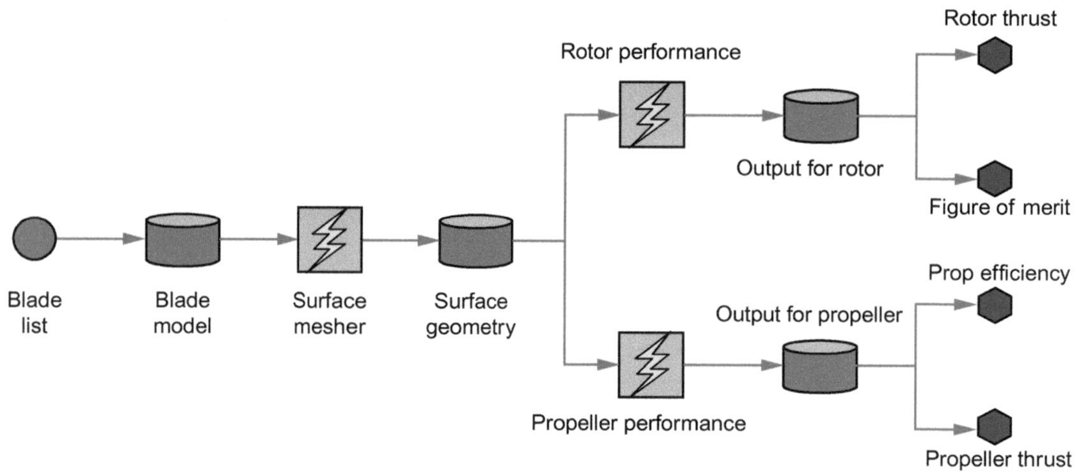

FIG. 34

Workflow for hover/cruise performance prediction.

After sorting the best design with improved propeller efficiency and figure of merit, its performance is compared with that of the baseline rotor in Fig. 36. This design, labelled 285, presents the following differences with respect to the baseline blade: the absence of airfoil A2, the replacement of airfoil A3 with A3_01, and the blade tip equipped by the A4 airfoil. It is important to emphasize that the blade is un-morphed when the rotor is in cruise flight, and that it is supposed to morph only in hover.

In order to find the optimal blade featuring a hover thrust 20% higher than the nominal value, a MOO is then performed by involving 10 further design variables which act on blade planform parameters, twist, blade segments length, airfoil distribution, and morphed shapes.

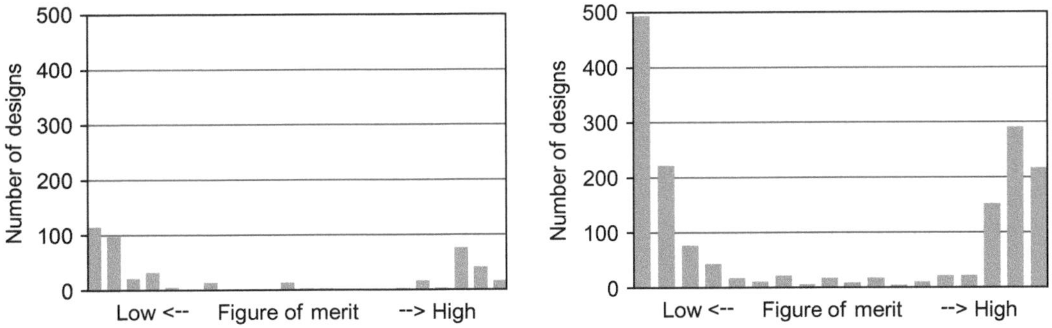

FIG. 35

Number of designs exceeding the nominal thrust in hover by 20% without morphing (left) and with morphing (right) for given ranges of figure of merit.

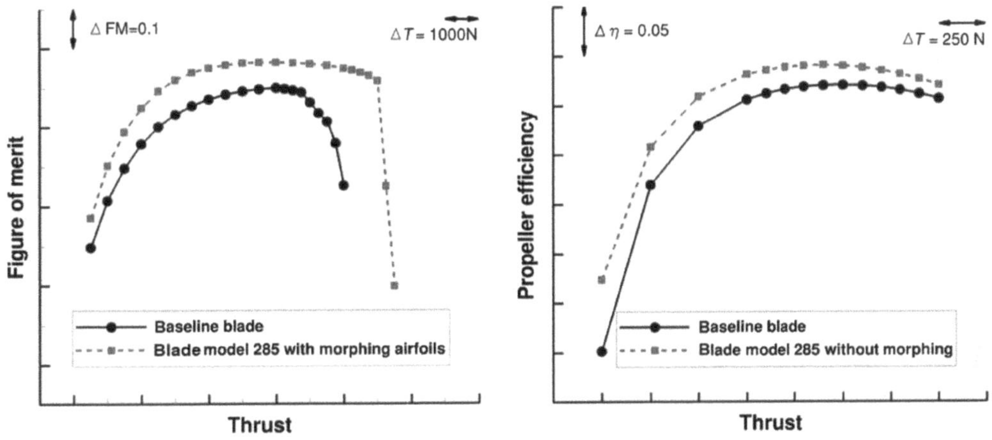

FIG. 36

Rotor performance of design 285 in hover (left) and cruise (right).

Low and high bounds are set for geometrical entities (sectional chord, twist, sweep angle, etc.). An evolutionary algorithm (NSEA+) [39] is applied on an initial population of 1649 individuals (already computed) and the evaluation of thousands of additional designs is needed.

Fig. 37 illustrates the designs into the objectives space showing that many designs are well beyond the baseline blade performance at nominal thrust conditions. In this case, the main interest is not in the design performance, but in the existence of blade models with their own airfoil distributions (including morphed shapes) fulfilling the thrust requirements.

Table 7 collects the top five designs with the highest propeller efficiency. Table 7 also includes the airfoil distribution from which it can be seen that the combination of LED and SETE generally results

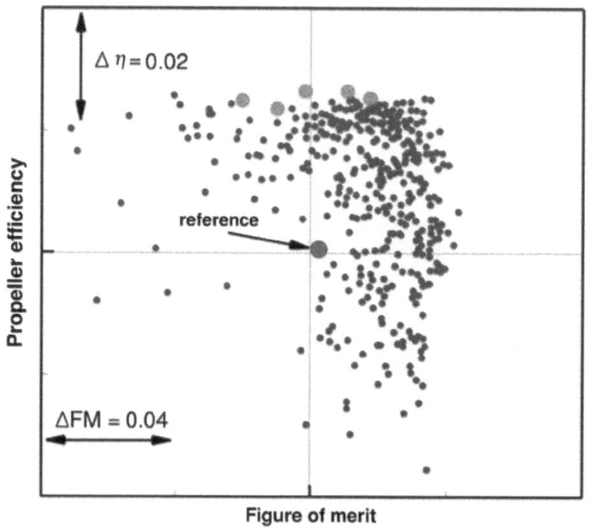

FIG. 37

Designs into the objectives plane. *Emphasized marks* are for designs selected in Table 7. The *reference symbol* indicates the baseline blade at thrust nominal conditions.

Table 7 Airfoil Distribution for Some Noticeable Designs

	Airfoil Position From Root to Tip					
Design	**1st**	**2nd**	**3rd**	**4th**	**5th**	**6th**
681	A1	A2	A2	A3_07	A3_07	A5
683	A1	A2	A2	A3_09	A3_09	A5
1557	A1	A2	A2	A3_12	A3_12	A4
582	A1	A2	A3_12	A3_12	A4	A5
891	A1	A3_09	A3_09	A4	A5	A5

in better performance. The chord and twist distributions of design 681 are compared with the geometric characteristics of the baseline blade in Fig. 38 and a 3D view, together with the airfoil selection and distribution, is depicted in Fig. 39. The performance of design 681 is opposed to that one of the baseline blade and design 285 both in terms of aerodynamic efficiency (Fig. 40) and in terms of power (Fig. 41).

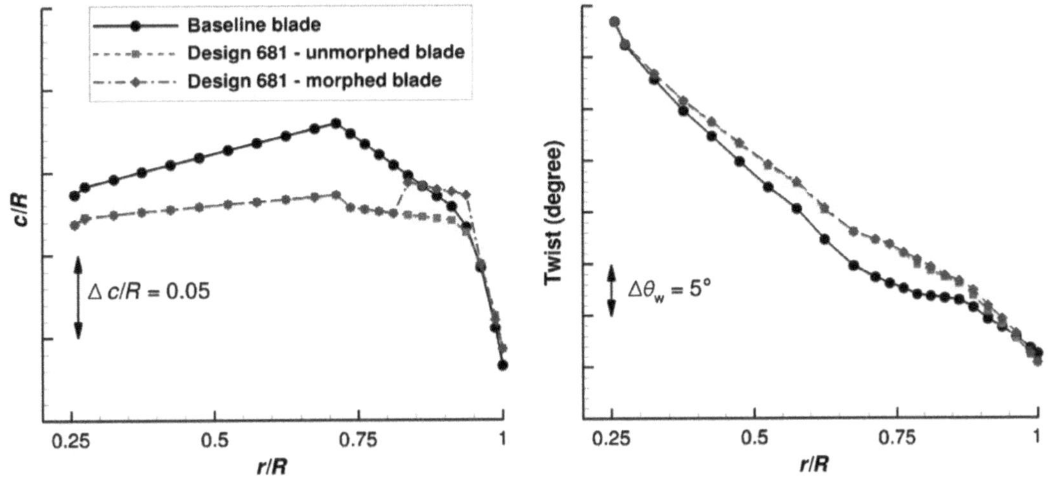

FIG. 38

Chord (left) and twist (right) distribution of design 681.

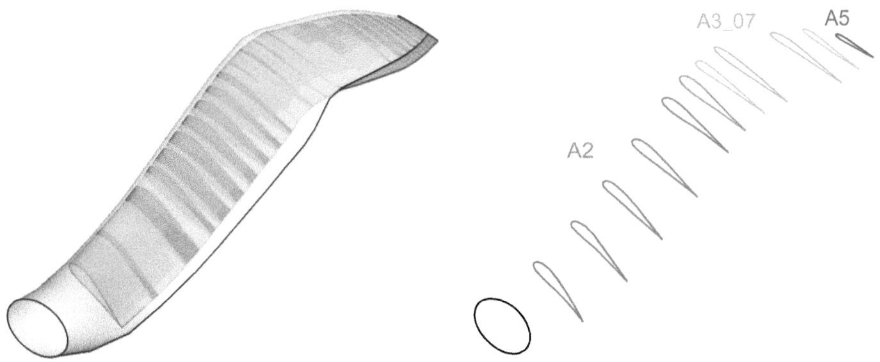

FIG. 39

3D view of design 681. On the left, the comparison with the baseline (gray shaded). On the right, the airfoil distribution.

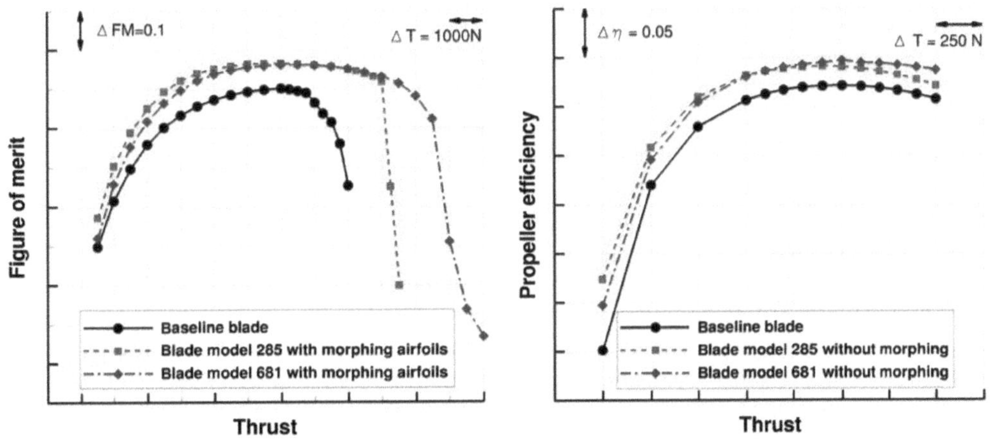

FIG. 40

Rotor performance of design 681 in terms of aerodynamic efficiency in hover (left) and cruise (right).

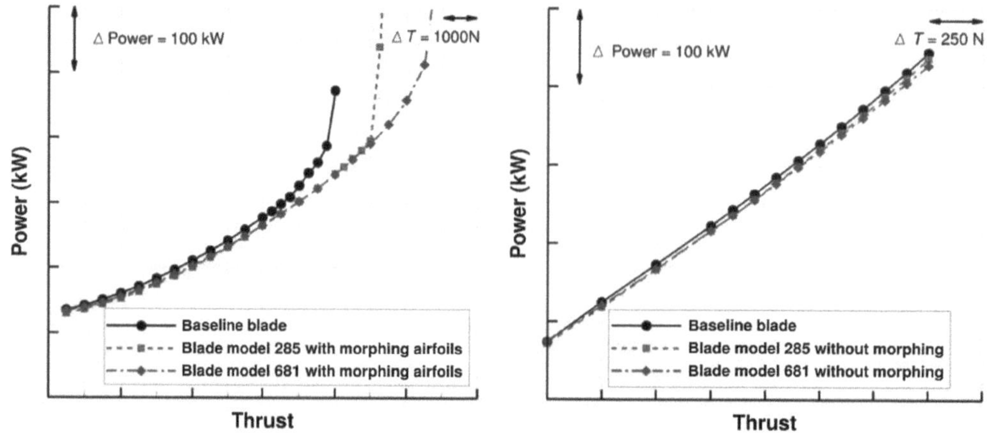

FIG. 41

Rotor performance of design 681 in terms of power in hover (left) and cruise (right).

9 CONCLUSIONS

The rotor of a convertiplane is a complex propulsion system component which must have good aerodynamic performance under varying operating conditions. In this chapter, the adoption of morphing devices on tiltrotor blades is explored by analyzing concepts allowing both spanwise and chordwise surface variations. The devices are supposed to be activated only in hover, letting the rotor provide a higher thrust. The research speculations are here associated with blade shape optimization. In this way, it is possible to simultaneously select those blades that are more aerodynamically efficient in both hover and in cruise, an operating condition in which the morphing devices are deactivated. The selection of morphing airfoils to be mounted on a blade and their spanwise position are fully investigated, and results are contrasted with those from optimized passive designs. In fact, the existence in the design space of solutions based exclusively on passive blades (e.g., from optimization of planform and airfoil selection and spanwise placement), even if characterized by a lower but still acceptable performance, may make the adoption of morphing sections less attractive.

A methodology for the aerodynamic shape optimization of tiltrotor morphing blades is illustrated. It is based on numerical processes executing aerodynamic and trim analysis procedures embedded within a commercial optimization environment. Several tools are used to morph the sections and to distribute selected airfoils on the blade planform. The fact that a blade is able to accommodate sections of different states (the un-morphed one plus several morphed ones) leads to the management of the design variables associated with those states.

The novelty of the present approach with respect to the state-of-the-art consists of a unique optimization process which meets the aerodynamic performance objectives by selecting the most suitable airfoil geometries, positioning them along the blade span, choosing the convenient morphing technology, and determining the amount of needed deformation. These features are added to the common optimization techniques based on planform related parameters and twist distribution. The potential of the methodology is discussed after performing both design space explorations and optimization tasks. The applications are accomplished on a realistic case even though the nature of the present study is essentially exploratory. In fact, the applications do not emanate from an industrial project, but are intended to assess the methodology and explore breakthrough technologies in light of forthcoming research projects.

The optimization of tiltrotor blades is fascinating, and the activities presented in this study are not exhaustive. Nevertheless, some useful insight can be drawn from them. Variable diameter rotors exhibit relevant performance improvements even though variable speed rotors are competing very closely. The influence of the airfoils (in terms of shape and spanwise position) is exploited so that their optimum selection leads to significant aerodynamic performance improvements. Indisputably, the design of advanced airfoils allows for better sectional aerodynamic performance, a basic element to further improve the rotor blade geometry. Finally, the incorporation of continuous morphing airfoils into the blades allows the nominal load conditions to be surpassed, making the design of new rotor configurations possible. Of course, the predicted benefits must be balanced with the complexity of the mechanisms and complemented by structural and power analyses.

These findings are originated by several simulation tools which, because of existing assumptions, may sometimes be subject to significant uncertainties, especially when nonconventional geometries are examined. Therefore, any result here shown must be considered as preliminary.

Since all the selected design variables affect the optimization objectives and mutually influence each other, it can be observed that deeper explorations of the design space requiring the evaluation of thousands (millions, in some cases) of designs make the whole activity unaffordable. Eventually a new generation of comprehensive rotor codes and Navier-Stokes solvers will be used for the optimization process. For the time being, the gradual assessment of less expensive methods provides numerical solutions that can be successfully applied for problems with highly intensive computation requirements.

REFERENCES

[1] F. Nannoni, G. Giancamilli, M. Cicalé, ERICA: the European Advanced Tiltrotor, Proceedings of 27th European Rotorcraft Forum, Moscow (RU), 2001.

[2] E. Porres, H. De Vries, M. Bauer, M. Da-Rold, F. De Nicola, E. Zoppitelli, DART tilt rotor program manufacturing and tests status, Proceedings of 32nd European Rotorcraft Forum, Maastricht (NL), 2006.

[3] T. Lefebvre, P. Beaumier, S. Canard, A. Pisoni, A. Pagano, A. Sorrentino, B. van der Wall, A. D'Alascio, C. Arzoumanian, S. Voutsinas, C. Hermans, Aerodynamic and aeroacoustic optimization of modern tilt-rotor blades within the ADYN project, Proceedings of the ECCOMAS-2004 conference, Jyvaskyla (FI), 2004.

[4] www.nicetrip.onera.fr.

[5] J. Decours, P. Beaumier, W. Khier, T. Kneisch, M. Valentini, L. Vigevano, Experimental validation of tilt-rotor aerodynamic prediction, Proceedings of 40th European Rotorcraft Forum, Southampton (UK), 2014.

[6] A. Le Pape, Numerical aerodynamic optimization of helicopter rotors: multi-objective optimization in hover and forward flight conditions, Proceedings of 31th European Rotorcraft Forum, Florence (IT), 2005.

[7] K. Collins, L. Sankar, Application of low and high fidelity simulation tools to helicopter rotor blade optimization, Proceedings of the American Helicopter Society 65th Annual Forum, Grapevine, TX (US), 2009.

[8] M. Imiela, High-fidelity optimization framework for helicopter rotors, Proceedings of 35th European Rotorcraft Forum, Hamburg (DE), 2009.

[9] D. Leusink, D. Alfano, P. Cinnella, Multi-fidelity optimization strategy for the industrial aerodynamic design of helicopter rotor blades, Aerosp. Sci. Technol. 42 (2015) 136–147.
[10] O. Gur, A. Rosen, Optimization of propeller based propulsion system, J. Aircr. 46 (1) (2009) 95–106.
[11] J. Dorfling, K. Rokhsaz, Constrained and unconstrained propeller blade optimization, J. Aircr. 52 (4) (2015) 1179–1188.
[12] T. Lefevbre, S. Canard, C. Le Tallec, P. Beaumier, F. David, ANIBAL: a new aero-acoustic optimized propeller for light aircraft applications, Proceedings of 27th International Congress of the Aeronautical Sciences (ICAS), Nice (FR), 2010.
[13] B.G. Marinus, M. Roger, R.A. Van den Braembussche, Aeroacoustic and aerodynamic optimization of aircraft propeller blades, Proceedings of 16th AIAA/CEAS Aeroacoustic Conference, Stockholm (SE), 2010.
[14] A. Pagano, L. Federico, M. Barbarino, F. Guida, M. Aversano, Multi-objective aeroacoustic optimization of an aircraft propeller, Proceedings of 12th AIAA/ISSMO Multidisciplinary Analysis and Optimization Conference, Victoria, BC (CA), 2008.
[15] P. Beaumier, J. Decours, T. Lefebvre, Aerodynamic and aero-acoustic design of modern tilt-rotors: the ONERA experience, Proceedings of the 26th Congress of International Council of the Aeronautical Sciences (ICAS), Anchorage, AK (US), 14–19 2008.
[16] J.G. Leishman, K.M. Rosen, Challenges in the aerodynamic optimization of high-efficiency proporotors, J. Am. Helicopter Soc. 56 (1) (2011), 012004 (1–21).
[17] C.W. Acree, Jr., Integration of rotor aerodynamic optimization with the conceptual design of a large civil tiltrotor, AHS Aeromechanics Conference, San Francisco, CA (US), 2010.
[18] S. Barbarino, O. Bilgen, R.M. Ajaj, M.I. Friswell, D.J. Inman, A review of morphing aircraft, J. Intell. Mater. Syst. Struct. 22 (9) (2011) 823–877.
[19] T. Weisshaar, Morphing aircraft systems: Historical perspectives and future challenges, J. Aircr. 50 (2) (2013) 337–353.
[20] E.A. Fradenburgh, Application of a variable diameter rotor system to advanced VTOL aircraft, Proceedings of the American Helicopter Society 31st Annual Forum, Washington, DC (US), 1975.
[21] D. Matuska, A. Dale, P. Lorber, Wind tunnel test of a variable-diameter (VDTR) model, NASA Contractor Report 177629 1994.
[22] J. Steiner, F. Gandhi, and Y. Yoshizaki, An investigation of variable rotor RPM on performance and trim, Proceedings of the American Helicopter Society 64th Annual Forum, Montréal (CA), April 29–May 1, 2008.
[23] M. Mistry, F. Gandhi, Helicopter performance improvement with variable rotor radius and RPM, J. Am. Helicopter Soc. 59 (2014), 042010 (1–19).
[24] E. Romander, 3-D Navier-Stokes analysis of blade root aerodynamics for a tiltrotor aircraft in cruise, Proceeding of the AHS Vertical Lift Aircraft Design Conference, San Francisco, CA (US), 2006.
[25] C.W. Acree, Performance optimization of the NASA large civil tiltrotor, Proceeding of the International Powered Lift Conference, London (UK), 2008.
[26] C. Stahlhut, J.G. Leishman, Aerodynamic design optimization of proprotors for convertible-rotor concepts, Proceedings of the American Helicopter Society 68th Annual Forum, Fort Worth, TX (US), 2012.
[27] R. Jain, H. Yeo, I. Chopra, Investigation of trailing-edge flap gap effects on rotor performance using high-fidelity analysis, J. Aircr. 50 (1) (2013) 140–151.
[28] T. Yokozeki, A. Sugiura, Y. Hirano, Development of variable camber morphing airfoil using corrugated structure, J. Aircr. 51 (3) (2014) 1023–1029.
[29] M. Khoshlahjeh, F. Gandhi, Extendable chord rotors for helicopter envelope expansion and performance improvement, J. Am. Helicopter Soc. 59 (1) (2014), 012007 (1–10).
[30] M.S. Murugan, B.K.S. Woods, M.I. Friswell, Morphing helicopter rotor blades with curvilinear fiber composites, Proceedings of the 38th European Rotorcraft Forum, Amsterdam (NL), 2012.
[31] R. Pecora, A. Concilio, I. Dimino, F. Amoroso, M. Ciminello, Structural design of an adaptive wing trailing edge for enhanced cruise performance, in: Proc. 24th AIAA/AHS Adaptive Structures Conference, January 4–8, 2016, San Diego, United States.

[32] I. Dimino, G. Diodati, A. Concilio, A. Volovick, L. Zivan, Distributed electromechanical actuation system design for a morphing trailing edge wing, SPIE Smart Structures/NDE, Las Vegas, Nevada (USA) March 2016. Proc. SPIE 9801, Industrial and Commercial Applications of Smart Structures Technologies 2016, 980108 (April 16, 2016). doi: 10.1117/12.2219223.
[33] E.S. Bae, F. Gandhi, CFD analysis of high-lift devices on the SC-1094R8 airfoil, Proceedings of the American Helicopter Society 67th Annual Forum, Virginia Beach, VA (US), 2011.
[34] R. Jain, H. Yeo, I. Chopra, Computational fluid dynamics–computational structural dynamics analysis of active control of helicopter rotor for performance improvement, J. Am. Helicopter Soc. 55 (4) (2010), 042004 (1–14).
[35] C. Garcia-Duffy, A. D'Andrea, S. Melone, Exploitation of active controls and morphing technologies to enhance rotor aerodynamic performance in hover conditions, Proceedings of AHS Aeromechanics Specialist's Conference, San Francisco, CA (US), 2010.
[36] A. Pagano, Advanced tiltrotor blade shapes by multi-objective optimization, Proceedings of Eurogen 2011 Conference, Capua (IT), 2011.
[37] A. Pagano, Aerofoil selection and spanwise placement in aerodynamic design and optimization of tiltrotor blades, Proceedings of the 39th European Rotorcraft Forum, Moscow (RU), 2013.
[38] A. Pagano, Aerodynamic shape optimization of tiltrotor blades equipped with continuous morphing aerofoils, Proceedings of 41st European Rotorcraft Forum, Munich (DE), September 1–4, 2015.
[39] Noesis Solutions (www.noesissolutions.com), Optimus rev. 10.15, 2014.
[40] A. Pagano, Code networking for multi-disciplinary analysis of rotor phenomena, Air Space Eur., Vol. 3, N. 3–4, pp 155–160, 2001.
[41] A. Pagano, L. Federico, L. Notarnicola, Process capturing and optimization for rotorcraft systems, Tech. Rep CIRA-TR-07-0152 and OPTIMUS Worldwide User Meeting, Leuven (BE), 2007.
[42] A. Pagano, S. Ameduri, V. Cokonaj, A. Prachař, Z. Zachariadis, D. Drikakis, Helicopter blade twist control through SMA technology: optimal shape prediction, twist actuator realisation and main rotor enhanced performance computation, Proceedings of 35th European Rotorcraft Forum, Hamburg (DE), 2009.
[43] A. Pagano, M. Barbarino, D. Casalino, L. Federico, Tonal and Broadband noise calculations for aeroacoustic optimization of propeller blades in a pusher configuration, J. Aircr. 47 (3) (2010) 835–848.
[44] J.G. Leishman, K.Q. Nguyen, State-space representation of unsteady airfoil behaviour, AIAA J. 28 (5) (1990) 836–844.
[45] H. Winarto, BEMT algorithm for the prediction of the performance of arbitrary propellers, Tech. Rep. CR CoE-AL 2004-HW3-01, Centre of Expertise in Aerodynamic Loads, Royal Melbourne Institute of Technology (AS), 2004.
[46] E. Bianchi, G. Preatoni, N. Dalla Rovere, Mach scaled wind tunnel tests on a 4-bladed half-span advanced Tilt-Rotor, Proceeding of the American Helicopter Society 63rd Annual Forum, Virginia Beach, VA (US), 2007.
[47] www.esposa-project.eu.
[48] A. Pagano, Investigations within the ADYN project: a year of activities, Tech. Rep. CIRA-TR-02-0503, CIRA (IT), 2003.
[49] M.H.L. Hounjet, A. Pagano, J.C. Le Balleur, D. Blaise, F. Salvatore, G. Cinquina, A. Kokkalis, Enhanced aerodynamic formulation of a European Rotorcraft Simulation Method, Tech. Rep. NLR-TR-98425, NLR (NL), 1998.
[50] M.H.L Hounjet, C.B. Allen, L. Gasparini, L. Vigevano, A. Pagano, GEROS: a european grid generator for rotorcraft simulation methods, Proceedings of 6th London Grid Generation Conference, London (UK), 1998.

CHAPTER

SYNERGIC EFFECTS OF PASSIVE AND ACTIVE ICE PROTECTION SYSTEMS 26

Tobias Strobl*, Stefan Storm†, Salvatore Ameduri‡

3DSE Management Consultants, Muenchen, Germany Airbus Group Innovations, Ottobrunn, Germany†*
The Italian Aerospace Research Centre, CIRA SCpA, Capua (CE), Italy‡

CHAPTER OUTLINE

NOMENCLATURE

AF	adhesive film
AGI	Airbus Group Innovations
AoA	angle of attack
$A_{\mathrm{surf\_ice}}$	surface area covered with ice
C	electrical current
°C	Celsius degree, temperature
C3D8E	eight-node three-dimensional linear piezoelectric brick elements

Morphing Wing Technologies. https://doi.org/10.1016/B978-0-08-100964-2.00026-5

C3D8R	eight-node three-dimensional linear brick elements
CAA	contact angle advancing
CAH	contact angle hysteresis
CAR	contact angle receding
CFD	computational fluid dynamics
DSLR	digital single-lens reflex
EDM	electro-discharge machining
FE	finite element
FEA	finite element analysis
FEM	finite element model
f_n	natural frequency
IPS	ice protection system
L	reference length to compute the *Re* number
LEWICE	Lewis Ice Accretion Program
LWC	liquid water content
m	meter, length
MVD	median volume diameter
NACA	National Advisory Committee for Aeronautics
P	electrical power
PSA	phosphoric sulfuric anodization
P_{d_tot}	total power density of the hybrid system
P_{heater}	power density of the heater system
P_{piezo}	power density of the piezoelectric system
q	power density of the heater system
Q_m	mechanical quality factor
R_a	surface roughness level
Re	Reynolds number
S	shear stress
T_∞	airstream temperature
Ton	ton
T_{tot}	stagantion temperature
V	voltage
V_∞	airstream velocity
WT	wind tunnel
W	watt, power
α	mass proportional Rayleigh damping coefficient
β	stiffness proportional Rayleigh damping coefficient
θ_{CB}	contact angle according to the Cassier-Baxter model
μ	dynamic viscosity
ρ	air density
τ_{adh}	adhesion shear

1 INTRODUCTION

Ice accretion events on general aerodynamic surfaces including aircraft empennages may represent a serious safety problem. The drastic modification of the shape close to the leading-edge zone causes a sudden lift decrease and drag arise with a consequent degradation of the airfoil performance.

In the worst scenario, such phenomenon can lead to the loss of the aircraft [1–3]. For this reason, aeronautical institutions are spending more and more efforts to develop methodologies to predict ice growth mechanisms and, from those, to develop more convenient and effective strategies to prevent ice accretion or remove formed particles before they accumulate beyond certain thresholds.

Generally, ice protection systems (IPS) can be classified under two families: anti-icing, aimed at preventing ice formation, and deicing devices, aimed at detaching already established ice accumulation. Among the most popular of these systems, thermal IPS are well-suited for aircraft surfaces prone to ice accretion. Energy demands are maximized when these systems operate in a running-wet, anti-icing mode [4]. This implies that the temperature of the contact area must be kept above 0°C, making the systems power consumption strongly dependent on the environmental temperature, which typically ranges from 0.6 to 7.2°C; this results in a very high power density requirement, spanning from 16.4 to 26.4 kW/m^2 [4,5]. This is generally associated with another drawback: the local action of this system is confined only to a small area around the heat source. These two shortcomings suggest that this equipment has limits, and should, for example, be flanked with other devices of a different nature. In this study, thermal heating was used only to inhibit ice accretion at the stagnation point, where the aerodynamics generate maximum pressure value (stagnation pressure) and minimum flow velocity value (stagnation, $v = 0$). Even in the presence of a hydrophobic coating, this situation makes it difficult to implement mechanical shear systems that would work only in conjunction with significant airflow.

The idea at the basis of this project is to melt ice on the leading edge, encourage the droplets to slide downward through the hydrophobic coating and, when ice starts reforming, remove it through piezoelectric actuators integrated inside the airfoil. Experiments utilizing a small-scale demonstrator in the laboratory-size icing wind tunnel of Airbus Group Innovations (AGI) supported this hybrid strategy, proving that the hybrid IPS can reduce power consumption up to 91% [6,7] as compared to a classical thermoelectric system, working on identical geometry.

The process of manufacturing the specific coating is described in detail, highlighting the different effects produced by the texture morphology. The optimal configuration is identified via an experimental test campaign, carried out on flat coupons excited by an electromagnetic inductor. Then, a NACA 0012 symmetric airfoil is machined to mirror finishing and coated with the icephobic film. Under certain conditions (wind speed, chamber temperature), the experimental tests proved the possibility of removing ice formation through only the action of aerodynamic forces. In another campaign, a piezoelectric actuator system is implemented to facilitate and anticipate detachment. The actuators were driven by a sweep signal, with a bandwidth able to detect the main resonant frequency of the airfoil structure, strongly dependent on the current ice shape and thickness. These last WT test results demonstrated the efficacy of combined use of the three IPSs and validated the design approach, the developed models, and the basic assumptions about the investigated phenomenon, with particular reference to the droplets behavior and the ice accretion dynamics.

2 PROS AND CONS OF CONSIDERED IPS

In this work, the combined use of three different IPS is examined. In detail, it is proposed to combine a classical thermal device with a suitable icephobic surface and an integrated and distributed mechanical exciter, based on piezoelectric actuators. Hereafter, a critical discussion on the advantages and the limitations of the three investigated IPS is provided.

2.1 THERMOELECTRIC IPS

Thermoelectric IPSs are more convenient than bleed air systems, which require a considerable amount of fuel. A practical approach is to utilize heater blankets in the regions that need protection from ice. These devices are usually activated in the running-wet mode; in other words, the protected zone is continuously kept at a temperature higher than that at which water freezes [8–10]. They have three main drawbacks:

(1) their energy supply is strongly affected by the outside temperature;
(2) they have a markedly localized effect, so that they must be materially extended to all the areas that have to be protected, with a consequent higher power supply;
(3) the running-wet mode produces ice accretions downstream.

These disadvantages impel the search for some alternative or cooperative solutions that could somehow improve the overall performance.

2.2 LOW-POWER CONSUMING PIEZOELECTRIC DEICING SYSTEMS

Electromechanical systems seem to be a valid way to lower the power consumption, even if specific zones like the leading edge stagnation point are difficult to treat this way because of the adverse action of aerodynamic loads which tends to press, and then preserve, the ice on the airfoil.

Like others of their class, electro-magnetic pulse actuators use high-intensity currents that are limited in duration to generate internal magnetic fields to deform the overlying skin and cause ice separation. However, the required elevation in amperage peak potentially leads to breakdown of the dielectric insulation, air electric arcs, and so on. Piezoelectric actuators may represent a suitable alternative [11–13]. The resulting apparatus is compact, reliable, and completely embedded in the reference architecture (smart structure). Its major limit is related to the low operational temperatures that have a remarkable impact upon the materials performance (authority degradation, fragility, bonding failure, etc.).

2.3 HYDROPHOBIC COATINGS

Passive systems do not need any power supply. This kind of protection can be then considered as the ideal solution to the problem.

The idea at the basis moves from considerations on the droplets surface tension. A droplet laying on a conventional aluminum alloy skin tends to spread and wet an increasingly larger portion of the surface as the liquid surface tension decreases. An illustration of liquids propensity to wet the surface is provided by the so-called contact angle (defined as the angle between the surface line and the tangent of the droplet profile) at the contact point, θ_{CB} in Fig. 1. By giving the surface a controlled nano-textured roughness, the situation sketched in the same Fig. 1 occurs: the droplet leans onto very small "pillars," but each other close enough to avoid any liquid penetration [16]. The main weakness of this system is in its functionality itself: because the aerodynamic forces are not able to induce the ice disconnection by themselves, the use of additional active protections is mandatory [7].

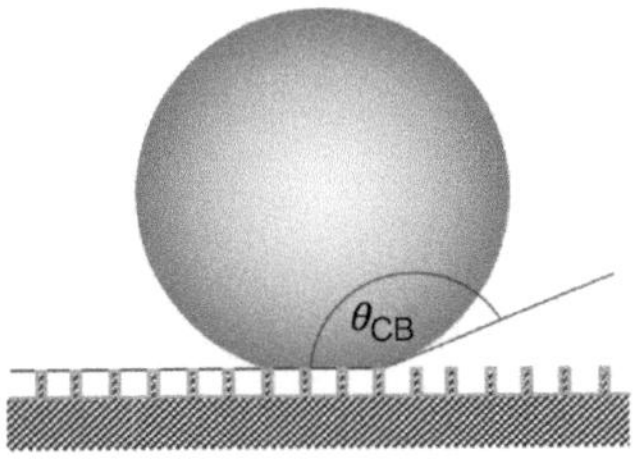

FIG. 1

Droplet adhesion and contact angle, according to the Cassie-Baxter behavior [14,15].

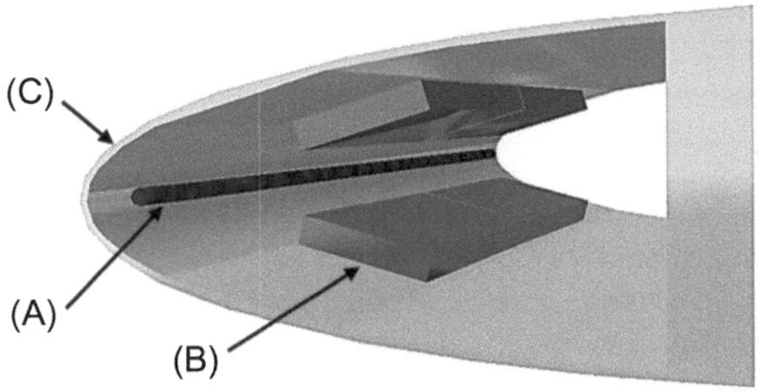

FIG. 2

Schematic of the integrated active-passive IPS: (A) thermoelectric heating wire, (B) piezoelectric actuators, and (C) icephobic surface (on the airfoil top surface—not depicted).

2.4 ALTERNATIVE STRATEGY BASED ON A HYBRID APPROACH

To overcome the weaknesses of the individual systems above, a combination of them is proposed. Because of its large energy demand and limited influence area, the thermoelectric device is used only at the most critical region of the wing, i.e., along the stagnation line. Here, neither the icephobic surface nor the piezoelectric-based device could be efficient because of the adverse flow pressure action that entraps the droplets and favors ice formation. The approach presented in this study differs from others presented in literature [4,5] exactly because heat is generated only at one specific zone. The melted beads then move fast downstream pushed by the aerodynamic flow, thanks to the reduced adhesion provided by the special coating. Away from the stagnation region, at the point where the ice should start forming, the action of the piezoelectric actuators produces a dynamic mechanical shear that definitively separates the accreted masses, aided by the peculiar properties of the prepared surface and the presence of the external air flow [17]. A sketch of the hybrid IPS layout is presented in Fig. 2.

3 DESIGN AND REALIZATION OF THE IPS

In this section, the processes and the design approach adopted to define the main features of the three IPS are presented.

3.1 HYDROPHOBIC COATING DESIGN AND PROCESS ASSESSMENT

The definition of the solid hosting support and the type of surface machining were fundamental steps in the design and the realization of an icephobic coating. In this case, the installation was performed on a 177.8 mm chord NACA 0012 airfoil, made from milled aluminum alloy 2024-T3. Wire-cut electro-discharge machining (EDM) was used to produce the hollow shape in the leading-edge area. The external surface was machined to obtain a mirror finish, as illustrated in Fig. 3 [7]. A phosphoric sulfuric anodization (PSA) process was then carried out to achieve a surface porosity level sufficient to fix the coating. The applied voltage and electrolyte temperature were the key parameters in this process. A number of different combinations of these variables are used to conduct tests on four different flat specimens (see Table 1 [7]). All of them were applied to a mirror finishing surface, Fig. 4 [7], to achieve a 0.9-μm-thick nanostructured aluminum oxide coating (minimum value), sufficient for icephobic purposes. Configuration (B) resulted the best porosity compromise between (A) and the two other cases, (C) and (D), leading respectively to a poor and an excessive density of nanopores, Fig. 4 [7]. In the first case, the result is due to the low chemical reactive level consequent to the application of low voltage and temperature values, generating few but larger pores. In the other arrangements, excessive voltages led instead to acceleration of the material accumulation process, while the effect of the increased temperature favored the chemical reactivity, which in turn generates a bigger distribution of tiny holes.

Indexes that helped evaluate the objective performance of a generic icephobic surface were the contact angles, measured downstream (CAA, contact angle advancing) and upstream (CAR, contact angle receding), together with its difference (CAH, contact angle hysteresis). These values were evaluated for the different products and reported in Table 2 [7]. They confirmed that texture (b) has better icephobic behavior. In the same table, the surface roughness (R_a) is also shown.

To enhance the icephobic layers operation, a special chemical agent (Episurf solution by Surfactis Technologies) was applied through a dip-coating process, assuring a more accurate leveling of the

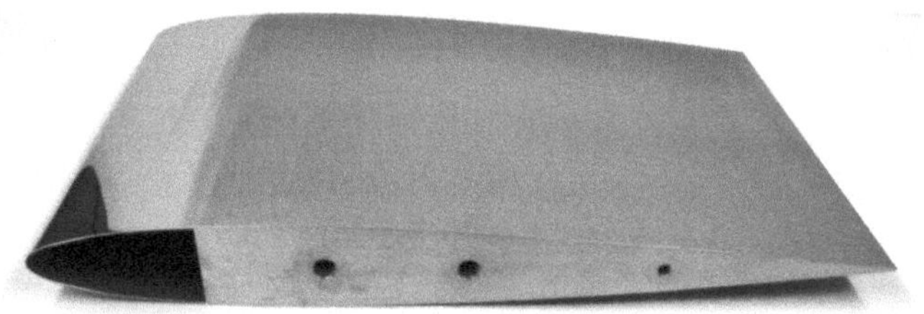

FIG. 3

NACA 0012: appearance of the airfoil surface after mirror polishing the leading-edge area [7].

Table 1 Voltage and Electrolyte Couples of Parameters Used for the Anodization [7]

Sample No.	(a)	(b)	(c)	(d)
Voltage (V)	18	18	18	22
Temperature (°C)	20	26	30	26

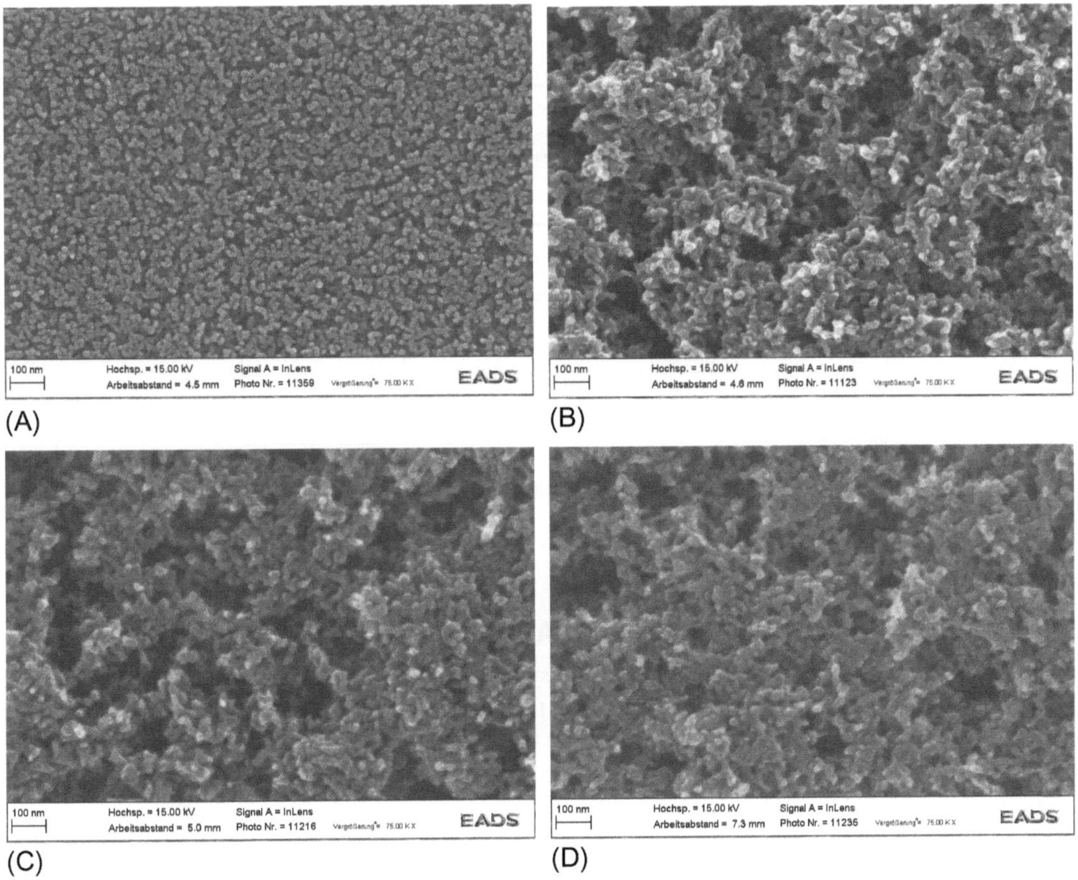

FIG. 4

Nanostructured patterns obtained for the voltage-temperature value couples [7].

Table 2 Contact Angles Versus Anodization Parameters (Voltage and Temperatures) [7]

Sample No.	(a)	(b)	(c)	(d)
R_a (μm)	0.020 ± 0.002	0.073 ± 0.005	0.077 ± 0.005	0.070 ± 0.007
CAA (°)	151.5 ± 1.21	160.6 ± 0.59	158.6 ± 0.56	160.0 ± 0.37
CAR (°)	136.3 ± 1.48	158.1 ± 0.14	155.8 ± 0.21	156.5 ± 0.47
CAH (°)	15.2	2.5	2.9	3.5

peaks of the rough, porous aluminum oxide cover. After this step, the ice adhesion strength was assessed through the use of an electromagnetic shaker [7,16,18]. As shown in Fig. 5 [7], the lowest shear necessary to remove the ice from the different samples (τ_{adh}) was observed for the texture (b), confirming the former attainments.

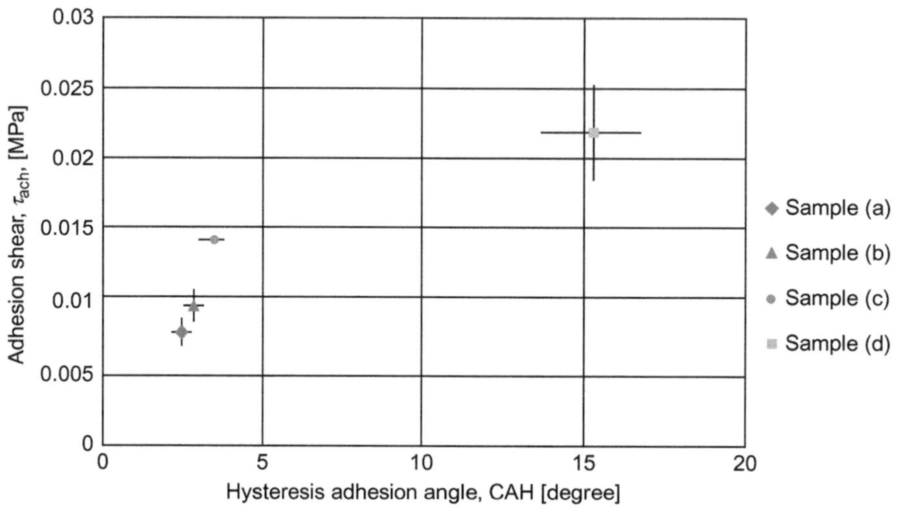

FIG. 5

Ice adhesion shear on nanostructured patterns [7].

3.2 THERMOELECTRIC SYSTEM DESIGN AND ICE SHEDDING PREDICTION

Under the combined action of the three proposed systems, the impinging water particles undergo the following evolution: the droplets move away from the heated zone, slide onto the icephobic treatment, are broken by the action of the mechanical exciters as they start slowing down and solidify, and finally are wiped away by the aerodynamic flow. This process can be simulated through three subsequent modeling steps:

- Simulation of the residual ice accretion through the LEWICE2D numerical tool;
- Computation of the aerodynamic loads generated by the ice distribution, by using proper CFD tool;
- Estimates of the shear at the interface between the ice and the airfoil surface by means of finite element analysis (FEA).

LEWICE (Lewis Ice Accretion Program) is a public software, realized by NASA that is widely used throughout the world for predicting the ice formation process and define anti-icing or deicing system requirements [19,20]. The code allows setting several environmental parameters; the results are generally used as inputs to CFD tools or for reference to wind tunnel or flight tests. In the specific application, to take into account the effect of the heat source, the IDEICE variable is set to 2. Setting the card ISH to 1, it is instead possible to simulate the ice shedding progression due to the delamination produced in zones where the temperature is higher than the freezing point. The ice adhesion strength is modeled by setting the ISTD entry to 1 (*section* mode) or 2 (*nodal* mode), representing high and low cohesion conditions, respectively.

Assuming the effects of the local curvature are negligible, a strong simplification of the model was achieved, and 2D schematics may be applied, Fig. 6 [6,21]. This is strictly valid only for the heat transfer part of the simulations.

The thermoelectric system implemented Joule heating of a carbon fiber cord, 95 mm long and 1.2 mm thick, bonded in correspondence of the stagnation line within the leading edge. To improve the adhesion quality and assure an adequate thermal continuity with the aluminum skin, the interface

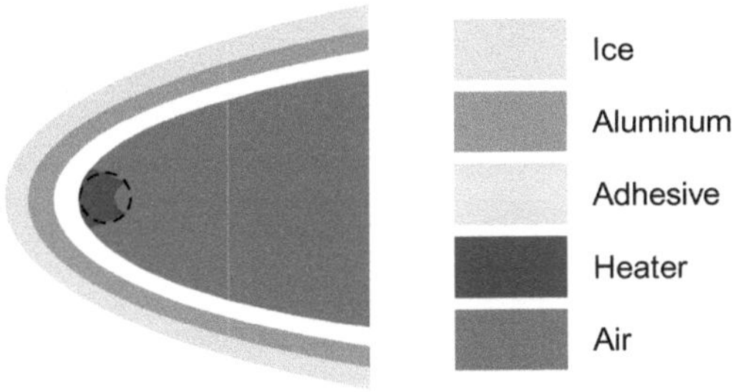

FIG. 6

Model for the LEWICE2D thermal deicing simulations [6,21].

was properly machined to guarantee the maximum contact surface between the elements. 3M Scotch-Weld Structural AF 3109-2, a high-temperature epoxy adhesive film, was used to connect the parts.

Table 3 collects the parameters used to model the heat transfer and to verify, in the design process, that the attained temperature was compatible with ice melting [6,21].

The air temperature in the cavity was assumed constant and equal to the static freestream temperature, according to the Dirichlet's boundary condition. The simulation included the external ice layer, since it affects the conduction mechanism; on the other side, the temperature at the airfoil interface influences the ice thickness itself. An iterative approach was then necessary: at first, a trial thickness outline was assumed for the ice layer; then, by using the three-step simulation defined above, an updated distribution was computed. Heater power needs were estimated to range between 200 and 325 kW/m^2. The LEWICE2D code runs covered periods of 120 s, with intervals of 5 s. As expected (Fig. 7) [6,21], this thermal system was able to produce a local ice-free region, very close to the energy source. Power densities greater than 200 kW/m^2 could shed the majority of the ice. However, excessive power densities favored ice accretion downstream. A good compromise between the need of pushing the ice from the leading edge as far as possible while keeping the power consumption reasonably low, was achieved with values of 275 kW/m^2.

The computation of the aerodynamic forces was performed through Loci/CHEM [25,26], a full featured flow solver based in the Loci framework [27,28]. Flow at 90 and 120 m/s velocity and Reynolds number (*Re*) in the ranges $1.25–1.37 \times 10^6$ and $1.70–1.87 \times 10^6$ were considered. In the formula for

Table 3 Properties of the Materials Involved in the Electrothermal Protection [6,21]

Layer	Density (kg/m^3)	Young Mod. (MPa)	Poisson Rat	Th. Cond. (W/(m K))	Sp. Heat. Cap. (J/ (kg K))	Th. diffusiv. (m^2/s)
Aluminum alloy 2024-T3 [21]	2780.0	73,100.0	0.33	121.0	875.0	4.974×10^{-5}
Heater (similar to FR4 [22])	1440.0	17,550.0	0.14	22.5×10^{-2}	900.0	1.736×10^{-7}
Adhesive [23,24]	1150.0	2689.0	0.31	0.170	1100.0	1.344×10^{-7}
Ice [20]	916.7	9000.0	0.33	23.2×10^{-1}	–	1.151×10^{-6}
Air [20]	133.5×10^{-2}	–	–	23.7×10^{-3}	1005.0	1.766×10^{-5}

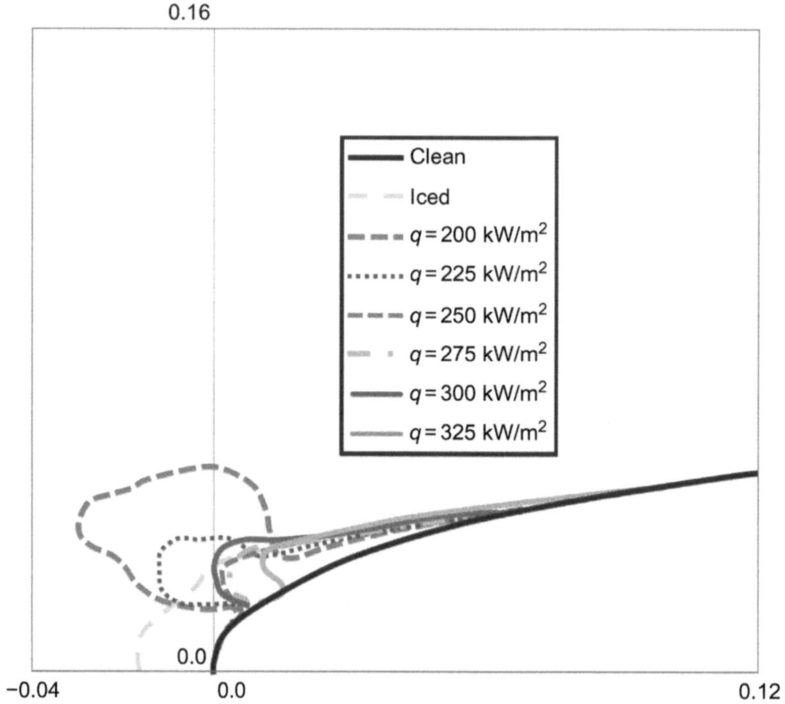

FIG. 7

Residual ice shapes versus heater density, in wet-running mode [6,21].

Re ($Re = \rho V_{\infty} L/\mu$), the reference length was the chord extension, while ρ and μ have the classical meaning of air density and viscosity, respectively.

The ice shedding phenomenon was modeled moving from the results of the thermal analyses (LEWICE2D) and CFD investigations (Loci/CHEM), respectively predicting the thermal and the aerodynamic force fields. FE analyses, implemented through Abaqus, were used to verify if the downstream ice deposits break under the occurring aerodynamic forces and moments. To reduce the computational effort, a symmetrical 2D problem was solved, taking into account the characteristics of the selected airfoil. A typical mesh, describing both the aerodynamic profile and the ice distribution is illustrated in Fig. 8 [6,21], corresponding to an exposition time of 120 s.

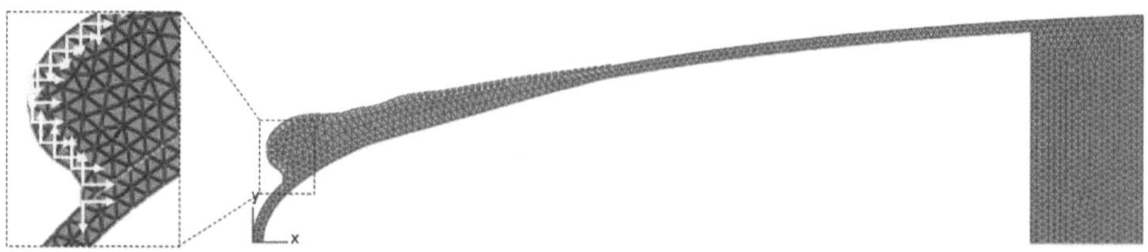

FIG. 8

2D Abaqus model of the leading edge upper half with accumulated ice and aerodynamic forces acting on the ice [6,21].

The stress field induced by the airflow loads was estimated at the interface between the ice and the structural skin. Shear values greater than the established threshold for the skin surface (evaluated for both conventional aluminum alloy and icephobic coating, τ_{adh}) indicate detachment events. The contour maps, reported in Fig. 9 [6,21], illustrate the estimated shear distributions every 30 s of exposure, up to a 150 s global observation time.

A shear threshold of 0.011 MPa was assumed and black and gray regions indicate the complete overshooting of this limit. The conservative assumption of considering the ice completely detached only when the threshold was overshot on the entire interface, was done. For this reason ice accretions were admitted passing from phase (A) to (E), even in presence of a prevalent detached region. Only for the last phase, when the shear at the interface was completely above the limit, the delamination was accepted.

3.3 PIEZOELECTRIC IPS SIZING AND PARAMETERS ASSESSMENT

As mentioned, the aim of the piezoelectric IPS is to generate a shear action at the interface between the structural skin and the ice, greater than the severance threshold to throw away the deposits that have the attitude to accumulate downstream.

To favor the bonding of the active elements on the structural body, flat zones were milled. In detail, two couples of multilayer piezoelectric ceramics (CeramTec Sonox P505, 30.0 mm long, 10.0 mm wide and 2.0 mm thick) were installed on the top and bottom internal surface of the aerodynamic profile. Each actuator consisted of approximately 310 layers with a thickness ranging between 0.09 and 0.092 mm.

The selected piezoelectric material is homogenous and transversely isotropic and presents the following characteristics [6]:

$$\begin{matrix}\text{Elasticity matrix}\\ \left(\text{stress coefficients, N/mm}^2\right):\end{matrix}\begin{bmatrix}156{,}000 & 119{,}000 & 94{,}000 & 0 & 0 & 0\\ 119{,}000 & 156{,}000 & 94{,}000 & 0 & 0 & 0\\ 94{,}000 & 94{,}000 & 106{,}000 & 0 & 0 & 0\\ 0 & 0 & 0 & 18{,}000 & 0 & 0\\ 0 & 0 & 0 & 0 & 18{,}000 & 0\\ 0 & 0 & 0 & 0 & 0 & 19{,}000\end{bmatrix} \quad (1)$$

$$\begin{matrix}\text{Piezoelectric coupling matrix}\\ \left(\text{converse piezoelectric effect constants, C/mm}^2\right):\end{matrix}\begin{bmatrix}0 & 0 & -4.9\times10^{-6}\\ 0 & 0 & -4.9\times10^{-6}\\ 0 & 0 & 1.69\times10^{-5}\\ 0 & 1.27\times10^{-5} & 0\\ 1.27\times10^{-5} & 0 & 0\\ 0 & 0 & 0\end{bmatrix} \quad (2)$$

$$\begin{matrix}\text{Dielectric matrix}\\ \left(\text{electrical permittivity, }\dfrac{\text{F}}{\text{mm}}\right):\end{matrix}\begin{bmatrix}7.97\times10^{-12} & 0 & 0\\ 0 & 7.97\times10^{-12} & 0\\ 0 & 0 & 6.91\times10^{-12}\end{bmatrix} \quad (3)$$

The characteristics of the other materials are reported in Table 4 [6]. The Abaqus code was used to predict the effectiveness of the action of the actuator system. The numerical model implemented different elements. Eight-node three-dimensional linear brick elements (C3D8R) were used to simulate the main structure, the reinforcements, the cord heater, and the residual ice.

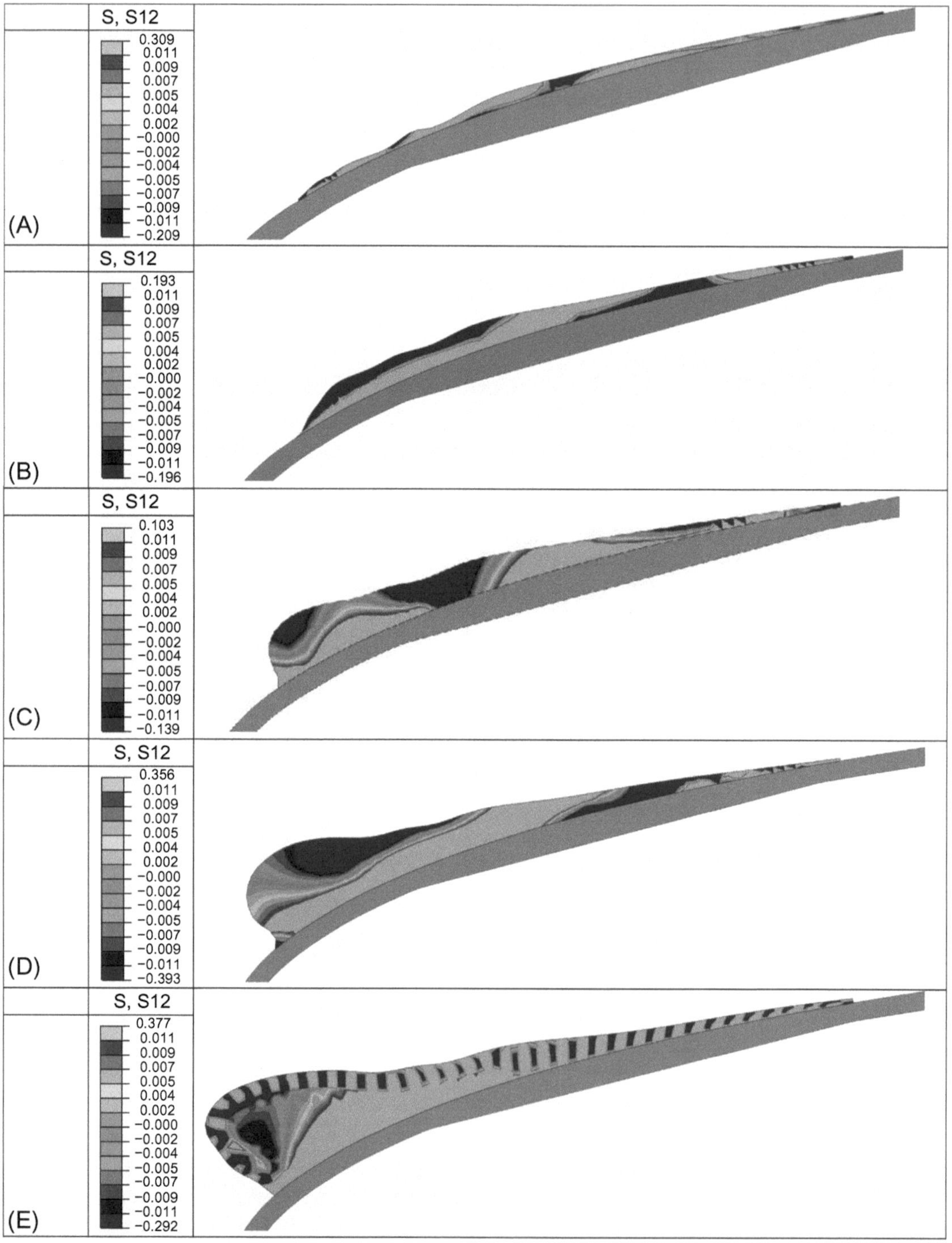

FIG. 9

Shear distribution vs. ice exposition times (30–150 s, step 30, top-down sequence) [6,21].

Table 4 Properties of the Materials of the Main Components [6]

Component	Material	Young's Modulus (N/mm^2)	Poisson's Ratio	Density (kg/mm^3)
Airfoil	Aluminum	73.100	0.33	$2{,}78 \times 10^{-9}$
Heater	Carbon fiber	17.550	0.14	$1{,}44 \times 10^{-9}$

Ceramics behavior, assumed to be perfectly bonded to the airfoil, was instead reproduced through eight-node three-dimensional linear piezoelectric brick elements (C3D8E). Proportional viscous damping was assigned to the actuators: the mass Rayleigh coefficient α was set to zero, while the stiffness coefficient β was computed through the classical equation [6,29,30]:

$$\beta = \frac{1}{2\pi f_n \cdot Q_m} \tag{4}$$

f_n and Q_m being, respectively, the natural frequency of the reference mode shape (the one that was selected for the simulation) and the mechanical quality factor of the piezo. Q_m was set to 80, according to the data sheet provided by the manufacturer.

Since the mechanical transmission can be magnified by exciting specific resonances, the first four natural frequencies of the structure were determined. They are equal to 3.445, 4.421, 11.473, and 15.578 kHz [6]. The first mode allowed achieving very high and well distributed shear levels (up to 116.8 MPa), Fig. 10 [6]. Therefore, only this one was excited.

The presence of the ice did significantly affect the structural dynamic response and its contribution was basically dependent on its mass and its volume. In the integration, such evolution was monitored at different intervals: 10, 20, 30, 40, 50, and 60 s. The interfacial surface and the ice thickness increase made the system stiffer, as demonstrated by the first frequency amplification, Table 5 [6].

Shear contour maps, taking into account the ice mass, were recorded every 10 s and shown in Fig. 11 [6]). The black and gray colors indicate that the absolute value of the shear level exceeded

FIG. 10

Shear distribution onto the skin; the dashed lines represent the location of the piezo [6].

Table 5 First Normal Frequency vs. Maximum Thickness of the Ice vs. Interfacial Area vs. Time [6]

Time Step	t_0	t_{10}	t_{20}	t_{30}	t_{40}	t_{50}	t_{60}
First system eigenfrequency (kHz)	3.445	3.488	3.493	3.577	3.662	3.731	3.819
Ice thickness (mm)	0	0.259	0.254	0.492	0.758	1.063	1.353
Interface area (mm^2)	0	2028.9	2450.4	2465.9	2604.6	2639.0	2709.7

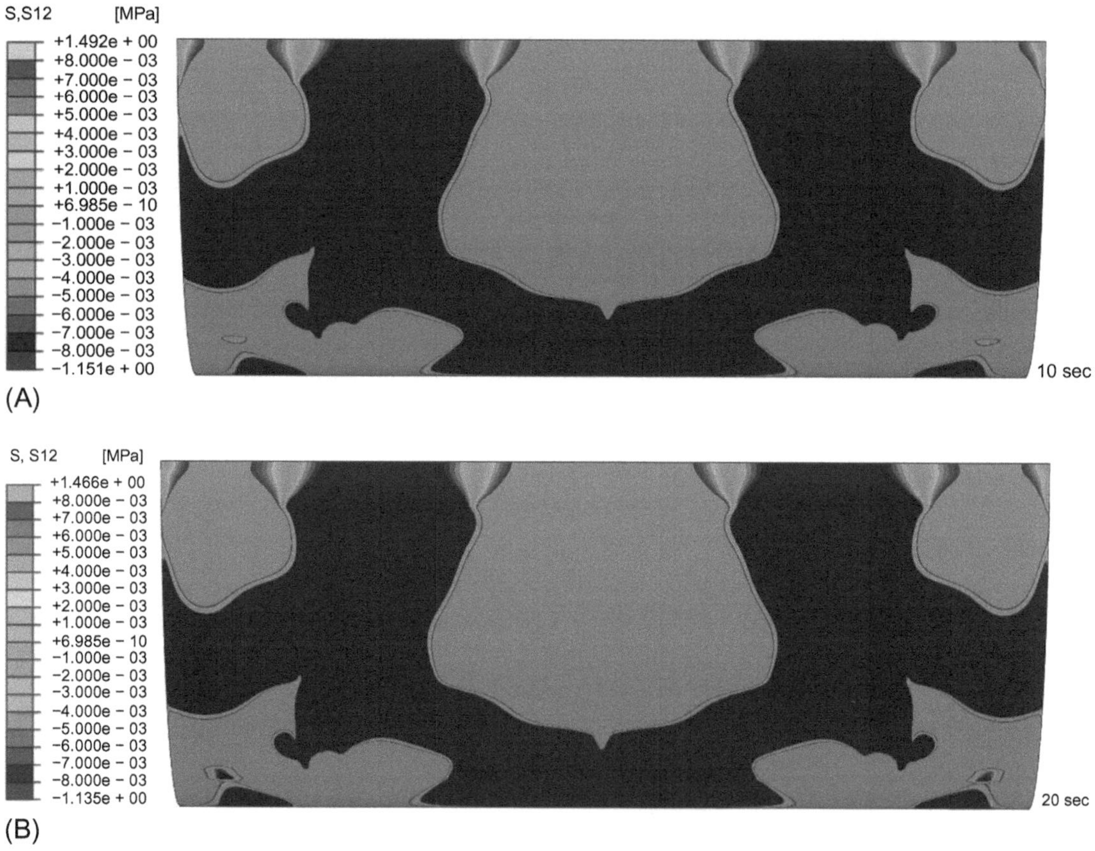

FIG. 11

Interface shear color maps for exposition times of 10–60 s, step 10 s, top-down [6].

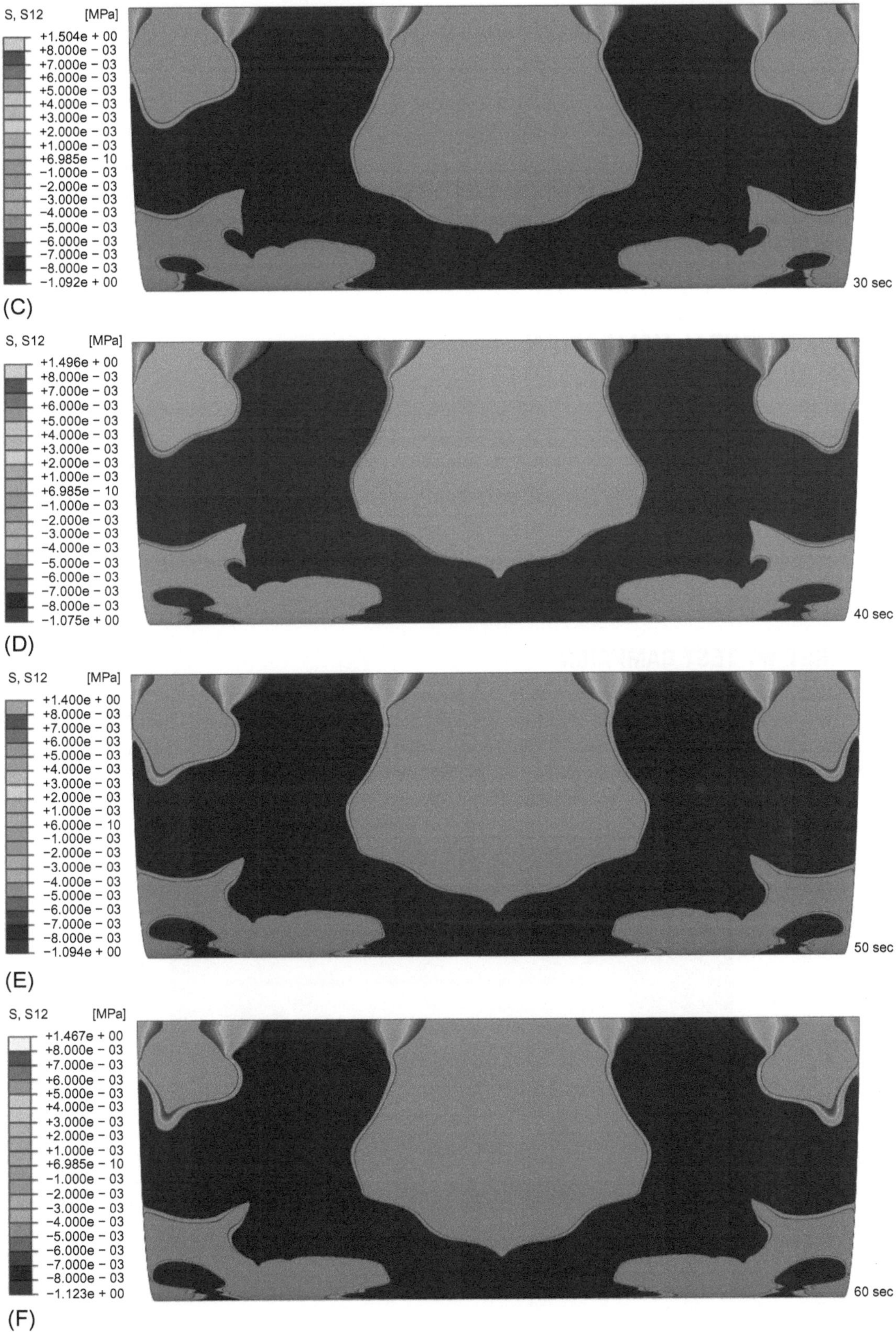
S, S12 [MPa]
+1.504e + 00
+8.000e − 03
+7.000e − 03
+6.000e − 03
+5.000e − 03
+4.000e − 03
+3.000e − 03
+2.000e − 03
+1.000e − 03
+6.985e − 10
−1.000e − 03
−2.000e − 03
−3.000e − 03
−4.000e − 03
−5.000e − 03
−6.000e − 03
−7.000e − 03
−8.000e − 03
−1.092e + 00
30 sec
(C)
S, S12 [MPa]
+1.496e + 00
+8.000e − 03
+7.000e − 03
+6.000e − 03
+5.000e − 03
+4.000e − 03
+3.000e − 03
+2.000e − 03
+1.000e − 03
+6.985e − 10
−1.000e − 03
−2.000e − 03
−3.000e − 03
−4.000e − 03
−5.000e − 03
−6.000e − 03
−7.000e − 03
−8.000e − 03
−1.075e + 00
40 sec
(D)
S, S12 [MPa]
+1.400e + 00
+8.000e − 03
+7.000e − 03
+6.000e − 03
+5.000e − 03
+4.000e − 03
+3.000e − 03
+2.000e − 03
+1.000e − 03
+6.000e − 10
−1.000e − 03
−2.000e − 03
−3.000e − 03
−4.000e − 03
−5.000e − 03
−6.000e − 03
−7.000e − 03
−8.000e − 03
−1.094e + 00
50 sec
(E)
S, S12 [MPa]
+1.467e + 00
+8.000e − 03
+7.000e − 03
+6.000e − 03
+5.000e − 03
+4.000e − 03
+3.000e − 03
+2.000e − 03
+1.000e − 03
+6.985e − 10
−1.000e − 03
−2.000e − 03
−3.000e − 03
−4.000e − 03
−5.000e − 03
−6.000e − 03
−7.000e − 03
−8.000e − 03
−1.123e + 00
60 sec
(F)

FIG. 11, CONT'D

the threshold necessary for shedding (experimentally determined to be 0.008 ± 0.001 MPa [7]). Phases A and B, corresponding to 10 and 20 s, respectively, show small colored zones at the bottom left and right of the map (upper view of the leading-edge zone), indicating that the ice remained attached to the surface in these zones. The following maps, on the contrary, show practically no color lines in the same zones, clear sign that the IPS was really effective everywhere. It was then effective only over a certain thickness of the ice.

4 EXPERIMENTAL VALIDATION

Two WT test campaigns were organized on a rectangular wing section, characterized by the NACA 0012 profile and by a 177.8 mm chord and a 100 mm span. They had the double aim of appreciating the actual effects of the combined IPS system and validating the numerical predictions. Both the experimental campaigns considered symmetric conditions (AoA = 0 degree) within the 100 mm × 150 mm test section of the facility at the Airbus Group Innovations laboratory, Fig. 12 [6,21]. In detail:

- The first campaign tested the wing model, integrated with the thermoelectric and the icephobic treatment;
- The second campaign tested the test article, integrated with all the three considered IPS.

4.1 FIRST WT TEST CAMPAIGN

The first test campaign was conducted with the presence of the thermoelectric and the hydrophobic IPS. It was mainly aimed at establishing the relation between the ice thickness (somehow related to the exposure time) and its propensity to remain attached. Glaze, mixed, and rime ice conditions were tested under two free airstream velocities ($V_{\infty} = 90$; 120 m/s). Table 6 collects the eight investigated cases, classified on the basis of wind speed, thermodynamic parameters, type of ice and needed electrical power for melting ice [6,21]. Both the median volume diameter, MVD, and the liquid water

FIG. 12

Airfoil mounted in the icing wind tunnel test section [6,21].

Table 6 Test Cases vs. Static and Total Thermodynamic Parameters, Ice Type, Droplet Generation Parameters and Electrical Voltage, Current and Power Needed for Melting Ice at the Stagnation Point [6,21]

Case	V_∞ (m/s)	T_{tot} (°C)	T_∞ (°C)	LWC (g/m³)	MVD (μm)	Ice Type	V (V)	C (A)	P (kW/m span)
1	90	−1.0	−4.0	0.55	20	Warm-glaze ice shape	6.6	0.8	0.05
2	90	−5.0	−9.0	0.55	20	Horn-glaze ice shape	11.4	1.5	0.17
3	90	−10.0	−14.0	0.55	20	Mixed-ice accretion	14.6	2.0	0.29
4	90	−15.0	−19.0	0.55	20	Rime-ice accretion	16.0	2.2	0.35
5	120	−1.0	−7.2	0.55	20	Warm-glaze ice shape	6.5	0.8	0.05
6	120	−5.0	−12.2	0.55	20	Horn-glaze ice shape	12.9	1.8	0.23
7	120	−10.0	−17.2	0.55	20	Mixed-ice accretion	16.0	2.2	0.35
8	120	−15.0	−22.2	0.55	20	Rime-ice accretion	17.9	2.6	0.47

content, LWC (basic parameters, describing the particles quality), were kept constant at 20 μm and 0.55 g/m^3, in that order. The carbon fiber cord heater worked in a running-wet anti-icing mode. It covered only a small area around the stagnation line, where it was installed.

The ice accretion on the airfoil surface was monitored with an original laser-scanning system, developed by AGI and including a laser diode and a detector. A digital single-lens reflex camera, DSLR, was used for system calibration by producing reference values for the ice thickness. The images taken by this camera were also used for verifying the results and checking the data post-processing. The accreted ice thickness is measured on 21 points on the upper airfoil surface, chord-wise, starting at a wrap distance of 5 mm from the stagnation site, Fig. 13, and spaced 1 and 2 mm (first 15 and the other 6 points, respectively). In this way, the entire region affected by the ice presence was monitored.

The IPS combined system was conceived to work continuously (wet-running modality). By contrast, per each test after achieving stable conditions in the WT section (V_∞ and T_∞), the heater and the laser scanner were switched on. After 5 s, supercooled water droplets with a MVD of 20.0 μm were supplied by a spray bar system; the ice was melted near the warm zone and shed downward until it is frozen again, far from the heater. Finally, it was removed by the aerodynamic loads that increase as the ice accretion goes ahead.

The ice accretion test runs took 3–10 min; during this time, ice shapes were recorded. The ice thickness was measured through the images acquired by the DSLR camera and the data recorded by the laser scanning system. An example of the acquired data is shown in Fig. 14 (case 2, Table 6) [6,21]. The graph describes the ice growing history at six stations: N.1, 3, 5, 10, 15, and 21.

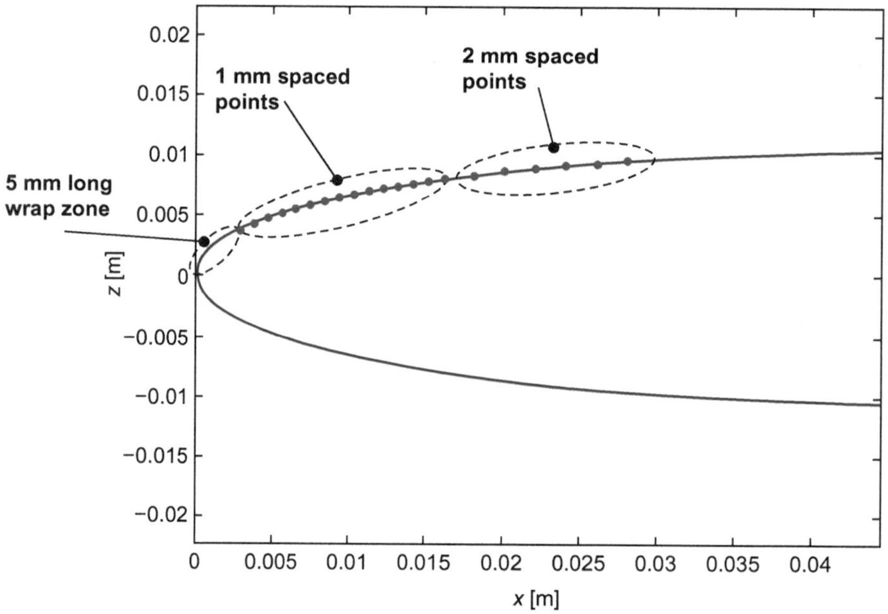

FIG. 13

Location of the 21 measurement points.

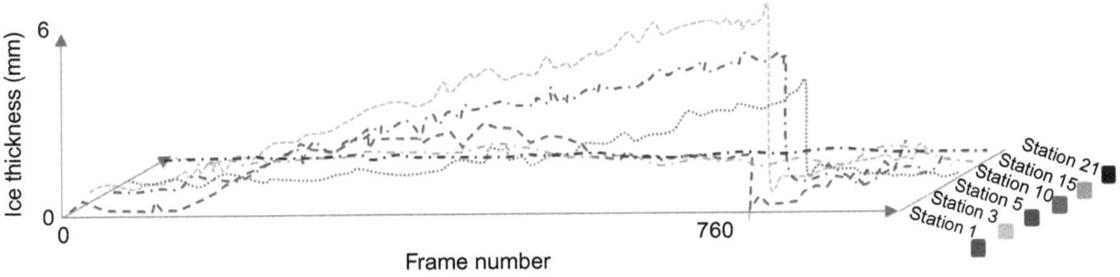

FIG. 14

Measurement of the ice growth and delamination for case 2 ($V_\infty = 90$ m/s and $T_{tot} = -5°C$) using the laser-scanning system [6,21].

Station N.3 shows the max growth level: the thickness reached 5.92 mm. The sharp fall of the curves at a frame of 760 ($T = 1.9$ s) highlights the detachment event. For case 3, the max ice thickness before delamination was 5.24 mm. For the cases 6, 7, and 8, it attained 6.79, 6.73, and 6.37 mm, respectively. No ice shedding was instead obtained for the warm and mushy shapes, corresponding to cases 1 and 5. Delamination was not recorded during the experiment N.4 due to a malfunction of the acquisition system. As a further result, the investigations showed that none of the residual ice shapes that accreted on the anticorrosion-prepared surface, separated under the only action of the aerodynamic forces.

The strongly hydrophilic character of the coated surface can explain the observed phenomenology. The interfacial forces of attraction between the airfoil metal surface and the ice were so strong that the stresses produced by the aerodynamic forces at V_∞ equal to 90 and 120 m/s were insufficient to initiate ice shedding.

A contribution to that strong adhesion was provided by the arithmetical mean roughness, R_a, of the coated aluminum skin, resulting from a milling process and equal to 1.05 μm. The droplets that impinged the leading edge could then penetrate the pores and crevices of the surface and be mechanically interlocked. In this way, a strong interfacial bond was generated and remained attached to the airfoil surface. It should be remarked that the mirror-polished skin of the icephobic aluminum airfoil had a value of R_a equal to 73 ± 5 nm (1–2 order of magnitude less). It could therefore prevent the phenomenon above, in principle. However, the anticorrosion primer that was sprayed onto to the airfoil, was about 2.5 μm thick and hence dramatically reduced the test article characteristics. Wind tunnel experiments confirmed that successful shedding of the ice deposits was only enabled by using the ultra-smooth, nanostructured super-hydrophobic layer. In this case, the max value of the ice thickness before disconnection (5.24 mm), was obtained for case 3 ($V_\infty = 90$ m/s; $T_{tot} = -10°C$).

4.2 SECOND WT TEST CAMPAIGN

In this test campaign, all the three Ice Protection Systems IPS were jointly considered: the electrothermal heater, the electromechanical system and the icephobic coating. An EA-PSI 8720-15 power supply fed both the active IPS. Three test runs investigated their behavior for a period of 60 s in the presence of warm-glaze ice, considered the most severe condition. A synthetic version of this WT test plan is reported in Table 7. As in the previous case, a sole condition for droplet generation was selected (MVD = 20.0 μm and LWC = 0.45 g/m^3), corresponding to the maximum intensity of atmospheric icing (CS-25, icing conditions [29]).

Table 7 Test Cases Conditions for the Second WT Campaign [6,21]

Test Run	Ice Type	T_{st} (°C)	T_{tot} (°C)	V_{air} (m/s)	MVD (μm)	LWC (g/m³)
1	Warm-glaze	−7.8	−0.8	118.9	20.0	0.45
2	Warm-glaze	−8.4	−1.2	120.1	20.0	0.45
3	Warm-glaze	−7.9	−0.9	118.7	20.0	0.45

According to the standards [30], the estimate of the actual LWC was performed through an icing blade method, before starting the deicing experiments with the hybrid IPS. Per each run, the atomization of super cooled water droplets started after having achieved WT stable conditions. At that time, the carbon fiber heater was switched on, operating in a wet running condition, and preventing the ice accretion at the stagnation line.

Along the tests, the piezoelectric systems were activated six times, by using a MAC audio amplifier, MPX 4500 type. An arbitrary waveform generator (Agilent 33220A) was used to provide a sinusoidal excitation signal with a frequency sweep ranging between 4.15 and 4.30 kHz and a 2 s overall cycle duration. The required total power density (P_{d_tot}) of the hybrid system includes the electrical power required by the carbon fiber cord heater (P_{heater}) and the power required by the piezoelectric mechanical exciter (P_{piezo}). It follows:

$$P_{d\_tot} = \frac{P_{heater} + P_{piezo}}{A_{surf\_ice}} \tag{5}$$

where A_{surf_ice} is the area covered with ice. The value of this area was pre-determined and corresponds to the wet surface in natural conditions, i.e., without using any IPS. It was determined as the product of the wrap chordwise times the spanwise extension of the impacted region. As for the absolute values of the specific experiment on the selected test article, the respective length resulted about 40 mm times 100 mm (the entire model extent). This led to an about 0.004 m^2 wet surface.

In Fig. 15 [6,21], a typical deicing sequence is shown, obtained during this second test campaign. It reports shot frames, taken at 2.5 ms step, by a Phantom v611 high-speed camera. The implemented Ice Protection System IPS demonstrated to produce deicing within 17.5 ms. The difference between the first and the second frame is remarkable. They show two different moments of the experiment, before (first) and after (second) the piezoelectric system was switched on. The third shot shows the dramatic moment when the ice was shed, following the excitation of the structural resonance (first mode) The further photograms give an idea of the ice breaking process and how the fragments were moved away by the airflow.

The carbon heater device was fed with 7.0 V and 0.8 A DC, leading to a global 5.6 W absorbed power, measured through an HP 4194A impedance analyzer. At the first structural resonance (frequency ranging between 4.15 and 4.30 kHz), the active mechanical system absorbed around 4.47 W. If a 20% safety coefficient is assumed, a final value of 5.36 W is obtained. These values give an estimate of the power density, equal to 2.74 kW/m^2. The experiments were carried out after a 30-s accretion time, to enable the growth of a sufficient ice volume.

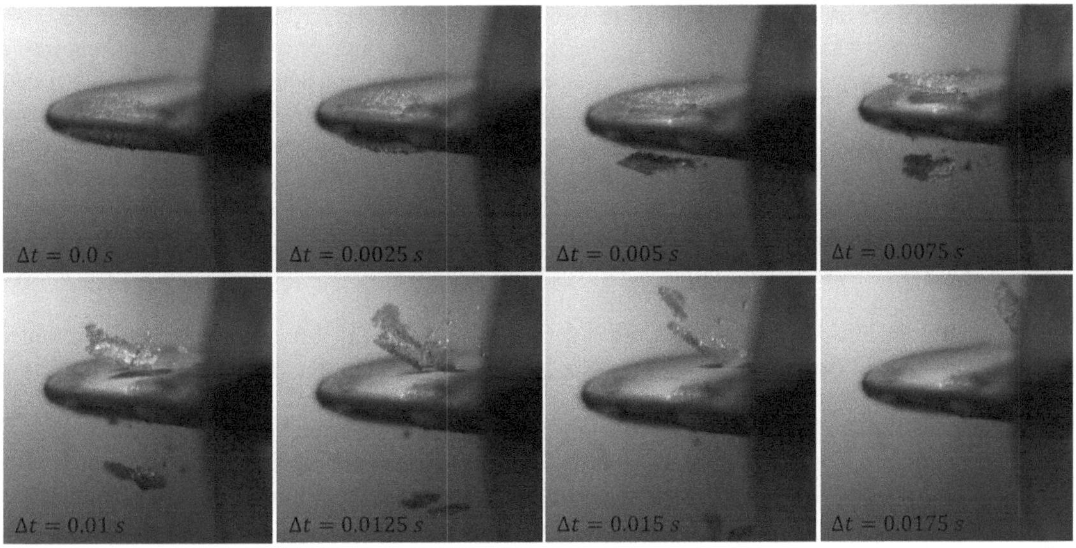

FIG. 15

Deicing sequence of the warm-glaze icing condition for 30 s of icing exposure [6,7,21].

These tests proved the effective potentiality of the proposed hybrid architecture to reduce the power consumption: in fact [4], conventional IPS energy absorption ranged between 16.4 and 62.0 kW/m^2. The piezoelectric system and the selected strategy of forcing the structural resonance mode allowed attaining an approximately 57% saving with respect to competing electromechanical subsystems [10]. The integrated IPS led also to benefits in terms of weight and compactness. The use of a single heater, installed at the stagnation line, represented in fact a valid alternative to heavier large-scale systems, extending all over the entire droplet impingement area (leading edge zone). Furthermore, the solid or quasisolid nature of the heater and piezo devices made them a profitable solution for the integration in narrow domains. Finally, their simplicity well matched the increasing demand for safer and safer life cycle design: the use of solid actuators and resistances represented a good response to this need. The combined performance of these devices was magnified by the presence of the selected and deployed icephobic treatment, allowing the ice formation to move further downstream, where the integrated smart actuators may suitably excite the structure.

5 CONCLUSIONS

In the work at hand, a hybrid Ice Protection System was investigated. Moving from common deicing and anti-icing approaches, characterized by relatively large energy consumption levels, the idea of combining some of them in a synergic operational mode was explored. The proposed strategy foresaw the combination of one passive device (a nano-textured hydrophobic coating) and two active devices (a carbon chord heater and a piezoelectric-based structural actuator).

The systems were used and integrated into an original architecture. In contrast to traditional heating apparatuses, usually widely distributed over the front side of the wing, only a small carbon chord was implemented in this case, integrated at the leading edge. Such equipment acted on a very limited zone to prevent ice formation at the stagnation line. Here, it worked alone. Aerodynamic actions pressed the droplet onto the surface with a near-zero flow velocity, hindering the actions of the hydrophobic coating, and then the water slides downstream along the airfoil skin, where general environmental conditions and low temperatures slowed it down and froze it. The action of the dynamic piezoelectric structural system generated a shear at the ice-structure interface, high enough to produce the desired shedding. The presence of the icephobic treatment was fundamental in lowering the separation stress to suitable values, attainable by the installed mechanics.

Pros and cons of each selected IPS were illustrated; then, moving from their working parameters and their specific peculiarities, the design phase was described. With reference to the hydrophobic system, both the materials and the fabrication processes were presented; the effect of some key parameters were introduced and discussed as the adopted temperature and voltage in the anodizing process. Their optimal combination, assuring the best icephobic performance, was found. It was finally proved by experimental measures like contact angles and the surface roughness. Different aspects of the thermoelectric system design were considered and numerically investigated. Heat transfer from the internally bonded carbon cable to the external surface of the airfoil was first computed. Then, the evolution and accumulation of the water as it moves downstream were simulated. Finally, the aerodynamic forces acting upon the ice-structure interface was estimated as a function of the curvilinear abscissa. The estimated ice shape and its position with respect to the stagnation point were used to define the chordwise location of the piezoelectric actuators working at the systems first structural resonance mode. An FE model gave essential input to the designers, from the structural eigenmodes determination to the actual stress distribution. The presence of the ice at different stages was taken into account because it strongly influences the systems dynamic response.

The results, achieved during two different WT tests at AGI, validated the numerical outcomes. The first experimental campaign focused on two main objectives: (1) the waters evolution, from its impact on the surface to ice formation far from the leading edge, in the absence of any piezoelectric actuation; (2) the quantification of the actual airflow shear, representing its capacity to detach ice masses. In this case, ice shedding was observed for certain conditions only (some specific ice typologies and when thicknesses overcome a certain threshold), demonstrating the necessity of deploying additional mechanical IPS. The second test campaign did instead address the possibility of including a piezoelectric system as a mechanical facilitator for ultimate ice removal. In the end, the needed power was estimated in the presence of warm-glaze ice (the most severe condition). Direct observation confirmed the systems expected functionality and physical process. It was equivalent to the former, but the final segment, wherein the action of the ceramics reaches critical shear values, was able to break the ice. This capability was enabled by the icephobic coating, which was equally important in the first impinging phase.

ACKNOWLEDGMENT

This work describes research activities carried out by Airbus Group Innovation Works, collected and edited by CIRA.

REFERENCES

[1] Civil Aviation Authority of New Zealand, Aircraft Icing Handbook: Meteorology, November 07, 2014, http://www.caa.govt.nz/safety_info/gaps/aircraft_icing_handbook.pdf.
[2] Federal Aviation Administration, Advisory Circular: Pilot Guide: Flight In Icing Conditions, November 07, 2014, http://www.faa.gov/documentLibrary/media/Advisory_Circular/AC_91-74A.pdf.
[3] Flight Operations Support—Customer Services Directorate, Getting to Grips With Cold Weather Operations, A Flight Operations View, November 07, 2014, http://www.smartcockpit.com/docs/Getting_To_Grips_With_Cold_Weather_Operations.pdf.
[4] K. Al-Khalil, Effect of Mixed Icing Conditions on Thermal Ice Protection Systems, November 07, 2014, http://www.coxandco.com/files/pdf/FAA-D9688.pdf.
[5] K. Al-Khalil, T. Ferguson, D. Phillips, in: A hybrid anti-icing ice protection system, 35th Aerospace Sciences Meeting and Exhibit, American Institute of Aeronautics and Astronautics, Roston, VA, USA, 1997.
[6] T. Strobl, S. Storm, D.S. Thompson, M. Hornung, Feasibility study of a hybrid ice protection system. in: 6th AIAA Atmospheric and Space Environments Conference, Atlanta, GA. 2014, https://doi.org/10.2514/6.2014-2060.
[7] T. Strobl, S. Storm, M. Kolb, J. Haag, M. Hornung, Development of a hybrid ice protection system based on nanostructured hydrophobic surfaces, in: in: 29th Congress of the International Council of the Aeronautical Sciences, 2014.
[8] N. Alegre, Full-Scale Runback Ice: Accretion and Aerodynamic Study, https://dspace.lib.cranfield.ac.uk/bitstream/18265547/1/Nathalie_Alegre_Thesis_2010.pdf.
[9] J. Fraga, AERO_Q407: 787 No-Bleed Systems: Saving Fuel and Enhancing Operational Efficiencies, http://www.boeing.com.
[10] K. Al-Khalil, Thermo-Mechanical Expulsion Deicing System – TMEDS, April 15, 2013, http://www.coxandco.com/files/pdf/AIAA-2007-0692.pdf.
[11] S. Struggl, J. Korak, C. Feyrer, M. Tomizuka, A basic approach for wing leading deicing by smart structures, in: Proc. SPIE 7981, Sensors and Smart Structures Technologies for Civil, Mechanical, and Aerospace Systems, April 18, 2011, pp. 79815L, https://doi.org/10.1117/12.880470.
[12] A. Overmeyer, J. Palacios, E. Smith, R. Royer, Rotating Testing of a Low-Power, Non-Thermal Ultrasonic De-icing System for Helicopter Rotor Blades, SAE Technical Paper 2011-38-0098, 2011.
[13] S.V. Venna, Y.-J. Lin, G. Botura, Piezoelectric transducer actuated leading edge de-icing with simultaneous shear and impulse forces, J. Aircr. 44 (2007) 509–515.
[14] W.M. Sigmund, S.-H. Hsu, Cassie–Baxter model, in: in: E. Drioli, L. Giorno (Eds.), Encyclopedia of Membranes, Springer-Verlag, Berlin, pp. 310–311. ISBN: 978-3-662-44323-1 (Print); ISBN: 978-3-662-44324-8 (Online).
[15] A. Marmur, Wetting of hydrophobic rough surfaces: to be heterogeneous or not to be?. Langmuir 19 (20) (2003) 8343–8348, https://doi.org/10.1021/la0344682.
[16] T. Strobl, D. Raps, M. Hornung, in: Evaluation of roughness effects on ice adhesion, 5th AIAA Atmospheric and Space Environments Conference, American Institute of Aeronautics and Astronautics, Roston, VA, USA.
[17] S. Struggl, J. Korak, C. Feyrer, in: A basic approach for wing leading deicing by smart structures, Proc. SPIE 7981, 2011, https://doi.org/10.1117/12.880470.
[18] W. Wright, User's Manual for LEWICE Version 3.2, NASA Report CR-2008-214255, 2008.
[19] T. Strobl, R. Adam, M. Tuschter, D. Thompson, M. Hornung, Feasibility study of a hybrid ice protection system based on passive removal of residual ice, in: in: M. Bauccio (Ed.), AIAA Science and Technology Forum 2015, American Institute of Aeronautics and Astronautics, 2015, third ed., ASM Metals Reference Book, ASM International, Materials Park, OH.

[20] Volture ProductsTM, Volture Products Spec Sheet & Material Properties: FR4 Spec Sheet & Material Properties, http://www.mide.com/pdfs/volture_specs_piezo_properties.pdf.
[21] 3M Aerospace Department Engineered Adhesives Division, 3M Scotch-WeldTM Structural Adhesive Film AF 3109-2: Technical Data, November 24, 2014, http://multimedia.3m.com/mws/media/241225O/3mtm-scotch-weldtm-structural-adhesive-film-af3109-2.pdf?fn=TDS_AF3109-2.pdf.
[22] http://www.substech.com/dokuwiki/doku.php?id=thermoset_epoxy_ep.
[23] http://www.seainc.com/productsLOCI-CHEM.php.
[24] E. Luke, in: On robust and accurate arbitrary polytope CFD solvers, 18th AIAA Computational Fluid Dynamics Conference, American Institute of Aeronautics and Astronautics, Roston, VA, USA, 2007, https://doi.org/10.2514/6.2007-3956.
[25] http://web.cse.msstate.edu/~luke/loci/index.html.
[26] E. Luke, T. George, Loci: a rule-based framework for parallel multi-disciplinary simulation synthesis. J. Funct. Program. 15 (3) (2005) 477–502, https://doi.org/10.1017/S0956796805005514.
[27] Nader, E.C.N. Silva, J.C. Adamowski, Effective Damping Value of Piezoelectric Transducer Determined by Experimental Techniques and Numerical Analysis, http://www.abcm.org.br/symposium-series/SSM_Vol1/Section_II_Control_Systems/SSM_II_14.pdf.
[28] J.E.S.M. Medina, F. Buiochi, J.C. Adamowski, Numerical Modeling of a Circular Piezoelectric Ultrasonic Transducer Radiating in Water, http://www.abcm.org.br/pt/wp-content/symposium-series/SSM_Vol2/Section_VI_Sensor_Actuators/SSM2_VI_11.pdf.
[29] Certification Specifications for Large Aeroplanes CS-25 – Flight in Icing Conditions, European Aviation Safety Agency.
[30] SAE Aerospace – An SAE International Group, Calibration and Acceptance of Icing Wind Tunnels: SAE ARP5905, Aerospace Recommended Practice, 2003, www.sae.org.

FURTHER READING

[1] Surfactis Technologies, EPISURF Solution Preparation and Surface Treatment: Technical Procedure, RD_120612_v01, (2012).
[2] Surfactis Technologies, EPISURF Solution Preparation and Surface Treatment: Technical Procedure (TP), Compound 08-202, RD_020909_v02, 2012.
[3] T. Strobl, D. Raps, M. Hornung, Comparative Evaluation of Ice Adhesion Behavior, World Acad. Sci. Eng. Technol. Int. J. Mech. Aerospace Ind. Mechatronic Manuf. Eng. 6 (8) (2012) 1622–1627.
[4] W. Wright, G. Addy, P. Struk, T. Bartkus, Recent Advances in the LEWICE Icing Model, NASA Glenn Research Center, Report N. GRC-E-DAA-TN23449, 2015.
[5] T. Strobl, Modellbasierter Entwurf eines hybriden Systemansatzes zur Eisentfernung von den Oberflächen einer Flugzeugkonfiguration (in German, "Model-based development of a hybrid approach for aircraft ice protection"), Master Thesis, Fakultät für Maschinenwesen, Technische Universitaet Muenchen, Muenchen, Germany, 2016.

CHAPTER

HELICOPTER VIBRATION REDUCTION

27

William A. Welsh

Sikorsky, A Lockheed Martin Company, Stratford, CT, United States

CHAPTER OUTLINE

Morphing Wing Technologies. https://doi.org/10.1016/B978-0-08-100964-2.00027-7

1 INTRODUCTION

Helicopter fuselage vibration degrades ride quality, causes crew fatigue, and damages components necessitating expensive part replacement. Average cabin floor vibration levels in the 1960s were often around 0.3 g resulting in a truly uncomfortable experience for crew and passengers and frequent replacement of damaged parts [1]. With a consistent demand from the helicopter-user community, levels have improved over the years as improved technology has been developed and applied. Now levels are almost always between 0.1 and 0.2 g at most places in the cockpit and cabin; still not a "Jet smooth ride," but a significant improvement.

In this chapter, existing approaches to helicopter vibration reduction as well as new technologies being pursued to achieve a jet-smooth ride are reviewed. (Note that the designations S-76 helicopter and S-92 helicopter are registered trademarks of Sikorsky a Lockheed Martin Company.)

2 NextGen VIBRATION LEVELS

A good "NextGen" rule is that helicopter vibration levels should be less than 0.03–0.05 g *throughout* the cockpit and cabin. These levels, would be virtually imperceptible [1] and, as shown in Fig. 1, helicopter reliability would be significantly improved at these vibration amplitudes. Such low levels are rarely achieved today in a consistent manner except in local areas in the helicopter cabin area (you can always find *somewhere* where vibrations are low). In the following paragraphs, the current state-of-the-art of helicopter vibration control will be summarized as well as some ideas on paths to achieving NextGen levels.

3 VIBRATION SPECIFICATIONS

Several civil and government specifications define vibration limits quantitatively. For military projects, the U.S. Army's ADS-27 [2] vibration specification is sometimes used where the intent is to minimize vibration experienced by the crew. In this context, the U.S. Army defines an "intrusion index" which is a

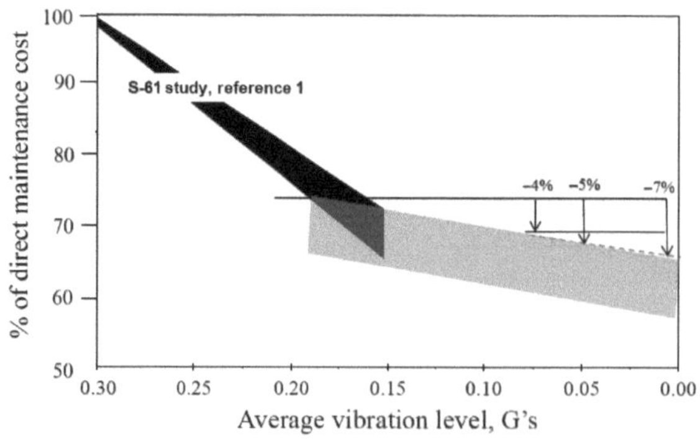

FIG. 1

Vibration reductions have potential to reduce helicopter operating cost.

parameter in units of velocity and is defined as a weighted square of the velocities in each of the orthogonal directions; vertical being weighted most importantly while lateral and longitudinal are weighted successively less, reflecting the relative tolerance of humans vibration in each direction. For civil applications, the ISO-2631 specification [3] is commonly used to define the vibration levels of comfort at various locations in an aircraft using a method not much different than that of ADS-27. The motive behind both ADS-27 and ISO-2631, in part, is to define vibration levels that will allow the crew to perform their tasks over some period without reduced proficiency. The FAA establishes the Federal Aviation Requirements or, more specifically, FAR29-251 [4] which states: "Each part of the rotorcraft must be free from excessive vibration under each appropriate speed and power condition." This somewhat vague guidance makes sense because the FAA is more focused upon safe operation and less concerned about passenger comfort or an advantage of one aircraft relative to its competitors.

4 SOURCE OF HELICOPTER VIBRATORY LOADS

A detailed description of the source of helicopter vibration is beyond the scope of the current summary but this very rich subject is covered in detail in a few excellent books by Bramwell et al. [5], Johnson [6], and Bielawa [7] in addition to many technical papers in the American Helicopter Society (AHS) journal and proceedings. It is sufficient for our purposes here that the vibratory loads originate with the main rotor and occur at the so-called blade passage frequency, that is, the rotor rotational speed times the number of blades. Typically this frequency is between 15 and 25 Hz. The term "blade passage frequency" is somewhat misleading because vibratory loads from the rotor are predominately transmitted to the fuselage structurally via the main rotor hub, rather than by aerodynamic pressure loading caused by blade passage over the fuselage. A secondary path is wake impingement on the vertical and horizontal tail aerodynamic surfaces and on the tail rotor itself. Where do main rotor vibratory loads come from? The vast change in apparent forward wind velocity over each blade as it progresses from the advancing side, where the tip velocity is often near Mach-1 to the retreating side of the rotor where much of the blade is stalled, causes sharp changes to loads versus blade azimuthal position during rotation. The largest aerodynamic loads on the blade are in the so-called "flatwise" direction, i.e., perpendicular to the chord and it is this abrupt loading change in this direction versus azimuth that causes the blade to vibrate at all harmonics of the rotor rotational speed. The usual nomenclature is that loads and motions that occur at a frequency of once per revolution of the rotor are termed "1P" loads and motions. Similarly, loads and motions that occur at the higher harmonics are termed "2P, 3P, etc." Luckily, and perhaps surprisingly to those unfamiliar with rotor dynamics, forces at all frequencies *except* the blade passage frequency "cancel out" at the rotor hub (where the blades are mounted). For example, for a 4-bladed helicopter, the only blade loads that eventually are transmitted to the fuselage are 3P, 4P, and 5P loads but all three of these "rotating system" loads result in only 4P loads on the "nonrotating" fuselage. In this case, the 3P blade loads originate on the rotating blade and in the rotating system. Some simple phase plots can show that the sum of 3P loads from all 4 blades but phase lagged 90 degrees from each other, result in a single vector of 3P force (or moment) that angularly travels in the same direction as the rotor rotation and progresses from 0 to 360 degrees at three times per revolution in the rotating frame of reference. As a result of this forward progression in the rotating frame, this identically sized force from each blade is transmitted to the fuselage as a 3P+1P=4P load. A similar phase plot would show that 2P loads on a 4-bladed rotor, "self-cancel."

This is a very fortunate outcome because the lower load harmonics are typically huge and would otherwise result in potentially damaging loads and high vibration being transmitted into the fuselage. Another nice "physics-fact" is that successive blade load harmonics become smaller with increasing frequency, i.e., 5P loads are almost always smaller than 2P loads provided no blade resonances aggravate the blade responses. Therefore, 7-bladed helicopters shake less than 2-bladed helicopters; for the former case, the only loads transmitted to the fuselage originate from higher frequency but smaller 6, 7, and 8P blade loads as opposed to much higher 1, 2, and 3P harmonic loads for the latter helicopter. The blade passage loads, generically referred to as "Np" loads, where N is the number of blades, are transmitted from the hub to the fuselage through the main rotor gearbox which is often rigidly attached to the top of the fuselage to provide a reliable and robust connection; an important feature, since the very large main rotor flight loads, i.e., the gross weight of the vehicle times any vehicle acceleration must pass through this same connection.

The fixed system Np loads can be viewed as six independent loads, three forces and three moments, as shown in Fig. 2. Any single one of these six loads can be large enough to cause undesirable fuselage vibration but superimposed together, especially for helicopters with six or fewer blades, some vibration mitigation is often required to make the vehicle practically useful. An alternative way of visualizing the vibratory hub loads is shown in Fig. 3. The three forces form a vector whose tip orbits an ellipsoid centered at the rotor hub while the three moments form a similar but differently shaped ellipsoid. Each ellipsoid changes shape vs. flight condition. In the discussion that follows, the reader might find both ways of viewing the Np loads useful when considering the methods for eliminating their effect. Suffice it say now that if a clever engineer could counter each of these six independent vibratory loads, preferably at the rotor hub, the helicopter dynamicist's task would be over!

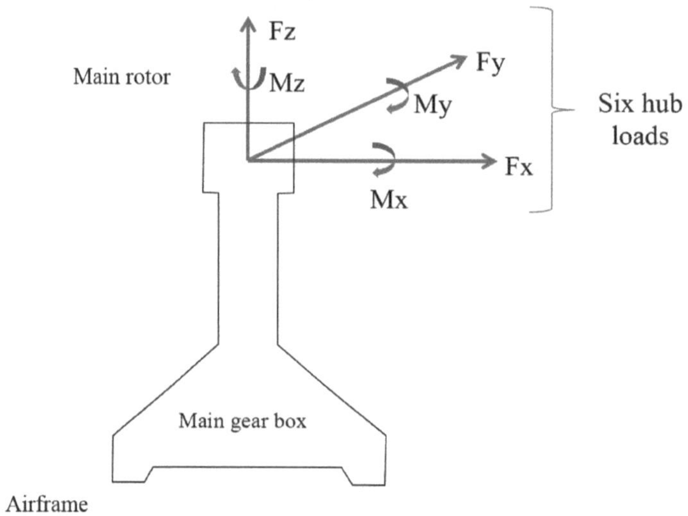

FIG. 2

Primary vibration path is the main gearbox which transmits six vibratory loads to the fuselage.

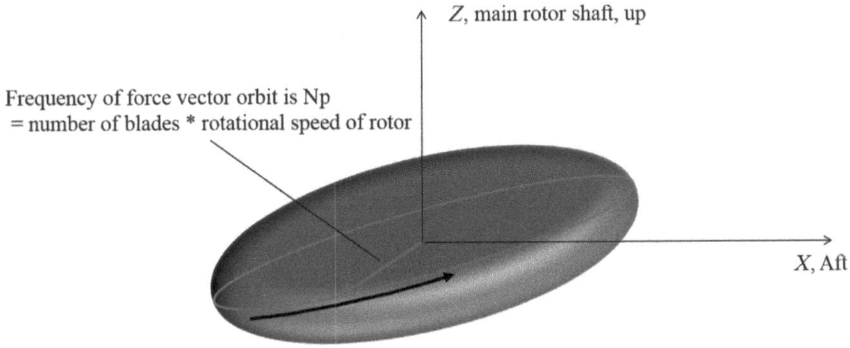

FIG. 3

Vibratory loads and movements produced by a rotor follow an ellipsoid.

5 HOW DO VIBRATORY LOADS GET INTO THE FUSELAGE?

In virtually all helicopters the Np hub loads, both forces and moments along with more-or-less quasisteady flight loads are directly transmitted to the fuselage through the main gearbox (MGB), which is, itself, very rigid and, of course, strong enough to withstand very large low-cycle flight loads and much smaller Np loads. But this nexus between the MGB and the helicopter fuselage is often where the big differences between antivibration approaches of various helicopter models become evident.

6 WHAT IS USED FOR VIBRATION CONTROL NOW?

The "fielded" approaches used for vibration control will be discussed first to be followed by approaches being studied for the future. To clarify the current state-of-the-art, various classes of antivibration approaches are shown in Table 1. The author notes that this table is limited by his knowledge of what is currently fielded. The entries that are, to the author's knowledge, not fielded on any production aircraft are labeled with the abbreviation "R&D" indicating that these have been studied in the research and development context only. In Table 1, note the use of the word "Frahm" named after Hermann Frahm who invented the "vibration absorber" in 1890 (US989958 A). A Frahm is a synonym for tuned vibration absorber; essentially a mass and spring tuned to the vibration forcing frequency. To tune a Frahm, the spring is anchored to a virtually immovable table and the spring-mass assembly is tuned to the exact forcing frequency that the engineer wants to suppress. Contradictory as it seems this process actually produces an antiresonance or "bucket" at the forcing frequency when the device is mounted in the helicopter fuselage. The result is that the Frahm produces very low vibration at its connection point on the fuselage, while the Frahm mass vibration amplitude is very large. Although beyond our scope for the current purposes to go into the physics of how this device works, suffice it to say that Frahms, or some variation of the idea, are used in a huge number of helicopters.

Table 1 Overview of Vibration-Reducing Methods Used in the Helicopter Industry

Company	Isolation		Fuselage Suppression		Rotor Suppression	
	Passive	Active	Frahm	Active	Passive	Active
Sikorsky	None	Zero-Vibe R&D	UH-60L	UH-60M	UH-60L, M (Bifilar)	HHC, IBC, active flap R&D
Bell	Main gearbox (MGB) isolation long springs, LIVE for vertical and roll	Active LIVE R&D	Bell-430 Bell-429	Bell-525 Bell-429 (R&D)	Bell-412 (pendular absorber) V-22 (pendular absorber)	None
Boeing	CH-47, floor isolation	None	CH-47	CH-47 self-tuning absorber		HHC and active flap R&D
Airbus	H125/H130/ AS365 (barbeque plate some with elastomer mounts instead of a metallic plate) Tiger, NH-90 SARIB	H135 (ARIS, hydraulically actuated DAVI principle)	H125 H120	H22 H130T2	H145 (out-of-plane pendulum) H120 (in-plane pendulum)	Active flap Blue Pulse, IBC R&D

6.1 WHY NOT ISOLATION?

The perceptive reader might wonder why simple elastomeric or metal isolation mounts, commonly used for isolating foundations from industrial equipment vibration, are not used on helicopters to *completely* isolate the fuselage from the MGB. The reason this is impractical is twofold. (1) Compared to industrial equipment, the MGBs of helicopters are designed to be very light for a given power rating; it is, after all, on a flying vehicle and every extra kilogram lessens the aircraft payload. (2) The Np frequency is commonly low being between 15 and 25 Hz. For good isolation, the natural frequency of the MGB on isolation mounts would need to be well below the Np frequency for all six possible translations and rotations of the MGB; say 1/2 of the Np frequency, to provide good isolation from the fuselage. This combination of low MGB mass and low required natural frequency, results in the need for very soft isolator spring rates. Low mounting stiffness would allow very large MGB motions under quasisteady flight loads. By themselves, large motions would not cause this approach to be impractical but one other fact ruins our hopes. In virtually all modern helicopters, gas-turbine engine(s), mounted to the fuselage, drive the MGB via high-speed drive shaft(s) running at speeds that vary from

6000 to 22,000 rpm. The MGB reduces this speed to a few hundred revolutions per minute to drive the main rotor. Couplings on the ends of the high-speed drive shafts are almost always designed to handle typical engine-to-MGB misalignments that occur due to installation tolerances but have limited ability to withstand relatively large quasisteady motions necessarily allowed by MGB isolation. Hence, to the author's knowledge, pure MGB isolation has never been successfully implemented in all six degrees-of-freedom on any helicopter.

Having discouraged the reader regarding six-degree-of-freedom isolation, we should not jump to the conclusion that isolation is never attempted. The use of the so-called "DAVI" or Dynamic Anti-resonant Vibration Isolator is a particularly interesting and widely used example which will be described later.

6.2 THE VENERABLE FRAHM

As mentioned earlier, most large helicopters are designed with a very stiff connection between the MGB and the fuselage. Aside from its obvious simplicity and weight efficiency, this approach also has the benefit of simplifying the connection of the MGB to the engine(s), i.e., excessive relative deflections between the MGB and engine are not a concern. Sikorsky and Boeing are two companies that generally follow this approach. Direct connection has the disadvantage of enabling the Np loads to easily get into the fuselage. Many companies attack this problem with a two-pronged strategy, by (1) countering the largest rotor in-plane vibratory loads at the rotor head (where the blades are connected) with special rotating Frahms of various styles and (2) suppressing the fuselage vibration caused by the remaining loads with Frahms in the airframe. In the next sections, we will discuss fuselage-based Frahms followed by a description of rotor-based Frahms.

6.3 FUSELAGE-BASED FRAHMS

From the early days of the helicopter industry, fuselage and rotor-based Frahms have been used to lessen vibration and are still in wide use today. The author's search for the earliest patents using Frahms on helicopters reached back to 1945 where pendular absorbers were used to lessen control vibrations in helicopters (US2489342). The imperative for vibration suppression on helicopters is revealed by the fact that this patent was granted only 5 years after the first flight of the Vought-Sikorsky VS-300; widely considered the first demonstration of a practical helicopter. One of the earliest uses of Frahms on production helicopters was in the Army's Chinook CH-47B helicopter in the early 1960s. This 40,500 lbm helicopter with two 3-bladed rotors, incorporated five absorbers each weighing 90 lbm; in total 1.1% of the aircraft gross weight. For the dynamics aficionado, the absorbers were arranged with one in the nose, one each under the pilot and copilot and two in the aft pylon. In the CH-47C, D, these fixed-tuned absorbers were replaced with three Self-Tuning Vibration Absorbers [8–10], or "STVAs" each weighing 170 lbm. About 10 years later, three Frahms were used on the UH-60A Black Hawk Helicopter. These absorbers were a clever design that used three nested leaf springs; the airframe-end was connected to the fuselage with a "simple" support while the other side of the leaf springs was cantilevered to the absorber mass. This approach was invented by John Marshall [11]. To lessen the effect of airframe deformations on damping, these were later replaced by self-contained "box frame" absorbers on the UH-60L as shown in Fig. 4. A perusal of this figure reveals three variations of fuselage absorbers; two of the designs are box-frames and one design is a single leaf spring type. Use of

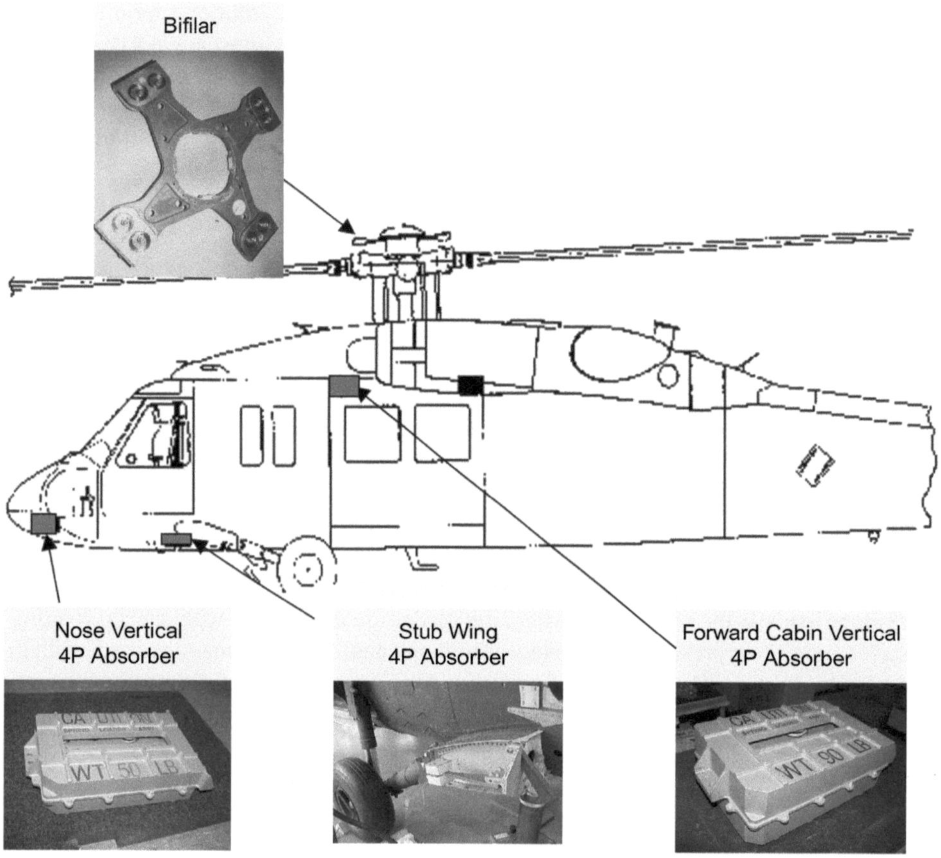

FIG. 4

Bifilar and Frahms on a UH-60L Black Hawk.

Taken from W.A. Welsh, Evolution of active vibration control technology, in: Proceedings of the American Helicopter Society, 4th Decennial Conference on Aeromechanics, January 21, 2004.

Frahms, in one form or another continues on currently produced helicopters but the fuselage-based versions are slowly being replaced by active systems. A perusal of the weights given above will reveals why absorbers are becoming obsolete; Frahms frequently make up from 1% to 2% of the helicopter gross weight. Given the relentless drive for more vehicle productivity, they are a prime target for weight reduction. The remaining Frahms in helicopters are on the helicopter rotors; a subject which will be covered next.

6.4 ROTOR-BASED FRAHMS

Clearly, to eliminate the vibratory loads emanating from the rotor, one would like to invent devices that produce loads with a similar ellipsoidal shape as previously discussed but in the "negative direction," then place these devices very near the source of rotor Np load generation, i.e., on the rotor head. Such counter-force devices capable of producing the negative ellipsoid do not yet exist but some approximations to these are in service. One example of a rotor-head mounted device is the bifilar. Many Sikorsky aircraft have one or two bifilars like the model shown in Fig. 4, which was invented by William Paul and Ken Mard of Sikorsky [12,13], in 1968 and 1969, respectively. A bifilar is a type of Frahm or pendular absorber where the Frahm mechanical springs are replaced with a "centrifugal" spring. Since the "spring" rate is now not fixed but rather, is proportional to rotor speed, the bifilar, as well as the more conventional pendular absorber, has the very nice feature of automatically remaining tuned to Np if the rotor speed changes. This can be seen by forming the natural frequency of a pendulum intended to reduce 3P rotating system loads as follows with: $L =$ pendulum length, $3\Omega =$ rotor speed times the harmonic to suppress, $m_{\text{pendular}} =$ the pendulum mass, and $g =$ gravitational constant. Notice that 3Ω is squared but is also inside of the square root, i.e., the natural frequency increases proportionately with rotor speed. Also, notice that the pendular mass term cancels, i.e., the natural frequency is not sensitive to the pendular mass.

$$\omega_n = \left(m_{\text{pendular}} * L * g * (3\Omega)^2 / m_{\text{pendular}}\right)^{1/2}$$

Bifilars as defined by the original inventors suffer from the same problem that tends to plague pendular absorbers, i.e., they tend to detune with increasing mass amplitude. To properly tune a conventional bifilar and pendulum to the Np − 1 or Np + 1 rotating system frequency, the c.g. of the somewhat large pendular mass must only be a short distance from the pivot point resulting in a very short pendular "arm." But as the mass oscillates to larger amplitudes, with an increasing forward speed of the helicopter, this large motion violates the small angle assumption resulting in a detuned natural frequency and decreased effectiveness. This detuning effect was solved by John Madden [14] by adding a special cycloidal shape to the bifilar bearings thus resulting in a natural frequency that is much less sensitive to mass amplitudes. The author finds it interesting that the peculiar bearing design used to retain the bifilar mass against centrifugal loading that was invented in 1969 also *enabled* the innovation of using cycloid bearings invented in 1980. It is doubtful (in the author's opinion) that the notion of using cycloidal bearings would have even occurred if conventional roller or ball bearings continued to be used to retain the mass as they are used in conventional pendular absorbers.

A bifilar (or pendulum) tuned to a single frequency creates a circular load (not an ellipse) that counters some of the in-plane rotor vibratory loads. Two bifilars, tuned to Np − 1 and Np + 1 frequencies can produce an ellipse in the plane of the rotor; but alas, not a tilted ellipsoid. Nevertheless, single or dual bifilars are a good compromise because this arrangement cancels one or two in-plane hub loads, respectively, while, regrettably, still allowing the vibratory vertical load, pitching and rolling moments to "leak" into the fuselage. Unfortunately, no currently available single device can cancel *all* three loads and three moments. The resulting residual fuselage vibration is, historically, lessened with one or more Frahms mounted in the airframe.

As previously alluded to, another example of a rotor-head mounted device is the pendular absorber which is used in the Bell 412 (Fig. 5) and the Boeing/Bell V-22 [15] (Fig. 6). These devices are simple masses on the end of a short pendulum arm that are mounted to the helicopter rotor head. As the rotor spins, the centrifugal force causes the mass to experience a virtual spring force in a manner very similar to that of the bifilar. The mass and pendulum arm are adjusted to produce a Frahm antiresonance effect either in or out of the rotor plane. For the case of the Bell 412, the pendular absorbers act

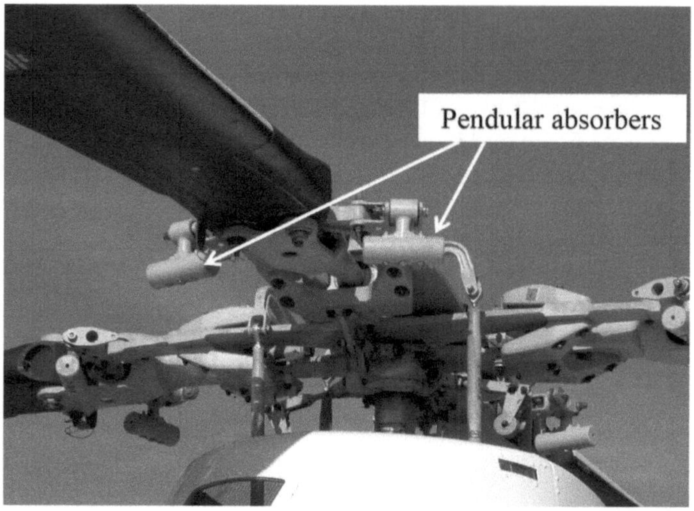

FIG. 5

Bell 412 pendular absorbers.

perpendicularly to the rotor plane (Fig. 5) and are intended to suppress Np vibration in both the rotating and fixed frames of reference while the V-22 pendular absorbers act in-plane (Fig. 6) and suppress Np − 1 vibration in the rotating frame of reference and Np in the fixed frame.

6.5 FRAHMS ARE HEAVY

Some weight estimates of Frahm-based systems can be gleaned from Fig. 4. The UH-60L system weights, per this figure, make up about 1.5% of the helicopters gross weight; or about 300 lbm for a 20,000 lb helicopter. The V-22 suffers from a smaller weight penalty as indicated by Rangacharyulu et al. [15], each pendulum mass weighing 40 lbm (there are six of them). Adding this to the weight of the fuselage based antivibration system, brings the total up to approximately 350 lbm. But the V-22 is a 40,000–65,000 lb aircraft so the Frahm weights are only 0.7% of the gross weight.

6.6 ACTIVE VIBRATION CONTROL

More recently, in the UH-60M, for example, fuselage mounted Frahms have been replaced with active vibration systems where active vibration control (AVC) actuators are placed where the previous Frahms resided. In this AVC system, the control computer acquires accelerations for 10 locations in the fuselage and optimally commands the AVC antivibration actuators to produce vibratory forces at Np to suppress these accelerations. The self-adaptive, real time, frequency domain controller implemented in the UH-60M AVC system is derived from the seminal work cited by Taylor et al. [16]. In this reference, the control methods developed by J. Molusis and F. Farrar are described in detail. In fact, the control algorithm, or derivatives of it, are probably being used in the vast majority of the fielded AVC systems and is generally the algorithm used for rotor-based active systems to be discussed later (IBC and HHC).

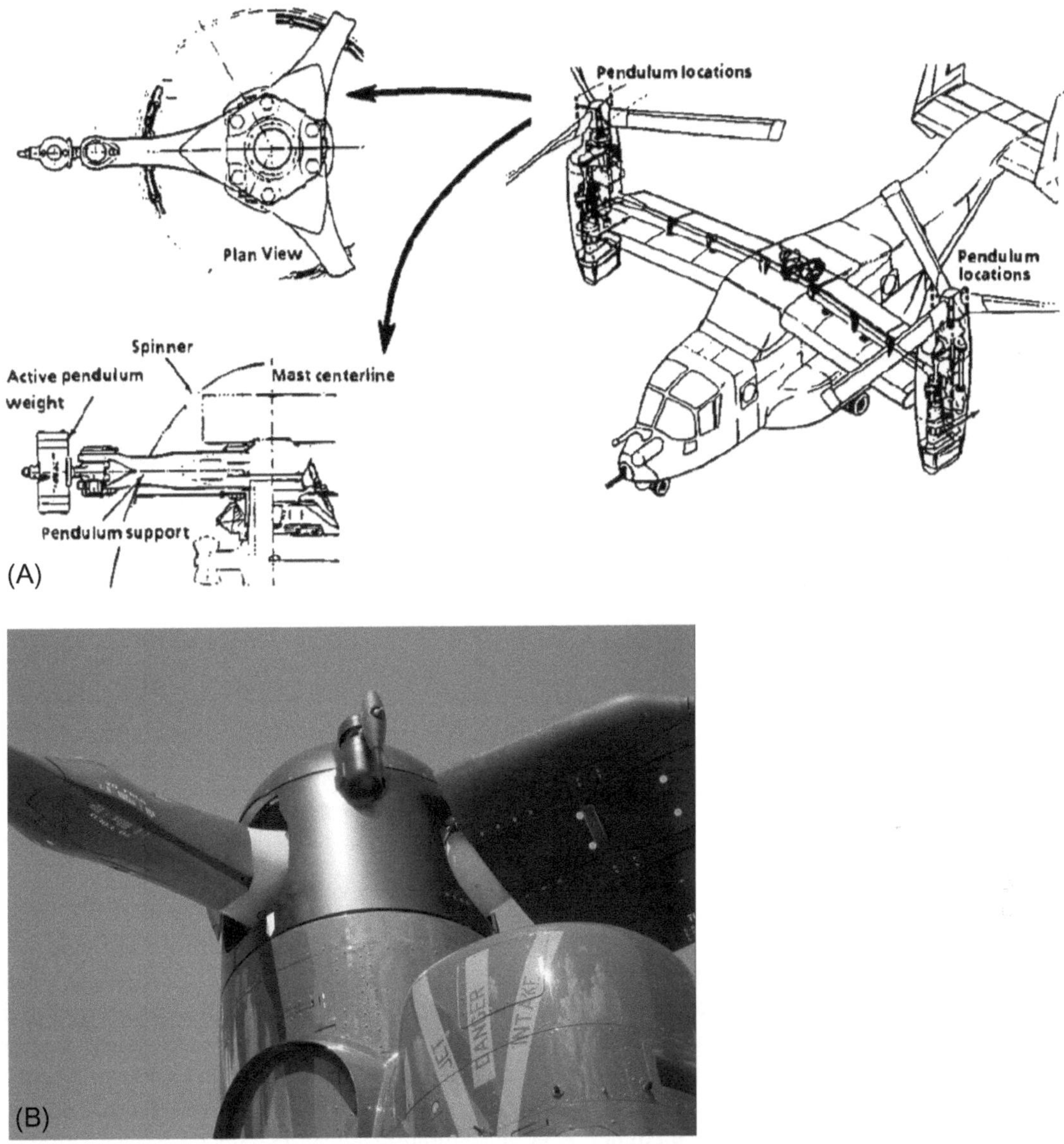

FIG. 6

(A) V-22 hub pendulum absorbers. (B) V-22 hub pendulum absorbers.

(A) Taken from M.A. Rangacharyulu, Bell Helicopter Textron Inc., M.J. Moore, Boeing Helicopter Company, Flight vibration testing of the V-22 tiltrotor aircraft, in: Proceedings of the 47th Annual Forum of the American Helicopter Society, Phoenix, AZ, May 6–8, 1991.

(B) Courtesy of the American Helicopter Society International.

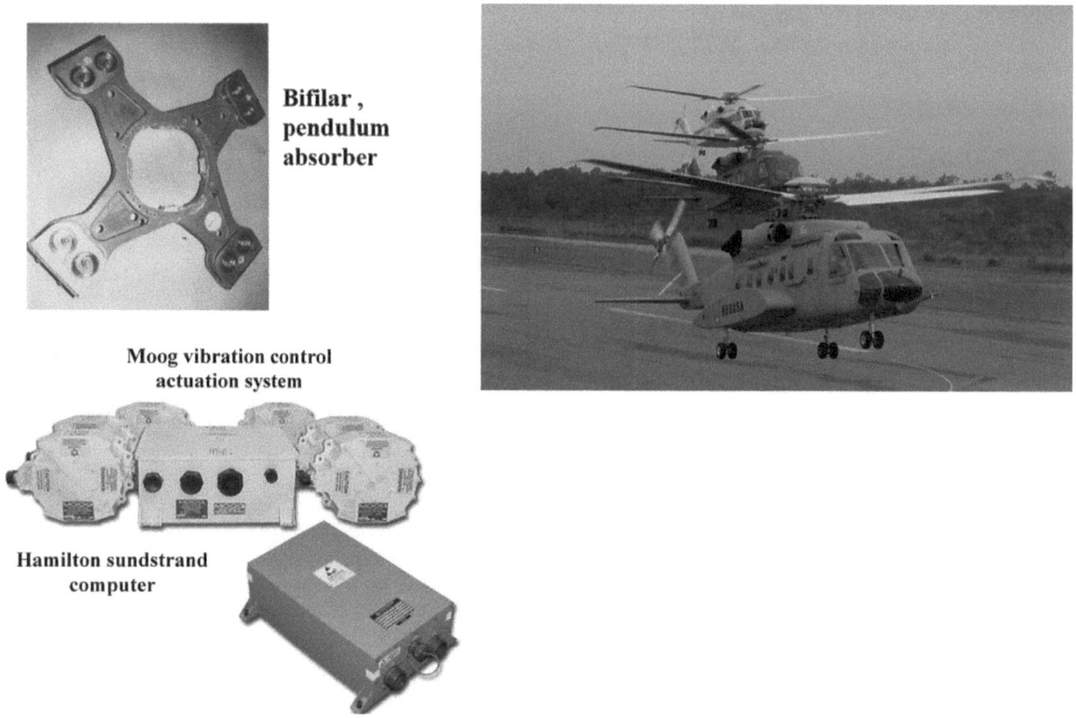

FIG. 7

Fuselage based active vibration system (AVC) on the Sikorsky S-92 helicopter.

Taken from W.A. Welsh, Evolution of active vibration control technology, in: Proceedings of the American Helicopter Society, 4th Decennial Conference on Aeromechanics, January 21, 2004.

The AVC system, one example of which is shown in Fig. 7, has significant advantages over fuselage based Frahms as shown by the author [17]; specifically, the AVC system can suppress vibration where the crew and passengers actually sit. The Frahms, being passive, can only suppress vibration at their respective mounting locations. This can be a serious limitation; if there is no physical room to mount Frahms near where the engineer would like to reduce vibration, they must be mounted at alternative locations, which are often less effective and less weight-efficient than desired. Thus, fuselage-based AVC is more weight efficient, i.e., lighter AVC antivibration actuators reduce fuselage vibration to the same or lower levels than Frahms systems with lighter and/or fewer AVC actuators (the heaviest components in AVC systems).

6.7 DYNAMIC ANTIRESONANT VIBRATION ISOLATOR

As stated earlier, no successful six-degree-of-freedom isolation system has ever been implemented in production helicopters. But very successful isolation systems *do* exist for eliminating the largest vibratory loads that would otherwise be transmitted into the fuselage. For example, some helicopter companies, like Bell Helicopters [18] position very clever "isolation" devices called LIVE (Liquid Inertia Vibration Eliminator) devices between the main transmission and the fuselage to interrupt the transmission of Np loads into the fuselage while allowing very reliable transfer of flight loads into the fuselage; obviously, a key aspect of safe helicopter operation. This idea was first developed by Kaman helicopter company and was called the "DAVI" or Dynamic Antiresonant Vibration Isolator and was invented by William Flannelly from Kaman Corporation [19]. LIVE is a derivative of this device, and is shown on a Bell-429 taken from Fig. 8 by Riedel [20]. It is placed in the main load path between the helicopter main transmission and the airframe and produces a "notch" or antiresonance in the load between the two. The DAVI was a purely mechanical device but subsequent refinements and advancements of the basic concept has resulted in the LIVE that is equivalent from a physics standpoint but replaces the DAVI's oscillatory mass with a heavy liquid and replaces the mechanical lever and bearings with pistons of differing areas. Presumably, this change improved reliability and certainly resulted in a more elegant, sealed, and compact device. On many Bell Helicopters, LIVE devices are used to lessen vertical and roll vibratory load transmission into the fuselage while passive isolation is used to lessen the effect of vibratory pitching moments and, finally, Frahms are sometimes used to lessen the residual vibration which still finds its way into the airframe.

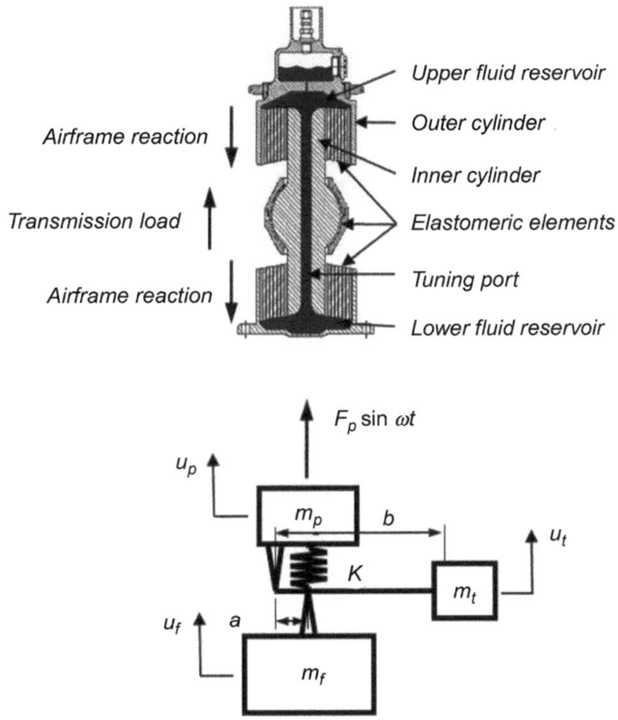

FIG. 8

Liquid Inertia Vibration Eliminator (LIVE) system.

Taken from K. Riedel, Vibration analysis and testing of Bell-429 helicopter, in: Proceedings of the American Helicopter Annual Forum, Phoenix, AZ, May 11–13, 2010.

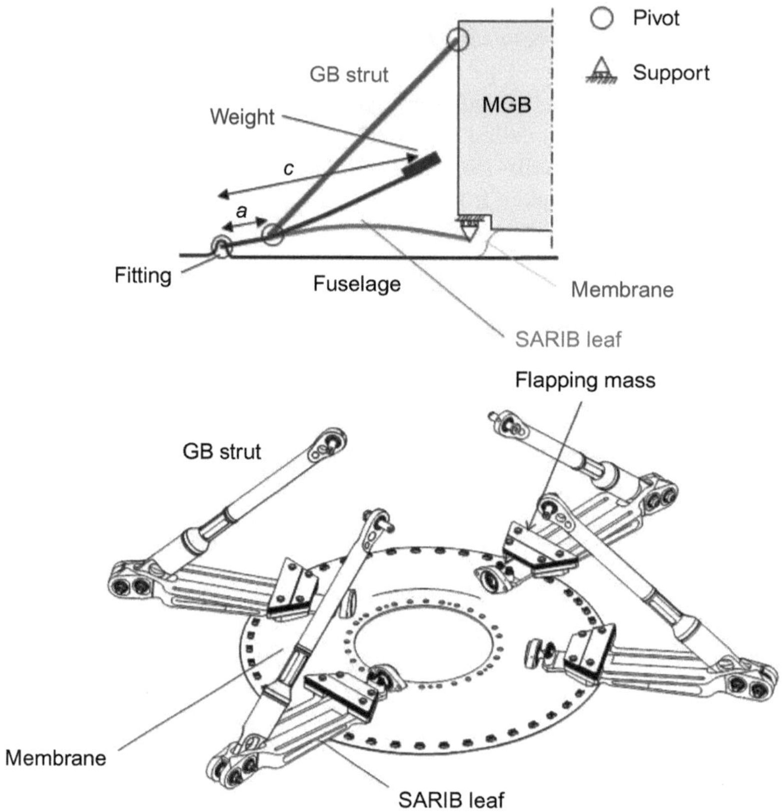

FIG. 9

Eurocopter's (now Airbus) SARIB gearbox isolation.

From P. Konstanzer, B. Enenkl, Recent advances in Eurocopter's passive and active vibration control, in: Proceedings of the American Helicopter Society 64th Annual Forum, Montreal, Canada, April 29–May 1, 2008.

In addition to LIVE, there are quite a few other variants of the original DAVI concept implementing innovations to adapt to the needs of a given helicopter. In the NH-90 Helicopter the so-called Système à Résonateurs Intégrés dans les Barres or "SARIB," shown in Fig. 9 taken from Konstanzer and Enenkl [21] is such a variant. Another variant is the "IRIS" (Improved Rotor Isolation System), developed by R. Desjardins and W.E. Hooper at Boeing [22,23].

7 MORE PROBLEMS WITH FRAHMS

As cited earlier, the venerable Frahm has been used (by the thousands) very successfully in helicopters for many years being distributed throughout the fuselage or, in their rotating-system version, as bifilars or pendulum absorbers on the rotor head. From the 1960s to now they have enabled vibration levels of 0.15 g or even lower, but we need to improve upon this for NextGen helicopters. Unfortunately, fuselage-mounted Frahms and AVC approaches both suffer from one common problem that ultimately limits their ability to reduce vibrations down to our target 0.03 g. Both approaches produce an effect called "modal spillover" and to understand this effect, it helps to view the fuselage dynamics from a "modal" and control system perspective. That is, fuselage vibration can be viewed as the superimposed response of an infinite number of natural frequencies and modes. Since the excitation is primarily due to rotor loads at Np, only a few of these modes really respond with significant amplitude; typically six or fewer modes are important. Consequently, the hopeful aircraft dynamicist/control engineer might conclude that these six modal degrees of freedom could be completely nullified by six "controls," which in this discussion, are represented by Frahms or AVC actuators distributed in the fuselage. Certainly, if six modes are responding to the hub loads, it is necessary for any control system (Frahms or AVC actuators) to have, at least, an equal number of controls to have any hope of nullifying the modal motion but, unfortunately this is not a *sufficient* condition. Focusing upon Frahms for a moment, six devices would surely produce virtually zero vibration but only at their six physical mounting locations. With some luck, low vibrations (0.1–0.15 g) can be obtained at locations other than at the Frahms mounting points. Unfortunately, some locations in the airframe will experience much higher levels to the point of making these seats the least favorite for passengers. This unfortunate effect occurs because while reducing vibration of the modes excited by rotor Np loads, Frahms also produce concentrated loads which excite modes that would *not* otherwise be excited by rotor Np loads in the first place. This effect is called "modal spillover" and the consequence is that vibrations everywhere in the fuselage cannot be reduced to the very low levels required for NextGen helicopters. A caveat to this; given a large enough number of antivibration devices mounted in the fuselage, i.e., Frahms or AVC actuators, NextGen vibrations can be achieved but experience demonstrates that the resulting weight increase would significantly lessen aircraft payload.

So, distributing antivibration devices in the airframe is destined to fail as a NextGen antivibration approach. But, following the same modal logic as previously indicated, it is also true that if antivibration loads are applied to the airframe in the same manner as the rotor Np loads, one would expect that the identical modes (to rotor Np loads) would be "excited," i.e., without modal spillover. The hoped-for result, would be global vibration reduction, i.e., one where the entire helicopter stops vibrating thus giving both passengers and equipment a smooth, damage-free ride.

8 ACTIVE COUNTER-FORCE

With the advent of rare earth magnets and high speed computer control, brushless electric motors have enabled some exciting developments in AVC. For example, it is now within the realm of possibility that all six hub loads comprising the elliptical figures previously described can be more or less directly suppressed by applying "equal and opposite" counter forces. This is still an area of active research as can be seen in Fig. 10, which was taken from Andrews [24]. In this Zero-Vibe system, active devices supplied by LORD Corporation are grouped around the main transmission in a manner that they can supply countering loads to every one of the six degrees-of-freedom. The system is comprised of a Hub Mounted Vibration Suppressor (HMVS) which is an active analog of two bifilars, i.e., it can create any

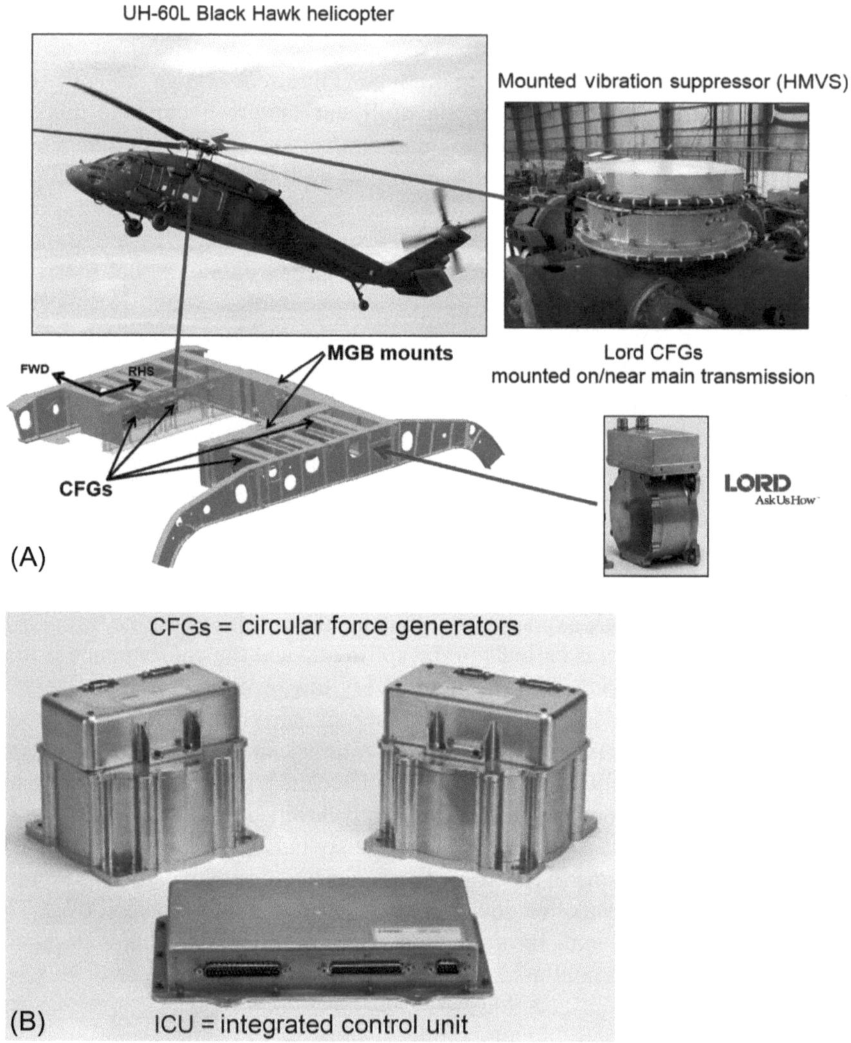

FIG. 10

(A) Sikorsky's Zero-Vibe system. (B) LORD Circular Force Generators and Integrated Control Unit.

(A) Taken from J. Andrews, W.A. Welsh, R. Altieri, J. DiOttavio, Ground and flight testing of a hub mounted vibration suppression system, in: Proceedings of the American Helicopter Society 70th Annual Forum, Montreal, Quebec, Canada, May 20–22, 2014. (B) Reprinted by permission of LORD Corporation.

elliptical in-plane counter-force to that produced by the main rotor. Other AVC actuators grouped around the main transmission supply the other counter-force and counter-moment loads. The Zero-Vibe intent is to avoid the modal spillover effects described in the Frahm discussion by applying counter-force loads at virtually the same location as does the main rotor thus exciting, or in this case, suppressing the same modal responses. A clarification is in order here; the newest actuators are still too large to literally place all six of them on the main rotor hub where the blade loads are applied, but this does not cause a problem because the main transmission, including the main rotor hub together, are a virtual rigid body. So, these six actuators can be placed either on the rotor hub or, alternatively, on or near the gearbox without suffering from being somewhat remote from the main rotor head. In fact, as a practical matter, actuators that produce a single directional counter-force must be placed a reasonable distance apart to act through a "moment arm" to produce a counter-moment thus necessitating being on opposite sides of the physically large part of the rigid body, i.e., the main transmission. A relatively recent addition to the helicopter world is the so-called "Circular Force Generators" (CFGs) by LORD Corporation shown in Fig. 10B. These novel devices produce a "circular" force of adjustable magnitude and phase. By adjusting these two parameters for six CFGs, it is possible to completely nullify the motion of a rigid body albeit the reader might find it hard to visualize how six whirling vectors, each with a controlled magnitude and phase, can add up to nullify the six hub loads (but it does work).

8.1 HIGHER HARMONIC CONTROL

A dream of helicopter dynamicists is to oscillate the main rotor blades in a manner to reduce the Np loads transmitted into the fuselage. The idea promises to achieve vibration reduction at the source (i.e., the rotor) and with low weight penalty since no extra actuators need be added to the aircrafts empty weight.

The first active rotor approach, tested in the 1960s (Warnicke and Drees [25]) on a 2-bladed UH-1 relied upon oscillating the existing main rotor blades through the rotor control system; a method known as Higher Harmonic Control (HHC). Typically, helicopters have three hydraulic servos that move the swashplate which, in turn, steers the main rotor according to the input from the pilot's flight controls. Normally, these servos move slowly in response to pilot inputs but, with HHC, the servos are oscillated at Np. The resulting rotor blade oscillations modify the Np vibratory loads that are transmitted to the fuselage. This approach has also been tried in many wind tunnel tests, for example by Shaw et al. [26] and some flight testing by Wood et al. [27] in 1985 and Straub and Byrnes [28] in 1986 on a 4-bladed OH-6A. In 1986, Miao et al. [29,30] tested HHC on an S-76 helicopter. This system is shown in Fig. 11. Again, the story is not an entirely happy one. In these references, it is shown that HHC is able to reduce vibrations in the fuselage but not to NextGen levels. Also, in some cases [29], at least for helicopters with articulated rotors as in this reference, the reduction came at the price of an extravagant expenditure of power in the form of hydraulic flow rates at high pressure (typically 3000 psi). This is a technical roadblock that has yet to be overcome and is probably responsible for the fact that no commercial or military helicopter has been equipped with HHC even 50 years after the approach was first attempted. What could overcome this roadblock? It is interesting to note that very little net energy is needed to oscillate a rotor blade. It surely takes considerable work to increase the pitch of a rotor blade because of the powerful "propeller" returning movement exerted by the spinning blade but, at least ideally, the blade will gladly return this work when pitch is reduced to zero. Unfortunately, because of their operating principle, typical hydraulic systems expend the same amount of work on both the up-stroke and down-stroke although the latter is, ideally, unnecessary. Perhaps a regenerative hydraulic system would overcome this problem. Another, more fundamental limitation of HHC is that, on a conventional helicopter, there are only three main rotor servos. Having only three controls necessarily limits the ability of HHC alone to suppress three hub loads or three modes or vibration at three locations. Clearly, for

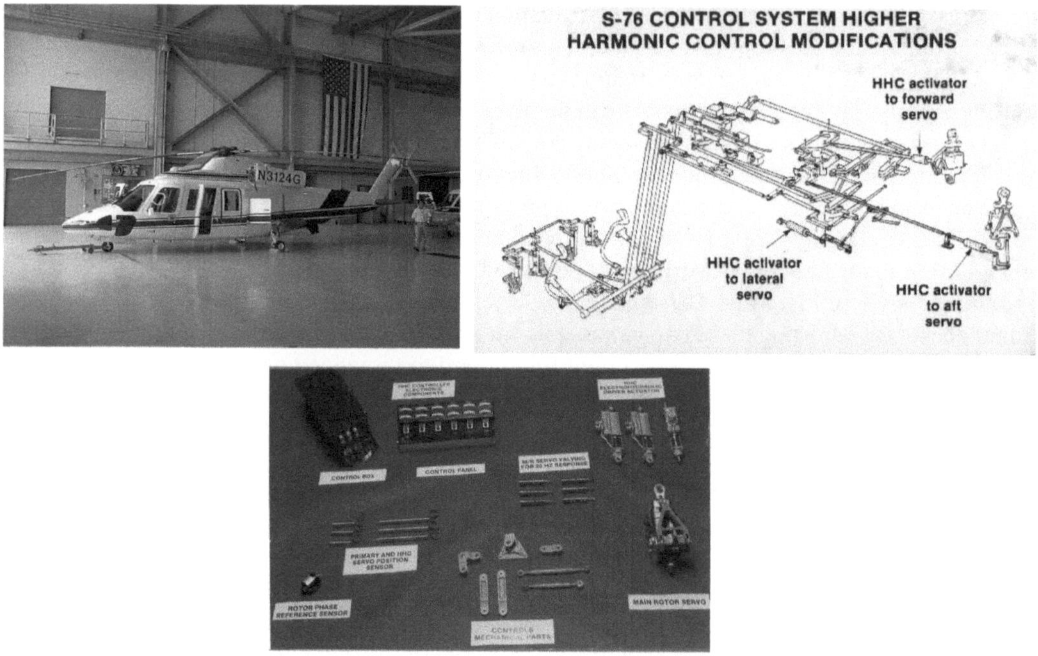

FIG. 11

Higher Harmonic Control (HHC) system.

Taken from W. Miao, S.B.R. Kottapalli, M.M. Frye, Flight demonstration of higher harmonic control (HHC) on S-76, in: Proceedings of the 42nd Annual Forum of the American Helicopter Society, Washington, DC, June 1986.

NextGen aircraft, we must suppress all six hub loads or hope that the helicopter only has three significant hub loads in the first place; not usually a good assumption. One potential solution to this controllability deficit is to use HHC to counter three of the hub loads, say vertical forces and pitching and rolling moments and augment HHC with some other actuation technology to attack the remaining loads, i.e., in-plane rotor loads: Fx, Fy, and Yaw moment: Mz. For example, the dual bifilar arrangement mentioned earlier could counter Fx and Fy. Another possibility is to augment HHC with a modern analog to the bifilar; the so-called Hub Mounted Vibration Suppressor [24] (HMVS) to be discussed later. The reader interested in further information regarding active rotor control should see the article by Dr. U. Arnold in the current book. Additionally, see Jacklin [31], for an excellent discussion and list of references regarding HHC and the next subject, IBC.

9 INDIVIDUAL BLADE CONTROL

An alternative approach to suppressing vibratory loads closer to the source was examined by Ham and Quackenbush [32] in the 1980s who proposed so-called "IBC" or Individual Blade Control for stall and vibration alleviation. The idea behind IBC, at least for vibration control, is to oscillate the entire blade at Np − 1, Np, and Np + 1 frequencies in the rotating frame of reference thus producing Np in the nonrotating frame of reference, i.e., the fuselage. IBC has an advantage over HHC because the blade pitches can also be oscillated at Np − 2 and Np + 2 which some researchers believe will improve rotor forward flight performance, i.e., L/De. This approach has, sporadically, seen about 40 years of experimentation in wind tunnels. See Jacklin et al. [33] for a BO-105 rotor IBC test, Jacklin et al. [34] (Fig. 12A) for a UH-60 rotor IBC test, at the Ames 80 × 120 wind tunnel. In the latter of these tests, the IBC actuators were designed and produced by ZF Luftfahrttechnik and are shown in Fig. 12B from Norman [35], and some results are given in Fig. 13 for a similar test at the Ames 40 × 80 wind tunnel taken from the same reference. Also, see IBC flight testing in Schimke et al. [36] on a BO-105 and Furst [37] for flight testing on a CH-53G. For readers who might want a more in-depth review of IBC technology and history, the author suggests the excellent paper by one of the pioneers in this area, P. Friedmann [38].

9.1 HYDRAULIC IBC

The approach in the testing [33,36,37] was to replace the usually rigid blade pitch control rods with hydraulic actuators; the latter is needed because the blade exhibits powerful propeller moments that require actuators with high-force capability. The motions, i.e., blade pitch inputs are also quite large, i.e., on the order of one degree; thus, eliminating any piezo-electric approach. The antivibration performance seen in the aforementioned testing was quite good albeit not to the desired NextGen standard of 0.03 g throughout the aircraft. Despite the partial success seen in wind tunnel and flight testing, the long history of work on this concept has shown that IBC, at least in the manner it was implemented, is impractical. The most intractable roadblock, which HHC did not have, is the need to get hydraulic power into the rotating system through some sort of hydraulic slip ring. Some researchers have had good luck with hydraulic slip rings during extended test programs but the authors experience has not been good and it is his opinion that no practical hydraulic slip ring has been invented which can withstand the combination of high pressure and high seal speeds and, at the same time, is reasonably reliable over a long period. IBC also suffers from the same potential for high power usage and controllability problems as does HHC. The power usage can be solved, just as for HHC, by regeneration and the controllability issue would yield to the same augmentation suggested for HHC, i.e., add more antivibration actuators to nullify the remaining hub loads.

9.2 ELECTRICAL IBC

Faced with the IBC and HHC road blocks, the research community has appeared to shift their focus to electrically powered IBC and on-blade devices to produce the same result, i.e., reduce fuselage vibrations by modifying Np loads transmitted by the rotor to the fuselage. Electrical IBC solves two of the problems exhibited by the hydraulic version: it eliminates the need for a hydraulic slip ring and also has the potential for lessening the power required. Electrical IBC actuators require an electrical slip ring but, unlike hydraulic slip rings, electrical slip rings are a proven technology, after all, many helicopters have de-icing systems that require large amounts of electrical power to be transmitted to the blades in order to heat them and preclude dangerous ice buildup. Also, as mentioned earlier, hydraulic actuators require high pressure fluid to both pitch the blade up but also to allow the blade to pitch down. The latter power requirement is unnecessary from a purely physics-based viewpoint because the blade propeller moments will cause the blade to return to flat-pitch with very little real need for any work to be expended. Electrical IBC solves this problem because they have the potential for "regeneration." That is, after work is done by the electrical actuator on the blade to pitch it up, the blade can then

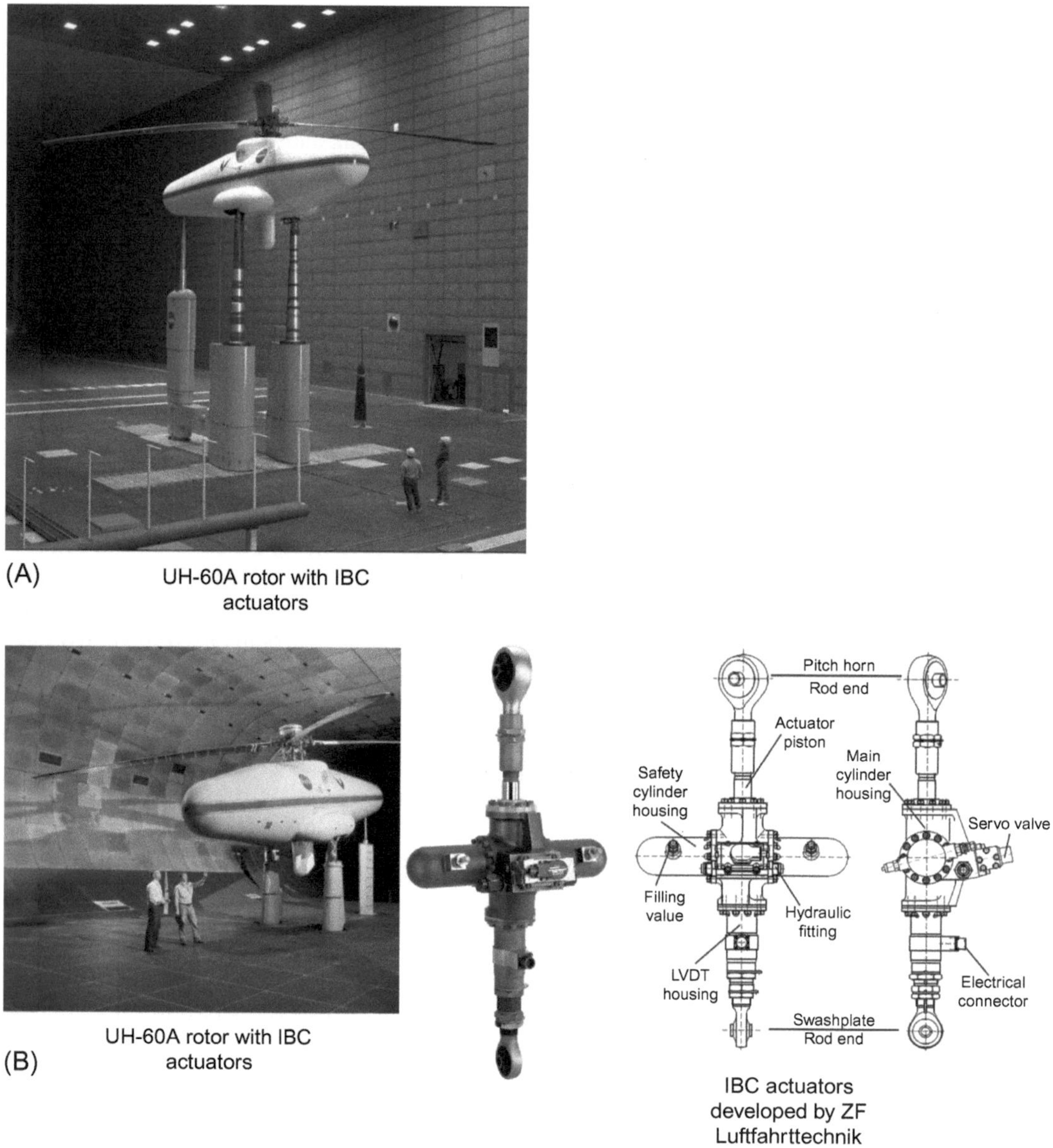

FIG. 12

(A) Individual Blade Control (IBC) system testing on the UH-60A rotor in the Ames 80 × 120 wind tunnel. (B) IBC actuators designed and produced by ZFL.

(A) Taken from S.A. Jacklin, A. Haber, G. de Simone, T.R. Norman, C. Kitaplioglu, P. Shinoda, Full-scale wind tunnel test of an individual blade control system for a Uh-60 helicopter, in: Proceedings of the American Helicopter 58th Annual Forum, Montréal, Canada, June 11–13, 2002.

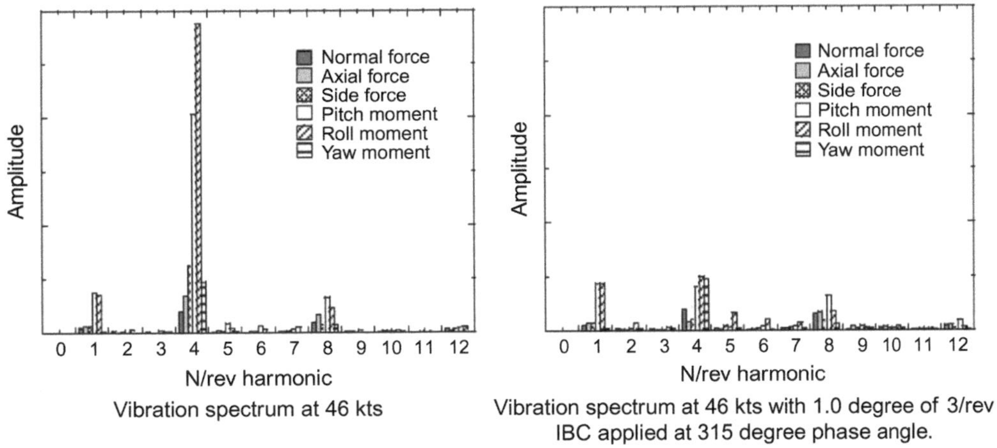

FIG. 13

IBC results in NFAC wind tunnel.

Taken from S.A. Jacklin, A. Haber, G. de Simone, T.R. Norman, C. Kitaplioglu, P. Shinoda, Full-scale wind tunnel test of an individual blade control system for a Uh-60 helicopter, in: Proceedings of the American Helicopter 58th Annual Forum, Montréal, Canada, June 11–13, 2002.

"back-drive" the actuator turning it into a virtual generator and thus recovering some of the power expended on the up-stroke. Various attempts at electrical IBC have been and still are being researched. Although no flight testing of this approach has yet been done, some notable research is being pursued, primarily by ZF Luftfahrttechnik GmbH [39,40].

9.3 ON-BLADE FLAPS

Another approach to electrical IBC is to activate a discrete on-blade trailing edge flap that morphs the blade instead of pitching the entire blade. One beauty of this method is that the actuators can be very small and light as compared to those required to pitch the entire blade. In fact, the Kaman company has always used trailing edge flaps which they call "servo-tabs" to provide *primary* pitch control for their helicopters. That is, they use the servo tabs to actually steer the helicopter in maneuvering flight. The servo-tab adds a second beauty to the on-blade flap approach because it is a so-called, balanced device, i.e., the servo-tab is hinged behind its leading edge as shown in Fig. 14. This clever arrangement, which is well-known to fixed wing airplane designers, enable servo-tabs to be actuated with very little effort, not only because they represent a small portion of the blade chord but also because they are subjected to only small aerodynamic pitching movements themselves. The latter is because the hinge is placed at the tab's quarter chord where the summation of the lift loads on the tab acts to produce pitching moments that are typically small. In fact, on Kaman's helicopters, the rotor is controlled via various mechanical control linkages, by the pilot's and copilot's efforts without the hydraulic boosting that is virtually universally used in the remainder of the helicopter world. On the negative side, that portion of the servo-tab which is ahead of its hinge, protrudes into the high-speed air traveling over the surface of the blade which, undoubtedly, introduces a drag penalty especially at very high speeds. For on-blade

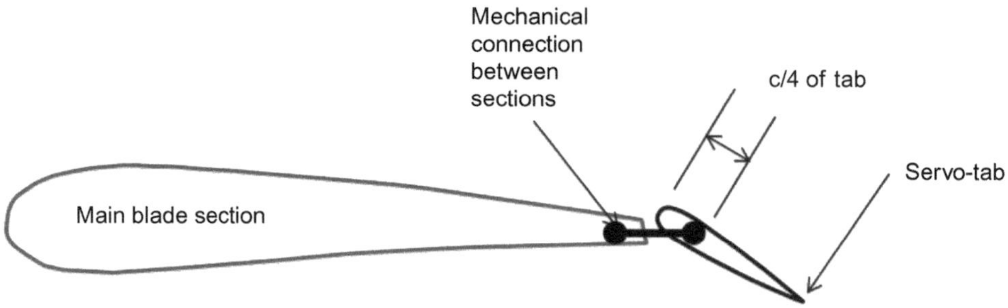

FIG. 14

Typical servo-tab.

active flaps intended for antivibration purposes, the servo-tab idea is often borrowed to lessen the required actuation effort. Typically, on-blade flaps used for antivibration are pitched only about ±5 degrees or less at the three frequencies we have discussed before, i.e., Np − 1, Np, and Np + 1. When the tab is pitched in the trailing-edge-down direction, for example, an upward load is exerted on the aft part of the chord which thus causes a pitching movement on that portion of the blade, which in turn, twists the blade nose-downward. The loss of lift thus causes the blade to "dive" downward. In the case of vibratory inputs, the blade thus oscillates in flap and generates the vibratory Np hub loads that can be part of the anti-load ellipse discussed earlier. Of course, the pitch stiffness, i.e., torsional stiffness of the blade must be low enough to allow this sort of deformation but this is the case on most helicopters. The new, commanded pitching of the blade at the three aforementioned frequencies is then modulated in phase and amplitude by an active controller to reduce fuselage Np vibrations. Much of the early whirl test and wind-tunnel work on this approach was performed Lemnios at Kaman [41,42]. Straub et al. [43,44] performed extensive whirl and wind tunnel testing using piezo-electrically powered trailing edge flaps as shown in Fig. 15. For the first time, flight testing of piezo-electrically controlled trailing edge flaps was performed by Konstanzer and Enenkl [21]. The configuration, for this test, called "Blue Pulse," is shown in Fig. 16. The latter two researchers used mechanically amplified piezo-electric actuators, aided by balanced servo-tabs. Lorber et al. [45] (Fig. 17) tested a nonbalanced-tab configuration using higher authority, on-blade actuators powered by brushless electric motors whose rotational motion was converted to linear motion by means of ball-screws. The nice feature of this design is that the aerodynamic drag penalty of the servo tab was eliminated and the authority of the actuator was much larger than that of the piezo-electric actuator thus opening the potential for primary flight control as well as vibration control. All the testing previously cited demonstrated very good vibration reductions albeit still limited by the same controllability problem exhibited by HHC and IBC. Of course, the same solutions to compensate for this shortcoming are equally applicable to blade morphing. One other nice feature of a discrete on-blade flap is that the entire mechanism including flap and actuator can be made accessible for routine maintenance. As the reader might surmise, the vibratory environment on a helicopter blade is hostile; steady acceleration levels can be almost 1000 g and vibratory levels can easily be 10–20 g. Clearly, any concept that requires maintenance due to failures caused by this high-vibration environment must be made accessible for replacement and/or repair. The on-blade flap equipment in the concept previously described, can fulfill this requirement by being placed under various sorts of access hatches designed into the blade.

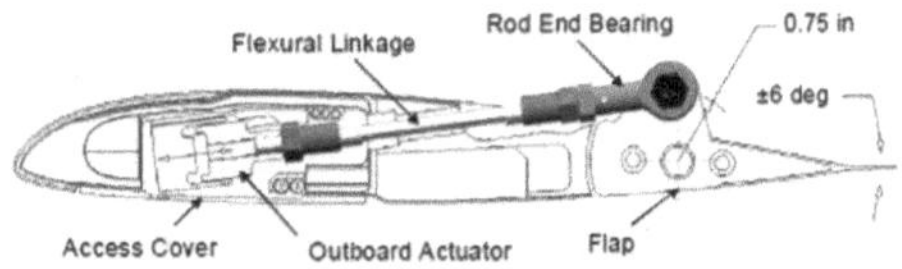

SMART Blade, flap, actuator cross-section

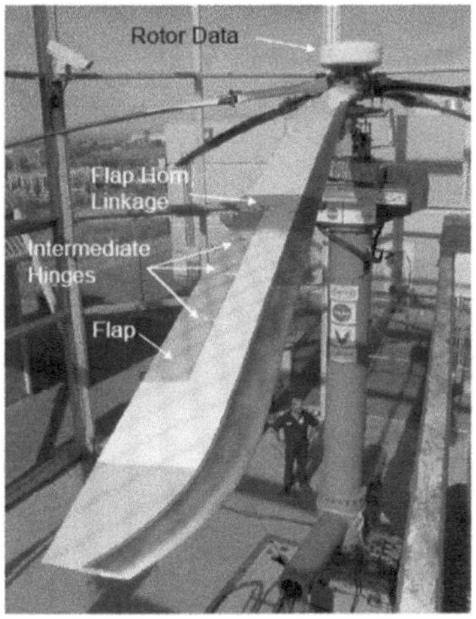

SMART rotor blade on whirl tower

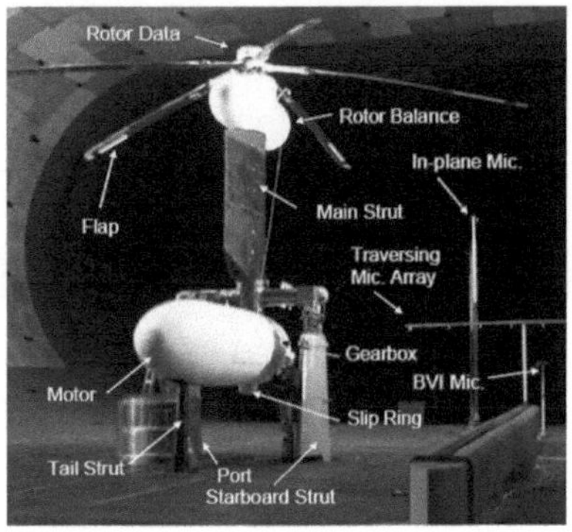

SMART rotor in the NFAC 40-by 80-foot wind tunnel (looking upstream)

FIG. 15

Boeing on-blade flap whirl and wind tunnel test (Refs. [43, 44]).

Close-up of deflected flap

BK117 helicopter with *Blue Pulse*™ active flap rotor

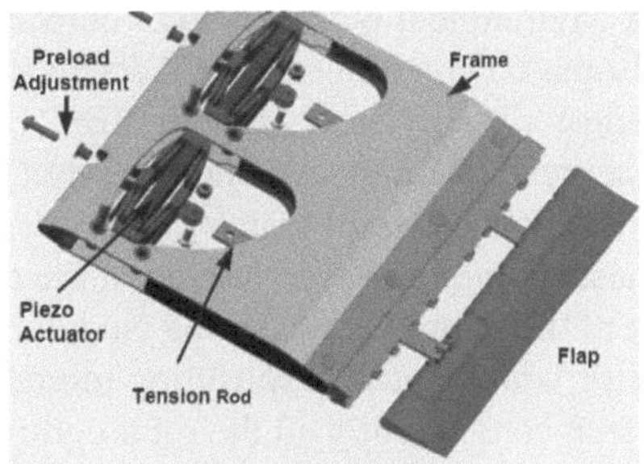

Flap unit assembly

FIG. 16

Eurocopter (now Airbus) active flap.

From P. Konstanzer, B. Enenkl, Recent advances in Eurocopter's passive and active vibration control, in: Proceedings of the American Helicopter Society 64th Annual Forum, Montreal, Canada, April 29–May 1, 2008.

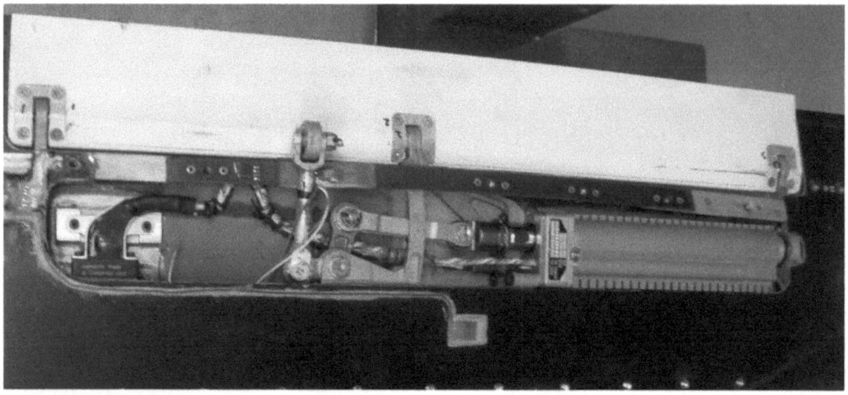

FIG. 17

Full authority active flap with electromechanical actuator (Ref. [46]).

Previously, we discussed discrete on-blade devices but some researchers have studied and tested more integrated morphing designs called "Active Twist Rotors" or ATRs. Cesnik [46] and Wilbur and coauthors [47], for example, have highly integrated piezo-electric actuators directly into the spar of the helicopter blade. This enables the blade twist to be varied via application of high-voltage to the embedded devices. ATRs have been wind-tunnel tested and have produced good vibration reductions, with limitations caused by the deflection capability of the piezo-electric material and again with the limitation caused by a limited number of control of degrees of freedom. The engineer might also consider that such highly integrated designs must also demand a high degree of reliability; it would be a very expensive failure if the piezo-electric device fails after years of service as it cannot be repaired or replaced, thus requiring the replacement of the entire blade and the failed blade to be discarded or re-built.

The results for both trailing edge flaps and active twist rotors are fairly impressive. So why are on-blade devices not universally adopted despite decades of development? The answer is complicated by the extreme challenges that must be overcome to make an on-blade "machine" work reliably in such a high-g environment but, the author considers that there is a more serious roadblock. Recall the previous discussion regarding the six independent loads or six degrees-of-freedom that must be eliminated to guarantee virtually zero vibration. IBC, HHC and on-blade flaps are all limited to essentially, three control degrees-of-freedom at the Np frequency. This can be visualized by imagining the commands in any of these active control systems to be one of three types (1) collective, where all the controls produce an in-phase or umbrella-type motion of the rotor disk, (2) pitch, where the controls produce a vibratory pitching of the rotor disk producing Np pitching moments, and (3) roll, where the controls yield rolling of the disk with resulting Np rolling moment. Three controls are not enough to guarantee complete suppression consequently all the rotor control methods must be augmented by additional anti-vibe devices to make up for the deficit. It is a "tough sell" to program managers if the technologist needs a lot of money to develop such a system and, at the same time, needs additional money for a second system. Consequently, the author's opinion is that this need for two systems is probably the primary reason why rotor-based methods have not been productionized.

10 THE PATH FORWARD

In this article, we have described some of the history of helicopter anti-vibration methods along with some that have been under development for many years and some approaches that are relatively recent. Which approach will make it to production is still a matter of speculation but some educated predictions can be made and some approaches can probably be eliminated. It seems certain that hydraulic HHC and IBC will not be making a resurgence anytime soon unless for very special applications; they just require too many inventions, i.e., better slip rings, better seals, and some ability to regenerate power expended. Thus, by elimination, electrical systems seem likely to be the wave of the future. How about on-blade devices vs. fuselage based devices? As we have suggested, on-blade devices suffer from an inability to do the entire task; they can reduce three degrees-of-freedom but not the six required to truly nullify all the vibratory loads created by the main rotor. The author considers that electrical IBC and electrical on-blade flaps have a real potential for "making the cut" to production. This notion is bolstered by the potential of such systems to compensate for the inevitable blade-to-blade differences that crop up in real aircraft. Such on-rotor systems, when augmented by three or more fuselage based antivibration actuators to produce controllability of six degrees-of-freedom seems a viable approach. Finally, the counter-force idea of having six controllable actuators on or near the gearbox is very alluring. Such a system is relatively simple and has solid physics-based reasons why it will get helicopter vibrations to near-zero levels. Current antivibration actuators seem just adequate for this job but there is a challenge here also. Antivibration actuators are heavy. Can the industry invent an actuator that is more weight efficient? Current units produce about 300 N of output load per kilogram of dead weight. The industry should pick up the challenge to double this value. If such a device were available, it could be used in the on-blade + fuselage systems as well as in the all-fuselage systems.

ACKNOWLEDGMENTS

The author thanks Robert Blackwell, Sikorsky Helicopter Co. (ret.), Lawrence Eastman, Sikorsky Helicopter Co. (ret.) and Christopher Sutton of Sikorsky a Lockheed Martin Company, Brahmananda Panda of Boeing, Troy Schank and Michael Smith of Bell Helicopter A Textron Company, Paul Cranga of Airbus Helicopters, Mark Jolly of Lord Corporation, Dr. Uwe Arnold of ZF Luftfahrttechnik and Michael Hirschberg of the American Helicopter Society International for their invaluable assistance in editing this article. The author also thanks their respective companies for kindly giving permission to reproduce the information from their excellent technical publications. All opinions, errors and omissions are solely the author's responsibility.

REFERENCES

[1] A.C. Veca, Vibration Effects on Helicopter Reliability and Maintainability, USAAMRDL Technical Report 73-11, April 1973.

[2] Aeronautical Design Standard, Standard Practice, Requirements for Rotorcraft Vibration Specifications, Modeling and Testing, ADS-27A-SP, May 2, 2006.

[3] International Standard, ISO-2631-1, Mechanical vibration and shock – Evaluation of human exposure to whole-body vibration.

[4] U.S. Federal Aviation Regulation, FAR-29.251.

[5] A.R.S. Bramwell, D. Balmford, G. Done, Bramwell's Helicopter Dynamics, second ed., AIAA (American Institute of Aeronautics & Ast); second ed., 2001.
[6] W. Johnson, Helicopter Theory, Revised ed., Dover Publications Inc., Mineola, NY, 1994.
[7] R.L. Bielawa, Rotary Wing Structural Dynamics and Aeroelasticity, second ed., American Institute of Aeronautics & Astronautics, Reston, VA, January 1, 2006.
[8] J.J. O'Leary, in: Reduction in vibration of the CH-47C helicopter using self-tuning vibration absorbers, Paper Proceedings of the Shock and Vibration Symposium, December 1969, pp. 191–202. (Bulletin No. 40, Part 5).
[9] R. Gabel, P. Lang, in: CH-47D airframe vibration reduction through airframe stiffening, Proceedings of the American Helicopter Society 51st Annual Forum, Fort Worth, TX, May 9–11, 1995.
[10] D. Richard Teal, D. McCorvey, in: Active vibration suppression for the CH-47D, Mallory, Proceedings of the American Helicopter Society 53rd Annual Forum, 1997.
[11] J. Marshall II, Tuned Spring Mass Vibration Absorber, U.S. Patent 4230291, October 28, 1980.
[12] W.F. Paul, K.C. Mard, Vibration Damped Helicopter Rotor, U.S. Patent 3540809, 1970.
[13] W.F. Paul, in: Development and evaluation of the main rotor bifilar absorber, Proceedings of the American Helicopter Society Annual Forum, May 14–16, 1969.
[14] J. Madden, Constant Frequency Bifilar Vibration Absorber, U.S. Patent 4218187, 1980.
[15] M.A. Rangacharyulu, Bell Helicopter Textron Inc., M.J. Moore, in: Flight vibration testing of the V-22 tiltrotor aircraft, Boeing Helicopter Company. Proceedings of the 47th Annual Forum of the American Helicopter Society, Phoenix, AZ, May 6–8, 1991.
[16] R.B. Taylor, P.E. Zwicke, P. Gold, W. Miao, Analytical Design And Evaluation Of An Active Control System For Helicopter Vibration Reduction And Gust Response Alleviation, July, 1980 (Prepared under Contract No. NAS2-10121 by United Technologies Research Center East Hartford, Connecticut and Sikorsky Aircraft for the Ames Research Center, NASA).
[17] W.A. Welsh, in: Evolution of active vibration control technology, Proceedings of the American Helicopter Society, 4th Decennial Conference on Aeromechanics, January 21, 2004.
[18] D.R. Halwes, in: LIVE—Liquid Inertial Vibration Eliminator, Proceedings of the American Helicopter Society 36th Annual Forum, Washington, DC, May 1980.
[19] W.G. Flannelly, in: The dynamic anti-resonant vibration absorber, Proceedings of the American Helicopter Annual Forum, 1966.
[20] K. Riedel, in: Vibration analysis and testing of Bell-429 helicopter, Proceedings of the American Helicopter Annual Forum, Phoenix, AZ, May 11–13, 2010.
[21] P. Konstanzer, B. Enenkl, in: Recent advances in eurocopter's passive and active vibration control, Proceedings of the American Helicopter Society 64th Annual Forum, Montreal, Canada, April 29–May 1, 2008.
[22] R.A. Desjardins, W.E. Hooper, in: Rotor isolation of the hingeless rotor BO-105 and YUH-61 helicopters, Proceedings of the Second European Rotorcraft and Powered Lift Aircraft Forum, September 1976.
[23] R. Desjardins, W.E. Hooper, Antiresonant rotor isolation for vibration reduction, J. Am. Helicopter Soc. 25 (3) (1980) 46–55.
[24] J. Andrews, W.A. Welsh, R. Altieri, J. DiOttavio, in: Ground and flight testing of a hub mounted vibration suppression system, Proceedings of the American Helicopter Society 70th Annual Forum, Montreal, Quebec, Canada, May 20–22, 2014.
[25] R.K. Wernicke, J.M. Drees, in: Second harmonic control, Proceedings of the 19th Annual National Forum of the American Helicopter Society, Washington, DC, May 1963.
[26] J. Shaw, N. Albion, E.J. Hanker, R.S. Teal, in: Higher harmonic control: wind tunnel demonstration of fully effective vibratory hub force suppression, Proceedings of the 41st Annual Forum of the American Helicopter Society, Fort Worth, TX, May 1985.
[27] E.R. Wood, R.W. Powers, J.H. Cline, C.E. Hammond, On developing and flight testing a higher harmonic control system, J. Am. Helicopter Soc. 30 (1) (1985) 3–20(18).

[28] F.K. Straub, E.V. Byrnes Jr, Application of Higher Harmonic Blade Feathering on the OH-6A Helicopter For Vibration Reduction, NASA-CR-4031, 1986.

[29] W. Miao, S.B.R. Kottapalli, M.M. Frye, in: Flight demonstration of higher harmonic control (HHC) on S-76, Proceedings of the 42nd Annual Forum of the American Helicopter Society, Washington, DC, June 1986.

[30] J. O'Leary, S. Kottapalli, M. Davis, in: Adaptation of a modern medium helicopter (Sikorsky S-76) to higher harmonic control, Proceedings of the 40th Annual Forum of the American Helicopter Society, 1984.

[31] S.A. Jacklin, in: Second test of a helicopter individual blade control system in the NASA AMES 40-by 80-foot wind tunnel, Proceedings of the 2nd International Aeromechanics Specialists' Conference, Bridgeport, CT, October 11–13, 1995.

[32] N.D. Ham, T.R. Quackenbush, in: A simple system for helicopter individual blade control and its application to stall induced vibration alleviation, Proceedings of the National Specialists' Meeting on Helicopter Vibration, Sponsored by the American Helicopter Society, Northeast Region, Hartford, CT, November 2–4, 1981.

[33] S.A. Jacklin, A. Blais, D. Teves, R. Kube, in: Reduction of helicopter BVI noise, vibration and power consumption through individual blade control, Proceedings of the American Helicopter 51st Annual Forum, Fort Worth, TX, May 9–11, 1995.

[34] S.A. Jacklin, A. Haber, G. de Simone, T.R. Norman, C. Kitaplioglu, P. Shinoda, in: Full-scale wind tunnel test of an individual blade control system for a UH-60 helicopter, Proceedings of the American Helicopter 58th Annual Forum, Montréal, Canada, June 11–13, 2002.

[35] T.R. Norman, C. Theodore, P. Shinoda, D. Fuerst, U.T.P. Arnold, S. Makinen, P. Lorber, J. O'Neill, in: Full-scale wind tunnel test of a UH-60 individual blade control system for performance improvement and vibration, loads, and noise control, Proceedings of the American Helicopter 65th Annual Forum, Grapevine, TX, 2009.

[36] D. Schimke, U.T.P. Arnold, R. Kube, in: Individual blade root control demonstration evaluation of recent flight tests, Proceedings of the American Helicopter Society 54th Annual Forum, Washington, DC, May 20–22, 1998.

[37] D. Fürst, C. Keßler, T. Auspitzer, M. Müller, A. Hausberg, H. Witte, in: Closed loop IBC-system and flight test results on the CH-53G helicopter, Proceedings of the American Helicopter Annual Forum, Baltimore, 2004.

[38] P.P. Friedmann, On blade control of rotor vibration, noise, and performance: just around the corner? The 33rd Nikolsky honorary lecture, J. Am. Helicopter Soc. 59 (4) (2014) 1–37.

[39] U.T.P. Arnold, D. Fuerst, T. Neuheuser, R. Bartels, in: Development of an integrated electrical swashplateless primary and individual blade control system, Cheeseman Award Paper Invited for Presentation at the American Helicopter Society International 63rd Annual Forum, Virginia Beach, VA, May 1–3, 2007.

[40] D. Fuerst, A. Hausberg, T. Neuheuser, in: Experimental verification of an electro-mechanical-actuator for a swashplateless primary and individual helicopter blade control system, Proceedings of the American Helicopter Society 64th Annual Forum, Montréal, Canada, April 29–May 1, 2008.

[41] A.Z. Lemnios, H.E. Howes. Wind Tunnel Investigation of the Controllable Twist Rotor Performance and Dynamic Behavior, JSAAMRDL-TR 77-10, June 1970.

[42] A.Z. Lemnios, A.F. Smith. An Analytical Evaluation of the Controllable Twist Rotor Performance and Dynamic Behavior, USAAMRDL-TR-72-16, 1972.

[43] F.K. Straub, et al., in: Development and whirl tower test of the SMART active flap rotor, SPIE Conference on Smart Materials and Structures, San Diego, 2004.

[44] F.K. Straub, V.R. Anand, T.S. Birchette, B.H. Lau, in: Wind tunnel test of the SMART active flap rotor, Proceedings of the American Helicopter Society 65th Annual Forum, Grapvine, TX, May 27–29, 2009.

[45] P.F. Lorber, B. Hein, J. Oneill, B. Isabella, J. Andrews, M. Brigley, J. Wong, P. LeMasurier, B.E. Wake, in: Whirl and wind tunnel testing of a sikorsky active flap demonstration rotor, Proceedings of the American Helicopter Society 67th Annual Forum, Virginia Beach, VA, May 3–5, 2011.

[46] C.E.S. Cesnik, S.J. Shin, W.K. Wilkie, M.L. Wilbur, P.H. Mirick, in: Modeling, design, and testing of the NASA/Army/MIT active twist rotor prototype blade, Proceedings of the American Helicopter Society 55th Annual Forum, Montreal, Canada, May 25–27, 1999.
[47] D. Fogarty, M.L. Wilbur, M.K. Sekula, D.D. Boyd Jr., in: Prediction of BVI noise for an active twist rotor using a loosely coupled CFD/CSD method and comparison to experimental data, Proceedings of the American Helicopter Society 68th Annual Forum, Fort Worth, Texas, May 1–3, 2012.

Afterword

There is no doubt today that the environmental impact of aviation remains a top priority for European research. The worldwide air traffic is expected to grow at a rate of 4%–5% per year, and higher growth rates are expected in the Middle East and Asia. Consequently, the need for efficient, sustainable, and competitive aviation, on one hand, and the resistance against noise, environmental pollution, and exhaustive usage of resources, on the other, could represent the new conflict of the industrialized society in the next years. In order to face such radical challenges and target durable progress in this changing world, a paradigm shift in the next generation aircraft design becomes fundamental.

Far from the day-to-day preoccupations of industry, often restraining even the transformation of emerging early-stage technologies into commercial products, this book offers a thorough experience of the current high-gain/high risk research on aircraft morphing devices ranging from multidisciplinary aerostructures design and manufacture to the description of design tools, prototyping guidelines, systems integration, and testing, through a bottom-up approach. Particular emphasis is given to the outstanding European achievements in this discipline of the SARISTU (acronym of Smart Intelligent Aircraft Structures) project. Coordinated by Airbus Germany and Leonardo Finmeccanica, SARISTU was surely the most advanced large-scale integrating program on morphing structures, targeting aircraft weight and operational costs reductions as well as enhanced aircraft aerodynamic performance. Ending in 2015, the project investigated the joint integration of different conformal morphing concepts in a laminar wing with the goal of improving aircraft performance through a 6% drag reduction inside the lift coefficient range of cruise, with a positive effect on fuel consumption. The project delivered the first full-scale, completely morphing wing prototype ever assembled in Europe, at Finmeccanica Headquarters (Pomigliano, Italy). The innovative seamless morphing wing incorporated a gapless morphing leading edge, a morphing trailing edge, and an adaptive winglet.

Turning emerging technologies into commercial products can be a lengthy, high-risk, and very complex process, especially in aeronautics. Matching technological capability to industrial standards is always difficult because any given technology, particularly if innovative, has to be combined with and integrated into existing technologies. However, although the research and technology results described in the book still need to be consolidated and flight tested, they are, nonetheless, a considerable advance. It has been more than a century since Wilbur Wright was casually twisting a cardboard inner-tube box in his bicycle shop and noticed how the deformation of the surface mimicked the profile of a pigeon's wing when it turned in the air. A few months later, in 1903, the first powered aircraft flyer could be controlled in three dimensions in the same way, with a system of levers. For many decades, then, rigid wing structures combined with hinged control surfaces, like flaps, ailerons, rudders, elevators, etc., have been the most effective means (and are still the standard today) to guarantee aircraft flight, stability, and controllability throughout the flight envelope. The combined use of multidisciplinary design tools, advanced composite materials, cost-effective manufacturing processes, rapid prototyping, and fully automated assembly for ground and flight-based installations, along with the latest advances in compact, high-authority, and lightweight electromechanical actuators, nowadays gives the unique chance to revisit the idea of shape-changing flight surfaces. New fuel-efficient wing concepts, radically contributing to reduce the environmental impact of aviation, may finally be realized.

The potential material instabilities throughout the relevant components' lifetime, the system weight penalties, the higher system complexity, and insufficient reliability have to date been the major obstacles to adoption of the morphing technology at an industrial scale. By removing some of these bottlenecks, this book provides a bridge in the maturation of such systems, which will bring the editors to focus on the remaining showstoppers, to enable the physical integration, per industrial standards, of a full-scale morphing wing device into a flying test-bed. The flight test demonstration of the morphing concepts detailed in this book is currently planned in 2020 in the framework of Clean Sky 2 Regional IADP. Sometime between 2025 and 2030, the editors predict a profound level of change in aircraft wing design and adaptation in the air. In the long term, the priority remains to demonstrate economic viability along with the technology's safety and reliability, remembering that gains in aircraft aerodynamic performance and fuel consumption should not be eclipsed by excessive manufacturing and maintenance costs.

There is a field, among several others, in which Europe excels. In these difficult times, when the confidence of European citizens in the European Union is being questioned, breakthrough research in aeronautics is the best possible response. It is not unreasonable to say that aeronautics has been a major contributor to the actual construction of Europe through a multitude of European research and demonstration activities built upon a vast European research network. Conformal morphing structures are part of this ambitious effort. The most advanced skills of several of Europe's public institutions and private firms are mobilized and (bio-)inspired to maintain the global competitiveness and prosperity of our union.

Alberto Milazzo
Full Professor, University of Palermo, Palermo, Italy

Index

Note: Page numbers followed by *f* indicate figures, and *t* indicate tables.

M

N

O

T

U

X

Z

Kolibri 360 GmbH
Ferdinand-Jühlke-Straße 7
99095 Erfurt, Deutschland
produktsicherheit@kolibri360.de